PacketCable Implementation

Jeff Riddel, CCIE No. 12798

Cisco Press
201 West 103rd Street
Indianapolis, IN 46290 USA

PacketCable Implementation

Jeff Riddel

Published by:
Cisco Press
800 East 96th Street
Indianapolis, IN 46240 USA

Printed in the United States of America 1 2 3 4 5 6 7 8 9 0

First Printing February 2007

Library of Congress Cataloging-in-Publication Number: 2003115151

ISBN-10: 1-58705-181-8

ISBN-13: 978-1-158705-181-4

Trademark Acknowledgments

Warning and Disclaimer

This book is designed to provide information about implementing multimedia services, such as voice telephony, on cable IP networks using PacketCable technology. Every effort has been made to make this book as complete and as accurate as possible, but no warranty or fitness is implied.

The information is provided on an “as is” basis. The authors, Cisco Press, and Cisco Systems, Inc., shall have neither liability nor responsibility to any person or entity with respect to any loss or damages arising from the information contained in this book or from the use of the discs or programs that may accompany it.

The opinions expressed in this book belong to the author and are not necessarily those of Cisco Systems, Inc.

Feedback Information

At Cisco Press, our goal is to create in-depth technical books of the highest quality and value. Each book is crafted with care and precision, undergoing rigorous development that involves the unique expertise of members from the professional technical community.

Readers' feedback is a natural continuation of this process. If you have any comments regarding how we could improve the quality of this book, or otherwise alter it to better suit your needs, you can contact us through email at feedback@ciscopress.com. Please make sure to include the book title and ISBN in your message.

We greatly appreciate your assistance.

Publisher	Paul Boger
Cisco Representative	Anthony Wolfenden
Cisco Press Program Manager	Jeff Brady
Managing Editor	Patrick Kanouse
Senior Development Editor	Christopher Cleveland
Project Editor	Tonya Simpson
Copy Editor	Kevin Kent
Technical Editors	Steve Brown, David Ensign
Team Coordinator	Vanessa Evans
Book Designer	Louisa Adair
Composition	Mark Shirar
Indexer	Ken Johnson

Americas Headquarters
Cisco Systems, Inc.
170 West Tasman Drive
San Jose, CA 95134-1706
USA
www.cisco.com
Tel: 408 526-4000
800 553-NETS (6387)
Fax: 408 527-0883

Asia Pacific Headquarters
Cisco Systems, Inc.
168 Robinson Road
#28-01 Capital Tower
Singapore 068912
www.cisco.com
Tel: +65 6317 7777
Fax: +65 6317 7799

Europe Headquarters
Cisco Systems International BV
Haarlerbergpark
Haarlerbergweg 13-19
1101 CH Amsterdam
The Netherlands
www-europe.cisco.com
Tel: +31 0 800 020 0791
Fax: +31 0 20 357 1100

Cisco has more than 200 offices worldwide. Addresses, phone numbers, and fax numbers are listed on the Cisco Website at **www.cisco.com/go/offices.**

About the Author

Jeff Riddel, CCIE No. 12798, is a network consulting engineer for Cisco Systems Broadband Advanced Services team. He has several years of experience providing broadband consulting and testing expertise to the major U.S. cable companies (Comcast, Time Warner Cable, Cox, Cablevision, and Charter) as well as several international cable companies. Jeff has also worked on various Cisco VoIP products and solutions and is a Cisco Cable Communications Specialist (CCCS) and a member of the Society of Cable Telecommunications Engineers (SCTE).

About the Technical Reviewers

Steve Brown is a consulting systems engineer for Cisco Systems. He has focused on systems integration, testing, and deployment of PacketCable telephony since its inception (only a slight exaggeration). In his role at Cisco, Steve has consulted with numerous North American MSOs to help them introduce VoIP services.

Dave Ensign is an engineering program manager in the Cable Video-IPTV Systems group at Cisco Systems, Inc. He has more than 20 years of experience in telecommunications and interactive broadband services and is currently the Cisco liaison to Cable Television Laboratories, Inc. (CableLabs).

Dedications

To my wonderful wife, Wanda, thank you for putting up with me during the nights and weekends spent working on this book. You are the best wife and friend a man could ask for.

Acknowledgments

First and foremost I would like to thank my technical reviewers: Dave Ensign and Steve Brown. Their feedback was instrumental in improving the readability and overall quality of this manuscript. I would also like to thank the "unofficial" technical reviewers who have helped with various aspects of the book: John Downey, Murry Gavin, Gene Keeling, and Tom Chang. Lastly, a big thank you goes out to the production team for this book. Kristin Weinberger, Christopher Cleveland, Tonya Simpson, and the rest of the Cisco Press folks have been incredibly professional and a pleasure to work with. I couldn't have asked for a finer team.

This Book Is Safari Enabled

The Safari® Enabled icon on the cover of your favorite technology book means the book is available through Safari Bookshelf. When you buy this book, you get free access to the online edition for 45 days.

Safari Bookshelf is an electronic reference library that lets you easily search thousands of technical books, find code samples, download chapters, and access technical information whenever and wherever you need it.

To gain 45-day Safari Enabled access to this book:

- Go to http://www.ciscopress.com/safarienabled
- Complete the brief registration form
- Enter the coupon code N8LB-4LLM-1KW4-UMMS-S1XU

If you have difficulty registering on Safari Bookshelf or accessing the online edition, please e-mail customer-service@safaribooksonline.com.

Contents at a Glance

Contents

Icons Used in This Book

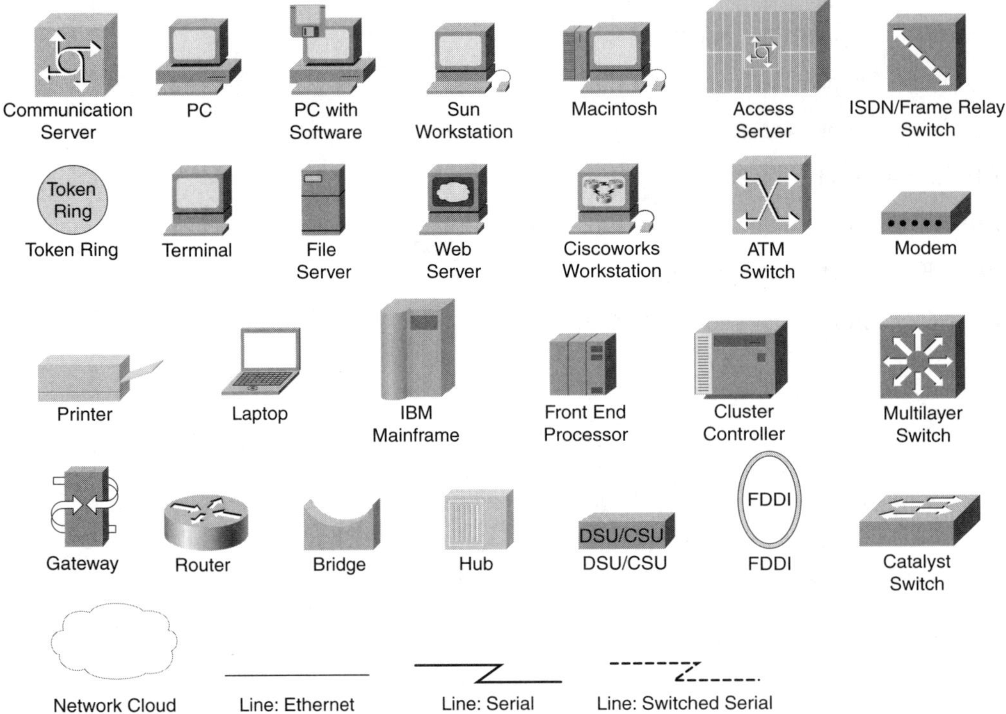

Command Syntax Conventions

The conventions used to present command syntax in this book are the same conventions used in the IOS Command Reference. The Command Reference describes these conventions as follows:

- **Boldface** indicates commands and keywords that are entered literally as shown. In actual configuration examples and output (not general command syntax), boldface indicates commands that are manually input by the user (such as a **show** command).
- *Italics* indicate arguments for which you supply actual values.
- Vertical bars (|) separate alternative, mutually exclusive elements.
- Square brackets [ ] indicate optional elements.
- Braces { } indicate a required choice.
- Braces within brackets [{ }] indicate a required choice within an optional element.

Introduction

In recent years, cable providers or Multiple Systems Operators (MSOs) have made a substantial investment in upgrading their existing infrastructure into a bidirectional Hybrid Fiber Coaxial (HFC) system capable of providing reliable high-speed Internet access and digital cable television. The bidirectionality component permits the transmission and receipt of IP packets, and it is this IP over cable network that enables the MSO to expand their competitive offerings in an effort to better serve their customers and increase revenue. High-speed data, video-on-demand, interactive gaming, videoconferencing, and telephony are some examples of these value-added services. PacketCable provides a framework MSOs can use to ensure the successful deployment of these services over a broadband cable network.

The PacketCable 1.x series of specifications define an architecture framework on how a cable MSO can provide a voice telephony service. PacketCable-based VoIP telephony services can be provided at a much smaller cost than that of the Incumbent Local Exchange Carriers (ILECs) legacy circuit switched networks. Cable MSOs becoming Competitive Local Exchange Carriers (CLECs) makes a lot of sense; in fact many of the Cable MSOs (including the five largest) are certified local exchange carriers. Telephony services are a natural progression for MSOs as they already have access to millions of potential subscribers. At of the end of 2005, major MSOs served 5.6 million cable telephony subscribers across the country (source: www.ncta.com). In fact MSOs are signing up 11,000 customers a day for VoIP telephony!! As you can see cable telephony is a key component of the cable industry's business growth strategy, and with the continued improvements in IP telephony, cable-delivered telephone service could evolve into a simple telecommunications afterthought for consumers, rather than a separate, independent service.

The PacketCable Multimedia specifications define a generic framework on how a cable MSO can provide any arbitrary multimedia service. These services can be voice, video, and/or data. By implementing PacketCable the MSO can have a single network to manage, which provides the backbone for all of these service types.

Goals and Methods

As a result of these expanded services, the industry need for knowledgeable engineering professionals is expected to increase dramatically. Cable's "triple play" offering (voice, video, and high-speed Internet access) has allowed cable to outpace DSL service in the U.S. as the more popular broadband "last mile" access solution.

PacketCable Implementation will supply IP networking information to those versed in cable video networks and help those deploying cable IP networks understand the ramifications of deploying PacketCable service on the cable network. A majority of this book deals with issues associated with implementing voice; however, you will also get some exposure into the more generic case of implementing any multimedia solution.

Who Should Read This Book?

This book is intended for engineers who have some experience in the area of cable modem technologies and VoIP. The primary audience includes:

- Cable network managers and deployment engineers who have experience with CATV but little experience with IP-based access networks or VoIP
- IP networking professionals wishing to learn more about cable as an access medium
- Engineers working with any of the Cisco products covered, such as the Cisco CMTS (uBR10012, uBR7246VXR), the Cisco call agent (BTS 10200), the Cisco multiservice media gateway (MGX8800 series with VXSM/VISM service modules), the Cisco IP Transfer Point (ITP) signaling gateway, and the Cisco Broadband Access Center for Cable (BAC) provisioning system
- Engineers working with any of the protocols covered, such as MGCP (including NCS and TGCP), SIP (including CMSS), RTP/RTCP, DOCSIS, COPS, RADIUS-based Event Messaging, the Sigtran protocol family, the IP Security protocol ESP, and the protocols involved with provisioning (DHCP, DNS, SNMP, Kerberos, etc.)
- This book might also be helpful to technical trainers and CMTS (Cisco cable modem termination system) equipment resellers

Readers should have knowledge of LAN and WAN technologies. Some experience with VoIP is also recommended. This book can be used by beginner, intermediate, advanced, and expert readers.

How This Book Is Organized

Useful as both a learning tool and reference, *PacketCable Implementation* introduces readers to the technologies associated with the implementation of PacketCable. Real-world case studies, tips, sample configurations, and sample network designs are included in this book. Tables and charts in every chapter serve as quick and easy references to key points, and each chapter closes with a summary section and chapter review questions, which assess the reader's understanding of the subject matter.

Although this book could be read cover-to-cover, it is designed to be flexible and allow you to easily move between chapters and sections of chapters to cover just the material that you need more work with. Chapter 1 provides an overview of the motivation behind PacketCable and contains high-level introductions to the various PacketCable specifications. Chapter 2 introduces the various components present in a PacketCable architecture and the corresponding Cisco products. Chapters 3 through 16 are the core chapters and can be covered in any order. If you do intend to read them all, the order in the book is an excellent sequence to use.

The core chapters, Chapters 3 through 16, cover the following topics:

Chapter 3, "Provisioning Overview"—This chapter provides information on the protocols involved in the provisioning process for PacketCable embedded multimedia termination adapters (eMTAs). It also covers in detail configuration steps and troubleshooting tips related to PacketCable provisioning on Cisco gear such as the Broadband Access Center for Cable (BAC) provisioning system.

Chapter 4, "Provisioning Flows"—This chapter covers the individual steps involved in provisioning a PacketCable eMTA in detail. It begins with a review of the steps in cable modem provisioning. Included with each step are traces and troubleshooting suggestions. Then eMTA provisioning is covered step by step. Likewise, for each step traces and logs describe the step.

Chapter 5, "The MTA Configuration File"—This chapter examines the configuration file used by PacketCable embedded MTAs in detail. The basic format of this file is covered, and examples are shown to reinforce key points. Also some tools that can be used to generate and view MTA configuration files are covered.

Chapter 6, "Signaling Interfaces and MGCP Overview"—This chapter introduces you to the Media Gateway Control Protocol (MGCP), the foundation of the Network-based Call Signaling (NCS) and Trunk Gateway Control Protocol (TGCP) used in PacketCable. Key aspects of the MGCP protocol are examined as is the use of the Session Description Protocol (SDP) for describing multimedia sessions.

Chapter 7, "NCS (Network-based Call Signaling)"—This chapter covers the NCS protocol used for call signaling between a PacketCable eMTA and call management server (CMS). How NCS varies from ordinary MGCP is examined, as well as how its messages are used to establish calls and provide feature functionality. Finally, configuring and troubleshooting NCS operation on the Cisco CMS, the BTS 10200, is covered. Because this is the first chapter where BTS configuration has been covered in any detail, a good portion of this chapter is dedicated to the BTS. This includes how to use the BTS command-line interface (CLI), some basic tips regarding BTS troubleshooting, and BTS configuration beyond setting up a PacketCable MTA, such as general system configuration, dial plan configuration, and call feature configuration.

Chapter 8, "TGCP and the PSTN Interconnect"—This chapter covers the protocols and PacketCable components used to interconnect to the legacy PSTN. This includes the Trunk Gateway Control Protocol (TGCP) used to communicate call signaling between a PacketCable Media Gateway (MG) and Media Gateway Controller (MGC). You will learn the differences between TGCP and pure MGCP preceded by a brief review of the SS7 and CAS protocols used in the traditional PSTN. Included with the SS7 discussion is an introduction to the Sigtran family of protocols used to communicate SS7 information over IP between the PacketCable Signaling Gateway (SG) and MGC components. Then call flows and traces illustrate how these protocols work and interact in PacketCable. Next, relevant configuration and troubleshooting on Cisco's MGC (the BTS 10200), Cisco's SG (the ITP), and two of Cisco's MGs (the VXSM and VISM) are covered.

Chapter 9, "Call Management Server Signaling Protcol (CMSS)"—This chapter covers the Call Management Server Signaling (CMSS) protocol used to communicate with remote VoIP endpoints. These endpoints reside in a different PacketCable network under the control of a different CMS. CMSS is modified version of the Session Initiation Protocol (SIP) and is used for communication between PacketCable call agents. The chapter begins with an overview of SIP followed by a look at the key differences between CMSS and SIP. Similar to the other call signaling chapters, messages and call flows are examined to help you understand the protocol. Finally, SIP and CMSS configuration and troubleshooting are addressed.

Chapter 10, "Audio CODECs"—This chapter examines how analog voice is transmitted across the digital PacketCable network. This involves the analysis of different types of CODECs that convert between analog voice streams and the digital IP packets used in PacketCable. Factors that can adversely affect call quality and techniques used to quantify call quality are examined. You will also learn how facsimile, voice-band modem, and telephony for the hearing impaired transmissions are handled in PacketCable.

Chapter 11, "RTP and RTCP"—This chapter covers the Real-time Transport Protocol (RTP) and Real-time Transport Control Protocol (RTCP) protocols used to transport and control the media streams discussed in Chapter 10. The RTP protocol is used to transport encoded speech, and the RTCP protocol is used to provide feedback on the quality of these RTP sessions. In this chapter you will see how functionality introduced in the previous chapter, such as dual-tone multiple frequency (DTMF) digit transport and facsimile transmission, apply to real call flows. Finally, pertinent configuration in Cisco PacketCable products is covered. This includes the BTS 10200 and Cisco media gateways.

Chapter 12, "DQoS Architecture and Framework"—A cable operator's ability to offer quality of service treatment for multimedia applications gives them a significant edge over other service providers that have no control over the MSO's transport network. This chapter covers dynamic quality of service (DQoS) in detail including how this functionality is provided and managed as well as the protocols that make it work. The DOCSIS MAC-layer protocol, the NCS protocol, and the Common Open Policy Service (COPS) protocol all contribute to PacketCable DQoS functionality.

Chapter 13, "Analyzing, Implementing, and Troubleshooting DQoS"—This chapter builds upon the information in the previous chapter by illustrating how these concepts are applied in the context of actual calls. How DQoS functionality is incorporated into NCS is given a closer look in this chapter. DQoS configuration on the Cisco CMTS and CMS is also examined. Finally, guidance on troubleshooting DQoS and troubleshooting the Cisco components are covered.

Chapter 14, "Multimedia Applications"—This chapter covers the architecture and protocols used to provide QoS for the generic type of multimedia service as defined under the PacketCable Multimedia (PCMM) specifications. It begins with an overview of how PacketCable Multimedia works. This includes an in-depth look at how PacketCable Multimedia varies from PacketCable telephony. Differences in the COPS and DOCSIS protocols and their messaging are compared to what you learned previously. Also call flows are examined to help illustrate PCMM functionality. PCMM configuration steps and troubleshooting tips on a Cisco CMTS are addressed as well.

Chapter 15, "Event Messaging and Lawful Intercept"—This chapter covers both the use of Event Messaging in a PacketCable Network as well as the lawful intercept requirement that is mandatory for service providers. The Event Messaging portion contains an overview of the transport layer protocol used for Event Messaging, the Remote Authentication Dial-In User Service (RADIUS). Then the different types of event messages are examined. The use of call flows and message traces is then covered to gain a better understanding on how these types of event messages are used. After this relevant configuration and troubleshooting tips on the Cisco BTS 10200 and Cisco CMTS platforms is addressed. After this point the chapter's focus turns to lawful intercept. Lawful intercept is the ability for a law enforcement agency such as the police department or the FBI to conduct a wiretap on a subscriber's phone calls. For a VoIP service provider, such as a PacketCable provider, special equipment and techniques are required to perform wiretapping services. This chapter provides an overview of this equipment and associated techniques. It also covers how this equipment interfaces with various PacketCable components such as the CMS, CMTS, and MG. As with other chapters the analysis of real call flows and configuration and troubleshooting relevant to lawful intercept are covered.

Chapter 16, "PacketCable Network Design Considerations"—This chapter covers design considerations not addressed in previous chapters. Examples of these design considerations include security, traffic engineering, QoS, and high availability requirements. The security portion covers the mechanisms used by PacketCable to provide data confidentiality and data integrity for the protocols used to communicate between components. The traffic engineering portion covers how to determine voice call capacity throughout a PacketCable network. Much of this focus is on the typical bottleneck in the network, the DOCSIS upstream channel. The QoS considerations address the treatment of all the different types of packets present in a PacketCable network. How these packets are treated can greatly affect the quality of service perceived by your customers. Similarly, ensuring your network is constantly available and operational is also vital in keeping your customers happy. Redundancy techniques used to ensure high availability throughout a PacketCable network address these concerns. Finally, at the end of this chapter you'll read a brief discussion about the future direction of PacketCable. This includes developments in DOCSIS as well as some of the key features that will be available with PacketCable Release 2.

Also three appendices are included at the end of the book. Appendix A lists the specifications and standards referenced throughout the book, Appendix B describes how the Ethereal packet sniffer utility can be used to troubleshoot a PacketCable network, and Appendix C contains complete call flow traces for the more common call types seen in a PacketCable network.

PART I

Introduction and Overview of PacketCable

CHAPTER 1

PacketCable Overview

This chapter introduces you to the PacketCable architecture and provides some background on the motivation driving the PacketCable initiative. The information presented here prepares you for the more technical information and case studies presented in the remainder of the book.

PacketCable Motivation

Cable providers or Multiple Systems Operators (MSOs) are expanding their competitive offerings in an effort to better serve their customers and increase revenue. High-speed data, video-on-demand, interactive gaming, videoconferencing, and telephony are some examples of these value-added services. PacketCable provides a framework MSOs can use to ensure the successful deployment of these services over a broadband cable network. Although many of the concepts and technologies presented in this book could be applied to implementing any one of these services, the primary focus is on telephony. As you will soon learn, telephony is the scope of PacketCable 1.x. PacketCable 2.0 and PacketCable Multimedia address a broader scope—the delivery of any type of voice, video, and/or data multimedia service over a broadband infrastructure.

In recent years, MSOs have made a substantial investment in upgrading their existing infrastructure into a bi-directional Hybrid Fiber Coaxial (HFC) system capable of providing reliable high-speed Internet access and digital cable television. The bi-directionality component permits the transmission and receipt of IP packets, and it is this IP network that provides high-speed data (HSD) services. Additionally, this same IP network can also be used to provide VoIP voice services at a much smaller cost than that of the Incumbent Local Exchange Carrier's (ILEC's) legacy circuit switched networks.

NOTE A voice service based upon an IP infrastructure is not the only way cable MSOs can provide a telephony service. Another way is to provide circuit switched voice over a dedicated portion of the radio frequency (RF) spectrum utilizing a traditional time-division multiplexing (TDM) switch. The drawback to this approach is that subscribers on the same HFC system cannot share this RF channel, therefore separate dedicated channels must be provided. The use of dedicated channels is very inefficient as this bandwidth pipe is wasted when subscribers are idle; a packet-based solution such as PacketCable allows for a substantially more efficient use of bandwidth. Also a circuit-switched implementation does not allow different services on the same network requiring the MSO to manage and provide separate networks for voice, video, and data.

Although still in the early stages, video over IP-based services can additionally be provided on the same Data-Over-Cable Service Interface Specification (DOCSIS) network as voice and data.

NOTE In today's networks cable television (both analog and digital) and premium video-on-demand programming are offered across the same HFC system but on different frequency carriers than the IP-based services are.

Imagine having one single network to manage that provides the backbone for all of these services. Cable MSOs becoming Competitive Local Exchange Carriers (CLEC) makes a lot of sense; in fact, many of the Cable MSOs (including the five largest) are already certified local exchange carriers. Telephony services are a natural progression for MSOs because they already have access to millions of potential subscribers. The capability of a single service provider to offer the triple play of video, data, and voice gives the Cable MSO a significant edge over Digital Subscriber Line (DSL) and Digital Satellite Broadcast (DSB) providers who typically aren't as well equipped to provide all of these services.

NOTE To compete with cable MSOs, DSL and DSB providers are often forced to have business relationships so that they, too, can give their customers all the same services possible with cable.

Financially, these different services make good business sense as well because they can be bundled together, resulting in a simple affordable option for end users.

Now that you know why supplying voice services makes sense for cable providers, the question turns to how to go about implementing it. Luckily, the MSOs have already figured this out. To help MSOs integrate emerging technologies into their business, a nonprofit research and development consortium called Cable Television Laboratories, Inc., or CableLabs, was formed.

CableLabs membership comprises the MSOs themselves. Because of this partnership between the research organization and the end service providers, CableLabs benefits the cable television industry by

- Researching and identifying new broadband technologies, helping cable operators deploy advanced services
- Authoring specifications and certifying products, enabling interoperability among different equipment vendors
- Providing an environment where equipment vendors can test their products for specification compliance and interoperability
- Disseminating information through publications such as technical newsletters and seminars

---

NOTE Remember that CableLabs itself does not create industry standards. However, many of the specifications authored by CableLabs have been adopted into standards by other organizations. For example, many of the PacketCable specifications have been adopted as standards by the Society of Telecommunications and Cable Engineers (SCTE) standards body in North America and by the International Telecommunication Union (ITU) standards body in Europe.

Once CableLabs create specifications, it is then up to the equipment vendors, such as Cisco, to create or enhance products adhering to these specifications.

You can obtain much more information about CableLabs and CableLabs projects via their website at http://www.cablelabs.com.

One of the broadband projects identified by CableLabs is PacketCable. According to the PacketCable project web page, "PacketCable is a CableLabs-led initiative aimed at developing interoperable interface specifications for delivering advanced, real-time multimedia services over a two-way cable plant." Its charter is to enable multimedia services by building upon the cable modem infrastructure using IP technology. As you will learn throughout this book, PacketCable leverages existing technologies, such as DOCSIS and existing VoIP protocols, to achieve its goal.

Understanding the PacketCable Big Picture

The PacketCable project defines specifications for the various components in the architecture. This architecture constitutes a baseline of what is needed to support broadband services over a DOCSIS/IP infrastructure. The creation of specifications enables you to have different vendors for each component, and as long as components are PacketCable certified or qualified, they can work together without issue.

The PacketCable architecture supports several end-to-end functions, such as signaling for services, quality of service (QoS)–enabled media transport, security, provisioning of the client device, billing, and other network management functions. The initial PacketCable charter was to design a network that provides telephony services with voice quality as good as the existing public switched telephone network (PSTN) service and with the same feature suite. By using an IP backbone, new features previously not possible with a circuit-switched architecture are now possible as well. A majority of this book deals with issues associated with implementing voice; however, you will also get a glimpse into the more generic case of implementing any multimedia solution. Figure 1-1 illustrates the PacketCable voice architecture.

Figure 1-1 *PacketCable Architecture*

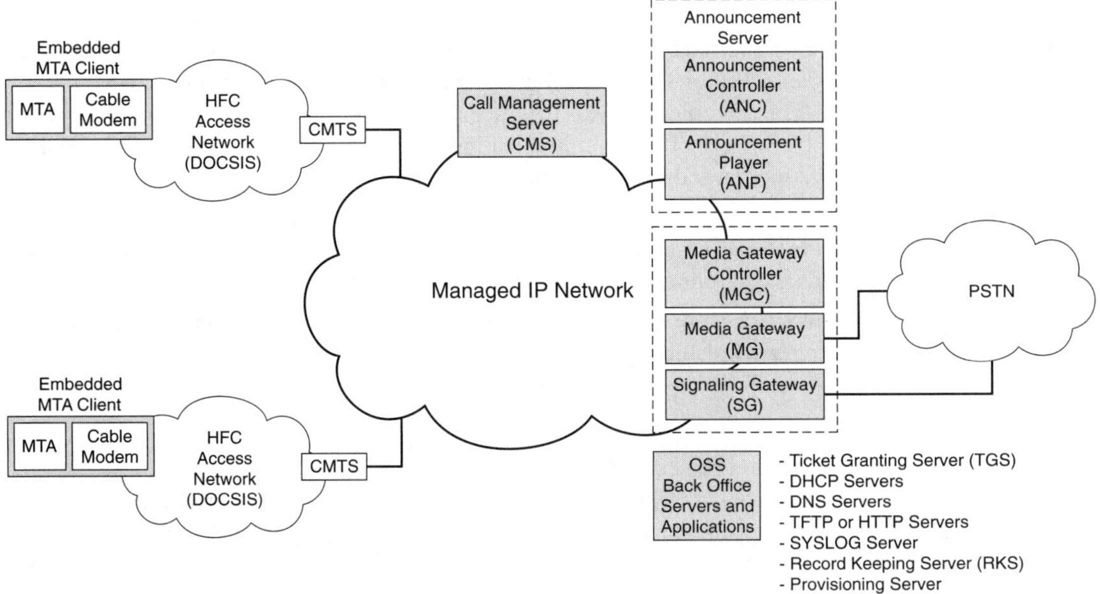

As you can see, several types of components comprise the PacketCable network, including the following:

- Provisioning components (such as Dynamic Host Configuration Protocol [DHCP], Domain Name System [DNS], and Trivial File Transfer Protocol [TFTP] servers)
- Billing components (such as Record Keeping Server [RKS])
- Signaling components (such as cable modem/multimedia terminal adapter [CM/MTA], cable modem termination system [CMTS], call management server [CMS], Media Gateway Controller [MGC], and Signaling Gateway [SG])
- Infrastructure components (such as HFC access network and Managed IP network)

Don't worry if you don't know what all these components are right now. Chapter 2, "PacketCable Functional Components," goes into detail about each of these components.

With this book you will learn how each component functions individually and how each component interacts with other components. By gaining a good understanding of how the pieces function and how to implement and troubleshoot the network to meet your business requirements, you are sure to succeed.

Cisco has been a critical player in the PacketCable initiative since its inception, making major contributions to the architecture and specifications. Cisco produces several of the architecture components; these are all covered in detail in this book. You will learn how these devices work, how to configure them, and how to troubleshoot issues that might come up. Please note, although many of the examples in the book are specific to Cisco products, the underlying concepts can be applied to non-Cisco equipment as well.

Standards

Many existing standards and specifications address various pieces of the PacketCable architecture. These standards include the following:

- Internet Engineering Task Force (IETF) Request For Comments (RFCs)
- American National Standards Institute (ANSI), Telcordia, and International Telecommunication Union (ITU) specifications
- CableLabs Data-Over-Cable Service Interface Specification (DOCSIS) specifications

PacketCable makes use of these standards whenever possible and tailors them to meet the needs of the cable industry.

IETF RFCs

The IETF has issued many RFCs, which define most of the protocols and technologies used in the Internet. The PacketCable network is built upon an IP backbone and makes use of many of these standards. Appendix A, "Standards and Specifications," provides a list of IETF RFCs referenced in the PacketCable specifications.

NOTE You can find all RFCs online at http://www.ietf.org/rfc/rfc*xxxx*.txt, where *xxxx* is the number of the RFC. If you do not know the number of the RFC, you can try searching by topic at http://www.rfc-editor.org/rfcsearch.html.

ANSI/Telcordia, ITU, and Other Specifications

In the early stages of deployment most of the voice calls made to and from PacketCable customers will be to non-PacketCable customers on the legacy PSTN. To provide transparent dialing to all telephony endpoints worldwide and because a large percentage of all telephony endpoints will continue to reside on the PSTN for the foreseeable future, some method of reaching these endpoints is required. Therefore, you will need to interact with the protocols (such as Signaling System 7 [SS7]) and switches used in the PSTN. Several ANSI, Telcordia, and ITU specifications are referenced in the PacketCable specifications for this reason. Appendix A provides a list of these specifications.

NOTE These standards bodies also author standards used in providing multimedia services over IP. Examples of these include CODEC specifications, facsimile and voice-band transmission specifications, and specifications for measuring call quality. These standards are also used in PacketCable and are included in Appendix A as well.

DOCSIS Specifications

DOCSIS defines technical specifications for equipment at both subscriber locations and the cable operator's headend or hub site. It defines the physical and data-link layer DOCSIS MAC protocols necessary to deliver IP connectivity over an HFC cable plant. Adoption of DOCSIS accelerates the deployment of data-over-cable services and ensures interoperability of equipment throughout system operators' infrastructures.

NOTE This book refers to the North American definition of DOCSIS. If you are implementing PacketCable in another part of the world, the HFC technical specifications can be different. For example, in Europe the Euro-DOCSIS specifications would be followed. The differences between U.S. DOCSIS and Euro-DOCSIS are very minor, however, so an overwhelming majority of the technical information presented in this book applies in both cases.

PacketCable networks require a DOCSIS 1.1 or later cable modem infrastructure as the basis for providing the IP connectivity and QoS needed across the HFC network. Consequently, you cannot enable PacketCable on DOCSIS 1.0 cable modems.

A solid understanding of what DOCSIS is and how it works is crucial to success in PacketCable. While it is beyond the scope of this text to cover all the details of DOCSIS, this chapter gives a high-level overview of the specifications, and the book references the DOCSIS standards throughout. Note that all the DOCSIS specifications can be found at http://www.cablemodem.com.

DOCSIS 1.0

The DOCSIS 1.0 Radio Frequency Interface (RFI) defines the specifications for communication needed to provide high-speed data service between the customer's location and the cable provider's headend. At the customer's location, a cable modem (CM) provides this service, and at the headend a cable modem termination system (CMTS) provides it. Thus, the CM and CMTS together enable the bidirectional exchange of IP packets over the HFC network. Figure 1-2 is from the DOCSIS RFI specification and illustrates this network.

Figure 1-2 *HFC Network for DOCSIS*

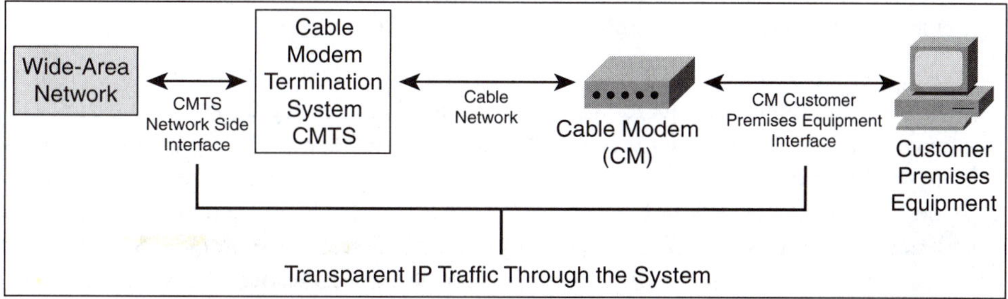

Separate parts of the RF spectrum are used for communication in each direction. In the downstream direction (CMTS to CM) a 6 MHz channel is used, and in the upstream direction a channel width of 200 KHz, 400 KHz, 800 KHz, 1.6 MHz, or 3.2 MHz can be used. In today's HFC networks, typically only the 1.6 MHz and 3.2 MHz upstream channel sizes are used.

NOTE In Euro-DOCSIS an 8 MHz channel is used in the downstream instead of a 6 MHz channel.

Signals in the downstream are modulated onto either 64-Quadrature Amplitude Modulation (QAM) or 256-QAM carriers, and signals in the upstream are modulated onto either Quadrature Phase-Shift Keying (QPSK) or 16QAM carriers.

The RF spectrum used in the upstream is in the lower end of the RF spectrum (5–42 MHz), which is typically much noisier than that of the downstream; therefore, modulation schemes with greater error tolerances are used.

These RF channels are shared bandwidth among all the cable modems wired to a particular Media Access Control (MAC) domain. A MAC domain is a collection of upstream and downstream channels available on the same physical HFC media. DOCSIS MAC domains can be configured in "sparse" mode, where a group of modems have access to only a single downstream channel and a single upstream channel, or they can be wired in "dense" mode, where they have access to multiple upstream channels and possibly multiple downstream channels. How DOCSIS MAC domains are configured is up to the MSO and can get rather complicated; you can find some more discussion relating to this topic in Chapter 16, "PacketCable Network Design Considerations."

The DOCSIS specifications define the requirements of these communication channels. The DOCSIS MAC protocol used for communication between the CM and CMTS is also defined, including the process by which a cable modem comes online with a CMTS and the process by which a cable modem transmits data. As one side of the CMTS is connected to the HFC network to reach the cable modems, the other side is connected to an IP backbone providing services such as Internet and email access.

Data sent to and from a CPE device such as a personal computer behind a cable modem reaches the Internet by DOCSIS encapsulation between the CM to the CMTS. DOCSIS encapsulation includes DOCSIS MAC layer overhead as well as physical overhead. The MAC layer overhead consists of headers containing information such as the source and destination of the packet. Upstream DOCSIS packets also have physical layer overhead to ensure they are understood correctly by the CMTS; downstream DOCSIS packets are split and encapsulated into 188-byte Motion Picture Experts Group (MPEG) frames.

NOTE The reason downstream packets are encapsulated into MPEG frames is so that a downstream channel could also be used for other services such as digital video; in this case the DOCSIS packets could be interleaved with the video packets. An important concept to remember about DOCSIS is that the downstream is broadcast to potentially hundreds of cable modems and that the upstream is contention-based between these same cable modems.

DOCSIS MAC frames are tracked using a Service Identifier (SID). The SID identifies the bi-directional traffic profile allocated to a particular cable modem. A class of service contains parameters quantifying the level of service given to that particular modem. Examples of these parameters include maximum downstream and upstream data rates, maximum burst sizes, and priority. The priority field can be used to give certain modems preferential treatment over others. The class of service is defined in the modem's configuration file; therefore, it is static and active for the duration the modem is online. A modem can have multiple classes of service; however, the mechanism that determines which packets use which class of service is not well defined and is up to the implementation of the specific vendor. As you will soon see, later versions of DOCSIS overcome this shortcoming using a well-defined means for mapping traffic to different service classes.

When a modem needs to send customer premises equipment (CPE) data in the upstream direction, it issues a request to the CMTS. The CMTS is periodically sending DOCSIS MAC messages called Bandwidth Allocation MAPs to the modems, defining when each modem can transmit data in the upstream direction; the SID is used for this identification. All data sent to and from the cable modems are sent in Best Effort fashion. That means each modem must compete against all others in the same MAC domain for bandwidth; modems are granted transmit opportunities as they become available by the CMTS. This transmission scheme works well for ordinary data; however, it does not work well for voice traffic where large delays and delay variations (also known as *jitter*) are extremely undesirable. Figure 1-3 is a graphical depiction of how a DOCSIS 1.0 cable modem transmits a packet.

Figure 1-3 *Sending an Upstream Packet in DOCSIS 1.0*

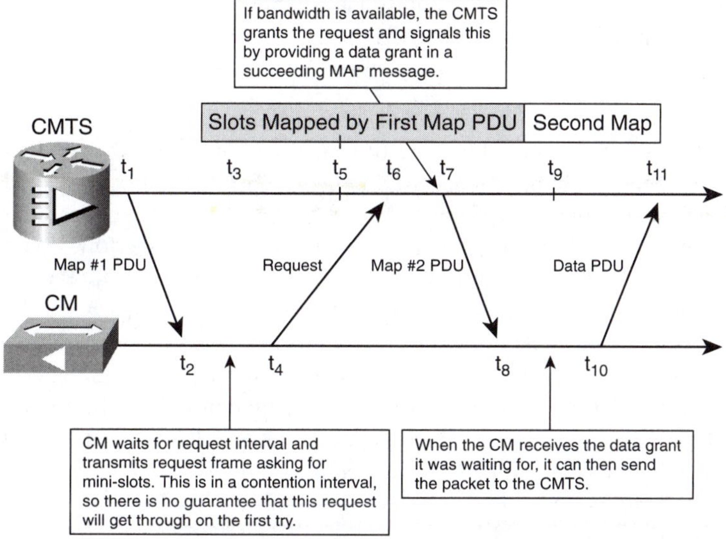

As you can see, first the CM uses a contention request interval to ask the CMTS for enough bandwidth to send the data it is waiting to transmit. If this request gets through to the CMTS and the CMTS has available bandwidth, it allocates an interval of bandwidth for this CM to send the data. Finally, the CM sees it has been granted bandwidth and transmits its data during this interval.

Other aspects of DOCSIS 1.0 worth a brief mention include the capability to concatenate multiple CPE packets together in a single upstream burst, the ability to piggyback bandwidth requests in data packets, and the encryption of DOCSIS packets using Baseline Privacy (BPI). Concatenation and piggybacking improve modem performance by reducing the amount of contention request packets that need to be sent. This reduces both packet transmission time and collision probability. BPI provides some measure of security on the shared HFC media by encrypting DOCSIS packets.

DOCSIS 1.1

DOCSIS 1.1 builds upon DOCSIS 1.0 by incorporating quality of service (QoS) functionality into the DOCSIS MAC protocol.

NOTE Integrating QoS into DOCSIS is not the only difference between DOCSIS 1.1 and DOCSIS 1.0. Other enhancements include improved security via BPI+, fragmentation, header suppression, and the support of additional Simple Network Management Protocol (SNMP) Management Information Bases (MIB).

DOCSIS 1.1 is better suited for real-time applications such as voice. In DOCSIS 1.0 traffic was governed by a bidirectional SID; in DOCSIS 1.1 traffic is governed by unidirectional service flows referenced by Service Flow IDs (SFID). So at a minimum each modem would have one upstream service flow and one downstream service flow. You can also have more than two service flows, so different types of traffic use different service flows and are subject to different traffic profiles. Packets are classified to a service flow before transmission using packet classifiers. Just as class of services are defined during the provisioning process for DOCSIS 1.0 modems, service flows and packet classifiers can be defined through provisioning for DOCSIS 1.1 modems. Also these service flows and classifiers can be dynamically created, modified, and deleted as needed to transmit data. So a service flow to carry voice would be created only when a call was in progress. The creation, modification, and deletion of dynamic service flows are done through additional DOCSIS MAC layer messages.

The manner in which traffic is serviced is different, too. Recall in DOCSIS 1.0 all data was transmitted in Best Effort fashion. DOCSIS 1.1 adds several scheduling services better suited for real-time traffic. In addition to the traditional Best Effort service, you also have

- Real-Time Polling Service (RTPS)
- Non Real-Time Polling Service (nRTPS)
- Unsolicited Grant Service (UGS)
- Unsolicited Grant Service with Activity Detection (UGS-AD)

The polling services provide modems with dedicated opportunities to request data not subject to collisions with other modem transmissions. This would work well for applications requiring regular transmission opportunities for time-sensitive data of variable packet sizes (such as real-time video). The last two scheduling services are of particular interest in PacketCable as they provide periodic uncontested opportunities to send fixed size data packets, which are characteristic of VoIP traffic. You can now see why DOCSIS 1.1 is a requirement for PacketCable. Chapter 12, "DQoS Architecture and Framework," and Chapter 13, "Analyzing, Implementing, and Troubleshooting DQoS," provide greater coverage of how DOCSIS 1.1 fits into PacketCable.

DOCSIS 2.0

DOCSIS 2.0 further builds upon earlier DOCSIS versions by expanding existing modulation schemes and introducing new ones used to allow for much greater bandwidth capacity in the same physical RF channels. Also upstream channel widths of up to 6.4 MHz are now possible for some of these modulation schemes, and network management objects to support these physical layer (PHY) changes have been added.

For applications such as Web browsing most of the data transfer is in the downstream direction, with the upstream traffic consisting of small acknowledgement packets. Applications such as video and voice require more symmetric types of traffic; consequently, larger upstream bandwidth is necessary. In DOCSIS 1.x, the only possible modulation scheme for use in the upstream direction is Time Division Multiple Access (TDMA) with either QPSK or 16-QAM modulation. With TDMA, bandwidth on a given frequency carrier is divided into timeslots, which is then assigned to modems by the CMTS. A 3.2-MHz channel using 16-QAM modulation provides around 10 Mbps of raw bandwidth, and once all the DOCSIS overhead is removed this leaves ~8 Mbps available for customer traffic.

NOTE To calculate the raw bandwidth, multiply the symbol rate by the number of bits in a symbol. For example, a 3.2-MHz channel provides a data rate of 2,560 kilosymbols per second, and 16-QAM modulation encoding uses 4 bits to represent each signal, equating to a raw bandwidth of roughly 10 Mbps.

As you might imagine, an application such as video would very quickly use up this bandwidth. DOCSIS 2.0 enhances upstream performance with the Advanced TDMA (A-TDMA) modulation scheme. In addition to supporting QPSK and 16-QAM, 8-QAM, 32-QAM, and 64-QAM modulation encodings are now possible. 64-QAM modulation uses 6 bits to represent each symbol, so in a 3.2-MHz channel you would have a raw bandwidth capacity of approximately 15 Mbps. If you used a 6.4-MHz upstream channel, the bandwidth capacity would double to approximately 30 Mbps.

DOCSIS 2.0 also introduces S-CDMA (Synchronous Code Division Multiple Access), allowing multiple CMs to transmit on the same upstream at the same time. Both A-TDMA and S-CDMA allow for much more data to be transmitted upstream in the same sized amount of RF spectrum.

DOCSIS 3.0

At the time of the writing of this book, DOCSIS 3.0 was still in the development stages, so please refer to the specifications for the most current information. DOCSIS 3.0 decouples the physical layer of DOCSIS from the MAC layer. This allows for channel bonding where multiple physical upstream and downstream channels can be combined to form a larger logical channel. This technology, also known as Modular CMTS (M-CMTS), results in very large bandwidth pipes in both the upstream and downstream channels, allowing DOCSIS to compete with other high bandwidth technologies such as fiber. DOCSIS 3.0 also supports the use of IP version 6 and the migration from IPv4.

PacketCable Specifications

PacketCable specifications are organized by release. The initial release of PacketCable defined only the baseline specifications needed for deployment. Subsequent versions of PacketCable addressed additional areas of the architecture building upon the previous version. For vendor equipment to become PacketCable certified (for multimedia terminal adapters [MTA]) or qualified (for everything else), it must conform to a subset of the specifications. This subset is defined by CableLabs, which is why equipment can become PacketCable certified or qualified without conforming to every PacketCable specification defined in that release.

PacketCable 1.0

PacketCable 1.0 is the first release of specifications defined by CableLabs. These specifications define the building blocks of what is required to provide a voice over IP (VoIP) telephony service across a broadband network in a single-zone architecture. A single-zone architecture is one in which all the PacketCable subscribers are under the control of the same call agent, and any call that needs to reach a subscriber not controlled by this call agent must be routed to the PSTN.

The topics covered in these specifications comprise much of the rest of the book. You can find the most current versions of the specifications at http://www.packetcable.com/specifications. Appendix A also provides a table that lists these specifications.

The specifications can be categorized into seven functional areas:

- **Device Provisioning (PKT-SP-PROV)**—Defines the manner in which an MTA is provisioned for it to be identified and authorized to place and receive calls. An MTA is the device providing telephony service the subscriber's phone is physically connected to. In most cases this is implemented in the cable modem.
- **Call Signaling**—Call signaling is needed to support three types of calls:
  - Calls originating from the PSTN and terminating to a cable user (OFFNET → ONNET)
  - Calls originating from a cable user and terminating to a cable user (ONNET → ONNET)
  - Calls originating from a cable user and terminating to the PSTN (ONNET → OFFNET)

  Call features such as 3-way calling, call waiting, call forwarding, and voice mail need to be supported as well.

  In PacketCable 1.0 you have three Call Signaling specifications:

  - **PKT-SP-EC-MGCP**—Defines the Network-based Call Signaling (NCS) protocol used by the MTA. The NCS protocol is based on the Media Gateway Control Protocol (MGCP), which is why MGCP is in the specification name.
  - **PKT-SP-TGCP**—Defines the Trunk Gateway Control Protocol (TGCP) protocol used to interface to the PSTN voice user.
  - **PKT-SP-ISTP**—Defines the Internet Signaling Transport Protocol (ISTP) also used to interface to the PSTN. Note, in many PacketCable architectures this protocol is not needed or used because the IETF Sigtran protocol has proved superior and sufficient for interfacing to the SS7 network. Hence, conforming to this specification is not a requirement for PacketCable qualification.
- **Media Stream Transport and Encoding (PKT-SP-CODEC)**—Requirements for the media stream or voice bearer traffic include
  - Minimizing the negative effects of latency, packet loss, and jitter and maximizing voice quality
  - Defining the CODECs responsible for converting analog voice signals into digital IP packets

- — Defining the requirements for echo cancellation, voice activity detection, dual-tone multiple frequency (DTMF) detection, fax detection, and modem detection
- — Providing mechanisms for CODEC switching during calls

- **Quality of service (PKT-SP-DQOS)**—Quality of service is required to ensure voice calls are of sufficient quality and the specification does the following:
  - — Provides a rich set of policy mechanisms to provide and manage the QoS services over the HFC network.
  - — Provides admission control for both upstream and downstream directions.
  - — Allows dynamic creation, modification, and deletion of QoS services.
  - — Mitigates potential abusive attacks such as theft and denial of service.
  - — Provides a priority mechanism for emergency 911 calls. All voice traffic always takes precedence over Best Effort traffic such as high-speed data traffic.
- **Event Messaging (PKT-SP-EM)**—Event Messaging defines the protocol and mechanism used to generate billing records for calls made in the PacketCable network. This requirement is now optional in PacketCable 1.0 implementations.
- **Security and Privacy (PKT-SP-SEC)**—Requirements perform the following functions:
  - — Enable residential voice capabilities with a security level at least as high as is currently provided by the PSTN.
  - — Protect the MSO from attacks such as denial of service and theft of service.
  - — Provide access control and authentication of voice subscribers.
  - — Security requirements cover several of the other functional areas such as provisioning, call signaling, and media transport.
- **Operational Support Systems**—The following four specifications describe how PacketCable devices can be managed by a Network Management System (NMS)
  - — PKT-SP-MIBS
  - — PKT-SP-MIB-MTA
  - — PKT-SP-MIB-SIG
  - — CL-SP-MIB-CLABDEF

In addition to these seven functional areas, a number of technical reports are also written that define how the components and protocols covered in the specifications relate to each other. You should exercise caution when referring to these technical reports, however, because some of this information is a bit outdated.

- **PKT-TR-ARCH**—Describes the overall system architecture and how everything fits together

- **PKT-TR-MGCP-PKG**—Further defines signaling packages relating to the NCS protocol
- **PKT-TR-CF-ON-ON**—Defines possible call flows for ONNET → ONNET call types
- **PKT-TR-CF-ON-PSTN**—Defines possible call flows for ONNET → OFFNET call types
- **PKT-TR-CF-PSTN-ON**—Defines possible call flows for OFFNET → ONNET call types
- **PKT-TR-OSS**—Defines a high-level framework for an end-to-end OSS/BSS (Operations Support System/Backoffice Support System) management system for a PacketCable network

PacketCable 1.5

PacketCable 1.5 builds upon the PacketCable 1.0 architecture by defining additional capabilities, one of which is the support for a multizone architecture. In this architecture VoIP subscribers under the control of different call agents can communicate without the need of the PSTN; instead, they communicate across the IP backbone.

For example, suppose a PacketCable user in New York wants to call a PacketCable user in Los Angeles. In PacketCable 1.0 this would be accomplished with an ONNET → OFFNET call in New York, traversing the PSTN across the country and an OFFNET → ONNET call in Los Angeles. With PacketCable 1.5 this call can be routed across an IP backbone, eliminating the PSTN completely, greatly reducing costs. Note that this type of scenario would require some sort of Service Level Agreement (SLA) between the two PacketCable providers.

The specification used to provide a multizone architecture is *CMS to CMS Signaling (PKT-SP-CMSS)*, which defines the Session Initiation Protocol (SIP)–based call signaling protocol used for communication between PacketCable zones.

Other capabilities defined in PacketCable 1.5 relate to providing primary line voice service, the requirements for Communications Assistance for Law Enforcement Act (CALEA) compliance, and the mechanism for PacketCable elements to report events:

- **Primary Line Support (PKT-SP-AIP)**—Defines the analog interface and powering (AIP) requirements for the MTA needed to qualify as primary line service. For a service provider to truly provide "primary" line service, more stringent requirements on reliability and emergency access exist than on a "secondary" line (sometimes referred to as a "teen" line).
- **Battery Backup MIBs (CL-SP-MIB-BB)**—Related to the previous specification, this specification defines the SNMP MIBs used to determine the availability and status of the MTA backup battery.

- **Lawful Intercept (PKT-SP-ESP)**—Defines the necessary requirements for a PacketCable network to comply with CALEA obligations. This allows law enforcement officials to collect data regarding a tapped subscriber's voice activity, including both call detail information and the call content.
- **Event Management Mechanism (PKT-SP-MEM)**—Defines how PacketCable elements can report asynchronous events for malfunction situations and other import events. The MIBs used to provide this functionality are described in the specification PKT-SP-EVEMIB.

Many of the specifications that exist in PacketCable 1.0 have also been enhanced in PacketCable 1.5. For example, the CODEC specification has been enhanced to support additional voice CODECs, T.38 fax relay, and DTMF digit relay.

Finally, a couple of PacketCable specifications address topics that have not been given much focus or attention in the industry. An example of such a specification is the Audio Server Protocol, which is a call signaling protocol designed for use with announcement server endpoints. In practice existing protocols such as MGCP or SIP are adequate for providing this functionality; thus, the ASP protocol is not needed.

The specifications in PacketCable 1.5 that have not been given much attention so far include the following:

- **Audio Server Protocol (PKT-SP-ASP)**—Defines the architecture and protocol for providing announcements in the PacketCable network
- **Call Management Server provisioning (PKT-SP-CMSPROV**—Defines the provisioning of the voice subscriber in the call agent or call management server (CMS)

As with PacketCable 1.0, PacketCable 1.5 includes a technical report (PKT-TR-ARCH1.5), which provides an architecture framework on how to implement a PacketCable 1.5 network and interdomain calls.

NOTE Previously, many of the PacketCable 1.5 specifications existed in the PacketCable 1.1, 1.2, and 1.3 releases. Those releases have now all been retired and rolled into PacketCable 1.5.

PacketCable Multimedia

Voice services are just the starting ground for PacketCable. PacketCable specifications also provide you with a framework for the generic multimedia case. PacketCable Multimedia (PCMM) expands the architecture into one that can support a wide variety of applications such as interactive gaming and videoconferencing. This does not mean voice will not be an application of PCMM. Quite the contrary—voice will probably be one of the key PCMM applications because now an embedded MTA is no longer required to provide voice services.

Figure 1-4 illustrates the PC Multimedia architecture.

Figure 1-4 *PacketCable Multimedia Architecture*

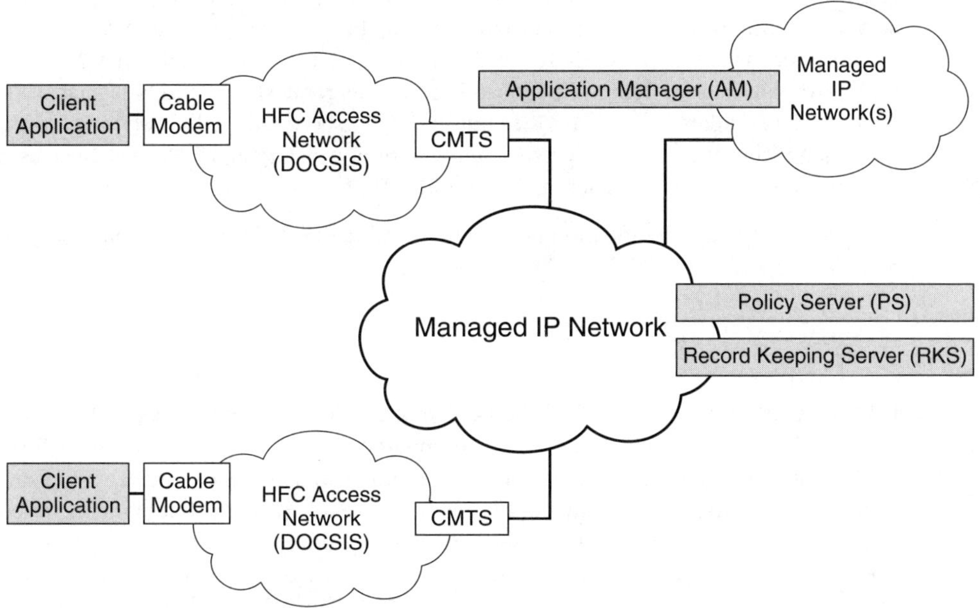

The PCMM architecture defines a new component known as the Policy Server (PS) that controls the creation and management of QoS across the HFC access network. Another new component is the Application Manager (AM), which is used as an interface between the PCMM application and the policy server.

The PCMM client is the device or application your customers are using for their multimedia services. This could be a telephone, a camera, a gaming console, or simply software running on a PC. The PCMM specification defines three types of clients:

- **Type I**—A client that is not QoS aware and relies on an Application Manager to set up quality of service
- **Type II**—A client that is QoS aware and interacts with an Application Manager
- **Type III**—A client that is QoS aware and is capable of setting up its QoS without the assistance of an Application Manager

So far, the PCMM specification covers only Type I clients—devices that are not QoS (that is, PCMM) aware—as these are what almost all of the existing clients are.

To help introduce PCMM to you, consider the example of a phone service provided by an IP phone residing on an Ethernet segment behind an ordinary DOCSIS 1.1 cable modem. You want phone calls from this device to have some measure of QoS across the cable network, but how? Well, first some call agent or Application Manager is signaled that a phone call is in progress. This Application Manager then signals the policy server controlling that particular client to authorize and set up the required QoS. After the policy server determines this Application Manager and client are permitted to have QoS enabled, it determines which CMTS this subscriber is connected to and signals the CMTS with the QoS information. Finally, the CMTS authorizes this request, and if this is successful and bandwidth is available, it initiates the creation of QoS across the HFC network. Don't worry too much if this doesn't make perfect sense right now. Chapter 15, "Event Messaging and Lawful Intercept," covers much more about PCMM.

Currently, only one specification deals with multimedia (PKT-SP-MM), and only one technical report (PKT-TR-MM-ARCH) exists.

PacketCable 2.0

Like DOCSIS 3.0, PacketCable 2.0 was also in the development stages at the time of the writing of this book. So please refer to the specifications for the most current information. PacketCable 2.0 provides some guidelines on how applications can utilize a PCMM framework for QoS. Some of the topics covered by PacketCable 2.0 include the integration of a cellular network with an HFC network for providing low-cost telephony, cross-platform integration (such as using a set-top box to provide caller ID on the television), the use of SIP-based endpoints, and the support for IPv6 in a PacketCable network.

Chapter Summary

Cable providers can increase their customer base and revenue by offering multimedia services in addition to basic data and television services. This fact is the driving motivation behind the inception of PacketCable. Through the implementation of PacketCable, MSOs can be true "triple-play" providers of voice, video, and data. All of these services can also be provided across a common infrastructure, greatly reducing both equipment and management costs.

CableLabs is a research and development organization benefiting the cable industry by accelerating the availability of new services and products. At CableLabs, new technologies are explored and designed leading to the creation of industry specifications. This in turn allows equipment vendors to design products to a common set of standards. The technologies relating to the deployment of multimedia services over a cable network are covered by CableLabs' PacketCable initiative. The first PacketCable release deals with the requirements needed for a cable MSO to offer residential telephony service to its customers. Specifically, the specifications detail the requirements for call signaling, QoS–enabled media transport, security, provisioning, billing, and network management. Cisco is heavily involved in PacketCable deployments and makes products related to many of the PacketCable architecture components.

One of the goals of PacketCable is to make use of existing standards and protocols whenever possible. Standards, specifications, and protocols developed by the IETF, ANSI, ITU, and CableLabs are all used in PacketCable.

The HFC portion of a PacketCable network requires, at minimum, a DOCSIS 1.1 capability. DOCSIS 1.0 uses a bi-directional class of service to categorize and provide QoS to a cable modem's traffic. These traffic flows are static as they are defined at the time a modem registers. Also the manner in which traffic is assigned to these flows is not very well defined in DOCSIS 1.0. These are some of the reasons why DOCSIS 1.1 is a requirement for PacketCable. DOCSIS 1.1 uses unidirectional service flows instead to categorize a cable modem's traffic. Classifiers are used in DOCSIS 1.1 to assign user traffic to specific service flows. These flows and classifiers can be statically provisioned in a modem's configuration file, or they can be dynamically created when needed. Thus, voice flows are created only when calls are active, and this bandwidth can be used for other traffic types when calls are not active. Also the way in which DOCSIS 1.1 flows are scheduled is enhanced compared to DOCSIS 1.0. In DOCSIS 1.0 all traffic was sent in a Best Effort fashion where upstream traffic competed against other upstream traffic. DOCSIS 1.1 has different scheduling types available that are better suited to real-time multimedia traffic such as voice. DOCSIS 2.0 allows for increased bandwidth capacity of upstream channels through enhanced and new modulation schemes.

The specifications included in the initial release of PacketCable, PacketCable 1.0, address the basic building blocks needed to offer a residential telephony service. These include device provisioning, call signaling, QoS for media packets, billing, security, and network management. The next release of PacketCable, PacketCable 1.5, extends the capabilities of PacketCable 1.0. Some of these extensions include the protocols needed for a multizone architecture, lawful intercept, and primary line considerations such as battery backup. PacketCable Multimedia (PCMM) is the next release of PacketCable, where the architecture is generalized to support a wide range of multimedia applications beyond residential telephony.

Chapter Review

1 List at least five examples of multimedia services possible with a PacketCable deployment.

Answer: Examples of services include residential telephony, enterprise telephony, interactive gaming, videoconferencing, and media streaming.

2 True or False: An HFC network can be used to provide circuit switched voice much like the legacy PSTN network.

Answer: True. However, this solution does not scale very well and can be costly compared to a packet switched voice solution.

3 True or False: CableLabs authors documents that are industry standards.

Answer: False. CableLabs creates specifications that are often adopted by other organizations into standards, but the CableLabs documents themselves are not industry standards.

4 A PacketCable 1.0 network can be categorized into provisioning components, billing components, signaling components, and infrastructure components. List an example of each.

Answer: Provisioning—DHCP,DNS, TFTP servers; Billing—RKS server; Signaling—CM/MTA, CMTS, CMS, MGC, SG; Infrastructure—HFC access network, Managed IP network

5 Which minimum DOCSIS version is needed for Implementing PacketCable?

Answer: DOCSIS 1.1

6 What is the main difference between DOCSIS 1.0 and DOCSIS 1.1?

Answer: DOCSIS 1.1 builds upon DOCSIS 1.0 by incorporating quality of service (QoS) functionality into the DOCSIS MAC protocol. In DOCSIS 1.0 traffic was governed by a bi-directional SID, in DOCSIS 1.1 traffic is governed by unidirectional service flows referenced by Service Flow ID (SFID).

7 Describe the process used by a DOCSIS 1.0 cable modem to send a packet upstream to the CMTS.

Answer: (1) The modem waits for a contention request opportunity. (2) It then issues a request for the bandwidth needed to send the packet. (3) The CMTS then authorizes the request and if it can be granted issues a data grant to the modem. (4) Upon receiving the data grant, the modem sends the packet in this interval.

8 What are the five types of scheduling services available in DOCSIS 1.1?

Answer: Best Effort (BE), Unsolicited Grant Service (UGS), Unsolicited Grant Service with Activity Detection (UGS-AD), Real-Time Polling Service (RTPS), Non Real-Time Polling Service (nRTPS).

9 What is the difference between DOCSIS 1.1 and DOCSIS 2.0?

Answer: DOCSIS 2.0 further builds upon earlier DOCSIS versions by expanding existing and introducing new modulation schemes used in the upstream channels to allow for much greater bandwidth efficiency. It also allows for larger U.S. bandwidth channels (up to 6.4 MHz) and adds additional SNMP MIBs to support these changes.

10 What are the seven functional areas that PacketCable 1.0 specs can be categorized into?

Answer: Device Provisioning, Call Signaling, Media Stream and Encoding, quality of service (QoS), Event Messaging, Security and Privacy, and Operational Support Systems (OSS).

11 In PacketCable 1.5, a multizone architecture is now supported. How is this done, and what is the benefit?

Answer: Calls between zones are now routed across an IP backbone using the SIP-based protocol CMSS. By doing this, the PSTN network is avoided entirely, greatly reducing long distance fees.

12 What new component introduced with PacketCable Multimedia is used to control and manage the creation of dynamic service flows on a CMTS?

Answer: A Policy Server

CHAPTER 2

PacketCable Functional Components

Now that you have a general idea of the motivation behind the PacketCable initiative and the architecture of a PacketCable network, it is time to delve deeper into the individual components comprising the architecture. This chapter introduces each component separately; the protocols and procedures required to communicate between components are covered in detail in further chapters.

Also in this chapter you learn how Cisco Systems is involved with each component type by learning what products are available and how to use them.

Figure 2-1 illustrates the overall PacketCable 1.0 architecture. Some of the components shown in this diagram aren't a requirement for PacketCable 1.0 but are included because they are referenced throughout this chapter. For example, the media servers are not explicitly a part of PacketCable 1.0 but are typically included in a PacketCable deployment. Also standalone multimedia terminal adapters (MTA) are not part of PacketCable 1.0, but they are included to illustrate their difference from embedded MTAs.

Figure 2-1 is referenced often in this chapter as the various sections are examined. In Chapter 1, "PacketCable Overview," you gained a high-level understanding of the PacketCable specifications and some background on DOCSIS and the importance of a DOCSIS 1.1/2.0 HFC plant. The sections that follow examine several of the hardware functional components in greater detail.

Figure 2-1 *PacketCable 1.0-Based Architecture*

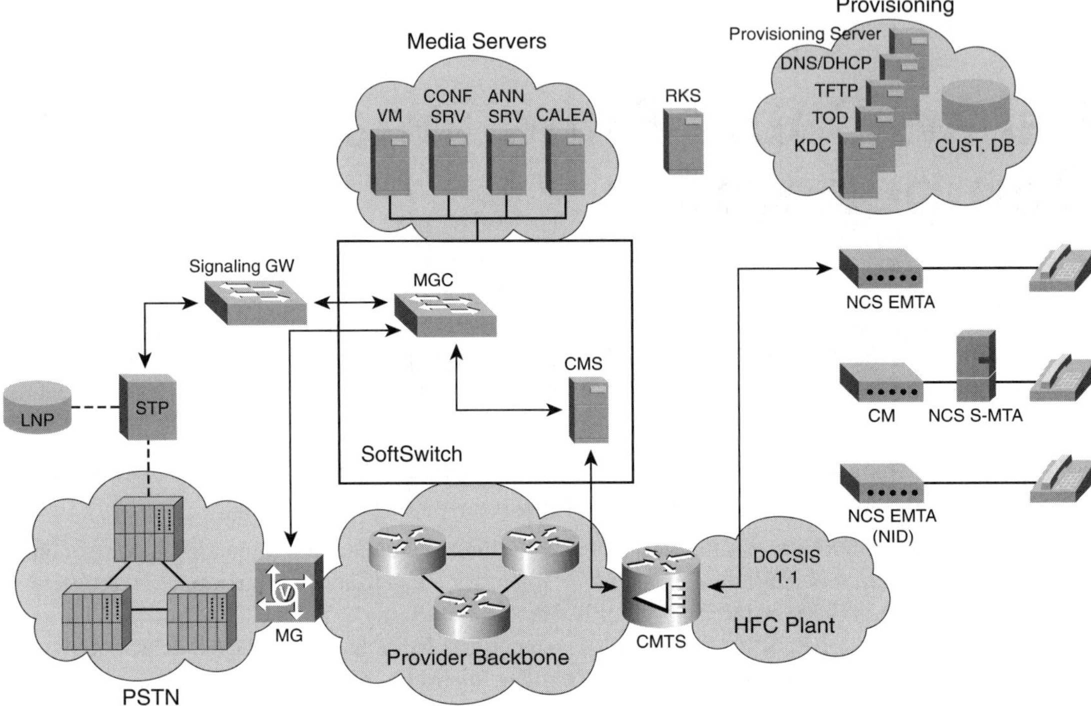

NOTE Another factor very crucial to a successful PacketCable deployment is the quality of the Hybrid Fiber Coaxial (HFC) plant. At a minimum the plant should meet the carrier-to-noise ratio and packet loss restrictions defined by Data-Over-Cable Service Interface Specification (DOCSIS). Just because a high-speed data Internet service can function fine on an HFC plant does not necessarily mean that a voice service will. Real-time applications such as voice are much more sensitive to packet loss caused by a noisy HFC plant than an Internet service would be. This topic is discussed more in the Part IV, "Media Stream," portion of this book.

Multimedia Terminal Adapter (MTA)

This section begins with the device your customers directly interface with to obtain telephony services, the multimedia terminal adapter (MTA). The MTA provides the telephony interface to the PacketCable network; thus, the MTA is where the subscriber's analog phone is connected either directly or via in-house wiring. These phones are the same analog phones used with the traditional PSTN service. An MTA has Foreign Exchange Station (FXS) interface ports for phone connectivity using standard RJ-11 cables. From a high level, the MTA simply takes the analog voice stream from the telephony interface, digitizes it, and then sends these samples out the network side interface. The manner in which this sampling occurs depends on the CODEC used by the MTA, covered in greater detail in Part IV of this book.

Two basic types of MTAs exist, the embedded MTA (eMTA) and the standalone MTA (sMTA). In the eMTA case, the MTA with FXS analog ports and the DOCSIS CM with the coaxial cable interface are integrated in the same physical device.

The embedded MTA is the only type of MTA required for a PacketCable 1.0 network and, therefore, has a greater focus in this book. A standalone MTA is exactly as it sounds—a standalone device, physically separate from the cable modem. The network side interface is typically Ethernet. Voice packets are digitized by the MTA and sent out as voice over IP (VoIP) packets out the Ethernet interface across a LAN until they reach the cable modem.

An MTA has the following responsibilities:

- **Call signaling with the CMS or call agent**—An MTA communicates with the Call Management Server (CMS) using the Network-based Call Signaling (NCS) protocol. The NCS protocol is a variant of the Media Gateway Control Protocol (MGCP). Both of these protocols are the focus of Part III, "Call Signaling," of this book.

NOTE The NCS and MGCP protocols use a centralized architecture. In a centralized architecture, most of the intelligence is the responsibility of one of the devices, allowing the other device to be relatively "dumb" and simple. It is this intelligence that is responsible for providing call features and functionality. Allowing one device to be simple reduces complexity on the device, making it cheaper and less prone to bugs and lessening the need to upgrade the software for feature enhancements. In the MGCP and NCS protocol architecture, the call agent/CMS component is the intelligent device, and the gateway (or MTA in the NCS case) is the unintelligent device. Also because the CMS is under total control of the MSO, the MSO has much tighter control over the services and functionality operating in its network. The opposite of a centralized architecture is a distributed architecture where both devices are required to have some intelligence. Session Initiation Protocol (SIP) and H.323 are examples of protocols utilizing this type of architecture.

- **Quality of service (QoS) signaling with the CMS and the cable modem termination system (CMTS)**—To ensure a high level of voice quality, PacketCable components need to establish some measure of QoS across the HFC and IP networks. Because an embedded MTA is located within the cable modem, the cable modem can take care of the HFC QoS on behalf of the MTA using the DOCSIS protocol to communicate with the CMTS. The NCS protocol is used to signal QoS information to the MTA, triggering the appropriate DOCSIS messaging.

A standalone MTA either needs to request QoS on its own or with the help of another device. Hopefully, this brings to your mind a topic addressed in the previous chapter, PacketCable Multimedia (PCMM). The standalone MTA can be thought of as a specialized case of any external device requiring QoS across a cable network, which is the focus of the PacketCable Multimedia specifications. Standalone MTAs are not addressed by the PacketCable 1.x specifications but will be fully supported in PacketCable 2.0.

The chapters in Part V, "Quality of Service," cover all these QoS topics in greater detail.

- **Authentication and security requirements with other PacketCable network entities**—Because an MTA resides at a subscriber's residence, it is an untrusted network entity and, hence, is vulnerable to security risks. All other PacketCable components are typically located within the MSO's facilities, making them more secure and consequently trusted network entities. A hacker could attempt to sabotage the MTA in an attempt to steal service or disrupt legitimate service. Also all MTA traffic traverses the HFC network, which is a shared media. With the right equipment this traffic could be intercepted by a third party. For these reasons, security and authentication are great concerns and methods of encrypting packets are required.

  To keep the MTA functionality controlled and portable it must be able to be provisioned remotely. An MTA dynamically obtains its configuration information just as a cable modem does. For an embedded MTA this might seem rather simple because the MTA and the cable modem are in the same physical device. However, because the cable provider and the telephony provider could be two different companies, it is a little more complex. The cable modem and MTA provisioning processes are separated. This requires that an embedded MTA have different MAC addresses, IP addresses, and Fully Qualified Domain Names (FQDN) for the MTA and CM components.

  The chapters in Part II, "MTA Provisioning," cover in more detail the provisioning process, which is a major portion of PacketCable.

- **Encoding/decoding of media streams**—As mentioned earlier, an MTA converts the analog stream from the telephone into digital VoIP packets that are sent out the network side interface. This is done using a CODEC, covered in more detail in Part IV of this book.

- **Providing standard PSTN analog services**—A phone in a PacketCable network must have at least the same functionality as a phone in the traditional PSTN. For example, ring tone, dial tone, and call-waiting tone must be available on the MTA, and functionality such as caller ID and voice mail indication must be available.

Some of these topics, as well as dual-tone multiple frequency (DTMF) digit transport, FAX, and modem transport, are covered in greater detail in Part IV of the book. Note that PacketCable 1.5 also has requirements for an embedded MTA to qualify as a primary line device. This includes things such as battery backup in the event of a power failure.

Overview of the CM2P2B

Through the Cisco acquisition of Linksys, Cisco now produces eMTAs. Linksys operates as a division of Cisco and continues to produce a wide range of VoIP, wireless, and Ethernet networking products and solutions for both home and small business users. Two Linksys MTA models are PacketCable 1.0 certified. Both products are also DOCSIS 1.0, 1.1, and 2.0 certified.

One is called the Linksys Cable Modem with 2 Phone Ports and Battery Backup (CM2P2B), and the other is called the Linksys Wireless-G Cable Gateway (CG2P2WB). Figure 2-2 provides an illustration of the Linksys CM2P2B.

Figure 2-2 *Linksys CM2P2B*

As you might guess, the CM2P2B has two analog phone jacks and supports two batteries for backup in the event of a power failure. It also has both a 10/100-Mbps Ethernet interface and a USB interface for PC or LAN connectivity.

Overview of the CG2P2WB

Figure 2-3 shows the Linksys CG2P2WB.

Figure 2-3 *Linksys CG2P2WB*

Like the CM2P2B, the CG2P2WB also has two analog phone jacks and supports two batteries for backup in the event of a power failure. For LAN connectivity, the CG2P2WB has a USB interface, a four port 10/100-Mbps Ethernet switch, and a built-in Wireless-G access point allowing wireless connectivity at a rate of 54 Mbps for up to 32 devices. It also has features such as firewall protection and parental control features like Internet access time limits and key word blocking.

Both directly and through the Linksys division, Cisco produces several devices that could be considered standalone MTAs. Pretty much any device with telephony ports and a LAN connection (Ethernet, USB, or wireless) qualifies. However, these devices are not certified by PacketCable. PacketCable poses additional functionality such as provisioning and security requirements. As alluded to earlier, these requirements will probably be a part of PacketCable 2.0.

Cable Modem (CM)

As you learned in the previous chapter's DOCSIS discussion, the cable modem is defined in the DOCSIS specifications and provides a means for sending IP packets across the RF or HFC network. The MSO's cable feed is connected to the RF interface of the cable modem. This feed is typically split using a simple two-way splitter with the MSO feed (from the wall outlet) as the input to the splitter and one of the outputs to the cable modem and the other to the television or cable set-top box (STB) for video service.

The customer connects his network to the cable modem via an Ethernet, USB, or wireless interface. This connection can be direct or through some networking device such as a router or switch. The modem takes packets from the LAN interface (Ethernet, USB, or wireless) and transfers them out the cable interface to the MSO. These IP packets are encapsulated in DOCSIS frames as they are sent across the HFC network. If a standalone MTA is connected via the LAN interface, these packets are treated just as any other customer premises equipment (CPE) device. If the CM is also an embedded MTA, packets from the telephony devices are also encapsulated in DOCSIS frames and sent out the cable interface. Typically, the modem functions as a simple bridge between these interfaces. In some vendor implementations the cable modem can function as a router as well, although the bridging functionality is all that is required by DOCSIS.

Keep in mind that for packets from the LAN interface to be forwarded, the cable modem must issue bandwidth requests. The CMTS controls how upstream bandwidth is allocated, so the CM is dependent upon the CMTS to provide the bandwidth it requires.

The cable modem is designed as a plug-and-play device; the CM dynamically determines its configuration via DOCSIS messaging and the provisioning process. This is to allow portability and service provider control of the cable modem, as it is an untrusted network entity. During cable modem provisioning HFC QoS parameters are assigned denoting the amount of upstream and downstream bandwidth allocated to the customer. Remember from Chapter 1 that this includes primary flows as well as optional secondary static flows.

For an embedded MTA, QoS is obtained via dynamically created Unsolicited Grant Service (UGS) service flows. The cable modem is responsible for classifying, marking, and policing the MTA voice packets onto these service flows. You'll learn much more about how this works in Part IV of this book. You'll find some coverage of the CM provisioning process and dynamic service flow creation process in this book; however, for more information please see the DOCSIS specifications.

Several companies, such as Cisco/Linksys, manufacture embedded MTAs that include a DOCSIS 1.1 or DOCSIS 2.0 cable modem. A standalone MTA can exist behind any DOCSIS 1.1/2.0–capable cable modem.

Cable Modem Termination System (CMTS)

As with the CM, the CMTS is defined in the DOCSIS specifications. The CMTS is typically located in the MSO's headend or distribution hub, making it the first network entity inside the trust boundary. The CMTS provides the DOCSIS connectivity to the cable modems and IP connectivity to the service provider's backbone via some sort of WAN interface.

The CMTS has the following responsibilities:

- **Allocating upstream bandwidth**—The CMTS must ensure that data grants are available to authorized CMs needing to send traffic. The term *CM* also refers to the cable modem portion of eMTAs. For telephony calls, data grants must be made available to the CMs so that voice packets can be sent at regular intervals and within acceptable jitter tolerances. In addition to the voice packets serviced by the CMTS, data packets are also granted upstream bandwidth. The CMTS can perform techniques such as fragmentation to allow data packets to be serviced without adversely affecting voice packets.
- **Providing QoS to the CMs based upon policy configuration**—Through communication with the CMS/Gate Controller the CMTS provides QoS only to CM/MTAs that are authorized to receive that QoS. PacketCable dynamic quality of service (DQoS) enables the CMS to request QoS from the CMTS on behalf of MTAs using the Common Open Policy Service (COPS) protocol (COPS is covered in detail in Part V of this book). Once the CMTS has agreed to provide the QoS, the CMS provides a Gate ID to the MTA, and the MTA then sets up the QoS directly over the DOCSIS interface with the CMTS. This is done using Dynamic Service messaging (defined in DOCSIS). The DQoS Gate parameters are converted into DOCSIS service flow and classifier parameters to achieve the required QoS. PacketCable DQoS prevents unauthorized devices from stealing bandwidth from authorized devices paying for QoS. In addition to providing QoS for the upstream packets across the HFC network, QoS can also be provided for the packets after they leave the CMTS out the WAN interface, typically done by manipulating IP Precedence or Differentiated Services Code Point (DSCP) bits in the IP header of the packets.
- **Classifying packets from the network side interface and assigning them to the appropriate QoS level**—Packets in the upstream direction are granted QoS from the CMTS allocating bandwidth. Packets in the downstream direction (from the WAN interface) must also be assigned to the appropriate QoS level as well. This is also specified in the COPS messaging from the Gate Controller. Downstream packets can be shaped to meet service agreements, and exceeding packets can be policed and dropped. Also voice packets can be given precedence to ordinary data packets, and then queuing techniques can be implemented, assuring timely delivery of voice packets destined for the same cable modem.

- **Recording usage of resources via Event Messaging**—The CMTS communicates with the Record Keeping Server (RKS) using Event Messaging to report usage of bandwidth resources. This information can be used for billing purposes and to notify the network administrator when resources are limited.

Cisco CMTS Products

Cisco Systems implementation of a CMTS is known as a Universal Broadband Router (uBR). Cisco produces two CMTSs that are PacketCable 1.0 qualified—the uBR7246VXR and the uBR10012. Both devices are also DOCSIS 1.1 and Euro-DOCSIS 1.1 qualified and NEBS 3 compliant. The uBR7246VXR is also PacketCable 1.1, PacketCable Multimedia, and Euro-DOCSIS 2.0 (for Advanced TDMA [A-TDMA]) qualified. In fact, the uBR7246VXR was one of the first two CMTSs to be PacketCable 1.0 qualified and the first CMTS to be PacketCable 1.1 qualified. A Cisco uBR integrates the CMTS functionality with a Cisco router. Thus, a rich set of IP features is available on the uBR.

Overview of the uBR7246VXR

The uBR7246VXR in essence is a Cisco 7200 series router integrated with a CMTS. Figure 2-4 illustrates an uBR7246VXR.

Figure 2-4 *Cisco uBR7246VXR*

The bottom four slots (numbered 3–6) are available for cable line cards. Each cable line card contains some number of downstream and upstream DOCSIS (or Euro-DOCSIS)

Radio Frequency Interfaces (RFIs). The following line cards are typically used on the uBR7246VXR:

- **MC16C**—Cable line card with one downstream IF output and six upstream RF inputs.
- **MC16E**—Similar to MC16C except for Euro-DOCSIS HFC plants.
- **MC16S**—Similar to the MC16C with enhanced spectrum management capabilities. Spectrum management works by continually monitoring the quality of the upstream frequency range (5–42 MHz) so the CMTS can make intelligent decisions when dynamically choosing or changing upstream frequencies.
- **MC28C**—Cable line card with two downstream IF outputs and eight upstream RF inputs (four upstream channels are associated with each downstream channel by default).
- **MC28U**—Cable line cards with two downstream RF outputs through the use of an integrated upconverter and eight upstream RF inputs with advanced RF capabilities (like DOCSIS 2.0 A-TDMA support) embedded. Like the MC28C card, here you have two DOCSIS MAC domains; by default the first four upstream channels are associated with the first downstream channel, and the last four upstream channels are associated with the second downstream channel. If desired, you can use a feature called "Virtual Interfaces" to change this default assignment.
- **MC28X**—Cable line card with the same advanced RF capabilities that the MC28U card has; however, it does not have an integrated upconverter.

Two port adapter slots on the chassis are for WAN interfaces (slots 1 and 2) for connecting to the provider's IP backbone. Types of WAN interfaces include

- Fast Ethernet
- Gigabit Ethernet
- Packet Over SONET (POS)
- Dynamic Packet Transport (DPT)

The processor of the uBR7246VXR is contained on a card called the Network Processing Engine (NPE) and is located on the back of the chassis.

The top slot (slot 0) in the front of the chassis is reserved for an I/O Controller card that contains management console and Ethernet ports as well as PCMCIA slots to be used for storing Cisco IOS images.

NOTE The latest NPE, the NPE-G1, provides the functionality of both the NPE and I/O Controller in addition to having a more powerful processor. Consequently, the I/O Controller is no longer needed. The NPE-G1 also has three Gigabit Ethernet WAN interfaces.

Other features of the uBR7246VXR include redundant power supplies, the ability to synchronize to an external clock, and the ability to provide N+1 redundancy for its cable line cards.

NOTE N+1 redundancy refers to the ability for one line card to take over for another in the event of a card failure. This functionality is accomplished through the use of a RF Switch and is covered in Chapter 16, "PacketCable Network Design Considerations."

Overview of the uBR10012

The uBR10012, in essence, is a CMTS integrated with a Cisco 10000 series aggregation router. Eight slots are available for cable line cards. These slots are numbered from left to right as follows: 8/0, 8/1, 7/0, 7/1, 6/0, 6/1, 5/0, and then 5/1, as illustrated in Figure 2-5.

Figure 2-5 *Cisco uBR10012*

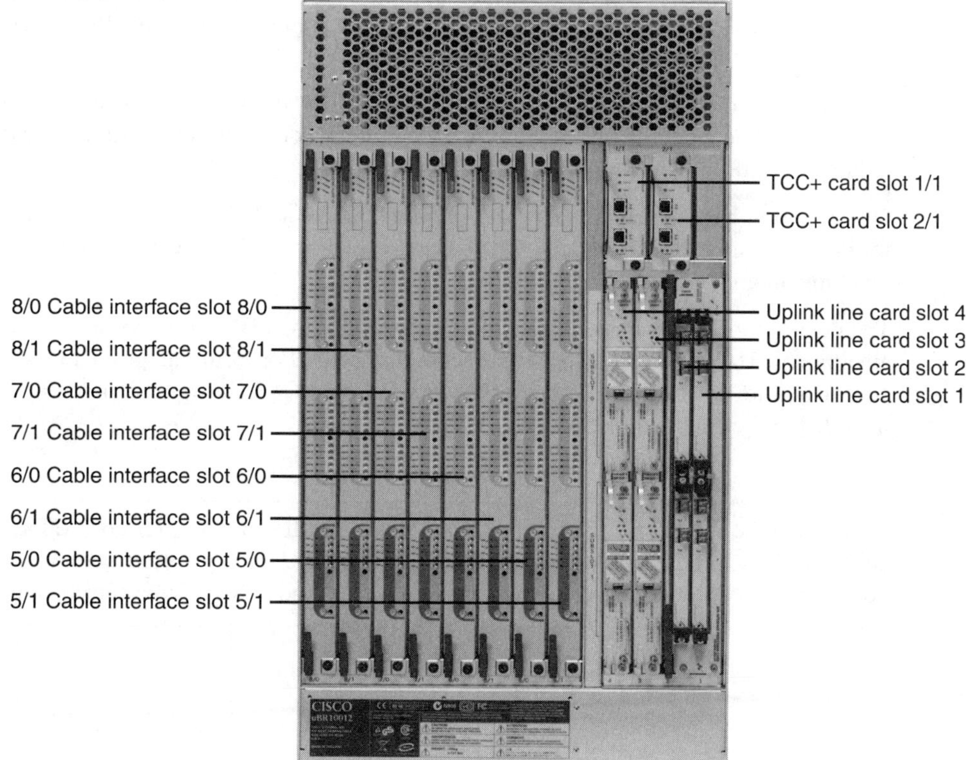

In addition to supporting all the cable line cards in the uBR7246VXR via an adapter card, the uBR10012 also supports the MC5x20S and MC5x20U cards. These cards have five downstream interfaces (thus five DOCSIS MAC domains) and 20 upstream interfaces. Like the MC28 cards, MAC domains are assigned by default with each downstream channel controlling four upstream channels, which can be changed by using the "Virtual Interfaces" feature. The difference between these two cards is that the MC5x20U card has an integrated upconverter for each downstream port and supports both DOCSIS and Euro-DOCSIS, whereas the MC5x20S card only supports DOCSIS. As you can see, the density of the uBR10012 is much larger than the uBR7246VXR.

Up to four WAN interfaces or uplink line cards are available here as well. These are located in slots 4/0, 3/0, 2/0, and 1/0. These cards can be Gigabit Ethernet, Packet over SONET, or Spatial Reuse Protocol (SRP).

A key difference from the VXR is that the uBR10012 has redundant processors. The processor cards are referred to as Performance Routing Engines (PREs). When the primary PRE module fails, the secondary PRE automatically takes over. If your network environment is one where high availability is a major concern, the uBR10012 would fit well. The PRE module also contains all the resources found on the I/O controller for the VXR. Flash disks, console ports, and Ethernet ports are available on the PRE itself, eliminating the need for an I/O controller.

The Timing, Communication, and Control Plus (TCC+) cards located in slots 1/1 and 2/1 can be used to connect to an external Stratum 3 clock source. Two cards are provided for redundancy purposes.

Other features such as redundant power supplies and the ability to provide N+1 redundancy for its cable line cards through an RF switch are available here as well. Chapter 16 covers this topic in greater detail.

The details concerning PacketCable configuration for the uBR7246VXR and uBR100012 are contained within the relevant chapters in this book. For example, provisioning commands are covered in the chapters in Part II, and DQoS configuration commands are included in the chapters in Part V.

NOTE It's beyond the scope of this book to go into detail on all aspects of Cisco CMTS configuration. Just the areas directly related with PacketCable operation are covered. Other books are available; you also can find information online at http://www.cisco.com.

Call Agent

The brain of a PacketCable network resides in the call agent. A call agent is responsible for things such as call control and signaling, maintaining call state, call features, reserving QoS, generating billing records, and initiating wiretapping. In PacketCable nomenclature call agent functionality is separated into several different devices; however, in implementation many of these functional devices are in the same physical device.

Call Management Server (CMS)

The CMS is responsible for providing call control and signaling for the MTAs using the NCS protocol. Recall from the MTA portion of the chapter that NCS is a centralized architecture and the CMS is the intelligent device whereas the MTA is the dumb device. The CMS maintains the call state for the MTAs. The CMS contains call routing information such as phone number and dial plan for each MTA. It provides call features (such as 3-way calling and call-waiting). The CMS also generates billing records using Event Messaging with the RKS.

Gate Controller (GC)

The GC is the logical portion of the CMS responsible for QoS authorization and control. Recall this is done using the COPS messaging with the CMTS. As you might expect, a GC controls gates in a PacketCable network, but what is a gate? A *gate* is a logical entity in a CMTS that controls access between the HFC network and the backbone IP network. The chapters in Part V cover the operation of the gate controller in detail.

Media Gateway Controller (MGC)

The MGC is similar to the CMS except that it provides call control and signaling for PSTN Media Gateways (MGs) instead of the MTAs. The MGC maintains call state for the MG endpoints. The MGC uses Trunk Gateway Control Protocol (TGCP) for communicating with the MGs. Just like NCS, TGCP is also a variant of the MGCP protocol and is covered in the chapters of Part III. When a PacketCable subscriber needs to communicate with a PSTN subscriber, the MGC is utilized. TGCP is used to set up bearer path connections between the MG and the PSTN.

The MGC with the help of another device, known as a Signaling Gateway (SG), also communicates call signaling to the PSTN using protocols such as Signaling System 7 (SS7).

The MGC also records usage via event messaging to the RKS.

Announcement Controller (ANC)

The Announcement Controller (ANC) is also similar to the CMS and MGC. It interfaces with the Announcement Player (ANP) to play network announcements. These announcements are used for call treatment scenarios to play announcements such as "Your call cannot be completed at this time, please try again later" to the call originator. If you remember from Chapter 1, PacketCable 1.5 contains a specification for the protocol used between the ANC and ANP; however, this specification has not been adopted by the industry. Generally communication to announcement devices is done using protocols such as SIP and MGCP.

Overview of the BTS 10200 Softswitch

The BTS 10200 is Cisco System's PacketCable call agent or softswitch. It incorporates the CMS, Gateway Controller (GC), MGC, and ANC functionalities. In older versions, it also handled the SG functionality (discussed in the next section). In later versions, this functionality is separated into another device.

The BTS 10200 is PacketCable 1.0 qualified both as a CMS/GC and as an MGC. In fact, it was one of the first two CMSs qualified and the first MGC qualified.

The BTS 10200 architecture is composed of four UNIX servers. Two of these servers perform the call agent functionality in a redundant configuration. These servers handle call control. In addition to the call agent functionality, these servers also act as feature servers. The feature servers provide functionality such as Centrex, Tandem, and Advanced Intelligent Networking (AIN) services. The other two servers handle the operations, maintenance, and configuration of MTAs, MGs, and CMTSs. These servers are referred to as an Element Management System (EMS) and also operate in a redundant fashion. Figure 2-6 illustrates one sample hardware configuration.

BTS 10200 provisioning can be executed directly on the EMS platform in a CLI session, using a Cisco application known as Extensible Provisioning and Operations Manager (EPOM) or by third-party applications that interface with the EMS API.

Figure 2-6 *Cisco System's BTS 10200*

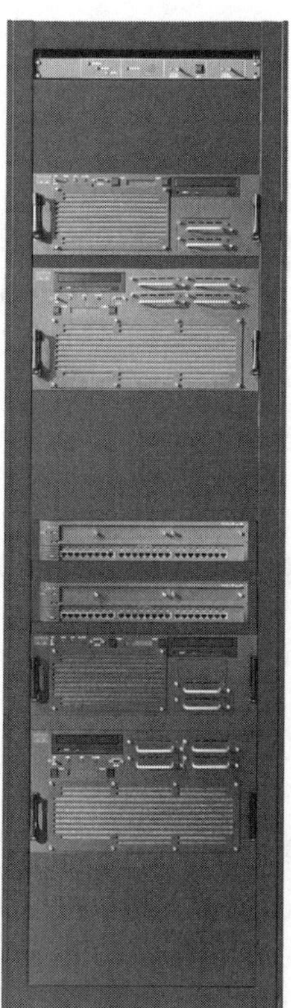

Among the services provided by the Cisco BTS 10200 Softswitch are

- Support for numerous traditional PSTN call signaling protocols (Integrated Services User Part [ISUP], Integrated Services Digital Network [ISDN], channel-associated signaling [CAS]) and variants)
- Support for numerous VoIP protocols (MGCP—including NCS and TGCP profiles—SIP, SIP-T, H.323)
- Operator services support
- 911 and 411 support
- Custom Local Area Signaling Service (CLASS) features such as call waiting, caller ID, 3-way calling, and Centrex features, such as call hold and call park.
- Call failure treatment handling with Announcement Players
- Communications Assistance for Law Enforcement Act (CALEA) support
- Call detail record (CDR) and Event Messaging billing generation

- Local number portability (LNP) support
- Voice mail support
- User activated call features (such as *69)
- Bulk provisioning
- Least cost and time-of-day routing

Just as with the CMTS products, the details concerning PacketCable configuration for the BTS10200 are contained within the relevant chapters in this book. For example, gateway configuration commands are included in the chapters in Part III, and DQoS configuration commands are included in the chapters in Part V.

In this book, BTS provisioning is shown from an EMS CLI session.

NOTE As you might imagine, the BTS10200 is a rather complex device, and it is beyond the scope of this book to go into the detail of all the provisioning required. Therefore, only the provisioning directly related to PacketCable and PacketCable endpoints is covered in this book. The provisioning described assumes a preexistent base configuration.

Signaling Gateway (SG)

The SG provides an interface between the SS7 call control messaging used in the PSTN and the PacketCable network. It uses either an SS7 A-link or D-link connected to an STP or an SS7 F-link connected to a SSP (Legacy PSTN Circuit Switch) or SCP (SS7 database). The MTP layer of the SS7 protocol is terminated on the SG, and the ISUP and TCAP layers are encapsulated in IP packets and sent to the MGC. This is a huge benefit because the MTP layers of the SS7 protocol can consume large amounts of bandwidth, wasting memory and CPU cycles if sent to the MGC. This traffic is now offloaded to the SG with only the upper layers of the SS7 protocol stack going to the MGC. The IP protocols used to transport the upper layer SS7 protocols are defined by the IETF Signaling Transport (Sigtran) working group. Don't worry if some of this terminology is unfamiliar to you; it is covered in greater detail in the PSTN call signaling chapter.

Overview of the ITP Signaling Gateway

The SG produced by Cisco Systems is called the Cisco IP Transfer Point (ITP). The ITP runs a special version of Cisco IOS Software on the following platforms:

- Cisco 2650 Router
- Cisco 2650XM Modular Access Router
- Cisco 2651 Router

- Cisco 2651XM Modular Access Router
- Cisco 7204 Router
- Cisco 7206 Router
- Cisco 7301 Router
- Cisco 7507 Router
- Cisco 7513 Router
- Versatile Interface Processor (VIP4-80)
- Versatile Interface Processor (VIP6-80)
- Cisco 7500 Route Switch Processor 16 (RSP8)
- Cisco 7500 Route Switch Processor 16 (RSP16)

ITP configuration and troubleshooting is covered in Chapter 8, "TGCP and the PSTN Interconnect," as well as much more information on how a signaling gateway works in a PacketCable network.

Media Gateway (MG)

The MG provides bearer connectivity between the PSTN and the PacketCable network. Typically this is done using SS7 Intermachine Trunks (IMT). Bearer circuits are set up using TGCP messaging with the MGC. Similar to the MTA, the MG is responsible for converting between the VoIP packets used in the PacketCable network and the circuit switched voice stream used in the PSTN. This is done using the CODECs described in the chapters in Part IV.

Overview of the MGX 8000 Series

The Cisco MGX 8800 series of carrier voice gateways are advanced Asynchronous Transfer Mode (ATM) multiservice switches and include the following models:

- MGX 8830
- MGX 8850
- MGX 8880

The MGX 8880 gateway is a newer platform based upon the MGX multiservice switches. The MGX 8880 has been specifically designed for voice service provides, like a PacketCable MSO. It has numerous QoS features and high-availability hardware and software.

The MGX supports both internal and external clocking. Both the MGX 8850 chassis and MGX 8880 chassis have 32 slots in the front for service modules and 32 slots in the back for network interface cards. The 32 slots are divided into two groups of 16, with one group existing on the top of the chassis and one group on the bottom. Some cards are of single height and fit into a single slot; other cards are double height and occupy both a top slot and the bottom slot directly below. The types of front cards typically used in the PacketCable architecture include the PXM, RPM, SRM, VISM, and VXSM. The MGX 8830 has 14 slots in each side with four reserved for PXM modules (double height), two slots reserved for an SRM module (double height), and eight single slots or four double slots for other service modules.

PXM Cards

The Processor Switch Module (PXM) card is the processor card for the MGX chassis. It is a double height card. For the MGX 8850 and MGC 8880, it must be located in the front slots 7/23 and/or 8/24. (These are double height cards; slots 23 and 24 are the bottom of slots 7 and 8. Consequently, these two slot locations are generally referred to as slots 7 and 8.) Utilizing both of these slots provides processor redundancy. The PXM-1E processor card can be used in the MGX 8830 and MGX 8850 platforms. In addition to being the processor card, it also provides the ATM switching fabric and the physical ATM interfaces. The back of the PXM card provides the ATM connection to the rest of the service provider's backbone. The back card can consist of DS-1, DS-3, OC-3, or OC-12 ATM interfaces. Typically, these interfaces are connected to a high-end router, such as the Cisco GSR 12000 series. The PXM-45 processor card can be used in the MGX 8850 and MGX 8880 platforms. This card provides 45 Gbps of non-blocking switching. When this processor card is used, RPM service modules are used to provide connectivity to upstream devices.

RPM Cards

The Route Processor Module (RPM) card is a high performance route processor module for the MGX chassis. It is a double height front card. In effect the RPM is a Cisco 7200 series router. Thus, features like Multiprotocol Label Switching (MPLS) and routing protocols such as Open Shortest Path First (OSPF) and Border Gateway Protocol (BGP) are available on this card. The back cards are single height and typically contain either Gigabit Ethernet or Packet Over SONET (POS) interfaces. It is these interfaces that provide connectivity to upstream devices when the more powerful PXM-45 processor card is used in the MGX chassis.

NOTE The most advanced version of this card is the RPM-XF. When the term *RPM* is used throughout this book, it refers to the RPM-XF.

SRM Cards

The Service Resource Module (SRM) card is a single height card and must be located in slots 15, 16, 31, or 32. The SRM card allows for DS-3, OC-3, and STM-1 connectivity to the PSTN. Other cards such as the VISM-PR card (described next) provide only DS-1 connectivity. Each SRM card would have a corresponding DS-3, OC-3, or STM-1 back card capable of supporting three DS-3 connections or a single OC-3 or STM-1 connection. For example in the DS-3 case, the SRM card performs much like an M13 MUX and cross-connects T1s off the DS-3 interface to modules such as the VISM. The SRM cards in slots 15 and 16 support distribution to the service modules in slots 1–6 and 9–14. The SRM cards in slots 31 and 32 support distribution to the service modules in slots 17–22 and 25–30. Just like the PXM cards, the SRM cards operate in a redundant manner with slots 15 and 16 working together and slots 31 and 32 working together in an active/standby configuration. In addition to the bulk distribution function, the SRM cards also enable the N+1 redundancy functionality of the service modules such as the VISM card. This allows certain VISM cards to serve as protection for other VISM cards in the event of a card failure. Finally, the SRM cards enable bit error rate testing (BERT) of T1 and E1 lines and ports.

VISM and VXSM Cards

Two types of voice enabling service modules are used in the MGX—the Voice Interworking Service Module (VISM) and the Voice Switch Service Module (VXSM). The most current version of the VISM card is the premium VISM card (VISM-PR); in the context of this book the term *VISM* actually refers to the VISM-PR service module. These cards are really the PacketCable MG. These are the cards that communicate with the MGC and convert between VoIP packets and circuit switched voice streams.

Each VISM card supports eight T1/E1 lines.

NOTE These lines can be connected either via a T1/E1 back card or through the distribution capability of the SRM cards. A fully loaded MGX 8850 or 8880 chassis supports 24 VISM cards, slots 1–6, 9–14, 17–22, and 25–30. By doing some math you can see this corresponds to 192 T1s per chassis (24 cards × 8 T1s per card). Recall that if you are doing DS-3 bulk distribution with the SRM cards, you can have only two active SRM cards (one on the top shelf and one on the bottom). Each SRM card can support up to three DS-3s, so six DS-3s total. This corresponds to 168 T1s per chassis (6 DS-3s × 28 T1s per DS-3). So with a SRM configuration you could support only 21 VISM cards (10.5 per shelf). This usually isn't a problem because typically N+1 redundancy would be implemented. This calls for some of the VISM cards to function as protection cards.

Currently, the VISM card supports the following CODECs:

- G.711 a-law and μ-law
- G.723.1
- G.726-16/24/32/40
- G.729A and G.729B

Other functionality of the VISM card includes echo cancellation, silence suppression, tone detection and generation, and CALEA functionality.

The VXSM is a higher density voice module and supports T1/E1, T3, and channelized OC-3/STM-1 interfaces. Each card supports up to 8,064 DS-0 trunks (8,064 equates to 4 OC-3 interfaces). The VXSM is a double height service module. Specifically, the available back cards include a 4-port OC-3/STM-1 interface card, a 6-port T3 interface card, and a 48 T1/E1 interface card.

Currently, the VXSM card supports the following CODECs:

- G.711 a-law and μ-law
- G.723.1
- G.726-16/24/32/40
- G.729A and G.729B

Like the VISM card, other functionality of the VXSM card includes echo cancellation, silence suppression, tone detection and generation, and CALEA functionality.

The MGX 8880 with the VXSM card is PacketCable 1.0 and 1.1 qualified, and the VXSM card also supports both RTP and call signaling encryption.

Configuration of the MGX service modules relevant to a PacketCable deployment is covered throughout this book, but specifically in Chapter 8.

Overview of the AS5000 Series

The MGX 8800 series of media gateways can handle a large number of trunks, over 3,840 DS-0s for a chassis with VISM cards and over 40,000 DS-0s for a chassis with VXSM cards.

For smaller deployments, you can use a media gateway from Cisco's AS5000 series of routers, specifically, the AS5400HPX (see Figure 2-7) and AS5350 routers.

Figure 2-7 *Cisco AS5400HPX Gateway*

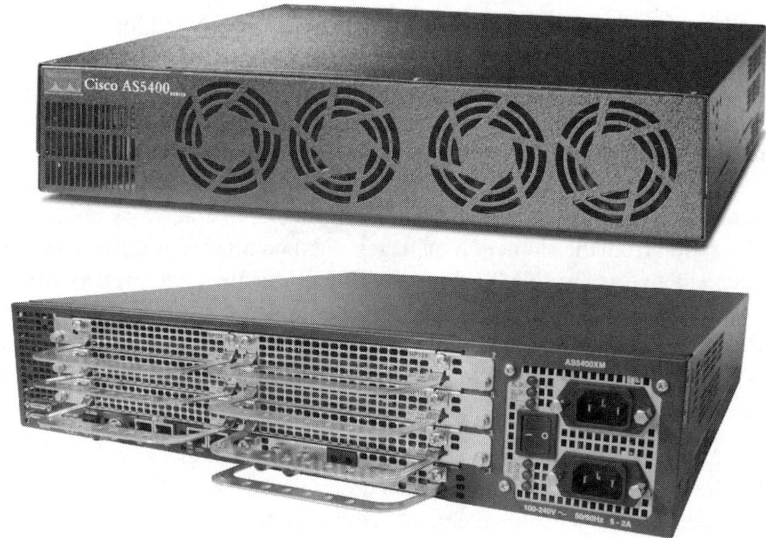

These routers also support the TGCP variant of MGCP. For PSTN connectivity, the AS5400HPX supports either an eight-port T1 interface or a single T3 interface, and the AS5350 supports a two-port, four-port, or eight-port T1 interface. Packets are routed to the PacketCable VoIP backbone via two 10/100 Fast Ethernet ports. Because they are Cisco routers, these products run Cisco IOS. Note that neither of these devices support PacketCable security (that is, call signaling and bearer packet encryption).

OSS/BSS BackOffice Components

Operational Support System (OSS) and Business Support System (BSS) components are responsible for functions such as the provisioning, billing, fault management, and security management of the PacketCable network. These components help the PacketCable service provider rapidly configure new customers, accurately bill for services, monitor network performance, and troubleshoot problems. A model commonly used to describe OSS/BSS functions is the ITU network management model FCAPS, which stands for Fault, Configuration, Accounting, Performance, and Security management.

Configuration management devices enable the service provider to quickly provision and enable new customers and services. Fault management and performance management devices ensure that the network remains operational and can provide the quality of service required by your customers. Accounting management devices lead directly to revenue by accurately measuring service and network usage on a per customer basis. Security management devices ensure the integrity of your network, and your customer's information is not compromised.

Another ITU model commonly used to describe OSS/NMS functions is the Telecommunications Management Network (TMN) model described in specification ITU-T M.3010. This model uses a layered approach to describe how OSS/NMS can be achieved for FCAPS management. Each component in the network is managed independently by an Element Management System (EMS); for example, in a PacketCable network you could have an EMS for the MTA, one for the CMS, one for the MG, etc. This is known as the Element Management Layer (EML). Each EMS in the network collects detailed information from the element it manages and then filters this information back to a Network Management System (NMS). This NMS manages the entire network by interfacing with the individual EMS devices. So for a PacketCable network one NMS would be controlling and communicating with the MTA EMS, CMS EMS, MG EMS, and so on. Because the NMS interfaces with each EMS, it can provide functionality such as "flow-through" provisioning for configuration management and event correlation for fault management.

Each EMS can communicate with the component it manages using either standardized interfaces or vendor-specific methods. For a PacketCable EMS, SNMPv3 is used to interface with PacketCable elements. However, the NMS device would typically interface using open standardized methods (such as Common Object Request Broker Architecture [CORBA]). This way a single NMS platform could manage a network with equipment from multiple vendors. In addition to the EML and network management layer (NML), the TMN model also includes the service management layer (SML) and business management layer (BML). The SML devices would control the NML devices, and the BML devices would control the SML devices. Examples of SML and BML devices include order entry systems, billing systems, and customer care systems.

The PacketCable OSS overview technical report provides a high-level discussion of how these models apply in a PacketCable network. Figure 2-8 is from this document and graphically illustrates how OSS components interact with the other PacketCable components and the TMN model. For more information please refer to this technical report.

Figure 2-8 *PacketCable OSS*

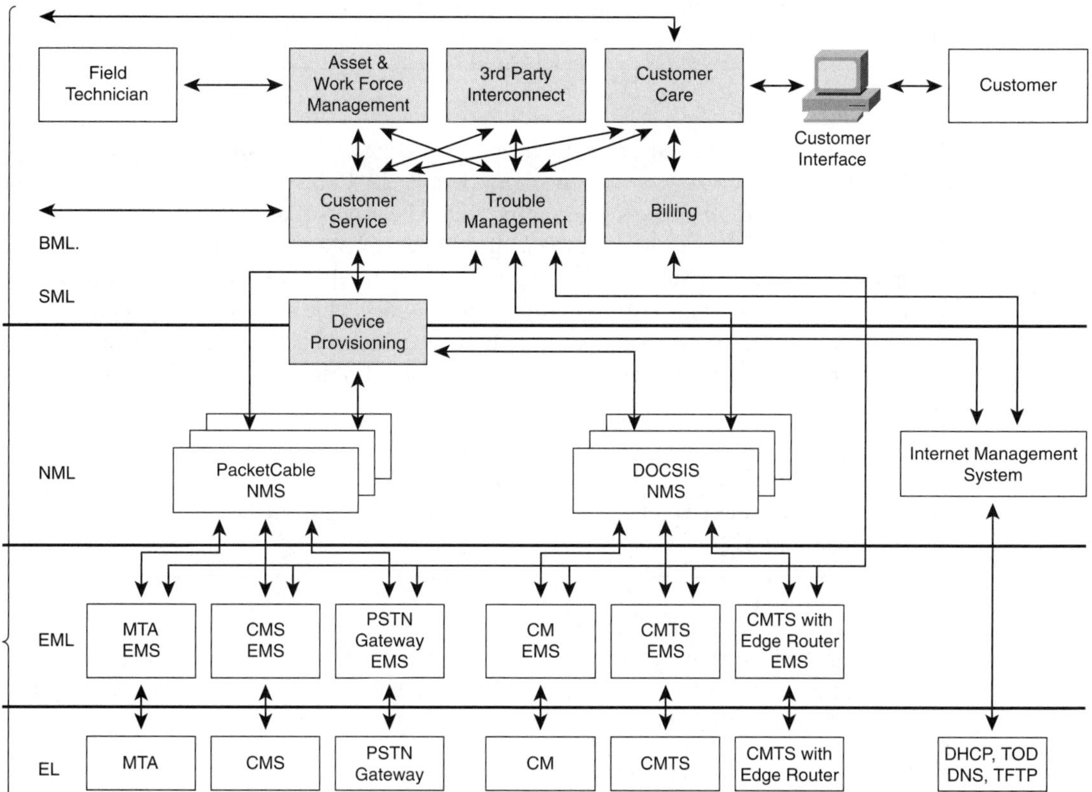

Overview of BAC

Cisco has a product that can be used for the configuration management of cable modems and PacketCable MTAs. This product is called Broadband Access Center (BAC). BAC automates the configuration and provisioning of network devices supported by a broadband service provider. It targets broadband service providers who want to deploy IP data, voice, and video on hybrid fiber and coaxial cable networks. BAC is designed with flexibility in mind; it can be adapted to handle a network of virtually any size. It supports both DOCSIS and PacketCable provisioning and provides redundancy and failover protection.

BAC is comprised of a number of different servers that can reside on the same or different physical machines.

NOTE In a live deployment these servers are usually installed on different machines, and the single machine configuration is used only in lab environments.

Typically, distributed servers are used close to the edge of the network to maximize availability and minimize provisioning delay. In a PacketCable network, the edge would be the group of CMTSs servicing a geographic area. The group of distributed servers handling the provisioning of devices in this area is also known as a *provisioning group*. Central servers are used that interface with these provisioning groups. These central servers provide a single place to manage and control the entire network. One set of central servers can support several sets of distributed servers. BAC is a PacketCable EMS for PacketCable MTAs, a DOCSIS EMS for cable modems, and also a CableHome EMS for CableHome devices.

NOTE CableHome is another CableLabs project designed for supporting home networks behind a cable infrastructure.

Upper-layer NMS systems can be integrated with BAC through the use of an application programming interface (API). Thus, third-party systems developed for customer support staff to configure customers can be integrated with BAC.

Figure 2-9 is a logical illustration of how BAC fits into a cable environment.

Regional Distribution Unit (RDU)

One of the central servers is the Regional Distribution Unit (RDU). This is the BAC server that provides a centralized point for the configuration and management of devices. Remember these devices include all cable modems, PacketCable MTAs, CableHome devices, and even regular CPE devices such as personal computers. A single RDU can be used to support an entire BAC deployment.

NOTE To maximize availability, you might find it desirable to deploy a redundant set of RDU servers using some sort of server load balancing (SLB) design.

Figure 2-9 *BAC*

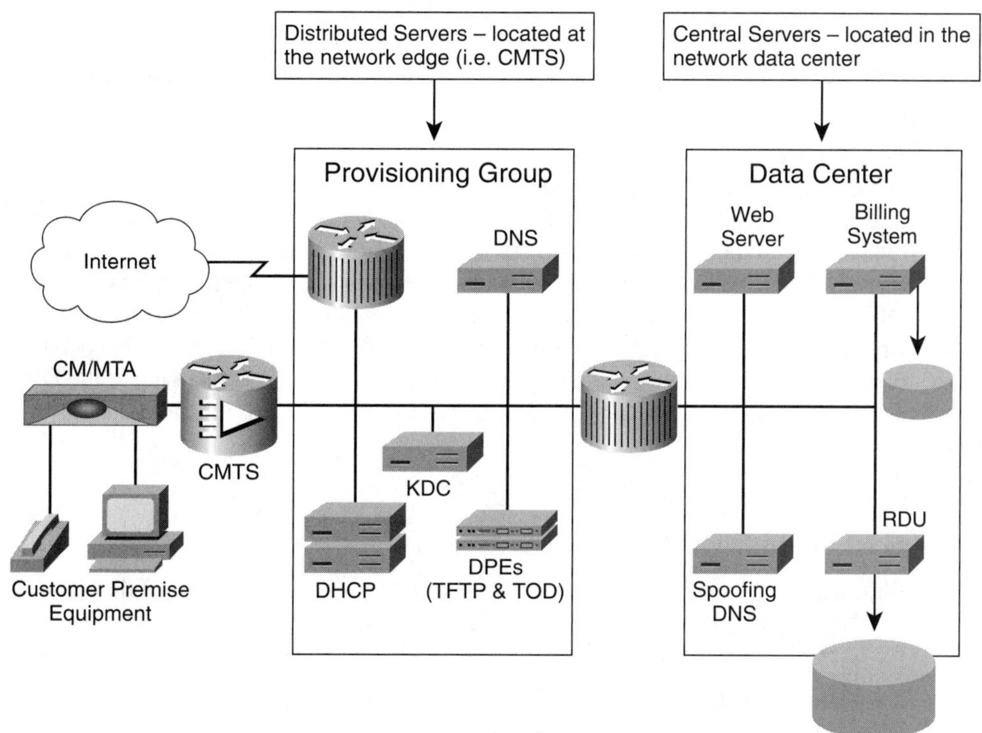

The RDU is required only for new device configuration; existing devices are provisioned by the provisioning group servers. Consequently, an RDU system failure will not prevent existing devices from operating. The RDU is also the BAC device that can be used to communicate with third-party applications through its API.

NOTE The spoofing DNS server shown in the central servers in Figure 2-9 can be used to redirect unprovisioned devices to a web server. This web server can then direct customers to self-provision these devices.

Device Provisioning Engines (DPEs)

Device configuration on the RDU is pushed to distributed BAC servers on the edge of the network where that device is connected. These servers are called Device Provisioning Engines (DPEs). DPEs cache device configuration, greatly increasing the speed in which devices become operational. A single DPE can cache configuration information for up to

1,000,000 devices. In addition to caching device configuration files, the DPE servers also function as the TFTP and Time-of-Day (ToD) servers needed in the provisioning process. As you will see in subsequent chapters, both cable modems and MTAs receive their configuration files using TFTP, and cable modems utilize ToD servers in their provisioning process.

BAC supports two platforms for the DPE functionality; one is using a Cisco appliance box, and the other is a software DPE running on a Sun Solaris machine. These DPE servers are typically deployed in pairs within a provisioning group. Not only does this provide redundancy in the event a single DPE fails, but load can also be shared between both servers. All DPEs in a provisioning group cache the same devices, so the failure of a single DPE is not service impacting. Also this enables other servers (such as DHCP) to contact any one of the DPEs to access device configuration. The DPE cache is also persistent, meaning a power failure does not result in the loss of data. The accuracy of the DPE cache is maintained by periodically synchronizing its cache with the information stored in the RDU.

Cisco Network Registrar (CNR)

Referring to Figure 2-9, you can see that the DPE is not the only type of distributed server. DHCP and DNS servers are also required for device provisioning. BAC works with another Cisco product that handles this functionality: the Cisco Network Registrar (CNR).

The DHCP servers work with the DPE servers to select the appropriate IP address for each device. Two DHCP servers are also typically installed in a provisioning group to provide redundancy in the event one server fails. With CNR, each DHCP server is allocated a portion of the IP address pool with the primary server getting a much larger portion. Lease updates are sent between servers so their databases are synchronized.

The DNS servers are also typically deployed in pairs in a provisioning group. After a DHCP server assigns a new lease to an MTA, it dynamically updates the DNS servers. Updates are also synchronized between the DNS servers. A good design is to use one server for making dynamic updates and the other server for resolving DNS queries. Much more about DNS and the entire PacketCable provisioning process is covered in the next three chapters.

Key Distribution Center (KDC)

One more type of distributed server is associated within a provisioning group, the Key Distribution Center (KDC). The KDC is used only for the provisioning of PacketCable MTAs. This server provides the Kerberos based security required for the PacketCable provisioning process. It authenticates MTAs and provides them with security "tickets" that permit secured access to the rest of the provisioning system. The KDC is loaded with a CableLabs certificate chain and a separate license file.

Greater detail concerning PacketCable configuration for the various BAC components is covered in Part II of the book.

Revisiting the Cisco BTS 10200

Before moving on to other PacketCable components, taking another look at the Cisco CMS, the BTS 10200, is useful after the discussion of OSS/NMS. Remember that part of the BTS architecture was the redundant pair of EMS machines. These machines are configuration element managers for the BTS 10200. The BTS 10200 database is populated through the BTS EMS machines, which are also performance and fault managers as they collect alarm and event information from the call agent and feature server machines. Finally, the BTS EMS is an accounting element manager as billing information is collected on this machine. Like BAC, the BTS EMS platform has an API that can be interfaced by third-party applications.

Record Keeping Server (RKS)

The RKS is the device used for the accounting management in PacketCable. It receives PacketCable Event Messages from the CMS, MGC, and CMTS. Event Messages contain information that identifies the calls and their resource usage. These packets can then be collected and correlated to create Call Detail Records (CDRs), which can then be made available to back office systems such as a billing server. Cisco Systems does not produce a RKS; it interfaces with third-party gear.

Lawful Intercept

A Communications Assistance for Law Enforcement Act (CALEA) server is the component used by law enforcement agencies to perform wiretapping of telephony subscribers. Just as CALEA support is required in the traditional PSTN, it is needed in the PacketCable network as well. The CALEA server serves as an interface to the law enforcement agency. The interfaces between a telecommunication service provider and law enforcement agency are defined in the ANSI standard, J-STD-025-A. It interfaces with the CMS, CMTS, MGC, and MG network entities to provide this information. There are two types of wiretapping:

- **Call Detail**—With a Call Detail wiretap only information about the call is sent to the CALEA server. This includes information such as the identities of persons the tapped subscriber called and received calls from.

- **Call Content**—A Call Content wiretap includes capturing the actual voice conversation. A key requirement for implementing CALEA services is that they must be transparent to the tapped subscriber. The subscriber must have no way of knowing they are under surveillance. Chapter 15, "Event Messaging and Lawful Intercept," deals with this topic in detail.

Cisco Systems does not produce a CALEA server; however, it does provide the required interfaces to it on other PacketCable network entities.

Chapter Summary

A PacketCable 1.0–based deployment has several different functional components:

- Multimedia terminal adapter (MTA)
- Cable modem (CM)
- Cable modem termination system (CMTS)
- Call management server (CMS)
- Media Gateway Controller (MGC)
- Media Gateway (MG)
- Signaling Gateway (SG)
- Media servers
- Operational Support System (OSS) components.

The MTA resides at the customer premises and provides the telephony interface to the PacketCable network. If the MTA functionality exists in the same hardware as the cable modem, it is known as an embedded MTA; if the MTA is a physically separate device, it is known as a standalone MTA. A PacketCable 1.x network supports only embedded MTAs. Some of the functionality of the MTA includes

- Communicating call signaling information with the CMS
- Communicating quality of service information with the CMTS
- Encoding/decoding media streams

Through Cisco's acquisition of Linksys, Cisco produces two PacketCable certified MTAs, the CM2P2B and the CG2P2WB.

The cable modem and the CMTS are the DOCSIS components in a PacketCable network. Together they provide the transport of voice and call signaling data across the HFC network. The CMTS also communicates QoS information with the CMS using the COPS protocol and reports usage information to the RKS using PacketCable event messaging. Cisco has two PacketCable qualified CMTS platforms—the uBR7246VXR and the uBR10012.

The call agent is the brain of a PacketCable network. It handles call control and signaling, maintains call state, provides call features, reserves QoS, generates billing records, and provides wiretapping control. The CMS is the call agent component responsible for providing call control and signaling for PacketCable MTAs. The gate controller interfaces with the CMTS to provide QoS for the MTA. The MGC is the call agent component responsible for providing call control and signaling for media gateways that interface with the PSTN. The ANC is responsible for providing call control and signaling for announcement player gateways. Cisco's BTS 10200 softswitch provides the CMS, GC, MGC, and ANC functionalities. It is PacketCable qualified as both a CMS and MGC.

The Signaling Gateway (SG) works with the MGC and MG components to interface with the PSTN. Specifically, it provides the SS7 interface. The lower layers of the SS7 protocol are terminated on the SG, and the upper layers are sent over IP to the MGC using the IETF Sigtran family of protocols. Several Cisco platforms known as IP Transfer Point (ITP) can function as a PacketCable SG.

The MG provides the bearer connectivity between the circuit-switched trunks connected to the PSTN and the PacketCable network. Cisco's MGX8800 family of multiservice switches can be used as PacketCable MGs. These platforms have a number of slots that support cards known as *service modules*. The service modules used to provide voice functionality are the VISM and VSXM. The MGX8880 gateway with VXSM service modules is PacketCable qualified. The AS5000 family of routers can also be used as an MG for smaller-scale deployments.

PacketCable OSS/NMS components are used for the provisioning, billing, fault management, and security management of the PacketCable network. The ITU FCAPS model and the TMN layered model are industry recognized methods for characterizing OSS/NMS components. The FCAPS model separates OSS/NMS management into five categories: fault, configuration, accounting, performance, and security. The TMN layered model discusses the hierarchy of management systems into the element layer, network layer, system layer, and business layer. Cisco's BAC solution is used in PacketCable deployments to handle MTA and cable modem management.

BAC is comprised of distributed servers near the network edge known as a provisioning group and central servers in a data center providing a single point of configuration and management. The central server is called the Regional Distribution Unit (RDU) and has an API that can be used to interface to third-party equipment. The RDU pushes device configurations to DPE servers in the provisioning group. These servers cache device configuration and interact with DHCP and DNS servers in the provisioning group. CNR is used as the DHCP and DNS servers. The KDC server is used to achieve the secure provisioning required by PacketCable.

Other components used in a PacketCable deployment include the RKS server for accounting/billing reporting and a CALEA server required by law enforcement agencies to provide wiretapping services.

Chapter Review

1 True or False: Standalone MTAs are used in a PacketCable 1.x deployment.

Answer: False. Only embedded MTAs are valid for PacketCable 1.x.

2 True or False: If your HFC network is currently supporting high-speed data, chances are it will have no problems supporting voice.

Answer: False. Dropped packets caused by noisy plants are a much bigger issue for voice packets than data packets.

3 True or False: Although a cable modem can function as either a bridge or a router, it is required by DOCSIS to function only as a bridge.

Answer: True

4 True or False: Not only can the CMTS be used to provide QoS for packets across the HFC network, but it can also influence QoS for packets across the IP network.

Answer: True

5 True or False: Both the Cisco uBR7246VXR and Cisco uBR10012 CMTS platforms support redundant processors.

Answer: False. Only the Cisco uBR10012 does.

6 True or False: To communicate to an announcement player, the announcement controller typically uses protocols such as SIP and MGCP rather than any PacketCable tailored protocol.

Answer: True

7 True or False: The VSXM service module supports TDM connectivity ranging from T1/E1 to channelized OC-3.

Answer: True

8 True or False: SNMPv2c is used by a PacketCable EMS to manage a PacketCable element.

Answer: False. SNMPv3 is used.

9 True or False: If the RDU in your network fails, you will not be able to provision new devices.

Answer: True

10 True or False: The KDC is needed to provision cable modems and MTAs.

Answer: False. It is needed only to provision MTAs.

11 What are the benefits in using a centralized call signaling protocol such as NCS between the MTA and CMS?

Answer: It reduces complexity on the MTA, making it less prone to bugs and the need for software upgrades to add feature functionality. Also it gives the MSO tighter control of services as the CMS is a much more secure component than the MTA.

12 What type of cable line card(s) on a Cisco uBR7246VXR support advanced RF capabilities like DOCSIS 2.0 A-TDMA support?

Answer: MC28U, MC28X

13 What are the three ways in which the BTS 10200 can be configured?

Answer: Via CLI session, Cisco's EPOM utility, or a third-party application via the API

14 What is the benefit of using a separate signaling gateway as opposed to running this functionality on the call agent?

Answer: The MTP layers of the SS7 protocol can be memory and CPU intensive, and this offloads this burden from the call agent.

15 What is the difference between the Cisco MGX 8880 media gateway and the Cisco MGX 8850 multiservice switch?

Answer: The MGX 8880 gateway has been optimized for voice with numerous QoS features and high-availability hardware and software.

16 What are three functions of the SRM service module?

Answer: T1/E1 bulk distribution, BERT testing, and enabling N+1 redundancy

17 What is some sample feature functionality that could be achieved by a PacketCable NMS?

Answer: "Flow through" provisioning, event correlation

18 What are the four types of servers found in a provisioning group?

Answer: DPE, DNS, DHCP, KDC

19 What are the two types of wiretaps? Describe them.

Answer: Call Detail, where identifying information such as called and calling number is recorded, and Call Content, where the actual voice conversations are recorded

PART II

MTA Provisioning

CHAPTER 3

Provisioning Overview

Hopefully, the previous two chapters have provided you with a solid background of what PacketCable is and what is involved with a PacketCable deployment. In these chapters you gained a high-level understanding of what technologies are addressed by the individual PacketCable specifications, as well as a detailed look at the functional components comprising a PacketCable deployment. You also learned how Cisco is involved with PacketCable and the Cisco equipment that can be used in a PacketCable network.

The remainder of the book covers these technologies and components in much greater detail. The next two chapters cover the provisioning aspects of PacketCable, and a majority of that focuses on the process used for provisioning embedded MTAs. Chapter 4, "Provisioning Flows," examines the individual provisioning steps one at a time. Included in each step is a look at the protocols and messaging involved, as well as troubleshooting tips. Chapter 5, "The MTA Configuration File," covers the configuration file used by embedded MTAs, including specifics on the individual parameters and how they work.

This chapter provides you with the background information needed to fully understand the concepts presented in Chapters 4 and 5. It begins with some overview information on the importance of having a well-defined dynamic method of provisioning multimedia terminal adapters (MTA). Then the protocols involved in the provisioning process are covered with a focus on the aspects related to PacketCable. Finally, configuration steps and troubleshooting tips related to PacketCable provisioning on Cisco gear are covered. The bulk of this discussion focuses on the Broadband Access Center (BAC) provisioning system.

Provisioning Overview

Provisioning is part of configuration management. It includes functions such as resource initialization and registration and attribute configuration. Two PacketCable specifications are related to provisioning, one for embedded MTAs (that first appears in the PacketCable 1.0 release) and one for the call management server (CMS) (that first appears in the PacketCable 1.5 release). Provisioning for embedded MTAs (eMTA) is a crucial part of PacketCable, and MTAs must conform to this specification for PacketCable certification. By having standardized guidelines for MTA provisioning, any vendor's MTA should be able to successfully provision with any vendor's Operational Support System (OSS). The

CMS provisioning specification has not been much of a focus and is more of a guideline than a requirement for CMS vendors.

Remember in the previous chapter you learned that the MTA portion of an embedded MTA has a separate MAC address, IP address, and Fully Qualified Domain Name (FQDN) than the cable modem has. This is because the MTA goes through its own provisioning process separate from the Data-Over-Cable Service Interface Specification (DOCSIS)–defined process the cable modem uses. The MTA has its own configuration file apart from the modem's DOCSIS configuration file. This permits different service providers to be in control of the DOCSIS and telephony portions of an eMTA. In other words, a single customer could have one service provider for voice services and a different service provider for high-speed data services.

NOTE As you'll see later on, some relationship exists between the cable modem and eMTA provisioning processes.

A PacketCable eMTA can be provisioned by three methods:

- Secure
- Basic
- Hybrid

The secure method is recommended because it places strict security requirements in the provisioning process. These security requirements help ensure only authorized devices are able to utilize the PacketCable network. Using the secure provisioning method makes it extremely difficult for unauthorized devices or hackers to steal service. The other two methods were added to the PacketCable specification later on because some Multiple Systems Operators (MSO) felt that the secure method was too complex and were willing to take on the security risks in return for a simpler solution. All three of these provisioning methods are covered in detail in Chapter 4.

Provisioning Protocols

As mentioned in Chapter 1, "PacketCable Overview," PacketCable attempts to use existing protocols whenever possible. This is definitely the case for PacketCable provisioning as IP protocols such as Dynamic Host Configuration Protocol (DHCP), Domain Name Server (DNS), Simple Network Management Protocol (SNMP), and Trivial File Transfer Protocol (TFTP) are used. You must be knowledgeable in these protocols to fully understand PacketCable provisioning.

DHCP

DHCP for IP version 4 is defined in the IETF RFC 2131. It provides a framework for passing configuration information to hosts on a TCP/IP network. DHCP is based on the Bootstrap Protocol (BOOTP) originally defined in the IETF RFC 0951. DHCP enhances the functionality of BOOTP with capabilities such as automatic allocation of reusable IP network addresses and additional configuration options.

With DHCP, a host can dynamically obtain an IP address and other configuration information from a server. The host can use this IP address for a period of time specified by the server known as a *lease*. The host can request renewal of this lease prior to expiry. Both DOCSIS cable modems and PacketCable MTAs obtain their IP address and other configuration parameters via DHCP.

A client device obtains its configuration information using a four-message exchange with the DHCP server, as shown in Figure 3-1.

Figure 3-1 *DHCP Message Exchange*

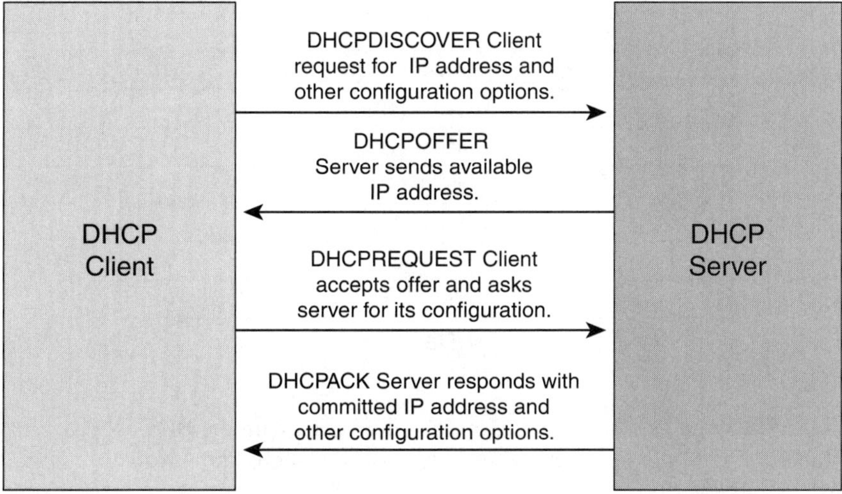

The DHCP Discover packet is broadcast by the client on its local subnet. If the DHCP server is on the same subnet, it sees this Discover packet; if the DHCP server is elsewhere in the network, a relay agent is needed to forward the packet to a DHCP server. The latter is what happens in a cable network. The CMTS acts as the DHCP relay agent; it forwards DHCP broadcast packets it receives to one or more preconfigured DHCP servers.

When a server receives the Discover packet, it responds to the client with a DHCP Offer packet. In this packet is an available IP address (in the yiaddr field) as well as other requested options. The Offer packet is either unicast back to the client or broadcast on the client's subnet, depending on the value of the Flags field in the Discover packet. The client might receive multiple offers; it selects the best one and signals this by broadcasting a DHCP Request packet identifying the selected server in the server identifier option (option #54). The servers not selected receive this packet, so they can reuse their offered IP addresses for other clients. When it receives the Request packet, the selected server commits the IP address binding and sends a DHCP Acknowledge packet to the client with all the requested configuration parameters. If the client wants to continue to use these configuration parameters past the lease expiry, it broadcasts a DHCP Request packet with its assigned address (in the ciaddr field). Servers with knowledge of this client's configuration extend the lease and respond with a DHCP Acknowledge packet.

The DHCP protocol uses the UDP protocol for transport; specifically, the host (that is, client) listens on port 68, and the DHCP server listens on port 67. Figure 3-2 illustrates the format of a DHCP packet.

Figure 3-2 *DHCP Packet Format*

op (1)	htype (1)	hlen (1)	hops (1)
xid (4)			
secs (2)		flags (2)	
ciaddr (4)			
yiaddr (4)			
siaddr (4)			
giaddr (4)			
chaddr (16)			
sname (64)			
file (128)			
options (312)			

All of these fields other than the options field exist with the BOOTP protocol as well. Table 3-1 provides descriptions of fields you'll want to know well.

Table 3-1 *Key BOOTP Fields*

BOOTP/DHCP Field	Description
flags	The leftmost bit is the broadcast bit and determines whether responses should be broadcast or unicast
siaddr	The server IP address used in the next step of the provisioning process (such as a TFTP server)
ciaddr	The client IP address, filled in by the client if renewing previously allocated parameters
yiaddr	Your (client) IP address, assigned by the server to the client

Table 3-1 *Key BOOTP Fields (Continued)*

BOOTP/DHCP Field	Description
giaddr	The gateway/relay agent IP address
chaddr	The client hardware address (that is, MAC address)

The options field consists of the DHCP configuration parameters. These options are defined in IETF RFC 2132 and are expressed in Tag/Length/Value (TLV) format. The first option is always option #53, which identifies the DHCP message type. Table 3-2 lists the DHCP options typically seen in cable modem and MTA provisioning.

Table 3-2 *DHCP Options*

DHCP Option #	Description/Comment
1	IP Subnet Mask. This option should be used for cable modems (CM) and must be used for MTAs. If it is absent, the default mask for the corresponding network class is assumed.
2	Time Offset. This option should be used for CMs.
3	Default Router. This option is required for both CMs and MTAs; it can also have more than one value.
4	Time Server. This option should be used for CMs. If it is absent, the default time is assumed to be midnight January 1, 1970.
6	Domain Name Server. This option is required for eMTAs. For eMTAs this option is only required to list one DNS server; however, it can have two.
7	Log Server (syslog). This option should be used for CMs and is required for eMTAs.
12	Host Name (with option #15 forms an FQDN). This option is required for eMTAs.
15	Domain Name (with option #12 forms an FQDN). This option is required for eMTAs.
43	Vendor Specific Information. This option is required for eMTAs and is discussed in more detail later.
50	Requested IP Address (similar to the ciaddr field).
51	IP Address Lease Time (in seconds).
53	DHCP Message Type (Discover, Offer, Request, Acknowledge, and so on). This option is required in all DHCP messages.
54	Server Identifier.

continues

Table 3-2 *DHCP Options (Continued)*

DHCP Option #	Description/Comment
55	Parameter Request List (indicates the DHCP options requested by the client device). This option is required for both CMs and MTAs.
57	Maximum DHCP Message Size (minimum size is 576 bytes).
60	Vendor Class Identifier (used to determine whether the device is a DOCSIS cable modem, PacketCable MTA, personal computer, and so on). This option is required for both CMs and MTAs; more on this option is discussed later.
61	Client identifier (indicates the client MAC address such as the chaddr field).
66	TFTP Server Name.
67	Configuration file name.
82	Relay Agent Information option (used by relay agents when forwarding requests). This option is often used by the CMTS when forwarding requests; more on this option is discussed later in the chapter.
122	CableLabs Client Configuration (holds PacketCable-specific options—discussed in detail later in the chapter). This option is required for both the CM and the MTA for PacketCable provisioning.
177	This option was used for PacketCable-specific options before the provisioning specification was finalized. If the eMTA is running old code, you might see this option requested, in which case you should upgrade the code on the eMTA to a more current version.

Option #43

Option #43, Vendor Specific Information, is used to transmit information specific to a certain vendor. It comprises a number of sub-options with sub-options #1–10, #31, and #32 specified by PacketCable. Table 3-3 lists the sub-options that are used by a PacketCable eMTA.

Table 3-3 *Option 43 Sub-Options Used in PacketCable*

Sub-Option #	Description
2	Identifies the component making the DHCP request. For PacketCable this is the MTA type (EMTA for embedded MTAs and SMTA for standalone MTAs). This sub-option is required for a PacketCable MTA.
3	Valid only for a cable modem and identifies all of its embedded components.
4	Device serial number. This sub-option is required for a PacketCable MTA.
5	Hardware version number. This sub-option is required for a PacketCable MTA.
6	Software version number. This sub-option is required for a PacketCable MTA.

Table 3-3 *Option 43 Sub-Options Used in PacketCable (Continued)*

Sub-Option #	Description
7	Boot ROM version. This sub-option is required for a PacketCable MTA.
8	Device's Organizational Unique Identifier (OUI). This sub-option is required for a PacketCable MTA.
9	Device's model number. This sub-option is required for a PacketCable MTA.
10	Vendor name. This sub-option is required for a PacketCable MTA.
31	MTA MAC address. This sub-option is required for a PacketCable MTA.
32	Correlation Identifier. This sub-option is required for a PacketCable MTA.

The Ethereal trace in Figure 3-3 shows an example of this DHCP option for an eMTA.

Figure 3-3 *DHCP Option #43 Example*

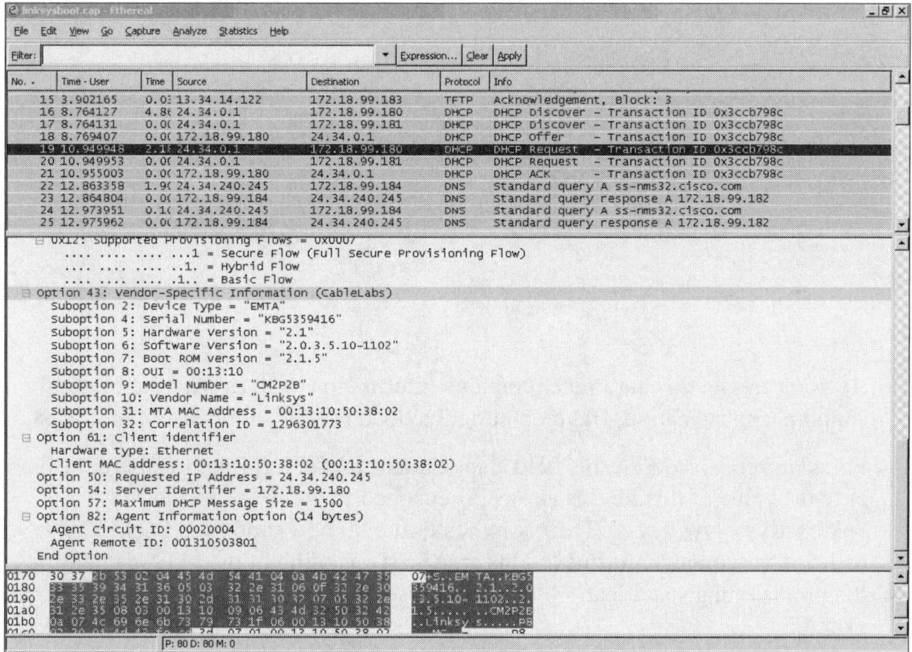

Option #60

Option #60, Vendor Class Identifier, identifies the device requesting configuration information. This parameter enables OSS/NMS devices to differentiate one device type from another. For a DOCSIS cable modem the value of this field is an ASCII encoded string of the format "docsisx.x:yyyyyyyyyyy". The x.x indicates the highest supported DOCSIS version: 1.0, 1.1, 2.0, or 3.0. The remainder of this field is an ASCII encoding of the Modem Capabilities parameter defined in DOCSIS. The trace in Figure 3-4 shows an example.

Figure 3-4 *DHCP Option #60 CM Example*

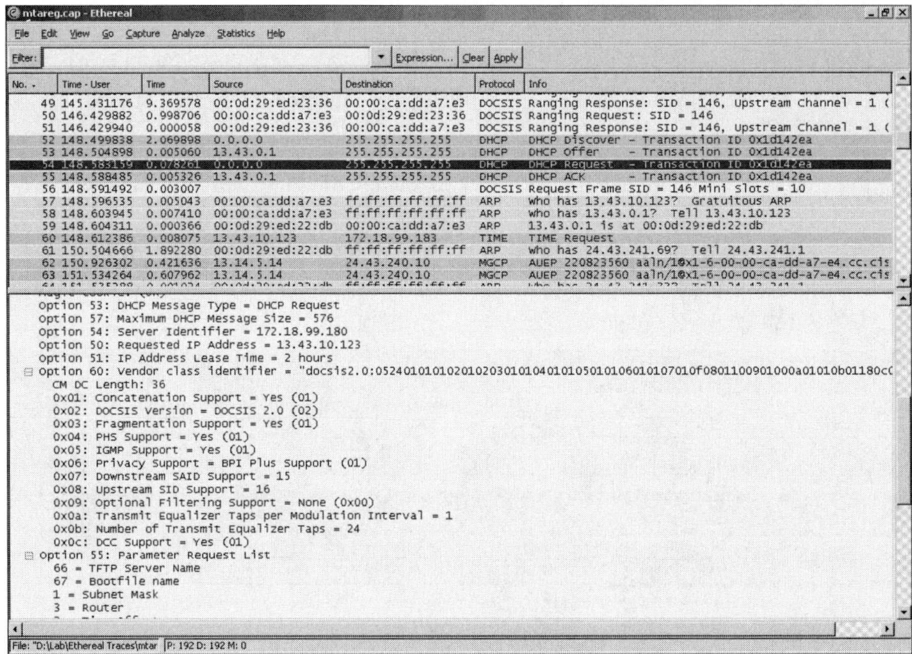

As you can see, this parameter contains information such as whether or not the modem supports concatenation, fragmentation, payload header suppression, multicast, and so on.

For a PacketCable MTA this field also contains useful information about the device. In this case the value of this field is an ASCII encoded string in the format "pktcx.x:yyyyyyyyyyy". The x.x indicates the highest supported PacketCable version: 1.0 or 1.5. The remainder of this field is an ASCII encoding of the MTA capabilities defined in the provisioning specification, specifically section 10. The trace in Figure 3-5 shows an example.

Figure 3-5 *DHCP Option #60 MTA Example*

```
mtareg.cap - Ethereal
File Edit View Go Capture Analyze Statistics Help
Filter:                                  Expression... Clear Apply
No. - Time - User   Time     Source             Destination        Protocol Info
  89 156.275214 0.229359 00:00:ca:dd:a7:e3 00:0d:29:ed:23:36 DOCSIS BPKM Request (Key Request)
  90 156.280855 0.005641 00:0d:29:ed:23:36 00:00:ca:dd:a7:e3 DOCSIS BPKM Response (Key Reply)
  91 156.370449 0.089594 13.14.5.14         24.43.240.10      MGCP   AUEP 220823560 aaln/1@x1-6-00-00-ca-dd-a7-e4.cc.cis
  92 156.783911 0.413462 13.14.5.14         24.43.240.10      MGCP   AUEP 220823564 aaln/1@x1-6-00-00-ca-dd-a7-e4.cc.cis
  93 158.299886 1.515975 0.0.0.0            255.255.255.255   DHCP   DHCP Discover - Transaction ID 0x1a12f58
  94 158.304095 0.004209 13.43.0.1          24.43.240.10      DHCP   DHCP Offer    - Transaction ID 0x1a12f58
  95 158.311934 0.007839 0.0.0.0            255.255.255.255   DHCP   DHCP Request  - Transaction ID 0x1a12f58
  96 158.315866 0.003932 13.43.0.1          24.43.240.10      DHCP   DHCP ACK      - Transaction ID 0x1a12f58
  97 158.324167 0.008301 00:00:ca:dd:a7:e4 ff:ff:ff:ff:ff:ff ARP    Who has 24.43.240.10?  Gratuitous ARP
  98 158.332585 0.008418 00:00:ca:dd:a7:e4 ff:ff:ff:ff:ff:ff ARP    Who has 24.43.240.1?  Tell 24.43.240.10

    Magic cookie: (OK)
    Option 53: DHCP Message Type = DHCP Request
    Option 57: Maximum DHCP Message Size = 576
    Option 54: Server Identifier = 172.18.99.180
    Option 50: Requested IP Address = 24.43.240.10
    Option 51: IP Address Lease Time = 4 days, 3 hours, 7 minutes, 4 seconds
  ⊟ Option 60: Vendor class identifier = "pktc1.0:051D0101000201020901010B09030506070809 0A0B0C0C01000D0101100109"
      MTA DC Length: 29
      0x01: PacketCable Version = PacketCable 1.0 (00)
      0x02: Number Of Telephony Endpoints = 02
      0x09: NVRAM Ticket/Session Keys Storage Support = Yes (01)
      0x0B: Supported CODEC(s) = G.729 (03), G.729E (05), PCMU (06), G.726-32 (07), G.728 (08), PCMA (09), G.726-16 (0A), G.7
      0x0C: Silence Suppression Support = No (00)
      0x0D: Echo Cancellation Support = Yes (01)
      0x10: MTA's "ifIndex" starting number in "ifTable" = 09
  ⊟ Option 55: Parameter Request List
      12 = Host Name
      15 = Domain Name
      1 = Subnet Mask
      3 = Router
      2 = Time Offset
      6 = Domain Name Server
      7 = Log Server
      122 = CableLabs Client Configuration
  ⊟ Option 61: Client identifier
      Hardware type: Ethernet
      Client MAC address: 00:00:ca:dd:a7:e4 (00:00:ca:dd:a7:e4)
    End Option
    Padding
File: "D:\Lab\Ethereal Traces\mtar  P: 192 D: 192 M: 0
```

As you can see, for an MTA this parameter contains information such as number of telephony ports/endpoints, supported CODECs, silence suppression support, and echo cancellation support.

Option #82

Option #82, Relay Agent Information, is defined in IETF RFC 3046. This option is inserted by DHCP relay agents such as a CMTS when forwarding requests to a server. DHCP servers can then use the information in this option when making IP address assignment decisions. The server sends this information back to the relay agent unchanged in the response, and then the relay agent removes this option before forwarding the response to the client.

Option #82 comprises a number of sub-options. In a cable network, sub-options #1 and #2 are typically the only ones used. Sub-option #1 is the agent circuit ID. Sub-option #2 is the agent remote ID—in a cable network it identifies the cable modem from where the DHCP request initiated.

The trace in Figure 3-6 shows an example.

Figure 3-6 *DHCP Option #82 Example*

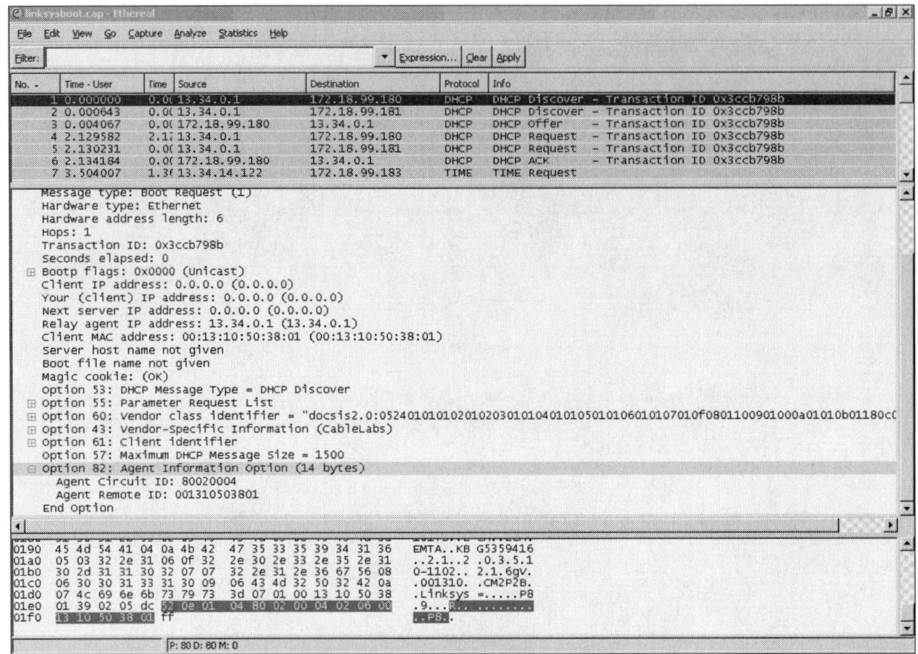

As you can tell from this example, sub-option #2 equates to the MAC address of the cable modem from where the DHCP packet originated. Thus, the provisioning server can use the information in option #82 to associate customer premises equipment (CPE) devices to cable modems.

Option #122

Option #122, CableLabs Client Configuration, consists of several sub-options that control the eMTA provisioning process. In addition to being defined in the PacketCable provisioning specification, option #122 is also defined in IETF RFC 3495. Option #122 is used for both the cable modem and eMTA provisioning processes. This is the option mentioned earlier that links these two provisioning processes together.

Sub-options #1 and #2 are the ones used in the cable modem provisioning process. They identify the DHCP servers the eMTA is authorized to use for its provisioning. Sub-option #1 is the primary DHCP server IP address and is required, while sub-option #2 is the secondary DHCP server IP address and is optional. Remember, DHCP Discover packets are broadcast so any DHCP server operating in the network could respond with a DHCP Offer. If an eMTA receives a DHCP Offer from a server other than the ones identified by sub-options #1 and #2 in its cable modem provisioning process, it ignores them. These options allow you to control what DHCP servers embedded MTAs can communicate with. If you want to bypass this security check, you can use a value of all ones (255.255.255.255) for sub-option #1. This permits the eMTA to communicate with any DHCP server. If you wanted to use eMTAs in your network only for high-speed data service, you can set sub-option #1 to a value of all zeros (0.0.0.0). This disables the eMTA provisioning process. An example use of this functionality is to disable telephony service for a non-paying customer.

Figure 3-7 shows an Ethereal trace of option #122 as used in the cable modem provisioning process.

Figure 3-7 *DHCP Option #122 CM Example*

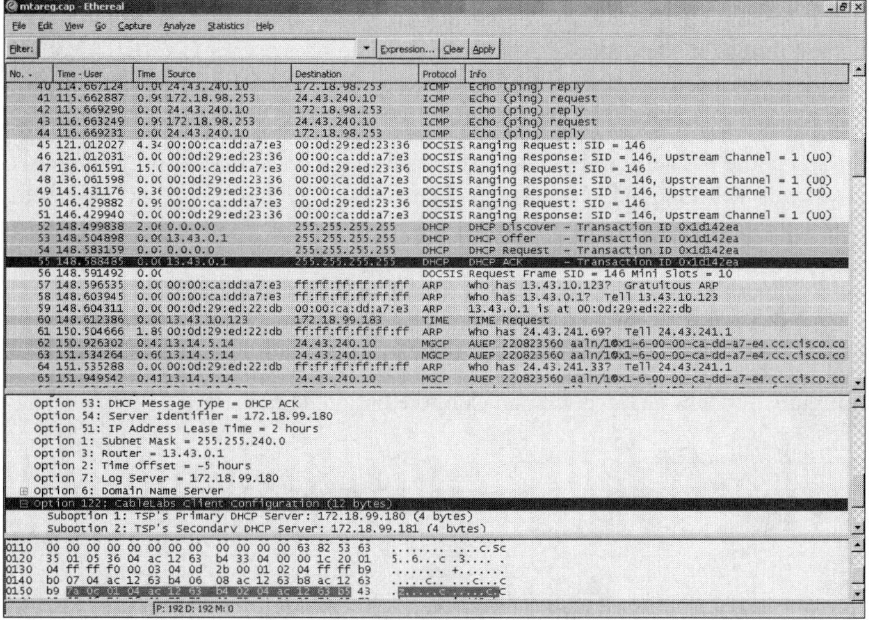

In this case both sub-option #1 and #2 are provided and equate to the DHCP server IP addresses the eMTA is authorized to communicate with.

The remaining sub-options are used during the eMTA provisioning process.

Sub-option #3 is required and identifies the SNMP provisioning entity used to configure the eMTA. Logically, this device uses the SNMP protocol to provision the eMTA; more on this device is covered in the SNMP portion of this chapter. This parameter can be supplied as a FQDN or an IP address. If this parameter is set to 0.0.0.0, the MTA stops its provisioning process. If this parameter is specified as a FQDN, the MTA issues a DNS query to the DNS server(s) specified by DHCP option #6 to resolve the FQDN into an IP address.

Sub-options #4 and #5 are optional and only valid for the secure provisioning method. In the secure method the Kerberos protocol is used for authenticating MTAs and establishing the security keys needed for the SNMPv3 protocol. More on Kerberos is covered later in this chapter. Sub-option #4 consists of variables used to control the Kerberos messaging back-off and retry mechanisms between the eMTA and Key Distribution Center (KDC).

Likewise, sub-option #5 consists of similar variables used to control the Kerberos messaging back-off and retry mechanisms between the eMTA and SNMP provisioning entity.

Sub-option #6 is required and defines the Kerberos realm if the secure provisioning method is used. If either the basic or hybrid provisioning methods are used, this parameter identifies this. An important fact to keep in mind is that this parameter is case-sensitive; it must be defined using all capital letters. If you want to use the basic provisioning method, sub-option #6 needs to be set to either BASIC.1 or BASIC.2. The difference between these values is that BASIC.2 requires the eMTA to send a SNMP provisioning complete notification whereas BASIC.1 does not. Likewise, if you want to use the hybrid provisioning method sub-option #6 needs to be set to either HYBRID.1 or HYBRID.2. Again the difference between these values is that HYBRID.2 requires the eMTA to send a SNMP provisioning complete notification whereas HYBRID.1 does not.

Sub-options #7, #8, and #9 are all optional. Sub-option #7 is valid only for the secure method and indicates if the MTA should first obtain a Ticket Granting Ticket (TGT) from a Kerberos Ticket Granting Server (TGS) before obtaining service tickets. As you'll learn about in the next chapter the use of a TGS is an optional part of the secure provisioning flow. Sub-option #8 defines the maximum amount of time an eMTA can take to fully provision. By default the value of this parameter is 10 minutes. Sub-option #9 defines the rules used to define security ticket invalidation. For more detail on this sub-option refer to IETF RFC 3594.

Figure 3-8 shows a trace with an example option #122 for an eMTA.

Figure 3-8 *DHCP Option #122 MTA Example*

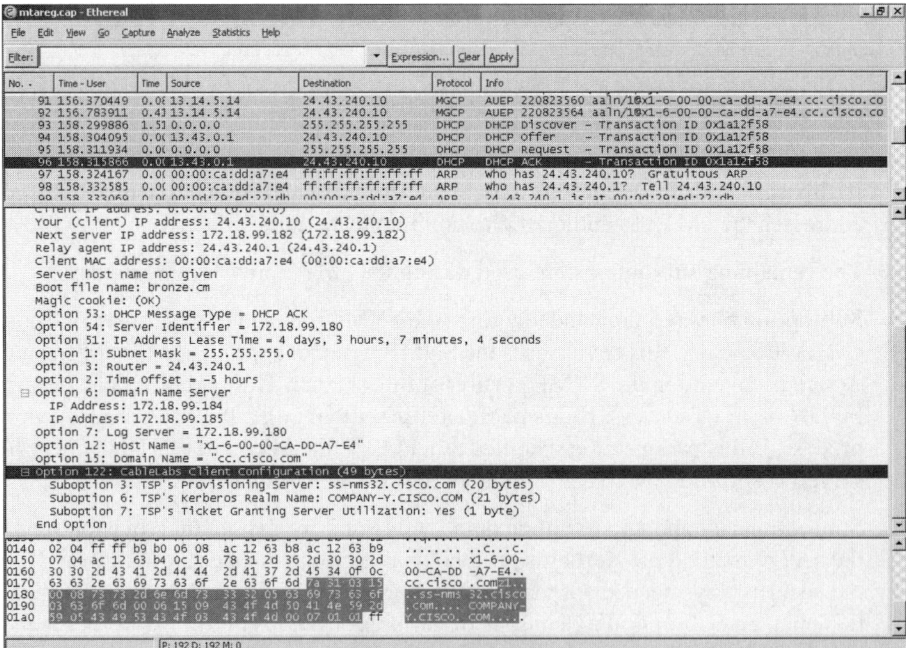

DNS

The Domain Name Server (DNS) for IP version 4 is defined in the IETF RFCs 1034 and 1035. DNS is a distributed database used by TCP/IP applications to map between IP addresses and domain names. DNS names are hierarchical with periods used to separate levels of hierarchy. The DNS database can be graphically represented as a tree where the top of the tree is the root represented by a dot or period. A DNS name ending with a period is known as an absolute domain name or FQDN. Put another way, a FQDN includes all the network domain names leading back to the root. A zone is a portion of the DNS hierarchical tree under a specific administration. Each zone has one primary DNS server and a number of secondary servers. These secondary servers populate their database by periodically copying the zone contents from the primary server. Because these servers have the zone information in their databases, they are said to be authoritative for that zone. DNS servers cache information learned by querying authoritative servers and can respond non-authoritatively to future queries for this data.

DNS can use either TCP or UDP for transport and uses port 53 in both cases.

NOTE Most of the time DNS uses UDP for transport; TCP is normally used only when large packets need to be sent such as for zone transfers.

DNS messages are either queries or responses. Figure 3-9 illustrates the format of a DNS packet.

Figure 3-9 *DNS Packet Format*

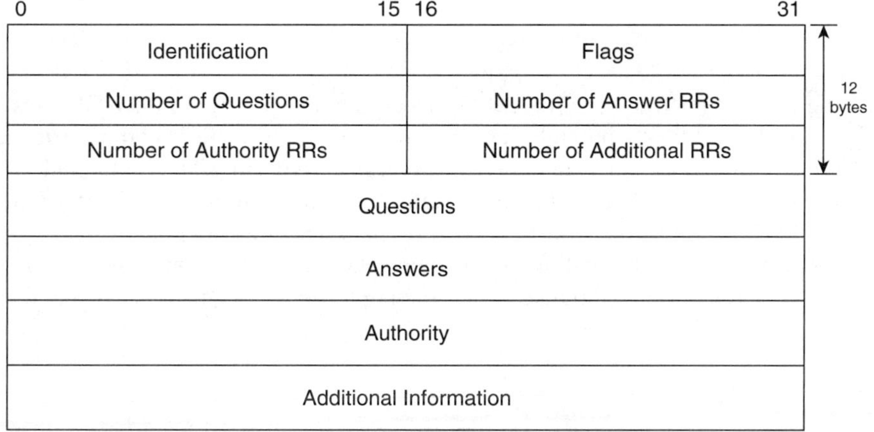

Queries are sent from a client to a DNS server, which resolves the query and sends a response back to the client.

If the DNS server that received the query cannot resolve the query directly, it can recursively forward the request to another DNS server. Part of the functionality of a DNS server is caching query/response information; thus, frequently queried records can be returned much more quickly.

After a PacketCable MTA goes through the DHCP steps previously covered, a DNS record also gets created. Specifically, the DHCP server sends a DNS update to the preconfigured DNS servers. This is known as dynamic DNS (DDNS). The capability of providing dynamic DNS updates is described in the IETF RFC 2136. The DHCP options #12 (hostname) and #15 (domain name) together form the FQDN of the MTA. The DHCP server requests the DNS servers to dynamically create an address record (known as an "A" record), mapping this FQDN to the IP address it assigned the MTA. Frequently a reverse mapping record is also created (known as a "PTR" record); by querying this record the IP address can be resolved into the FQDN.

A PacketCable MTA also uses DNS to resolve the names of other servers it interacts with. For example the CMS it communicates with is defined as a FQDN in the MTA configuration file. The MTA contacts a DNS server to resolve this FQDN into one or more IP addresses. DHCP option #6 is mandatory for a PacketCable MTA and contains the IP address of one or more DNS servers. An eMTA also uses DNS before it can communicate with the KDC. Remember DHCP option #122 used for PacketCable provisioning? Sub-option #6 is the Kerberos realm name. Before the eMTA can begin the process of obtaining security keys, it needs to contact the DNS server to figure out how to contact the KDC. First the eMTA requests a server (SRV) lookup to find out the FQDN of the KDC in that realm. Then it contacts the DNS server again to resolve that FQDN into an IP address.

TFTP

The Trivial File Transfer Protocol (TFTP) is defined in RFC 1350. TFTP uses UDP for transport, specifically port 69. Cable modem configuration files are transferred using TFTP. MTA configuration files are also typically transferred using TFTP, although they can optionally use HTTP as well. Additionally, cable modem firmware is transferred using TFTP. TFTP operation is fairly simple. A client makes a read request to a server for a file. The server then starts sending the file to the client in data packets. The receipt of each data packet is acknowledged by the client sending acknowledgement packets to the server.

HTTP

The Hypertext Transfer Protocol (HTTP), defined in RFC 2616, is also a request/response protocol. A client sends a request to the server in the form of a request method, Universal Resource Identifier (URI), and protocol version, followed by a message containing request modifiers, client information, and possible body content over a connection with a server.

The server responds with a status line, including the message's protocol version and a success or error code, followed by a message containing server information, entity information, and possible entity-body content. As previously stated, a PacketCable MTA can obtain its configuration file using HTTP instead of TFTP.

Time-of-Day (ToD) Protocol

The Time-of-Day (ToD) protocol provides a mechanism for client devices to obtain the current time and date. It can use either UDP or TCP for transport and in both cases uses port 37. A client sends a request to a server for the current data and time. The server responds with the number of seconds since midnight January 1, 1900, GMT. DHCP option #4 is used to provide the identity of the time server to a client. This option is used during the cable modem provisioning process. A related option is DHCP option #2, which is time offset. This identifies the time zone of the device by specifying the offset from GMT.

System Logging (syslog)

The system logging (syslog) protocol described in RFC 3164 is used by a device to convey information such as event notifications that typically require no acknowledgement. Each device is configured to send log messages to a particular logging server. Multiple devices can send logs to the same or different locations. The syslog protocol uses UDP for transport, specifically port 514.

A syslog message has three parts—priority, header, and the message itself. The first part (the priority) consists of a number inside of angle brackets, for example <132>. The priority value represents both the facility and severity of the message. The message facility helps identify the source of the message such as the device kernel, a mail system, a system daemon, etc. The values 16–23 are for processes or daemons that have no explicitly defined facility. These are known as "local use" facilities where 16 represents local use 0 (local0), 17 represents local use 1 (local1), 18 represents local use 2 (local2), and so on. Table 3-4 lists the message severity codes.

Table 3-4 *Syslog Message Severity*

Code	Severity
0	Emergency
1	Alert
2	Critical
3	Error
4	Warning
5	Notice
6	Informational
7	Debug

The priority value is calculated by multiplying the facility value by 8 and then adding the severity value. For a PacketCable MTA the message facility is always local use 0; thus, the priority is always between 128 ($16 \times 8 + 0$) and 135 ($16 \times 8 + 7$).

The next part of the message is known as the header and consists of a timestamp followed by an indication of the hostname or IP address of the device.

The third part of the message is the message itself. This usually contains additional information of the process that generated the message and the text of the message.

The identity of the syslog server can be sent to a device via DHCP using DHCP option #7. This option is mandatory for a PacketCable MTA. If it is set to all zeros (0.0.0.0) or all ones (FF.FF.FF.FF), syslog logging is turned off. Otherwise, this field is the IP address of the syslog server.

Figure 3-10 shows a syslog message.

Figure 3-10 *Example PacketCable Syslog Packet*

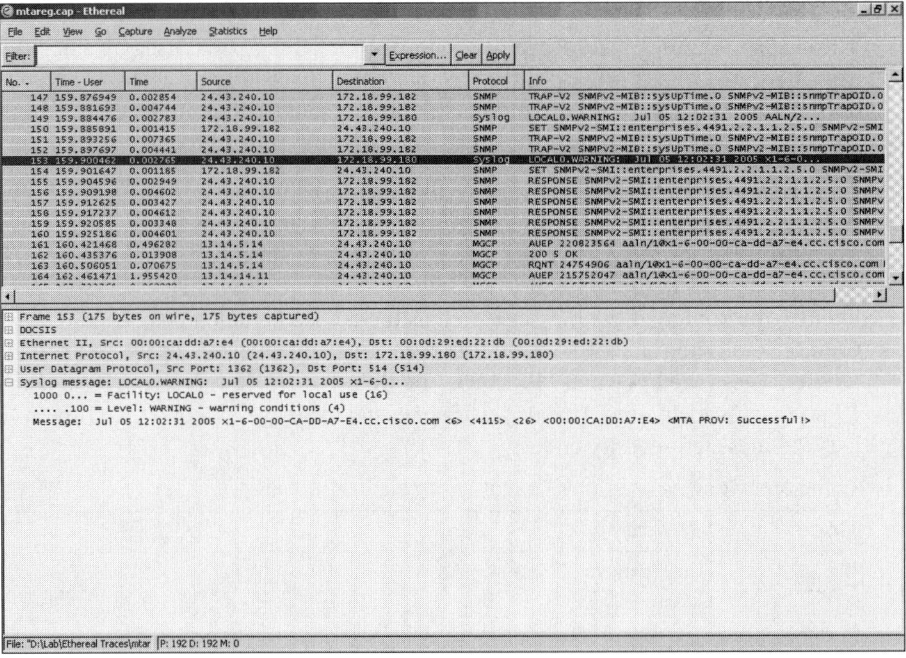

SNMP

The Simple Network Management Protocol (SNMP) was originally defined in RFC 1157. SNMP can be used for several of the FCAPS management functions described in Chapter 2, "PacketCable Functional Components." SNMP functionality comprises managers and agents. The SNMP manager compiles one or more Management Information Bases (MIBs). A MIB is a virtual database that identifies each manageable object by its name, syntax, accessibility, status, a text description, and a unique manageable Object Identification Number (OID). Three types of MIBs exist—public, experimental and private enterprise. OIDs represent each manageable object with a unique sequence of numbers and names. An SNMP Agent is running in most networked devices; its function is to constantly gather information so that the manager can request and respond to requests from one or more managers with the appropriate data.

SNMPv1 defines five types of messages also called Protocol Data Units (PDUs):

- **Get Request**—Used by a manager to request individual non-tabled objects
- **Get Next Request**—Used by a manager to request tabled objects
- **Set Request**—Used by a manager to write new values to objects
- **Get Response**—Response to the previous three messages
- **Trap Response**—Used to send a pre-programmed event notification to the manager

SNMPv2c added three additional PDU types:

- **Get Bulk Request**—A more efficient Get Next Request to get multiple objects simultaneously
- **Inform Request**—Used for manager-to-manager transfers
- **SNMPv2 Trap Response**—A new format for sending traps; obsoletes the previous Trap Response PDU type

SNMPv3 adds security to SNMP. In PacketCable, the Kerberos protocol, described next, is used to exchange security keys between the manager and agent. These security keys are then used to authenticate and optionally encrypt the SNMP packets. A parameter known as a transform identifier is used to negotiate an encryption algorithm (currently, only DES encryption or NULL encryption [no encryption] are supported). The HMAC MD5 and SHA-1 authentication algorithms are supported for SNMPv3 message authentication in PacketCable.

In PacketCable, SNMP is used to provision the MTA. One of the sub-options in the PacketCable DHCP option (option #122) is the provisioning SNMP entity (sub-option #3). This entity handles the SNMP aspects of the provisioning process, such as communicating the name and location of the MTA configuration file. Because SNMPv3 adds security to the provisioning process, it is the version required for the secure method of PacketCable provisioning. For the basic and hybrid methods of provisioning, SNMPv2c is used instead.

If you are using the secure method for PacketCable provisioning, it might be desirable to manage or monitor the eMTA via SNMP without the security requirement. A capability in PacketCable known as *co-existence* can meet this requirement. Co-existence means that SNMPv2c can be used for network management and/or monitoring of eMTA MIBs with secure SNMPv3 required only for the provisioning process.

Kerberos

Kerberos is the protocol used to provide security to the provisioning process. It allows clients and services to securely identify themselves to each other. Kerberos was initially developed at the Massachusetts Institute of Technology and is specified in RFC 4120. Specifically, Kerberos is used to establish the security keys needed for SNMPv3 messaging. Thus, Kerberos is needed only in the secure method of PacketCable eMTA provisioning. However, Kerberos does have uses in PacketCable other than for eMTA provisioning; it is also used to provide security for the call signaling data exchanged between a MTA and a CMS by negotiating the security keys used for IPSec.

It is conceivable that a rogue MTA could steal telephony service by misrepresenting itself as another legitimate and authorized MTA. This type of identity theft is analogous to someone stealing your driver's license and credit cards and assuming your identity. Additionally, this MTA clone could not only get phone service for free but could also intercept and manipulate the service of the original MTA. Kerberos protects against this possibility by providing a mechanism in which MTA devices and PacketCable services are authenticated and then granted keys used to secure future communication.

At this point it is probably worth mentioning how Kerberos differs from other key management protocols such the Internet Key Exchange (IKE) protocol described in IETF RFC 4306. With this type of protocol, key management is handled directly between the parties involved. This is possible because the parties know the identity of each other well in advance. Also typically a small number of security associations are required on each party, permitting the computationally intense security algorithms to reside directly in the devices involved. In PacketCable several MTAs need to establish security associations with the provisioning entity and CMS. Also, because the MTA obtains its configuration dynamically, its information is typically not well known in advance; therefore, Kerberos is better suited for PacketCable.

Kerberos provides security keys to clients and servers through the use of an authentication server (AS) and a TGS. Think of these servers as the official "government" agencies responsible for issuing licenses to these clients for a particular service. In a PacketCable network, the MTAs are the clients, the KDC is both the Kerberos AS and TGS, and the services are secure provisioning and secure MTA-CMS call signaling. The AS uses public key cryptology to authenticate clients and if successful issues these clients a ticket known as a TGT. This TGT permits the client to contact the TGS. The TGT is presented to the TGS, which returns a key or ticket for a particular service. In PacketCable the TGS step is optional and often not used; in this case the AS provides the service tickets directly. For PacketCable provisioning the service is SNMPv3 secure communication with the provisioning server. For secure call signaling the service would be the security keys needed for the IPSec messaging between the MTA and CMS. Remember that not only do the user clients (that is, MTAs) need to have keys registered with the Kerberos AS, but the application servers (that is, NMS/OSS [PacketCable provisioning] CMS [call signaling]) need to as well.

The collection of clients and servers under the domain of a single KDC is known as a Kerberos *realm*. The KDC handles the authentication and key distribution for all clients and servers in its realm. In standard Kerberos secret pre-shared keys are used for the client to server authentication. Separate keys are needed between each client/server pair. As you might imagine, in a realm with many clients and servers this need for separate keys does not scale well. An enhancement to Kerberos known as the Public Key Cryptology for Initial Authentication (PKINIT) extension allows digital signatures and certificates to be used for this authentication. In this case only public keys are shared between users. In public key cryptology, two keys are used to encrypt and decrypt data. These keys are known as the *public key* and the *private key*. When one key is used to encrypt data, the other key is required to decrypt the data. The public key can be known by multiple entities whereas the private key is assumed to be secret. So if a device or entity encrypts a message with the private key that only it knows, and another device is able to decrypt this message with the public key, it is in essence verifying the identity of the device that sent the message. With this process the private key never needs to be communicated; this process is the essence on how digital signatures work.

NOTE This discussion is very high-level; if you are interested in the details, I would encourage you to refer to the relevant standards and specifications.

Digital certificates are defined under the ITU X.509 standard. These certificates are typically structured in a hierarchical fashion so that organizations can issue certificates that are related or "chained" to more authoritative organizations.

Two X.509 certificate hierarchies are associated with the KDC and PacketCable—the MTA device hierarchy and the CableLabs service provider hierarchy. In the MTA device hierarchy each MTA has a certificate (unique to the MTA) that is chained to a manufacturer MTA certificate (unique to the MTA vendor), which is in turn chained to the MTA root certificate (common for all PacketCable MTAs). In the CableLabs service provider hierarchy, the KDC has a certificate (unique to the KDC) that is chained to the Local System Operator certificate (unique to the service provider division—this level of hierarchy is optionally implemented), which is chained to the Service Provider certificate (unique to the service provider), which is chained to the CableLabs service provider root certificate (common for all PacketCable service providers).

Consequently, the KDC is typically loaded with the following five certificates:

- CableLabs Service Provider Root
- Service Provider
- Local System
- KDC
- MTA Root

Logically, the MTA Root certificate contains the root certificate used by the KDC to validate certificates sent by the MTA. The KDC actually validates the whole MTA certificate chain. The signature on the MTA manufacturer certificate is validated against the MTA Root certificate, and the signature on the MTA device certificate is validated against the MTA manufacturer certificate.

An important fact to remember with Kerberos is that both the client and the server are authenticated. That is in PacketCable provisioning, the KDC authenticates the identity of the MTA, and the MTA authenticates the identity of the KDC. The messages used for this authentication are known as the Authentication Service (AS) Request and Response.

NOTE In PacketCable, all Kerberos messages use UDP for transport.

The KDC can return either the provisioning service ticket directly in the AS-RSP, or it can return TGT. If a TGT is returned, the MTA would then make another request to the KDC known as a TGS Request. The KDC responds to this message with a TGS Response containing the provisioning service ticket. As alluded to earlier, in addition to the provisioning service ticket, these messages can contain a secure call signaling service ticket used for the creation of IPSec security associations between the MTA and CMS components. So far the use of call signaling encryption has not been widely deployed in the field.

An important fact to remember is that the security tickets obtained from the KDC are stored in persistent storage, meaning they will not be lost on a MTA power cycle. This helps reduce the amount of processing load on the KDC by letting it focus on issuing new keys and key renewal instead of on dealing with MTAs that already have valid keys.

After the MTA has the security keys it needs, it obtains the application-specific security information via another Kerberos message exchange, this time with the application server. These messages are known as the Application Server Request and Response (AP-REQ/AP-RSP). In the case of secure provisioning this message exchange is with the SNMP provisioning entity, and SNMPv3 security keys are exchanged. In the case of secure call signaling this message exchange is with the CMS and IPSec security keys are exchanged.

CMTS Configuration

Most of the provisioning related configuration takes place on the OSS/NMS servers. However, some configuration is required on the CMTS component such as the location of the DHCP server(s).

Recall that the CMTS acts as a relay agent that forwards the broadcast DHCP requests it receives to one or more preconfigured DHCP servers. For a Cisco CMTS this is configured using the following command:

```
CMTS(config-if)#cable helper-address DHCP server IP
```

You can have this command configured multiple times on an interface. In fact, redundant pairs of helper addresses are typically defined. Also keep in mind that these commands need to be on the same interface that the IP address ranges are defined. For example, if you are bundling cable interfaces together, this command must be on the virtual bundle interface instead of the physical cable interfaces. Also with the command configured, the CMTS uses this helper address for DHCP broadcast packets received from cable modems as well as their CPE devices. The optional keywords **cable-modem** and **host** can be appended to this command to tell the CMTS to forward broadcasts only for these device types. In PacketCable, if you are using different DHCP servers for cable modem provisioning versus MTA provisioning, you would use these keywords to identify the separate DHCP servers.

You can also define how the CMTS populates the **giaddr** field in DHCP packets using the following command.

```
CMTS(config-if)#cable dhcp-giaddr [policy | primary]
```

The **giaddr** field is used to identify the relay agent to the DHCP server(s). In most deployments, different IP networks are used for cable modems and MTAs. Thus, multiple IP addresses are defined on the Layer 3 interface. If this configuration command is set to **primary**, the CMTS uses the main interface IP address in this field for all requests regardless of where they originate. If it is set to **policy**, the CMTS uses the main interface IP address for requests coming from cable modems but uses the first secondary interface IP address for requests coming from CPE devices such as computers and PacketCable MTAs.

Another way the CMTS can tell the DHCP server about where the request came from is by appending DHCP option #82 to requests before forwarding them. Servers can then use this information when making IP address assignment decisions. On a Cisco CMTS you can configure this option to be included using the following global configuration command:

```
CMTS(config)#ip dhcp relay information option
```

By default BAC uses DHCP option #82 to detect devices. If you don't want the CMTS to validate option 82 in messages it receives from DHCP servers, the following global configuration command can be used:

```
CMTS(config)#no ip dhcp relay information check
```

NOTE Obviously, other configuration needs to exist on the CMTS, such as interface IP addresses for the cable, MTA, and CPE subnets.

BAC Configuration

Chapter 2 introduced Cisco's solution for provisioning PacketCable eMTAs, BAC, and its various components. This section takes a look at the configuration steps needed to get a BAC solution operational. Many of the key concepts involved with configuring BAC for a PacketCable deployment are covered here.

NOTE For additional information please consult the Cisco Broadband Access Center for Cable Installation Guide, the Cisco Broadband Access Center for Cable Administration Guide, and the Cisco Broadband Access Center for Cable Command Line Interface Reference.

Recall that the BAC solution includes CNR DHCP and DNS servers working with BAC servers such as the RDU for centralized configuration and management, the DPE devices used to cache client configuration, and the KDC for eMTA security ticket distribution. These servers run on Solaris machines and securely communicate with each other through the use of a common preconfigured password. Consequently, if a mismatch occurs in this shared secret between BAC servers, they are unable to communicate.

CNR Configuration

In the BAC solution, Cisco Network Registrar (CNR) is used as the DHCP servers in a provisioning group. It can also be used for the DNS servers. The CNR software can run on Sun, Windows, or Linux servers. Logically, for a server provider network, high-end servers are recommended. CNR can be configured three ways—on the server directly via a CLI interface, via a Windows GUI interface, or remotely via a standard web browser. Obviously, the Windows GUI interface applies only to Windows-based installations. In this section, all configuration examples are shown using the web interface.

After logging in to the web interface, five functions are available:

- Administration
- Zone
- Host
- Address Space
- DHCP

The Administration function is used for tasks such as defining users, passwords, and privilege levels. The Zone and Host functions are applicable to DNS configuration. The Address Space and DHCP functions are applicable to DHCP configuration.

DHCP Configuration

Figure 3-11 shows an example of the web interface seen after logging into CNR and selecting the DHCP function.

Figure 3-11 *Main CNR DHCP Menu*

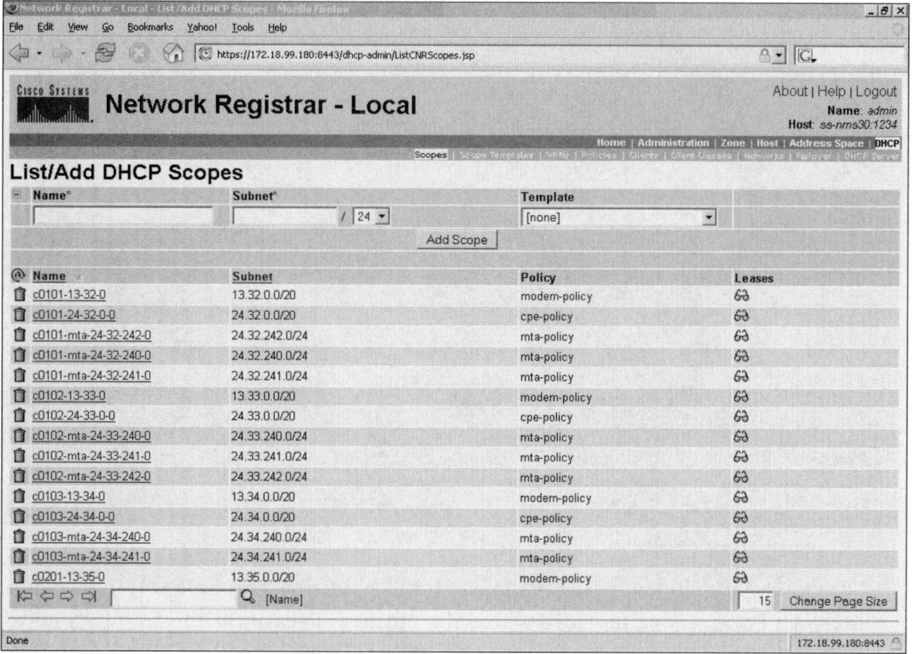

The default menu after selecting the DHCP function relates to the configuration management of scopes. *Scopes* are groups of one or more ranges of IP address that the DHCP server uses to assign an address to a client. To figure out what scope to use when assigning a client an IP address, CNR first uses the giaddr field in the incoming DHCP packet to determine what scope a particular client belongs to. Next CNR uses a parameter known as a selection tag. The *selection tag* determines what clients are authorized to receive address assignments from a specific scope. The scope also typically has an associated policy that defines DHCP options to be used for its clients. Selection tags and policies are covered in more detail later. Figure 3-12 shows an example scope configuration.

Figure 3-12 *Example Scope Configuration*

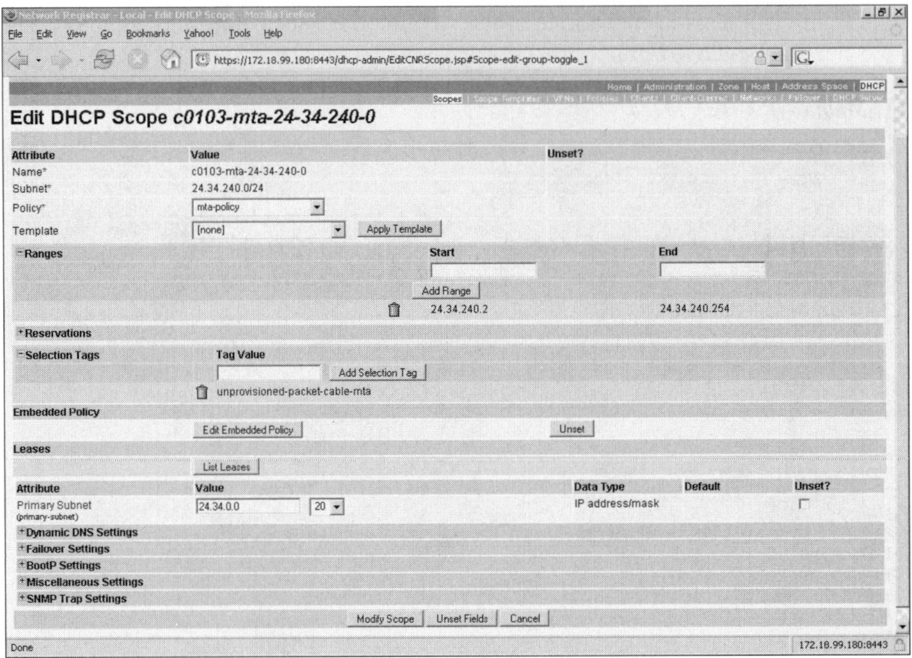

Often, multiple logical networks exist on the same physical network. This is typically the case in a PacketCable network where the physical network is the DOCSIS network with both ordinary data modems and PacketCable MTAs connected. Usually separate logical networks are used for cable modem IP addresses, MTA IP addresses, and ordinary CPE device (that is, personal computer) IP addresses. To configure this in CNR, multiple scopes would need to be configured. But how does CNR know that these scopes reside on the same physical network? The answer is to define some of these scopes as secondary and assign them to a primary subnet. If you look at the scope in Figure 3-12, you can see this is in fact a secondary scope. Towards the bottom of the screen, you can see an attribute called "Primary Subnet"; this defines the primary subnet. In this case the primary subnet is used for normal CPE address, and this secondary subnet is used for PacketCable MTAs.

As you learned earlier in the chapter, after a PacketCable MTA is assigned an IP address, the DHCP server dynamically contacts the DNS server to create a resource record. This functionality is called dynamic DNS and is also a part of the scope configuration. Referring to Figure 3-12, you can see below the Primary Subnet attribute an expandable box called Dynamic DNS Settings. If you expand this box, you see something similar to Figure 3-13.

Figure 3-13 *Scope DDNS Configuration*

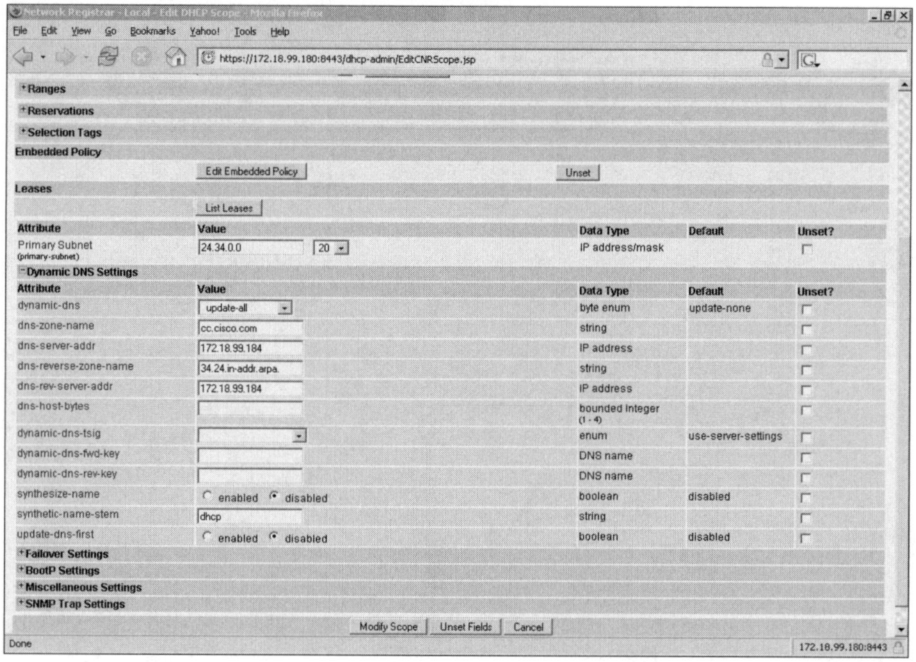

In this case DDNS is set to **update-all** meaning a dynamic update is sent for both the forward "A" record and the reverse record. The DNS server IP address and zone names for both the forward and reverse record are also specified. Notice, the forward DNS zone name is the same as the client's domain name (DHCP option #15). Some configuration is also required on the DNS server, which is covered in that portion of this chapter.

In CNR, clients can be grouped together into a client-class. This way common information can be defined in a single place that applies to all of these clients. A client-class is created under the Client-Classes menu as demonstrated in Figure 3-14.

Figure 3-14 *Client Class Configuration Example*

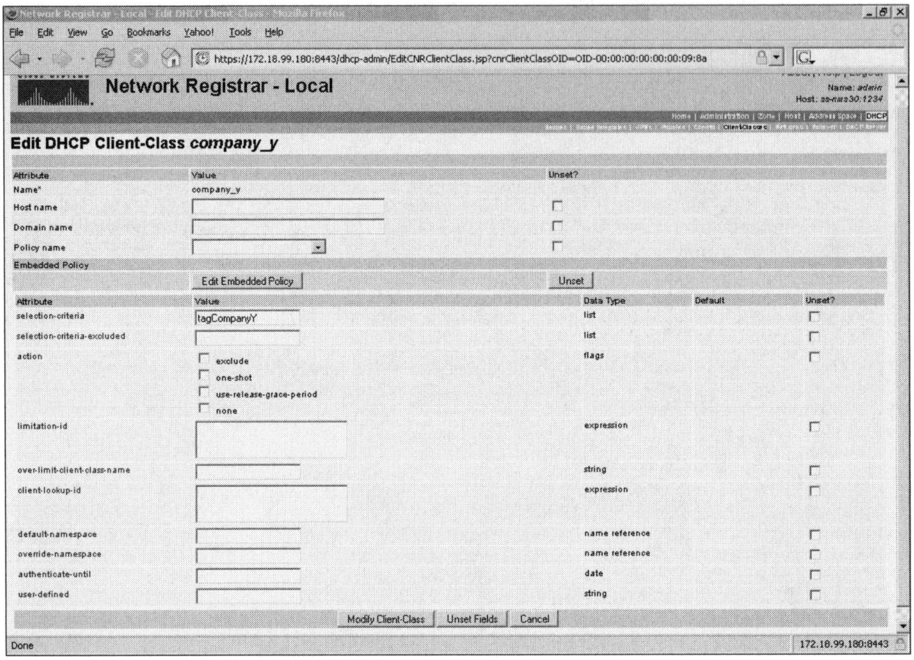

As with scope configuration, named policies and embedded policies can be associated with a client-class. A client-class attribute important in a BAC deployment is the selection-criteria attribute. This is the scope selection tag mentioned earlier in the chapter. All clients in this client-class have a selection tag of "tagCompanyY" and are permitted to get addresses only from scopes that have this selection tag included.

CNR assigns DHCP options to clients based upon configured polices. You can implement policies in four ways in CNR:

- Attached to a particular client
- Attached to a particular client-class
- Attached to a particular scope
- System default policies

The system default policy should be used to define options that are to be the same for all clients across all scopes. Policies can be attached to clients, client-classes, and scopes using either named policies or embedded policies. Named policies can be shared in different places, whereas an embedded policy is part of the scope, client-class, or client. If the client requests an option that is found in multiple policies, the following rules apply, listed in order of preference.

1 Client embedded policy
2 Client assigned/named policy
3 Client-class embedded policy
4 Client-class assigned/named policy
5 Scope embedded policy
6 Scope assigned/named policy
7 System default policy

You can create a named policy by selecting the Add Policy button under the Policies menu under the DHCP function, as shown in Figure 3-15.

Figure 3-15 *Named Policy Example*

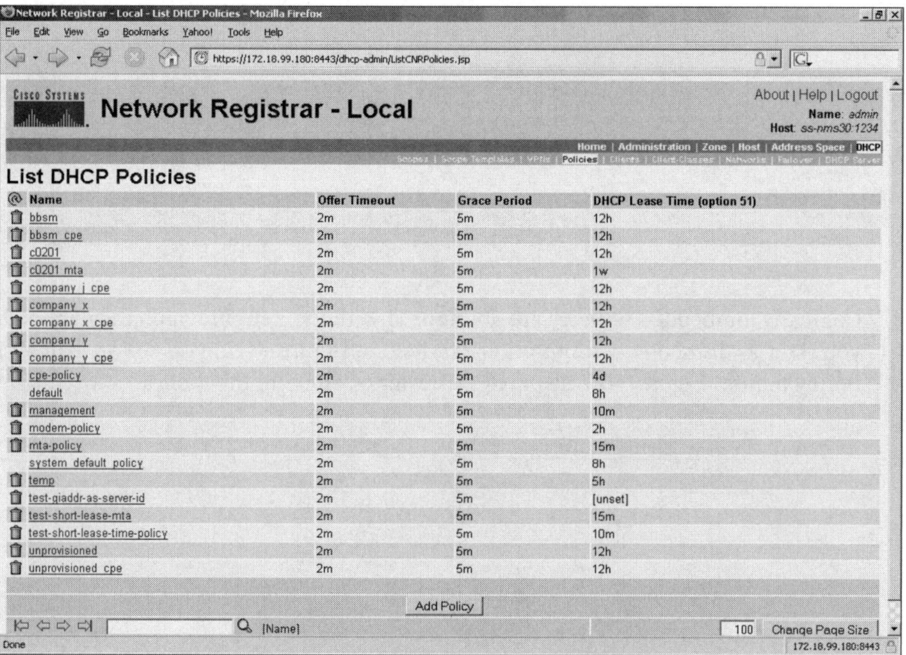

As you can see, the Add Policy button is at the bottom of the screen. After clicking this you have a window where you can fill in the various fields you want to be included in your policy, as shown in Figure 3-16. The DHCP specific options are listed under the Options attribute. After selecting the option and entering its value, make sure to select the Add

Option button. Once you have entered all the parameters for your policy, select the Add Policy button at the bottom of the screen.

Figure 3-16 *Configuring Policy Options*

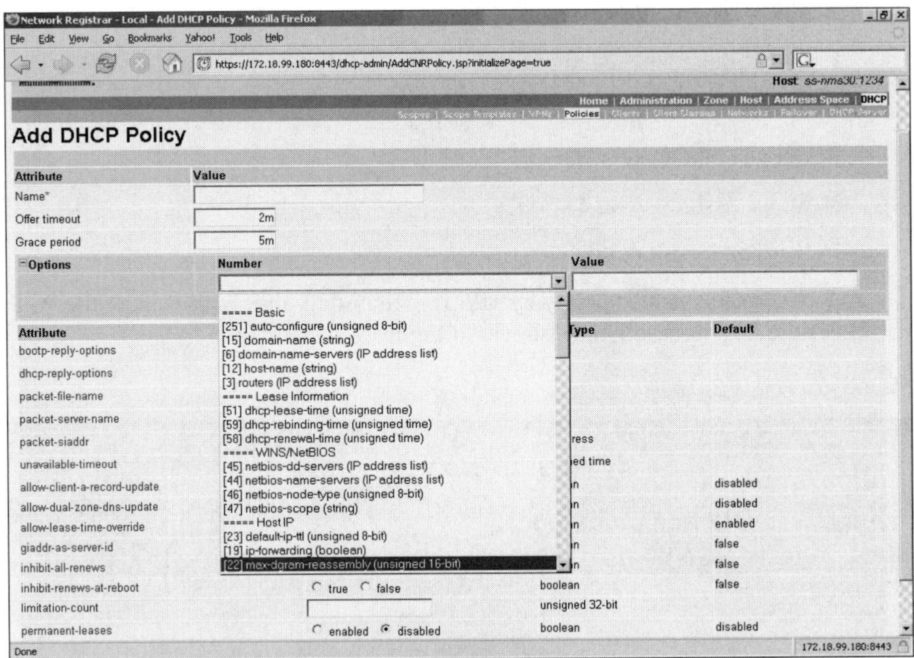

Logically, embedded policies are configured with the item in which they are embedded. If you refer to the scope configuration example in Figure 3-12, you see an Edit Embedded Policy button. Figure 3-17 displays the result of clicking this button.

Figure 3-17 *Embedded Policy Example*

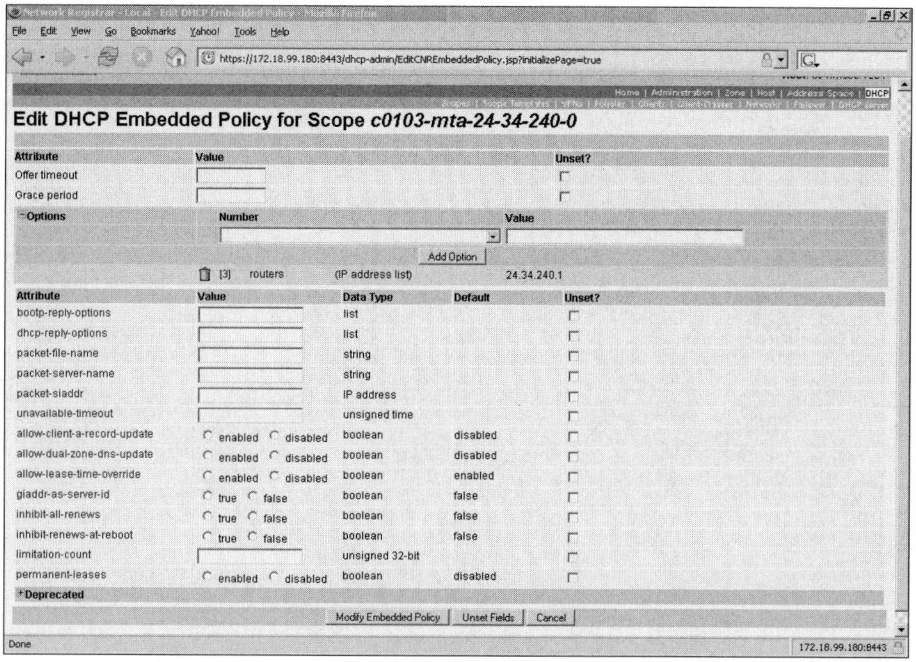

A scope embedded policy is a good place to define DHCP options that would be specific to that particular scope, such as default router IP address.

The operation of CNR can be altered or customized through the use of extension scripts that can be written in C/C++ or Toolkit Command Language (TCL). CNR utilizes extensions to communicate with outside devices such as BAC servers. The BAC extensions are installed on the CNR DHCP servers after CNR is installed.

During this installation process, PacketCable DHCP options are configured as well. Among these are the IP addresses of the primary and secondary DHCP servers (option #122, sub-options #1, #2); the IP addresses of the primary and secondary DNS servers (option #6); the Kerberos realm name (option #122, sub-option #6); and the shared secret used for the authenticated communication with the other BAC servers. These parameters are stored in a file called cnr_ep.properties, as illustrated in Figure 3-18.

Figure 3-18 *CNR Extension Point Properties Example*

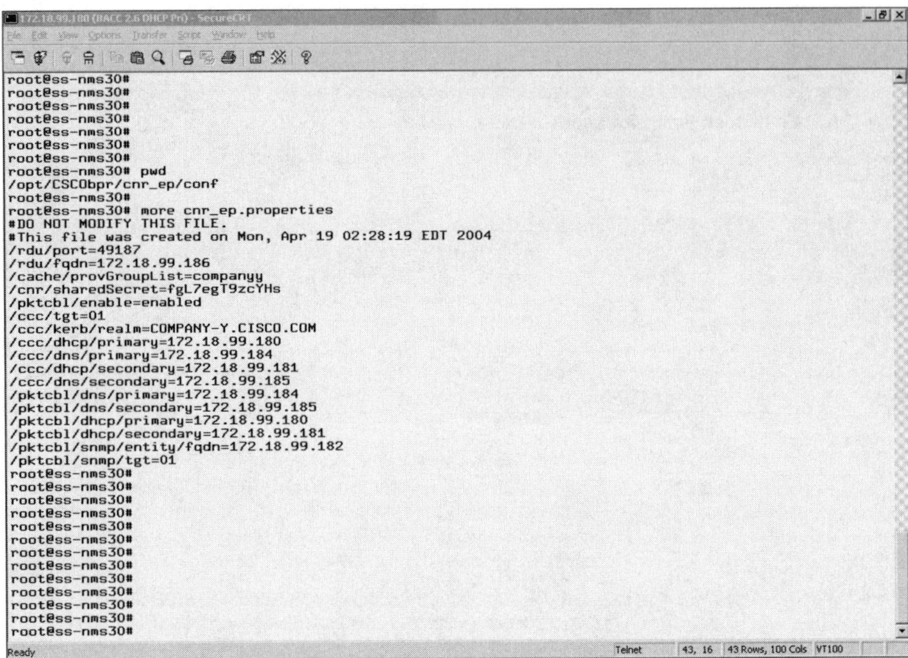

The properties beginning with /ccc/ (CableLabs Client Configuration) represent the DHCP option #122 sub-options and the properties beginning with /pktcbl/ represent the deprecated DHCP option #177 sub-options. Note, the exceptions are the /ccc/dns/ properties, which denote DNS server DHCP option #6.

Once the BAC extensions have been installed, they must be configured in CNR, and then the DHCP process needs to be reloaded. A utility called changeNRProperties.sh enables you to change these parameters if necessary. This utility is located in the <BAC HOME>/cnr_ep/bin directory (by default the <BAC HOME> path is /opt/CSCObpr). Figure 3-19 illustrates the syntax of this command.

Figure 3-19 *Change Extension Properties Utility*

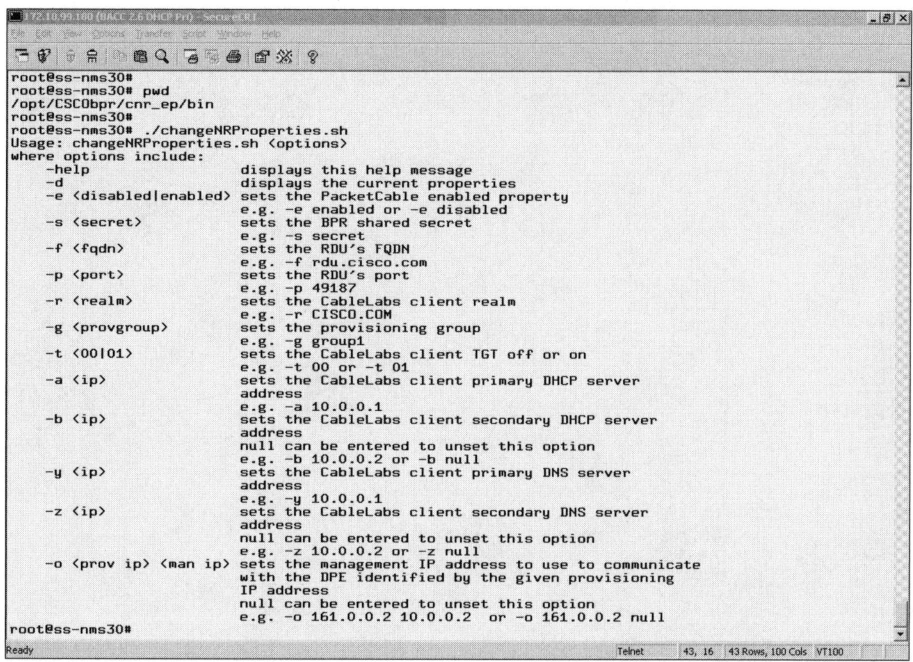

For DHCP failover between servers to be possible, both must have identical scope, policy, lease, and client class configuration. The primary server updates the secondary with every client operation. In normal operation, the secondary DHCP server ignores all DHCP Discover messages. Remember DHCP failover is provided by having the primary server reserve a portion of the address space for the secondary server. If the secondary server detects the primary server has failed, it can use this address allocation.

NOTE The secondary server's use of a pre-identified address space eliminates the possibility of duplicate address assignment by both servers in the event server-to-server communication is lost.

In CNR, the failover configuration is done under the DHCP Server menu, as shown in Figure 3-20.

Figure 3-20 *DHCP Server Failover Configuration*

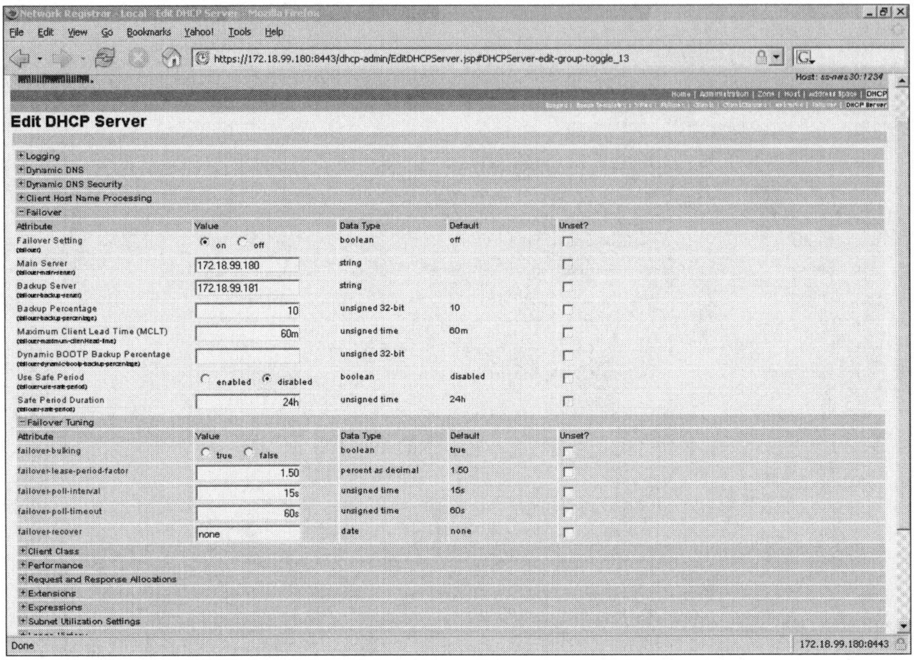

As you can see, both the main server and backup server IP addresses are defined as well as the percentage of IP addresses allocated to the backup server (in this example 10 percent is allocated).

DNS Configuration

At this point your DHCP servers should be ready to go; next the configuration of the DNS servers is examined. Figure 3-21 shows an example of the web interface seen after logging into CNR and selecting the Zone function.

New zones on a primary DNS server can be added from this page. Figure 3-22 shows an example configuration of one of the existing zones, cc.cisco.com.

Figure 3-21 *Main CNR Zone Menu*

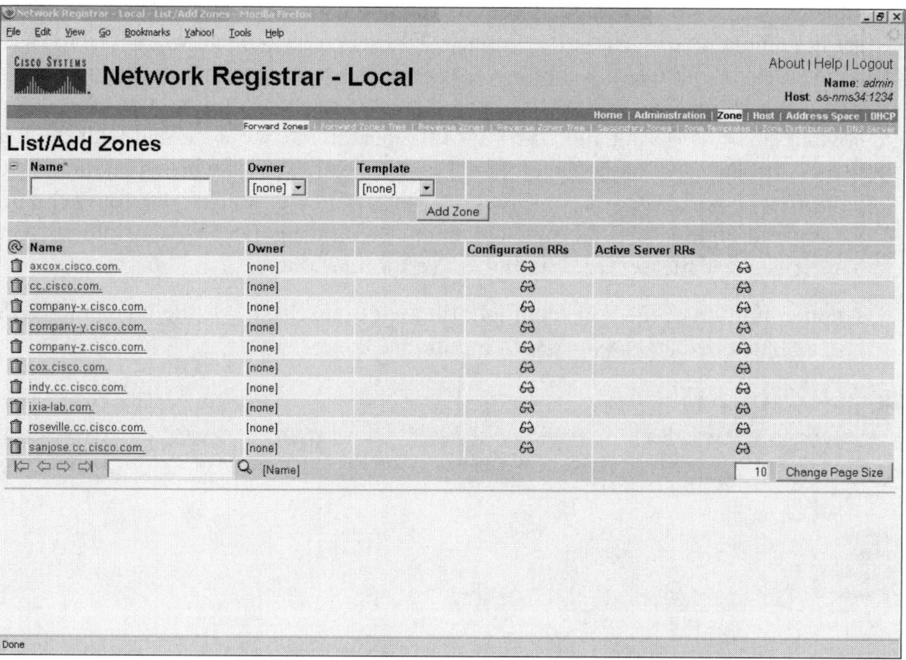

Figure 3-22 *Example Zone Configuration*

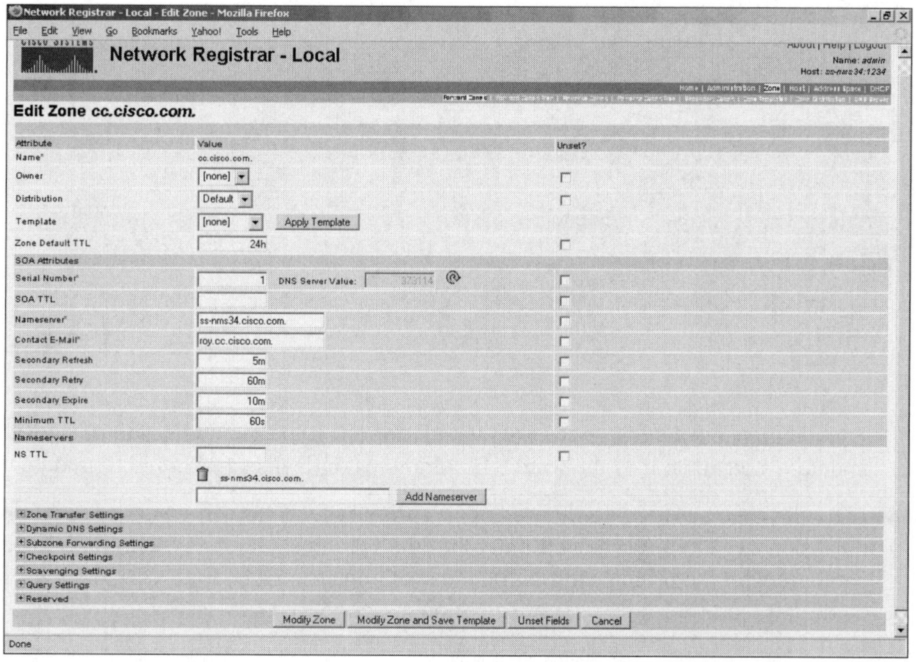

Network Registrar - Local - Edit Zone - Mozilla Firefox
File Edit View Go Bookmarks Yahoo! Tools Help
Network Registrar - Local
Name: admin
Host: ss-nms34:1234
Edit Zone cc.cisco.com.
Attribute
Value
Unset?
Name* cc.cisco.com.
Owner [none]
Distribution Default
Template [none] Apply Template
Zone Default TTL 24h
SOA Attributes
Serial Number* 1 DNS Server Value:
SOA TTL
Nameserver* ss-nms34.cisco.com.
Contact E-Mail* roy.cc.cisco.com.
Secondary Refresh 5m
Secondary Retry 60m
Secondary Expire 10m
Minimum TTL 60s
Nameservers
NS TTL
ss-nms34.cisco.com.
Add Nameserver
+Zone Transfer Settings
+Dynamic DNS Settings
+Subzone Forwarding Settings
+Checkpoint Settings
+Scavenging Settings
+Query Settings
+Reserved
Modify Zone
Modify Zone and Save Template
Unset Fields
Cancel
Done

The first thing you need to do when creating a zone is define the Start of Authority (SOA) Attributes. The serial number, name server, and contact email fields are mandatory. The serial number field is used by a primary DNS server to indicate when its database changes; an increase in value of this parameter triggers a zone transfer to the secondary server(s). The name server field is used to define the value for the Name Server (NS) record. The contact email field is an altered form of the person in charge's email address. Below the SOA Attributes section is one entitled Nameservers. This is where you identify the authoritative name servers for this zone. You must enter at least one for each zone; in the example in Figure 3-22 the same server as listed for the NS SOA attribute is configured. This is because this server is authoritative for this zone.

Another portion of the zone configuration worth discussing is the Dynamic DNS Settings tab, as illustrated and expanded in Figure 3-23.

Figure 3-23 *Zone DDNS Configuration*

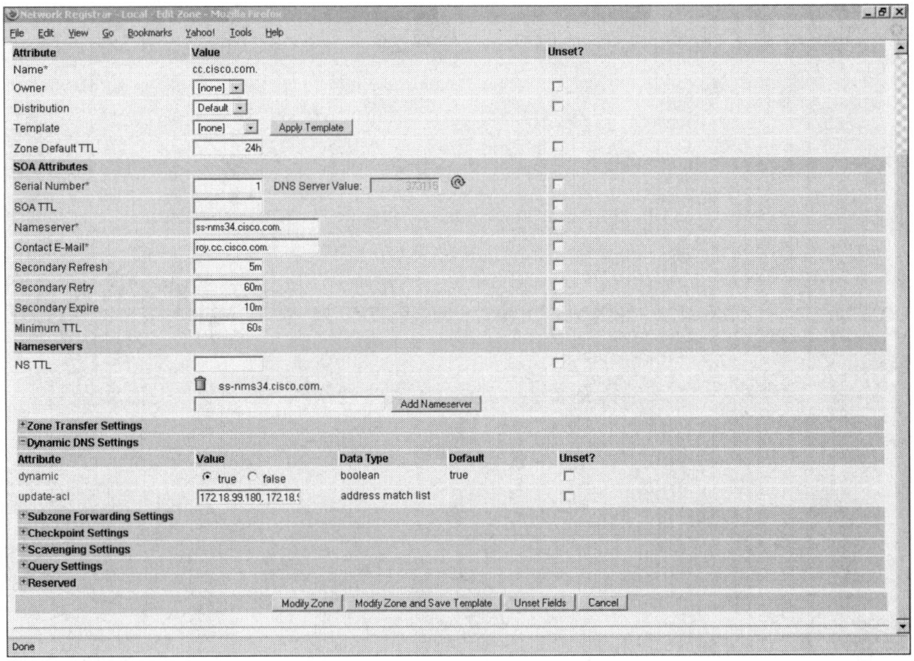

The dynamic attribute is set to true, which permits dynamic updates to this zone. The update-acl attribute lists the source IP addresses of devices permitted to update this zone.

After the zone has been configured on the primary DNS server, you need to configure it on the secondary DNS server(s). Remember the secondary server synchronizes its zone database with the primary through the use of zone transfers. Figure 3-24 shows the configuration of the cc.cisco.com zone on a secondary DNS server.

Figure 3-24 *Secondary Zone Configuration*

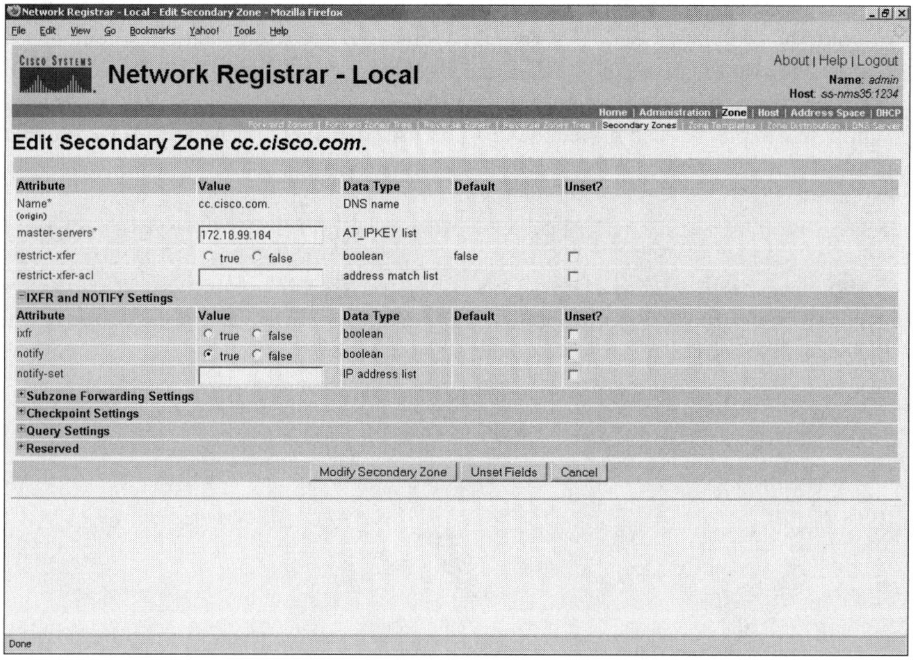

The master-servers attribute is required and identifies the primary server for this zone. The restrict-xfer options can be used to limit the resource records in this zone learned by the secondary server.

NOTE It is also a good practice to create reverse zones for each network you use. A *reverse zone* allows you to convert IP addresses back to hostnames. Reverse zones use the in-addr.arpa domain. Reverse zones are not a requirement for PacketCable but are still worthwhile to configure. Please consult Cisco's CNR documentation for more details on how to configure reverse zones.

Address resource records for PacketCable MTAs are dynamically created when they provision. But some records you must statically define, such as those for the KDC and CMS components. These records would all be created on the authoritative primary DNS server. Simple hostname-to-IP address or "A" resource records can be created under the Host function. Simply click on the zone in which you want to add the record and then add the hostname and IP address in the dialog boxes that appear on the top of the screen.

Resource records of any type can also be created by clicking the Configuration RRs option on the zone of interest in the forward zones menu. This is how the "SRV" record of the KDC would be added. Figure 3-25 shows the web page used to add static resource records. In this example you see three static RRs, and the first one is a KDC "SRV" record. Pay attention to this record because it is required for PacketCable provisioning. You'll see more about how in Chapter 4. Also notice the third record, which is the "A" record for the KDC FQDN.

Figure 3-25 *Static Resource Record Configuration*

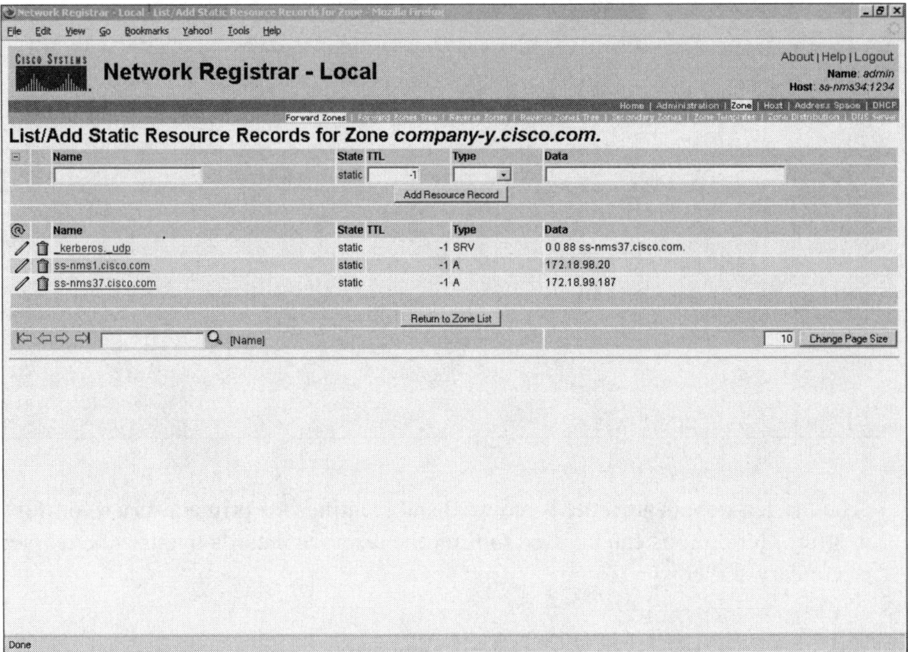

RDU Configuration

Remember from the previous chapter that the Regional Distribution Unit (RDU) is the centralized server for device configuration and management. The RDU pushes device configuration to the distributed DPE devices throughout the MSO's network. The RDU also has an API allowing third-party NMS/OMS devices to interface with BAC.

Like CNR the RDU can be configured via a web interface (specifically, port 8100 is used). After you log into the RDU the web page looks similar to the one shown in Figure 3-26.

Figure 3-26 *Main RDU Menu*

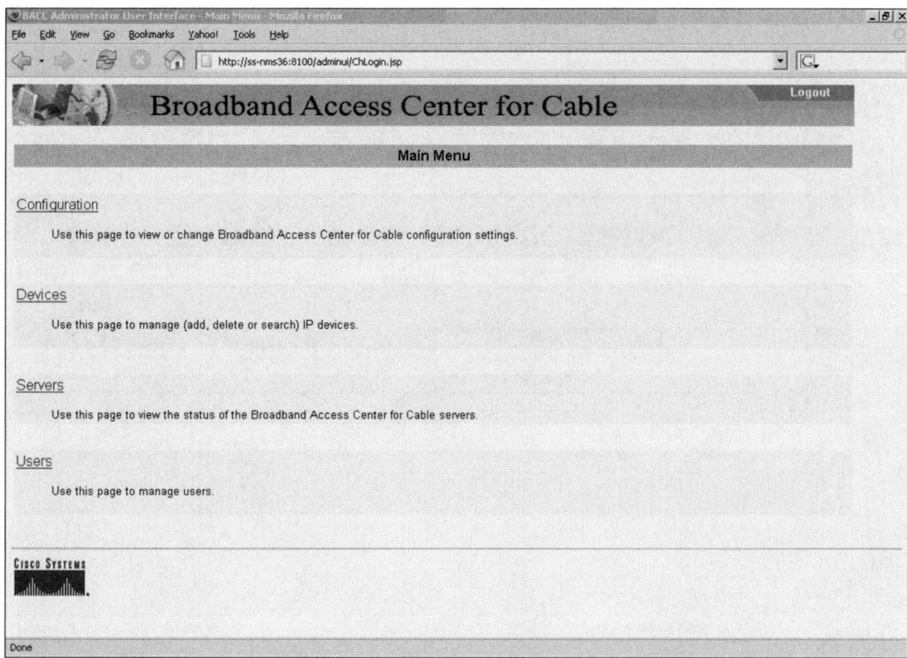

As you can see, you have four options on the main menu: Configuration, Devices, Servers, and Users. The Configuration option is used for:

- Defining DHCP criteria (which map to CNR selection tags)
- Defining classes of service (used to map configuration data to clients)
- Uploading/downloading external files (such as modem and MTA configuration files)

The Devices option is used to configure and manage devices such as cable modems and MTAs. Here configuration information is associated to individual devices. The Servers option is used to monitor the status of DPEs and DHCP servers. The Users option is used to add, delete, and modify users.

To help you understand how RDU configuration works, consider the following example. From the Devices option you can search for a device based upon its MAC address, IP address, FQDN, associated DHCP criteria, or associated class of service. In the example shown in Figure 3-27 the search was performed based upon the device MAC address.

Figure 3-27 *Device Management Example*

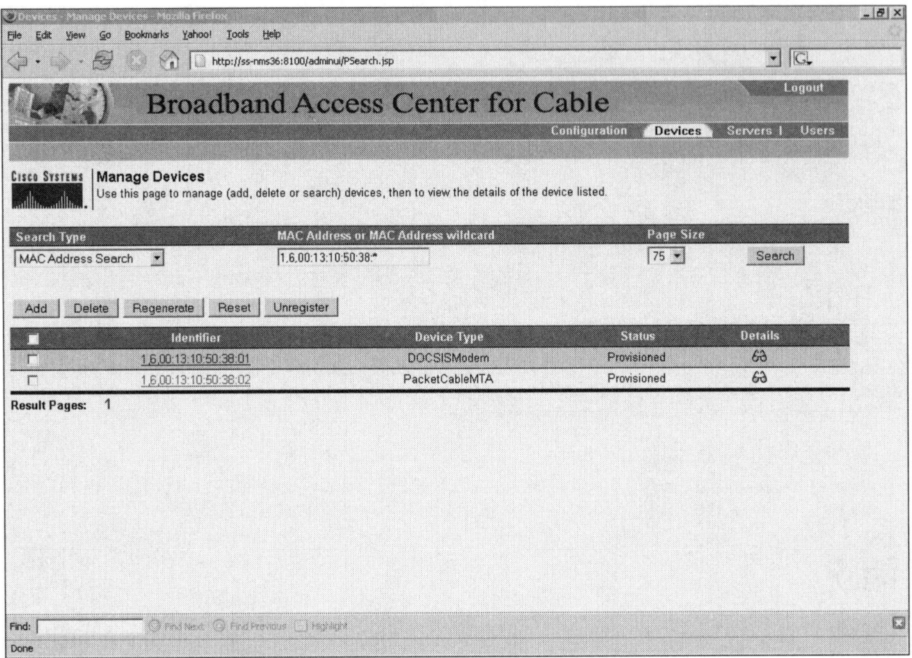

Notice the use of the * character as a wildcard in the MAC address search. In this case, the final byte is wildcarded so all devices beginning with the MAC address 00:13:10:50:38 are listed. As you can see, in this case this resulted in a single DOCSIS cable modem and its associated PacketCable eMTA. By clicking on the Identifier link you can see the configuration details of a particular device. Figure 3-28 shows this information for the eMTA.

TIP You can also add a device from this menu. This works well if you have only a few devices to add, but if you have several to add, this method can be quite tedious. A better way to provision large numbers of devices is through the API interface.

As you can see, the host name and the domain name that make up the MTA FQDN are defined here. Also defined are the Class Of Service and the DHCP Criteria configuration options. In this example the Class Of Service is set to **LinksysMTA-CC44** and the DHCP Criteria is set to **unprovisioned-packet-cable-mta**. Recall that the DHCP criteria is used to associate clients to CNR DHCP selection tags. Under the Configuration option in the main menu you can see how this DHCP criteria is defined, as shown in Figure 3-29.

Figure 3-28 *MTA Configuration Example*

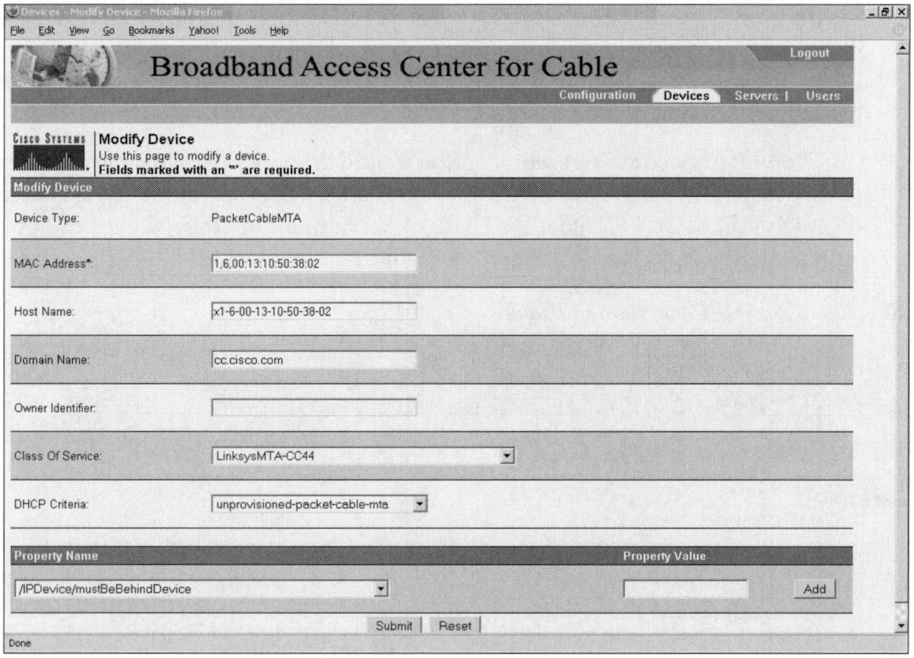

Figure 3-29 *DHCP Criteria Configuration Example*

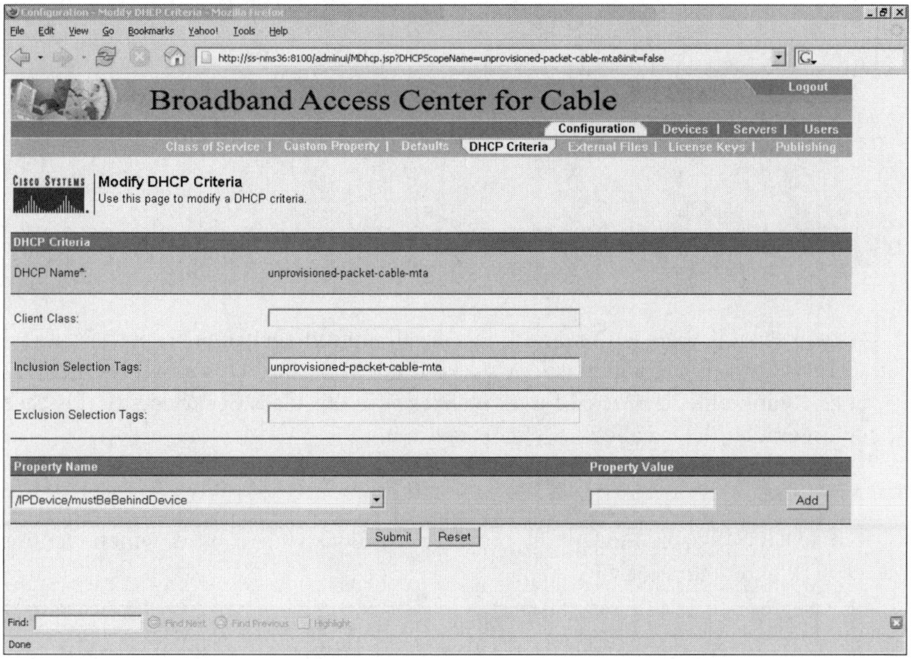

In this case the Inclusion Selection Tag has the same value as the DHCP criteria, **unprovisioned-packet-cable-mta**. If you refer to Figure 3-12 in the CNR DHCP discussion concerning scopes, you can see the field Selection Tags for that scope was set to **unprovisioned-packet-cable-mta**. Thus, this client is authorized to get an IP address from that scope.

The Class Of Service configuration option is used primarily to define the configuration file the device gets. In the previous example, the Class Of Service for the eMTA device was set to **LinksysMTA-CC44**. Under the Class of Service sub-option you can see how this particular class of service is defined, as Figure 3-30 illustrates.

Figure 3-30 *Class of Service Configuration Example*

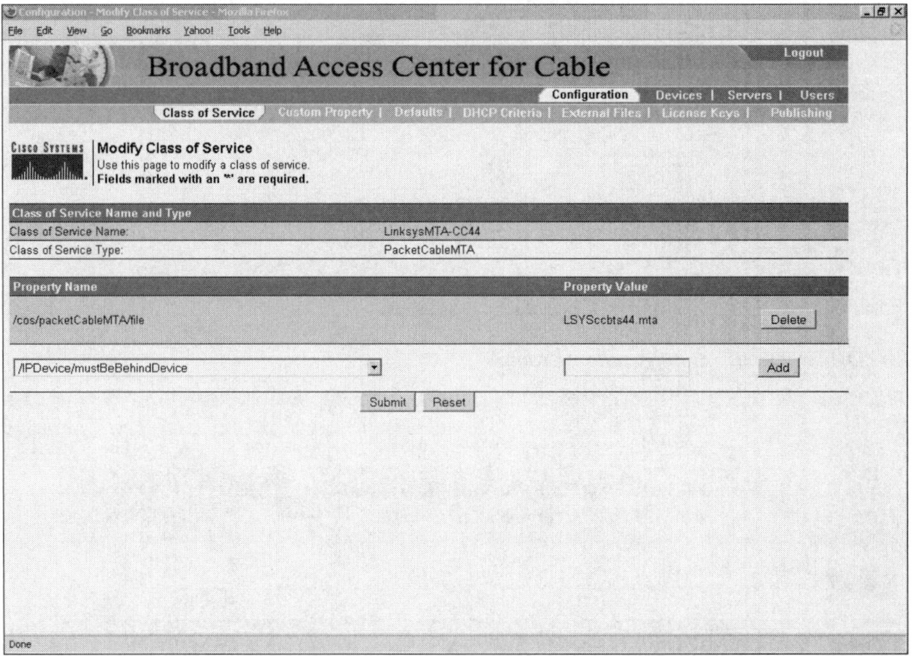

As you can tell, there isn't much to this configuration. Only one property is associated with this class of service, which is /cos/packetCableMTA/file. This identifies the MTA configuration file to be used for all devices with this class of service. In this case, the configuration file is called LSYSccbts44.mta.

NOTE For a DOCSIS cable modem the property /cos/docsis/file is used, which identifies a DOCSIS configuration file.

These configuration files are initially stored on the RDU server. Configuration files are later pushed to the DPE cache engines during the device's initial provisioning attempt (Chapter 4 covers this in more detail). To upload and download configuration files on the RDU the External Files Configuration option is used. From this window you can either use the default * wildcard to list all configuration files or you can enter the specific file name, as shown in Figure 3-31.

Figure 3-31 *Managing External Files Example*

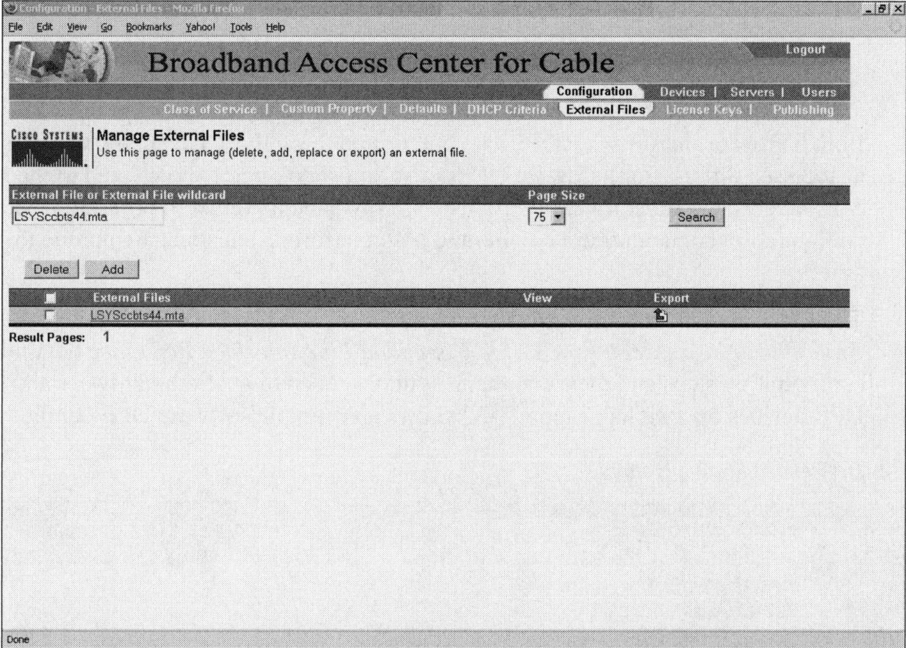

By clicking the filename, you can replace the file with one on your local machine. By clicking the export link you can download a copy of that file to your local machine. You can also delete that file and add a new file using the respective buttons. Both binary files and text template files can be manipulated in the External Files menu.

DPE Configuration

Also from Chapter 2, you learned about the Device Provisioning Engine (DPE) devices that cache client configuration received from the RDU. In addition to communicating with the RDU, DPE devices communicate with the CNR DHCP servers to assign IP addresses to clients. The DPE devices also function as the TFTP server used to transfer DOCSIS and PacketCable configuration files and the TOD server required by DOCSIS cable modems for provisioning. Typically, a redundant pair of DPEs and a redundant pair of DHCP servers are deployed in an area forming a provisioning group capable of supporting up to 1,000,000 devices.

Recall from Chapter 2 that you find two types of DPEs:

- Hardware DPEs (that is, DPE-590 or DPE-2115)
- Software DPEs that run on Solaris machines

The configuration of a DPE is fairly simple. Basic tasks include:

- Configuring IP interfaces
- Assigning the DPE to a provisioning group
- Configuring the IP address and port number of the RDU

Both hardware and software DPEs are configured through a CLI, but the way this interface is accessed differs. For hardware DPEs, a PC must be directly connected to the serial console port and for software DPEs, you simply need to Telnet to port 2323. Many of the configuration commands are common to both platforms, but some are unique to one or the other.

The *Cisco Broadband Access Center for Cable Administrative Guide* and the *Cisco Broadband Access Center for Cable Command Line Interface Reference* detail the configuration steps and their command syntax needed to get both hardware and software DPE devices operational. Figure 3-32 shows an example software DPE configuration.

Figure 3-32 *DPE Configuration Example*

```
172.18.99.182 (BACC 2.6 DPE #1) - SecureCRT
File Edit View Options Transfer Script Window Help
ss-nms32# show ?
  bundles          Shows the archived bundles.
  clock            Shows the current system time.
  commands         Shows the full command hierarchy.
  cpu              Shows the current CPU usage.
  device-config    Show device configuration.
  disk             Shows the current disk usage.
  dpe              Shows the status of the DPE process if started.
  hostname         Shows the system hostname.
  interface        Shows all of the interfaces.
  ip               Shows IP configuration details.
  log              Shows recent log entries.
  memory           Shows the current memory usage.
  packetcable      Shows PacketCable.
  running-config   Shows the appliance configuration.
  tftp             Shows TFTP details.
  version          Shows DPE version.

ss-nms32# show running-config
  dpe port 49186
  dpe provisioning-group primary companyy
  dpe rdu-server 172.18.99.186 49187
  dpe shared-secret fgL7egT9zcYHs
  interface ethernet bge0 provisioning enabled true
  interface ethernet bge0 provisioning fqdn ss-nms32.cisco.com
  log level 5-notification
  no debug
  no debug dpe cache
  no debug dpe connection
  no debug dpe dpe-server
  no debug dpe event-manager
  no debug dpe events
  no debug dpe exceptions
  no debug dpe framework
  no debug dpe messaging
  no debug dpe netsnmp
  no debug dpe registration
  no debug dpe snmp
  no debug dpe tftp
  no dpe provisioning-group secondary
  no tftp allow-create-dirs
  no tftp allow-override
  no tftp allow-read-access
  no tftp allow-write-access
  no tftp verify-ip
  packetcable enable
  packetcable registration kdc-service-key <value is set>
  packetcable snmp key-material <value is set>
ss-nms32#
Ready                                    Telnet   49, 11   49 Rows, 132 Cols   VT100
```

Notice the location of the RDU is specified as well as the name of the provisioning group the DPE is assigned to. The DPE **shared-secret** command sets the key used for secure communication with the RDU; logically if you have a mismatch between the DPE and the RDU, they can't communicate. The **interface** commands identify the Ethernet ports authorized to communicate with the DHCP servers. Pay close attention to the bottom part of the configuration. As you can see, the PacketCable service is enabled, and the KDC service key is defined. Without these commands you would be unable to provision PacketCable MTAs. Of course, the KDC service key needs to match the key configured on the KDC.

KDC Configuration

The final BAC component to learn about is the Key Distribution Center (KDC) used to authenticate MTAs and grant security tickets. Recommended practice is to deploy one KDC with each provisioning group to achieve maximum performance. The KDC is made up of a KDC agent and KDC certificates. The KDC agent monitors the status of the KDC process and can restart this process should it ever go down.

Much of the KDC configuration happens during the installation. A file named kdc.ini contains properties such as the KDC realm name, KDC FQDN, and the interface address used to send and receive Kerberos packets. Remember a KDC realm comprises all the clients and servers interacting with the KDC and that the KDC realm name is case-sensitive and must be defined only in uppercase characters. In the kdc.ini file, you can also adjust the maximum and minimum ticket durations. You must take extreme care when modifying these parameters as this can result in a much heavier load on the KDC. The default minimum of 6 days and the default maximum of 7 days are recommended.

Recall from the Kerberos discussion that the KDC is typically loaded with the following five certificates:

- CableLabs Service Provider Root
- Service Provider
- Local System
- KDC
- MTA Root

The KDC certificates used for MTA authentication must be obtained from CableLabs. A tool in the KDC called PKCert is used to convert the certificates into a format that can be used by the KDC. This tool is located on the KDC in the <BAC_HOME>/kdc directory (by default the <BAC HOME> path is /opt/CSCObpr). Figure 3-33 illustrates the syntax of this tool.

Figure 3-33 *KDC Certificate Utility*

```
root@ss-nms37#
root@ss-nms37#
root@ss-nms37# pwd
/opt/CSCObpr/kdc
root@ss-nms37#
root@ss-nms37#
root@ss-nms37# ./PKCert.sh
Usage: Pkcert [-v|-c|-?] <-z>

    -c                    Create KDC Certificate
    -v                    Verify and Normalize PacketCable Certificate Set
    -z <error|info|debug> Print much debug output

 DESCRIPTION:
    This utility is intended to manage the X.509 certificates
    required for proper operation of the PacketCable Authentication
    Server (KDC)
root@ss-nms37#
root@ss-nms37#
root@ss-nms37# ./PKCert.sh -c
Pkcert Version 1.0
Logging to pkcert.log
Usage: -s <dir> -d <dir> -c <cert> -r <realm> -a <name> -k <keyFile> [-n <serial>] [-o]
      -a <KDC DNS Name> Specify DNS Name of KDC
      -c <Cert File> Use Service Provider Certificate (DER Encoded)
      -d <directory> Specify destination directory
      -k <Key File>  Use Service Provider Private key (DER Encoded)
      -n <Serial#>   Set Certificate Serial Number
      -o             Overwrite existing files
      -r <Realm>     Specify Kerberos Realm for KDC Certificate
      -s <directory> Specify source directory

DESCRIPTION:
      Create and install PacketCable KDC certificate.
      The new certificate contains the Realm in the subject alternate name field.
      The new certificate is unique to the current environment in that it contains:
        1 - The KDC Realm
        2 - The DNS name associated with this KDC which will be used by the MTA.
root@ss-nms37#
root@ss-nms37#
root@ss-nms37#
root@ss-nms37#
root@ss-nms37#
```

The KDC can also be used in a Euro-PacketCable deployment. In this case the certificates need to be obtained from tComLabs instead of CableLabs as the root certificates are different.

Make sure that the MTAs in your network are configured to use the real PacketCable root certificate. Also a test PacketCable root certificate is available for lab environments. Some MTAs have the capability to switch their certificates between these two; any MTA loaded with the test root attempting to authenticate a KDC loaded with the real root (or vice-versa) fails.

In addition to creating the certificates you also need to create the service keys used for KDC to DPE communication. This tool is called *keygen* and is also on the KDC in the <BAC_HOME>/kdc directory. Figure 3-34 shows the syntax of this command, as well as an example of the keys used in a BAC configuration.

Figure 3-34 *KDC Service Key Utility*

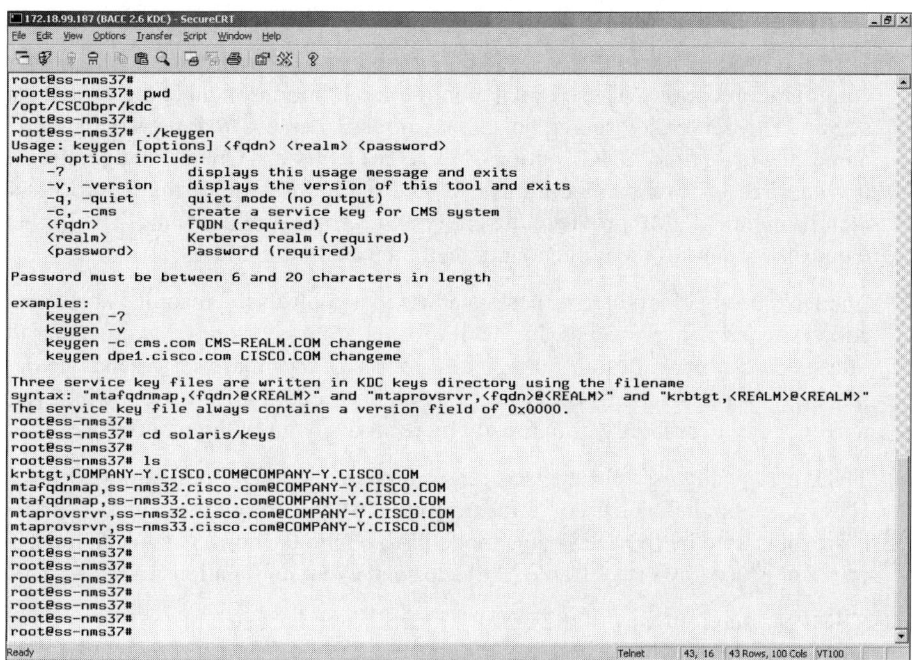

In this example, the KDC communicates with two DPEs. Their FQDNs are ss-nms32.cisco.com and ss-nms33.cisco.com, and the KDC realm name is COMPANY-Y.CISCO.COM. The mtafqdnmap keys are used for the FQDN exchange between the KDC and DPE. You'll learn more next chapter. The mtaprovsrvr keys are used for the AS exchange between the KDC and a MTA, also discussed next chapter. The krbtgt key is used when the optional Ticket Granting Server (TGS) step is used; this is specified when sub-option #7 of DHCP option #122 is set.

NOTE If you wanted to use Kerberozied IPSec for MTA to CMS communication, you would also have to create service keys for the CMS FQDN.

Chapter Summary

PacketCable specifications exist that cover the provisioning aspects for both the MTA and CMS components. The MTA provisioning specification details the mechanisms and steps required to get a MTA operational so that it can provide telephony service. Pre-existent protocols such as DHCP, DNS, SNMP, TFTP, and Kerberos are used for this process. PacketCable MTAs must conform to the provisioning specification to achieve certification.

DHCP is used by both the CM and MTA to obtain an IP address as well as other configuration options. These DHCP configuration options include the default router, time server, DNS server, log server, host name, domain name, TFTP server, configuration file name, and lease time. DHCP option #122 is the CableLabs Client Configuration option and is comprised of several sub-options that control the MTA provisioning process such as identifying the SNMP provisioning entity and Kerberos realm name. The Kerberos realm name also identifies the provisioning method used by the MTA.

The DNS protocol is a hierarchical database that controls the mappings between device IP addresses and domain names. Inside of this hierarchy are zones that are under the control of a specific administration and typically consist of a primary server and one or more secondary servers. Clients query these servers for record information, which then attempt to respond either directly or indirectly by recursively querying other servers.

TFTP is typically used for the transfer of modem and MTA configuration files, although HTTP can optionally be used for the transfer of MTA configuration files. The Time-of-Day protocol is used by DOCSIS cable modems to obtain the current data and time. The syslog protocol is used by PacketCable MTAs to send event notifications to log servers.

SNMP is used for many of the FCAPS functions on PacketCable MTAs. It is used to provide the MTA with its configuration file name and location as well as by the MTA to convey capabilities and configuration status. Version 3 of the protocol adds security to SNMP via the support of authentication and encryption algorithms. For the secure method of MTA provisioning, SNMPv3 is a requirement.

The security keys used by SNMPv3 are provided using the Kerberos protocol. Kerberos protects against unauthorized devices using the PacketCable network by authenticating both PacketCable MTAs and the KDC. The KDC and all the clients and servers in the PacketCable network reside in a Kerberos realm. The KDC controls the security key establishment between MTAs and the SNMP provisioning entity as well as optionally between the MTA and CMS. ITU X.509 digital certificates are used to authenticate PacketCable devices. Once a MTA successfully authenticates with the KDC, it is assigned security keys that have a limited lifetime and must not be lost by the MTA in the event of a power cycle.

Much of the configuration relating to MTA provisioning in a PacketCable network resides on the NMS/OMS devices; however, some configuration is required on the CMTS. This configuration consists of the location of the DHCP servers and related options.

Configuration of Cisco's BAC solution involves CNR DHCP server, DNS server, RDU server, DPE server, and KDC server configuration.

CNR DHCP configuration steps consist of defining IP scopes, client classes, and policies. Policies are where DHCP options are specified. Policies can be defined in several places and can either be embedded to the specific location where defined or given a name so that they can be referenced in multiple places. A CNR extension script is used to define how CNR DHCP servers communicate with other BAC devices such as the RDU and DPE. This extension is installed after CNR is installed and also contains the definition of PacketCable DHCP options. The same configuration must exist on both the primary and secondary DHCP server. The secondary server monitors the health of the primary server and is allocated a portion of each scope's address space to use in the event the primary fails.

CNR DNS configuration involves defining zones, enabling dynamic DNS updates, and adding static records.

The BAC RDU is the centralized server for device configuration and management. This is where CMs and MTAs and their configuration parameters are defined. For each device you define parameters such as its DHCP criteria (used to map to CNR selection tags) and its class of service. The class of service associates groups of devices to parameters such as their configuration file. This configuration can be manually entered from the RDU web interface, or it can be added via the RDU API interface.

Device configuration is transferred from the RDU to DPE cache engines when needed. The DPE works with the RDU and CNR DHCP servers to determine DHCP information; it also functions as the TFTP server for sending configuration files, the TOD server needed in cable modem provisioning, and the SNMP provisioning entity for MTA provisioning. DPE configuration tasks consist of configuring IP interfaces, assigning the DPE to a provisioning group, and configuring the location of the RDU.

A majority of the configuration required on the BAC KDC happens during installation. During installation the KDC realm name, FQDN, X.509 certificates, and service keys are all set up. The BAC KDC is equipped with utilities to assist in generating these certificates and service keys.

Chapter Review

1 True or False: The MTA MAC address is equal to the MAC address of the DOCSIS cable modem.

Answer: False. Different MAC addresses and IP addresses are required for the MTA and CM components of an embedded MTA.

2 True or False: DHCP option #43 is the vendor-specific option and is used in PacketCable to convey information about the MTA hardware and software versions.

Answer: True

3 True or False: Option #122 sub-options #1 and #2 are sent in the DHCP Offer to a cable modem to specify the DHCP server(s) the embedded MTA is authorized to communicate with.

Answer: True

4 True or False: The Kerberos realm name is case-sensitive and must be specified in all lowercase letters.

Answer: False. It is case-sensitive but must be specified in all uppercase letters.

5 True or False: If **cable dhcp-giaddr primary** is configured, the CMTS uses the primary interface address in the giaddr field regardless of where the DHCP packet originated.

Answer: True

6 True or False: For Dynamic DNS updates to work in CNR not only does the DHCP scope need to be configured to update DNS, but the DNS Zone needs to be configured to accept these updates.

Answer: True

7 True or False: The BAC KDC can operate in both the PacketCable and Euro-PacketCable environments with the same X.509 certificates.

Answer: False. Euro-PacketCable certificates need to be chained to the tComLabs Service Provider root as opposed to the CableLabs Service Provider root used in PacketCable.

8 What are the three methods of MTA provisioning, and which one is generally preferred and why?

Answer: Secure, Basic, and Hybrid. The Secure method is generally preferred because it is the only method that provides security to the provisioning process via device authentication and Kerberos.

9 What is the role of the CMTS in the DHCP process?

Answer: The CMTS is the DHCP relay agent that forwards packets broadcast by DHCP clients to preconfigured servers.

10 How are MTA FQDNs formed and how are they populated in DNS?

Answer: By combining DHCP options #12 and #15, the DHCP server then sends dynamic DNS updates to the configured DNS server(s).

11 What message facility is used for syslog messages in PacketCable?

Answer: Local 0

12 Why is the Kerberos protocol better suited for providing security keys than other protocols, such as the Internet Key Exchange (IKE) protocol?

Answer: Because IKE is better suited when the parties involved know each other well in advance and only need to support a small number of secure connections. Kerberos uses a separate device known as a KDC that is better suited for the PacketCable environment.

13 What are the two X.509 certificate hierarchies used in PacketCable?

Answer: The MTA device hierarchy (comprised of the MTA Device, MTA Manufacturer, and MTA Root certificates) and the CableLabs Service Provider hierarchy (comprised of the KDC, Local System Operator, Service Provider, and Service Provider Root certificates)

14 How does CNR determine what scope a device is granted an IP address from?

Answer: Through the use of scope selection tags and by the value of the giaddr field in received DHCP packets. A device's scope selection tag is determined by its DHCP criteria configuration on the BAC RDU.

15 List four places where CNR DHCP policies can be defined:

Answer: Client configuration, client class configuration, scope configuration, system defaults

16 How do you load a new or modified configuration file to a BAC system?

Answer: By using the external file configuration option on the RDU

17 How is the configuration CLI accessed for both hardware and software DPEs?

Answer: For hardware DPEs a direct connection to the serial interface is needed, and for software DPEs you must telnet to port 2323.

CHAPTER 4

Provisioning Flows

This chapter covers the individual steps involved in provisioning a PacketCable embedded multimedia terminal adapter (eMTA) in detail. Chapter 3, "Provisioning Overview," covered the protocols used for PacketCable provisioning as well as the basic configuration required on a Cisco cable modem termination system (CMTS) and on the Cisco Broadband Access Center (BAC) provisioning solution. This information serves as the foundation for the information presented in this chapter. The chapter begins with a review of the steps in cable modem provisioning. Included with each step are traces and troubleshooting suggestions. Then eMTA provisioning is covered step by step. Likewise, for each step, traces and logs describe the step. The provisioning steps outlined in this chapter assume the necessary configuration already exists in the Network Management System/Operational Support System (NMS/OSS) systems used for MTA provisioning. Typically, a third-party application interfaces with the application programming interface (API) of these systems to populate the necessary configuration information. Chapter 3 covered much of the configuration related to the Cisco BAC solution, so please refer to that chapter as necessary.

DOCSIS Cable Modem Provisioning

As you know, an eMTA is physically co-located with a cable modem, and before the eMTA can successfully provision, the cable modem must successfully provision. This process is defined in section 9 (Cable Modem—CMTS Interaction) of the Data-Over-Cable Service Interface Specification (DOCSIS) Radio Frequency Interface (RFI) specification.

Specifically the cable modem initialization process is broken down into the following stages:

- Scanning and Downstream Synchronization
- Obtaining Upstream Parameters
- Ranging
- IP Connectivity Establishment
- Time of Day Establishment
- Transfer of Configuration Information

- Registration
- Baseline Privacy Initialization

NOTE Most cable modems have LED indicators that provide a hint as to what stage in their provisioning process they are at currently.

Figure 4-1 illustrates the first two stages.

Figure 4-1 *Cable Modem Provisioning - Stages 1 and 2*

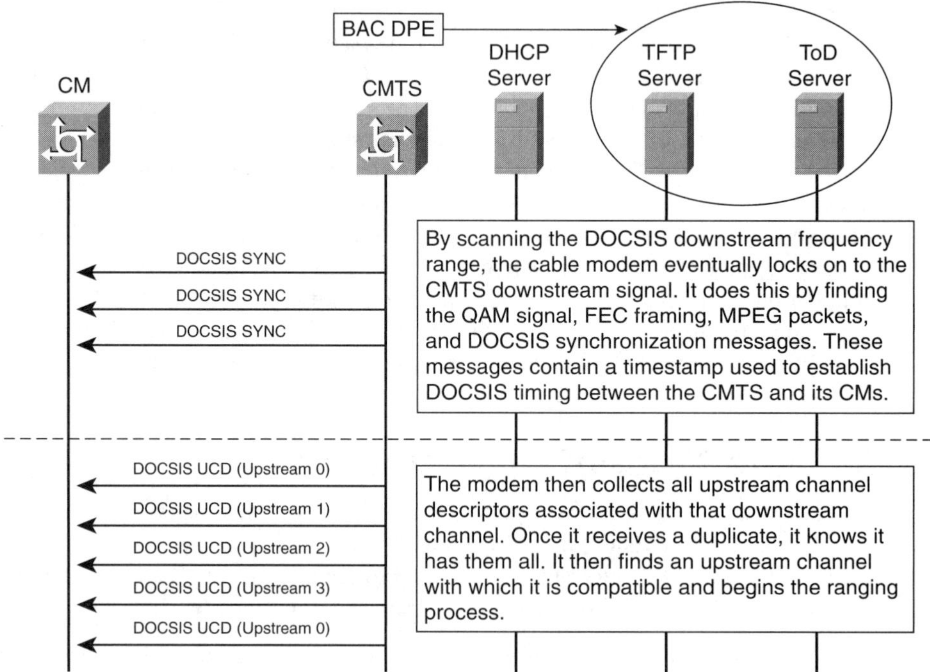

Scanning and Downstream Synchronization

When a cable modem is initialized or reinitialized, the first thing it does is look for a downstream DOCSIS channel. This initialization could be triggered by a number of things including a power cycle, a software reset, a loss of communication with the CMTS, or a Simple Network Management Protocol (SNMP) reset.

A cable modem searches or scans the radio frequency (RF) spectrum for a valid channel. Remember with standard DOCSIS this would be a 6-MHz channel in the 88 to 860-MHz range. A cable modem "locks" on a downstream channel via a four-step synchronization:

1 The modem looks for the Quadrature Amplitude Modulation (QAM) symbol timing (either 64 QAM or 256 QAM).

2 The modem looks for Forward Error Correction (FEC) encoded frames.

3 The modem looks for Moving Picture Experts Group (MPEG) packets within those frames.

4 The modem looks for DOCSIS SYNC messages within those MPEG packets. These SYNC messages synchronize the DOCSIS timing between the CMTS and CM.

A failure in this stage most likely indicates a physical layer problem such as faulty cabling, a faulty cable modem, or a problem with the CMTS and/or upconverter. This could include CMTS configuration. Figure 4-2 is a configuration example on a Cisco uBR10012 CMTS using a MC5x20U cable line card.

Figure 4-2 *Cisco CMTS Downstream Configuration Example*

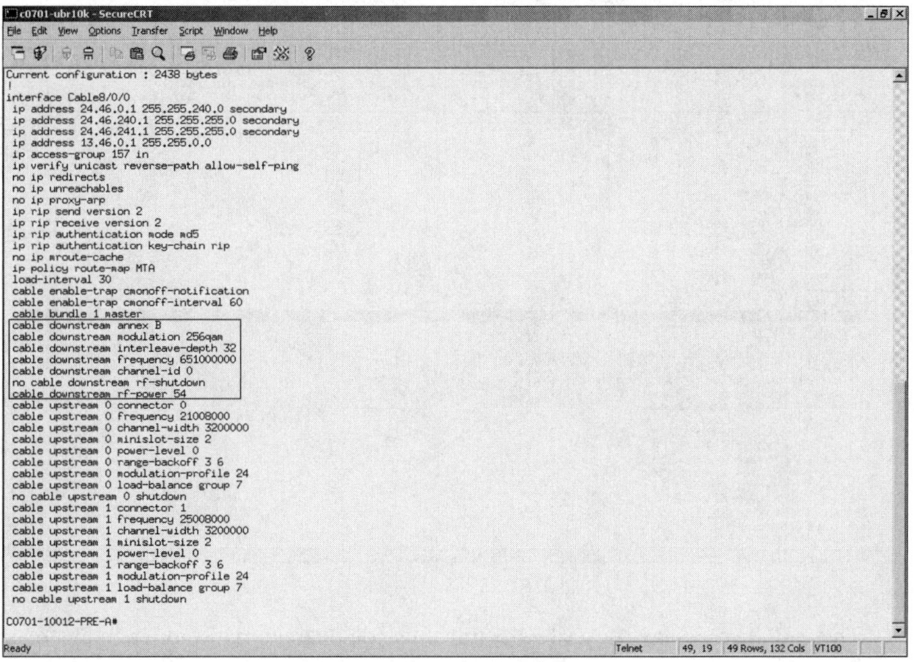

In Figure 4-2, notice the seven cable downstream configuration commands. Annex B indicates this is a 6-MHz DOCSIS downstream channel; Annex A would indicate an 8-MHz Euro-DOCSIS channel. The QAM modulation used in this case is 256 QAM. 256 QAM provides greater capacity than 64 QAM but is more sensitive to noise; hence, a larger SNR is required. The Interleave Depth indicates how the FEC encoding is done; higher depths provide better protection from impulse noise but increase latency. The downstream center frequency is set to 651 MHz and is assigned an identifier of 0. Finally, because this card has an integrated upconverter, the RF output is enabled and configured to transmit at a power level of 54 dBmV.

Obtaining Upstream Parameters

Now that the cable modem has successfully found a valid downstream DOCSIS channel, it needs to find an upstream DOCSIS channel. The cable modem does this by collecting Upstream Channel Descriptor (UCD) DOCSIS messages sent downstream by the CMTS, as illustrated by Figure 4-3.

Figure 4-3 *Example DOCSIS UCD Message*

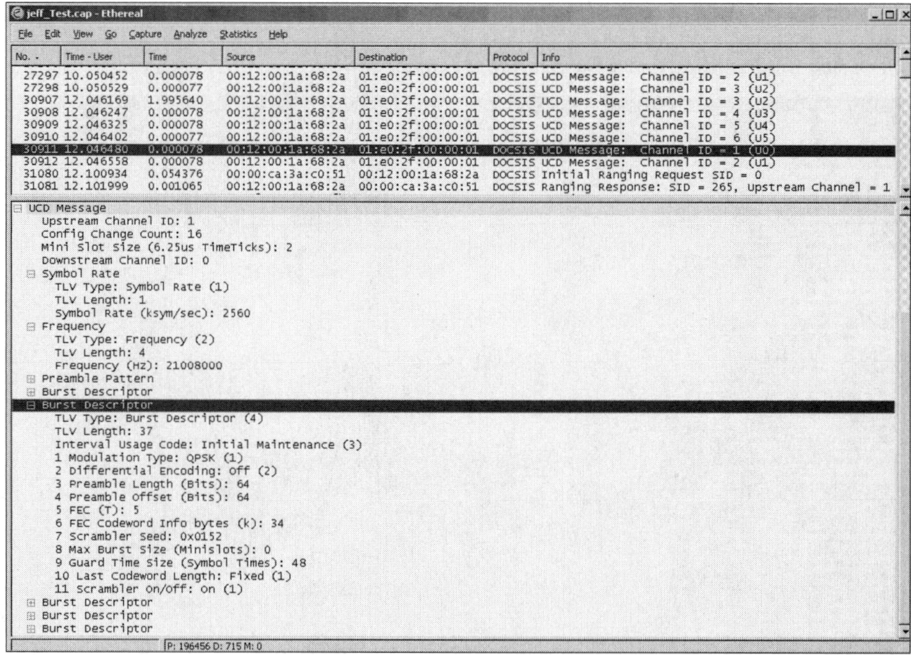

These UCD messages describe the transmission characteristics of each upstream channel. From Figure 4-3 you can tell that this upstream (upstream channel ID 0) has a center frequency of 21 MHz and a channel width of 3.2 MHz (indicated by the symbol rate of 2,560 kilosymbols per second). Also included are burst descriptors, which define how the modem sends different types of DOCSIS packets. In this example five different burst types are available on this upstream channel: Initial Maintenance, Station Maintenance, Request, Short Data Grant, and Long Data Grant.

On a Cisco CMTS, the parameters for these burst types are defined in a modulation profile. Figure 4-4 is an example configuration of the upstream parameters for a MC5x20U cable line card on an uBR10012 CMTS.

Figure 4-4 *Cisco CMTS Upstream Configuration Example*

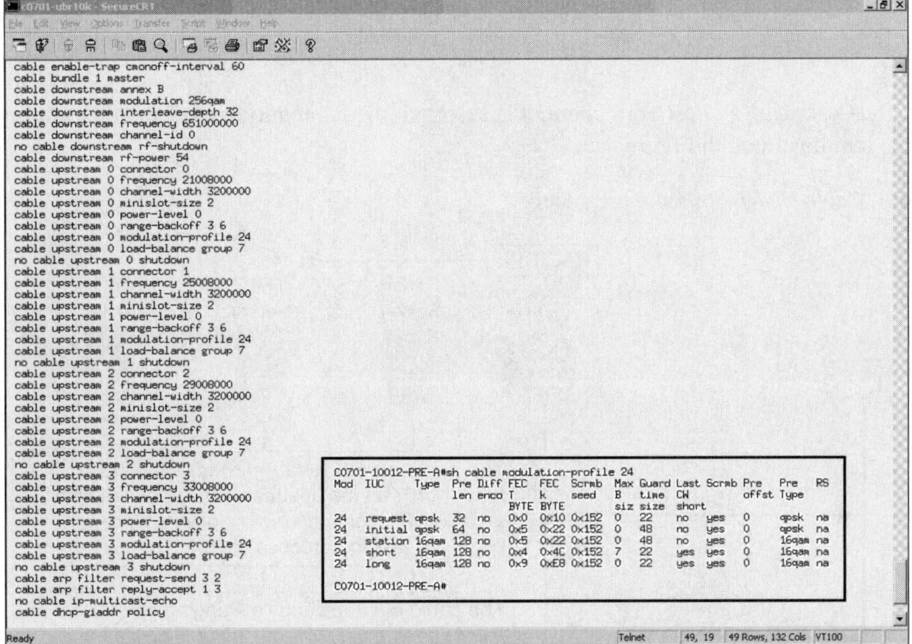

In Figure 4-4, four upstream channels are associated with this downstream channel. For each one, the center frequency and channel width are defined. Also defined is the minislot size, which relates to the capacity of each minislot allocated in the DOCSIS MAP messages. The power level states that cable modem transmissions should be received at a level of 0 dBmV. The ranging back-off parameters define how the modem should behave if its ranging transmissions encounter collisions. The modulation profile defines the characteristics of the various burst types; the parameters of modulation profile 24 are shown in the box in the lower right of Figure 4-4. The load balance configuration is used to move modems around on the upstream channels to achieve a relatively uniform load.

Because a single downstream typically controls multiple upstream channels, the cable modem collects UCDs for all upstreams associated with that downstream. It then determines which ones it is compatible with and selects one for upstream transmission. If none of these upstream channels are acceptable, it reverts to the previous step and looks for another downstream channel.

NOTE The cable modem configuration file can instruct a CM to use a particular downstream frequency and/or a particular upstream channel. In this case, the modem moves to the specified downstream and/or upstream after it has received its configuration file.

Ranging

As soon as an upstream channel is selected, the cable modem begins the ranging process, as illustrated in Figure 4-5.

Figure 4-5 *Cable Modem Provisioning - Stage 3*

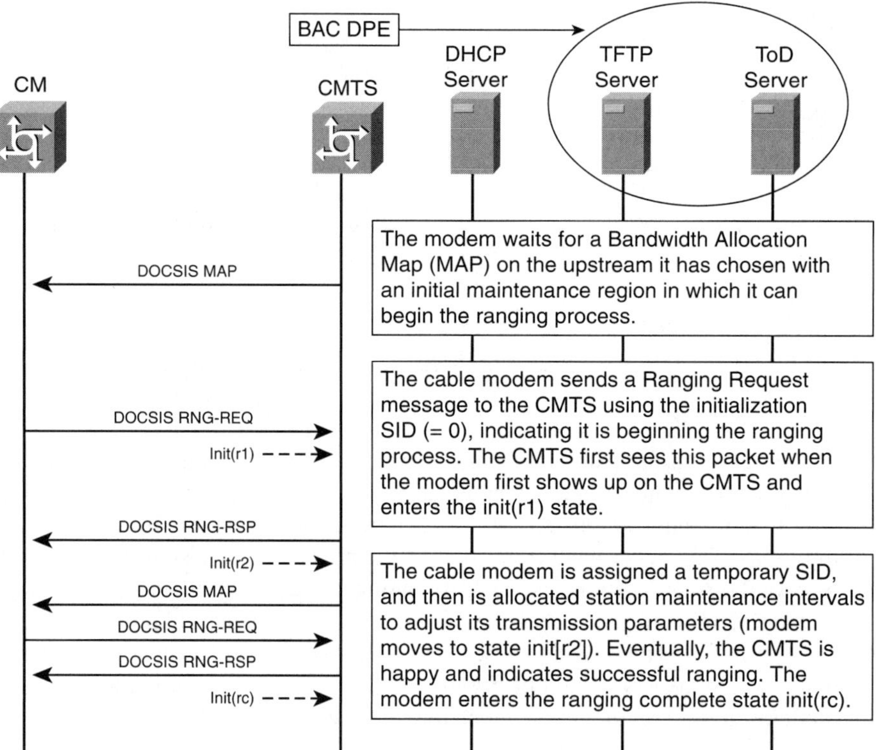

If you recall from Chapter 1 "PacketCable Overview," the CMTS sends messages called Bandwidth Allocation Maps (MAPs) to describe what devices are authorized to transmit data in what time slots. Each of the time intervals specified in the MAP messages is associated with a Service Identifier (SID) identifying the devices authorized to use the interval and an Interval Usage Code (IUC) describing what type of data it can transmit. These IUCs correspond to the burst types mentioned in the previous step and include data grants for sending ordinary packets, request grants for making bandwidth requests, and ranging intervals.

You have two types of ranging intervals:

- **Initial maintenance**—The initial maintenance interval is the IUC used in the next step in cable modem provisioning. These intervals are sent to the broadcast SID, meaning any device can use them and are long enough to accommodate the largest anticipated timing offset between a cable modem and the CMTS.
- **Station maintenance**—Station maintenance ranging intervals are used as a keepalive mechanism to ensure online devices are still online and also to request the device make adjustments to its transmit power, frequency, or timing.

The cable modem uses one of these intervals to transmit a Ranging Request (RNG-REQ) message using the initialization SID. The value "0" is used as the initialization SID value. The cable modem continues to transmit ranging messages in these intervals until it receives acknowledgement from the CMTS or it exhausts the number of tries. The modem starts out transmitting at some specific power level, and if it does not receive acknowledgement, it gradually increases its output power level until it reaches its maximum.

Also because these messages are sent in a broadcast interval, they might collide with the transmissions of other modems. The failure in receiving a ranging acknowledgment is also known as a *T3 timeout*, as the DOCSIS timer T3 defines the amount of time the modem waits for a response to a RNG-REQ. If all retries have been exhausted and still no acknowledgement is received, the cable modem repeats this ranging process on another upstream channel if available. If one is not, the cable modem scans for another downstream channel.

The acknowledgement message the cable modem expects to see is a Ranging Response (RNG-RSP) message from the CMTS. The CMTS sends this message as soon as it is able to decipher a RNG-REQ sent by the cable modem, meaning the cable modem is transmitting at a large enough power level for the CMTS to detect the transmission. In fact, it is at this point that you first see the cable modem on the CMTS. If you are using a Cisco CMTS, you can use the **show cable modem** command for this purpose:

```
C0701-10012-PRE-A#show cable modem 0013.1050.3801
MAC Address     IP Address      I/F        MAC         Prim RxPwr  Timing  Num BPI
                                           State       Sid  (dB)   Offset  CPE Enb
0013.1050.3801 13.46.12.205    C7/0/0/U1 init(r1)    30   0.00   1284    0   N
```

As you can see at this point, the cable modem is in the init(r1) state. In addition to acknowledging the receipt of the RNG-REQ message the RNG-RSP message servers other purposes. The CMTS assigns a temporary SID to the cable modem, which is sent in this message.

NOTE This temporary SID typically becomes the permanent SID for the cable modem's primary upstream service flow; SIDs for secondary upstream flows are assigned later in the provisioning process.

The RNG-RSP message is also used to adjust the cable modem's transmission parameters. If needed, a frequency adjust is included if the CMTS detects the cable modem transmission is outside of a predefined acceptable threshold. Also a timing adjust is included if the cable modem transmissions are received earlier or later than expected according to the bandwidth allocation MAPs. When a cable modem has properly adjusted its timing, it appears to the CMTS as if it is physically next to it. Finally, a power adjust is included so that the received power from the cable modem is at some predefined value. This value is typically 0 dBmV, plus or minus some tolerable variance. Figure 4-6 is an Ethereal trace of an initial RNG-REQ and its associated RNG-RSP.

Figure 4-6 *Example Initial Ranging*

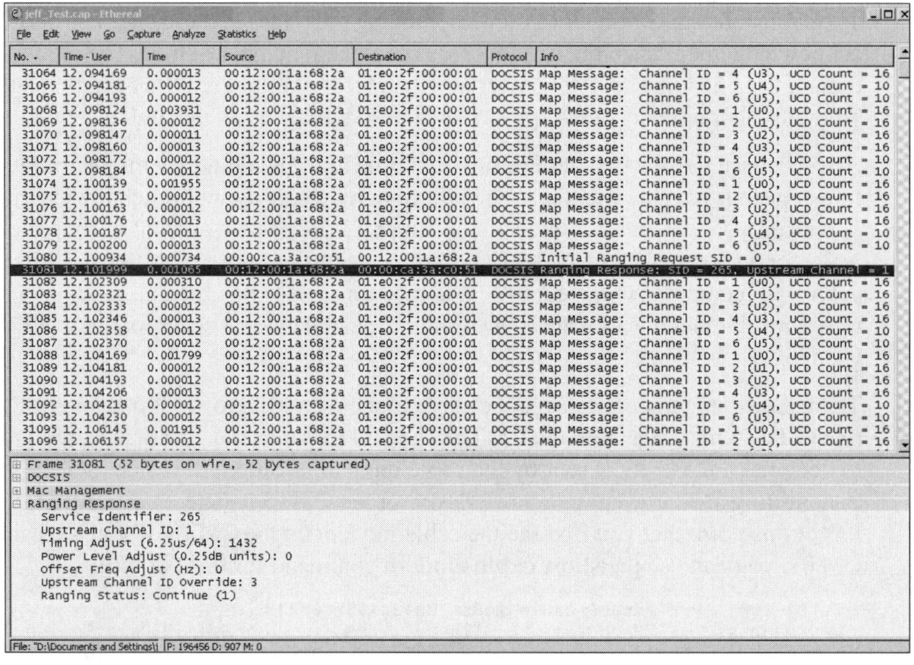

In this case, the modem is told to adjust its timing.

NOTE Another parameter worth mentioning in this figure is the upstream channel identifier. Often cable plants are wired in sparse mode where only a single upstream is on the same physical cable as the downstream. In these cases the same frequency is typically used for all upstream channels associated with the downstream. This parameter lets the cable modem know which upstream it is actually using. If this identifier is different from what the modem thinks it's using, it should immediately adjust its transmission parameters to the UCD associated with this upstream channel identifier.

In addition to sending the RNG-RSP message to adjust the cable modem's transmission parameters, the CMTS also adds this modem to its polling list. The polling list consists of the SIDs of all operational devices. This is where the station maintenance intervals discussed earlier come into play. Each one of these devices is periodically allocated a unicast station maintenance interval to transmit a RNG-REQ message; this serves as a keepalive mechanism between the CMTS and its associated devices. During the cable modem provisioning process, these station maintenance intervals are also used to provide the cable modem with an opportunity to adjust its transmission parameters and send additional RNG-REQ messages, as shown in Figure 4-7.

Figure 4-7 *Ranging Example*

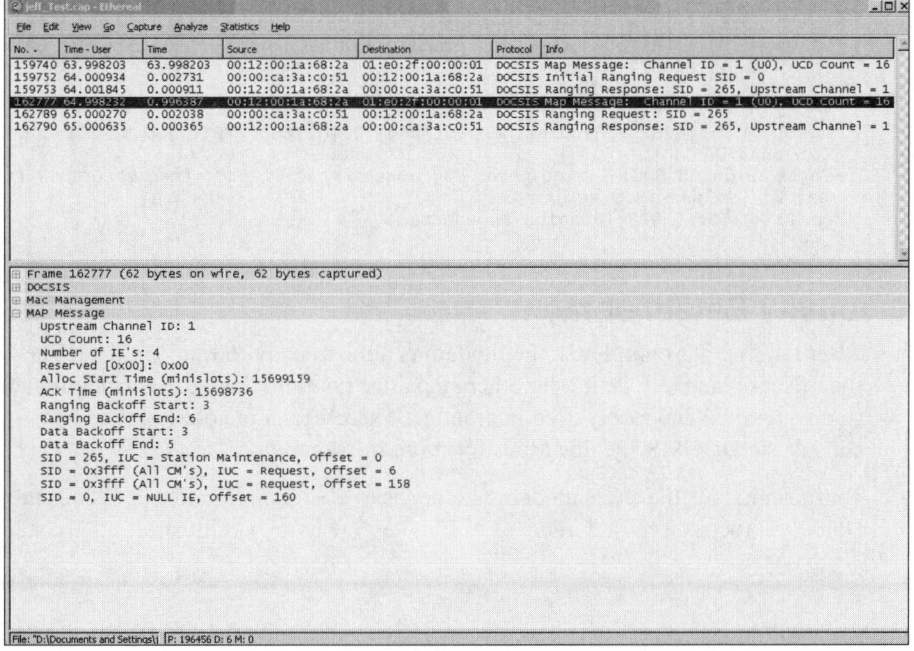

In this case, the modem is allocated temporary SID 265. On a Cisco CMTS the cable modem enters the init(r2) ranging stage.

Eventually the cable modem should have adjusted its transmission parameters such that they are all acceptable by the CMTS. The CMTS then sends a RNG-RSP message indicating that the ranging process has been successful. On a Cisco CMTS, the cable modem transitions from the init(r2) state to the init(rc) (Ranging Complete) state:

```
C0701-10012-PRE-A#show cable modem 0013.1050.3801
MAC Address     IP Address      I/F        MAC          Prim RxPwr  Timing  Num BPI
                                           State        Sid  (dB)   Offset  CPE Enb
0013.1050.3801 13.46.12.205     C7/0/0/U1 init(rc)      30   0.00   1284    0   N
```

NOTE If the modem gets stuck in the init(r2) state, this most likely means that the cable modem is unable to adhere to the transmission requirements of the CMTS—for example, if you have too much loss in the Hybrid Fiber Coaxial (HFC) network so that the minimum power setting is unattainable.

The ranging process on a Cisco CMTS can be troubleshot using the **debug cable range** command in conjunction with the **debug cable mac-address** *CM_MAC_Address* **verbose** command for a particular modem:

```
Feb 13 20:55:55.618: Initial Ranging: Downstream channel ID is 0 (configured 0)
Feb 13 20:55:55.622: Reusing old primary sid 58 for modem 0013.1050.3801
Feb 13 20:55:55.622: Timing error 1134, power error 0.00dB, freq error 156 (thres
  6400 adj 0) [sm per 13.0 sec]
Feb 13 20:55:56.622: Ranging CM 0013.1050.3801, SID 58 on I/f Cable3/0/U4
Feb 13 20:55:56.622: Ranging request from 0013.1050.3801, SID 58 [58] on Interface
  Cable3/0/U4
Feb 13 20:55:56.622: Timing error -1, power error 0.25dB, freq error 390 (thres 6400
  adj 0) [sm per 13.0 sec]
Feb 13 20:55:56.622: Ranging successful.
```

IP Connectivity Establishment

After ranging has completed, the modem is almost ready to transmit data. The DOCSIS specification actually defines another step after ranging, device class identification. This step is optional and rarely used in practice. Therefore, it is not discussed here—please consult the DOCSIS specifications for more information.

To transmit data, the cable modem first needs to obtain an IP address. It does this using the DHCP protocol discussed in detail in Chapter 3. Figure 4-8 illustrates these steps.

Figure 4-8 *Cable Modem Provisioning - Stage 4*

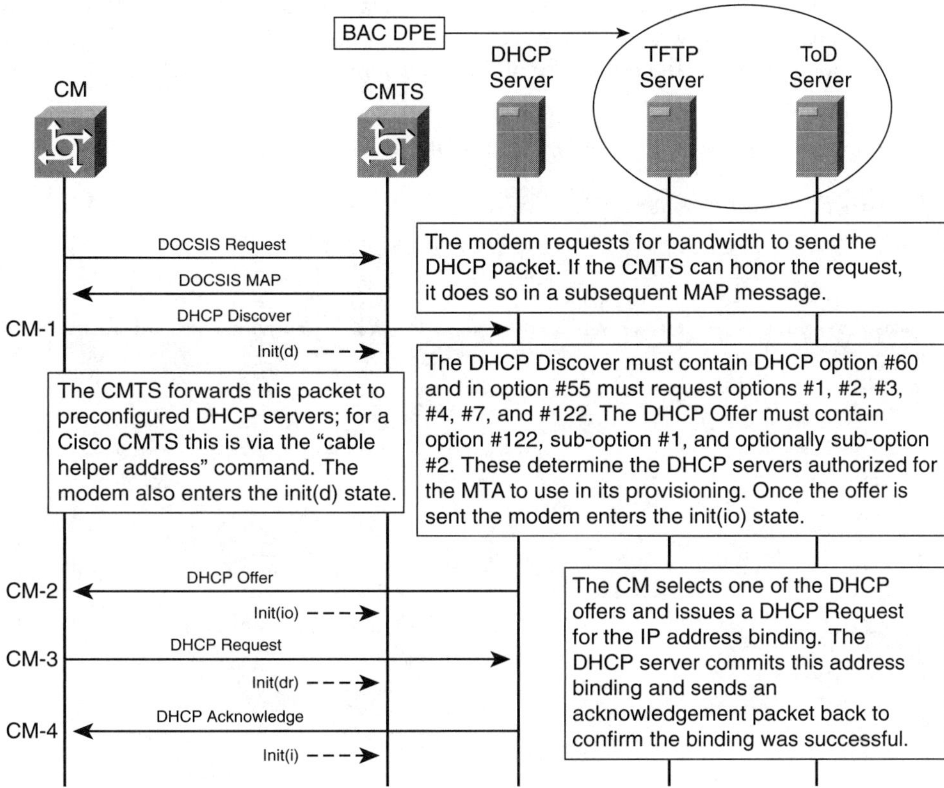

First, the cable modem broadcasts a DHCP Discover packet. This step is also the first one covered in the PacketCable provisioning specification and is referred to as CM-1 (Cable Modem Step #1). But before the modem can send this packet, it needs a data grant from the CMTS. The cable modem transmits a DOCSIS request packet asking for the amount of minislots needed to send the DHCP packet. Remember this request interval is contention-based, so it could collide with request packets from other devices and need to be retransmitted. If the CMTS successfully receives the packet and has available bandwidth, it allocates the modem a unicast data grant in an upcoming MAP message. The cable modem can then send the DHCP Discover packet in this interval. When a Cisco CMTS receives the DHCP Discover, it transitions the cable modem to the init(d) state, initialization DHCP:

```
C0701-10012-PRE-A#show cable modem 0013.1050.3801
MAC Address     IP Address      I/F        MAC          Prim RxPwr  Timing  Num BPI
                                           State        Sid  (dB)   Offset  CPE Enb
0013.1050.3801 13.46.12.205    C7/0/0/U1  init(d)      30   0.00   1284    0   N
```

If the modem is stuck in the init(rc) state and unable to transition to the init(d) state, this could mean that the upstream channel is over capacity, and the modem is not given the data grant needed to transmit the DHCP Discover packet. Please keep in mind the DOCSIS steps that need to occur whenever a modem needs to transmit data. For simplicity reasons they are not included in the remainder of the provisioning discussion but obviously still need to occur.

Figure 4-9 shows an Ethereal trace of a DHCP Discover packet.

Figure 4-9 *DHCP Discover Example (CM)*

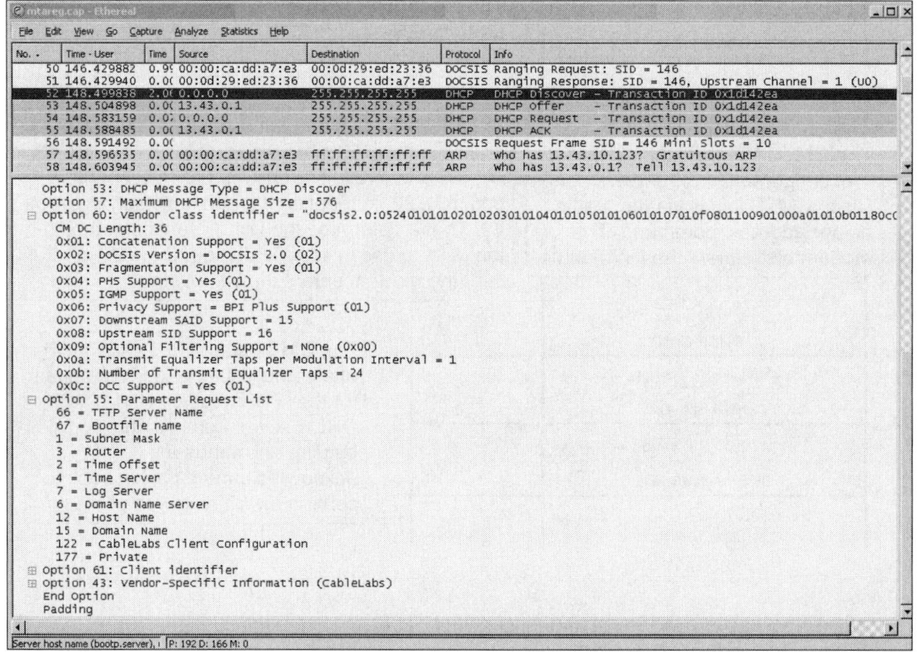

Pay close attention the DHCP options present in the Discover packet and to the DHCP options requested. Notice Option #60, which identifies the device as a DOCSIS 2.0 cable modem and details many of the modem's capabilities. Note the options requested in Option #55; if your provisioning system cannot return all of these options, you will most likely run into problems, and if the PacketCable option, option #122, is not available, the MTA will not provision.

The CMTS acts a relay agent for DHCP. When it receives the DHCP packet, it fills in the "giaddr" field, optionally adds the relay agent option (DHCP Option #82), and forwards these packets to the preconfigured DHCP servers (with a Cisco CMTS this is identified by the **cable helper address** *ip-address* interface command).

Stuck in the init(d) State

Getting stuck in the init(d) state can signify a number of things. The DHCP servers could be down or unreachable. The DHCP server might not have any available leases (that is, IP addresses). The particular client might not be configured properly, or some other problem could exist with the configuration of the OSS/NMS servers.

If you are using the Cisco BAC solution, the DHCP process is handled in the following manner:

1 The DHCP Discover is sent to the Cisco Network Registrar (CNR) DHCP servers.

2 The extension script configured on CNR tells it to contact the Device Provisioning Engine (DPE) for additional information before processing the packet.

3 If the modem's configuration is already stored in the DPE cache, it sends this information back to the CNR. This is the case with a previously configured and operational modem that reboots, for example. If this is the first time the device boots or the DPE cache has timed-out or been cleared, the DPE drops the request from the DHCP server; so the DHCP process fails, and the modem needs to restart the provisioning flow.

   Meanwhile, the DHCP server contacts the Regional Distribution Unit (RDU) and requests it to build the device configuration and send it to the DPE. After the DPE has the required modem information, it sends this back to CNR.

4 CNR can then determine the IP address and other parameters the modem is to receive and sends the DHCP Offer.

If any configuration errors cause any of the previously mentioned steps to fail, the modem gets stuck in the init state. Obviously, if the device is not configured or is misconfigured on the RDU, it cannot receive an Offer packet. One example of a misconfiguration is an improperly formatted template file. Chapter 5, "The MTA Configuration File," covers the use of templates in BAC. Remember how the RDU DHCP criteria relate to CNR scope selection tags? A mismatch here causes CNR to be unable to find the appropriate IP address range for the device. Also if you have a mismatch in the shared secret configuration between the DPE and RDU, they cannot communicate and pass device information. Logically, each of these devices must have the proper IP address and port number of the other device configured correctly. Not only does the DPE-to-RDU communication need to be operational, but the DPE-to-CNR DHCP server communication needs to be operational as well, which means IP connectivity and proper configuration of the BAC extension script on CNR.

Stuck in the init(io) State

Assuming the Discover packet makes it to a DHCP server and that server can honor that request, it returns a DHCP Offer to the cable modem (step CM-2). On a Cisco CMTS, the cable modem transitions from the init state to the init(io) state when it forwards the Offer packet to the cable modem:

```
C0701-10012-PRE-A#show cable modem 0013.1050.3801
MAC Address     IP Address      I/F       MAC          Prim RxPwr  Timing  Num BPI
                                          State        Sid  (dB)   Offset  CPE Enb
0013.1050.3801  13.46.12.205    C7/0/0/U1 init(io)      30   0.00   1284    0   N
```

NOTE The init(io) and init(dr) cable modem states were added in Cisco IOS Software Release 12.3(13a)BC. Previous to this release, these states would show up as init(i).

Figure 4-10 shows a trace of the DHCP Offer packet.

Figure 4-10 *DHCP Offer Example (CM)*

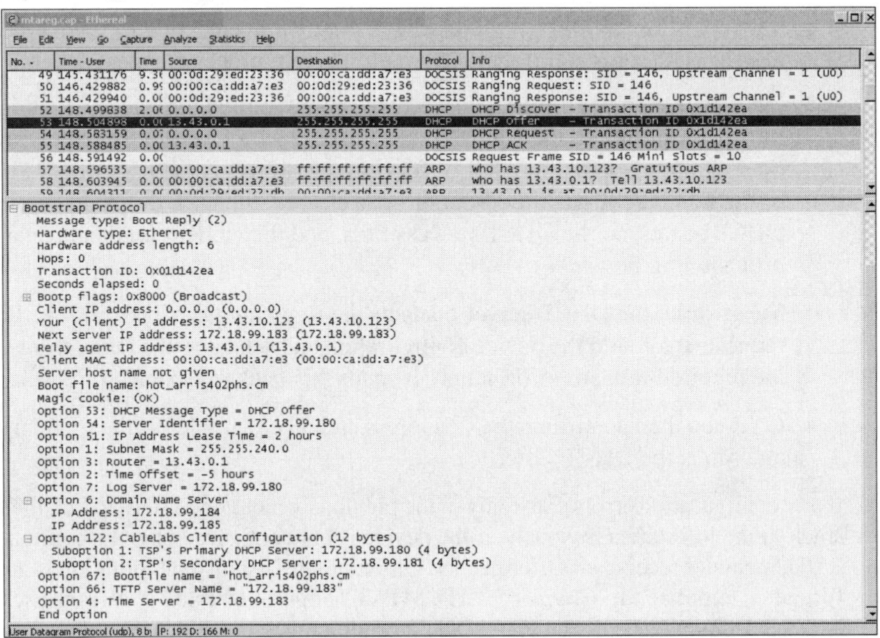

In this packet, option #4 and option #66 are included, which identify the time server and TFTP server (both the BAC DPE in this case) used later in the CM provisioning process. Notice in this figure option #122 is also present and identifies the primary and secondary DHCP servers authorized to communicate with the MTA. The cable modem waits some time for a DHCP Offer with DHCP option #122; however, if none is received, the modem eventually timeouts and accepts one without this option. In this case, the MTA does not attempt to provision.

The cable modem then chooses one of the offers it receives and sends a DHCP Request packet (step CM-3). Again, this packet is broadcast by the cable modem and forwarded by the CMTS. After the CMTS receives the DHCP Request packet, it transitions the modem to the init(dr) state. If the cable modem fails to choose any of the received offers, it gets stuck in the init(io) state. This typically indicates at least one of the options requested by the cable modem was not returned by the DHCP servers.

After the DHCP server receives the Request packet, it adds this IP address binding to its database and sends a DHCP Acknowledgement packet back to the cable modem (step CM-4). At this point, the Cisco CMTS transitions the modem to the init(i) state.

Troubleshooting the DHCP Process

On a Cisco CMTS you can debug the DHCP process using the **debug cable dhcp** command in conjunction with the **debug cable mac-address** *CM_MAC_Address* command for a particular modem, as shown in Figure 4-11.

Figure 4-11 *CMTS DHCP Debugging (CM)*

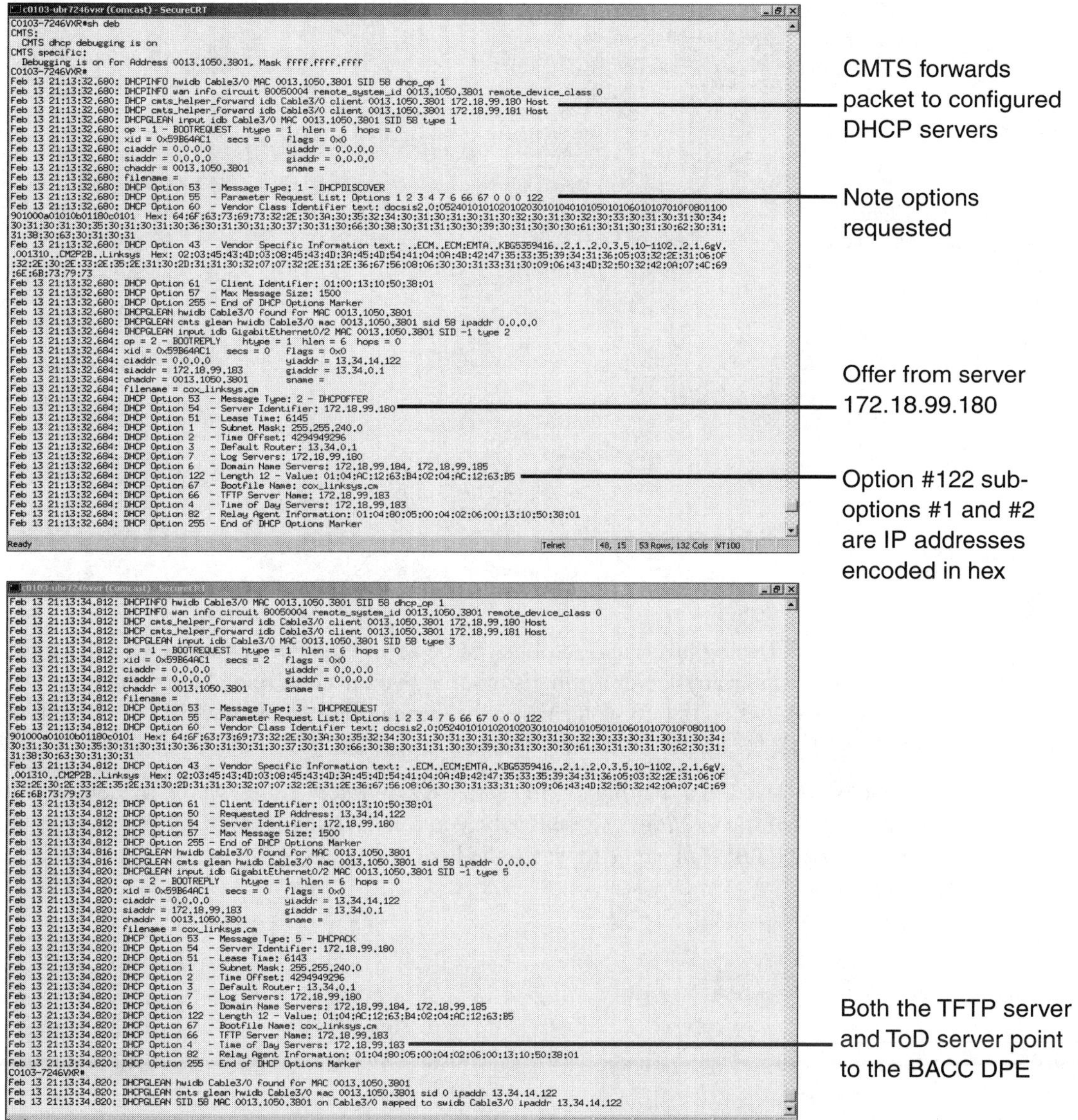

Another place that can help in troubleshooting these steps is the CNR DHCP logs. These logs can be viewed either from the web interface or the CLI. If using the web interface, simply go to the DHCP Server submenu under the DHCP menu, as shown in Figure 4-12.

Figure 4-12 *CNR DHCP Logs*

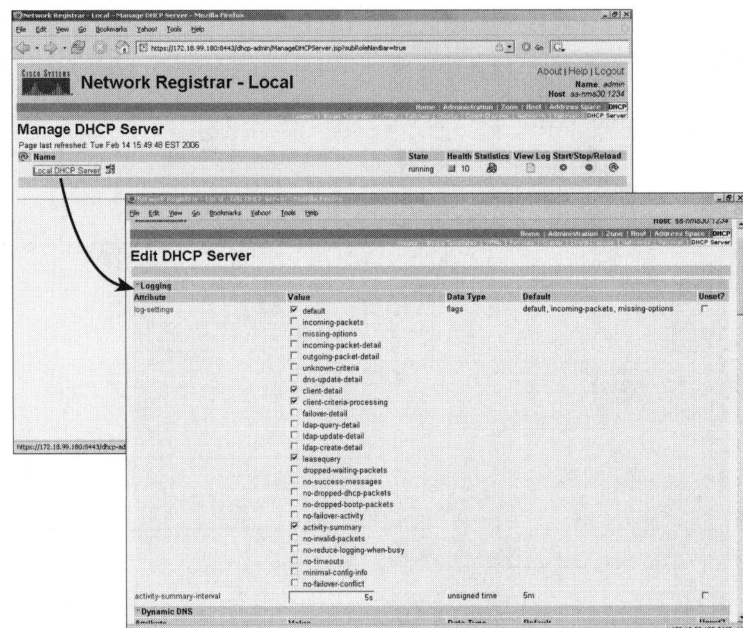

Select the server to adjust logging options

As you can see, the log levels and settings can be adjusted by selecting the DHCP server. Remember that after any server setting is modified you need to reload the server. By selecting the View Log option, you can see the current log file and navigate through its contents looking for the logs related to the device you are troubleshooting.

These logs are saved in a rotating log file called *name_dhcp_1_log*, which by default is in the directory */var/nwreg2/logs*. Alternatively, you can navigate to this directory and use the UNIX command **tail -f** *filename* to watch the log contents in real-time, as Figure 4-13 demonstrates.

Figure 4-13 *CNR DHCP Log Example*

```
02/14/2006 16:05:49 name/dhcp/1 Info Protocol 0 04935 R959160:   ->     chaddr = 0:13:10:50:38:1
02/14/2006 16:05:49 name/dhcp/1 Info Protocol 0 04935 R959160:   ->       dhcp-message-type = 1 discover
02/14/2006 16:05:49 name/dhcp/1 Info Protocol 0 04935 R959160:   ->       dhcp-parameter-request-list =
02/14/2006 16:05:49 name/dhcp/1 Info Protocol 0 04935 R959160:   ->           1    subnet-mask
02/14/2006 16:05:49 name/dhcp/1 Info Protocol 0 04935 R959160:   ->           2    time-offset
02/14/2006 16:05:49 name/dhcp/1 Info Protocol 0 04935 R959160:   ->           3    routers
02/14/2006 16:05:49 name/dhcp/1 Info Protocol 0 04935 R959160:   ->           4    time-servers
02/14/2006 16:05:49 name/dhcp/1 Info Protocol 0 04935 R959160:   ->           7    log-servers
02/14/2006 16:05:49 name/dhcp/1 Info Protocol 0 04935 R959160:   ->           6    domain-name-servers
02/14/2006 16:05:49 name/dhcp/1 Info Protocol 0 04935 R959160:   ->           66   tftp-server
02/14/2006 16:05:49 name/dhcp/1 Info Protocol 0 04935 R959160:   ->           67   boot-file
02/14/2006 16:05:49 name/dhcp/1 Info Protocol 0 04935 R959160:   ->           0    pad
02/14/2006 16:05:49 name/dhcp/1 Info Protocol 0 04935 R959160:   ->           0    pad
02/14/2006 16:05:49 name/dhcp/1 Info Protocol 0 04935 R959160:   ->           0    pad
02/14/2006 16:05:49 name/dhcp/1 Info Protocol 0 04935 R959160:   ->           122  cablelabs-client-configuration
02/14/2006 16:05:49 name/dhcp/1 Info Protocol 0 04935 R959160:   ->       dhcp-class-identifier = "docsis2.0:0524010101020102
3010104010105010106010107010f0801100901000a01010b01180c0101"
02/14/2006 16:05:49 name/dhcp/1 Info Protocol 0 04935 R959160:   ->       vendor-encapsulated-options = 83 octets
SubOption 2 (3 octets) : (String)"ECM"
SubOption 3 (8 octets) : (String)"ECM:EMTA"
SubOption 4 (10 octets) : (String)"KBG5359416"
SubOption 5 (3 octets) : (String)"2.1"
SubOption 6 (15 octets) : (String)"2.0.3.5.10-1102"
SubOption 7 (7 octets) : (String)"2.1.6gV"
SubOption 8 (6 octets) : (String)"001310"
SubOption 9 (6 octets) : (String)"CM2P2B"
SubOption 10 (7 octets) : (String)"Linksys"

02/14/2006 16:05:49 name/dhcp/1 Info Protocol 0 04935 R959160:   ->       dhcp-client-identifier = 01:00:13:10:50:38:01
02/14/2006 16:05:49 name/dhcp/1 Info Protocol 0 04935 R959160:   ->         ethernet? = 0:13:10:50:38:1
02/14/2006 16:05:49 name/dhcp/1 Info Protocol 0 04935 R959160:   ->       dhcp-max-message-size = 1500
02/14/2006 16:05:49 name/dhcp/1 Info Protocol 0 04935 R959160:   ->       relay-agent-info = 01:04:80:01:00:04:02:06:00:13:10:
50:38:01
02/14/2006 16:05:49 name/dhcp/1 Info Protocol 0 04935 R959160:   ->         circuit-id = 80:01:00:04
02/14/2006 16:05:49 name/dhcp/1 Info Protocol 0 04935 R959160:   ->         remote-id = 00:13:10:50:38:01
02/14/2006 16:05:49 name/dhcp/1 Info Protocol 0 04935 R959160:   ->       end
02/14/2006 16:05:49 name/dhcp/1 Info Protocol 0 04935 R959160:   ->     sname = ""
02/14/2006 16:05:49 name/dhcp/1 Info Protocol 0 04935 R959160:   ->     file = ""
02/14/2006 16:05:49 name/dhcp/1 Info Protocol 0 04935 R959160: ----- END OF RECEIVED -- R959160 -----
02/14/2006 16:05:49 name/dhcp/1 Info Configuration 0 04488 Client detail for Client CID: 01:00:13:10:50:38:01  from Request '
R959160'  Client entry name: '' Client Class Name = 'none', Specified host name: no, Host Name = 'none', Specified domain nam
e: no, Domain Name = '(null)', Client Policy = 'none', Client Class Policy = 'none', Action = 'none', InclusionCriteria = 'ta
gCompanyY', ExclusionCriteria = '', over-limit-client-class = 'none', Client Class Name Derived From: no client-entry, given
no default-client-class-name & no default entry client class-name.
02/14/2006 16:05:49 name/dhcp/1 Info Configuration 0 04400 When processing Request: R959160 for Client: CID: 01:00:13:10:50:3
8:01  the Lease: 13.34.14.122 was found to be acceptable.
02/14/2006 16:05:49 name/dhcp/1 Info Protocol 0 04935 X964434: ----- TRANSMITTED -- X964434 -----
02/14/2006 16:05:49 name/dhcp/1 Info Protocol 0 04935 X964434: <-       port = 67    transmitted to = 13.34.0.1
02/14/2006 16:05:49 name/dhcp/1 Info Protocol 0 04935 X964434: <-       packet length = 357
02/14/2006 16:05:49 name/dhcp/1 Info Protocol 0 04935 X964434: <-       op = 2 reply
02/14/2006 16:05:49 name/dhcp/1 Info Protocol 0 04935 X964434: <-       htype = 1  ethernet    hlen = 6
02/14/2006 16:05:49 name/dhcp/1 Info Protocol 0 04935 X964434: <-       hops = 0
Ready                                                  Telnet   49, 16   52 Rows, 125 Cols   VT100
```

In this example, you can see the DHCP Discover packet and corresponding DHCP Offer for a cable modem. If some of the requested options could not be provided, you would see that information here. As you might imagine, if you have a large network, these log files can fill up very quickly. Consequently, you must also exercise caution when choosing log levels. For example, you would not want to enable the **incoming-packet-detail** and **outgoing-packet-detail** levels for very long in a production system.

Likewise, for a CNR DNS server, the rotating log file is called name_dns_1_log in the same directory. This is a good place to look for help when diagnosing problems relating to dynamic DNS updates. When these log files reach their maximum size, an incremental version number is appended to the file, and a new file is started. For example, the first time a new DHCP log file is needed, the old file is renamed name_dhcp_1_log.01. This rotating and renaming continues until the maximum number of log files is reached; at this point the oldest file is deleted, and the next oldest assumes its name and so on. The maximum file size and maximum number of log file parameters are configurable.

As you know, in BAC the DHCP process is handled via communication between the DHCP server and the DPE cache engine; thus, the DPE log is also a useful source of information when debugging these steps. The DPE log is contained in the file dpe.log, which by default is in the directory /var/CSCObpr/dpe/logs. The DPE log can be viewed simply by going to this directory and looking at the file contents or by using the command **show log** from the

DPE CLI. With this command you can either look at the entire log, the last x amount of lines, or the running log. If you refer back to the discussion on DPE configuration in Chapter 3, you can see the command **log level 5-notification** in the running configuration; this command is setting the log level. Based upon this configuration, events from the log level you specify and all lower number levels are written to the log file. Figure 4-14 demonstrates an example of this log output with the level set as 6-info.

Figure 4-14 *DPE Log Example*

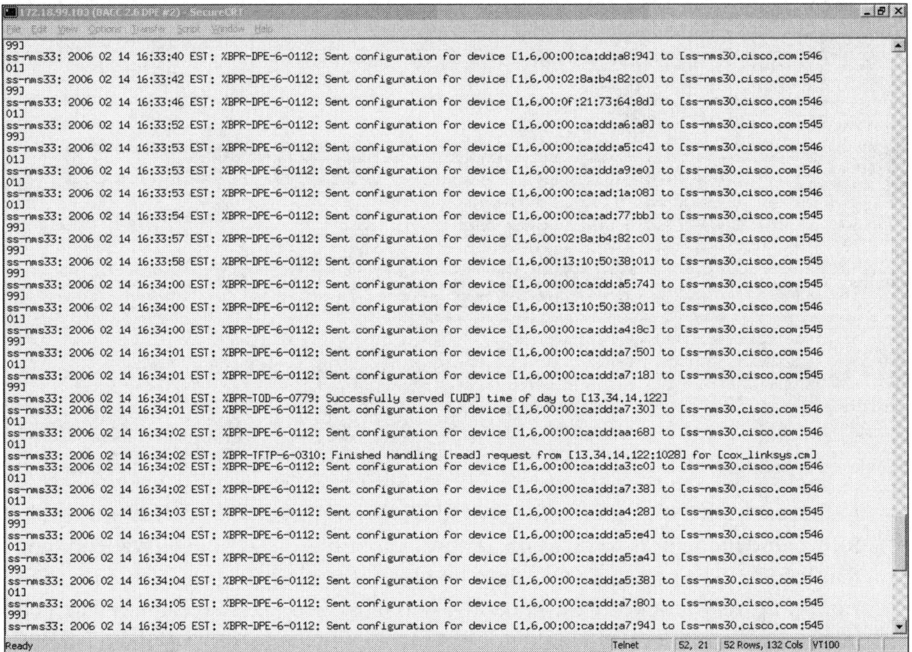

In this figure you can see logs as the result of the DPE sending device configuration to the DHCP servers (these have a label beginning with **%BPR-DPE**), from the DPE sending time of day responses to cable modems (these have a label beginning with **%BPR-TOD**), and from the DPE sending configuration files to cable modems (these have a label beginning with **%BPR-TFTP**).

If you are having problems with a device after initial configuration, another good place to help diagnose problems is the RDU log. The RDU log is contained in the file rdu.log, which by default is in the directory /var/CSCObpr/rdu/logs. The following is a sample log found in this file:

```
ss-nms36: 2006 02 14 10:38:29 EST: %BPR-EXT-4-0258: [PacketCableMTAExtension]:
  [Device has requested deprecated CableLabs DHC
P option 177; Encountered while attempting to generate a configuration for device
  [1,6,00:00:39:07:02:20] via relay agent [24.38.0.1]]
```

In this case, the MTA is running old code and requesting DHCP option #177 instead of option #122. The RDU log levels are also adjustable via the utility setLogLevel.sh located by default in the /opt/CSCObpr/rdu/bin, as shown in Figure 4-15.

Figure 4-15 *Setting RDU Log Levels*

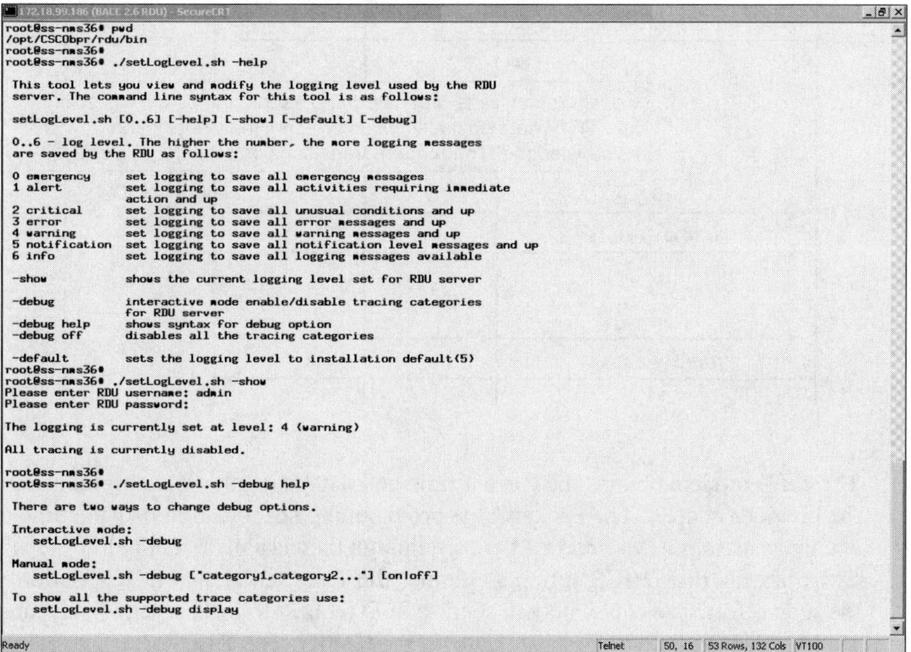

Time of Day Establishment

Figure 4-16 illustrates the next two cable modem provisioning stages.

Figure 4-16 *Cable Modem Provisioning - Stages 5 and 6*

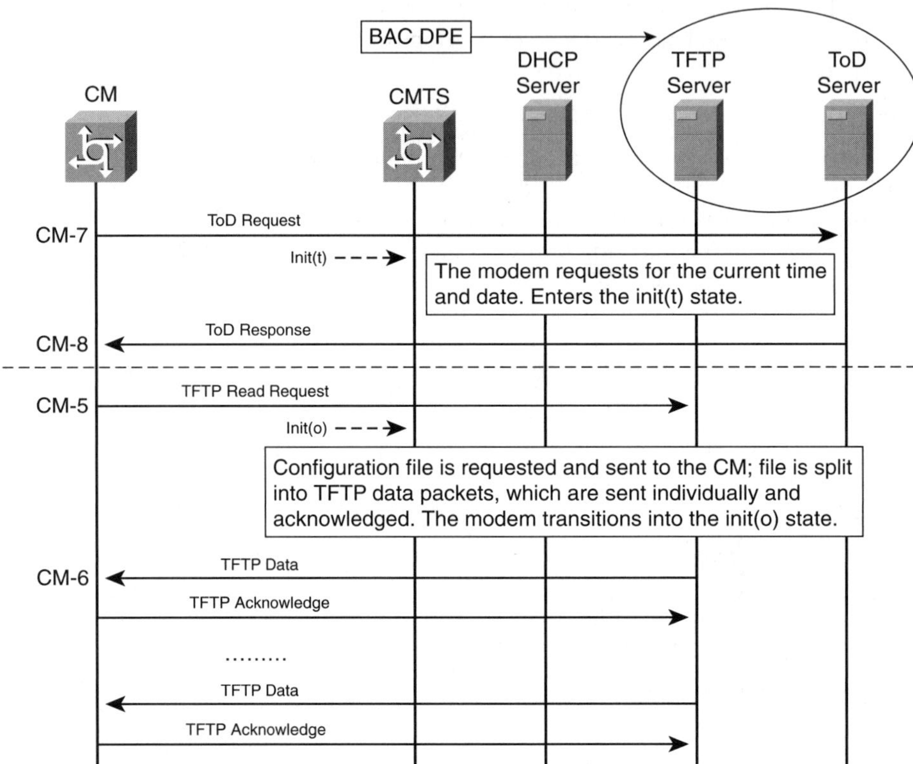

The cable modem obtains the current time and date using the Time protocol described in the previous chapter. The PacketCable provisioning specification lists the time of day exchange as steps CM-7 and CM-8 even though they usually occur before steps CM-5 and CM-6. Recall that DHCP option #4 defines the Time server, and DHCP option #2 defines the time offset. The cable modem sends a ToD request to this server, which causes the modem to transition to the init(t) state on the CMTS. The Time server returns the current time and date in the ToD response.

When using the Cisco BAC solution, the DPE acts as the time server, so if an IP connectivity issue exists between the DPE and CMTS, the modem stays in the init(t) state until it times out. Refer back to the section "Troubleshooting the DHCP Process" earlier in the chapter to resolve this problem.

NOTE If no ToD response is received, this does not prevent the cable modem from provisioning; however, this can prevent the baseline privacy (BPI+) phase mentioned later from completing. BPI+ requires the cable modem and CMTS to verify the timestamps on their digital certificates used for authentication. If these timestamps are not synchronized, causing BPI+ authentication to fail, the modem enters the reject(pk) state.

Transfer of Configuration Information

The next step is the transfer of configuration information (steps CM-5 and CM-6). The DOCSIS configuration file is transferred to the cable modem using the TFTP protocol. The cable modem initiates a TFTP request for its configuration file. The TFTP server IP address and the configuration file name are specified in the DHCP messages received earlier. The TFTP server then locates this file and starts transferring it in TFTP data packets.

When a Cisco CMTS detects the modem is transferring its configuration file, it transitions to the init(o) state:

```
C0701-10012-PRE-A#show cable modem 0013.1050.3801
MAC Address     IP Address      I/F       MAC          Prim RxPwr  Timing  Num BPI
                                          State        Sid  (dB)   Offset  CPE Enb
0013.1050.3801 13.46.12.205    C7/0/0/U1 init(o)      30   0.00   1284    0   N
```

If the modem gets stuck in this state, this typically indicates the configuration file is not on the TFTP server or the modem determines that the configuration file it receives is invalid. Also if the TFTP server is not operational or is unreachable, this also causes the modem to "hang" in the init(o) state.

Remember when you are using BAC, the DPE is the TFTP server. So if you have a routing issue between the DPE and CMTS or the configuration file does not exist on the DPE, the modem gets stuck in the init(o) state. Refer back to the section "Troubleshooting the DHCP Process" earlier in the chapter for how to debug on the DPE.

Registration

After the cable modem has received its configuration file, it first determines its validity. The cable modem makes sure all mandatory items are present in the configuration file and also performs a message integrity check (MIC) on the file contents and validates it. If either of these fails, the modem stays in the init(o) state as previously mentioned. The CM MIC check protects against another device modifying or corrupting the configuration file contents before reaching the cable modem.

If the configuration file passes these checks, the cable modem follows the steps shown in Figure 4-17.

Figure 4-17 *Cable Modem Provisioning - Stages 7 and 8*

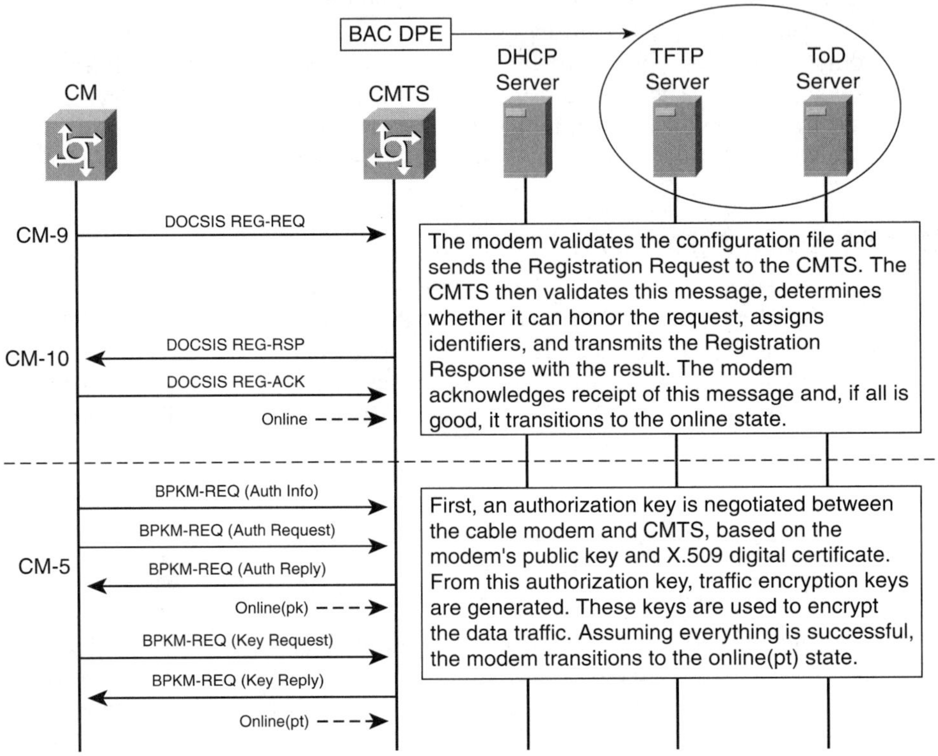

BAC DPE

CM

CMTS

DHCP Server

TFTP Server

ToD Server

CM-9 DOCSIS REG-REQ

CM-10 DOCSIS REG-RSP

DOCSIS REG-ACK

Online

The modem validates the configuration file and sends the Registration Request to the CMTS. The CMTS then validates this message, determines whether it can honor the request, assigns identifiers, and transmits the Registration Response with the result. The modem acknowledges receipt of this message and, if all is good, it transitions to the online state.

CM-5

BPKM-REQ (Auth Info)

BPKM-REQ (Auth Request)

BPKM-REQ (Auth Reply)

Online(pk)

BPKM-REQ (Key Request)

BPKM-REQ (Key Reply)

Online(pt)

First, an authorization key is negotiated between the cable modem and CMTS, based on the modem's public key and X.509 digital certificate. From this authorization key, traffic encryption keys are generated. These keys are used to encrypt the data traffic. Assuming everything is successful, the modem transitions to the online(pt) state.

First, the cable modem uses the configuration file to formulate the DOCSIS Registration Request (REG-REQ) message it sends to the CMTS (Step CM-9). The REG-REQ message contains all service flows and classifiers defined in the configuration file, as demonstrated in Figure 4-18.

Figure 4-18 *Registration Request Example*

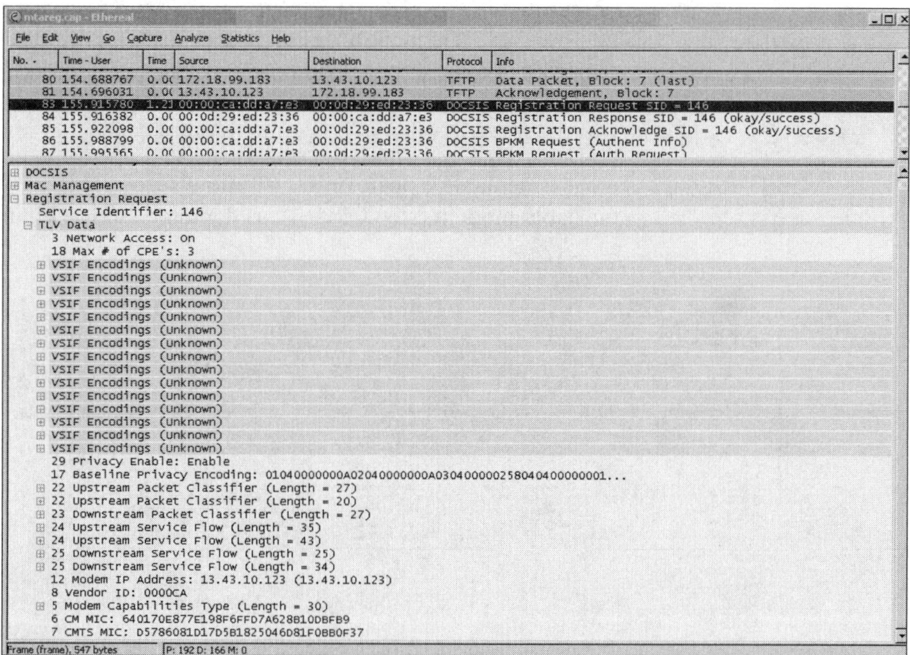

Upon receipt of the REG-REQ, the CMTS also performs a message integrity check. The CMTS has a password or shared secret it uses to calculate this value. This shared secret is also known to the provisioning system but unknown to the cable modem; therefore, the CMTS can determine if the cable modem attempted to alter or "hack" their configuration. The provisioning system calculates the CMTS MIC and includes it in the configuration file to be passed transparently by the modem to the CMTS; you can see this parameter at the bottom of Figure 4-18. The CMTS uses its configured shared secret and the parameters in the REG-REQ to calculate the CMTS MIC. If the REG-REQ were altered, a mismatch in the CMTS MIC values occurs and the REG-REQ is rejected; on a Cisco CMTS the cable modem transitions into the reject(m) state.

On a Cisco CMTS, the basic way to configure the shared secret is through the **cable shared-secret** *authorization key* command under the cable interface. Cisco also offers the dynamic shared secret or DMIC feature to provide additional security. This feature is configured using the **cable dynamic-secret** cable interface command. When this feature is used, the CMTS intercepts the configuration file during the TFTP process and dynamically creates a secret to generate the CMTS MIC sent with the configuration file. The CMTS then remembers this secret when verifying the REG-REQ sent by the cable modem. With this feature, each shared secret is valid only for a certain time and only for a particular modem; that is, if the modem re-registers, a different shared secret would be used.

Assuming the CMTS MIC check passes, the CMTS then checks to see if it can honor the service flow parameters requested in the REG-REQ. If it can, it assigns service flow identifiers to each service flow and SIDs to each upstream service flow. Similarly, the CMTS also assigns classifier identifiers and payload header indices as appropriate. If the CMTS cannot honor one of the service flow requests, it rejects the REG-REQ message. On a Cisco CMTS, this causes the cable modem to transition to the reject state.

If the REG-REQ message is acceptable a positive Registration Response (REG-RSP) message is returned to the cable modem (step CM-10), as demonstrated in Figure 4-19.

Figure 4-19 *Registration Response Example*

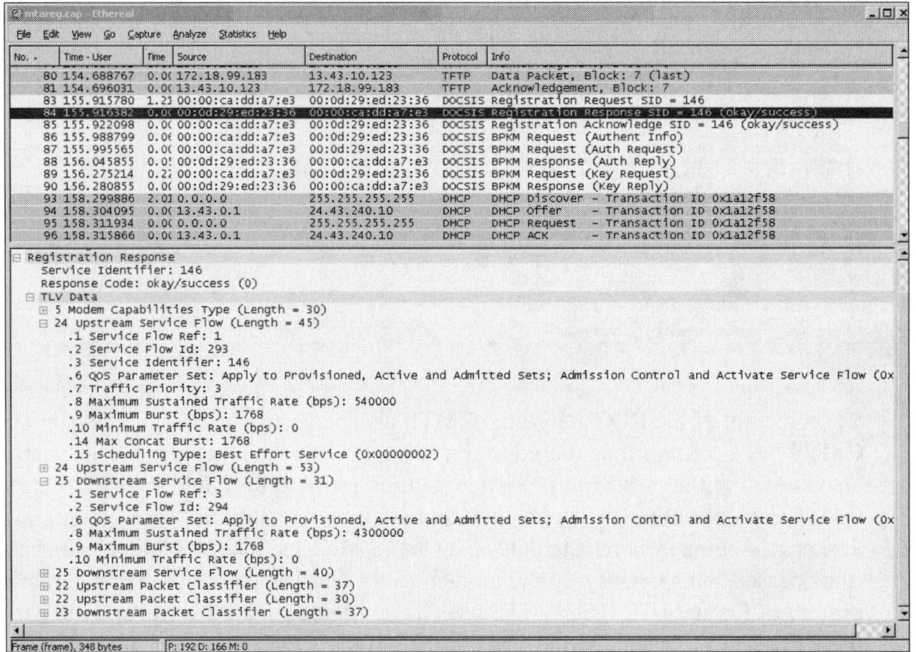

As you can see, in this case a response code of okay/success is returned and identifiers have been assigned to the service flows and classifiers. The cable modem then acknowledges the receipt of this message by sending a Registration Acknowledgement (REG-ACK) message to the CMTS. At this point the cable modem should have transitioned to the online state on the CMTS:

```
C0701-10012-PRE-A#show cable modem
    MAC Address    IP Address      I/F        MAC      Prim  RxPwr  Timing  Num BPI
                                              State    Sid   (dB)   Offset  CPE Enb
   0013.1050.3801 13.46.12.205    C7/0/0/U1  online   30    0.00   1284     1   Y
```

Another state you might see on a Cisco CMTS is reject(na). A modem enters this state if a Registration Negative Acknowledgement (REG-NACK) message is received in response to the Registration Response or if the modem fails to send the Registration Acknowledgement within the acceptable amount time.

On a Cisco CMTS, you can debug the registration process using the **debug cable registration** command in conjunction with the **debug cable mac-address** *CM_MAC_Address* **verbose** command for a particular modem. Additionally, if you include the **debug cable tlvs** command, the parameters in the registration messages are decoded in the debug output, as shown in Figure 4-20.

Figure 4-20 *CMTS CM Registration Debugging*

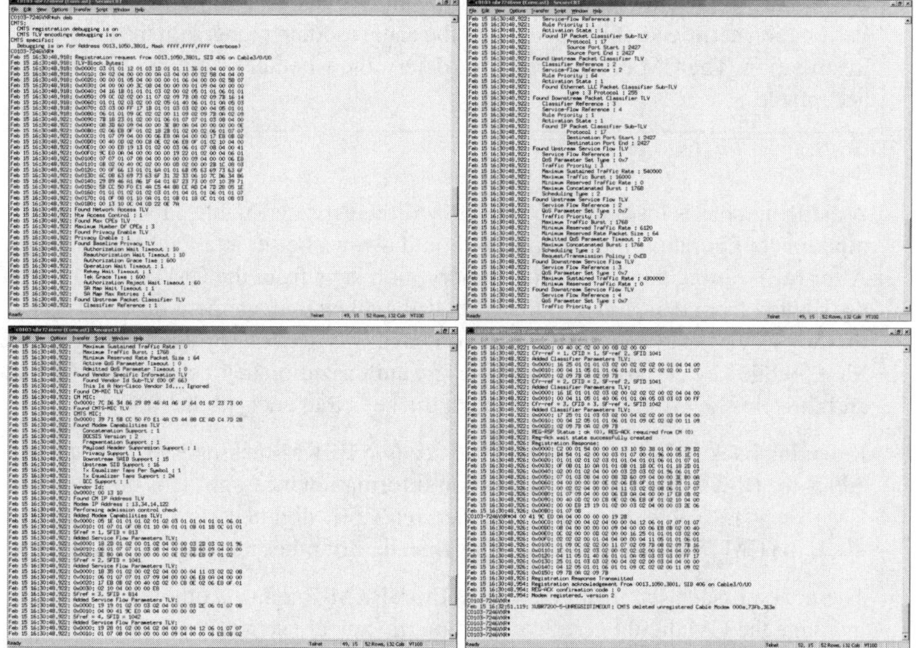

As you can see, these debugs generate a great deal of information that can help you diagnose issues with modem registration. Also if you are implementing the dynamic shared secret feature, the command **debug cable dynamic-secret** can help provide additional information.

Baseline Privacy Initialization

Although optional for ordinary cable modems, the security provided by Baseline Privacy Initialization Plus (BPI+) is mandatory for PacketCable MTAs. PacketCable still works with BPI disabled; however, this is not recommended.

BPI+ provides encryption for data packets across the DOCSIS network. In addition to providing packet encryption, BPI+ uses digital certificates to authenticate cable modems with the CMTS, similar to how digital certificates are used to authenticate PacketCable MTAs with the KDC. The cable modem has an ITU X.509 digital certificate, which is signed by the cable modem manufacturer; it also has the manufacturer X.509 certificate, which is signed by the DOCSIS root authority.

NOTE X.509 digital certificates are also used to provide a secure mechanism for upgrading and downgrading cable modem firmware files. In this case, a certificate known as a code verification certificate (CVC) is sent to the cable modem along with the upgrade/downgrade information. The CM uses the CVC to determine whether or not the new image is acceptable.

A cable modem is instructed to run BPI+ if the Privacy Enable TLV is set in the cable modem configuration file. This TLV should always be set in a PacketCable environment. After registration, the CM requests encryption keys from the CMTS. Encryption keys are distributed from the CMTS to the CMs using a key management protocol called Baseline Privacy Key Management (BPKM). You have two tiers of key exchanging with BPKM. First, public key cryptology establishes an authorization key between the CM and CMTS, and then this key is utilized to establish the keys that encrypt the modem's data traffic.

Referring back to Figure 4-17, you can see five BPKM messages: The first one is optional and is the BPKM-Request Authorization Information message. This message is used by the CM to send the cable modem's manufacturer X.509 digital certificate. If this message is not used, the CMTS needs to learn this certificate thru other means.

The next message sent from the CM is the BPKM-Request Authorization Request. In this message the CM digital certificate and public key are sent to the CMTS along with information such as the CM MAC address, manufacturer identifier, and serial number. Also included in this message are the modem's security capabilities and primary SID value. The primary SID is equal to the primary Security Association Identifier (SAID). Figure 4-21 shows an example.

Figure 4-21 *BPKM Authentication Request Example*

```
mtareg.cap - Ethereal
File  Edit  View  Go  Capture  Analyze  Statistics  Help
No. - Time - User    Time  Source               Destination          Protocol Info
  78 154.681152  0.0( 172.18.99.183       13.43.10.123         TFTP     Data Packet, Block: 6
  79 154.687952  0.0( 13.43.10.123        172.18.99.183        TFTP     Acknowledgement, Block: 6
  80 154.688767  0.0( 172.18.99.183       13.43.10.123         TFTP     Data Packet, Block: 7 (last)
  81 154.696031  0.0( 13.43.10.123        172.18.99.183        TFTP     Acknowledgement, Block: 7
  82 155.155840  0.4! 00:0d:29:ed:22:db   ff:ff:ff:ff:ff:ff    ARP      Who has 24.43.241.33?  Tell 24.43.241.1
  83 155.915780  0.7! 00:00:ca:dd:a7:e3   00:0d:29:ed:23:36    DOCSIS   Registration Request SID = 146
  84 155.916382  0.0( 00:0d:29:ed:23:36   00:00:ca:dd:a7:e3    DOCSIS   Registration Response SID = 146 (okay/success)
  85 155.922098  0.0( 00:00:ca:dd:a7:e3   00:0d:29:ed:23:36    DOCSIS   Registration Acknowledge SID = 146 (okay/success)
  86 155.988799  0.0( 00:00:ca:dd:a7:e3   00:0d:29:ed:23:36    DOCSIS   BPKM Request (Authent Info)
  87 155.995565  0.0( 00:00:ca:dd:a7:e3   00:0d:29:ed:23:36    DOCSIS   BPKM Request (Auth Request)
  88 156.045855  0.0! 00:0d:29:ed:23:36   00:00:ca:dd:a7:e3    DOCSIS   BPKM Response (Auth Reply)
  89 156.275214  0.2: 00:00:ca:dd:a7:e3   00:0d:29:ed:23:36    DOCSIS   BPKM Request (Key Request)
  90 156.280855  0.0( 00:0d:29:ed:23:36   00:00:ca:dd:a7:e3    DOCSIS   BPKM Response (Key Reply)
  91 156.370449  0.0{ 13.14.5.14          24.43.240.10         MGCP     AUEP 220823560 aaln/1@x1-6-00-00-ca-dd-a7-e4.cc.cisco.co
  92 156.783911  0.4] 13.14.5.14          24.43.240.10         MGCP     AUEP 220823564 aaln/1@x1-6-00-00-ca-dd-a7-e4.cc.cisco.co
⊞ Frame 87 (1003 bytes on wire, 1003 bytes captured)
⊟ DOCSIS
    11.. .... = FCType: MAC Specific (0x03)
    ..00 001. = FCParm: Mac Management Message (0x01)
    .... ...0 = EHDRON: Extended Header Absent
    MacParm: 0x00
    Length after HCS (bytes): 1001
    Header check sequence: 0xd6ae
⊞ Mac Management
⊟ BPKM Request Message
    BPKM Code: Auth Request (4)
    BPKM Identifier: 24
    BPKM Length: 973
⊟ BPKM Attributes
  ⊟ 5 CM Identification
      1 Serial Number: 4AEBGC21B00745E
      2 Manufacturer Id: 0000CA
      3 Mac Address: 00:00:ca:dd:a7:e3 (00:00:ca:dd:a7:e3)
      4 RSA Public Key: 30818902818100B1219697615F5EFEE5884FDDE1C9FA533A...
    18 CM Certificate: 30820302308201EAA003020102021100F55084036478B209...
  ⊟ 19 Security Capabilities
      21 Cryptographic Suite List: 0100
      22 BPI Version: BPI+ (1)
    12 SAID: 146
Frame (frame), 1003 bytes    P: 192 D: 192 M: 0
```

The CMTS validates the cable modem's certificate and then uses the contained public key to encrypt the authorization key. Remember the CM certificate is validated using the CM manufacturer certificate, and the manufacturer certificate is validated against the DOCSIS root certificate. The authorization key is sent to the cable modem in the BPKM-RSP Authorization Reply message. Then the cable modem uses this key to securely initiate the negotiation of traffic encryption keys (TEKs). The request for TEKs is sent in the BPKM-REQ Key Request message and the TEKs are returned in the BPKM-RSP Key Reply message. These TEKs encrypt data traffic; they have a limited lifetime and, therefore, must be periodically regenerated. The authorization key also has a limited lifetime and is periodically regenerated.

On a Cisco CMTS, if you have a failure in the authorization key process, the cable modem enters the reject(pk) state, and if you have a failure in the traffic encryption key process, the cable modem enters the reject(pt) state. A modem that successfully completes the BPI+ process enters the online(pt) state. A failure here is most likely caused by a bad certificate on the modem, a missing or invalid root certificate on the CMTS, or a problem in the timestamp synchronization between the cable modem and the CMTS. On a Cisco CMTS, the root certificate is saved in the boot flash under the name **root-cert**. The DOCSIS root certificate (and PacketCable root certificate) can be obtained from VeriSign at the following

URL: https://www.verisign.com/products/cable/root.html. A timing problem is probably because of a failure in the Time of Day establishment phase mentioned earlier.

As soon as all the keys are successfully negotiated, the modem enters the online(pt) state:

```
C0701-10012-PRE-A#show cable modem 0013.1050.3801
MAC Address     IP Address      I/F         MAC           Prim RxPwr  Timing  Num BPI
                                            State         Sid  (dB)   Offset  CPE Enb
0013.1050.3801  13.46.12.205    C7/0/0/U1   online(pt)    30   0.00   1284    0   Y
```

On a Cisco CMTS, if the cable modem fails to renew a traffic key after it expires, it enters the expire(pt) state, and if it fails to renew the authorization key, it enters the expire(pk) state.

Using the **debug cable privacy** command in conjunction with the **debug cable mac-address** *CM_MAC_Address* **verbose** command for a particular modem can help troubleshoot the BPI+ process, as shown in Figure 4-22.

Figure 4-22 *CMTS Baseline Privacy Debugging*

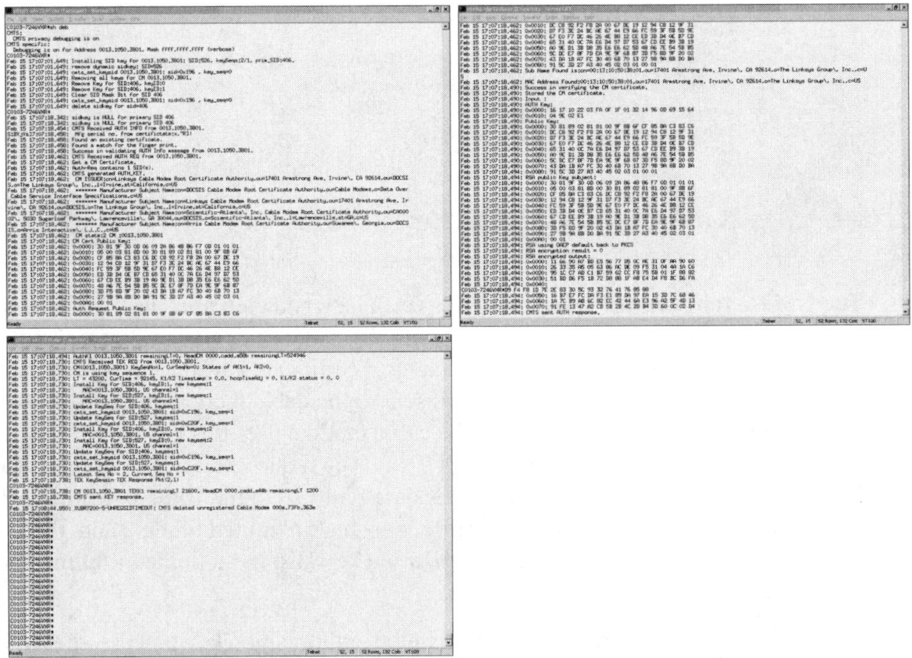

In these debugs you can see all the BPKM messages, certificate validation, and key assignments.

For more details on how BPI+ works please consult the DOCSIS BPI+ specification.

PacketCable MTA Provisioning

Now that the cable modem has successfully provisioned, the MTA can now begin its provisioning process. Remember that a PacketCable MTA has a separate MAC address, IP address, and FQDN from the cable modem, allowing these two entities to be managed separately. Also remember that you have three methods for MTA provisioning:

- The Secure method
- The Basic method
- The Hybrid method

The Secure Method

Because the secure method is the recommended way of provisioning MTAs, it is discussed first.

As with the cable modem, MTA provisioning has an IP connectivity establishment phase, a security phase, a transfer of configuration phase, and a registration phase.

IP Connectivity Establishment

As you know, the MTA also uses the DHCP protocol to obtain an IP address, and you learned that this is a four-step process, as illustrated in Figure 4-23.

Figure 4-23 *MTA Provisioning - IP Connectivity*

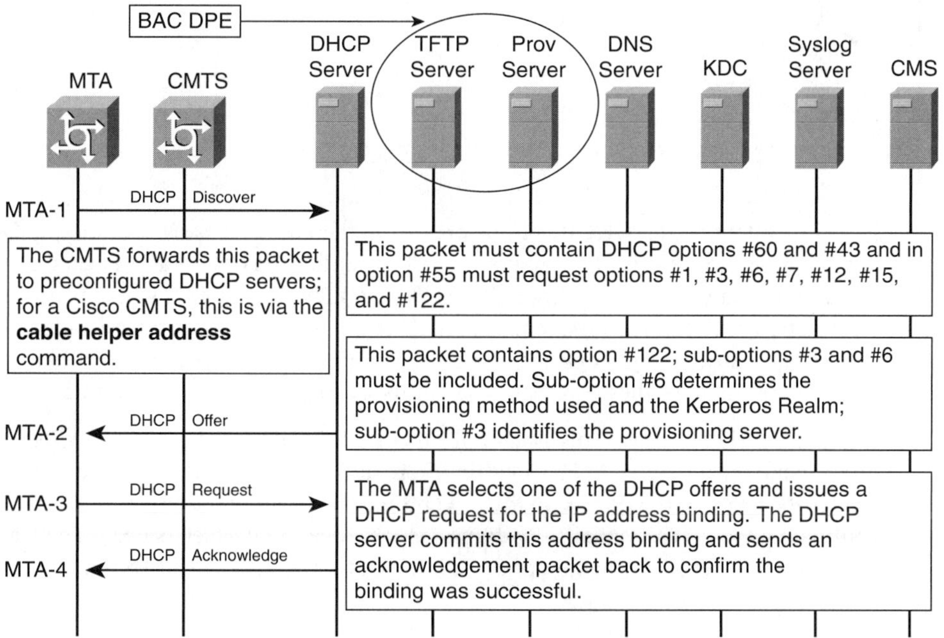

The PacketCable provisioning specification refers to these steps as MTA-1 through MTA-4, and these steps are common to all three methods of provisioning. The first step, MTA-1, is where the MTA broadcasts a DHCP discover packet, which is forwarded by the CMTS. If you are not seeing this DHCP discover packet, this probably means you did not return DHCP option #122 to the cable modem, or it was returned with a value of 0.0.0.0 for sub-option #1. Figure 4-24 shows an Ethereal trace of provisioning step MTA-1.

Figure 4-24 *DHCP Discover Example (MTA-1)*

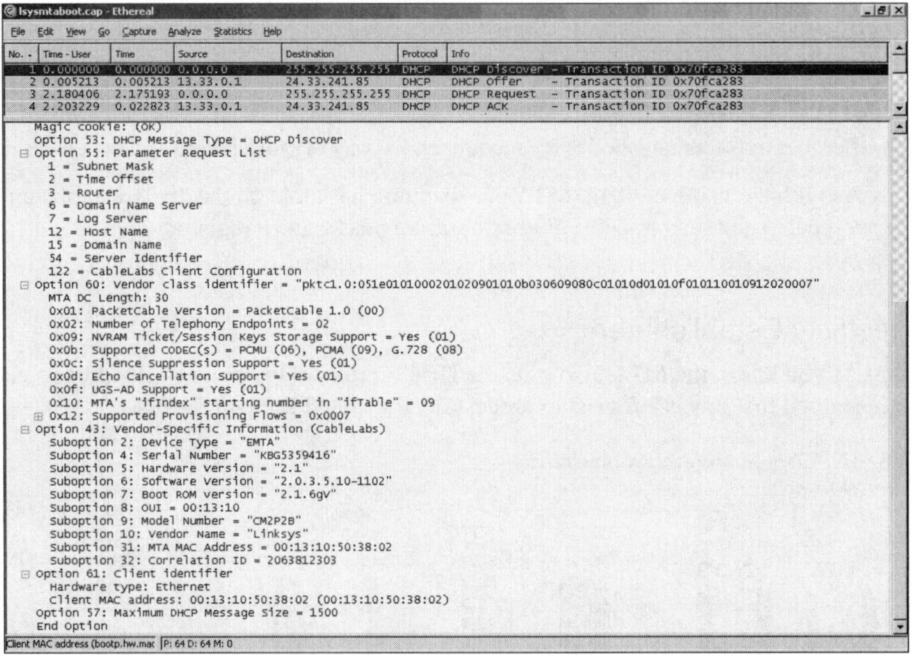

Pay close attention the DHCP options present in the Discover packet and to the DHCP options requested. Notice Option #60, which identifies the device as a PacketCable MTA and details many of the MTA capabilities. Also notice Option #43, which provides information such as the MTA vendor name, serial number, and hardware and software versions. Finally, note the options requested in Option #55; if your provisioning system does not return all of these options, you are likely to get "stuck" in this state.

When the CMTS receives this packet, it fills in the "giaddr" field to identify itself as the relay agent. On a Cisco CMTS, if **cable dhcp giaddr primary** is configured, the "giaddr" field is the primary IP address on the Layer 3 interface, and if **cable dhcp giaddr policy** is configured, the first secondary IP address on the Layer 3 interface is used. The CMTS also optionally adds the relay agent option (DHCP Option #82). The CMTS then unicasts these packets to the preconfigured DHCP servers. Thus if you have more than one server configured, the CMTS unicasts the packet to each server.

A failure in this state has many of the same probable causes as a failure in the cable modem state, CM-1. For example, if the DHCP servers are down or unreachable, make sure the CMTS and DHCP servers can ping each other. Remember it is definitely possible for the CM and MTA to use different DHCP servers. On a Cisco CMTS this distinction can be made by using the keywords cable-modem or host at the end of the helper address configuration. Also make sure the cable dhcp-giaddr option is set properly as either **policy** or **primary**. Remember this effects how scopes are selected by the DHCP server; so if it's configured wrong, the DHCP server is unable to find a matching scope. As in step CM-1, you might also have a problem with the configuration of the OSS/NMS servers, or the DHCP server might not have any available leases.

Remember the steps involved if using the Cisco BAC solution. First, the DHCP Discover is sent to the CNR DHCP servers and passed to the DPE via the BAC extension script. If the MTA configuration is already stored in the DPE cache, it sends this information back to CNR; otherwise, the RDU needs to be contacted and told to provide the DPE with the MTA information. As soon as the DPE has the required MTA information, it sends this back to CNR, which determines the IP address and other parameters needed. Finally, the DHCP Offer is built and sent to the MTA.

Just as with the cable modem case earlier, if the device is not configured or is misconfigured on the RDU, it cannot receive an Offer packet.

Assuming the Discover packet makes it to a DHCP server and that server can honor that request, the server returns a DHCP Offer to the MTA (step MTA-2). Figure 4-25 is an Ethereal trace of the DHCP Offer packet.

The cable modem waits some time for a DHCP Offer with DHCP option #122; however, if none is received, the modem eventually timeouts and accepts one without this option. In this case, the MTA does not attempt to provision. In the DHCP Offer the DHCP server identifies itself by filling in its IP address in DHCP option #54. The use of this parameter is very important because it relates to whether or not the MTA accepts the offer. Recall that in the cable modem portion of the provisioning process sub-options #1 and #2 of DCHP option #122 identify the primary and secondary servers authorized to communicate with the MTA. Therefore, if the server IP address does not equal one of these values, the MTA rejects the DHCP Offer. Remember, if you have an implementation that causes the server IP address to differ from the option #122 values, you can bypass this security check by using a value of 255.255.255.255 for the option #122, sub-option #1 sent to the cable modem.

Figure 4-25 *DHCP Offer Example (MTA-2)*

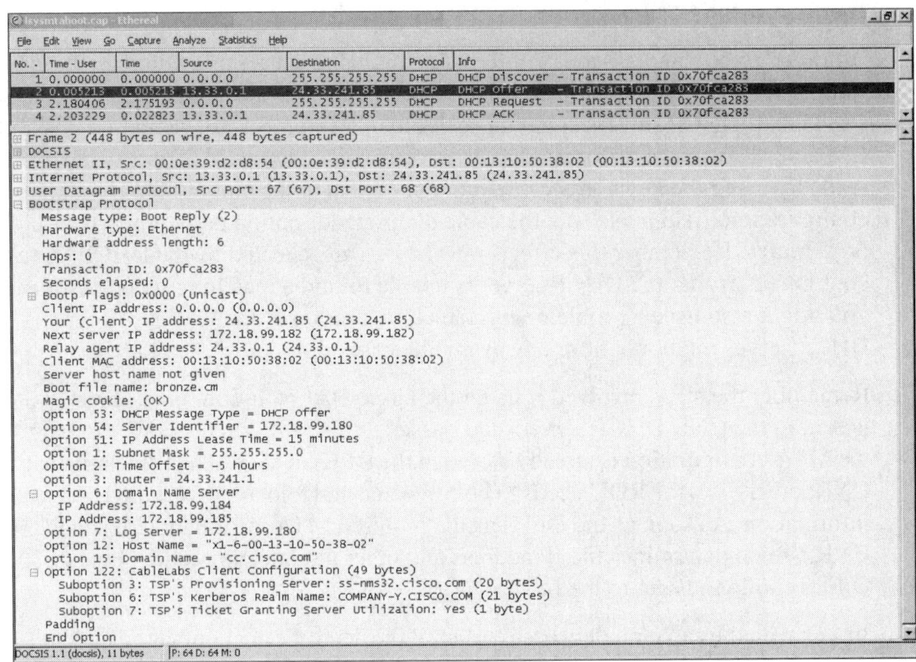

In the DHCP Offer sent to the MTA, sub-options #3 and #6 are required in option #122. Recall from Chapter 3 that sub-option #3 identifies the SNMP provisioning entity and sub-option #6 identifies the Kerberos realm. The Kerberos realm in secure provisioning defines the Kerberos domain and is in a DNS domain name format. In the non-secure provisioning flows, Kerberos is bypassed, so two special values in sub-option #6 trigger these flows.

If this parameter has a value of "BASIC.1" or "BASIC.2," the basic method is used; if it has a value of "HYBRID.1" or "HYBRID.2," the hybrid method is used; otherwise, the secure method is used. As you can see from Figure 4-25, in this example the Kerberos realm is equal to "COMPANY-Y"; therefore, the secure method of provisioning is used.

NOTE Recall the Kerberos realm name must be specified in all capital letters.

Assuming the DHCP Offer is sent from an authorized server, and all the required DHCP options are present and valid, the MTA then chooses one of the valid offers it receives and sends a DHCP Request packet (step MTA-3). Again, this packet is broadcast by the MTA and forwarded by the CMTS. Remember, if the MTA fails to choose any of the received offers, this typically indicates at least one of the options requested was not returned by the DHCP servers or was deemed invalid.

Figure 4-26 shows an example of the DHCP Request packet.

Figure 4-26 *DHCP Request Example (MTA-3)*

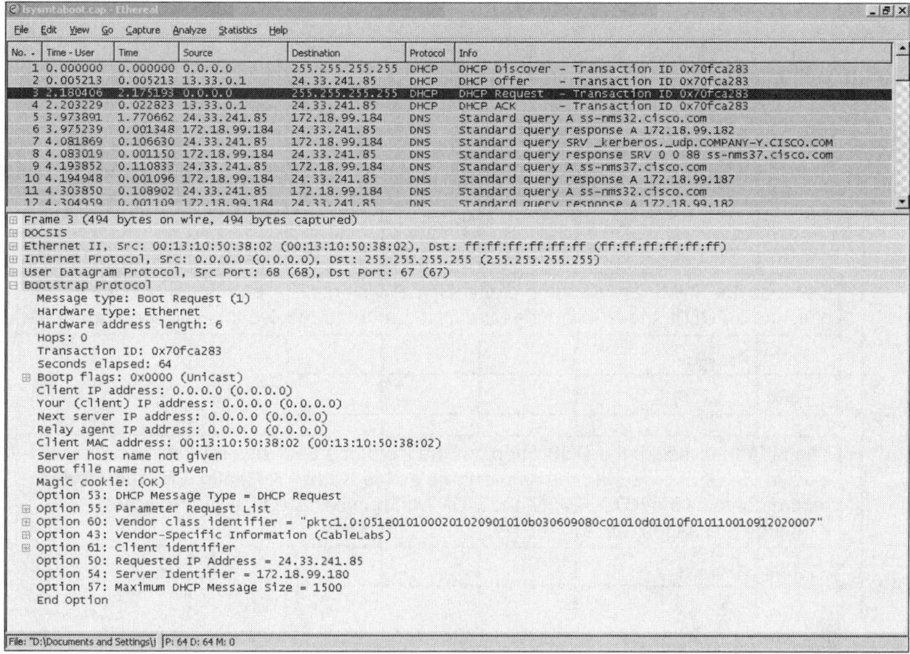

After the DHCP server receives the Request packet, it adds this IP address binding to its database and sends a DHCP Acknowledgement packet back to the MTA (step MTA-4).

At this point on a Cisco CMTS you should see the MTA show up as a host entry behind the cable modem:

```
C0103-7246VXR#show cable modem 13.34.14.122 cpe
IP address          MAC address
24.34.240.245       0013.1050.3802
```

Security Establishment

Now that the DHCP process has been successfully completed, the MTA needs to establish its security keys for communication with the provisioning system as well as with the CMS. The steps involved in this process are split into three phases with the first phase illustrated in Figure 4-27.

NOTE If the MTA already has valid security keys stored in nonvolatile random access memory (NVRAM), many of these steps are skipped (specifically steps MTA-5 through MTA-12).

Figure 4-27 *MTA Provisioning - Security - Part I*

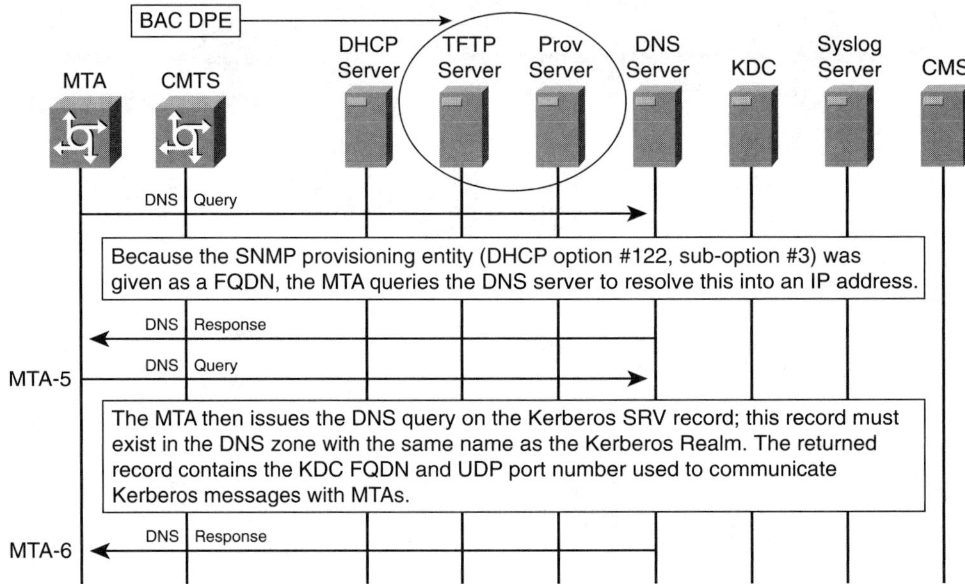

Although not defined as a step in the provisioning specification, the MTA at some point needs to resolve the SNMP provisioning entity FQDN into an IP address. In this example this is done next, as you can see in Figure 4-27. Remember DNS queries are sent to the DNS server(s) specified in DHCP option #6. Logically, if you do not have an "A" record configured in the DNS server, this step fails, and you are unable to provision PacketCable MTAs.

The next official step, MTA-5, is where the MTA queries the DNS server for the KDC server in the Kerberos realm specified by option #122, sub-option #6. This type of DNS record is a SRV record, and the format of the query is **_kerberos._udp**. followed by the Kerberos realm name; so in this case the SRV query is for **_kerberos._udp.COMPANY-Y.CISCO.COM**. Figure 4-28 illustrates an example of this query.

Figure 4-28 *DNS Query for KDC SRV*

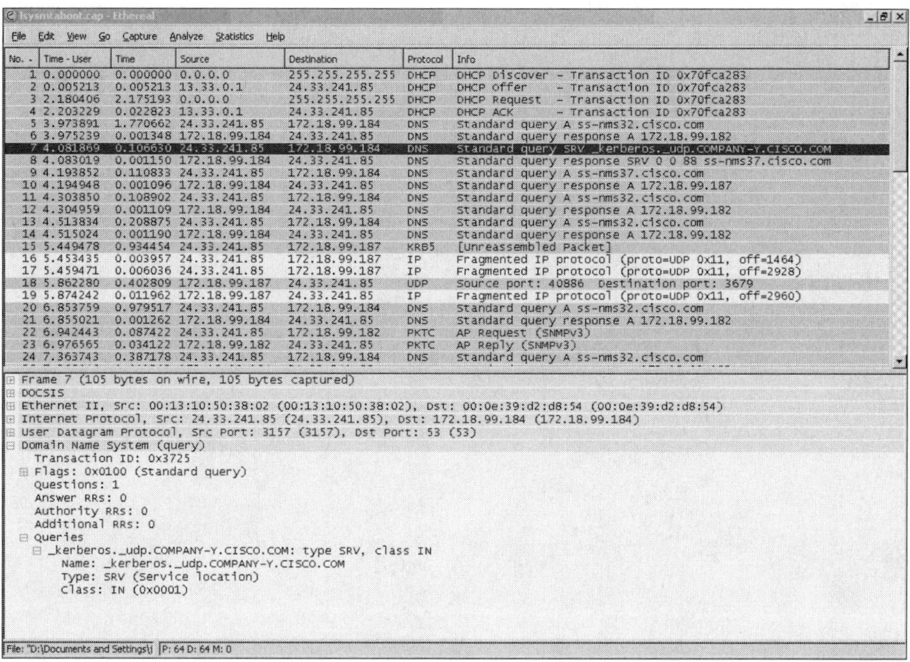

Figure 3-25 in Chapter 3 showed the format of this record using CNR for the DNS server; this record is shown again here as Figure 4-29.

The data in this record identifies the FQDN of the KDC (in this case ss-nms37.cisco.com) as well as the UDP port number used on the KDC (in this case UDP port 88). The DNS server responds to this query (step MTA-6) by sending this information to the MTA, as shown in Figure 4-30. From the previous figure you can also tell the last record is the "A" record for the KDC FQDN. If either this record or the "SRV" record is missing, the provisioning process is not able to pass the security phase.

Figure 4-29 *Static Resource Record Configuration*

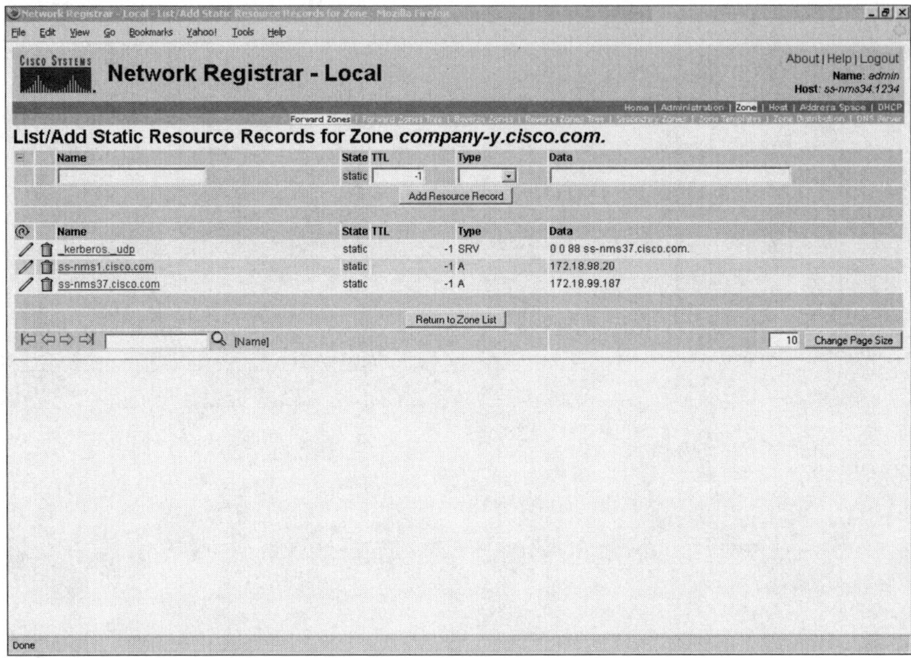

Figure 4-30 *DNS Response for KDC SRV*

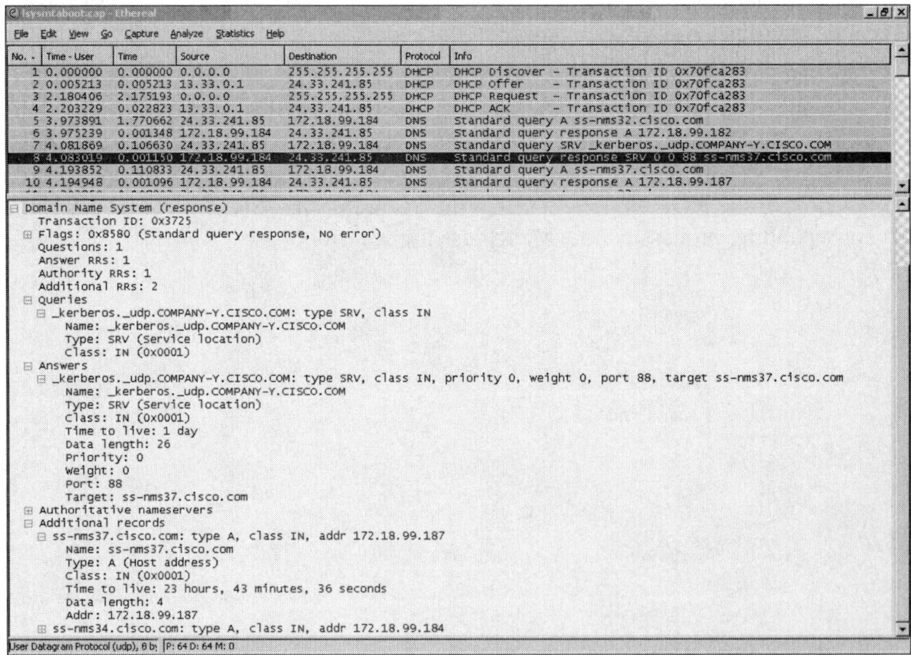

Upon receiving the "SRV" response, the MTA then queries the DNS server again to resolve the KDC FQDN into an IP address. This query and response are steps MTA-7 and MTA-8. Optionally, the DNS can go ahead and return the KDC IP address in the response to the SRV lookup (step MTA-6). In this case steps MTA-7 and MTA-8 are not needed. Figure 4-31 illustrates the logic for steps MTA-7 through MTA-10.

Figure 4-31 *MTA Provisioning - Security - Part II*

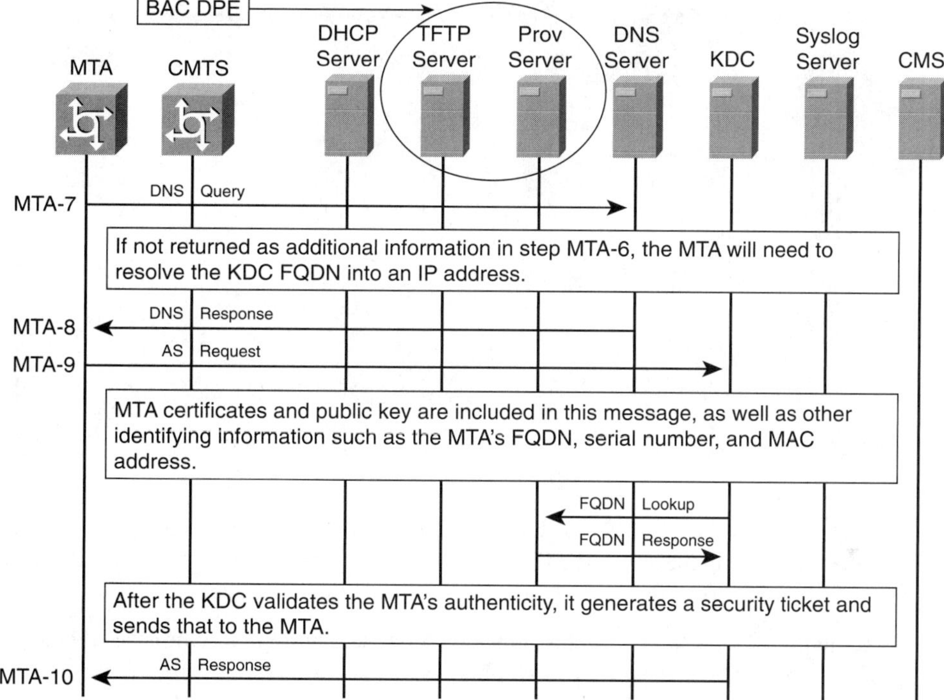

After the MTA knows the location of the KDC, it initiates the request for a Kerberos ticket by sending an authentication server request (AS request) message to the KDC. Figure 4-32 shows this message (step MTA-9).

Figure 4-32 *AS Request*

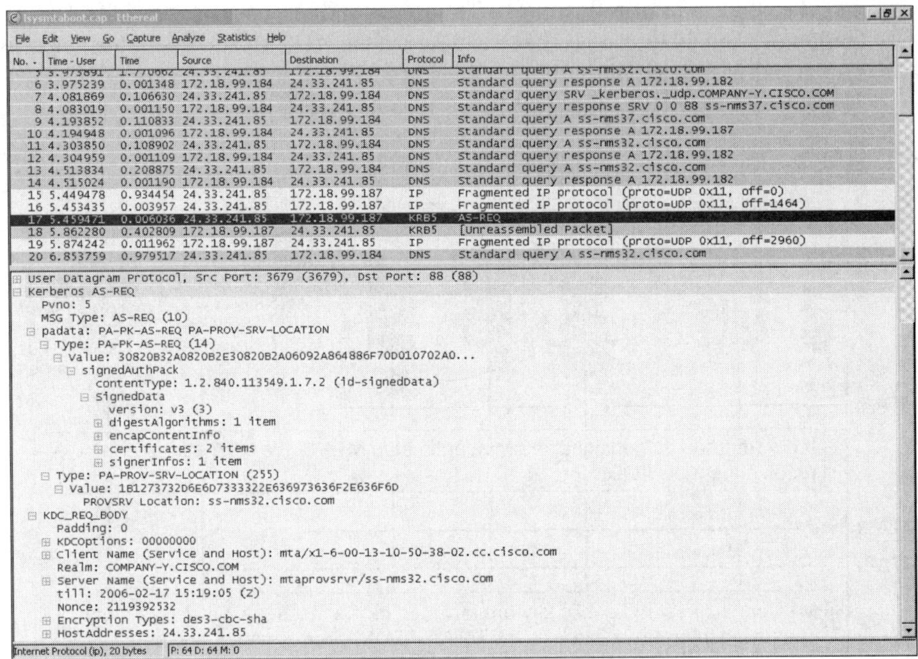

Because of all the information sent in this message, rather large packets are needed; therefore, fragmentation at the IP layer is required. Fortunately, Ethereal can reassemble IP fragments, so you can still see the AS-REQ packet in the trace.

Remember this message contains the MTA device and manufacturer certificates, digital signature, public key, and serial number; as you can see from the figure, this information is in a field known as the PKINIT pre-authenticator field, PA-PK-AS-REQ. The KDC validates the received MTA Manufacturer certificate chains to the MTA Root certificate and the received MTA Device certificate chains to the MTA Manufacturer certificate.

NOTE The encapContentInfo sub-field contains the MTA public key as well as a checksum covering the contents of the KDC_REQ_BODY field.

As you can see from Figure 4-32, the FQDN of the SNMP provisioning entity is sent in the PA-PROV-SRV-LOCATION field. If you look at the KDC_REQ_BODY field you can see the MTA FQDN, IP address, and the services for which the MTA requests security keys. In this case, only the PacketCable provisioning service (mtaprovsrvr) is requested. The MTA device certificate only contains the MTA MAC address, not the MTA FQDN, so for the KDC to validate this MAC address, it needs the FQDN to MAC mapping. The KDC verifies this mapping by contacting the provisioning server specified in the PA-PROV-SRV-LOCATION field. With a BAC implementation the KDC contacts the DPE to validate the FQDN to MAC mapping. This step is commonly referred to as the *FQDN query to the provisioning server.*

The protocol used for the FQDN exchange is Kerberos-based; specifically, UDP is used, and the provisioning server listens on port 2246. For more details on this process refer to the PacketCable Security specification.

If these checks are successful the KDC returns a key to the MTA in a Kerberos authentication response (AS response) message. This key is encrypted using the MTA public key so that only an MTA with the matching private key is able to access this returned key.

This encrypted key could be the provisioning service ticket or a Ticket Granting Ticket (TGT). Which of these is sent depends whether or not the optional Ticket Granting Server (TGS) was used; more on this is discussed later.

The AS-RSP is step MTA-10 in the provisioning specification. If you have a problem receiving the AS-RSP message, this typically indicates the KDC is not operational. Or you could have a mismatch between the service key configured on the DPE and the one configured on the KDC. These keys must have the same name and content for the DPE and KDC to communicate successfully with each other. If you do have a mismatch, the DPE cannot respond to the KDC FQDN query, and hence, no AS-RSP is sent to the AS-REQ. Another possible problem affecting this step is clock skew; all BAC components need to have their clocks synchronized via NTP. If the clock skew is too large, you see Kerberos error messages sent to the MTA in response to the AS-REQ. Lastly a routing error between the DPE and KDC or between the MTA and KDC results in no AS-RSP message being received by the MTA.

Besides the ticket, the AS-RSP packet also includes the KDC certificate, Service Provider certificate, and optional Local System Operator certificate. Remember not only does the KDC authenticate the MTA, but the MTA also authenticates the KDC. This is different from the BPI+ flow where the CMTS authenticates the CM, but the reverse does not happen. The MTA verifies the received certificates all chain to its stored Service Provider Root certificate. A failure to move beyond this stage indicates the MTA failed to validate these certificates probably because of certificates rooted to the real Service Provider Root on one device and certificates rooted to the test Service Provider Root on the other. The test root is commonly used in lab environments.

In addition to verifying the certificates, the MTA validates the KDC digital signature, the integrity checksum, and that the received ticket has not already expired.

Figure 4-33 illustrates the final steps in the security phase.

Figure 4-33 *MTA Provisioning - Security - Part III*

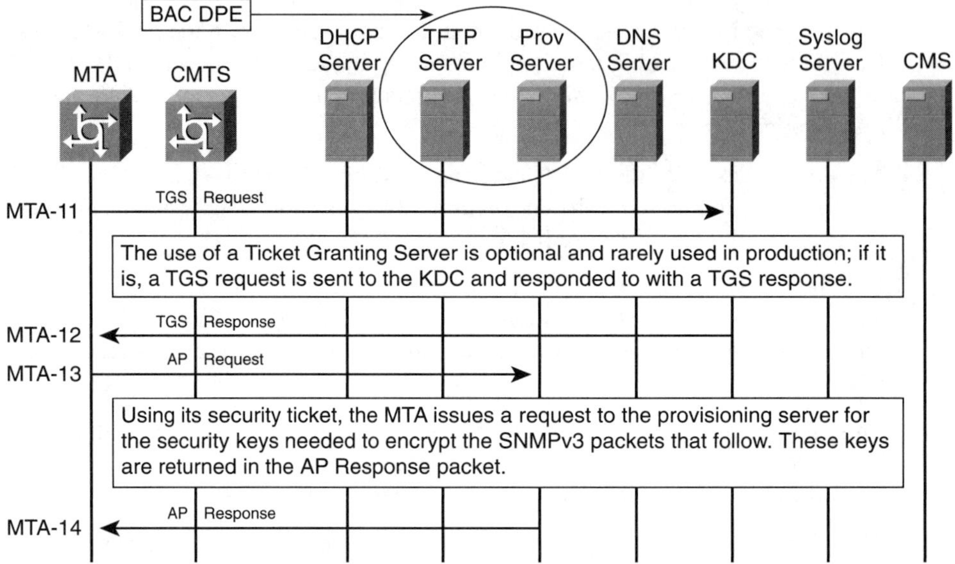

Sub-option #7 of option #122 specifies whether or not a TGS is used in the provisioning process. This is an optional parameter that has a default value of FALSE, meaning no TGS server is used. In most production systems this is the case. If this parameter is set to a value of TRUE and a TGT is returned in step MTA-10, then the following two steps (MTA-11 and MTA-12) would now take place. In these steps a TGS request is sent to the KDC (MTA-11), which then returns the secure provisioning service ticket in a TGS response message (MTA-12).

At this point the MTA uses its provisioning service ticket to obtain the security keys needed for the secure SNMPv3 packets used later in the provisioning process. The MTA sends an AP request (step MTA-13) message to the provisioning server containing an encrypted form of the service ticket. For a BAC deployment the provisioning server is the DPE device. Figure 4-34 illustrates this step.

Figure 4-34 *AP Request Example*

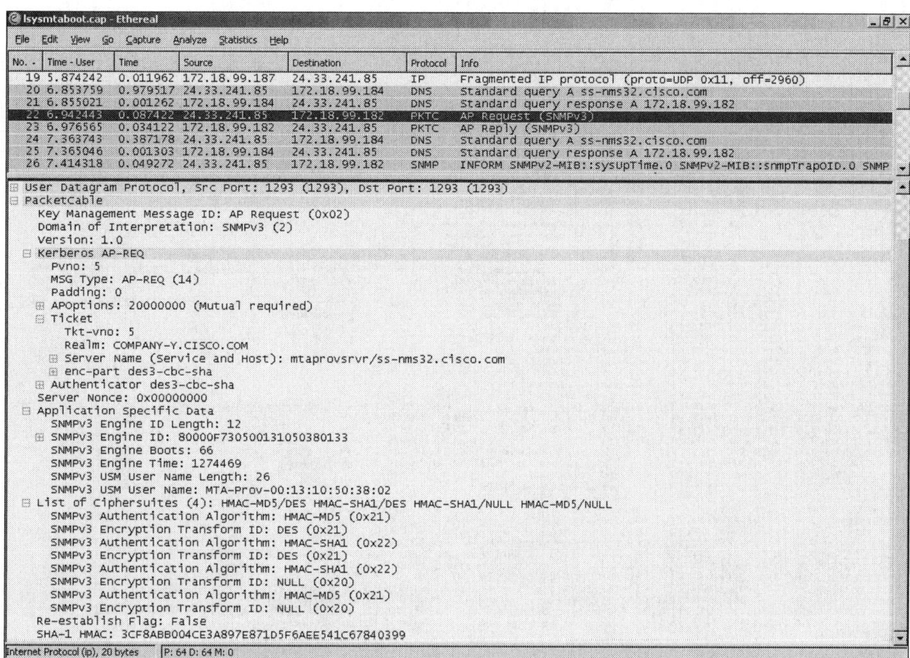

In this figure, you can see the provisioning service ticket as well as a list of supported authentication and encryption algorithms. The application server then decrypts the service ticket, validates it, and sends the SNMP keying information back to the MTA in an AP response (step MTA-14) message. The SNMPv3 packets that follow are now authenticated and optionally encrypted.

As you might expect the KDC log is a good place to start when troubleshooting issues in this stage of the provisioning process. By default the KDC log is stored in the directory /opt/CSCObpr/kdc/logs. Like with the other BAC components this log file is rotated, and the most recent entries are in the file KDC.log.0. Figure 4-35 illustrates an example of the type of information in this file.

Figure 4-35 *KDC Log Example*

```
20051228/083647 [MAIN]{9:bffffffff} -- Incoming message sent to thread 0 for processing
20051228/083647 [0]{9:bffffffff} -- AS-REQ received
20051228/083647 [0]{9:bffffffff} --   AS-REQ contains PKINIT
20051228/083647 [0]{9:bffffffff} --   Client is MTA
20051228/083647 [0]{9:bffffffff} -- Apparent clock skew: 1135740981; client_time: 19700101050026
20051228/083647 [0]{9:bffffffff} -- KRB_ERR message sent to 24.32.241.111:1103: Client/KDC clock skew too large
20051228/083658 [MAIN]{10:ffffffffff} -- Received datagram from 24.32.241.111, port # 1103; datagram # 391156; length = 2981
20051228/083658 [MAIN]{9:bffffffff} -- Incoming message sent to thread 0 for processing
20051228/083658 [0]{9:bffffffff} -- AS-REQ received
20051228/083658 [0]{9:bffffffff} --   AS-REQ contains PKINIT
20051228/083658 [0]{9:bffffffff} --   Client is MTA
20051228/083658 [0]{9:bffffffff} -- Signer Identifier is issuer and serial number
20051228/083658 [0]{9:bffffffff} -- Performing MTA FQDN exchange with PS
20051228/083658 [0]{9:bffffffff} -- Response received from PS
20051228/083658 [0]{9:bffffffff} -- Received non-KRB-ERROR response from PS
20051228/083658 [0]{9:bffffffff} -- Request is for ticket for service: mtaprovsrvr/ss-nms32.cisco.com
20051228/083658 [0]{9:bffffffff} -- Creating ticket that expires: 20060104080258
20051228/083658 [0]{9:bffffffff} -- AS-REP sent to 24.32.241.111:1103
20051228/083659 [DATABASE MANAGER]{10:ffffffffff} -- Processing command: ADD 24.32.241.111 20060104080258
20051228/084651 [MAIN]{10:ffffffffff} -- Received datagram from 24.32.240.216, port # 1103; datagram # 391157; length = 2981
20051228/084651 [MAIN]{9:bffffffff} -- Incoming message sent to thread 0 for processing
20051228/084651 [0]{9:bffffffff} -- Cached response sent to 24.32.240.216:1103
20051228/084702 [MAIN]{10:ffffffffff} -- Received datagram from 24.32.240.216, port # 1103; datagram # 391158; length = 2981
20051228/084702 [MAIN]{9:bffffffff} -- Incoming message sent to thread 0 for processing
20051228/084702 [0]{9:bffffffff} -- AS-REQ received
20051228/084702 [0]{9:bffffffff} --   AS-REQ contains PKINIT
20051228/084702 [0]{9:bffffffff} --   Client is MTA
20051228/084702 [0]{9:bffffffff} -- Signer Identifier is issuer and serial number
20051228/084702 [0]{9:bffffffff} -- Performing MTA FQDN exchange with PS
20051228/084702 [0]{9:bffffffff} -- Response received from PS
20051228/084702 [0]{9:bffffffff} -- Received non-KRB-ERROR response from PS
20051228/084702 [0]{9:bffffffff} -- Request is for ticket for service: mtaprovsrvr/ss-nms32.cisco.com
20051228/084702 [0]{9:bffffffff} -- Creating ticket that expires: 20060104072902
20051228/084703 [0]{9:bffffffff} -- AS-REP sent to 24.32.240.216:1103
20051228/084729 [DATABASE MANAGER]{10:ffffffffff} -- Processing command: ADD 24.32.240.216 20060104072902
20051228/085209 [MAIN]{10:ffffffffff} -- Received datagram from 24.47.241.27, port # 1828; datagram # 391159; length = 2974
20051228/085209 [MAIN]{9:bffffffff} -- Incoming message sent to thread 0 for processing
20051228/085209 [0]{9:bffffffff} -- AS-REQ received
20051228/085209 [0]{9:bffffffff} --   AS-REQ contains PKINIT
20051228/085209 [0]{9:bffffffff} --   Client is MTA
20051228/085209 [0]{9:bffffffff} -- Signer Identifier is issuer and serial number
20051228/085209 [0]{9:bffffffff} -- Performing MTA FQDN exchange with PS
20051228/085209 [0]{9:bffffffff} -- Response received from PS
20051228/085209 [0]{9:bffffffff} -- Received non-KRB-ERROR response from PS
20051228/085210 [0]{9:bffffffff} -- Request is for ticket for service: mtaprovsrvr/ss-nms32.cisco.com
20051228/085210 [0]{9:bffffffff} -- Creating ticket that expires: 20060104073010
20051228/085210 [0]{9:bffffffff} -- AS-REP sent to 24.47.241.27:1828
20051228/085229 [DATABASE MANAGER]{10:ffffffffff} -- Processing command: ADD 24.47.241.27 20060104073010
20051228/090029 [DATABASE MANAGER]{10:ffffffffff} -- Processing command: PRUNE 20051228090027
20051228/090800 [MAIN]{10:ffffffffff} -- Received datagram from 24.47.241.26, port # 1960; datagram # 391160; length = 2972
20051228/090800 [MAIN]{9:bffffffff} -- Incoming message sent to thread 0 for processing
--More--(6%)
```

In this example you can see AS-REQ messages come into the KDC; the KDC identifies the client as a PacketCable MTA performs the FQDN lookup with the provisioning server and then generates the security ticket for MTA provisioning. At the top of the example you can see an exchange that fails because of excessive clock skew between the MTA and KDC.

Transfer of Configuration Information

At this point the MTA can communicate securely with the provisioning system; therefore, it is ready to be configured. The logical steps behind this process are split in two phases, the first of which Figure 4-36 illustrates.

Figure 4-36 *MTA Provisioning - Configuration - Part I*

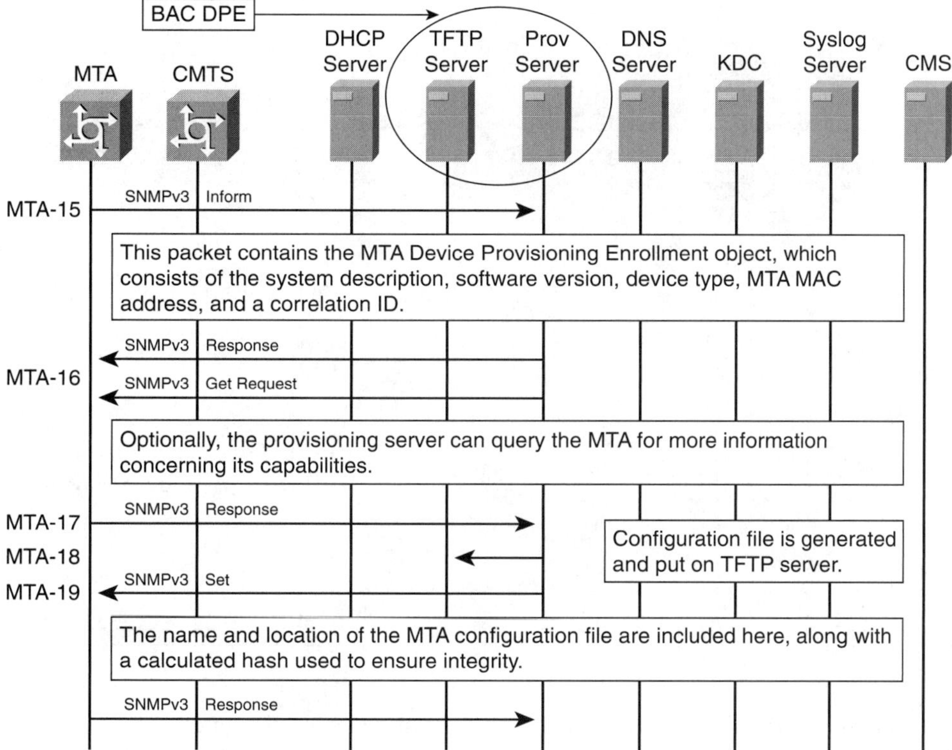

First the MTA sends a SNMPv3 Inform packet to the provisioning server. This is the SNMP entity specified in sub-option #3 of DHCP option #122 returned to the MTA. In a Cisco BAC implementation this would be to the DPE device. This message lets the DPE know that an MTA is ready to be configured. Figure 4-37 illustrates the MTA-15 step.

Figure 4-37 *SNMPv3 Inform Example - Enrollment*

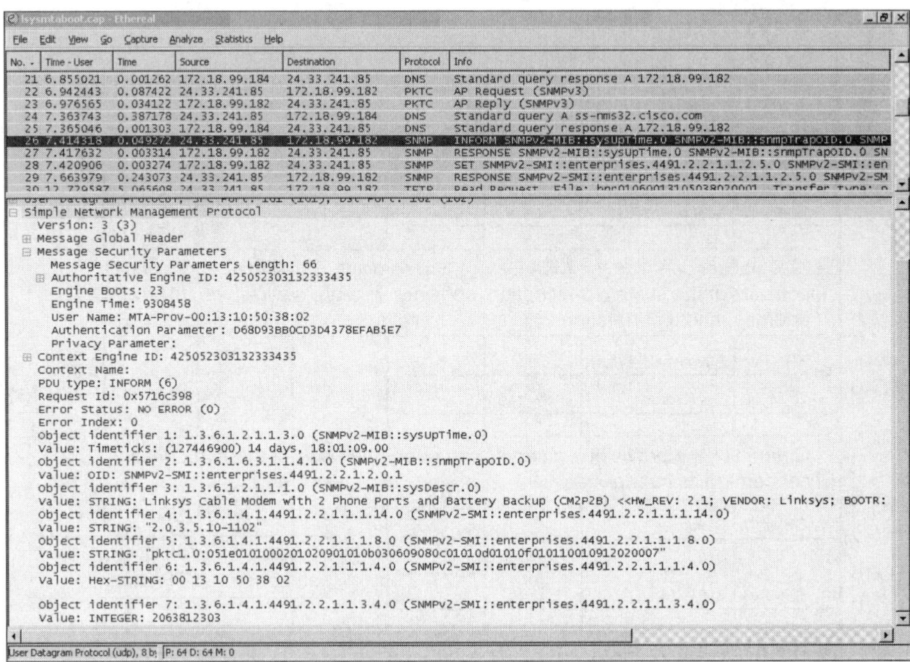

This SNMP Inform contains the MTA provisioning enrollment object, pktcMtaDevProvisioningEnrollment, which is defined in the MTA Device MIB specification. This MIB object consists of five other MIB objects: the system description, current software version, MTA device type, MTA MAC address, and a correlation identifier. You can see an example of this object and its contents in Figure 4-37. In this figure you can also tell that the provisioning server acknowledges receipt of this packet by sending an SNMP Response packet.

The next two steps, MTA-16 and MTA-17, are optional. Step MTA-16 is used by the provisioning server to find out more information about the capabilities of the MTA. If this step is used, one or more SNMPv3 Get Request packets are sent to the MTA to find out this information. Step MTA-17 consists of the response packets to these requests. Notice in the provisioning example shown in this chapter these optional steps are not used.

When the provisioning server has all the information it needs from the MTA, it is able to construct the MTA configuration file. What an MTA configuration file looks like and how it is structured is covered in detail in Chapter 5, "The MTA Configuration File." The mechanics detailing how the provisioning server generates the configuration file is not explicitly stated in the provisioning specification. It is simply referred to as step MTA-18.

Now that the configuration file is created, it needs to be sent to the MTA. An SNMPv3 Set packet is sent from the provisioning server to the MTA with the name and location of the configuration file. Figure 4-38 illustrates an example of this packet (step MTA-19).

Figure 4-38 *SNMPv3 Set Packet*

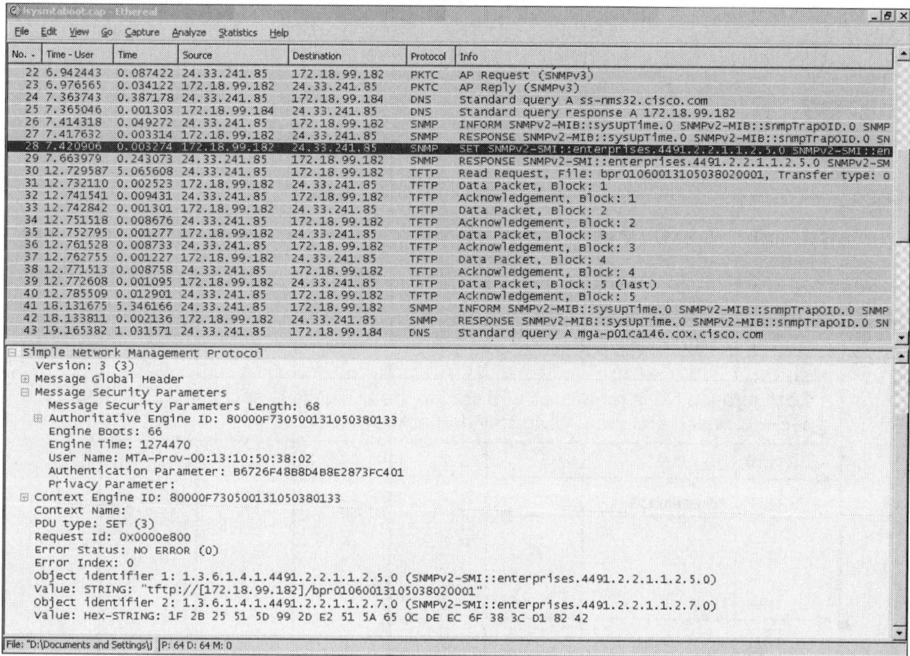

As you can see from Figure 4-38, the location of the configuration file is specified as a URL. Either the TFTP or HTTP protocol is used to transfer the configuration file. In this example the TFTP protocol is used. Notice that the provisioning server also calculates a hash of the configuration file that is sent in the SNMPv3 set packet so that the authenticity of the file can be verified. In Figure 4-38 the first Object Identification Number (OID) specifies the configuration file location and name, and the second OID is the authenticity hash. Optionally, the configuration file can also be encrypted. In this case an encryption key is sent in this step as well.

NOTE The encryption key that could have been included in the SNMPv3 Set message is for encryption of the MTA configuration file sent in the TFTP data packets. The SNMPv3 messages themselves could have also been encrypted; in these examples they are not. The encryption key in this case would have been returned in the AP-RSP message (step MTA-14).

The MTA acknowledges receipt of this packet by sending a SNMP Response packet.

Figure 4-39 illustrates the remaining configuration phase steps.

Figure 4-39 *MTA Provisioning – Configuration - Part II*

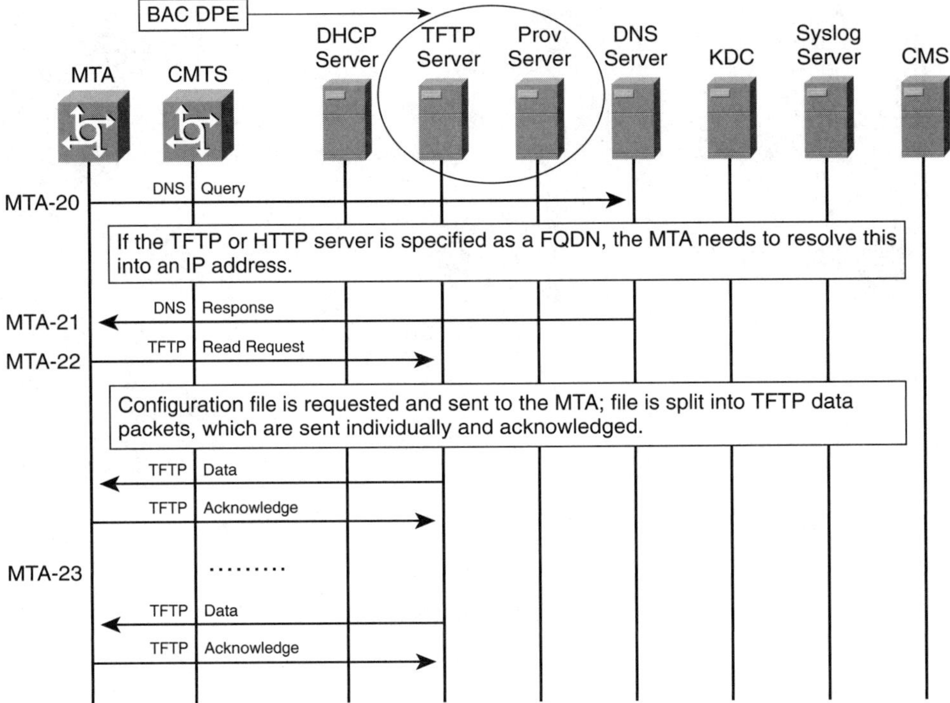

If in the configuration file URL the TFTP or HTTP server is specified as an FQDN, the MTA needs to resolve this into an IP address. The DNS query and response used to accomplish this task are referred to as steps MTA-20 and MTA-21. Notice in the example the TFTP server is expressed as an IP address; therefore, steps MTA-20 and MTA-21 are not required.

After the IP address of the TFTP or HTTP server is known by the MTA, it then initiates the request for the configuration file. Figure 4-40 is an example of this step (MTA-22) using the TFTP protocol to transfer the configuration file.

Figure 4-40 *MTA Configuration File Transfer Example*

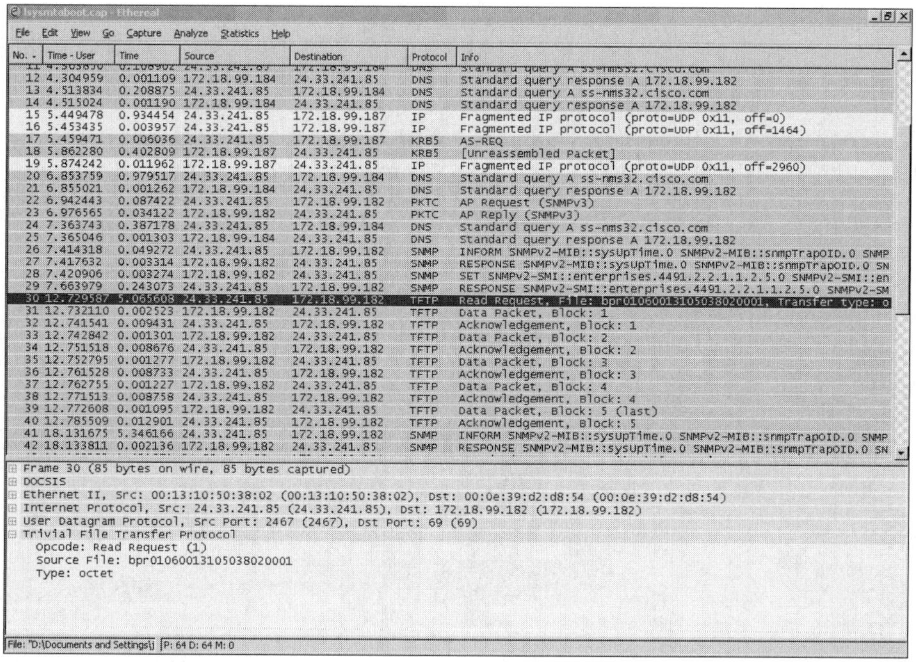

```
File  Edit  View  Go  Capture  Analyze  Statistics  Help
No. - Time - User  Time      Source          Destination     Protocol Info
   12 4.304959  0.001109 172.18.99.184   24.33.241.85    DNS      Standard query response A 172.18.99.182
   13 4.513834  0.208875 24.33.241.85    172.18.99.184   DNS      Standard query A ss-nms32.cisco.com
   14 4.515024  0.001190 172.18.99.184   24.33.241.85    DNS      Standard query response A 172.18.99.182
   15 5.449478  0.934454 24.33.241.85    172.18.99.187   IP       Fragmented IP protocol (proto=UDP 0x11, off=0)
   16 5.453435  0.003957 24.33.241.85    172.18.99.187   IP       Fragmented IP protocol (proto=UDP 0x11, off=1464)
   17 5.459471  0.006036 24.33.241.85    172.18.99.187   KRB5     AS-REQ
   18 5.862280  0.402809 172.18.99.187   24.33.241.85    KRB5     [Unreassembled Packet]
   19 5.874242  0.011962 172.18.99.187   24.33.241.85    IP       Fragmented IP protocol (proto=UDP 0x11, off=2960)
   20 6.853759  0.979517 24.33.241.85    172.18.99.184   DNS      Standard query A ss-nms32.cisco.com
   21 6.855021  0.001262 172.18.99.184   24.33.241.85    DNS      Standard query response A 172.18.99.182
   22 6.942443  0.087422 24.33.241.85    172.18.99.182   PKTC     AP Request (SNMPv3)
   23 6.976565  0.034122 172.18.99.182   24.33.241.85    PKTC     AP Reply (SNMPv3)
   24 7.363743  0.387178 24.33.241.85    172.18.99.184   DNS      Standard query A ss-nms32.cisco.com
   25 7.365046  0.001303 172.18.99.184   24.33.241.85    DNS      Standard query response A 172.18.99.182
   26 7.414318  0.049272 24.33.241.85    172.18.99.182   SNMP     INFORM SNMPv2-MIB::sysUpTime.0 SNMPv2-MIB::snmpTrapOID.0 SNMP
   27 7.417632  0.003314 172.18.99.182   24.33.241.85    SNMP     RESPONSE SNMPv2-MIB::sysUpTime.0 SNMPv2-MIB::snmpTrapOID.0 SN
   28 7.420906  0.003274 172.18.99.182   24.33.241.85    SNMP     SET SNMPv2-SMI::enterprises.4491.2.2.1.1.2.5.0 SNMPv2-SMI::en
   29 7.663979  0.243073 24.33.241.85    172.18.99.182   SNMP     RESPONSE SNMPv2-SMI::enterprises.4491.2.2.1.1.2.5.0 SNMPv2-SM
   30 12.729587 5.065608 24.33.241.85    172.18.99.182   TFTP     Read Request, File: bpr01060013105038020001, Transfer type: o
   31 12.732110 0.002523 172.18.99.182   24.33.241.85    TFTP     Data Packet, Block: 1
   32 12.741541 0.009431 24.33.241.85    172.18.99.182   TFTP     Acknowledgement, Block: 1
   33 12.742842 0.001301 172.18.99.182   24.33.241.85    TFTP     Data Packet, Block: 2
   34 12.751518 0.008676 24.33.241.85    172.18.99.182   TFTP     Acknowledgement, Block: 2
   35 12.752795 0.001277 172.18.99.182   24.33.241.85    TFTP     Data Packet, Block: 3
   36 12.761528 0.008733 24.33.241.85    172.18.99.182   TFTP     Acknowledgement, Block: 3
   37 12.762755 0.001227 172.18.99.182   24.33.241.85    TFTP     Data Packet, Block: 4
   38 12.771513 0.008758 24.33.241.85    172.18.99.182   TFTP     Acknowledgement, Block: 4
   39 12.772608 0.001095 172.18.99.182   24.33.241.85    TFTP     Data Packet, Block: 5 (last)
   40 12.785509 0.012901 24.33.241.85    172.18.99.182   TFTP     Acknowledgement, Block: 5
   41 18.131675 5.346166 24.33.241.85    172.18.99.182   SNMP     INFORM SNMPv2-MIB::sysUpTime.0 SNMPv2-MIB::snmpTrapOID.0 SNMP
   42 18.133811 0.002136 172.18.99.182   24.33.241.85    SNMP     RESPONSE SNMPv2-MIB::sysUpTime.0 SNMPv2-MIB::snmpTrapOID.0 SN

Frame 30 (85 bytes on wire, 85 bytes captured)
DOCSIS
Ethernet II, Src: 00:13:10:50:38:02 (00:13:10:50:38:02), Dst: 00:0e:39:d2:d8:54 (00:0e:39:d2:d8:54)
Internet Protocol, Src: 24.33.241.85 (24.33.241.85), Dst: 172.18.99.182 (172.18.99.182)
User Datagram Protocol, Src Port: 2467 (2467), Dst Port: 69 (69)
Trivial File Transfer Protocol
    Opcode: Read Request (1)
    Source File: bpr01060013105038020001
    Type: octet

File: "D:\Documents and Settings\j  P: 64 D: 64 M: 0
```

In response to this request the TFTP or HTTP server sends the configuration file to the MTA (step MTA-23). Upon receiving the configuration file the MTA validates the configuration file hash it received earlier by calculating the hash on the received file and comparing the results. If applicable the MTA also decrypts the configuration file using the encryption key it received in step MTA-19.

If you encounter a failure in this step, the most likely cause is a bad configuration file—most likely a required parameter is missing or invalid. Chapter 5 covers configuration files in more depth. Also if the computed hash differs from the hash received from the provisioning entity, then this causes a failure in this step. This type of error is indicative of a denial or theft of service attempt as something has corrupted the configuration file contents.

The DPE logs can help troubleshoot issues with MTA provisioning in this phase. Figure 4-41 shows an example of these logs (the log level was set to 6-Info).

Figure 4-41 *DPE Log - MTA*

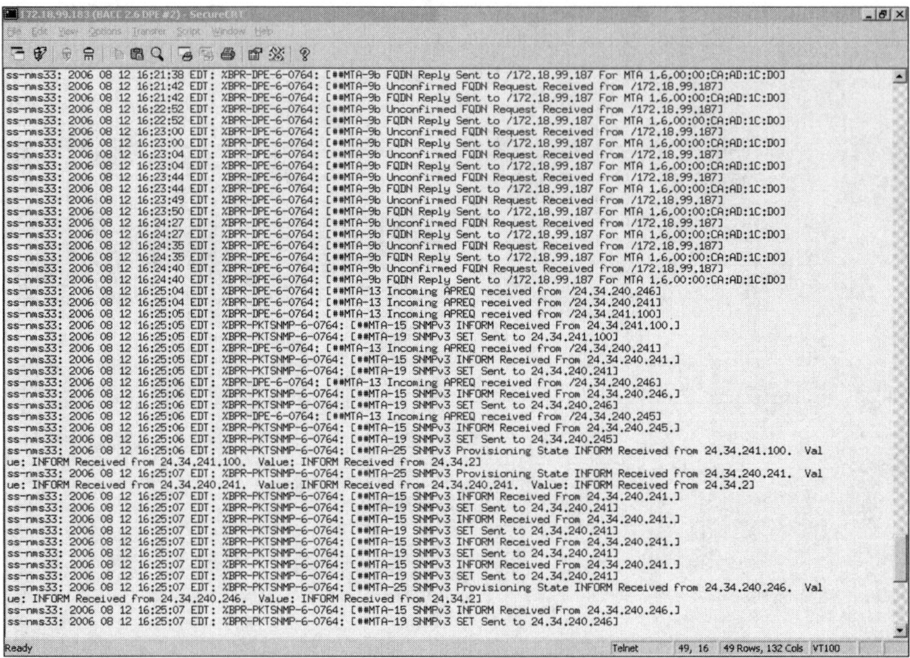

The command **grep '##MTA' dpe.log** was used to filter out just the MTA provisioning related logs. In this log you can see messages relating to the FQDN look-up exchange with the KDC (step MTA-9), incoming AP requests (MTA-13), incoming SNMPv3 Inform messages for enrollment (MTA-15), outgoing SNMPv3 Set messages (MTA-19), and SNMPv3 Inform messages for provisioning status (MTA-25).

Registration

The final phase in the MTA provisioning process involves registration and logging as Figure 4-42 illustrates.

Figure 4-42 *MTA Provisioning - Registration*

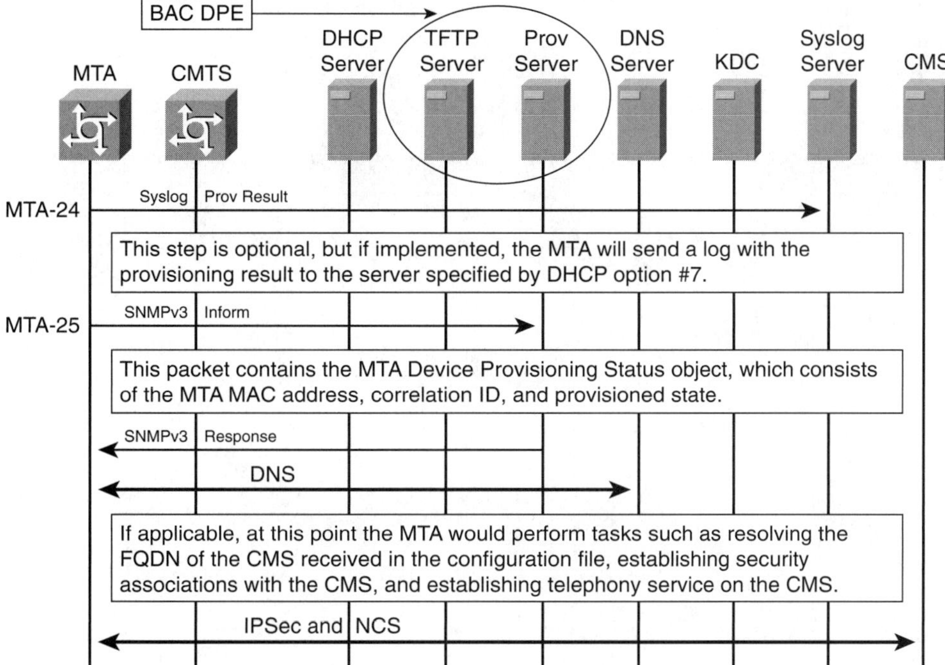

At this point, the MTA has received and validated its configuration file. The MTA then updates the logging server, specified in DHCP option #7, with the results of this validation using the system logging (syslog) protocol (step MTA-24). This step is optional and not used in the provisioning example shown. A sample of what this step looks like is included in the Chapter 3 discussion about system logging; specifically refer to Figure 3-10.

In addition to updating the logging server, the MTA also notifies the provisioning server with the result (step MTA-25). If everything is successful, it sends an SNMPv3 Inform packet to the provisioning server, as shown in Figure 4-43.

Figure 4-43 *SNMPv3 Inform Example - Provisioning Status*

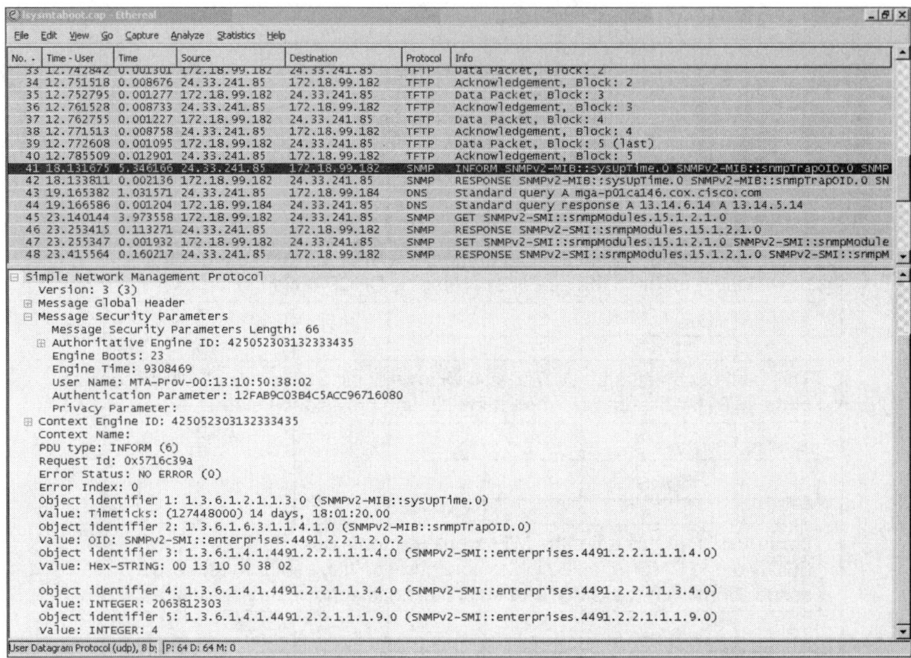

This SNMP Inform packet contains the MTA provisioning status object, pktcMtaDevProvisioningStatus, which is defined in the MTA Device MIB specification. This MIB object consists of three other MIB objects:

- The MTA MAC address
- A correlation identifier
- The current provisioning state

You can see an example of this object and its contents in Figure 4-43. The last MIB object is the provisioning state, pktcMtaDevProvisioningState, which is expressed as an integer. It currently has seven possible values according to Table 4-1.

Table 4-1 *pktcMtaDevProvisioningState Values*

Value	Meaning
1	Pass
2	In progress
3	Failed (Configuration File Error)
4	Pass with warnings
5	Pass with incomplete parsing

Table 4-1 *pktcMtaDevProvisioningState Values (Continued)*

Value	Meaning
6	Failed (Internal Error)
7	Failed (Other Reason)

Notice in the example this object has a value of 4, indicating "pass with warnings." If there is a failure, an SNMPv3 Inform packet is also sent by the MTA, and most likely manual intervention is required to correct the problem. For example, if a value of 3 is received indicating a problem with the configuration file, the file most likely is missing or has invalid mandatory parameters.

From Figure 4-43, you can also tell that the provisioning server acknowledges receipt of this packet by sending a SNMP Response packet.

Step #25 completes the provisioning process as defined in the provisioning specification. However, additional steps occur before the MTA is able to process phone calls. An explanation of these steps is covered after the discussion on the Basic and Hybrid provisioning methods.

The Basic Method

The Basic Method is a much simpler way to provision MTAs. In this method, the MTA configuration file transfer is very similar to how the cable modem obtains its configuration file. As a trade-off to this simplicity the Basic Method does not provide any sort of security to the provisioning process. Therefore, it is recommended you use the secure method discussed previously if possible.

Recall that sub-option #6 of DHCP option #122 signifies which method of PacketCable provisioning is used. This parameter is the Kerberos Realm; because the Basic Method does not use Kerberos, a value of BASIC.1 or BASIC.2 is used instead. Figure 4-44 shows an example.

Figure 4-44 *DHCP - Basic Method*

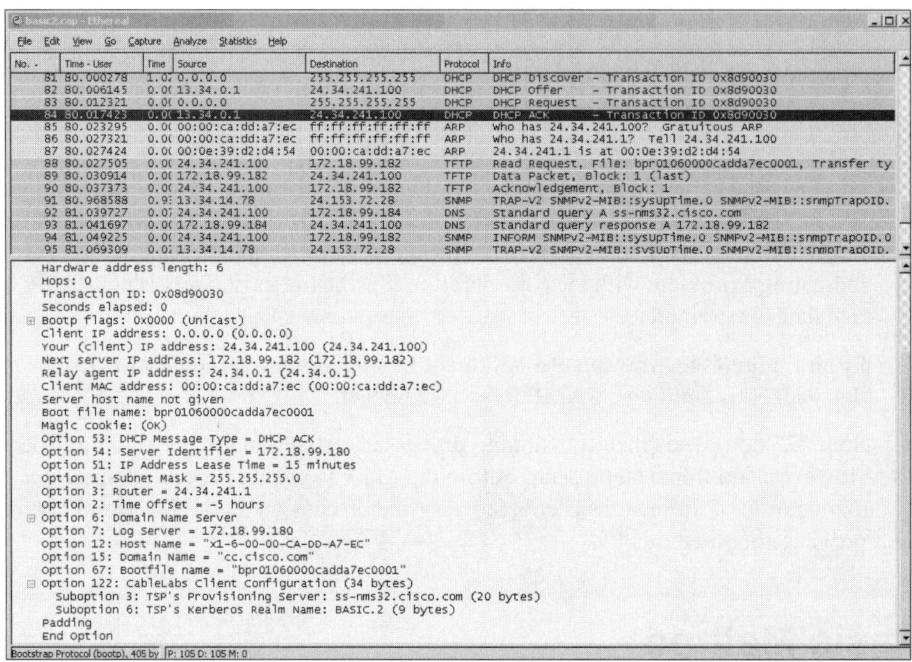

The basic method consists of four additional steps following the DHCP steps common to all three provisioning methods (steps MTA-1 through MTA-4). Figure 4-45 illustrates these steps.

The first step is known as B-MTA-22. This step is the MTAs request for its configuration file. The DHCP Acknowledgement packet in the previous step tells the MTA the name and location of the configuration file in the "siaddr" and "file" fields. This is the same way in which a cable modem learns the name and location of its DOCSIS configuration file. The MTA simply initiates a TFTP request to this server for the specified file. If you refer back to Figure 4-44, you can see that the MTA requested the file specified by these fields. The TFTP server then responds with TFTP data packets containing the configuration file (step B-MTA-23).

In addition to standard configuration information the configuration file also contains a hash that is used by the MTA to authenticate the file contents. The hash is sent in the SNMP variable pktcMtaDevProvConfigHash defined in the PacketCable MTA MIB specification.

Figure 4-45 *MTA Provisioning - Basic Method*

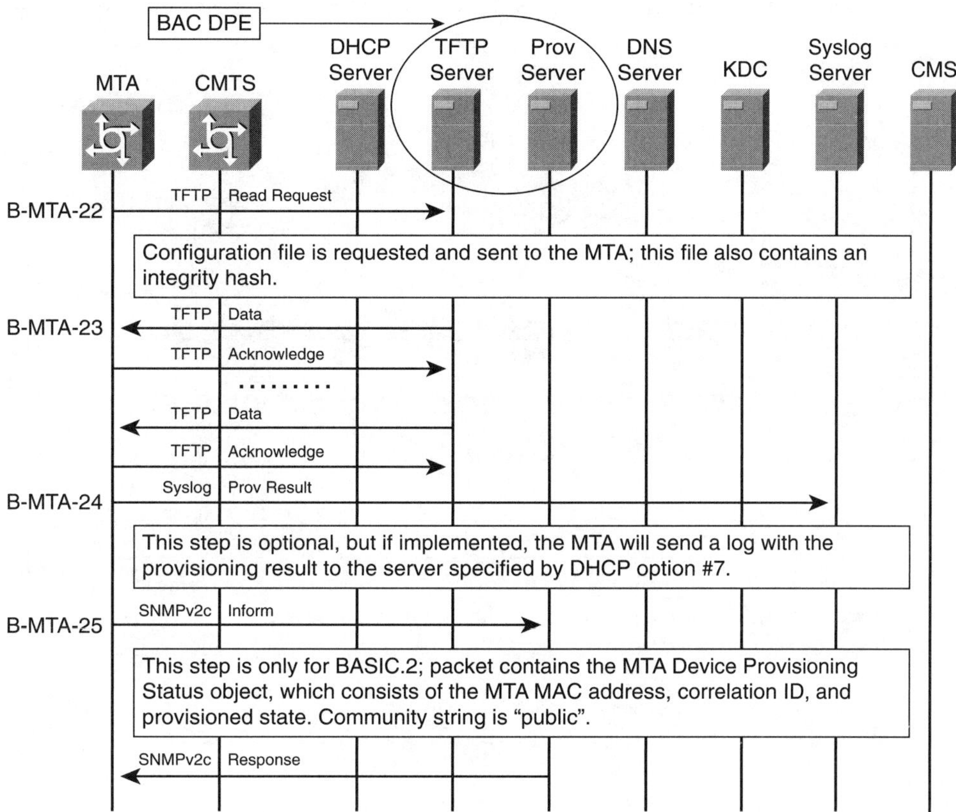

NOTE This is different from how this hash is transferred to the MTA in the other provisioning methods. In those cases the hash is included in the SNMP Set packet that also has the configuration file name and location.

The MTA can optionally send a syslog packet to the logging server with the results of the configuration file validation (step B-MTA-24) just as with the secure flow. And similar to the secure flow, the final step (B-MTA-25) is an SNMP Inform packet sent to the provisioning server, identified in sub-option #3 of option #122, with the results of the provisioning process. In this case SNMPv2c is used instead of SNMPv3, and the community string is always the value "public". You might be wondering what the difference is between BASIC.1 and BASIC.2? The difference is whether or not this final step is performed; with BASIC.2 this final step is included, and with BASIC.1 it is not.

If you are using BAC, the easiest way to set it up to support the basic method of provisioning is to define a dedicated MTA Class of Service, as shown in Figure 4-46.

Figure 4-46 *MTA Class of Service - Basic Method*

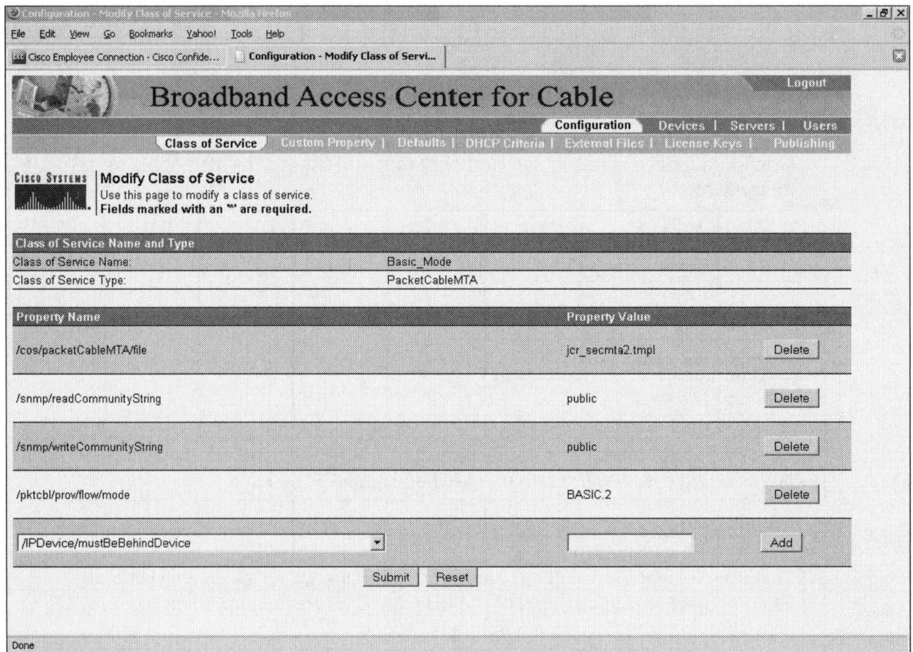

In this case the configuration file references a BAC template. If a static configuration file is used, it needs to be specially generated such that it includes the hash variable. This file can be generated using the BAC configuration utility discussed in Chapter 5. The provisioning flow mode variable tells BAC that MTAs using this class of service are provisioning under the BASIC.2 method. Lastly the SNMP community strings are set to the value "public," which is required for BASIC method provisioning.

The Hybrid Method

Like the basic method, the hybrid method simplifies MTA provisioning by removing the security steps. However, the flow of the hybrid method is much more like the secure method than the basic method. Basically, the steps are the same as in the secure method except that the Kerberos steps don't exist, and SNMPv2c is used instead of SNMPv3.

In this case, sub-option #6 of DHCP option #122 would have a value of HYBRID.1 or HYBRID.2 to signify use of the hybrid method. Cisco's BAC does not currently support the hybrid method for PacketCable provisioning; therefore, no examples are available.

After completion of the common DHCP steps (steps MTA-1 through MTA-4) a SNMPv2c Inform packet is sent to the provisioning server specified in sub-option #3 of DHCP option #122. Illustrated in Figure 4-47, this is referred to as step H-MTA-15.

Figure 4-47 *MTA Provisioning - Hybrid Method - Part I*

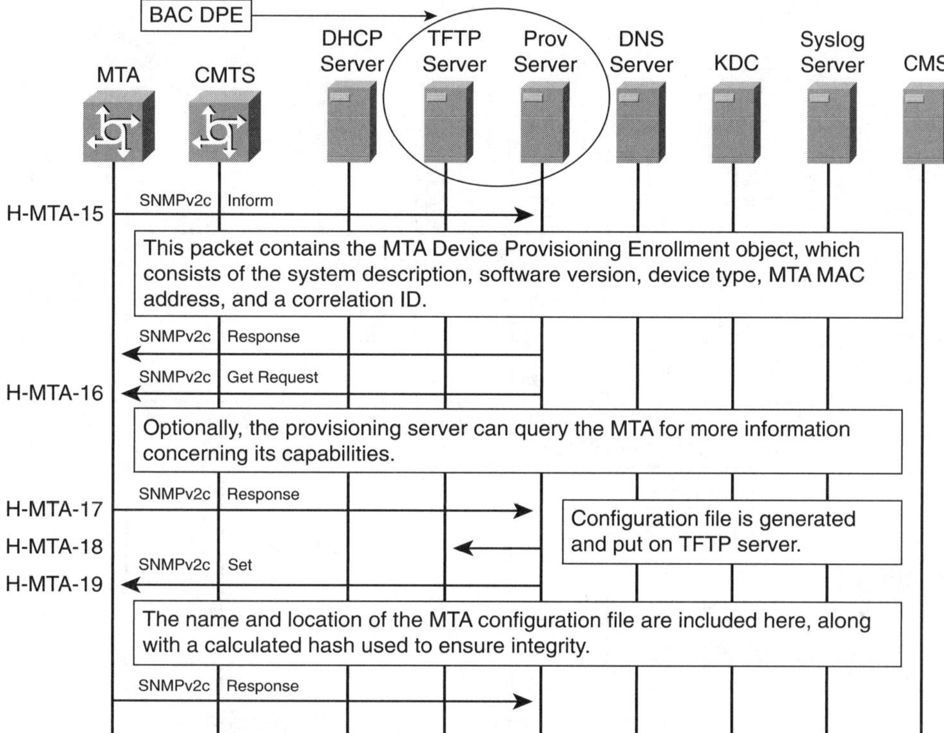

The next two steps, H-MTA-16 and H-MTA-17, are optional and are used by the provisioning server to find out more information about the capabilities of the MTA. If used, one or more SNMPv2c Get Request packets are sent to the MTA to find out this information, and SNMPv2c Get Response packets are returned.

The provisioning server then uses this information to construct the MTA configuration file (step H-MTA-18). Again, the mechanics detailing how this configuration file is generated is not explicitly stated in the provisioning specification.

Now that the configuration file is created, it needs to be sent to the MTA (step H-MTA-19). A SNMPv2c Set packet is sent from the provisioning server to the MTA with the URL encoded name and location of the configuration file. A hash of the configuration file is also sent so that the authenticity of the file can be verified.

If in this URL the TFTP or HTTP server is specified as an FQDN, the MTA needs to resolve this into an IP address. The DNS query and response used to accomplish this task are referred to as steps H-MTA-20 and H-MTA-21. Figure 4-48 illustrates these steps.

Figure 4-48 *MTA Provisioning: Hybrid Method - Part II*

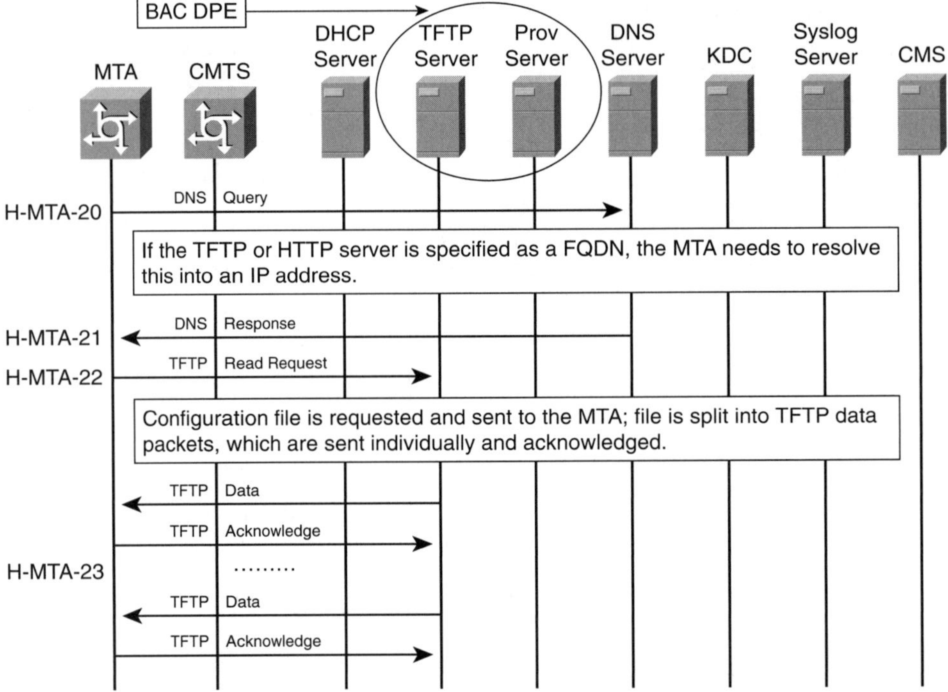

After the IP address of the TFTP or HTTP server is known by the MTA, it then initiates the request for the configuration file (step H-MTA-22). In response to this request the TFTP or HTTP server sends the configuration file to the MTA (step H-MTA-23). Upon receiving the configuration file the MTA validates the configuration file hash it received earlier by calculating the hash on the received file and comparing the results.

At this point the MTA has received and validated its configuration file. The MTA then updates the logging server, specified in DHCP option #7, with the results of this validation using the system logging (syslog) protocol (step H-MTA-24). Figure 4-49 illustrates these final steps.

Figure 4-49 *MTA Provisioning: Hybrid Method - Part III*

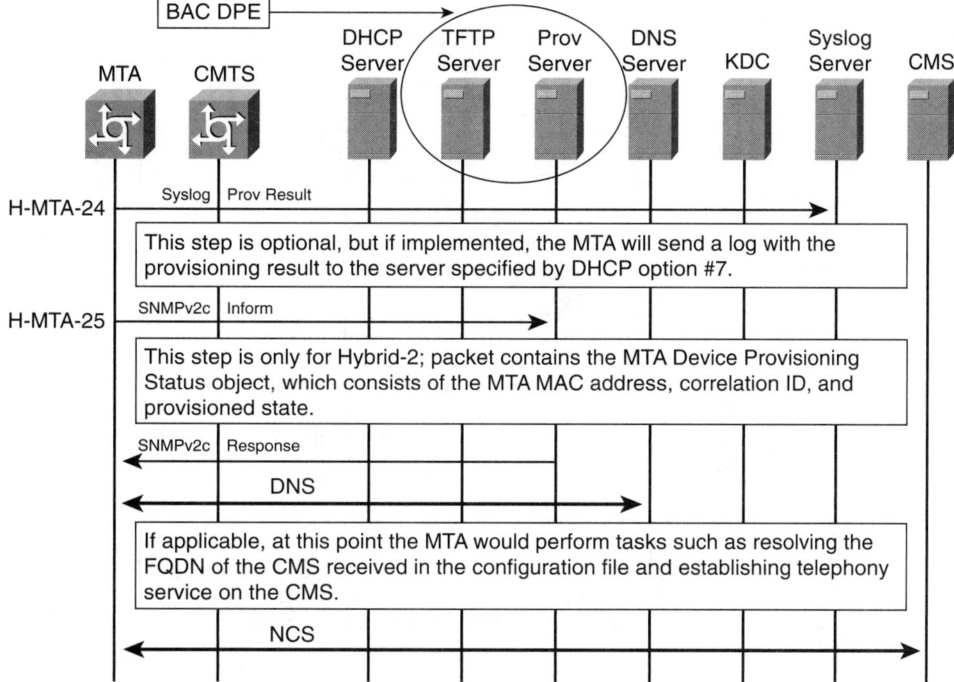

In addition to updating the logging server the MTA also notifies the provisioning server with the result (step H-MTA-25). If everything is successful, it sends a SNMPv2c Inform packet to the provisioning server. Similar to the basic method, the difference between HYBRID.1 and HYBRID.2 relates to the inclusion of a step. With HYBRID.2 this final step is included, and with HYBRID.1 it is not.

MTA to CMS Initialization

As you'll learn about more in the next chapter, the MTA configuration file contains the FQDN of the CMS with which the MTA communicates. The CMS FQDN is associated with a Kerberos Realm, which might or might not be the same realm used in the provisioning process. Another parameter included in the MTA configuration file specifies whether or not secure call signaling (using IPsec) needs to be used with this CMS. If secure call signaling is disabled for the CMS, then the MTA sends call signaling messages in the clear. If secure call signaling is enabled for the CMS, then the steps (labeled SEC-1 through SEC-10) described in Figure 4-50 and 4-51 occur next (these steps are defined in the PacketCable Security Specification—refer to this for more detail on these steps).

Figure 4-50 *MTA to CMS Initialization - Secure Method - Part I*

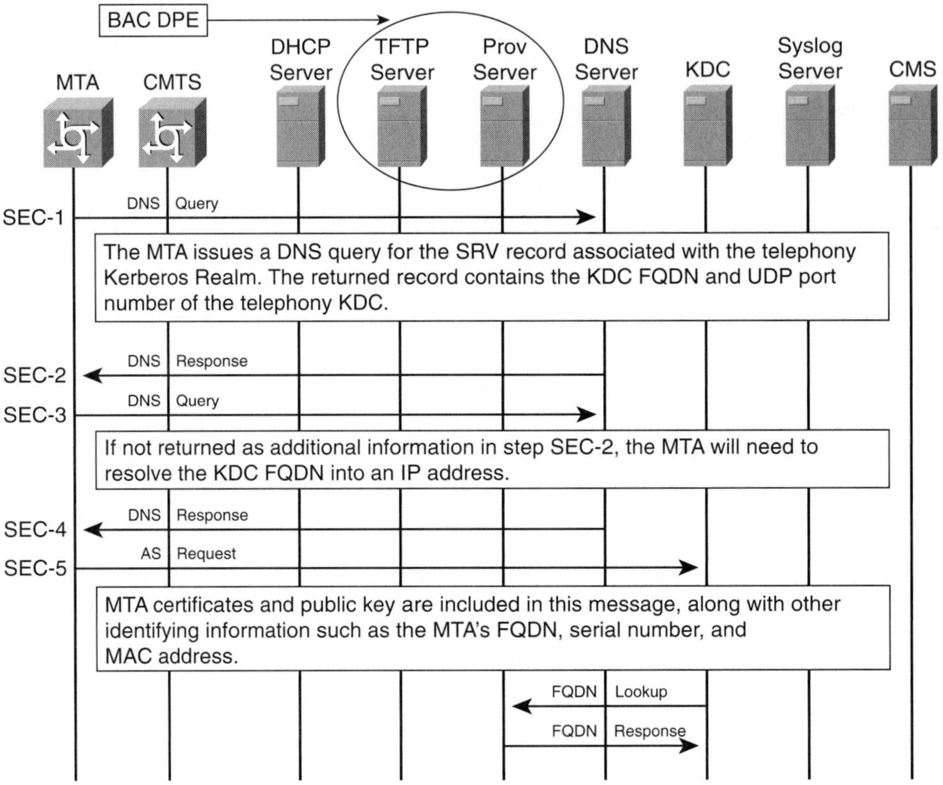

BAC DPE

SEC-1	MTA → DNS Server	DNS Query
		The MTA issues a DNS query for the SRV record associated with the telephony Kerberos Realm. The returned record contains the KDC FQDN and UDP port number of the telephony KDC.
SEC-2	DNS Server → MTA	DNS Response
SEC-3	MTA → DNS Server	DNS Query
		If not returned as additional information in step SEC-2, the MTA will need to resolve the KDC FQDN into an IP address.
SEC-4	DNS Server → MTA	DNS Response
SEC-5	MTA → KDC	AS Request
		MTA certificates and public key are included in this message, along with other identifying information such as the MTA's FQDN, serial number, and MAC address.
	KDC → Prov Server	FQDN Lookup
	Prov Server → KDC	FQDN Response

Columns: MTA, CMTS, DHCP Server, TFTP Server, Prov Server, DNS Server, KDC, Syslog Server, CMS (BAC DPE encompasses TFTP Server and Prov Server).

Figure 4-51 *MTA to CMS Initialization - Secure Method - Part II*

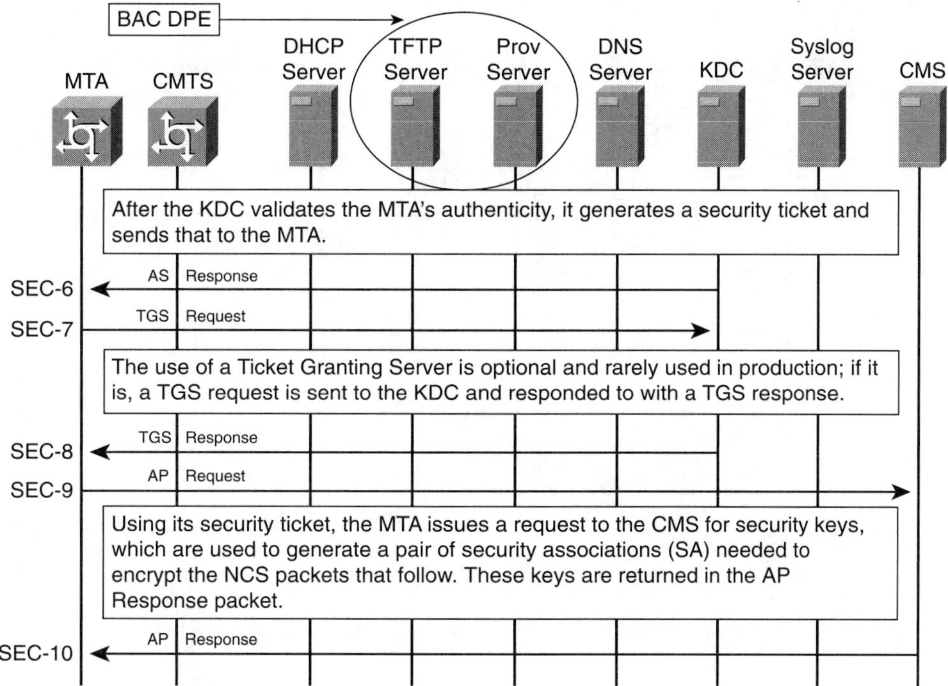

As you can see, these steps are very similar to the security steps used in the MTA provisioning process. Troubleshooting techniques described in this section generally apply here as well. Figure 4-52 shows a trace of the AP-REQ packet sent to the CMS.

Figure 4-52 *MTA to CMS AP-REQ Trace*

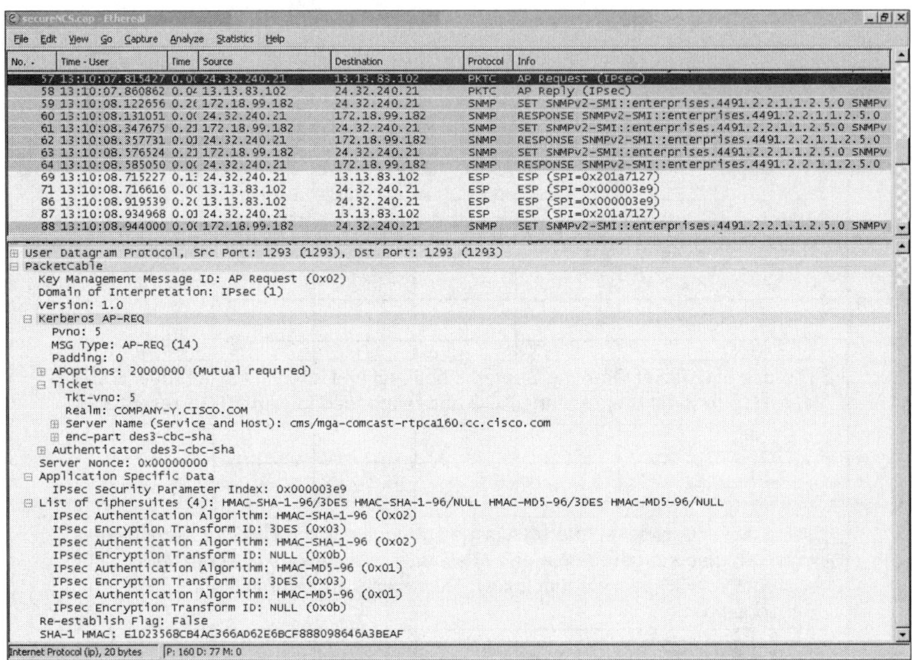

Notice the Encapsulation Security Payload (ESP) packets in the figure; these are the encrypted NCS call signaling packets.

Whether call signaling is encrypted or not, the MTA is now ready to begin communication with the CMS. First, the MTA needs to send a DNS query to resolve the CMS FQDN. The CMS FQDN typically resolves to multiple IP addresses. After this is done the MTA sends a NCS packet known as a "ReStart In Progress" or RSIP to the CMS to signal it is ready to process phone calls (see Figure 4-53). The specific type of RSIP message is known as a RSIP restart. To protect against flooding the CMS with RSIP messages in the event of a catastrophic outage the MTA waits a random delay before sending the message. After receiving the RSIP restart the CMS then provides information to the MTA such as how to process dialed digits and where to send them. More details on how the NCS protocol works is covered in the Call Signaling portion of this book. The MTA also sends notification to the SNMP provisioning entity concerning the status of each of its endpoints. An SNMP trap known as a "Link Up" trap is sent.

NOTE These messages are sent for each endpoint on the MTA. Thus, if both phone lines of the MTA are enabled in the configuration file, then an SNMP Link Up trap is sent for each endpoint.

Figure 4-53 *CMS Initialization Example*

File Edit View Go Capture Analyze Statistics Help

No.	Time - User	Time	Source	Destination	Protocol	Info
33	1.362299	0.0(	24.34.240.241	172.18.99.182	SNMP	INFORM SNMPv2-MIB::sysUpTime.0 SNMPv2-MIB::snmpTrapOID.
34	1.365373	0.0(	172.18.99.182	24.34.240.241	SNMP	RESPONSE SNMPv2-MIB::sysUpTime.0 SNMPv2-MIB::snmpTrapOI
35	1.366423	0.0(	24.34.240.241	172.18.99.182	SNMP	TRAP-V2 SNMPv2-MIB::sysUpTime.0 SNMPv2-MIB::snmpTrapOID
36	1.370570	0.0(	24.34.240.241	172.18.99.182	SNMP	TRAP-V2 SNMPv2-MIB::sysUpTime.0 SNMPv2-MIB::snmpTrapOID
37	1.374446	0.0(	24.34.240.241	172.18.99.180	Syslog	LOCAL0.WARNING: Feb 15 14:43:50 2006 x1-6-0...
38	1.376720	0.0(	172.18.99.182	24.34.240.241	SNMP	SET SNMPv2-SMI::enterprises.4491.2.2.1.1.2.5.0 SNMPv2-S
39	1.379766	0.0(	24.34.240.241	172.18.99.182	SNMP	TRAP-V2 SNMPv2-MIB::sysUpTime.0 SNMPv2-MIB::snmpTrapOID
40	1.382670	0.0(	24.34.240.241	172.18.99.182	SNMP	TRAP-V2 SNMPv2-MIB::sysUpTime.0 SNMPv2-MIB::snmpTrapOID
41	1.386404	0.0(	24.34.240.241	172.18.99.180	Syslog	LOCAL0.WARNING: Feb 15 14:43:50 2006 AALN/1...
42	1.388954	0.0(	172.18.99.182	24.34.240.241	SNMP	SET SNMPv2-SMI::enterprises.4491.2.2.1.1.2.5.0 SNMPv2-S
43	1.390616	0.0(	24.34.240.241	172.18.99.182	SNMP	TRAP-V2 SNMPv2-MIB::sysUpTime.0 SNMPv2-MIB::snmpTrapOID
44	1.394719	0.0(	24.34.240.241	172.18.99.182	SNMP	TRAP-V2 SNMPv2-MIB::sysUpTime.0 SNMPv2-MIB::snmpTrapOID
45	1.398479	0.0(	24.34.240.241	172.18.99.180	Syslog	LOCAL0.WARNING: Feb 15 14:43:50 2006 AALN/2...
46	1.400886	0.0(	172.18.99.182	24.34.240.241	SNMP	SET SNMPv2-SMI::enterprises.4491.2.2.1.1.2.5.0 SNMPv2-S
47	1.402595	0.0(	24.34.240.241	172.18.99.182	SNMP	TRAP-V2 SNMPv2-MIB::sysUpTime.0 SNMPv2-MIB::snmpTrapOID
48	1.406748	0.0(	24.34.240.241	172.18.99.182	SNMP	TRAP-V2 SNMPv2-MIB::sysUpTime.0 SNMPv2-MIB::snmpTrapOID
49	1.410445	0.0(	24.34.240.241	172.18.99.180	Syslog	LOCAL0.WARNING: Feb 15 14:43:50 2006 x1-6-0...
50	1.412271	0.0(	172.18.99.182	24.34.240.241	SNMP	SET SNMPv2-SMI::enterprises.4491.2.2.1.1.2.5.0 SNMPv2-S
51	1.414373	0.0(	24.34.240.241	172.18.99.184	DNS	Standard query A mga-comcast-rtpca160.cc.cisco.com
52	1.416317	0.0(	172.18.99.184	24.34.240.241	DNS	Standard query response A 13.13.83.118 A 13.13.83.102
53	1.418608	0.0(	24.34.240.241	172.18.99.182	SNMP	RESPONSE SNMPv2-SMI::enterprises.4491.2.2.1.1.2.5.0 SNM
54	1.422633	0.0(	24.34.240.241	172.18.99.182	SNMP	RESPONSE SNMPv2-SMI::enterprises.4491.2.2.1.1.2.5.0 SNM
55	1.426631	0.0(	24.34.240.241	172.18.99.182	SNMP	RESPONSE SNMPv2-SMI::enterprises.4491.2.2.1.1.2.5.0 SNM
56	1.430659	0.0(	24.34.240.241	172.18.99.182	SNMP	RESPONSE SNMPv2-SMI::enterprises.4491.2.2.1.1.2.5.0 SNM
57	1.894323	0.4(	13.14.13.11	24.34.240.241	MGCP	AUEP 216620479 aaln/1@x1-6-00-00-ca-dd-a7-e4.cc.cisco.c
58	2.337428	0.44	24.34.240.241	13.13.83.102	MGCP	RSIP 1 aaln/*@x1-6-00-00-CA-DD-A7-E4.cc.cisco.com MGCP
59	2.339876	0.0(	13.13.83.102	24.34.240.241	MGCP	200 1 OK
60	2.454015	0.11	13.13.83.102	24.34.240.241	MGCP	RQNT 108468703 aaln/1@x1-6-00-00-ca-dd-a7-e4.cc.cisco.c
61	2.466019	0.01	24.34.240.241	13.13.83.102	MGCP	200 108468703 OK
62	2.664319	0.1(	13.13.83.102	24.34.240.241	MGCP	RQNT 108468704 aaln/2@x1-6-00-00-ca-dd-a7-e4.cc.cisco.c
63	2.677175	0.01	24.34.240.241	13.13.83.102	MGCP	200 108468704 OK

```
Ethernet II, Src: 00:00:ca:dd:a7:e4 (00:00:ca:dd:a7:e4), Dst: 00:0e:39:d2:d4:54 (00:0e:39:d2:d4:54)
Internet Protocol, Src: 24.34.240.241 (24.34.240.241), Dst: 13.13.83.102 (13.13.83.102)
User Datagram Protocol, Src Port: 2427 (2427), Dst Port: 2727 (2727)
Media Gateway Control Protocol
  RSIP (RestartInProgress)
  Transaction ID: 1
  Endpoint: aaln/*@x1-6-00-00-CA-DD-A7-E4.cc.cisco.com
  Version: MGCP 1.0 NCS 1.0
  [The response to this request is in frame 59]
  Parameters
    RestartMethod (RM): restart
```

File: "D:\Documents and Settings\ P: 88 D: 88 M: 0

Ready Telnet 52, 16 52 Rows, 132 Cols VT100

As you can see in the figure, packet #51 and #52 are the DNS query and response for the CMS IP addresses, and packet #58 is the RSIP restart message sent from the MTA to the CMS letting it know it's now operational. The Notification Request (RQNT) NCS messages that follow tell the MTA where to send future packets and to be on the lookout for an off-hook event. The SNMP trap messages in this trace signify the successful transfer of the configuration file, the successful provisioning of line #1, the successful provisioning of line #2, and the successful overall provisioning. In this case the MTA also generates syslog messages for these events.

Examples of problems that would prevent the RSIP message acknowledgement include an IP routing failure between the MTA and the CMS, a failure in the CMS NCS process, a failure in the security negotiation, or a problem in the MTA configuration on the CMS. For example, if the CMS does not recognize the MTA attempting to register, it sends a negative NCS response acknowledgement back to the MTA. Specifically, the "500 - MGW not configured on CA" message is used. Also if no NCS messages are seen after the RSIP exchange, this typically indicates the CMS is unable to resolve the configured MTA FQDN. This could be because of an OSS and CMS configuration mismatch or a failure in the dynamic DNS process during MTA provisioning. Again, much more on NCS messaging is covered in the chapters in Part III of this book.

MTA to CMS Communication Loss

Similar messaging occurs when the MTA loses communication with the CMS. The NCS aspects of MTA to CMS communication loss are covered in detail in the Call Signaling portion of the book. From a high level, after the MTA detects the CMS is not receiving its signaling packets (via the lack of acknowledgements), it enters the RSIP disconnected state. This is a different type of a RSIP message that signifies that the MTA is no longer in communication with the CMS.

When the MTA enters this state, it also sends an SNMP "Link Down" trap to the SNMP provisioning entity. Figure 4-54 illustrates examples of these messages.

Figure 4-54 *CMS-MTA Communication Loss Example*

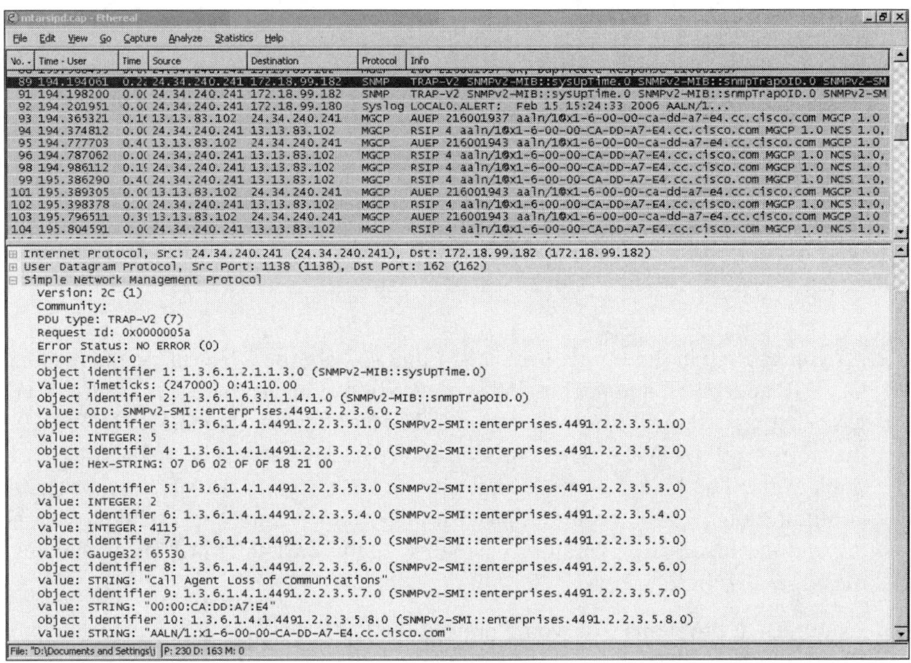

From this figure you can see the SNMP Trap indicating loss of communication with the CMS; RSIP disconnected messages are shown in this trace as well as a syslog message that also convey the lack of CMS communication.

Chapter Summary

The cable modem provisioning steps are broken down into eight stages.

In the first stage the modem scans the frequency range for a downstream DOCSIS carrier. After it has "locked" on to the downstream, it collects the information about associated upstream channels and determines which ones it can use. Upstream information is sent in DOCSIS UCD messages. These messages contain parameters such as frequency, symbol rate, and burst descriptors. On a Cisco CMTS, the characteristics of the different burst types are configured in a modulation profile. If the modem finds an upstream channel it can use, it begins the ranging process. As soon as it transmits at sufficient power to be heard by the CMTS, the CMTS fine tunes the modem's transmissions by requesting it to adjust parameters like output power, time offset, and frequency offset. During this process the modem is also assigned a temporary Service Identifier (SID).

After the ranging process successfully completes, the modem attempts to establish IP connectivity via the DHCP protocol. The DHCP packets sent by the modem are forwarded by the CMTS to the DHCP server(s). In the list of requested DHCP options sent by the modem is option #122, which is used for PacketCable MTA provisioning. With the Cisco BAC provisioning system, the DHCP server contacts the DPE for device configuration. If the DPE has this information in its cache, it is sent to the DHCP server; otherwise, the DHCP server needs to request the RDU to build this configuration and send it to the DPE. If the modem receives an offer packet with all the information it asks for, it then requests the DHCP server for this lease.

Next, the modem requests for the current date and time. Then the modem gets its configuration file using the TFTP protocol. After the file is received, the modem verifies its integrity and validity, and if all is well, it begins the registration process with the CMTS. The registration request contains information such as modem capabilities and service flow and classifier definitions. The CMTS verifies the integrity and validity of this message, and if it can honor the request, it assigns identifiers and sends a positive registration response message to the CM. In the final stage the modem and CMTS negotiate the security keys used to encrypt traffic over the DOCSIS network. This is accomplished with the Baseline Privacy Plus (BPI+) feature, which makes use of ITU X.509 digital certificates.

At this point, the CM has completed its provisioning, and the MTA can begin its provisioning process. This begins with the MTA establishing IP connectivity using the DHCP protocol. The MTA sends a DHCP Discover packet requesting the following options: subnet mask, default router, host name, domain name, DNS server, log server, and the CableLabs Client Configuration option. If using BAC, the DHCP server contacts the

DPE for MTA configuration information; if it exists, the DPE returns this information. Otherwise, the DHCP server requests the RDU to build the configuration and send it to the DPE devices. After the MTA receives an authorized offer packet with all the information it asks for, it requests the DHCP server for the lease.

If the secure method is used, the following steps occur next. Assuming the MTA does not already have valid security keys, it begins the process of obtaining them from the KDC. First it queries the DNS server to get the FQDN of the KDC in the realm specified by DHCP option #122, sub-option #6. Then it queries again to get the IP address associated with this FQDN. After it knows how to contact the KDC, the MTA sends an Authentication Server (AS) Request message to the KDC. This packet contains the necessary certificates and MTA identification information for the KDC to authenticate the MTA. If the KDC is successful in this authentication, it sends an AS Response message to the MTA containing the security key as well as X.509 certificates used by the MTA to authenticate the KDC. The MTA then uses the provisioning service key in the AP message exchange with the SNMP provisioning server. It is from this exchange that the security keys needed for SNMPv3 are obtained.

Secure SNMPv3 packets provide the MTA with its configuration parameters. First an SNMP Inform packet is sent with the enrollment MIB object. The provisioning server uses this information to construct the MTA configuration file. The name and location of this configuration file as well as an integrity hash are then sent to the MTA in a SNMP Set packet. The MTA then obtains this file using either the TFTP or HTTP protocols. After the file transfer has completed the MTA validates and verifies the integrity of the file. If these checks succeed, the MTA sends a SNMP Inform packet to the provisioning server indicating this and optionally generates a system logging indication as well.

In addition to the secure method two other methods exist for MTA provisioning, the basic method and the hybrid method. These methods are insecure; therefore, it is recommended to use the secure method whenever possible. For the basic method the MTA configuration file name and location are given in the DHCP exchange, and an integrity hash is included in the file. For the hybrid method SNMP is still used for MTA configuration, but SNMPv2c is used instead of SNMPv3.

No matter which method is used, the MTA initializes communication with the CMS after it successfully provisions. These steps involve resolving the CMS FQDN, optionally establishing secure communication with the CMS, exchanging initialization NCS messaging with the CMS, and notifying the provisioning entity with the status of this initialization exchange. If the MTA detects a loss of communication with the CMS, this information is also sent to the provisioning entity and NCS messaging is sent to the CMS indicating this disconnected state.

Chapter Review

1 Correlate the described problem with the cable modem provisioning stage the modem will be stuck in.

Provisioning Stages

1) Scanning and Downstream Synch
2) Obtaining Upstream Parameters
3) Ranging
4) IP Connectivity
5) ToD Exchange
6) Configuration Transfer
7) Registration
8) BPI+

Problems

a) The DPE is unable to ping the cable modem subnets.
b) All of the upstream carriers support only DOCSIS 2.0 and the MTA is only DOCSIS 1.1 capable.
c) No DOCSIS root certificate exists on the CMTS.
d) The modem is unable to adjust its power level to meet the requirements of the CMTS; thus, it never transitions from the init(r1) state to the init(r2) state.
e) The modem configuration file is missing a downstream service flow definition.
f) The CMTS does not have the available bandwidth to honor all the requested QoS parameters from a modem.
g) The DHCP criteria selection tags configured on the RDU don't match any of the scope selection tags on CNR.
h) On the CMTS no downstream frequency is specified.

Answer: 1h, 2b, 3d, 4g, 5a, 6e, 7f, 8c

2 Correlate the following problems and symptoms related to the MTA provisioning process.

Problems

1) There is a mismatch in the service keys configured on the DPE and KDC.
2) DNS SRV record for the Kerberos realm name missing.
3) Server identifier in DHCP Offer differs from the values in option #122, sub-option #1 and #2.
4) MTA loaded with the wrong Service Provider Root certificate.

5) The MTA configuration file is missing a mandatory parameter.

6) Option #122 not returned during cable modem provisioning.

Symptoms

a) MTA never sends DHCP Discover packet.

b) MTA never sends the AS Request packet.

c) The AS Response is sent to the MTA, but the MTA never moves past this step.

d) DHCP Offer has all requested options, but MTA never sends DHCP Request

e) KDC never sends the AS Response packet to the MTA.

f) Final SNMPv3 Inform packet indicates a provisioning failure.

Answer: 1e, 2b, 3d, 4c, 5f, 6a

3 What four things does a DOCSIS cable look for before it can lock onto a downstream carrier?

Answer: QAM signal, FEC framing, MPEG packets, DOCSIS Synchronization (SYNC) messages

4 What is the difference between an initial maintenance burst interval and a station maintenance burst interval?

Answer: Initial maintenance intervals are broadcast and used by a modem the first time it attempts to range, whereas station maintenance intervals are unicast and used to give a modem an opportunity to fine tune transmission parameters or as a keep-alive mechanism.

5 Before a modem can send a Best Effort data packet, what needs to occur?

Answer: The modem sends a bandwidth request packet in a contention interval denoted in a MAP message. If this request gets through, the CMTS determines if it has available bandwidth and if the modem is authorized to transmit this data. Assuming it is, the CMTS issues a data grant to the modem in an upcoming MAP message. The modem then uses this data grant to send the packet.

6 What needs to be configured to permit a MTA to consider DHCP offers from any DHCP server?

Answer: Option #122, sub-option #1 needs to be set to the broadcast address 255.255.255.255.

7 Assuming all network components are Cisco products, identify four ways to help isolate problems in the DHCP process.

Answer: CMTS DHCP debugs, CNR DHCP logs, DPE logs, RDU logs

8 What is the benefit of using the Cisco DMIC feature over specifying the shared secret using the **cable shared secret** *authorization key* command?

Answer: With DMIC, secrets are dynamically generated, making each valid only for a limited time, and different for each device; consequently, a compromised secret is much less damaging.

9 What X.509 certificates are needed on the CM and CMTS for BPI+ to be successful?

Answer: The CM needs the CM Device and CM Manufacturer certificates, and the CMTS needs the CM Root certificate.

10 What DHCP option #122 sub-options must be returned to a MTA, and what do they represent?

Answer: Sub-option #3, which identifies the SNMP provisioning entity, and sub-option #6, which identifies the Kerberos Realm name/provisioning method.

11 What is the difference between the BASIC.1 and BASIC.2 provisioning methods?

Answer: The SNMP Inform packet signifying the provisioning status is included only in the BASIC.2 method.

12 What X.509 certificates are included in the AS Request message sent from the MTA to the KDC?

Answer: The MTA Device and MTA Manufacturer certificates

13 What X.509 certificates are included in the AS Response message sent from the KDC to the MTA?

Answer: The KDC, Local System Operator (optionally implemented), and Service Provider certificates

14 List the information sent by the MTA to the provisioning entity in the enrollment MIB object.

Answer: System description, Software version, MTA device type, MTA MAC address, correlation identifier

15 How do the basic and hybrid provisioning methods differ from the secure method?

Answer: These methods are insecure: the basic method flow is similar to the modem provisioning flow, and the hybrid method flow is like the secure method, except no Kerberos and SNMPv2c is used instead of SNMPv3.

16 What type of call signaling message is sent to the CMS when a MTA initializes?

Answer: RSIP restart

17 What type of call signaling message is sent to the CMS when a MTA detects a loss of communication?

Answer: RSIP disconnected

18 If no NCS messaging is seen from the CMS to the MTA after the RSIP exchange, what does this probably indicate?

Answer: The CMS is unable to resolve the configured MTA FQDN probably because of a failure in DDNS or misconfiguration.

CHAPTER 5

The MTA Configuration File

This chapter examines the configuration file used by PacketCable embedded multimedia terminal adapters (eMTA). The chapter covers the basic format of this file and provides examples to reinforce key points. In addition, the chapter covers some tools that can be used to generate and view MTA configuration files, one of which is Cisco Broadband Configurator (CBC). CBC is an application that can be used to create and edit Data-Over-Cable Service Interface Specification (DOCSIS) and PacketCable configuration files. Additionally, the template capability within the Cisco Broadband Access Center (BAC) can be used to dynamically generate configuration files. A utility within BAC can also be used to convert between binary and text versions of configuration files, as well as to add functionality.

As you know by now, a PacketCable MTA is configured using Simple Network Management Protocol (SNMP). This can be both through the transfer of an initial configuration file and incrementally, by an external device. You can think of the MTA configuration file as simply a collection of SNMP sets of parameters identified by their Management Information Base (MIB) value.

NOTE A MIB is a virtual database that identifies each manageable object by its name, syntax, accessibility, status, a text description, and a unique manageable Object Identification Number (OID). OIDs represent each manageable object with a unique sequence of numbers and names and are defined using Abstract Syntax Notation One (ASN.1). ASN.1 utilizes a Tag/Length/Value (TLV) format that allows complex data types to be defined out of a small number of predefined simple data types, other complex data types, or both. Types of MIBs include public (MIB II), experimental, and private enterprise; PacketCable MIBs are private enterprise MIBs.

The MTA configuration file can be either fully defined ahead of time or dynamically generated based on a template. In either case, the configuration file or template must be created ahead of time.

Refer to the PacketCable MTA MIB, Signaling MIB, and Management Event MIB specifications, as well as to the PacketCable MIBs Framework specification, for details on the MIBs used by a PacketCable MTA and their structure.

Cisco Broadband Configurator

Cable operators can either create their own applications for generating and editing configuration files or use ones obtained from vendors. Because DOCSIS cable modem and PacketCable MTA configuration files relate directly to their products, cable modem and MTA vendors typically have built these applications. Cisco is no exception; specifically, the CBC product can be used for this purpose.

CBC is a Java-based application that can run on Windows, Solaris, and Linux platforms and can create and modify DOCSIS and PacketCable configuration files. You can open and save files both in their binary form (the version used in production) and in text form (for simplicity). A configuration wizard setting exists that provides default values for mandatory parameters. CBC also comes with several sample configuration files that you can modify to meet your needs and provides a streamlined way of modifying multiple files simultaneously. CBC has both a graphical user interface (GUI) and a command-line interface (CLI); more focus in this chapter is spent on the GUI.

Using the CBC GUI

After you install and launch Cisco Broadband Configuration, you see a GUI, as shown in Figure 5-1.

Figure 5-1 *Initial CBC GUI*

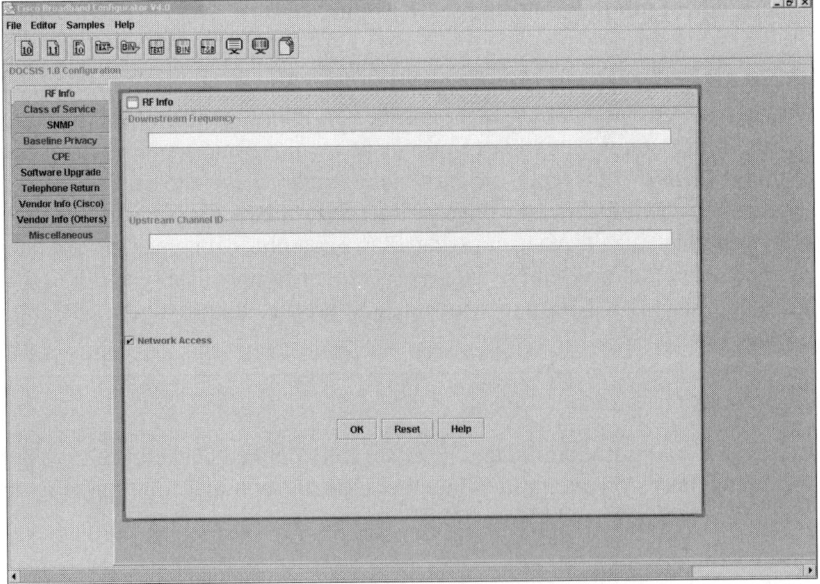

The screen shown is for creating DOCSIS 1.0 style cable modem configuration files. The menu tab at the left of the screen enables you to navigate to various configuration variables based on their category. For example, to configure the upstream and downstream rates and parameters you would need to navigate to the Class of Service tab. As you know, PacketCable requires the cable modem to be configured using DOCSIS 1.1 parameters, so how would you use CBC to accomplish this?

You actually have two ways:

- By selecting the appropriate option under the File menu at the top of the screen.
- By selecting the icons near the top of the screen. The leftmost three icons are for selecting a DOCSIS 1.0 configuration GUI, a DOCSIS 1.1 configuration GUI, or a PacketCable 1.0 configuration GUI. The next two icons are for opening a configuration file in either text format or binary format. Similarly, the next three icons are for saving the current GUI configuration data in text format, binary format, or both text and binary format. The following two icons are for changing the view between the binary and text formats. The rightmost icon is for modifying multiple files simultaneously, which is discussed later.

As previously mentioned, CBC provides many sample configuration files that can help you get started; Figure 5-2 is an illustration of the Samples menu.

Figure 5-2 *Sample CBC Configuration Files*

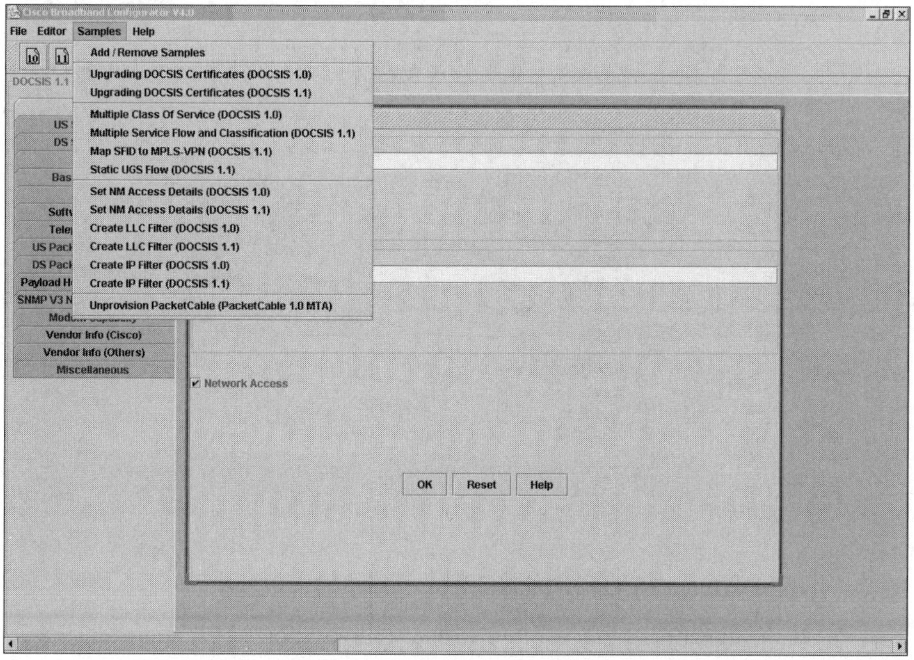

The types of sample configuration files are pretty self explanatory from their descriptions.

NOTE As you can see from Figure 5-2, it is also possible to remove these samples or to add your own sample configuration files to this menu.

The Help menu can be used to launch a web browser with detailed assistance on how to create configuration files and specifics on what the individual configuration parameters mean, as well as their possible values.

As mentioned earlier, CBC provides the capability to create or modify multiple configuration files simultaneously. Figure 5-3 shows an example of the GUI used in this case.

Figure 5-3 *Using CBC to Modify Multiple Files*

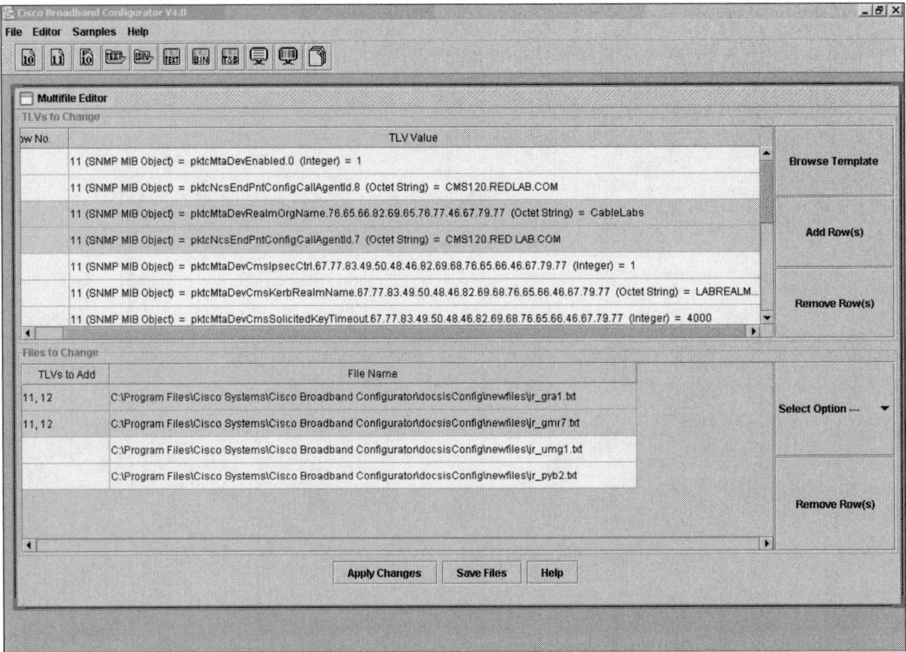

As you can see, you find two panes in this window:

- In the upper pane you identify all the configuration TLV parameters to change.
- In the lower pane you identify all the configuration files to which these changes apply.

You can add TLVs either manually or by importing a template file. A template file is simply a text file with a .tmpl extension that contains the TLVs. Much more on using template files is discussed later in this chapter. Two sample template files are supplied with CBC; you can use these as references for creating template files. The configuration files can preexist, or they can be created from a base configuration file. After the TLVs and configuration files have been identified, you simply select a group of TLVs and a group of configuration files to which you want these TLVs applied, and select Apply Changes.

Using the CBC CLI

CBC also has a CLI interface that can be used to convert configuration files between the binary and text formats. The CLI can be used to automate configuration file tasks in a script. For example, a script could be written to generate the text version of a configuration file, and then the CBC CLI could be used to generate the binary version.

For more information on CBC, please consult the user guide available through the Help menu, or for online documentation, access http://www.cisco.com/univercd/home/home.htm.

MTA Configuration File Format

As with a DOCSIS cable modem configuration file, the format of the MTA configuration file is a series of TLV entries. You have five valid tag types, as shown in Table 5-1.

Table 5-1 *Valid Tags for the MTA Configuration File TLV Entries*

Tag	Description
11	DOCSIS-specified TLVs
38	PacketCable-specified TLVs
43	Vendor-specific TLVs
64	PacketCable-specified TLVs
254	Configuration file delimiter (Start/End)

DOCSIS tag type 11 is defined in Appendix C of the DOCSIS RFI specification. It is used for SNMP MIB objects that can be set in the TFTP process during provisioning. The length field for tag type 11 is a single byte; consequently, the tag value can be no greater than 254 bytes. The PacketCable tag type 64 has a 2-byte length field; therefore, it is needed for tag values greater than 254 bytes. Tag type 38 defines SNMP notification receivers and is actually defined both in Appendix C of the DOCSIS specification and in the PacketCable provisioning specification. The configuration file delimiter, tag type 254, is one byte in length and is either 0x01, denoting the beginning of the file, or 0xFF (255), denoting the end of the file. Vendor-specific information can also be added to the configuration file; in this case, tag type 43 is used. MTA configuration MIBs are defined using either DOCSIS tag type 11 or the PacketCable tag types 38 or 64. As you will see, most parameters are defined under DOCSIS tag type 11, and this is the preferred implementation.

You find several types of data in the MTA configuration file, specifically:

- Device-level configuration data
- Device-level service data
- Per-endpoint configuration data
- Per-realm configuration data
- Per-CMS configuration data

Device-Level Configuration Data

Device-level configuration data can be set only in the MTA configuration file. Figure 5-4 is an example of an MTA configuration file where the required device level configuration attributes are pointed out. The textual view option within CBC displays the configuration file.

Figure 5-4 *Device-Level Configuration Data*

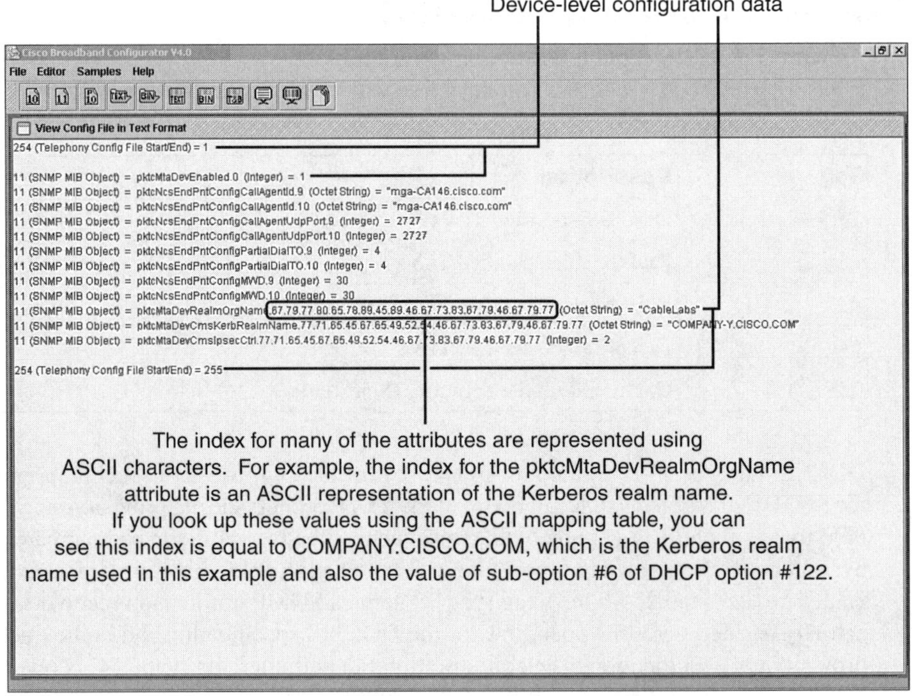

The ASCI character set mapping used in this figure is as follows:

0 NUL	1 SOH	2 STX	3 ETX	4 EOT	5 ENQ	6 ACK	7 BEL
8 BS	9 HT	10 NL	11 VT	12 NP	13 CR	14 SO	15 SI
16 DLE	17 DC1	18 DC2	19 DC3	20 DC4	21 NAK	22 SYN	23 ETB
24 CAN	25 EM	26 SUB	27 ESC	28 FS	29 GS	30 RS	31 US
32 SP	33 !	34 "	35 #	36 $	37 %	38 &	39 '
40 (	41)	42 *	43 +	44 ,	45 -	46 .	47 /
48 0	49 1	50 2	51 3	52 4	53 5	54 6	55 7
56 8	57 9	58 :	59 ;	60 <	61 =	62 >	63 ?
64 @	65 A	66 B	67 C	68 D	69 E	70 F	71 G
72 H	73 I	74 J	75 K	76 L	77 M	78 N	79 O
80 P	81 Q	82 R	83 S	84 T	85 U	86 V	87 W
88 X	89 Y	90 Z	91 [	92 \	93]	94 ^	95 _
96 `	97 a	98 b	99 c	100 d	101 e	102 f	103 g
104 h	105 i	106 j	107 k	108 l	109 m	110 n	111 o
112 p	113 q	114 r	115 s	116 t	117 u	118 v	119 w
120 x	121 y	122 z	123 {	124 |	125 }	126 ~	127 DEL

As you can see, the configuration file begins and ends with type 254 delimiter TLVs. Pay attention to the pktcMtaDevEnabled attribute, which must be sent to a value of 1 (equal to "true") if voice services are to be provided on the MTA. The other attribute identified here is the Realm Organization Name, which must equal the value specified in the service provider certificate on the Key Distribution Center (KDC).

For more information on attributes in this category, please refer to the provisioning specification and the MTA Device MIB specification.

Device-Level Service Data

Device-level service data can be set either remotely via SNMP or in the MTA configuration file. All these attributes are optional, and many of them have default values if not explicitly specified. These attributes are all defined in the MTA Signaling MIB specification. Examples of these attributes include ones used to set IP Type of Service or Differentiated Services Code Point (DSCP) codepoint values for Network-based Call Signaling (NCS) signaling packets and Real-time Transport Protocol (RTP) bearer packets. Other examples are attributes used to define MTA ring tone cadences.

Per-Endpoint Configuration Data

Per-endpoint configuration data is sent to the MTA at the time voice services are enabled. Therefore, if voice services are enabled during the initialization phase, this data needs to be present in the configuration file. If voice servers are activated later via SNMP, then per-endpoint configuration data must be set at this point.

Logically, these attributes apply to only a single MTA endpoint; therefore, if you want to configure both lines on an MTA, each of these attributes needs to be specified twice.

The same configuration file shown previously in Figure 5-4 is shown again in Figure 5-5, only this time the per-endpoint configuration data is highlighted. In the figure, note that for endpoint attributes, the index 9 corresponds to the first line on the MTA, and the index 10 corresponds to the second line.

Figure 5-5 *Per-Endpoint Configuration Data*

The only mandatory attribute in this class is the CMS name attribute pktcNcsEndPntConfigCallAgentId. This is the FQDN of the CMS with which the endpoint initially communicates (where the endpoint is sent). The CMS specified here also must be defined in the per-CMS configuration data attributes.

Per-endpoint configuration data

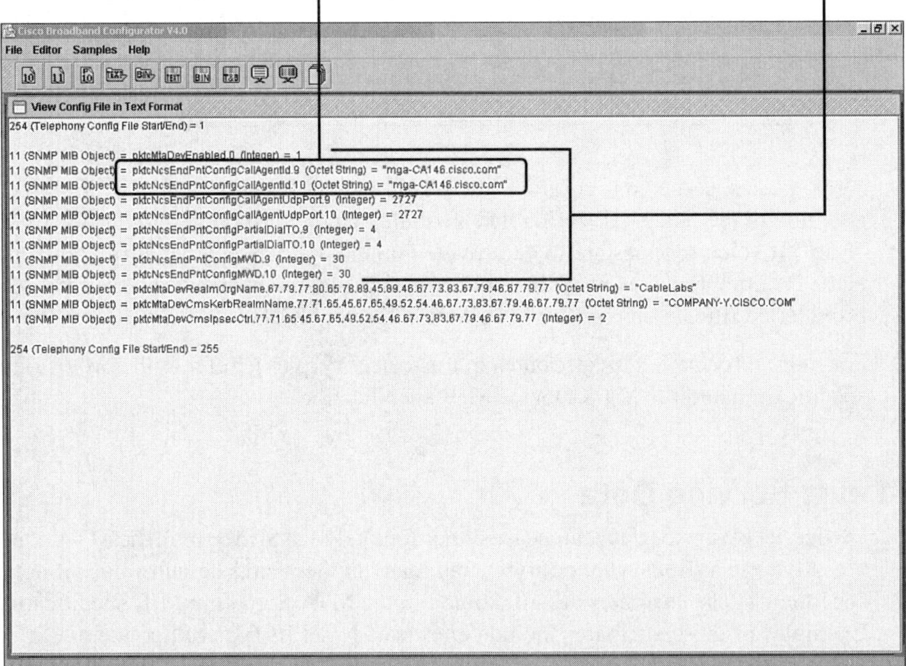

As mentioned in Figure 5-5, the first endpoint or line on the MTA is referenced by index 9, and the second endpoint or line is referenced by index 10. As the figure points out, only the call management server (CMS) name attribute is mandatory. An important fact about this attribute is that it only defines the initial CMS with which an endpoint attempts to communicate; it is definitely possible to have multiple CMS devices defined in the configuration file with which the endpoint could communicate.

Optional attributes in this class include definitions of call signaling timers, such as those used to define how long to wait for dial tone and ring back tone. Timers and variables used to control the operation of the NCS protocol itself exist here as well. Please take extreme caution if you are altering these timer values from their default values because it can drastically affect MTA operation. In Figure 5-5, the attribute used to alter the partial dial time out and the attribute used to manipulate how the MTA behaves when a restart is received are included as an example.

Most of the per-endpoint configuration data attributes are defined in the MTA Signaling MIB specification. The exception is the port administration state attribute, ifAdminStatus, which specifies whether an endpoint or line is enabled or disabled. If this attribute is set to a value of 2, meaning "down," this line is not operational regardless of the configuration on the CMS. This attribute is defined in IETF RFC 2863, which defines the Interfaces Group (IF) MIB.

Per-Realm Configuration Data

The attributes for per-realm configuration data can be set only in the MTA configuration file. They are all defined in the MTA Device MIB specification. These attributes relate to the Kerberos realm identified by sub-option #6 of DHCP option #122.

The only required attribute in this category is the Realm Organization Name attribute (refer to Figure 5-4). This attribute must equal the value specified in the service provider certificate on the KDC. Notice how the index of this attribute is specified. This index is the ASCII representation of the Kerberos realm name; if you look at the ASCII character set mapping table in that figure, you can tell this index is equal to the Kerberos realm name identified in sub-option #6 of DHCP option #122.

The optional attributes in this category relate to timers and retry values used in obtaining Kerberos certificates.

Per-CMS Configuration Data

Per-CMS configuration data attributes are also defined in the MTA Device MIB specification. For an endpoint to be able to communicate with a CMS, at least one CMS entry must be defined. Consequently, if you want the MTA to be able to process calls on a CMS after initialization, one CMS entry is required in the MTA configuration file. You can have multiple CMS entries defined if you want to have different endpoints use different CMS devices or if you want a CMS to have the ability to move between CMS devices.

These attributes define how security associations are managed between the MTA and CMS. If you are not using secure call signaling in your network, most of these attributes do not apply.

Figure 5-6 is the same MTA configuration file example shown previously in the chapter, this time with the per-CMS configuration data highlighted.

Figure 5-6 *Per-CMS Configuration Data*

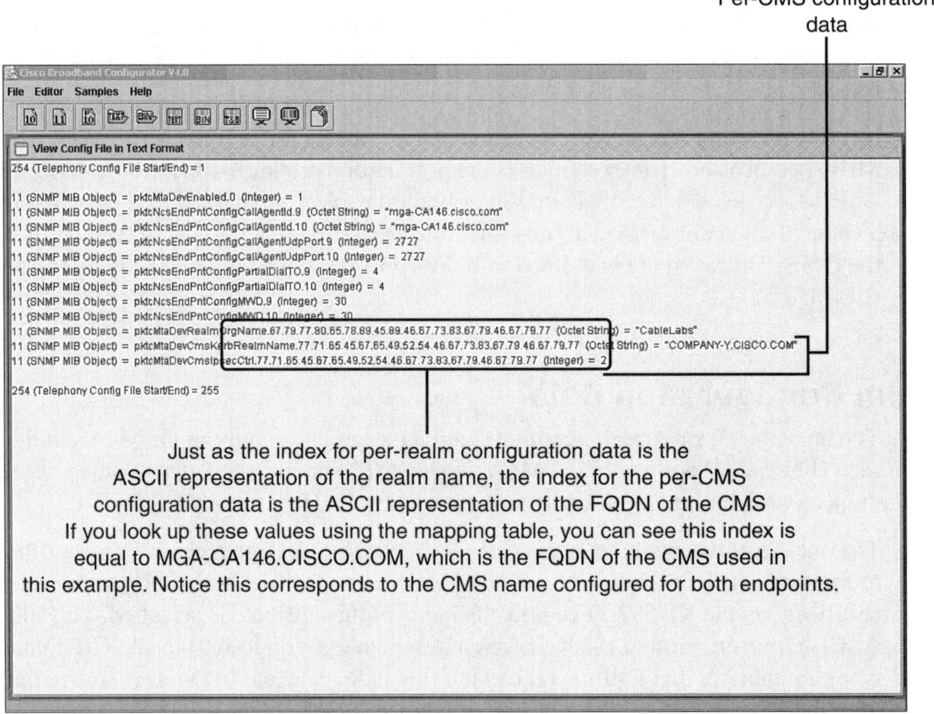

The Kerberos realm name attribute, pktcMtaDevCmsKerbRealmName, is the only one required that associates the CMS to a particular Kerberos realm. This realm name must also be defined in a Realm Organization Name attribute. As you can tell in this example, the pktcMtaDevCmsKerbRealmName attribute is equal to COMPANY-Y.CISCO.COM, which is the index of the pktcMtaDevRealmOrgName attribute. The remaining attributes in this class are all optional and relate to the security establishment between the MTA and CMS. In this example only the IPsec Control attribute, pktcMtaDevCmsIpsecCtrl, is included, which defines how IP security is established. The value 2 in this example signifies that no IPsec encapsulation is to exist for call signaling packets between the MTA and CMS; a value of 1 enables NCS signaling security.

Notification Receivers

Other than the configuration file delimiter TLVs, perhaps you've noticed that all the TLVs shown in the examples have been DOCSIS tag type 11. So you might be wondering, "When are the other PacketCable tag types used?"

One use is for defining NMS devices that need to be notified when an MTA completes the provisioning process. In this case PacketCable tag type 38 is used. This tag type comprises a number of sub-options. Some examples include sub-type 1, which identifies the IP address of the NMS station, and sub-type 3, which identifies the message (Trap or Inform) and SNMP protocol version the MTA should use when sending packets to the NMS station. For more on defining notification receivers refer to section 11 of the provisioning specification.

BAC Configuration File Capabilities

Within the BAC provisioning solution, several utilities exist that help with the setup and maintenance of a PacketCable deployment. In Chapter 3, "Provisioning Overview," you learned about the utility on the KDC used to create the necessary digital certificates and security keys for PacketCable operation. In this chapter you were also introduced to a BAC utility on the CNR DHCP servers that lets you modify the BAC extension script configuration. BAC also offers some features that help with the use and management of configuration files—specifically, the use of templates to dynamically build configuration files and a utility that lets you view and manipulate these files.

Using Templates

The Cisco BAC provisioning system has the capability to use templates to dynamically generate modem and MTA configuration files instead of static files. Templates are created as basic ASCII text files, making them very easy to view and edit.

Template files comprise four different entry types:

- **Comments**—Logically, comments let you document the template file; the # character indicates that a comment follows.
- **Includes**—Includes let you create a hierarchy of template files; in other words, you can have a template made up of other templates. For example, if you have several options that are common to several classes of devices, you could have one template file for these common options and then include this template in a template for each device class.

- **Options**—Options let you define DOCSIS and PacketCable configuration parameters. The format of an option is the word *option* followed by an identifier and value pair. For custom options it is also necessary to include the option encoding type (hex, ASCII, IP, etc.). For SNMP-style options such as DOCSIS type 11 and PacketCable type 64 the OID must be specified as well.
- **Instance modifiers**—Instance modifiers are needed when you want to group compound options into specific TLVs.

Sample template files are included on the BAC Regional Distribution Unit (RDU), as shown in Figure 5-7.

Figure 5-7 *Sample BAC Template Files*

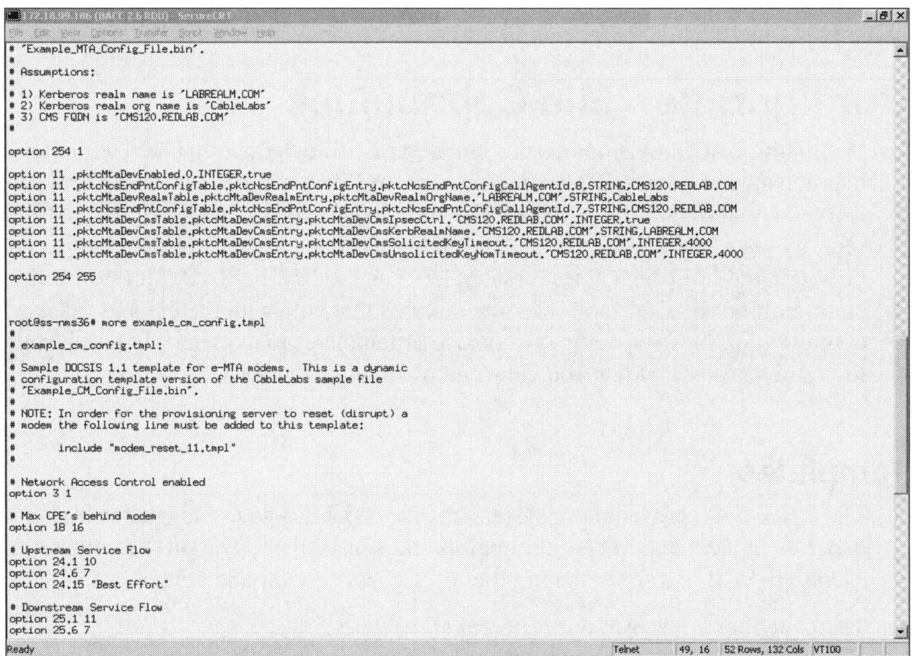

As you can see, at the top of this figure is a PacketCable MTA template, and at the bottom is a DOCSIS CM template. Notice that in the MTA example the DOCSIS type 11 options have the OID specified, followed by a comma and the format (string, integer, and so on), followed by a comma and the value.

In the previous example, all the options were specified as static values. The power and usefulness of templates really come into play when you use macro variables. With macro variables, options can be changed or added depending on where the variable is set. There are five places on the BAC RDU where macro variables can be defined:

1 System defaults
2 Technology defaults (that is, DOCSIS, PacketCable)
3 DHCP criteria
4 Class of Service
5 Device configuration

The system defaults and technology defaults are accessible from the Defaults tab under the Configuration menu on the RDU. This is also the priority flow for setting variables. If the same variable is defined to different values in all five places, the technology defaults take precedence over the system defaults, the DHCP criteria takes precedence over the technology defaults, and so forth.

Probably the easiest way to illustrate how macro variables can be used is through an example. In this example, a cable modem template is used with a variable for the maximum downstream rate for the primary downstream service flow. Figure 5-8 illustrates this template.

Figure 5-8 *Macro Variable Template Example*

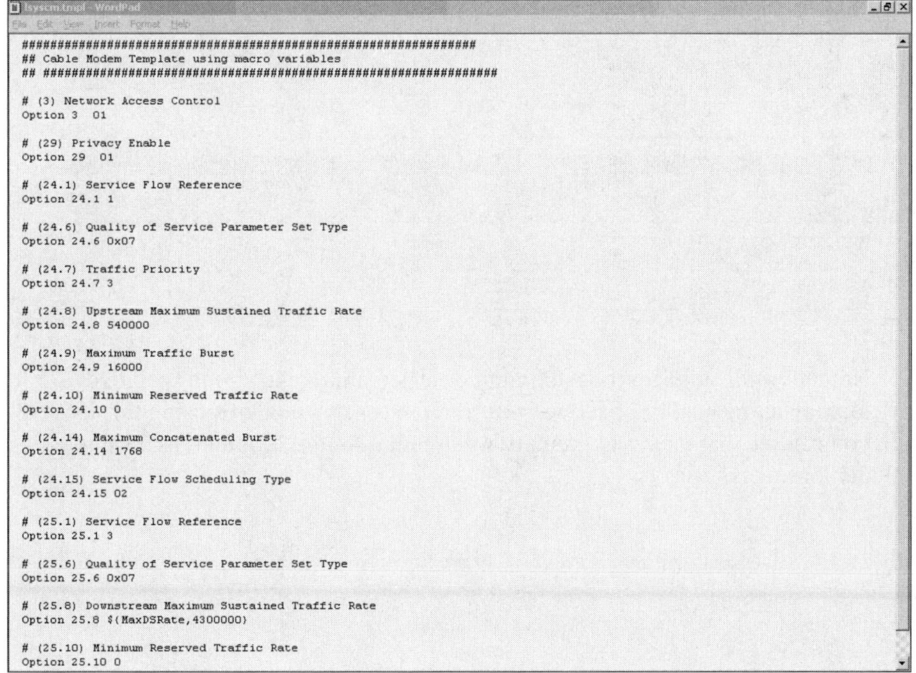

```
#####################################################################
## Cable Modem Template using macro variables
## ##################################################################

# (3) Network Access Control
Option 3  01

# (29) Privacy Enable
Option 29  01

# (24.1) Service Flow Reference
Option 24.1 1

# (24.6) Quality of Service Parameter Set Type
Option 24.6 0x07

# (24.7) Traffic Priority
Option 24.7 3

# (24.8) Upstream Maximum Sustained Traffic Rate
Option 24.8 540000

# (24.9) Maximum Traffic Burst
Option 24.9 16000

# (24.10) Minimum Reserved Traffic Rate
Option 24.10 0

# (24.14) Maximum Concatenated Burst
Option 24.14 1768

# (24.15) Service Flow Scheduling Type
Option 24.15 02

# (25.1) Service Flow Reference
Option 25.1 3

# (25.6) Quality of Service Parameter Set Type
Option 25.6 0x07

# (25.8) Downstream Maximum Sustained Traffic Rate
Option 25.8 ${MaxDSRate,4300000}

# (25.10) Minimum Reserved Traffic Rate
Option 25.10 0
```

Notice near the bottom of the figure, a macro variable has been inserted for the value of the maximum downstream rate option. A default value of 4,300,000 is also shown in case the macro variable is not defined in any of the five places mentioned earlier for a particular device. Also note that before this template can be used, this macro variable, MaxDSRate, must be defined as a custom property on the RDU. The custom property menu exists below the configuration menu. These variables must be defined as DataType.STRING.

So take a look at what happens when this variable is defined for a particular modem, as shown in Figure 5-9.

Figure 5-9 *Macro Variable Setting Example*

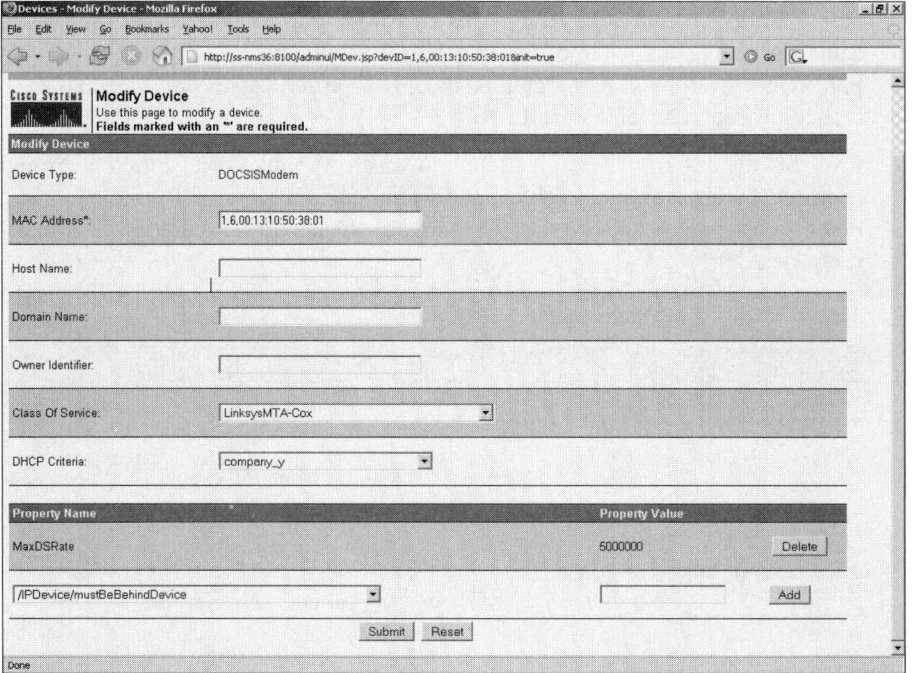

Notice for this modem the maximum downstream rate has been specified as 5 Mbps. By looking at the quality of service information for two modems using this class of service, you can see that their maximum downstream rates are different, as Example 5-1 demonstrates.

Example 5-1 *Macro Variable Example*

```
C
0103-7246VXR#show cable modem 13.34.14.63
MAC Address     IP Address      I/F        MAC          Prim RxPwr  Timing  Num BPI
                                           State        Sid  (dB)   Offset  CPE Enb
0013.1050.37f5 13.34.14.63     C3/0/U0    online(pt)   2    0.50   1131    1   Y

C0103-7246VXR#show cable modem 13.34.14.63 qos
Sfid  Dir Curr  Sid   Sched  Prio MaxSusRate  MaxBrst   MinRsvRate  Throughput
          State       Type
5     US  act   2     BE     3    540000      16000     0           0
6     DS  act   N/A   BE     0    4300000     3044      0           0

C0103-7246VXR#show cable modem 13.34.14.122
MAC Address     IP Address      I/F        MAC          Prim RxPwr  Timing  Num BPI
                                           State        Sid  (dB)   Offset  CPE Enb
0013.1050.3801 13.34.14.122    C3/0/U0    online(pt)   406  0.00   1134    1   Y

C0103-7246VXR#show cable modem 13.34.14.122 qos
Sfid  Dir Curr  Sid   Sched  Prio MaxSusRate  MaxBrst   MinRsvRate  Throughput
          State       Type
813   US  act   406   BE     3    540000      16000     0           0
814   DS  act   N/A   BE     0    5000000     3044      0           0
```

As you can see, the maximum sustained rate for the specified modem was changed as expected to 5 Mbps.

Configuration File Utility

On the BAC RDU, a configuration file utility exists that can be used to convert configuration files between their binary form and templates. This utility can also be used to add shared secrets to CM configuration files, to test template variables, and to create basic method MTA configuration files. The utility is called runCfgUtil.sh and is located in the <BAC HOME>/rdu/bin directory (by default, the <BAC HOME> path is /opt/CSCObpr).

Figure 5-10 shows the command syntax of this utility.

Figure 5-10 *BAC Configuration Utility*

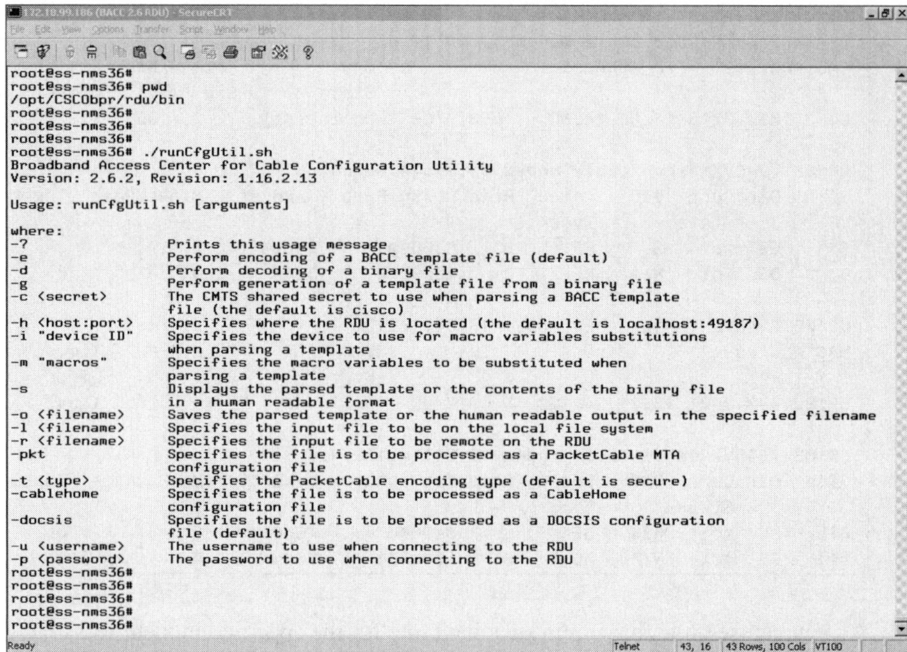

Figure 5-11 shows how this utility could be used to view the resulting binary file for the modem configured with the specific maximum downstream rate in the template example used earlier in the chapter.

Figure 5-11 *BAC Configuration Utility Example*

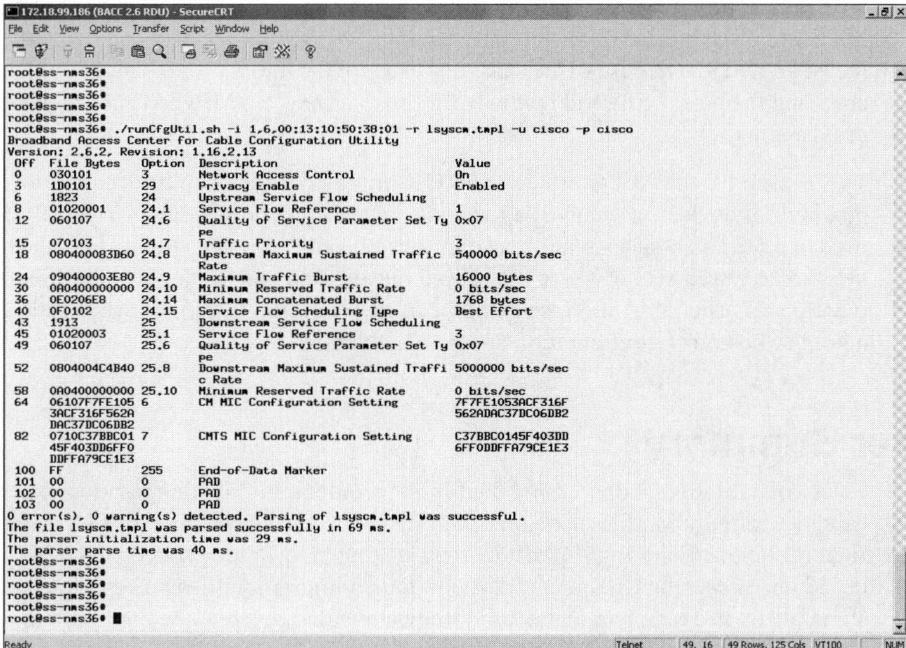

MTA Configuration After Initialization

Obviously, changes to MTA configuration can be made by modifying the MTA configuration file in the provisioning system and reloading the MTA. For example, if you want to change the CMS FQDN authorized to communicate with the MTA telephony endpoints, this is typically how you would accomplish this task.

However, suppose you want to enable the second line on an MTA that already has the first line active. If you followed the previous procedure, you would have to temporarily take the first line out of service while you reload the MTA to bring up the second line. To eliminate this loss of service, Post Initialization Incremental Provisioning (PIIP) can be used. As another example, suppose that one of your customers has cancelled their service or has not paid their bill; in this case, you could use PIIP to disable their service.

As you might have guessed, PIIP uses the SNMP protocol. SNMP sets are applied to the same MIB variables that you can set in the MTA configuration file. Please refer to the PacketCable specifications for the MIB variable definitions. These SNMP sets are generated on an interface from an NMS/OMS device, such as the BAC API Scripter tools suite discussed next.

If you are using the secure method for provisioning, you need to use SNMPv3 for incremental provisioning. This means that a security association must exist between the MTA and the NMS/OMS device. The ability to create this security can be a little tricky if the NMS/OMS device is not the same one used for the initial provisioning. Logically, if you are using the basic or hybrid methods for provisioning, SNMPv2c is used for incremental provisioning.

Java is used for the API interface to BAC. Simple Java code can be used to provision hundreds of devices in seconds or to handle incremental provisioning. This API can also be used to write a Java script to redirect unprovisioned devices that attempt to come up on your network to a web server where they could register and sign up for service. Finally, as mentioned earlier, this interface could be used to develop third-party applications that automate customer configuration and management.

Chapter Summary

MTA configuration is done using the SNMP protocol. Initial configuration is performed using a configuration file made up of SNMP MIB objects specified in TLV parameters. Incremental provisioning is performed by sending SNMP set packets to the MTA with the definition of specific MIB objects. The initial configuration file can be defined either as a static file or as a template that is used to dynamically generate the configuration file. Incremental provisioning has many applications, such as enabling or disabling an MTA or an individual line.

The Cisco tool used to build DOCSIS and PacketCable configuration files is called Cisco Broadband Configurator (CBC). This tool is Java-based and can be used to view, edit, and save configuration files in both their binary and text forms. CBC is equipped with sample files as well as a configuration wizard to ease configuration file generation. CBC also has the capability to modify TLV parameters in multiple configuration files simultaneously. The CBC command-line interface can also be used to generate binary configuration files from text configuration files.

You find five categories of MIB-based TLVs in an MTA configuration file:

- Device-level configuration
- Device-level service
- Per-endpoint configuration
- Per-realm configuration
- Per-CMS configuration

Most of these MIBs are defined in the PacketCable MTA MIB, Signaling MIB, and Management Event MIB specifications.

A sample device-level configuration MIB is pktcMtaDevEnabled, which must be set to a value of true for the MTA to be operational. Device-level service MIBs are optional and can be set either during initialization or incrementally. Per-endpoint configuration MIB attributes need to be set if voice services are to be enabled. These attributes identify the CMS to which the NCS RSIP is sent and can also be used to change NCS timers and variables from their default values. Logically, per-realm configuration attributes relate to the Kerberos realm name. These attributes can only be set in the configuration file, and the only mandatory one is pktcMtaDevRealmOrgName, which must equal the organization name in the KDC service-provider certificate. Per-CMS attributes denote data associated with the CMS, such as its Kerberos realm name and whether the MTA needs to use secure signaling to communicate with it.

PacketCable type 38 TLVs define SNMP notification receivers. These are devices that need to be notified of various MTA events such as provisioning status and CMS communication status.

The Cisco BAC solution supports the use of templates to dynamically generate configuration files, which can greatly reduce the number of configuration files a service provider needs to maintain. BAC template files consist of comments used to document the file, includes allowing template files to be nested, options to define DOCSIS and PacketCable TLVs, and instance modifiers used to group options. Options can be defined statically or by using macro variables so that TLVs can be changed based on the specific device, class of service, DHCP criteria, or technology configuration.

The Cisco BAC solution also provides a number of helpful utilities, one of which is the configuration utility located on the RDU. This utility can be used for a number of applications, such as converting between template files and binary configuration files. BAC also has an open Java-based API interface allowing the capability for services such as bulk configuration, self-provisioning, and flow through provisioning

Chapter Review

1 True or False: You can customize CBC by creating your own sample files and removing unwanted samples.

Answer: True

2 True or False: Most PacketCable TLVs are specified using tag type 38.

Answer: False. DOCSIS tag type 11 is preferred and typically used.

3 True or False: The index for realm attributes such as pktcMtaDevRealmOrgName is the ASCII encoding of the MTA FQDN.

Answer: False. This is a per-realm attribute, so the index is the ASCII representation of the Kerberos realm name.

4 True or False: The indices for PacketCable MTA endpoints are 7 for line #1 and 8 for line #2.

Answer: False. They are 9 for line #1 and 10 for line #2.

5 True or False: The CMS specified for an individual endpoint must also have an entry in the CMS table.

Answer: True

6 True or False: Even if the MTA has been provisioned using the secure method, SNMP notification receivers can be defined that communicate with the MTA using SNMPv2c.

Answer: True

7 List an example of a device-level service attribute.

Answer: Signaling DSCP value, Bearer DSCP value, ring tone cadence

8 What is an example of a MIB that can be specified in an MTA configuration file that isn't defined by a PacketCable specification?

Answer: ifAdminStatus, interface MIB used to set port administrative status

9 What is the index for per-CMS configuration attributes?

Answer: The ASCII representation of the CMS FQDN

10 Describe how you could use templates in a network with three different cable modem vendors that all get the same configuration parameters except the upgrade file name and CVC.

Answer: You could have a template containing all the common parameters, and then include this template in each vendor-specific template with the specific upgrade file name and CVC.

11 What are the five places in BAC where macro variables can be defined? List them in order of increasing preference.

Answer: 1. System defaults, 2. Technology defaults, 3. DHCP criteria, 4. Class of Service, 5. Device configuration

12 Identify three applications of the BAC configuration file utility.

Answer: Adding shared secrets to CM configuration files, testing templates, and creating basic method MTA configuration files

13 Why can configuring an MTA via post initialization incremental provisioning be difficult?

Answer: Because if the secure method was used, the NMS/OSS application needs to communicate with the MTA via SNMPv3.

Call Signaling

CHAPTER 6

Signaling Interfaces and MGCP Overview

Chapter 2, "PacketCable Functional Components," introduced the importance of call signaling. Calls are set up, modified, and torn down in a PacketCable VoIP network by call signaling. Because the PacketCable architecture consists of multiple components, they need to communicate call information using some sort of protocol. This chapter introduces you to Media Gateway Control Protocol (MGCP), the foundation of the Network-based Call Signaling (NCS) and Trunk Gateway Control Protocol (TGCP) protocols used in PacketCable. In Chapter 1, "PacketCable Overview," you learned that one of the goals when developing the PacketCable architecture was to use existing protocols and standards where possible. NCS and TGCP are perfect examples of this because they are modified versions of the MGCP protocol tailored to the PacketCable application.

This chapter examines key aspects of MGCP, as they are the building blocks of the NCS and TGCP protocols. This chapter also introduces use of the Session Description Protocol (SDP) by MGCP. Chapter 7, "NCS (Network-based Call Signaling)," and Chapter 8, "TGCP and the PSTN Interconnect," cover specifics relating to NCS and TGCP as well as implementation and troubleshooting.

NOTE It is beyond the scope of this text to cover every detail of the MGCP and SDP protocols. The current RFC or standard should always be referenced. At the time of the writing of this book, RFC 3435—Media Gateway Control Protocol (MGCP) Version 1.0 is the latest and is the basis of the current version of the NCS and TGCP PacketCable specifications. RFC 2327—SDP: Session Description Protocol is the latest version of the SDP specification, and RFC 3661 contains information on MGCP Return Code usage.

Remember, IETF RFCs were introduced in Chapter 1 and can be found online at http://www.ietf.org/rfc/rfc*xxxx*.txt, where *xxxx* is the number of the RFC. If you do not know the number of the RFC, you can try searching by topic at http://www.rfc-editor.org/rfcsearch.html.

Device Relationship

Chapter 2 touched on the difference between signaling protocols that utilize a distributed architecture and those that use a centralized architecture. This concept is very important to fully understand before we get into too much detail with any of the protocols, so it is covered again here for emphasis. In a distributed architecture, both devices are required to have some intelligence or capability to perform tasks such as providing feature functionality and maintaining call state. Session Initiation Protocol (SIP) and H.323 are examples of protocols utilizing this type of architecture. Another example is the Call Management Server Signaling (CMSS) protocol used to communicate between PacketCable call agents or call management servers (CMS). The CMSS protocol is a variant of SIP and is covered in Chapter 9, "Call Management Server Signaling Protocol (CMSS)."

In a centralized architecture most of the intelligence is the responsibility of one of the devices allowing the other device to be relatively "dumb" or simple. This intelligence is responsible for providing call features and functionality. Allowing one device to be simple reduces complexity on the device, making it cheaper and less prone to bugs, and lessens the need to upgrade the software for feature enhancements. MGCP, NCS, and TGCP are all centralized architecture protocols where the call agent component is the intelligent device and the gateway component is the unintelligent device. Also because the call agent component is under total control of the Multiple Systems Operator (MSO), the MSO has much tighter control over the services and functionality operating in its network.

The call agent/gateway relationship is often referred to as a master/slave relationship, where the slave (gateway) relies on the master (call agent) for all its instruction. The following are some of the functions of the call agent:

- Control call routing
- Maintain call state
- Provide call features
- Set up and tear down connections on the gateways
- Instruct the gateways to generate signals and detect events

The gateway simply does what the call agent instructs it to do. The gateway sets up connections between the voice over IP (VoIP) network and the outside time-division multiplexing (TDM) world as instructed by the call agent. A gateway also generates signals, such as ring tone, and detects events, such as a hook flash.

NOTE In Chapter 2, you learned about the individual PacketCable components. Remember, the CMS is the call agent responsible for providing call control and signaling for PacketCable MTA gateways. Specifically, the NCS variant of MGCP, covered in Chapter 7, is used. Similarly, the MGC is the call agent responsible for providing call control and signaling for the media gateways that interface to the legacy public switched telephone network (PSTN). The TGCP variant of MGCP, covered in Chapter 8, is used for the call agent-gateway signaling in this case.

Endpoints and Connections

The two basic elements of MGCP are *endpoints* and *connections*. An endpoint is what an MGCP call agent controls.

Endpoints exist on gateways and typically correlate to a physical entity outside the VoIP network, although as you'll soon see virtual endpoints are also possible. The MGCP specification defines endpoints as the "sources and/or sinks of data," such as VoIP media streams. Types of endpoints in a PacketCable network include

- Physical interfaces to PSTN trunk circuits (that is, digital channels [DS0])
- Physical interfaces to multimedia terminal adapter (MTA) analog lines
- Virtual interfaces to announcement server audio content
- Virtual interfaces to Interactive Voice Response (IVR) audio content

A digital channel (DS0) or trunk is a communication channel in the legacy circuit switched PSTN. Therefore, these endpoints exist on PSTN trunking gateways. This is the Media Gateway (MG) in the PacketCable architecture.

NOTE In the legacy PSTN, voice is digitized and transmitted from switch to switch in a 64-Kbps communication channel known as a *trunk*. TDM allows for multiple trunks to be carried on the same physical media. For example, the T1 carrier used in North America contains 24 DS0 trunks and has a total bandwidth of 1.544 Mbps (8 Kbps of framing overhead are sent in addition to the 24 DS0 circuits). Each 64-Kbps DS0 is capable of supporting one uncompressed voice call.

An analog line is a phone line just like you have in your home. Analog line endpoints exist on residential gateways or MTAs in the PacketCable architecture.

Announcement server access points are virtual endpoints where the interface is to software-based content rather than to a physical entity. They refer to the point on an announcement server supplying the voice message to the VoIP network. These endpoints typically operate in a half-duplex manner; the announcement media stream travels to the person on the other end of the call, and no media stream is sent in the opposite direction. In the PacketCable architecture, announcement server access points exist on the Announcement Player (ANP).

IVR endpoints provide an interactive menu-driven command/response dialogue to provide and collect information. These virtual endpoints are also typically implemented in software. Examples of IVR endpoints include voice mail systems and an application such as Movie Phone.

The other basic MGCP element is a connection. Connections exist between endpoints and can be either point to point or multipoint. MGCP defines connections over various types of media such as IP, ATM, and TDM networks.

Connections permit the data transfer between gateway endpoints. The MGCP Call Agent dictates the creation, modification, and deletion of connections on these endpoints. This information is signaled to the gateways in MGCP messages. As you might have guessed, a typical call consists of at least two connections:

- One connection on the endpoint associated to the originating device
- One connection on the endpoint associated with the terminating device

MGCP Message Format

The format of MGCP messages is simple. MGCP messages are encapsulated in UDP packets. By default, packets destined for gateways are sent to port 2427, and packets destined for call agents are sent to port 2727. The MGCP message itself is encoded in plain ASCII text. The fact that MGCP messages are encoded this way has its positives and negatives. Troubleshooting is very easy, as any protocol sniffer, such as the UNIX **snoop** command, can display these messages without the need for any complex interpretation. This capability is also a potential security vulnerability and enforces the need for packet encryption.

Recall that UDP is a non-reliable protocol; you have no guarantee a packet will reach its destination in UDP itself. This is handled by the upper-layer protocol, in this case MGCP, which uses a command/response model. You have two types of MGCP messages, commands and responses. Both consist of two parts, a header and optional parameters. First, the format of MGCP commands is covered in this chapter, followed by some examples; then the format of MGCP responses is covered, likewise followed by some examples.

MGCP Commands

An MGCP command header is comprised of a four-field command line followed by optional parameter lines. Unless otherwise specified, all the values in the command line and the parameter lines are case insensitive. A blank line precedes the optional SDP information. Figure 6-1 shows a breakdown of the command header.

Figure 6-1 *MGCP Command Header Syntax*

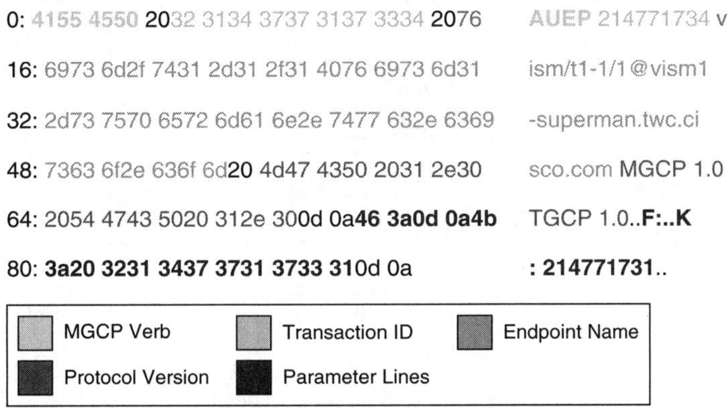

Notice the four fields in the command line (MGCP verb, transaction ID, endpoint name, and protocol version) and the optional K parameter.

In this figure, the UNIX **snoop** command was used to display the MGCP messaging. By using the **-x 42** option, IP packet content from byte offset 42 is displayed in both hex format (on the left) and ASCII format (on the right). The 42 means the first 42 bytes of the packet—which consist of the Ethernet header (14 bytes), IP header (20 bytes), and UDP header (8 bytes)—are not shown. Consequently, only the MGCP layer is shown and decoded. The **snoop** utility syntax and usage manual pages can be found by typing **man snoop** on a UNIX machine. You might find that an external protocol sniffer such as Ethereal is easier to use than **snoop**. Ethereal is covered in detail in Appendix C, "Complete Call Flows," and is also used heavily throughout the book. However, familiarity with **snoop** is beneficial for situations in which it might not be feasible to hook up an external protocol analyzer. Also because most call agents run on a UNIX platform (such as the Cisco BTS 10200), **snoop** is usually readily available. Thus, **snoop** is used exclusively in this chapter; however, in future messages the hex portion of the output has been removed to make the messages easier to read.

Verbs

MGCP command verbs are encoded as four case-insensitive characters that determine whether connections are created, modified, or deleted, or whether some other action occurs. As you can see in Table 6-1, most messages are sent from the call agent to the gateway (GW); this makes sense because the call agent is the brain in the protocol.

Table 6-1 *MGCP Verbs*

MGCP Verb Code	MGCP Verb	Message Flow
CRCX	Create Connection	CA → GW
MDCX	Modify Connection	CA → GW
DLCX	Delete Connection	CA ↔ GW
RQNT	Notification Request	CA → GW
NTFY	Notify	GW → CA
AUEP	Audit Endpoint	CA → GW
AUCX	Audit Connection	CA → GW
RSIP	Restart In Progress	GW → CA
EPCX[1]	Endpoint Configuration	CA → GW

[1]Although the Endpoint Configuration verb is defined by MGCP, it is not used in PacketCable.

Transaction IDs

Transaction IDs (TID) are used to correlate commands and responses. The TID is a string of up to nine decimal digits. Each gateway supports two separate TID name spaces (one for sending transactions and one for receiving transactions). A TID must not be reused more quickly than three minutes from the completion of its previous command; this applies to both call agents and gateways.

Endpoint Name

As discussed earlier in the chapter, endpoints are what the MGCP call agent controls, and they refer to the interface to the VoIP network. An endpoint name consists of two parts, the local name and the domain name. Note that the endpoint name is used in the message regardless of whether the message is going from the call agent to the gateway or from the gateway to the call agent. Both portions of the endpoint name are case-insensitive and can be up to 255 characters in length. Endpoint names are encoded like email addresses, with the @ symbol separating the local name from the domain name. The domain name is typically the FQDN of the gateway under control. The network's DNS server resolves this FQDN into one or more IP addresses.

NOTE The domain name can also be expressed as an IP address, but this is generally discouraged. Not only is this not possible if the gateway has multiple IP addresses, but if the gateway IP address changes, then it must also be changed on all call agents that communicate with the gateway, as opposed to simply updating a DNS entry. Remember, you learned in Chapters 3, 4, and 5 that MTA DNS entries are dynamically updated, so nothing needs to be changed on the call agent if the FQDN is used and an MTA IP address changes.

The endpoint's local name is hierarchical and identifies an endpoint on that gateway. Hierarchical levels are delineated by a slash (/), with the leftmost value the least specific and the rightmost value the most specific. For example, suppose you had a trunking gateway with an OC3 interface, divided into DS3 interfaces, further divided into DS1 interfaces, and finally divided into DS0 endpoints. You could possibly have a local name something like OC3-#/DS3-#/DS1-#/DS0#. Endpoints can be specified using wildcards; the * character indicates an "all" specification, and the $ character indicates an "any" specification.

For example, suppose you have a Voice Interworking Service Module (VISM) card as a trunking gateway with FQDN vism1-mgx1.company-x.com. The endpoint name, *@vism1-mgx1.company-x.com, would refer to all endpoints on that gateway. The endpoint name, $@vism1-mgx1.company-x.com, would refer to any endpoint on that gateway.

Protocol Version

The protocol version is the keyword MGCP followed by a version number (in this case 1.0). An optional MGCP profile name can exist after the version number. This is used in PacketCable for the NCS and TGCP profiles of MGCP.

MGCP Parameters

Each MGCP parameter line is contained on a single line and is of the format *<Parameter code>*: *<Parameter value>*. Mandatory and optional parameters are determined based on the MGCP verb. The use of several of these parameters is covered later in the chapter. Table 6-2 lists the MGCP parameters and their codes.

Table 6-2 *MGCP Parameters, Codes, and Descriptions*

Parameter Code	Parameter Name	Parameter Description
B	Bearer Information	Attribute list describing media stream
C	Call Id	Hex string assigned by call agent; used to identify calls
A	Capabilities	Parameter list of gateway capabilities; used in audit
I	Connection ID	Hex string assigned by gateway; used to identify connections
M	Connection Mode	Connection operation mode
P	Connection Parameters	Connection statistic parameters
T	Detect Events	List of events to detect
D	Digit Map	Text encoding of digit map
ES	Event States	List of events
L	Local Connection Parameters	Call agent specified connection operational parameters
MD	Maximum MGCP Datagram Size	Maximum MGCP datagram size
N	Notified Entity	Identifier of device the gateway needs to notify
O	Observed Events	List of events observed by gateway
PL	Package List	List of packages supported by gateway
Q	Quarantine Handling	Description of how gateway should process quarantined and observed events
E	Reason Code	Numerical reason code
R	Requested Events	List of events the call agent requests the gateway to watch for
F	Requested Information	Identifies call agent requested information in an audit
X	Request Identifier	Identifier used to associate a NTFY and RQNT commands
K	Response Acknowledgement	Code used to confirm the receipt of responses
RD	Restart Delay	Denotes the time before restart
RM	Restart Method	Denotes the type of restart
I2	Second Connection Id	Hex identifier of second connection
Z2	Second Endpoint Id	Identifier of second endpoint

Table 6-2 *MGCP Parameters, Codes, and Descriptions (Continued)*

Parameter Code	Parameter Name	Parameter Description
S	Signal Requests	Signals the gateway should generate
Z	Specific Endpoint Id	Used to specify specific endpoints
RC[1]	Remote Connection Descriptor	Session Description Protocol information of remote endpoint
LC[1]	Local Connection Descriptor	Session Description Protocol information of local endpoint

1 Connection descriptors are encoded using the Session Description Protocol and not in the same format as the other parameters; the parameter codes RC and LC are not actually used. The Session Description Protocol is covered toward the end of this chapter.

MGCP Commands Not Directly Related to Call Control

Not all MGCP commands are used for call control purposes. Some commands are used for informational purposes and for changing the operational state of an endpoint. In fact, these are the commands first used when an MTA initializes.

RSIP Command

In Chapter 4, "Provisioning Flows," you learned about the individual steps involved in MTA provisioning. Remember, the last part of this process is MTA-to-CMS initialization. During this process an MGCP ReStart In Progress (RSIP) command is sent from the gateway to alert the call agent that an endpoint or group of endpoints have been put in service. This command can also be used by a gateway to tell the call agent that an endpoint or group of endpoints have been taken out of service or lost call agent communication. Table 6-3 lists the possible parameters of the RSIP command.

Table 6-3 *RSIP Parameters*

Parameter Name	Code	Mandatory/ Optional	Usage
Reason Code	E	Optional	Code identifying why connection is restarted
Response Acknowledgement	K	Optional	Acknowledges transaction IDs of responses received
Restart Delay	RD	Optional	Delay in seconds; default is null or zero
Restart Method	RM	Mandatory	Type of restart occurring

Values of the Restart Method include

- **graceful**—Endpoint(s) to be taken out of service after the value specified by the Restart Delay parameter. Any established connection(s) should be taken down gracefully. No new calls should be permitted on these endpoints. A null RD parameter indicates the endpoint(s) should be taken out of service after all active calls have naturally completed.
- **forced**—Endpoint(s) abruptly taken out of service; any established connection will be dropped.
- **restart**—Endpoint(s) put back in service after the Restart Delay parameter value.
- **disconnected**—Endpoint has lost communication with the call agent and is trying to reestablish connectivity. The Restart Delay parameter specifies the number of seconds the endpoint has been disconnected. Refer to the "MTA to CMS Communication Loss" section in Chapter 4 for more information.

A sample RSIP command is as follows:

```
RSIP 311482023 aaln/*@x1-6-00-13-10-50-37-F6.cc. cisco.com MGCP 1.0 NCS 1.0
RM: restart
```

The only parameter in this message is the RM: restart parameter, which indicates that all analog line endpoints on this gateway are now back in service. Because no restart delay is included, the endpoints are to be put back into service immediately. Notice the use of the endpoint wildcard to refer to all analog line endpoints.

AUEP Command

The Audit Endpoint (AUEP) command is used to find out information about endpoints. This information includes the current values of MGCP parameters such as Requested Events (R), Signal Request (S), Notified Entity (N), and Digit Map and is requested using the Requested Info parameter. One of the more useful parameters to query an endpoint about is Capabilities (A). The values returned are similar to the local connection options and indicate what the endpoint supports; a list of supported packages and connection modes are also returned. This type of message exchange typically occurs after an MTA initializes so the call agent can learn its status and capabilities. If no information is requested in the AUEP, the gateway simply returns a positive acknowledgement, assuming the endpoint is valid regardless of its operational state. This type of message is often used as a "keepalive" mechanism between the call agent and gateway. Table 6-4 lists the possible parameters of the AUEP command.

Table 6-4 *AUEP Parameters*

Parameter Name	Code	Mandatory/ Optional	Usage
Requested Information	F	Optional	Comma-separated list of parameter codes to be audited
Response Acknowledgement	K	Optional	Acknowledges transaction IDs of responses received

A sample AUEP command is as follows:

```
AUEP 214954705 aaln/1@x1-6-00-13-10-50-37-f6.cc.cisco.com MGCP 1.0 NCS 1.0
F: A
K: 214954685
```

Two parameters are in this message:

- **F: A**—The Requested Information parameter identifies queried parameter codes; in this case the Capabilities parameter is queried, requesting the endpoint's capabilities such as supported CODECs, packetization periods, connection modes, packages, dynamic quality of service (DQoS) support, and packet encryption support.
- **K: 214954685**—The Response Acknowledgement parameter indicating the call agent has received the response with transaction ID 214954685.

AUCX Command

Information pertaining to a connection can be obtained using the Audit Connection (AUCX) command. Like the AUEP command, the Requested Info field is used. The call agent can use this command to find out information such as the current Local and Remote Connection Descriptors (LC and RC), Connection Mode (M), and Connection Parameters (P). Table 6-5 lists the possible parameters of the AUCX command.

Table 6-5 *AUCX Parameters*

Parameter Name	Code	Mandatory / Optional	Usage
Connection ID	I	Mandatory	Identifies the connection being audited
Requested Information	F	Mandatory	Comma-separated list of parameter codes to be audited
Response Acknowledgement	K	Optional	Acknowledges transaction IDs of responses received

Packages, Events, and Signals

Before moving on to the specifics of call control commands, this section visits some other important aspects of MGCP packages, events, and signals.

MGCP is a very modular and extensible protocol; as new functionality is required it can be added through the use of packages. Inside an MGCP package, well-defined extensions, such as events and signals, exist. An event is something a call agent requests a gateway to detect, such as a phone line going off hook or dialed digits. Call agents also request gateways to generate signals such as ring-tone or ring-back tone to be provided to either the endpoint or the VoIP connection.

NOTE Events and signals aren't the only extensions that are defined in packages. Other types of extensions refer to values of MGCP parameters, such as Bearer Information, Local Connection Options, and Connection Modes. Refer to RFC 3435 for more information.

Events and signals share the same name space inside a package. For example, the Line package contains the off-hook and on-hook events denoted by "hd" (hook down) and "hu" (hook up), respectively. The package name is an optional component listed with events and signals; if it is absent, the default package is assumed.

By default, events and signals are detected and generated on endpoints (that is, the analog line itself or the trunk DS0). However, they can be applied to a connection if followed by the @ symbol and the connection identifier.

What to do when an event occurs is denoted by the action options specified in parentheses after the event. Action options and their codes are as follows:

- **N**—Notify immediately (default action)
- **S**—Swap audio (used with call waiting)
- **A**—Accumulate the event in a buffer
- **D**—Accumulate according to digit map
- **K**—Keep signal active
- **E**—Process an embedded RQNT
- **I**—Ignore the event

Signals are categorized into three types:

- **On/Off (OO)**—Once applied, these signals last until they are turned off. An example is the visual message-waiting indicator (vmwi). The values (+) or (-) can be used after the signal to turn signals on/off. If nothing is specified, the default is on.

- **Time-out (TO)**—Signals last for a period of time but can be explicitly turned off as well. Ringback tone is an example of this type of signal. When a TO signal times out, it generates an operation complete (oc) event, which can be notified to the call agent. Time-out signals have a default time-out period that can be altered via provisioning or set as a parameter.
- **Brief (BR)**—The duration of these signals is so short that they stop on their own. For example, the generation of a dual-tone multiple frequency (DTMF) digit.

Making Calls and Call Control Commands

Now that you know what MGCP messages look like, it's time to go into detail about how they are used in the context of actual calls. In PacketCable, calls are originated either ONNET (via an MTA) or OFFNET (via the PSTN). This chapter looks only at calls originated ONNET. Chapter 8 examines OFFNET calls. For analog lines, signaling information is sent across the same pair of wires as the actual voice stream, so MGCP is used to convey call signaling information as well as to manage connections.

RQNT Command

Remember, gateways are dumb devices. Without the call agent they do not know what to look for on their endpoints or how to process the information they receive. The call agent conveys this information to the gateways in the RQNT (Notification Request) message. The Notification Request command instructs the gateway to send notifications based on the occurrence of specified events on an endpoint. Examples of these events include off-hook, on-hook line transitions; digit detection; and tone detection. Table 6-6 lists some key parameters of the RQNT command.

Table 6-6 *RQNT Parameters*

Parameter Name	Code	Mandatory/ Optional	Usage
Digit Map	D	Optional	Defines the format of dialed digits to be detected by the gateway
Notified Entity	N	Optional	FQDN of the device to be notified of the event occurrence
Requested Events	R	Optional	List of events the gateway is expected to detect and report
Request Identifier	X	Mandatory	Used to correlate this request with the Notify command it triggers
Response Acknowledgement	K	Optional	Acknowledges transaction IDs of responses received

continues

Table 6-6 *RQNT Parameters (Continued)*

Parameter Name	Code	Mandatory/ Optional	Usage
Signal Requests	S	Optional	Signals to be applied by the gateway, such as tones
Quarantine Handling	Q	Optional	Specifies how events detected before the arrival of this command are to be processed

Consider the following capture of an RQNT command:

```
RQNT 161101944 aaln/1@x1-6-00-13-10-50-37-f6.cc.cisco.com MGCP 1.0 NCS 1.0
N: 104@mgcp-104CA104.roseville.cc.cisco.com:2728
X: 1F300000001
R: L/hd(N)
S:
Q: discard
K: 161101943
```

Six parameters are in this message:

- **N: 104@mgcp-104CA104.roseville.cc.cisco.com:2728**—The Notified Entity parameter specifies where the Notify command is sent when the requested events are detected. This is the FQDN of the call agent that controls the endpoint. Usually, the call agent FQDN consists of a local name and a domain name separated by the @ symbol. In this example, the local name is 104. Also notice the use of the port number at the end of the FQDN to override the default call agent port number of 2727. Remember that at any point in time an endpoint has exactly one Notified Entity (NE) to where the gateway sends commands. Typically, the gateway is provisioned with an initial NE parameter; this is the case in PacketCable. Remember in Chapter 5, "The MTA Configuration File," the MTA configuration file parameter, pktcNcsEndPntConfigCallAgentId, which identifies the CMS FQDN with which the MTA initially communicates. If no NE value is provisioned or set in the RQNT, the parameter defaults to the source address of the last successful non-audit command. For security reasons, this is strongly discouraged.

NOTE The gateway can still receive commands from other entities; this is an important security issue to keep in mind. Another call agent could potentially "steal" the gateway by sending a RQNT message with an invalid NE. If you secure call signaling through encryption, the chances of such an attack can be greatly reduced. Also PacketCable MTAs generally have an added measure of security by only accepting notified entities from CMS FQDNs that are defined in the MTA configuration file.

- **X: 1F300000001**—The Request ID parameter correlates the future Notify command with the RQNT message.
- **R: L/hd(N)**—The Requested Events parameter is a list of events to be detected by the gateway. The action to take upon event detection is in parentheses after the event. In this case, the event is from the Line Package, denoted by the L package name, and the event is a hook-detect (that is, off-hook). The action to be taken upon detection is to Notify (N) the call agent specified by the NE parameter.
- **S:**—The Signal Requests parameter is a list of signals to be generated by the endpoint. An empty list indicates all currently playing time-out signals will be stopped, and no new ones will be generated.
- **Q: discard**—The Quarantine Handling parameter tells the endpoint what to do about events detected before this RQNT was received. In this case the endpoint is told to discard these events.
- **K: 161101943**—The Response Acknowledgement parameter indicates that the call agent has received the response with transaction ID 161101943. In this case this transaction is an AUEP command issued by the call agent immediately before this RQNT was issued.

NTFY Command

When a gateway detects an event requested by a call agent, it issues an NTFY command to the call agent to let it know. This message is sent to the current Notified Entity for that endpoint. Table 6-7 lists the key parameters of the NTFY command.

Table 6-7 *NTFY Parameters*

Parameter Name	Code	Mandatory/ Optional	Usage
Notified Entity	N	Optional	FQDN of the device to be notified of the event occurrence
Observed Events	O	Mandatory	Events detected on that endpoint
Request Identifier	X	Mandatory	Used to correlate this request with the Notify command it triggers
Response Acknowledgement	K	Optional	Acknowledges transaction IDs of responses received

Consider the following capture of the NTFY command relating to the preceding RQNT command:

```
NTFY 311482028 aaln/1@x1-6-00-13-10-50-37-F6.cc.cisco.com MGCP 1.0 NCS 1.0
N: 104@mgcp-104ca104.roseville.cc.cisco.com:2728
X: 1F300000001
O: hd
```

Three parameters are in this message:

- **N: 104@mgcp-104ca104.roseville.cc.cisco.com:2728** —The Notified Entity parameter identifying the device requesting the notification. Notice it is equal to the parameter in the RQNT if the Notified Entity parameter is in the corresponding RQNT message. If the Notified Entity parameter is not in the corresponding RQNT message, then this parameter is not sent in the NTFY message either.
- **X: 1F300000001**—The Request Identifier parameter correlating the Notify command to the RQNT command.
- **O: hd**—The list of observed events, in this case a hook detect (off-hook). Notice that no package name is included, indicating this event is from the default package, the line package (L) in this case.

Before moving on to the CRCX, look at another RQNT that occurs later in this call:

```
RQNT 161101945 aaln/1@x1-6-00-13-10-50-37-f6.cc.cisco.com MGCP 1.0 NCS 1.0
X: 1F300000002
R: L/hu(N),L/hf(N),L/[0-9#*T],L/oc(N)
D: (0T|00|[2-9]11|[2-9]xx[2-9]xxxxxx|1[2-9]xx[2-9]xxxxxx|0 [2-9]xx[2-
  9]xxxxxx|011xxxxxx.T|01xxxxxx.T|101xxxx|#|*xx|11xx|xxxxxxxxxxxxxxxxxx)
S: L/dl(to=16000)
K: 161101944
```

Amazingly enough you find only five parameters in this message:

- **X: 1F300000002**—The Request Identifier parameter used in the Notify message to correlate it to the RQNT message.
- **R: L/hu(N),L/hf(N),L/[0-9#*T],L/oc(N)**—A list of four Requested Events, all from the Line package. The first two events are for notification of the on-hook (hang up) and hook-flash events. The third event asks for all the digits on a standard telephone (0 through 9, * and #) and a timer expiry (T) to be processed according to the digit map. The digit map is the next parameter in the command. The last event is for notification of the operation complete event. This occurs as a result of the timeout of certain signals, in this case the dial-tone signal.
- **D: (0T|00|[2-9]11|[2-9]xx[2-9]xxxxxx|1[2-9]xx[2-9]xxxxxx|0 [2-9]xx[2-9]xxxxxx|011xxxxxx.T|01xxxxxx.T|101xxxx|#|*xx|11xx|xxxxxxxxxxxxxxxxxx)**— The digit map parameter dictates how the endpoint processes or accumulates digits. We will not decode the entire digit map as surely that would put you to sleep, but we will decode the first five strings. The syntax of the digit map is derived from the UNIX command egrep. It is a list of alternative strings made up of digits, digit ranges, and timer expiries. The gateway attempts to find a shortest possible match from the list. The gateway continues to detect digit events until one of the strings is matched or until it is impossible to match any of the strings. When either of these occurs, the call agent is notified. Please refer to the MGCP RFC 3435 or the PacketCable NCS specification for more detail on digit maps. As with the Notified Entity parameter, the gateway

remembers the last digit map it received; consequently, it is typically not sent for each call. Usually it is provided only after the MTA initializes or after a change in the digit map defined on the call agent.

Getting back to decoding the first five strings in this digit map:

— **0T**—Maps a 0 followed by a timer expiry corresponding to an operator call

— **00**—Maps two 0s in a row, which typically corresponds to an external operator (while the preceding case is an internal operator)

— **[2-9]11**—Matches 211,311,411,...,911 operator call types

— **[2-9]xx[2-9]xxxxxxx**—Matches 10-digit dialing

— **1[2-9]xx[2-9]xxxxxxx**—Matches toll calls where a 1 is dialed followed by a 10-digit number

---

NOTE The Timer expiry value (T) corresponds to one of two timers, either the partial dial timer Tpar or the critical dial timer Tcrit. If at least one more digit is required, the timer Tpar is used; if all that's required is the timer expiry, Tcrit is used. For example, suppose you have dialed the digits 555-123. In this case at least one more digit is required; therefore, Tpar is used. If you had dialed 555-1234, then Tcrit would start because a valid number of digits has been dialed. Both timers are provisionable with default values of 16 seconds for Tpar and 4 seconds for Tcrit. These values can be altered from their defaults by defining the MTA signaling MIBs pktcNcsEndPntConfigPartialDialTO and pktcNcsEndPntConfigCriticialDialTO in the MTA configuration file.

- **S: L/dl(to=16000)**—The Signal Requests parameter indicates dial tone is to be generated out the endpoint; this signal times out after 16 seconds.
- **K: 161101944**—The Response Acknowledgement parameter indicates the call agent has received the response with transaction ID 161101944.

CRCX Command

Skip ahead now a couple of messages to the point where a connection can be created. Don't worry about missing some of the call flow; that information is covered in detail in the next two chapters. Here, just focus on the format of the messages and the parameters inside them.

A connection needs to be created for a media stream to pass between endpoints across the VoIP network. Connections are created on gateways as requested by a call agent. Table 6-8 lists the key parameters of the CRCX command.

Table 6-8 *CRCX Parameters*

Parameter Name	Code	Mandatory/ Optional	Usage
Call ID	C	Mandatory	Identifies the call to which this connection belongs
Connection Mode	M	Mandatory	Mode of operation for this side of the connection
Digit Map	D	Optional	Defines the format of dialed digits to be detected by the gateway
Local Connection Options	L	Optional	Used to direct handling of the connection by the gateway
Notified Entity	N	Optional	FQDN of the device to be notified of the event occurrence
Requested Events	R	Optional	List of events the gateway is expected to detect and report
Request Identifier	X	Optional	Used to correlate this request with the Notify command it triggers
Response Acknowledgement	K	Optional	Acknowledges transaction IDs of responses received
Signal Requests	S	Optional	Signals to be applied by the gateway, such as tones
Remote Connection Descriptor	RC	Optional	SDP information about the endpoint at the other end of the call

Some of the parameters have already been covered with the RQNT and NTFY commands; however, several have not yet been seen and warrant some discussion. Specifically, notice the Requested Events and digit map parameters can also be included in the CRCX. Some call agents include these parameters in this message instead of in the RQNT when establishing calls.

Call ID Parameter

MGCP deals with connections; a call is really the next layer up. In other words, a call is a collection of connections. Consequently, call identifiers have little meaning in MGCP itself. A call ID is assigned by the call agent for each connection it creates and should be the same for all connections in a call. The call ID can be used to correlate connections to a given call for accounting purposes. Note: Connection IDs are assigned by the gateways and sent to the call agent in the response to the CRCX command. For example, in a basic call between two endpoints, each endpoint connection has its own connection ID but a common call ID.

Connection Mode Parameter

Whenever a connection is created, the mode or manner in which data is sent between the IP part of the connection and the endpoint or TDM part of the connection needs to be defined. This is defined by the Connection Mode (M) parameter. Key values include

- **sendonly**—Data is sent from the endpoint out the IP network.
- **recvonly**—Data is sent from the IP network to the endpoint.
- **sendrecv**—Data is sent in both directions.
- **inactive**—No data is sent in either direction.
- **loopback**—Used in maintenance mode for Signaling System 7 (SS7) continuity tests, which are covered in Chapter 8.
- **conttest**—Used in maintenance mode for SS7 continuity tests, which are covered in Chapter 8.

Local Connection Options

Local Connection Options (LCO) allow the call agent to direct the gateway's handling of a connection by influencing parameters such as the following (parameter code included in parentheses):

- (a) CODEC compression algorithm (listed in order of preference).
- (p) Packetization period in milliseconds (a single ms value or range).
- (b) Allowable bandwidth in kilobits per second; this covers payload and header from the IP layer up.
- (t) Type of Service byte value to use in the IP header Precedence/DiffServ field; if not included the gateway uses a default value of 0 unless provisioned otherwise.
- Echo cancellation "on" or "off"; it is recommended the gateway handle this parameter rather than the call agent.
- (s) Silence suppression (Voice Activity Detection) "on" or "off".
- (r) Resource reservation can be "g" (guaranteed), "cl" (controlled load), or "be" (best effort); if set to "g" or "cl", the gateway starts emitting RSVP PATH messages.

In the next two chapters you will see that LCO options differ slightly for the NCS and TGCP variants of MGCP.

Now, consider an actual capture of a CRCX command:

```
CRCX 161101947 aaln/1@x1-6-00-13-10-50-37-f6.cc.cisco.com MGCP 1.0 NCS 1.0
C: 293
L: mp:20;20; 20,  a:PCMU;PCMA;G728,  e:off,  s:off,  t:a0
M: inactive
K: 161101946
```

You find four parameters in this message:

- **C: 293**—The Call Identifier parameter identifying the call of which this connection is a part.
- **L: mp:20;20; 20, a:PCMU;PCMA;G728, e:off, s:off, t:a0**—The Local Connection Options parameter directing the gateway how to handle the connection. In this case, the call agent is instructing the gateway about packetization periods to use, CODECs to use, what to do about echo and silence suppression, and how to mark IP Type of Service bits.
- **M: inactive**—This Connection Mode parameter instructs the gateway how to handle bearer data between the VoIP connection and the analog endpoint. Inactive mode means that no data is sent between the IP connection and the analog endpoint.
- **K: 161101946**—The Response Acknowledgement parameter indicating the call agent has received the response with transaction ID 161101946.

MDCX Command

A Modify Connection (MDCX) message is used to do exactly what it sounds like it will do—modify an existing connection. This command is also sent from the call agent to the gateway. Typical uses for the MDCX command include notifying one endpoint about the endpoint at the other leg of the call enabling the exchange of media packets, modifying the connection mode permitting these packets to be exchanged, and modifying the Local Connection Options. Table 6-9 lists the key parameters of the MDCX command.

Table 6-9 *MDCX Parameters*

Parameter Name	Code	Mandatory/ Optional	Usage
Call ID	C	Mandatory	Identifies the call to which this connection belongs
Connection ID	I	Mandatory	Identifies the connection being modified
Connection Mode	M	Optional	Mode of operation for this side of the connection
Digit Map	D	Optional	Defines the format of dialed digits to be detected by the gateway

Table 6-9 *MDCX Parameters (Continued)*

Parameter Name	Code	Mandatory/ Optional	Usage
Local Connection Options	L	Optional	Used to direct handling of the connection by the gateway
Notified Entity	N	Optional	FQDN of the device to be notified of the event occurrence
Requested Events	R	Optional	List of events the gateway is expected to detect and report
Request Identifier	X	Optional	Used to correlate this request with the Notify command it triggers
Response Acknowledgement	K	Optional	Acknowledges transaction IDs of responses received
Signal Requests	S	Optional	Signals to be applied by the gateway, such as tones
Remote Connection Descriptor	RC	Optional	SDP information about the endpoint at the other end of the call

So moving forward in the call example, you can consider a couple messages and analyze an actual capture of an MDCX:

```
MDCX 161101949 aaln/1@x1-6-00-13-10-50-37-f6.cc.cisco.com MGCP 1.0 NCS 1.0
I: BE8C
C: 293
L: dq-gi:d423,  dq-rr:snrcresv
M: recvonly
K: 161101947

v=0
o=- 161101948 29088 IN IP4 24.34.240.247
c=IN IP4 24.34.240.247
b=AS:82
t=0 0
m=audio 53456 RTP/AVP 0 8 15
a=X-pc-secret:base64:<hex omitted>
a=X-pc-csuites-rtp:62/51 64/51 60/51 60/50
a=X-pc-csuites-rtcp:81/71 81/70 80/70
a=X-pc-nrekey:0
a=mptime:20 20 20
a=ptime:20
```

The following are the parameters in this message:

- **I: BE8C**—The Connection Identifier parameter uniquely identifies this connection on the gateway. Connection IDs are assigned by gateway. This is sent to the call agent in the response to the CRCX command.
- **C: 293**—The Call Identifier parameter identifies the call of which this connection is a part. Notice it's the same call of which the previous CRCX is part.

- **L: dq-gi:d423, dq-rr:snrcresv**—These are LCO options specific to the NCS protocol and PacketCable DQoS; they are covered in detail in the next chapter.
- **M: recvonly**—The Connection Mode parameter instructs the gateway how to handle bearer data between the VoIP connection and the analog endpoint. Notice that this has changed from the CRCX. Receive only dictates that data is sent from the IP connection out the endpoint but not the other way around.
- **K: 161101947**—The Response Acknowledgement parameter indicates the call agent has received the Response with transaction ID 161101947.

Another interesting thing about this message is that it includes the optional SDP information talked about earlier. This is the information about the endpoint at the other end of the call, the Remote Connection Descriptor parameter. This information needs to be exchanged between endpoints so they know how and where to send the media stream. SDP is covered later in this chapter.

DLCX Command

The Delete Connection (DLCX) command is used to terminate a connection. It is the only command that can be sent from the call agent to the gateway or from the gateway to the call agent. A gateway could send a DLCX as a result of losing connectivity to an endpoint. However, sending a DLCX from the gateway is discouraged and should be done only as a last resort. The decision to delete a connection should be at the discretion of the call agent. When a DLCX is sent from the call agent, the response from the gateway contains statistics about the connection. See the "MGCP Responses" section of this chapter for more information. Table 6-10 lists the key parameters of the DLCX command.

Table 6-10 *DLCX Parameters*

Parameter Name	Code	Mandatory/ Optional	Usage
Call ID	C	Optional	Identifies the call to which this connection belongs
Connection ID	I	Optional	Identifies the connection being deleted
Connection Parameters	P	Optional	Call statistic information such as packets sent/received, jitter, and latency
Digit Map	D	Optional	Defines the format of dialed digits to be detected by the gateway
Notified Entity	N	Optional	FQDN of the device to be notified of the event occurrence
Reason Code	E	Optional	Code identifying why connection is to be deleted, used in a DLCX sent from a gateway

Table 6-10 *DLCX Parameters (Continued)*

Parameter Name	Code	Mandatory/ Optional	Usage
Requested Events	R	Optional	List of events the gateway is expected to detect and report
Request Identifier	X	Optional	Used to correlate this request with the Notify command it triggers
Response Acknowledgement	K	Optional	Acknowledges transaction IDs of responses received
Signal Requests	S	Optional	Signals to be applied by the gateway, such as tones

Consider the following capture of a DLCX command:

```
DLCX 161101960 aaln/1@x1-6-00-13-10-50-37-f6.cc.cisco.com MGCP 1.0 NCS 1.0
I: BE8C
C: 293
K: 161101958
```

You can see three parameters in this message:

- **I: BE8C**—The Connection Identifier parameter identifies the connection to be deleted.
- **C: 293**—The Call Identifier parameter identifies the call of which this connection is part. Notice it's the same call that the previous CRCX and MDCX are a part of.
- **K: 161101958**—The Response Acknowledgement parameter indicates the call agent has received the Response with transaction ID 161101958.

MGCP Responses

Recall that MGCP is a command/response protocol. All MGCP commands are acknowledged with responses consisting of two parts, a header and optional response parameters.

A response header is comprised of a two- or three-field response line consisting of the following information:

- Response code
- Transaction ID tying the response to the associated command
- Optional ASCII commentary

A blank line precedes the optional response parameters. Figure 6-2 illustrates a breakdown of the response header.

Figure 6-2 *MGCP Response Header Syntax*

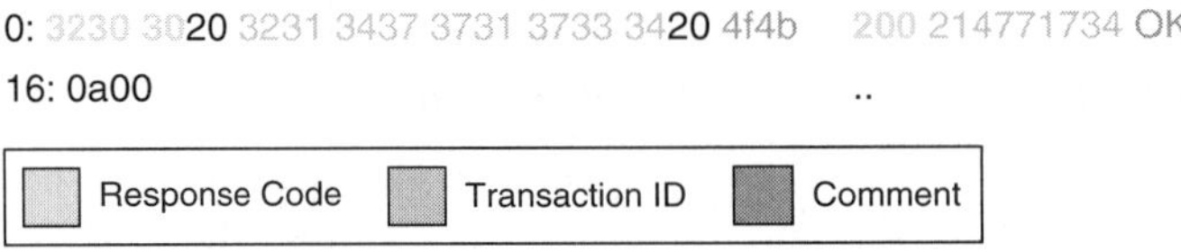

Response Codes and Reason Codes

Response codes (also referred to as Return codes) are three-digit integers that indicate the status of a command. The following ranges of values are defined:

- **000–099**—Values in this range indicate an acknowledgment of a previously received response.
- **100–199**—Values in this range indicate a provisional response, meaning the command is still in progress.
- **200–299**—Values in this range indicate the successful completion of the command; the value 200 is typically seen for all commands other than the DLCX, where the value 250 is used.
- **400–499**—Values in this range indicate a transient error, which is expected to be temporary, such as 401 (already off hook) and 404 (insufficient bandwidth).
- **500–599**—Values in this range indicate a permanent error such as a protocol violation or error.
- **800–899**—Values in this range are specific to a package and defined there.

Reason codes are used in DLCX messages originating from gateways. Remember, normally the call agent should be responsible for tearing down connections. When the gateway is tearing down the connection, this usually corresponds to a problem. Reason codes can also be sent in RSIP messages to convey the reason for the RSIP. Only a small number of reason codes have been defined, as shown in Table 6-11.

Table 6-11 *Reason Codes*

Reason Code	Description
000	Endpoint state is normal
900	Endpoint malfunctioning
901	Endpoint taken out-of-service
902	Loss of lower layer connectivity
903	QoS resource reservation lost
904	Manual intervention
905	Facility failure

NOTE Several of the response codes and reason codes have predefined values and can be found in Section 2.4 of RFC 3435 and RFC 3661—Media Gateway Control Protocol (MGCP) Return Code Usage. Sections 4.5 and 4.6 of the PacketCable NCS specification also contain descriptions of many of the codes used in a PacketCable deployment.

Consider the following captured MGCP response:

```
250 161101960 Connection was deleted
P: PS=321, OS=52002, PR=461, OR=73760, PL=0, JI=0, LA=42, PC/RPS=461, PC/ROS=74682,
  PC/RPL=2, PC/RJI=0
```

The response code is 250, indicating this is a positive response to a DLCX command. The optional comment "Connection was deleted" is included. The optional parameter Connection Parameters (P), containing statistics about the connection, is also included. These statistics include

- Number of packets sent (PS)
- Octets sent (OS)
- Packets received (PR)
- Octets received (OR)
- Packets lost (PL)
- Average packet jitter (JI) in milliseconds
- Average packet latency (PL) in milliseconds

This example also contains additional statistic information specifically related to PacketCable. These additions are covered more in Chapter 7.

NOTE Be aware that these statistics originate from gateways, which are typically untrusted network entities (that is, MTAs). Caution should be taken with the accuracy of these values, and the statistics should be used only to provide an overall picture of network performance and not for accounting purposes.

Additional MGCP Information

This chapter provides an overview of the MGCP protocol, but several topics that have not been covered might still be of interest. RFC 3435 should be referenced for further information.

Some of these topics include

- **The ability to piggyback or concatenate multiple commands and responses in a single message**—Message concatenation improves bandwidth efficiency by reducing the number of packet headers that need to be sent, and it reduces the delay in sending packets. Remember also that MGCP uses UDP, a connectionless transport protocol that does not guarantee packet delivery or order. By piggybacking messages you are also specifying the order in which these messages are processed. Also if the messages are such that it would be undesirable for one to be received without first receiving the other, piggybacking ensures this cannot happen—that is, if one message is lost, they are both lost.
- **The use of provisional responses notifying MGCP entities that commands have been received but not yet executed**—As you'll see more later, this is especially useful in PacketCable. Because it can take some time for DOCSIS resources to be reserved and committed, the MTA can send a provisional response to the CMS to let it know this transaction is still in progress.
- **Buffering of responses and the use of response acknowledgements ensuring the successful delivery of commands and responses**—In the earlier examples you saw the Response Acknowledgement (K) parameter included in several messages. This parameter is used a great deal in PacketCable to let the remote device know it has successfully received the message(s) identified by the transaction ID(s) after the "K" parameter. It is also used to request the remote device to send an acknowledgement of the successful receipt of a response packet; in this case an empty "K" parameter is sent in the response packet, and the remote device responds with another response packet with a response code in the 000 to 099 range (000 is normally used).
- **Considerations and handling of call agent failovers**—In most PacketCable deployments the CMS consists of multiple machines with multiple interfaces capable of processing call signaling traffic. Therefore, the Notified Entity (N) FQDN typically resolves to multiple IP addresses so that if one interface is lost, gateways can still communicate with the call agent. The BTS 10200 uses this type of implementation, although to avoid gateways trying to contact the unoperational standby call agent, virtual IP addresses are used that always exist on the operational active call agent. An alternative to using active and standby call agents is using primary and backup call agents. In the later case, either call agent is capable of servicing gateways, whereas in the previous case only the active call agent is capable. The primary and backup call agents can then use the Notified Entity (N) parameter to redirect gateways to one another to optimally share the load between machines.
- **Quarantine event handling**—Covers how a gateway should process events that have not explicitly been requested by a call agent.
- **Use of experimental packages, events, and signals**—You can find a list of MGCP packages registered with IANA at http://www.iana.org/assignments/mgcp-packages.

SDP Protocol Overview

The MGCP Remote Connection Descriptor (RC) and Local Connection Descriptor (LC) parameters exist at the end of the MGCP parameter list and define the media streams associated with the remote and local endpoints. These media streams are encoded using the Session Description Protocol (SDP) defined in RFC 2327. The SDP is defined for the general multimedia case; this chapter focuses on its use in PacketCable voice.

Like MGCP, SDP is also encoded in ASCII. All SDP parameters are described on a single line using a single case-significant character code, an = sign, and its associated value. Session descriptions consist of two parts, the session-level description, which applies to all included media streams, and optionally several media-level descriptors, each applying to a single media stream.

The session-level descriptor begins with the SDP version parameter (v=), which is always equal to 0. The session-level descriptor continues to the first media-level descriptor, which begins with the media announcement parameter (m=). Additional parameters can then follow that override session-level descriptors if a conflict exists. In other words, if multiple media streams are defined, common parameters can be defined in the session-level descriptor, and specific parameters can be defined in the relevant media-level descriptor. The end of the media-level descriptor is marked by a new media level descriptor, m=, or the end of the session description.

Three mandatory session-level SDP parameters must be included, although they are pretty useless in this application. These parameters have recommended values that are identified in the MGCP RFC. These parameters are

- Owner/creator (o=)
- Session name (s=)
- Time active (t=)

Optional parameters that you might encounter in the session-level descriptor and/or media-level descriptors include the following:

- **Connection Data (c=)**—This parameter must be present in either the session-level descriptor or the media-level descriptor for each media stream. This field contains the network address for the media stream and is of the following format:

  ```
  c=<network type> <address type> <connection address>
  ```

  For PacketCable, *<network type>* is "IN" meaning "Internet" and *<address type>* is "IP4" meaning IP version 4. The *<connection address>* field can be either an IP address or a FQDN.

- **Bandwidth (b=)**—This parameter specifies the bandwidth used by the session or specific media stream in kilobits per second and is of the following format:

  ```
  b=<modifier>:<bandwidth-value>
  ```

In PacketCable the *<modifier>* is always equal to AS, meaning Application Specific maximum.

- **Session Attribute lines (a=)**—Attributes are the primary means for extending SDP to an application. Media descriptors can have any number of attribute lines. Attribute lines are of two possible formats:

  ```
  a=<flag>
  a=<attribute>:<value>
  ```

  The first format is for property attributes, and its presence indicates the attribute is a property of the session (that is, a=recvonly). The second format is for value attributes that set the value of an attribute. A common use of this format is for dynamic RTP payload type definitions.

Media descriptors start with the media announcement parameter, the format for which is as follows:

```
m=<media> <port> <transport> <fmt list>
```

For PacketCable voice, the *<media>* field is always the value "audio." The *<port>* field is the RTP port number used for the bearer stream and should be an even port number in the range 1024 to 65535. The *<transport>* field identifies the transport protocol, which is RTP in this case. The value of this field is "RTP/AVP" (RTP using the Audio/Video profile). The last field is the media format, or RTP payload type in this case. This can be a static defined payload, such as 0 indicating PCM μ-law, or a dynamic payload type requiring a defining attribute line. Refer to the chapters in Part IV, "Media Stream," for more information or the RTP specifications.

Revisiting the MDCX example decoded earlier, you can see a session descriptor to analyze:

```
MDCX 161101949 aaln/1@x1-6-00-13-10-50-37-f6.cc.cisco.com MGCP 1.0 NCS 1.0
I: BE8C
C: 293
L: dq-gi:d423,  dq-rr:snrcresv
M: recvonly
K: 161101947

v=0
o=- 161101948 29088 IN IP4 24.34.240.247
s=-
c=IN IP4 24.34.240.247
b=AS:82
t=0 0
m=audio 53456 RTP/AVP 0 8 15
a=X-pc-secret:base64:<hex omitted>
a=X-pc-csuites-rtp:62/51 64/51 60/51 60/50
a=X-pc-csuites-rtcp:81/71 81/70 80/70
a=X-pc-nrekey:0
a=mptime:20 20 20
a=ptime:20
```

After the MDCX command parameters, there is a blank line followed by the SDP information. The list that follows decodes this line by line.

NOTE This SDP information is the Remote Connection Descriptor describing how to reach the endpoint at the other end of the call.

- **v=0**—The SDP version parameter indicating the start of the session-level descriptor.
- **o=- 161101948 29088 IN IP4 24.34.240.247**—The mandatory owner parameter not used.
- **s=-**—The mandatory session name parameter not used.
- **c=IN IP4 24.34.240.247**—The connection information parameter indicating all media streams are reached using the IP address 24.34.240.247.
- **b=AS:82**—The bandwidth parameter indicating an application-specific maximum amount of bandwidth needed for this session (in this case, 82 kbps).
- **t=0 0**—The mandatory time active parameter not used.
- **m=audio 53456 RTP/AVP 0 8 15**—The media-level descriptor indicating that the remote media stream is accessed using RTP port number 43456 and payload types 0, 8, and 15 corresponding to PCMU (G.711 µ-law), PCMA (G.711 a-law), and G.728.
- **a=X-pc-secret:base64:<hex omitted>**—Attribute parameter specifically used for PacketCable security; this is covered in the next chapter.
- **a=X-pc-csuites-rtp**—Attribute parameter specifically used for PacketCable security; this is covered in the next chapter.
- **a=X-pc-csuites-rtcp**—Attribute parameter specifically used for PacketCable security; this is covered in the next chapter.
- **a=X-pc-nrekey**—Attribute parameter specifically used for PacketCable security; this is covered in the next chapter.
- **a=mptime:20 20 20**—Attribute parameter used in PacketCable to define multiple packetization periods; this is covered in the next chapter.
- **a=ptime:20**—Attribute parameter used to define supported packetization period.

Chapter Summary

PacketCable components such as the Call Management Server (CMS), multimedia terminal adapter (MTA), Media Gateway Controller (MGC), and Media Gateway (MG) are all involved in call processing and, therefore, must communicate call signaling information with each other. The CMS and MTA use the NCS variant of the MGCP protocol for call signaling. The MGC and MG use the TGCP variant of the MGCP protocol for call signaling. A solid understanding of the MGCP protocol is crucial to understanding these two PacketCable protocol variants.

MGCP is a centralized architecture wherein the call agent is the intelligent device and the gateway is the unintelligent device. The call agent handles the bulk of call signaling functionality, such as call setup and teardown, as well as feature implementation. The gateway is dependent upon the call agent to provide guidance. This master/slave relationship allows the slave device to be simple and relatively cheap to operate and maintain. It also permits the MSO to be in tighter control of its network because the call agent is a trusted network element and MTA gateways are generally untrusted network elements.

The basic constructs of MGCP are endpoints and connections. Endpoints refer to the interfaces to external call elements such as analog lines and digital trunks. Connections permit the transfer of media streams between endpoints across the VoIP network used by PacketCable. Two types of MGCP messages exist, commands and responses, both of which are simple to decode and analyze. MGCP commands include call signaling commands such as CRCX (Create Connection), MDCX (Modify Connection), DLCX (Delete Connection), RQNT (Notification Request), and NTFY (Notify). These commands are used to set up, modify, and tear down connections to endpoints, which form call legs. Events and signals are grouped into MGCP packages; they describe things such as digit detection and tone generation. MGCP parameters are used to instruct gateways on what events and signals to detect and generate as well as how to process them.

Not all MGCP commands are used for call signaling; some are used to gather information such as the AUEP (Audit Endpoint) and AUCX (Audit Connection) commands. The RSIP (ReStart In Progress) command is used to notify the call agent of changes in endpoint operational status.

All MGCP commands are positively or negatively acknowledged using MGCP responses. In the response messages are return codes that indicate whether a command was successful and, if not, why it might have failed.

The Session Description Protocol is used to convey information about the media sessions supported by MGCP. Currently in PacketCable only voice media stream support is defined. The format of the SDP protocol also consists of a small number of parameters that describe the RTP audio stream. Session descriptors are divided into two parts: a session-level descriptor that applies to all included media streams and optional media-level descriptors that apply to a single media stream.

Chapter Review

1 How are the PacketCable signaling protocols NCS and TGCP related to MGCP?

Answer: NCS and TGCP are variants of MGCP optimized for PacketCable MTAs and PacketCable MGs, respectively.

2 What are the benefits of a centralized signaling protocol architecture over a distributed architecture from a gateway perspective?

Answer: A centralized architecture allows the gateway component to be simpler and easier to maintain and makes upgrades less necessary when new functionality is introduced as a majority of the complexity is the responsibility of the call agent component.

3 List some functions of a MGCP call agent.

Answer: Control call routing, maintain call state, provide call features, set up and tear down connections on the gateways, and instruct the gateways to generate and detect events

4 What types of MGCP endpoints exist in the PacketCable architecture?

Answer: Analog line endpoints on MTAs, trunk/DS0 endpoints on MGs, and virtual announcement and IVR endpoints on announcement servers

5 What do MGCP connections provide a connection between?

Answer: They facilitate the transfer of media streams between endpoints.

6 Why is packet encryption an important aspect to consider with MGCP?

Answer: Because MGCP is encoded in plain text, which makes it easy for a third party to decode it.

7 Identify the following commands as flowing from call agent to gateway, gateway to call agent, or both:

CRCX

NTFY

RQNT

DLCX

RSIP

Answer: call agent → gateway: CRCX, RQNT

gateway → call agent: NTFY, RSIP

both: DLCX

8 What are the two parts of an endpoint name and how are they formatted?

Answer: The local name, which uses a hierarchical convention to denote where the endpoint exists on the gateway, and a domain name, which is fully qualified and refers to the gateway itself.

9 List some examples of events and signals.

Answer: Example events include an off hook, on hook, flash hook, DTMF digit detection, fax/modem tone detection. Example of signals are dial tone, ring tone, ring back tone, call waiting tone.

10 True or False: An MGCP endpoint can be associated with multiple notified entities.

Answer: False

11 True or False: Transaction identifiers are unique values for each NTFY and RQNT command.

Answer: True

12 True or False: The default action to take when a requested event occurs is to notify the call agent.

Answer: True

13 True or False: An empty signal requests parameter indicates no change be made to the currently playing signals.

Answer: False. This indicates to stop playing any previously started signals.

14 True or False: The digit map string, x.T, matches any number of DTMF digits followed by a timer expiry.

Answer: True

15 True or False: The Local Connection parameter can be used to influence the way a gateway handles a connection.

Answer: True

16 True or False: The Audit Endpoint (AUEP) command can be used to find out whether any existing connections are on an endpoint.

Answer: True

17 True or False: An RSIP with a Restart Method parameter of "graceful" and a Restart Delay parameter of "0" indicates the endpoint should be immediately taken out of service, even if an active connection exists on it.

Answer: False. This means the endpoint should be taken out of service after the active connection is naturally released.

18 True or False: Receiving an MGCP response with a code in the 5xx range could be an indication of a protocol violation.

Answer: True

19 True or False: The SDP parameters Owner "o," Session "s," and Time "t" convey very useful information for PacketCable calls.

Answer: False. These parameters exist in PacketCable calls only to maintain compliance with the SDP specification.

CHAPTER 7

NCS (Network-based Call Signaling)

In Chapter 6, "Signaling Interfaces and MGCP Overview," you learned the importance of call signaling and its place in the PacketCable architecture. You also gained a solid understanding of Media Gateway Control Protocol (MGCP). This chapter provides you with more real-world exposure to call signaling between PacketCable components; specifically, the multimedia terminal adapter (MTA) and call management server (CMS). These components communicate using a profile of MGCP called Network-based Call Signaling (NCS). This chapter examines how NCS varies from ordinary MGCP, as well as how its messages establish calls and provide feature functionality. Finally, this chapter covers how to configure and troubleshoot NCS operation on the Cisco CMS—the BTS 10200. Because this is the first chapter where BTS configuration is covered in any detail, a good portion of this chapter is dedicated to the BTS. This includes how to use the BTS command-line interface (CLI), some basic tips regarding BTS troubleshooting, and BTS configuration beyond setting up a PacketCable MTA, such as general system configuration, dial plan configuration, and call feature configuration.

NCS Variations from MGCP

Recall from Chapter 6 that MGCP supports several different types of endpoints: analog lines, trunking gateways, and announcement servers, to name a few. MGCP is also very flexible because it supports a wide variety of events and signals. The NCS protocol is a "cleaned-up" version of MGCP, supporting only PacketCable embedded MTAs (currently, only voice MTAs).

Aspects of MGCP not needed for supporting analog voice lines have been removed in the NCS profile. Also some extensions to MGCP exist in NCS to support additional PacketCable functionality such as quality of service (QoS) and security.

This section examines the following variations of NCS from standard MGCP:

- Format of the NCS command header.
- NCS endpoint naming convention.
- NCS packages, events, and signals. This includes the FAX package and voice over IP (VoIP) metrics package defined in the PacketCable 1.5 version of the NCS specification.

- Changes to the Local Connection Option (L) parameters relating to CODECs and packetization periods.
- Additions to the Local Connection Option parameters for the support of dynamic quality of service (DQoS).
- Additions to the Local Connection Option parameters for the support of secure media.
- Additions to the Session Description Protocol (SDP) attributes.
- Other modifications.

Command Header Format

The only change to the command header is the inclusion of the NCS keyword and version number after the MGCP keyword and version number. This indicates that NCS is a profile of MGCP.

NOTE The version number for both NCS and MGCP is 1.0.

Endpoint Naming Convention

As you know from what was covered in Chapter 6, an endpoint name consists of two parts, a local name and a domain name in the following format:

<local name>@<domain name>

The domain name is typically the Fully Qualified Domain Name (FQDN) of the MTA, and the local name is a series of hierarchical terms separated by slashes (/), where the leftmost value is the least specific, and the rightmost value is the most specific.

For NCS, the only type of endpoint that exists at this time is analog voice line.

NOTE Support for video endpoints is planned in the future.

The TGCP profile of MGCP supports trunk endpoints; Chapter 8, "TGCP and the PSTN Interconnect," covers TGCP in further detail.

Consequently, unless wildcard characters are used, the leftmost term is always aaln, indicating an analog access line. The local name is made up of at most two terms; the second term indicates the port number of the analog line on the MTA.

NOTE MTA port numbers start from 1 and increment from there; this is different from pure MGCP, where port numbers can start from 0.

So a typical local name of an MTA endpoint is aaln/1, representing line #1 on the MTA.

The wildcard characters $, meaning "any," and *, meaning "all," can also be used in the local name. For example, the local names $, $/$, aaln/$, and aaln refer to any endpoint on the MTA. The local names *, */*, and aaln/* refer to all endpoints on the MTA.

NOTE A common practice in PacketCable deployments is to use a form of the MTA MAC address as the MTA FQDN. For example, suppose you (PacketCable voice provider PC-company) are deploying an MTA with the following MAC address: 000F:66F9:48EA. Then, the MTA FQDN could be x1-6-00-0F-66-F9-48-EA.PC-company.com, and its telephony endpoints would be aaln/1@x1-6-00-0F-66-F9-48-EA.PC-company.com and aaln/2@x1-6-00-0F-66-F9-48-EA.PC-company.com. The endpoint name *@x1-6-00-0F-66-F9-48-EA.PC-company.com refers to all endpoints on the MTA and might be used when the MTA goes in to or out of service.

Packages, Events, and Signals

In PacketCable 1.0 only one package is available in NCS, the Line package denoted by the package name L. The Line package supports a majority of the events, signals, and parameters needed to handle most types of phone calls. Two additional NCS packages are available in PacketCable 1.5: the Fax package, denoted by the package name FXR, and the VoIP metrics package, denoted by the package name XRM.

Logically, the Fax package supports facsimile transmission to and from an MTA over a PacketCable network. By default, fax transmissions are sampled and transferred across a VoIP network in the same manner as ordinary voice. This method is not very efficient or robust; a few lost packets in a voice call in all likelihood go unnoticed, but a few lost packets in a facsimile transmission can have a significant impact on fax quality and transmission speed. The NCS Fax package defines two methods of fax transmission better suited for the PacketCable VoIP environment. One uses the ITU T.38 recommendation for call agent–controlled fax relay, and the other allows the gateways to independently determine the method of fax transmission without involving the call agent. Chapter 10, "Audio CODECs," goes into more detail on fax transmission in a PacketCable network.

The VoIP metrics package provides a means for collecting and reporting statistical information on PacketCable calls. Remember from the previous chapter that when an MGCP connection is deleted, call metric information is returned by the gateway in the DLCX 250 response. This information includes packets sent and received, octets sent and received, the number of packets lost, and jitter and latency measurements. The VoIP metrics package supports additional call metrics that can be used to better gauge call quality. This package is based on the Real-Time Control Protocol Extended Report (RTCP-XR) packet type defined in IETF RFC 3611. RTCP is covered in detail in Chapter 11, "RTP and RTCP."

As you'll learn in this chapter, RTCP is used to send call information between gateways during the life of a call. The type of information includes packets sent and received, percent packet loss, and a measurement of the inter-arrival jitter between received packets. The additional information available with the VoIP metrics package includes jitter buffer overflows, statistics during periods of packet bursts and gaps, delay measurements, signal level measurements, and overall call quality (Mean Opinion Score) calculations.

The Line package is the default package for both PacketCable 1.0 and 1.5 NCS endpoints. Remember that the package name does not need to be specified for events and signals in the default package; for example, the event L/hu is equivalent to the event hu.

Table 7-1 describes some of the more common events and signals used in the Line package; for a full list please refer to Appendix A of the PacketCable NCS specification.

Table 7-1 *Line Package Events and Signals*

Event/Signal Code	Description
0–9,*,#	Dual-tone multiple frequency (DTMF) tones representing the digits on a standard telephone
bz	Busy tone (default timeout is 30 seconds)
ci(ti, nu, na)	Caller ID (ti = time, nu = number, na = name)
dl	Dial tone (default timeout is 16 seconds)
ft	Fax tone
hd	Off hook transition (hook down)
hf	Flash hook
hu	On hook transition (hook up)
mt	Modem tone
mwi	Message waiting indicator (default timeout is 16 seconds)
oc	Operation complete; indicates a signal has timed out
of	Operation failure; typically indicates a failure in a signal generation
ot	Off hook warning tone
rg	Ring tone (default timeout 180 seconds)
ro	Reorder tone (default timeout 30 seconds)
rt	Ring back tone (default timeout 180 seconds)
sl	Slutter tone (default timeout 16 seconds)
vmwi	Visual message waiting indicator
wt1, wt2, wt3, wt4	Call waiting tones

Local Connection Option Parameters for CODECs and Packetization

Recall that one of the Local Connection Option (LCO) parameters is the CODEC compression algorithm (a), which is an ordered list of supported CODECs to be used by the gateway. CODECs describe how voice is converted from the analog stream into digital packets. This topic is covered in detail in Chapter 10. For PacketCable MTAs the supported CODECs for this parameter are restricted to the values outlined in the PacketCable CODEC specifications. For PacketCable 1.0, the only mandatory CODEC is G.711 (Pulse Code Modulation) (a: PCMU/PCMA), and the compression CODECs G.728 (a: G728) and G.729E (a: G729E) are recommended. For PacketCable 1.5, the iLBC (a: iLBC) and BV16 (a: BV16) CODECs are mandatory, in addition to G.711; the G.728 and G.729E CODECs remain recommended. Again, refer to Chapter 10 for much more information on CODECs.

Another LCO parameter closely related to the CODEC used is *packetization period*. This defines the length of time a CODEC is applied to a voice stream to construct a single digital packet. Packetization is also covered in Chapter 10. The way this parameter is presented is slightly different than in pure MGCP.

Chapter 6 introduced the LCO parameter Packetization Period (p). This was defined as either a single value or a range of values both specified in milliseconds. In NCS, this functionality is split into two parameters:

- The Packetization Period (p) that exists in MGCP. However, this parameter now has a restriction that it can be only a single value. This single value would be used for all CODECs specified in the a parameter.
- If different packetization periods are required for different CODECs, a new parameter must be used. This parameter is the Multiple Packetization Period (mp).

These two packetization parameters are mutually exclusive—one or the other is used but never both. The value of the mp parameter is an ordered list of packetization periods in milliseconds corresponding to the list of CODECs denoted in the a parameter. A hyphen (-) can be used in the list to indicate that this value is the same as another in the list.

For example, suppose a CMS instructs an MTA to use either the G.711 or the G.729E CODECs with a packetization period of 20 milliseconds to set up a call, and use of the compression CODEC G.729E is preferred—what would the LCO parameters sent by the call agent look like? The answer is L: p:20, a:G729E;PCMU. Because the same packetization period is used, the p parameter can be sent; also note that the G.729E CODEC is specified before the PCMU (G.711u) CODEC because it is preferred.

Now suppose that, instead, a 10 millisecond packetization period is to be used with the G.729E CODEC, but a 20 millisecond packetization period should still be used for the G.711 CODEC—what would the LCO parameters now look like? In this case, the mp parameter must be sent so the resulting LCO would look like L: mp:10; 20, a:G729E;PCMU. The packetization period value 10 corresponds with the CODEC value G729E, and the packetization period value 20 corresponds to the CODEC value PCMU.

Local Connection Option Parameters for DQoS

Chapter 1, "PacketCable Overview," introduced the PacketCable requirement for a QoS mechanism ensuring voice calls are of a sufficient quality. This requirement includes the capability to create, manage, and control access to QoS services over the shared, and sometimes congested, Data-Over-Cable Service Interface Specification (DOCSIS) network.

In Chapter 2, "PacketCable Functional Components," you learned that the CMTS provides this dynamic quality of service (DQoS) under the direction of the gate controller functionality provided by the CMS.

The DQoS information is relayed to an embedded MTA by the CMS in the NCS messaging, where it is then used to request quality of service. For an embedded MTA, the cable modem requests access to DQoS services on behalf of the MTA from information received in the NCS signaling.

NOTE The managing and policing of DQoS services is performed through communication between the CMS and CMTS and is the topic of Part V of this book, "Quality of Service."

DQoS information is incorporated into the NCS protocol with the Local Connection Option parameters dq-gi, dq-rr, dq-ri, and dq-rd. The following sections explain the use of these parameters.

DQoS Gate ID (dq-gi)

A DQoS gate is an access point to QoS across the DOCSIS network. How gates work will be given much more detail in Chapter 12, "DQoS Architecture and Framework," but for now think of a gate as the key to a premium service between the cable modem (CM) and the CMTS. A 4-byte gate identifier references a gate.

The presence of the gate identifier parameter in an NCS CRCX or MDCX command indicates that DQoS services are to be reserved or committed for the NCS connection. However, this parameter is not mandatory if DQoS has already been performed for the connection in question. For example, if DQoS services are fully committed in a CRCX command, then this parameter would not be needed in subsequent MDCX commands for the same connection. Quality of service resources that are in the reserved state are not yet available for use by the MTA, but they are put aside by the CMTS so that they are available when needed. QoS resources in the committed state are available for traffic.

If this parameter is absent, the actions taken by the MTA depend on that vendor's particular implementation. The MTA could attempt to set up unauthorized QoS resources on the

CMTS (which might or might not permit this request). To permit this functionality on a Cisco CMTS, PacketCable functionality must be disabled, or the command **packetcable authorize vanilla-docsis-mta** must be configured. Another option is that the MTA sends voice traffic in a Best Effort fashion just like ordinary CPE traffic.

DQoS Resource Reservation (dq-rr)

An optional parameter that is typically included along with the gate ID is the resource reservation parameter (dq-rr). This parameter specifies whether the Hybrid Fiber Coaxial (HFC) resources should be reserved and/or committed in both the upstream (send) and downstream (receive) directions. Committed resources are available for immediate use, whereas reserved resources must be committed before they can be used. The parameter has the following values:

- Reserve upstream resources (sendresv)
- Reserve downstream resources (recvresv)
- Reserve resources in both directions (snrcresv)
- Commit upstream resources (sendcomt)
- Commit downstream resources (recvcomt)
- Commit resources in both directions (snrccomt)

The parameter can have multiple values separated by commas. Also if this parameter is absent and the gate ID parameter is present, the default reservation action is to reserve in both directions. The default commit action is determined based on the NCS Connection Mode (M) parameter according to Table 7-2.

Table 7-2 *Default Commit Actions per Connection Mode*

Connection Mode	Commit Action
"inactive"	Do not commit resource in either direction
"send only" or "replicate"	Commit resources in the send/upstream direction
"receive only"	Commit resources in the receive/downstream direction
all others	Commit resources in both directions

NOTE If a reserve action is specified but no commit action is specified, then no resources are committed. If a commit action is specified and the resources have not yet been reserved, the reservation occurs automatically.

DQoS Resource ID (dq-ri)

The optional resource identifier parameter identifies resources reserved or committed at the CMTS. The Resource ID is also a 4-byte field. This parameter can be used in scenarios where resources can be shared between multiple endpoints, such as a Call Waiting scenario. This parameter is assigned by the CMTS when resources are reserved and sent to the MTA in the DOCSIS messaging; the MTA then forwards this parameter to the CMS in the 200 OK response to the CRCX/MDCX command that instructed the MTA to reserve/commit DQoS resources.

DQoS Reserve Destination (dq-rd)

The optional reserve destination parameter contains an IP address and possibly a User Datagram Protocol (UDP) port number to be used when the terminating endpoint is not yet known. The IP address and RTP port number of the terminating endpoint is sent to the originating endpoint using SDP in the Remote Connection Descriptor parameter. If this parameter has not yet been received by the originating endpoint, the reserve destination parameter can be used.

Local Connection Option Parameters for Security

Another PacketCable requirement introduced in Chapter 1 was security. A PacketCable customer's voice conversations traverse a shared DOCSIS network as well as some type of IP network, necessitating a means for ensuring that these conversations are private.

A function of the MTA is to provide this security. The MTA does this by encrypting the bearer packets before sending them across the DOCSIS network. How the MTA does this encrypting is conveyed in NCS signaling messages from the CMS.

Just like DQoS functionality, security functionality is incorporated into NCS using new Local Connection Option parameters. These parameters are as follows:

- **sc-rtp (RTP Ciphersuite)**—A ciphersuite dictates how data is to be authenticated and encrypted; it consists of two terms separated by a slash (/). The first term represents the authentication algorithm, and the second represents the encryption algorithm. Permissible values for both of these terms are covered in the PacketCable security specification.

  The RTP Ciphersuite parameter contains a list of ciphersuites in order of preference to be applied on the RTP bearer packets.

- **sc-rtcp (RTCP Ciphersuite)**—The Real-time Transport Control Protocol (RTCP) conveys statistical information about its associated RTP session. Chapter 11 covers both RTP and RTCP in detail. The RTCP Ciphersuite parameter contains a list of ciphersuites in order of preference to be applied on the RTCP packets.

SDP Attributes

The use of the Session Description Protocol in NCS is, for the most part, the same as its use with MGCP discussed in Chapter 6. The use of the version (v), origin (o), session (s), connection data , bandwidth (b), time (t), and media announcement (m) parameters is as was described with MGCP. However, some additional attributes are used with the attribute (a) parameter:

- **X-pc-codecs**—Contains a list of PacketCable CODECs the endpoint is capable of using.
- **mptime**—Contains a list of support packetization periods. You have one value for each CODEC defined in the <fmt list> entry in the media announcement parameter.
- **X-pc-secret**—Contains the end-to-end encryption key information used to establish RTP and RTCP security.
- **X-pc-nrekey**—Contains a counter for the number of rekey events, used with RTP security.
- **X-pc-csuites-rtp**—Contains the list of supported ciphersuites for RTP security.
- **X-pc-csuites-rtcp**—Contains the list of supported ciphersuites for RTCP security.

NOTE The character X at the beginning of an MGCP, NCS, or SDP parameter, package, event, or signal indicates the value is experimental. The second character indicates whether the message is dropped if this parameter is not understood. An X+ means the message is dropped if the parameter is not known and a negative response is returned. An X- means the parameter should just be ignored and the message should still be processed.

Other Notable Changes

At this point, the majority of the differences between MGCP and NCS have been covered. A couple of other things are worth mentioning, though.

For example, attributes associated with DQoS and security can be audited using the AUEP and AUCX commands, just like any other type of attribute.

Call statistic information obtained from RTCP packets is returned in the 250 response to a call agent–initiated DLCX. Recall that in an MGCP 250 response, information associated with bytes and octets sent and received, as well packet jitter and latency, is returned. In an NCS 250 response, the information obtained from received RTCP reports is returned as well. This information contains packets sent (RPS), octets sent (ROS), packet loss (RPL), and packet jitter information (RJI) as seen by the remote end of the conversation. These metrics are preceded by a PC/ in the response message, indicating they are PacketCable variables.

Examining Call Flows

The best way to understand PacketCable voice is to look at real call flows. You will do this throughout the book, each time focusing on a different aspect of the call. Right now, the NCS signaling used to establish calls between MTAs is examined. These are known as ONNET-to-ONNET calls because both the originating and the terminating party are "on" the PacketCable "network." In Chapter 8, you will examine calls that originate or terminate to OFFNET or PSTN subscribers.

The following are some points to keep in mind about these sample call flows:

- These traces are for calls between MTAs controlled by the same call agent or CMS. Calls between endpoints controlled by different CMSs require the use of the CMS signaling protocol (CMSS) covered in Chapter 9, "Call Management Server Signaling Protocol (CMSS)."
- All the MTAs involved are connected to the same CMTS; as far as NCS is concerned, it doesn't matter whether the MTAs are on the same CMTS or on different ones.
- The PacketCable specifications do not dictate how messages formulate calls; this is up to the vendors of the various components. As long as the specifications are not violated, you might have many ways to accomplish the same task. All the examples in this text utilize the Cisco BTS 10200, running version 4.5, as the softswitch or CMS.
- The BTS configuration used to make these calls possible is covered in the section "Configuring the BTS 10200."

Basic ONNET Call

The first type of call examined is a basic ONNET-to-ONNET call between Oscar (the originating party) and Tina (the terminating party. Both Oscar and Tina are PacketCable subscribers, and their MTAs are referred to as MTA_O and MTA_T, respectively. The examination of their phone call is broken up into three stages:

1. Stage one details the messaging used from when Oscar picks up his phone until the point where Tina's phone is ringing.
2. Stage two shows the messaging from there up to the point where Oscar and Tina have bidirectional voice communication.
3. Stage 3 shows the messaging resulting from Oscar and Tina hanging up their phones.

In addition to the message flows, the details of each packet are shown. These traces were obtained using the UNIX **snoop** command on the active BTS call agent. To make the messages easier to read, the hex portion of the **snoop** output was removed; for more information on **snoop**, refer to Chapter 6.

Basic ONNET Call: Stage One

Figure 7-1 illustrates the NCS signaling used for stage one.

Figure 7-1 *Basic ONNET Call Flow (Stage One)*

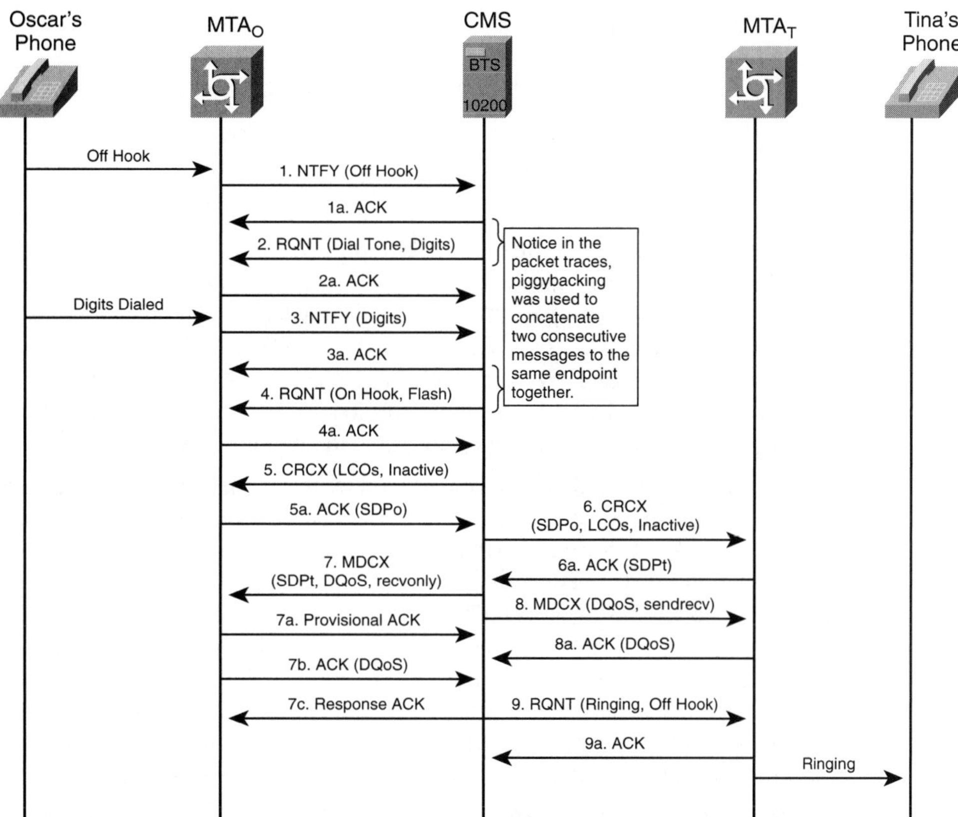

Notice that a number identifies each command and its associated responses. The following examines these one at a time:

1 Oscar picks up his phone, which causes a NTFY message to be sent from his MTA:

```
NTFY 890324736 aaln/1@x1-6-00-13-10-50-38-02.cc.cisco.com MGCP 1.0 NCS
  1.0
N: 104@mgcp-104ca104.roseville.cc.cisco.com:2728
X: 1F100000009
O: hd
```

The off hook is identified by the Observed Event hook-detect (O: hd). Also note the presence of the Notified Entity parameter (N) identifying the CMS device requesting this notification. At some point in the past this endpoint received an RQNT message requesting notification of the off hook event. In that RQNT message was the Request Identifier (X) parameter, which was equal to the Request Identifier parameter in this message. Remember, this parameter correlates NTFY and RQNT messages.

2 The next MGCP message is as follows:

```
200 890324736 OK
.
RQNT 161096006 aaln/1@x1-6-00-13-10-50-38-02.cc.cisco.com MGCP 1.0 NCS
 1.0
X: 1F10000000A
R: L/hu(N),L/hf(N),L/[0-9#*T],L/oc(N)
S: L/dl(to=16000)
```

This message shows an example of MGCP piggybacking as you actually see two MGCP messages concatenated together; a single line consisting of a period separates the two messages. First is the 200 response acknowledgement indicating the previous NTFY message was processed successfully by the CMS. This is message "1a" in the previous call flow figure. In the next message, the CMS requests the MTA to provide Oscar with dial tone and requests to be notified after digits are dialed, if Oscar hangs up his phone or presses the flash button, or if the dial tone timer expires.

The request to provide dial tone is identified by the Signal Event parameter, S: L/dl(to=16000). Remember, the L indicates that the dial tone signal, dl, is part of the NCS Line Package. A timeout value (in milliseconds) is included, so dial tone plays for a maximum of 16 seconds. If this occurs, an operation complete (oc) event is generated; notice this event is included in the Requested Events parameter (R). The Requested Event, L/[0-9#*T], indicates that all DTMF digits, as well as a digit timer expiry, are to be processed according to the endpoint's digit map. The digit map is kept in memory by the endpoint; it was sent by the CMS in a previous transaction. This avoids making MGCP messages unnecessarily larger by having to resend the same digit map repeatedly; of course, if it is changed, it is resent to the endpoint.

The Requested Events, L/hu(N) and L/hf(N), request notification of the on hook and flash hook events, respectively. The other parameter is the Request Identifier (X) to be used in the resulting Notify message if one of these events occurs. The receipt and successful processing of this RQNT message is then positively acknowledged by Oscar's MTA as shown:

```
200 161096006 OK
```

3 Oscar dials Tina's number, triggering another NTFY message to be sent to the CMS:

```
NTFY 890324737 aaln/1@x1-6-00-13-10-50-38-02.cc.cisco.com MGCP 1.0 NCS
 1.0
N: 104@mgcp-104ca104.roseville.cc.cisco.com:2728
X: 1F10000000A
O: 3,2,1,5,0,1,2,0,1,0
```

Notice the format of the digits in the Observed Event parameter. Each digit is specified in the order dialed, one at a time. Had some other requested event occurred during the digit dialing, this would be included here as well.

4 The response for this NTFY and the next RQNT command sent to this endpoint are piggybacked together:

```
200 890324737 OK.
RQNT 161096007 aaln/1@x1-6-00-13-10-50-38-02.cc.cisco.com MGCP 1.0 NCS
 1.0
X: 1F10000000B
R: L/hu(N),L/hf(N)
```

In this RQNT, the CMS issues a new list of requested events consisting of only the on hook and the flash hook events. The acknowledgement to this RQNT is as follows:

```
200 161096007 OK
```

5 Now that the CMS has all the digits it needs to route the call, a connection is initiated on Oscar's MTA:

```
CRCX 161096008 aaln/1@x1-6-00-13-10-50-38-02.cc.cisco.com MGCP 1.0 NCS
 1.0
C: 284
L: mp:20;20;20,  a:PCMU;PCMA;G728,  e:off,  s:off,  t:a0
M: inactive
K: 161095965
```

The CMS creates a Call ID (C: 284) to reference this call that is used for both call legs. The Connection Mode parameter (M: inactive) indicates that no packets are to pass between Oscar's phone and the VoIP network at this time. The CMS also includes the Local Connection Parameter option to influence how Oscar's MTA handles the call. The mp: parameter identifies what packetization periods are to be used for each of the CODECs referenced by the a: parameter. Remember, these parameters are in order of preference, and a one-to-one mapping exists between the two. In this case, the CMS is giving Oscar's MTA three CODECs to choose from with G.711μ (PCMU) being preferred; all CODECs would use a 20-ms packetization period. The other three LCO parameters indicate that echo cancellation (e:off) and silence suppression (s:off) are not to be used for this connection and that the MTA should set the IP ToS byte for voice packets to 0xa0(t:a0).

Oscar's MTA responds to the CRCX with a Connection ID (I: 575) used to reference this newly created connection:

```
200 161096008 OK
I: 575

v=0
o=- 161096008 1397 IN IP4 24.34.240.245
s=-
c=IN IP4 24.34.240.245
```

```
b=AS:82
t=0 0
m=audio 53456 RTP/AVP 0 8 15
a=X-pc-secret:base64:<hex digit string omitted>
a=X-pc-csuites-rtp:62/51 64/51 60/51 60/50
a=X-pc-csuites-rtcp:81/71 81/70 80/70
a=X-pc-nrekey:0
a=mptime:20 20 20
a=ptime:20
```

Also in the response is the SDP information that serves as the Local Connection Descriptor (LCD) and defines the media stream associated with Oscar's endpoint. The SDP information begins with the version field (v), which is always equal to 0. This is followed by the owner field (o) and session name field (s). Remember, these are two of the three fields that are not that useful for PacketCable but are included because they are mandatory in the SDP specification (RFC 2327).

Next is the connection data field , which identifies the IP address of Oscar's MTA, in this case 24.34.240.245. The bandwidth field (b) is next, which specifies that this audio stream requires 82 kbps. The third mandatory but not too useful field follows, which is the time active field (t).

All the SDP information up to this point comprises the session-level descriptor portion. Next is the media-level descriptor, which begins with the media announcement parameter (m). This parameter identifies the media stream as audio using RTP port number 53456. The last part of this parameter contains the supported RTP payload types. Recall that these can be well-defined static values or dynamic values that require a defining attribute parameter. In this case, they are all well-defined static values representing voice CODECs: 0 (G.711μ), 8 (G.711a), and 15 (G.728). The remaining parameters are all attribute parameters (a) that define PacketCable-specific data. The attributes X-pc-secret, X-pc-csuites-rtp, X-pc-csuites-rtcp, and X-pc-nreky all define encryption capabilities. The attribute ptime defines a common supported packetization period for all supported CODECs, whereas the mptime attribute defines a supported packetization period for each CODEC defined in the media announcement parameter.

6 The CMS then requests a connection to be created on Tina's MTA and provides Tina's MTA with information on how to reach Oscar's MTA:

```
CRCX 161096009 aaln/1@x1-6-00-0a-73-fa-be-8c.cc.cisco.com MGCP 1.0 NCS
  1.0
C: 284
L: mp:20;20;20,  a:PCMU;PCMA;G728,  e:off,  s:off,  t:a0
M: inactive
K: 161095980

v=0
o=- 161096008 1397 IN IP4 24.34.240.245
```

```
c=IN IP4 24.34.240.245
b=AS:82
t=0 0
m=audio 53456 RTP/AVP 0 8 15
a=X-pc-secret:base64:<hex digit string omitted>
a=X-pc-csuites-rtp:62/51 64/51 60/51 60/50
a=X-pc-csuites-rtcp:81/71 81/70 80/70
a=X-pc-nrekey:0
a=mptime:20 20 20
a=ptime:20
```

Parameters in this CRCX are similar to the parameters in Oscar's CRCX. Notice that the call ID is the same because this is the terminating leg of the same call. Oscar's SDP information is also forwarded to Tina's MTA so that it knows how and where to forward voice packets. The response to this CRCX is as follows:

```
200 161096009 OK
I: 719E

v=0
o=- 161096009 29086 IN IP4 24.34.240.247
s=-
c=IN IP4 24.34.240.247
b=AS:82
t=0 0
m=audio 53456 RTP/AVP 0 8 15
a=X-pc-secret:base64:<hex digit string omitted>
a=X-pc-csuites-rtp:62/51 64/51 60/51 60/50
a=X-pc-csuites-rtcp:81/71 81/70 80/70
a=X-pc-nrekey:0
a=mptime:20 20 20
a=ptime:20
```

Tina's MTA responds with a Connection ID referring to the connection for this leg of the call. This response also includes the SDP information detailing how to reach Tina's endpoint. For example, the connection data field specifies the IP address of Tina's MTA, in this case 24.34.240.247; the media announcement parameter (m) defines the RTP port used on Tina's MTA, in this case 53456, as well as a list of supported RTP payload types. Notice this list is identical to the one used by Oscar's MTA.

7 The CMS then provides Oscar's MTA with Tina's information by sending an MDCX:

```
MDCX 161096010 aaln/1@x1-6-00-13-10-50-38-02.cc.cisco.com MGCP 1.0 NCS
  1.0
I: 575
C: 284
L: dq-gi:1463,  dq-rr:snrcresv
M: recvonly
```

```
K: 161096008

v=0
o=- 161096009 29086 IN IP4 24.34.240.247
s=-
c=IN IP4 24.34.240.247
b=AS:82
t=0 0
m=audio 53456 RTP/AVP 0 8 15
a=X-pc-secret:base64:<hex digit string omitted>
a=X-pc-csuites-rtp:62/51 64/51 60/51 60/50
a=X-pc-csuites-rtcp:81/71 81/70 80/70
a=X-pc-nrekey:0
a=mptime:20 20 20
a=ptime:20
```

In the MDCX, you see Tina's SDP information is included so Oscar's MTA knows how and where to forward voice packets. Also the mode of the connection has been changed to receive only, allowing packets from the VoIP network to be heard by Oscar. This is primarily done so that in case ring back tone is generated across the PacketCable VoIP network, it can be heard by Oscar. Another key thing to note in this message is the inclusion of LCO parameters relating to DQoS. Before this step, the CMS has communicated with the CMTS to obtain gate information. The gate ID parameter (dq-gi:1463) references the PacketCable gate at the CMTS to be used by Oscar's MTA to obtain quality of service across the DOCSIS network. The resource reservation parameter (dq-rr:snrcresv) indicates resources should be reserved only by the MTA at this point. If Tina is not able to answer the call, tying up valuable resources that could be used by other customers is pointless. This means that if ring back tone were being provided across the VoIP network, these packets could not use the premium DQoS resources and would have to compete with other types of traffic.

Notice in the call flow in Figure 7-1 you have multiple response packets. The first is as follows:

```
100 161096010
```

Recall that response code 100 indicates a provisional response. In the previous MDCX command, the MTA was instructed to initiate the request for DQoS resource reservation on the CMTS. This process can take some time, so in the meantime the MTA lets the CMS know that the MDCX was received and is currently in progress.

As soon as the MTA gets acknowledgement from the CMTS that DQoS resource reservation was successful, it sends a positive response message to the CMTS:

```
200 161096010 OK
K: DQ-RI: 2BF
```

In this response, the LCO DQoS resource ID parameter (DQ-RI: 2BF) identifies the quality of service resources on the CMTS. The inclusion of the response acknowledgement parameter (K) without any value tells the CMS that the MTA requests an acknowledgement of the successful receipt of this response packet. This response acknowledgment packet is as follows:

```
000 161096010
```

As you can see, the use of response acknowledgements provides reliability and insurance that DQoS information is properly transferred between the MTA and CMS.

8 Similarly, the CMS modifies the connection on Tina's MTA:

```
MDCX 161096011 aaln/1@x1-6-00-0a-73-fa-be-8c.cc.cisco.com MGCP 1.0 NCS
 1.0
I: 719E
C: 284
L: dq-gi:5443,  dq-rr:snrcresv
M: sendrecv
K: 161096009
```

In this MDCX, the mode of the connection has been changed to send/receive, permitting the exchange of bearer information between the VoIP network and Tina's phone. DQoS information for Tina's end of the call is included as well. The gate ID (dq-gi:5443) references the gate at the CMTS to be used by Tina's MTA to obtain quality of service. The resource reservation (dq-rr:snrcresv) indicates Tina's MTA should reserve these DOCSIS resources only at this point. So at this point packets could traverse from Tina's MTA to Oscar's MTA, although without DQoS, but packets could not traverse the other direction because Oscar's MTA is in receive-only mode.

The associated response from Tina's MTA is as follows:

```
200 161096011 OK
DQ-RI: 2C1
```

The response parameter (DQ-RI: 2C1) references the quality of service resources on the CMTS to be used by Tina.

9 Now that the CMS knows quality of service resources can be obtained on both ends of the call and connections can be created on both Oscar's and Tina's MTA, it requests Tina's MTA to ring her phone:

```
RQNT 161096012 aaln/1@x1-6-00-0a-73-fa-be-8c.cc.cisco.com MGCP 1.0 NCS
 1.0
X: 1EB0000000C
R: L/hd(N)
S: L/rg(to=0)
K: 161096011
```

Ring tone is requested using the Signal Event parameter, S: L/rg(to=0). The timeout value of 0 indicates that the MTA generates ring tone until instructed not to or until the phone is answered (specified by the R: L/hd(N) parameter). The corresponding response to this message is as follows:

```
200 161096012 OK
```

Basic ONNET Call: Stage Two

Figure 7-2 illustrates the NCS signaling for stage two.

Figure 7-2 *Basic ONNET Call Flow (Stage Two)*

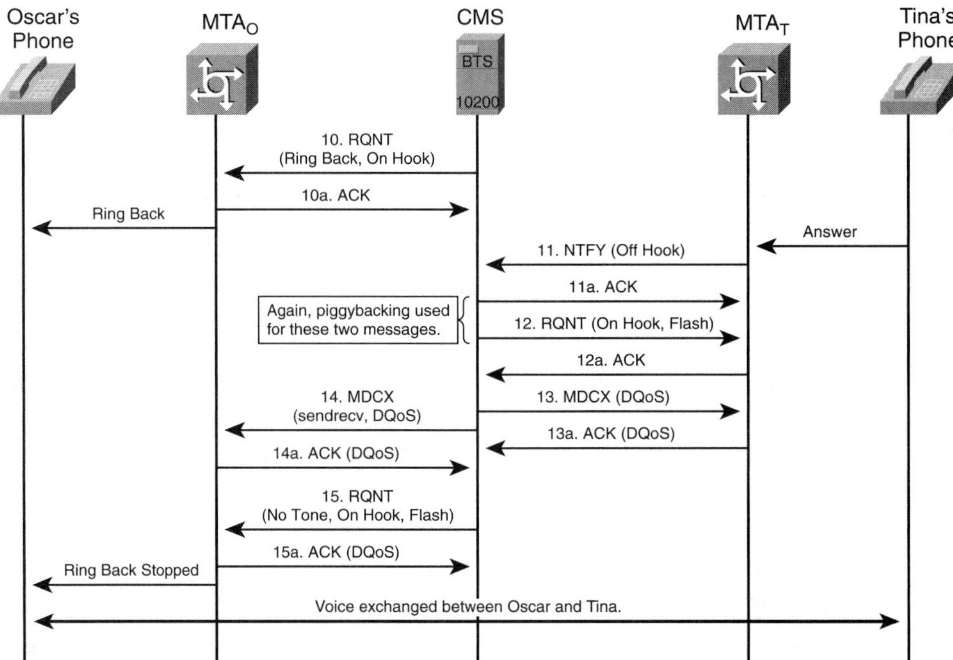

10 The CMS alerts Oscar that Tina's phone is ringing by requesting Oscar's MTA to provide ring back tone:

```
RQNT 161096013 aaln/1@x1-6-00-13-10-50-38-02.cc.cisco.com MGCP 1.0 NCS
 1.0
X: 1F10000000C
R: L/hu(N),L/hf(K,I)
S: L/rt(to=0)
K: 161096010
```

Ring back tone is requested using the Signal Event parameter, S: L/rt(to=0). This tone plays until Oscar's MTA is instructed to stop it or Oscar hangs up his phone (specified by the Requested Event L/hu(N)). Notice the parameters specified with the hook flash event, L/hf(K,I). The combination of the Keep Signal Active action (K) and the Ignore action (I) means that if Oscar presses the flash button on his phone, the CMS is not notified and the ring back tone signal continues to play.

The response to this packet is as follows:

```
200 161096013 OK
```

11 Eventually, Tina answers her phone, triggering a NTFY to be sent from her MTA to the CMS:

```
NTFY 10708 aaln/1@x1-6-00-0a-73-fa-be-8c.cc.cisco.com MGCP 1.0 NCS 1.0
N: 104@mgcp-104ca104.roseville.cc.cisco.com:2728
X: 1EB0000000C
O: hd
```

The answering is indicated by the Observed Event hook detect (O: hd).

12 The response acknowledgement to this NTFY is piggybacked with the CMS request to be notified if Tina hangs up her phone or performs a hook flash:

```
200 10708 OK
.
RQNT 161096014 aaln/1@x1-6-00-0a-73-fa-be-8c.cc.cisco.com MGCP 1.0 NCS
 1.0
X: 1EB0000000D
R: L/hu(N),L/hf(N)
```

The response to this packet is as follows:

```
200 161096014 OK
```

13 Now that Tina has answered her phone, the quality of service resources on the HFC network can be authorized for use. First, take a look at how this is done on Tina's MTA:

```
MDCX 161096015 aaln/1@x1-6-00-0a-73-fa-be-8c.cc.cisco.com MGCP 1.0 NCS
 1.0
I: 719E
C: 284
L: dq-gi:5443,  dq-rr:snrccomt,  dq-ri:2C1
K: 161096012
```

Notice the DQoS parameters in the MDCX. The gate ID and resource ID reference the quality of service previously reserved back in Step 8. The resource reservation parameter is now set to snrccomt, indicating resources can be committed in both the send and receive directions.

The response includes the DQoS resource identifier indicating these DQoS resources are now available for use:

```
200 161096015 OK
DQ-RI: 2C1
```

14 Similarly, QoS resources are modified on Oscar's MTA:

```
MDCX 161096016 aaln/1@x1-6-00-13-10-50-38-02.cc.cisco.com MGCP 1.0 NCS
 1.0
I: 575
C: 284
L: dq-gi:1463,  dq-rr:snrccomt,  dq-ri:2BF
M: sendrecv
K: 161096013
```

The gate and resource previously reserved in Step 7 are referenced, and the resource reservation parameter is now set to snrccomt, committing the DOCSIS DQoS resources in both directions. Also the NCS connection mode has been modified to send/receive, permitting the exchange of voice between Oscar's phone and the PacketCable network. The corresponding response, indicating the DQoS resources now available for use, is as follows:

```
200 161096016 OK
DQ-RI: 2BF
```

15 Because Tina has answered the call, the CMS requests Oscar's MTA to stop providing ring back tone:

```
RQNT 161096017 aaln/1@x1-6-00-13-10-50-38-02.cc.cisco.com MGCP 1.0 NCS
 1.0
X: 1F10000000D
R: L/hu(N),L/hf(N)
S:
K: 161096016
```

The empty Signal Events parameter stops all signals that were previously applied, in this case ring back tone. Also notification of an on hook or flash hook event is requested by the CMS. The corresponding response is as follows.

```
200 161096017 OK
```

At this point, Tina and Oscar are engaged in an active voice call with guaranteed quality of service across the DOCSIS network.

Basic ONNET Call: Stage Three

Figure 7-3 illustrates stage three of the call, which examines the NCS signaling used to disconnect it.

Figure 7-3 *Basic ONNET Call Flow (Stage Three)*

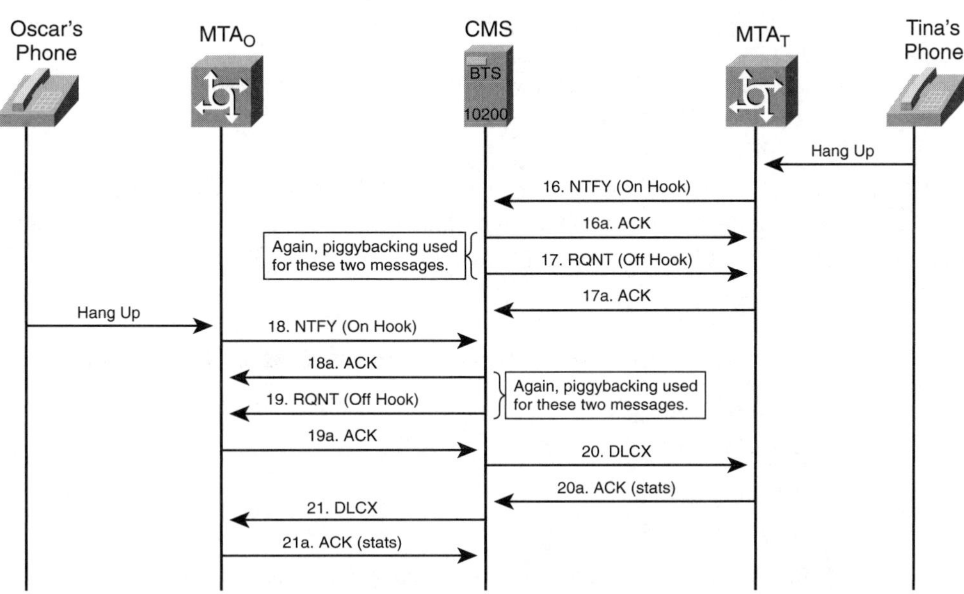

16 Tina decides she is finished talking to Oscar and hangs up her phone, triggering a notification to the CMS:

```
NTFY 10709 aaln/1@x1-6-00-0a-73-fa-be-8c.cc.cisco.com MGCP 1.0 NCS 1.0
N: 104@mgcp-104ca104.roseville.cc.cisco.com:2728
X: 1EB0000000D
O: hu
```

The Observed Event, O: hu, identifies the hang-up event.

17 The response to this NTFY message is piggybacked with the CMS request for Tina's MTA to notify it if Tina indicates she wants to make another phone call by going off hook:

```
200 10709 OK
.
RQNT 161096018 aaln/1@x1-6-00-0a-73-fa-be-8c.cc.cisco.com MGCP 1.0 NCS
  1.0
X: 1EB0000000E
R: L/hd(N)
```

The corresponding response is as follows:

```
200 161096018 OK
```

18 Realizing that Tina has hung up her phone, Oscar hangs up his phone as well:

```
NTFY 890324738 aaln/1@x1-6-00-13-10-50-38-02.cc. cisco.com MGCP 1.0 NCS
 1.0
N: 104@mgcp-104ca104.roseville.cc.cisco.com:2728
X: 1F10000000D
O: hu
```

19 Similar to Step 17, the CM requests Oscar's MTA to notify it if Oscar takes his phone off hook. Also similar to that step, this message is piggybacked to the response message for the previous step:

```
200 890324738 OK
.
RQNT 161096019 aaln/1@x1-6-00-13-10-50-38-02.cc.cisco.com MGCP 1.0 NCS
 1.0
X: 1F10000000E
R: L/hd(N)
```

The corresponding response is as follows:

```
200 161096019 OK
```

20 Now that the CMS knows both parties have disconnected their phones, it deletes the connections on the MTAs. First, the connection on Tina's MTA is deleted:

```
DLCX 161096020 aaln/1@x1-6-00-0a-73-fa-be-8c.cc.cisco.com MGCP 1.0 NCS
 1.0
I: 719E
C: 284
K: 161096015
```

In the DLCX command, the identifier of the call and the connection on Tina's MTA are included. In the response to the DLCX, the Connection Parameters (P) is included, providing the CMS with statistics about the call:

```
250 161096020 Connection was deleted
P: PS=707, OS=114534, PR=575, OR=92000, PL=2, JI=0, LA=49, PC/RPS=480,
PC/ROS=77760, PC/RPL=1, PC/RJI=0
```

This response includes packets sent/received, jitter, and latency information. Remember, the information received from Oscar's MTA in RTCP reports is included in the response as well.

21 Finally, the connection on Oscar's MTA is deleted:

```
DLCX 161096021 aaln/1@x1-6-00-13-10-50-38-02.cc.cisco.com MGCP 1.0 NCS
 1.0
I: 575
C: 284
K: 161096017
```

Call statistic information, as seen from Oscar's MTA, is reported to the CMS, as well as information received from RTCP reports originated from Tina's MTA:

```
250 161096021 Connection was deleted
P: PS=579, OS=93798, PR=706, OR=112960, PL=1, JI=197, LA=49, PC/RPS=609,
PC/ROS=98658, PC/RPL=2, PC/RJI=0
```

Feature Calls

Now that you have a good understanding of how a basic ONNET-to-ONNET call works, you can examine some calls that involve a bit more complexity. The level of detail covered in these examples won't be as much as in the previous example; instead, the focus is on aspects of NCS not previously seen.

Three PacketCable subscribers are involved in these examples: Alan, Bob, and Cathy. Their MTAs are denoted as MTA_A, MTA_B, and MTA_C, respectively. Just as in the previous example, the call is broken up into stages. Because more feature functionality is involved, you find more stages. Also instead of looking at the details of the packets with the UNIX **snoop** command, you are introduced to another method of packet sniffing—the Ethereal packet sniffer utility. In these examples, the WAN or upstream interface on the CMTS is connected to a Cisco 7600 router via a Gigabit Ethernet link. A PC running Ethereal is also connected to this 7600 router; the span monitoring feature is enabled on the 7600, which allows all packets to and from the CMTS to be replicated to the PC. For more information on the span feature, refer to the Cisco online documentation.

NOTE Appendix B, "Cable Monitor and Ethereal," covers Ethereal as well as the Cisco CMTS cable monitor feature in more detail.

The Cable Monitor feature enables you to see DOCSIS packets as seen by a cable interface and copies these packets out an Ethernet port to a protocol sniffer for analysis. In Ethereal, a filter was then applied to show only the MGCP/NCS messages for these three MTAs.

Feature Call #1: Caller ID and Call Waiting

The first feature call is made up of seven stages. The specific features in this case are caller ID and call waiting. Figure 7-4 illustrates the NCS call flow for the first stage.

Figure 7-4 *Feature Call #1: Caller ID and Call Waiting (Stage One)*

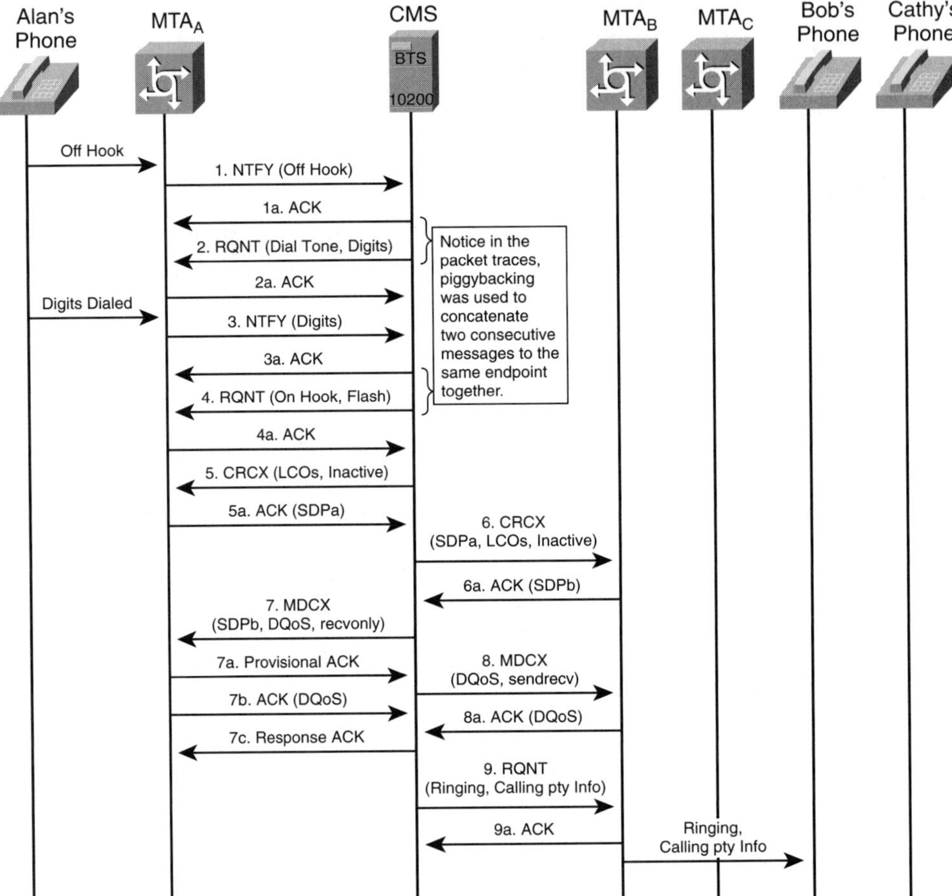

The first eight steps are really no different than the first eight steps in the basic call example:

1. Alan takes his phone off hook, triggering a NTFY message to be sent to the CMS.
2. The CMS then requests dial tone be provided to Alan and asks to be notified when digits are collected according to the MTA's associated digit map for that endpoint. The CMS also requests to be notified if Alan goes on hook or presses the flash button or if the dial tone signal times out.
3. As soon as Alan hears dial tone, he dials Bob's phone number, which is forwarded to the CMS in another NTFY message.
4. The CMS then issues a RQNT to Alan's MTA asking to be notified if Alan hangs up his phone or issues a flash hook.

5 The CMS realizes it can route the call based on those digits and creates an inactive connection on Alan's MTA. The response from the CRCX contains SDP information such as the IP address and RTP port number used to reach this endpoint, as well as other bearer-related information such as supported CODECs and packetization periods.

6 The CMS does the same thing to Bob's MTA, except this CRCX also contains Alan's SDP information. Logically, the response to this CRCX contains SDP information about the endpoint on Bob's MTA.

7 After consulting the CMTS to obtain gate information, the CMS forwards this information to the MTAs using MDCX messages. First, the connection on Alan's MTA is modified. The MTAs then contact the CMTS to create the quality of service pipe. At this time, the DQoS resources are only in the reserved state. In the response to the MDCX messages, the DQoS resource identifier returned by the CMTS is passed along to the CMS. As you can see in Figure 7-4, provisional responses and response acknowledgements ensure this DQoS information is accurately transferred between the MTA and CMS. Remember, this MDCX is also used to relay Bob's SDP information to Alan's MTA.

8 Similarly, the connection on Bob's MTA is modified.

9 Step 9 is where some additional feature functionality is seen; Figure 7-5 shows the Ethereal capture of these first nine steps, with Step 9 highlighted.

Figure 7-5 *Ethereal Capture of Steps 1–9*

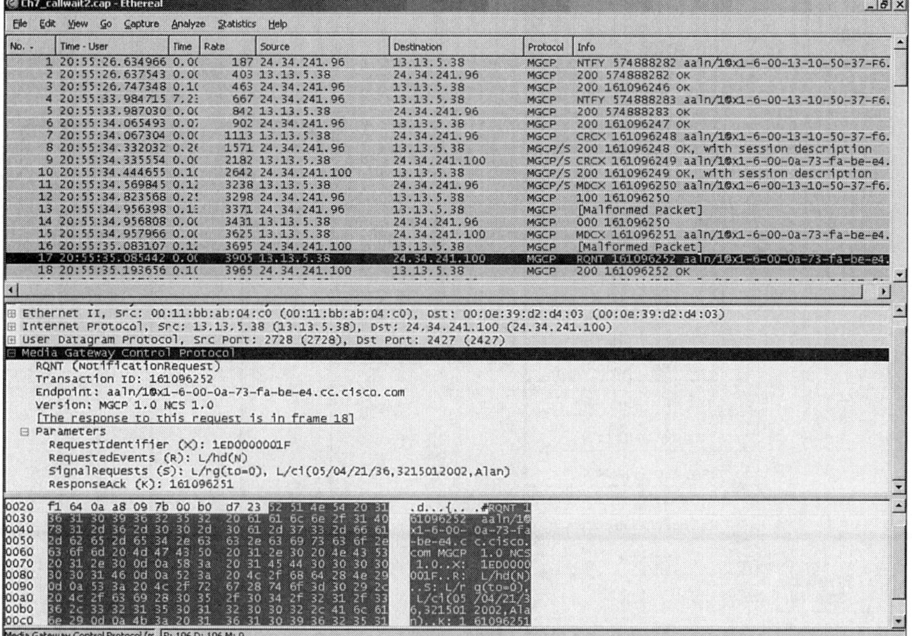

In this step, the CMS sends a RQNT to Bob's MTA to ring Bob's phone. The following is a closer look at this RQNT message:

```
RQNT 161096252 aaln/1@x1-6-00-0a-73-fa-be-e4.cc.cisco.com MGCP 1.0 NCS
 1.0
X: 1ED0000001F
R: L/hd(N)
S: L/rg(to=0), L/ci(05/04/21/36,3215012002,Alan)
K: 161096251
```

In addition to the ring tone signal you find another signal, L/ci(05/04/21/36,3215012002,Alan). This is the caller ID signal. It has three parameters: time, number, and name. The time parameter denotes the time of the call; in this case, May 4th at 9:36 p.m. (2136). The number parameter identifies Alan's phone number, in this case 321-501-2002. The name field is set to a value Alan, which indicates the name of the calling party. Note: Because the switch used in this case is in a lab environment, the phone numbers and names were chosen arbitrarily. In the real world, the name would typically indicate both the first and last name of the calling party.

Figure 7-6 illustrates the NCS call flow for the second stage. Notice these messages are almost identical to the basic call example as well.

Figure 7-6 *Feature Call #1: Caller ID and Call Waiting (Stage Two)*

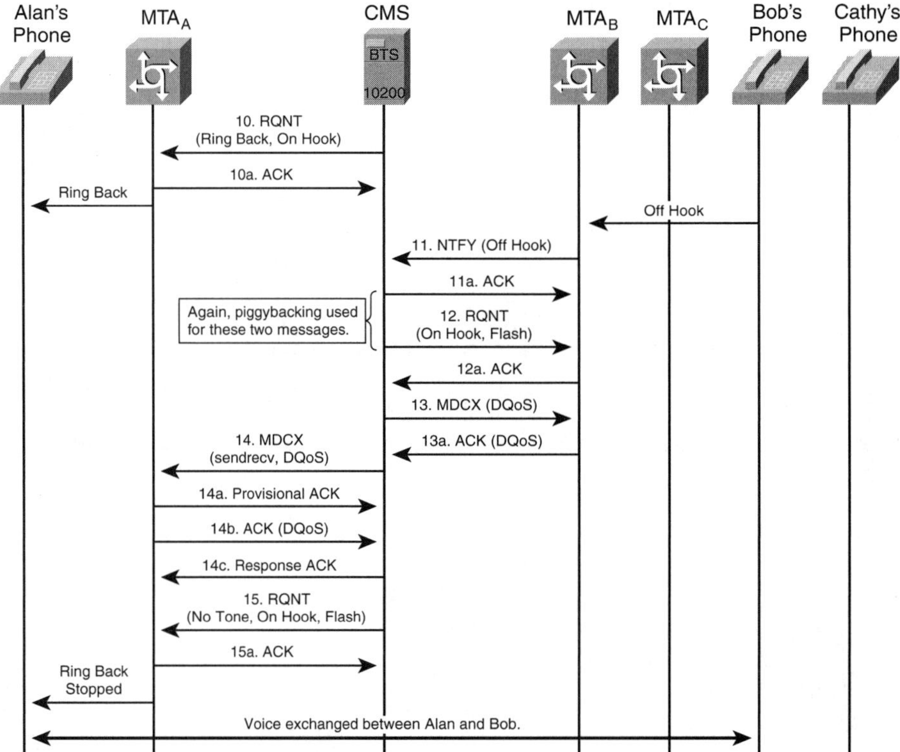

10 The CMS requests Alan's MTA to provide ring back tone and to be notified if Alan hangs up.

11 Bob answers his phone, triggering a NTFY with the hook detect observed event to be sent to the CMS.

12 The CMS then requests Bob's MTA to notify it if Bob hangs up his phone or does a hook flash.

13 Knowing that the call has been answered, the CMS instructs both MTAs to enable the exchange of voice packets. First, Bob's MTA is instructed to activate its DQoS resources. Again, Bob's MTA requests a response acknowledgment to the MDCX response with the DQoS resource identifier.

14 Similarly, the CMS instructs Alan's MTA to activate its DQoS resources. The mode of the NCS connection is also modified to send/receive. Notice again the use of provisional responses and response acknowledgements to ensure DQoS information is accurately transferred between the MTA and CMS.

15 Next, the CMS requests Alan's MTA to stop providing ring back tone and to notify it if Alan hangs up his phone or does a hook flash.

At this point Alan and Bob are involved in a normal phone conversation. As seen in the third stage of this call, Alan is not the only one interested in talking to Bob. Cathy decides to make a call of her own, as seen in Figure 7-7.

Figure 7-7 *Feature Call #1: Caller ID and Call Waiting (Stage Three)*

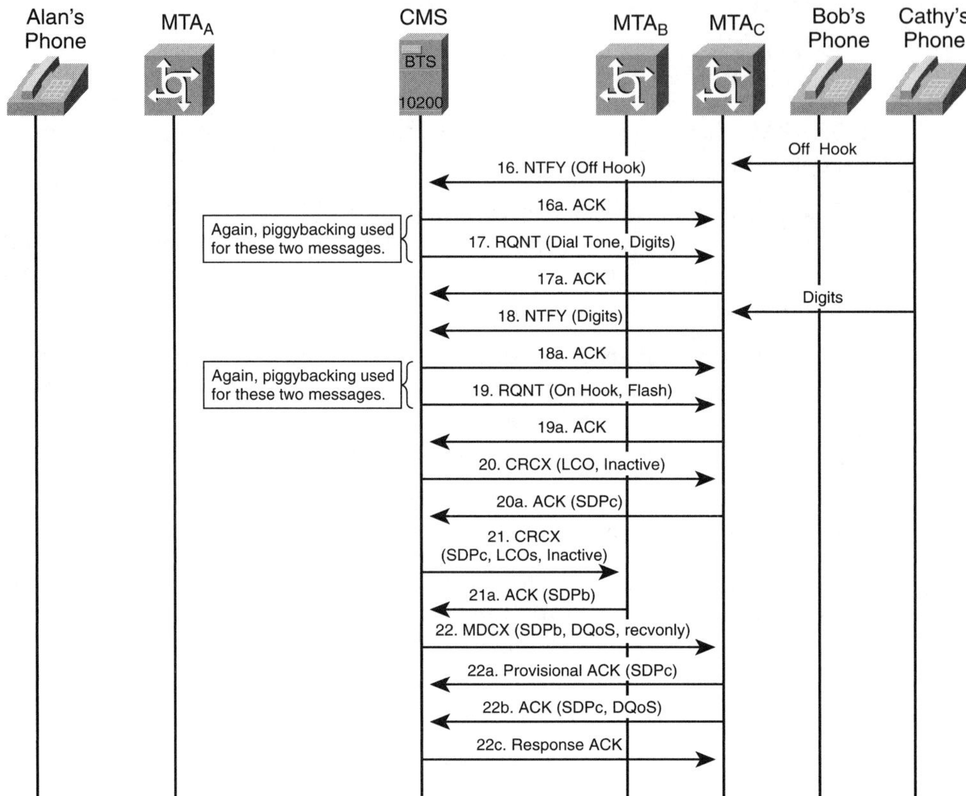

16 Cathy picks up her phone, triggering a NTFY to be sent to the CMS.

17 The CMS instructs Cathy's MTA to provide dial tone and to collect digits.

18 Cathy dials Bob's number, triggering another NTFY to the CMS.

19 CMS requests to be notified if Cathy hangs up or does a hook flash.

20 The CMS realizes it can route the call, so it creates an inactive connection on Cathy's MTA and provides the connection options to be used. The SDP information about how to reach Cathy is returned in the acknowledgement response.

21 An inactive connection for this conversation is also created on Bob's MTA. Cathy's SDP information is included, along with the connection options. Remember, Bob already has an active connection he is using to talk to Alan. In the response to the CRCX, Bob's SDP information to be used by Cathy is returned.

22 After consulting the CMTS to obtain gate information for Cathy's QoS, the CMS modifies her connection with this information and Bob's SDP information and changes the connection mode to receive only. Cathy's MTA first responds with a provisional response letting the CMS know that the DQoS reservation is currently in progress. Her MTA also resends her SDP information in this response. As soon as her MTA gets the DQoS confirmation from the CMTS, it sends a 200 positive response; the DQoS resource identifier is returned, along with the SDP information. Then the CMS sends a response acknowledgement confirming the successful receipt of this 200 response message.

Recapping, Alan and Bob are still involved in a call, and Cathy is in the process of calling Bob as well. As you might have guessed, this call includes call-waiting functionality illustrated in Figure 7-8.

Figure 7-8 *Feature Call #1 : Caller ID and Call Waiting (Stage Four)*

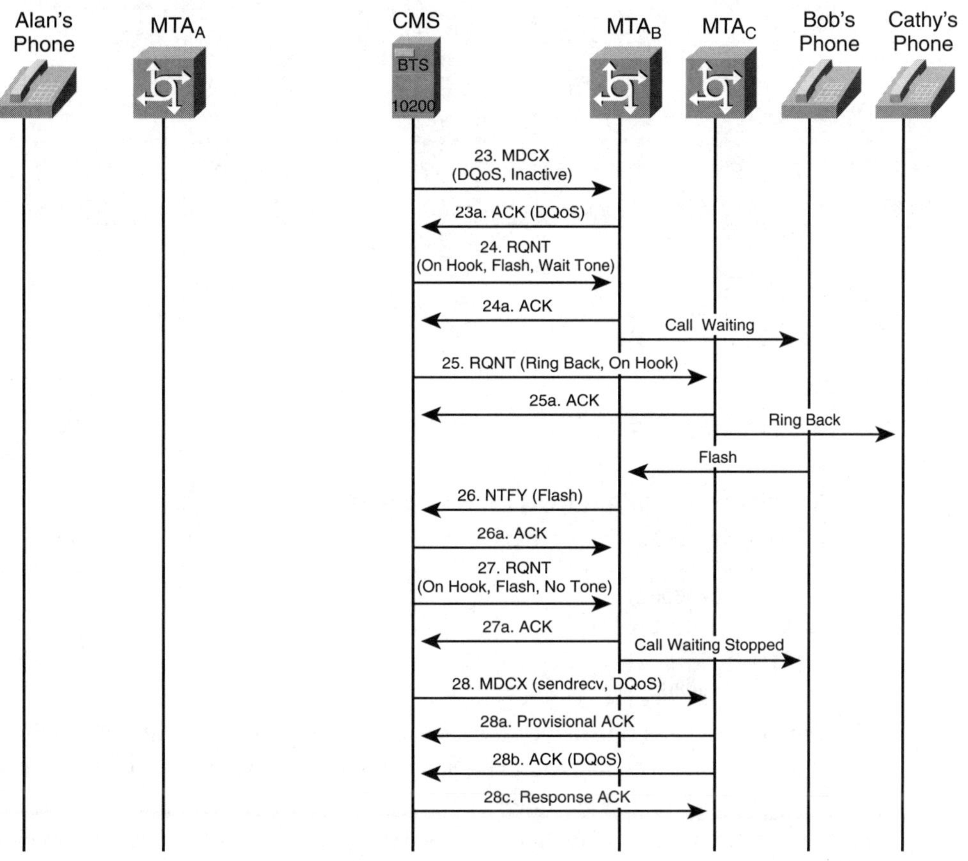

23 Similar to Step 22, the CMS passes along DQoS information to Bob to be used for the call with Cathy. The DQoS resource identifier is returned in the response.

24 This step is where you see the call-waiting functionality. The CMS sends a wait tone to Bob to alert him someone else is trying to reach him. You can see this in the Ethereal capture in Figure 7-9.

Figure 7-9 *Ethereal Capture of Call Waiting*

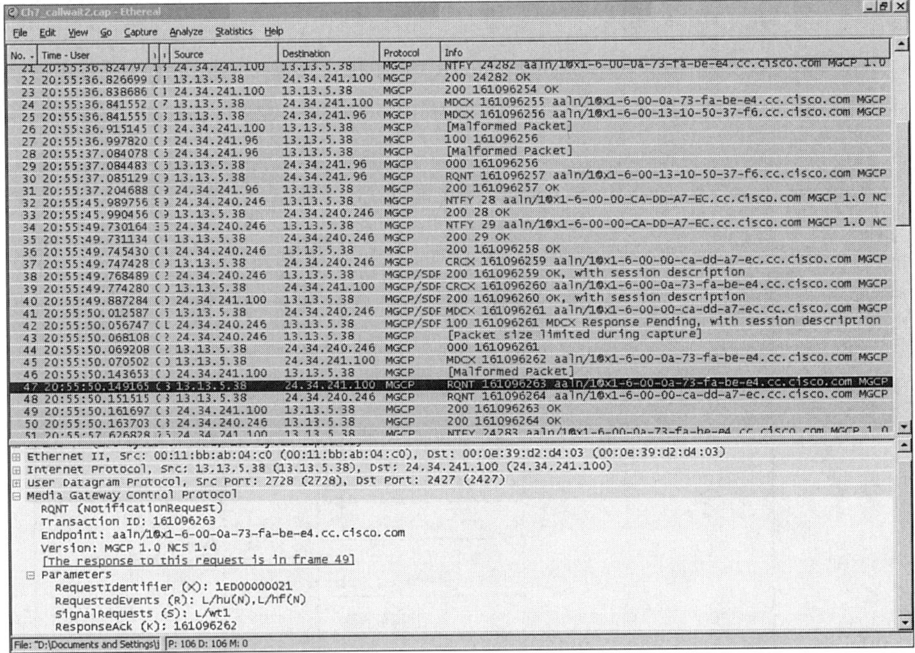

A close-up view of this RQNT message is as follows:

```
RQNT 161096263 aaln/1@x1-6-00-0a-73-fa-be-e4.cc.cisco.com MGCP 1.0 NCS
 1.0
X: 1ED00000021
R: L/hu(N),L/hf(N)
S: L/wt1
K: 161096262
```

The Line package signal, wt1, is requested, which is the call-waiting tone signal. Also notice the requested events: on hook and hook flash. As you'll see in a little bit, the hook flash is how a user toggles between calls.

25 The CMS informs Cathy that Bob has been made aware of her call by providing her with ring back tone.

26 Upon hearing the call-waiting tone, Bob decides to find out who it is and presses the flash button to change calls. This triggers a NTFY to the CMS with the hook flash observed event:

```
NTFY 24283 aaln/1@x1-6-00-0a-73-fa-be-e4.cc.cisco.com MGCP 1.0 NCS 1.0
N: 104@mgcp-104ca104.roseville.cc.cisco.com:2728
X: 1ED00000021
O: hf
```

27 Knowing that Bob has issued a flash hook, the CMS turns off the call waiting tone provided by Bob's MTA.

28 The CMS then instructs Cathy's MTA to modify its connection to enable bidirectional communication and tells the MTA to activate its DQoS connection. Again, provisional responses and response acknowledgements are used by Cathy's MTA.

So now that the connection on Cathy's MTA is ready, the CMS needs to modify the connections on Bob's MTA, as illustrated in Figure 7-10.

Figure 7-10 *Feature Call #1: Caller ID and Call Waiting (Stage Five)*

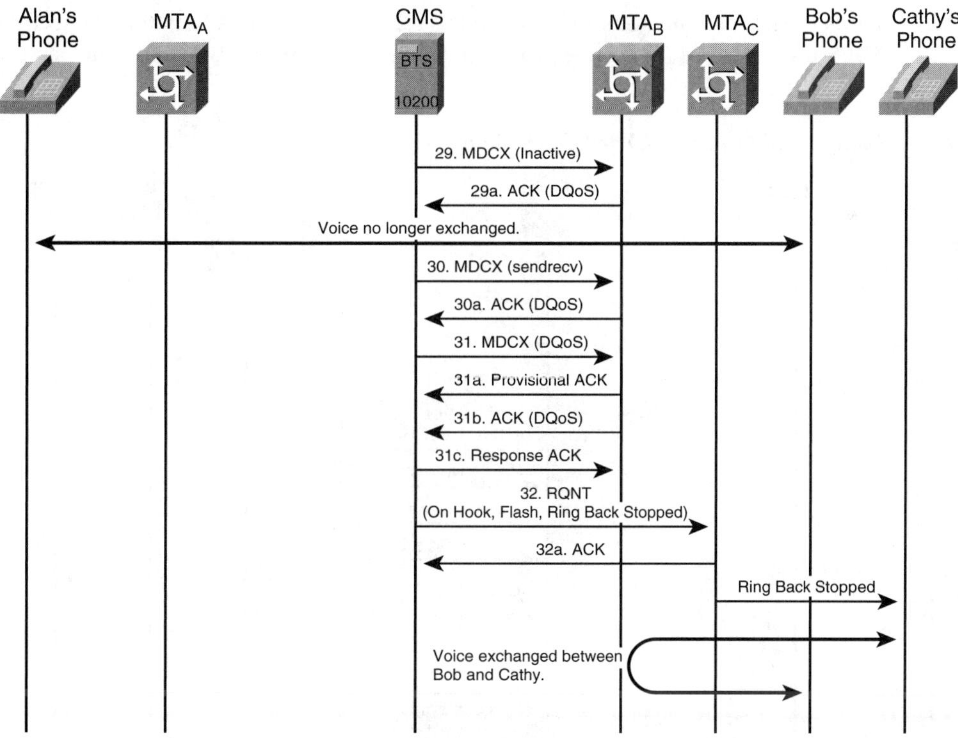

29 First, the CMS deactivates the connection on Bob's MTA used to communicate with Alan. After this point, Bob and Alan can no longer communicate. To do this, an MDCX is sent with the connection mode changed to the "inactive" state. This not only deactivates the MGCP connection on Bob's MTA used to communicate with Alan, but also triggers DOCSIS messaging to the CMTS to change these DQoS resources from the active/committed state back to the admitted/reserved state.

30 The CMS then activates the connection on Bob's MTA used to communicate with Cathy.

31 Then the CMS instructs Bob to activate the DQoS resources to be used for this conversation. Provisional responses and response acknowledgements are used by Bob's MTA.

32 Now that all the resources are set up for Bob and Cathy to communicate, the CMS instructs Cathy's MTA to stop providing ring back tone, which alerts her that Bob has answered the phone and is ready to talk.

Notice that the connection between Bob and Alan still exists—it was just put into inactive mode on Bob's MTA, and DQoS resources for this connection on Bob's MTA were changed back into the admitted/reserved state. On Alan's MTA both the connection and DQoS resources are still active. Thus the connection between Bob and Alan can easily be reestablished, as shown in Figure 7-11.

Figure 7-11 *Feature Call #1: Caller ID and Call Waiting (Stage Six)*

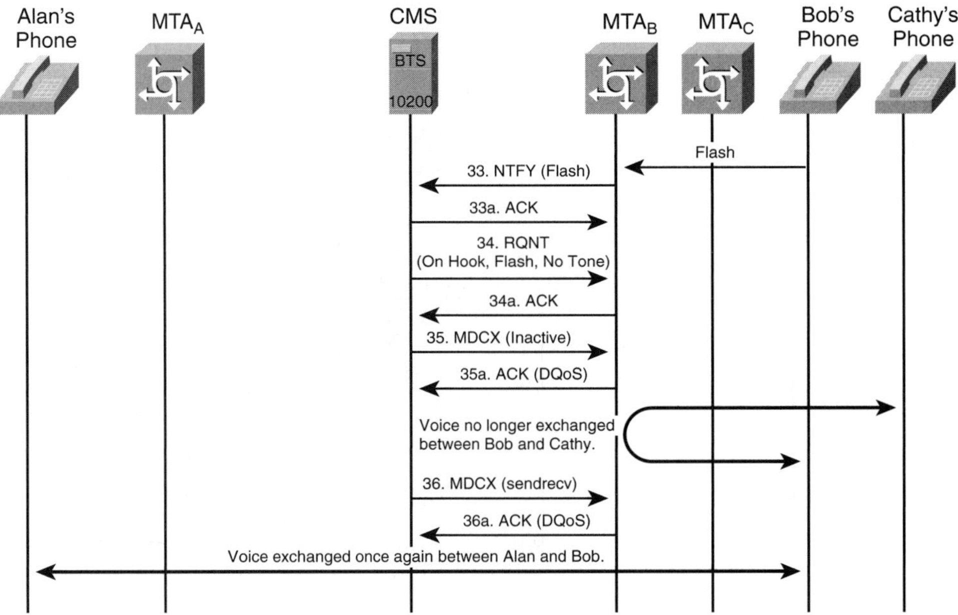

33 Bob presses the flash button on his phone again to stop talking to Cathy and resume his conversation with Alan. Thus a NTFY with the hook flash observed event is sent to the CMS.

34 The CMS again asks to be notified if Bob hangs up or does another flash hook.

35 This time, the connection used to talk with Cathy is put into inactive state on Bob's MTA. So Bob and Cathy can no longer communicate. Likewise, Bob's DQoS resources used to communicate with Cindy are put in the admitted/reserved state.

36 The connection used to talk with Alan is put back into send/receive mode so that Bob and Alan can once again communicate. This triggers the activation of Bob's DQoS resources used to communicate with Alan.

Figure 7-12 illustrates the final stage of this call, which looks at how the parties disconnect.

Figure 7-12 *Feature Call #1: Caller ID and Call Waiting (Stage Seven)*

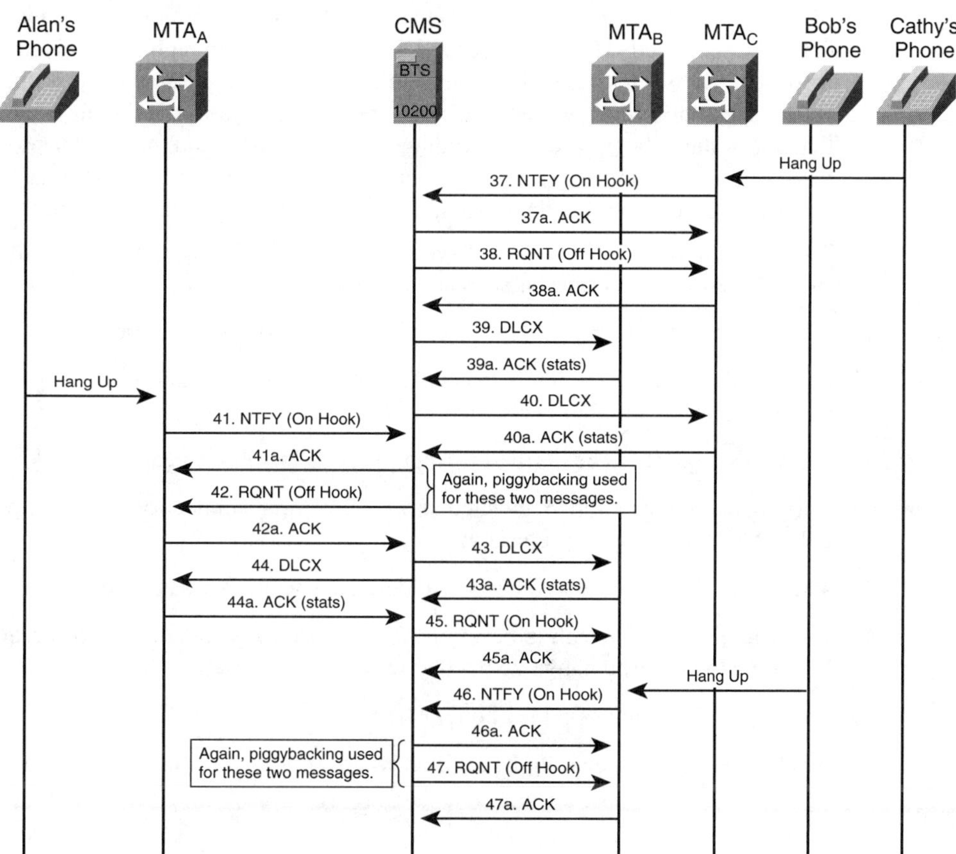

37 Because Bob decided to stop talking to Cathy in favor of Alan, Cathy hangs up her phone, triggering a NTFY to the CMS.

38 The CMS asks Cathy's MTA to let it know if she takes her phone off hook again. The RQNT sent in this case is as follows:

```
RQNT 161096277 aaln/1@x1-6-00-00-ca-dd-a7-ec.cc.cisco.com MGCP 1.0 NCS
 1.0
N: 104@mgcp-104CA104.roseville.cc.cisco.com:2728
X: 1FB0000001C
R: L/hd(N,E(R(L/[0-9#*T],L/hu(N),L/hf(N),L/oc(N),L/of(N)),S(L/dl)))
Q: process, loop
K: 161096276
```

Notice the Requested Events parameter in this message. It starts simple enough with a request to be notified if Cathy takes her phone off hook, L/hd(N), but then the E action is specified. This is the Embedded Notification Request action and is symbolic of a separate RQNT message. It tells Cathy's MTA that after an off hook event is detected, send the NTFY message, and then proceed as if you have received the embedded RQNT message. The embedded RQNT message shown is just like the RQNT messages you have seen earlier. It requests generation of the dial tone signal; that dialed digits be processed according to the digit map; and that the on hook, hook flash, operation complete (used for timer expiry), and operation failure (used in case a problem generating the dial tone signal is encountered) events be notified.

39 The connection Bob's MTA used to communicate with Cathy is deleted by the CMS. Bob's MTA returns some statistics about that connection.

40 The connection to Cathy's MTA is also deleted and statistics are returned.

41 Eventually, the conversation between Alan and Bob is over, and Alan hangs up his phone, triggering another NTFY to the CMS.

42 The CMS requests to be notified if Alan picks up his phone again.

43 This connection can now be deleted as well. First, the connection on Bob's MTA is deleted, and statistics are returned.

44 The connection on Alan's MTA is also deleted, and statistics are returned.

45 Knowing that the connection between Alan and Bob has ended, the CMS requests to be notified when Bob hangs up his phone.

46 Eventually, Bob hangs up his phone.

47 Finally, the CMS requests to be notified when Bob takes his phone off hook again.

Feature Call #2: Three-Way Calling

The second feature call shown also involves Alan, Bob, and Cathy. This example does not go into as much detail on each message because much of this is repetitive. An Ethereal capture was also done for this call, and some of the key parts are shown.

This call comprises eight stages. Figure 7-13 illustrates the NCS call flow for the first stage.

Figure 7-13 *Feature Call #2: Three-Way Calling (Stage One)*

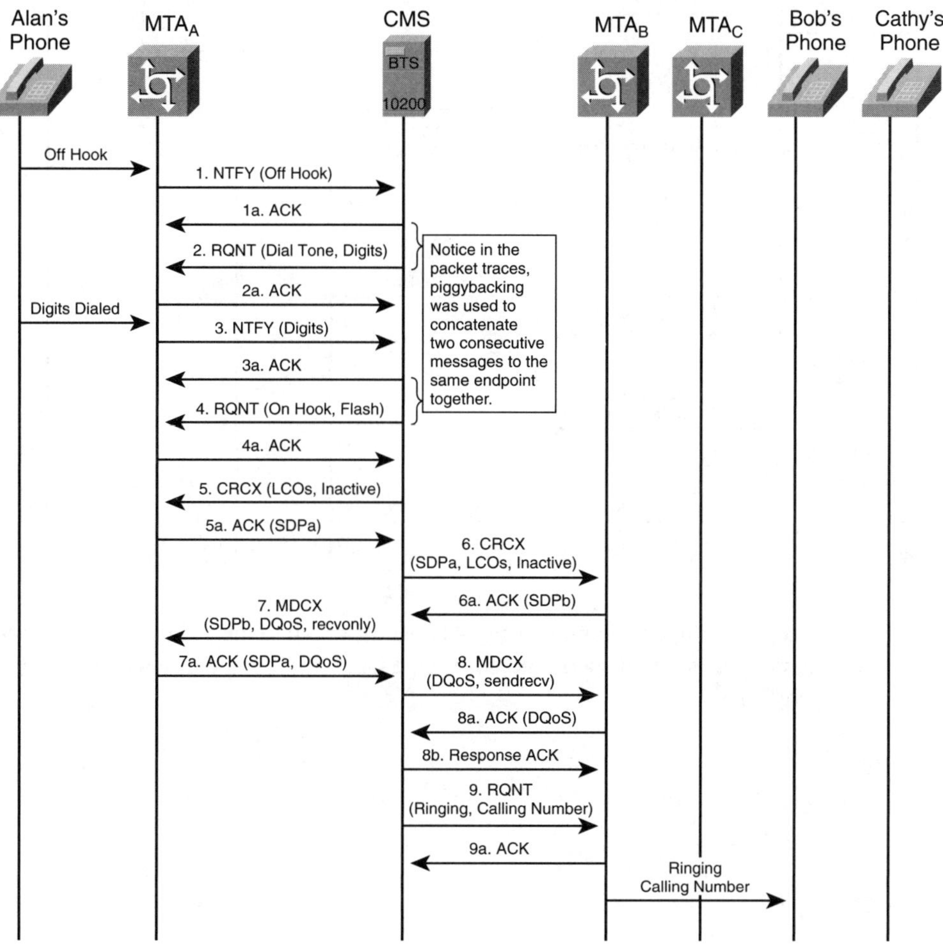

The first nine steps are identical to the previous example and involve Alan making a call to Bob up to the ringing stage. The second stage is also identical to the previous example and is illustrated in Figure 7-14.

Figure 7-14 *Feature Call #2: Three-Way Calling (Stage Two)*

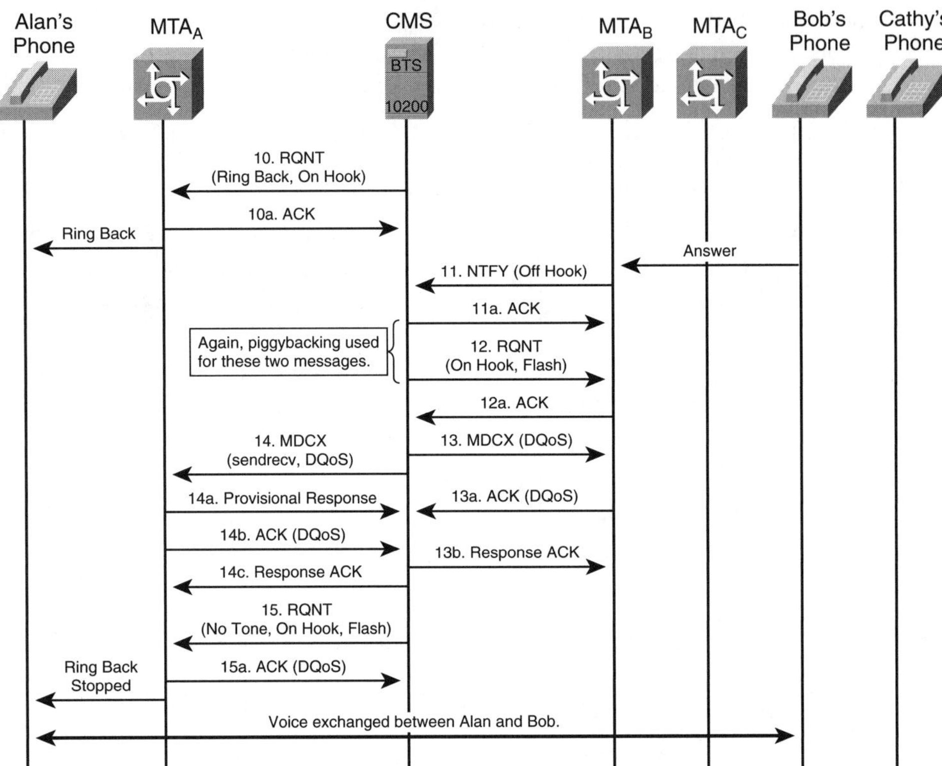

These steps continue the call between Alan and Bob up to the point where bidirectional communication is established. The third stage is where things start to change, as illustrated in Figure 7-15.

Figure 7-15 *Feature Call #2: Three-Way Calling (Stage Three)*

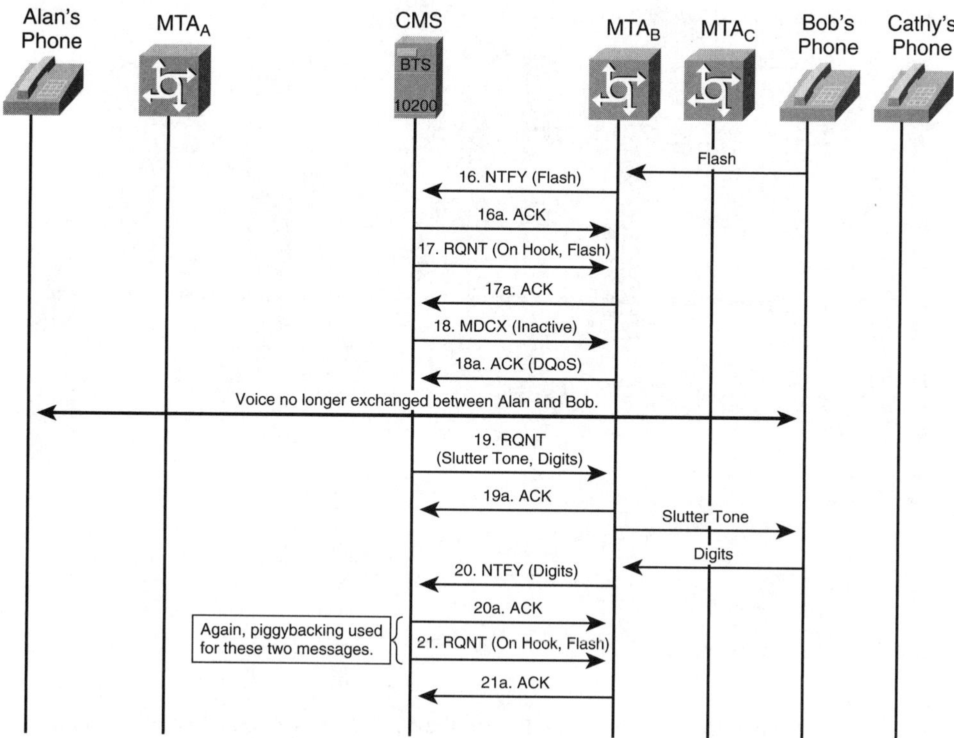

16 While talking to Alan, Bob decides he would also like to bring Cathy into the conversation. So he issues a hook flash triggering a NTFY to be sent to the CMS.

17 CMS asks to be notified if Bob hangs up his phone or does another hook flash.

18 The CMS deactivates the connection between Bob and Alan for the moment. It does this by sending an MDCX with connection mode "inactive" to Bob's MTA, which causes Bob's DQoS resources to return to the admitted state.

19 The CMS instructs Bob's MTA to provide slutter dial tone (L/sl) and to collect digits according to the digit map. This is how three-way call functionality is implemented. Figure 7-16 shows the RQNT used to do this.

Figure 7-16 *Ethereal Capture of Three-Way Calling at Stage Three*

```
Ch7_twc.cap - Ethereal
File Edit View Go Capture Analyze Statistics Help
No. Time - User        Source          Destination     Protocol Info
25 16:42:26.085514  24.34.241.100   13.13.5.38      MGCP   NTFY 24292 aaln/1@x1-6-00-0a-73-fa-be-e4.cc.cisco.com MGCP 1.0 N
26 16:42:26.087397  13.13.5.38      24.34.241.100   MGCP   200 24292 OK
27 16:42:26.099410  24.34.241.100   13.13.5.38      MGCP   200 161096952 OK
28 16:42:26.101660  13.13.5.38      24.34.241.100   MGCP   MDCX 161096953 aaln/1@x1-6-00-0a-73-fa-be-e4.cc.cisco.com MGCP 1
29 16:42:26.102849  13.13.5.38      24.34.241.96    MGCP   MDCX 161096954 aaln/1@x1-6-00-13-10-50-37-f6.cc.cisco.com MGCP 1
31 16:42:26.145906  24.34.241.100   13.13.5.38      MGCP   100 161096953
32 16:42:26.175883  24.34.241.100   13.13.5.38      MGCP   [Malformed Packet]
33 16:42:26.176294  13.13.5.38      24.34.241.100   MGCP   000 161096953
35 16:42:26.352824  24.34.241.96    13.13.5.38      MGCP   [Malformed Packet]
36 16:42:26.353321  13.13.5.38      24.34.241.96    MGCP   000 161096954
37 16:42:26.354536  13.13.5.38      24.34.241.96    MGCP   RQNT 161096955 aaln/1@x1-6-00-13-10-50-37-f6.cc.cisco.com MGCP 1
38 16:42:26.483491  24.34.241.96    13.13.5.38      MGCP   200 161096955 OK
39 16:42:47.716686  24.34.241.100   13.13.5.38      MGCP   NTFY 24293 aaln/1@x1-6-00-0a-73-fa-be-e4.cc.cisco.com MGCP 1.0 N
40 16:42:47.717260  13.13.5.38      24.34.241.100   MGCP   200 24293 OK
41 16:42:47.718647  13.13.5.38      24.34.241.100   MGCP   RQNT 161096956 aaln/1@x1-6-00-0a-73-fa-be-e4.cc.cisco.com MGCP 1
42 16:42:47.731946  24.34.241.100   13.13.5.38      MGCP   200 161096956 OK
43 16:42:47.733639  13.13.5.38      24.34.241.100   MGCP   MDCX 161096957 aaln/1@x1-6-00-0a-73-fa-be-e4.cc.cisco.com MGCP 1
44 16:42:47.806375  24.34.241.100   13.13.5.38      MGCP   [Malformed Packet]
45 16:42:47.808474  13.13.5.38      24.34.241.100   MGCP   RQNT 161096958 aaln/1@x1-6-00-0a-73-fa-be-e4.cc.cisco.com MGCP 1
46 16:42:47.820458  24.34.241.100   13.13.5.38      MGCP   200 161096958 OK
47 16:42:55.626317  24.34.241.100   13.13.5.38      MGCP   NTFY 24294 aaln/1@x1-6-00-0a-73-fa-be-e4.cc.cisco.com MGCP 1.0 N
48 16:42:55.627378  13.13.5.38      24.34.241.100   MGCP   200 24294 OK
49 16:42:55.639296  24.34.241.100   13.13.5.38      MGCP   200 161096959 OK
50 16:42:55.640238  13.13.5.38      24.34.241.100   MGCP   CRCX 161096960 aaln/1@x1-6-00-0a-73-fa-be-e4.cc.cisco.com MGCP 1
51 16:42:55.736936  24.34.241.100   13.13.5.38      MGCP/S 200 161096960 OK, with session description

Internet Protocol, Src: 13.13.5.38 (13.13.5.38), Dst: 24.34.241.100 (24.34.241.100)
User Datagram Protocol, Src Port: 2728 (2728), Dst Port: 2427 (2427)
Media Gateway Control Protocol
  RQNT (NotificationRequest)
  Transaction ID: 161096958
  Endpoint: aaln/1@x1-6-00-0a-73-fa-be-e4.cc.cisco.com
  Version: MGCP 1.0 NCS 1.0
  [The response to this request is in frame 46]
  Parameters
    RequestIdentifier (X): 1ED00000034
    RequestedEvents (R): L/hu(N),L/hf(N),L/[0-9#*T](D)
    SignalRequests (S): L/sl
    ResponseAck (K): 161096957
  RQNT 161096958 aaln/1@x1-6-00-0a-73-fa-be-e4.cc.cisco.com MGCP 1.0 NCS 1.0\r\n
  X: 1ED00000034\r\n
  R: L/hu(N),L/hf(N),L/[0-9#*T](D)\r\n
  S: L/sl\r\n
  K: 161096957\r\n
File: "D:\Documents and Settings\ P: 130 D: 114 M: 0
```

Notice at the bottom of the figure, the MGCP message is displayed in plain text, similar to the output you've seen previously in this chapter. This is an optional feature in Ethereal that can be enabled by choosing the Display Raw Text for MGCP Messages option under Edit > Preferences > Protocols > MGCP. The \r\n at the end of each line represents the carriage return and newline characters included at the end of each line in an MGCP message. Remember, Ethereal is covered in detail in Appendix B.

20 Upon receiving this tone, Bob dials Cathy's number, triggering a NTFY to the CMS.

21 The CMS requests to be notified if Bob hangs up his phone or does another hook flash.

Figure 7-17 illustrates stage four, where a connection is set up between Bob and Cathy.

Figure 7-17 *Feature Call #2: Three-Way Calling (Stage Four)*

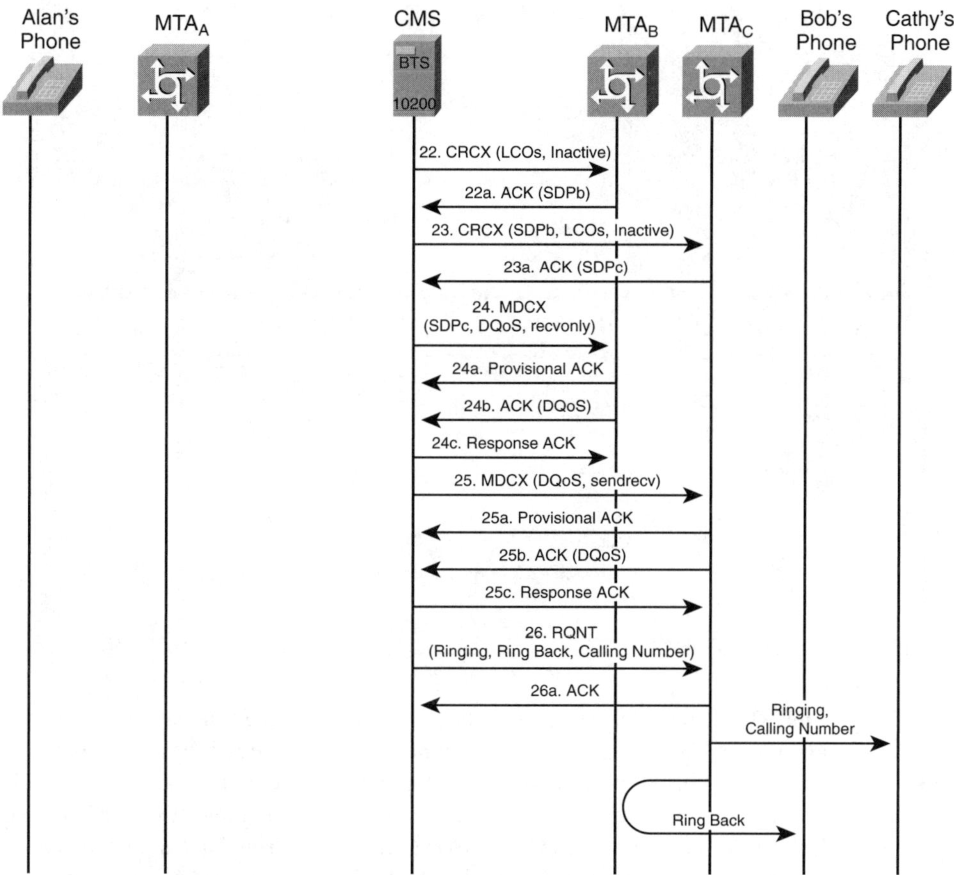

22 An inactive connection is created on Bob's MTA.

23 Similarly, an inactive connection is created on Cathy's MTA, and her MTA is given Bob's SDP information.

24 The connection on Bob's MTA is modified with DQoS information and Cathy's SDP info, and the connection is put into receive-only mode.

25 Cathy's MTA connection is modified to the send/receive state, and DQoS information is provided.

26 The final step in this stage is somewhat interesting. This RQNT message is illustrated in Figure 7-18.

Figure 7-18 *Ethereal Capture of Three-Way Calling at Stage 4*

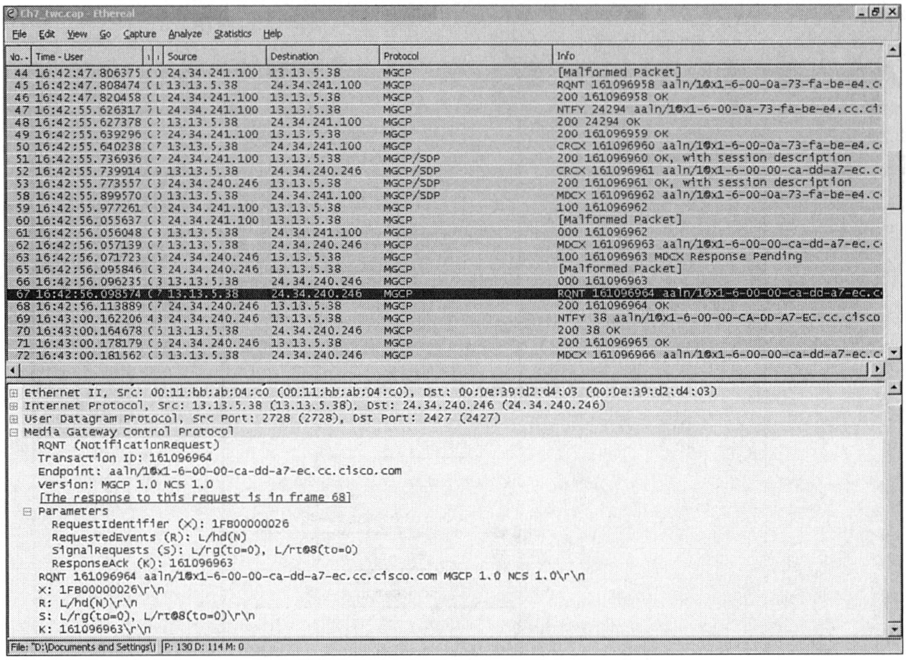

This RQNT requests two signals be generated: the ring tone signal to Cathy's phone and the ring back signal to Bob's phone. The ring tone is specified as you've seen before; it's simply played out the analog port connected to Cathy's phone. The way ring back tone is supplied is different, however. Remember this RQNT is sent to Cathy's MTA. So how can Cathy's MTA provide the ring back signal to Bob's phone? The answer is the ring back signal generated on Cathy's MTA is played out the connection to the VoIP network instead of the analog port. This is known as *inband ringback*. The character @ and the connection identifier 8 after the ring back signal name identify the connection to the VoIP network where ring back tone is applied.

The call continues to the point where Cathy answers her phone and is ready to talk to Bob and Alan, which begins in stage 5 (see Figure 7-19).

Figure 7-19 *Feature Call #2: Three-Way Calling (Stage Five)*

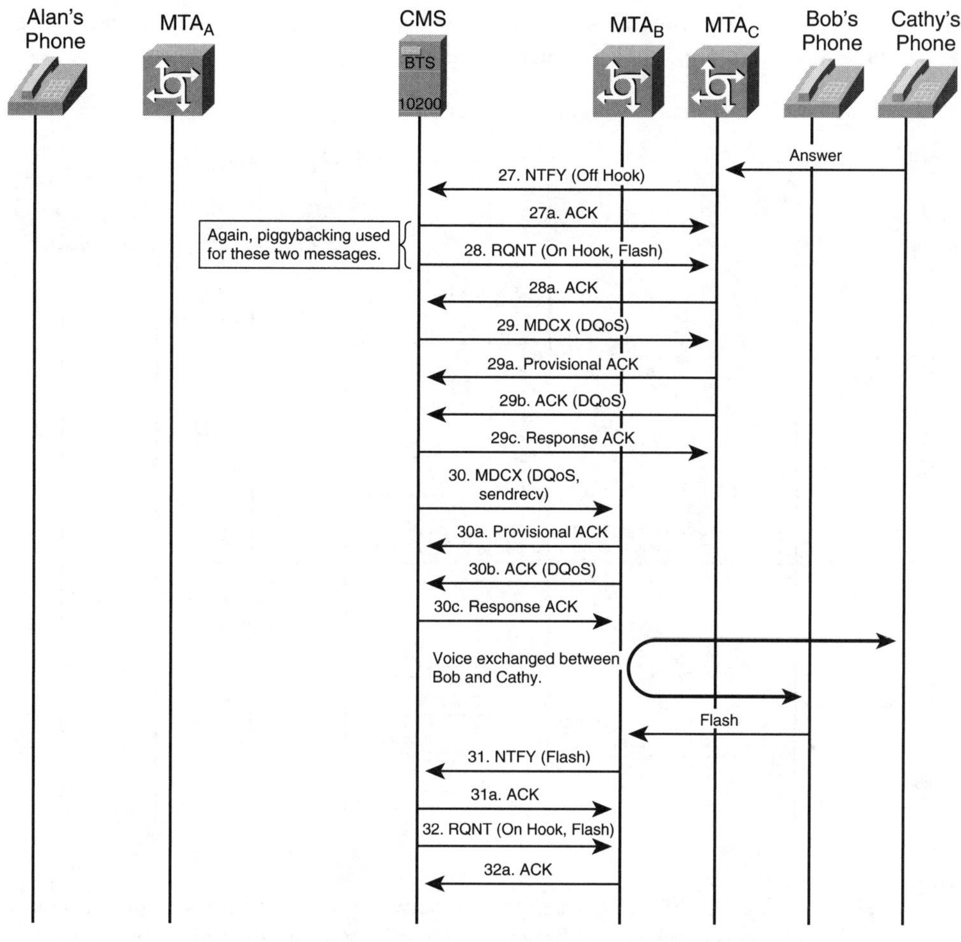

27 Cathy answers her phone, triggering a NTFY to the CMS.

28 CMS requests notification of Cathy's hanging up her phone or doing a hook flash.

29 CMS first modifies the connection to Bob on Cathy's MTA by requesting the activation of DQoS resources.

30 CMS then modifies the connection to Cindy on Bob's MTA by setting the connection mode to send/receive and requests Bob to activate DQoS resources. At this point, Bob and Cathy can communicate.

31 Bob wants to bring Alan back into the call and performs a hook flash on his phone that is relayed to the CMS.

32 CMS again requests notification of an on hook or hook flash from Bob's MTA.

At this point, you might be wondering how the voice packets are exchanged between all three MTAs. Figure 7-20 illustrates how this is done.

Figure 7-20 *Feature Call #2: Three-Way Calling (Stage Six)*

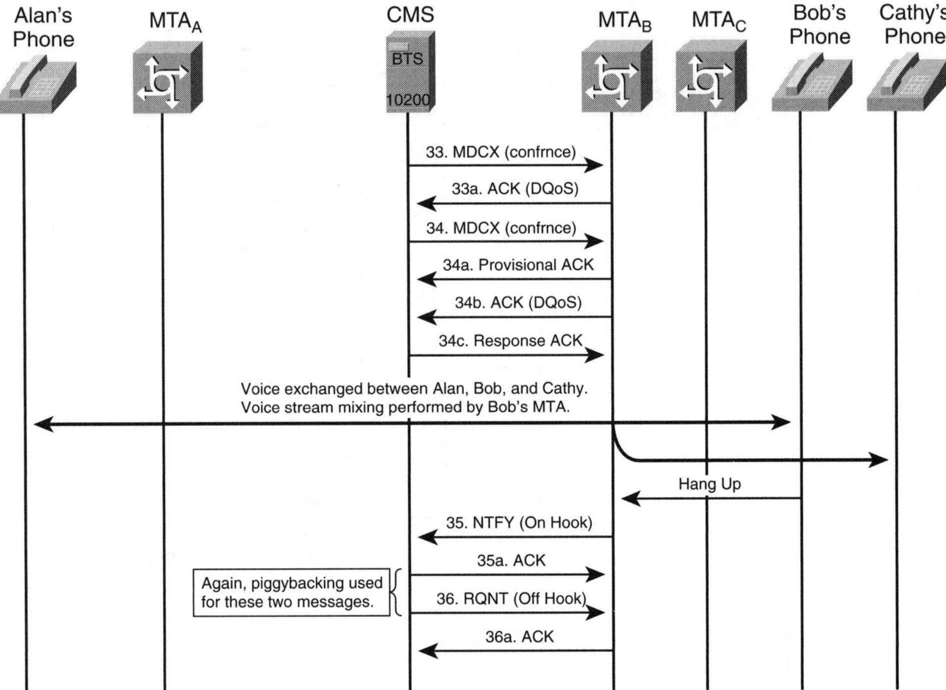

33 Way back in Step 18 the connection between Bob and Alan was put into inactive mode. This connection is reactivated in this step. But instead of the connection being put in send/receive mode, it is put in conference mode. Figure 7-21 shows the Ethereal capture of the MDCX.

Figure 7-21 *Ethereal Capture of Three-Way Calling at Stage Six*

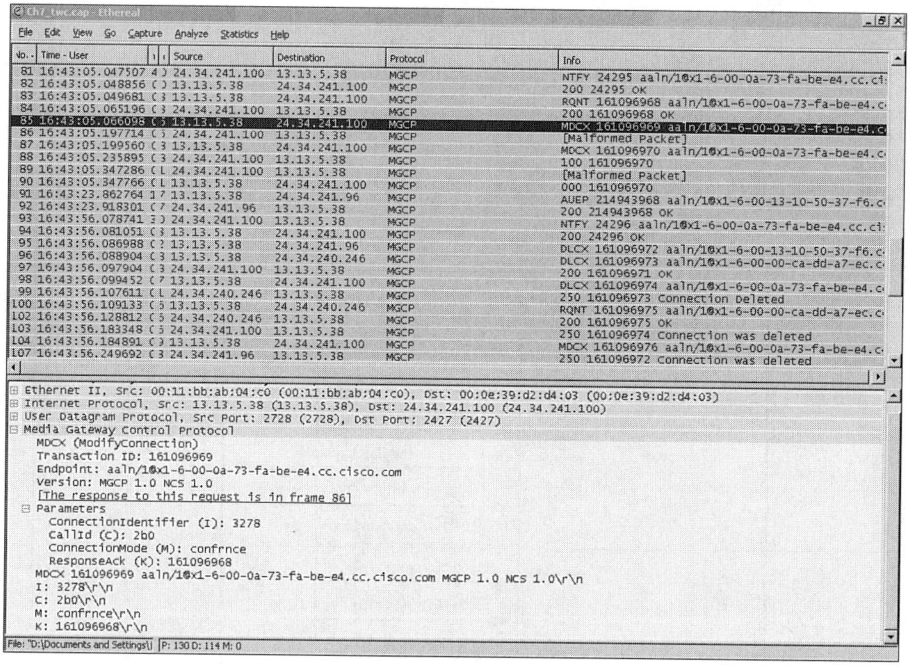

34 In conference mode, not only is the media stream exchanged between the endpoints involved in the connection, but it is also sent to all other connections that are in conference mode. Also because this connection was changed from the inactive mode, Bob's DQoS resources used for communication with Alan are reactivated as well.

So in this step you see the connection on Bob's MTA that communicates with Cathy's MTA is also modified and put into conference mode. Now the media stream exchanged between Bob and Cathy is also sent to Alan. Similarly, the media stream exchanged between Bob and Alan is also sent to Cathy. Thus, Alan, Bob, and Cathy can all now communicate with each other. Bob's MTA acts as a media stream mixer by taking the media stream it receives from one party and combining it with the media stream it is sending. This mixed audio is then put into RTP packets and sent to the other party.

35 This call then proceeds to the disconnect stage; first, Bob hangs up his phone, triggering a NTFY to the CMS.

36 The CMS asks to be notified if Bob picks up his phone again.

Because Bob decided to hang up his phone first, this created an interesting situation. Alan and Cathy did not have any connections with each other; all of their communication was occurring via Bob's MTA. So how can Alan and Cathy still talk if Bob is out of the picture? By looking at Figure 7-22 you see they can't.

Figure 7-22 *Feature Call #2: Three-Way Calling (Stage Seven)*

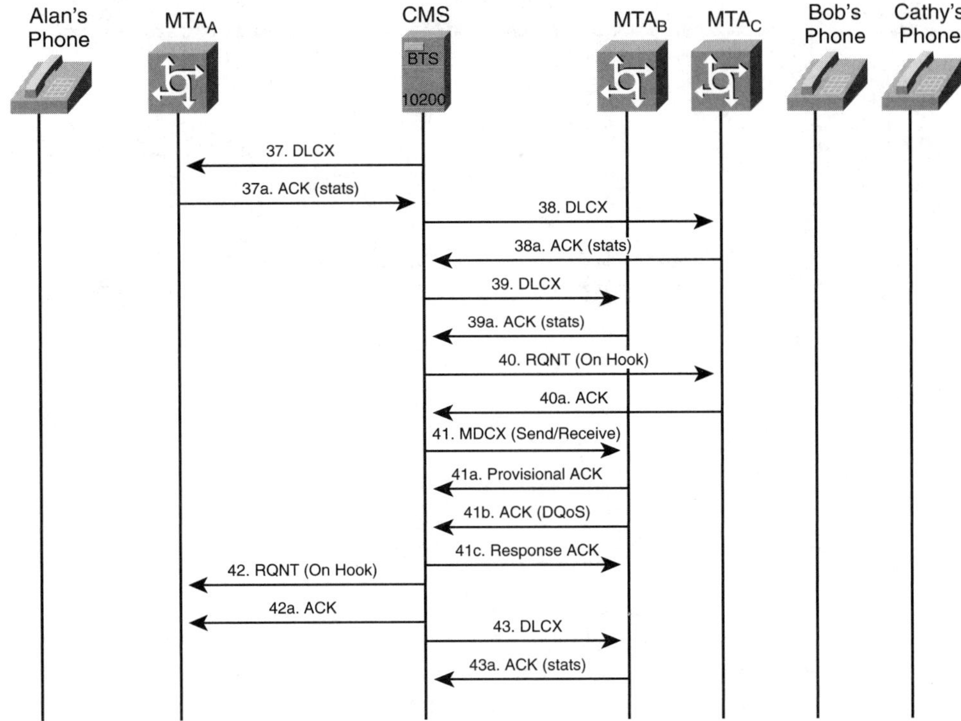

37 The connection between Bob and Alan is deleted on Alan's MTA and statistics are returned.

38 Similarly, the connection between Bob and Cathy is deleted on Cathy's MTA and statistics are returned.

39 CMS then deletes the connection on Bob's MTA used to communicate with Alan and collects connection statistics.

40 As soon as the CMS has confirmation that Cathy's connection has been deleted, it requests to be notified when Cathy hangs up her phone.

41 The next step is a little interesting as well. Both sides of the connection between Bob and Alan have been deleted, but the one between Bob and Cathy on Bob's MTA hasn't totally been deleted yet. So the CMS knows this connection no longer needs to be in conference mode because Alan is gone, so it changes this connection back to send/receive mode.

42 Similar to Step 40, as soon as the CMS has confirmation that Alan's connection has been deleted, it requests to be notified when Alan hangs up his phone.

43 At last the connection between Bob and Cathy on Bob's MTA is deleted, and statistics are returned.

The remainder of the disconnect phase is shown in stage eight of the call in Figure 7-23.

Figure 7-23 *Feature Call #2: Three-Way Calling (Stage Eight)*

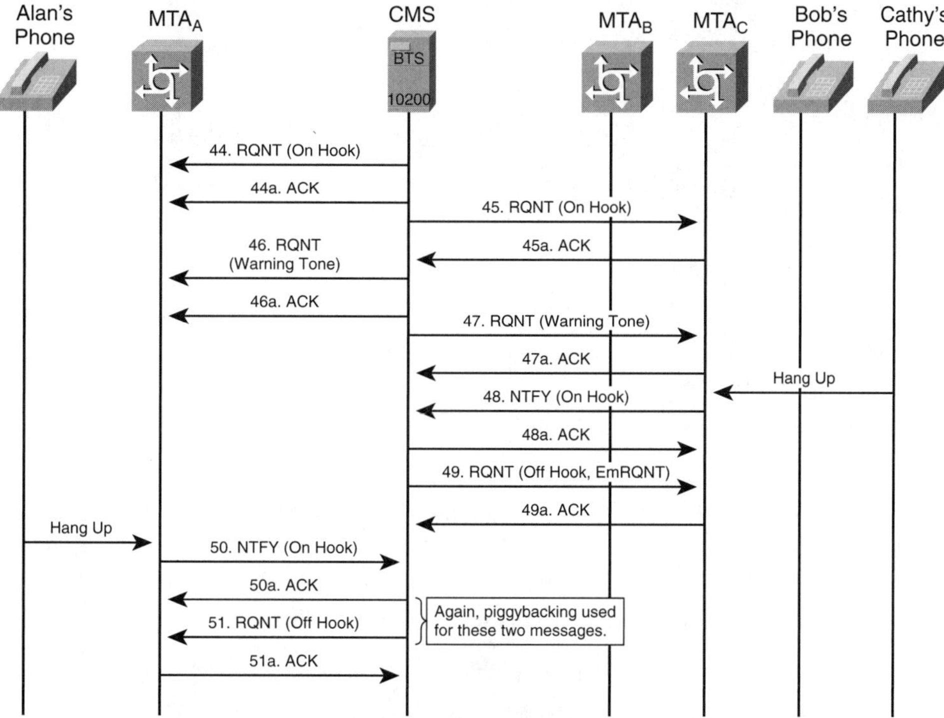

44 The CMS is still waiting for Alan and Cathy to hang up their phones, so it sends another request for the on hook event—first to Alan.

45 Similarly, another RQNT for the on hook event is sent to Cathy.

46 With still no on hook reported, the CMS requests the off hook warning signal be generated by Alan's MTA, as shown in Figure 7-24.

Figure 7-24 *Ethereal Capture of Three-Way Calling at Stage Eight*

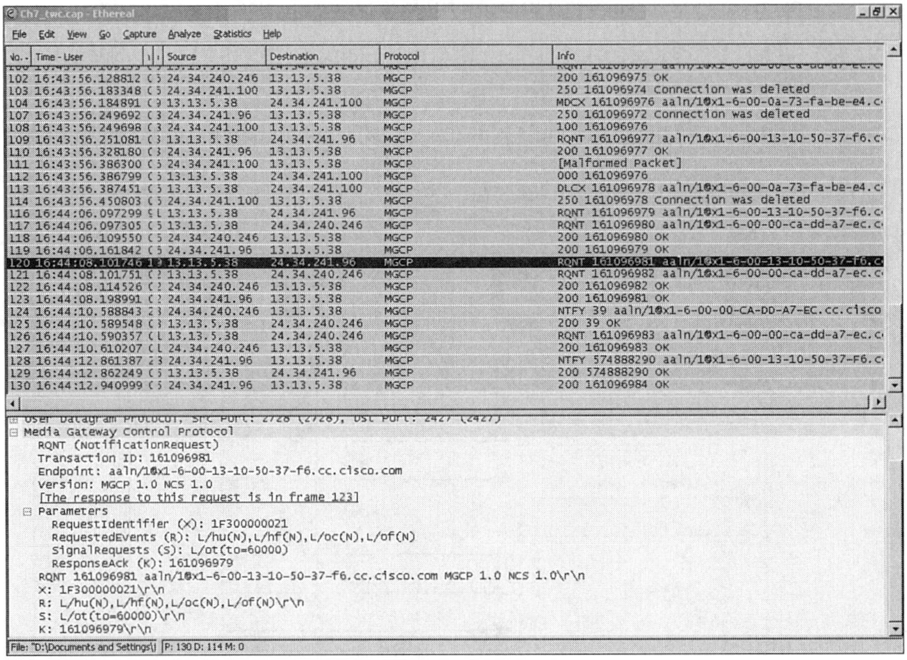

```
Ch7_twc.cap - Ethereal
File  Edit  View  Go  Capture  Analyze  Statistics  Help

No. Time - User      Source          Destination     Protocol  Info
102 16:43:56.128812  24.34.240.246   13.13.5.38      MGCP      200 161096975 OK
103 16:43:56.183348  24.34.241.100   13.13.5.38      MGCP      250 161096974 Connection was deleted
104 16:43:56.184891  13.13.5.38      24.34.241.100   MGCP      MDCX 161096976 aaln/1@x1-6-00-0a-73-fa-be-e4.c
107 16:43:56.249692  24.34.241.96    13.13.5.38      MGCP      250 161096972 Connection was deleted
108 16:43:56.249698  24.34.241.100   13.13.5.38      MGCP      100 161096976
109 16:43:56.251081  13.13.5.38      24.34.241.96    MGCP      RQNT 161096977 aaln/1@x1-6-00-13-10-50-37-f6.c
110 16:43:56.328180  24.34.241.96    13.13.5.38      MGCP      200 161096977 OK
111 16:43:56.386300  24.34.241.100   13.13.5.38      MGCP      [Malformed Packet]
112 16:43:56.386799  13.13.5.38      24.34.241.100   MGCP      000 161096976
113 16:43:56.387451  13.13.5.38      24.34.241.100   MGCP      DLCX 161096978 aaln/1@x1-6-00-0a-73-fa-be-e4.c
114 16:43:56.450803  24.34.241.100   13.13.5.38      MGCP      250 161096978 Connection was deleted
116 16:44:06.097299  13.13.5.38      24.34.241.96    MGCP      RQNT 161096979 aaln/1@x1-6-00-13-10-50-37-f6.c
117 16:44:06.097305  13.13.5.38      24.34.240.246   MGCP      RQNT 161096980 aaln/1@x1-6-00-00-ca-dd-a7-ec.c
118 16:44:06.109550  24.34.240.246   13.13.5.38      MGCP      200 161096980 OK
119 16:44:06.161842  24.34.241.96    13.13.5.38      MGCP      200 161096979 OK
120 16:44:08.101746  13.13.5.38      24.34.241.96    MGCP      RQNT 161096981 aaln/1@x1-6-00-13-10-50-37-f6.c
121 16:44:08.101751  13.13.5.38      24.34.240.246   MGCP      RQNT 161096982 aaln/1@x1-6-00-00-ca-dd-a7-ec.c
122 16:44:08.114526  24.34.240.246   13.13.5.38      MGCP      200 161096982 OK
123 16:44:08.198991  24.34.241.96    13.13.5.38      MGCP      200 161096981 OK
124 16:44:10.588843  24.34.240.246   13.13.5.38      MGCP      NTFY 39 aaln/1@x1-6-00-00-CA-DD-A7-EC.cc.cisco
125 16:44:10.589548  13.13.5.38      24.34.240.246   MGCP      200 39 OK
126 16:44:10.590357  13.13.5.38      24.34.240.246   MGCP      RQNT 161096983 aaln/1@x1-6-00-00-ca-dd-a7-ec.c
127 16:44:10.610207  24.34.240.246   13.13.5.38      MGCP      200 161096983 OK
128 16:44:12.861387  24.34.241.96    13.13.5.38      MGCP      NTFY 574888290 aaln/1@x1-6-00-13-10-50-37-F6.c
129 16:44:12.862249  13.13.5.38      24.34.241.96    MGCP      200 574888290 OK
130 16:44:12.940999  24.34.241.96    13.13.5.38      MGCP      200 161096984 OK

User Datagram Protocol, Src Port: 2728 (2728), Dst Port: 2427 (2427)
Media Gateway Control Protocol
  RQNT (NotificationRequest)
  Transaction ID: 161096981
  Endpoint: aaln/1@x1-6-00-13-10-50-37-f6.cc.cisco.com
  Version: MGCP 1.0 NCS 1.0
  [The response to this request is in frame 123]
  Parameters
    RequestIdentifier (X): 1F300000021
    RequestedEvents (R): L/hu(N),L/hf(N),L/oc(N),L/of(N)
    SignalRequests (S): L/ot(to=60000)
    ResponseAck (K): 161096979
  RQNT 161096981 aaln/1@x1-6-00-13-10-50-37-f6.cc.cisco.com MGCP 1.0 NCS 1.0\r\n
  X: 1F300000021\r\n
  R: L/hu(N),L/hf(N),L/oc(N),L/of(N)\r\n
  S: L/ot(to=60000)\r\n
  K: 161096979\r\n
File: "D:\Documents and Settings\  P: 130 D: 114 M: 0
```

The Signal Requests parameter, L/ot(to=60000), indicates the off hook warning signal should play for 60 seconds. The Requested Events parameter includes the on hook event, hook flash event, operation complete event, and operation failure event.

47 Likewise, a similar message is sent to Cathy's MTA.

48 Finally, Cathy hangs up her phone.

49 The CMS then sends an RQNT message to Cathy's MTA asking for notification if Cathy takes her phone off hook again. This message also contains an embedded RQNT similar to the one shown in the previous example.

50 Alan finally hangs up his phone as well.

51 Finally, the CMS requests notification should Alan take his phone off hook again.

Configuring the BTS 10200

By now, you should have a pretty good understanding of the NCS protocol and its use in providing call feature functionality. The next part of this chapter covers configuring MTAs on the Cisco BTS 10200. Some other types of BTS configuration, such as basic functionality, dial plan configuration, and call feature configuration, are covered as well. Additionally, you will learn the configuration required to make the call examples in this chapter. Be aware that BTS configuration tables and fields can change from one version of BTS software to another; these configuration examples are taken from version 4.5 of the BTS software.

NOTE It's beyond the scope of this book to cover all functional areas and aspects of BTS configuration. The book's focus is on the parts directly related to a PacketCable deployment. If you are interested in other areas of BTS configuration, please consult BTS-specific documentation.

Back in Chapter 2, you were introduced to the BTS 10200. You learned that its architecture typically consists of four UNIX servers, with two of them functioning as a redundant pair of call agent and feature server machines and the other two functioning as a redundant pair of Element Management System (EMS) machines responsible for BTS configuration, fault, performance, and accounting management. You learned that BTS 10200 configuration can be executed directly on the EMS platform in a CLI session, using a Cisco application known as Extensible Provisioning and Operations Manager (EPOM) or using third-party applications that interface with the EMS API. In this chapter and throughout the book, provisioning examples are shown via an EMS CLI session. Upon completed configuration on the active EMS machine, this information is replicated to the standby EMS machine and forwarded to the active call agent machine. After receiving the data, the call agent machine sends the data to the appropriate application (call agent or feature server), sends an acknowledgement to the EMS, and updates the standby call agent machine. It is important to note that the call agent machines maintain only the data necessary for call processing; this expedites the configuration process and lessens the processing burden on the call agent machines.

To access the EMS CLI, you need to log in to an EMS machine under a user with CLI privileges; by default the username optiuser has these privileges. The basic syntax of the CLI is a verb followed by a noun followed by a series of name=value pairs separated by semicolons. BTS configuration data is organized in a series of tables; the verb identifies the action to be taken on the table. Sample verb values include **show**, **add**, **change**, **delete**, **status**, and **control**. The noun identifies the table name, and the name=value pairs identify the individual fields in the table and their associated values. The field names are case insensitive, whereas the field values are case sensitive. The wildcard character % can be used in **show**, **add**, **change**, and **delete** commands; for example, the command **show mgw id=PC%** displays all media gateway table entries beginning with the characters PC.

The wildcard character * can be used in status and control commands; for example, the command **status mgw id=*** displays the status for all media gateways. As of version 4.5, the BTS is also equipped with command completion help; if you don't remember the exact syntax of a command, type as much of the command as you do know, and then press Tab. The BTS displays all the valid options for the next part of the command. This is somewhat similar to the use of the ? help in Cisco IOS. Also several control commands exist that can aid in CLI session navigation. For example, the up and down arrows can be used to scroll through the history of entered commands; Ctrl-A and Ctrl-E move the cursor to the beginning and end of the command line, respectively; Ctrl-K deletes all characters to the right of the cursor; Ctrl-I toggles between insert and overwrite modes; and Ctrl-L redisplays the current line.

If you have a large amount of configuration data to add and still want to do this directly from the EMS, you can create a bulk configuration script instead of manually entering each command one at a time. The script is simply a text file with a configuration CLI command on each line. After this file is created, transfer it to the /opt/ems/ftp/deposit directory on the active EMS machine using secure FTP. The system periodically checks this directory (every 20 seconds) for a script to execute. A report file is generated, showing the results of this script execution, which can be viewed by pointing a web browser to https://*EMS FQDN or IP address*, as demonstrated in Figure 7-25.

Figure 7-25 *BTS Bulk Configuration*

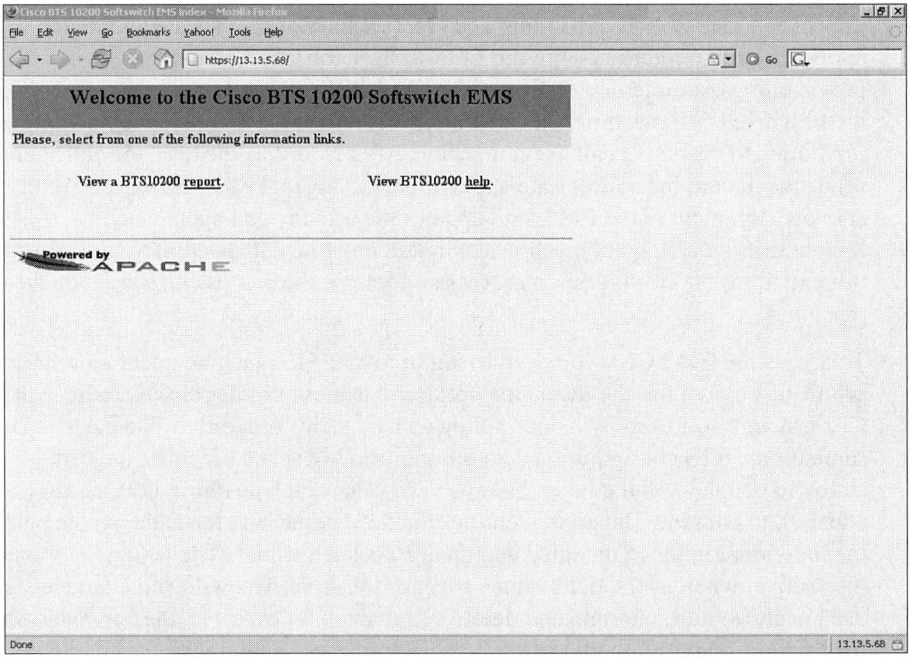

To avoid trying to control an entity in service that has not yet been provisioned, recommended practice is to not have configuration commands (Add, Change, and so on) in the same script as control commands (Control, Status, and so on).

Basic or Core Configuration Tasks

This section begins with some of the basic configuration tasks required for PacketCable operation. These tables are typically defined at installation time, and usually no modification is necessary. Each of these tasks includes recommended configuration settings; these setting should be fine for the general case but might require some adjusting depending on the environment.

The following are the basic BTS configuration steps:

Step 1 Define the call agent characteristics (call agent profile table).

Step 2 Define the call agent (call agent table).

Step 3 Configure call agent variables (call agent Configuration table).

Step 4 Define points of presence (point of presence table).

Step 1: Define the Call Agent Characteristics

As you'll see throughout BTS configuration, a profile table often exists where common characteristics can be defined that multiple entities can reference. For example, you have a media gateway profile table that multiple media gateways can refer to, and you have trunk group profile tables that multiple trunk groups can refer to. Even though you have only a single call agent defined, you also have a call agent profile table, where call agent characteristics are defined. This table includes several fields pertinent to a PacketCable deployment, many of which must be changed from their default values. The CLI syntax for defining a call agent profile is as follows:

```
CLI>add call_agent_profile id=call agent id; cms_id=cms id; mgc_id=mgc id;
  feid=financial entity id; dqos_supp=y; cdb_billing_supp=n; em_billing_supp=y;
  pri_rks_profile_id=primary RKS id; [sec_rks_profile_id=secondary RKS id];
```

Table 7-3 defines the fields for the call agent profile table.

Table 7-3 *Call Agent Profile Table Field Descriptions*

Field	Type and Range	Default Value and Comment	Description
CMS-ID	Integer: 00000–99999	99999	Five-digit element ID of the CMS; used to denote CMS in Event Messaging
MGC-ID	Integer: 00000–99999	99999	Five-digit element ID of the MGC; used to denote MGC in Event Messaging
FEID	Text	33333	Financial Entity ID of the BTS used in event messages for billing purposes
DQOS-SUPP	Boolean	N (No) In a PacketCable deployment this must be set to Y because DQoS is a part of PacketCable.	DQoS supported by this call agent (CMS)
CDB-BILLING-SUPP	Boolean	Y (Yes) If you are using PacketCable Event Messaging for billing, this field should be set to N, and the EM-BILLING-SUPP field must be set to Y.	Generate Call Detail Blocks (CDB) for billing
EM-BILLING-SUPP	Boolean	N (No) If you are using PacketCable Event Messaging for billing, this field must be set to Y, and the CDB-BILLING-SUPP field should be set to N.	Generate PacketCable event messages for billing
FORCED	Character (Y or N)	N (No) If you want to enable both types of billing record generation, this field must be set to Y. This is strongly discouraged because of the processing load it creates on the EMS.	Allows the user to enable both CDB and Event Messaging billing record generation

Table 7-3 *Call Agent Profile Table Field Descriptions (Continued)*

Field	Type and Range	Default Value and Comment	Description
PCMM-SUPP	Character (Y or N)	N (No) Set this field to Y if you are using the BTS for PacketCable Multimedia.	Specifies whether PacketCable Multimedia is supported
PRI-RKS-PROFILE-ID	Foreign Key to table radius-profile	N/A	Refers to the Primary RKS used for Event Messaging
SEC-RKS-PROFILE-ID	Foreign Key to table radius-profile	N/A	Refers to the Secondary RKS used for Event Messaging
DESCRIPTION	Text	N/A	User-defined description

The configuration of this table used for the call flow examples is

```
CLI>show call_agent_profile
ID=CA104
CMS_ID=104
MGC_ID=104
FEID=CA104
DQOS_SUPP=Y
CDB_BILLING_SUPP=Y
EM_BILLING_SUPP=N
GTD_SUPP=N
PCMM_SUPP=N
AMID=99999
PCMM_VERSION_MAJOR=1
PCMM_VERSION_MINOR=0
Reply : Success: Entry 1 of 1 returned.
```

Step 2: Define the Call Agent

Now that the call agent profile has been defined, you can define the call agent. The entry in this field has the same index as the entry in the call agent profile table:

```
CLI>add call_agent id=call agent id; tsap_addr=FQDN and port number of the call agent
  application; mgw_monitoring_enabled=y; clli=common language location identifer;
```

The TSAP_ADDR (Transport Service Access Point address) is a common field seen in several BTS tables; it is usually a Fully Qualified Domain Name (FQDN) and optional port number that identify the access point of the specific entity, in this case the call agent application. A Common Language Location Identifier (CLLI) is an 11-character string traditionally used in the PSTN that uniquely identifies a telephony switch. Setting the media gateway monitoring enabled flag to yes means the call agent uses a keepalive mechanism with its media gateways using ICMP pings and/or MGCP AUEPs to keep track of their operational states. It is recommended to keep this functionality enabled so the BTS doesn't try to route calls to nonoperational gateways.

The configuration of this table used for the call flow examples is

```
CLI>show call-agent
ID=CA104
TSAP_ADDR=sim-CA104.roseville.cc.cisco.com:9104
MGW_MONITORING_ENABLED=Y
CLLI=RSVLMNFYPS0
Reply : Success: Entry 1 of 1 returned.
```

Step 3: Configure Call Agent Variables

The call agent configuration table contains variables such as timers that affect how calls are handled. Most of these should never need changing from their defaults; variables especially pertinent to a PacketCable deployment are shown in Table 7-4. The syntax is as follows:

```
CLI>add ca-config type=variable name; datatype=boolean¦integer¦string¦digits;
  value=variable value;
```

Table 7-4 *CA-CONFIG Table Variables*

Variable Name	Data Type and Value Range	Default/ Recommended Value	Description
BEST-EFFORT-ON-QOS-FAIL	Boolean	Y (Yes)	Determines whether calls are allowed to complete using best-effort DOCSIS flows if DQoS fails.
CODEC-G728-PTIME	Integer	10–30 milliseconds	Packetization period used for the G.728 CODEC.
CODEC-G729E-PTIME	Integer	10–30 milliseconds	Packetization period used for the G.729E CODEC.
CODEC-MOD-DURING-CALL	Boolean	Y (Yes)	Determines whether the CODEC used for the first leg of a call is modified to G.711 for multiparty calls.
CODEC-PCMA-PTIME	Integer	10–30 milliseconds	Packetization period used for the G.711 (PCM) CODEC with a-law encoding.
CODEC-PCMU-PTIME	Integer	10–30 milliseconds	Packetization period used for the G.711 (PCM) CODEC with μ-law encoding.
CODEC-T38-PTIME	Integer	10–30 milliseconds	Packetization period used for T.38 fax relay.

Table 7-4 *CA-CONFIG Table Variables (Continued)*

Variable Name	Data Type and Value Range	Default/ Recommended Value	Description
COPS-DSCP-TOS	Integer	96	Differentiated Services Codepoint value/IP TOS byte value for outgoing COPS packets.
DEFAULT-CODEC-TYPE	String	PCMU	Default CODEC to be used.
DEFAULT-OFFICE-SERVICE-ID	String	N/A	Specifies the default services.
DEFAULT-POP	Integer	255	Specifies the default Point of Presence (POP).
DEFAULT-QOS-ID	String	Default	Specifies the default Quality of Service (QoS).
DIAL-TONE-TMR	Integer 0–32	16	Default timer for the dial-tone signal.
DIGIT-TO-TMR	Integer 30–300	60	Default timer for digit timeout.
DQOS-DS-SLACK-TERM-TIMER	Integer 0–60000 microseconds	0 microseconds	Tolerated latency for downstream service flows (0 indicates no restriction).
DQOS-GATE-TIMER	Integer 0–10 seconds	2 seconds	CMTS gate operation response timer.
DQOS-T1-TIMER	Integer 0–500 seconds	200 seconds	Amount of time a gate can remain in the authorized state.
DQOS-T5-TIMER	Integer 1–60 seconds	5 seconds	Amount of time the CMTS must hold resources in the Admitted QoS set in excess of the Active QoS set.
DQOS-T7-TIMER	Integer 0–300 seconds	200 seconds	Amount of time the CMTS must hold resources in the Admitted/Reserved QoS set in excess of the Active/ Committed QoS set.

continues

Table 7-4 *CA-CONFIG Table Variables (Continued)*

Variable Name	Data Type and Value Range	Default/ Recommended Value	Description
DQOS-T8-TIMER	Integer 0–300 seconds	0 seconds	Amount of time Committed/ Active CMTS resources can remain unused (0 indicates forever) Keeping this parameter at the default of 0 seconds provides no protection from "hung" dynamic service flows that can tie up bandwidth. It is recommended to set this to a nonzero value; 200–300 seconds should be adequate.
DQOS-US-SLACK-TERM-TIMER	Integer 0–60000 microseconds	800 microseconds	Tolerated grant jitter for upstream service flows
INACTIVE-CONN-MODE-BEFORE-ANSWER	Boolean	N(No)	Specifies the use of the inactive connection mode for calls before they are answered.
LOCAL-NO-ANSWER-TMR	Integer 0–300	185 seconds	No answer timer value for local calls.
LOCAL-RINGBACK	Boolean	Y (Yes)	Determines whether ring back tone is provided locally for ONNET-to-ONNET calls.
MAX-MGCP-DATAGRAM	Integer 2000–8000 bytes	4000 bytes	Maximum size of a MGCP datagram.
MGCP-ICMP-PING-RETRANSMIT-DURATION	Integer 2–10	2 seconds	Time the BTS waits for a response from a gateway after sending an MGCP AUEP or ICMP ping.
MGCP-ICMP-PING-RETRY-COUNT	Integer 1–6	2	Number of MGCP AUEP or ICMP ping retries sent by the BTS before declaring the gateway unreachable.
MGCP-INIT-DURATION	Integer 1–10	1 second	Duration after which a media gateway starts initializing terminations.

Table 7-4 *CA-CONFIG Table Variables (Continued)*

Variable Name	Data Type and Value Range	Default/ Recommended Value	Description
MGCP-INIT-TERMS	Integer 100–5000	160	Number of terminations initialized in every MGCP-INIT-DURATION.
MGCP-MAX-FAULT-COUNT	Integer 0–10	2	Maximum number of fault counts.
MGCP-MAX-KEEPALIVE-AUEP	Integer 1000–100000	40000	Maximum number of gateways to be AUEP pinged in a 10-second interval.
MGCP-MAX-KEEPALIVE-ICMP	Integer 100–10000	4000	Maximum number of gateways to be ICMP pinged in a 10-second interval.
MGCP-MAX-NO-OF-ADM-RESP	Integer 1–100	100	Maximum number of ADM responses sent in one message.
MGCP-MAX-UNREACH-COUNT	Integer 0–10	0	Maximum number of unreachable counts.
MGCP-RECOVERY-DURATION	Integer 1–10	5 seconds	Time before an attempt to recover faulty/unreachable endpoints occurs.
MGCP-RTO-MAX	Integer 1–30	4 seconds	Maximum time between two successive MGCP retransmissions.
MGCP-SIG-TOS-LOWDELAY	Boolean	Y	Whether the IP TOS bit for low delay is set in outgoing MGCP packets.
MGCP-SIG-TOS-PRECEDENCE	Integer 0–7	3	Value of the IP precedence bits in outgoing MGCP packets.
MGCP-SIG-TOS-RELIABILITY	Boolean	N	Whether the IP TOS bit for reliability is set in outgoing MGCP packets.
MGCP-SIG-TOS-THROUGHPUT	Boolean	N	Whether the IP TOS bit for maximum throughput is set in outgoing MGCP packets.
MGCP-T-HIST	Integer 3–120	30 seconds	Maximum time allowed before the copy of an MGCP Response packet is destroyed.

continues

Table 7-4 *CA-CONFIG Table Variables (Continued)*

Variable Name	Data Type and Value Range	Default/ Recommended Value	Description
MGCP-T-MAX	Integer 10–60	20 seconds	Maximum time before stopping retransmissions after sending an initial MGCP packet.
MGCP-T38-FAX-MODE-PREF1	String	T38_FXR_LOOSE	First preferred method for handling fax transmissions.
MGCP-T38-FAX-MODE-PREF2	String	T38_FXR_STRICT	Second preferred method for handling fax transmissions.
MGCP-T38-FAX-MODE-PREF3	String	T38_FXR_GW	Third preferred method for handling fax transmissions.
NO-ANSWER-TMR	Integer 1–300	185 seconds	Value of the no answer timer.
NO-RESP-TMR	Integer 1–300	20 seconds	Length of time MGCP and SIP endpoints are allowed to stay in the "alerting" phase before the call is torn down.
RADIUS-DSCP-TOS	Integer 0–255	96	DSCP codepoint value used for outgoing RADIUS (Event Messaging) packets.
SCTP-DSCP	String	CS3	DSCP codepoint value used for outgoing SCTP (Sigtran) packets.
START-NO-ANSWER-TMR	Boolean	N	Whether the no answer time is stared.
SUSP-TMR	Integer 0–300	10 seconds	If the called party hangs up and then picks the phone back up before this timer expires, the call is reconnected.
T38-MAX-BIT-RATE	Integer	14400	Default bit rate for a T.38 fax.
T38-MAX-BUFFER-SIZE	Integer	200	Default maximum buffer size for a T.38 fax.
T38-MAX-DATAGRAM-SIZE	Integer	72	Default maximum datagram size for a T.38 fax.
TONE-TIMER	Integer 0–180	60 seconds	Length of time tone is played to a subscriber before the call is disconnected.

NOTE To see the default values of these variables, look at the call agent configuration base table (ca-config-base).

Step 4: Define Points of Presence (POP)

The BTS 10200 can support subscribers in multiple geographic regions. In fact, this is a common deployment where one BTS system services several regions or points of presence (POP). Each POP can have its dialing characteristics, so the POP identifier provides a means of distinguishing subscribers—that is, each subscriber is assigned to a particular POP. The syntax for adding a POP is as follows:

```
CLI>add pop id=POP identifier; country=integer; timezone=timer value;
```

Logically, the time zone field identifies the time zone of that particular POP; the default value is the local time zone of the BTS.

The configuration of this table used for the call flow examples is

```
CLI>show pop id=PacketCable_PoP
ID=PacketCable_PoP
TIMEZONE=LOCAL
LATA=99999
LOCAL_7D_DIALING=Y
ITP=N
ZERO_MINUS=LEC
BLOCK_EAWOPIC=Y
CNAM_OPTION=LOCAL_OR_LIDB
PIC2_REQD=N
TREAT_IMS_ANONYMOUS=N
ZERO_PLUS_LOCAL=N
TEMP_DISC_SERVICE_ALLOWED=N
SENSOR_ID=000000
Reply : Success: Entry 1 of 1 returned.
```

One field in this table is of particular interest to one of the feature call examples. The field CNAM_OPTION specifies whether internal entries (LOCAL) are used for Caller ID names or an external database (EXT_LIDB) is consulted. The value LOCAL_OR_LIDB means the internal entry is used if it exists; otherwise, an external lookup is made.

Dial Plan Configuration Tasks

The BTS is a very feature-rich platform with support for much Class IV and Class V telephony switch functionality. Among this functionality is the support for a wide range of call types. You can display a list of valid call types on the BTS by looking at table CALL_TYPE:

```
CLI>show call_type
CALL_TYPE=500
DESCRIPTION=Service Access Code 500, use carrier to route the call
CALL_TYPE=700
DESCRIPTION=700 SAC call, route via PIC or dialed CAC
```

```
CALL_TYPE=900
DESCRIPTION=Service Access Code 900, use carrier to route the call
CALL_TYPE=976
DESCRIPTION=Information Services Calls
CALL_TYPE=AIRLINES
DESCRIPTION=Airlines Information and Reservation
CALL_TYPE=AMBULANCE
DESCRIPTION=Ambulance
CALL_TYPE=ANA
DESCRIPTION=Automatic Number Announcement
CALL_TYPE=BUSINESS
DESCRIPTION=811- calls to business office
CALL_TYPE=CUT_THRU
DESCRIPTION=Cut-thru calls (calls dialed as 101XXXX+#)
CALL_TYPE=DA
DESCRIPTION=411, NPA-555-1212 calls
CALL_TYPE=DA_TOLL
DESCRIPTION=1+411, 1+NPA-555-1212 calls
CALL_TYPE=EMG
DESCRIPTION=911 calls
CALL_TYPE=FIRE
DESCRIPTION=Fire
CALL_TYPE=INFO
DESCRIPTION=Information Services Calls
CALL_TYPE=INTERLATA
DESCRIPTION=InterLATA call (uses PIC or dialed CAC)
CALL_TYPE=INTL
DESCRIPTION=International call type
CALL_TYPE=INTL_OPR
DESCRIPTION=International Operator
CALL_TYPE=INTL_WZ1
DESCRIPTION=International Call Within World Zone 1
CALL_TYPE=LB_TEST
DESCRIPTION=LoopBack Test Call (108 Test Line)
CALL_TYPE=LOCAL
DESCRIPTION=Local Call
CALL_TYPE=LRN
DESCRIPTION=Location Routing Number
CALL_TYPE=MOBILE
DESCRIPTION=Calls to Mobile Network
CALL_TYPE=NAS
DESCRIPTION=Network Access Server (same as DYNAMIC)
CALL_TYPE=NATIONAL
DESCRIPTION=National Number. Use LSA and LATA tables to determine if local, toll or
  interLATA call
CALL_TYPE=NAT_OPR
DESCRIPTION=National Operator Assisted Calls
CALL_TYPE=NON_EMG
DESCRIPTION=311 calls
CALL_TYPE=OPERATOR
DESCRIPTION=Operator Call
CALL_TYPE=PCS
DESCRIPTION=Service Access Code 500, use carrier to route the call
CALL_TYPE=POLICE
DESCRIPTION=Police
CALL_TYPE=PREMIUM
DESCRIPTION=Service Access Code 900, use carrier to route the call
CALL_TYPE=RAILWAYS
DESCRIPTION=Railways Information and Reservation
CALL_TYPE=RELAY
DESCRIPTION=711 calls
CALL_TYPE=REPAIR
DESCRIPTION=611 calls
CALL_TYPE=SVC_CODE
```

```
DESCRIPTION=Generic Service Code Call Type. Used when none of the call types apply
CALL_TYPE=TANDEM
DESCRIPTION=Tandem call between Call Agent and the next switch or Call Agent
CALL_TYPE=TEST_CALL
DESCRIPTION=Test call dialed as: 958/959-xxxx or 1xx
CALL_TYPE=TIME
DESCRIPTION=Time
CALL_TYPE=TOLL
DESCRIPTION=IntraLATA toll call
CALL_TYPE=TOLL_FREE
DESCRIPTION=Toll free call (800, 888, 877, 866, 855, 844...)
CALL_TYPE=TRAFFIC
DESCRIPTION=Traffic Accident Report
CALL_TYPE=TW
DESCRIPTION=Time and Temperature Service
CALL_TYPE=UAN
DESCRIPTION=Universal Access Number
CALL_TYPE=VACANT
DESCRIPTION=Vacant code
CALL_TYPE=WEATHER
DESCRIPTION=Weather Report
Reply : Success:  Entries 1-44 of 44 returned.
```

In this chapter, only the local ONNET call type is covered; the basic steps to configure a dial plan for local routing on the BTS are as follows:

Step 1 Define national destination codes (national destination code table).

Step 2 Define exchange codes (exchange code table).

Step 3 Define office codes (office code table).

Step 4 Define destinations (destination table).

Step 5 Define digit map(s) (digit map table).

Step 6 Define dial plan characteristics (dial plan profile table).

Step 7 Define dial plan (dial plan table).

Step 1: Define National Destination Codes

The national destination codes are the home area codes serviced by this particular call agent. This configuration is very simple—just add the 3-digit area codes supported by this BTS:

```
CLI>add ndc digit_string=area code;
```

The configuration of this table used for the call flow examples is

```
CLI>show ndc
DIGIT_STRING=321
DIGIT_STRING=322
DIGIT_STRING=987
DIGIT_STRING=988
Reply : Success:  Entries 1-4 of 4 returned.
```

Step 2: Define Exchange Codes

Exchange codes are the three-digit codes in a phone number after the area code. Each exchange code within a home area code serviced by this particular call agent comprises an office code serviced by the call agent. This configuration is also very simple—just add the three-digit area codes supported by this BTS:

```
CLI>add exchange_code ndc=ndc from previous step; ec=exchange code;
```

The configuration of this table used for the call flow examples is

```
CLI>show exchange_code limit=1
DIGIT_STRING=321501
NDC=321
EC=501
OFFICE_CODE_INDEX=1
MIN_DN_LENGTH=10
MAX_DN_LENGTH=10
Reply : Success: Entry 1 of 12 returned.
```

Step 3: Define Office Codes

This table defines the office codes serviced by the call agent; an office code is basically the combination of the national destination code and exchange code. The syntax to define office codes is as follows:

```
CLI>add office_code call-agent_id=call agent identifier; ndc=ndc from previous
  steps; ec=exchange code from previous step; dn_group=xxxx;
```

The configuration of this table used for the call flow examples is

```
CLI>show office_code limit=1
DIGIT_STRING=321501
OFFICE_CODE_INDEX=1
DID=N
CALL_AGENT_ID=CA104
DIALABLE=Y
NDC=321
EC=501
DN_GROUP=xxxx
Reply : Success: Entry 1 of 12 returned.
```

Step 4: Define Digit Map(s)

Remember the MGCP digit map parameter described in the previous chapter that tells a gateway how to process dialed digits? The digit map table is where this parameter is defined on the BTS:

```
CLI>add digit_map id=digit map identifier; digit_pattern=digit map expression list;
  description=digit map description;
```

Note that a default digit map is automatically built in to the BTS, and it is sent if you don't specify an alternative digit map.

For the call flow examples in this book the default digit map is used; for reference, here is an alternative digit map configuration:

```
CLI>show digit_map limit=1
ID=NPA321
DIGIT_PATTERN=0T¦00¦[2-9]11¦1[2-9]11¦[2-9]xx[2-9]xxxxxx¦¦1[2-9]xx[2-9]xxxxxx¦0[2-
  9]xx[2-9]xxxxxx¦011xxxxxx.T¦01xxxxxx.T¦101xxxx¦#¦*xx¦[2-9]#¦[2-4]x#¦[2-9]T¦[2-4]xT
Reply : Success: Entry 1 of 2 returned.
```

Step 5: Define Destinations

The destination table defines the call type and routing information for dialed digits; because at this point only ONNET-to-ONNET or local calls have been discussed, this is the only call type covered. Other call types are covered in the next chapter. The syntax to define destinations is as follows:

```
CLI>add destination dest-id=destination identifier; call_type=local;
  route_type=sub;
```

The route type of subscriber means that the office code index and last four digits of the dialed number are used to find the subscriber; specifically the DN2Subscriber table is used. The call type **local** means this destination is for 7- or 10-digit nontoll calls.

The configuration of this table used for the call flow examples is

```
CLI>show destination dest-id=ONNET_LC
DEST_ID=ONNET_LC
CALL_TYPE=LOCAL
ROUTE_TYPE=SUB
ZERO_PLUS=Y
INTRA_STATE=Y
GAP_ROUTING=N
CLDPTY_CTRL_REL_ALWD=N
CALL_SUBTYPE=NONE
ACQ_LNP_QUERY=NA
Reply : Success: Entry 1 of 1 returned.
```

Step 6: Define Dial Plan Characteristics

A dial plan defines how dialed digits are processed. The dial plan profile table really just defines the dial plan; it must be added before entries to the dial plan can be added. It is the dial plan profile identifier that links a dial plan to subscribers and trunk groups. The syntax to define dial plan characteristics is as follows:

```
CLI>add dial_plan_profile id=dial plan identifier; description=dial plan
  description;
```

The configuration of this table used for the call flow examples is

```
CLI>show dial_plan_profile id=INCOMING
ID=INCOMING
DESCRIPTION=dialing plan profile for incoming calls
NANP_DIAL_PLAN=Y
OVERDECADIC_DIGITS_SUPP=N
```

```
NOA_BASED_ROUTING=N
Reply : Success: Entry 1 of 1 returned.
```

The field NANP_DIAL_PLAN= Y indicates that this is a North American dial plan. This means dialed digits follow the format NXX-NXX-XXXX, where X is a digit from 0–9, and N is a digit from 2–9.

Step 7: Define the Dial Plan

As soon as the dial plan profile has been defined, the individual entries in the dial plan can be added using the following syntax:

```
CLI>add dial_plan id=dial plan identifier from previous step; digit-string=numerical
  string of dialed digits; dest-id=destination identifier previously defined;
```

The configuration of this table used for the call flow examples is (notice the entries are tied to the ONNET local call destination defined earlier)

```
CLI>show dial-plan id=INCOMING
ID=INCOMING
DIGIT_STRING=321
DEST_ID=ONNET_LC
SPLIT_NPA=NONE
DEL_DIGITS=0
MIN_DIGITS=10
MAX_DIGITS=10
NOA=NATIONAL
ID=INCOMING
DIGIT_STRING=322
DEST_ID=ONNET_LC
SPLIT_NPA=NONE
DEL_DIGITS=0
MIN_DIGITS=10
MAX_DIGITS=10
NOA=NATIONAL
Reply : Success:  Entries 1-2 of 2 returned.
```

MTA Configuration Tasks

The following are the basic steps to configure an MTA media gateway on the BTS and associated provisioning tables:

Step 1 Configure the CMTS to which the MTA is connected (aggregation table).

Step 2 Configure the MTA characteristics (media gateway profile table).

Step 3 Configure the MTA (media gateway table).

Step 4 Configure the NCS endpoints (termination table).

Step 5 Optionally, configure the subscriber quality of service parameters (QoS table).

Step 6 Configure the subscriber characteristics (subscriber profile table).

Step 7 Configure the PacketCable subscriber (subscriber table).

Step 8 Bring the MTA into service.

Step 9 Prepare the endpoints for service.

Step 10 Bring the endpoints into service.

Step 1: Configure the CMTS to Which the MTA Is Connected

For the BTS to provide DQoS information to the MTA, it needs to know what CMTS the MTA is connected to. The CMTS is defined in the aggregation table:

```
CLI>add aggr id=CMTS name; tsap-addr=CMTS ip address | FQDN; dqos-supp=y; es-event-
  supp=y; es-supp=y;
```

Keep in mind that for DQoS to work properly, the CMTS also must support DQoS messaging and be configured to do so; this is accomplished on the Cisco CMTS with the global configuration command **packetcable**. Much more on CMTS configuration and DQoS is covered in Chapter 13, "Analyzing, Implementing, and Troubleshooting DQoS."

Table 7-5 lists the key fields and fields relating to call signaling.

Table 7-5 *AGGR Table*

Field	Default Value	Recommended Value	Description
TSAP-ADDR	N/A	IP address or FQDN of a loopback interface on the CMTS.	IP address or FQDN of the CMTS.
DQOS-SUPP	N	Y	Enable DQoS (must be Y for PacketCable).
ES-EVENT-SUPP	N	Set this to Y if the CMTS supports PacketCable lawful intercept and you intend to use it.	CMTS supports Event Messaging for PacketCable lawful intercept.
ES-SUPP	N	Set this to Y if the CMTS supports PacketCable lawful intercept and you intend to use it.	CMTS supports packet replication for PacketCable lawful intercept.

A CMTS loopback interface is recommended over a physical interface because typically at least two wide area network (WAN) interfaces on the CMTS are connected to the IP backbone. If you define one of these interfaces for CMS communication and this interface goes down, the CMS connection goes down even though an alternative path exists. Also it is recommended that this IP address not be reachable to external users for security reasons.

Obviously, IP connectivity to this interface is required.

The aggregate configuration used in the examples in this chapter is

```
CLI>show aggr
ID=PC-CMTS
TSAP_ADDR=PC-CMTS.cc.cisco.com
ES_SUPP=Y
DQOS_SUPP=Y
KA_TIMER=2
ACK_TIMEOUT=1000
ES_EVENT_SUPP=Y
IPSEC_SA_ESP_CS=3DES-MD5,3DES-SHA1,NULL-MD5,NULL-SHA1
IPSEC_SA_LIFETIME=86400
IPSEC_SA_GRACE_PERIOD=21600
IPSEC_ULP_NAME=IP
IKE_GROUP=2
IKE_SA_LIFETIME=86400
IKE_CS=3DES-MD5,3DES-SHA1
TYPE=CMTS
Reply : Success: Entry 1 of 1 returned.
```

As soon as a CMTS aggregation device is defined, the BTS attempts to establish the TCP connection to the CMTS on port 2126. You can verify the status of this connection using the **status** verb, as demonstrated here:

```
CLI>status aggr id=PC-CMTS
AGGR ID -> PC-CMTS
OPER STATE -> AGGR IN Service
RESULT -> ADM configure result in success
REASON -> ADM executed successfully
Reply : Success:
```

If everything is working, you see an operational state of AGGR IN Service. A state of Connecting means the BTS is in the process of establishing the TCP connection, and a state of Out of Service means this connection is not operational. For more information on CMTS-to-CMS communication, refer to Chapters 12 and 13.

Step 2: Configure the MTA Characteristics

Instead of having to define all the characteristics of every MTA serviced by the BTS, you define a template that the MTA references. This template is referred to as a media gateway (abbreviated *mgw* in BTS nomenclature) profile. Among the characteristics defined here are several that relate to call signaling. The following syntax shows how to add a media gateway profile:

```
CLI>add mgw-profile id=mta mgw profile name; vendor=MTA vendor name; mgcp-
  version=MGCP_1_0; mgcp-variant=NCS_1_0; description=MTA description;
```

Table 7-6 lists key fields in this table relating to a PacketCable deployment; a majority of these fields are related to call signaling.

Table 7-6 *Key MGW-PROFILE Table Fields*

Field	Default Value	Recommended Value	Note
VENDOR	N/A	MTA vendor name.	MTA vendor name.
DESCRIPTION	N/A	Defining descriptions is a good practice and should be added when possible.	Textual description of the media gateway profile.
Fields relating to the command header and parameters as well as general protocol operation			
MGCP_VERSION	MGCP_0_1	MGCP_1_0	MGCP protocol version used.
MGCP_VARIANT	None	NCS_1_0 for MTAs TGCP_1_0 for trunking media gateways	MGCP protocol variant.
MGCP_NE_LOCAL NAME_SUPP	N	Y	CMS Notified Entity parameter expressed as <local>@FQDN.
MGCP_ERQNT_ SUPP	N	Y Embedded RQNT use can reduce the number and size of MGCP messages that need to be sent, improving network speed and efficiency; however, some MTAs might not handle this functionality very well, so set this field to N for these gateways.	MTA supports embedded RQNT. Recall this is when an RQNT message is in effect embedded or included in another message such as a CRCX, an MDCX, or even another RQNT. This in effect allows the BTS to send multiple commands at once. For example, instead of having to send one RQNT message to detect an off hook event and then another to detect digits and provide dial tone, a single RQNT command can be used.
MGCP_3WAY_ HSHAKE_SUPP	Y	Y	MTA can handle Response Acks.
MGCP_CONN_ID_ AT_GW_SUPP	Y	Y Set this parameter to Y so the BTS can delete hung connections on a gateway.	Delete only connections returned from AUEP.

continues

Table 7-6 *Key MGW-PROFILE Table Fields (Continued)*

Field	Default Value	Recommended Value	Note
MGCP_HAIRPIN_ SUPP	Y	N Hairpinning is not permitted in PacketCable.	MTA can do TDM hairpinning, meaning if both endpoints are on the same gateway, then the network type parameter local is used (nt:local) instead of IP.
MGCP_HAIRPIN_ Z2_SUPP	N	N Hairpinning is not permitted in PacketCable.	MTA supports second endpoint; parameter applies only if MGCP_HAIRPIN_SUPP is Y.
MGCP_XDLCX_ SUPP	N	N	Disables restrictions on the X: parameter in a DLCX.
CONN_MODE_ REQUIRED_IN_ MDCX	N	N	The MTA should need the connection mode parameter only if the mode has changed.
MGCP_PIGGY BACK_MSG_SUPP	Y	Y	BTS can piggyback MGCP messages in response packets sent to a gateway.
MGCP_REQ_ID_ SUPP	N	Y	Gateway supports correlation of request identifier from RQNT to NTFY messages.
MGCP_RSIPSTAR_ SUPP	Y	Y	An RSIP restart message using the * wildcard for the endpoint name implies all endpoints are available.
MGCP_TERM_ INIT_LEVEL	0	0 For trunking media gateways; implies endpoints initialized on the DS-0 level.	Applicable for trunking media gateways; defines initialization level.
MGCP_TEST_ CONN_SUPP	Y	Y	Permits the network loopback (netwloop) and continuity test (netwtest) connection modes on the media gateway.

Table 7-6 *Key MGW-PROFILE Table Fields (Continued)*

Field	Default Value	Recommended Value	Note
PARALLEL_TEST_ CONN_SUPP	Y	Y	Whether a network loopback or continuity test connection is allowed at the same time as another inactive, recvonly, or sendrecv mode connection.
MGCP_T_TRAN	400 ms	400 ms	Initial timeout after which a command is retransmitted; future timeouts increase exponentially until a limit is reached.
MGCP_T_ LONGTRAN	5 ms	5 ms	MGCP transaction timeout after receiving a 100 provisional response.
REFRESH_DIGIT_ MAP	N	N A PacketCable MTA remembers its last digit map.	Digit map is refreshed on the gateway for every call.
MGCP_EP_ SPECIFIC_CAP_ SUPP	N	N	If the media gateway supports endpoint-specific queries.
Fields relating to packages, events, and signals			
MGCP_DEFAULT_ PKG	NONE	LINE Set to LINE for MTAs; for SS7 media gateways the value TRUNK is used; for CAS E911 media gateways the value MO is used.	Default package name.
MGCP_VMWI_ SUPP	N	Y	MTA can set visual message waiting indicator. S: vmwi
MGCP_MWI_SUPP	Y	Y The user hears an MWA tone if they have a message; if this is set to N, a slutter tone is played instead.	MTA can set message waiting indicator. S: mwi

continues

Table 7-6 *Key MGW-PROFILE Table Fields (Continued)*

Field	Default Value	Recommended Value	Note
MGCP_DIALTONE_TO_SUPP	Y	Y	Gateway supports the timeout parameter for the dial tone signal; value is set in the CA_CONFIG table.
MGCP_TO_SUPP	Y	Y	Gateway supports the timeout parameter for signals; values set in the CA_CONFIG table.
RBK_ON_CONN_SUPP	Y	Y	MTA can send ring back on connection. S: rt@<conn>
RBK_ON_INACTIVE_CONN_SUPP	N	N	MTA cannot send ring back on connection in inactive mode.
MGCP_QLOOP_SUPP	Y	Y	MTA supports quarantine looping.
MGCP_QDISCARD_SUPP	Y	Y	The discard method for quarantine handling is supported.
DTMF_OOB_SUPP	N	N	Whether or not the gateway supports mid-call notification of DTMF digits by sending a NTFY command.
Fields relating to the LCO parameter and options			
CODEC_NEG_SUPP	Y	Y	Negotiate call CODEC; works with QOS table.
PC_MPTIME_SUPP	Y	Y	LCO parm mp: supported.
MGCP_CAP_NEG_REQ	Y	Y	MTA can negotiate capabilities.
EC_SUPP	N	N Normally you want the gateway to decide on its own when echo cancellers should be enabled and disabled.	Echo cancellation. LCO parm e: supported.

Table 7-6 *Key MGW-PROFILE Table Fields (Continued)*

Field	Default Value	Recommended Value	Note
IPTOS_RTP_SUPP	N	Y For the value of the IP ToS byte see the QoS table description.	Set TOS byte: LCO parm t: supported.
MGCP_RSVP_ SUPP	N	N	Media gateway supports the Resource Reservation Protocol (RSVP).
Fields relating to the SDP parameters			
SDP_ORIGFIELD_ SUPP	Y	Y	SDP parm o: supported.
SDP_SESSNAME_ SUPP	Y	Y	SDP parm s: supported.
SDP_EMAIL_SUPP	Y	N	SDP parm e: supported.
SDP_PHONE_ SUPP	Y	N	SDP parm p: supported.
SDP_URI_SUPP	Y	N	SDP parm u: supported.
SDP_BANDWIDTH _SUPP	Y	Y	SDP parm b: supported.
SDP_INFO_SUPP	Y	N	SDP parm i: supported.
SDP_TIME_SUPP	Y	Y	SDP parm t: supported.
SDP_CAP_ ENCODE_TYPE	AUTO	AUTO Values encoded as received from the remote endpoint.	Specified SDP capabilities encoding.
SDP_ATTRIB_ SUPP	Y	Y	SDP parm a: supported.
Fields relating to gateway keepalive methods			
KEEPALIVE_ METHOD	AUEP	AUEP It is important for the BTS to know the current operational state of its gateways; if it does not, it can try to route calls to gateways that are not operational. Therefore, it is recommended to keep this field set to its default value of AUEP.	Defines the keepalive method between the BTS and media gateway. AUEP means MGCP AUEP messages are used; other values include AUEP-ICMP, which means AUEP is tried first, and if that fails, ICMP pings are used, and NONE, which means no heartbeat mechanism is used.

continues

Table 7-6 *Key MGW-PROFILE Table Fields (Continued)*

Field	Default Value	Recommended Value	Note
MGCP_KEEPALIVE_INTERVAL	60 seconds	60 seconds	MTA AUEP interval when no activity detected between BTS and gateway.
MGCP_KEEPALIVE_RETRIES	3	3	Number of keepalive retries if the first one fails.
MGCP_MAX_KEEPALIVE_INTERVAL	600 seconds	600 seconds	Maximum MGCP keepalive interval.
Fields relating to DNS			
MGCP_MAX1_RETRIES	2	2 Because PacketCable gateways typically have one only IP address, this field isn't really pertinent.	Number of retries for each gateway IP address.
MGCP_MAX2_RETRIES	3	3	Total number of retries for a gateway before declared unreachable.
DOMAIN_NAME_CACHING_SUPP	Y	Y Disabling DNS caching can cause an unnecessary high load on the BTS under high traffic conditions.	(Y) Cache the MTA IP address so a DNS lookup is needed only for the first message to a gateway, or (N) do a DNS lookup for every message that needs to be sent.
Fields related to facsimile transmission			
FAX_INBAND_METHOD	GW_SPECIFIED	GW_SPECIFIED This field is meaningful only if the field FAX_T38_ENABLED is set to N in the QoS table.	If set to GW_SPECIFIED, the BTS is not involved in how the gateways handle in-band fax transmissions; if set to T38_FXR_OFF, the BTS explicitly instructs the gateways not to do T38 fax relay.
T38_FXR_LOOSE_SUPP	AUTO	AUTO Support determined by gateway auditing.	Gateway supports call agent controlled T38 Loose mode fax.

Table 7-6 *Key MGW-PROFILE Table Fields (Continued)*

Field	Default Value	Recommended Value	Note
Fields related to security			
KRB_RESET_FLAG	Y	Y	Kerberos SA is automatically reestablished upon expiration.
IKE_CS	3DES-MD5, 3DES-SHA1	3DES-MD5, 3DES-SHA1	List of supported ciphersuites for authentication and encryption in priority order.
IKE_GROUP	2	2	Diffie-Helman exchange group; valid only for trunking gateways.
IKE_KEY	N/A	N/A	Preshared key for security association between BTS and trunking gateway.
IKE_SA_LIFETIME	86400 seconds	86400 seconds	Lifetime of IKE security association.
IPSEC_SA_ESP_CS	3DES-MD5, 3DES-SHA1, NULL-MD5, NULL-SAH1	3DES-MD5, 3DES-SHA1, NULL-MD5, NULL-SAH1s	List of supported IPsec ciphersuites for authentication and encryption in priority order.
IPSEC_SA_GRACE_PERIOD	21600 seconds	21600 seconds	IPsec key expiration grace time; must be less than or equal to 25 percent of the IPSEC_SA_LIFETIME value.
IPSEC_SA_LIFETIME	86400 seconds	86400 seconds	Lifetime of IPsec security association.
IPSEC_ULP_NAME	IP	IP	IPsec SA upper-layer protocol name.

Three types of MTAs are used for the examples in this chapter: Linksys CM2P2B, Scientific Atlanta DPX2203, and Arris TTM402. A media gateway profile has been created for each one, and that configuration is

```
CLI>show mgw-profile id=Linksys_CM2P2B
ID=Linksys_CM2P2B
VENDOR=Cisco_Linksys
MGCP_VARIANT=NCS_1_0
RBK_ON_CONN_SUPP=Y
PACKET_TYPE=IP
```

```
AAL1=N
AAL2=N
AAL5=N
PVC=N
SVC=N
SPVC=N
EC_SUPP=N
SDP_ORIGFIELD_SUPP=Y
SDP_SESSNAME_SUPP=Y
SDP_EMAIL_SUPP=N
SDP_PHONE_SUPP=N
SDP_URI_SUPP=N
SDP_BANDWIDTH_SUPP=Y
SDP_INFO_SUPP=N
SDP_TIME_SUPP=Y
SDP_ATTRIB_SUPP=Y
MGCP_ERQNT_SUPP=N
MGCP_HAIRPIN_SUPP=N
MGCP_QLOOP_SUPP=Y
MGCP_3WAY_HSHAKE_SUPP=Y
MGCP_CONN_ID_AT_GW_SUPP=Y
MGCP_VMWI_SUPP=Y
PORT_START=1
DESCRIPTION=Linksys CM2P2B eMTA
MGCP_VERSION=MGCP_1_0
MGCP_RSVP_SUPP=N
MGCP_RSIPSTAR_SUPP=Y
MGCP_TERM_INIT_LEVEL=0
MGCP_HAIRPIN_Z2_SUPP=N
DTMF_OOB_SUPP=N
MGW_TYPE=UNSPECIFIED
OSI_SUPP=N
MGCP_DIALTONE_TO_SUPP=Y
MGCP_MWI_SUPP=Y
SPARE1_SUPP=N
SPARE2_SUPP=N
IPTOS_RTP_SUPP=Y
USE_STATIC_PROFILE=N
REFRESH_DIGIT_MAP=N
MGCP_XDLCX_SUPP=N
MGCP_CAS_BLOCK_SUPP=N
CODEC_NEG_SUPP=Y
MGCP_DEFAULT_PKG=LINE
MGCP_KEEPALIVE_INTERVAL=60
MGCP_KEEPALIVE_RETRIES=3
MGCP_T_TRAN=400
MGCP_MAX1_RETRIES=2
MGCP_MAX2_RETRIES=3
MGCP_T_LONGTRAN=5
PC_MPTIME_SUPP=Y
MGCP_CAP_NEG_REQ=Y
RBK_ON_INACTIVE_CONN_SUPP=N
MGCP_NAS_SUPP=Y
DOMAIN_NAME_CACHING_SUPP=Y
BTXML_SUPP=N
CONN_MODE_REQUIRED_IN_MDCX=N
MGCP_NE_LOCALNAME_SUPP=Y
KRB_REEST_FLAG=Y
IPSEC_SA_ESP_CS=3DES-MD5,3DES-SHA1,NULL-MD5,NULL-SHA1
IPSEC_SA_LIFETIME=86400
IPSEC_SA_GRACE_PERIOD=21600
IPSEC_ULP_NAME=IP
IKE_GROUP=2
IKE_SA_LIFETIME=86400
```

```
IKE_CS=3DES-MD5,3DES-SHA1
MGCP_EP_SPECIFIC_CAP_SUPP=N
KEEPALIVE_METHOD=AUEP
MGCP_MAX_KEEPALIVE_INTERVAL=600
MGCP_REQ_ID_SUPP=Y
MGCP_PIGGYBACK_MSG_SUPP=Y
MGCP_QDISCARD_SUPP=Y
MGCP_TO_SUPP=Y
MGCP_TEST_CONN_SUPP=Y
PARALLEL_TEST_CONN_SUPP=Y
T38_FXR_LOOSE_SUPP=AUTO
FAX_INBAND_METHOD=GW_SPECIFIED
SDP_CAP_ENCODE_TYPE=AUTO
Reply : Success: Entry 1 of 1 returned.
CLI>show mgw-profile id=SA_DPX2203
ID=SA_DPX2203
VENDOR=Cisco_SA
MGCP_VARIANT=NCS_1_0
RBK_ON_CONN_SUPP=Y
PACKET_TYPE=IP
AAL1=N
AAL2=N
AAL5=N
PVC=N
SVC=N
SPVC=N
EC_SUPP=N
SDP_ORIGFIELD_SUPP=Y
SDP_SESSNAME_SUPP=Y
SDP_EMAIL_SUPP=N
SDP_PHONE_SUPP=N
SDP_URI_SUPP=N
SDP_BANDWIDTH_SUPP=Y
SDP_INFO_SUPP=N
SDP_TIME_SUPP=Y
SDP_ATTRIB_SUPP=Y
MGCP_ERQNT_SUPP=N
MGCP_HAIRPIN_SUPP=N
MGCP_QLOOP_SUPP=Y
MGCP_3WAY_HSHAKE_SUPP=Y
MGCP_CONN_ID_AT_GW_SUPP=Y
MGCP_VMWI_SUPP=Y
PORT_START=1
DESCRIPTION=Scientific Atlanta DPX2203 eMTA
MGCP_VERSION=MGCP_1_0
MGCP_RSVP_SUPP=N
MGCP_RSIPSTAR_SUPP=Y
MGCP_TERM_INIT_LEVEL=0
MGCP_HAIRPIN_Z2_SUPP=N
DTMF_OOB_SUPP=N
MGW_TYPE=UNSPECIFIED
OSI_SUPP=N
MGCP_DIALTONE_TO_SUPP=Y
MGCP_MWI_SUPP=Y
SPARE1_SUPP=N
SPARE2_SUPP=N
IPTOS_RTP_SUPP=Y
USE_STATIC_PROFILE=N
REFRESH_DIGIT_MAP=N
MGCP_XDLCX_SUPP=N
MGCP_CAS_BLOCK_SUPP=N
CODEC_NEG_SUPP=Y
MGCP_DEFAULT_PKG=LINE
MGCP_KEEPALIVE_INTERVAL=60
```

```
MGCP_KEEPALIVE_RETRIES=3
MGCP_T_TRAN=400
MGCP_MAX1_RETRIES=2
MGCP_MAX2_RETRIES=3
MGCP_T_LONGTRAN=5
PC_MPTIME_SUPP=Y
MGCP_CAP_NEG_REQ=Y
RBK_ON_INACTIVE_CONN_SUPP=N
MGCP_NAS_SUPP=Y
DOMAIN_NAME_CACHING_SUPP=Y
BTXML_SUPP=N
CONN_MODE_REQUIRED_IN_MDCX=N
MGCP_NE_LOCALNAME_SUPP=Y
KRB_REEST_FLAG=Y
IPSEC_SA_ESP_CS=3DES-MD5,3DES-SHA1,NULL-MD5,NULL-SHA1
IPSEC_SA_LIFETIME=86400
IPSEC_SA_GRACE_PERIOD=21600
IPSEC_ULP_NAME=IP
IKE_GROUP=2
IKE_SA_LIFETIME=86400
IKE_CS=3DES-MD5,3DES-SHA1
MGCP_EP_SPECIFIC_CAP_SUPP=N
KEEPALIVE_METHOD=AUEP
MGCP_MAX_KEEPALIVE_INTERVAL=600
MGCP_REQ_ID_SUPP=Y
MGCP_PIGGYBACK_MSG_SUPP=Y
MGCP_QDISCARD_SUPP=Y
MGCP_TO_SUPP=Y
MGCP_TEST_CONN_SUPP=Y
PARALLEL_TEST_CONN_SUPP=Y
T38_FXR_LOOSE_SUPP=AUTO
FAX_INBAND_METHOD=GW_SPECIFIED
SDP_CAP_ENCODE_TYPE=AUTO
Reply : Success: Entry 1 of 1 returned.
CLI>show mgw-profile id=Arris_TTM402
ID=Arris_TTM402
VENDOR=Arris
MGCP_VARIANT=NCS_1_0
RBK_ON_CONN_SUPP=Y
PACKET_TYPE=IP
AAL1=N
AAL2=N
AAL5=N
PVC=N
SVC=N
SPVC=N
EC_SUPP=N
SDP_ORIGFIELD_SUPP=Y
SDP_SESSNAME_SUPP=Y
SDP_EMAIL_SUPP=N
SDP_PHONE_SUPP=N
SDP_URI_SUPP=N
SDP_BANDWIDTH_SUPP=Y
SDP_INFO_SUPP=N
SDP_TIME_SUPP=Y
SDP_ATTRIB_SUPP=Y
MGCP_ERQNT_SUPP=Y
MGCP_HAIRPIN_SUPP=N
MGCP_QLOOP_SUPP=Y
MGCP_3WAY_HSHAKE_SUPP=Y
MGCP_CONN_ID_AT_GW_SUPP=Y
MGCP_VMWI_SUPP=Y
PORT_START=1
DESCRIPTION=Arris TTM402 eMTA
```

```
MGCP_VERSION=MGCP_1_0
MGCP_RSVP_SUPP=N
MGCP_RSIPSTAR_SUPP=Y
MGCP_TERM_INIT_LEVEL=0
MGCP_HAIRPIN_Z2_SUPP=N
DTMF_OOB_SUPP=N
MGW_TYPE=UNSPECIFIED
OSI_SUPP=N
MGCP_DIALTONE_TO_SUPP=Y
MGCP_MWI_SUPP=Y
SPARE1_SUPP=N
SPARE2_SUPP=N
IPTOS_RTP_SUPP=Y
USE_STATIC_PROFILE=N
REFRESH_DIGIT_MAP=N
MGCP_XDLCX_SUPP=N
MGCP_CAS_BLOCK_SUPP=N
CODEC_NEG_SUPP=Y
MGCP_DEFAULT_PKG=LINE
MGCP_KEEPALIVE_INTERVAL=60
MGCP_KEEPALIVE_RETRIES=3
MGCP_T_TRAN=400
MGCP_MAX1_RETRIES=2
MGCP_MAX2_RETRIES=3
MGCP_T_LONGTRAN=5
PC_MPTIME_SUPP=Y
MGCP_CAP_NEG_REQ=Y
RBK_ON_INACTIVE_CONN_SUPP=N
MGCP_NAS_SUPP=Y
DOMAIN_NAME_CACHING_SUPP=Y
BTXML_SUPP=N
CONN_MODE_REQUIRED_IN_MDCX=N
MGCP_NE_LOCALNAME_SUPP=Y
KRB_REEST_FLAG=Y
IPSEC_SA_ESP_CS=3DES-MD5,3DES-SHA1,NULL-MD5,NULL-SHA1
IPSEC_SA_LIFETIME=86400
IPSEC_SA_GRACE_PERIOD=21600
IPSEC_ULP_NAME=IP
IKE_GROUP=2
IKE_SA_LIFETIME=86400
IKE_CS=3DES-MD5,3DES-SHA1
MGCP_EP_SPECIFIC_CAP_SUPP=N
KEEPALIVE_METHOD=AUEP
MGCP_MAX_KEEPALIVE_INTERVAL=600
MGCP_REQ_ID_SUPP=Y
MGCP_PIGGYBACK_MSG_SUPP=Y
MGCP_QDISCARD_SUPP=Y
MGCP_TO_SUPP=Y
MGCP_TEST_CONN_SUPP=Y
PARALLEL_TEST_CONN_SUPP=Y
T38_FXR_LOOSE_SUPP=AUTO
FAX_INBAND_METHOD=GW_SPECIFIED
SDP_CAP_ENCODE_TYPE=AUTO
Reply : Success: Entry 1 of 1 returned.
```

Step 3: Configure the MTA

After the MTA characteristics are defined, you need to define the MTA itself and associate it to a media gateway profile and a CMTS, using the following syntax:

```
CLI>add mgw id=mta mgw name; tsap-addr=mta FQDN; call-agent-id=call agent name; mgw-
  profile-id=mta mgw profile name; type=rgw; aggr-id=CMTS name;
```

Table 7-7 lists some of the key fields and fields relating to call signaling.

Table 7-7 *Gateway Table Field Descriptions*

Field	Default Value	Recommended Value	Description
TSAP-ADDR	N/A	MTA FQDN	IP address or FQDN of the MTA.
CALL-AGENT-ID	N/A	Key for the call agent table.	Identifier that the call agent subscribers of this mgw are assigned to.
MGW-PROFILE-ID	N/A	Key for the media gateway profile table describing this MTA.	Identifier of the media gateway profile used.
AGGR-ID	N/A	Key for the aggregation table referencing the CMTS the MTA is connected to.	Identifier of the CMTS used.
NODE	N/A	Descriptive name for the HFC node this MTA is physically connected to.	Defines the HFC node the MTA is connected to.
Type	N/A	RGW	A MTA is a residential gateway; in the next chapter you'll see trunking media gateways are defined as TGW.
CALL-AGENT-CONTROL-PORT	0 (Call agent automatically assigns it.)	0	Call agent MGCP listening port for this media gateway.

NOTE The aggregation ID is an optional field for a media gateway; however, without it the BTS does not try to provide DQoS with the CMTS and MTA. In terms of NCS this means that CRCX and MDCX commands do not have the DQoS Local Connection Parameters included, in turn causing the MTA not to signal the CMTS for quality of service.

In the examples you saw five PacketCable subscribers: Oscar, Tina, Alan, Bob, and Cathy. Oscar and Alan used the Linksys MTAs jriddel_lsys1 and jriddel_lsys2, respectively; Tina and Bob used the Scientific Atlanta MTAs jriddel_sa1 and jriddel_sa2, respectively; and

Cathy used the Arris MTA jriddel_arris1. The media gateway table configuration for these MTAs was as follows:

```
CLI>show mgw id=jriddel%
ID=jriddel_arris1
TSAP_ADDR=x1-6-00-00-ca-dd-a7-e4.cc.cisco.com
CALL_AGENT_ID=CA104
MGW_PROFILE_ID=Arris_TTM402
STATUS=INS
AGGR_ID=PC-CMTS
TYPE=RGW
ID=jriddel_lsys1
TSAP_ADDR=x1-6-00-13-10-50-38-02.cc.cisco.com
CALL_AGENT_ID=CA104
MGW_PROFILE_ID=Linksys_CM2P2B
STATUS=INS
AGGR_ID=PC-CMTS
TYPE=RGW
ID=jriddel_lsys2
TSAP_ADDR=x1-6-00-13-10-50-37-f6.cc.cisco.com
CALL_AGENT_ID=CA104
MGW_PROFILE_ID=Linksys_CM2P2B
STATUS=INS
AGGR_ID=PC-CMTS
TYPE=RGW
ID=jriddel_sa1
TSAP_ADDR=x1-6-00-0a-73-fa-be-8c.cc.cisco.com
CALL_AGENT_ID=CA104
MGW_PROFILE_ID=SA_DPX2203
STATUS=INS
AGGR_ID=PC-CMTS
TYPE=RGW
ID=jriddel_sa2
TSAP_ADDR=x1-6-00-0a-73-fa-be-e4.cc.cisco.com
CALL_AGENT_ID=CA104
MGW_PROFILE_ID=SA_DPX2203
STATUS=INS
AGGR_ID=PC-CMTS
TYPE=RGW
Reply : Success:  Entries 1-5 of 5 returned.
```

Step 4: Configure the NCS Endpoints

Now that you have the MTA configured, you need to configure the NCS endpoints on that MTA. This is done in the termination table:

```
CLI>add termination prefix=local endpoint name prefix; port-start=starting port;
  port-end=ending port; type=endpoint type; mgw-id= mgw name;
```

Table 7-8 lists the key fields relating to PacketCable configuration.

Table 7-8 *Termination Table Field Descriptions*

Field	Default Value	Recommended Value	Description
PREFIX	N/A	aaln/ For NCS endpoints; you'll see the value for other types of endpoints later.	Endpoint prefix
PORT-START	1	1	Initial port number
PORT-END	24	2 This range covers both endpoints of a PacketCable MTA.	Ending port number
TYPE	N/A	LINE For MTAs; for trunking gateways use the value TRUNK.	Termination type
MGW-ID	N/A	Key for the media gateway.	Media gateway name

The following is the configuration for the terminations on Oscar's MTA media gateway (jriddel_lsys1). Notice at this point they show a status of unequipped (UEQP):

```
CLI>show termination mgw-id=jriddel_lsys1
ID=aaln/1
MGW_ID=jriddel_lsys1
TYPE=LINE
STATUS=UEQP
ID=aaln/2
MGW_ID=jriddel_lsys1
TYPE=LINE
STATUS=UEQP
Reply : Success:  Entries 1-2 of 2 returned.
```

Step 5: Configure the Subscriber QoS Parameters

In the media gateway profile, you saw that some of the fields influence the Local Connection Option parameters sent to MTA endpoints. However, you probably noticed that not all of the LCO parameters could be specified here. The reason is they are defined in another table, the QoS table, with the following syntax:

```
CLI>add qos id=QoS profile name; codec-type=preferred CODEC; port-end=ending port;
  type=LINE; mgcp-pkg-type=LINE_NCS; mgw-id=mta mgw name;
```

Table 7-9 lists some of the key fields relating to PacketCable deployment.

Table 7-9 *QoS Table Fields*

Field	Default Value	Recommended Value	Description
CLIENT_TYPE	NONE	DQOS For a PacketCable Multimedia subscriber this should be set to a value of MM_COPS.	Identifies type of client requiring QoS
CODEC-TYPE	N/A	PCMU Optionally you might want to create an entry for a compression CODEC such as G.728 or G.729E	Preferred CODEC (see discussion after this table for more information on CODEC negotiation).
DOCSIS_DSCP_TOS	160 (10100000 in binary)	160 (10100000 in binary)	DSCP/IP ToS byte value used for matching packets to the DQoS flow; this is the value sent in the NCS LCO attribute t:.
DOCSIS_DSCP_TOS_BITMASK	224 (11100000 in binary)	224 (11100000 in binary)	Identifies the bits in the DSCP/IP ToS byte that must be matched when classifying packets.
DQOS_CMTS_DSCP_TOS	160 (10100000 in binary)	160 (10100000 in binary)	DSCP/IP ToS byte value sent to the CMTS in COPS Gate Set messages; the CMTS overwrites RTP packets with this value before sending them out the WAN interface.
DQOS_DSCP_TOS_BITMASK	224 (11100000 in binary)	224 (11100000 in binary)	Identifies the bits in the DSCP/IP ToS byte that must be matched when classifying packets.
DTMF-PREF-MODE	DTMF-INBOUND	DTMF-INBOUND	Preferred DTMF relay method.
FAX_T38_ENABLED	Y	Y	Use T.38 for facsimile transmission.
HBW	0 (gateway chosen)	0	Higher bandwidth in kilobits/second.

continues

Table 7-9 *QoS Table Fields (Continued)*

Field	Default Value	Recommended Value	Description
HPTIME	10 ms	20 ms	Higher packetization period.
LBW	0 (gateway chosen)	0	Lower bandwidth in kilobits/ second
LPTIME	10 ms	20 ms	Lower packetization period.
IPTOS_RTP_ LOWDELAY	N	N	Whether or not the Delay bit in the IP ToS byte is set.
IPTOS_RTP_ PRECEDENCE	CRITICAL(5)	CRITICAL (5)	Value of the three IP Precedence bits in the IP ToS byte.
IPTOS_RTP_ RELIABILITY	N	N	Whether or not the Reliability bit in the IP ToS byte is set.
IPTOS_RTP_ THROUGHPUT	N	N	Whether or not the Throughput bit in the IP ToS byte is set
MAX_DQOS_ AUTH_ BANDWIDTH	HIGH	HIGH	Maximum authorized amount of DQoS bandwidth (High = G.711, Low = G.72x).
MAX_DQOS_ SESSIONS	5	5	Maximum number of DQoS session permitted for a subscriber.
RESOURCE_ RESERVATION	BE (Best Effort)	BE (Best Effort)	PacketCable 1.x devices do not use RSVP.
RSVP_ REQUIRED	N	N	PacketCable 1.x devices do not use RSVP.
SILENT_ SUPPRESSION	N	N If you want to utilize the UGS-AD service to free up DOCSIS bandwidth, set this parameter to Y.	Enable silence suppression (a.k.a. VAD). LCO parm s:.

If CODEC negotiation is enabled for the MTA in the media gateway profile, the CODEC and packetization periods defined in the QOS table identify the highest priorities. Values obtained from an audit of the endpoint's capabilities are included after that. If CODEC negotiation is disabled, the CODEC and packetization period defined here are used for the MTA endpoint.

The QoS profile used by the subscribers in the call flow examples is

```
CLI>show qos id=G711u_20ms
ID=G711u_20ms
CODEC_TYPE=PCMU
RSVP_REQUIRED=N
RESOURCE_RESERVATION=BE
IPTOS_RTP_LOWDELAY=N
IPTOS_RTP_THROUGHPUT=N
IPTOS_RTP_RELIABILITY=N
IPTOS_RTP_PRECEDENCE=CRITICAL
MAX_DQOS_SESSIONS=5
MAX_DQOS_AUTH_BANDWIDTH=HIGH
DTMF_PREF_MODE=DTMF_INBAND
DQOS_CMTS_DSCP_TOS=160
LBW=0
HBW=0
AUTO_GAIN_CONTROL=N
GAIN_VALUE=0
SILENT_SUPPRESSION=N
LPTIME=20
HPTIME=20
VIDEO_CODEC_TYPE=NONE
CLIENT_TYPE=DQOS
DQOS_DSCP_TOS_BITMASK=224
DOCSIS_DSCP_TOS=160
DOCSIS_DSCP_TOS_BITMASK=224
FAX_T38_ENABLED=Y
Reply : Success: Entry 1 of 1 returned.
```

Step 6: Configure Subscriber Characteristics

Some mechanism needs to exist to correlate information such as dial plans to subscribers; this is the subscriber profile table. This table is also used to associate POPs, digit maps, and QoS profiles to a subscriber or group of subscribers:

```
CLI>add subscriber_profile id=subscriber profile identifier; dial_plan_id=dial plan
  identifier; pop_id=Point of Presence (POP) identifier; digit_map_id=digit map
  identifier; qos_id=QoS table identifier;
```

The following is the configuration of the subscriber profile used for the examples in this chapter:

```
CLI>show subscriber_profile id=PacketCable_subs
ID=PacketCable_subs
DIAL_PLAN_ID=INCOMING
LOCAL_PFX1_OPT=NR
TOLL_PFX1_OPT=RQ
POP_ID=PacketCable_PoP
OLI=0
EA_USE_PIC1=Y
QOS_ID=G711u_20ms
INTERLATA_PFX1_OPT=RQ
Reply : Success: Entry 1 of 1 returned.
```

Step 7: Configure the PacketCable Subscriber

The last configuration step is to configure the parameters associated with the subscriber using the NCS endpoint such as phone number:

```
CLI>add subscriber id=subscriber identifier; name=subscriber name;
  category=INDIVIDUAL; dn1=sub phone number; term-id=termination identifier; mgw-
  id=mgw name; sub-profile-id=sub profile name;
```

Table 7-10 lists some of the key fields of the subscriber table.

Table 7-10 *Subscriber Table Key Fields and Descriptions*

Field	Default Value	Recommended Value	Description
NAME	N/A	Subscriber name; typically LAST NAME, FIRST NAME Avoid using the ",',-, and _ characters in this field.	Subscriber name used for CNAM (Caller ID) feature.
PRIVACY	NONE (Name and number)	None Other valid values are FULL (neither name nor number displayed), NAME (name not displayed), and USER (use user-provided information).	Specifies how calling party information (name, #) is displayed to the called party.
CATEGORY	INDIVIDUAL	INDIVIDUAL	Indicates an individual subscriber.
DN1	N/A	Subscriber telephone number	Subscriber Directory Number.
TERM_TYPE	TERM (MGCP termination)	TERM (MGCP termination)	Specifies termination type.
TERM-ID	N/A	aaln/1 or aaln/2	NCS endpoint identifier the subscriber is using.
MGW-ID	N/A	Key for the MTA media gateway	MTA media gateway name.
SUB-PROFILE-ID	N/A	Key for the subscriber profile	Subscriber profile used to associate POP and dial plan to subscriber.
QOS-ID	N/A	Key for QoS	QoS profile used for the subscriber.

The following is the configuration of the key fields for the subscribers used in the examples:

```
CLI>show subscriber display=id,category,name,dn1,privacy,term-id,mgw-id,
  sub-profile-id,term-type
ID=sub1@jriddel_arris2
CATEGORY=INDIVIDUAL
NAME=Cathy
DN1=3215012022
PRIVACY=NONE
TERM_ID=aaln/1
MGW_ID=jriddel_arris2
SUB_PROFILE_ID=PacketCable_subs
TERM_TYPE=TERM
ID=sub1@jriddel_lsys1
CATEGORY=INDIVIDUAL
NAME=Oscar
DN1=3215012000
PRIVACY=NONE
TERM_ID=aaln/1
MGW_ID=jriddel_lsys1
SUB_PROFILE_ID=PacketCable_subs
TERM_TYPE=TERM
ID=sub1@jriddel_lsys2
CATEGORY=INDIVIDUAL
NAME=Alan
DN1=3215012002
PRIVACY=NONE
TERM_ID=aaln/1
MGW_ID=jriddel_lsys2
SUB_PROFILE_ID=PacketCable_subs
TERM_TYPE=TERM
ID=sub1@jriddel_sa1
CATEGORY=INDIVIDUAL
NAME=Tina
DN1=3215012010
PRIVACY=NONE
TERM_ID=aaln/1
MGW_ID=jriddel_sa1
SUB_PROFILE_ID=PacketCable_subs
TERM_TYPE=TERM
ID=sub1@jriddel_sa2
CATEGORY=INDIVIDUAL
NAME=Bob
DN1=3215012011
PRIVACY=NONE
TERM_ID=aaln/1
MGW_ID=jriddel_sa2
SUB_PROFILE_ID=PacketCable_subs
TERM_TYPE=TERM
Reply : Success:  Entries 1-5 of 5 returned.
```

As soon as a subscriber is defined with a directory number (DN), this also triggers population of the DN2Subscriber table. As you learned in the section "Dial Plan Configuration Tasks," it is this table that is referenced for local call routing. This table also holds the administrative state for a particular DN. When a person's phone number is changed or disconnected or ported out to another provider, it is this table that needs to be updated. An example of this table is as follows:

```
CLI>show dn2subscriber office_code_index=1
Please wait....
OFFICE_CODE_INDEX=1
```

```
DN=2000
STATUS=ASSIGNED
RING_TYPE=1
LNP_TRIGGER=N
NP_RESERVED=N
SUB_ID=sub1@jriddel_lsys1
LAST_CHANGED=2006-05-03 17:32:35
ADMIN_DN=N
PORTED_IN=N
OFFICE_CODE_INDEX=1
DN=2010
STATUS=ASSIGNED
RING_TYPE=1
LNP_TRIGGER=N
NP_RESERVED=N
SUB_ID=sub1@jriddel_sa1
LAST_CHANGED=2006-05-04 10:31:38
ADMIN_DN=N
PORTED_IN=N
```

Step 8: Put the MTA into Service

Before MTA endpoints can be used to place and receive calls you need to perform a few administrative tasks on the BTS. The first one is to enable the MTA or put it into service:

```
CLI>control mgw id=MTA mgw identifier; target-state=INS; mode=FORCED;
```

The result of this operation can be verified with the **status** command:

```
CLI>status mgw id=MTA mgw identifier
```

The MTAs must be in the working state before they can be used. For example:

```
CLI>status mgw id=jriddel_lsys1
MGW_ID -> jriddel_lsys1
RESULT -> ADM configure result in success
REASON -> ADM executed successfully
ADMIN_STATE -> ADMIN_INS
OPER_STATE -> Media gateway in working status
Reply : Success: Entry 1 of 1 returned.
```

The other possible operational states are unknown, down, and cannot be reached. Logically, the cannot be reached state indicates a problem in IP connectivity; the down state typically indicates the receipt of a negative MGCP response message; and the unknown state can be the result of disabling media gateway monitoring.

Step 9: Prepare the Endpoints for Service

Now that the MTA is in a working state, the endpoints need to be put into a working state as well. The first step to doing this is to **equip** the termination:

```
CLI>equip subscriber-termination id=MTA subscriber endpoint identifier;
```

Step 10: Bring the Endpoints into Service

Finally, the MTA endpoints need to be enabled just as the MTA itself was:

```
CLI>control subscriber-termination id=MTA mgw identifier; target-state=INS;
  mode=FORCED;
```

Just like for the MTA, the result of this operation can be verified with the **status** command:

```
CLI>status subscriber-termination id=MTA subscriber identifier
```

The following is an example of a subscriber in an operational state:

```
CLI>status subscriber-termination id=sub1@jriddel_lsys1
SUBSCRIBER_ID -> sub1@jriddel_lsys1
RESULT -> ADM configure result in success
REASON -> ADM executed successfully
ADMIN_STATE -> ADMIN_INS
OPER_STATE -> Term is available for new calls
FAULT_REASON -> No fault reason available
Reply : Success: Entry 1 of 1 returned.
```

Notice the operational state (OPER_STATE) is Term is available for new calls, meaning the endpoint is ready to process calls.

Call Feature Configuration Tasks

The configuration steps discussed so far allow for basic ONNET-to-ONNET calls, but in all likelihood you intend to offer your customers a variety of call features such as the CLASS features call waiting, call forwarding, three-way calling, and so on. The steps and table used for providing call features are as follows (Note: These tables are also typically populated at the time of installation and shouldn't require modification with the exception of the subscriber service profile table):

Step 1 Define feature servers (feature server table).

Step 2 Define features (feature table).

Step 3 Define vertical service codes (vertical service code table).

Step 4 Define services (service table).

Step 5 Associating subscribers with services (subscriber service profile table).

Step 6 Associating subscribers with features (subscriber feature data table).

Step 1: Define the Feature Servers

Remember that a BTS call agent handles feature functionality in addition to providing basic call control functionality using different internal processes. You have already defined the call agent process when you added the call agent; you must also define the feature servers. Two feature servers are normally defined. One is known as the Advanced Intelligent Networking (AIN) feature server, and the other is known as the plain old telephone service

(POTS) feature server. The POTS feature server handles POTS features as well as Centrex features and Tandem features. With release 4.5 of the BTS you can also have external feature servers (type = 3PTY). The syntax to define the feature server is as follows:

```
CLI>add feature_server id=Feature Server identifier; tsap_addr=FQDN and port number
  used to reach the feature server; type=AIN¦POTS¦3PTY;
```

The configuration of this table used for the call flow examples is

```
CLI>show feature_server
ID=FSAIN304
TSAP_ADDR=asm-AIN304.roseville.cc.cisco.com:11205
TYPE=AIN
EXTERNAL_FEATURE_SERVER=N
ID=FSPTC504
TSAP_ADDR=pots-PTC504.roseville.cc.cisco.com:11235
TYPE=POTS
EXTERNAL_FEATURE_SERVER=N
Reply : Success:  Entries 1-2 of 2 returned.
```

Step 2: Define Features

The characteristics of these call features are defined in the Feature table.

In the call flow examples, the features calling number delivery (CND), calling name delivery (CNAM), call waiting (CW), and three-way calling (TWC) were used. As an example, the configuration in this table for these features is as follows:

```
CLI>show feature fname=CND
FNAME=CND
TDP1=FACILITY_SELECTED_AND_AVAILABLE
TID1=TERMINATION_RESOURCE_AVAILABLE
TTYPE1=R
FEATURE_SERVER_ID=FSPTC504
DESCRIPTION=Calling number delivery
GRP_FEATURE=N
Reply : Success: Entry 1 of 1 returned.
CLI>show feature fname=CNAM
FNAME=CNAM
TDP1=FACILITY_SELECTED_AND_AVAILABLE
TID1=TERMINATION_RESOURCE_AVAILABLE
TTYPE1=R
FEATURE_SERVER_ID=FSPTC504
DESCRIPTION=Calling Name
GRP_FEATURE=N
Reply : Success: Entry 1 of 1 returned.
CLI>show feature fname=CW
FNAME=CW
TDP1=T_BUSY
TID1=T_BUSY
TTYPE1=R
FEATURE_SERVER_ID=FSPTC504
DESCRIPTION=Call Waiting
GRP_FEATURE=N
Reply : Success: Entry 1 of 1 returned.
CLI>show feature fname=TWC
FNAME=TWC
TDP1=O_MID_CALL
TID1=O_SWITCH_HOOK_FLASH_IMMEDIATE
```

```
TTYPE1=R
TDP2=T_MID_CALL
TID2=T_SWITCH_HOOK_FLASH_IMMEDIATE
TTYPE2=R
FEATURE_SERVER_ID=FSPTC504
DESCRIPTION=Three way call
GRP_FEATURE=N
Reply : Success: Entry 1 of 1 returned.
```

The TDP fields refer to the Trigger Detection Points of the individual features, and the TID fields refer to the Trigger Identifiers. When one of these triggers occurs, the call agent communicates with the specified feature server. The trigger type fields (TTYPE) specify whether a response is required (R) from the feature server.

Step 3: Define Vertical Service Codes

As you are probably aware, when subscribers want to activate or deactivate certain call features, typically an asterisk (*) followed by a two-digit code is dialed on their phone. This star code is also known as a vertical service code (VSC). The VSC table on the BTS defines the mappings between these codes and the feature they activate or deactivate:

```
CLI>add vsc digit_string=VSC code (*xx); fname=feature name;
```

Many of the VSC table entries are preprovisioned:

```
CLI>show vsc
DIGIT_STRING=#76#
FNAME=HOTVD
DESCRIPTION=Hot Line Variable Activation
DIGIT_STRING=*57
FNAME=COT
DIGIT_STRING=*60
FNAME=SCR_ACT
DIGIT_STRING=*61
FNAME=DRCW_ACT
DIGIT_STRING=*63
FNAME=SCF_ACT
DIGIT_STRING=*64
FNAME=SCA_ACT
DIGIT_STRING=*66
FNAME=AC_ACT
DIGIT_STRING=*67
FNAME=CNDB
DIGIT_STRING=*69
FNAME=AR_ACT
DIGIT_STRING=*70
FNAME=CCW
DIGIT_STRING=*72
FNAME=CFUA
DIGIT_STRING=*73
FNAME=CFUD
DIGIT_STRING=*74
FNAME=SC1D_ACT
DIGIT_STRING=*75
FNAME=SC2D_ACT
DIGIT_STRING=*76*
FNAME=HOTVA
DESCRIPTION=Hot Line Variable Activation
DIGIT_STRING=*77
```

```
FNAME=ACR_ACT
DIGIT_STRING=*78
FNAME=DND_ACT
DIGIT_STRING=*79
FNAME=DND_DEACT
DIGIT_STRING=*80
FNAME=SCR_ACT
DIGIT_STRING=*81
FNAME=DRCW_ACT
DIGIT_STRING=*82
FNAME=CIDSD
DIGIT_STRING=*83
FNAME=SCF_ACT
DIGIT_STRING=*84
FNAME=SCA_ACT
DIGIT_STRING=*86
FNAME=AC_DEACT
DIGIT_STRING=*87
FNAME=ACR_DEACT
DIGIT_STRING=*89
FNAME=AR_DEACT
DIGIT_STRING=*90
FNAME=CFBVA
DIGIT_STRING=*91
FNAME=CFBVD
DIGIT_STRING=*92
FNAME=CFNAVA
DIGIT_STRING=*93
FNAME=CFNAVD
DIGIT_STRING=*95
FNAME=CNAB
DIGIT_STRING=*98
FNAME=RACF_PIN
Reply : Success:  Entries 1-32 of 32 returned.
```

Step 4: Define Services

Telephony service providers typically bundle call features into packages for their subscribers. On the BTS these feature bundles are referred to as *services*. A service comprises features that are defined in the service table. Up to 10 features can be bundled together in a single service using the following syntax:

```
CLI>add service id=service identifier fname1=feature name; fname2=feature name; ...
```

The following three services are used in the call flow examples:

```
CLI>show service id=7
ID=7
FNAME1=CW
FNAME2=CCW
DESCRIPTION=Call Waiting and Cancel Call waiting
Reply : Success: Entry 1 of 1 returned.
CLI>show service id=9
ID=9
FNAME1=CND
FNAME2=CNAM
DESCRIPTION=Calling number delivery and CNAM
Reply : Success: Entry 1 of 1 returned.
CLI>show service id=13
ID=13
```

```
FNAME1=TWC
DESCRIPTION=Three way call
Reply : Success: Entry 1 of 1 returned.
```

Step 5: Associate Subscribers with Services

Now that you have your services defined, you need to assign subscribers to these services; the table subscriber service profile is used for this purpose. Until this table is populated, the subscriber cannot use the features defined in these services:

```
CLI>add subscriber_service_profile sub_id=subscriber identifier service_id=
  service name;
```

Up to 50 service identifiers can be assigned to a single subscriber. As you might recall, all the call feature functionality shown in this chapter was used by Bob; therefore, Bob's subscriber was assigned to these services as shown:

```
CLI>show subscriber-service-profile sub-id=sub1@jriddel_sa2
SUB_ID=sub1@jriddel_sa2
SERVICE_ID=13
SUB_ID=sub1@jriddel_sa2
SERVICE_ID=7
SUB_ID=sub1@jriddel_sa2
SERVICE_ID=9
Reply : Success:  Entries 1-3 of 3 returned.
```

Step 6: Associate Subscribers to Features

Another way to assign features to a subscriber is by using the subscriber feature data table. It is recommended, however, to use the subscriber service profile table for this purpose instead. The main use of this table is to see what the current activation state is for the VSC-controlled features. The syntax to do this is as follows:

```
CLI>show subscriber_feature_data sub_id=subscriber identifier
```

Troubleshooting Tips

The last part of this chapter contains some troubleshooting tips that can help you diagnose and fix problems relating to NCS.

General BTS Troubleshooting Tips

Before getting into suggestions on troubleshooting specific types of problems you might encounter, some basic tips specific to BTS troubleshooting are covered. The topics covered in this section are covered at a very high level. For more details please refer to BTS-specific documentation.

Displaying Running Processes on the BTS

On each BTS machine, you can obtain the status of the running processes with the **nodestat** command from the UNIX prompt. This command also tells you the BTS version numbers as well as the status of the individual interfaces on that machine. Figure 7-26 shows an example of this output on a call agent machine.

Figure 7-26 *BTS **nodestat** Command*

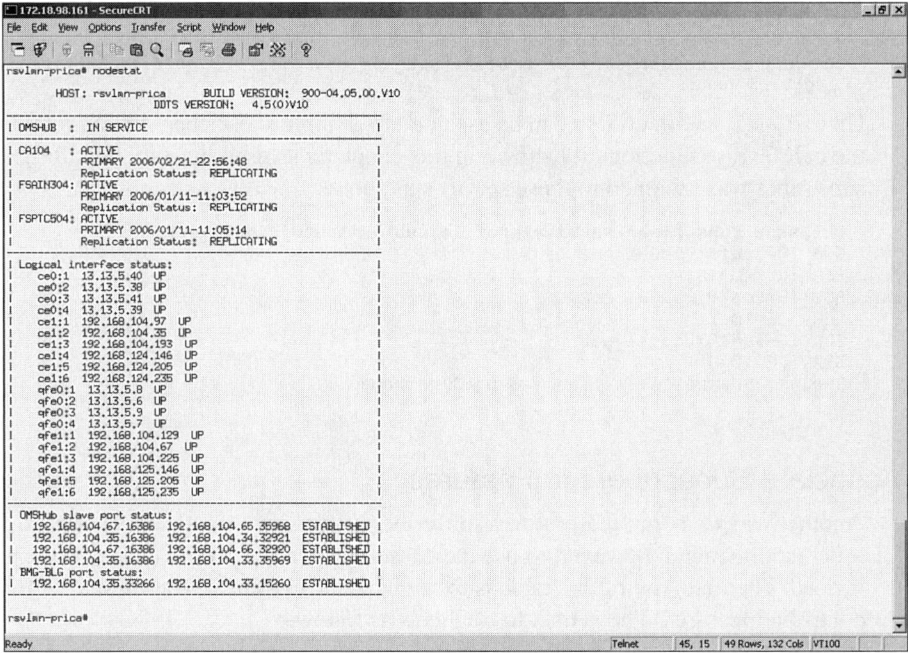

Each process should show a status of ACTIVE, indicating the process is running properly, and a replication status of REPLICATING, indicating the process is properly synchronizing with the standby machine. Notice that because this is a call agent machine, you have a call agent process as well as a process for each feature server. On an EMS machine you would see a general EMS process and a billing (BDMS) process. Processes can be stopped and started using the platform utility from the UNIX command line. The syntax of this command on a call agent machine is as follows:

```
rsvlmn-prica# platform
Usage: /opt/bts/bin/platform {start¦stop¦-h} [-noping¦-noirdp¦-bypass¦-nocopy]
  [-f¦-i PROCESS]
        start      : start all BTS 10200 PROCESS.
        stop       : stop BTS 10200 application instances.
        -bypass    : start application without error checking.
        -nocopy    : not saving shared memory DB on start.
        -noping    : start application without mate pinging.
        -noirdp    : start application using static routing.
```

```
        -notimecheck: start application without time check between the mate.
        -nooracheck : do not check system configurations for Oracle DB.
        -reboot     : run this program under reboot. Skip several check.
        -h  help    : print this usage information.
        -f  force   : terminate all BTS 10200 PROCESS.
        -i  instance: BTS 10200 application instance.
Valid PROCESS:
        CA104
        FSAIN304
        FSPTC504
```

NOTE Root access is required on the machine to stop and start processes.

Displaying BTS Events and Alarms

The BTS generates events and alarms for the real-time notification of network conditions. These alarms and events can be retrieved either via SNMP or through CLI commands. The difference between events and alarms is their severity level; alarms are classified as minor, major, or critical, whereas events are classified as either informational or warning. Alarms and events are also classified by category type; the values are audit, billing, callp, config, database, maintenance, oss, security, signaling, statistics, and system. The following is an example alarm:

```
CLI>show alarm COMPONENT_ID=jriddel_lsys1
ID=5567
TYPE=SIGNALING
NUMBER=79
DESCRIPTION=Media Gateway unreachable
SEVERITY=MAJOR
ALARM_STATUS=ON
TIMESTAMP=2006-05-07 17:47:33
ORIGIN=MGA.A.CA104
THREAD=MGA.A
COMPONENT_ID=jriddel_lsys1
ENTITY NAME=x1-6-00-13-10-50-38-02.cc.cisco.com
GENERAL CONTEXT=ICMP ping status success
SPECIFC CONTEXT=MGW_RGW
FAILURE CONTEXT=MGCP connectivity not available
Reply : Success: Entry 1 of 1 returned.
```

The number field identifies a specific type of alarm or event in that category. The BTS troubleshooting guide contains detailed descriptions of each alarm and event as well as probable causes and suggested resolution procedures. For example, the documentation for signaling alarm number 79 suggests the probable cause is an MGCP interoperability issue between the media gateway and call agent.

Accessing BTS Log Files

On each BTS machine, running log files are kept that periodically rotate. Separate log files are kept for each process running on the machine. For example, on a call agent machine you have a separate running log file for the basic call agent process and each feature server process. On a call agent machine, these log files are stored in the directory /opt/OptiCall/<process>/bin/logs by default, where <process> indicates the name of the call agent or feature server application. The name of the log files contain a timestamp indicating the period of time covered in the log file. By typing **tailt** in the /opt/OptiCall/<process>/bin directory, you can see the running log of that process. On an EMS machine the log files are kept by default in the directories /opt/ems/bin/logs and /opt/bdms/bin/logs for the EMS and BDMS applications; similarly, the command **tailt** in the bin directory can be used to get a running log of that process.

Displaying the EMS Database Transaction Queue

After a command is entered on the EMS to modify the database, it is placed into the transaction queue, where it remains until acknowledgement is received from the target call agent or feature server process that the database change was successful. Entries should remain in the transaction queue for only a matter of seconds; if items remain, this indicates an error condition. To check to see whether any entries exist in the transaction queue execute the following command: **show transaction-queue**.

Auditing the Media Gateway

To ensure that the database on the EMS and on the call agent and feature servers is consistent, you can use the **audit** command. This command can be executed on the entire database or on individual tables. For example, the following example shows an audit on the media gateway table:

```
CLI>audit mgw
Auditing table MGW on [CA104]
..............
Reply : Success: Request was successfully completed
```

Running BTS Diagnostics

Diagnostic tests can be run on several types of BTS entities, which can help in troubleshooting. To perform diagnostic tests, you must first put that entity into maintenance state. The following example shows putting an MTA media gateway into maintenance mode:

```
CLI>control mgw id=jriddel_lsys1; mode=forced; target-state=maint
MGW ID -> jriddel_lsys1
INITIAL STATE -> ADMIN_INS
REQUEST STATE -> ADMIN_MAINT
RESULT STATE -> ADMIN_MAINT
FAIL REASON -> ADM found no failure
REASON -> ADM executed successfully
RESULT -> ADM configure result in success
Reply : Success: CLI change successfully
```

For a media gateway, two types of diagnostic tests can be performed:

- MGW Network Connectivity Test
- MGW MGCP Connectivity Test

The first test verifies IP connectivity to the media gateway by performing an ICMP ping test; the second test verifies MGCP is operational on the device by sending an AUEP message for the gateway capabilities. The following is an example of running the second test:

```
CLI>diag mgw id=jriddel_lsys1; test=2
REASON -> PASSED
RESULT -> ADM ran test succeessfully
MGW DN -> jriddel_lsys1
TEST TYPE -> ADM running MGCP connectivity test on MGW
TEST DURATION -> 63
Reply : Success:
```

Similarly, diagnostic tests also exist for subscriber terminations; just as with media gateways, subscriber terminations must first be put into maintenance mode before tests can be executed. Three types of diagnostic tests are available:

- Subscriber MGCP connectivity test
- Subscriber termination connection test
- Subscriber termination ring test

The first test sends an AUEP to the subscriber endpoint for its capabilities and active connections; the second test verifies further MGCP operation by creating, modifying, and deleting a connection to the subscriber endpoint; and the third test sends an MGCP RQNT message to the endpoint requesting the generation of ring tone. The following is an example showing the diagnostic tests available for a subscriber termination:

```
CLI>diag subscriber-termination id=sub1@jriddel_lsys1
===
(1)  Subscriber MGCP Connectivity Test
(2)  Subscriber Termination Connection Test
(3)  Subscriber Termination Ring Test
(4)  ALL
Reply : Success: Diagnostic Subscriber Menu.
```

Validating Call Routing

Call routing can be validated on the BTS using the Translation Verification Tool (TVT) from the EMS CLI interface. This tool simulates a call from an originator based on dialed digits. The following is an example of the translations for the basic call flow example from Oscar (phone #: 321-501-2000) to Tina (phone #: 321-501-2010):

```
CLI>translate line calling_dn=3215012000;called_dn=3215012010
TRANSLATE ON CALL AGENT CA104 IS... ->
TABLEINFO ->
TABLE: SUBSCRIBER
```

```
ID=sub1@jriddel_lsys1;CATEGORY=INDIVIDUAL;NAME=Oscar;STATUS=ACTIVE;DN1=3215012000;
  PRIVACY=NONE;RING_TYPE_DN1=1;TERM_ID=aaln/
  1;MGW_ID=jriddel_lsys1;PIC1=NONE;PIC2=NONE;PIC3=NONE;GRP=N;USAGE_SENS=Y;
  SUB_PROFILE_ID=PacketCable_subs;TERM_TYPE=TERM;IMMEDIATE_RELEASE=N;TERMINATING_
  IMMEDIATE_REL=N;SEND_BDN_AS_CPN=N;SEND_BDN_FOR_EMG=N;PORTED_IN=N;BILLING_
  TYPE=NONE;VMWI=Y;SDT_MWI=Y;
TABLE: SUBSCRIBER_PROFILE
ID=PacketCable_subs;DIAL_PLAN_ID=INCOMING;LOCAL_PFX1_OPT=NR;TOLL_PFX1_OPT=RQ;
  POP_ID=PacketCable_PoP;OLI=0;EA_USE_PIC1=Y;QOS_ID=G711u_20ms;INTERLATA_PFX1_
  OPT=RQ;
TABLE: DIAL_PLAN_PROFILE
ID=INCOMING;DESCRIPTION=dialing plan profile for incoming
  calls;NANP_DIAL_PLAN=Y;OVERDECADIC_DIGITS_SUPP=N;NOA_BASED_ROUTING=N;
TABLE: DIAL_PLAN
ID=INCOMING;DIGIT_STRING=321;DEST_ID=ONNET_LC;SPLIT_NPA=NONE;DEL_DIGITS=0;MIN_DIGI
  TS=10;MAX_DIGITS=10;NOA=NATIONAL;
TABLE: DESTINATION
DEST_ID=ONNET_LC;CALL_TYPE=LOCAL;ROUTE_TYPE=SUB;ZERO_PLUS=Y;INTRA_STATE=Y;
  GAP_ROUTING=N;CLDPTY_CTRL_REL_ALWD=N;CALL_SUBTYPE=NONE;ACQ_LNP_QUERY=NA;
TABLE: OFFICE_CODE
DIGIT_STRING=321501;OFFICE_CODE_INDEX=1;DID=N;CALL_AGENT_ID=CA104;DIALABLE=Y;NDC=3
  21;EC=501;DN_GROUP=xxxx;EC_DIGIT_STRING=321501;
TABLE: DN2SUBSCRIBER
OFFICE_CODE_INDEX=1;DN=2010;STATUS=ASSIGNED;RING_TYPE=1;LNP_TRIGGER=N;NP_RESERVED=
  N;SUB_ID=sub1@jriddel_sa1;LAST_CHANGED=2006-05-04 10:31:38.0;VIRTUAL_DN=N;
  PORTED_IN=N;
TABLE: SUBSCRIBER
ID=sub1@jriddel_sa1;CATEGORY=INDIVIDUAL;NAME=Tina;STATUS=ACTIVE;DN1=3215012010;PRI
  VACY=NONE;RING_TYPE_DN1=1;TERM_ID=aaln/
  1;MGW_ID=jriddel_sa1;PIC1=NONE;PIC2=NONE;PIC3=NONE;GRP=N;USAGE_SENS=Y;
  SUB_PROFILE_ID=PacketCable_subs;TERM_TYPE=TERM;IMMEDIATE_RELEASE=N;TERMINATING_
  IMMEDIATE_REL=N;SEND_BDN_AS_CPN=N;SEND_BDN_FOR_EMG=N;PORTED_IN=N;BILLING_
  TYPE=NONE;VMWI=Y;SDT_MWI=Y;
Reply : Success:
```

As you can see, the output shows all the tables involved in routing this call. First, Oscar's entry in the subscriber table is referenced to get his subscriber profile name. Then this subscriber profile is referenced to get the dial plan name. At this point, the entry in the dial plan for the dialed digits (called #) is identified. This entry points to the ONNET local destination, which is of route type subscriber. Recall that route type subscriber means that a lookup in the DN2SUBSCRIBER table occurs for that particular office code index and DN number (the office code index can be found by looking up the first six digits of the called # in the office code table; the last four digits is the DN). This matching entry is Tina's entry in the subscriber table; hence, a reply code of success is returned.

Collecting Traffic Statistics

The BTS regularly collects statistical measurements for various types of traffic. Individual types of measurements are collected in 5-, 15-, 30-, or 60-minute intervals. The table MEASUREMENT_PROV defines the measurement interval for each traffic category. The following example shows these traffic categories:

```
CLI>report measurement_
measurement_ain_tools_summary
measurement_ainsvc_summary
measurement_anm_summary
```

```
measurement_audit_summary
measurement_billing_summary
measurement_call_tools_summary
measurement_callp_summary
measurement_dqos_summary
measurement_em_summary
measurement_h323_summary
measurement_isdn_summary
measurement_isup_summary
measurement_m3ua_summary
measurement_mgcp_summary
measurement_pct_tools_summary
measurement_pots_acar_summary
measurement_pots_cos_summary
measurement_pots_cot_summary
measurement_pots_local_summary
measurement_pots_misc_summary
measurement_pots_sle_summary
measurement_sctp_summary
measurement_sia_summary
measurement_sim_summary
measurement_snmp_summary
measurement_sua_summary
measurement_tcap_summary
measurement_tg_usage_summary
```

Of particular interest to the call types covered thus far are the CALLP, DQOS, MGCP, and POTS_LOCAL categories. Start and stop times can be specified to retrieve the measurements for a particular time period. The following example shows the MGCP measurements from 6:00 to 6:15 p.m. (1800 to 1815) on May 7, 2006:

```
CLI>report measurement_mgcp_summary start_time=2006-05-07 18:00:00;end_time=2006-
  05-07 18:15:00
TIMESTAMP                               2006-05-07 18:00:00
CALL_AGENT_ID                           CA104
CONDITION                               Normal
MGCP_DECODE_ERROR                       0
MGCP_ENCODE_ERROR                       0
MGCP_UNREACHABLE                        0
MGCP_SEND_FAILED                        0
MGCP_CRCX_ACK_RX                        2
MGCP_CRCX_NACK_RX                       0
MGCP_CRCX_TX                            2
MGCP_MDCX_ACK_RX                        2
MGCP_MDCX_NACK_RX                       0
MGCP_MDCX_TX                            2
MGCP_DLCX_RX                            0
MGCP_DLCX_TX                            2
MGCP_DLCX_ACK_RX                        2
MGCP_DLCX_NACK_RX                       0
MGCP_RQNT_ACK_RX                        7
MGCP_RQNT_NACK_RX                       0
MGCP_RQNT_TX                            7
MGCP_AUEP_ACK_RX                        3
MGCP_AUEP_NACK_RX                       0
MGCP_AUEP_TX                            3
MGCP_NTFY_RX                            0
MGCP_RSIP_RX                            3
MGCP_RSIP_ACK_TX                        3
TIMESTAMP                               2006-05-07 18:15:00
CALL_AGENT_ID                           CA104
CONDITION                               Normal
MGCP_DECODE_ERROR                       0
```

```
MGCP_ENCODE_ERROR                  0
MGCP_UNREACHABLE                   0
MGCP_SEND_FAILED                   0
MGCP_CRCX_ACK_RX                   1
MGCP_CRCX_NACK_RX                  0
MGCP_CRCX_TX                       1
MGCP_MDCX_ACK_RX                   1
MGCP_MDCX_NACK_RX                  0
MGCP_MDCX_TX                       1
MGCP_DLCX_RX                       0
MGCP_DLCX_TX                       1
MGCP_DLCX_ACK_RX                   1
MGCP_DLCX_NACK_RX                  0
MGCP_RQNT_ACK_RX                   5
MGCP_RQNT_NACK_RX                  0
MGCP_RQNT_TX                       5
MGCP_AUEP_ACK_RX                   2
MGCP_AUEP_NACK_RX                  24
MGCP_AUEP_TX                       26
MGCP_NTFY_RX                       0
MGCP_RSIP_RX                       4
MGCP_RSIP_ACK_TX                   4
Reply : Success:  Entries 1-2 of 2 returned.
```

Displaying Billing Records

If you have configured the BTS to generate call detail records (CDRs) for billing, these records can help in troubleshooting as well. Billing records are stored in ASCII files in the directory specified by the BILLING_DIRECTORY field in the BILLING_ACCT_ADDR table; the following is an example:

```
CLI>show billing_acct_addr
BILLING_DIRECTORY = /opt/bms/ftp/billing
BILLING_FILE_PREFIX = bil
BILLING_SERVER_DIRECTORY = /dev/null
POLLING_INTERVAL = 15
SFTP_SUPP = N
DEPOSIT_CONFIRMATION_FILE = N
Reply : Success: Request was successful.
```

As with the log files, billing files are periodically rotated, and the filename contains a timestamp indicating the time period of the contained records. Typically, third-party applications are used to parse the data in these files; consult the BTS Billing Guide reference for more detail on how to parse these files. Billing records can also be seen using the EMS CLI; specifically, the command **report billing-record** is used. The following is an example that returns the most recently generated record:

```
CLI>report billing-record
..
CALLTYPE=LOCAL
SIGSTARTTIME=2006-05-04 16:04:49
SIGSTOPTIME=2006-05-04 16:05:05
CALLCONNECTTIME=2006-05-04 16:04:53
CALLANSWERTIME=2006-05-04 16:04:53
CALLDISCONNECTTIME=2006-05-04 16:05:05
CALLELAPSEDTIME=00:00:12
INTERCONNECTELAPSEDTIME=00:00:00
ORIGNUMBER=3215012000
TERMNUMBER=3215012010
```

```
CHARGENUMBER=3215012000
DIALEDDIGITS=3215012010
OFFHOOKINDICATOR=1
SHORTOFFHOOKINDICATOR=0
CALLTERMINATIONCAUSE=NORMAL_CALL_CLEARING
OPERATORACTION=0
ORIGSIGNALINGTYPE=0
TERMSIGNALINGTYPE=0
ORIGTRUNKGROUPNUMBER=0
TERMTRUNKGROUPNUMBER=0
OUTGOINGTRUNKNUMBER=0
ORIGCIRCUITID=0
TERMCIRCUITID=0
ORIGQOSTIME=2006-05-04 16:05:05
ORIGQOSPACKETSSENT=579
ORIGQOSPACKETSRECD=706
ORIGQOSOCTETSSENT=93798
ORIGQOSOCTETSRECD=112960
ORIGQOSPACKETSLOST=1
ORIGQOSJITTER=197
ORIGQOSAVGLATENCY=49
TERMQOSTIME=2006-05-04 16:05:05
TERMQOSPACKETSSENT=707
TERMQOSPACKETSRECD=575
TERMQOSOCTETSSENT=114534
TERMQOSOCTETSRECD=92000
TERMQOSPACKETSLOST=2
TERMQOSJITTER=0
TERMQOSAVGLATENCY=49
PACKETIZATIONTIME=20
SILENCESUPPRESSION=0
ECHOCANCELLATION=0
CODECTYPE=PCMU
CONNECTIONTYPE=IP
OPERATORINVOLVED=0
CASUALCALL=0
INTERSTATEINDICATOR=0
OVERALLCORRELATIONID=CA104626
TIMERINDICATOR=0
RECORDTYPE=NORMAL RECORD
CALLAGENTID=CA104
ORIGPOPTIMEZONE=GMT
ORIGTYPE=ON NET
TERMTYPE=ON NET
NASERRORCODE=0
NASDLCXREASON=0
FAXINDICATOR=NOT A FAX
FAXPAGESSENT=0
FAXPAGESRECEIVED=0
ORIGPOPID=PacketCable_PoP
TERMPOPID=PacketCable_PoP
TERMPOPTIMEZONE=GMT
DIALPLANID=INCOMING
CALLINGPARTYCATEGORY=Ordinary Subscriber
CALLEDPARTYINDICATOR=No Indication
CALLEDPARTYPORTEDIN=No
CALLINGPARTYPORTEDIN=No
BILLINGRATEINDICATOR=None
ORIGENDPOINTADDR=x1-6-00-13-10-50-38-02.cc.cisco.com
TERMENDPOINTADDR=x1-6-00-0a-73-fa-be-8c.cc.cisco.com
ORIGCMTSID=PC-CMTS.cc.cisco.com
TERMCMTSID=PC-CMTS.cc.cisco.com
SENSORID=000000
ORIGPRIVACYINDICATOR=NONE
Reply : Success: Entry 1 of 1 returned from host: rsvlmn-priems
```

Many of the fields of the record are self-explanatory; for detailed explanations refer to the BTS Billing Guide. Specific records can be retrieved by providing additional arguments to the **report billing-record** command. These arguments include specifying specific call types (local, toll, and so on), originating and terminating phone numbers, and specific time periods.

Troubleshooting an MTA Not In Service

You can verify the status of an MTA on the BTS using the **status** commands for both the media gateway name and the subscriber name. Both should be administratively in service and operationally in service. The media gateway/MTA is considered in service by the BTS if it responds to a heartbeat AUEP command. Note, even if the AUEP response is negative, the media gateway still shows in service.

If the media gateway is operational and the subscriber endpoint is not, this typically indicates a problem with the NCS messaging. The endpoint name could be incorrect, the protocol version could be wrong, or you could have a compatibility problem with the NCS parameters (such as embedded RQNT support). A packet trace is the easiest way to diagnose this type of problem. Command responses in the 5xx range indicate this type of problem. Remember, command responses in the 4xx range indicate transient/temporary problems. One exception to this is one of the more common problems seen in a PacketCable deployment, the failure to reserve DQoS resources. The MGCP response code used by most MTA vendors in this case is 526 - No bandwidth. This problem might just be temporary, or it could be indicative of a persistent problem relating to DOCSIS resources.

If the MTA media gateway is also down according to the BTS, you can check several other things. Verify the BTS can resolve the MTA FQDN specified in the TSAP_ADDR parameter for the media gateway to the correct IP address. Verify the BTS can ping this MTA IP address; if not, you might have a routing issue. Remember that you can perform ping and MGCP tests using the diagnostic utilities described earlier; alternatively, you can simply issue a ping from the UNIX prompt on the active call agent machine. To do these two verifications, you would need access to the active call agent machine of the BTS. Recall that you can determine the status of a BTS machine by typing **nodestat** from the UNIX command line.

If the MTA is not pingable and you are sure you do not have a routing issue, the problem is most likely with the MTA and/or CM provisioning.

Make sure the CM is online on the CMTS and also look to see whether the MTA shows up as a host entry behind the cable modem. On a Cisco CMTS the modem status can be found by the **show cable modem** *mac-address* command, and the MTA status can be found by the **show cable modem** *mac-address* **cpe** command. If the CM is online but the MTA does not show it, that indicates you might have a problem with the MTA provisioning process. The problem could also be bad or missing CM PacketCable DHCP options. Please refer to

Chapter 4, "Provisioning Flows," for more assistance with debugging the MTA provisioning steps.

If both the MTA and the CM are online and the MTA is pingable but still shows down on the BTS, this probably indicates a problem with the MTA configuration file. For example, an incorrect value of the CMS FQDN in the configuration file can cause the MTA to not respond to NCS commands from the CMS. Or this type of problem would occur if either the MTA or the CM is provisioned for secure signaling but the other is not. Also, this type of problem would occur if secure signaling is configured on both devices and you have a problem in the encryption/decryption process.

Assuming the CMS FQDN entry in the MTA configuration file is correct and resolvable, the MTA should generate an RSIP message to the CMS. Note: The MTA waits a random time before sending its RSIP command. By default, this random time is between 0 and 600 seconds. So you might have to wait up to 10 minutes before the MTA tries to contact the CMS. This maximum waiting delay is configurable. The act of taking a phone off hook bypasses this delay and forces the RSIP message to be sent immediately by the MTA. A packet sniffer can be used to verify RSIP commands as well. Another good place to look for problems with the MTA configuration and/or provisioning process is with a logging server. Check these logs for entries from the MTA that can help you identify problems.

DNS Issues

As mentioned in the previous section, if the MTA configuration file has a problem with the CMS FQDN, the MTA cannot come online. Verify the MTA is able to resolve the CMS FQDN. Note that the CMS FQDN typically resolves into multiple IP addresses; thus, it could take the MTA a long time to get to the working IP address. The NCS timers Max1 and Max2 can be adjusted to speed up this process.

Also mentioned in the previous section was the possibility of the CMS encountering DNS issues when resolving the MTA FQDN. Verify that the MTA FQDN is resolvable and resolving to the correct IP address. It could be possible that the CMS is caching an old IP address for the MTA. If you are using the BTS call agent, DNS servers are defined in the file /etc/resolv.conf; verify that the configuration specified in this file is correct and that the servers specified are reachable and operational.

No Dial Tone or Ring Tone

If everything looks good on the BTS but no dial tone is received when the subscriber picks up the phone, this could be caused by a couple of things. You could have an impedance issue on the analog line causing the MTA not to be able to detect the off hook event. This type of problem could also cause false off hook events to be reported.

Also the MTA must be told to play dial tone by the CMS. This is usually the result of the MTA detecting the off hook event. So by sniffing the NCS packets, you should see a NTFY of the off hook event to the CMS followed by an RQNT requesting the dial tone signal event to be played. Verify that these packets are exchanged without error. If you never see the off hook NTFY, verify that the CMS requests this event after the endpoint sends its RSIP restart command.

The same issues for dial tone can also cause problems with other signals such as ring tone. If you are not receiving ring tone, verify that this signal is requested by the CMS. If you are not seeing any RQNT or any other NCS commands on your terminating MTA, verify the call routing on the CMS. Also examining billing records for release reasons is another good way to track down problems.

No Ring Back Tone

If you look back to the fields defined on the media gateway profile, you'll see one of them was RBK_ON_CONN_SUPP. This field defines whether a media gateway is capable of sending ring back tone out its NCS connection over the IP network. If a media gateway is not capable of doing this, you must set this parameter to N. Also verify the ring back tone is permitted to reach the originating endpoint. This means that the connection on the originating endpoint must be in a mode that allows packets to be received. Also if the CMTS aggregation device (AGGR_ID) is misconfigured or down, this is likely where you notice a problem. The MTA still gets dial tone because NCS connectivity is fine, but after digits are dialed the CMS tries to create gates on the CMTS specified by the AGGR-ID. This fails, and you likely hear reorder tone on the phone.

No Voice Path and One Way Voice

Related to the ring back case are problems with no voice path. Both NCS connections must be in at least the send/receive mode. If you have issues with no voice or one-way voice, check the NCS connection modes.

Also recall that the voice packets travel between MTAs using the RTP/UDP/IP protocols. If the MTAs do not have IP connectivity between each other or the RTP ports are blocked by a firewall, you have problems with voice path.

Problems with voice quality are typically related to the exchange of RTP packets and are covered in the section "Voice Path Problems" of Chapter 10.

DQoS Not Established or DQoS Information Missing

If the media gateway is not provisioned with an aggregation device or if the aggregation device indicates that DQoS is not supported, you are not able to generate HFC quality of service because the DQoS Local Connection Options is not included in the NCS signaling. Also if DQoS is not supported or enabled on the CMTS, you have this problem. Remember, you can check the status of the CMTS on the BTS using the **status aggr** command.

Large Post Dial Delay Issues

If users are complaining about the length of time it takes for a phone to ring after dialing, you have an issue with post dial delay. The most likely cause is dropped or delayed NCS packets; ensure QoS treatment for call signaling packets is in place. For more information on this topic refer to Chapter 16, "PacketCable Network Design Considerations." This problem could also be caused by digit timers set too long or a poorly designed digit map. Both of these are configurable and should be adjusted as needed. The digit timers are set in the MTA configuration file, and the digit map is defined on the CMS (table DIGIT_MAP on the BTS).

Parameter Negotiation Problems

The two endpoints involved in a PacketCable network negotiate the CODEC and packetization parameters with each other. The CMS provides these options to the originating endpoint in the initial CRCX command, specifically in the LCO parameter. The BTS determines what these LCO parameters are by auditing the capabilities of the endpoint and by the values specified in the QoS table. Remember, the BTS uses the value in the QoS table as the preferred value if CODEC negotiation is enabled and appends the other values. If negotiation is not enabled, the QoS value is the only value sent. The originating endpoint then takes the intersection of this list and what it supports or what it is configured to support and sends it to the CMS in the CRCX response, specifically in the SDP information. Because the BTS determines the LCO parameters by auditing the gateway, the SDP parameters are typically equal to the LCO parameters.

The originating endpoint's SDP along with LCO parameters for the terminating endpoint is sent in the CRCX to the terminating endpoint. Based on these values and the supported values for the terminating endpoint, the terminating endpoint identifies the acceptable list of CODEC and packetization period pairs that can be used for the call. This information is then relayed back to the originating gateway in an MDCX command.

If the originating and terminating endpoints cannot agree on a CODEC and packetization period, negotiation fails, and the call fails. Make sure the two endpoints you are using are compatible and they are configured with the correct preferences.

SDP Problems

Problems can also exist within SDP. Remember, the SDP information conveys bearer information between the two endpoints. Often the CMS does not validate the SDP information and just transparently relays this information between the endpoints. If you have a problem with the SDP parameters, such as bad values, missing parameters, or incorrect parameter order, the MTA gateways should report this. A protocol sniffer can diagnose this type of problem as well.

Chapter Summary

In the PacketCable architecture the CMS and the MTA communicate using the Network-based Call Signaling (NCS) protocol. NCS is a modified version of MGCP aimed at supporting only PacketCable embedded voice MTAs. Aspects of MGCP not needed for this type of endpoint are removed from NCS. The NCS profile of MGCP contains extensions to support PacketCable functions like dynamic quality of service and media security.

MTA endpoints are analog telephone lines and are represented by a local name of aaln/x, where x is the physical port number on the MTA. The Line package is the only package for NCS in PacketCable 1.0; in PacketCable 1.5 the Fax package and VoIP metrics package are added. The Line package is the default package for both PacketCable 1.x releases and consists of DTMF tones, signals such as dial tone and ring tone, and events such as on hook, off hook, and hook flash. In addition to new Local Connection Options for the support of DQoS and security, the multiple packetization parameter, mp, is also added in NCS. This option allows for different packetization periods for different voice CODECs. If this parameter is used, you have a one-to-one mapping with the values specified by the CODEC parameter.

DOCSIS quality of service resources are provided by the DQoS function of PacketCable. This information is communicated between the CMS and MTA using new NCS Local Connection Option parameters. Among these parameters are the gate ID referencing the PacketCable gate on the CMTS, the resource reservation parameter identifying the status of this gate, and the resource ID referring to specific CMTS resources. LCO parameters used to encrypt media streams are introduced with NCS as well.

The best way to understand how NCS is used is to examine call flows. Assuming the signaling is not encrypted, NCS messages can easily be examined using a protocol sniffer. The cable monitor utility within the Cisco CMTS software can be used in conjunction with the Ethereal sniffer to decode packets off the DOCSIS network. By examining NCS call flows and messages, you can see how various call features are accomplished and obtain information to help diagnose any problems.

Before an MTA can provide voice service it must be configured to do so on the CMS. On the BTS 10200 CMS, a number of provisioning steps are required covering the MTA, the MTA endpoints, and the CMTS. Several of the fields in these steps relate to how the NCS protocol is used. Other types of BTS configuration are also needed before calls can be processed, such as general system configuration, dial plan configuration, and call feature configuration. Some tips to help in BTS troubleshooting include examining alarms, log files, and billing records; performing diagnostic tests; and validating call routing translations.

NCS signaling problems between the CMS and the MTA can prevent the MTA from being usable or in an in service state. IP connectivity problems, the ability to resolve DNS names, invalid MTA provisioning information, and problems with the NCS messages and parameters themselves all prevent an MTA from being functional. Skill in troubleshooting these areas is invaluable in diagnosing problems. Issues with dial tone, ring tone, and voice path can be caused by problems in these areas as well. Finally, recall some communication parameters are negotiated between gateways, and if you encounter a failure in this process, problems occur.

Chapter Review

1 Which of the following are valid NCS endpoint names?

   a aaln/0

   b $

   c aaln/1/1

   d aaln/*

   e aaln

   Answer: $, aaln/*, and aaln are valid.

2 What is the Line package event oc used for?

   Answer: The operation complete event is used for the notification of signal timeouts.

3 For bidirectional voice to be possible between two MTA endpoints, what conditions must be met?

   Answer: T NCS connections modes must be in at least the send/receive state for both endpoints.

4 What additional information is provided when an NCS connection is deleted compared to when a pure MGCP connection is deleted?

   Answer: Statistics from received RTCP reports are sent in the delete connection acknowledgement.

5 Assuming no encryption, what are some ways to look at the NCS signaling?

   Answer: UNIX snoop utility, Cable Monitor, and Ethereal

6. What does the response code, 000, indicate?

   Answer: A response acknowledgement

7. Before an MTA endpoint can be controlled in service on the BTS, what step needs to occur assuming the MTA itself is in service?

   Answer: The endpoints need to be equipped.

8. If you notice no DQoS resources are created when placing calls, what should you check?

   Answer: PacketCable enabled on the CMTS, aggregation device defined for the MTA, and DQoS enabled for the aggregation device

9. What tables on the BTS 10200 contain fields that define NCS signaling properties?

   Answer: Media Gateway Profile, QOS

10. If the BTS reports both the MTA endpoint and the MTA itself as out of service, what are some things to check?

    Answer: Cable modem and MTA online, IP connectivity to the MTA, that BTS can resolve MTA FQDN, and MTA configuration file

11. True or False: The multiple packetization parameter, mp, identifies a range of supported packetization periods for a given CODEC.

    Answer: False. This parameter defines different packetization periods for different CODECs.

12. True or False: Either the mp or p packetization parameter can be used to define a connection but not both.

    Answer: True

13. True or False: Both the termination and originating endpoints in a call must use the same packetization period when encoding voice packets.

    Answer: True

14. True or False: The LCO parameter dq-rr describes the current state of the DQoS resources identified by the LCO parameter dq-gi.

    Answer: False. This parameter tells the embedded MTA what state to request the HFC resources to be put in by the CMTS.

15. True or False: NCS has LCO parameters that determine whether or not the NCS messages are encrypted.

Answer: False. The NCS encryption parameters are for RTP/RTCP message encryption, not signaling encryption.

16 True or False: NCS negative acknowledgements to heartbeat AUEP commands from the BTS cause the BTS to mark the MTA as out of service.

Answer: False. As long as a response is received by the BTS, the media gateway shows in service.

17 True or False: Missing mandatory SDP parameters and invalid SDP parameters are typically not detected by the CMS.

Answer: True

CHAPTER 8

TGCP and the PSTN Interconnect

In the early stages of a PacketCable deployment it is more than likely that a majority of the calls made are going to be between PacketCable customers and customers still using the legacy public switched telephone network (PSTN).

Consequently, PSTN interconnection devices and methods are required. The Media Gateway Controller (MGC), Media Gateway (MG), and Signaling Gateway (SG) PacketCable components provide this interconnection. The previous chapter focused on the Network-based Call Signaling (NCS) profile of Media Gateway Control Protocol (MGCP) used for call signaling communication between a PacketCable call management server (CMS) and multimedia terminal adapter (MTA). This chapter is concerned with the PSTN or trunking equivalent, namely the Trunk Gateway Control Protocol (TGCP). The TGCP profile of MGCP is used for call signaling between a PacketCable MGC and an MG. You will learn the differences between TGCP and pure MGCP, preceded by a brief review of the Signaling System 7 (SS7) and channel-associated signaling (CAS) protocols used in the traditional PSTN. Included with the SS7 discussion is an introduction to the Sigtran family of protocols used to communicate SS7 information over IP between the SG and MGC. The chapter then uses call flows and traces to illustrate how these protocols work and interact in PacketCable. Next, you will see relevant configuration and troubleshooting on the Cisco MGC (the BTS 10200), the Cisco SG (the IP Transfer Point [ITP]), and two Cisco MGs (the Voice Switch Service Module [VXSM] and Voice Interworking Service Module [VISM]).

PSTN Call Signaling Overview

Figure 8-1 illustrates the devices and protocols typically involved in connecting a PacketCable voice over IP (VoIP) network to the traditional circuit-switched PSTN.

Figure 8-1 *Sample PSTN Interconnect Architecture*

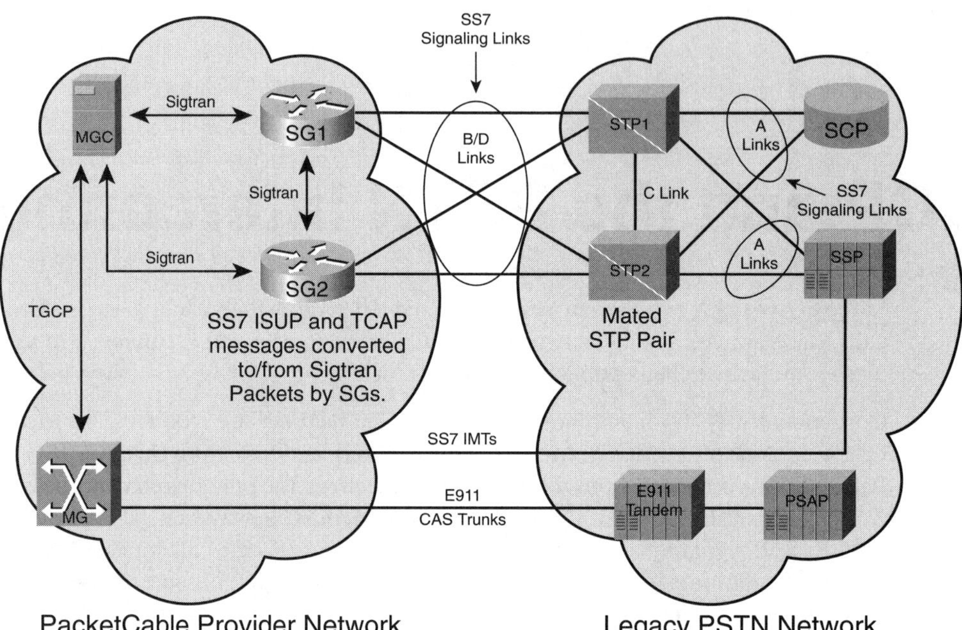

The cloud on the left side of this figure depicts the portion of the PacketCable network involved in the PSTN interconnect, and on the right side you can see the PSTN components involved. All these components, as well as the protocols they use to communicate, are covered in this chapter.

In this example, you can see four places where the PacketCable and PSTN networks interconnect:

- Two pairs of SS7 signaling links between the PacketCable Signaling Gateways (SGs) and PSTN Signaling Transfer Points (STPs)
- SS7 inter-machine trunks (IMTs) between the PacketCable MG and PSTN Signaling Service Point (SSP), which is typically a legacy Class V telephony switch
- CAS trunks used for emergency 911 services between the PacketCable MG and PSTN E911 tandem switch, which has connectivity to the Public Safety Answering Point (PSAP)

Recall from Chapter 6, "Signaling Interfaces and MGCP Overview," that a trunk is a communication channel in the circuit-switched PSTN capable of supporting one uncompressed voice call. A trunk is also known as a DS0 and has a capacity of 64 kbps.

The trunks between these networks contain the voice streams of active calls between subscribers on the two networks. But how is a call set up between the networks to begin with? Logically, you must also have a way of communicating call routing information such as called and calling party phone numbers. This call signaling information can be communicated to the PSTN either in-band or out-of-band.

In-band communication is when the call signaling information travels the same path as the bearer or voice information. In a PacketCable network, in-band call signaling would travel between the MGC and the PSTN switch through the MG. The MG gateway would have to be able to extract the call signaling information from the PSTN trunk and send it to the MGC. Also, the MG must be able to generate call signaling information toward the PSTN as directed by the MGC. If in-band signaling is embedded within the DS0 trunk or channel containing the voice stream, it is known as channel-associated signaling (CAS). Trunks used for emergency 911 access and operator services often use CAS. The sample architecture in Figure 8-1 shows CAS trunks for connectivity to an Enhanced 911 (E911) tandem switch, which has access to a PSTN Public Safety Answering Point (PSAP).

NOTE In Chapter 6, you also learned that multiple trunks simultaneously utilize the same physical carrier via time-division multiplexing (TDM). In North America, the T1 is the smallest carrier and consists of 24 DS0 trunks. Another common type of PSTN signaling is where one of these circuits is used for the call control of the other 23 circuits. This type of signaling is known as Common Channel Signaling (CCS), and this particular variety is known as an Integrated Services Digital Network (ISDN) Primary Rate Interface (PRI). Similarly, in Europe, the smallest carrier is an E1 that consists of 32 DS0 circuits, with the first one dedicated for framing and synchronization. If PRI signaling is used, one additional channel handles the call control of the other 30 channels used for voice bearer streams. This call control data channel is also known as the D-channel, whereas the bearer circuits are referred to as B-channels. Please refer to standards such as the ITU Q.921 and Q.931 specifications for more information on ISDN PRI.

Out-of-band signaling is exactly as it sounds; in this case, call signaling information is transmitted to the PSTN through a separate path than that for the bearer or voice information. In a PacketCable network, out-of-band signaling is not seen by the media gateway; instead, the signaling gateway is used. An example is the Signaling System 7 (SS7) protocol used for most OFFNET call types in PacketCable. In this case, the full SS7 protocol stack from the PSTN terminates on the signaling gateway, which in turn strips off the upper layers of the protocol and sends those to the MGC.

Referring to Figure 8-1, you can see the SS7 signaling information is sent in the SS7 links between the SGs and STPs. An SS7 signaling link typically operates at a speed of either 56 kbps or 64 kbps on a single DS0 channel within a T1 or E1 carrier. For capacity and redundancy reasons, multiple signaling links are typically used, as shown in Figure 8-1. The corresponding bearer circuits these signaling links control reside in the SS7 IMTs between the MGC and PSTN switch (SSP). Like ISDN PRI, SS7 is also a Common Channel Signaling (CCS) protocol as a single signaling channel services many bearer channels.

The following sections provide a short overview of SS7 and CAS as well as their use in PacketCable. This helps you understand the big picture of how all these protocols and components relate to each other in the call flow portion of the chapter.

SS7 Overview

Entire books can be written covering the SS7 protocol; this is not one of them. However, because of its importance in a PacketCable network's PSTN interconnect, a quick overview is warranted.

You have three types or categories of SS7 devices:

- **Service Switching Point (SSP)**—An example of an SSP is the traditional circuit-switched PSTN switch to which a MG's trunks are connected. These PSTN switches are further categorized into Class 5 or end-office switches and Class 4 or tandem switches. Historically, Class 5 switches are the switches operated by your local telephone carrier, and Class 4 switches are the switches operated by your long-distance carrier. An SSP is responsible for routing calls and setting up voice circuits. An SSP can originate, terminate, or route SS7 signaling packets.
- **Service Control Point (SCP)**—The SCP is simply a database of call information. For example, when a 1-800 or toll-free number is dialed, an SCP can be used to translate the toll-free number into a local number providing a certain service.
- **Signaling Transfer Point (STP)**—The STP is responsible for routing SS7 messages between SSPs, SCPs, and other STPs. STPs are usually distributed throughout the SS7 network, providing regionalized connectivity for other SS7 devices. This eliminates the need for every SSP and SCP in the SS7 network to have a direct connection to each other. In North America, a distributed STP network is typically used for the connectivity between SSPs and SCPs. STPs can also be organized in a hierarchical network topology.

Figure 8-2 illustrates how these devices are used in the traditional PSTN.

Figure 8-2 *SS7 Signaling Devices and Links*

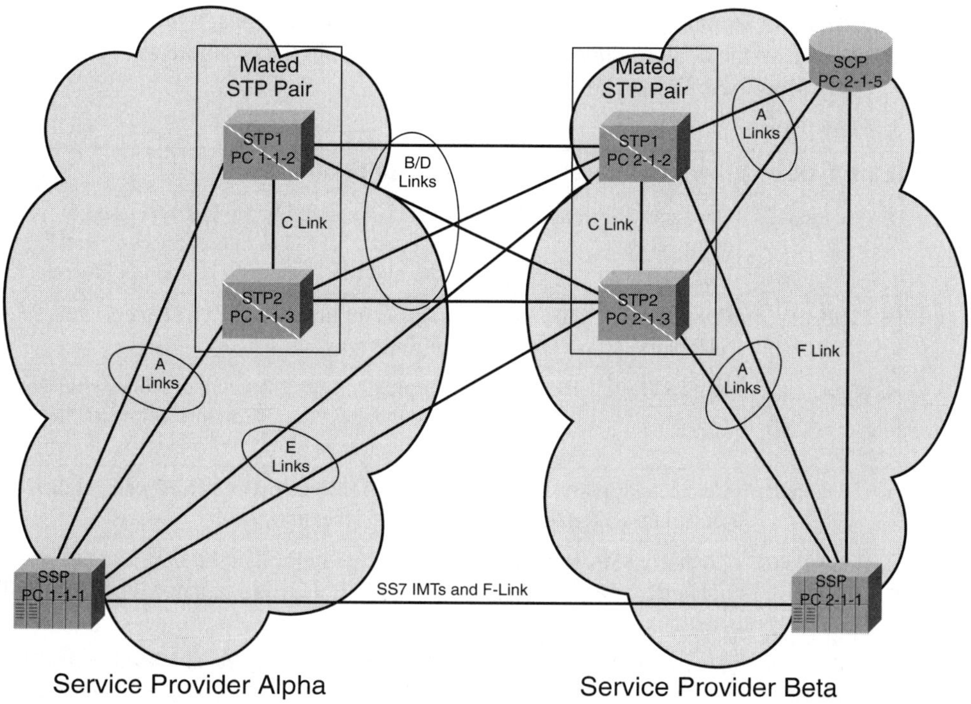

As you can see in Figure 8-2, there are two SS7 service provider networks: Alpha and Beta. Both provider Alpha and Beta have an SSP and a pair of STPs; service provider Beta also has an SCP.

Each one of these SS7 devices is identified by a construct known as a *point code*. You can think of a point code as the SS7 equivalent to an IP address. In North America, the American National Standards Institute (ANSI) point code structure is used (defined as a 24-bit value broken down into three subfields: a network identifier [8 bits], a network cluster identifier [8 bits], and a network cluster member identifier [8 bits]). In Europe, the International Telegraph and Telephone Consultative Committee (CCITT) point code structure is used (defined as a 14-bit value also broken down into three subfields: a zone identifier [3 bits], an area/network identifier [8 bits], and a signal point identifier [3 bits]).

Links, Linksets, and Routes

SS7 devices communicate with each other using signaling links, which are categorized depending on the devices they interconnect as shown in the example in Figure 8-2. Table 8-1 describes these link types.

Table 8-1 *SS7 Link Types*

Link Type	Devices Interconnected	Description
A (Access)	Connects SSPs and SCPs to home STPs	This is most likely the link type used by your PacketCable SG to interconnect to an STP if your SG and MGC share a common point code.
B (Bridge)	Connects STP pairs to other STP pairs	This is for links between STP pairs at the same level of hierarchy.
C (Cross)	Joins STPs together	Typically, two STPs connect to a switch for redundancy; the C link interconnects these two STPs.
D (Diagonal)	Connects primary and secondary STP pairs	This is for links between STP pairs at different levels of hierarchy.
E (Extended)	Connects SSPs and SCPs to nonhome STPs	E-links are typically used only as a last resort; for example, if there is a failure with the local STP pair A links.
F (Fully Associated)	Connect SSPs and SCPs to each other directly	This type of connection is often used in Europe; in this case, the signaling DS0 uses one of the timeslots on the same circuit as the voice DS0s. For an E1 carrier this is usually timeslot 16.

NOTE Often, STP hierarchy levels aren't well defined; therefore, the distinction between B links and D links isn't well defined either. Consequently, these links are generally referred to as B/D links.

Referring to Figure 8-2, you can see both provider Alpha and Beta have an SSP and a mated pair of home STPs. The link between an SSP and a home STP is classified as an Access (A) link. A mated pair of STPs is used by each provider for redundancy reasons in case one of these STPs fail; the link between a mated STP pair is known as a Cross link. Because Alpha's customers and Beta's customers must be able to make telephone calls with each other, the Alpha and Beta STP pairs are interconnected. Each STP in Alpha's network has a connection to each STP in Beta's network; hence, you see four links in total. Because STP hierarchy levels aren't very well defined, these links can be classified as Bridge (B) links, Diagonal links, or B/D links as shown in Figure 8-2. The SSPs in the two networks also have a direct SS7 signaling link. Because this type of link traverses the same path as the bearer circuits, it is categorized as a Fully Associated link. Service Provider Alpha wants to be very certain that it can communicate with Service Provider Beta, so it also has direct links from its SSP to Beta's mated STP pair. These links would typically be used only as a last resort and are classified as Extended links. Service Provider Beta's STP pair also has connectivity to an SCP database it uses for services such as number translation. These links are also classified as A links.

Multiple links between a pair of SS7 nodes are often used to increase capacity and provide redundancy. This collection of links is referred to as a linkset. Each link in the linkset is identified by a unique Signaling Link Code (SLC). A single linkset can have up to 16 links; thus, the SLC is a value between 0 and 15. The point code of the remote SS7 node is known as an Adjacent Point Code (APC). SS7 nodes load share traffic across the links in a linkset.

An SS7 path between an SSP and another SSP or SCP is known as a *route*. A route can traverse several linksets to reach the destination point code of the remote SSP or SCP. The collection of routes that can be used between an SSP and another SSP or SCP is referred to as a *routeset*.

So how many routes exist in the routeset between Alpha's SSP and Beta's SSP? The answer is 21:

1. Alpha SSP -> Alpha STP #1 -> Beta STP #1 -> Beta SSP
2. Alpha SSP -> Alpha STP #1 -> Beta STP #1 -> Beta STP #2 -> Beta SSP
3. Alpha SSP -> Alpha STP #1 -> Beta STP #2 -> Beta SSP
4. Alpha SSP -> Alpha STP #1 -> Beta STP #2 -> Beta STP #1 -> Beta SSP
5. Alpha SSP -> Alpha STP #2 -> Beta STP #1 -> Beta SSP
6. Alpha SSP -> Alpha STP #2 -> Beta STP #1 -> Beta STP #2 -> Beta SSP
7. Alpha SSP -> Alpha STP #2 -> Beta STP #2 -> Beta SSP
8. Alpha SSP -> Alpha STP #2 -> Beta STP #2 -> Beta STP #1 -> Beta SSP
9. Alpha SSP -> Alpha STP #1 -> Alpha STP #2 -> Beta STP #1 -> Beta SSP
10. Alpha SSP -> Alpha STP #1 -> Alpha STP #2 -> Beta STP #1 -> Beta STP #2 -> Beta SSP
11. Alpha SSP -> Alpha STP #1 -> Alpha STP #2 -> Beta STP #2 -> Beta SSP
12. Alpha SSP -> Alpha STP #1 -> Alpha STP #2 -> Beta STP #2 -> Beta STP #1 -> Beta SSP
13. Alpha SSP -> Alpha STP #2 -> Alpha STP #1 -> Beta STP #1 -> Beta SSP
14. Alpha SSP -> Alpha STP #2 -> Alpha STP #1 -> Beta STP #1 -> Beta STP #2 -> Beta SSP

15 Alpha SSP -> Alpha STP #2 -> Alpha STP #1 -> Beta STP #2 -> Beta SSP

16 Alpha SSP -> Alpha STP #2 -> Alpha STP #1 -> Beta STP #2 -> Beta STP #1 -> Beta SSP

17 Alpha SSP -> Beta STP #1 -> Beta SSP

18 Alpha SSP -> Beta STP #1 -> Beta STP #2 -> Beta SSP

19 Alpha SSP -> Beta STP #2 -> Beta SSP

20 Alpha SSP -> Beta STP #2 -> Beta STP #1 -> Beta SSP

21 Alpha SSP -> Beta SSP

The first 16 of these are the paths that utilize the A links between Alpha's SSP and STP pair. The next eight are the paths that utilize the E links between Alpha's SSP and Beta's STP pair, and the final one is the F link connecting Alpha's SSP directly to Beta's SSP.

The MTP Layers

Like the TCP/IP protocol suite, the SS7 protocol consists of multiple layers and can be compared to the Open System Interconnection (OSI) model. Figure 8-3 illustrates the SS7 protocol stack compared to the OSI model; it also shows the SS7 protocol layers that operate on each type of SS7 device.

Figure 8-3 *SS7 Protocol Stack*

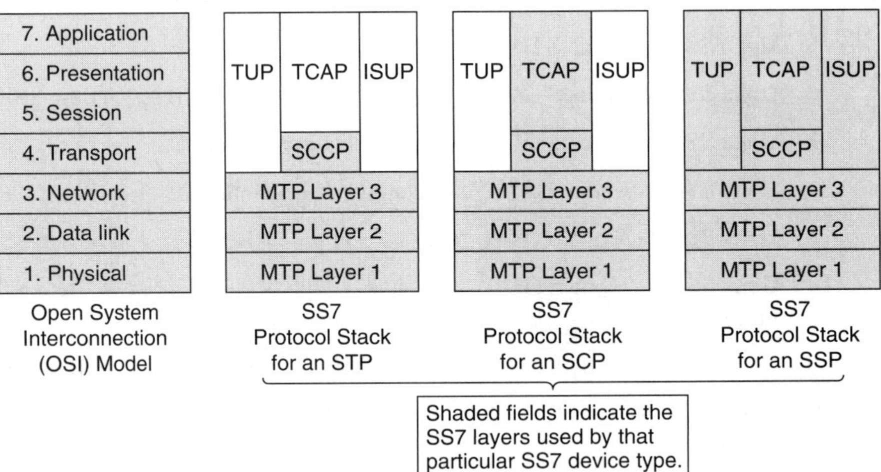

The Message Transfer Part (MTP) layers are the lowest three layers and control the transport system between SS7 nodes. The MTP layers are somewhat analogous to the IP network layer protocol and lower layers in the TCP/IP stack or the physical, data link, and

network layers in the OSI model. The MTP layers are responsible for providing an interface to the physical media, detecting and removing transmission errors, and routing data between SS7 devices. SSPs, SCPs, and STPs all utilize the MTP layers to create SS7 signaling links between them. The higher-layer SS7 protocols traverse these links. The MTP layers serve the following functionality:

- MTP Layer 1 defines the access to the physical media, which is typically a 64-kbps channel.
- MTP Layer 2 defines three basic types of signal units: Message Signal Units (MSUs), in which the upper-layer protocols are encapsulated; Link Status Signal Units (LSSUs), used to bring the link into service; and Fill In Signal Units (FISUs), used when no other messages exist to send. All these messages have flag fields used to delineate message boundaries, sequence numbers, and indicator bits used for error detection and flow control. MSUs are the only type of signal units that travel from link to link and, consequently, are the only signal units that contain routing information such as origination and destination point codes.
- MTP Layer 3 routes messages between SS7 nodes, distributes upper-layer protocols to the appropriate places, and defines messages used to ensure the reliable transmission of messages. For example, when an STP detects that an SSP is no longer reachable, it uses MTP Layer 3 messages to notify other SS7 devices.

NOTE SS7 links are synchronous in nature, meaning messages are constantly exchanged. This is the main reason why the entire SS7 protocol stack is not implemented directly on the PacketCable MGC. A PacketCable SG can handle the lower-layer SS7 protocols by itself and pass only the upper-layer protocols to the MGC. This greatly reduces the burden on the MGC and provides it with only the messaging relevant to call control.

The SCCP and TCAP Layers

The Signaling Connection Control Part (SCCP) layer enhances the routing capabilities of the MTP layer. This is somewhat akin to how the User Datagram Protocol (UDP) and TCP transport layer protocols work with the IP protocol. Instead of ports, fields such as the subsystem number and translation type determine how the upper-layer protocol is handled. Again, all types of SS7 devices utilize this layer as well. The MTP and SCCP layers are the only layers that an STP understands.

The Transaction Capabilities Application Part (TCAP) layer is the application layer protocol used in transaction type applications. The TCAP layer lies above the SCCP layer. For example, database lookups utilize the TCAP layer, so when an SSP needs to translate a 1-800 number, it issues a TCAP query to an SCP, which returns the actual phone number. Another common use for the TCAP layer is with local number portability (LNP).

At this point, it is worth talking a little bit about Global Title Translation (GTT). A global title is an application address such as a 1-800 toll-free number or calling card number. These numbers are global so that the same phone number can be dialed regardless of your geographic location. These phone numbers are not the actual phone numbers of the destination and must be translated into their actual numbers. This translation is usually facilitated by an STP. The STP is configured to look for these global titles and translate them into the point code and SCCP subsystem number of the SS7 device that can further process the call. The STP initiates a TCAP query to the database specified by the translated SCP destination point code and SCCP subsystem number. Because STPs are typically deployed in mated pairs, both members of the pair are capable of GTT. Thus, a common point code called a capability point code is defined on the STP pair, which is used for GTT.

The ISUP Layer

The final SS7 protocol layer covered is the one you need to understand the best. This is the Integrated Services Digital Network (ISDN) User Part (ISUP) layer. This is the protocol used for call signaling between SSPs. Inside ISUP messages, you find information such as called number and calling number. ISUP messages are the SS7 messages depicted in the call flows later in the chapter.

Each ISUP message also references a DS0 trunk where the corresponding bearer circuit resides. Usually, the bearer circuits are physically separated from the signaling circuit, as you learned earlier. The TDM circuit on which the bearer circuits reside is known as an inter-machine trunk (IMT).

DS0s on an IMT are identified in the ISUP messages by a Circuit Identification Code (CIC). Each switch maintains a mapping of ISUP CICs to physical timeslots or DS0s.

NOTE Two switches verify only the existence of CICs between each other, not its location. Consequently, if you have a mismatch between the two switches, you do not have voice path. The continuity test mechanisms described later are methods to protect against this possibility.

Next, the more common ISUP messages are covered. Please note most countries in the world have their own version of ISUP that often varies slightly from each other. However, most of the concepts presented here are universal to all ISUP variants. In North America, the ANSI variation of ISUP is used, which is also the variant used in all the examples in this book. The following ISUP messages are used in a completed call:

- **Initial Address Message (IAM)**—This is the call setup message and contains parameters such as the called or dialed number and the calling or originating number.

- **Acknowledgement Message (ACM)**—This message indicates that the IAM has been received and the call is being processed. Often, the receipt of this message infers that the voice or bearer path is now open.
- **Answer Message (ANM)**—Logically, this message indicates the call has been answered.
- **Release Message (REL)**—This message indicates call teardown and contains a cause value parameter that specifies the reason the call is torn down. If the call is torn down abnormally, this cause value identifies the reason so the switch can play the appropriate announcement, such as "All Circuits are Busy."
- **Release Complete Message (RLC)**—This message acknowledges the successful receipt and processing of the release message.
- **Suspend Message (SUS)/Resume Message (RES)**—For calls from an OFFNET user to an ONNET user, the ONNET user might decide to switch to another phone. In this case, as soon as the user hangs up, the call is not immediately torn down. Instead, the call is "suspended" and an ISUP SUS message is sent to the PSTN letting it know the call is in the suspend state. A timer is associated with this process. If this timer expires, the call is released and a REL message is sent to the PSTN. If the ONNET user picks up the phone again before this timer expires, the call is "resumed," and an ISUP RES message is sent to the PSTN letting it know that the call is active again.

In addition to call processing ISUP messages, you have ISUP messages to manage circuits. Some of the more common messages include

- **Blocking (BLO)/Blocking Acknowledge (BLA)**—The BLO message tells the remote switch that a circuit is removed from service. This could be because of an equipment failure or because of some sort of maintenance. The remote switch acknowledges the successful receipt and processing of the BLO by sending a BLA.
- **Circuit Group Block (CGB)/Circuit Group Block Acknowledge (CGA)**—The BLO and BLA messages can be used only for a single circuit. The CGB and CGA messages are the equivalent for a range of circuits (such as an entire T1 or E1).
- **Unblock (UBL)/Unblock Acknowledge (UBA)**—As soon as a previously blocked circuit becomes available, the remote switch is notified with a UBL message. The successful receipt and processing of a UBL is acknowledged by sending a UBA.
- **Circuit Group Unblock (CGU)/Circuit Group Unblock Acknowledge (CGUA)**—The equivalent to the UBL/UBA messages for a range of circuits.
- **Reset Circuit (RSC)**—This message synchronizes the state of a circuit between two switches. It is acknowledged with an RLC message.
- **Circuit Group Reset (GRS)/Circuit Group Reset Acknowledge (GRA)**—The equivalent to the RSC message for a range of circuits.

NOTE Telephony User Part (TUP) is another SS7 layer used for call control that has messages somewhat similar to ISUP. In some parts of the world TUP is used instead of ISUP; however, in most countries ISUP has replaced TUP.

By now, you should have a good understanding of the SS7 protocols. In the section "Examining Call Flows" you see how SS7 signaling interacts with TGCP signaling for OFFNET call types. For more information relating to SS7, please consult the specifications—for ANSI SS7, this would be the Bellcore GR246 specification, and for ITU/CCITT SS7, this would be the ITU G.76x specification series.

Signaling Gateway Overview and Sigtran

So far in this chapter, the signaling gateway hasn't been covered in great detail. You know that the signaling gateway communicates with the PSTN on all layers of the SS7 protocol, and you know that the SG relays only the upper-layer protocols to the MGC such as ISUP and TCAP. This is done to dump some of the SS7 processing burden to the SG. The protocols used between the MGC and the SG in the Cisco Solution are the IETF Sigtran protocols. This chapter provides an overview of Sigtran; more information can be obtained on the Internet at the following address: http://www.ietf.org/html.charters/sigtran-charter.html.

These Sigtran protocols consist of two adaptation layer protocols running over the Stream Transport Control Protocol (SCTP) transport layer protocol, which runs over IP.

NOTE SCTP is a reliable transport layer protocol that is more efficient than TCP, making it more suitable for message-oriented data. It typically works by creating a session using two networks between the MGC and SG, with one network acting as a secondary path if the other path fails. For more information on SCTP refer to the IETF RFCs 2960 and 3286.

The two adaptation layer protocols, M3UA (defined in RFC 3332) and SUA (defined in RFC 3868), are adapted versions of the SS7 MTP Layer 3 and SCCP protocol layers. Therefore, ISUP messages are carried over the M3UA protocol and TCAP messages are carried over the SUA protocol. Figure 8-4 depicts this relationship.

Figure 8-4 *Sigtran Protocol Family*

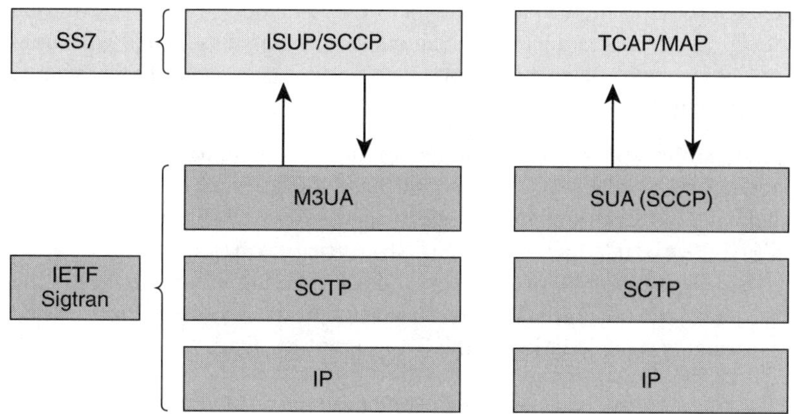

If you refer to Figure 8-3, which shows the SS7 protocol stack, you see that the TUP, SCCP, and ISUP layers all reside above MTP Layer 3; therefore, in a Sigtran implementation, any of these could be sent over the M3UA protocol. Consequently, TCAP could be sent over SCCP over M3UA. However, because the SUA protocol is design explicitly for the transfer of SCCP messages in an IP environment, TCAP messages are sent over SUA instead.

You should be aware of some additional Sigtran terminology. This can be especially helpful in understanding the configuration portion of this chapter. An *application server* is a logical entity serving a specific routing key. A *routing key* is identified by a specific routing context (which is simply an integer). A routing key is the set of SS7 parameters that uniquely identify the range of signaling traffic handled by a particular application server. An example of a routing key would be the ISUP messaging between an MGC and remote SSP. In this example, the application server is the portion of the MGC responsible for processing these ISUP messages. An application server is typically associated with multiple physical entities performing either an active or a backup role. Each of the processes on these entities is referred to as an Application Server Process (ASP). In other words, the active MGC has the active ASP, and the standby MGC has the backup ASP.

NOTE For more on this topic, see RFC 3332, "Signaling System 7 (SS7) Message Transfer Part 3 (MTP3) User Adaptation Layer (M3UA)," Section 1.2.

CAS Overview

Before out-of-band signaling protocols were used in the PSTN, in-band protocols such as CAS were the norm. In today's PSTN, the CAS protocol is still often used for connections to operator services switches and emergency access switches; consequently, CAS support is necessary on PacketCable media gateways. The reason CAS is still used for these applications is that the increased cost and complexity associated with implementing more advanced protocols such as SS7 on these devices is deemed by many to outweigh any potential benefits.

CAS signaling works by utilizing bits in the voice DS0 to convey signaling information. Specifically, the least significant bit in each DS0 of every sixth DS1 frame is used for signaling. This technique is also referred to as *Robbed Bit Signaling* for obvious reasons. For T1s, the extended super frame (ESF) format is often used, which assigns consecutive DS1 frames into groups of 24. Thus, four of these signaling bits, the A, B, C, and D bits, robbed from the 6th, 12th, 18th, and 24th frames, respectively. Figure 8-5 illustrates the T1 ESF format and the location of the CAS signaling bits.

Figure 8-5 *T1 ESF Frame Format and CAS Signaling Bit Locations*

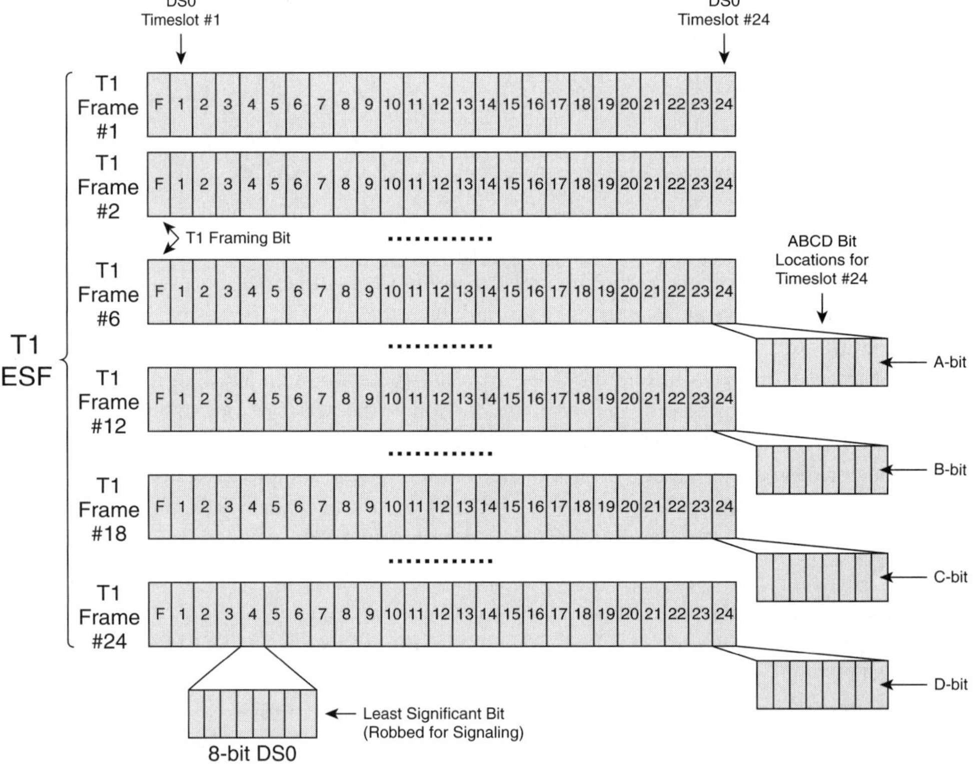

The values of these four bits convey information about the state of the channel, such as line seizure and answer status. For example, when the DS0 channel is idle, the four bits all have the value 0, and when the DS0 channel contains an active call, the four bits all have the value 1. So you might be wondering how information such as the called and calling number is passed from switch to switch using CAS signaling. These are sent in-band using multi-frequency (MF) tone pairs that have predefined mappings to digits.

TGCP Overview

Remember how NCS can be categorized as a "cleaned up" version of MGCP that supports only PacketCable MTA endpoints? Well, TGCP is a "cleaned up" version of MGCP that supports only the PSTN endpoints that exist on PacketCable Media Gateways (MG). These gateways connect to the traditional PSTN via a physical device known as a trunk; hence, these gateways are often referred to as *trunking gateways*.

So, what is a trunk? A *trunk* is a time-division multiplexing (TDM) circuit capable of carrying digitally encoded voice traffic. Chapter 10, "Audio CODECs," details how this encoding process works, so refer to that chapter for more specifics. In Chapter 10, you will learn that, to digitally represent analog speech at an acceptable quality, 8-bit samples are required at a rate of 8,000 samples per second. Thus, in the PSTN a single voice call requires a 64-kbps (8 bits/sample * 8,000 samples/second) channel, commonly referred to as a DS0. Multiple voice calls or DS0s can exist on a single physical connection by giving each call dedicated periodic intervals in which to transmit. This multiplexing of voice calls into dedicated time slots is time-division multiplexing.

In North America, the smallest TDM trunk type is the T1 or DS1. A DS1 frame consists of 24 DS0s multiplexed together plus a framing bit, yielding a total frame size of 193 bits. So, there are 24 DS0 channels or timeslots in a T1, each capable of supporting a single uncompressed voice call. This frame is sent at the same rate of 8,000 per second, equating to a bit rate of 1.544 Mbps. In this book the terms T1 and DS1 are used interchangeably; however, technically, a DS1 refers to the circuit framing, and a T1 refers to the carrier containing the DS1 frame.

In Europe, an E1 carrier is the smallest TDM trunk type consisting of 32 DS0s, resulting in a bit rate of 2.048 Mbps. For an E1 carrier one of the DS0s is used for framing bits, so 31 voice calls can be carried within a single E1.

As you might imagine, a 24- or 31-call capacity trunk is not much when interconnecting telephony switches; therefore, higher-capacity trunks are typically used. These circuits are the DS3/T3 carrier in North America and the E3 carrier in Europe. A DS3 frame contains 28 DS1 frames plus some additional overhead, yielding a total bit rate of 44.736 Mbps. Similarly, an E3 frame can hold 16 E1 frames and has a total bit rate of 34.368 Mbps.

In North America, if larger-capacity circuits are required, the Synchronous Optical Network (SONET) technology is used. Just as with DS1/T1 and DS3/T3, there is a framing component and a carrier component to these circuits. The smallest SONET frame type is the Synchronous Transport Signal - Level 1 (STS-1) frame, with a size of 810 bytes. The corresponding carrier is known as an Optical Carrier - Level 1 (OC-1), with a bit rate of

51.84 Mbps, which is enough to carry a single DS3/T3. An STS-3 frame consists of three STS-1 frames capable of transporting three DS3s/T3s or 84 DS1s/T1s. This is an OC-3 carrier and is also commonly used to interconnect to the PSTN. To better carry T1 and E1 frames within an STS-1 frame, the STS-1 frame is broken into seven virtual tributary groups (VTG). Each VTG is comprised of several virtual tributaries (VT), each used to carry a specific type of frame payload. Different sizes of VTs are used for carrying different frames types; T1 frames are carried in VT type 1.5 (VT1.5), which has a capacity of 1.728 Mbps (remember, a T1 carrier is 1.544 Mbps), and E1 frames are carried in VT type 2 (VT2), which has a capacity of 2.304 Mbps (remember, an E1 carrier is 2.048 Mbps). One VTG can hold 4 VT1.5s or 3 VT2s.

As you can see, TDM circuits are hierarchical in nature, which fits nicely with the endpoint naming convention used in TGCP. For example, in an OC-3 carrier, a specific DS0 trunk would first be referenced by its STS-1 frame (one OC-3 carries three STS-1 frames), then its VTG group number (one STS-1 frame has seven VTGs), then the VT number (four VT1.5s or three VT2.0s in a VTG), and finally by its DS0 timeslot within that T1 or E1 frame in the virtual tributary. This is covered in greater detail later in the chapter.

Because TGCP is a variation of MGCP, it uses a centralized architecture. The MGC or call agent handles most of the complexity, whereas the media gateway just performs basic tasks such as creating, modifying, and deleting connections between the PSTN world and the PacketCable world.

TGCP Variations from MGCP

The fact that TGCP supports only trunk endpoints is one of the differences between TGCP and "pure" MGCP. A few other topics are worth mentioning concerning how these protocols vary. For example, TGCP has some extensions to MGCP required to support PacketCable functionality such as security and lawful intercept. In the previous chapter you learned how NCS varies from standard MGCP; many of these differences also exist between TGCP and MGCP. However, as a whole, these differences are relatively minor (no DQoS component to TGCP is arguably the most critical difference between NCS and pure MGCP). In fact, many Multiple Systems Operators (MSO) deploy standard MGCP instead of TGCP for MGC control of the MGs in their PacketCable networks.

The TGCP variations from MGCP include

- Format of the TGCP command header
- TGCP endpoint naming convention
- TGCP packages, events, and signals
- Changes to the Local Connection Option (L) parameters relating to CODECs and packetization periods
- Additions to the Local Connection Option parameters for the support of secure media

- Additions to the Local Connection Option parameters for the support of lawful intercept
- Additions to the Session Description Protocol (SDP) attributes
- Other modifications

The sections that follow cover these variations in greater detail.

Command Header Format

Similar to NCS, the only change to the command header is the inclusion of the TGCP keyword and version number after the MGCP keyword and version number. This indicates that TGCP is a profile of MGCP. Note: For both TGCP and MGCP the version number is 1.0.

TGCP Endpoint Naming Convention

Just like an NCS endpoint name, a TGCP endpoint name consists of two parts, a local name and a domain name in the following format:

<local name>@<domain name>

The domain name is typically the Fully Qualified Domain Name (FQDN) of the MG, and the local name is a series of hierarchical terms separated by slashes (/), where the leftmost value is the least specific and the rightmost value is the most specific.

In TGCP, the only valid type of endpoint is a trunk or DS0. Consequently, unless wildcard characters are used, the leftmost term of the local name is always ds, indicating a single trunk or 64-kbps DS0 circuit. Unless wildcard characters are used, the rightmost term is always a number identifying the timeslot or channel number of the DS0 (1–24 if carried in a T1, 1–31 if carried in an E1).

The terms in between describe the hierarchy of the trunking gateway. All intermediate terms are comprised of two parameters, a *unit type* and a *unit number* separated by a hyphen. Some examples of unit type are s, su, oc3, ds3, e3, ds1, and e1. From the earlier discussion about trunks, you can figure out what most of these parameters represent, although the s and su unit types might be unclear. Trunking gateways can be comprised of multiple slots, and these slots can have multiple subunits. This is where the s and su identifiers are used.

For example, the Cisco AS5350 and AS5400 gateways have multiple slots that could each have a wide-area network (WAN) interface card (WIC) with T1 daughter cards. The slot or s unit type would be used in this case to identify the WIC, and the subunit or su unit type would be used to identify the daughter card. In the case of the VXSM and VISM gateways within the MGX series, each VXSM or VISM card can be thought of as a trunking gateway by itself as each card has a unique domain name and IP address. A VISM card can interface only to T1 or E1 lines; therefore, the endpoint name in this case is ds/ds1-*x*/*y*@<*FQDN*>, where *x* indicates the T1 or E1 number and *y* indicates the DS0 number.

NOTE This is the case even if the DS3/E3 interfaces on the SRM module back card are used for the physical connection to the PSTN instead of the T1/E1 interfaces on the VISM back cards. Remember that the SRM module can be used to bulk distribute T1s/E1s from its DS3/E3 interface to other service modules in the MGX chassis. Please refer to the section "Configuring a Trunking Media Gateway," later in this chapter or to the "Media Gateway (MG)" section in Chapter 2, "PacketCable Functional Components," for more information.

A VXSM card can utilize DS1, DS3, or OC-3 lines to connect to the PSTN. If DS1 lines are used, there are two DS1 back cards (each supporting 24 DS1 lines) in the VXSM back card slots. If DS3 lines are used, there are two DS3 back cards (each supporting three DS3 lines) in the VXSM back card slots. If OC-3 lines are used, there can be two OC-3 back cards in the VXSM slot (each supporting four OC-3 lines). However, unlike the other cases, these back cards provide redundancy for each other and cannot be combined to provide additional capacity. Regardless of the type of back card used, a common naming scheme is used for the endpoint local names. This format is ds/s-0/ds1-*x*/*y*, where *x* indicates the T1 or E1 number and can have a value from 0–335, and *y* indicates the DS0 number. The full range of 0–335 is needed only if you are using the highest capacity back card, the OC-3 variety. Recall that one OC-3 can carry 84 T1s worth of traffic; therefore, four OC-3s can hold 336 T1s worth of traffic. The following bulleted lists show how to figure out the DS1 numbers based on the associated OC-3 and DS3 carriers.

If OC-3 back cards are used:

- **OC-3 #1 carries the first 84 T1s numbered from 0–83**—Remember, each OC-3 carrier supports three STS-1 frames, and a single STS-1 frame can consist of a single DS3 frame or seven virtual tributary groups (one VTG can hold 4 1.5VTs, each used for carrying a single T1).

  If the STS-1 frames are broken out into DS3s, then DS3 #1 holds T1s 0–27, DS3 #2 holds T1s 28–55, and DS3 #3 holds T1s 56–83.

  If the STS-1 frames are divided into 1.5VT VTGs, then STS-1 #1 holds VTGs 1–7 where VTG #1 holds T1s 0–3, VTG #2 holds T1s 4–7, VTG #3 holds T1s 8–11, VTG #4 holds T1s 12–15, VTG #5 holds T1s 16–19, VTG #6 holds T1s 20–23, and VTG #7 holds T1s 24–27.

  Likewise for STS-1 #2: VTG #1 holds T1s 28–31, VTG #2 holds T1s 32–35, VTG #3 holds T1s 36–39, VTG #4 holds T1s 40–43, VTG #5 holds T1s 44–47, VTG #6 holds T1s 48–51, and VTG #7 holds T1s 52–55.

Finally, for STS-1 #3: VTG #1 holds T1s 56–59, VTG #2 holds T1s 60–63, VTG #3 holds T1s 64–67, VTG #4 holds T1s 68–71, VTG #5 holds T1s 72–75, VTG #6 holds T1s 76–79, and VTG #7 holds T1s 80–83.

- **OC-3 #2 carries the next 84 T1s numbered from 84–167**—If the STS-1 frames are broken out into DS3s, then DS3 #1 holds T1s 84–111, DS3 #2 holds T1s 112–139, and DS3 #3 holds T1s 140–167.

  If the STS-1 frames are divided into 1.5VT VTGs, then STS-1 #1 holds VTGs 1–7 where VTG #1 holds T1s 84–87, VTG #2 holds T1s 88–90, VTG #3 holds T1s 91–94, VTG #4 holds T1s 95–98, VTG #5 holds T1s 99–103, VTG #6 holds T1s 104–107, and VTG #7 holds T1s 108–111.

  Likewise for STS-1 #2: VTG #1 holds T1s 112–115, VTG #2 holds T1s 116–119, VTG #3 holds T1s 120–123, VTG #4 holds T1s 124–127, VTG #5 holds T1s 128–131, VTG #6 holds T1s 132–135, and VTG #7 holds T1s 136–139.

  Finally, for STS-1 #3: VTG #1 holds T1s 140–143, VTG #2 holds T1s 144–147, VTG #3 holds T1s 148–151, VTG #4 holds T1s 152–155, VTG #5 holds T1s 156–159, VTG #6 holds T1s 160–163, and VTG #7 holds T1s 164–167.

- **OC-3 #3 carries the next 84 T1s numbered from 168–251**—If the STS-1 frames are broken out into DS3s, then DS3 #1 holds T1s 168–195, DS3 #2 holds T1s 196–223, and DS3 #3 holds T1s 224–251.

  If the STS-1 frames are divided into 1.5VT VTGs, then STS-1 #1 holds VTGs 1–7 where VTG #1 holds T1s 168–171, VTG #2 holds T1s 172–175, VTG #3 holds T1s 176–179, VTG #4 holds T1s 180–183, VTG #5 holds T1s 184–187, VTG #6 holds T1s 188–191, and VTG #7 holds T1s 192–195.

  Likewise for STS-1 #2: VTG #1 holds T1s 196–199, VTG #2 holds T1s 200–203, VTG #3 holds T1s 204–207, VTG #4 holds T1s 208–211, VTG #5 holds T1s 212–215, VTG #6 holds T1s 216–219, and VTG #7 holds T1s 220–223.

  Finally, for STS-1 #3: VTG #1 holds T1s 224–227, VTG #2 holds T1s 228–231, VTG #3 holds T1s 232–235, VTG #4 holds T1s 236–239, VTG #5 holds T1s 240–243, VTG #6 holds T1s 244–247, and VTG #7 holds T1s 248–251.

- **OC-3 #4 carries the last 84 T1s numbered from 252–335**—If the STS-1 frames are broken out into DS3s, then DS3 #1 holds T1s 252–279, DS3 #2 holds T1s 280–307, and DS3 #3 holds T1s 308–335.

  If the STS-1 frames are divided into 1.5VT VTGs, then STS-1 #1 holds VTGs 1–7 where VTG #1 holds T1s 252–255, VTG #2 holds T1s 256–259, VTG #3 holds T1s 260–263, VTG #4 holds T1s 264–267, VTG #5 holds T1s 268–271, VTG #6 holds T1s 272–275, and VTG #7 holds T1s 276–279.

Likewise for STS-1 #2: VTG #1 holds T1s 280–283, VTG #2 holds T1s 284–287, VTG #3 holds T1s 288–291, VTG #4 holds T1s 292–295, VTG #5 holds T1s 296–299, VTG #6 holds T1s 300–303, and VTG #7 holds T1s 304–307.

Finally, for STS-1 #3: VTG #1 holds T1s 308–311, VTG #2 holds T1s 312–315, VTG #3 holds T1s 316–319, VTG #4 holds T1s 320–323, VTG #5 holds T1s 324–327, VTG #6 holds T1s 328–331, and VTG #7 holds T1s 332–335.

If DS3 back cards are used (DS3s #1–3 are in the top back card slot and DS3s #4–6 are in the bottom back card slot):

- DS3 #1 carries the first 28 T1s numbered from 0–27.
- DS3 #2 carries the next 28 T1s numbered from 28–55.
- DS3 #3 carries the next 28 T1s numbered from 56–83.
- DS3 #4 carries the next 28 T1s numbered from 84–111.
- DS3 #5 carries the next 28 T1s numbered from 112–139.
- DS3 #6 carries the last 28 T1s numbered from 140–167.

If DS1 back cards are used, DS1s 0–24 are in the top back card slot and DS1s 25–47 are in the bottom back card slot.

In the examples in this chapter, OC-3 back cards are used with the VXSM cards. Specifically, the OFFNET trunk group used the first T1 on the second STS-1 of the third OC-3 interface. This equates to T1 #196, so in this case the local endpoint names are ds/s-0/ds1-196/#.

The use of the wildcard characters $, meaning any, and *, meaning all, is the same as for NCS. But in TGCP, wildcard ranges can also be used in the format [*N-M*], where *N* is the lower bound of the range and *M* is the upper bound. For example, if you wanted to specify that only the first five DS0s on T1 number 2 can be used, the endpoint local name would be ds/ds1-2/[1-5] .

Packages, Events, and Signals

PacketCable 1.0 defines three packages for TGCP:

- ISUP trunk (IT) package
- MF Operator Services Signaling (OSS) (MO) package
- MF Terminating (MT) package

Support for all three packages is required for PacketCable 1.0 MGCs and MGs. The one you will use most of the time is the IT package. This package is used when the endpoint is an SS7 ISUP trunk connected to the PSTN. Consequently, it is this package that is used for basic OFFNET calls. If no package name is specified, the IT package is assumed because this is the default package for TGCP.

The MO and MT packages are used with MF CAS trunk endpoints.

The MO package is used for one-way outgoing MF Feature Group D (FGD) trunks. These trunks are used for connecting to an operator switch for emergency 911 access.

The MT package is used for one-way incoming MF FGD trunks used for Busy Line Verification (BLV) and Operator Interrupt (OI) functionality.

Just as with NCS, in PacketCable 1.5 two additional packages are available: the Fax package, denoted by the package name FXR, and the VoIP metrics package, denoted by the package name XRM. Please refer to Chapter 7, "NCS (Network-based Call Signaling)," for more information on these two packages. Support for all five packages is required for PacketCable 1.5 MGCs, and support for the IT, FXR, and XRM packages are required for PacketCable 1.5 MGs with support for the MO and MT packages recommended.

Table 8-2 describes the events and signals used in the IT package.

Table 8-2 *IT Package Events and Signals*

Event/Signal Code	Description
co1	Continuity tone 1 (2010 Hz tone; default timeout is 3 seconds)
co2	Continuity tone 2 (1780 Hz tone; default timeout is 3 seconds)
ft	Fax tone
ld	Long duration connection; used when a connection is up for an extended time (default value is 1 hour)
ma	Media start; occurs when first valid RTP packet is received
mt	Modem tone
oc	Operation complete; indicates a signal has timed out
of	Operation failure
ro	Reorder tone (dual tones of 480 Hz and 620 Hz; default timeout 30 seconds)
rt	Ring back tone (dual tones of 440 Hz and 480 Hz; default timeout 180 seconds)
tdd	Telecom devices for the deaf tone

Table 8-3 describes the events and signals used in the MO package.

Table 8-3 *MO Package Events and Signals*

Event/Signal Code	Description
ans	Call answer; used after operator answers the call, and after this point only the operator can terminate the call
ft	Fax tone

continues

Table 8-3 *MO Package Events and Signals (Continued)*

Event/Signal Code	Description
ld	Long duration connection; used when a connection is up for extended time (default value is 1 hour)
mt	Modem tone
oc	Operation complete; indicates a signal has timed out
of	Operation failure
orbk	Operator ring back; used to alert the calling party
rbz	Reverse make busy; used to mark the trunk as busy
rcl	Operator recall; used to bring the operator back after an event such as a customer hook flash
rel	Release call; might include a cause code parameter indicating the reason for the release
res	Resume call
rlc	Release complete
sup	Call setup; the address and ID are included as parameters in the format sup(addr(<*MF digits*>,id(<*MF digits*>) where the <*MF digits*> identify the called address (*addr/DNIS*) and calling address (*id/ANI*)
sus	Suspend call
swk	Wink start
TDD	Telecom devices for the deaf tone

Table 8-4 describes the events and signals used in the MT package.

Table 8-4 *MT Package Events and Signals*

Event/Signal Code	Description
ans	Call answer; used after operator answers the call, and after this point only the operator can terminate the call
bl	Call block; used to indicate no further call is accepted
bz	Busy tone (dual tones of 480 Hz and 620 Hz)
hf	Hook flash
inf	Information digits; includes as a parameter all digits received in the format of a comma-separated list of MF digits
oc	Operation complete; indicates a signal has timed out

Table 8-4 *MT Package Events and Signals (Continued)*

Event/Signal Code	Description
of	Operation failure
oi	Operator interrupt; occurs when the operator attempts to interrupt the call
pst	Permanent signal tone
rel	Release call; can include a cause code parameter indicating the reason for the release
res	Resume call
rlc	Release complete
ro	Reorder tone
sup	Call setup; the address and ID are included as parameters in the format sup(addr(<*MF digits*>,id(<*MF digits*>) where the <*MF digits*> identify the called address (*addr/DNIS*) and calling address (*id/ANI*)
sus	Suspend call

Local Connection Option Parameters for CODECs and Packetization

In Chapter 7, you learned about the relationship between the LCO CODEC compression algorithm parameter (a) and the LCO packetization parameters (p) and (mp). This same relationship and these same rules exist for TGCP. So refer to Chapter 7 for more detail on these parameters. Also remember that the same CODECs used with NCS must be supported with TGCP as well.

Local Connection Option Parameters for Security

Local Connection Option (LCO) parameters for the support of secure voice are also needed for TGCP just as they are with NCS. This includes the Real-time Transport Protocol (RTP) and Real-time Transport Control Protocol (RTCP) ciphersuite parameters. Just as with the packetization and CODEC parameters, these work the same way as in the NCS case. So refer to Chapter 7 for more information.

Local Connection Option Parameters for Lawful Intercept

TGCP endpoints must also be able to interact with lawful intercept devices if a PacketCable customer is under electronic surveillance. Normally, this functionality is handled by the CMTS, but scenarios do exist where the only place a customer's voice conversation can be tapped is at the trunking gateway. A good example of this is when PacketCable subscribers have forwarded their calls to an OFFNET destination. Inbound PSTN calls would be forwarded right back out to the PSTN, leaving the MG as the only place where the call could be tapped.

Thus, additional LCO parameters are used in TGCP to handle this requirement. These parameters are the call content connection identifier (es-cci) and the call content destination (es-ccd). The use of these parameters is as follows:

- **Call Content Connection Identifier (es-cci)**—The CMS, CMTS, and media gateway can all contribute to providing electronic surveillance information, and multiple customers can be under surveillance at the same time. The Call Content Connection Identifier is a 32-bit value used to correlate all this information. This field is added to the header of all intercepted voice packets before they are sent to the electronic surveillance delivery function.
- **Call Content Destination (es-ccd)**—This field identifies the IP address and UDP port number of the delivery function that is collecting the intercepted voice packets.

NOTE Chapter 15, "Event Messaging and Lawful Intercept," covers how electronic surveillance or lawful intercept works in PacketCable.

SDP Attributes

The use of the Session Description Protocol (SDP) in TGCP is nearly identical to its use in NCS. The use of the version (v), origin (o), session (s), connection data , bandwidth (b), time (t), and media announcement (m) parameters is as was described with NCS and MGCP. Also the additional attributes for CODECs and security defined for NCS also apply for TGCP. Attributes identifying whether or not the endpoint supports electronic surveillance also exist with TGCP.

Other Notable Changes

Again, these changes are the same as with NCS. All the previously mentioned additions to the Local Connection Options can also be audited using the AUEP and AUCX commands. Finally, call quality information obtained via RTCP is returned after an MGC-initiated DLCX.

Examining Call Flows

Now that you know the basics of the protocols involved when interworking with the PSTN, this section examines the way they all fit together. In Chapter 7, you saw the details of how ONNET-to-ONNET calls work. In this chapter, you will see the details of calls to OFFNET subscribers. Remember, an ONNET subscriber is someone who has their phone service on the PacketCable network, whereas an OFFNET subscriber is someone whose phone service is still handled by the PSTN.

These call flows are from actual calls and serve as an example of the messaging and parameters involved in OFFNET calls. Remember, though, the PacketCable specifications do not dictate the specifics of how the protocols are used to formulate calls; this is up to the vendors of the various components. As long as the specifications are not violated, you have many ways to accomplish the same task. All the examples in this text use the Cisco BTS 10200, running version 4.5, as the softswitch or MGC/CMS, Cisco 7301 and 7206VXR routers as the SG, and either the Cisco VXSM or VISM card as the MG.

Also keep in mind that you can have the MGC and CMS PacketCable components be separate devices instead of consolidated into a single device as with the Cisco BTS.

Basic ONNET-to-OFFNET Call Flow

The first example is a basic call between Oscar (the originating party) and Tina (the terminating party) much like you saw in Chapter 7. The only difference in this case is that Tina still has her phone service from the legacy PSTN, whereas Oscar is a PacketCable subscriber. The examination of their phone call is broken up into two stages:

- Stage one details the messaging used from when Oscar picks up his phone until the point where Tina's phone is ringing.
- Stage two shows the messaging from here up to the point where Oscar and Tina have bidirectional voice communication and eventually hang up their phones, ending the call.

In addition to the message flows, you will see the details of key packets.

Stage One: Basic ONNET-to-OFFNET Call Flow

Figure 8-6 illustrates the call signaling used for stage one of a basic ONNET-to-OFFNET call.

Figure 8-6 *Basic ONNET-to-OFFNET Call Flow: Stage One*

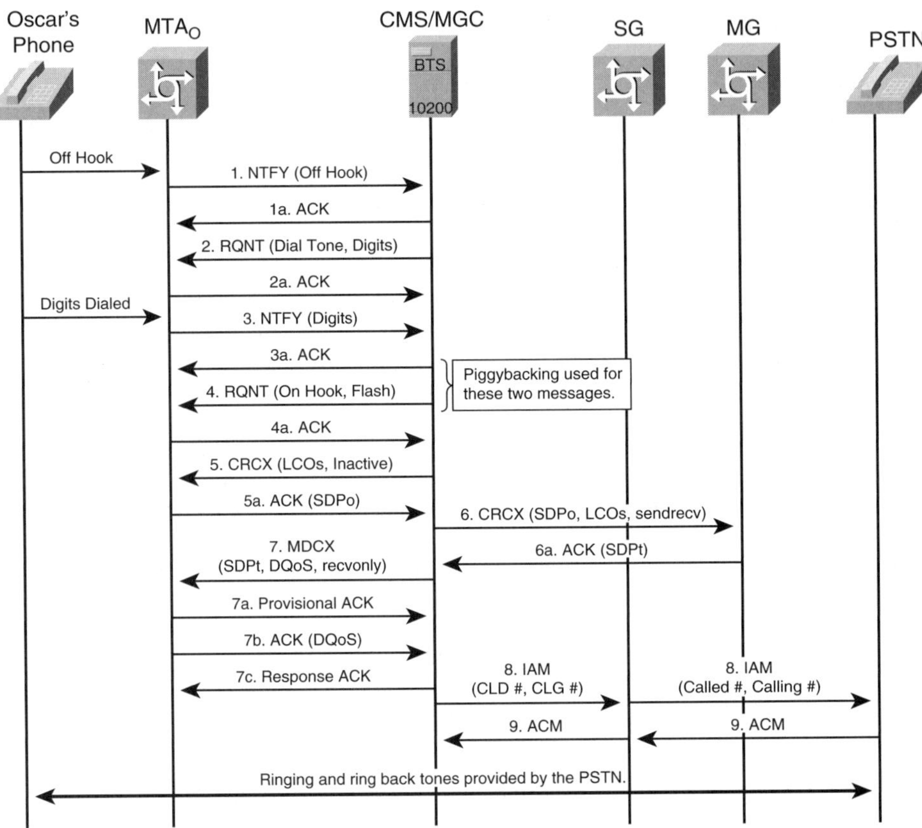

The first five steps are no different than the first five steps in the basic ONNET-to-ONNET call example in Chapter 7, so more focus is devoted to Steps 6 through 9.

1. Oscar takes his phone off hook, triggering a NTFY message to be sent to the CMS. This message contains the Observed Event hook detect.
2. The CMS then requests dial tone be provided to Oscar and asks to be notified when digits are collected according to the MTA's associated digit map for that endpoint.
3. As soon as Oscar hears dial tone, he dials Tina's phone number, which is forwarded to the CMS in another NTFY message.
4. The CMS then issues an RQNT to Oscar's MTA asking to be notified if Oscar hangs up his phone or issues a flash hook.

5 The CMS realizes it can route the call based on those digits and creates an inactive connection on Oscar's MTA. The response from the CRCX contains SDP information containing the IP address, RTP port number, and other bearer-related information about the endpoint.

6 In this step, the MGC component of the BTS initiates the creation of a connection on the MG interfacing to the PSTN. In this CRCX are local connection options influencing how the connection is handled. Information such as CODEC and packetization period is included here. Notice that the mode of the connection is send/receive, meaning the MG immediately starts forwarding packets between the trunk DS-0 and the PacketCable network. For OFFNET calls, ring back tone is typically provided from the PSTN, so if you call a destination (such as another country) that uses a different type of ring back signal, you hear this unique signal. Thus, this tone needs to pass across the PacketCable VoIP network to the originating endpoint. For this to happen, the connection on the MG must be in send mode when the PSTN is providing ring back. This is why the BTS immediately puts the connection in send/receive mode for OFFNET calls. Finally, the SDP information about Oscar's MTA is also included so that the MG knows how and where to send the bearer stream. The following is a detailed look at this message:

```
CRCX 72093 DS/S-0/DS1-196/1@tgw01-vxsm-5.cisco.com MGCP 1.0 TGCP 1.0
C: 32d
L: p:20,  a:PCMU;PCMA,  e:on,  s:off,  t:a0
M: sendrecv
K: 72092

v=0
o=- 105003104 105003104 IN IP4 24.34.240.241
c=IN IP4 24.34.240.241
t=0 0
m=audio 54042 RTP/AVP 0 8
b=AS:81
a=mptime:20 20
a=rtpmap:0 PCMU/8000/1
a=rtpmap:8 PCMA/8000/1
a=X-pc-secret:base64:<hex digit string omitted>
a=X-pc-csuites-rtp:62/51 64/51 60/51 60/50
a=X-pc-csuites-rtcp:81/70 81/71 82/70 82/71 80/70
```

From the trace you can see the TGCP endpoint name is ds/s-0/ds1-196/1@tgw01.-cisco.com, where tgw01.vxsm-5.cisco.com is the FQDN of the MG. The local name indicates that the PSTN trunk used is on the second DS0 on the first DS1 on the MG. The rest of the message is similar to what you have seen with NCS.

7 Next, the CMS passes the media gateway's SDP information to Oscar's MTA so that it knows how to communicate with the PSTN. Also DQoS information is passed to the MTA, and the connection mode is modified to receive only. This triggers the MTA to reserve the radio frequency (RF) quality of service (QoS) pipe to the CMTS for the upstream direction (MTA to CMTS) and reserve and commit the QoS pipe for the downstream direction (CMTS to MTA). This aspect of PacketCable calls is covered much more in the DQoS chapters of this book (Chapter 12, "DQoS Architecture and Framework," and Chapter 13, "Analyzing, Implementing, and Troubleshooting DQoS"). Also notice the use of provisional responses and response acknowledgments, which was covered in detail in Chapter 7.

8 In this step you see the first SS7 message. An ISUP IAM is sent to the PSTN, instructing it to set up the call to Tina's phone. Remember, SS7 messages are sent as Sigtran messages for the communication between the MGC and SG. Ethereal has the capability to decode the Sigtran family of protocols. Figure 8-7 shows the Sigtran form of this ISUP IAM message.

Figure 8-7 *Ethereal Capture of ISUP IAM*

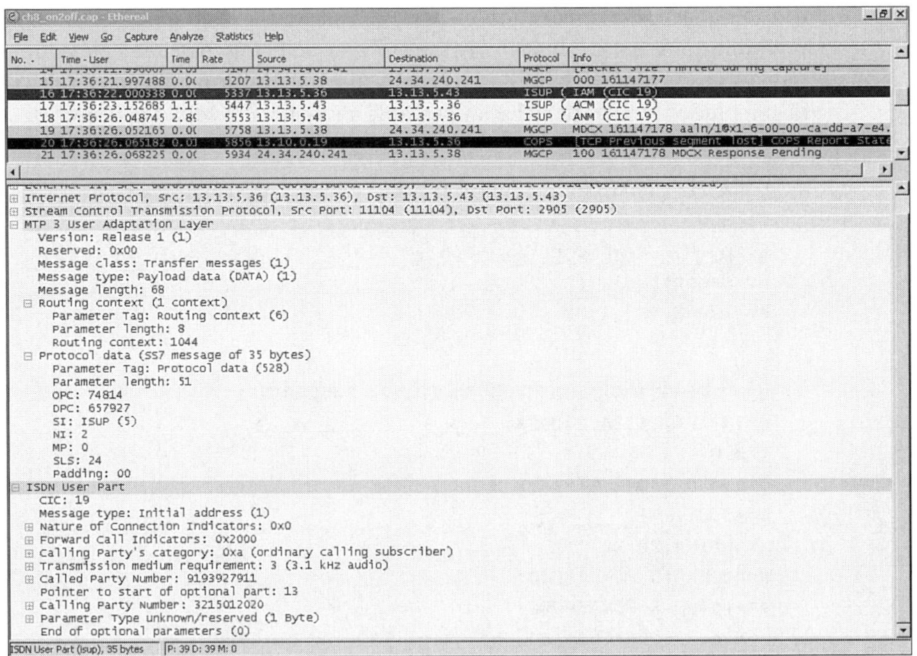

You can see much of the Sigtran protocol family in Figure 8-7. An SCTP header indicates the SCTP source and destination port numbers used for communication between the MGC and SG. An M3UA header defines the SS7 MTP3 type parameters. You can see the origination and destination point codes of the message in the OPC and DPC protocol data fields, as well as the SI field denoting the upper-layer protocol (in this case ISUP).

NOTE These point codes are in decimal format. To convert them into the three-field format, first convert the decimal value to a six-digit hexadecimal value (for example, 74814 equates to 0x01243E). Then, separate the six-digit field into three two-digit pairs (01-24-3E). Finally, convert these values back into decimal (1-32-62).

The rest of the message is the ISUP portion. You can see this message is an IAM for CIC #19. Parameters such as called and calling party number are included in this message.

9. The last message in this stage of the call is an ACM received from the PSTN. This indicates that the call has been completed, and one-way voice path (from the PSTN to Oscar's MTA) is now open. Consequently, Oscar now hears ring back tone on his phone indicating Tina's phone is ringing. Keep in mind that if the call failed to route somewhere in the PSTN, you would also receive an ACM—but in that case a tone or announcement indicating the reason for failure instead of ring back tone.

Stage Two: Basic ONNET-to-OFFNET Call Flow

Figure 8-8 illustrates the call flow for the second stage.

Figure 8-8 *Basic ONNET-to-OFFNET Call Flow: Stage Two*

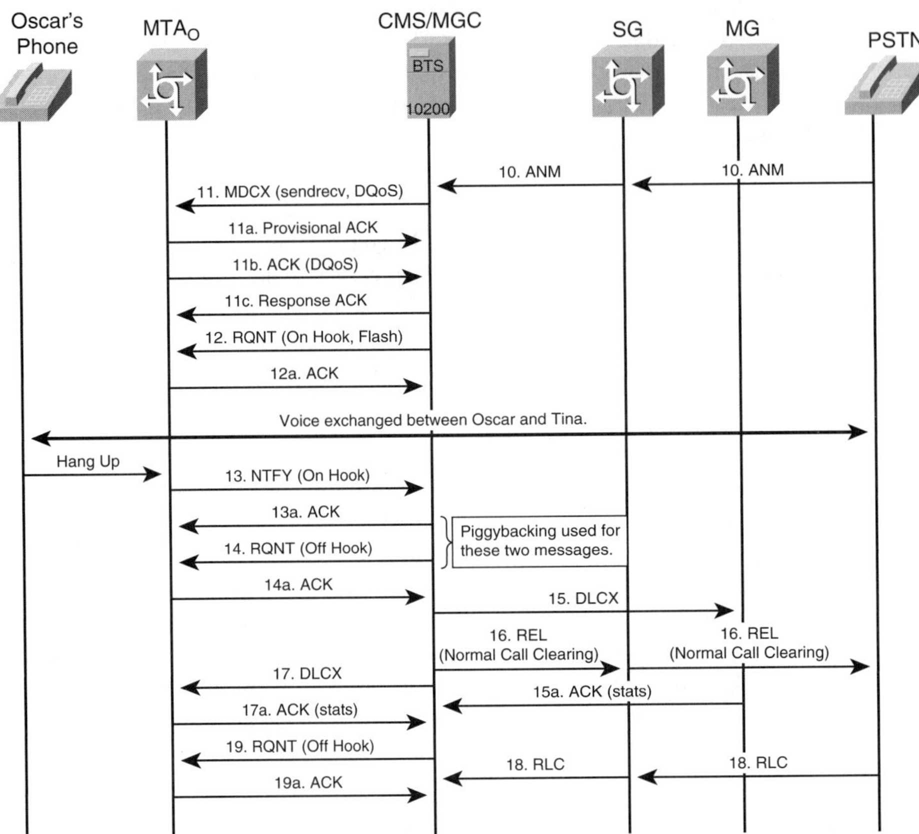

10 Eventually, Tina answers her phone, triggering an ISUP ANM message from the PSTN toward the MGC. Figure 8-9 shows the Sigtran form of this ISUP ANM message.

Figure 8-9 *Ethereal Capture of ISUP ANM*

```
ch8_on2off.cap - Ethereal
File Edit View Go Capture Analyze Statistics Help
No. . Time - User          Time  Rate Source          Destination     Protocol Info
   16 17:36:22.000338 0.0(  5337 13.13.5.36     13.13.5.43      ISUP (  IAM (CIC 19)
   17 17:36:23.152685 1.1!  5447 13.13.5.43     13.13.5.36      ISUP (  ACM (CIC 19)
   18 17:36:26.048745 2.8!  5553 13.13.5.43     13.13.5.36      ISUP (  ANM (CIC 19)
   19 17:36:26.052165 0.0(  5758 13.13.5.38     24.34.240.241   MGCP    MDCX 161147178 aaln/1@x1-6-00-00-ca-dd-a7-e4.
   20 17:36:26.065182 0.01  5856 13.10.0.19     13.13.5.36      COPS    [TCP Previous segment lost] COPS Report State
   21 17:36:26.068225 0.0(  5934 24.34.240.241  13.13.5.38      MGCP    100 161147178 MDCX Response Pending
   22 17:36:26.076276 0.0(  6008 24.34.240.241  13.13.5.38      MGCP    [Malformed Packet]
   23 17:36:26.076528 0.0(  6068 13.13.5.38     24.34.240.241   MGCP    000 161147178

Internet Protocol, Src: 13.13.5.43 (13.13.5.43), Dst: 13.13.5.36 (13.13.5.36)
Stream Control Transmission Protocol, Src Port: 2905 (2905), Dst Port: 11104 (11104)
MTP 3 User Adaptation Layer
  Version: Release 1 (1)
  Reserved: 0x00
  Message class: Transfer messages (1)
  Message type: Payload data (DATA) (1)
  Message length: 44
  Routing context (1 context)
    Parameter Tag: Routing context (6)
    Parameter length: 8
    Routing context: 1044
  Protocol data (SS7 message of 9 bytes)
ISDN User Part
  CIC: 19
  Message type: Answer (9)
  Pointer to start of optional part: 1
  Backward Call Indicators: 0x614
    Optional Parameter: 17 (Backward call indicators)
    Parameter length: 2
    .... ..10 .... .... = Charge indicator: Charge (0x0002)
    .... 01.. .... .... = Called party's status indicator: Subscriber free (0x0001)
    ..00 .... .... .... = Called party's category indicator: No indication (0x0000)
    00.. .... .... .... = End-to-end method indicator: No End-to-end method available (only link-by-link method available)
    .... .... .... ...0 = Interworking indicator: no interworking encountered (No.7 signalling all the way)
    .... .... .... ..0. = End-to-end information indicator: no end-to-end information available
    .... .... .... .1.. = ISDN user part indicator: ISDN user part used all the way
    .... .... .... 0... = Holding indicator: holding not requested
    .... .... ...1 .... = ISDN access indicator: terminating access ISDN
    .... .... ..0. .... = Echo Control Device Indicator: Echo control device not included
    .... .... 00.. .... = SCCP method indicator: No indication (0x0000)
  End of optional parameters (0)
P: 39 D: 39 M: 0
```

The only parameters to this message are the Backward Call Indicators. The receipt of this message indicates the call has been answered and voice path can be cut through in both directions.

11 To accomplish this, the CMS instructs Oscar's MTA to modify its connection to send/receive mode and to commit its DQoS resources for the upstream direction (remember, downstream DQoS resources have already been committed). After this point bidirectional voice communication between Oscar and Tina is available.

12 The CMS then requests to be notified if Oscar hangs up his phone or initiates a hook flash. Notice, this is not necessary on the MG because ISUP messages are received for events occurring in the PSTN.

13 At some point later, Oscar hangs up his phone, triggering a NTFY message to be sent to the CMS.

14 The CMS then requests to be notified if Oscar takes his phone off hook again.

15 The result of Oscar's hanging up his phone causes both of the MGCP connections to be deleted. First, the MGC sends a DLCX to the MG to delete the PSTN connection.

16 The MGC must also notify the PSTN that this call is now over, so it sends an ISUP REL message to the PSTN via the SG. This message is depicted in Figure 8-10.

Figure 8-10 *Ethereal Capture of ISUP REL*

```
ch8_on2off.cap - Ethereal
File  Edit  View  Go  Capture  Analyze  Statistics  Help
No. - Time - User        Time     Rate Source          Destination     Protocol Info
   28 17:36:39.428042 13.3357   6432 24.34.240.241   13.13.5.38      MGCP     NTFY 81 aaln/1@x1-6-00-00-CA-DD-A7-E4.cc.
   29 17:36:39.430261 0.00221   6592 13.13.5.38      24.34.240.241   MGCP     200 81 OK
   30 17:36:39.431886 0.00162   6733 13.13.5.38      13.13.9.129     MGCP     DLCX 72094 DS/S-0/DS1-196/1@tgw01-vxsm-5.
   31 17:36:39.432662 0.00077   6835 13.13.5.36      13.13.5.43      ISUP (   REL (CIC 19)
   32 17:36:39.445459 0.01279   6895 24.34.240.241   13.13.5.38      MGCP     200 161147180 OK
   33 17:36:39.446262 0.00080   7042 13.13.5.38      24.34.240.241   MGCP     DLCX 161147181 aaln/1@x1-6-00-00-ca-dd-a7
   34 17:36:39.459013 0.01275   7156 13.13.9.129     13.13.5.38      MGCP     250 72094 OK

Internet Protocol, Src: 13.13.5.36 (13.13.5.36), Dst: 13.13.5.43 (13.13.5.43)
Stream Control Transmission Protocol, Src Port: 11104 (11104), Dst Port: 2905 (2905)
MTP 3 User Adaptation Layer
  Version: Release 1 (1)
  Reserved: 0x00
  Message class: Transfer messages (1)
  Message type: Payload data (DATA) (1)
  Message length: 40
  Routing context (1 context)
    Parameter Tag: Routing context (6)
    Parameter length: 8
    Routing context: 1044
  Protocol data (SS7 message of 8 bytes)
    Parameter Tag: Protocol data (528)
    Parameter length: 24
    OPC: 74814
    DPC: 657927
    SI: ISUP (5)
    NI: 2
    MP: 1
    SLS: 24
ISDN User Part
  CIC: 19
  Message type: Release (12)
  Cause indicators, see Q.850 (2 bytes length)
    Mandatory Parameter: 18 (Cause indicators)
    Pointer to Parameter: 2
    Parameter length: 2
    Cause indicators (-> Q.850)
    .... 0010 = Cause location: Public network serving the local user (LN) (2)
    .00. .... = Coding standard: ITU-T standardized coding (0x00)
    1... .... = Extension indicator: last octet
    .001 0000 = Cause indicator: Normal call clearing (16)
    1... .... = Extension indicator: last octet
  No optional parameter present (Pointer: 0)
File: "D:\Documents and Settings\j  P: 39 D: 37 M: 0
```

In this message, you can see the cause value indicating the reason the MGC is ending the call. In this case, the cause value is 16, meaning "normal call clearing."

17 The CMS then instructs Oscar's MTA to delete its NCS connection.

18 The PSTN then signals the MGC that the call has been successfully released by sending an ISUP RLC.

19 Finally, the CMS requests notification if Oscar takes his phone off hook again.

Basic OFFNET-to-ONNET Call Flow

The second example is the reverse case. Here, Tina is the PacketCable subscriber, and Oscar's phone service is from the legacy PSTN. Again, the call is broken into two stages:

- Stage one details the messaging used from when Oscar dials Tina's number, triggering the PSTN to contact the PacketCable network up to the point where Tina answers her phone.
- Stage two covers the remainder of the call.

As with the previous examples, traces show the details of key packets.

Stage One: Basic OFFNET-to-ONNET Call Flow

Figure 8-11 illustrates the call signaling used for stage one.

Figure 8-11 *Basic OFFNET-to-ONNET Call Flow: Stage One*

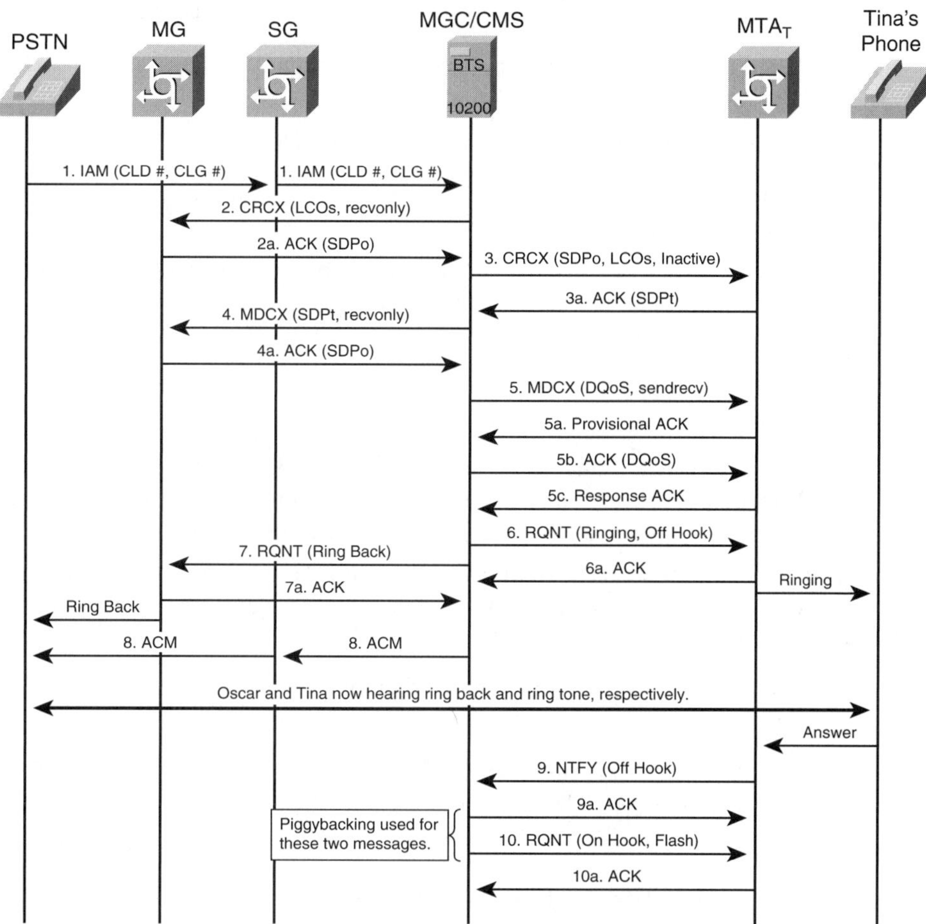

1 First, the PSTN sends an ISUP IAM to the MGC. This message is similar to the IAM from the previous example, except in this case Tina's phone number is the called phone number, and Oscar's phone number is the calling phone number.

2 Next, a TGCP connection is created in the receive-only state on the MG endpoint referenced by the CIC in the IAM message. Local Connection Options are included in the CRCX as well, and the endpoint's bearer information is returned using the SDP protocol.

3 The CMS then passes Oscar's SDP information to Tina's MTA in an NCS CRCX. This connection is put into the inactive state. As with the previous step, Local Connection Options are included in the CRCX, and the MTA endpoint's bearer information is returned using the SDP protocol.

4 The MGC relays Tina's SDP information to the MG in a TGCP MDCX command.

5 The CMS modifies the MTA connection by putting it in send/receive mode and instructing the reservation of DQoS resources.

6 Next, the CMS requests the MTA to ring Tina's phone.

7 Similarly, the MGC requests the MG to provide ring back tone to Oscar's phone in the PSTN. This message is as follows:

```
RQNT 72089 DS/S-0/DS1-196/1@tgw01-vxsm-5.cisco.com MGCP 1.0 TGCP 1.0
N: mgcp-104CA104.cisco.com:2427
X: 21900000013
R:
S: IT/rt(to=0)
K: 72088
```

Notice the TGCP Signal Requests parameter (S); it is instructing the MG to play the IT package ring back tone (rt) out the DS0 endpoint. The (to=0) parameter means this signal is to continue playing until the MGC instructs it to stop.

8 The MGC then sends an ISUP ACM to the PSTN to indicate the call has successfully been set up and to open up the voice path. This is necessary so Oscar can hear the ring back tone generated by the MG.

9 Eventually, Tina answers her phone, triggering her MTA to send an NCS NTFY to the CMS.

10 In response to this message, the CMS requests to be notified if Tina hangs up her phone or presses the flash button.

Stage Two: Basic OFFNET-to-ONNET Call Flow

Figure 8-12 illustrates the call flow for the second stage.

Figure 8-12 *Basic OFFNET-to-ONNET Call Flow: Stage Two*

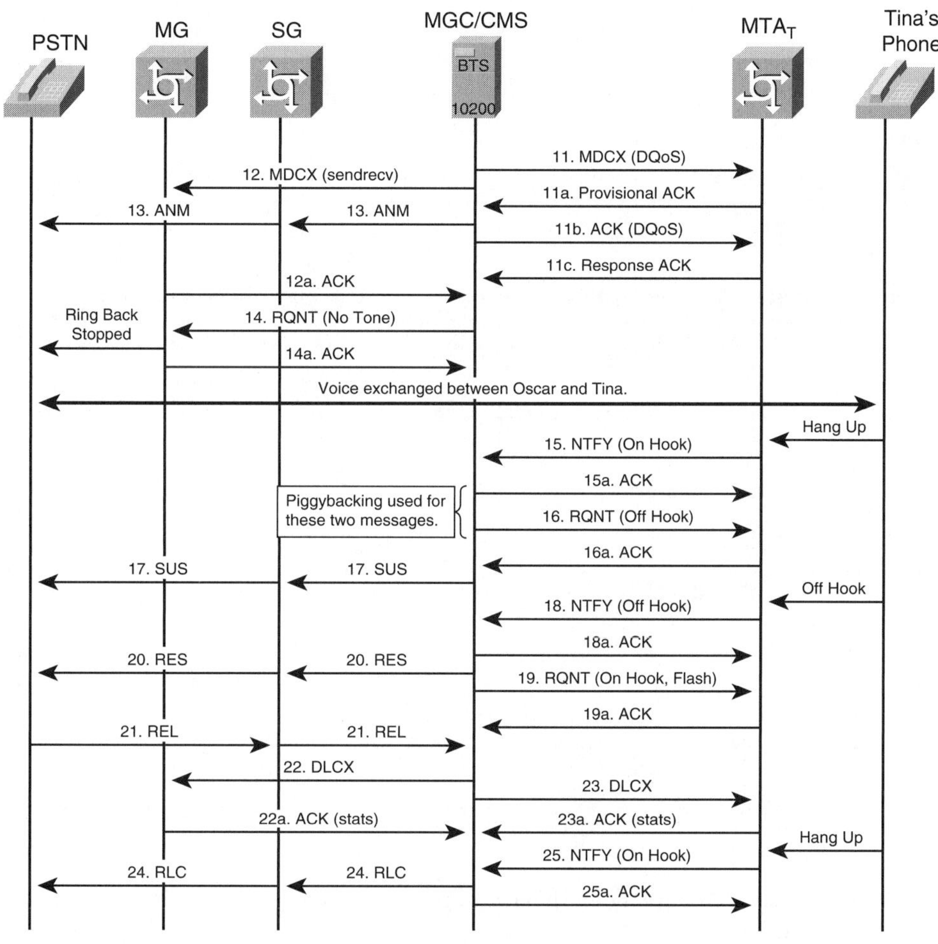

11 Next, the CMS instructs Tina's MTA to commit or activate its DQoS resources.

12 The MGC then opens up the voice path in both directions by modifying the connection on the MG to send/receive mode.

13 The PSTN is notified of Tina answering her phone by the receipt of an ISUP ANM message. After this point voice path is available in both directions end to end.

14 Because Tina has answered her phone, the MGC instructs the MG to stop playing ring back tone to the PSTN. Remember this was set up to continuously play in Step 7.

15 Eventually, Tina hangs up her phone, triggering an NCS NTFY message to be sent to the CMS.

16 The CMS then requests to be notified if Tina picks up her phone again.

17 This step might be interesting to you. Instead of an ISUP REL message, an ISUP SUS message is sent to the PSTN. This means if Tina picks up her phone again before the suspend timer expires, the call is resumed. Figure 8-13 depicts the SUS message.

Figure 8-13 *Ethereal Capture of ISUP SUS*

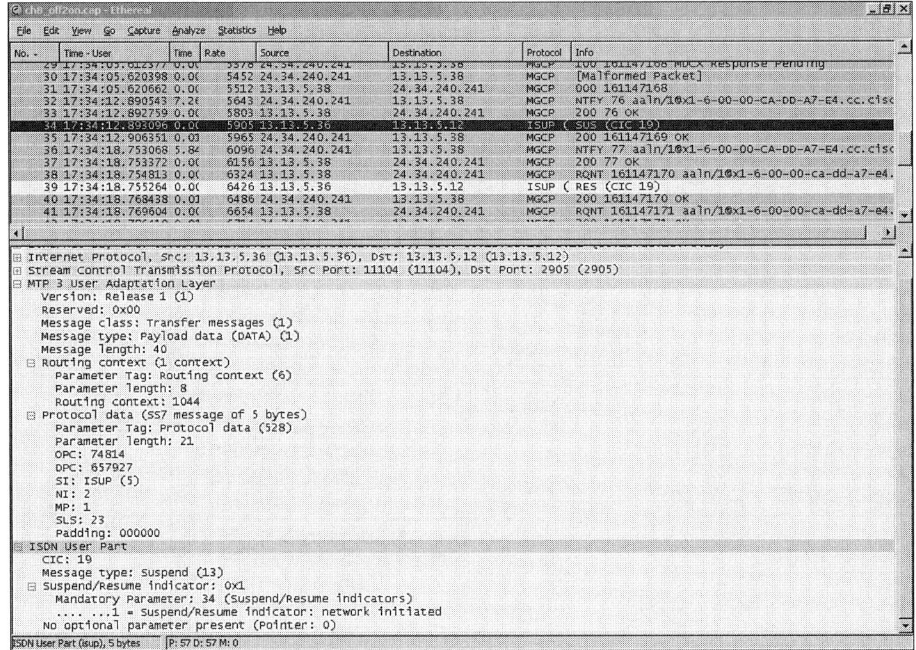

18 Tina does re-answer her phone, triggering another NTFY message to be sent to the CMS.

19 Again, the CMS then requests to be notified if Tina picks up her phone again.

20 The receipt of the NTFY off hook event means that the call should be resumed, so an ISUP RES message is sent to the PSTN. Figure 8-14 shows this message.

Figure 8-14 *Ethereal Capture of ISUP RES*

```
No.  Time - User              Time  Rate  Source          Destination     Protocol  Info
31   17:34:05.620662 0.0(  5512 13.13.5.38      24.34.240.241   MGCP      000 161147168
32   17:34:12.890543 7.2(  5643 24.34.240.241   13.13.5.38      MGCP      NTFY 76 aaln/1@x1-6-00-00-CA-DD-A7-E4.cc.cis(
33   17:34:12.892759 0.0(  5803 13.13.5.38      24.34.240.241   MGCP      200 76 OK
34   17:34:12.893096 0.0(  5905 13.13.5.36      13.13.5.12      ISUP (    SUS (CIC 19)
35   17:34:12.906351 0.0]  5965 24.34.240.241   13.13.5.38      MGCP      200 161147169 OK
36   17:34:18.753068 5.8(  6096 24.34.240.241   13.13.5.38      MGCP      NTFY 77 aaln/1@x1-6-00-00-CA-DD-A7-E4.cc.cis(
37   17:34:18.753372 0.0(  6156 13.13.5.38      24.34.240.241   MGCP      200 77 OK
38   17:34:18.754813 0.0(  6324 13.13.5.38      24.34.240.241   MGCP      RQNT 161147170 aaln/1@x1-6-00-00-ca-dd-a7-e4.
39   17:34:18.755264 0.0(  6426 13.13.5.36      13.13.5.12      ISUP (    RES (CIC 19)
40   17:34:18.768438 0.0]  6486 24.34.240.241   13.13.5.38      MGCP      200 161147170 OK
41   17:34:18.769604 0.0(  6654 13.13.5.38      24.34.240.241   MGCP      RQNT 161147171 aaln/1@x1-6-00-00-ca-dd-a7-e4.
42   17:34:18.786419 0.0]  6714 24.34.240.241   13.13.5.38      MGCP      200 161147171 OK
43   17:34:26.206264 7.41  6816 13.13.5.43      13.13.5.36      ISUP (    REL (CIC 19)

Internet Protocol, Src: 13.13.5.36 (13.13.5.36), Dst: 13.13.5.12 (13.13.5.12)
Stream Control Transmission Protocol, Src Port: 11104 (11104), Dst Port: 2905 (2905)
MTP 3 User Adaptation Layer
  Version: Release 1 (1)
  Reserved: 0x00
  Message class: Transfer messages (1)
  Message type: Payload data (DATA) (1)
  Message length: 40
  Routing context (1 context)
    Parameter Tag: Routing context (6)
    Parameter length: 8
    Routing context: 1044
  Protocol data (SS7 message of 5 bytes)
    Parameter Tag: Protocol data (528)
    Parameter length: 21
    OPC: 74814
    DPC: 657927
    SI: ISUP (5)
    NI: 2
    MP: 1
    SLS: 23
    Padding: 000000
ISDN User Part
  CIC: 19
  Message type: Resume (14)
  Suspend/Resume indicator: 0x1
    Mandatory Parameter: 34 (Suspend/Resume indicators)
    .... ...1 = Suspend/Resume indicator: network initiated
  No optional parameter present (Pointer: 0)
File: "D:\Documents and Settings\j  P: 57 D: 57 M: 0
```

21 At some point Oscar's hangs up his phone. This triggers the PSTN to send an ISUP REL message to the MGC.

22 This, in turn, causes the MGC to instruct the MG to delete its TGCP connection.

23 Likewise, the CMS instructs Tina's MTA to delete its NCS connection.

24 The MGC alerts the PSTN of the successful call teardown by sending an ISUP RLC message.

25 Finally, Tina hangs up her phone, triggering a NTFY message to be sent to the CMS.

Emergency 911 Call Flow

In this example, the PacketCable customer Oscar needs emergency assistance, so he dials 911, routing him to an operator services switch. This switch connects to the PacketCable MGC via the CAS signaling protocol. Remember that with the CAS protocol, no Signaling Gateway is involved in the call. Instead, all call signaling information is included in the TGCP protocol, as you'll soon see.

As with any ONNET-to-OFFNET call, the first message sent to the MG is a CRCX:

```
CRCX 107374406 ds/ds1-2/1@vism14.cisco.com MGCP 1.0 TGCP 1.0
C: 2a
L: p:10,  a:PCMU;PCMA,  e:off,  s:on
M: sendrecv
K: 107374405

v=0
o=- 15469687 15469687 IN IP4 24.34.240.87
s=-
c=IN IP4 24.34.240.87
t=0 0
m=audio 52458 RTP/AVP 0 8
b=AS:96
a=mptime:10 10
a=rtpmap:0 PCMU/8000/1
a=rtpmap:8 PCMA/8000/1
```

The TGCP CRCX is similar to the CRCX in the previous examples. The connection is placed in the send/receive state, and Oscar's SDP information is included. The MG's TGCP endpoint SDP information is returned in the acknowledgement to this CRCX:

```
200 107374406 OK
I: 3

v=0
o=- 0 0 IN IP4 13.137.20.18
s=-
c=IN IP4 13.137.20.18
t=0 0
m=audio 49156 RTP/AVP 0
```

The next message, a TGCP RQNT, is where things start to look different:

```
RQNT 107374407 ds/ds1-2/1@vism14.cisco.com MGCP 1.0 TGCP 1.0
N: mga-rtpCA160.cisco.com:2428
X: 34F00000002
R: MO/rel(N),MO/ans(N),MO/oc(N),MO/of(N)
S: MO/sup(addr(K0,9,1,1,S0),id(K0,0,9,9,7,0,0,0,1,S1))
K: 107374406
200 107374407 OK
```

Recall that PacketCable operator services are provided using the TGCP MO package. Because the IT package is the default for TGCP, all events and signals need to be prefixed by the MO package name. If you look at the TGCP Signal Requests parameter, you see the supervision/call setup signal. This is how the called and calling number parameters are sent in the CAS protocol. First, you see the MF digits representing the dialed number (or DNIS), 911, followed by the MF digits representing the calling number (or ANI), 997-0001. The K# and S# MF digits denote the beginning and end of a digit string, and the 0 preceding the ANI is a CAS information digit. The TGCP Requested Events parameter includes notification if the operator services switch answers the call (ans) or releases the call (rel).

Almost immediately, an emergency operator answers the call, triggering a NTFY message to be sent to the MGC:

```
NTFY 307390 ds/ds1-2/1@vism14.cisco.com MGCP 1.0 TGCP 1.0
X: 34F00000002
O: mo/ans
200 307390 OK
```

The MGC sends down a new list of requested events, which consists of the release event and the operator ring back event:

```
RQNT 107374408 ds/ds1-2/1@vism14.cisco.com MGCP 1.0 TGCP 1.0
X: 34F00000003
R: MO/rel(N),MO/orbk(N)
K: 107374407
200 107374408 OK
```

Eventually, the connection is deleted, triggering a DLCX sent from the MGC to the MG:

```
DLCX 107374409 ds/ds1-2/1@vism14.cisco.com MGCP 1.0 TGCP 1.0
I: 3
C: 2a
K: 107374408
250 107374409 OK
P: PS=716, OS=57201, PR=558, OR=44640, PL=0, JI=24, LA=0
```

The MGC then requests the MG to generate the release event to the operator services switch and to detect the release complete event:

```
RQNT 107374410 ds/ds1-2/1@vism14.cisco.com MGCP 1.0 TGCP 1.0
X: 34F00000004
R: MO/rlc(N)
S: MO/rel
K: 107374409
200 107374410 OK
```

You can see in the next and final message that the MG does, in fact, detect the release complete event from the operator services switch:

```
NTFY 308144 ds/ds1-2/1@vism14.cisco.com MGCP 1.0 TGCP 1.0
X: 34F00000004
O: mo/rlc
200 308144 OK
```

COT Testing

In the SS7 overview section you learned that it is possible for the two switches between which an SS7 IMT trunk is connected to have mismatched mappings of CICs to DS0s. This results in a lack of bearer connectivity on this trunk. A procedure known as a *continuity test* (COT) protects against this by testing the reception of predefined tones across the bearer path.

You can do COT tests using two methods:

- One method is where one switch simply "loops" its receive path to its transmit path. The other switch then generates a tone and expects to receive that same tone back. This is known as the "single tone" or "loop back" method and is defined by ITU.
- One method uses two different tones, a go tone and a return tone. Switch A sends the go tone; as soon as switch B detects the go tone, it sends the return tone. If switch A detects the return tone, the test has passed. This method is used in the United States and is also known as the "dual-tone" or "transponder" method.

COT tests can be performed either as part of call processing or as a maintenance act. On the BTS, you can initiate a COT test from the CLI using the following command:

```
CLI>diag ss7-trunk-termination tgn-id=8502; cic=648; test=3
TGN ID -> 8502
REASON -> Continuity test done
RESULT -> ADM ran test succeessfully
START CIC -> 648
END CIC -> 648
TEST TYPE -> ADM running SS7 continuity test
TEST DURATION -> 0
Reply : Success:
```

NOTE Before this command can be executed, the trunk termination must first be put in the "maintenance" state; more on using BTS diagnostic commands is covered in the "Troubleshooting Tips" portion of this chapter.

As you can see, this test was successful. You might be wondering what the messaging involved in this test looked like. In this example, the "transponder" method of continuity testing was configured. When this diagnostic command is entered, first the switch instructs the remote switch to set up this circuit for a continuity test. This is done by sending an ISUP CCR (Continuity Check Request) message. When the remote switch has completed this task, it sends an ISUP LPA (Loopback Acknowledge) message back to let the originating switch know.

Now that the switches are prepared for the test, the MG endpoint is instructed to generate the "go" tone and detect the "return" tone. The following RQNT message accomplishes this:

```
RQNT 72432 DS/S-0/DS1-84/24@tgw01-vxsm-5.com MGCP 1.0 TGCP 1.0
N: mgcp-104CA104.cisco.com:2427
X: 21800000006
R: IT/co1(N),IT/of(N)
S: IT/co2
K: 72431.
200 72432 OK
```

The "go" tone is the signal IT/co2, and the "return" tone is the signal IT/co1. Assuming the MG successfully detects the "return" tone, it sends a NTFY message to the MGC:

```
NTFY 318038610 DS/S-0/DS1-84/24@tgw01-vxsm-5.cisco.com MGCP 1.0
N: mgcp-104CA104.cisco.com:2427
X: 21800000006
O: IT/co1
200 318038610 OK
```

The switch then informs the remote switch that the test was successful by sending an ISUP COT message to the remote switch. It then informs the remote switch that the test is done and sends an ISUP REL message to the remote switch for the SS7 CIC on which the test was performed.

If the continuity test is initiated from the remote switch, the MGC creates a connection on the MG endpoint in "conttest" mode after receiving the ISUP CCR message:

```
CRCX 22075 DS/S-0/DS1-84/24@tgw01-vxsm-5.cisco.com MGCP 1.0 TGCP 1.0
C: 16
M: conttest
K: 22074
```

This instructs the MG endpoint to look for the "go" tone and generate the "return" tone. As soon as the MGC receives the ISUP REL message, it sends a DLCX to the MG to remove this connection.

Another form of COT testing known as a *network continuity test* exists. In this case, the bearer path between the two VoIP endpoints is tested. This ensures that IP connectivity (that is, routing) exists between the two endpoints.

Component Configuration

The last part of this chapter covers configuration of parameters relevant to the topics of this chapter on the various PacketCable components. These components are the MGC, MG, and the SG.

Signaling Gateway Configuration

In Chapter 2 you learned the SG produced by Cisco Systems is called the Cisco IP Transfer Point (ITP). The ITP runs a special version of IOS on the following platforms:

- Cisco 2650 Router
- Cisco 2650XM Modular Access Router
- Cisco 2651 Router
- Cisco 2651XM Modular Access Router
- Cisco 7204 Router
- Cisco 7206 Router
- Cisco 7301 Router
- Cisco 7507 Router[1]
- Cisco 7513 Router[1]

[1] For the 75xx router platforms, they must be equipped with both a VIP4-80 or VIP6-80 versatile interface processor (VIP) and an RSP8 or RSP16 route switch processor (RSP).

NOTE Which platform you choose to implement depends on your performance requirements.

The 2650/2651 routers are at the low end of the scale; they can support at most four signaling links and do not have redundant processors or hot-swappable line cards. The 7204/6VXR and 7301 routers are considered in the middle range of the scale; they can support 24–48 signaling links and have hot-swappable line cards but do not have redundant processors. The 7507/13 routers are the high end of the scale; they support up to 800 signaling links, have hot-swappable line cards, and have redundant processors.

Remember, in North America SS7 telephony switches (SSP) usually connect to each other as well as to SS7 databases (SCP) through an STP network. A PacketCable domain acts just like one of these switches; hence, it also interconnects through an STP network via its signaling gateways. Usually, a pair of SGs is used for redundancy reasons just like a pair of STPs is used in the PSTN. You typically use either A links or B/D links to interconnect PacketCable SGs to a mated pair of STPs. This depends on whether the signaling gateways share a common point code with the MGC or if they have their own point codes. If they share a point code, the combination of the signaling gateways and media gateway controller constitute a SSP; therefore, the links from the SGs are A links. If different point codes are used, the connectivity between the MGC and SGs are really A links because the MGC is still considered an SSP, but the SGs are now viewed as STPs; remember, the connectivity between STP mated pairs is via B/D links. In Europe, SSPs often connect to other SSPs and SCPs directly using F links. So for a Euro-PacketCable deployment, the SS7 links from the signaling gateway are likely F links as well.

If B/D links connect an ITP pair to a mated STP pair, the ITP pair is referred to as an SG mated pair; if A links are used instead, then the ITPs are said to be connected as an ITP group. Both configuration options are covered in this chapter with examples.

Figure 8-15 is an example of the ITP group implementation. In this example, Cisco 7301 routers are used for the ITPs.

Figure 8-15 *Sample ITP Group Implementation*

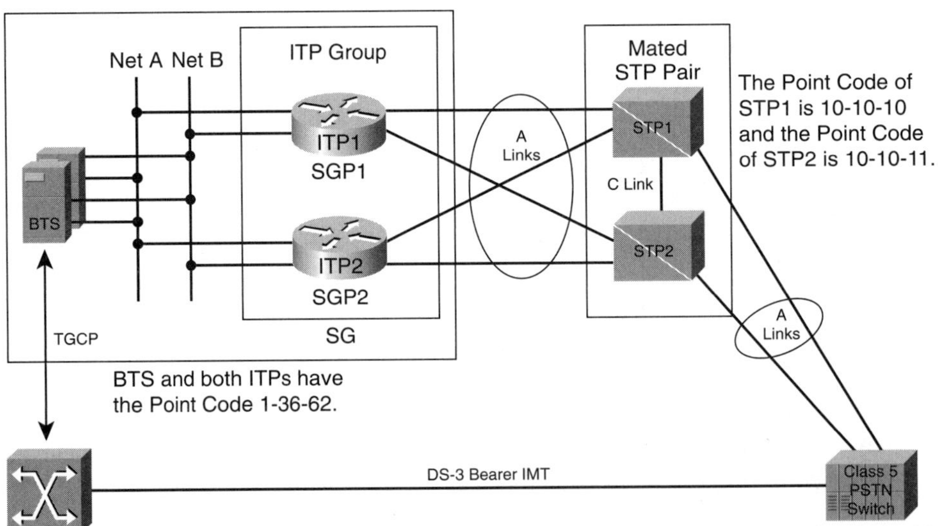

As you can see from Figure 8-15, the BTS and the two ITPs share the point code 1-36-62; therefore, the SS7 links between the ITPs and STPs are A links. In fact, the two physical ITPs logically form a single SG. The STPs are not aware that they are connected to different physical machines because they are represented by a common point code. Each ITP runs its own SG process (SGP) and communicates information with its peer; thus, the two SGPs together form one logical SG. You'll see more of this terminology in the "BTS 10200 Configuration" section.

One of the ITPs operates in active mode and processes traffic, while the other one operates in standby mode. You can use the **show cs7 group state** command on the ITP to view its current status.

The configuration needed for this example is shown later. Some key aspects of using the ITP group implementation to keep in mind are as follows:

- A separate pair of ITPs is needed to support each MGC origination point code.
- The ITPs must be physically close to the MGC because they must exchange real-time state information.
- If the MGC is down, the ITPs are not usable either; they send user part unusable messages to the STPs.
- Global Title Translation (GTT) is not supported on the ITPs because they are working with the MGC to act as an SSP instead of an STP mated pair.

Figure 8-16 is an example of the SG mated pair implementation. In this example, Cisco 7206VXR routers are used for the ITPs.

Figure 8-16 *Sample SG Mated Pair Implementation*

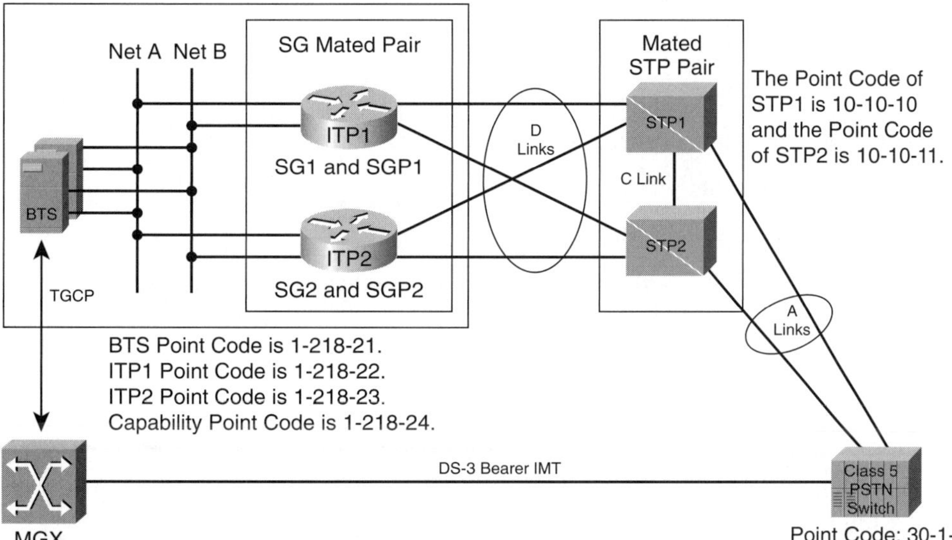

As you can see from Figure 8-16, different point codes are used for each ITP and the BTS. The ITPs function as a mated pair of STPs; therefore, the SS7 links between the ITPs and STPs are B/D links. For simplicity reasons, they are referred to as D links in the diagram. Each ITP runs an SG process and is considered an SG. Unlike the ITP group implementation, traffic can be load shared between the ITPs.

The configuration needed for this example is also shown later. Some key aspects of using the SG mated pair implementation to keep in mind are as follows:

- A single SG mated pair can work with multiple MGC origination point codes. In fact, a single SG mated pair could be used to connect to multiple MGCs.
- The ITPs can be geographically separated from each other and the MGC.
- If an MGC is down, the ITPs can still route SS7 traffic and support any other MGCs to which they are connected. SS7 Transfer Prohibit (TFP) messages are sent to the STPs to identify the MGC point code that is no longer reachable.
- Because the SG mated pair functions similar to an STP mated pair, GTT is supported on the ITPs, and a common capability point code can be defined on the ITPs.

One possible downside to the SG mated pair implementation compared to the ITP group implementation is that SS7 service providers usually charge more for D link interconnects than they do for A link interconnects. This is because multiple point codes can be supported across a D link, whereas only a single point code is supported across an A link.

The configuration of an ITP can be logically split into two parts: one part for the SS7 connectivity to the PSTN network and the other part for the Sigtran connectivity to the PacketCable VoIP network.

The basic steps for configuring an ITP on the PSTN side are as follows:

Step 1 ITP point code configuration.

Step 2 Controller configuration.

Step 3 Serial interface configuration.

Step 4 Define linksets and add links.

Step 5 Define route table.

The basic steps for configuring an ITP on the Sigtran side are as follows:

Step 1 Define Sigtran interfaces.

Step 2 Define M3UA (local SCTP info).

Step 3 Define SUA (local SCTP info).

Step 4 Define ASPs (remote SCTP info).

Step 5 Define ASs with routing keys.

Step 6 SG mated pair specific configuration.

Step 7 ITP group specific configuration.

After all the configuration steps are covered, the configuration for the examples in Figure 8-15 and Figure 8-16 is shown.

ITP Point Code Configuration (PSTN Side)

The point code of the ITP must be defined as well as the SS7 variant. Currently, the supported variant types are ANSI, China, ITU, and TTC; in North America the ANSI variant would be used. The syntax for this configuration is as follows:

```
Router(config)#cs7 variant {ansi ¦ china ¦ itu ¦ ttc}
Router(config)#cs7 point-code network.cluster.number
```

By default, the point code delimiter is a period (**.**); this can be changed to a hyphen (**-**) using the configuration command **cs7 point-code delimiter dash.** If you are using an SG mated pair configuration, you might also want to define an SS7 capability point code. This point code is the same on both SGs and identifies the mated pair for GTT. This configuration is as follows:

```
Router(config)#cs7 capability-pc network.cluster.number
```

Controller Configuration (PSTN Side)

The interface on the ITP where the SS7 signaling link exists is typically a DS0 channel on a T1 or E1 controller. You need to define this DS0 by creating a "channel group" under the controller interface; you might also need to specify whether the WAN interface card (WIC) on the Cisco router is for T1 or E1 connectivity. The syntax for this configuration is as follows:

```
Router(config)#card type {e1 ¦ t1} card_slot_number
Router(config)#controller {T1 ¦ E1} controller slot #/controller port #
```

This last configuration command takes you into the controller specific configuration; here you need to define parameters such as the T1/E1 framing format, clock source, line coding, and cable length. The syntax for this configuration is

```
Router(config-controller)#framing {esf ¦ sf}
Router(config-controller)#clock source {bits {primary ¦ secondary [priority]} ¦
  internal ¦ line {primary ¦ secondary [priority]}}
Router(config-controller)#linecode {ami ¦ b8zs}
Router(config-controller)#cablelength {long ¦ short} length parameter
```

The T1 framing format typically used is ESF, and the line coding typically used is Binary 8 Zero Substitution (B8ZS). For an E1, CRC4 framing and High-Density Bipolar 3 (HDB3) line coding is typically used. Internal clocking logically means the clock source is internally generated. If line clocking is used, the ITP clock is derived from the connected node on that line, and the Building Integrated Timing Supply (BITS) option means that the clocking source is externally connected through a dedicated controller port. A controller dedicated for BITS clocking cannot be used to carry data as well. The cable length parameter is configured in terms of length or attenuation and is used to adjust power levels accordingly. Finally, you need to define channel groups for the DS0 timeslots containing SS7 signaling links. The syntax is

```
Router(config-controller)#channel-group group # timeslots signaling link timeslot
  [speed 56 ¦ 64]
```

You must define a separate channel group for each signaling link on the particular T1 or E1. In the channel group definition, you define which DS0 timeslot is used for the SS7 signaling link; remember that a T1 is comprised of 24 DS0s, so the valid range is 1–24 in this case. Optionally, you can also define the signaling link transmission speed (by default, 64 kbps is assumed).

Serial Interface Configuration (PSTN Side)

After a channel group is defined, a serial interface is created with the following name: **serial** *controller slot #/controller port #:channel group #*. Usually, the only configuration you have to modify here is the encapsulation type. This must be set to MTP2, indicating this serial interface is used for SS7 communication. The syntax for this configuration is as follows:

```
Router(config)#interface Serial controller slot #/controller port #:channel group #
Router(config-if)#encapsulation mtp2
```

Define Linksets and Add Links (PSTN Side)

At this point, the SS7 signaling link has been created; next, you need to define what it connects to. Remember that the collection of links between a pair of SS7 nodes is known as a linkset. The syntax for configuring a linkset is as follows:

```
Router(config)#cs7 linkset linkset name adjacent point code
```

The adjacent point code is the point code of the SS7 device at the other end of the link. This is typically an STP unless an F-link is used. The previous command takes you into the linkset configuration mode, where you define the individual links within that linkset. The syntax for configuring a link is as follows:

```
Router(config-cs7-ls)#link SLC [Group name] serial controller slot #/controller port
  #:channel group # ¦ sctp remote IP address remote SCTP port local SCTP port
  [passive]
```

The SLC is the Signaling Link Code and has a value from 0–15; remember, the most links a linkset can have is 16. If you are using the ITP group configuration, this group name is specified next. Two types of links can be defined. One is a physical link to a legacy SS7 node. This is the link type discussed so far, which is defined by a serial interface. The other type is a Sigtran-based link to another SG. This type of link is used for the C link between

ITPs in the mated SG pair configuration; refer to the "Mated SG Pair Specific Configuration" section for more detail.

Define Route Table (PSTN Side)

Recall that the series of linksets used to get from one SS7 node to another is known as a route. On a Cisco ITP, the collection of routes is stored in a route table. Routes that are one hop away are automatically put into the route table, but routes that are more than one hop away must be manually entered. To manually enter routes you must first go into route table configuration mode:

```
Router(config)#cs7 route-table [system]
```

The **system** keyword is optional because only one route table is currently supported on the ITP. After you are in this configuration mode either you can manually add routes one at a time using the **update** command, or you can load a saved table configuration from a file. This file is stored in binary form. The syntax for manually adding route table entries is as follows:

```
Router(config-cs7-rt)#update route Destination Point Code and Mask linkset Linkset
  name [priority 1-9] [qos-class 1-7 | default]
```

The lower priority number indicates a higher priority route; the default route priority is 5. The highest priority route is used for traffic if available; if a tie exists between multiple routes, traffic is load-shared. A sample use of the priority field is in making routes traversing a C link the lowest priority. After you have manually created the route table you can create the binary file using the following global command:

```
Router#cs7 save route-table device:filename
```

After a route table has been saved, all update entries are removed from the running configuration because these entries are now contained in this file. To remove entries from the route table, you can use the **remove** option from route table configuration mode as shown:

```
Router(config-cs7-rt)#remove route Destination Point Code and Mask linkset Linkset
  name
```

Define Sigtran Interfaces (Sigtran Side)

At this point the SS7 configuration is complete, and the Sigtran part of the ITP needs to be defined. The first step is to configure the Ethernet interfaces on the ITP connected to the Sigtran networks. You usually have two Sigtran networks, and the connection from the ITP is typically Fast Ethernet.

Define M3UA (Sigtran Side)

Next, you need to configure the M3UA protocol on the ITP; remember, on top of this protocol ISUP messages are sent between the SG and MGC. First, the process needs to be defined along with the local SCTP port number:

```
Router(config)#cs7 m3ua local port # [offload]
```

The **offload** parameter can be used on the 75xx platform to offload M3UA SCTP processing to the VIP. The default SCTP port number for the M3UA protocol is 2905; this port number must match up with the MGC configuration, so be careful if you are using a nondefault value. After this command is entered, you go into the M3UA configuration mode. Here you can specify several options that define how the SCTP protocol operates; the only required configuration is to define the local IP address(es) used. Typically, the two Ethernet interface IP addresses defined in the previous step are used.

```
Router(config-cs7-m3ua)#local-ip Ethernet IP #1
Router(config-cs7-m3ua)#local-ip Ethernet IP #2
```

Define SUA (Sigtran Side)

Similarly, you need to configure the SUA protocol on the ITP; remember, on top of this protocol TCAP messages are sent between the SG and the MGC. As with the M3UA definition, the process first needs to be defined along with the local SCTP port number:

```
Router(config)#cs7 sua local port # [offload]
```

The **offload** parameter can be used on the 75xx platform to offload SUA SCTP processing to the VIP. The default SCTP port number for the SUA protocol is 14001; this port number must match up with the MGC configuration, so be careful if using a nondefault value. After this command is entered you go into the SUA configuration mode. Here you can specify several options that define how the SCTP protocol operates; the only required configuration is to define the local IP address(es) used. Typically, the two Ethernet interface IP addresses defined earlier are used.

```
Router(config-cs7-sua)#local-ip SG Ethernet IP #1
Router(config-cs7-sua)#local-ip SG Ethernet IP #2
```

Define ASPs (Sigtran Side)

The next step is to define the Application Server Processes (ASP). Remember, these are the processes on the various physical MGC machines responsible for providing a certain type of SS7 functionality. Because the MGC in this case is the BTS 10200, which uses a redundant pair of MGC machines, typically a pair of ASPs is defined to handle ISUP functionality controlled by the BTS call agent process, and a pair is defined for each feature server process. One of these ASP pairs handles TCAP functionality controlled by the BTS PTC (POTS/Tandem/Centrex) feature server, and the other handles TCAP functionality controlled by the BTS AIN (Advanced Intelligent Network) feature server. The syntax for configuring an ASP is as follows:

```
Router(config)#cs7 asp ASP name Remote SCTP port # Local SCTP port # [m3ua ¦ sua]
```

Again, the port numbers must match with what is configured on the MGC. After defining the ASP, you go into ASP configuration mode where you can define various ASP and SCTP parameters. The only parameters you likely need to configure are the remote MGC IP address(es). Because you should have a pair of Sigtran networks, the MGC should have an IP address on each network.

```
Router(config-cs7-asp)#remote-ip MGC Ethernet IP #1
Router(config-cs7-asp)#remote-ip MGC Ethernet IP #2
```

Define Application Servers and Routing Keys (Sigtran Side)

After the ASPs have been defined, you must define the individual application servers. Recall that each application server is associated with a unique routing key that defines the SS7 service provided by that application server. You can think of the routing key as a filter that defines the traffic permitted to be sent to the application server. If multiple routing keys are defined that match the incoming traffic, then the more specific match is used. For example, if one routing key matches only on destination point code, and another matches on both origination and destination point codes, the latter routing key is used. The syntax for configuring an application server is as follows:

```
Router(config)#cs7 as AS name [m3ua ¦ sua]
```

This puts you into Application Server configuration mode, where you need to define the routing key as well as associated ASPs. The syntax for routing key configuration is

```
Router(config-cs7-as)#routing-key routing context {gtt ¦ MGC point code [opc Point
  Code and Mask] [si {isup [cic Min value Max value] ¦ sccp [ssn Subsystem #] ¦ tup
  [cic Min value Max value]} ¦ ssn Subsystem #]}
```

At first, this command might be a bit intimidating; so break it down bit by bit. First, define the routing context, which is just an integer that references this routing key. Then, you need to specify whether or not this application server is associated to a particular point code or is used for GTT. The GTT option is valid only if the mated SG architecture is used. The GTT option requires additional configuration that is outside the scope of this book.

If the routing key is associated to a particular point code, you need to define the routing key for a particular origination point code, a particular upper-layer SS7 service—ISUP, SCCP, or TUP—or a particular subsystem number. If an origination point code is specified, you can optionally also specify an SS7 service or subsystem number. If the routing key is for the ISUP or TUP services, you can optionally specify specific CIC ranges, and if the routing key is for the SCCP service, you can optionally specify a subsystem number.

The syntax for associating the ASPs to the application server is

```
Router(config-cs7-as)#asp ASP name
```

Other options you might see in a configuration are network appearance and traffic mode.

```
Router(config-cs7-as)#network-appearance value
Router(config-cs7-as)#traffic-mode broadcast | loadshare [bindings | roundrobin] |
  override
```

If you are using the BTS as the MGC, you want to set the **traffic-mode** to the value **override**. This means either one of the ASPs processes the traffic but not both. The BTS call agent machines function as an active/standby pair and not in a load sharing fashion. The **network appearance** option should be set to a value of 1 if you are using BTS version 4.4 or earlier; if you are using BTS version 4.5 or later, you should not configure this option.

Mated SG Pair Specific Configuration (Sigtran Side)

If you are using the mated SG pair architecture, some additional configuration is required specific to this architecture.

To enable the exchange of SS7 messaging between the ITPs in a mated SG pair, you need to create a C link Sigtran connection. The first step is to define the local IP address and SCTP port number on each ITP. This is done with the following command:

```
Router(config)#cs7 local-peer local SCTP port # [offload VIP slot #]
```

The **offload** option can be used to offload peer link management to the specified VIP. The previous command puts you into local peer configuration mode, where you can define a number of parameters. One of these you need to define is the IP addresses of the local SG.

```
Router(config-cs7-lp)#local-ip SG Ethernet IP #1
Router(config-cs7-lp)#local-ip SG Ethernet IP #2
```

The information for the remote SG is specified by creating a linkset to the remote SG consisting of an SCTP link using the following configuration commands:

```
Router(config)#cs7 linkset linkset name remote SG point code
Router(config-cs7-ls)#link SLC sctp remote IP address remote SCTP port local SCTP
  port [passive]
```

If the optional **passive** keyword is used, this means the remote SG must initiate the SCTP connection.

The previous configuration allows Sigtran user traffic to pass between the ITPs; you must also configure a Sigtran connection that passes state information between the ITPs. This is done using the Signaling Gateway Mate Protocol (SGMP). First, define the local parameters using the following command:

```
Router(config)#cs7 sgmp local SCTP port #
```

This puts you into SGMP configuration mode, where you can define a number of parameters. One of these you need to define is the IP addresses of the local SG.

```
Router(config-cs7-sgmp)#local-ip SG Ethernet IP #1
Router(config-cs7-sgmp)#local-ip SG Ethernet IP #2
```

The information for the remote SG is specified using the following configuration command:

```
Router(config)#cs7 mated-sg mated SG name [remote SCTP port # [passive]]
```

If the **passive** keyword is used, this means the remote SG has to initiate this connection. This puts you into mated SG configuration mode, where you can define a number of parameters. One of these you need to define is the IP addresses of the remote SG:

```
Router(config-cs7-mated-sg)#remote-ip SG Ethernet IP #1
Router(config-cs7-mated-sg)#remote-ip SG Ethernet IP #2
```

ITP Group-Specific Configuration (Sigtran Side)

If you are using the ITP group architecture, some additional configuration is required specific to this architecture.

Remember that in the ITP group architecture, the two ITPs together form a logic SG; thus, redundancy is provided inter-device. Consequently, you need to enable inter-device redundancy using the following command:

```
Router(config)#redundancy inter-device
```

You then need to define a mechanism to allow members of the ITP group to monitor the status and operation of its peer members. For redundancy reasons this is done in two places. First, redundancy state transport is configured by defining an IPC zone, which contains the IP addresses and SCTP port information for communication between ITPs. The configuration commands needed are as follows:

```
Router(config)#ipc zone default
Router(config-ipczone)#association <association id
Router(config-ipczone-assoc)#no shutdown
Router(config-ipczone-assoc)#protocol sctp
Router(config-ipc-protocol-sctp)#local-port local SG SCTP port #>
Router(config-ipc-local-sctp)#local-ip local SG IP #1
Router(config-ipc-local-sctp)#local-ip local SG IP #2
Router(config-ipc-protocol-sctp)#remote-port remote SG SCTP port #
Router(config-ipc-remote-sctp)#remote-ip remote SG IP #1
Router(config-ipc-remote-sctp)#remote-ip remote SG IP #2
```

The other transport defined is the group control transport; this is done by configuring a CS7 group. Like the IPC zone configuration, this consists of defining IP addresses and SCTP ports for communication between ITPs. The configuration commands needed are as follows:

```
Router(config)#cs7 group [group name] [local port #]
Router(config-cs7-group)#local-ip local SG IP #1
Router(config-cs7-group)#local-ip local SG IP #2
Router(config-cs7-group)#peer group peer name [remote port #]
Router(config-cs7-group-peer)#remote-ip remote SG IP #1
Router(config-cs7-group-peer)#remote-ip remote SG IP #2
```

Final ITP Configurations

Now that you know the steps needed to configure an ITP, look back at the examples in Figure 8-15 and Figure 8-16. First, take a look back at the configuration for the ITP group implementation in Figure 8-15. Figure 8-17 shows the first part of this configuration. In these configuration figures, the configuration of ITP #1 is shown on the left, and the corresponding configuration of ITP #2 is shown on the right.

Figure 8-17 *ITP Group Configuration Example: Part I*

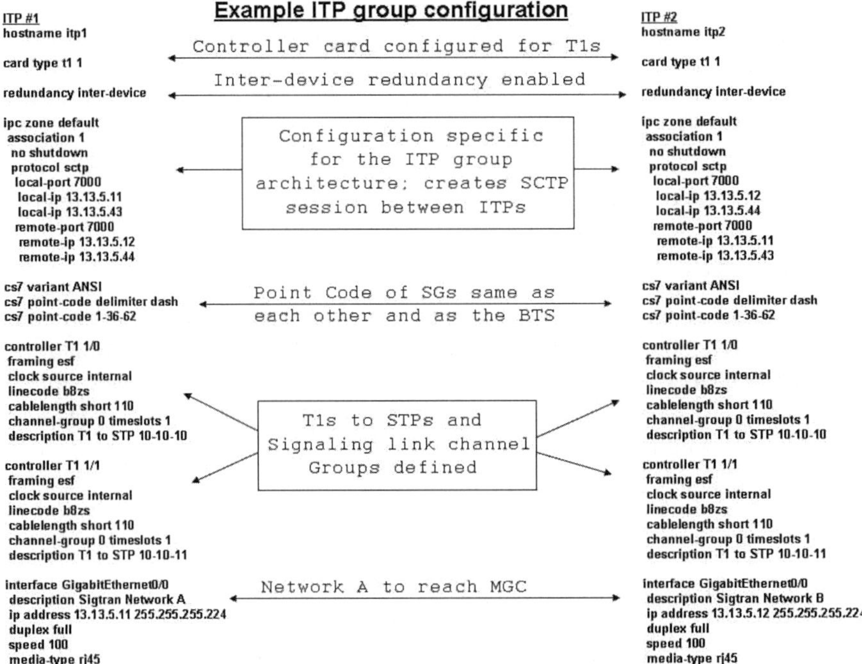

The figure should be relatively self-explanatory. Notice the same point code is defined on both ITPs. Also notice the IPC zone configuration used to enable the communication between the ITPs in the ITP group. Figure 8-18 shows the next part of the configuration.

Figure 8-18 *ITP Group Configuration Example: Part II*

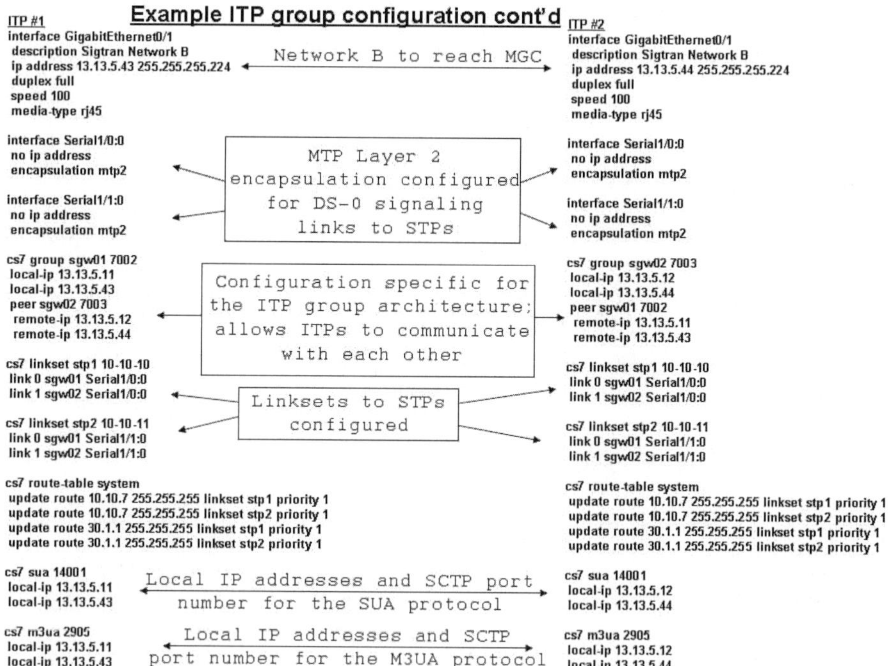

In this figure, the remaining ITP group-specific configuration is shown, the CS7 Group configuration. Also defined in this figure are the A links and corresponding linksets to the mated STP pair. Notice you have a linkset to each STP consisting of two links—one on each physical ITP. Also notice the configuration for links on both ITPs is defined on both ITPs. Two destination point codes are defined in the route table definition, which represent remote SSPs. Finally, the local parameters for the Sigtran M3UA and SUA protocols are defined. Figure 8-19 shows the rest of the configuration.

Figure 8-19 *ITP Group Configuration Example: Part III*

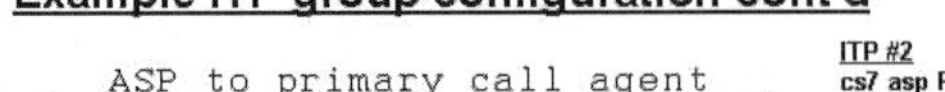

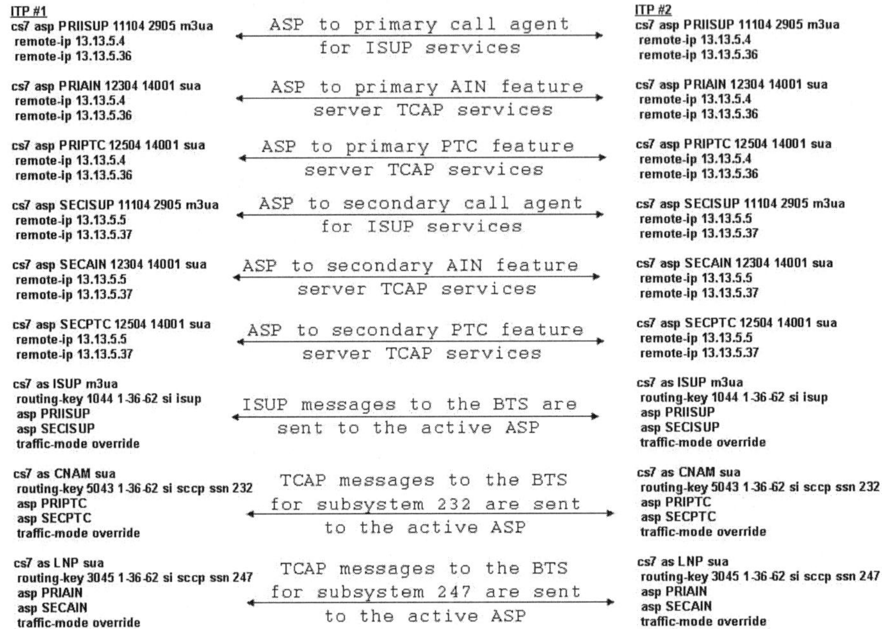

In this figure, the ASPs for the ISUP service and the two TCAP feature services are defined for both BTS machines. Then application servers are defined that reference these ASPs. In the AS definitions, routing keys are also defined that describe the traffic that is routed to the AS. For example, the routing key under the ISUP AS indicates all ISUP messages with a destination point code of 1-36-62 are sent to either the PRIISUP or SECISUP ASP (whichever is active). As another example, the calling name (CNAM) AS has a routing key that directs all TCAP messages destined for point code 1-36-62 for subsystem 232 to the active BTS PTC feature server, and the local number portability (LNP) AS has a routing key that directs all TCAP messages destined for point code 1-36-62 for subsystem 247 to the active BTS AIN feature server.

Next, examine the configuration for the SG mated pair implementation shown in Figure 8-16. Figure 8-20 shows the first part of this configuration. In these configuration figures the configuration of ITP #1 is shown on the left, and the corresponding configuration of ITP #2 is shown on the right.

Figure 8-20 *SG Mated Pair Configuration Example: Part I*

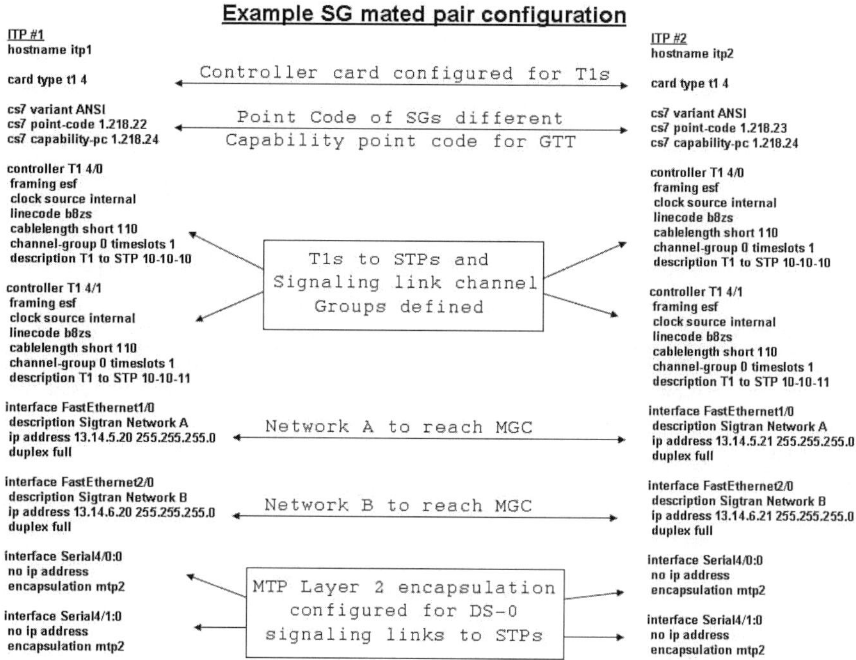

The figure should be relatively self-explanatory. The only thing really different from the ITP group configuration is the definition of a capability point code. This can be used to define a destination point code for GTT. Again, additional configuration is required for GTT; please refer to Cisco ITP–specific documentation for more information. Figure 8-21 shows the next part of the configuration.

Figure 8-21 *SG Mated Pair Configuration Example: Part II*

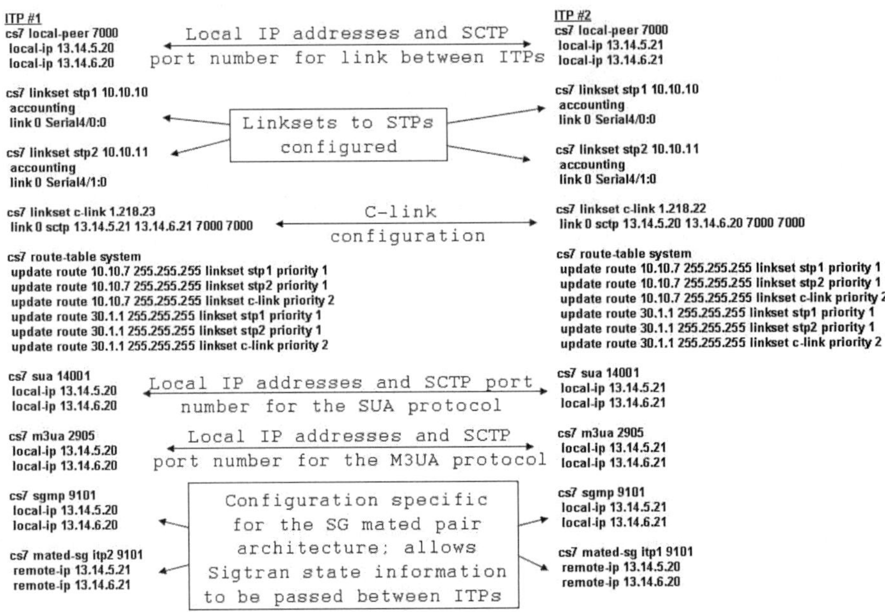

This figure shows the configuration specific to the SG mated pair implementation. This includes the local peer and C link configuration needed to pass SS7 traffic between the ITPs and the SGMP and mated SG configuration required to pass Sigtran information between the ITPS. Also the linksets and links to each STP are defined, as well as the destination routes and the local parameters for the Sigtran M3UA and SUA protocols. Figure 8-22 shows the rest of the configuration.

Figure 8-22 *SG Mated Pair Configuration Example: Part III*

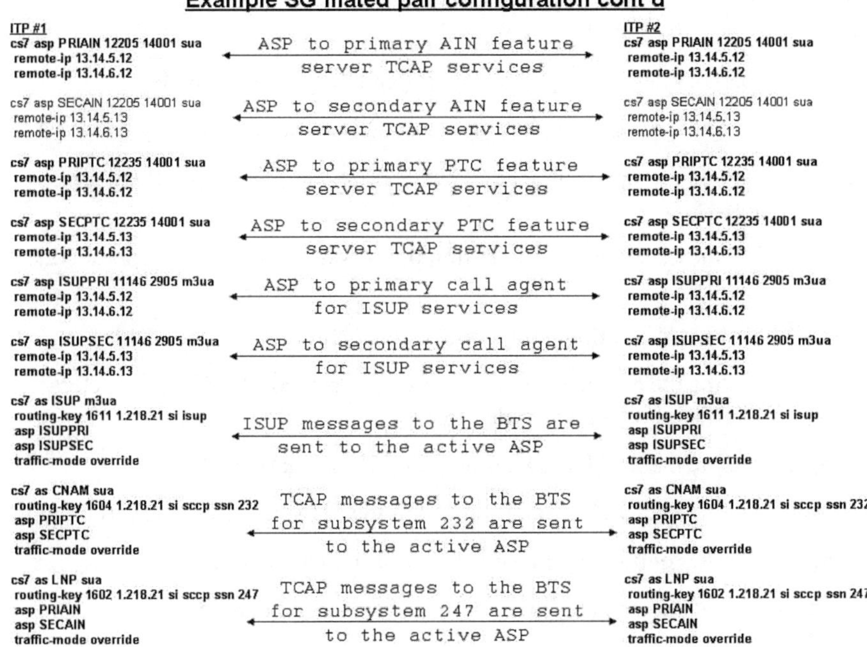

This part of the configuration is similar to the final part of the ITP group configuration, so please refer to that for additional information.

A great deal of ITP functionality is not covered in this chapter; for more information about Cisco ITPs, please refer to the following URLs:

http://www.cisco.com/univercd/cc/td/doc/product/wireless/moblwrls/itp/
http://www.cisco.com/en/US/products/sw/wirelssw/ps1862/index.html

BTS 10200 Configuration

In Chapter 7, you learned about MTA configuration on the BTS 10200 as well as basic system configuration, ONNET or local dial plan configuration, and call feature configuration. This chapter builds on this knowledge base by covering three new aspects related to the PSTN interconnect. You will learn how SS7 and signaling gateways are configured, how trunking media gateways and their trunks are configured, and how to configure dial plans for OFFNET call routing.

SS7 and Signaling Gateway Configuration Tasks

You perform 11 required configuration steps to configure BTS to signaling gateway communication for ISUP services:

Step 1 Define the signaling gateway(s) (signaling gateway table).

Step 2 Define the signaling gateway group(s) (signaling gateway group table).

Step 3 Define the signaling gateway process(es) (signaling gateway process table).

Step 4 Define the switch's origination point code(s) (originating point code table).

Step 5 Define destination point code(s) (destination point code table).

Step 6 Define the Service Control Transport Protocol (SCTP) parameters (SCTP association profile table).

Step 7 Define the SCTP associations (SCTP association table).

Step 8 Define the ISUP variant(s) used in the network (user part variant table).

Step 9 Define routing key(s) for ISUP (routing key table).

Step 10 Define the call control route (call control route table).

Step 11 Enable the SCTP association.

Step 1: Defining the Signaling Gateways

In this step, all you do is define your signaling gateways by assigning them internal identifiers. The only additional optional field is a priority field where the use of one SG can be given preference over another. If the priorities are the same, traffic is load shared between the SGs:

```
CLI>add sg id=SG name;
```

If you are using the ITP group architecture, you configure only one SG. For the ITP group configuration the two physical ITPs constitute a single logical SG; redundancy is achieved in the signaling group process step. The following is the SG configuration from the ITP group implementation:

```
CLI>show sg
ID=SG
PRIORITY=1
DESCRIPTION=Signaling Gateway 1
```

If you are using the mated SG pair architecture, you configure two SGs, one for each physical ITP; here is the SG configuration from the SG mated pair implementation example:

```
CLI>show sg
ID=ITP1
PRIORITY=1
ID=ITP2
PRIORITY=1
```

Step 2: Defining the Signaling Gateway Groups

Next, associations between signaling gateways that are members of the same group are made (that is, a mated pair of SGs). Again, for the mated SG pair architecture two SGs are in the SG group, and if you are using the ITP group configuration, only one SG is in the SG group:

```
CLI>add sg-grp id=SG group name; sg1-id=SG#1 name; sg2-id=SG#2 name;
```

Here is the SG group configuration from the ITP group implementation example:

```
CLI>show sg_grp
ID=SGG
SG1_ID=SG
DESCRIPTION=SG Group 1
```

Here is the SG group configuration from the SG mated pair implementation example:

```
CLI>show sg-grp
ID=ITP-PAIR
SG1_ID=ITP1
SG2_ID=ITP2
```

Step 3: Defining the Signaling Gateway Processes

In this step, a process is defined for each physical signaling gateway. Because in the SG mated pair architecture, each physical ITP is a separate signaling gateway, you would define a separate SG processes for each. Because the physical ITPs together form a logical SG in the ITP group architecture, you would define separate SG processes for a common SG.

```
CLI>add sgp id=SG process name; sg-id=SG name;
```

Here is the SG process configuration from the ITP group implementation example:

```
CLI>show sgp
ID=SGP1
SG_ID=SG
DESCRIPTION=SG Process on ITP#1
ID=SGP2
SG_ID=SG
DESCRIPTION=SG Process on ITP#2
```

Here is the SG process configuration from the SG mated pair implementation example:

```
CLI>show sgp
ID=SGP1
SG_ID=ITP1
DESCRIPTION=SG process on ITP#1
ID=SGP2
SG_ID=ITP2
DESCRIPTION=SG process on ITP#2
```

Step 4: Defining the Switch's Origination Point Code

Now you need to define the SS7 point code used by the BTS.

```
CLI>add opc id=OPC name; point-code=point code;
```

NOTE It is perfectly valid to have more than one origination point code defined. This is sometimes the case when you are connected to two different SS7 service provider networks; you would use an OPC provided by provider "A" for their links and a different OPC provided by provider "B" for their links.

The optional field **point-code-type** can also be included. It has the values **ANSI_CHINA** and **ITU**, which defines the point code format. By default, **ANSI_CHINA** is assumed, in which the point code is expressed in the following notation:

network id-network cluster id-network cluster member id

Here is the OPC configuration from the ITP group implementation example:

```
CLI>show opc
ID=OPC_001-036-062
POINT_CODE=1-36-62
POINT_CODE_TYPE=ANSI_CHINA
```

Here is the OPC configuration from the SG mated pair implementation example:

```
CLI>show opc
ID=opc
POINT_CODE=1-218-21
POINT_CODE_TYPE=ANSI_CHINA
```

Step 5: Define the Destination Point Codes

Similarly, you need to define the SS7 point codes of all the SSPs and SCPs the BTS intends to communicate with

```
CLI>add dpc id=DPC name; point-code=point code;
```

NOTE You do not need to add the point codes of the ITPs or the STPs to which they connect; the exception to this is if any of these devices are doing GTT, in which case you would typically define the capability point codes of these devices as a DPC.

The optional field **point-code-type** also exists here with the same default. The SS7 network indicator field **net-ind** can also be included; it has the values **national** (default), **international**, **reserved**, and **spare**.

The DPC configuration from the ITP group implementation example and the SG mated pair implementation are identical and as follows:

```
CLI>show dpc
ID=DPC_10-10-7
POINT_CODE=10-10-7
POINT_CODE_TYPE=ANSI_CHINA
NET_IND=NATIONAL
NET_AP=0
ID=DPC_30-30-1
POINT_CODE=30-30-1
POINT_CODE_TYPE=ANSI_CHINA
NET_IND=NATIONAL
NET_AP=0
```

Step 6: Define the SCTP Parameters

In this step, you define the attributes of the SCTP connection to your SGs. This same profile can then be used for multiple SCTP associations.

```
CLI>add sctp-assoc-profile id=SCTP association profile name;
```

A number of optional parameters are available here, such as timer and retry values.

Here is the SCTP association profile configuration from the ITP group implementation example:

```
CLI>show sctp_assoc_profile
ID=SCTP_PF0
BUNDLE_TIMEOUT=100
MAX_ASSOC_RETRANS=4
MAX_PATH_RETRANS=3
RETRIEVE_FLAG=Y
MAX_RTO=2000
MIN_RTO=300
SACK_TIMEOUT=200
HB_TIMEOUT=2000
```

Here is the SCTP association profile configuration from the SG mated pair implementation example:

```
CLI>show sctp-assoc-profile
ID=sctp-prof
BUNDLE_TIMEOUT=500
MAX_ASSOC_RETRANS=5
```

```
MAX_PATH_RETRANS=5
RETRIEVE_FLAG=Y
MAX_RTO=6000
MIN_RTO=301
SACK_TIMEOUT=101
HB_TIMEOUT=1000
```

Step 7: Define the SCTP Associations

Next, each SCTP session or association to an SG process is defined; this includes parameters such as the IP addresses and port numbers of the SGs.

```
CLI>add sctp-assoc id=SCTP association name; sgp-id=SG process name from Step 3;
  sctp-assoc-profile-id=SCTP association profile name from Step 6; platform-id=BTS
  platform using this SCTP association; remote-port=SG port number; remote-tsap-
  addr=IP address #1 of the SG; remote-tsap-addr2=IP address #2 of the SG;
```

Each entry in this table defines the SCTP parameters for a single ITP for a single application service. Because you typically have three application services (one for ISUP, one for the PTC feature server, and one for the AIN feature server) and you also typically have a pair of ITPs, usually you have six entries in this table.

Here are the SCTP associations configured for the ITP group implementation example:

```
CLI>show sctp-assoc
ID=SCTP-ISUP-ITP1
SGP_ID=SGP1
SCTP_ASSOC_PROFILE_ID=SCTP_PF0
REMOTE_PORT=2905
REMOTE_TSAP_ADDR1=13.13.5.11
REMOTE_TSAP_ADDR2=13.13.5.43
PLATFORM_ID=CA104
LOCAL_RCVWIN=3000
MAX_INIT_RETRANS=3
MAX_INIT_RTO=1000
STATUS=INS
ULP=XUA
DESCRIPTION=For ISUP over M3UA communication to ITP#1
ID=SCTP-AIN-ITP1
SGP_ID=SGP1
SCTP_ASSOC_PROFILE_ID=SCTP_PF0
REMOTE_PORT=14001
REMOTE_TSAP_ADDR1=13.13.5.11
REMOTE_TSAP_ADDR2=13.13.5.43
PLATFORM_ID=FSAIN304
LOCAL_RCVWIN=3000
MAX_INIT_RETRANS=3
MAX_INIT_RTO=1000
STATUS=INS
ULP=XUA
DESCRIPTION=For TCAP over SUA communication to ITP#1 for AIN FS
ID=SCTP-PTC-ITP1
SGP_ID=SGP1
SCTP_ASSOC_PROFILE_ID=SCTP_PF0
REMOTE_PORT=14001
REMOTE_TSAP_ADDR1=13.13.5.11
REMOTE_TSAP_ADDR2=13.13.5.43
PLATFORM_ID=FSPTC504
LOCAL_RCVWIN=3000
```

```
MAX_INIT_RETRANS=3
MAX_INIT_RTO=1000
STATUS=INS
ULP=XUA
DESCRIPTION=For TCAP over SUA communication to ITP#1 for PTC FS
ID=SCTP-ISUP-ITP2
SGP_ID=SGP2
SCTP_ASSOC_PROFILE_ID=SCTP_PF0
REMOTE_PORT=2905
REMOTE_TSAP_ADDR1=13.13.5.12
REMOTE_TSAP_ADDR2=13.13.5.44
PLATFORM_ID=CA104
LOCAL_RCVWIN=3000
MAX_INIT_RETRANS=3
MAX_INIT_RTO=1000
STATUS=INS
ULP=XUA
DESCRIPTION=For ISUP over M3UA communication to ITP#2
ID=SCTP-AIN-ITP2
SGP_ID=SGP2
SCTP_ASSOC_PROFILE_ID=SCTP_PF0
REMOTE_PORT=14001
REMOTE_TSAP_ADDR1=13.13.5.12
REMOTE_TSAP_ADDR2=13.13.5.44
PLATFORM_ID=FSAIN304
LOCAL_RCVWIN=3000
MAX_INIT_RETRANS=3
MAX_INIT_RTO=1000
STATUS=INS
ULP=XUA
DESCRIPTION=For TCAP over SUA communication to ITP#2 for AIN FS
ID=SCTP-PTC-ITP2
SGP_ID=SGP2
SCTP_ASSOC_PROFILE_ID=SCTP_PF0
REMOTE_PORT=14001
REMOTE_TSAP_ADDR1=13.13.5.12
REMOTE_TSAP_ADDR2=13.13.5.44
PLATFORM_ID=FSPTC504
LOCAL_RCVWIN=3000
MAX_INIT_RETRANS=3
MAX_INIT_RTO=1000
STATUS=INS
ULP=XUA
DESCRIPTION=For TCAP over SUA communication to ITP#2 for PTC FS
```

Similarly, here are the SCTP associations configured for the SG mated pair implementation example:

```
CLI>show sctp-assoc
ID=SCTP-AIN-ITP1
SGP_ID=SGP1
SCTP_ASSOC_PROFILE_ID=sctp-prof
REMOTE_PORT=14001
REMOTE_TSAP_ADDR1=13.14.5.20
REMOTE_TSAP_ADDR2=13.14.6.20
PLATFORM_ID=FSAIN205
DSCP=NONE
IP_TOS_PRECEDENCE=FLASH
LOCAL_RCVWIN=3000
MAX_INIT_RETRANS=3
MAX_INIT_RTO=500
STATUS=INS
ULP=XUA
DESCRIPTION=For TCAP over SUA communication to ITP#1 for AIN FS
```

```
ID=SCTP-ISUP-ITP1
SGP_ID=SGP1
SCTP_ASSOC_PROFILE_ID=sctp-prof
REMOTE_PORT=2905
REMOTE_TSAP_ADDR1=13.14.5.20
REMOTE_TSAP_ADDR2=13.14.6.20
PLATFORM_ID=CA146
DSCP=AF11
IP_TOS_PRECEDENCE=ROUTINE
LOCAL_RCVWIN=3000
MAX_INIT_RETRANS=3
MAX_INIT_RTO=500
STATUS=INS
ULP=XUA
DESCRIPTION=For ISUP over M3UA communication to ITP#1
ID=SCTP-PTC-ITP1
SGP_ID=SGP1
SCTP_ASSOC_PROFILE_ID=sctp-prof
REMOTE_PORT=14001
REMOTE_TSAP_ADDR1=13.14.5.20
REMOTE_TSAP_ADDR2=13.14.6.20
PLATFORM_ID=FSPTC235
DSCP=NONE
IP_TOS_PRECEDENCE=FLASH
LOCAL_RCVWIN=3000
MAX_INIT_RETRANS=3
MAX_INIT_RTO=500
STATUS=INS
ULP=XUA
DESCRIPTION=For TCAP over SUA communication to ITP#1 for PTC FS
ID=SCTP-AIN-ITP2
SGP_ID=SGP2
SCTP_ASSOC_PROFILE_ID=sctp-prof
REMOTE_PORT=14001
REMOTE_TSAP_ADDR1=13.14.5.21
REMOTE_TSAP_ADDR2=13.14.6.21
PLATFORM_ID=FSAIN205
DSCP=NONE
IP_TOS_PRECEDENCE=FLASH
LOCAL_RCVWIN=3000
MAX_INIT_RETRANS=3
MAX_INIT_RTO=500
STATUS=INS
ULP=XUA
DESCRIPTION=For TCAP over SUA communication to ITP#2 for AIN FS
ID=SCTP-ISUP-ITP2
SGP_ID=SGP2
SCTP_ASSOC_PROFILE_ID=sctp-prof
REMOTE_PORT=2905
REMOTE_TSAP_ADDR1=13.14.5.21
REMOTE_TSAP_ADDR2=13.14.6.21
PLATFORM_ID=CA146
DSCP=AF11
IP_TOS_PRECEDENCE=ROUTINE
LOCAL_RCVWIN=3000
MAX_INIT_RETRANS=3
MAX_INIT_RTO=500
STATUS=INS
ULP=XUA
DESCRIPTION=For ISUP over M3UA communication to ITP#2
ID=SCTP-PTC-ITP2
SGP_ID=SGP2
SCTP_ASSOC_PROFILE_ID=sctp-prof
REMOTE_PORT=14001
```

```
REMOTE_TSAP_ADDR1=13.14.5.21
REMOTE_TSAP_ADDR2=13.14.6.21
PLATFORM_ID=FSPTC235
DSCP=NONE
IP_TOS_PRECEDENCE=FLASH
LOCAL_RCVWIN=3000
MAX_INIT_RETRANS=3
MAX_INIT_RTO=500
STATUS=INS
ULP=XUA
DESCRIPTION=For TCAP over SUA communication to ITP#2 for PTC FS
```

Step 8: Define the ISUP Variant Used in the Network

Here, each variant of ISUP used by the BTS is defined; remember, typically each country has its own ISUP variant.

```
CLI>add user-part-variant id=ISUP variant name;
```

You can find currently supported ISUP variants by looking at the table user-part-variant-base. Here is an example of this table:

```
CLI>show user_part_variant_base; display=id
ID=ANSISS7_GR317
ID=Q761_ARGENTINA
ID=Q761_ARGENTINA_C2
ID=Q761_AUSTRALIAN
ID=Q761_CHILE
ID=Q761_CHINA
ID=Q761_ETSIV2
ID=Q761_ETSIV2_ISRAEL
ID=Q761_ETSIV3
ID=Q761_ETSIV3_HUNGARY
ID=Q761_HONGKONG
ID=Q761_STANDARD
ID=Q761_STANDARD97
ID=Q761_THAILAND
ID=Q767_BRAZIL
ID=Q767_MEXICO
ID=Q767_STANDARD
```

In both the ITP group implementation example and the SG mated pair implementation the ANSI SS7 variant is used, so the configuration of this table is the same and as follows:

```
CLI>show user_part_variant
ID=ANSISS7_GR317
PROTO_FAM=SS7_ANSI
RESET_SUPPORTED=Y
MAX_RESET_MASTER_RANGE=1
MAX_RESET_SLAVE_RANGE=23
MAX_BLOCK_MASTER_RANGE=1
MAX_BLOCK_SLAVE_RANGE=23
MAX_PARALLEL_JOBS=10
```

Step 9: Define the Routing Keys for ISUP

Recall that the routing key defines the traffic sent to a particular application service:

```
CLI>add routing-key id=Routing Key name; opc-id=OPC name from Step 4; sg-grp-id=SG
  group name from Step 2; si=ISUP; rc=Routing Key number; platform-id=BTS call agent
  platform identifier;
```

Here is the routing key configuration from the ITP group implementation example:

```
CLI>show routing_key
ID=ISUP
OPC_ID=OPC_001-036-062
SG_GRP_ID=SGG
SI=ISUP
RC=1044
PLATFORM_ID=CA104
```

Here is the routing key configuration from the SG mated pair implementation example:

```
CLI>show routing-key
ID=rk_isup
OPC_ID=opc
SG_GRP_ID=ITP-PAIR
SI=ISUP
RC=1611
PLATFORM_ID=CA146
```

Step 10: Define the Call Control Route

This is the step where everything comes together. When you define a call control route, you specify which destinations are reachable with which routing keys and ISUP variant.

```
CLI>add call-ctrl-route id=Name of route to destination; dpc-id=DPC name from Step
  5; routing-key-id=Routing Key name from Step 9; si=ISUP; user-part-variant-id=ISUP
  variant name from Step 8;
```

Here is the call control route configuration from the ITP group implementation example:

```
CLI>show call_ctrl_route
ID=DPC_10-10-7
DPC_ID=DPC_10-10-7
USER_PART_VARIANT_ID=ANSISS7_GR317
SI=ISUP
ROUTING_KEY_ID=ISUP
ID=DPC_30-30-1
DPC_ID=DPC_30-30-1
USER_PART_VARIANT_ID=ANSISS7_GR317
SI=ISUP
ROUTING_KEY_ID=ISUP
```

Here is the call control route configuration from the SG mated pair implementation example:

```
CLI>show call_ctrl_route
ID=DPC_10-10-7
DPC_ID=DPC_10-10-7
USER_PART_VARIANT_ID=ANSISS7_GR317
SI=ISUP
ROUTING_KEY_ID=rk_isup
ID=DPC_30-30-1
DPC_ID=DPC_30-30-1
USER_PART_VARIANT_ID=ANSISS7_GR317
```

```
SI=ISUP
ROUTING_KEY_ID=rk_isup
```

Step 11: Enable the SCTP Association

The final step is to bring the SCTP association into service:

```
CLI>control sctp-assoc id=SCTP Association identifier; target-state=INS;
 mode=FORCED;
```

NOTE Before you can make SS7 calls, you also need to do tasks such as defining the Media Gateway; these steps are covered in the "Trunking Media Gateway Configuration Tasks" section later in the chapter.

The result of this operation can be verified with the **status** command, as shown:

```
CLI>status sctp-assoc ID=SCTP-ISUP-ITP1
SCTP ASSOC ID -> SCTP-ISUP-ITP1
ADMIN STATE -> ADMIN_INS
OPER STATE -> SCTP-ASSOC in service
REASON -> ADM executed successfully
RESULT -> ADM configure result in success
Reply : Success:
```

TCAP Configuration Tasks

The previous steps only completed configuring ISUP services. To fully configure TCAP services on the BTS a few additional steps are required and are as follows:

Step 12 Define a SCCP network (SCCP network table).

Step 13 Define subsystem characteristics (subsystem group table).

Step 14 Define the subsystems (subsystem table).

Step 15 Define routing key(s) for TCAP/SCCP (routing key table).

Step 16 Define individual TCAP/SCCP routes (SCCP route table).

Step 17 Define service logic host route characteristics (service logic host route profile table).

Step 18 Define service logic host route (service logic host route table).

Step 19 Modify call agent configuration (call agent configuration table).

Step 20 Enable the SCTP associations.

Step 21 Enable the subsystem groups or subsystems.

Step 12: Define an SCCP Network

The SCCP network table contains SCCP attributes associated with the SS7 network. A one-to-one mapping exists between the entries in this table and the entries in the origination point code table:

```
CLI>add sccp_nw id=integer that references the network; net-ind=INTERNATIONAL ¦
 SPARE ¦ NATIONAL ¦ RESERVED; sub_svc= INTERNATIONAL¦SPARE¦NATIONAL¦RESERVED;
 hop_count=1-15;
```

The following is a example configuration of this table:

```
CLI>show sccp_nw
ID=1
NET_IND=NATIONAL
SUB_SVC=NATIONAL
HOP_COUNT=3
```

Step 13: Define Subsystem Characteristics

The subsystem group table defines all valid SSN (Subsystem Number) IDs at a global level. Previous to the 4.5 release, the subsystem profile table was used for this purpose. These settings apply for a given subsystem regardless of the OPC it uses; per OPC settings are configured in the next step.

```
CLI>add subsystem-grp id=subsystem identifier; platform-id=identifier of the Feature
 Server responsible for this subsystem; tcap-version=identifies the TCAP
 specification used by the subsystem;
```

An example configuration of this table is as follows:

```
CLI>show subsystem_grp
ID=CNAM_SSN
PLATFORM_ID=FSPTC504
STATUS=UOS
TCAP_VERSION=ANS96
ID=LNP_SSN
PLATFORM_ID=FSAIN304
STATUS=UOS
TCAP_VERSION=ANS96
```

Step 14: Define the Subsystems

The subsystem table defines all subsystems using SCCP. It defines all local and remote SSNs at the OPC level.

```
CLI>add subsystem id=subsystem identifier from Step 13; opc-id=origination point
 code from Step 4; local-ssn=local subsystem #, unique per OPC; remote-ssn=remote
 subsystem #; sccp-nw-id=SCCP network ID from Step 12; application-version=TCAP
 protocol version used in the query message either AIN01 or IN1;
```

The following is a sample configuration of this table:

```
CLI>show subsystem
ID=CNAM_SSN
LOCAL_SSN=232
```

```
REMOTE_SSN=232
SCCP_NW_ID=1
OPC_ID=OPC_001-036-062
SCCP_VERSION=ANS92
APPLICATION_VERSION=IN1
STATUS=UOS
ID=LNP_SSN
LOCAL_SSN=247
REMOTE_SSN=247
SCCP_NW_ID=1
OPC_ID=OPC_001-036-062
SCCP_VERSION=ANS92
APPLICATION_VERSION=AIN01
STATUS=UOS
```

The identifier in this table must also exist in the subsystem group table defined in Step 13.

Step 15: Define the Routing Key for TCAP/SCCP Services

Similar to what was done for ISUP services in Step 9, you must also define routing keys for the TCAP/SCCP services.

```
CLI>add routing-key id=Routing Key name; opc-id=OPC name from Step 4; sg-grp-id=SG
  group name from Step 2; si=SCCP; subsystem-grp-id=Subsystem group identifier from
  Step 13; rc=Routing Key number; platform-id=BTS feature server platform identifier;
```

The following is a sample configuration of this table:

```
CLI>show routing-key
ID=CNAM
OPC_ID=OPC_001-036-062
SG_GRP_ID=SGG
SI=SCCP
SUBSYSTEM_GRP_ID=CNAM_SSN
RC=5043
PLATFORM_ID=FSPTC504
ID=LNP
OPC_ID=OPC_001-036-062
SG_GRP_ID=SGG
SI=SCCP
SUBSYSTEM_GRP_ID=LNP_SSN
RC=3045
PLATFORM_ID=FSAIN304
```

Step 16: Define Individual TCAP/SCCP Routes

The SCCP route table defines the end-to-end connectivity of a subsystem in the BTS to a subsystem in the SS7 network.

```
CLI>add sccp-route subsystem-grp-id=Subsystem group identifier from Step 13; opc-
  id=OPC name from Step 4; rk-id=Routing Key identifier from Step 15; dpc-id=DPC name
  from Step 5; bpc-id=Optional backup DPC also defined in Step 5;
```

The following is a sample configuration of this table:

```
CLI>show sccp-route
SUBSYSTEM_GRP_ID=LNP_SSN
OPC_ID=OPC_001-036-062
```

```
DPC_ID=STP1
RK_ID=LNP
BPC_ID=STP2
```

Step 17: Define Service Logic Host Route Characteristics

The service logic host route (SLHR) profile table needs to be defined before the SLHR table in Step 18 can be configured.

```
CLI>add slhr-profile id=SLHR identifier;
```

The following is a sample configuration of this table:

```
CLI>show slhr_profile
ID=slhr_cnam
DESCRIPTION=Service Logic Host Routing Table for IN1 CNAM Service
ID=slhr_lnp
DESCRIPTION=Service Logic Host Routing Table for AIN LNP Service
```

Step 18: Define Service Logic Host Route

The service logic host route (SLHR) table contains the information necessary to make the MGC generate a TCAP query to an SCP. The specific events that can cause these queries to be generated are known as Trigger Detection Points (TDP).

```
CLI>add slhr id=SLHR identifier from Step 17; opc-id=OPC name from Step 4; dpc-id=DPC
 name from Step 5; subsystem-grp-id=Subsystem group identifier from Step 13;
 tt=translation type; gtt-req=Y¦N; gtt-addr-type=CDPN¦CLGN¦DN; gtt-addr=GTT
 addresss;
```

The following is a sample configuration of this table:

```
CLI>show slhr
ID=slhr_lnp
OPC_ID=OPC_001-036-062
DPC_ID=STP1
SUBSYSTEM_GRP_ID=LNP_SSN
GTT_REQ=Y
TT=11
GTT_ADDR_TYPE=CDPN
GTT_ADDR=3
```

Step 19: Modify Call Agent Configuration

Now that you have created the SLHR, you might have to modify the call agent configuration table to set this entry as the default.

The following example shows how to see the current configuration:

```
CLI>show ca_config type=DEFAULT-LNP-SLHR-ID;
TYPE=DEFAULT-LNP-SLHR-ID
DATATYPE=STRING
VALUE=SLHR_LNP
```

Step 20: Enable the SCTP Association

Just as you did in the ISUP configuration steps, you need to enable the SCTP associations for TCAP services.

```
CLI>control sctp-assoc id=SCTP Association identifier; target-state=INS;
 mode=FORCED;
```

The result of this operation can be verified with the **status** command as shown:

```
CLI>status sctp-assoc id=SCTP-AIN-ITP1
SCTP ASSOC ID -> SCTP-AIN-ITP1
ADMIN STATE -> ADMIN_INS
OPER STATE -> SCTP-ASSOC in service
REASON -> ADM executed successfully
RESULT -> ADM configure result in success
Reply : Success:
```

Step 21: Enable the Subsystem Groups or Subsystems

You must also control the subsystems into service. You can do this by controlling the subsystem in-service or by controlling the subsystem group in-service. Controlling the subsystem group in and out of service has the effect of controlling all subsystems in the group in and out of service.

```
CLI>control subsystem_grp id=Subsystem group identifier; target-state=UIS;
 mode=FORCED;
```

The result of this operation can be verified with the **status** command as shown:

```
CLI>status subsystem_grp id=LNP_SSN;
ADMIN STATE -> User in service
OPER STATE -> Subsystem allowed
REASON -> ADM executed successfully
RESULT -> ADM configure result in success
SUBSYSTEM GRP ID -> LNP_SSN
Reply : Success:
```

You can also verify the status of the individual subsystems as shown in the following command:

```
CLI>status subsystem id=LNP_SSN; OPC_ID=OPC_001-036-062
ADMIN STATE -> User in service
OPER STATE -> Subsystem allowed
REASON -> ADM executed successfully
RESULT -> ADM configure result in success
SUBSYSTEM ID -> LNP_SSN
OPC ID -> OPC_001-036-062
Reply : Success:
```

Trunking Media Gateway Configuration Tasks

Now that the basic steps required to provision SS7 are done, you need to configure the MG where the bearer circuits are located. You might notice that many of these steps are similar to the steps involved in configuring an MTA outlined in the previous chapter. In each of these steps are recommended configuration settings; these settings should be fine in general but might require some adjusting depending on the environment. The basic tasks to configure a trunking media gateway on the BTS and their associated provisioning tables are

Step 1 Configure the MG characteristics (media gateway profile table).

Step 2 Configure the MG (media gateway table).

Step 3 Configure the TGCP endpoints (termination table).

Step 4 Optionally, configure the trunk group quality of service parameters (QoS table).

Step 5 Configure the trunk group characteristics (SS7 or CAS trunk group profile table).

Step 6 Configure the trunk group (trunk group table).

Step 7 Configure the trunks (trunk table).

Step 8 Bring the MG into service.

Step 9 Bring the trunk group into service.

Step 10 Prepare the endpoints for service.

Step 11 Bring the endpoints into service.

Step 1: Configure the MG Characteristics

Instead of having to define all the characteristics of every MG serviced by the BTS, a template can be defined that the MG references. This template is referred to as a *media gateway profile*. Among the characteristics defined here are several that relate to call signaling. Remember in Chapter 7 that this table was also used to define MTA profiles; the way in which MG profiles are defined is similar:

```
CLI>add mgw-profile id=mg mgw profile name; vendor=vendor name; mgcp-
  version=MGCP_1_0; mgcp-variant=TGCP_1_0; description=MG description;
```

Table 8-5 lists some of the key fields and fields relating to a PacketCable deployment, a majority of which are related to call signaling. Notice, this table is nearly identical to Table 7-4 in Chapter 7.

Table 8-5 *Key MGW-PROFILE Table Fields*

Field	Default Value	Recommended Value	Note
VENDOR	N/A	MG vendor name	MG vendor name
DESCRIPTION	N/A	Defining descriptions is a good practice and should be added when possible.	Textual description of the media gateway profile

Table 8-5 *Key MGW-PROFILE Table Fields (Continued)*

Field	Default Value	Recommended Value	Note
Fields relating to the command header and parameters, as well as general protocol operation			
MGCP_VERSION	MGCP_0_1	MGCP_1_0	MGCP protocol version used.
MGCP_VARIANT	NONE	TGCP_1_0 for trunking media gateways. NCS_1_0 for MTAs	MGCP protocol variant.
MGCP_NE_LOCAL NAME_SUPP	N	Y	CMS Notified Entity parameter expressed as <local>@FQDN.
MGCP_ERQNT_ SUPP	N	N Embedded RQNT use can reduce the number and size of MGCP messages that need to be sent, improving network speed and efficiency; however, some MGs might not handle this functionality very well, so set this field to N for these gateways. Also typically for an MG this message is not used that often compared to NCS messaging with an MTA.	MG supports embedded RQNT. Recall this is when an RQNT message is in effect embedded or included in another message such as a CRCX, an MDCX, or even another RQNT. This in effect allows the BTS to send multiple commands at once.
MGCP_3WAY_ HSHAKE_SUPP	Y	Y	MG can handle Response Acks.
MGCP_CONN_ID_ AT_GW_SUPP	Y	Y Set this parameter to Y so the BTS can delete hung connections on a gateway.	Delete only connections returned from AUEP.
MGCP_HAIRPIN_ SUPP	Y	N Hairpinning is not permitted in PacketCable.	MG can do TDM hairpinning, meaning if both endpoints are on the same gateway then the network type parameter local is used (nt:local) instead of IP.

continues

Table 8-5 *Key MGW-PROFILE Table Fields (Continued)*

Field	Default Value	Recommended Value	Note
MGCP_HAIRPIN_ Z2_SUPP	N	N Hairpinning is not permitted in PacketCable.	MG supports second endpoint parameter; applies only if MGCP_HAIRPIN_SUPP is Y.
MGCP_XDLCX_ SUPP	N	N	Disables restrictions on the X: parameter in a DLCX.
CONN_MODE_ REQUIRED_IN_ MDCX	N	N	The MG should need the connection mode parameter only if the mode has changed.
MGCP_PIGGYBACK _MSG_SUPP	Y	Y	BTS can piggyback MGCP messages in response packets sent to a gateway.
MGCP_REQ_ID_ SUPP	N	Y	Gateway supports correlation of request identifier from RQNT to NTFY messages.
MGCP_RSIPSTAR_ SUPP	Y	Y	An RSIP restart message using the * wildcard for the endpoint name implies all endpoints are available.
MGCP_TERM_INIT _LEVEL	0	0 For trunking media gateways; implies endpoints initialized on the DS0 level.	Applicable for trunking media gateways; defines initialization level.
MGCP_TEST_ CONN_SUPP	Y	Y	Permits the network loopback (netwloop) and continuity test (netwtest) connection modes on the media gateway.
PARALLEL_TEST_ CONN_SUPP	Y	Y	Whether a network loopback or continuity test connection is allowed at the same time as another inactive, recvonly, or sendrecv mode connection.

Table 8-5 *Key MGW-PROFILE Table Fields (Continued)*

Field	Default Value	Recommended Value	Note
MGCP_T_TRAN	400 ms	400 ms	Initial timeout after which a command is retransmitted; future timeouts increase exponentially until a limit is reached.
MGCP_T_LONGTRAN	5 ms	5 ms	MGCP transaction timeout after receiving a 100 provisional response.
MGCP_EP_SPECIFIC_CAP_SUPP	N	N	If the media gateway supports endpoint specific queries.
Fields relating to packages, events, and signals			
MGCP_DEFAULT_PKG	NONE	TRUNK For SS7 media gateways the value TRUNK would be used; for CAS E911 media gateways the value MO would be used. This is set to LINE for PacketCable MTAs.	Default package name.
MGCP_TO_SUPP	Y	Y	Gateway support the timeout parameter for signals; values set in the CA_CONFIG table.
RBK_ON_CONN_SUPP	Y	Y	MG can send ring back on connection. S: rt@<conn>
RBK_ON_INACTIVE_CONN_SUPP	N	N	MG cannot send ring back on connection in inactive mode.
MGCP_QLOOP_SUPP	Y	Y	MTA supports Quarantine Looping.
MGCP_QDISCARD_SUPP	Y	Y	The discard method for quarantine handling is supported.

continues

Table 8-5 *Key MGW-PROFILE Table Fields (Continued)*

Field	Default Value	Recommended Value	Note
DTMF_OOB_SUPP	N	N	Whether or not the gateway supports mid-call notification of DTMF digits by sending a NTFY command.
Fields relating to the LCO parameter and options			
CODEC_NEG_SUPP	Y	Y	Negotiate call CODEC; works with QOS table.
PC_MPTIME_SUPP	Y	Y	LCO parameter "mp:" supported.
MGCP_CAP_NEG_REQ	Y	Y	MG can negotiate capabilities.
EC_SUPP	N	N Normally you want the gateway to decide on its own when echo cancellers should be enabled and disabled.	Echo cancellation LCO parameter "e:" supported.
IPTOS_RTP_SUPP	N	Y For the value of the IP ToS byte see the QoS table description.	Set TOS byte: LCO parameter "t:" supported.
MGCP_RSVP_SUPP	N	N	Media gateway supports the Resource Reservation Protocol (RSVP).
Fields relating to the SDP parameters			
SDP_ORIGFIELD_SUPP	Y	Y	SDP parameter o: supported.
SDP_SESSNAME_SUPP	Y	Y	SDP parameter s: supported.
SDP_EMAIL_SUPP	Y	N	SDP parameter e: supported.
SDP_PHONE_SUPP	Y	N	SDP parameter p: supported.
SDP_URI_SUPP	Y	N	SDP parameter u: supported.
SDP_BANDWIDTH_SUPP	Y	Y	SDP parameter b: supported.

Table 8-5 *Key MGW-PROFILE Table Fields (Continued)*

Field	Default Value	Recommended Value	Note
SDP_INFO_SUPP	Y	N	SDP parameter i: supported.
SDP_TIME_SUPP	Y	Y	SDP parameter t: supported.
SDP_CAP_ENCODE_TYPE	AUTO	AUTO Values encoded as received from the remote endpoint	Specified SDP capabilities encoding.
SDP_ATTRIB_SUPP	Y	Y	SDP parameter a: supported.
Fields relating to gateway keepalive methods			
KEEPALIVE_METHOD	AUEP	AUEP It is important for the BTS to know the current operational state of its gateways; if it does not it can try to route calls to gateways that are not operational; therefore, it is recommended to keep this field set to its default value of AUEP.	Defines the keepalive method between the BTS and media gateway. AUEP means MGCP AUEP messages are used; other values include AUEP-ICMP, which means AUEP is tried first, and if that fails, ICMP pings are used, and NONE, which means no heartbeat mechanism is used.
MGCP_KEEPALIVE_INTERVAL	60 seconds	60 seconds	MG AUEP interval when no activity detected between BTS and gateway.
MGCP_KEEPALIVE_RETRIES	3	3	Number of keepalive retries if the first one fails.
MGCP_MAX_KEEPALIVE_INTERVAL	600 seconds	600 seconds	Maximum MGCP keepalive interval.
Fields relating to DNS			
MGCP_MAX1_RETRIES	2	2	Number of retries for each gateway IP address.
MGCP_MAX2_RETRIES	3	3	Total number of retries for a gateway before declared unreachable.
DOMAIN_NAME_CACHING_SUPP	Y	Y Disabling DNS caching can cause an unnecessary high load on the BTS under high traffic conditions.	(Y) Cache the MG IP address so a DNS lookup is needed only for the first message to a gateway, or (N) do a DNS lookup for every message that needs to be sent.

continues

Table 8-5 *Key MGW-PROFILE Table Fields (Continued)*

Field	Default Value	Recommended Value	Note
Fields related to Facsimile transmission			
FAX_INBAND_METHOD	GW_SPECIFIED	GW_SPECIFIED This field is meaningful only if the field FAX_T38_ENABLED is set to N in the QoS table.	If set to GW_SPECIFIED, the BTS is not involved in how the gateways handle in-band fax transmissions; if set to T38_FXR_OFF, the BTS explicitly instructs the gateways not to do T38 fax relay.
T38_FXR_LOOSE_SUPP	AUTO	AUTO Support determined by gateway auditing	Gateway supports call agent controlled T38 Loose mode fax.
Fields related to security			
KRB_RESET_FLAG	Y	Y	Kerberos SA is automatically reestablished upon expiration.
IKE_CS	3DES-MD5, 3DES-SHA1	3DES-MD5, 3DES-SHA1	List of supported ciphersuites for authentication and encryption in priority order.
IKE_GROUP	2	2	Diffie-Helman exchange group; valid only for trunking gateways.
IKE_KEY	N/A	N/A	Preshared key for security association between BTS and trunking gateway.
IKE_SA_LIFETIME	86400 seconds	86400 seconds	Lifetime of IKE security association.
IPSEC_SA_ESP_CS	3DES-MD5, 3DES-SHA1, NULL-MD5, NULL-SAH1	3DES-MD5, 3DES-SHA1, NULL-MD5, NULL-SAH1s	List of supported IPsec ciphersuites for authentication and encryption in priority order.
IPSEC_SA_GRACE_PERIOD	21600 seconds	21600 seconds	IPsec key expiration grace time; must be less than or equal to 25 percent of the IPSEC_SA_LIFETIME value.

Table 8-5 *Key MGW-PROFILE Table Fields (Continued)*

Field	Default Value	Recommended Value	Note
IPSEC_SA_ LIFETIME	86400 seconds	86400 seconds	Lifetime of IPsec security association.
IPSEC_ULP_NAME	IP	IP	IPsec SA upper layer protocol name.

The MGW profile used for the VXSM and VISM cards in the call flow examples is as follows:

```
CLI>show mgw-profile id=Cisco_VXSM_VISM ID=Cisco_VXSM_VISM
MGCP_VARIANT=TGCP_1_0
RBK_ON_CONN_SUPP=Y
PACKET_TYPE=IP
AAL1=N
AAL2=N
AAL5=N
PVC=N
SVC=N
SPVC=N
EC_SUPP=N
SDP_ORIGFIELD_SUPP=Y
SDP_SESSNAME_SUPP=Y
SDP_EMAIL_SUPP=N
SDP_PHONE_SUPP=N
SDP_URI_SUPP=N
SDP_BANDWIDTH_SUPP=Y
SDP_INFO_SUPP=N
SDP_TIME_SUPP=Y
SDP_ATTRIB_SUPP=Y
MGCP_ERQNT_SUPP=N
MGCP_HAIRPIN_SUPP=N
MGCP_QLOOP_SUPP=Y
MGCP_3WAY_HSHAKE_SUPP=Y
MGCP_CONN_ID_AT_GW_SUPP=Y
MGCP_VMWI_SUPP=N
PORT_START=1
MGCP_VERSION=MGCP_1_0
MGCP_RSVP_SUPP=N
MGCP_RSIPSTAR_SUPP=Y
MGCP_TERM_INIT_LEVEL=0
MGCP_HAIRPIN_Z2_SUPP=N
DTMF_OOB_SUPP=N
MGW_TYPE=MGX8850
OSI_SUPP=N
MGCP_DIALTONE_TO_SUPP=Y
MGCP_MWI_SUPP=Y
SPARE1_SUPP=N
SPARE2_SUPP=N
IPTOS_RTP_SUPP=Y
USE_STATIC_PROFILE=N
REFRESH_DIGIT_MAP=N
MGCP_XDLCX_SUPP=N
MGCP_CAS_BLOCK_SUPP=N
CODEC_NEG_SUPP=Y
MGCP_DEFAULT_PKG=TRUNK
MGCP_KEEPALIVE_INTERVAL=60
MGCP_KEEPALIVE_RETRIES=3
```

```
MGCP_T_TRAN=400
MGCP_MAX1_RETRIES=2
MGCP_MAX2_RETRIES=3
MGCP_T_LONGTRAN=5
PC_MPTIME_SUPP=Y
MGCP_CAP_NEG_REQ=Y
RBK_ON_INACTIVE_CONN_SUPP=N
MGCP_NAS_SUPP=Y
DOMAIN_NAME_CACHING_SUPP=Y
BTXML_SUPP=N
CONN_MODE_REQUIRED_IN_MDCX=N
MGCP_NE_LOCALNAME_SUPP=N
KRB_REEST_FLAG=Y
IPSEC_SA_ESP_CS=3DES-MD5,3DES-SHA1,NULL-MD5,NULL-SHA1
IPSEC_SA_LIFETIME=86400
IPSEC_SA_GRACE_PERIOD=21600
IPSEC_ULP_NAME=IP
IKE_GROUP=2
IKE_SA_LIFETIME=86400
IKE_CS=3DES-MD5,3DES-SHA1
MGCP_EP_SPECIFIC_CAP_SUPP=N
KEEPALIVE_METHOD=AUEP
MGCP_MAX_KEEPALIVE_INTERVAL=600
MGCP_REQ_ID_SUPP=Y
MGCP_PIGGYBACK_MSG_SUPP=Y
MGCP_QDISCARD_SUPP=Y
MGCP_TO_SUPP=Y
MGCP_TEST_CONN_SUPP=Y
PARALLEL_TEST_CONN_SUPP=N
T38_FXR_LOOSE_SUPP=AUTO
FAX_INBAND_METHOD=GW_SPECIFIED
SDP_CAP_ENCODE_TYPE=AUTO
```

Step 2: Configure the MG

After the MG characteristics are defined, you need to define the MG itself and associate it to a media gateway profile:

```
CLI>add mgw id=mg mgw name; tsap-addr=mg FQDN; call-agent-id=call agent name; mgw-
  profile-id=mg mgw profile name from Step 1; type=tgw;
```

Table 8-6 lists some of the key fields relating to a PacketCable deployment; again, this table is nearly identical to Table 7-5 in Chapter 7.

Table 8-6 *Media Gateway Table Field Descriptions*

Field	Default Value	Recommended Value	Description
TSAP-ADDR	N/A	MG FQDN	IP address or FQDN of the MG
CALL-AGENT-ID	N/A	Key for the call agent table	Identifier that the call agent trunks of this mgw are assigned to
MGW-PROFILE-ID	N/A	Key for the media gateway profile table describing this MG	Identifier of the media gateway profile used

Table 8-6 *Media Gateway Table Field Descriptions (Continued)*

Field	Default Value	Recommended Value	Description
Type	N/A	TGW	An MG is a trunking gateway
CALL-AGENT-CONTROL-PORT	0 (call agent automatically assigns it)	0	Call agent MGCP listening port for this media gateway

The following is the MGW configuration for the VXSM card used in the call flow examples:

```
CLI>show mgw type=tgw
ID=VXSM_05
TSAP_ADDR=tgw01-vxsm-5.cisco.com
CALL_AGENT_ID=CA104
MGW_PROFILE_ID=Cisco_VXSM_VISM
STATUS=INS
TYPE=TGW
```

Step 3: Configure the TGCP Endpoints

Now that you have the MG configured, you need to configure the TGCP endpoints on that MG. This is done in the termination table. Here you can define a range of endpoints; typically, these are added one T1 or E1 at a time:

```
CLI>add termination prefix=prefix format; port-start=starting port; port-end=ending
  port; type=TRUNK; mgw-id=mg mgw name from Step 2;
```

Table 8-7 lists some of the key fields relating to PacketCable configuration; this table is similar to Table 7-6 in Chapter 7.

Table 8-7 *Termination Table Field Descriptions*

Field	Default Value	Recommended Value	Description
PREFIX	N/A	Hierarchical local name endpoint format (i.e., ds/ds1-1/)	Endpoint prefix for TGCP
PORT-START	1	1	Initial port number
PORT-END	24	24 (for a T1)	Ending port number
TYPE	N/A	TRUNK	Termination type
MGW-ID	N/A	Key for the MG media gateway	Media gateway name

Here is the configuration for three of the terminations on the VXSM card:

```
CLI>show termination tgn-id=1234; limit=2;
ID=DS/S-0/DS1-196/1
MGW_ID=VXSM_05
TYPE=TRUNK
STATUS=INS
TGN_ID=1234
TRUNK_ID=19
ID=DS/S-0/DS1-196/2
MGW_ID=VXSM_05
TYPE=TRUNK
STATUS=INS
TGN_ID=1234
TRUNK_ID=20
Reply : Success:  Entries 1-2 of 5 returned.
```

The termination identifier represents the local part of the endpoint name; if you combine this with the TSAP-ADDR in the MGW table configuration, you have the full endpoint name. So in the example the endpoint names are DS/S-0/DS1-196/1@tgw01-vxsm-5.cisco.com and DS/S-0/DS1-196/2@tgw01-vxsm-5.cisco.com, which represent the first two DS0s on DS1 #196 on VXSM card 5.

When you initially configure the terminations, they have a status of unequipped (UEQP), and the trunk group number identifier (TGN_ID) and trunk identifier (TRUNK_ID) fields do not show up. These fields are automatically populated when you correlate the terminations to trunks, as you'll see in Step 7.

Step 4: Configure the Trunk Group QoS Parameters

In the media gateway profile, you saw that some of the fields influence the Local Connection Option parameters sent to MG endpoints. However, you probably noticed that not all of the LCO parameters could be specified there. The reason is they are defined in another table, the QOS table. You learned about this same table in Chapter 7 (Table 7-7) for defining these characteristics for MTA subscribers:

```
CLI>add qos id=QoS profile name; codec-type=preferred CODEC;
```

Table 8-8 lists some of the key fields and fields relating to a PacketCable deployment.

Table 8-8 *QoS Table Fields*

Field	Default Value	Recommended Value	Description
CODEC-TYPE	N/A	PCMU Optionally, you might want to create an entry for a compression CODEC such as G.728 or G.729E.	Preferred CODEC (see discussion after this table for more information on CODEC negotiation)

Table 8-8 *QoS Table Fields (Continued)*

Field	Default Value	Recommended Value	Description
DTMF-PREF-MODE	DTMF-INBOUND	DTMF-INBOUND	Preferred DTMF relay method
FAX_T38_ENABLED	Y	Y	Use T.38 for facsimile transmission
HBW	0 (gateway chosen)	0	Higher bandwidth in kilobits/second
HPTIME	10 ms	20 ms	Higher packetization period
LBW	0 (gateway chosen)	0	Lower bandwidth in kilobits/second
LPTIME	10 ms	20 ms	Lower packetization period
IPTOS_RTP_LOWDELAY	N	N	Whether or not the Delay bit in the IP ToS byte is set
IPTOS_RTP_PRECEDENCE	CRITICAL(5)	CRITICAL (5)	Value of the three IP Precedence bits in the IP ToS byte
IPTOS_RTP_RELIABILITY	N	N	Whether or not the Reliability bit in the IP ToS byte is set
IPTOS_RTP_THROUGHPUT	N	N	Whether or not the Throughput bit in the IP ToS byte is set
RESOURCE_RESERVATION	BE (Best Effort)	BE (Best Effort)	PacketCable 1.x devices do not use RSVP
RSVP_REQUIRED	N	N	PacketCable 1.x devices do not use RSVP
SILENT_SUPPRESSION	N	N If you want to utilize the UGS-AD service to free up DOCSIS bandwidth, set this parameter to Y.	Enable silence suppression (a.k.a. VAD) LCO parm s:

If CODEC negotiation is enabled for the MG in the media gateway profile, the CODEC and packetization periods defined in the QOS table identify the highest priorities. Values obtained from an audit of the endpoints capabilities are included after that. If CODEC negotiation is disabled, the CODEC and packetization period defined here are used for the MG endpoint.

The following is the QoS profile used by the trunk groups in the call flow examples:

```
CLI>show qos id=G711u_20ms
ID=G711u_20ms
CODEC_TYPE=PCMU
RSVP_REQUIRED=N
RESOURCE_RESERVATION=BE
IPTOS_RTP_LOWDELAY=N
IPTOS_RTP_THROUGHPUT=N
IPTOS_RTP_RELIABILITY=N
IPTOS_RTP_PRECEDENCE=CRITICAL
MAX_DQOS_SESSIONS=5
MAX_DQOS_AUTH_BANDWIDTH=HIGH
DTMF_PREF_MODE=DTMF_INBAND
DQOS_CMTS_DSCP_TOS=160
LBW=0
HBW=0
AUTO_GAIN_CONTROL=N
GAIN_VALUE=0
SILENT_SUPPRESSION=N
LPTIME=20
HPTIME=20
VIDEO_CODEC_TYPE=NONE
CLIENT_TYPE=DQOS
DQOS_DSCP_TOS_BITMASK=224
DOCSIS_DSCP_TOS=160
DOCSIS_DSCP_TOS_BITMASK=224
FAX_T38_ENABLED=Y
Reply : Success: Entry 1 of 1 returned.
```

Step 5: Configure the Trunk Group Characteristics

In this table you configure common characteristics for a collection of trunk groups. For example, if you have 10 T1s interconnecting to the local PSTN, they probably all have some common parameters. Different tables are used for this, depending on the type of signaling the trunk group uses. For CAS signaling, the cas-tg-profile table is used:

```
CLI>add cas-tg-profile id=CAS trunk group profile name; test-line=y¦n; no-test-
  trunk=y ¦ n; e911=y ¦ n; oss-sig-type=MOSS ¦ NBEC ¦ EAOSS ¦ NONE; sig-type=MF ¦
  DTMF ¦ MF-OSS ¦ MF-TERM ¦ LINE; MF-OSS-TYPE=denotes how ANI and information digits
  are signaled;
```

For SS7 signaling, different tables are used, depending on the SS7 variant. For North American ANSI SS7 the ss7-ansi-tg-profile table is used (other SS7 profile tables include the ss7-q761-tg-profile and ss7-q767-tg-profile table for the ITU Q.761 and Q.767 variants):

```
CLI>add ss7-ansi-tg-profile id=SS7 trunk group profile name; type=a7; cot-orig=y¦n;
  cot-duration=length of continuity test in seconds; cot-freq=frequency of cot tests
  on outgoing calls; cot-tone=type of cot performed
```

NOTE There are several other configurable SS7 parameters, most of which relate to timers and the support of optional parameters; please consult your BTS documentation for more information.

Table 8-9 lists some of the key fields and fields relating to a PacketCable deployment.

Table 8-9 *Key Trunk Group Profile Fields*

Field	Default Value	Recommended Value	Description
SS7-x-TG-PROFILE Fields			
TYPE	A7	A7 for ANSI	Denotes ANSI variant of SS7.
COT-ORIG	Y	Y	Whether or not COT tests are supported, this field must be Y for the other COT fields to be meaningful.
COT-DURATION	1 second	1 second	Length of COT test in seconds.
COT-FREQ	7	7	Percentage of outgoing calls requesting a COT test on a per-trunk basis. This frequency is 1/x; for example, the default of 7 means 1 out of every 7 calls requests COT on a given trunk.
COT-TONE	4W-2W	4W-2W for transponder mode; 4W-4W for loop back mode	Type of COT test performed (4W-2W is Tx of 1780 Hz and Rx of 2010 Hz; 4W-4W is Tx and Rx of 2010 Hz).
CAS-TG-PROFILE Fields			
TEST-LINE	N	N	Whether or not the trunk group is a test line trunk group
NO-TEST-TRUNK	N	Y (for busy line verification trunk groups) N (else)	Identifies whether or not the trunk group is used for busy line verification.

continues

Table 8-9 *Key Trunk Group Profile Fields (Continued)*

Field	Default Value	Recommended Value	Description
E911	N	Y (for Emergency 911 trunk groups) N (else)	Identifies whether or not the trunk group is used for emergency 911 operator services.
OSS-SIG-TYPE	N	NONE (Default—no operator signaling); MOSS (Modified operator services signaling); NBEC (Non Bell Exchange Carrier); EAOSS (Exchange Access operator services signaling)	Identifies whether or not the trunk group is used for operator services and if so what type of operator services signaling (OSS) is used.
SIG-TYPE	N/A	MF-OSS (for operator services) MF-TERM (for BLV) MF (MF signaling using the MS package) DTMF (DTMF signaling using the DT package) LINE (Regular Subscriber Line)	CAS signaling type.
MF-OSS-TYPE	N/A	N/A	Identifies how ANI and information digits are signaled on operator services trunk groups.

Here are same sample configurations for a standard SS7 trunk group profile and a CAS E911 trunk group profile:

```
CLI>show ss7-ansi-tg-profile ID=SS7_PF
ID=SS7_PF
HOP_COUNTER=20
CFN_SUPP=N
COT_ORIG=Y
COT_FREQ=7
COT_TONE=4W_TO_2W
COT_DURATION=1
INBAND_INFO=N
SEND_CIP=Y
SEND_CPN=Y
SEND_OCN=Y
SEND_REDIR_NUM=Y
SEND_ATP=Y
SEND_GAP=Y
SEND_GN=N
```

```
SEND_JIP=Y
T_IAM=30
T_REL=6
T_RSC=15
T_GRS=15
T_CGB=15
T_BLO=6
T_8=15
T_CCR_R=12
T_COT_R=20
T_COT_L=300
FAST_ANSWER_SUPP=N
SEND_HOPCOUNTER=Y
SEND_CIP_NONGEO=N
SEND_CPN_PRES=N
SEND_REDIRINFO=Y
SEND_REDIRCAP=N
SEND_REDIRCOUNTER=N
REDIR_MAX=5
T_9=255
ALARM_CARRIER=UNKNOWN
ALLOW_EXM=Y
ALLOW_CRMCRA=Y
CCT_GRP_CARRIER=DIGITAL
DEFAULT_BC=SPEECH
DEFAULT_CHGNOA=NOTUSED
DEFAULT_CHGNPI=NOTUSED
DEFAULT_OLI=NOTUSED
SEND_CHNOLIP=Y
SEND_CHN_NONGEO=Y
T_14=6
T_18=15
T_20=15
CLI>show cas-tg-profile id=CAS_E911
ID=CAS_E911
OSS_SIG=Y
TEST_LINE=N
NO_TEST_TRUNK=N
E911=Y
SIG_TYPE=MF_OSS
MF_OSS_TYPE=MO_10I
PLAY_DIAL_TONE=N
TEST_LINE_TYPE=NONE
OSS_SIG_TYPE=NONE
INBAND_INFO=N
NO_ANSWER_TIMER=0
NO_ANSWER_ACTION=NA
```

Step 6: Configure the Trunk Group

Now that a trunk group profile has been defined, you need to define your trunk groups:

```
CLI>add trunk-grp id=Trunk group number; call-agent-id=call agent name; tg-
  type=SS7¦CAS; tg-profile=Trunk group profile name from Step 5; call-ctrl-route-
  id=Call control route name; dial-plan-id=Dial plan name; mgcp_pkg_type=MGCP/TGCP
  package name;
```

Table 8-10 lists some of the key fields and fields relating to PacketCable.

Table 8-10 *Key Trunk Group Fields*

Field	Default Value	Recommended Value	Description
CALL-AGENT-ID	N/A	Key for the call agent table	Identifier that the call agent trunks of this trunk group are assigned to
TG-TYPE		SS7 for SS7 trunk groups; CAS for CAS trunk groups	Identifies the type of trunks in this trunk group
TG-PROFILE	N/A	Key for appropriate trunk group profile table	Identifier for the trunk group profile characterizing general common trunk group parameters
CALL-CTRL-ROUTE-ID	N/A	Key for call control route table	Only valid for SS7 trunk groups; identifier for call control route specifying SS7 parameters needed to reach the destination
DIAL-PLAN-ID	N/A	Key for dial plan table	Dial plan used for calls originating from this trunk group
MGCP-PKG-TYPE	N/A	IT for SS7 trunks MO or MT for CAS trunks	MGCP package used by the trunk group
QOS-ID	N/A	Key for QoS	QoS profile used for the trunk group
SEL-POLICY	ASC	Opposite of remote end	Trunk selection policy

The following is the trunk group configuration used in the call flow examples:

```
CLI>show trunk-grp id=1234
ID=1234
CALL_AGENT_ID=CA104
TG_TYPE=SS7
NUM_OF_TRUNKS=5
DPC=10-10-7
TG_PROFILE_ID=SS7_PF
STATUS=INS
DIRECTION=BOTH
SEL_POLICY=ASC
GLARE=SLAVE
ALT_ROUTE_ON_CONG=N
```

```
SIGNAL_PORTED_NUMBER=N
DIAL_PLAN_ID=INCOMING
DEL_DIGITS=0
QOS_ID=G711u_20ms
TRAFFIC_TYPE=LOCAL
ANI_BASED_ROUTING=N
CALL_CTRL_ROUTE_ID=DPC_10-10-7
MGCP_PKG_TYPE=IT
ANI_SCREENING=N
SEND_RDN_AS_CPN=N
STATUS_MONITORING=N
SEND_EARLY_BKWD_MSG=N
EARLY_BKWD_MSG_TMR=5
SCRIPT_SUPP=N
VOICE_LAYER1_USERINFO=AUTO
VOICE_INFO_TRANSFER_CAP=AUTO
PERFORM_LNP_QUERY=N
```

Notice in this example the trunk selection policy is set to ASC or ascending; this indicates that when this trunk group is used for outgoing calls the lowest available circuit (CIC) is used first. Thus, the remote switch should use a descending selection policy and select the highest available circuit (CIC) first. This minimizes the chance both switches will try to place a call on the same circuit at the same time, a condition known as "glare."

Step 7: Configure the Trunks

Next, configure the individual DS0s on your trunk groups. It is with this table that SS7 CICs are mapped to TGCP endpoints. So if this is not configured correctly, you will have voice patch problems. Also note the term CIC is used even if the trunk type is MF CAS; in this case, CIC refers to DS0 channel or timeslot number. Just as with terminations, ranges of trunks can be added at once. Thus, the specified CIC range must be the same as the termination range:

```
CLI>add trunk cic-start=First SS7 CIC number; cic-end=Last SS7 CIC number; tgn-
  id=Trunk Group number from Step 6; mgw-id=Media Gateway name from Step 2;
  termination-prefix=Endpoint local name prefix; termination-port-start=First
  endpoint number; termination-port-end=Last endpoint number;
```

Table 8-11 lists some of the key fields and fields relating to PacketCable.

Table 8-11 *Key Trunk Fields*

Field	Default Value	Recommended Value	Description
CIC-START	N/A		First SS7 CIC or DS0 timeslot of trunks to be added
CIC-END			Last SS7 CIC or DS0 timeslot of trunks to be added
TGN-ID	N/A	Key for trunk group table	Number of the trunk group these trunks are associated with

continues

Table 8-11 *Key Trunk Fields (Continued)*

Field	Default Value	Recommended Value	Description
MGW-ID	N/A	Key for media gateway table	Identifier of the media gateway these trunks reside on
TERMINATION-PREFIX	N/A		Local endpoint name except for the rightmost term
TERMINATION-PORT-START	N/A		Endpoint number (rightmost term) for the first SS7 CIC or DS0 timeslot of trunks to be added
TERMINATION-PORT-END	N/A		Endpoint number (rightmost term) for the last SS7 CIC or DS0 timeslot of trunks to be added

The following is a sample of the trunk configuration used in the call flow examples (the ID number is the CIC # of the trunk) :

```
CLI>show trunk tgn-id=1234; limit=2
ID=19
TGN_ID=1234
TERM_ID=DS/S-0/DS1-196/1
MGW_ID=VXSM_05
ID=20
TGN_ID=1234
TERM_ID=DS/S-0/DS1-196/2
MGW_ID=VXSM_05
```

Step 8: Bring the MG into Service

Before MG endpoints can be used to place and receive calls, you need to perform a few administrative tasks on the BTS. The first one is to enable the MG or put it into service:

```
CLI>control mgw id=MG mgw identifier; target-state=INS; mode=FORCED;
```

The result of this operation can be verified with the following command:

```
CLI>status mgw id=MG mgw identifier
```

The MGs must be in the working state before they can be used, for example:

```
CLI>status mgw id=VXSM_05
MGW_ID -> VXSM_05
RESULT -> ADM configure result in success
REASON -> ADM executed successfully
ADMIN_STATE -> ADMIN_INS
OPER_STATE -> Media gateway in working status
```

The other possible operational states are unknown, down, and cannot be reached. Logically, the cannot be reached state indicates a problem in IP connectivity; the down state typically indicates the receipt of a negative MGCP response message; and the unknown state can be the result of disabling media gateway monitoring.

Step 9: Bring the Trunk Group into Service

Similarly, the trunk group needs to be enabled:

```
CLI>control trunk-grp id=Trunk group number; target-state=INS; mode=FORCED;
```

The result of this operation can be verified with the **status** command as shown:

```
CLI>status trunk-grp id=Trunk group number
```

Again, the trunk group needs to have an operational status of in-service before it can be used, as shown in the following example:

```
CLI>status trunk-grp id=1234
TGN_ID -> 1234
RESULT -> ADM configure result in success
REASON -> ADM executed successfully
ADMIN_STATE -> ADMIN_INS
OPER_STATE -> Trunk group in-service
PRIMARY_OPER_STATE -> NOT USED
BACKUP_OPER_STATE -> NOT USED
```

Step 10: Prepare the Endpoints for Service

Now that the MG and trunk group are in a working state, the endpoints need to be put into a working state as well. The first step to doing this is to **equip** the terminations:

```
CLI>equip trunk-termination tgn-id=Trunk group number; cic=all;
```

Step 11: Bring the Endpoints into Service

Finally, the MG endpoints need to be enabled just as the MG itself was.

```
CLI>control trunk-termination tgn-id=Trunk group number; cic=all; target-state=INS;
 mode=FORCED;
```

Just as was the case for the MG, the result of this operation can be verified with the **status** command:

```
CLI>status tt tgn-id=Trunk group number; cic=all;
```

The keyword **tt** displays a condensed form of the trunk termination status output, as demonstrated here:

```
CLI>status tt tgn-id=1234; cic=all
 1234  19   ADMIN_INS     TERM_ACTIVE_IDLE      ACTV    IDLE    NON_FAULTY
 1234  20   ADMIN_INS     TERM_ACTIVE_IDLE      ACTV    IDLE    NON_FAULTY
 1234  21   ADMIN_INS     TERM_ACTIVE_IDLE      ACTV    IDLE    NON_FAULTY
 1234  22   ADMIN_INS     TERM_ACTIVE_IDLE      ACTV    IDLE    NON_FAULTY
 1234  23   ADMIN_INS     TERM_ACTIVE_IDLE      ACTV    IDLE    NON_FAULTY
Reply : Success:
```

Looking at the trunks used in the examples, you see the operational state (OPER_STATE) is TERM ACTIVE IDLE, meaning the endpoints are ready to process calls. Alternatively, you can use the **status trunk-termination** command, which has a more verbose output for the same data.

Dial Plan Configuration Tasks for OFFNET Call Routing

In Chapter 7 you learned how to create a dial plan on the BTS 10200 and how to configure it for local or ONNET call routing. Now you will learn how to configure OFFNET routing.

The basic steps to configure OFFNET routing are as follows:

Step 1 Define routes for trunk groups (route table).

Step 2 Optionally, define route guides for policy-based routing (route guide table).

Step 3 Optionally, define carriers (carrier table).

Step 4 Define destinations (destination table).

Step 5 Define dial plan entries (dial plan table).

Step 1: Define Routes for Trunk Groups

A route consists of a list of trunk groups where calls are routed. Up to 10 trunk groups can exist in a single route. The syntax for adding an entry to this table is as follows:

```
CLI>add route id=route identifier; tgn1-id=trunk group #1 identifier; tgn2-id=trunk
 group #2 identifer; …; tgn10-id=trunk group #10 identifer; tg_selection=LCR ¦ RR
 ¦SEQ; alt-route-id=alternate route identifier;
```

The trunk group selection field, tg_selection, specifies the priority for choosing trunk groups within a route. The priority method can be least cost routing (LCR), round robin (RR), or sequential (SEQ).

The following shows the configuration of this table used for the call flow examples:

```
CLI>show route ID=OffNet
ID=OffNet
TGN1_ID=1234
DEL_DIGITS1=0
DEL_DIGITS2=0
DEL_DIGITS3=0
DEL_DIGITS4=0
DEL_DIGITS5=0
DEL_DIGITS6=0
DEL_DIGITS7=0
DEL_DIGITS8=0
DEL_DIGITS9=0
DEL_DIGITS10=0
TG_SELECTION=SEQ
NEXT_ACTION=NONE
```

As you can see, in this case only one trunk group is defined for this route, so the trunk group selection field doesn't matter. In this table you can also manipulate the called party number and calling party number digits sent to the next switch.

Step 2: Define Route Guides for Policy-Based Routing (Optional)

The BTS can be configured to route calls based on certain predefined policies, which can be especially useful when you have multiple routes or trunk groups to a particular destination.

For example, calls can be routed so that each trunk group receives a certain percentage of the call traffic volume, or calls can be routed depending on the phone number of the person placing the call. Another possibility is time- or date-dependent routing; for example, suppose it's cheaper to route traffic on one carrier during the business day but on another carrier at night.

Policy routing is configured using the ROUTE_GUIDE table; the syntax is as follows:

```
CLI>add route_guide id=route guide identifier; policy_id=index into policy table;
  policy_type=ODR ¦ TOD ¦ PERCENT ¦ PREFIX ¦ OLI ¦ POP ¦ ROUTE ¦ NXX ¦ REGION;
```

Table 8-12 describes the individual policy types as well as the next table indexed by the policy_id field.

Table 8-12 *Route Guide Fields*

Policy Type	Next Table	Description
ODR (Origination Dependent Routing)	POLICY_ODR	Call is routed based on the NDC or Exchange Code of the originating/calling party.
TOD (Time Of Day)	POLICY_TOD	Call is routed based on the current time and/or date.
PERCENT	POLICY_PERCENT	Call is routed to multiple destinations based on predefined percentages of call volume.
PREFIX	POLICY_PREFIX	Call is routed based on the type of call (i.e., national, international, operator, etc.).
OLI (Originating Line Information)	POLICY_OLI	Call is routed based on the OLI associated with the calling party number.
POP (point of presence)	POLICY_POP	Call is routed based on the POP associated with the calling party.
ROUTE	ROUTE	Call is routed based on the ROUTE table described in the previous step.
NXX	POLICY_NXX	Used for routing number services calls.
REGION	POLICY_REGION	Call is routed based on the region, which is determined from the calling party number.

In the examples in this chapter, only basic routing was used so this table was not needed. Please refer to BTS documentation for more details on more complicated call routing.

Step 3: Define Carriers (Optional)

The carrier table defines the characteristics of interLATA, intraLATA, and international carriers. A Local Access Transport Area (LATA) is a geographic region in the United States assigned to one or more telephony service providers (TSP). Some U.S. states are made up of only one or two LATAs, whereas other states have several. Because these TSPs are providing local telephony service, they are also known as Local Exchange Carriers (LEC). LECs are further classified as the Incumbent LEC (ILEC) and Competitive LECs (CLEC). A call between LECs is an intraLATA call, whereas a call that crosses LATA boundaries must be handed off to an Inter Exchange Carrier (IXC) and is classified as an interLATA (that is, long-distance) call.

In the carrier table, you can specify the various call types supported by a particular carrier, as demonstrated here:

```
CLI>show carrier
ID=0386
STATUS=INS
INTER=Y
INTRA=Y
INTL=Y
CASUAL=N
CUT_THRU=N
OP_SERVICES=Y
SEND_CN=N
SEND_CSP=Y
USE_DIAL_PLAN=Y
LNP_QUERY=N
```

This table was not needed for the basic call types in this chapter; please refer to the BTS specific documentation if you would like more information.

Step 4: Define Destinations

Recall from Chapter 7 that the destination table defines the call type and routing information for dialed digits. In that chapter, the call type local and the route type subscriber were covered because they are used for local ONNET call routing. Now it is time to look at some other call types and route types.

```
CLI>add destination dest-id=destination identifier; call_type=call type identifier;
  route_type=route type identifier; route_guide_id=index into next table;
```

Table 8-13 describes the possible values for the route type field.

Table 8-13 *Route Types*

Route Type	Description
SUB	Call is routed to a local subscriber (this is the routing discussed in Chapter 7).
CARRIER	Call is routed based on information specified in the CARRIER table.
ROUTE	Call is routed based on information in the ROUTE GUIDE table.
ANNC	Call is routed to a specified announcement.
RID	Call is routed based on information in the ROUTE table.
DP	Call is retranslated based on a new dial plan identifier.
NONE	No route required.

The possible values of the call type field were discussed in Chapter 7; they can also be found by looking at the CALL_TYPE table.

The destination entry used in the call flow examples is

```
CLI>show destination DEST_ID=OffNet
DEST_ID=OffNet
CALL_TYPE=TOLL
ROUTE_TYPE=RID
ZERO_PLUS=N
INTRA_STATE=Y
ROUTE_ID=OffNet
GAP_ROUTING=N
CLDPTY_CTRL_REL_ALWD=N
CALL_SUBTYPE=NONE
ACQ_LNP_QUERY=NA
```

Notice that the RID route type is specified, which means the ROUTE table is used next, specifically the index **OffNet**, which as you can see was defined in Step 1. The call type field is set to the value Toll, which means this destination is for interLATA calls.

Step 5: Define Dial Plan Entries

So the destination defined in the example in the previous step is for calls to subscribers within the same LATA as the PacketCable subscribers but handled by a different Local Exchange Carrier (LEC). The next step is to define the phone numbers reachable by this destination in the dial plan:

```
CLI>add dial_plan id=dial plan identifier from previous step; digit-string=numerical
  string of dialed digits; dest-id=destination identifier previously defined;
```

The configuration used in the call flow examples is

```
CLI>show dial-plan dest-id=OffNet
ID=INCOMING
DIGIT_STRING=919
DEST_ID=OffNet
SPLIT_NPA=NONE
DEL_DIGITS=0
```

```
MIN_DIGITS=10
MAX_DIGITS=10
NOA=NATIONAL
```

In the "Troubleshooting Tips" portion of this chapter, you'll see how to verify the BTS translations for OFFNET call types, similar to the ONNET example shown in the previous chapter.

Configuring a Trunking Media Gateway

At this point, you know how to configure Cisco ITPs and the BTS to get call signaling operational to the PSTN. You also know how to configure a trunking media gateway on the BTS and configure basic OFFNET call routing.

This section looks at configuring the MG itself—specifically, the configuration required for basic PacketCable operation on the Cisco VXSM and VISM gateways. Only the basic steps and the steps directly related to TGCP signaling are covered here. In Chapter 11, "RTP and RTCP," you will learn more about configuration options related to the bearer stream, and configuration related to security and lawful intercept is also covered later in the book.

Remember from Chapter 2 that the VXSM and VISM gateways are service modules inside an MGX 8800 series platform. Configuration is accomplished by first logging in to the active PXM processor card (typically using Secure Shell). After you are logged in, you can use the display cards command, **dspcds,** to see the individual service module cards inside the chassis. The following is an example from an MGX 8880:

```
MGX8880.8.PXM.a > dspcds
MGX8880                         System Rev: 05.02   May. 19, 2006 10:43:28 EDT
Chassis Serial No:   SCA0917004A Chassis Rev: C0     GMT Offset: -4
                                                     Node Alarm: CRITICAL
Card Front/Back         Card              Alarm      Redundant    Redundancy
Slot Card State         Type              Status     Slot         Type
01   Empty              ---               ---        ---          ---
02   Empty              ---               ---        ---          ---
03   Standby/Empty      VISM_PR_8T1       MAJOR      13           PRIMARY SLOT
04   Active/Empty       VISM_PR_8T1       NONE       NA           NO REDUNDANCY
05   Active/Active      VXSM_155          MAJOR      NA           NO REDUNDANCY
06   Active/Active      VXSM_155          MAJOR      NA           NO REDUNDANCY
07   Standby/Active     PXM45C            NONE       08           PRIMARY SLOT
08   Active/Active      PXM45C            NONE       07           SECONDARY SLOT
09   Active/Active      RPM_XF            CRITICAL   NA           NO REDUNDANCY
10   Active/Active      RPM_XF            MAJOR      NA           NO REDUNDANCY
11   Active/Empty       VISM_PR_8T1       NONE       NA           NO REDUNDANCY
12   Active/Empty       VISM_PR_8T1       NONE       NA           NO REDUNDANCY
13   Active/Empty       VISM_PR_8T1       NONE       03           SECONDARY SLOT
14   Active/Active      VISM_PR_8T1       MAJOR      NA           NO REDUNDANCY
15   Standby/Active     SRMEB_3T3         NONE       16           PRIMARY SLOT
16   Active/Active      SRMEB_3T3         NONE       15           SECONDARY SLOT
17   Empty              ---               ---        ---          ---
18   Empty              ---               ---        ---          ---
19   Active/Empty       VISM_PR_8T1       NONE       NA           NO REDUNDANCY
20   Active/Empty       VISM_PR_8T1       NONE       NA           NO REDUNDANCY
27   Active/Empty       VISM_PR_8T1       NONE       NA           NO REDUNDANCY
28   Active/Empty       VISM_PR_8T1       NONE       NA           NO REDUNDANCY
29   Empty              ---               ---        ---          ---
```

```
30   Active/Empty      VISM_PR_8T1     NONE      NA        NO REDUNDANCY
31   Standby/Active    SRMEB_3T3       NONE      32        PRIMARY SLOT
32   Active/Active     SRMEB_3T3       NONE      31        SECONDARY SLOT
```

First, the cards in the top half of the chassis (slots 1–15) are shown, then the bottom half of the chassis (slots 16–32) are shown. The redundant processor PXM-45 cards are in the required slots 7 and 8. The router RPM-XF cards for connectivity to the rest of the VoIP network are in slots 9 and 10. The SRM cards used for bulk distribution, redundancy, and BERT testing are in the required slots 15, 16, 31, and 32. In this case, you also have VISM-PR cards in slots 3, 4, 11, 12, 13, 14, 19, 20, 27, 28, and 30, and VXSM cards are in slots 5 and 6 (the 155 number indicates these cards are to be used with OC-3 back cards—the bandwidth of an OC-3 is ~155 Mbps). You might be wondering why slots 21–26 do not show up; this is because these are the bottom slots in places where double height cards are used (PXM, RPM, and VXSM).

For each of these 32 slots in the front, there is a corresponding slot in the back of the chassis. What type of card is used in the back slot depends on the associated card in the front slot. Also all cards for the back slots are single height; therefore, a double height front card can have two associated back cards. Many of the back cards provide TDM connectivity; in the previous example the SRM back cards each contain three T3 coaxial interfaces, the VISM back cards each contain eight T1 RJ-45 copper interfaces, and the VXSM back cards support four OC-3 optical interfaces. Often, the T1s configured on a VISM card are physically connected to the PSTN inside a DS-3 interface on the SRM back cards instead of the T1 interfaces on a VISM back card. Currently, there are three options for VXSM back cards:

- A pair of 24-port T1/E1 cards used to support a total of 48 T1/E1 lines
- A pair of 3-port T3/E3 cards used to support a total of 6 T3/E3 lines
- A pair of 4-port OC-3 cards

Unlike the first two options, the OC-3 back cards provide redundancy for each other and are not combined for additional capacity. If you do not want to use this additional redundancy feature, you can use only a single OC-3 back card. Staying on the redundancy topic, a VISM or VXSM card can be configured as a backup for other VISM or VXSM cards. For VXSM cards, the redundancy technique is 1:1 where an adjacent card operates in standby mode for an active card. The PSTN TDM lines are often split using Y-cables with an output going to back card interface for each VXSM card. This type of redundancy protects against a back card failure as well as a front card failure. Optionally, you might choose to provide only front card redundancy; in this case, a special type of back card is used for the redundancy slot, which has no physical connections. Much more about redundancy is covered in Chapter 16, "PacketCable Network Design Considerations." The PXM back cards consist of ports for management console and Ethernet connectivity, and the RPM-XF back cards have Gigabit Ethernet or Packet over SONET ports for WAN uplink connectivity as well as management console and Ethernet ports.

Figure 8-23 show the MGX8800 implementation used for the examples in this chapter.

Figure 8-23 *Sample MGX880 Implementation*

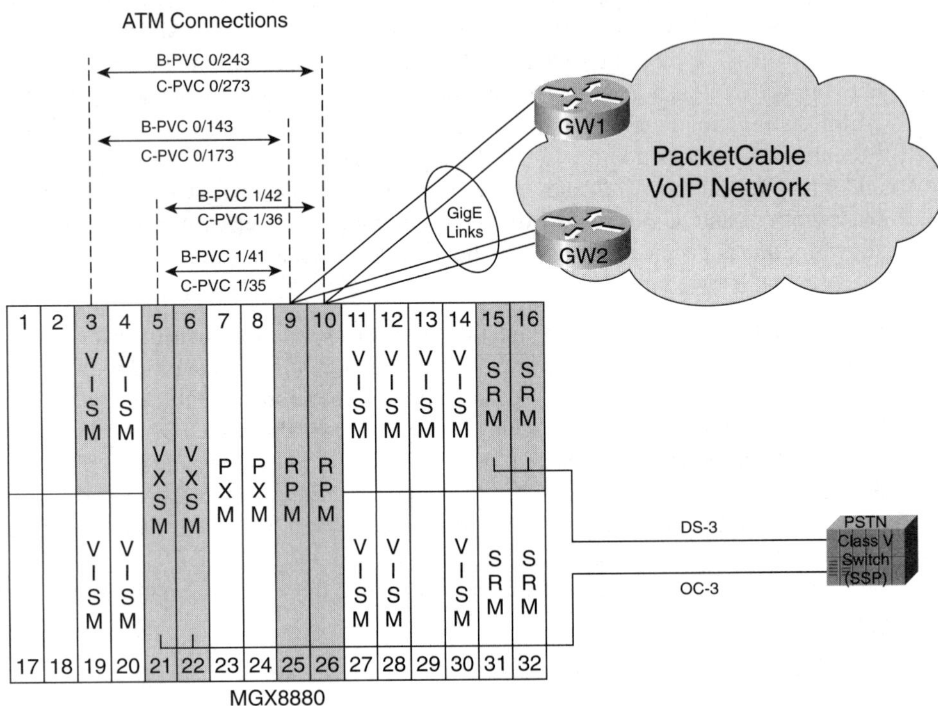

To configure one of the VISM or VXSM cards you must change configuration to the slot number of the particular card you want to configure. For example, to configure the VXSM card in slot 5 you would use the **cc 5** command as demonstrated here:

```
MGX8800.8.PXM.a > cc ?
ERR: syntax error near "?"
Syntax: cc "slotNumber"
        slotNumber -- a slot number between 1 and 32
MGX8800.8.PXM.a > cc 5
(session redirected)
MGX8800.5.VXSM.a >
```

As you can see from this example, the question mark can be used to display command syntax; just type as much of the command as you know, and then type **?** to display the syntax. If you enter only a **?**, this results in a display of the entire list of available commands. You can narrow the list of commands shown by typing **?** followed by as much of the command as you do know; for example, to see all MGC configuration commands available on a VXSM card, you might enter the following:

```
MGX8800.5.VXSM.a > ? cnfmgc
    Available commands
    ------------------
    cnfmgc                      Config MGC
    cnfmgcgrpmgc                Modify port & pref of a MGC within a MGC group
    cnfmgcip                    Config MGC IP
```

Keep in mind that any part of the command you know can be used with the **?**; if you were to type **? mgc**, you would get all commands containing the letters **mgc**.

The format of the card CLI prompt tells you some information about the card and its state. The format is

```
MGX name.slot #.card type.card state
```

For example, the CLI prompt for the VXSM card in slot 5 used in this chapter is

```
MGX8880.8.PXM.a >
```

The state a indicates the card is active; other possible state values are i (initialization) and s (standby).

The steps for configuring a trunking media gateway can be logically broken out into three areas:

- PXM configuration for basic gateway setup
- RPM configuration
- VXSM and VISM configuration

PXM Configuration for Basic Gateway Setup

The first thing to do is to define the node name of the MGX chassis using the **cnfname** command, as shown in the following example:

```
MGX8880.7.PXM.a > cnfname MGX8880
This node name will be changed to MGX8880. Please Confirm
cnfname: Do you want to proceed (Yes/No)? yes
cnfname: Configured this node name to MGX8880 Successfully.
MGX8880.7.PXM.a >
```

Then, configure the date and time for the MGX using the **cnfdate** command to set the data, the **cnftmzn** command to set the time zone, the **cnftmzngmt** command to set the time offset from GMT, and the **cnftime** command to set the time.

Next, use the **addcontroller** command to define the PNNI controller used to control the ATM connectivity between cards. The controller identifier must be 2 for a PNNI controller, the location is **internal**, the type is PNNI, and the slot is either 7 or 8 to specify the location of the PXM. The controller configuration used in the examples is as follows:

```
MGX8800.7.PXM.a > dspcontrollers
MGX8800                        System Rev: 05.02   Jun. 26, 2006 12:57:03 EDT
MGX8880-CH                                          Node Alarm: CRITICAL
Number of Controllers:        1
Controller Name:
Controller Id:                2
Controller Location:          Internal
Controller Type:              PNNI
Controller Logical Slot:      7
Controller Bay Number:        0
Controller Line Number:       0
Controller VPI:               0
Controller VCI:               0
Controller In Alarm:          NO
Controller Error:
```

Next, use the **ipifconfig** command to define the IP parameters for the management interface lnPci0.

The last step in this task is probably the most important one, especially if you are using VXSM cards. It is to use the **setrev** command to define the firmware images used for the various slots in the chassis. With this command, you also define the call signaling protocol used for VXSM cards. The options are **H.248**, **MGCP**, and **TGCP** with **H.248** the default.

NOTE If you neglect to set this value properly and need to change it later on, you lose the entire configuration on the card.

The syntax of the **setrev** command is as follows:

```
MGX8800.7.PXM.a > setrev
ERR: Syntax: setrev <slot> <primary revision> [-ccp <CallControlProtocol>]
 [-codectmp <CodecTemplateNumber>] [-service <ServiceType>]
         slot -- <1-32> except 15,16,31,32
          revision -- revision number. E.g.,
                      2.0(1)
                      2.0(1.255)
                      2.0(0)I  or 2.0(0)A
                      2.0(0)P1 or 2.0(0)P2
                      2.0(0)P3 or 2.0(0)P4
                      2.0(0)D
                      2.0(1.166)I  or 2.0(1.166)A
                      2.0(1.166)P1 or 2.0(1.166)P2
                      2.0(1.166)P3 or 2.0(1.166)P4
                      2.0(1.166)D
CallControlProtocol -- Call control protocol (supported only on VXSM)
                       1.H.248 (default)
                       2.TGCP
                       3.MGCP
CodecTemplateNumber -- Codec Template Number (supported only for VXSM)
                       1.TGW/Wireline (default)
                       2.Cable
ServiceType -- Service to be used for this card (supported only on MPSM16T1E1/
 MPSM155)
                       0 = ATM/FR (Default)
                       1 = MLPPP
```

RPM Configuration

Remember that the RPM blade in an MGX chassis is really a Cisco router running a version of Cisco IOS appropriate for an MGX ATM switch. Redundant RPM modules typically are used, and ATM connections are built between each RPM and each VISM/VXSM card. In the example used in this chapter, redundant RPM-XF modules exist in slots 9 and 10. Probably the easiest way to learn RPM configuration is to look at the configuration for this example. This configuration is broken into three parts, the first of which is depicted in Figure 8-24.

Figure 8-24 *Sample RPM Configuration: Part I*

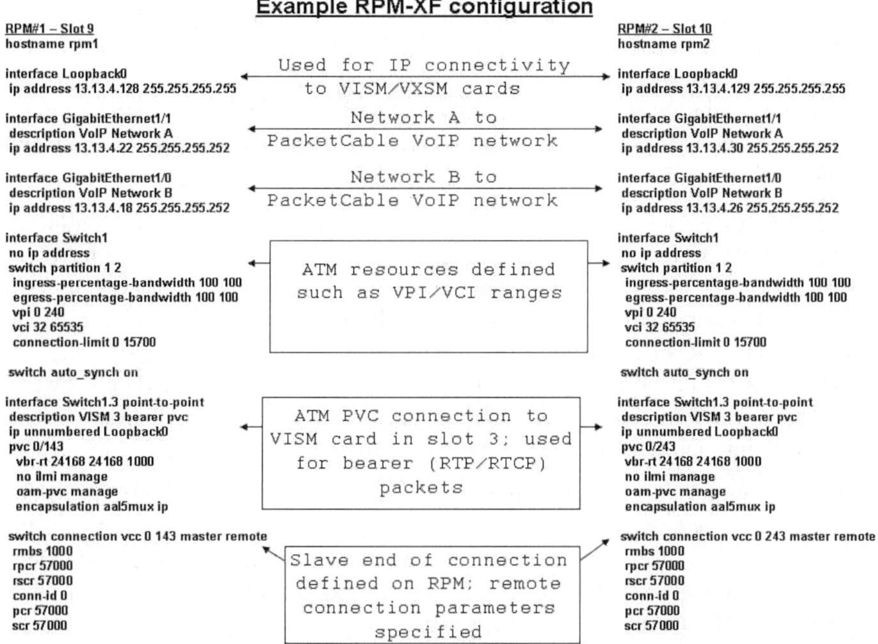

Figure 8-24 shows the configuration for the RPM-XF card in slot 9 on the left and the configuration for the RPM-XF card in slot 10 on the right. First, loopback interfaces are defined for communicating with the VXSM and VISM cards in the chassis. To preserve IP addresses, the ATM interface connections to each VXSM and VISM card all reference these loopback interfaces for their IP connectivity instead of using separate IP addresses on each ATM subinterface. Two Gigabit Ethernet interfaces are used on each RPM card to connect the MGX switch to the rest of the PacketCable network. Both interfaces should be used to connect to different upstream routers to provide protection against a single point of failure.

The next step is to define the ATM resources used by the RPM when establishing permanent virtual circuits (PVCs) with other service modules by defining a switch interface. The switch interface configuration involves defining the allowable VPI and VCI ranges as well as the minimum and maximum connection limits.

As soon as this is done, the individual ATM connections to the other service modules (that is, VXSM and VISM cards) are defined by creating logical switch subinterfaces. Because each interface is for the communication between another card and the RPM, point-to-point subinterfaces are used. As mentioned earlier, the loopback IP address avoids needing a separate subnet for communication to each service module. To avoid call signaling and bearer traffic competing on a common ATM PVC, separate ones are defined for each VXSM/VISM card. In the subinterface configuration are the ATM parameters such as local and remote VPI/VCI numbers and traffic descriptors. This configuration needs to agree with the configuration on the remote VXSM/VISM card for the connection to operate properly. The first subinterface defined is the one for bearer traffic to and from the VISM card in slot 3.

All RTP voice packets to and from the 192 (8 T1s * 24 DS0/T1) DS0 endpoints are transferred across this connection. In this case, voice is digitized using the G.711 CODEC, and an RTP/UDP/IP packet is generated every 20 milliseconds. The result of this is 50 200-byte packets sent and received every second for every active DS0 endpoint. How voice is digitized into RTP/UDP/IP packets is covered in detail in Chapters 10 and 11. Because this is an ATM connection, each outgoing packet is converted into ATM cells by the RPM card, and received ATM cells are reassembled back into packets before being sent out the Gigabit Ethernet interfaces to the rest of the PacketCable VoIP network.

Notice from the PVC configuration, AAL5 (ATM Adaptation Layer 5) is used to encapsulate each IP packet into ATM cells. The result of this encapsulation is that each RTP packet requires five ATM cells; thus, for each call 250 cells are sent and received each second (5 cells/packet * 50 pps). If all 192 endpoints are active, a total of 48,000 cells per second of RTP traffic are sent and received across this connection. Because this type of voice traffic is sent and received at a constant rate, you would think a CBR (constant bit rate) service would be used for the ATM connection. However, the RPM supports only the UBR, VBR-nrt, and VBR-rt services. By setting the peak cell rate and sustained cell rate to the same value in a VBR-rt service, you have, in effect, a CBR service.

As you can see from the bottom of Figure 8-24, the remote PCR and SCR are all set to a value of 57,000 cells per second. The reason 57,000 is used instead of 48,000 is to allow overhead for RTCP packets, ATM OAM management cells, and possible CALEA replicated streams. Please refer to the section "Capacity Planning on the Trunking Gateway" in Chapter 16 for more information. If you look at the VBR-rt configuration line immediately below the PVC configuration, you might notice the PCR and SCR values are set to 24,168. These are the local PCR and SCR values and are specified in kilobits per second instead of cells per second; if you convert 57,000 cells per second into kbps, this value is the result. Finally, notice the master is specified as remote under the switch connection command, indicating that the RPM is the slave end and the VISM card is the master end.

NOTE ATM is not covered in detail in this book because its use in a PacketCable network is very specific to using the Cisco MGX series of switches. For more information refer to the Cisco URL http://www.cisco.com/univercd/cc/td/doc/cisintwk/ito_doc/atm.htm. You can get the ATM specifications at the following URL: http://www.mfaforum.org/tech/atm_specs.shtml.

Figure 8-25 shows the definition of additional ATM subinterfaces and PVCs.

Figure 8-25 *Sample RPM Configuration: Part II*

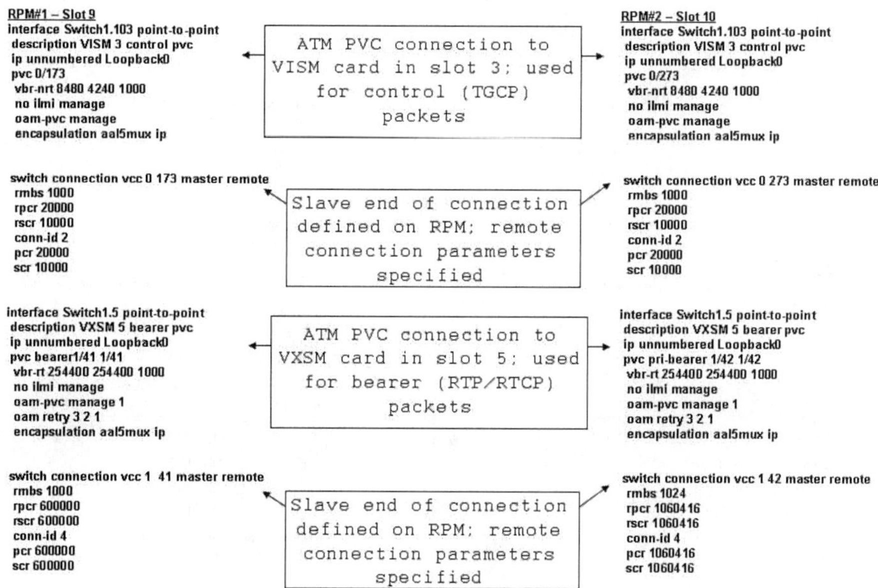

The top part of the configuration contains the definition of the PVCs used for control (that is, TGCP) traffic to the VISM card in slot 3. In this case, the VBR non real-time service is used because, unlike bearer traffic signaling, traffic is not overly sensitive to jitter. The Peak Cell Rate is set to 8,480 kbps (20,000 cells per second), the Sustained Cell Rate is set to 4,240 kbps (10,000 cells per second), and the Maximum Burst Size is set to 1,000 cells. If all 192 DS0 endpoints were active at once, this would leave an SCR of ~22 kbps and a PCR of ~44 kbps per endpoint, which should be more than enough.

The bottom part of the configuration contains the definition of the PVCs used for bearer traffic to the VXSM card in slot 5. As with the VISM card, the VBR-rt service is used. However, because the number of DS0 endpoints is much bigger with a VXSM card the traffic parameters have increased as well. In this example, it is assumed at most one OC-3 worth of traffic is active (2,016 DS0s). Again, some overhead for RTCP packets, ATM OAM management cells, and possible CALEA replicated streams has been included. The result is a PCR and SCR of 600,000 cells per second (254,400 kbps).

Figure 8-26 shows the final part of the RPM configuration.

Figure 8-26 *Sample RPM Configuration: Part III*

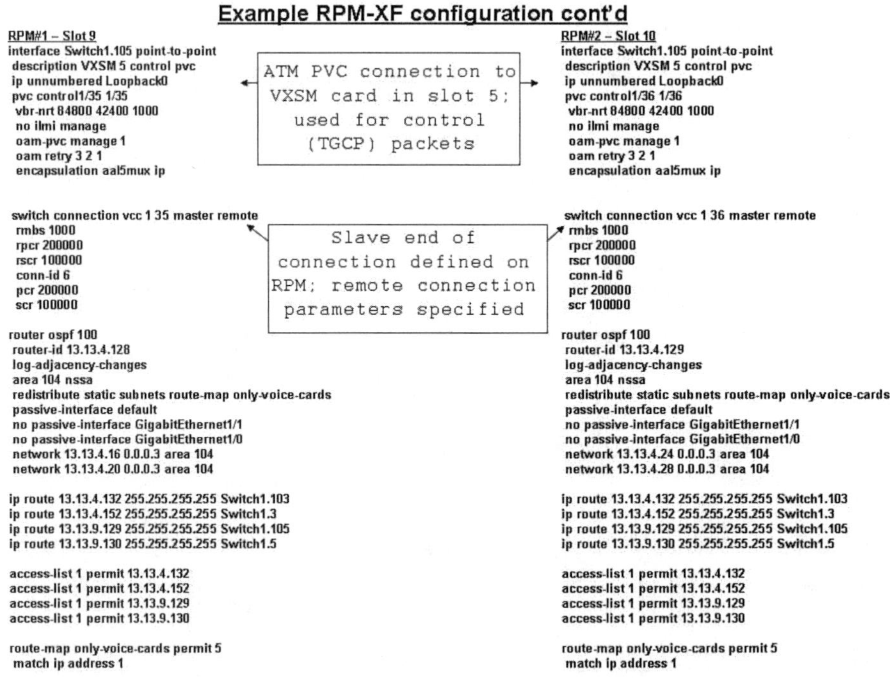

Here, one additional ATM communication channel is defined for the TGCP call signaling traffic to the VXSM card in slot 5. As with the VISM case, a VBR-nrt service is used. Because the number of DS0 endpoints has increased by roughly a factor of 10, the traffic parameters have increased by a factor of 10. The PCR is set to 84,800 kbps (200,000 cells per second), the SCR is set to 42,400 kbps (100,000 cells per second), and the MBS remains 1,000 cells.

Also a routing protocol is configured to communicate the VXSM/VISM networks to the rest of the PacketCable network. In this case, the Open Shortest Path First (OSPF) protocol is used, and the MGX is configured as a Not So Stubby Area (NSSA). This means a default route is used by the MGX to reach the rest of the network, and the external redistributed static routes for the VXSM/VISM cards are advertised.

VXSM and VISM Configuration

Now that the RPM cards are configured, you can configure the VXSM and VISM cards. The steps required for this can be divided into three areas:

- For ATM connectivity to the RPM cards (Steps 1–4)
- For TDM connectivity to the PSTN (Steps 5–12)
- For call signaling connectivity to the MGC (Steps 13–19)

These steps are as follows:

Step 1 Verify the VISM operating mode (VISM only).

Step 2 Enable ATM port (VISM only).

Step 3 Enable ATM resources.

Step 4 Configure PVC on the VISM/VXSM card.

Step 5 Configure CAS variants (Optional).

Step 6 Enable TDM lines.

Step 7 Enable STS-1 paths (VXSM OC-3 cards only).

Step 8 Configure line/path parameters (Optional).

Step 9 Bring the lines into service.

Step 10 Define Voice Interfaces (VXSM only).

Step 11 Add DS0 endpoints to the line (VISM only).

Step 12 Configure CAS endpoints (VISM only—Optional).

Step 13 Configure the gateway domain name.

Step 14 Configure gateway IP addresses.

Step 15 Configure MGC Domain names and Domain Name Servers.

Step 16 Configure Domain Name IP addresses.

Step 17 Add the MGCs to a redundancy group.

Step 18 Configure a signaling protocol to this MGC redundancy group (VISM only).

Step 19 Adjust xGCP parameters (Optional).

Step 1: Verify VISM Operating Mode (VISM Only)

Actually, the first thing you should do is verify whether the VISM card is configured for the right mode of operation. For PacketCable this should be voipSwitching/voipTrunking. You can check the current configuration using the **dspvismparam** command, and you can change it using the **cnfvismmode** command:

```
MGX8800.1.3.VISM8.a > dspvismparam
```

Step 2: Enable the ATM Port (VISM Only)

Next, you need to add an ATM port to the card using the following command (this command has no argument):

```
MGX8800.1.3.VISM8.a > addport
```

Step 3: Enable ATM Resources

The same command, **addrscprtn**, is used on both the VXSM and the VISM cards; however, the parameters are different. An example for the VXSM card is as follows:

```
MGX8800.5.VXSM.a > addrscprtn 1 1 2 10000 1000000 10000 1000000 0 240 32 65535 1 810
```

The first three parameters are always **1 1 2**; the remaining parameters specify the minimum and maximum values for VPI/VCI, bandwidth, connections, and so on. Please consult the Cisco VXSM Command Reference Guide for more details.

On a VISM card the command is much simpler, for example:

```
MGX8800.1.3.VISM8.a > addrscprtn 1
```

Step 4: Configure PVC on the VISM/VXSM Card

When PVCs are created between two cards, one card is the master end, and the other card is the slave end. In the examples in this chapter the RPM is always functioning as the slave end, and the VISM/VXSM is always functioning as the master end. The slave ends need to be created first, which causes the MGX to generate an ATM address for the connection. That ATM address is needed when creating the master end. To see what this ATM address is, use the **display connection** command from the active PXM card, as demonstrated here:

```
MGX8880.8.PXM.a > dspcon 9:1.2:2 1 41
Port                     Vpi Vci             Owner      State      Persistency
-------------------------------------------------------------------------------
Local  9:1.2:2           1.41                SLAVE      FAIL       Persistent
```

```
        Address: 47.00918100000000059a3fa172.000001091802.00
        Node name: MGX8880
Remote Routed          0.0                  MASTER      --          Persistent
        Address: 00.000000000000000000000000.000000000000.00
        Node name:
MGX8880.8.PXM.a > dspcon 9:1.2:2 1 35
Port                   Vpi Vci              Owner       State       Persistency
-------------------------------------------------------------------------------
Local  9:1.2:2         1.35                 SLAVE       FAIL        Persistent
        Address: 47.00918100000000059a3fa172.000001091802.00
        Node name: MGX8880
Remote Routed          0.0                  MASTER      --          Persistent
        Address: 00.000000000000000000000000.000000000000.00
        Node name:
```

The arguments for this command are the port identifier, followed by the VPI and VCI. The port identifier in these examples is **9:1.2:2**, which refers to the RPM in slot 9, subslot 1, using the PNNI port (2:2). You can use the **dsppnports** command to view the port identifiers on the chassis. The two connections shown are for the bearer and control PVCs to the VXSM card in slot 5. Notice that the remote port is not specified and that the local slave connections are in the FAIL state; this is because the master side has not been defined.

The following example shows how to create the master end of the ATM connections for the VXSM card in slot 5:

```
MGX8880.5.VXSM.a > addcon 1 1 41 4 1 2 1 -slave
 47009181000000000059a3fa17200000109180200.1.41 -lpcr 600000 -rpcr 600000 -lscr
 600000 -rscr 600000 -lmbs 1000
master endpoint added successfully
master endpoint id : 47009181000000000059A3FA17200000105180100.1.41
MGX8880.5.VXSM.a > addcon 1 1 35 4 1 1 1 -slave
 47009181000000000059a3fa17200000109180200.1.35 -lpcr 200000 -rpcr 200000 -lscr
 100000 -rscr 100000 -lmbs 1000
master endpoint added successfully
master endpoint id : 47009181000000000059A3FA17200000105180100.1.35
```

Arguments in this command include the VPI/VCI of the connection, the slave ATM address, and the traffic characteristics of the ATM connection. Remember, these should match what was configured on the slave end.

If the **dspcon** command is entered again, it now shows the information for the remote master end of the connection, and the status of the connection indicates **OK**, meaning operational:

```
MGX8880.8.PXM.a > dspcon 9:1.2:2 1 41
Port                   Vpi Vci              Owner       State       Persistency
-------------------------------------------------------------------------------
Local  9:1.2:2         1.41                 SLAVE       OK          Persistent
        Address: 47.00918100000000059a3fa172.000001091802.00
        Node name: MGX8880
Remote 5:1.1:1         1.41                 MASTER      OK          Persistent
        Address: 47.00918100000000059a3fa172.000001051801.00
        Node name: MGX8880
MGX8880.8.PXM.a > dspcon 9:1.2:2 1 35
Port                   Vpi Vci              Owner       State       Persistency
-------------------------------------------------------------------------------
Local  9:1.2:2         1.35                 SLAVE       OK          Persistent
        Address: 47.00918100000000059a3fa172.000001091802.00
        Node name: MGX8880
```

```
Remote 5:1.1:1          1.35                  MASTER     OK         Persistent
       Address: 47.00918100000000059a3fa172.000001051801.00
       Node name: MGX8880
```

You can also verify the operation of these ATM connections using the **test connection** command:

```
MGX8800.5.VXSM.a > tstcon 1 1 41
test is in progress ..
Connection Id    Test Type    Direction    Result     Round Trip Delay
=============    =========    =========    =======    ================
01.0001.00041     OAM Lpbk     ingress     Success         96 microsec
MGX8800.5.VXSM.a > tstcon 1 1 35
test is in progress ..
Connection Id    Test Type    Direction    Result     Round Trip Delay
=============    =========    =========    =======    ================
01.0001.00035     OAM Lpbk     ingress     Success         96 microsec
```

The syntax of the **addcon** command for creating the master connection end on VISM cards is slightly different:

```
MGX8800.1.3.VISM8.a > addcon localVCI preference pvcType application PCR mastership
  [remoteConnId servicetype [scr] [mbs]]
```

NOTE If you are using an MGX82xx series chassis, you most likely build PVCs from the VISM to the PXM module, which in turn connects to another device with ATM interfaces such as a Cisco GSR. The *localVCI* is the VCI identifier of the PVC; the VPI is always 0 in the MGX82xx case and the card slot # in the MGX88xx case. The *preference* parameter enables you to have multiple PVCs for redundancy; the values are 1 for the primary PVC and 2 for the secondary PVC. In a PacketCable application the *pvcType* is always 1, indicating AAL5 ATM adaptation layer encapsulation. The *application* parameter is 3 for a signaling (MGCP/TGCP) PVC and 2 for a bearer or combined bearer/signaling PVC. The optional *remoteConnId* parameter applies only to the master side; it's the identifier for the slave side of the PVC. The remaining parameters define the ATM service such as constant or variable bit rate as that services associated parameters.

Step 5: Configure CAS Variants (Optional)

If you are using a VXSM or VISM card to connect to an E911 tandem switch using the CAS protocol, a number of configuration steps are required.

First, you must define the CAS variant. On a VXSM card a number of built-in variants are predefined, which you can display by using the **dspcasbuiltinvars** command:

```
MGX8880.5.VXSM.a > dspcasbuiltinvars
=======================================================
=======================================================
    Built-in CAS Variants (dspcasbuiltinvars)
=======================================================
 File Name:    fgd_os_e911
 File Name:    fgd_os_psap
 File Name:    mf_wink_start_incoming
```

```
File Name:     mf_wink_start_outgoing
File Name:     dtmf_wink_start_incoming
File Name:     dtmf_wink_start_outgoing
```

For E911 you want to use the **fgd_os_e911** variant. To add this variant, use the following command on the VXSM card:

```
MGX8800.5.VXSM.a > addcasvar CasVarIndex file_name [-source Source File]
```

The index is a value from 1 to 25 that uniquely identifies the CAS variant, and the *Source File* parameter is 1 if the file is internal and 2 if it is external. So the following command would add the internal E911 variant:

```
MGX.5.VXSM.a > addcasvar 1 fgd_os_e911 -source 1
```

Use the **dspcasvars** command to verify that the internal E911 variant is successfully added:

```
MGX8880.5.VXSM.a > dspcasvars
============================================================================
                              CAS Variants
============================================================================
VarIndx  Variant File Name  Source File  Variant State     Num of Associated
                                                           DS0 Group
=======  =================  ===========  ================  =================
   1     fgd_os_e911        internal     initSuccessfully          0
```

For a VISM card, the **addcasvar** command is also used, but its syntax is slightly different:

```
MGX8800.1.3.VISM8.a > addcasvar variant_name file_name [cas_var_source]
```

The variant name is a local name used to reference this CAS variant, the file name can be either a predefined CAS variant file or one loaded via TFTP. One predefined variant that might be of interest is *fgd_os_e911.o*, which would be used for emergency 911 operator services trunks.

Step 6: Enable TDM Lines

On a VXSM card each TDM line needs to be enabled using the **up line** command:

```
MGX8800.5.VXSM.a > upln bay.line
```

The *bay* number is 1 if the TDM line is connected to the top back card and 2 if the TDM line is on the bottom back card. The *line* number identifies the line on that card, and its value depends on the type of VXSM card; for an OC-3 the value is from 1 to 4, for a T3/E3 the value is from 1 to 3, and for a T1/E1 the value is from 1 to 24.

As soon as the line has been enabled, you can verify its status using the **display line** command:

```
MGX8800.5.VXSM.a > dspln 1.3
=====================================================
            Sonet Line Information
=====================================================
Sonet Line              :                1.3
Medium Type             :                sonet
Interface Name          :                sonet.5.1.3
Admin Status            :                up
Operational Status      :                up
```

```
Time Elapsed(seconds)  :                    693
Valid Intervals        :                    96
Invalid Intervals      :                    0
Line Type              :                    sonetSts3
Loopback Config        :                    noLoopback
Transmit Clock Source  :                    localTiming
Frame Scramble         :                    enabled
RDIV Type              :                    onebit
RDIP Type              :                    onebit
Section Transmit Trace :
 00000000000000000000000000000000000000000000000000000000000000000000000000000000
 000000000000000000000000000000000000000000000d0a
Section Expect Trace   :
 00000000000000000000000000000000000000000000000000000000000000000000000000000000
 000000000000000000000000000000000000000000000d0a
Section Trace Failure  :                    true
Section Received Trace :
 55555555555555555555555555555555555555555555555555555555555555555555555555555555
 555555555555555555555555555555555555555555555555
```

You should see an operational status of up. If you have problems, you can get more information using the **display line alarm** command:

```
MGX8800.5.VXSM.a > dsplnalm 1.3
=======================================================
           Sonet Line Alarm
=======================================================
Sonet Line       :          1.3
Section LOSs     :          0
Section LOFs     :          0
Line AISs        :          79
Line RFIs        :          0
Path AISs        :          0
Path RFIs        :          1
Section Stats    :          No Defect
Line Stats       :          No Defect
```

On a VISM card you enable a TDM line using the **add line** command:

```
MGX8800.1.3.VISM8.a > addln line_number
```

There are eight T1 or E1s on a VISM card. The line number here refers to the T1 or E1 number to add (1–8). This must be repeated for each T1/E1. This command must be done even if the physical connectivity is provided via the SRM card instead of the VISM back card. You can display the status of these lines with the **display lines** command:

```
MGX8800.1.3.VISM8.a > dsplns
  Line  Conn       Type     Status/Coding       Length       XmtClock Alarm
        Type                                                 Source
  ----  ----- ----------- ------ -------- ------------ -------- -----
  3.1   RJ-48 dsx1ESF        Ena/dsx1B8ZS 0-131 ft      LocalTim  No
  3.2   RJ-48 dsx1ESF        Dis/dsx1B8ZS 0-131 ft      LocalTim
  3.3   RJ-48 dsx1ESF        Dis/dsx1B8ZS 0-131 ft      LocalTim
  3.4   RJ-48 dsx1ESF        Dis/dsx1B8ZS 0-131 ft      LocalTim
  3.5   RJ-48 dsx1ESF        Dis/dsx1B8ZS 0-131 ft      LocalTim
  3.6   RJ-48 dsx1ESF        Dis/dsx1B8ZS 0-131 ft      LocalTim
  3.7   RJ-48 dsx1ESF        Ena/dsx1B8ZS 0-131 ft      LocalTim  No
  3.8   RJ-48 dsx1ESF        Dis/dsx1B8ZS 0-131 ft      LocalTim
  LineNumOfValidEntries: 8
```

You can display the detailed configuration of a line with the **dspln** *line#* command. And if you have problems, you can check the line alarm status with the **display alarms** command as follows:

```
MGX8800.1.3.VISM8.a > dspalms -ds1
   Line  AlarmState    StatisticalAlarmState
   ----  -----------   ---------------------
   3.1   No Alarms     No Statistical Alarms
   3.7   No Alarms     No Statistical Alarms
```

Step 7: Enable STS-1 Paths (VXSM OC-3 Cards Only)

An OC-3 optical carrier is comprised of three Synchronous Transport Signal–Level 1 (STS-1) frames. If you are using OC-3 VXSM cards, the individual STS-1 frames must be enabled as well.

```
MGX8800.5.VXSM.a > uppath -sts bay.line.path
```

The *path* number identifies the STS-1 frame on that particular OC-3; it has a value from 1 to 3. You can verify the status of the path as demonstrated here:

```
MGX8800.5.VXSM.a > dsppath -sts 1.3.2
=====================================================
          Sonet Path Information
=====================================================
Sonet Path                 :             1.3.2
Interface Name             :             sonetPath.5.1.3.OC-3.2
Admin Status               :             up
Operational Status         :             down
Path Payload               :             ds3
Tributary Mapping Type     :             asynchronous
Signalling Transport Mode  :             notApplicable
Tributary Grouping Type    :             notApplicable
Transmit Trace             :
 000000000000000000000000000000000000000000000000000000000000000000000000000000000
 00000000000000000000000000000000000000000000d0a
Expect Trace               :
 000000000000000000000000000000000000000000000000000000000000000000000000000000000
 00000000000000000000000000000000000000000000d0a
Trace Failure              :             true
Received Trace             :
 fffffffffffffffffffffffffffffffffffffffffffffffffffffffffffffffffffffffffffffffff
 fffffffffffffffffffffffffffffffffffffffffffffff
```

After the STS-1 frame has been enabled, you might need to change the configuration of its payload. Remember that an STS-1 frame can contain a single DS3 (default), or it can be made up of seven virtual tributary groups. A VTG can comprise four 1.5 Virtual Tributaries (VT 1.5), each capable of supporting a single T1, or three 2.0 Virtual Tributaries (VT 2.0), each capable of supporting a single E1. To change the format of an STS-1 frame, use the **payload** option in the **cnfpath -sts** command:

```
MGX8800.5.VXSM.a > cnfpath -sts
Err: Not enough arguments.
Syntax: cnfpath -sts <bay.line.path>
    [-payload <Path Payload>][-tm <Tributary Mapping Type>]
    [-tg <Tributary Grouping>][-txtrace <Transmit Trace>]
    [-exptrace <Expect Trace>]
```

```
--
        bay: 1(upper) line: 1 - 4 path: 1 - 3
  payload -- Path Payload: 3(ds3, NA for SDH) 4(vt15vc11) 5(vt2vc12)
  tm -- Tributary Mapping Type: 1(asyn) 2 (byteSync, NA for ds3)
  tg -- Tributary Grouping(NA for Sonet): 2(au3) 3(au4)
  txtrace -- Transmit Trace(ASCII string): size is 0-15(SDH) or 0-62(Sonet)
  exptrace -- Expect Trace(ASCII string): size is 0-15(SDH) or 0-62(Sonet)
MGX8800.5.VXSM.a > cnfpath -sts 1.3.2 -payload 4
```

In the examples in this chapter, the STS-1 frame is divided up into 1.5 virtual tributaries.

The next step is to enable the various DS1s in the STS-1 frame. You can specify these DS1s using either the DS3 format or the VTG format:

```
MGX8800.5.VXSM.a > uppath -ds1 bay.line.path.vtg.vt
```

The *vtg* number identifies the VT group; thus, it has a value from 1 to 7, and the *vt* number identifies the virtual tributary within that group. For the examples in this chapter, the command **uppath -ds1 1.3.2.1.1** was used.

At this point, you can verify the status of these paths using the **dsppath** command; again, modifiers are used for the various path types. The output for both the STS-1 and DS1 used in the examples is as follows:

```
MGX8800.5.VXSM.a > dsppath -sts 1.3.2
=======================================================
          Sonet Path Information
=======================================================
Sonet Path                 :           1.3.2
Interface Name             :           sonetPath.5.1.3.OC-3.2
Admin Status               :           up
Operational Status         :           up
Path Payload               :           vt15vc11
Tributary Mapping Type     :           asynchronous
Signalling Transport Mode :            notApplicable
Tributary Grouping Type    :           notApplicable
Transmit Trace             :

00000000000000000000000000000000000000000000000000000000000000000000000000000000
 0000000000000000000000

0000000000000000000000d0a
Expect Trace               :

00000000000000000000000000000000000000000000000000000000000000000000000000000000
 0000000000000000000000

0000000000000000000000d0a
Trace Failure              :           false
Received Trace             :

00000000000000000000000000000000000000000000000000000000000000000000000000000000
 0000000000000000000000

000000000000000000000000
MGX8800.5.VXSM.a > dsppath -ds1 1.3.2.1.1
=======================================================
               DS1 Path Information
=======================================================
DS1                      :      1.3.2.1.1
Interface Name           :      ds1VTpath.5.1.3.OC-3.2.1.1
```

```
Admin Status            :      up
Operational Status      :      up
Valid Intervals         :      96
Invalid Intervals       :      0
Time Elapsed(seconds)   :      83
Type                    :      dsx1ESF
Send Code               :      dsx1SendNoCode
Loopback Config         :      dsx1NoLoop
Signal Mode             :      none
Transmit Clock Source   :      localTiming
Loopback State          :      NoLpbk
Loopback Code Detection:       false
Trunk Condition         :      disable
Status                  :      NoAlarm
```

If you have problems, the **dsppathalm** command can be used to check alarm status.

Step 8: Configure Line/Path Parameters (Optional)

DS1 lines on a VXSM card default to no call signaling; if the line is used for CAS signaling, you need to modify this. If you are using the DS1 or DS3 variety, the **cnfln** command is used; if you have the OC-3 variety, the **cnfpath -ds1** command is used:

```
MGX8800.5.VXSM.a > cnfpath -ds1 bay.line.path.vtg.vt -signal 2
```

Signal mode 1 is no call signaling (SS7), and signal mode 2 is for CAS signaling.

For a VISM card the line defaults to ESF framing, B8ZS line coding, and local timing. If you need to change any of the parameters, you can do so with the **cnfln** *line_number* command.

The line also defaults to no call signaling (option 3), which is used for SS7 IMT trunks, but if the line is used for CAS signaling, you need to modify its signaling using the **cnflnsig** *line_number* **1** command.

The configuration of a line can be verified with the **dspln** *line_number* command.

Step 9: Bring the Lines into Service

As soon as each line is configured, it needs to be enabled using the **cnflnis** command:

```
MGX8800.5.VXSM.a > cnflnis line_number
```

Individual lines can be taken out of service using the **cnflnoos** command. The status of each line can be seen with the **dsplnstate** command, and if you have alarms, they can be viewed with the **dsplnalm** command.

Step 10: Define Voice Interfaces (VXSM Only)

A voice interface (VIF) is a group of configurable parameters that apply to a group of DS0 trunks within a DS1. The syntax for defining a VIF varies slightly, depending on the type of VXSM card. The syntax for an OC-3 card is as follows:

```
MGX8800.5.VXSM.a > addvif bay.line.path.vtg.vt DS0group index DS0BitMap service type
  BulkProvisionNumber
```

The *DS0groupindex* is a value from 0–23 for a T1 line (0–31 for an E1 line) and identifies the group of DS0s within that T1 or E1. The next parameter, *DS0BitMap*, specifies the individual DS0s in the group. Typically, all DS0s are in the same group. The service type should be *xgcp* for a PacketCable deployment.

```
MGX8800.5.VXSM.a > addvif 1.3.2.1.1 0 1-5 11
MGX8800.5.VXSM.a > dspvif 1.3.2.1.1 0
=========================================================
       Configuration of DS0 Group in DS1/E1
=========================================================
DS1/E1                   :                1.3.2.1.1
DS0 Group                :                0
Service Type             :                xgcp
DS0 Bit Map              :                1-5
```

In the chapter example, a VIF was defined for the first five DS0s on the DS1 enabled in the previous step.

If the VIF is for CAS signaling, you also need to associate the VIF with a CAS variant using the **cnfvifcas** command:

```
MGX8800.5.VXSM.a > cnfvifcas bay.line.path.vtg.vt DS0group index CasVariant
  CasProfile
```

Step 11: Add DS0 Endpoints to the Line (VISM Only)

To add DS0 endpoints to the line, use the following command:

```
MGX8800.1.3.VISM8.a > addendpts endpt_num ds1_num ds0_num endpts_num
```

The field *endpt_num* is the starting endpoint number (1–192 for a T1 card), the field *endpts_num* indicates the number of endpoints to add, and *ds0_num* is the DS0 or timeslot used for the first endpoint. Note: Endpoints can also be added individually using the **addendpt** command.

Step 12: Configure CAS Endpoints (VISM Only: Optional)

To configure CAS endpoints, use the following command:

```
MGX8800.1.3.VISM8.a > cnfcasendpt endpt_number casVariantName
```

If any of these endpoints are CAS endpoints, you need to configure them as such with the **cnfcasendpt** command.

Step 13: Configure the Gateway Domain Name

The gateway name is the endpoint domain name (DN) and needs to match with the configuration on the MGC for call signaling to work. On a VXSM card, the gateway name is set with the following command:

```
MGX8800.5.VXSM8.a > cnfgwdn domain_name
```

On a VISM card, the gateway name is set with the following command:

```
MGX8800.1.3.VISM8.a > cnfvismdn domain_name
```

In both cases, the FQDN can be up to 64 characters in length. On a VXSM card, the value can be displayed using the **dspdns** command, and on a VISM card the value can be displayed using the **dspvismdn** command. The VXSM configuration in the examples is as follows:

```
MGX8800.5.VXSM.a > dspdns
=====================================================
            All Domain Name
=====================================================
Index      Entity        Domain Name
=====      ==========    ====================
  1               mgc    mgcp-104CA104.cisco.com
 13           gateway    tgw01-vxsm-5.cisco.com
```

Step 14: Configure Gateway IP Addresses

For a VXSM card, an IP address is assigned to each ATM connection using the **addconip** command as shown:

```
MGX8800.5.VXSM.a > addconip Index PortNum VPI VCI IpAddr PrefixLength DefaultGwIp
```

The *Index* is a value from 1–16, the *PortNum* is always 1, the *VPI* and *VCI* values are for the connection this IP address is assigned to, and the *DefaultGwIP* is 1 if this IP address is to be used as a default gateway and 2 if it isn't.

This step would be repeated for every bearer and control connection. Here is the configuration on the VXSM card used in the call flow examples:

```
MGX8800.5.VXSM.a > dspconips
============================================================
                  Gateway IPs
============================================================
Index Interface VPI   VCI      IP Address    Prefix Length Default GW IP
===== ========= ===== ===== =============== ============= =============
 1        1         1    35 13.13.9.129          32        true
 2        1         1    41 13.13.9.130          32        false
```

For a VISM card, an IP address is assigned using the following command:

```
MGX8800.1.3.VISM8.a > cnfvismip vismIpAddr netMask [vismBearerIpAddr bearerNetMask]
```

NOTE If only one IP address is specified, this is used for both signaling packets and bearer packets. If two IP addresses are used, the first one is used for call signaling, and the second one is used for bearer packets. Chapter 16 describes the advantage of the latter configuration.

Step 15: Configure Domain Names and Domain Name Servers

For VISM cards, Domain Name Server and Media Gateway Controller (MGC) FQDNs are defined using the same command:

```
MGX8800.1.3.VISM8.a > adddn domain_num mgc_domain_name [Resolution_Type]
```

The *Resolution Type* describes how the IP addresses for the domain name are resolved. It has four values:

- 1 = Internal Only (Default). IP addresses must be statically defined on the VISM card.
- 2 = External Only. IP addresses are resolved via an external DNS server.
- 3 = Internal First. Local IP addresses are tried first; if they fail, resolved IP addresses are tried.
- 4 = External First. Resolved IP addresses are tried first; if they fail, local IP addresses are tried.

If the domain name (DN) is to be used as a DNS server, you need to configure this using the **cnfdnssrvr** *DN* command.

For a VXSM card, MGC entities and Domain Name Servers are defined using different commands. The syntax for defining a DNS server is as follows:

```
MGX8800.5.VXSM.a > adddnsdn Domain Name
```

MGC DNs are first defined by specifying a MGC domain name using the following command:

```
MGX8800.5.VXSM.a > addmgcdn MGC Index> Domain Name
```

The *MGC Index* is a value from 1–4, and the *Domain Name* is up to 64 characters in length. As was the case with VISM card configuration, the DN can be resolved internally by manually associating the DN with IP addresses or externally using a DNS server. To configure the resolution method, use the **cnfmgc** command:

```
MGX8800.5.VXSM.a > cnfmgc MGC Index Resolution
```

The *Resolution* value can be 1 or 2. A value of 1 indicates internal resolution, which implies IP address mapping for that DN must be configured using the **addmgcip** command, shown next. A value of 2 indicates external resolution, which implies a DNS server has been configured as shown in Step 10.

In addition to the **dspdns** command shown in Step 13, you can also use the **dspmgcs** command to see this configuration:

```
MGX8800.5.VXSM.a > dspmgcs
============================================================
                  All MGC Information
============================================================
MGC      Num Grp    Num IP     Domain Name
=====    =======    =======    ================
  1        1          2        mgcp-104CA104.cisco.com
```

NOTE If you define the domain name to use external resolution, you must have a DNS server defined. Unless your MGC is regularly changing IP addresses, it is recommended to just use internal resolution.

Step 16: Configure Domain Name IP Addresses

If you are using a DNS server, you must specify at least one IP address for it, and if you are using internal resolution for the MGC FQDN, you must specify at least one IP address for that as well. The same command is used for both purposes on a VISM card:

```
MGX8800.1.3.VISM8.a > adddnip Resolution_index domain_name IP_address preference
```

The *preference* is the resolution preference of the IP address where the lowest number is the highest preference.

As soon as the MGC FQDN, its resolution method and any manually entered IP addresses are configured, the DN needs to be identified as an MGC:

```
MGX8800.1.3.VISM8.a > addmgc mgcNumber DomainName
```

For a VXSM card, DNS server IP addresses are defined using the following command:

```
MGX8800.5.VXSM.a > adddnssrvrip Index IP Address
```

Up to three IP addresses can be defined; hence, the index is a value from 1–3.

MGC IP addresses are defined using the following command:

```
MGX8800.5.VXSM.a > addmgcip MGC Index MGC IP Index MGC IP Address Preference
```

Up to four IP address(es) can be associated with an MGC DN; therefore, the *MGC IP Index* parameter has a value from 1 to 4. Logically, the *Preference* parameter indicates the order of preference or priority of the specified IP address; it has a value from 1 to 8, and a lower value indicates a higher preference.

The following command shows the configuration for the VXSM card used in the examples:

```
MGX8800.5.VXSM.a > dspmgcips
======================================================
              All MGC IP Information
======================================================
 MGC       Index          IP Address        Type  Preference
=====      =====     =================     ====== =========
   1         1            13.13.5.6         ipv4  1
   1         2           13.13.5.38         ipv4  2
```

NOTE Because a single VXSM card can support such a large number of TDM endpoints (possibly four OC-3s worth), it can be configured for use by multiple call agent groups. In this case the VXSM card is divided into a number of virtual gateways (VGW). Up to 12 VGWs can exist on a single physical VXSM card. Each virtual gateway is configured to communicate with a single call agent group. A call agent group is the collection of MGC machines. This book assumes that each VXSM card is controlled by a single MGC group; therefore, only one VGW is defined.

Step 17: Add the MGCs to a Redundancy Group

Depending on the MGC implementation, you might have multiple MGCs controlling the gateway. In a typical BTS implementation you have only one MGC controlling the gateway. Even though you have a redundant pair of call agent machines with the BTS, the IP addresses and FQDN used for call signaling stay with the active machine. Regardless of the implementation, the individual MGC entries need to be associated together to form an MGC redundancy group even if the group consists of a single MGC entry.

On a VXSM card this is done using the following command:

```
MGX8800.5.VXSM.a > addmgcgrpmgc MGC Group Index MGC Index Preference TCP/UDP port
```

The *Preference* is a value from 1–4, where the higher the value the more preferred the MGC. The *TCP/UDP port* parameter is optional and indicates the port used to communicate with the MGC; if a value of 0 is used, the port specified by the protocol table is used. Use the **dspprotocolport** command to see these port values and the **cnfprotocolport** command to change them.

The following is the configuration used for the call flow examples:

```
MGX8800.5.VXSM.a > dspmgcgrpmgcs
=====================================================
                 All MGC on MGC Group
=====================================================
 Group        MGC          Prefe         TCP/UDP Port
=======      ======       =======       =============
    1           1            1               2727
```

On a VISM card, MGC redundancy groups are configured using the following command:

```
MGX8800.1.3.VISM8.a > addmgcgrpentry Red_Group_Num mgcRedGrpMgcNum mgcRedGrpPref
```

The preference is the preference of the MGC within the group; the lowest number is the highest preference.

Step 18: Configure a Signaling Protocol to this MGC Redundancy Group (VISM Only)

To configure a signaling protocol for an MGC redundancy group, enter the following command:

```
MGX8800.1.3.VISM8.a > addmgcprotocol Red_Group_Num protocol_number ]
```

The protocol number defines the signaling protocol used; 5 indicates MGCP 1.0, and 6 indicates the PacketCable TGCP 1.0 protocol variant of MGCP. Also a number of optional parameters are used with this command to tweak the way the signaling protocol behaves. Usually, the defaults are preferred, but they can be changed if need be; consult the VISM documentation for more details.

Remember with a VXSM card that the call signaling protocol is set on the PXM with the **setrev** command during the initial setup. You can verify the call signaling protocol used by a VXSM card using the following command:

```
MGX8800.5.VXSM.a > dspprotocolport
======================================================
            Media Gateway Protocol
======================================================
Protocol Index        :                  1
Protocol              :                  tgcp
Version               :                  1.0
MG Protocol port      :                  2427
MGC Protocol port     :                  2727
```

Step 19: Adjust xGCP Parameters (Optional)

For a VXSM card, a number of parameters associated with the xGCP call signaling protocol can be customized through the creation of a profile using the **addxgcpprof** command. As soon as a profile is created, there are several commands available to customize parameters as demonstrated here:

```
MGX8800.5.VXSM.a > ? cnfxgcpprof
    Available commands
    ------------------
    cnfxgcpprofdtmf            Configure DTMF Parameters in Profile
    cnfxgcpprofretry           Configure Retry Parameters in Profile
    cnfxgcpproftcot            Configure Continuity Tone Timers in Profile
    cnfxgcpproftd              Configure Disconnect Timers in Profile
    cnfxgcpproftdmap           Configure Digit-Map Timers in Profile
    cnfxgcpprofthist           Configure Transition Kept Timer in Profile
    cnfxgcpproftsmax           Configure Tsmax Timer in Profile
    cnfxgcpprofttone           Configure Tone Timers in Profile
```

Other commands exist on both the VXSM and VISM cards for adjusting call signaling parameters; use the**? cnfxgcp** command to see a full list of these cards.

Troubleshooting Tips

The last part of this chapter contains some troubleshooting tips that can help you in diagnosing and fixing problems relating to TGCP and the PSTN interconnect.

Troubleshooting the Cisco ITP

Just as was the case with the configuration, troubleshooting on a Cisco ITP can be separated into the IP Sigtran side and the TDM side.

As far as the Sigtran side goes, you want to verify that your ASPs and Application Servers are in service. Examples of the commands to do this are as follows:

```
itp1#sh cs7 asp
                                                          Effect Primary
ASP Name       AS Name        State      Type  Rmt Port Remote IP Addr  SCTP
------------   ------------   --------   ----  -------- --------------  ----------
PRIAIN         cox800A        active     SUA   12205    13.14.5.12      0x00980112
PRIAIN         BTS4-LNP       active     SUA   12205    13.14.5.12      0x00980112
PRIAIN         BTS4-800T      active     SUA   12205    13.14.5.12      0x00980112
PRIPTC         ACAR           active     SUA   12235    13.14.5.12      0x0098012F
PRIPTC         coxCNAM        active     SUA   12235    13.14.5.12      0x0098012F
SECAIN         cox800A        inactive   SUA   12205    13.14.5.13      0x00DB0072
SECAIN         BTS4-LNP       inactive   SUA   12205    13.14.5.13      0x00DB0072
SECAIN         BTS4-800T      inactive   SUA   12205    13.14.5.13      0x00DB0072
SECPTC         ACAR           inactive   SUA   12235    13.14.5.13      0x00580291
SECPTC         coxCNAM        inactive   SUA   12235    13.14.5.13      0x00580291
ISUPPRI        ISUP           active     M3UA  11146    13.14.5.12      0x00980161
ISUPSEC        ISUP           inactive   M3UA  11146    13.14.5.13      0x0058023A
itp1#sh cs7 as
                        Routing     Routing Key                                Cic   Cic
AS Name        State    Context     Dpc           Si    Opc              Ssn Min   Max
------------   ------   ---------   ------------  ----  ------------     --- ----- -----
ACAR           active   1605        1.218.21      sccp                   251
cox800A        active   1603        1.218.21      sccp                   248
coxCNAM        active   1604        1.218.21      sccp                   232
BTS4-LNP       active   1602        1.218.21      sccp                   247
BTS4-800T      active   1601        1.218.21      sccp                   254
ISUP           active   1611        1.218.21      isup
```

You can tell that the ASPs designated for the primary call agent are active and operational, whereas their standby counterparts are inactive. The Application Servers using these ASP pairs are also active. If you are having problems getting ASPs and Application Servers working, start with the lower-layer protocols and then work your way up. First, verify IP connectivity between the designated SG and MGC IP addresses. Next, verify whether the SCTP port numbers match up correctly. After this is done, your ASPs should come up; if they don't, you probably have an MGC misconfiguration. If your ASPs are okay but the Application Server does not become active, check parameters such as routing key, point code, and service indicator to make sure they are configured the same on both the SG and the MGC.

As for troubleshooting the TDM side of things, start out by verifying the T1 controller is up without alarms. Check to make sure framing, line coding, and clocking are properly configured. If this is good, verify the serial interface (DS0) is operational. Make sure the right timeslot is identified, it is configured with MTP2 encapsulation, and has the correct speed. If this is okay, verify your links and linksets are in-service:

```
itp1#sh cs7 linkset
lsn=stp1            apc=10.10.10     state=avail      avail/links=1/1
  SLC  Interface                   Service    PeerState        Inhib
  00   Serial4/0:0                 avail      ---------        -----
lsn=stp2            apc=10.10.11     state=avail      avail/links=1/1
  SLC  Interface                   Service    PeerState        Inhib
  00   Serial4/1:0                 avail      ---------        -----
lsn=c-link          apc=1.218.23     state=avail      avail/links=1/1
  SLC  Interface                   Service    PeerState        Inhib
  00   13.14.5.21 7000 7000        avail      InService        -----
```

At least one link in each linkset must be available for the linkset to be operational. If your SS7 links do not show available, the most likely cause is a mismatched SLC value. Another helpful command in diagnosing problems with bringing links up is **show cs7 mtp2 state**, which logically shows the status of the MTP2 layer. The **debug cs7 mtp2 lsc serial** *interface* command shows the LSSU messages exchanged across the link and can help in troubleshooting, as demonstrated here:

```
*Jun 28 18:06:05.738: %LINEPROTO-5-UPDOWN: Line protocol on Interface Serial4/0:0,
 changed state to down
*Jun 28 18:06:05.890: itu2LSC_SetEmergency(Serial4/0:0): MTP2LSC_OOS
*Jun 28 18:06:05.890: itu2LSC_Start(Serial4/0:0): MTP2LSC_OOS
*Jun 28 18:06:16.722: itu2LSC_Rcvd_SIOS(Serial4/0:0): MTP2LSC_INITIAL_ALIGNMENT
*Jun 28 18:06:17.106: itu2LSC_Rcvd_SIO(Serial4/0:0): MTP2LSC_INITIAL_ALIGNMENT
*Jun 28 18:06:17.106: itu2LSC_Rcvd_SIE(Serial4/0:0): MTP2LSC_INITIAL_ALIGNMENT
*Jun 28 18:06:17.106: itu2LSC_Rcvd_SIE(Serial4/0:0): MTP2LSC_INITIAL_ALIGNMENT
*Jun 28 18:06:17.166: itu2LSC_Rcvd_SIE(Serial4/0:0): MTP2LSC_INITIAL_ALIGNMENT
*Jun 28 18:06:17.226: itu2LSC_Rcvd_SIE(Serial4/0:0): MTP2LSC_INITIAL_ALIGNMENT
*Jun 28 18:06:17.458: itu2LinkEvent(Serial4/0:0): MTP1 AERM failure
*Jun 28 18:06:17.462: itu2LSC_Rcvd_SIE(Serial4/0:0): MTP2LSC_INITIAL_ALIGNMENT
*Jun 28 18:06:17.462: itu2LSC_Rcvd_SIE(Serial4/0:0): MTP2LSC_INITIAL_ALIGNMENT
*Jun 28 18:06:17.522: itu2LSC_Rcvd_SIE(Serial4/0:0): MTP2LSC_INITIAL_ALIGNMENT
*Jun 28 18:06:18.062: itu2LSC_AlignmentComplete(Serial4/0:0):
 MTP2LSC_INITIAL_ALIGNMENT
*Jun 28 18:06:18.062: %CS7MTP3-5-ADJRBEGN: Adjacent SP restart beginning for linkset
 stp1
*Jun 28 18:06:18.082: %CS7MTP3-5-LINKUPDOWN: Link 0 in linkset stp1 is available
*Jun 28 18:06:18.110: %CS7MTP3-5-ADJRENDS: Adjacent SP restart ending for linkset
 stp1
*Jun 28 18:06:18.110: %CS7MTP3-3-LINKSETSTATE: Linkset stp1 is available
*Jun 28 18:06:19.062: %LINEPROTO-5-UPDOWN: Line protocol on Interface Serial4/0:0,
 changed state to up
```

If you have multiple links in a linkset and want to stop traffic on one of the links to perform some kind of maintenance, you can gracefully take it out of service using the global command **cs7 inhibit**. If you have only one link operational in a linkset, you cannot inhibit it because it makes the linkset unusable. To restore service on an inhibited link, use the global command **cs7 uninhibit**.

To route SS7 traffic, the route sets also need to be operational. The following example shows how you can verify this:

```
itp1#sh cs7 route
Dynamic Routes 0 of 1000
Routing table = system Destinations = 9 Routes = 21
```

```
Destination            Prio Linkset Name        Route
---------------------- ---- ------------------- -------
0.0.1/24        acces    1  stp1                avail
                         1  stp2                avail
                         2  c-link              avail
1.218.23/24     acces    1  c-link              avail
10.5.1/24       acces    1  stp1                avail
                         1  stp2                avail
                         2  c-link              avail
10.10.7/24      acces    1  stp1                avail
                         1  stp2                avail
                         2  c-link              avail
10.10.10/24     acces    1  stp1                avail
10.10.11/24     acces    1  stp2                avail
30.1.1/24       acces    1  stp1                avail
                         1  stp2                avail
                         2  c-link              avail
100.60.44/24    acces    1  stp1                avail
                         1  stp2                avail
                         2  c-link              avail
101.101.101/24  INACC    1  stp1                UNAVAIL
                         1  stp2                UNAVAIL
                         2  c-link              UNAVAIL
Routing table = XUA
Destination          Type
-------------------- --------
1.218.21/24     acces AS
```

If your SS7 routes are not accessible, likely causes include problems on the far end and those particular destinations not being permitted. Use the following command to check SS7 route connectivity:

```
ping cs7 [flags] destination
  -duration                          Test duration (default 1 second)
  -ni                                Network Indicator (default 2)
  -rate                              Test message rate (default 1 pps)
  -size                              Test message size (default 40 bytes)
  -sls                               Signaling Link Selector (default 0)
  SS7 name or network.cluster.member Destination CS7  Point Code
  stop                               stop a running test
```

Finally, you can use a few different commands to verify ITP operation is dependent on the implementation type. For example, you can use the **show cs7 group state** command to verify the status of the ITP group implementation and use the **show cs7 mated-sg** command to verify the status of the SG mated pair implementation. The following shows examples of both of these commands:

```
itp1#sh cs7 mated-sg
                                                       Effect Primary
Mate Name     State            Passive  Remote Port Remote IP Addr   SCTP
------------  --------------   -------  ----------- ---------------  ----------
itp2          active           no/alt   9101        13.14.5.21       0x00DB00BA
itp11#sh cs7 group state
ITP Group state information:
  Redundancy Facility state:
    RF State:                ACTIVE
    RF Peer State            STANDBY HOT
    Inter-device transport:  Up
  Group role information:
    Current State:    Manager
    Previous State:   ManagerNegotiating
```

```
    Latest event:      NEGO_MANAGER
    Peer present:      TRUE
    Peer name:         sgw02
    Peer comm:         UP
  Group timers:
    Peer wait timeout:    60 seconds
    Active timer:         none
  Checkpointing state:
    Last seq # sent:           1183
    Last seq # rcvd:           0
    Congested:                 FALSE
    Current send queue depth: 0
```

Troubleshooting the Cisco BTS 10200

Much of the BTS troubleshooting covered in the previous chapter applies here as well. Looking at log files, alarms and events, and measurements can provide helpful information. Monitoring the transaction queue and auditing key tables can help diagnose potential database problems.

As far as troubleshooting the Sigtran connection to the SG goes, a similar procedure to the one described on the SG should be followed. First, verify IP connectivity. Then, verify whether SCTP ports are properly configured. The BTS command **status sctp-assoc id=***sctp assoc name* covered earlier can be used and should result in an operational state of OPER STATE -> SCTP-ASSOC in service. If this is not the case, you most likely have a configuration problem or mismatch.

You can verify end-to-end SS7 connectivity using the BTS command **status dpc id=***dpc name*. This should yield in an operational state of OPER STATE -> DPC available. If it doesn't, there could be a far end problem or an issue with the permitted routes on the SG.

Procedures for troubleshooting the MG are very similar to those for troubleshooting the MTA described in Chapter 7. So refer to that chapter for issues relating to IP connectivity, MGCP port numbers, and MGCP endpoint name problems. Remember that you can get a pretty good idea of the problem based on the MGCP response code returned by the MG. Again, like the MTA a common problem is the MG configured TSAP-ADDR FQDN not DNS resolvable.

As mentioned in the BTS configuration section, the BTS command **status trunk-termination tgn-id=***trunk group number*; **cic=***cic | cic range* should indicate ADMIN_INS TERM_ACTIVE_IDLE   ACTV   IDLE   NON_FAULTY. If it doesn't indicate TERM_ACTIVE_IDLE, you probably have a TGCP problem; if it is not IDLE and NON_FAULTY, you probably have a signaling (SS7 or CAS) problem. Refer to the BTS documentation for more information on the output of this command.

Another good way to verify trunks is using the BTS diagnostic commands:

```
CLI>diag cas-trunk-termination tgn-id=9401; cic=121
===
(1)  CAS MGCP Connectivity Test
(2)  CAS Termination Connection Test
```

```
(3)  ALL
Reply : Success: Diagnostic CAS Menu.
CLI>diag ss7-trunk-termination tgn-id=1234; cic=19
===
(1)  SS7 MGCP Connectivity Test
(2)  SS7 Termination Connection Test
(3)  SS7 COT Test
(4)  SS7 CQM Test
(5)  SS7 CVT Test
(6)  SS7 CIC Audit
(7)  ALL
Reply : Success: Diagnostic SS7 Menu.
```

Remember, you learned about performing SS7 COT tests (SS7 test #3) earlier in this chapter. To execute these commands, you must first put the individual trunk into maintenance state with the following BTS command:

```
control trunk-termination tgn-id=trunk group number; cic=cic number; mode=forced;
 target-state=maint
```

Tests 1 and 2 in both cases verify the MGCP/TGCP connectivity between the MGC and MG. SS7 test number 3 verifies the voice path by doing a continuity test. The other SS7 tests can be used to verify the remote switch's configuration and view of the CIC states.

OFFNET call routing can also be verified using the BTS Translation Verification Tool (TVT) introduced in the previous chapter. For example, consider the translations for the ONNET-to-OFFNET call example:

```
CLI>translate line calling-dn=3215012020; called-dn=19193927911
TRANSLATE ON CALL AGENT CA104 IS... ->
TABLEINFO ->
TABLE: SUBSCRIBER
ID=sub1@jriddel_arris1;CATEGORY=INDIVIDUAL;NAME=Jeff;STATUS=ACTIVE;DN1=3215012020;
 PRIVACY=NONE;RING_TYPE_DN1=1;TERM_ID=aaln/
 1;MGW_ID=jriddel_arris1;PIC1=NONE;PIC2=NONE;PIC3=NONE;GRP=N;USAGE_SENS=Y;SUB_
 PROFILE_ID=PacketCable_subs;TERM_TYPE=TERM;IMMEDIATE_RELEASE=N;TERMINATING_
 IMMEDIATE_REL=N;SEND_BDN_AS_CPN=N;SEND_BDN_FOR_EMG=N;PORTED_IN=N;BILLING_
 TYPE=NONE;VMWI=Y;SDT_MWI=Y;
TABLE: SUBSCRIBER_PROFILE
ID=PacketCable_subs;DIAL_PLAN_ID=INCOMING;LOCAL_PFX1_OPT=NR;TOLL_PFX1_OPT=RQ;POP_ID=P
 acketCable_PoP;OLI=0;EA_USE_PIC1=Y;QOS_ID=G711u_20ms;INTERLATA_PFX1_OPT=RQ;
TABLE: DIAL_PLAN_PROFILE
ID=INCOMING;DESCRIPTION=dialing plan profile for incoming
 calls;NANP_DIAL_PLAN=Y;OVERDECADIC_DIGITS_SUPP=N;NOA_BASED_ROUTING=N;
TABLE: DIAL_PLAN
ID=INCOMING;DIGIT_STRING=919;DEST_ID=OffNet;SPLIT_NPA=NONE;DEL_DIGITS=0;MIN_DIGITS
 =10;MAX_DIGITS=10;NOA=NATIONAL;
TABLE: DESTINATION
DEST_ID=OffNet;CALL_TYPE=TOLL;ROUTE_TYPE=RID;ZERO_PLUS=N;INTRA_STATE=Y;ROUTE_ID=
 OffNet;GAP_ROUTING=N;CLDPTY_CTRL_REL_ALWD=N;CALL_SUBTYPE=NONE;ACQ_LNP_QUERY=NA;
TABLE:  ROUTE
ID=OffNet;TGN1_ID=1234;DEL_DIGITS1=0;DEL_DIGITS2=0;DEL_DIGITS3=0;DEL_DIGITS4=0;DEL
 _DIGITS5=0;DEL_DIGITS6=0;DEL_DIGITS7=0;DEL_DIGITS8=0;DEL_DIGITS9=0;DEL_DIGITS10=0
 ;TG_SELECTION=SEQ;NEXT_ACTION=NONE;
TABLE: TRUNK_GRP
ID=1234;CALL_AGENT_ID=CA104;TG_TYPE=SS7;NUM_OF_TRUNKS=5;DPC=10-10-
 7;TG_PROFILE_ID=SS7_PF_ATT;STATUS=INS;DIRECTION=BOTH;SEL_POLICY=ASC;GLARE=SLAVE;
 ALT_ROUTE_ON_CONG=N;SIGNAL_PORTED_NUMBER=N;DIAL_PLAN_ID=INCOMING;DEL_DIGITS=0;
 QOS_ID=G711u_20ms;TRAFFIC_TYPE=LOCAL;ANI_BASED_ROUTING=N;CALL_CTRL_ROUTE_ID=
 DPC_10-10-
 7;MGCP_PKG_TYPE=IT;ANI_SCREENING=N;SEND_RDN_AS_CPN=N;STATUS_MONITORING=N;
 SEND_EARLY_BKWD_MSG=N;EARLY_BKWD_MSG_TMR=5;SCRIPT_SUPP=N;VOICE_LAYER1_USERINFO=
 AUTO;VOICE_INFO_TRANSFER_CAP=AUTO;ACCESS_TYPE=COMBINED;POI=INTER_ENDOFFICE;
 PERFORM_LNP_QUERY=N;
Reply : Success:
```

As you can see, the output shows all the tables involved in routing this call. First, Oscar's entry in the subscriber table is referenced to get his subscriber profile name. Then this subscriber profile is referenced to get the dial plan name. At this point, the entry in the dial plan for the dialed digits (called #) is identified. This entry points to the OFFNET local destination, which is of route type RID with route ID OFFNET. Recall that route type RID means that a lookup in the route table on the index specified by the route ID is used to route the call. This route points to the SS7 trunk group which connects to the PSTN.

Similarly, you can use this tool to verify the OFFNET-to-ONNET example. Notice in this case the originating trunk group is specified instead of the originating subscriber's phone number:

```
CLI>translate trunk tgn-id=1234; called_dn=3215012020
TRANSLATE ON CALL AGENT CA104 IS... ->
TABLEINFO ->
TABLE: TRUNK_GRP
ID=1234;CALL_AGENT_ID=CA104;TG_TYPE=SS7;NUM_OF_TRUNKS=5;DPC=10-10-
 7;TG_PROFILE_ID=SS7_PF_ATT;STATUS=INS;DIRECTION=BOTH;SEL_POLICY=ASC;GLARE=SLAVE;
 ALT_ROUTE_ON_CONG=N;SIGNAL_PORTED_NUMBER=N;DIAL_PLAN_ID=INCOMING;DEL_DIGITS=0;
 QOS_ID=G711u_20ms;TRAFFIC_TYPE=LOCAL;ANI_BASED_ROUTING=N;CALL_CTRL_ROUTE_ID=DPC_
 10-107;MGCP_PKG_TYPE=IT;ANI_SCREENING=N;SEND_RDN_AS_CPN=N;STATUS_MONITORING=N;
 SEND_EARLY_BKWD_MSG=N;EARLY_BKWD_MSG_TMR=5;SCRIPT_SUPP=N;VOICE_LAYER1_USERINFO=
 AUTO;VOICE_INFO_TRANSFER_CAP=AUTO;ACCESS_TYPE=COMBINED;POI=INTER_ENDOFFICE;
 PERFORM_LNP_QUERY=N;
TABLE: DIAL_PLAN_PROFILE
ID=INCOMING;DESCRIPTION=dialing plan profile for incoming
 calls;NANP_DIAL_PLAN=Y;OVERDECADIC_DIGITS_SUPP=N;NOA_BASED_ROUTING=N;
TABLE: DIAL_PLAN
ID=INCOMING;DIGIT_STRING=321;DEST_ID=ONNET_LC;SPLIT_NPA=NONE;DEL_DIGITS=0;MIN_DIGI
 TS=10;MAX_DIGITS=10;NOA=NATIONAL;
TABLE: DESTINATION
DEST_ID=ONNET_LC;CALL_TYPE=LOCAL;ROUTE_TYPE=SUB;ZERO_PLUS=Y;INTRA_STATE=Y;
 GAP_ROUTING=N;CLDPTY_CTRL_REL_ALWD=N;CALL_SUBTYPE=NONE;ACQ_LNP_QUERY=NA;
TABLE: OFFICE_CODE
DIGIT_STRING=321501;OFFICE_CODE_INDEX=1;DID=N;CALL_AGENT_ID=CA104;DIALABLE=Y;NDC=3
 21;EC=501;DN_GROUP=xxxx;EC_DIGIT_STRING=321501;
TABLE: DN2SUBSCRIBER
OFFICE_CODE_INDEX=1;DN=2020;STATUS=ASSIGNED;RING_TYPE=1;LNP_TRIGGER=N;NP_RESERVED=
 N;SUB_ID=sub1@jriddel_arris1;LAST_CHANGED=2006-05-04
 11:16:14.0;VIRTUAL_DN=N;PORTED_IN=N;
TABLE: SUBSCRIBER
ID=sub1@jriddel_arris1;CATEGORY=INDIVIDUAL;NAME=Jeff;STATUS=ACTIVE;DN1=3215012020;
 PRIVACY=NONE;RING_TYPE_DN1=1;TERM_ID=aaln/
 1;MGW_ID=jriddel_arris1;PIC1=NONE;PIC2=NONE;PIC3=NONE;GRP=N;USAGE_SENS=Y;
 SUB_PROFILE_ID=PacketCable_subs;TERM_TYPE=TERM;IMMEDIATE_RELEASE=N;TERMINATING_
 IMMEDIATE_REL=N;SEND_BDN_AS_CPN=N;SEND_BDN_FOR_EMG=N;PORTED_IN=N;BILLING_
 TYPE=NONE;VMWI=Y;SDT_MWI=Y;
Reply : Success:
```

The translations in this case are similar to the ONNET-to-ONNET translations in the previous chapter except for the way the dial plan is determined. In this case, it is identified from the dial plan profile associated with the trunk group.

Again, examining billing records is another good place for troubleshooting. The following are sample billing records for an ONNET-to-OFFNET toll call followed by an OFFNET-to-ONNET call:

```
CLI>report billing-record tail=2
..
```

```
CALLTYPE=TOLL
SIGSTARTTIME=2006-06-28 18:16:00
SIGSTOPTIME=2006-06-28 18:16:16
ICSTARTTIME=2006-06-28 18:16:00
ICSTOPTIME=2006-06-28 18:16:16
CALLCONNECTTIME=2006-06-28 18:16:06
CALLANSWERTIME=2006-06-28 18:16:06
CALLDISCONNECTTIME=2006-06-28 18:16:16
CALLELAPSEDTIME=00:00:10
INTERCONNECTELAPSEDTIME=00:00:16
ORIGNUMBER=3215012020
TERMNUMBER=9193927911
CHARGENUMBER=3215012020
DIALEDDIGITS=19193927911
OFFHOOKINDICATOR=1
SHORTOFFHOOKINDICATOR=0
CALLTERMINATIONCAUSE=NORMAL_CALL_CLEARING
OPERATORACTION=0
ORIGSIGNALINGTYPE=0
TERMSIGNALINGTYPE=1
ORIGTRUNKGROUPNUMBER=0
TERMTRUNKGROUPNUMBER=1234
OUTGOINGTRUNKNUMBER=0
ORIGCIRCUITID=0
TERMCIRCUITID=19
PICSOURCE=2
ICINCIND=1
ICINCEVENTSTATUSIND=20
ICINCRTIND=0
ORIGQOSTIME=2006-06-28 18:16:16
ORIGQOSPACKETSSENT=496
ORIGQOSPACKETSRECD=780
ORIGQOSOCTETSSENT=79360
ORIGQOSOCTETSRECD=124800
ORIGQOSPACKETSLOST=0
ORIGQOSJITTER=0
ORIGQOSAVGLATENCY=0
TERMQOSTIME=2006-06-28 18:16:16
TERMQOSPACKETSSENT=780
TERMQOSPACKETSRECD=494
TERMQOSOCTETSSENT=124800
TERMQOSOCTETSRECD=79040
TERMQOSPACKETSLOST=0
TERMQOSJITTER=0
TERMQOSAVGLATENCY=20
PACKETIZATIONTIME=20
SILENCESUPPRESSION=0
ECHOCANCELLATION=0
CODECTYPE=PCMU
CONNECTIONTYPE=IP
OPERATORINVOLVED=0
CASUALCALL=0
INTERSTATEINDICATOR=0
OVERALLCORRELATIONID=CA104744
TIMERINDICATOR=0
RECORDTYPE=NORMAL RECORD
CALLAGENTID=CA104
ORIGPOPTIMEZONE=GMT
ORIGTYPE=ON NET
TERMTYPE=OFF NET
NASERRORCODE=0
NASDLCXREASON=0
FAXINDICATOR=NOT A FAX
FAXPAGESSENT=0
```

```
FAXPAGESRECEIVED=0
ORIGPOPID=PacketCable_PoP
TERMPOPTIMEZONE=GMT
DIALPLANID=INCOMING
CALLINGPARTYCATEGORY=Ordinary Subscriber
CALLEDPARTYINDICATOR=No Indication
CALLEDPARTYPORTEDIN=No
CALLINGPARTYPORTEDIN=No
BILLINGRATEINDICATOR=None
ORIGENDPOINTADDR=x1-6-00-00-ca-dd-a7-e4.cc.cisco.com
TERMENDPOINTADDR=tgw01-vxsm-5.cisco.com
ORIGCMTSID=PC-CMTS.cc.cisco.com
SENSORID=000000
ORIGPRIVACYINDICATOR=NONE
CALLTYPE=LOCAL
SIGSTARTTIME=2006-06-28 19:20:41
SIGSTOPTIME=2006-06-28 19:20:48
CALLCONNECTTIME=2006-06-28 19:20:44
CALLANSWERTIME=2006-06-28 19:20:44
CALLDISCONNECTTIME=2006-06-28 19:20:48
CALLELAPSEDTIME=00:00:04
INTERCONNECTELAPSEDTIME=00:00:00
ORIGNUMBER=9193927911
TERMNUMBER=3215012020
CHARGENUMBER=9193927911
DIALEDDIGITS=3215012020
OFFHOOKINDICATOR=1
SHORTOFFHOOKINDICATOR=0
CALLTERMINATIONCAUSE=NORMAL_CALL_CLEARING
OPERATORACTION=0
ORIGSIGNALINGTYPE=1
TERMSIGNALINGTYPE=0
ORIGTRUNKGROUPNUMBER=1234
TERMTRUNKGROUPNUMBER=0
OUTGOINGTRUNKNUMBER=0
ORIGCIRCUITID=19
TERMCIRCUITID=0
ORIGQOSTIME=2006-06-28 19:20:49
ORIGQOSPACKETSSENT=200
ORIGQOSPACKETSRECD=199
ORIGQOSOCTETSSENT=32000
ORIGQOSOCTETSRECD=31840
ORIGQOSPACKETSLOST=0
ORIGQOSJITTER=0
ORIGQOSAVGLATENCY=0
TERMQOSTIME=2006-06-28 19:20:49
TERMQOSPACKETSSENT=201
TERMQOSPACKETSRECD=201
TERMQOSOCTETSSENT=32160
TERMQOSOCTETSRECD=32160
TERMQOSPACKETSLOST=0
TERMQOSJITTER=0
TERMQOSAVGLATENCY=0
PACKETIZATIONTIME=0
SILENCESUPPRESSION=0
ECHOCANCELLATION=1
CODECTYPE=PCMU
CONNECTIONTYPE=IP
OPERATORINVOLVED=0
CASUALCALL=0
INTERSTATEINDICATOR=0
OVERALLCORRELATIONID=CA104745
TIMERINDICATOR=0
RECORDTYPE=NORMAL RECORD
```

```
CALLAGENTID=CA104
ORIGPOPTIMEZONE=GMT
ORIGTYPE=OFF NET
TERMTYPE=ON NET
NASERRORCODE=0
NASDLCXREASON=0
FAXINDICATOR=NOT A FAX
FAXPAGESSENT=0
FAXPAGESRECEIVED=0
TERMPOPID=PacketCable_PoP
TERMPOPTIMEZONE=GMT
DIALPLANID=INCOMING
CALLINGPARTYCATEGORY=Ordinary Subscriber
CALLEDPARTYINDICATOR=No Indication
CALLEDPARTYPORTEDIN=No
CALLINGPARTYPORTEDIN=No
BILLINGRATEINDICATOR=None
ORIGENDPOINTADDR=tgw01-vxsm-5.cisco.com
TERMENDPOINTADDR=x1-6-00-00-ca-dd-a7-e4.cc.cisco.com
TERMCMTSID=PC-CMTS.cc.cisco.com
ORIGINTERNATIONALINDICATOR=NO
ORIGPRIVACYINDICATOR=NONE
```

Troubleshooting the Cisco VXSM and VISM Media Gateways

From the active PXM card you can look at the log messages on a per-slot basis using the **dsplog -sl** *Slot#* command. You can also look at the alarms on a per-slot basis using the **dspcdalm -sl** *Slot#* command.

As for troubleshooting commands on the VXSM and VISM cards, the best way to start troubleshooting is from the bottom up. Start with verifying the status of the card itself using the **dspcd** command:

```
MGX8800.5.VXSM.a > dspcd
                     Front Card           Upper Card           Lower Card
                     ----------           ----------          ------------
Card Type:           VXSM_155             VXSM_SMFIR_155       ---
State:               Active               Active               Empty
Serial Number:       SAD082007L8          SAD09430642          ---
Boot FW Rev:         5.2(10.205)P1        ---                  ---
DCAP Boot FW         5                    ---                  ---
SW Rev: swtools      5.52(10.205)P1       ---                  ---
800-level Rev:       B0                   A0                   ---
Orderable Part#:     800-15121-06         800-21428-07         ---
PCA Part#:           73-6604-6            73-8062-7            ---
CLEI Code:           BA6A790FAB           BAUIAD7AAA           ---
Line FRU Type            Vendor Name      Part #
---- ---------------- ---------------- ----------------
1.1  NOT INSTALLED       OCP              ---
1.2  SMFIR_155           OCP              TRP-03L3I1BCS
1.3  SMFIR_155           OCP              TRP-03L3I1BCS
1.4  NOT INSTALLED       OCP              ---
Reset Reason:        Reset from PXM
```

From this output, you can verify that the back cards are properly identified.

Remember, the operation of these cards relies on the underlying ATM PVCs used to connect the card with the rest of the network. The **dspcon** and **tstcon** commands covered in the configuration section can be used to verify these PVCs.

Calls can be rejected because of not enough ATM bandwidth being available. Remember, when the PVCs were defined, information such as peak cell rate and maximum burst size were defined. To verify whether calls are failing because of connection admission control (CAC) use the **dspconcacs** command as demonstrated here:

```
MGX8800.5.VXSM.a > dspconcacs
==============================================================================
                               CAC statistics
==============================================================================
if  VPI  VCI   CAC    total     total   utilized     VAD        CAC        CAC
Num            master calls   cell rate cell rate   calls    succeeded    failed
=== === ===== ====== ======= ========= ========= ========= ========= =========
 1  1   35    master n/a     n/a        n/a        n/a        n/a        n/a
 1  1   41    master 1       600000     273        0          33         0
```

If this is okay, verify the IP and MGCP/TGCP connectivity. Logically, IP connectivity can be verified with ICMP pings. A common mistake is not enabling the gateway itself with the **cnfgwis** command. You can obtain the current status, which should be **inService**, as demonstrated in the following example:

```
MGX8800.5.VXSM.a > dspgwstate
=====================================================
            XGCP Gateway State
=====================================================
Gateway Id       :                      1
Service State    :                 inService
```

For a VISM card, commands such as the Display MGC groups and Display DN IPs also tell you the status of these entities, as demonstrated here:

```
MGX8800.1.3.VISM8.a > dspmgcgrps
mgcRedgrpNum     mgcNum     Preference     State
------------     ------     ----------     -----
     1             1             1       Active
MGX8800.1.3.VISM8.a > dspdnallips
 ResolutionIndex     DomainName                       IP            State    Pref
 ---------------   ------------                  ------------   ---------  ----
       1          mgcp-104CA104.cisco.com         13.13.5.6      Active      1
       2          mgcp-104CA104.cisco.com         13.13.5.38     Inactive    2
```

You can display call signaling protocol statistics on either a VXSM or VISM card using the following command:

```
MGX8800.5.VXSM.a > dspxgcpdetailcnts
MGCP Messages from Call Agent to GW:
         ACK total received             : 144
         ACK total retransmitted        : 0
         NACK total received            : 0
         NACK total retransmitted       : 0
         CRCX total received    *       : 35
         CRCX total successful *        : 33
         CRCX total failed      *       : 2
         CRCX total prov. successful    : 0
         CRCX total retransmissions     : 0
         MDCX total received    *       : 29
```

```
          MDCX total successful *        : 29
          MDCX total failed     *        : 0
          MDCX total prov. successful    : 0
          MDCX total retransmissions     : 0
          DLCX total received   *        : 32
          DLCX total successful *        : 32
          DLCX total failed     *        : 0
          DLCX total retransmissions     : 0
          RQNT total received   *        : 50
          RQNT total successful *        : 50
          RQNT total failed     *        : 0
          RQNT total retransmissions     : 0
          AUCX total received   *        : 0
          AUCX total successful *        : 0
          AUCX total failed     *        : 0
          AUCX total retransmissions     : 0
          AUEP total received   *        : 63931
          AUEP total successful *        : 13318
          AUEP total failed     *        : 50613
          AUEP total retransmissions     : 0
 MGCP Messages from GW to Call Agent:
          ACK (100) total transmissions  : 0
          ACK (200/250) total transmissions       : 13462
          ACK total retransmissions               : 143
          NACK total transmissions                : 50615
          NACK total retransmissions              : 0
          RSIP total transmitted                  : 139
          RSIP total successful transmissions     : 139
          RSIP total failed transmsissions        : 0
          RSIP total retransmissions              : 0
          RSIP total successful retransmissions   : 0
          RSIP total failed retransmissions       : 0
          DLCX total transmissions                : 0
          DLCX total successful transmissions     : 0
          DLCX total failed transmissions         : 0
          DLCX total retransmissions              : 0
          DLCX total successful retransmissions   : 0
          DLCX total failed retransmissions       : 0
          NTFY total transmissions                : 5
          NTFY total successful transmissions     : 5
          NTFY total failed transmissions         : 0
          NTFY total retransmissions              : 0
          NTFY total successful retransmissions   : 0
          NTFY total failed retransmissions       : 0
```

The verification of the operational status of TDM lines and paths was covered in the configuration section; please refer there for assistance.

On a VXSM card, you can display the operational status of endpoints using the following command:

```
MGX8800.5.VXSM.a > dspxgcpendpts DS/S-0/DS1-196/*
Total number of matching end-point(s) : 5
====================================================================
                    Display List of End-Point(s)
====================================================================
Endpoint              Admin State  Oper State  Accept Call  Connection
====================  ===========  ==========  ===========  ==========
DS/S-0/DS1-196/1      inService    up          Yes          Yes
DS/S-0/DS1-196/2      inService    up          Yes          No
DS/S-0/DS1-196/3      inService    up          Yes          No
DS/S-0/DS1-196/4      inService    up          Yes          No
DS/S-0/DS1-196/5      inService    up          Yes          No
```

Notice five endpoints are shown that are all operational and able to accept calls; the first endpoint actually has an active call on it.

You can find a lot of information for active calls using the **dspxgcpendptcons** and **dspxgcpendptcon** commands, as demonstrated in the following example from a VXSM card:

```
MGX8800.5.VXSM.a > dspxgcpendptcons ds/s-0/ds1-196/*
====================================================================
          Display All Connections on the given End-Point(s)
====================================================================
Index  Endpoint               Call ID / Conn ID     Conn Mode  Call Mode
=====  ====================   ====================  =========  =========
1      DS/S-0/DS1-196/1       352 / 22              sendrecv   Voice
Total number of regular connections for the given end-point(s) : 1
MGX8800.5.VXSM.a > dspxgcpendptcon ds/s-0/ds1-196/1
 Endpoint Name                    : DS/S-0/DS1-196/1
 DSX1 / DS0                       : 196 / 1
 Connection Status                : Connection exists
========= Connection 1 Information ===========
------------- Overall Information ------------
 Protocol version                 : MGCP 1.0
 Call ID                          : 352
 Connection ID                    : 22
 Connection Mode                  : sendrecv
 Connection Type                  : RTP
 Network Type                     : IN
 Call Mode                        : Voice
 MGCPApp FSM state                : CALL_ACTIVE
 Connection is UP for             : 0:11:59 (hr:min:sec)
 - NOTE: The above connection time is not preserved across switchovers;
         it is independently generated on each card.
----------------------------------------------
------- Local Connection Options (LCO) -------
 Last Specified LCO Codec         : PCMU;PCMA;G.726-32
 Echo Cancellation                : on
 Silence Suppression              : off
 Packet Period                    : 20 (ms)
 Type of Service (TOS)            : 160
 RSVP (bitmap)                    : 0x1
 Network Type                     : IN
 NAS Bearer Type                  : DIGITAL (0)
 NAS Bearer Capability            : 0
 NAS CDN                          :
 NAS CGN                          :
----------------------------------------------
---- Telephony Leg End-point Information -----
 Quarantine Mode                  : process, step
 Quarantine State                 : Process
 Request ID (X:)                  : 21900000014
 Persistent Event List            :
 Admin State                      : inService
 Oper State                       : up
 Accept Call                      : Yes
----------------------------------------------
-------- Peer Packet Leg Information ---------
 Connection Mode                : sendrecv
 Codec                          : PCMU
 Packetization Period           : 20 (ms)
 Local Bearer IP / Port         : 13.13.9.130 / 16448
 Remote Bearer IP / Port        : 24.34.240.241 / 56654
----------------------------------------------
```

```
--- Peer Packet Leg End-point Information ----
Bearer Security Information
------------------------------
----------------------------------------------
==============================================
============= Bearer Information =============
     Connection Mode      : 'Send-recv'
     Transmit Codec       : 'G.711 U'
     VBD Upspeed Codec    : 'G.711 U'
     TTY Upspeed Codec    : 'G.711 U'
     Packetization Period : '20 ms'
     VAD Mode             : 'Disabled'
     VAD Threshold        : -38
     VAD Hangover Time    : 250 ms
     ECAN                 : Enabled
     ECAN Tail Length     : 128 ms
     Tx Gain              : 0
     Rx Gain              : 0
     Jitter Mode          : 'Fixed'
     Jitter Min. Delay    : 0 ms
     Jitter Max. Delay    : 40 ms
     Jitter Nom. Delay    : 2 ms
     Detect Digit Type    : 'DTMF'
     Digit Relay Method   : 'Send as Voice'
     RTP Encapsulation Parameters ...
     SSRC                 : 31852930 (0x1e60982)
     RTCP                 : Enabled
     NSE Payload Type     : 100
     NTE Tx Payload Type  : 101
     NTE Rx Payload Type  : 101
```

Some other VISM commands that are useful in looking at active calls and MGCP/TGCP related information are **dspxgcpcons**, **dspxgcpdetailcnts**, **dspxgcplncons**, and **dspxgcpcalls**. Remember that if you do not know the exact syntax of a command, you can display help by typing a question mark (**?**) followed by any part of the command you do remember.

Chapter Summary

In the initial stages of a PacketCable deployment, most of your customers' phone calls are with customers still using the traditional PSTN. These ONNET-to-OFFNET calls require the use of the Media Gateway Controller (MGC), Media Gateway (MG), and Signaling Gateway (SG) PacketCable components. The SG offloads some of the SS7 complexity from the MGC by interworking at all SS7 layers with the PSTN and passing along only the upper layers to the MGC. The MG handles the bearer portion of the PSTN interconnection and communicates with the MGC using the TGCP variant of MGCP. TGCP is similar to NCS in that it is a streamlined version of MGCP aimed at supporting a certain type of gateway. Because of this optimization, some notable differences exist between TGCP and basic MGCP. However, many MSOs consider this optimization small enough that they deploy basic MGCP in the field.

This optimization includes eliminating support for endpoints other than trunk endpoints and the addition of parameters needed for the support of PacketCable functionality, such as secure media and lawful intercept. TGCP endpoints are trunk DS0s; thus, the local part of their endpoint name is of the form ds/.../x where x is the channel number of the DS0. The middle terms describe the hierarchy of the MG, such as whether a DS1, DS3, or OC-3 interface is used for PSTN connectivity. TGCP has five defined packages: the IT package used for standard SS7 trunks, the MO package used for applications such as emergency 911 operator access, the MT package used for applications such as busy line verification, the FXR package for T.38 facsimile support, and the XRM for extended VoIP metrics. The FXR and XRM packages were added in PacketCable 1.5.

Call signaling communication with the PSTN is done using legacy protocols such as SS7 and CAS. SS7 is an out-of-band protocol and is comprised of multiple layers used to interconnect three types of PSTN devices. These are the telephone switches themselves (SSPs), databases (SCPs), and transit devices (STPs). All three devices operate on the lower MTP layers of SS7. The MTP layer is responsible for interfacing upper layers to the physical media, ensuring a reliable connection between devices, and routing between devices. The SCCP layer also exists on all device types and enhances the routing capabilities of the MTP layer. The TCAP layer is an upper layer used in database transactions between an SSP and an SCP. The ISUP layer is the upper layer used for call signaling. ISUP messages such as IAM, ACM, ANM, REL, and RLC are used to establish calls in the PSTN. The SG PacketCable component passes along the ISUP and TCAP layers to the MGC. The Sigtran protocol suite is used for this communication and includes the SCTP transport layer protocol and the M3UA and SUA adaptation layer protocols. Cisco's SG is known as an ITP or IP Transfer Point and runs on routers using a special version of Cisco IOS. Cisco ITPs are usually deployed in pairs in either an ITP group implementation or an SG mated pair implementation.

CAS utilizes in-band signaling and is often used in telephony service provider networks for emergency operator access and busy line verification.

Just as is the case with any other part of PacketCable, the best way to understand how the PSTN interconnect works is by examining the call flows and their messages and parameters.

The configuration of a Cisco ITP has five basic steps for configuring the connectivity to the PSTN and seven basic steps for configuring the Sigtran connectivity to the MGC.

The configuration of an SG on the Cisco BTS 10200 consists of 11 steps to get ISUP services operational and an additional 10 steps to get TCAP services operational. The configuration of an MG on the Cisco BTS 10200 also consists of 11 steps that include parameters concerning the operation of the TGCP protocol. Basic OFFNET call routing can be configured in three steps, although many options are available for more sophisticated routing features.

The configuration of a Cisco MGX 88000 series gateway includes PXM configuration, RPM configuration, and VXSM/VISM configuration. ATM PVCs are typically built between the RPM modules and the individual VXSM and VISM cards. The VXSM and

VISM card configuration can be broken down into three areas: ATM configuration for connectivity to the PacketCable network, TDM configuration for connectivity to the PSTN, and TGCP/MGCP configuration for connectivity to the PacketCable MGC.

Chapter Review

1 What is the speed or bit-rate of the following TDM circuits?

DS0

DS1

DS3

OC3

Answer: DS0 (64 kbps), DS1 (1.544 Mbps), DS3 (44.736 Mbps), OC-3 (155.52 Mbps)

2 Suppose an MG has six slots each, containing a channelized DS3 connection to the PSTN. What would be the local endpoint name be for the first DS0 on the tenth T1 on DS3 in slot 5?

Answer: ds/s-5/ds3-1/ds1-10/1

3 If the MGC wants to create a connection only on timeslots 20–24 on this T1, what would the local endpoint name be?

Answer: ds/s-5/ds3-1/ds1-10/[20-24]

4 What TGCP package is used for one-way incoming MF FGD CAS trunks?

Answer: MT (MF Terminating)

5 What TGCP parameters are sent from an MGC to an MG when a wire tap is needed?

Answer: Call content connection identifier (es-cci) and call content destination (es-ccd)

6 True or False: An MG must be able to extract and generate call signaling information if an in-band PSTN signaling protocol is used.

Answer: True. No SG is used in this case.

7 What is the benefit of using STPs in an SS7 network?

Answer: Provides a distributed location for SSPs and SCPs to connect to, eliminating the need for a full mesh of links

8 What type of SS7 signaling link will you most likely use in a PacketCable network if you want to use the same point code on the MGC and SG?

Answer: A link

9 What are the three types of MTP layer 2 messages?

Answer: FISU (Fill In Signal Unit), LSSU (Link Status Signal Unit), and MSU (Message Signal Unit)

10 How are SS7 devices identified?

Answer: By their point code

11 If there is no voice path across an SS7 IMT, what is the probable cause and how can that be verified?

Answer: CIC mismatch; COT testing

12 If a group of SS7 circuits are taken out of service for maintenance, how does the remote switch know it should no longer pass traffic to them?

Answer: By the receipt of an ISUP CGB (Circuit Group Block) message

13 Why is CAS signaling also known as "Robbed Bit Signaling?"

Answer: Because bits from the bearer circuit are used for call signaling information

14 Why is the SCTP transport layer protocol used in Sigtran instead of UDP or TCP?

Answer: SCTP provides reliability unlike UDP and yet is more efficient for message-oriented data than TCP.

15 On a Cisco ITP, what nomenclature is given to the MGC and its Sigtran processes?

Answer: Application Server Process (ASP) and Application Server (AS)

16 For ONNET-to-OFFNET calls, why is ring back tone supplied by the PSTN instead of locally by the MTA?

Answer: So different ring tones from different parts of the PSTN can be conveyed to the originating party

17 List at least three limitations of an ITP group implementation compared to an SG mated pair implementation.

Answer: 1) Separate ITPs are needed for each MGC OPC. 2) The ITPs must be geographically close. 3) If the MGC is down the ITPs are not usable. 4) Global Title Translation is not supported.

18 What command can you use (and from what card) to find the ATM address of the slave side of a PVC in a MGX gateway?

Answer: **dspcon** (Display Connection) from the active PXM

19 What command can you use on a Cisco ITP to test the connectivity to an SS7 destination point code?

Answer: ping cs7 *Destination*

20 On the Cisco BTS 10200, how can you see what ISUP variants are available?

Answer: Supported ISUP variants are listed in the user-part-variant-base table.

21 On the Cisco BTS 10200, where is COT support configured?

Answer: The SS7 trunk group profile table (that is, ss7-ansi-tg-profile)

22 On a Cisco VISM card how do you change a DS1 from an SS7 IMT without call signaling to a CAS trunk with signaling?

Answer: The command **cnflnsig**

23 On an MGX gateway, how do you see log messages for a specific slot in the chassis?

Answer: dsplog -sl *Slot #* from the active PXM

CHAPTER 9

Call Management Server Signaling Protocol (CMSS)

In the previous three chapters, you learned all about the Media Gateway Control Protocol (MGCP) and its PacketCable variants, Network-based Call Signaling (NCS) and Trunk Gateway Control Protocol (TGCP). These protocols enable a PacketCable call agent to interface with its local subscribers and subscribers off the legacy public switched telephone network (PSTN). In this chapter, you will gain an understanding of the Call Management Server Signaling (CMSS) protocol used to communicate with remote voice over IP (VoIP) endpoints. These endpoints reside in a different PacketCable network under the control of a different call management server (CMS).

CMSS is a modified version of the Session Initiation Protocol (SIP) and is used for communication between PacketCable call agents. The chapter begins with an overview of SIP followed by a look at the key differences between CMSS and SIP. Similar to the other call signaling chapters, messages and call flows are examined to help you understand the protocol. Finally, SIP and CMSS configuration and troubleshooting are addressed.

SIP Overview

Recall from Chapter 6, "Signaling Interfaces and MGCP Overview," that call signaling protocols can be classified based on their architecture. Protocols such as MGCP, NCS, and TGCP utilize a *centralized architecture* where a majority of the complexity lies within the call agent component (CMS or Media Gateway Controller [MGC]). This permits the gateway component (multimedia terminal adapter [MTA] or Media Gateway [MG]) to be relatively simple. This, in turn, allows the gateways to be easier and cheaper to maintain and makes software upgrades because of the creation of new feature functionality less necessary.

A *distributed architecture* is the other type, wherein all the components speaking the protocol are required to have some intelligence. Because call agent-to-call agent communication by nature is between intelligent devices, a distributed architecture protocol is the logical choice. As you know by now, PacketCable attempts to reuse existing protocols whenever possible, and this case is no different. The IETF Session Initiation Protocol (SIP) is used as the basis for PacketCable call agent-to-call agent communication. Just as NCS and TGCP are versions of MGCP tailored for PacketCable, the modified version of SIP is called Call Management Server Signaling (CMSS).

CMSS is used between PacketCable call agents whether they are CMSs, MGCs, or, as is often the case, CMS/MGC combinations. In today's networks, the CMS and MGC components are typically bundled into the same physical devices; however, it is anticipated that in the future these components will be unbundled, allowing them to scale independently.

NOTE If you want to support calls between PacketCable devices and non-PacketCable devices on different call agents, you can use standard SIP.

Another variant of SIP is commonly used when interconnecting a VoIP network to the PSTN; this variation is known as SIP for Telephones (SIP-T). SIP-T is defined in IETF RFC 3372 and is better suited for PSTN gateway interconnects because it supports the transfer of Signaling System 7 (SS7)–specific parameters across a VoIP network. Not only does SIP-T support the encapsulation of Integrated Services User Part (ISUP) parameters inside SIP messages, but it also supports the translation of critical ISUP parameters into SIP parameters. This capability allows SIP servers to make decisions based on these translated parameters.

NOTE At press time, the CMSS specification was still evolving, and consequently many CMS devices have yet to fully support the protocol. As a result of this, in today's networks either basic SIP or SIP-T is often used instead of CMSS. The drawback to using these protocols is that some functionality unique to a PacketCable environment is lost. You'll learn more on this topic later in the chapter.

With the use of CMSS (or SIP or SIP-T), an MTA controlled by one CMS can talk to an MTA or MG controlled by a different CMS or MGC. This allows users in different parts of the country to communicate solely over IP, bypassing the PSTN and its costly connection fees entirely. As you might imagine, this functionality is very attractive to a company deploying PacketCable.

Another important note is that the two CMSs communicating using CMSS might or might not be under the domain of the same service provider. This factor leads to privacy and security considerations as you'll see later; it also assumes some level of trust between service providers.

CMSS/SIP in a PacketCable Network

At this point a diagram illustrating how CMSS or SIP fits into a PacketCable network might help in understanding its uses. Figure 9-1 is such an example between two PacketCable service providers.

Figure 9-1 *Sample SIP/CMSS Architecture*

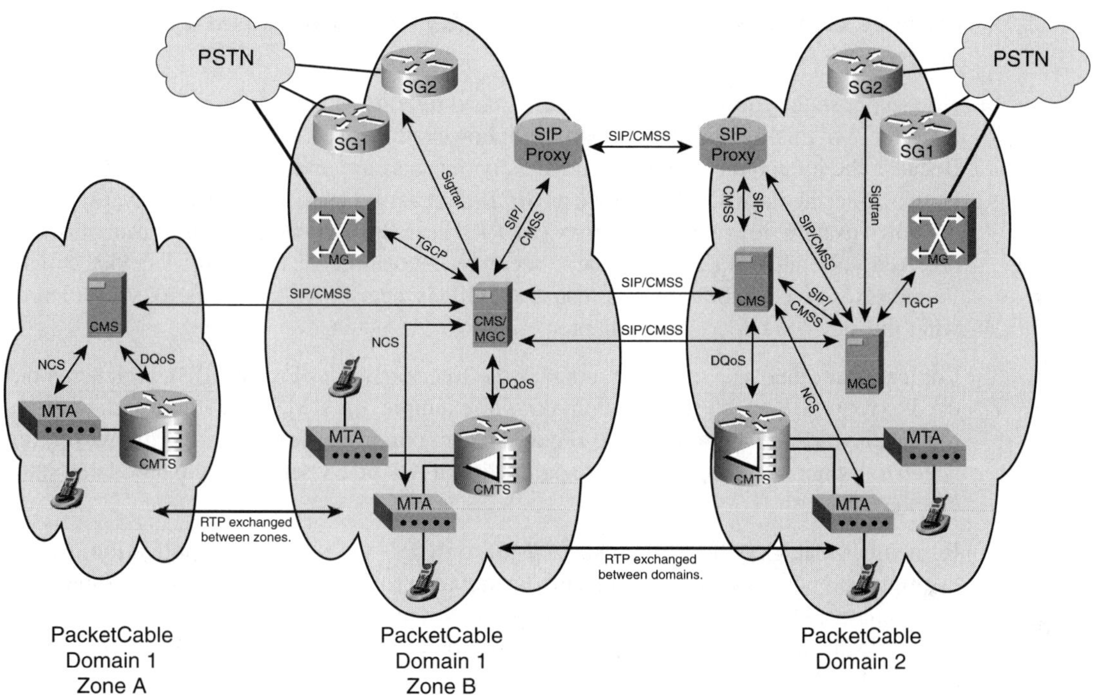

Notice the two domains in this example correspond to two different service providers. Domain 1 is further split into two separate zones; you can think of a zone as corresponding to a specific geographic area. Specifically, a PacketCable zone is defined as the set of MTAs in one or more DOCSIS networks under the control of a single CMS system. As you can see, Zone A in Domain 1 supports only ONNET call types because there is no PSTN connectivity in that zone. Conversely, Zone B in Domain 1 supports its own ONNET subscribers and also has PSTN connectivity through the devices and protocols you learned about in Chapter 8, "TGCP and the PSTN Interconnect." PacketCable Domain 2 consists of a single zone supporting its ONNET subscribers and is also equipped with PSTN connectivity.

The PacketCable service provider in Domain 2 has chosen to use separate physical devices for the CMS and MGC components. So how would OFFNET calls from its subscribers reach the PSTN? The answer is the CMS and MGC components would be configured to route calls between each other over the VoIP network using the CMSS protocol. Keep in mind that basic SIP or SIP-T could be used instead of CMSS; however, for simplicity reasons the term CMSS is used and implies CMSS, SIP, or SIP-T could be used.

Similarly, CMSS is also used for the communication between the CMS in Domain 1 Zone A and the CMS/MGC in Domain 1 Zone B. This enables all the subscribers in Domain 1 to reach each other as well as the PSTN. Because all the devices involved are in the same domain, security is not a concern, so no intermediary devices are used between the CMS systems.

Now suppose the subscribers in Domain 1 need to communicate with the subscriber in Domain 2. In this case, CMSS is also used; however, the way it's used is slightly different. Because the interconnecting CMS devices in this case are under different administrative control, intermediary devices known as *SIP proxy servers* are used. Specifically, this device is known as an *Exterior Border Proxy (EBP)*. These devices can be used to route traffic between VoIP networks and to provide security and resource accounting. A device called a Session Border Controller (SBC) would also likely be used in this scenario; you'll learn what this device is later in this chapter.

Logically, another type of proxy server is the Interior Border Proxy (IBP), which can be used to route SIP traffic inside a domain. For example, the CMS in Domain 1 Zone A and the CMS/MGC in Domain 1 Zone B could have connected through an IBP; however, this is generally not a requirement. More on the use of SIP proxy servers is covered later in this chapter.

Remember that a key benefit of using CMSS or SIP is the toll bypass application. As an example, suppose Domain 1 represented a service provider in New York and Domain 2 represented a service provider in Los Angeles. Not only can a call between their customers utilize CMSS to bypass the costly PSTN, but a call from one service provider subscriber to a legacy PSTN in the region of the other service provider can utilize CMSS to bypass much of the PSTN.

SIP RFCs

The PacketCable CMSS specification is based on SIP version 2.0. Several Internet Engineering Task Force (IETF) Request For Comments (RFC) are associated with SIP, with RFC 3261 being the main one. This RFC is used as the basis for this overview. Table 9-1 lists the SIP RFCs referenced by the PacketCable CMSS specification; high-level overviews of several of these RFCs are also covered in this chapter.

Table 9-1 *SIP RFCs Referenced by PacketCable CMSS*

RFC Number	Title	Description
RFC 3261	SIP: Session Initiation Protocol	Core RFC defining SIP version 2.0
RFC 3262	Reliability of Provisional Responses in Session Initiation Protocol (SIP)	The "100rel" extension used to ensure the reliability of 1xx provisional responses using the PRACK method
RFC 3263	Session Initiation Protocol (SIP): Locating SIP Servers	Defines the DNS procedures for converting an SIP URI into an IP address and port number of the next device to contact
RFC 3265	Session Initiation Protocol (SIP) - Specific Event Notification	Enables SIP nodes to request notifications from other nodes of specific events
RFC 3311	The Session Initiation Protocol (SIP) UPDATE Method	Enables the parameters of a session to be updated without affecting the state of a dialog
RFC 3312	Integration of Resource Management and Session Initiation Protocol (SIP)	Defines the framework for using "preconditions," which are used in CMSS to ensure distributed quality of service (DQoS) resources are available before users are alerted (i.e., ring tone and ring back tone provided)
RFC 3323	A Privacy Mechanism for the Session Initiation Protocol (SIP)	Defines guidelines for the creation of messages that do not compromise personal identity information
RFC 3325	Privacy Extensions to the Session Initiation Protocol (SIP) for Asserted Identity within Trusted Networks	Describes privacy extensions that enable trusted SIP proxy servers to assert the identity of authenticated users
RFC 3398	Integrated Services Digital Network (ISDN) User Part (ISUP) to Session Initiation Protocol (SIP) Mapping	Defines how ISUP parameters and messages are mapped into SIP parameters and messages and vice-versa
RFC 3420	Internet Media Type message/sipfrag	Permits the secure transfer of SIP messages end to end without revealing content to intermediary devices
RFC 3515	The Session Initiation Protocol (SIP) Refer Method	Defines the REFER method used to redirect a request. Needed for applications such as call transfer
RFC 3603	Private Session Initiation Protocol (SIP) Proxy-to-Proxy Extensions for Supporting the PacketCable Distributed Call Signaling Architecture	Describes extensions for the exchange of customer information and billing information between PacketCable CMS systems

continues

Table 9-1 *SIP RFCs Referenced by PacketCable CMSS (Continued)*

RFC Number	Title	Description
RFC 3842	A Message Summary and Message Waiting Indication Event Package for the Session Initiation Protocol (SIP)	Describes a method to carry message waiting status and message summary from a message waiting system to an interested SIP user agent
RFC 3966	The tel URI for Telephone Numbers	Specifies the "tel" URI used to represent telephone numbers

SIP User Agents and Sessions

Remember, in MGCP the key constructs are connections and endpoints. The endpoint is the demarcation point between the VoIP world and the time-division multiplexing (TDM) world, and connections are controlled by the call agent to establish communication between an endpoint and the VoIP network. The call agent then establishes calls by creating and managing multiple connections. Thus, two or more endpoints are able to communicate across the VoIP network.

In SIP, the key constructs are user agents and sessions. User agents are simply the devices speaking SIP, and you have two types—user agent clients (UAC) and user agent servers (UAS). Keep in mind that SIP utilizes a distributed intelligence, so at first glance figuring out what is categorized as a client versus a server might be somewhat unclear. Basically, the client is the device that makes the request, and the server is the device that accepts or rejects this request. This distinction between server and client is still a bit confusing because it changes on a per-request basis. For example, suppose you have an IP phone that communicates to an IP-based Private Branch Exchange (PBX) using SIP—what type of user agent is the phone? Well, if the IP phone is the originator of the call, it would signal the PBX to create a session. So in this case it is the UAC, and the PBX is the UAS. However, if the far end tears down the call, it initiates a tear-down request; so, the PBX would be the UAC, and the IP phone receiving this request would be considered the UAS. Therefore, in the life of a call, a user agent typically behaves as both a UAC and a UAS. In a PacketCable network, the CMS is an SIP user agent.

Now that you know what user agents are, you can learn a bit more about sessions. Simply put, a session is the dynamically created exchange of multimedia between user agents. So in the previous example the session is the voice call itself. And just like with MGCP, sessions are defined using another protocol, such as the Session Description Protocol (SDP).

Okay, so how is an SIP user agent identified? The answer is with a Uniform Resource Identifier (URI). Like email addresses and MGCP endpoint names, SIP URIs consist of a local name (or username) and a domain name. The general syntax for URIs is defined in IETF RFC 2396. A sample URI could be sip:oscar@packetcablenet.com, where oscar is the username and packetcablenet.com is the domain name. This example is an SIP URI and is noted by the sip: syntax. Secure URIs can also exist and have the syntax sips: instead of sip:. The Transaction Layer Security (TLS) protocol secures SIP signaling in this case. When the SIP user agent is a telephone, a telephone URI is commonly used that has the syntax tel:, and the username represents the telephone number. The format for telephone URIs is defined in IETF RFC 3966. Telephone user agents can also be specified in the sip: and sips: URI formats; in these cases, the qualifying parameter user=phone is added to the URI. In PacketCable the sip:, sips:, and tel: URI formats can all be used.

SIP Proxy Servers

In the sample architecture in Figure 9-1 you were introduced to SIP proxy servers. SIP proxies receive SIP request messages from clients and forward them on behalf of the sender to servers. Their primary role is to route SIP traffic. They can do this based on either preconfigured static mappings or dynamic methods. An SIP user agent client can blindly send requests to a certain SIP proxy server. This proxy server can then discover where the request should be sent by performing some kind of lookup. This capability to perform routing lookups eliminates the need for all call agents that need to intercommunicate from having to know and connect to one another directly.

Proxy servers can also be used as aggregation points for security associations. For example, instead of security associations needing to be established between all CMS pairs across domain boundaries, security associations can just be established between their proxy servers. Proxy servers can also be used to authorize users, translate IP addresses, and collect billing information to track network resources.

Finally, proxy servers can be used to provide call feature functionality. For example, if they receive a response to a call setup request indicating a subscriber is unavailable, the call could then be forwarded to another number (such as a cell phone) or a voice mail service.

Proxy servers can be either *stateful* or *stateless*. A *stateful* proxy server has the ability to correlate requests and their associated responses and can handle functions such as detecting timeouts and generating retransmissions. A *stateless* proxy server is a simple forwarding element; as soon as a packet is forwarded, it is forgotten. If a PacketCable CMS is used as an intermediary device routing SIP traffic between two other CMS systems, it is operating as a stateless proxy server.

Other types of SIP servers you should be aware of are *registrar servers*, which collect the location of user agents dynamically, and *redirect servers*, which help requests reach their intended destinations. Typically, register and redirect servers are collocated with proxy servers.

SIP Security

Security between SIP-based networks is typically handled by a Session Border Controller (SBC). An SBC is not a proxy server, but instead another type of SIP device known as a back-to-back user agent (B2BUA). It receives requests from a user agent client in network 1 and processes them as a user agent server; then it uses this information to act like a user agent client in network 2 and generates new requests. This enables the network topologies (such as IP addresses) in the two networks to be hidden from each other.

Another key point about SIP is that it is independent of the underlying transport layer protocol. That means it is perfectly valid to run SIP over User Datagram Protocol (UDP), Transmission Control Protocol (TCP), or even Stream Control Transmission Protocol (SCTP). Remember from Chapter 8 that SCTP is the transport protocol used with Sigtran. Secure SIP using the Transport Layer Security (TLS) protocol can be provided over the TCP and SCTP transport layer protocols. TLS requires a connection-oriented transport layer protocol; therefore, it is not supported over UDP. Please refer to IETF RFC 4346, "The Transport Layer Security (TLS) Protocol Version 1.1"; IETF RFC 4366, "Transport Layer Security (TLS) Extensions"; and IETF RFC 3436, "Transport Layer Security over Stream Control Transmission Protocol," for more information about TLS.

Also it is up to the application to choose the port(s) it uses for SIP, although port 5060 is the standard default for UDP, TCP, and SCTP. Also note that SIP works with both IPv4 and IPv6.

SIP Protocol Format

Now that you know the basics of SIP, this section covers the details of the messages and their associated parameters.

Like MGCP, SIP messages consist of requests and responses encoded in plaintext. The combination of an SIP request and its associated responses are more commonly referred to as an SIP method.

NOTE Because SIP messages are encoded in plaintext, debugging is relatively easy if the messages are not encrypted. Conversely, a lack of encryption introduces a number of potential security threats.

SIP messages consist of three parts:

- **The start line**—Either defines the method type or identifies the message as a response
- **The message header**—Defines the SIP parameters

- **The message body**—Contains the multimedia session information, which is defined using another protocol such as SDP

Start Line

If the start line begins with the term SIP/2.0, identifying the version of SIP used (currently 2.0), the message is a response, and this line is also known as the status line; otherwise, this term identifies the SIP method, and this line is also known as the request line.

Six SIP methods are defined in RFC 3261: INVITE, REGISTER, ACK, BYE, CANCEL, and OPTIONS.

- **INVITE**—The main method used to establish a session between user agents. In PacketCable, this method establishes calls between CMSs.
- **REGISTER**—Used to register SIP devices to a server. This method is not used in PacketCable 1.x.
- **ACK**—Used by an originating user agent to acknowledge the receipt of the final response to an INVITE request. Completes the three-way handshake used in establishing a session.
- **BYE**—Used for the teardown of sessions.
- **CANCEL**—Used by an originating user agent to cancel a pending INVITE request.
- **OPTIONS**—Used to query user agents or proxy servers for their capabilities; this is somewhat like the auditing function of MGCP.

Several other SIP methods are defined in other RFCs. The ones that are most relevant to a PacketCable deployment are

- **PRACK**—Used to confirm the successful end-to-end receipt of a provisional response.
- **UPDATE**—Used to convey the status of QoS resource reservation from CMS to CMS.
- **SUBSCRIBE**—Used to request notification of changes in resource or call status in the network.
- **NOTIFY**—Used to notify a device about the occurrence of an event such as a resource or call status change.
- **REFER**—Used to redirect an SIP request to another resource; a typical application is a call transfer from one party to another.

After the SIP method, the identifier of the destination or called party is included. This is typically the same URI as in the "To" field, covered in the next section.

If the message is a response, the next parameter is a three-digit code describing the current status of the SIP method. This parameter is very similar to MGCP response codes and is usually followed by a short text description of the code. Values in the 1xx range indicate a provisional response meaning the user agent is still processing the SIP method. Examples of this are

- **100**—Trying
- **183**—Session Progress
- **180**—Ringing

Values in the 2xx range indicate the SIP method was successfully processed and executed.

The 3xx range indicates a redirection. This is used in scenarios such as call forwarding, where one user agent reroutes a call to another user agent.

Values in the 4xx range indicate a client error such as a syntax error in the request or an inability for the server to fulfill the request. These error codes indicate a definite problem, and the requests should not be retransmitted unmodified.

Values in the 5xx range indicate a server error and usually reflect the server's inability to process and grant the request. In most cases nothing is wrong with the request; the server just can't currently fulfill the request. This could be because of a lack of resources. Because this condition can be temporary, these requests can be retransmitted.

Finally, the 6xx range signifies a global failure.

Message Header

The message header contains the SIP parameters or header fields characterizing the message. Each parameter is defined by a name followed by a ":" and then the parameter value. This part of the message is analogous to the MGCP parameter list. Table 9-2 shows the SIP headers you are most likely to see in a PacketCable environment.

Table 9-2 *SIP Parameters Encountered in a PacketCable Environment*

Parameter	Description	Mandatory
Via	Address (IP address or Fully Qualified Domain Name [FQDN]) the originator is expecting to receive responses; also contains the branch parameter that identifies the transaction. Proxy servers append an additional Via parameter with their own address before they forward a request message. They also remove their own Via parameter from responses before forwarding them.	Mandatory in a request; for CMSS if the originator requested privacy, no FQDNs should be included. IP addresses used instead.

Table 9-2 *SIP Parameters Encountered in a PacketCable Environment (Continued)*

Parameter	Description	Mandatory
From	Sender of the request. Contains the URI and display name of the SIP user agent (UA) associated with the calling party. A tag parameter is also included, which is a random string generated by the originating user agent used for identification purposes. If multiple dialogs exist between two parties (such as one for voice and one for video), this tag is used to differentiate the dialogs.	Mandatory in a request; for CMSS if the originator requested privacy, the display name must be "Anonymous."
To	Recipient of the request. Contains the URI and display name of the SIP UA associated with the called party. A tag parameter is appended to this parameter in future messages. As with the From parameter, this is a random string generated by the terminating user agent used for identification purposes.	Mandatory in a request; for CMSS if the originator requested privacy, the display name must be absent.
Call-ID	Globally unique identifier for the call (consists of random string and originator hostname or IP address). For the INVITE method, the combination of the To, From, and Call-ID fields completely defines the peer-to-peer relationship between user agents and is known as a dialog.	Mandatory in a request; for CMSS when the originating party requests privacy, this field must be a random single word and cannot contain the "@" symbol.
CSeq	Command Sequence; consists of integer and method name. This integer is incremented for each new request within a dialog.	Mandatory in a request.
Max-Forwards	Maximum number of hops a request can traverse. This is typically set to 70, which is large enough to guarantee delivery and protect against loops.	Mandatory in a request.
Supported	Used to identify SIP extensions the UAC supports. This lets the UAS know what extensions it can use.	Should be in an INVITE request.
Contact	URI representing direct route; this is where future requests should be sent. This is different from the Via header, which identifies where responses should be sent. If the proxy wants to receive future requests, it can utilize the Record-Route header described later. Also some UACs can be specifically programmed to ignore this header and continue to use the proxy server for future requests.	Mandatory in an INVITE request.
Require	Used to identify SIP extensions that the UAC uses, and the UAS must understand.	Can be in an INVITE request.

continues

Table 9-2 *SIP Parameters Encountered in a PacketCable Environment (Continued)*

Parameter	Description	Mandatory
Allow	Used to indicate what methods can be invoked within a dialog.	Should be in an INVITE request.
Accept	Used to indicate what content types are supported.	Can be in an INVITE request.
Record-Route	Used by proxies so that future requests will be received. It allows the proxy to know the entire call flow so that it could perform mid-call features.	Optionally added by proxy servers.
User-Agent	Contains information, such as software version, about the user agent making the request.	Can be in an INVITE request.
Content-Length	Byte count of message body	Can be in an INVITE request.
Content-Type	Description of content message body.	Can be in an INVITE request.

In a UAS's response to an INVITE request the Via, From, Call-ID, and CSeq fields are copied from the request unchanged. The To field is also copied, and if no tag was in it, the UAS appends a randomly generated tag used to identify the UAS. Remember these two tags along with the Call-ID uniquely identify the dialog.

Message Body

As stated before, the message body is defined through some other protocol. For PacketCable this protocol is the Session Description Protocol defined in RFC 2327. Please refer to Chapter 6 or RFC 2327 for more information. Also if the transfer of ISUP messages is needed between the call agent machines through the use of the CMSS or SIP-T protocols, then an Integrated Services User Part (ISUP) message body can be included as well.

Examining Call Flows

Okay, now that you have been exposed to the basics of SIP, the analysis of some real call flows should help reinforce what you've learned. In Chapter 8 you learned how the call flows look when one of the subscribers is still using the legacy PSTN. This time you see what these flows look like when the two subscribers reside in different PacketCable networks.

These call flows are from actual calls and serve as an example of the messaging and parameters involved. However, remember that the PacketCable specifications do not dictate the specifics of how the protocols formulate calls; this is up to the vendors of the various components. As long as the specifications are not violated, you have many ways to accomplish the same task. All the examples in this text utilize the Cisco BTS 10200 as the softswitch or MGC/CMS. Also the two CMSs in these examples interconnect directly; no proxy servers are used. Consequently, the two CMSs are configured like two different zones in a common domain. One of the BTS systems is running version 4.5 and the other version 4.4, neither of which supports CMSS; therefore, regular SIP is used to interconnect the PacketCable networks.

Basic Call

The first example is a basic call between Oscar (the originating party) and Tina (the terminating party), much like you saw in previous chapters. The only difference is that in this case Tina has her phone service from a different PacketCable service provider than Oscar.

Oscar's MTA is referred to as MTA_o, and the CMS his MTA communicates with is referred to as CMS_o; similarly, Tina's MTA is referred to as MTA_t, and the CMS her MTA communicates with is referred to as CMS_t. The examination of their phone call is broken up into three stages. The first stage details the messaging used from when Oscar picks up his phone until the point where Tina's phone is ringing. Stage two shows the messaging from here up to the point where Oscar and Tina have bidirectional voice communication. Stage three shows the messaging from this point until both parties eventually hang up their phones, ending the call. In addition to the message flows, the details of key packets are shown.

One last comment before getting to the analysis—this example can be thought of as an NCS-to-SIP call as well as an SIP-to-NCS call, depending on which CMS you are focusing.

The call signaling used for Stage 1 is illustrated in Figure 9-2.

Figure 9-2 *Basic SIP Call Flow (Stage 1)*

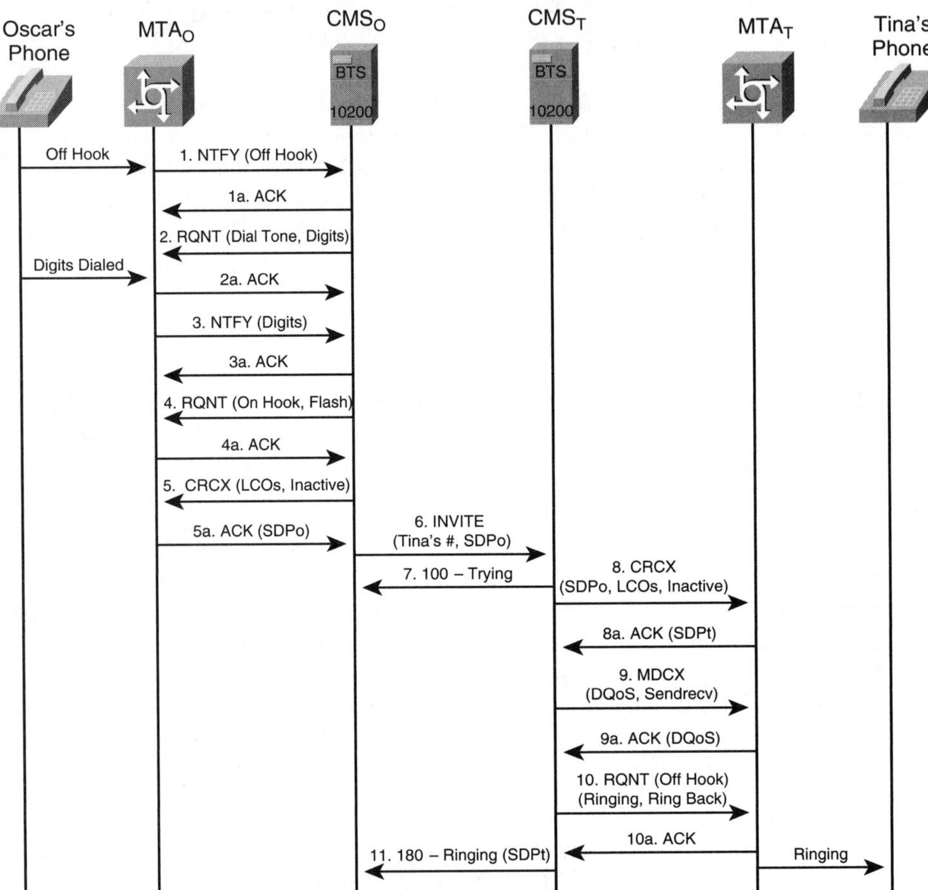

The first five steps are no different than the first five steps in the basic ONNET-to-OFFNET call example in Chapter 8, so little focus is devoted to those.

1. Oscar takes his phone off hook, triggering a NTFY message to be sent to the CMS.
2. The CMS then requests dial tone be provided to Oscar and asks to be notified when digits are collected according to the associated digit map for that MTA endpoint.
3. As soon as Oscar hears a dial tone, he dials Tina's phone number, which is forwarded to the CMS in another NTFY message.
4. The CMS then issues an RQNT to Oscar's MTA asking to be notified if Oscar hangs up his phone or issues a flash hook.

5 The CMS realizes it can route the call based on those digits and creates an inactive connection on Oscar's MTA. The response from the CRCX houses SDP information containing the IP address, Real-time Transport Protocol (RTP) port number, and other bearer-related information about the endpoint.

6 In this step, you see the first SIP message. The CMS realizes that to reach Tina it needs to communicate with another CMS. So it initiates the creation of a session by sending an INVITE request to Tina's CMS. Figure 9-3 is an Ethereal capture of this message.

Figure 9-3 *Ethereal Capture of SIP INVITE*

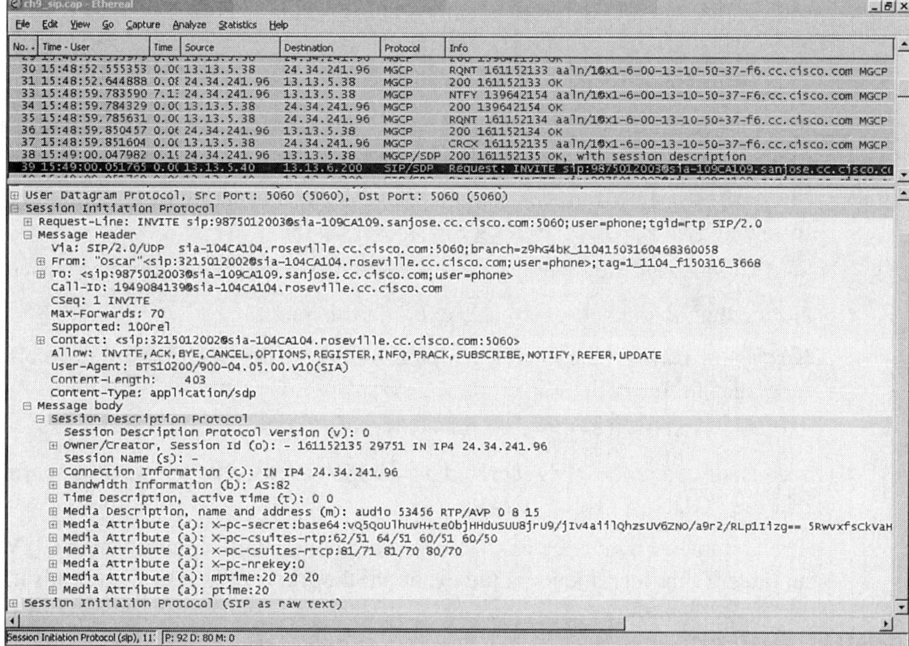

As you can see, Ethereal can fully decode SIP. Also remember that, like MGCP messages, SIP messages are encoded in plain text. Because this is your first SIP message, it is probably worthwhile to examine it in detail, piece by piece. As you see, UDP is being used as the transport layer protocol with the default port of 5060.

The start line shows that the message is a request for the INVITE method:

```
INVITE sip:9875012003@sia-
  109CA109.sanjose.cc.cisco.com:5060;user=phone;tgid=rtp SIP/2.0
```

Notice that the destination is an SIP URI referring to Tina's phone. The host name is the FQDN used by SIP on Tina's CMS, in this case sia-109CA109.sanjose.cc.cisco.com, and the local name is Tina's phone number. In PacketCable the local name is always the called party's phone number.

Next are the SIP parameters in the Message Header portion of the request. The first parameter (Via) identifies the path the request is taking.

```
Via: SIP/2.0/UDP  sia-104CA104.roseville.cc.cisco.com:5060;branch=
  z9hG4bK_11041503160468360058
```

In this case, only one Via parameter is equal to the FQDN used by SIP on Oscar's CMS. Notice it identifies the transport layer protocol SIP is using; in this case UDP. Also included within the Via parameter is the branch sub-parameter, which identifies the transaction and associate responses with requests.

The From parameter contains Oscar's URI, which consists of his phone number followed by the FQDN used by SIP on his CMS.

```
From: "Oscar"<sip:3215012002@sia-104CA104.roseville.cc.cisco.com;
  user=phone>;tag=1_1104_f150316_3668
```

It also contains his display name "Oscar" and the tag sub-parameter, which identifies the UAC and provides message integrity.

Similarly, the To parameter contains Tina's URI consisting of her phone numbered followed by the FQDN used by SIP on her CMS.

```
To: <sip:9875012003@sia-109CA109.sanjose.cc.cisco.com;user=phone>
```

Notice that no tag sub-parameter exists at this point.

Next is the Call-ID parameter, which is a unique identifier that exists in all SIP messages in this call.

```
Call-ID: 1949084139@sia-104CA104.roseville.cc.cisco.com
```

It consists of a randomly generated part, which is globally unique over space and time, and the FQDN of Oscar's CMS.

The Command Sequence parameter, CSeq, denotes the method name (INVITE) and an integer that increments in the event another request is sent within this dialog.

```
CSeq: 1 INVITE
```

Next is the Max-Forwards parameter, which is there to guard against loops.

```
Max-Forwards: 70
```

The Supported parameter is next and has the value 100rel, indicating the UAC supports this extension. This extension enables the reliable delivery of provisional response messages; more on this extension is covered in the "CMSS Extensions of SIP" section later in this chapter.

```
Supported: 100rel
```

The Contact parameter contains the URI for Oscar's MTA where all future SIP requests should be sent.

```
Contact: <sip:3215012002@sia-104CA104.roseville.cc.cisco.com:5060>
```

The Allow parameter identifies what methods are valid during the time this SIP dialog is active. As you can see, the six methods defined in RFC 3261 are listed here as well as six other methods.

```
Allow: INVITE,ACK,BYE,CANCEL,OPTIONS,REGISTER,INFO,PRACK,SUBSCRIBE,
  NOTIFY,REFER,UPDATE
```

The User-Agent parameter contains information about Oscar's CMS; in this case it indicates the CMS is a Cisco BTS 10200 running software version 4.5.0 (v10).

```
User-Agent: BTS 10200/900-04.05.00.V10(SIA)
```

The last two parameters describe the message body. The Content-Type parameter identifies the message body as being SDP (RFC 2327), and the Content-Length parameter gives the message length of 403 bytes.

```
Content-Length:    403
Content-Type: application/sdp
```

After these parameters is the SDP message body, which was obtained from the response to the NCS CRCX in the previous step.

```
v=0
o=- 161152135 29751 IN IP4 24.34.241.96
s=-
c=IN IP4 24.34.241.96
b=AS:82
t=0 0
m=audio 53456 RTP/AVP 0 8 15
a=X-pc-secret:base64:<hex digit string omitted>
a=X-pc-csuites-rtp:62/51 64/51 60/51 60/50
a=X-pc-csuites-rtcp:81/71 81/70 80/70
a=X-pc-nrekey:0
a=mptime:20 20 20
a=ptime:20
```

7 Hopefully, SIP makes a little more sense now. If not, more messages should help. So after Tina's CMS receives the INVITE, it lets Oscar's CMS know this by sending a 100 Trying provisional response. It's going to take some time for Oscar's CMS to fully process the INVITE request, so in the meantime it sends this message just to let Tina's CMS know the INVITE was received. Figure 9-4 is a capture of this message.

Figure 9-4 *Ethereal Capture of SIP 100 Trying*

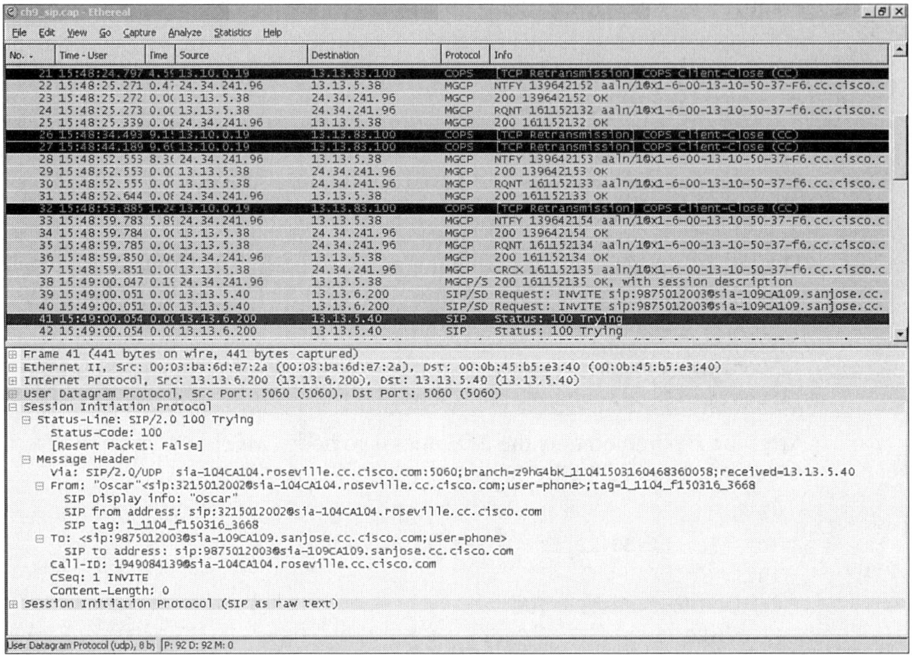

As you can see, the method line indicates that this is a 100 Trying response. Notice the Via, From, Call-ID, CSeq, and To parameters are virtually unchanged from the INVITE request. The received sub-parameter has been added to the Via field, which indicates the source IP address of the previous INVITE request. Finally, the Content-Length parameter is equal to zero, indicating no Message Data is associated with this response.

8 Now that Tina's CMS has been informed about the call, it creates an inactive NCS connection to her MTA. Included is the SDP information about Oscar's MTA, which was forwarded from Oscar's CMS in the SIP INVITE request. The response from the CRCX contains SDP information containing the IP address, RTP port number, and other bearer-related information about the NCS endpoint on Tina's MTA. This information is then used by Tina's CMS to begin the process of setting up DQoS resources on the CMTS for her MTA.

9 Tina's CMS then modifies the connection on her MTA by putting it into send/receive mode and relaying DQoS gate information. This triggers the MTA to reserve and commit the RF quality of service pipe to the CMTS; this aspect of PacketCable calls is covered much more in Chapters 12 and 13.

10 Next, Tina's CMS instructs her MTA to ring her phone and to play ring back tone through the PacketCable network back to Oscar's phone. This is the reason DQoS resources were committed instead of just reserved in the previous step—to allow the ring back tone to use these premium resources.

11 The last step in Stage 1 is another SIP response. Tina's CMS alerts Oscar's CMS that her phone is now ringing by sending another provisional response. In this case a 180 Ringing is sent, as shown in Figure 9-5.

Figure 9-5 *Ethereal Capture of SIP 180 Ringing*

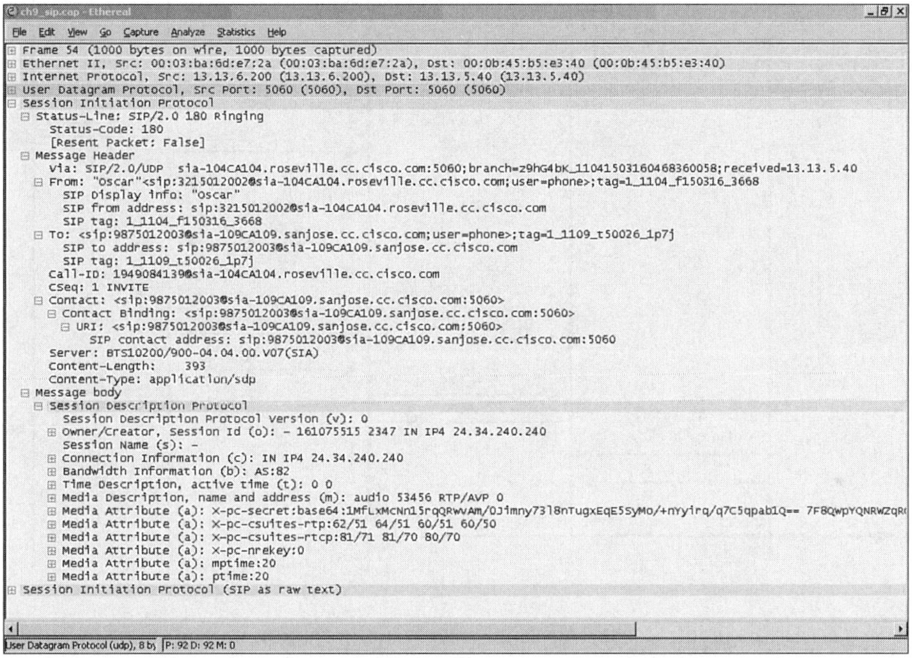

```
ch9_sip.cap - Ethereal
File Edit View Go Capture Analyze Statistics Help
⊞ Frame 54 (1000 bytes on wire, 1000 bytes captured)
⊞ Ethernet II, Src: 00:03:ba:6d:e7:2a (00:03:ba:6d:e7:2a), Dst: 00:0b:45:b5:e3:40 (00:0b:45:b5:e3:40)
⊞ Internet Protocol, Src: 13.13.6.200 (13.13.6.200), Dst: 13.13.5.40 (13.13.5.40)
⊞ User Datagram Protocol, Src Port: 5060 (5060), Dst Port: 5060 (5060)
⊟ Session Initiation Protocol
  ⊟ Status-Line: SIP/2.0 180 Ringing
      Status-Code: 180
      [Resent Packet: False]
  ⊟ Message Header
      Via: SIP/2.0/UDP  sia-104CA104.roseville.cc.cisco.com:5060;branch=z9hG4bK_11041503160468360058;received=13.13.5.40
    ⊟ From: "Oscar"<sip:3215012002@sia-104CA104.roseville.cc.cisco.com;user=phone>;tag=1_1104_f150316_3668
        SIP Display info: "Oscar"
        SIP from address: sip:3215012002@sia-104CA104.roseville.cc.cisco.com
        SIP tag: 1_1104_f150316_3668
    ⊟ To: <sip:9875012003@sia-109CA109.sanjose.cc.cisco.com;user=phone>;tag=1_1109_t50026_1p7j
        SIP to address: sip:9875012003@sia-109CA109.sanjose.cc.cisco.com
        SIP tag: 1_1109_t50026_1p7j
      Call-ID: 1949084139@sia-104CA104.roseville.cc.cisco.com
      CSeq: 1 INVITE
    ⊟ Contact: <sip:9875012003@sia-109CA109.sanjose.cc.cisco.com:5060>
      ⊟ Contact Binding: <sip:9875012003@sia-109CA109.sanjose.cc.cisco.com:5060>
        ⊟ URI: <sip:9875012003@sia-109CA109.sanjose.cc.cisco.com:5060>
            SIP contact address: sip:9875012003@sia-109CA109.sanjose.cc.cisco.com:5060
      Server: BTS10200/900-04.04.00.V07(SIA)
      Content-Length:    393
      Content-Type: application/sdp
  ⊟ Message body
    ⊟ Session Description Protocol
        Session Description Protocol Version (v): 0
      ⊞ Owner/Creator, Session Id (o): - 161075515 2347 IN IP4 24.34.240.240
        Session Name (s): -
      ⊞ Connection Information (c): IN IP4 24.34.240.240
      ⊞ Bandwidth Information (b): AS:82
      ⊞ Time Description, active time (t): 0 0
      ⊞ Media Description, name and address (m): audio 53456 RTP/AVP 0
      ⊞ Media Attribute (a): X-pc-secret:base64:1MfLxMcNn15rqQRwvAm/0Jimny73l8nTugxEqE5SyMo/+nYyirq/q7C5qpab1Q== 7F8QwpYQNRWZqR
      ⊞ Media Attribute (a): X-pc-csuites-rtp:62/51 64/51 60/51 60/50
      ⊞ Media Attribute (a): X-pc-csuites-rtcp:81/71 81/70 80/70
      ⊞ Media Attribute (a): X-pc-nrekey:0
      ⊞ Media Attribute (a): mptime:20
      ⊞ Media Attribute (a): ptime:20
⊞ Session Initiation Protocol (SIP as raw text)
User Datagram Protocol (udp), 8 by  P: 92 D: 92 M: 0
```

Notice the Via, From, Call-ID, and CSeq parameters are unchanged from the previous response. The To parameter is also copied, and a tag sub-parameter has been appended. This identifies the UAS (Tina's CMS) and provides message integrity. Remember, in SIP it is possible for more than one UAS to respond to the INVITE request. The Contact parameter has been added, which contains the URI for Tina's MTA where all future SIP requests should be sent. The User-Agent parameter has also been added, which contains information about Tina's CMS—in this case, it indicates the CMS is a Cisco BTS 10200 running software version 4.4.0 (v07). The last two parameters describe the message body. The Content-Type parameter identifies the message body as being SDP (RFC 2327), and the Content-Length parameter gives the message length of 393 bytes. After these parameters is the SDP message body, which was obtained from the response to the NCS CRCX in Step 8.

The call flow for the second stage is included in Figure 9-6.

Figure 9-6 *Basic SIP Call Flow (Stage 2)*

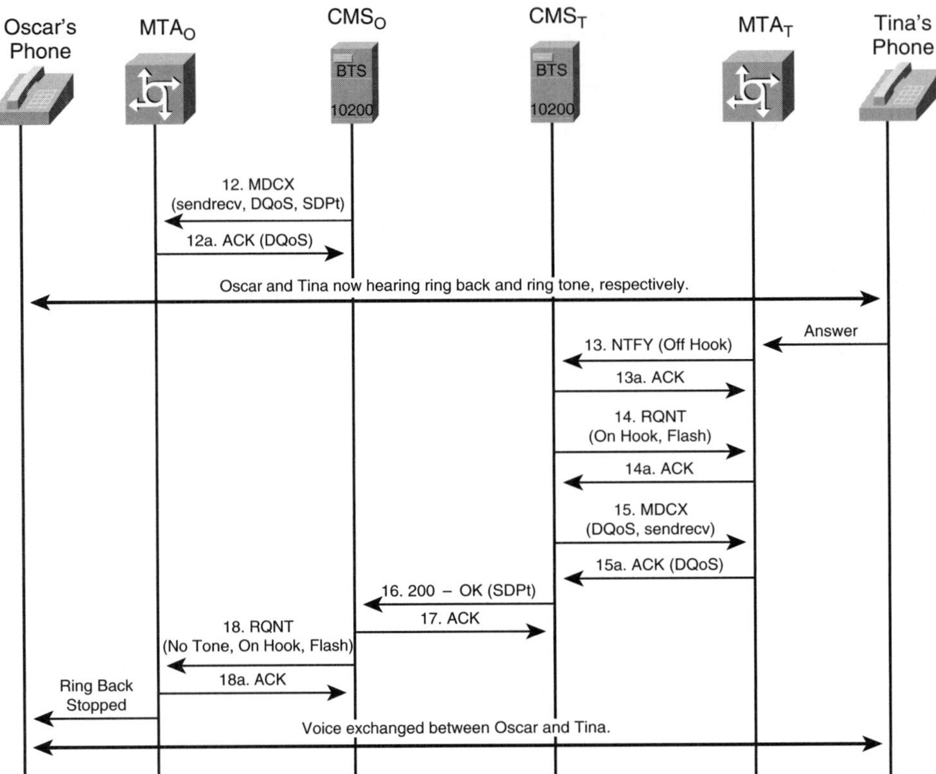

12 At this point, Oscar's CMS has received the ringing indication from Tina's CMS as well as the SDP information for Tina's MTA. So Oscar's CMS passes this information along to his MTA. It also instructs Oscar's MTA to change its connection mode to send/receive and reserve and commit DQoS resources. This is so Oscar can hear the ring back tone sent from Tina's MTA.

13 Eventually, Tina answers her phone, triggering an NCS NTFY message to be sent to her CMS.

14 Tina's CMS then requests to be notified if she hangs up her phone or hits her flash button.

15 Tina's CMS then modifies the connection on her MTA to send/receive and instructs the commitment of DQoS resources. In this case this step is not really needed because these tasks were already done in Step 9. However, if ring back was not provided across the PacketCable network, resources would have only been reserved in Step 9 and committed in this step.

16 Now that Tina's CMS knows the call has been answered, it can notify Oscar's CMS that the SIP session is now active. Thus, a final 200 OK response can be sent for the initial INVITE message. This message is illustrated in Figure 9-7.

Notice that the Via, From, To, Call-ID, and CSeq fields are again unchanged from the previous SIP packets. The Supported header is included, identifying the SIP methods supported by Tina's CMS, and the Allow header is included, identifying the methods supported by Tina's CMS in this dialog. Tina's SDP information is passed in the message body of this response.

Figure 9-7 *Ethereal Capture of 200 OK*

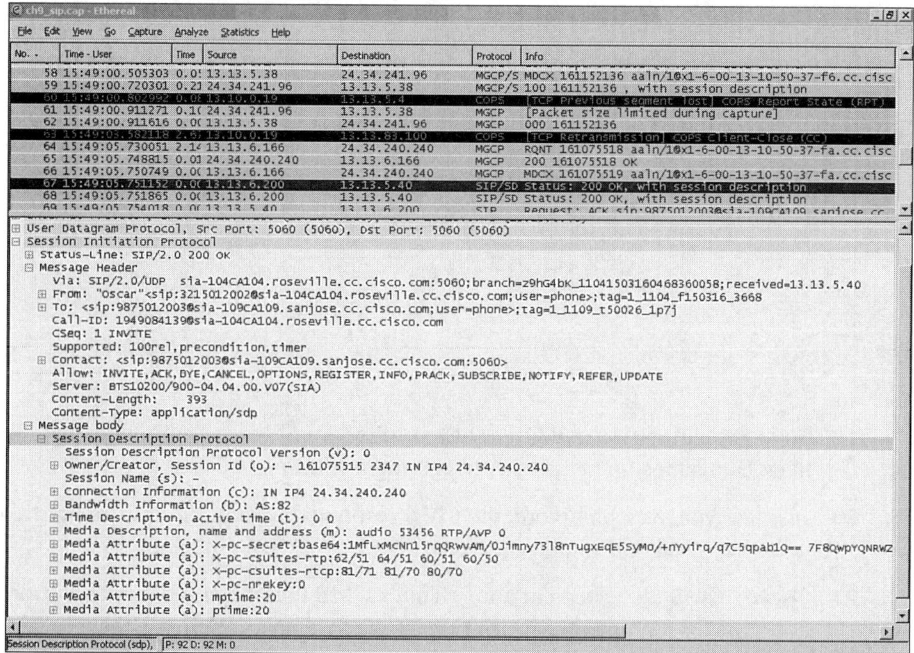

17 As part of the three-way handshake procedure used with the SIP INVITE method, Oscar's CMS acknowledges the receipt of the final INVITE response by sending an ACK message to Tina's CMS. This message is shown in Figure 9-8.

Notice that in this message the branch sub-parameter has changed because this message begins a new transaction; also, the CSeq header has changed to 1 ACK to reflect the new SIP method.

18 In the last step of this stage Oscar's MTA is instructed to stop playing ring back tone and to be notified if Oscar hangs up or presses his flash button.

The call flow for the final stage is included in Figure 9-9.

Figure 9-8 *Ethereal Capture of ACK Message*

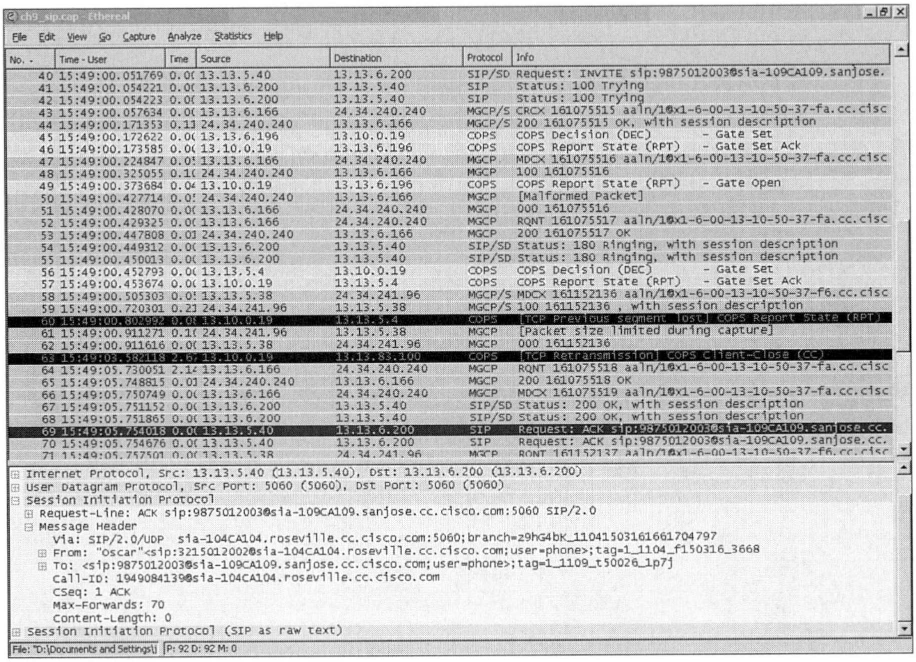

19 So at this point Tina and Oscar have bidirectional voice communication. Some time later Tina hangs up her phone, triggering a NTFY to be sent to her CMS.

20 Just like you've seen before, the CMS responds by requesting to be notified if Tina takes her phone off hook.

21 Oscar's CMS also needs to notify Tina's CMS that the call has been disconnected. It does this by sending an SIP BYE message to Tina's CMS, as shown in Figure 9-10.

Again the branch sub-parameter has changed, indicating this is a new transaction. Also, the CSeq value has changed to 2 BYE, identifying the new method.

Figure 9-9 *Basic SIP Call Flow (Stage 3)*

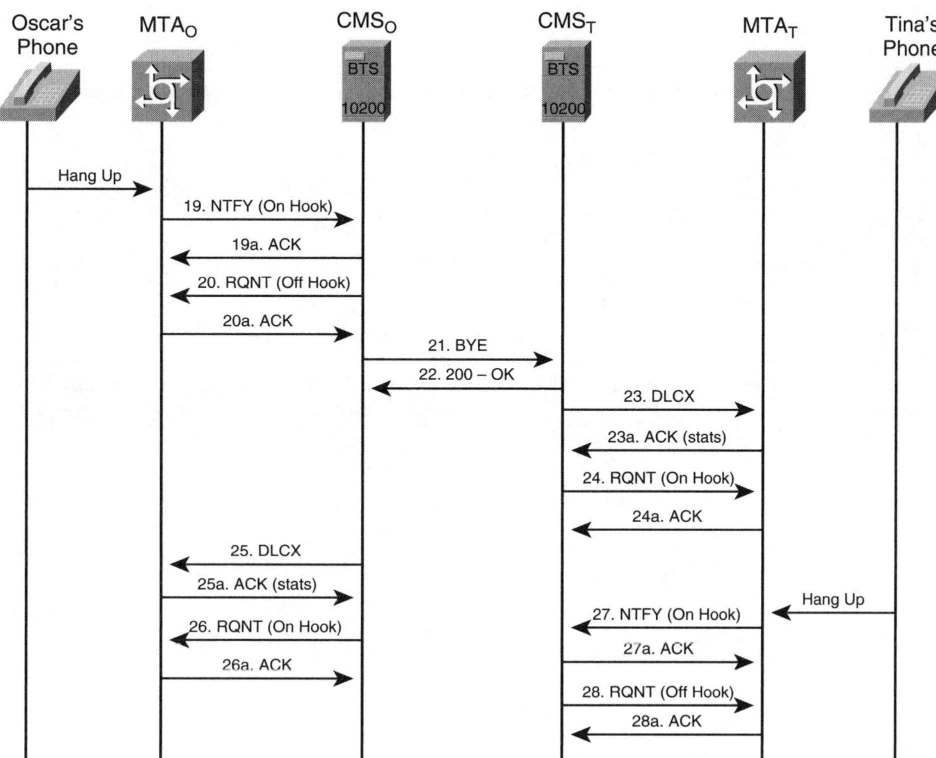

22 After receiving this message, Oscar's CMS acknowledges it by sending a 200 OK response.

23 Tina's CMS then deletes the NCS connection it created on her MTA. The MTA returns call statistic information in the acknowledgement.

24 Next, Tina's CMS requests to be notified when Tina hangs up her phone.

25 In this step the connection on Oscar's MTA is deleted and call statistics are returned.

26 Next, Oscar's CMS requests to be notified when Oscar picks up his phone again.

27 Eventually, Tina does hang up her phone, triggering a NTFY to be sent to his CMS.

28 Finally, the CMS requests to be notified when Oscar picks up his phone again.

Figure 9-10 *Ethereal Capture of BYE Message*

```
ch9_sip.cap - Ethereal
File  Edit  View  Go  Capture  Analyze  Statistics  Help
No. - Time - User        Time  Source         Destination    Protocol Info
   62 15:49:00.911616 0.0( 13.13.5.38     24.34.241.96   MGCP     000 161152136
   63 15:49:03.582118 2.6; 13.10.0.19     13.13.83.100   COPS     [TCP Retransmission] COPS Client-Close (CC)
   64 15:49:05.730051 2.1< 13.13.6.166    24.34.240.240  MGCP     RQNT 161075518 aaln/1@x1-6-00-13-10-50-37-fa.cc.cisc
   65 15:49:05.748815 0.01 24.34.240.240  13.13.6.166    MGCP     200 161075518 OK
   66 15:49:05.750749 0.0( 13.13.6.166    24.34.240.240  MGCP     MDCX 161075519 aaln/1@x1-6-00-13-10-50-37-fa.cc.cisc
   67 15:49:05.751152 0.0( 13.13.6.200    13.13.5.40     SIP/SD   Status: 200 OK, with session description
   68 15:49:05.751865 0.0( 13.13.6.200    13.13.5.40     SIP/SD   Status: 200 OK, with session description
   69 15:49:05.754018 0.0( 13.13.5.40     13.13.6.200    SIP      Request: ACK sip:9875012003@sia-109CA109.sanjose.cc.
   70 15:49:05.754676 0.0( 13.13.5.40     13.13.6.200    SIP      Request: ACK sip:9875012003@sia-109CA109.sanjose.cc.
   71 15:49:05.757501 0.0( 13.13.5.38     24.34.241.96   MGCP     RQNT 161152137 aaln/1@x1-6-00-13-10-50-37-f6.cc.cisc
   72 15:49:05.799261 0.0< 24.34.240.240  13.13.6.166    MGCP     [Malformed Packet]
   73 15:49:05.801078 0.0( 24.34.241.96   13.13.5.38     MGCP     200 161152137 OK
   74 15:49:12.466449 6.6( 24.34.241.96   13.13.5.38     MGCP     NTFY 139642155 aaln/1@x1-6-00-13-10-50-37-F6.cc.cisc
   75 15:49:12.467216 0.0( 13.13.5.38     24.34.241.96   MGCP     200 139642155 OK
   76 15:49:12.468403 0.0( 13.13.5.38     24.34.241.96   MGCP     RQNT 161152138 aaln/1@x1-6-00-13-10-50-37-f6.cc.cisc
   77 15:49:12.470535 0.0( 13.13.5.40     13.13.6.200    SIP      Request: BYE sip:9875012003@sia-109CA109.sanjose.cc.
   78 15:49:12.471195 0.0( 13.13.5.40     13.13.6.200    SIP      Request: BYE sip:9875012003@sia-109CA109.sanjose.cc.
   79 15:49:12.471719 0.0( 13.13.6.200    13.13.5.40     SIP      Status: 200 OK
   80 15:49:12.472376 0.0( 13.13.6.200    13.13.5.40     SIP      Status: 200 OK
   81 15:49:12.472870 0.0( 13.13.6.166    24.34.240.240  MGCP     DLCX 161075520 aaln/1@x1-6-00-13-10-50-37-fa.cc.cisc
   82 15:49:12.520323 0.0< 24.34.240.240  13.13.6.166    MGCP     250 161075520 Connection was deleted
   83 15:49:12.521486 0.0( 13.13.6.166    24.34.240.240  MGCP     RQNT 161075521 aaln/1@x1-6-00-13-10-50-37-fa.cc.cisc
   84 15:49:12.536282 0.01 24.34.241.96   13.13.5.38     MGCP     200 161152138 OK
   85 15:49:12.536431 0.0( 24.34.240.240  13.13.6.166    MGCP     200 161075521 OK
   86 15:49:12.537098 0.0( 13.13.5.38     24.34.241.96   MGCP     DLCX 161152139 aaln/1@x1-6-00-13-10-50-37-f6.cc.cisc
   87 15:49:12.563810 0.0: 13.10.0.19     13.13.6.196    COPS     [TCP Previous segment lost] COPS Report State (RPT)
   88 15:49:12.673609 0.1( 24.34.241.96   13.13.5.38     MGCP     250 161152139 Connection was deleted
   89 15:49:12.677331 0.0( 13.10.0.19     13.13.5.4      COPS     [TCP Previous segment lost] COPS Report State (RPT)
   90 15:49:13.278357 0.6( 13.10.0.19     13.13.83.100   COPS     [TCP Retransmission] COPS Client-Close (CC)
   91 15:49:14.092485 0.81 13.13.6.166    24.34.240.240  MGCP     RQNT 161075522 aaln/1@x1-6-00-13-10-50-37-fa.cc.cisc
   92 15:49:14.108693 0.01 24.34.240.240  13.13.6.166    MGCP     200 161075522 OK

⊞ Internet Protocol, Src: 13.13.5.40 (13.13.5.40), Dst: 13.13.6.200 (13.13.6.200)
⊞ User Datagram Protocol, Src Port: 5060 (5060), Dst Port: 5060 (5060)
⊟ Session Initiation Protocol
  ⊞ Request-Line: BYE sip:9875012003@sia-109CA109.sanjose.cc.cisco.com:5060 SIP/2.0
  ⊟ Message Header
     Via: SIP/2.0/UDP  sia-104CA104.roseville.cc.cisco.com:5060;branch=z9hG4bK_1104150316067726073l
   ⊞ From: "Oscar"<sip:3215012002@sia-104CA104.roseville.cc.cisco.com;user=phone>;tag=1_1104_f150316_3668
   ⊞ To: <sip:9875012003@sia-109CA109.sanjose.cc.cisco.com;user=phone>;tag=1_1109_t50026_1p7j
     Call-ID: 1949084139@sia-104CA104.roseville.cc.cisco.com
     CSeq: 2 BYE
     Max-Forwards: 70
     Content-Length: 0
⊞ Session Initiation Protocol (SIP as raw text)
File: "D:\Documents and Settings\j  P: 92 D: 92 M: 0
```

Error Traces

Now that you've seen what a good call looks like, this section examines some SIP messaging from unsuccessful calls.

Call Not Answered

This example demonstrates what happens when Tina does not answer Oscar's call. As you might imagine, the first three messages are the same as you've seen before. An INVITE is sent from Oscar's CMS to Tina's CMS, Tina's CMS acknowledges the receipt of this message with a 100 Trying response, and at some point later Tina's CMS tells Oscar's CMS that her phone is ringing by sending a 180 Ringing response.

Eventually, Oscar gets tired of waiting for Tina to answer her phone, so he hangs up. Oscar's CMS needs to let Tina's CMS know this, but because the INVITE three-way handshake has not yet been completed, it can't send the BYE request. Instead, a CANCEL request is sent, and the following message results:

```
CANCEL sip:3215012002@sia-104CA104.roseville.cc.cisco.com:5060;user=phone;
  tgid=rtp SIP/2.0
```

```
Via: SIP/2.0/UDP  sia-109CA109.sanjose.cc.cisco.com:5060;branch=z9hG4bK_
  1109500952019169415
From: "Tina"<sip:9875012003@sia-109CA109.sanjose.cc.cisco.com;user=phone>;
  tag=1_1109_f50095_2u41
To: <sip:3215012002@sia-104CA104.roseville.cc.cisco.com;user=phone>
Call-ID: 0034814788@sia-109CA109.sanjose.cc.cisco.com
CSeq: 1 CANCEL
Max-Forwards: 70
Content-Length: 0
```

Logically, this message cancels the previous INVITE request. Tina's CMS acknowledges the receipt of this message with a 200 OK response. Then the call attempt to Tina's MTA is terminated, and Oscar's CMS is notified of this by the receipt of a final response message for the INVITE request. This response has a response code of 487 and a description of Request Terminated. Recall that response codes in the 4xx range indicate a client error. This response message is as follows:

```
487 Request Terminated
Via: SIP/2.0/UDP  sia-109CA109.sanjose.cc.cisco.com:5060;branch=z9hG4bK_
  1109500952019169415;received=13.13.6.16
From: "Tina"<sip:9875012003@sia-109CA109.sanjose.cc.cisco.com;user=phone>;
  tag=1_1109_f50095_2u41
To: <sip:3215012002@sia-104CA104.roseville.cc.cisco.com;user=phone>;tag=1_1104_t
  150328_262
Call-ID: 0034814788@sia-109CA109.sanjose.cc.cisco.com
CSeq: 1 INVITE
Server: BTS 10200/900-04.05.00.V10(SIA)
Content-Length: 0
```

As with any other final response to an INVITE request, the originating CMS acknowledges the receipt of this response message with an ACK request to complete the three-way handshake.

Do Not Disturb

In this example, Oscar has activated the "Do Not Disturb" feature indicating he is not willing to accept any calls at this point.

The first two messages are nothing new. An INVITE is sent between the two CMSs, and the terminating CMS responds with a provisional response of 100 Trying. The next message is as follows:

```
SIP/2.0 403 Forbidden
Via: SIP/2.0/UDP  sia-109CA109.sanjose.cc.cisco.com:5060;branch=z9hG4bK_
  1109500911360468000;received=13.13.6.168
From: "Tina"<sip:9875012003@sia-109CA109.sanjose.cc.cisco.com;user=phone>;
  tag=1_1109_f50091_9587
To: <sip:3215012002@sia-104CA104.roseville.cc.cisco.com;user=phone>;
  tag=1_1104_t150327_t1sm
Call-ID: 1720760918@sia-109CA109.sanjose.cc.cisco.com
Seq: 1 INVITE
Server: BTS 10200/900-04.05.00.V10(SIA)
Content-Length: 0
```

The response has a response code of 403 and a description of Forbidden. In this case, the request is denied because the terminating CMS cannot fulfill the INVITE request.

Specifically, the reason for the Forbidden response is because the terminating MTA has activated the "Do Not Disturb" feature.

The originating CMS acknowledges the receipt of this response message to complete the three-way handshake.

DQoS Resource Reservation Failure

This example examines what happens when you have a failure in DQoS resource reservation. Just as with the previous example, the first two steps are pretty normal, and the problem is seen with the third message:

```
502 Bad Gateway
Via: SIP/2.0/UDP  sia-109CA109.sanjose.cc.cisco.com:5060;branch=z9hG4bK_
  1109501040399175816;received=13.13.6.168
From: "Tina"<sip:9875012003@sia-109CA109.sanjose.cc.cisco.com;user=phone>;
  tag=1_1109_f50104_0f2s
To: <sip:3215012002@sia-104CA104.roseville.cc.cisco.com;user=phone>;
  tag=1_1104_t150329_5w65
Call-ID: 1799943527@sia-109CA109.sanjose.cc.cisco.com
CSeq: 1 INVITE
Server: BTS 10200/900-04.05.00.V10(SIA)
Content-Length: 0
```

Again, a negative response is seen with a response code of 502 and a description of Bad Gateway. Remember, responses in the 5xx range are server errors and typically indicate the CMS is simply unable to fulfill a perfectly valid request. In this case the problem is that PacketCable DQoS failed for the terminating MTA. Because there is a chance that this problem is only temporary, the 502 response code is used. Like the previous example, the last message is an ACK request confirming the receipt of the 502 response.

CMSS Extensions of SIP

Now that you have a good handle on the basics of the SIP protocol, this section examines the differences between CMSS and SIP. The CMSS protocol is defined under the PacketCable 1.5 series of specifications. CMSS can really be thought of as an extension to SIP. Like MGCP, SIP is very flexible and has mechanisms for the addition of fields needed to support capabilities for a PacketCable environment.

These extensions include

- DQoS extensions
- Privacy extensions
- Extensions to pass additional information
- Other extensions
- Other modifications

DQoS Extensions

In NCS, additional parameters exist to support the PacketCable DQoS functionality covered in Chapter 12, "DQoS Architecture and Framework," and Chapter 13, "Analyzing, Implementing, and Troubleshooting DQoS." This is also required in CMSS. As you will learn, part of DQoS involves reserving and committing DOCSIS resources in separate steps. This information needs to be communicated between CMSs when the CMSS protocol is used. This is done with an additional SIP request/response exchange during the session setup.

Recall that SIP utilizes a three-way handshake when initiating a session. First, the UAC sends an INVITE message requesting a session be established. At some point later the UAS confirms that this session can be established and sends a 2xx response code to the UAC. Finally, the UAC acknowledges the receipt of this response by sending an ACK request message.

Now you will learn how session setup works in CMSS. Recall that the Require message header parameter identifies SIP extensions the UAC uses and that the UAS needs to understand. The option precondition is one of these extensions used in CMSS. This is used for each media flow described in the SDP in the INVITE request, and the values are appended in the SDP using the attribute parameter (a). Because this is a bidirectional end-to-end QoS requirement, the values in the INVITE need to be precondition type="qos", desired status type="e2e", and direction-tag="sendrecv". This means the called party should not be alerted until DQoS resources are reserved in both directions end-to-end.

NOTE In addition to the precondition option, the 100rel option must also be included in the Require parameter in every CMSS INVITE request. This option specifies that provisional responses need to be acknowledged to ensure they are successfully received. As you'll soon see, the PRACK method is used for this purpose.

When the UAS receives an INVITE with the precondition option, it generates the provisional response 183 Session Progress containing the SDP information for the terminating endpoint. Included in the response is the precondition option as well, with the same values as in the INVITE, except the direction tag is set to recv. This means the UAS handles QoS reservation for the UAS-to-UAC direction. The terminating CMS also requests that the originating CMS notify it when the QoS reservation is complete by adding the a=conf to the SDP field.

The receipt of the 183 Session Progress response is acknowledged by sending a PRACK message to the terminating CMS. The originating CMS conveys the status of the QoS reservation by sending an UPDATE request message. In this message is a success/failure indication of each precondition. Assuming the UPDATE message indicates the resource reservation was successful and resource reservation was also successful on the UAS, the UPDATE message is positively acknowledged with a 200 OK response.

After DQoS resources have been reserved by both MTAs, the terminating CMS notifies the originating CMS by sending another provisional response. A 180 Ringing message is used if the originating CMS is to instruct the originating MTA to provide ring back tone, and a 183 Session Progress message is used if the ring back tone is provided from the terminating MTA. Because this is another provisional response, the PRACK/200-OK exchange is used again to acknowledge the receipt of the provisional response.

At some point later the call is answered, triggering DQoS resources to be committed and triggering the final 200 OK response to the initial INVITE request to be sent. The call is now up and active; at the end of this section you'll see a call flow of how the messaging might look.

The precondition extension details are outlined in RFC 3312, "Integration of Resource Management and Session Initiation Protocol (SIP)." The PRACK method for ensuring the reliable delivery of response methods is detailed in RFC 3262. The UPDATE method used to convey the status of QoS resources is detailed in RFC 3311.

Privacy Extensions

Privacy extensions enable calls to be made between CMS machines without compromising the identity of the calling party when they have requested privacy. These extensions also make the Customer Originated Trace feature possible. This feature allows a destination party to report a harassing call even if the originator requested privacy.

When a trusted proxy receives a message from a node it does not trust, it authenticates the identity of the node. As a result of this authentication it ascertains the identity of the node and inserts this identity in a message header called P-Asserted-Identity. This header looks similar to the From header in that it contains the URI of the originator and an optional display name. If this message is to be forwarded to an untrusted proxy or UA and the calling party requested privacy, this header's values are removed. The presence or absence of this parameter and the "trust" state of adjacent nodes dictate how this parameter is used throughout the network.

If privacy is enabled, other parameters in the CMSS messages need to be modified as well before sensitive data can traverse an untrusted network. Specifically, a device known as an *anonymizer* is used. The anonymizer handles three different types of privacy: user, name, and IP address; and it provides three different privacy functions: signaling content privacy, signaling IP address privacy, and media IP address privacy.

The signaling content privacy function is what you've learned about so far regarding using the Anonymous tag when a subscriber's name is private. This function is typically built in to the CMS. IP address privacy is normally handled by routing the messages through a separate device such as the Session Border Controller (that is, SIP Back-to-Back User

Agent) talked about earlier. This involves modifying fields such as the SDP c: parameter, which is adjusted to point to this device instead of conveying an endpoint's IP address.

The privacy extensions are outlined in RFC 3323, "A Privacy Mechanism for the Session Initiation Protocol (SIP)," and RFC 3325, "Privacy Extensions to the Session Initiation Protocol (SIP) for Asserted Identity within Trusted Networks."

Extensions to Pass Additional Information

CMSS also supports extensions used to support the transport of additional information between CMSs. This information includes authorization, billing, and lawful intercept data. Much of this information is defined in RFC 3603, "Private Session Initiation Protocol (SIP) Proxy-to-Proxy Extension for Supporting the PacketCable Distributed Call Signaling Architecture."

These headers include

- **P-DCS-TRACE-PARTY-ID**—Used to enable the Customer Originated Trace feature for identifying the source of harassing calls even when features such as privacy have been activated by the originating party. Support for this header is not mandatory for CMSS compliance.
- **P-DCS-OSPS**—Used for Operator Services calls between CMS systems, such as Busy Line Verification (BLV) and Emergency Interrupt (EI). For example, this header is used if an operator reachable through a particular PacketCable MGC needs to interrupt the line of a subscriber on a particular PacketCable CMS and the MGC and CMS are interconnected using CMSS.
- **P-DCS-BILLING-INFO**—Used to transfer billing-related information between CMSs. Among this information is an identifier used to correlate billing information between the CMSs and a charge parameter indicating the party billed.
- **P-DCS-LAES and P-DCS-REDIRECT**—The P-DCS-LAES extension contains information needed to support lawful intercept for CMS-to-CMS call types. Basically, it is used when the responsibility for lawful intercept moves from one CMS to another. This is required in situations where a subscriber under surveillance has forwarded his or her phone. The P-DCS-REDIRECT extension contains information specific to redirected or forwarded calls. For lawful intercept the REFER method transports information. Support of these headers is required for CMSS compliance.

Other Extensions

Other extensions provided with CMSS include the capability for one CMS to tell another to issue a new INVITE request, and the capability for one CMS to update the parameters of a session without changing its state. The update of parameters can be accomplished using the SIP UPDATE method; remember this method was also used to update the "preconditions" used to signify DQoS resource status.

The SIP REFER method can be used by an SIP device to redirect a request to another resource. For example, it can be used in a call transfer situation. In this case, the REFER method would contain the URI identifying the user agent to which the session should be redirected. The SIP NOTFIY method would then be used to convey the status of the referral or redirection.

The SIP Replaces header is defined in Appendix C of the PacketCable CMSS specification and can be used to logically replace an existing SIP dialog (remember, a dialog comprises the Call-ID header and the tags from the From and To headers) with a new one. This can be used for call features such as "Attendant Transfer" and "Call Park." This header can be used only with the INVITE method.

SIP devices can "subscribe" or request to be notified when the status of a particular resource or call in their network changes using the SIP SUBSCRIBE method. When an entity detects a change in state of one of these resources or calls, it then uses the SIP NOTIFY message to alert the subscribed SIP device. This is somewhat similar to how the RQNT and NTFY messages in MGCP-based protocols request and detect events.

The extensions enabling the CMS to send a new INVITE request are outlined in RFC 3515, "The Session Initiation Protocol (SIP) Refer Method"; RFC 3265, "Session Initiation Protocol (SIP) - Specific Event Notification"; and Appendix C of the PacketCable CMSS specification, "The Session Initiation Protocol (SIP) 'Replaces' Header."

Other Modifications

The tel URI specified in RFC 3966, "The tel URI for Telephone Numbers," is often used in CMSS to describe resources identified by telephone numbers. Telephone numbers are specified either in global notation (ITU E.164) or local notation. Global numbers are unique worldwide and begin with the "+" character followed by the country code (CC) and national significant number (NSN). Conversely, local numbers are unique only within a certain part of the telephone network. CMSS supports additional extensions to the tel URI relating to services such as number portability and freephone.

As mentioned earlier, because CMSS is used only for inter-call-agent communication and not directly with the end gateways, the REGISTER method is not required. The REGISTER method is often used by gateways such as IP phones to dynamically register with a softswitch. Typically, a CMS is statically configured with the address of either the other CMSs it needs to communicate with or the address of a proxy server.

CMSS Timers

In SIP, retransmission is handled based on two timers, T1 and T2. The retransmission interval starts at T1 seconds and is doubled for each iteration (as long as this is below the maximum of T2 seconds) until a maximum number of retransmissions is reached.

CMSS adds a new timer, T3, to protect against "hung" sessions because of an application layer failure. This timer is different depending on whether you are referring to the originating CMS or the terminating CMS. On the originating CMS, this timer is known as *T-setup* and should have a value of 5 to 6 minutes. It is the time between receiving the first provisional response to an INVITE request and the final 200 OK. If this timer expires, the CMS sends out a CANCEL request aborting the call. On the terminating CMS this timer is known as T-ringing and should have a value of 3 to 4 minutes. It is the time between the receiving of an INVITE request and the ACK request completing the three-way handshake procedure. If this timer expires, the CMS sends out a 480 Temporary Unavailable response.

CMSS Call Flow

Now we'll revisit the basic call flow between Oscar and Tina, except this time full-fledged CMSS is used between their call agents. No packet traces are shown because a CMS with CMSS support was not available; instead, only the high-level call flow is shown. Figure 9-11 shows the messaging for the first stage of the call.

Figure 9-11 *Basic CMSS Call (Stage 1)*

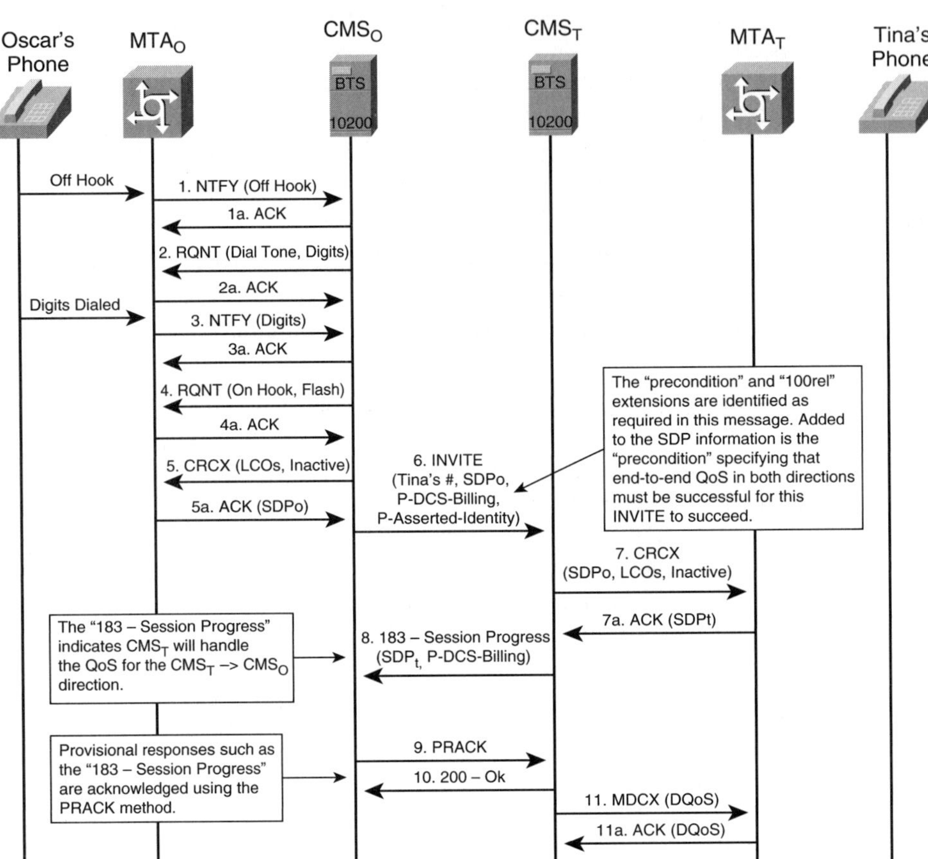

First, look at the INVITE message in Step 6; notice the P-DCS-billing message header is included to provide billing information, and the P-Asserted-Identity message header is there to provide the Caller ID feature. The INVITE also indicates that the 100rel and precondition extensions are required and contains the precondition requesting that QoS resources be reserved in both directions end-to-end before ring tone or ring back tone is provided.

The following is an example of what this INVITE might look like:

```
INVITE sip:9875012003@sia-109CA109.sanjose.cc.cisco.com:5060;user=phone;
  tgid=rtp SIP/2.0
Via: SIP/2.0/UDP  sia-104CA104.roseville.cc.cisco.com:5060;branch=z9hG4bK_
  1104150322142038661 3
From: "Oscar"<sip:3215012002@sia-104CA104.roseville.cc.cisco.com;user=phone>;
  tag=1_1104_f150322_7195
To: <sip:9875012003@sia-109CA109.sanjose.cc.cisco.com;user=phone>
Call-ID: 1700461475<long random value used instead of UAC FQDN>
```

```
CSeq: 1 INVITE
Max-Forwards: 70
Supported: 100rel,precondition,P-DCS
Contact: <sip:3215012002@sia-104CA104.roseville.cc.cisco.com:5060>
Require: 100rel,precondition
Allow: INVITE,ACK,BYE,CANCEL,OPTIONS,REGISTER,INFO,PRACK,SUBSCRIBE,
  NOTIFY,REFER,UPDATE
P-Asserted-Identity: "Oscar"<sip:3215012002@sia-104CA104.roseville.cc.cisco.com;
  user=phone>
P-DCS-Billing: <contains charging number, calling jurisdiction information, and
  location routing number>
Content-Length:    403
Content-Type: application/sdp

v=0
o=- 161152184 29757 IN IP4 24.34.241.96
s=-
c=IN IP4 24.34.241.96
b=AS:82
t=0 0
m=audio 53456 RTP/AVP 0 8 15
a=X-pc-secret:base64:<hex string omitted>
a=X-pc-csuites-rtp:62/51 64/51 60/51 60/50
a=X-pc-csuites-rtcp:81/71 81/70 80/70
a=X-pc-nrekey:0
a=mptime:20 20 20
a=ptime:20
a=des:qos mandatory e2e sendrecv      <- QoS precondition
```

In this case, Oscar has not requested privacy, so his identity is not hidden.

Tina's CMS acknowledges the INVITE with the provisional response of 183 Session Progress and includes Tina's SDP information as well as its own billing information. This message also indicates that Tina's CMS handles the QoS resource precondition for the receive direction, and it requests confirmation when Oscar's CMS has successfully reserved QoS resources on its end. The following is an example of what this message might look like:

```
SIP/2.0 183 Session Progress
Via: SIP/2.0/UDP  sia-104CA104.roseville.cc.cisco.com:5060;branch=z9hG4bK_
  11041503221420386613;received=13.13.5.40
From: "Oscar"<sip:3215012002@sia-104CA104.roseville.cc.cisco.com;user=phone>;
  tag=1_1104_f150322_7195
To: <sip:9875012003@sia-109CA109.sanjose.cc.cisco.com;user=phone>
Call-ID: 1700461475<long random value used instead of UAC FQDN>
Contact: <sip:9875012003@sia-109CA109.sanjose.cc.cisco.com:5060>
Require: 100rel
Allow: INVITE,ACK,BYE,CANCEL,OPTIONS,REGISTER,INFO,PRACK,SUBSCRIBE,
  NOTIFY,REFER,UPDATE
P-Asserted-Identity: "Oscar" <sip:3215012002@sia-
  104CA104.roseville.cc.cisco.com;user=phone>;
Content-Length:    393
Content-Type: application/sdp

v=0
o=- 161075574 2353 IN IP4 24.34.240.240
c=IN IP4 24.34.240.240
b=AS:82
t=0 0
```

```
m=audio 53456 RTP/AVP 0
a=X-pc-secret:base64:<hex string omitted>
a=X-pc-csuites-rtp:62/51 64/51 60/51 60/50
a=X-pc-csuites-rtcp:81/71 81/70 80/70
a=X-pc-nrekey:0
a=mptime:20
a=ptime:20
a=des:qos mandatory e2e sendrecv
a=conf:qos e2e recv
```

Oscar's CMS then acknowledges the receipt of the provisional response by sending a provisional acknowledge (PRACK) message. This message is then acknowledged by Tina's CMS. At this point, Tina's CMS also initiates the reservation of DQoS resources on the CMTS connected to Tina's MTA.

Figure 9-12 shows the messaging for the second stage of the call.

Figure 9-12 *Basic CMSS Call (Stage 2)*

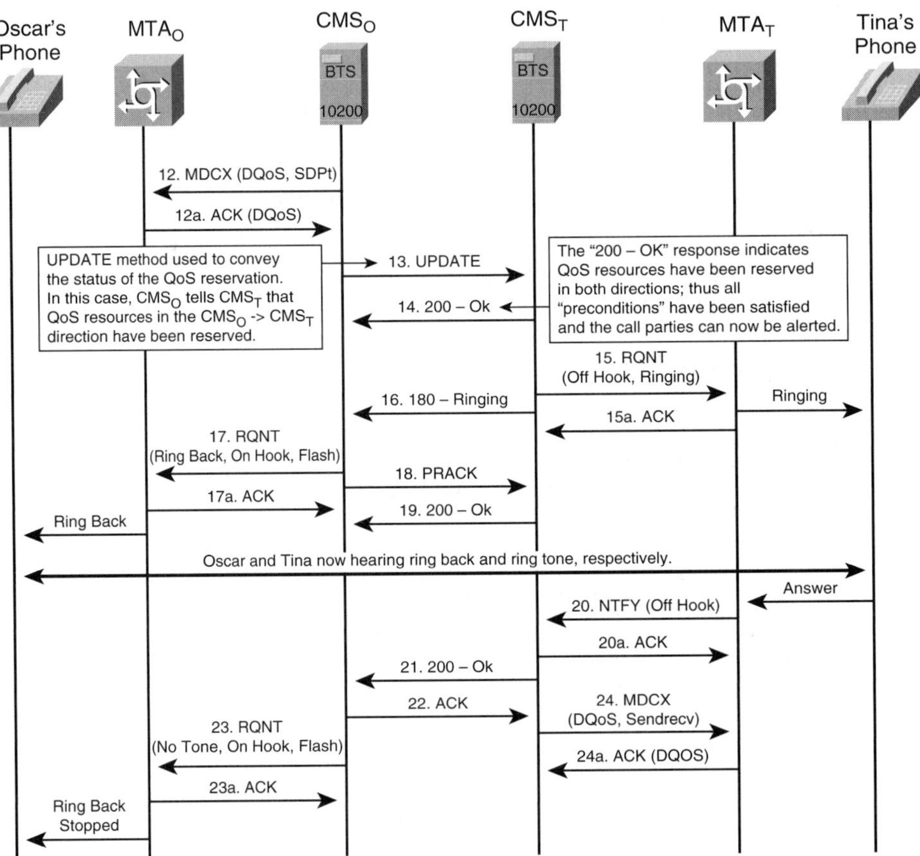

After Oscar's MTA has successfully reserved DQoS resources, his CMS conveys this information to Tina's CMS by sending an UPDATE message.

Knowing that DQoS resources are successfully reserved for both call legs, Tina's CMS instructs her MTA to ring her phone. Tina's CMS also instructs Oscar's CMS to provide local ring back tone to Oscar's phone. Again, because the 180 Ringing message is a provisional response, Oscar's CMS sends a PRACK message to alert Tina's CMS that the message was received.

As you saw in the SIP example, after Tina answers the call, the final 200 OK response to the INVITE is sent and DQoS resources are committed. And to complete the three-way handshake, Oscar's CMS sends an ACK message acknowledging the receipt of the final response message.

Figure 9-13 shows the messaging for the final stage of the call.

Figure 9-13 *Basic CMSS Call (Stage 3)*

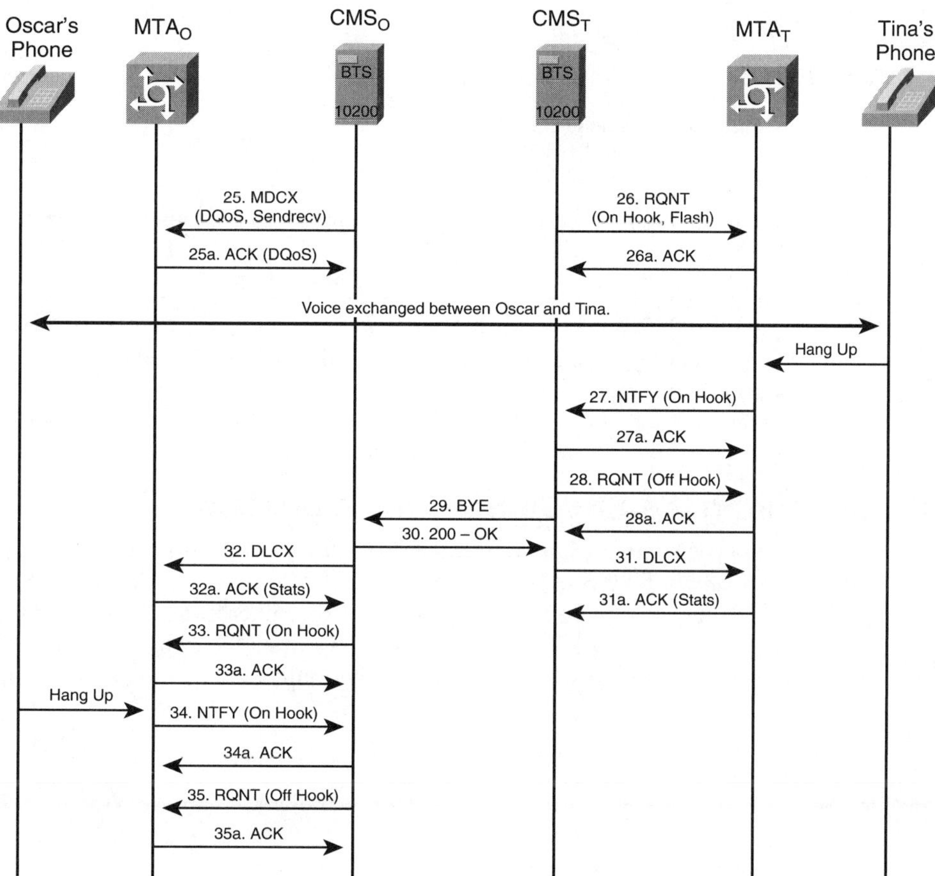

This call flow is no different than you saw earlier with the basic SIP call.

Configuring SIP and CMSS on the BTS 10200

By now you should have a pretty good understanding of the SIP and CMSS protocols and how they are used in PacketCable. The next part of this chapter covers the configuration on the Cisco BTS 10200 related to SIP and CMSS signaling.

In other chapters you learned how to configure MTAs, CMTSs, SGs, and MGs, so please refer to those if needed. Only a couple steps are needed to configure inter-call-agent communication, as you will soon see. As always, keep in mind that each of these tasks has recommended configuration settings; these setting should be fine for the general case but might require some fine tuning depending on the environment. The following are the basic steps to configure SIP, SIP-T, and/or CMSS on the BTS:

Step 1 Configure the SIP/CMSS characteristics that are to be associated to the softswitch trunk group (Softswitch trunk group profile table).

Step 2 Configure the SIP/CMSS trunk group (Trunk group table).

Step 3 Bring the trunk group into service.

Step 4 Define routes for trunk groups (Route table).

Step 5 Define destinations (Destination table).

Step 6 Define dial plan entries (Dial Plan table).

Keep in mind that similar configuration is also required on the remote softswitch.

NOTE The Cisco BTS 10200 also supports SIP gateways such as Cisco IP phones; the steps required for this configuration are not covered in this book because they are outside the scope of PacketCable 1.x.

Step 1: Configuring the SIP/CMSS Characteristics

In this section you configure common characteristics for a collection of trunk groups. If you are using SIP and/or CMSS to connect to multiple CMSs and they all share the same characteristics, they can share a common profile. Remember from Chapter 8 that different tables are used for this, depending on the type of signaling the trunk group uses. SIP and CMSS trunks are categorized as softswitch trunk groups, so they use the softsw-tg-profile table:

```
CLI>add softsw-tg-profile id=Softswitch trunk group profile name; protocol-type=CMSS
  ¦ SIP ¦ SIP-T;
```

Table 9-3 lists the key fields relating to SIP/CMSS.

Table 9-3 *Key Trunk Group Profile Fields*

Field	Default Value	Recommended Value	Description
PROTOCOL-TYPE	N/A	CMSS if the remote user agent is also a PacketCable CMS and CMSS supported on both switches SIP if the remote user agent is non-PacketCable SIP-T if ISUP parameters need to be transported	Denotes the protocol used to communicate between devices. For a PacketCable CMS you should use CMSS. For non-PacketCable you should use SIP (or SIP-T if you need to be able to transport ISUP parameters).
APPLY-USER-PRIVACY	N	Y	If set to "Y" and the originator requested privacy, information such as the calling # and name is hidden in the "From" header in the initial outbound INVITE message.
DIVERSION-HEADER-SUPP	N	N	Whether or not the SIP Diversion header is supported, which can be used for applications such as call forwarding.
DTMF-RELAY-METHOD	NONE	N/A	When out-of-band DTMF relay is used, this specifies what method conveys DTMF digits; valid values are NOTIFY and INFO for those methods.
ES-SUPP	N	N/A	Whether or not electronic surveillance is supported on the SIP/CMSS interface.
MAX-FORWARDS	70	70	Defines the value to use in the MAX-FORWARDS parameter in outbound INVITE messages.

continues

Table 9-3 *Key Trunk Group Profile Fields (Continued)*

Field	Default Value	Recommended Value	Description
NON-SRV-TRANSPORT	UDP	UDP	Defines the transport layer protocol used; valid values are UDP, TCP, and UDP_ONLY. If set to UDP_ONLY, TCP is not attempted even if the maximum UDP message size is reached.
PRACK-FLAG	N	Y	Specifies whether INVITE messages require provisional responses.
REDIRECT-SUPPORTED	VALID-DOMAINS-ONLY	VALID-DOMAINS-ONLY	Defines what the BTS should do in the event a 3xx redirect response is received.
SIP-SIG-LOWDELAY	Y	Y	Whether or not to set the IP TOS delay bit in SIP/CMSS packets.
SIP-SIG-PRECEDENCE	3 (Flash)	3 (Flash)	The IP Precedence value used for SIP/CMSS packets.
SIP-SIG-RELIABILITY	N	N	Whether or not to set the IP TOS reliability bit in SIP/CMSS packets.
SIP-SIG-THROUGHPUT	N	N	Whether or not to set the IP TOS throughput bit in SIP/CMSS packets.
SIP-TIMER-PROFILE-ID	N/A	Key into SIP-TIMER-PROFILE table	This table can be used to set the values for various protocol timers.
SIPT-ISUP-VER	N/A	N/A	Only valid for the SIP-T protocol; defines the SIP-T version.
USE-PAI-HDER-FOR-ANI	N	N/A	Controls the P-Asserted-ID (PAI) header; used to send and receive calling party information.

The configuration used for the call flow examples is as follows:

```
CLI>show softsw_tg_profile
ID=8503
PROTOCOL_TYPE=SIP
INBAND_TONE_AVAILABLE=Y
DESCRIPTION=Softswitch TG to SNJS BTS
PRACK_FLAG=N
REDIRECT_SUPPORTED=VALID_DOMAINS_ONLY
```

```
DTMF_RELAY_METHOD=NONE
SESSION_TIMER_ALLOWED=N
REFER_ALLOWED=N
DNS_SRV_SUPP=NONE
AUTHORIZATION_REQUIRED=N
SIP_SIG_LOWDELAY=Y
SIP_SIG_THROUGHPUT=N
SIP_SIG_RELIABILITY=N
SIP_SIG_PRECEDENCE=FLASH
DIVERSION_HEADER_SUPP=N
NON_SRV_TRANSPORT=UDP_ONLY
VOICE_MAIL_TRUNK_GRP=N
TRUNK_SUB_GRP_TYPE=TGID
USE_SIPT_ISUP_BASE=Y
ES_SUPP=N
AUDIT_THRESHOLD=3
GTD_PARMS=ALL
GTD_MODE=COMPACT
MAX_FORWARDS=70
SCALE_FACTOR=1
HOP_COUNTER_SUPP=Y
HOP_COUNTER_MAX=20
SEND_FULL_E164=N
SEND_CIC_PARAM=Y
USE_PAI_HDR_FOR_ANI=N
APPLY_USER_PRIVACY=N
DNS_SRV_ADV_ON_RETRANS_TIMEOUT=Y
```

Notice basic SIP is used as the CMS-to-CMS protocol in this case.

Step 2: Configuring the SIP/CMSS Trunk Group

Now that a trunk group profile has been defined, you need to define your trunk groups:

```
CLI>add trunk-grp id=Trunk group number; call-agent-id=call agent name; tg-
  type=SOFTSW; tg-profile-id=Trunk group profile name from previous Step; softsw-
  tsap-addr=FQDN of remote CMS; dial-plan-id=Dial plan name;
```

Table 9-4 lists the key fields relating to SIP/CMSS.

Table 9-4 *Key Trunk Group Fields*

Field	Default Value	Recommended Value	Description
CALL-AGENT-ID	N/A	Key for the Call Agent table	Identifier for the call agent to which trunks off this trunk group are assigned.
TG-TYPE		SOFTSW	Identifies the trunk group as a softswitch type (i.e., SIP/CMSS).
TG-PROFILE	N/A	Key for appropriate trunk group profile table	Identifier for the trunk group profile characterizing general common trunk group parameters; defined in previous step.

continues

Table 9-4 *Key Trunk Group Fields (Continued)*

Field	Default Value	Recommended Value	Description
SOFTSW-TSAP-ADDR	N/A	FQDN of the remote CMS	Identifier of the remote CMS; a port number can be added to the end if the default of 5060 is not used. This is also needed if multiple trunk groups to the same switch are needed.
DIAL-PLAN-ID	N/A	Key for Dial Plan table	Dial plan used for calls originating from this trunk group.

Here is the trunk group configuration used in the call flow examples:

```
CLI>show trunk-grp tg-type=SOFTSW
ID=8503
CALL_AGENT_ID=CA104
TG_TYPE=SOFTSW
SOFTSW_TSAP_ADDR=sia-109CA109.sanjose.cc.cisco.com:5060
TG_PROFILE_ID=8503
STATUS=INS
DIRECTION=BOTH
SEL_POLICY=ASC
GLARE=SLAVE
ALT_ROUTE_ON_CONG=N
SIGNAL_PORTED_NUMBER=N
POP_ID=S218954_HAS_DAKO
DIAL_PLAN_ID=INCOMING
DESCRIPTION=Softswitch Trunk Group to SNJS BTS
DEL_DIGITS=0
TRAFFIC_TYPE=LOCAL
ANI_BASED_ROUTING=N
CLLI=RSVLMNFYPS0
MGCP_PKG_TYPE=NA
TRUNK_SUB_GRP=rtp
ANI_SCREENING=N
SEND_RDN_AS_CPN=N
STATUS_MONITORING=N
SEND_EARLY_BKWD_MSG=N
EARLY_BKWD_MSG_TMR=5
SCRIPT_SUPP=N
VOICE_LAYER1_USERINFO=AUTO
VOICE_INFO_TRANSFER_CAP=AUTO
PERFORM_LNP_QUERY=N
```

Step 3: Putting the Trunk Group into Service

Because this trunk group runs over IP you have no physical trunks or terminations to create. You just need to enable the trunk group and configure routing. Enabling the trunk group works the same way as for SS7 and channel-associated signaling (CAS) trunk groups as shown in the previous chapter:

```
CLI>control trunk-grp id=Trunk group number; target-state=INS; mode=FORCED;
```

The result of this operation can be verified with the **status** command, as shown:

```
CLI>status trunk-grp id=Trunk group number
```

Again, the trunk group needs to have an operational status of in-service before it can be used as shown in the following example:

```
CLI>status trunk-grp id=8503
TGN_ID -> 8503
RESULT -> ADM configure result in success
REASON -> ADM executed successfully
ADMIN_STATE -> ADMIN_INS
OPER_STATE -> Trunk group in-service
PRIMARY_OPER_STATE -> NOT USED
BACKUP_OPER_STATE -> NOT USED
```

Step 4: Defining Routes for Trunk Groups

At this point, you can configure routing for this trunk group. Call routing was covered in detail in Chapter 8, so less detail is shown here. For basic routing you first need to define a route for the SIP/CMSS trunk group. The syntax for adding an entry to this table is as follows:

```
CLI>add route id=route identifier; tgn1-id=trunk group #1 identifier; tgn2-id=trunk
  group #2 identifer; …; tgn10-id=trunk group #10 identifer;
```

The configuration of this table used for the call flow examples is

```
CLI>show route tgn1-id=8503
ID=SNJSBAPS0_8503
TGN1_ID=8503
DEL_DIGITS1=0
DEL_DIGITS2=0
DEL_DIGITS3=0
DEL_DIGITS4=0
DEL_DIGITS5=0
DEL_DIGITS6=0
DEL_DIGITS7=0
DEL_DIGITS8=0
DEL_DIGITS9=0
DEL_DIGITS10=0
TG_SELECTION=SEQ
NEXT_ACTION=NONE
```

Step 5: Defining Destinations

Next, create a destination associating this route with a call type and route type:

```
CLI>add destination dest-id=destination identifier; call_type=call type identifier;
  route_type=route type identifier; route_guide_id=index into next table;
```

The destination entry used in the call flow examples is as follows:

```
CLI>show destination route_id=SNJSBAPS0_8503
DEST_ID=SNJSBAPS0IA_8503
CALL_TYPE=TOLL
ROUTE_TYPE=RID
ZERO_PLUS=Y
```

```
INTRA_STATE=Y
ROUTE_ID=SNJSBAPS0_8503
GAP_ROUTING=N
CLDPTY_CTRL_REL_ALWD=N
CALL_SUBTYPE=NONE
ACQ_LNP_QUERY=NA
```

Step 6: Defining the Dial Plan

Finally, create entries in the dial plan for this destination:

```
CLI>add dial_plan id=dial plan identifier from previous step; digit-string=numerical
  string of dialed digits; dest-id=destination identifier previously defined;
```

The configuration used in the call flow examples is as follows:

```
CLI>show dial-plan dest-id=SNJSBAPS0IA_8503
ID=INCOMING
DIGIT_STRING=987503
DEST_ID=SNJSBAPS0IA_8503
SPLIT_NPA=NONE
DEL_DIGITS=0
MIN_DIGITS=10
MAX_DIGITS=10
NOA=NATIONAL
```

In the troubleshooting portion of this chapter, you'll see how to verify the BTS translations similar to the examples shown in Chapter 8.

Troubleshooting Tips

The last part of this chapter contains some troubleshooting tips that might help you in diagnosing and fixing problems relating to CMS-to-CMS communication.

As you saw earlier, one of the best ways to troubleshoot an SIP or CMSS problem is to simply look at the messaging. The response messages give great insight into the probable cause of a problem with the response code and description they return. Also the fact that SIP messages are encoded in plaintext means you can easily comprehend the packets by inserting a packet sniffer. If your CMS is UNIX-based, the **snoop** command can be used to view the messages. Of course, if these messages are encrypted, you have to find some other means to troubleshoot, such as examining call agent log files.

If you aren't seeing any messages between CMSs, you most likely have a problem with IP routing or DNS resolution. Check items such as routing tables, DNS configuration, and firewalls. Remember that on the Cisco BTS 10200, the status of the trunk group should have an operation state of in-service. If not, make sure the TSAP-ADDR of the remote CMS is resolvable and pingable.

Also any of the problems that can exist with NCS, TGCP, and DQoS can occur in a CMSS call as well. If you have problems with CODEC negotiation or reserving DQoS resources, those might appear as a CMSS failure. Refer to Chapter 10, "Audio CODECs," and Chapter 13 for troubleshooting tips as well.

Chapters 7 and 8 contain a lot of information on BTS troubleshooting that can be applied here as well. SIP OFFNET call routing can be verified with the BTS Translation Verification Tool (TVT) the same way as ISUP OFFNET call routing. For example, the translations for the ONNET-to-OFFNET call example are as follows:

```
CLI>translate line calling_dn=321-501-2002;called_dn=19875012003
TRANSLATE ON CALL AGENT CA104 IS... ->
TABLEINFO ->
TABLE: SUBSCRIBER
ID=sub1@jriddel_lsys2;CATEGORY=INDIVIDUAL;NAME=Oscar;STATUS=ACTIVE;DN1=3215012002;
  PRIVACY=NONE;RING_TYPE_DN1=1;TERM_ID=aaln/1;MGW_ID=jriddel_lsys2;
  PIC1=NONE;PIC2=NONE;PIC3=NONE;GRP=N;USAGE_SENS=Y;
  SUB_PROFILE_ID=PacketCable_subs;TERM_TYPE=TERM;IMMEDIATE_RELEASE=N;
  TERMINATING_IMMEDIATE_REL=N;SEND_BDN_AS_CPN=N;SEND_BDN_FOR_EMG=N;
  PORTED_IN=N;BILLING_TYPE=NONE;VMWI=Y;SDT_MWI=Y;
TABLE: SUBSCRIBER_PROFILE
ID=PacketCable_subs;DIAL_PLAN_ID=INCOMING;LOCAL_PFX1_OPT=NR;TOLL_PFX1_OPT=RQ;
  POP_ID=PacketCable_PoP;OLI=0;EA_USE_PIC1=Y;QOS_ID=G711u_20ms;
  INTERLATA_PFX1_OPT=RQ;
TABLE: DIAL_PLAN_PROFILE
ID=INCOMING;DESCRIPTION=dialing plan profile for incoming calls;
  NANP_DIAL_PLAN=Y;OVERDECADIC_DIGITS_SUPP=N;NOA_BASED_ROUTING=N;
TABLE: DIAL_PLAN
ID=INCOMING;DIGIT_STRING=987501;DEST_ID=SNJSBAPS0IA_8503;SPLIT_NPA=NONE;
  DEL_DIGITS=0;
  MIN_DIGITS=10;MAX_DIGITS=10;NOA=NATIONAL;
TABLE: DESTINATION
DEST_ID=SNJSBAPS0IA_8503;CALL_TYPE=TOLL;ROUTE_TYPE=RID;ZERO_PLUS=Y;
  INTRA_STATE=Y;ROUTE_ID=SNJSBAPS0_8503;GAP_ROUTING=N;CLDPTY_CTRL_REL_ALWD=N;
  CALL_SUBTYPE=NONE;ACQ_LNP_QUERY=NA;
TABLE:  ROUTE
ID=SNJSBAPS0_8503;TGN1_ID=8503;DEL_DIGITS1=0;DEL_DIGITS2=0;DEL_DIGITS3=0;
  DEL_DIGITS4=0;DEL_DIGITS5=0;DEL_DIGITS6=0;DEL_DIGITS7=0;DEL_DIGITS8=0;
  DEL_DIGITS9=0;DEL_DIGITS10=0;TG_SELECTION=SEQ;NEXT_ACTION=NONE;
TABLE: TRUNK_GRP
ID=8503;CALL_AGENT_ID=CA104;TG_TYPE=SOFTSW;
  SOFTSW_TSAP_ADDR=sia-109CA109.sanjose.cc.cisco.com:5060;TG_PROFILE_ID=8503;
  STATUS=INS;DIRECTION=BOTH;SEL_POLICY=ASC;GLARE=SLAVE;
  ALT_ROUTE_ON_CONG=N;SIGNAL_PORTED_NUMBER=N;POP_ID=S218954_HAS_DAKO;
DIAL_PLAN_ID=INCOMING;DESCRIPTION=Softswitch Trunk Group to SNJS BTS;
  DEL_DIGITS=0;TRAFFIC_TYPE=LOCAL;ANI_BASED_ROUTING=N;CLLI=RSVLMNFYPS0;
  MGCP_PKG_TYPE=NA;TRUNK_SUB_GRP=rtp;ANI_SCREENING=N;SEND_RDN_AS_CPN=N;
  STATUS_MONITORING=N;SEND_EARLY_BKWD_MSG=N;EARLY_BKWD_MSG_TMR=5;SCRIPT_SUPP=N;
  VOICE_LAYER1_USERINFO=AUTO;VOICE_INFO_TRANSFER_CAP=AUTO;ACCESS_TYPE=COMBINED;
  POI=INTER_ENDOFFICE;PERFORM_LNP_QUERY=N;
Reply : Success:
```

And as was the case in these previous chapters, examining billing records is another good place for troubleshooting. Following are sample billing records for an ONNET-to-OFFNET toll call using SIP, followed by an OFFNET-to-ONNET call using SIP:

```
CLI>report billing-record
..
CALLTYPE=TOLL
SIGSTARTTIME=2006-07-14 18:36:25
SIGSTOPTIME=2006-07-14 18:36:33
ICSTARTTIME=2006-07-14 18:36:25
```

```
ICSTOPTIME=2006-07-14 18:36:33
CALLCONNECTTIME=2006-07-14 18:36:29
CALLANSWERTIME=2006-07-14 18:36:29
CALLDISCONNECTTIME=2006-07-14 18:36:33
CALLELAPSEDTIME=00:00:04
INTERCONNECTELAPSEDTIME=00:00:08
ORIGNUMBER=3215012002
TERMNUMBER=9875012003
CHARGENUMBER=3215012002
DIALEDDIGITS=19875012003
OFFHOOKINDICATOR=1
SHORTOFFHOOKINDICATOR=0
CALLTERMINATIONCAUSE=NORMAL_CALL_CLEARING
OPERATORACTION=0
ORIGSIGNALINGTYPE=0
TERMSIGNALINGTYPE=5
ORIGTRUNKGROUPNUMBER=0
TERMTRUNKGROUPNUMBER=8503
OUTGOINGTRUNKNUMBER=0
ORIGCIRCUITID=0
TERMCIRCUITID=0
PICSOURCE=2
ICINCIND=1
ICINCEVENTSTATUSIND=20
ICINCRTIND=0
ORIGQOSTIME=2006-07-14 18:36:33
ORIGQOSPACKETSSENT=396
ORIGQOSPACKETSRECD=361
ORIGQOSOCTETSSENT=64152
ORIGQOSOCTETSRECD=57760
ORIGQOSPACKETSLOST=0
ORIGQOSJITTER=0
ORIGQOSAVGLATENCY=0
TERMQOSTIME=2006-07-14 18:36:33
TERMQOSPACKETSSENT=0
TERMQOSPACKETSRECD=0
TERMQOSOCTETSSENT=0
TERMQOSOCTETSRECD=0
TERMQOSPACKETSLOST=0
TERMQOSJITTER=0
TERMQOSAVGLATENCY=0
PACKETIZATIONTIME=20
SILENCESUPPRESSION=0
ECHOCANCELLATION=0
CODECTYPE=PCMU
CONNECTIONTYPE=IP
OPERATORINVOLVED=0
CASUALCALL=0
INTERSTATEINDICATOR=0
OVERALLCORRELATIONID=CA104796
TIMERINDICATOR=0
RECORDTYPE=NORMAL RECORD
TERMCLLI=RSVLMNFYPS0
CALLAGENTID=CA104
ORIGPOPTIMEZONE=GMT
ORIGTYPE=ON NET
TERMTYPE=OFF NET
NASERRORCODE=0
NASDLCXREASON=0
FAXINDICATOR=NOT A FAX
FAXPAGESSENT=0
FAXPAGESRECEIVED=0
ORIGPOPID=PacketCable_PoP
TERMPOPID=S218954_HAS_DAKO
```

```
TERMPOPTIMEZONE=EST
DIALPLANID=INCOMING
TERMINATINGSIPUSERNAME=3215012002
TERMINATINGSIPCALLID=1060718038@sia-104CA104.roseville.cc.cisco.com
TERMINATINGSIPADJACENTHOPADDR=13.13.6.200
TERMINATINGSIPCALLTYPE=SIP
CALLINGPARTYCATEGORY=Ordinary Subscriber
CALLEDPARTYINDICATOR=Ordinary Subscriber
CALLEDPARTYPORTEDIN=No
CALLINGPARTYPORTEDIN=No
BILLINGRATEINDICATOR=None
ORIGENDPOINTADDR=x1-6-00-13-10-50-37-f6.cc.cisco.com
TERMENDPOINTADDR=sia-109CA109.sanjose.cc.cisco.com:5060
ORIGCMTSID=PC-CMTS.cc.cisco.com
SENSORID=000000
ORIGPRIVACYINDICATOR=NONE
CALLTYPE=LOCAL
SIGSTARTTIME=2006-07-14 14:37:03
SIGSTOPTIME=2006-07-14 14:37:12
CALLCONNECTTIME=2006-07-14 14:37:06
CALLANSWERTIME=2006-07-14 14:37:06
CALLDISCONNECTTIME=2006-07-14 14:37:12
CALLELAPSEDTIME=00:00:06
INTERCONNECTELAPSEDTIME=00:00:00
ORIGNUMBER=9875012003
TERMNUMBER=3215012002
CHARGENUMBER=9875012003
DIALEDDIGITS=3215012002
OFFHOOKINDICATOR=1
SHORTOFFHOOKINDICATOR=0
CALLTERMINATIONCAUSE=NORMAL_CALL_CLEARING
OPERATORACTION=0
ORIGSIGNALINGTYPE=5
TERMSIGNALINGTYPE=0
ORIGTRUNKGROUPNUMBER=8503
TERMTRUNKGROUPNUMBER=0
OUTGOINGTRUNKNUMBER=0
ORIGCIRCUITID=0
TERMCIRCUITID=0
ORIGQOSTIME=2006-07-14 14:37:12
ORIGQOSPACKETSSENT=0
ORIGQOSPACKETSRECD=0
ORIGQOSOCTETSSENT=0
ORIGQOSOCTETSRECD=0
ORIGQOSPACKETSLOST=0
ORIGQOSJITTER=0
ORIGQOSAVGLATENCY=0
TERMQOSTIME=2006-07-14 14:37:12
TERMQOSPACKETSSENT=439
TERMQOSPACKETSRECD=400
TERMQOSOCTETSSENT=71118
TERMQOSOCTETSRECD=64000
TERMQOSPACKETSLOST=0
TERMQOSJITTER=0
TERMQOSAVGLATENCY=24
PACKETIZATIONTIME=20
SILENCESUPPRESSION=0
ECHOCANCELLATION=0
CODECTYPE=PCMU
CONNECTIONTYPE=IP
OPERATORINVOLVED=0
CASUALCALL=0
INTERSTATEINDICATOR=0
OVERALLCORRELATIONID=CA104797
```

```
TIMERINDICATOR=0
RECORDTYPE=NORMAL RECORD
ORIGCLLI=RSVLMNFYPS0
CALLAGENTID=CA104
ORIGPOPTIMEZONE=EST
ORIGTYPE=OFF NET
TERMTYPE=ON NET
ORIGINATINGSIPUSERNAME=9875012003
ORIGINATINGSIPCALLID=0722393924@sia-109CA109.sanjose.cc.cisco.com
ORIGINATINGSIPADJACENTHOPADDR=13.13.6.168
NASERRORCODE=0
NASDLCXREASON=0
FAXINDICATOR=NOT A FAX
FAXPAGESSENT=0
FAXPAGESRECEIVED=0
ORIGPOPID=S218954_HAS_DAKO
TERMPOPID=PacketCable_PoP
TERMPOPTIMEZONE=GMT
DIALPLANID=INCOMING
ORIGINATINGSIPCALLTYPE=SIP
CALLINGPARTYCATEGORY=Ordinary Subscriber
CALLEDPARTYINDICATOR=No Indication
CALLEDPARTYPORTEDIN=No
CALLINGPARTYPORTEDIN=No
BILLINGRATEINDICATOR=None
ORIGENDPOINTADDR=sia-109CA109.sanjose.cc.cisco.com:5060
TERMENDPOINTADDR=x1-6-00-13-10-50-37-f6.cc.cisco.com
TERMCMTSID=PC-CMTS.cc.cisco.com
SENSORID=000000
ORIGPRIVACYINDICATOR=NONE
```

Notice in these billing records the inclusion of SIP specific parameters. For example, the username (ORIGINATINGSIPUSERNAME/TERMINATINGSIPUSERNAME), Call-ID (ORIGINATINGSIPCALLID/TERMINATINGSIPCALLID), and adjacent IP address (ORIGINATINGSIPADJACENTHOPADDR/ TERMINATINGSIPADJACENTHOPADDR) of the SIP leg of the calls are included. Also notice that the call statistic information (Packets Sent/Received, Octets Sent/Received, Packets Lost, Packet Jitter, and Packet Latency) are not included for the SIP leg of the calls. This information is collected by the local BTS the MTA communicates with in MGCP DLCX response messages and is not conveyed in SIP messages between the BTS systems.

Finally, SIP-related measurements are also available on the BTS (specifically, the SIA and SIM categories), as shown in the following example:

```
CLI>report measurement_sia_summary
TIMESTAMP                               2006-07-14 14:45:00
CALL_AGENT_ID                           CA104
CONDITION                               Normal
SIS_TOTAL_INCOM_MSG                     7
SIS_TOTAL_SUCC_INCOM_MSG                7
SIS_TOTAL_OUTG_MSG_ATTMP                7
SIS_TOTAL_SUCC_OUTG_MSG                 7
SIS_REQ_RETRAN_RX                       0
SIS_REQ_RETRAN_TX                       0
SIS_RSP_RETRAN_RX                       0
SIS_RSP_RETRAN_TX                       0
SIS_T1_TIMER_EXPIRED                    0
SIS_T2_TIMER_REACHED                    0
SIS_INVITE_RX                           1
SIS_INVITE_TX                           1
```

```
SIS_CANCEL_RX                           0
SIS_CANCEL_TX                           0
SIS_BYE_RX                              1
SIS_BYE_TX                              1
SIS_ACK_RX                              1
SIS_ACK_TX                              1
SIS_OPTIONS_RX                          0
SIS_OPTIONS_TX                          0
SIS_REGISTER_RX                         0
SIS_REGISTER_TX                         0
SIS_INFO_RX                             0
SIS_INFO_TX                             0
SIS_NOTIFY_RX                           0
SIS_NOTIFY_TX                           0
SIS_100_RX                              1
SIS_100_TX                              1
SIS_18X_RX                              1
SIS_18X_TX                              1
SIS_200_RX                              2
SIS_200_TX                              2
SIS_3XX_RX                              0
SIS_3XX_TX                              0
SIS_4XX_RX                              0
SIS_4XX_TX                              0
SIS_5XX_RX                              0
SIS_5XX_TX                              0
SIS_6XX_RX                              0
SIS_6XX_TX                              0
SIS_7XX_RX                              0
SIS_7XX_TX                              0
SIS_PROV_RSP_RETRAN_RX                  0
SIS_PROV_RSP_RETRAN_TX                  0
SIS_PRACK_RX                            0
SIS_PRACK_TX                            0
SIS_SUBSCRIBE_RX                        0
SIS_SUBSCRIBE_TX                        0
SIS_REFER_RX                            0
SIS_REFER_TX                            0
SIS_REFER_W_REPLACES_RX                 0
SIS_INVITE_REPLACES_TX                  0
SIS_INVITE_REPLACES_RX                  0
SIS_REL100_RX                           0
SIS_REL100_TX                           0
SIS_UNSUPPORTED_RX                      0
SIS_UPDATE_RX                           0
SIS_UPDATE_TX                           0
SIA_OUTG_INIT                           1
SIA_OUTG_SUCC                           1
SIA_OUTG_FAIL                           0
SIA_INCOM_INIT                          1
SIA_INCOM_SUCC                          1
SIA_INCOM_FAIL                          0
SIA_TOTAL_SUCC                          0
SIA_TOTAL_FAIL                          0
SIA_TOTAL_OUTG_MSG_FAIL                 0
SIA_TOTAL_INCOM_MSG_FAIL                0
SIA_REFRESHES_TX                        0
SIA_TOTAL_SESS_TIMER_FAIL               0
SIA_CALL_FAIL_BY_EXPIRED_REG            0
SIA_MWI_NOTIFY_TX                       0
SIA_MWI_NOTIFY_TX_FAIL                  0
SIA_MWI_NOTIFY_RX                       0
SIA_AUDIT_CCB_FREED                     0
SIA_AUDIT_CALL_RELEASED                 0
```

```
SIA_AUDIT_BCM_CALL_RELEASED           0
SIA_AUDIT_REGCONTACT_FREED            0
SIA_SECURE_FQDN_VIOLATION_REQ         0
SIA_SECURE_FQDN_VIOLATION_RESP        0
Reply : Success: Entry 1 of 1 returned.
CLI>report measurement_sim_summary
TIMESTAMP                             2006-07-14 14:45:00
CALL_AGENT_ID                         CA104
CONDITION                             Normal
SIS_TOTAL_INCOM_MSG                   22
SIS_TOTAL_SUCC_INCOM_MSG              22
SIS_TOTAL_OUTG_MSG_ATTMP              22
SIS_TOTAL_SUCC_OUTG_MSG               22
SIS_REQ_RETRAN_RX                     0
SIS_REQ_RETRAN_TX                     0
SIS_RSP_RETRAN_RX                     0
SIS_RSP_RETRAN_TX                     0
SIS_T1_TIMER_EXPIRED                  0
SIS_T2_TIMER_REACHED                  0
SIS_INVITE_RX                         0
SIS_INVITE_TX                         10
SIS_CANCEL_RX                         0
SIS_CANCEL_TX                         0
SIS_BYE_RX                            1
SIS_BYE_TX                            0
SIS_ACK_RX                            0
SIS_ACK_TX                            10
SIS_OPTIONS_RX                        0
SIS_OPTIONS_TX                        0
SIS_REGISTER_RX                       0
SIS_REGISTER_TX                       0
SIS_INFO_RX                           1
SIS_INFO_TX                           0
SIS_NOTIFY_RX                         0
SIS_NOTIFY_TX                         0
SIS_100_RX                            10
SIS_100_TX                            0
SIS_18X_RX                            0
SIS_18X_TX                            0
SIS_200_RX                            1
SIS_200_TX                            2
SIS_3XX_RX                            0
SIS_3XX_TX                            0
SIS_4XX_RX                            0
SIS_4XX_TX                            0
SIS_5XX_RX                            0
SIS_5XX_TX                            0
SIS_6XX_RX                            0
SIS_6XX_TX                            0
SIS_7XX_RX                            9
SIS_7XX_TX                            0
SIS_PROV_RSP_RETRAN_RX                0
SIS_PROV_RSP_RETRAN_TX                0
SIS_PRACK_RX                          0
SIS_PRACK_TX                          0
SIS_SUBSCRIBE_RX                      0
SIS_SUBSCRIBE_TX                      0
SIS_REFER_RX                          0
SIS_REFER_TX                          0
SIS_REFER_W_REPLACES_RX               0
SIS_INVITE_REPLACES_TX                0
SIS_INVITE_REPLACES_RX                0
SIS_REL100_RX                         0
SIS_REL100_TX                         0
```

```
SIS_UNSUPPORTED_RX                          0
SIS_UPDATE_RX                               0
SIS_UPDATE_TX                               0
SIM_BCM_MSG                                 3
SIM_RELATIONS                               2
SIM_TDP_N                                   0
SIM_TDP_R                                   2
SIM_EDP_N                                   0
SIM_EDP_R                                   0
SIM_INSTRUCT                                2
SIM_INSTRUCT_RSP                            1
SIM_TERMINATE_RX                            0
SIM_TERMINATE_TX                            2
SIM_FS_MSG_TX                               14
SIM_FS_MSG_RX                               22
SIM_FS_PING_MSG_TX                          8
SIM_FS_PING_NO_RSP_FAULTY                   0
SIM_FS_RESTART_MSG_TX                       0
SIM_FS_ASYNC_MSG_TX                         0
SIM_AUDIT_CCB_FREED                         0
SIM_AUDIT_SIP_CCB_FREED                     0
Reply : Success: Entry 1 of 1 returned.
```

Chapter Summary

A major benefit of a VoIP network is the capability to bypass the PSTN when your subscribers make long-distance or even inter-LATA telephone calls. By avoiding the PSTN you also avoid having to pay for any toll access charges; thus, telephone service can be provided at a much lower cost. The SIP protocol can be used for this purpose by communicating call signaling information between two or more different softswitches. The Call Management Server Signaling (CMSS) protocol is a variant of SIP used for this purpose in a PacketCable environment.

The SIP and CMSS signaling protocols utilize a distributed architecture model. This means that they rely on both devices involved in the call to have some intelligence. This is in contrast to protocols such as MGCP and NCS that use a centralized architecture.

The main constructs of SIP are user agents and sessions. SIP user agents are simply the devices speaking SIP, and sessions refer to the multimedia exchange of data between two or more user agents. SIP user agents are identified by a Uniform Resource Identifier (URI), which is similar to an MGCP endpoint name in that it consists of a local name portion and a domain name. Another type of SIP device is a proxy server, which can perform many functions such as routing SIP traffic, authorizing clients, accounting, and providing an aggregation point for security.

SIP messages consist of requests and responses, are encoded in plaintext, and consist of three parts: the start line, which identifies the message type; the message header, which defines the SIP parameters; and the message body, which contains a description of the multimedia session. The combination of an SIP request and its responses is known as a method; some SIP methods are INVITE, REGISTER, ACK, BYE, CANCEL, and OPTIONS. Key message headers are Via, From, To, and Call-ID.

Just as was the case with many other parts of PacketCable, examining message traces is an excellent way to gain an understanding of how the protocol works.

CMSS is a variation of SIP tailored for PacketCable call agents. It utilizes some SIP extensions to provide all the necessary functionality mandated by PacketCable. Among these extensions are methods of conveying DOCSIS resource reservation status (DQoS), providing telephony CLASS feature functionality such as call privacy, and passing additional information from CMS to CMS.

The configuration of SIP and CMSS on the Cisco BTS 10200 is a fairly simple and straightforward task; only a few tables are involved. Troubleshooting is best handled by looking at message traces and analyzing response codes. Verification of basic stuff like IP connectivity and DNS resolution is also a good place to start. Also many of the BTS troubleshooting techniques discussed in previous chapters apply here as well.

Chapter Review

1 True or False: You can use CMSS to place a call between a PacketCable MTA on a PacketCable CMS and a non-PacketCable IP phone.

Answer: False. CMSS can be used only between PacketCable CMSs; however, plain SIP could be used in this situation.

2 True or False: You can use CMSS to place a call between a PacketCable MTA on a PacketCable CMS and a PacketCable MG on a PacketCable MGC.

Answer: True. CMSS is for a CMS, MGC, or a CMS/MGC combination.

3 True or False: In the course of a call a CMS acts as both a user agent client (UAC) and a user agent server (UAS).

Answer: True. The UAC is the device sending a request, and the UAS is the device receiving it.

4 True or False: In CMSS it is valid to use a subscriber's name instead of their phone number in their URI.

Answer: False. The local name in the URI must represent the subscriber's phone number.

5 True or False: The CANCEL message is sent when an active call is abnormally disconnected.

Answer: False. The CANCEL message is used only when the session is not yet established.

6 Under what situation would you want to use TCP as the transport layer protocol for SIP/CMSS rather than UDP?

Answer: For messages larger than the maximum UDP supports and also if you want security via TLS.

7 What is the purpose of the Contact message header?

Answer: It tells the receiving SIP user agent where to send future requests.

8 How can an SIP proxy make sure it receives all the requests during a call so that it can provide mid-call feature functionality?

Answer: By using the Record-Route message header telling the SIP user agents to route all requests through the proxy server

9 Illustrate the message flow for the session setup in SIP.

Answer:

Orig CMS		Term CMS
-----	INVITE	-----→
←---	100 Trying	--
←---	180 Ringing	--
←----	200 Ok	-------
------	ACK	-----→

10 Illustrate the message flow for the session setup in CMSS.

Answer:

Orig CMS		Term CMS
-----	INVITE	-----→
←---	183 Session Progress	--
-----	PRACK	-----→
←----	200 Ok	-------
------	UPDATE	---→
←-----	200 Ok	-------
←-----	180 Ringing	---
-----	PRACK	-----→
←----	200 Ok	-------
←-----	200 Ok	-------
-------	ACK	------→

11 What is the fundamental difference between a response code in the 4xx range and one in the 5xx range?

Answer: Responses in the 5xx could indicate a temporary problem that might be fixed in the future, whereas responses in the 4xx indicate a permanent problem.

12 An SIP dialog comprises what three SIP message headers?

Answer: From, To, and Call-ID.

13 In CMSS, what triggers the UPDATE message to be sent during session setup?

Answer: The reservation of DQoS resources on the originating leg.

14 What does an SIP anonymizer do?

Answer: Provides privacy between service provider networks.

15 What safeguard in CMSS exists to protect against "hung" sessions?

Answer: The use of timer T3 during session setup.

16 How do you configure the BTS 10200 to use TCP for SIP transport instead of the default UDP?

Answer: By changing the value of the NON-SRV-TRANSPORT parameter in the softsw-tg-profile table.

17 What would be the reason for using a Session Border Controller between PacketCable VoIP networks?

Answer: To provide security by protecting the topology of one network from the other.

18 Why isn't call statistic data available for both sides of the call when SIP is used?

Answer: Call statistic information is returned in MGCP DLCX 250 response messages and is not relayed from CMS to CMS in SIP messaging.

PART IV

Media Stream

CHAPTER 10

Audio CODECs

Part III of this book covered PacketCable call signaling in detail. Specifically, Media Gateway Control Protocol (MGCP) and its PacketCable variants Network-based Call Signaling (NCS) and Trunk Gateway Control Protocol (TGCP), as well as the PacketCable Session Initiation Protocol (SIP) variant (Call Management Server Signaling [CMSS]) were examined. Now you will learn more about the technologies related to the bearer portion of PacketCable calls. You will gain a solid understanding of how analog voice is transmitted across the digital PacketCable network. This involves the analysis of different types of CODECs that convert between analog voice streams and the digital IP packets used in PacketCable. Factors that can adversely affect call quality and techniques used to quantify call quality are examined. You will also learn how facsimile, voice-band modem, and telephony for the hearing impaired transmissions are handled in PacketCable. Chapter 11, "RTP and RTCP," builds upon the information presented here by introducing the Real-time Transport Protocol (RTP) and Real-time Transport Control Protocol (RTCP) protocols and then showing how the concepts in this chapter apply to the real world using Cisco PacketCable components. RTP is used to transport encoded speech, and RTCP is used to provide feedback on the quality of these RTP sessions.

How CODECs Work

The first concept covered is the most fundamental one—how is voice transported from one party's mouth to the other party's ear in a voice over IP (VoIP) network? The microphone and headset in a phone convert between the sound waves coming from your mouth or heard by your ear and the electric signals used by the gateway to propagate speech across the telephony network. For a PacketCable network, it is then the gateway's responsibility to convert this electric analog signal into digital IP packets. So one gateway (such as the multimedia terminal adapter [MTA]) encodes the analog signal into IP packets, and the gateway at the other end of the call decodes the IP packets back into an analog signal. This coding and decoding is accomplished via a CODEC.

You will see how the CODECs used in PacketCable work and their differences later in the chapter, but first we examine greater detail on the general principles behind coding and decoding voice. Although the frequency range of human speech reaches 20 KHz, most human speech falls in the frequency range from 300 Hz to about 3,400 Hz. To make the analog to digital conversion simpler (and cheaper), other unneeded frequencies are removed or filtered out. Just to be safe, a high-pass filter is used to remove all frequencies greater than 4,000 Hz or 4 KHz.

NOTE This 4-KHz frequency range is often referred to as *narrowband voice*.

Now that only the portion of the analog signal needed for voice is obtained, the next step is to obtain samples of this signal to use for digital transmission. But how often do samples need to be taken? A scientist named Harry Nyquist proved that for the receiving end of the call to be fully reconstructed, the original analog signal samples must be taken at a rate of at least twice the maximum frequency in the original signal. In this case, the maximum frequency is 4 KHz, so sampling must occur at a rate of 8,000 samples per second. After this step you have analog samples. Next, these samples need to be expressed digitally. This is done by quantization. Figure 10-1 illustrates this sampling process, also known as Pulse Amplitude Modulation (PAM).

Figure 10-1 *The Sampling Process*

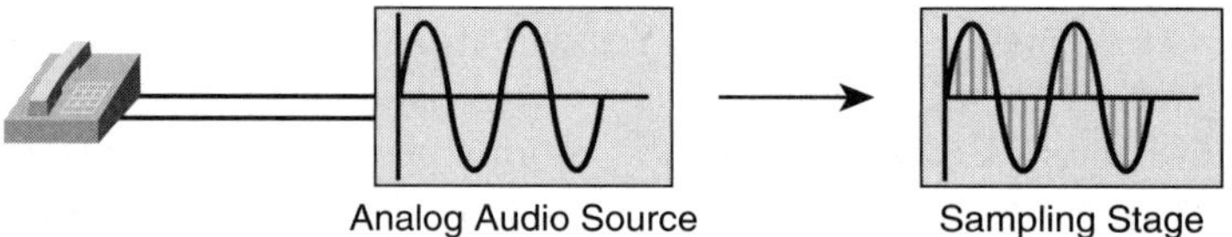

Quantization is the process of mapping ranges of analog values to discrete digital values. These digital values can then comprise the payload portion of the voice packets sent across the IP network. Some number of bits represents these digital values. Obviously, the more bits you use, the greater your granularity and the closer the digital value is to the actual value. This difference, or round-off error, is referred to as *quantization noise*. Logically, you want this quantization noise to be as small as possible to make your voice sample as true to the original as possible. Quantization noise increases the noise floor in the voice signal affecting its overall Signal to Noise Ratio (SNR). Higher SNR values result in better voice quality.

One way to define quantization levels is uniformly across the entire range of the input analog signal. This is the technique used in the "clear channel" CODEC. Using this CODEC, you need 12 bits to create enough levels so that the quantization noise is small enough that voice quality does not suffer.

NOTE 12 bits correlates to 4,096 quantization levels ($2^{12} = 4{,}096$). Therefore, 12 bits per sample combined with a sampling rate of 8,000 samples per second yields a bit-rate of 96 kilobits/second.

As it turns out, linear quantization levels are not very optimal. In this case, the possible distortion or quantization noise is constant throughout the entire signal range. Thus, in areas where the audio or signal level is lower, the SNR is smaller. This problem is compounded by the fact that a majority of human speech is in the lower signal levels. A better approach is to try to keep the signal-to-ratio level constant. The G.711 CODEC addresses this deficiency, as you will see in the next section.

PacketCable CODECs

Currently, the mandatory PacketCable CODECs are the ITU standard G.711, Global IP Sound's Internet Low Bitrate Codec (iLBC), and Broadcom's BV16. The ITU standard G.728 and G.729E CODECs are identified as recommended optional CODECs. An overview of all these CODECs is included in this section.

NOTE The addition of iLBC and BV16 to the list of mandatory CODECs is part of PacketCable 1.5. In PacketCable 1.0, only G.711 is mandatory.

G.711

The G.711 CODEC (also known as PCM) provides roughly the same voice quality as "clear channel" using only 8 bits simply by altering the quantization technique. Instead of using a linear quantization scale, a technique called *companding* is used. Companding uses a logarithmic scale for quantization levels that results in keeping signal-to-quantization noise levels or Signal to Noise Ratio (SNR) constant.

In North America and Japan the μ-law companding standard is used (hence, this CODEC is commonly referred to as G.711μ or PCMU), whereas in Europe the a-law companding standard is used (hence, this CODEC is commonly referred to as G.711a or PCMA). Both of these variations are a requirement in PacketCable. Both of these techniques are logarithmic, but enough differences exist between them that if you encode with μ-law and decode with a-law, voice quality will suffer. Basically, the quantization levels are divided up into 16 level segments. But the segments at the lower signal levels are of a smaller size, thus the quantization noise is smaller because of the greater granularity. If you use these companding standards, 8 bits provide roughly the same quality as 12 bits using linear companding. So the bit rate for the G.711 CODEC is 64 Kbps (8,000 samples/second * 8 bits/sample).

NOTE This is the bandwidth consumption of the actual voice stream only and does not take into account any protocol layer overhead (this is covered in the next chapter).

Figure 10-2 depicts an illustration of the quantization process for G.711/PCM encoding.

Figure 10-2 *PCM Quantization*

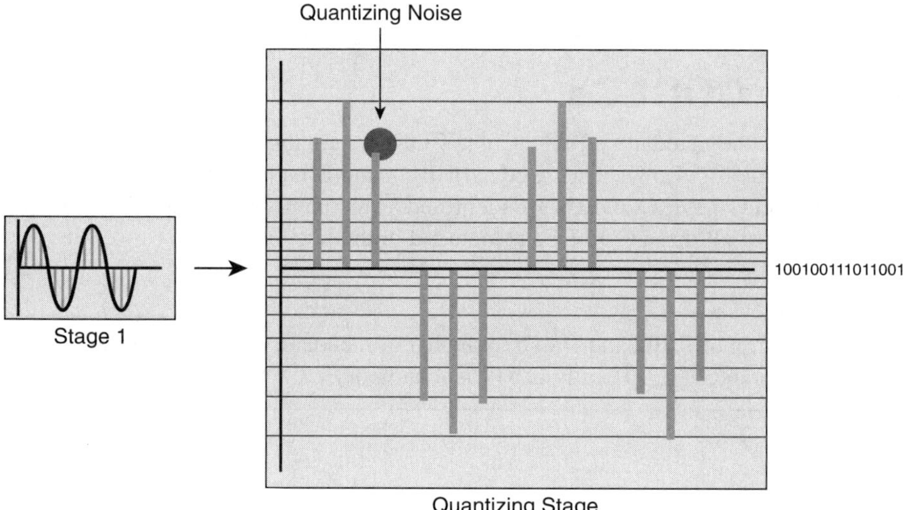

Now that the voice stream has been converted into digital samples, these samples need to be encapsulated into IP packets so they can be sent across the VoIP network. The length of time samples are collected before encapsulated into IP packets determines the amount of voice payload in each packet. This time is known as the *packetization period* because this process repeats at periodic intervals.

PacketCable has three supported packetization periods for use with the G.711 CODEC: 10 milliseconds (ms), 20 ms, and 30 ms. Using smaller packetization periods lowers voice delay (because packets are sent more frequently), but it also adds more packet overhead and thus reduces bandwidth efficiency. Also note that the smaller the packetization period, the less harmful a dropped packet is because less data is in that packet. The combination of this fact, along with the delay consideration, is why a 10-ms packetization period is typically used for Voice Band Data (VBD) calls. What VBD calls are and how they work is covered later in this chapter. For normal voice, a 20-ms packetization period is typically used in today's PacketCable deployments. Chapter 11 discusses the format of these voice packets in more detail.

Before moving on to other CODECs, the topic of what happens when packets are lost is another important consideration. Some causes for packet loss include RF noise in the cable network and congestion on network interfaces. Obviously, a direct correlation exists between packet loss and voice quality. The PacketCable CODEC specification states that a 3 percent packet loss is the absolute maximum tolerable amount. Where packet drops occur is also critical; ten dropped packets in a row is much worse than one drop every once in a while. You will see throughout this book how to minimize the chances of packet loss, but for now focus on what happens when this does occur.

Because voice is a real-time application, there is no point in having any type of retransmission scheme because it would introduce additional delay, which would impact voice quality. Instead, Packet Loss Concealment (PLC) algorithms are used to mitigate the effects of missing packets. These algorithms range from simply repeating the preceding packet to more advanced interpolation and extrapolation techniques that make intelligent guesses on what the packet is based on past and future packets. The G.711 Appendix 1 specification addresses PLC for G.711; many of the other CODECs have PLC considerations included within the CODEC specification itself.

G.728

You might decide that a 64-Kbps CODEC still consumes too much bandwidth and would like to reduce that further if possible. This is where other CODECs that utilize additional compression techniques come into the picture.

CODECs are categorized into two types:

- **Waveform encoders**—A waveform encoder works by attempting to reproduce the analog waveform as much as possible. These are generally of lower complexity but produce relatively high bit rates. G.711 and G.726 are waveform encoders.
- **Vocoders**—Vocoders work by outputting characteristics of the analog waveform instead of the waveform itself. These produce much lower bit rates at the expense of higher complexity. G.728 and G.729E are vocoders.

One technique a waveform encoder can use is to send only encoded versions of the differences between successive samples instead of sending the entire encoded samples themselves. This is beneficial because successive samples tend to be similar, and thus, their differences are small. Consequently, a smaller number of bits are needed to encode them. This compression technique is referred to as Adaptive Differential Pulse Code Modulation (ADPCM). ADPCM is the method used by the ITU G.726 specification.

Another compression technique is to send "codes" in the payload instead of the samples themselves. These codes are well defined in the CODEC definition and map to one or more of these samples. This compression technique is referred to as Code Excited Linear Predictive (CELP). As it turns out, CELP achieves very high quality using a minimal number of bits. Of course, nothing is free—the drawback of CELP CODECs is that they are mathematically complex and require the use of more Digital Signal Processor (DSP) resources than other CODECs.

The G.728 and G.729E CODECs used in PacketCable utilize the CELP type of compression. The basics of how these CODECs work are as follows; if you are interested in the full details, please refer to the ITU specifications.

NOTE Another way to reduce the bandwidth utilization of a CODEC is with the implementation of voice activity detection (VAD). This is a type of variable rate transmission, as very few or no data packets are sent at times when the speaker is silent. The functionality and implementation of VAD is covered in detail in Chapter 11 and Chapter 13, "Analyzing, Implementing, and Troubleshooting DQoS."

In the G.728 CODEC, five consecutive samples are processed at a time. From these samples a construct known as a *codepoint* is generated. This codepoint is then compared against the 1,024 known codepoints defined in G.728. The one that is closest to the actual value (that is, the one with the minimal error) is then used in the payload portion of the packet to represent these five samples. As you might guess, because 1,024 codepoints are possible, 10 bits are needed to identify a codepoint ($2^{10} = 1{,}024$). Four successive codepoints are then concatenated to form a single G.728 frame, where the oldest codepoint is the most significant. Thus, the size of a G.728 frame is

5 bytes (10 bits/codepoint * 4 codepoints/frame = 40 bits/frame)

The gateway at the other end of the call would then do a reverse lookup on the received codepoints and convert them back to samples.

This is a very high-level description of the CODEC. Parameters used to figure out the codepoint such as excitation gain, synthesis filter coefficients, and weighting filter coefficients are also used and are dynamically determined based on previous codepoints. The math behind these computations occurs relatively quickly, and no "look-ahead" delay is involved, which is why this CODEC is also referred to as Low Delay CELP (LD-CELP). Again, if you are interested in more details see the specification.

So what are the resulting bandwidth requirements of G.728? This is fairly easy to compute. Recall that voice is sampled at a rate of 8,000 samples per second, and one codepoint represents five samples. So codepoints must be sent at a rate of 1,600 per second (8,000/5), and because each codepoint corresponds to 10 bits, the resulting bit rate is 16,000 bits per second or 16 Kbps.

G.729E

The G.729 CODEC family uses a different type of CELP known as Conjugate Structure Algebraic CELP (CS-ACELP). This CODEC operates on speech every 10 ms. During each one of these intervals 80 samples (8,000 samples/sec * 10 milliseconds) are collected and processed. G.729 is a Linear Predictor (LP) for the forward direction. An LP is a mathematical model where future values of a signal are estimated as a linear function from previous samples.

G.729E is different from ordinary G.729 in that the algebraic codeword size is increased, and it has an LP for the backward direction as well; this makes the CODEC well-suited for music transmission as well as plain speech.

Two distinct codebooks exist, one for the forward direction that is typically used for speech transmission and one in the backward direction used for music transmission. The CODEC figures out the frame each way, and then selects the one it thinks will result in the best quality. In the packet payload is an LP mode indicator bit identifying whether the forward or backward LP was used. Other transmitted parameters include pitch delay values, gain parameters, and the codebook vectors themselves. The total size of these parameters is 118 bits. So, 118 bits are transmitted every 10 ms, yielding a bit rate of 11.8 Kbps.

One additional comment about the G.729E CODEC is that it also uses a look-ahead delay of 5 ms before packet transmission. This delay is also referred to as an algorithmic delay because the CODEC uses some knowledge of future voice samples before encoding the current samples.

Figure 10-3 shows a depiction of this process.

Figure 10-3 *The G.729 CODEC*

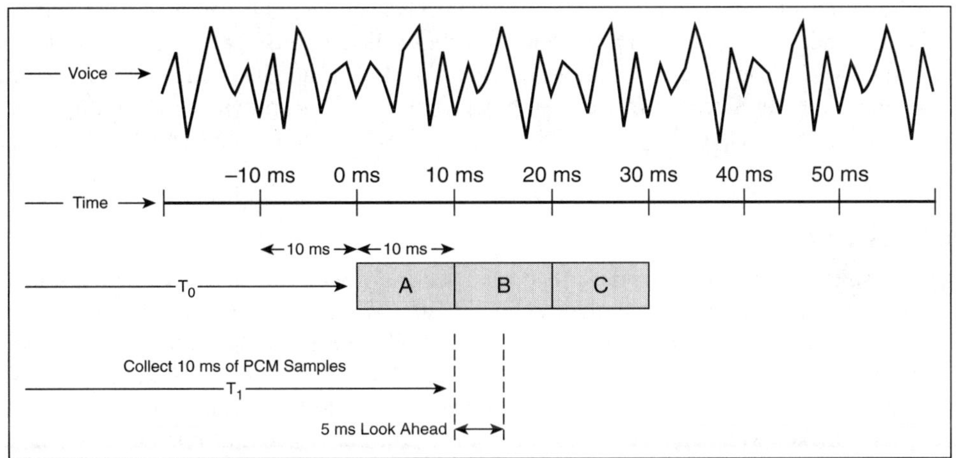

One additional thing to keep in mind about the G.728 and G.729E CODECs is that they require more DSP resources than the G.711 CODEC to perform the computational aspects of the CODECs. If these DSP resources are not currently available for a call wanting to use one of these CODECs, the call should fall back to using a simpler CODEC such as G.711.

iLBC

The Global IP Sound (GIPS) Internet Low Bitrate Codec (iLBC) CODEC is a royalty free narrowband CODEC that has been standardized by the IETF. It uses a linear predictive algorithm and has two supported frame or block lengths. When using a block length of 20 ms, it produces 304 bits per block, yielding a bit rate of 15.2 Kbps (304 bits/20 ms). When using a block length of 30 ms, it produces 400 bits per block, yielding a bit rate of 13.33 Kbps (400 bits/30 ms). A look-ahead of 5 ms is used with the 20 ms block length, and a look-ahead of 10 ms is used with the 30 ms block length.

The robustness to packet loss of the iLBC CODEC is similar to that of G.711, yet as you can see, the bandwidth requirement is much smaller. The reason iLBC handles packet loss better than other low bit rate CODECs is that it does not exploit dependencies between successive packets. Thus, a single packet can be lost without affecting the usefulness of other packets. The voice quality achieved with iLBC is on par with that achieved with the G.729E CODEC with much better resiliency to packet loss.

BV16

Broadcom's BroadVoice 16 (BV16) CODEC is another narrowband CODEC that provides toll quality service with a bit rate of 16 Kbps. BV16 is royalty free for PacketCable deployments. A document describing BV16 is available in the PacketCable 1.5 specification series. BV16 uses an algorithm called Two-Stage Noise Feedback Coding (TSNFC). You have no look-ahead delay with BV16, and its frame size is only 5 ms; therefore, the size of a frame is 80 bits (10 bytes). BV16 provides toll-quality speech exceeding or equivalent to G.728 and G.729E while using a very small coding delay.

Wideband CODECs

Thus far all the CODECs described have been narrowband CODECS. As mentioned earlier, most human speech lies in the frequency range from 300 Hz to 3400 Hz. However, some speech is outside this range. Wideband CODECs take advantage of this fact by doubling the sampling frequency range. Frequencies from 0–8,000 Hz are now sampled; consequently, the sampling frequency needs to be increased to 16 KHz. This increased frequency range also makes wideband CODECs better suited for non-voice transport, such as music. The use of wideband CODECs is currently optional in PacketCable.

CODEC Negotiation

Some final comments about CODECs; recall from Chapter 7, "NCS (Network-based Call Signaling)," that CODECs are negotiated between gateways. That is, for a CODEC to be used, both gateways must support it. Also remember that the CMS gives gateways a list of CODECs to use in order of preference. So if the CODEC you are trying to use is not currently being used, verify that both gateways do support it, and it is at the beginning of the list.

As an example, revisit the basic ONNET-to-ONNET call between Oscar and Tina from Chapter 7. The CRCX sent to Oscar's MTA is as follows:

```
CRCX 161096008 aaln/1@x1-6-00-13-10-50-38-02.cc.cisco.com MGCP 1.0 NCS 1.0
C: 284
L: mp:20;20;20,  a:PCMU;PCMA;G728,  e:off,  s:off,  t:a0
M: inactive
K: 161095965
```

In the Local Connection Parameters you see three CODECs listed: G.711 μ-law (PCMU), G.711 a-law (PCMA), and G.728—with G.711 μ-law being the preferred CODEC. In the SDP parameters in the response to this message, Oscar's MTA tells the CMS which of these CODECs it supports:

```
200 161096008 OK
I: 575

v=0
o=- 161096008 1397 IN IP4 24.34.240.245
s=-
c=IN IP4 24.34.240.245
b=AS:82
t=0 0
m=audio 53456 RTP/AVP 0 8 15     ← Prioritized list of possible CODECs
a=X-pc-secret:base64:<hex digit string omitted>
a=X-pc-csuites-rtp:62/51 64/51 60/51 60/50
a=X-pc-csuites-rtcp:81/71 81/70 80/70
a=X-pc-nrekey:0
a=mptime:20 20 20
a=ptime:20
```

In this case all three of these CODECs are supported and are specified here in the media announcement (m) parameter; 0 represents PCMU, 8 represents PCMA, and 15 represents G.728. Chapter 11 goes into detail on these mappings. The CMS then passes this list of supported CODECs in the CRCX to Tina's MTA:

```
CRCX 161096009 aaln/1@x1-6-00-0a-73-fa-be-8c.cc.cisco.com MGCP 1.0 NCS 1.0
C: 284
L: mp:20;20;20,  a:PCMU;PCMA;G728,  e:off,  s:off,  t:a0
M: inactive
K: 161095980

v=0
o=- 161096008 1397 IN IP4 24.34.240.245
c=IN IP4 24.34.240.245
b=AS:82
t=0 0
m=audio 53456 RTP/AVP 0 8 15
a=X-pc-secret:base64:<hex digit string omitted>
```

```
a=X-pc-csuites-rtp:62/51 64/51 60/51 60/50
a=X-pc-csuites-rtcp:81/71 81/70 80/70
a=X-pc-nrekey:0
a=mptime:20 20 20
a=ptime:20
```

Also included in this message is the prioritized list of CODECs for Tina's MTA (Local Connection Options), which is the same as the list that was sent for Oscar's MTA. Tina's MTA is capable of supporting all of these CODECs, so it indicates this in the SDP in its response to the CRCX:

```
200 161096009 OK
I: 719E

v=0
o=- 161096009 29086 IN IP4 24.34.240.247
s=-
c=IN IP4 24.34.240.247
b=AS:82
t=0 0
m=audio 53456 RTP/AVP 0 8 15    <- Prioritized list of possible CODECs
a=X-pc-secret:base64:<hex digit string omitted>
a=X-pc-csuites-rtp:62/51 64/51 60/51 60/50
a=X-pc-csuites-rtcp:81/71 81/70 80/70
a=X-pc-nrekey:0
a=mptime:20 20 20
a=ptime:20
```

This information is then passed along to Oscar's MTA in an MDCX command:

```
MDCX 161096010 aaln/1@x1-6-00-13-10-50-38-02.cc.cisco.com MGCP 1.0 NCS 1.0
I: 575
C: 284
L: dq-gi:1463,  dq-rr:snrcresv
M: recvonly
K: 161096008

v=0
o=- 161096009 29086 IN IP4 24.34.240.247
s=-
c=IN IP4 24.34.240.247
b=AS:82
t=0 0
m=audio 53456 RTP/AVP 0 8 15
a=X-pc-secret:base64:<hex digit string omitted>
a=X-pc-csuites-rtp:62/51 64/51 60/51 60/50
a=X-pc-csuites-rtcp:81/71 81/70 80/70
a=X-pc-nrekey:0
a=mptime:20 20 20
a=ptime:20
```

So at this point, both Oscar's MTA and Tina's MTA know what the other one is capable of and are assured that if one MTA starts sending RTP packets using any of these CODECs, the other MTA is able to understand and decode them.

CODEC Metrics

Now that you are familiar with the possible CODECs to use, you can take a look at some of the metrics to help you differentiate them. The first and most obvious one is voice quality. Three primary areas are associated with voice quality:

- **Clarity**—Clarity refers to the listener's ability to easily comprehend the received audio stream. Poor voice clarity will undoubtedly not sit well with your customers.
- **Delay**—Delay refers to the time between when one person speaks and when the other party hears it. If delay is too large, voice communication can be very frustrating even if it's perfectly clear.
- **Echo**—Echo is when persons hear themselves talk, and this, too, can be annoying.

Proactively making sure all three of these factors are as optimal as possible is crucial to keeping customers happy.

The other major metric is *bit rate*. The whole purpose of using a more advanced CODEC is to reduce the bandwidth requirements per call, permitting more calls to be placed in the same amount of bandwidth. Without taking into consideration packet overhead, you saw that a G.711 call takes up 64 Kbps, a G.728 call takes up 16 Kbps, a G.729E call takes up 11.8 Kbps, an iLBC call takes up either 13.33 Kbps or 15.2 Kbps, and a BV16 call takes up 16 Kbps. So, for a circuit-switched network (such as the legacy PSTN) where there is no packet overhead, you can see that four G.728 or BV16 calls or five G.729E or iLBC calls consume roughly the same amount of bandwidth as a single G.711 call. The impact of packet overhead (IP, UDP, and RTP) is examined in the next chapter.

Voice Clarity

Probably the voice quality measurement you are most familiar with is Mean Opinion Score (MOS). MOS works on a 5-point scale, with a grade of 5 being the best and 1 being the worst. Table 10-1 lists the associated speech quality and distortion level for each value.

Table 10-1 *Voice Clarity MOS Ratings*

Rating	Speech Quality	Distortion Level
5	Excellent	Imperceptible
4	Good	Just perceptible, not annoying
3	Fair	Perceptible, slightly annoying
2	Poor	Annoying, but not objectionable
1	Unsatisfactory	Very annoying, objectionable

A MOS score of 4.0 is considered "toll quality," meaning it's the quality associated with traditional PSTN service. Historically, MOS scores were purely subjective. They were based on people listening to voice samples and grading them. Consequently, external factors such as background noise and a person's mood could influence their grading. Using live human subjects to quantify voice quality is time-consuming and expensive. More scientific ways of quantifying voice quality were later developed. Many of these methods can be mapped directly into traditional MOS type scores.

NOTE Historical research suggests that a 3 percent packet loss results in the MOS score dropping by 0.5.

PAMS

One of these methods is the Perceptual Analysis/Measurement System (PAMS) developed by British Telecom. This method uses a mathematical model of human hearing. A PAMS measurement reports two scores—a Listening Quality (LQ) score and a Listening Effort (LE) score. These are both on a five-point scale very similar to the MOS score. In fact, research has proven that PAMS scores average within one-half of a point compared to the traditional MOS score. Another point to note about PAMS is that errors in the input signal are taken into consideration.

PSQM and PSQM+

Another method is the Perceptual Speech Quality Measure (PSQM) developed by ITU (standard P.861). PSQM scores are on a scale of 0 to 6.5, where 0 indicates no distortion and a score of 6.5 indicates much distortion. Remember, this method is the opposite of all the others in that the lower the number, the better the voice quality. There is no direct correlation between traditional MOS scores and PSQM. PSQM was designed to measure compression's affect on quality; it does not account for lost packets. Also PSQM does not take into account the quality of the input signal. In other words, if the input signal and the output signal were the same, the PSQM score would indicate perfect quality even if the input signal were bad.

The PSQM developers addressed some of these problems by developing PSQM+. PSQM+ does additional post-processing on PSQM values. This value correlates more directly to traditional MOS scores. However, PSQM+ still has a problem with lost packets.

PESQ

The most advanced and best method for measuring voice quality is Perceptual Evaluation of Speech Quality (PESQ). PESQ was designed as the successor to PSQM and is defined in the ITU specification P.862. PESQ combines the techniques used in PSQM+ and PAMS. Distortion, errors, packet loss, and delay are all accounted for in PESQ measurements. PESQ uses a sensory model and creates a quality score analogous to MOS. The best PESQ score that can be achieved is 4.5.

ITU E-model and R-factors (ITU-T G.107)

The voice quality methods discussed so far are determined based on the voice streams themselves. There is another type of voice quality method that is better suited for quantifying the effects of the VoIP environment. The ITU E-model defined in the G.107 specification is an example of one of these methods. The E-model estimates voice quality based on transmission parameters (such as packet loss, delay, jitter, and echo).

In this case, voice quality is expressed by a transmission quality rating also known as an R-factor (Rating factor). For narrowband voice the range of the R-factor is from 0 to 100 (for wideband voice values greater than 100 are possible). A value of 94 corresponds to "toll quality," and values of 50 or less are considered unusable. The G.107 specification also defines an equation for converting R-factors into MOS scores. Specifically, a MOS Listening Quality (LQ) score and a MOS Conversation Quality (CQ) score are generated. The difference between these is that the MOS-LQ does not take into account the effects of delay and echo, whereas the MOS-CQ does.

Delay

The voice quality component of delay is very important in providing quality services. Some amount of delay is unavoidable, but care should be taken to minimize delay as much as possible. End-to-end delays less than 100 ms are imperceptible in normal conversation, and delays greater than 300 ms degrade voice quality to the point where it is unacceptable. An end-to-end delay no greater than 150 ms is generally regarded as the maximum allowable. When the delay is greater than this, the call parties begin to talk over each other. The PacketCable 1.x architecture framework technical reports specify that the one-way delay for local IP access and IP egress (meaning the end-to-end delay excluding the IP backbone) is less than 45 ms.

End-to-end delay consists of all components in between the speaker's mouth and the listener's ear. Therefore, this delay is sometimes referred to as mouth-to-ear delay. Delay components can also be categorized as being static (the same for all packets) or dynamic (different for each packet).

For this discussion end-to-end delay is separated into three networks:

1. Delay from the source of the speech until it is turned into IP packets
2. Delay transmitting these packets across the backbone network
3. Delay from the terminating edge of the backbone until these packets are turned back into the speech heard by the receiver

Networks 1 and 3 are the IP access and IP egress networks specified by PacketCable as having a combined delay less than 45 ms.

Figure 10-4 depicts a very high-level illustration of this delay.

Figure 10-4 *End-to-End Delay Model*

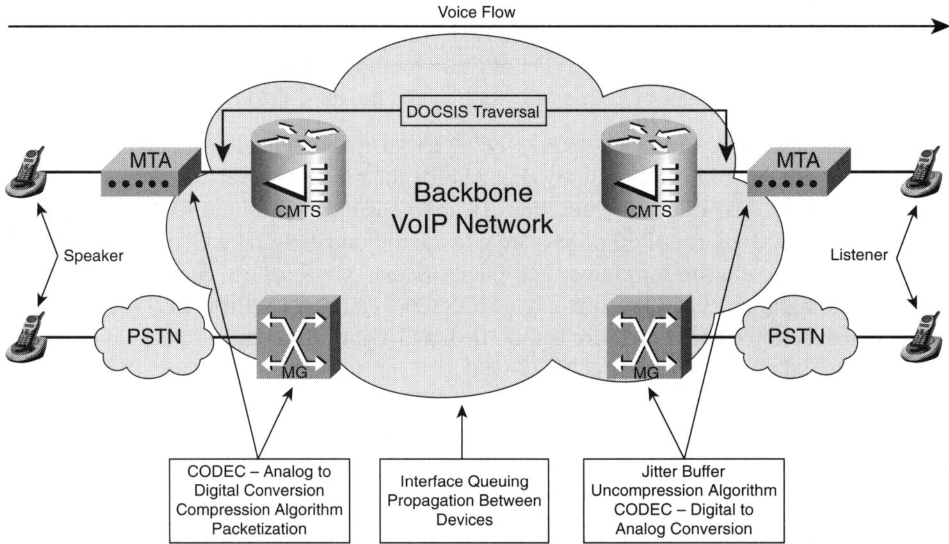

Speaker to Backbone Delay

The first category is the delay from the source of the speech until it is turned into IP packets. For a PacketCable customer this is the delay up to the point where IP packets are created by the telephony interface of the MTA. For a call coming into the PacketCable network from the PSTN, this is the delay up to the point where IP packets are created by the telephony interface of the MG.

First, the PacketCable customer case is examined. The first thing the MTA must do when it receives voice from an analog port is to encode the voice stream into packets or frames using a CODEC.

NOTE Some delay also exists in getting the voice through the telephone to the MTA; however, this delay is assumed negligible.

The MTA encoding delay is static and includes the process of sampling the analog signal and performing quantization until PCM samples are obtained. For compression CODECs, these samples are then further processed, and frames are constructed. The PacketCable CODEC specification references an ITU formula for calculating CODEC delay that is equal to the look-ahead delay plus twice the frame size. The theory behind this calculation is that a CODEC must be able to encode and decode frames at a rate of at least the frame size to be usable.

Because voice is sampled at a rate of 8,000 samples/sec, the frame size of one sample is the inverse of this, or 0.125 ms. This is the encoding delay for a single digital PCM and G.711 sample. Recall that G.728 frames are created from four codepoints, each of which is generated from five PCM samples. So the encoding delay for a G.728 frame is 2.5 ms:

4 codepoints/frame * 5 samples/codepoint * 0.125 ms/sample

The encoding delay for a G.729E frame is 15 ms (a frame is generated every 10 ms, but you also have a 5 ms look-ahead delay).

The encoding delay for an iLBC frame depends on the block size; for a 20 ms block size this delay is 25 ms (after adding the 5 ms look-ahead delay), and for a 30 ms block size this delay is 40 ms (after adding the 10 ms look-ahead delay).

Finally, BV16 frames are generated every 5 ms, and you have no look-ahead delay. So, the encoding delay is only 5 ms.

Next, the MTA must put these digital samples or frames into the payload of IP packets. The delay for this process depends on the rate at which IP packets are generated (the packetization period). Recall that for PacketCable packetization periods of 10 ms, 20 ms, or 30 ms are supported. The packetization process can occur as soon as the digital samples or frames are ready.

For the G.711 CODEC, all the digital samples that are collected during the packetization period comprise the payload portion of the IP packets.

For the G.728 CODEC, frames are ready every 2.5 ms. So, if packets are generated every 10 ms, the MTA puts 4 frames in every packet; if packets are generated every 20 ms, the MTA puts 8 frames in a packet; and if packets are generated every 30 ms, the MTA puts 12 frames in every packet.

For the G.729E CODEC, frames are ready every 10 ms. Thus, with a 10 ms packet rate, one frame is in every packet; with a 20 ms packet rate, two frames are in every packet; and with a 30 ms packet rate, three frames are in every packet.

For iLBC, frames are ready either every 20 ms or every 30 ms, depending on the block size. Consequently, you can use only a 20 ms packetization with the 20 ms block size and only a 30 ms packetization with the 30 ms block size.

BV16 frames are generated every 5 ms, so you can use 10 ms packetization (resulting in 2 frames per packet), 20 ms packetization (resulting in 4 frames per packet), and 30 ms packetization (resulting in 6 frames per packet).

Packetization delay maps to the length of the packetization period. This delay is also classified as static, and logically larger packetization periods induce larger delays.

Next, delay for when calls originate from the PSTN interconnect part of the PacketCable network is examined. The delay through the PSTN network is completely out of your control. However, this delay should not be too large because the PSTN network is circuit switched. One exception to this case is if a satellite circuit is involved in the call, as the propagation delay here is rather significant. If no satellite circuits are involved in the call, PSTN delay is typically on the order of tens of milliseconds.

Call bearer information in the TDM circuits connecting to the MG is already formatted as digital samples, typically one of the G.711 variants. If a CODEC other than G.711 is used, the CODECs compression algorithm needs to convert these samples into the frame format used by that CODEC.

The packetization process then occurs on the MG as it did on the MTA in the previous example.

It is assumed that as soon as enough samples or frames are available to fill the IP packet payload, the IP packets can be generated almost immediately. Therefore, the speaker to backbone delay for G.711, G.728, and BV16 is roughly equal to the packetization period (10 ms, 20 ms, or 30 ms).

The G.729E and iLBC CODECs have look-ahead delays that delay the generation of IP packets. Therefore, the sum of the CODEC look-ahead delay and the packetization delay roughly equals the speaker to backbone delay. For G.729E this would be 15 ms, 25 ms, or 35 ms. For iLBC this would be 25 ms or 40 ms.

Backbone Delay

The second delay category is for backbone delay. You have really only two components here. One is the propagation delay between backbone devices, and the other is the queuing delay on the egress interfaces of the backbone devices.

You can't do much about the propagation delay other than to put high-speed links in the backbone and ensure that your backbone routing protocol is picking the fastest routes across the backbone network. As for queuing delay, a Differentiated Services Code Point (DSCP)–aware network can help here by giving voice packets preferential treatment of backbone resources. This works by identifying delay sensitive packets, such as voice, and applying queuing techniques, such as Low Latency Queuing (LLQ), where these packets are given priority for transmission. If your backbone network is Multiprotocol Label Switching (MPLS)–based, experimental bits in the MPLS header can provide quality of service (QoS) in a manner similar to DSCP. Refer to Chapter 16, "PacketCable Network Design Considerations," for more information on backbone QoS.

A special case of backbone QoS is for traversal of the DOCSIS network between an embedded MTA and CMTS. In this case, IP packets are encapsulated into DOCSIS frames. Framing delay should be very small; the only possible exception is if Payload Header Suppression (PHS) is enabled, which suppresses some portion of the header overhead with an index value to save bandwidth. Refer to the DQoS chapters (Chapters 12 and 13) for more information about PHS.

Recall that for the upstream direction (MTA → CMTS) these frames are sent using a DOCSIS Unsolicited Grant Service (UGS). This eliminates the need to issue DOCSIS data grant requests and the possibility of these data grants not being available. PacketCable dynamic quality of service (DQoS) ensures this bandwidth is available. Without these features, the delay here would be considerably longer. Assuming the MTA CODEC clock is synchronized to the CMTS DOCSIS clock, the only delay here is the propagation delay across the DOCSIS network. Synchronization refers to the UGS grants being ready for use as soon as the MTA has them to send. Note, without proper synchronization, delay can be an issue especially for voice-band transmissions such as fax.

For the downstream direction (CMTS → MTA) a high priority best-effort service flow is set up in the downstream direction for voice. This service flow is set up dynamically the same time the UGS flow for upstream voice traffic is set up. It has reserved bandwidth for downstream voice packets and is given priority over other types of downstream traffic. Refer to the chapters in Part V of this book for more information on these DOCSIS service flows.

The sum of the queuing delays on the egress interfaces on all devices on the path from the talker to the listener and the propagation delays between devices comprise the backbone delay.

Backbone to Listener Delay

Finally, this section examines the delay from the terminating edge of the backbone until it is turned back into the speech heard by the receiver. If the listening party is a PacketCable customer, this is the delay from the CMTS down to the MTA and then through the MTA. If the listening party is connected via the PSTN, this delay encompasses the delay thru the trunking media gateway and the PSTN delay.

The main delay component here is the jitter or playout buffer. A jitter buffer delays the playing of packets for a while by holding them in memory. This ensures that even if packets arrive at variable intervals, they can be played out at regular intervals corresponding to the packetization period. The drawback of a jitter buffer is that it adds delay. Typically, jitter buffers are adaptive in that they start out at some value and then adjust themselves to a value optimal for the network. Figure 10-5 shows a graphical depiction of how a jitter buffer works.

Figure 10-5 *Jitter Buffer*

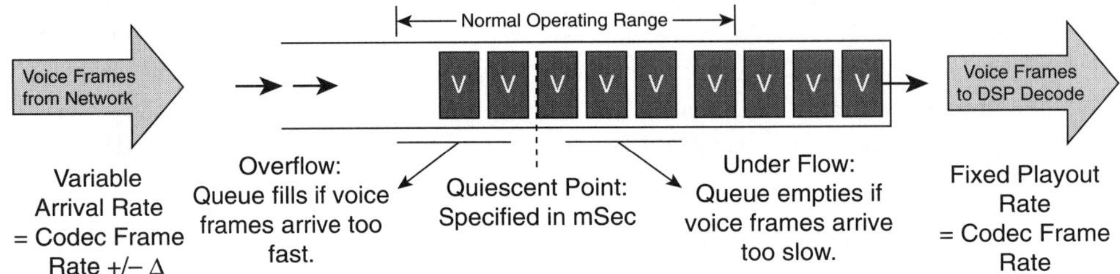

If the length of time packets are held in the jitter buffer is too short, you face a possibility that no packets will be available when a packet needs to be played out. If the length of time packets are held in the jitter buffer is too long, not only does this add unnecessary delay, but you also face the risk that the buffer will overflow and packets will be dropped. A common implementation is to use adaptive jitter buffers with a relatively high initial delay value (40–60 ms) and let them adapt to a more optimal value.

After a packet leaves the jitter buffer, the bearer payload is extracted and sent to the CODEC. If a compression CODEC is used, the payload is uncompressed and converted back into PCM samples. If the terminating gateway is an MTA, the decoder functionality of the CODEC then converts the digital PCM samples back into an analog stream. If the terminating gateway is an MG, these PCM samples are sent in the payload of the TDM circuits connecting to the PSTN. Just as with the PSTN talking party case, unavoidable delay is associated with the PSTN network itself. The CODEC delay in either case is estimated as the CODEC frame size. Recall that this is 0.125 ms for G.711, 2.5 ms for G.728, 10 ms for G.729E, 20 ms or 30 ms for iLBC, and 5 ms for BV16.

Therefore, the backbone to listener delay is estimated to be the jitter buffer delay plus the decoding delay of the CODEC.

In summary, the biggest contributor to end-to-end delay is typically backbone delay where the backbone comprises several devices. However, the biggest single delay sources are usually packetization delay and jitter buffer playout delay.

Echo

The final voice quality metric you will learn about is echo. *Echo* is when the talking party hears his or her own speech. This occurs when the signal in the transmission path is "leaked" onto the receive path, as illustrated in the example in Figure 10-6.

Figure 10-6 *Echo Example*

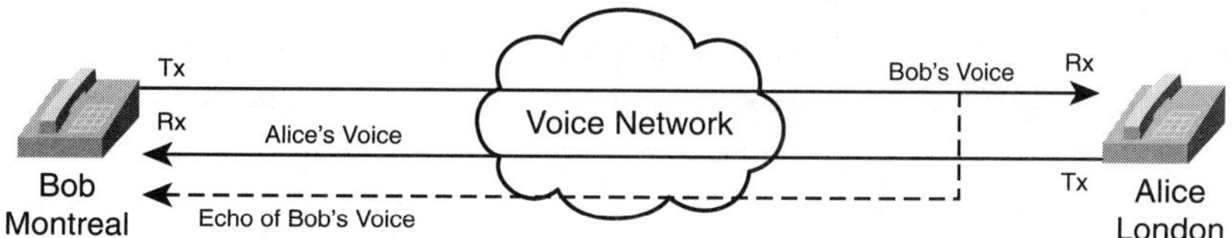

If the time between the original speech and the returning speech is very small (less than 25 or 30 milliseconds), echo is not a problem. In fact, this type of echo (also known as *sidetone*) is actually beneficial because it often reassures the talker to hear his or her own voice. Also, if the level of echo is very low, it should not be a problem. An echo attenuation level of at least –50 dB is desired.

So what causes echo? It is important to remember that the cause of echo is almost always at the far end. That is, in an ONNET-to-OFFNET call or an OFFNET-to-ONNET call if the PacketCable customer is hearing echo, the cause is most likely in the PSTN. The most common cause is from the 2- to-4-wire conversion on the PSTN interface in the PSTN gateway or the telephony interface in the MTA. This 2- to-4-wire conversion is also known as a *hybrid* because it is physically a hybrid coil that converts the separate audio streams in the 4-wire interface to the combined 2-wire interface. There is an impedance mismatch between the 2-wire side and the 4-wire side that causes signal leakage from the transmit pair to the receive pair. A key thing to keep in mind is that no echo is introduced on digital links. Bits inside packets do not leak. So if echo is caused by telephony interfaces, why is it more problematic in a PSTN-to-VoIP call than in a PSTN-to-PSTN call? The answer is delay. Delay is typically much shorter in a traditional PSTN call; therefore, echo problems are not noticeable. The additional delays imposed by a VoIP network such as a PacketCable network cause the echo problem to be much more noticeable.

If you are experiencing echo with your trunking gateway, you can try adding some attenuation toward the PSTN. Hopefully, this reduces the reflected energy or echo to the point where it is tolerable. You could also add attenuation to the receive interface from the PSTN. However, be cautious when doing this so you don't reduce the regular speech level so much that it is difficult to hear anything.

The best solution to echo problems is the use of echo cancellers. Echo cancellers work by examining the analog voice stream they receive and removing the attenuated form of it from the analog voice stream they send. Another device can be used to combat echo, the echo suppressor. This device simply disables the receive path when it detects the transmit path is active. The problem with these devices is that it takes some time to detect the transmit path is no longer active so the receive path can be disabled when it has valid speech to transmit. Thus, the beginning parts of a conversation can be lost; this condition is commonly referred to as *clipping*.

Therefore, the use of echo cancellers is preferable to echo suppressors.

NOTE With echo suppressors, both parties cannot be talking at the same time; echo cancellers allow this to occur. This is known as *double-talk* and must be permitted to occur. The ITU specifications G.165 and G.168 describe the use of echo cancellers.

Echo cancellers are applied toward the telephony circuit; this is also referred to as a tail circuit. The amount of time the echo canceller looks for echo from the received signal in the transmit signal is known as *echo tail* and is usually configurable. A typical value for this parameter is 32 ms. As a general rule, you can set this parameter to the maximum value on the trunking media gateway. Although less of a problem, echo can also be generated on the PacketCable MTA. In this case, the PSTN customer would detect the echo. An echo tail length of around 8 ms is usually sufficient on the MTA.

NOTE If the MTA is connected to a non-standard telephone with a 4-wire microphone and headset, echo cancellation should not be needed at all.

If the call is carrying voice band data such as a facsimile transmission, echo cancellation must be disabled to ensure the quality of the fax transmission. This is done by the gateway sending a tone to disable the echo canceller upon detection of fax tone. This tone is a 2100 Hz tone with phase reversals.

NOTE Echo cancellation should also be disabled when performing SS7 continuity tests on the trunking gateway.

The type of echo discussed so far is known as *electrical echo* or *line echo*; another type of echo is known as *acoustic echo*. Acoustic echo occurs because of coupling between the speaker and microphone through the air. Typically, this happens from using a speakerphone with the volume set too high.

Voice Path Problems

This section contains a description of some of the voice problems you might encounter and ideas on the probable causes. Problems caused explicitly by echo and delay are not covered here because they were covered in detail earlier.

No Voice and One-way Voice

As mentioned in Chapter 7, problems relating to no voice are most likely related to IP connectivity issues. These could be routing issues where the two gateways do not have IP routes to each other or issues that arise from access lists or firewalls preventing the passing of RTP packets between gateways. Firewalls and access lists are also the most likely cause of one-way voice where RTP traffic is permitted to pass in one direction but not the other. Another possibility is that at least one of the gateways is not operating in "Send/Receive" mode; this can be verified by examining the MGCP-related messaging.

Garbled Voice

Garbled voice is characterized by breaks in the audio stream such as the loss of syllables within a word. The most likely causes of this are packet drops and/or jitter problems.

Other Voice Problems

Other voice problems include clipping, crackling, and robotized voice. Remember, clipping is when the beginning of words or syllables is lost. This is most likely because of voice activity detection. If VAD is enabled, it can take some time for the presence of actual voice to be detected by the gateway so the beginning of words or syllables can be lost.

Crackling is when static is heard in the conversation. This is most likely caused by electrical interference or a bad power supply in the MTA.

Robotized voice is exactly what it sounds like, when the person you are talking to sounds like a robot. The most likely cause of this is when a companding mismatch exists when using the G.711 CODEC. For example, if voice is encoded using μ-law companding and decoded using a-law companding, this can occur. This should not occur in a PacketCable network as the CODEC is negotiated between the gateways; if this does occur, it indicates a malfunctioning gateway.

DTMF Digit Transport

Certain types of calls require the transmission of dual-tone multiple frequency (DTMF) digits in the voice path. These are the 0–9, *, and # keys on a standard telephone. These digits are used for applications such as accessing voicemail, making credit card calls, and Interactive Voice Response (IVR) call types. IVR applications are when the user is given a menu type voice script such as when checking movie times.

By default, DTMF digit tones are sent just like normal voice. The devices in the PacketCable network are completely unaware that they are transporting these tones. Consequently, no mechanism ensures they are delivered reliably to the far end. And if packets are lost, the capability for the far end to properly identify the tones becomes difficult. Also because the goal of these tones is to convey the pressing of telephone digits, you can find much easier ways to do this in an IP environment.

A better way to transport DTMF digits is described in IETF RFC 2833. This technique uses a separate type of RTP packet called a Named Telephony Event (NTE). RTP is covered in more detail in Chapter 11. Basically, within the RTP header is a field that identifies the payload type contained in the packet. Some of the payload types include identifiers for the CODEC used to carry the normal voice stream, G.711, G.729, etc. With RFC 2833 a separate payload type is used for the transmission of DTMF digits. This technique is also referred to as *DTMF relay*. The capability to use RFC 2833 for DTMF relay is a requirement in PacketCable 1.5.

You have two ways to handle this type of DTMF transmission:

- **Gateway controlled**—In gateway controlled mode the two gateways negotiate this payload type in the exchanged SDP parameters included in the call signaling packets. Remember from the previous chapters that information such as gateway IP address, RTP port number, CODEC, and packetization period are sent in the SDP parameters. The payload type used for RFC 2833 DTMF transmission can also be negotiated here. In this case the call agent or CMS does not influence this negotiation; it is completely controlled by the gateways.
- **Call agent controlled**—In call agent controlled mode, the DTMF relay parameters are sent to the gateway in the call signaling Local Connection parameters. In both of these modes the gateway must be able to detect and identify DTMF digits.

An important thing to keep in mind with DTMF relay is that it takes some time for the gateway to detect a DTMF digit. Therefore, it is not uncommon for the beginning of a DTMF digit to be sent in the normal voice stream. In this case, you face the danger that a DTMF digit can be generated twice at the receiving gateway—once in the regular voice stream and once via the DTMF relay operation. If you experience problems where DTMF digits are being repeated, this is probably the cause. One way this problem can be alleviated is if the gateway blocks the sending of normal audio packets when it is detecting DTMF digits. This capability is a requirement in PacketCable 1.5.

Voice Band Data (VBD) Transport

Voice Band Data (VBD) refers to the transmission of data signals over a packet network voice channel using an appropriate encoding method. These data signals include those used by data modems, fax machines, and text telephones used by the hearing-impaired community.

Using CODECs to turn analog voice streams into digital samples is needed to transport voice across the digital PacketCable VoIP network. Using CODECs to turn VBD (which is already a digitally modulated analog signal) into digital samples can cause problems, however.

First, this process can distort the modulated signal so that the receiving end is unable to recognize the modulated waveforms. Currently, only G.711 is an acceptable CODEC for VBD transmission because it imposes the minimal amount of distortion. If your call is initially using a lower bit-rate CODEC, it needs to change to G.711 upon detection of VBD. Also the high-pass filter in the CODEC must be disabled, and voice activity detection must be turned off at this point.

Second, even a small amount of packet loss can be devastating for VBD. In a normal voice call, the listener does not in all likelihood even notice the occasional dropped packet, especially if packet loss concealment techniques are in place. For VBD call types, dropped packets are a much bigger problem. Packet loss in fax calls result in line errors that can cause transmission failures, packet loss in modem calls cause modem retraining to much lower transfer speeds, and packet loss in a text telephony call results in dropped characters, making text unreadable. Therefore, you must make sure jitter buffers are large enough to avoid overruns; in fact, you can often configure gateways to use larger jitter buffers when VBD is detected.

A third issue is the requirement for the end-to-end delay to be relatively constant. This is to ensure that synchronization between the two endpoints is maintained.

The ITU V.152 specification defines some mechanisms gateways can use to provide a more reliable method of passing VBD in-band after it is detected. Some of these mechanisms include

- Capability to negotiate a CODEC (such as G.711) to use explicitly for VBD
- Capability to ensure constant end-to-end delay
- Capability to automatically disable VAD and echo cancellations
- Implementation of Forward Error Correction (FEC) techniques to automatically fix corrupted packets (for example, RFC 2377)
- Implementation of packet redundancy techniques to protect against lost packets (for example, RFC 2198)
- Implementation of packet loss concealment techniques more suitable for VBD

An alternative to in-band transmission is the use of packet relay mechanisms that employ retransmission and/or redundancy schemes.

Analog Modem Transport

One question that you might have is if a PacketCable subscriber has a cable modem with connectivity to the Internet, why would he or she want to hook up an analog modem to it? In this case, you are converting digital data to analog form with the traditional modem and then converting it back into digital form with the MTA. At first glance it doesn't make much sense. A couple of situations where this scenario would make sense include if the MSO were supplying only voice service to a customer without high-speed data service. In this case, they would still need to dial in to gain Internet access. Another scenario is if the customer has a home security system that requires PSTN access via an analog modem within the security system. Yet another scenario is the case of a customer needing to connect to a private intranet that supports only dial-up connectivity.

So when a gateway detects modem tone, it does four things:

- Up-speeds to the G.711 CODEC if needed
- Disables the high-pass filter
- Disables echo cancellation
- Disables VAD

Another thing that can happen at this time is that jitter buffers change from a dynamic adaptive size to a fairly large fixed size. This accomplishes two things:

- Protects against clocking differences between the two gateways
- Prevents the modems from retraining, which happens when the jitter buffer is adapting

You have two ways to transport modem traffic in a VoIP network:

- **Modem passthrough**—In modem passthrough mode, modem traffic is carried using the G.711 CODEC just like normal voice traffic. This method has the problems of packet loss, jitter, and latency, which are mitigated with modem relay. If modem relay is not an option, the performance of modem passthrough can still be improved. Performance problems because of packet loss can be alleviated simply by sending each packet twice. The drawback to this is that bandwidth usage also doubles for each modem call.
- **Modem relay**—In modem relay operation, the modem signals are demodulated at one gateway, converted to digital form, and then sent to the other gateway. These packets are sent using the Simple Packet Relay Transport (SPRT) protocol that runs over UDP. The receiving gateway then converts these packets back into modem signals that are sent to the receiving modem. In addition to reducing the problems caused by packet loss, jitter, and latency, bandwidth utilization is also improved with modem relay.

Facsimile Transport

Just as with modem detection, when the G.711 CODEC is used and fax transmission is detected, echo cancellation needs to be turned off so it won't adversely affect fax quality. This is done using a 2100 Hz tone with phase reversals to notify the far end that echo needs to be disabled. If a CODEC using compression is used and fax transmission is detected, the gateway needs to up-speed to the G.711 CODEC in addition to disabling its echo cancellers. Also the high pass filter needs to be disabled, and VAD needs to be disabled.

T.30 and T.4 are the ITU standards for fax transmission used in the PSTN by traditional Group 3 category fax machines. The T.30 specification defines the format of non-page data that is used for tasks such as capabilities negotiation. The T.4 specification defines the format of page data (that is, the fax pages themselves). You have two parts to successful fax transmission: the negotiation phase, ensuring that the receiving fax machine understands the received data, and the confirmation phase, where the receiving fax machine informs the sender of the successful receipt.

If fax relay is not an option for fax calls, the performance of fax passthrough can still be improved. Simply sending fax control packets twice can alleviate performance problems because of packet loss. The drawback to this is that bandwidth usage also doubles for each fax call.

In-band fax transmission might work okay, but different modes can be used to improve the reliability of fax transmission. Just as modem relay can be used to improve the transmission of modem signals, fax relay can be used for reliable facsimile transmission.

A third type of fax transmission exists called *fax store-and-forward*; it is defined in the T.37 standard. In this technology the originating gateway converts the fax transmission into an email message and sends it to the other gateway across the IP network. The terminating gateway then converts this email message back into a traditional fax transmission to the terminating fax machine.

Similar to T.37, T.38 fax relay is also handled on the two gateways. And like T.37, the two fax machines involved in the call have no idea T.38 fax relay is occurring. What T.38 does is demodulates the normal T.30 fax transmission and sends the control and data information over encapsulated IP packets. The T.38 IP packets exchanged between the gateways are referred to as Internet Fax Protocol (IFP) packets. IFP packets can utilize either the TCP or UDP transport layer protocols; the UDP option is commonly referred to as UDPTL. If UDP is used, fax transmission is fast; however, it can be unreliable. T.38 adds reliability through redundancy of the image data and a Forward Error Correction (FEC) scheme. T.38 is the preferred method of fax transmission in PacketCable networks and is a requirement in PacketCable 1.5. Figure 10-7 shows a graphical representation of T.38 fax relay.

Figure 10-7 *FAX Relay*

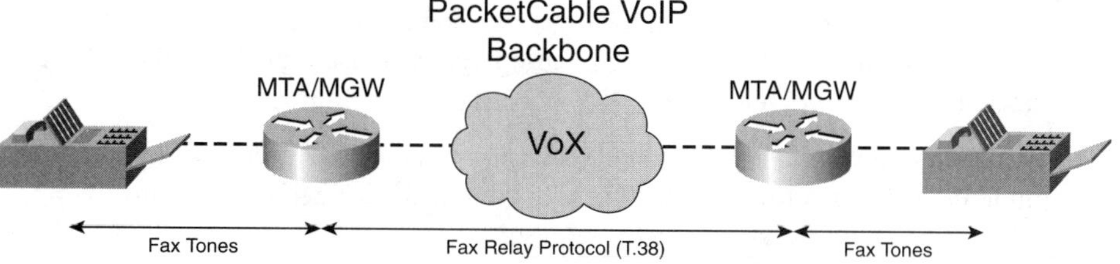

Just as is the case with DTMF relay, the fax relay method is negotiated between the gateways. And just as with DTMF relay, you have two methods in which this can be done:

- **Gateway controlled**—In gateway controlled mode, fax handling parameters are negotiated in the SDP parameters transferred between gateways without call agent involvement.
- **Call agent controlled**—In call agent controlled mode, the way fax is handled is identified in the Local Connection parameter within the NCS or TGCP signaling sent to the gateways.

In these modes, when a gateway detects a fax tone, it switches to T.38 mode. Call agent controlled mode is the preferred method in PacketCable. The main reason for this is that the call agent controls QoS resource manipulation, so it needs to be aware of changing QoS requirements because of fax.

Teletype Technology Transport

The last type of Voice Band Data to consider in a PacketCable network is Teletype Technology (TTY). TTY is the protocol used to enable telephony services for the hearing impaired. Text telephones are used that utilize modem technologies. The ITU standard V.18 details this technology.

NOTE In the PSTN today TTY calls cannot be made between countries because many countries are using their own TTY protocols.

A text telephone consists of text input/output devices combined with a voice-band modem. Therefore, only the G.711 CODEC can be used for the same reason as in the modem and fax cases. The MTA needs to be able to detect V.18 signals and signal them to the CMS in the NCS Requested Events parameter. This event is referred to as a Telecom Devices for the Deaf, or TDD, event. Also when an MTA detects an answer signal from a TTY device, it must notify the CMS in the NCS Requested Events parameter. Just as with fax and other modem calls, echo cancellation must be disabled upon detection of a TTY call. TTY data is transmitted in the RTP stream using a separate payload type defined in RFC 2793 for the transmission of text characters.

TTY is an evolving technology, and emerging standards are anticipated to redefine how it's implemented—specifically, the ITU V.151 and IETF RFC 4103 standards, which are likely to be a part of the PacketCable 2.0 release.

Chapter Summary

Analog voice is converted into digital samples by the use of a CODEC. Mandatory PacketCable CODECs include the ITU standard G.711 and the royalty free CODECs iLBC and BV16, although the royalty free CODECs are mandatory only for PacketCable 1.5. Recommended optional PacketCable CODECs include the ITU standards G.728 and G.729 Appendix E. The narrowband G.711 CODEC works by sampling the analog audio source at rate of 8,000 samples per second. These samples are then quantized using μ-law or a-law companding. At this point this data is inserted into the payload portion of RTP packets. Not including IP packet encapsulation overhead, the G.711 CODEC has a bandwidth requirement of 64 Kbps. The G.728 and G.729E vocoders utilize compression techniques to lessen this bandwidth requirement to 16 Kbps for G.728 and 11.8 Kbps for G.729E. The G.728 CODEC is a low-delay CODEC, and the G.729E CODEC is well suited for music as well as voice. The iLBC CODEC also uses a low bandwidth, either 13.33 Kbps or 15.2 Kbps, and is much better at dealing with dropped packets. The BV16 CODEC provides toll quality voice with a bandwidth requirement of 16 Kbps and a coding delay of only 5 ms.

Ways of quantifying CODEC performance include voice quality and bandwidth utilization. Voice quality consists of clarity, delay, and echo. The PESQ method is the preferred way of measuring voice clarity and correlates to traditional MOS scores. Another recommended method of measuring the effects of a VoIP network on voice clarity is the ITU E-model. Delay can be broken down into three areas: from the speaker to the backbone, across the backbone, and from the backbone to the listener. Delays in these areas come from the CODEC themselves, the IP packetization process, jitter buffers, queuing on intermediary devices, and propagation time between devices. An end-to-end delay of 150 ms is generally regarded as the maximum acceptable. Electrical or line echo occurs because of an impedance mismatch in the 2-to-4-wire conversion used in telephony interfaces. Echo cancellers can be used to reduce echo by remembering the received audio stream and removing the degraded portion of it from the transmit stream.

Voice problems, such as no voice or one-way voice, are probably caused by routing issues or firewalls. Problems associated with garbled voice are indicative of lost packets or jitter problems.

The transmission of DTMF digits between gateways is needed to support applications such as IVR and voicemail. DTMF digits can be transmitted either in-band or via DTMF relay using the techniques described in RFC 2833. DTMF relay provides a more reliable method for transmitting DTMF digits and is a requirement in PacketCable 1.5.

Voice Band Data transmission refers to the ability to support analog modem, facsimile, and telephony for the hearing impaired signals across the PacketCable network. If no relay method is used when VBD is detected by a gateway, the CODEC needs to up-speed to G.711, echo cancellation needs to be disabled, voice activity detection needs to be disabled, and the high-pass filter in the CODEC should also be disabled. The V.152 specification defines additional techniques for improving in-band VBD transmission.

Relay techniques provide an alternative method of transmitting VBD. The T.38 recommendation for fax relay is an example of one of these methods. T.38 supports error-correction techniques and redundancy schemes, making it the preferred method of fax transmission for a PacketCable network.

Chapter Review

1 What CODECs were added to the PacketCable list of mandatory CODECs in PacketCable 1.5?

Answer: Global IP Sound's iLBC and Broadcom's BLV16 were added to G.711 as mandatory CODECs.

2 What frequency range does most human speech fall between?

Answer: 300 Hz to 3,400 Hz (this is narrowband speech)

3 Why is linear quantization not a very good technique?

Answer: Because quantization noise is a much bigger problem at lower audio levels and greater granularity is needed in this range

4 What are the bandwidth requirements for each of the following CODECs used in PacketCable?

G.711

G.728

G.729E

iLBC

BV16

Answer: G.711: 64 Kbps; G.728: 16 Kbps; G.729E: 11.8 Kbps; iLBC: 13.33 Kbps and 15.2 Kbps; BV16: 16 Kbps

5 What are some methods CODECs use to lessen the problem of dropped packets?

Answer: Repetition, extrapolation, interpolation

6 What CODEC is very robust with respect to lost packets?

Answer: iLBC

7 What are the three components of voice quality?

Answer: Clarity, delay, echo

8 What is the MOS rating of PSTN toll quality voice?

Answer: 4.0

9 What is regarded as the maximum tolerable end-to-end delay for voice packets in a PacketCable call?

Answer: 150 milliseconds

10 What are the benefits and drawbacks to larger jitter buffers versus smaller jitter buffers?

Answer: With larger jitter buffers interpacket delays can be bigger but delay suffers.

11 What are the two types of echo and how are they caused?

Answer: Electrical echo caused by the 2- to 4-wire hybrid in telephony circuits and acoustic echo, typically caused by speakerphones

12 How do echo cancellers work, and what is the main adjustable parameter associated with them?

Answer: They remove the degraded portion of the receive signal from the transmit signal; tail length is the main adjustable parameter.

13 What possible voice path problems could be caused by a firewall?

Answer: No voice and one-way voice

14 If you notice the beginning of sentences is being cut off, what is the probable cause?

Answer: VAD (Voice Activity Detection)

15 If a voicemail server cannot detect DTMF digits in the voice stream, how can this problem be fixed?

Answer: By implementing DTMF relay

16 When a Voice Band Data transmission such as analog modem or fax is detected, what steps does the gateway need to perform?

Answer: Disable echo, disable VAD, disable high-pass filter, and up-speed to G.711

17 If fax or modem relay is not an option, how can a gateway improve performance?

Answer: By sending packets multiple times or using error correction schemes as described in V.152

18 Why is telephony for the hearing impaired not possible between countries?

Answer: Because they all use different protocols

19 In the ITU E-model, what is the R-factor value that corresponds to toll quality voice?

Answer: 94

CHAPTER 11

RTP and RTCP

Now that you are familiar with how CODECs function, this chapter builds on this knowledge by examining the resulting packet streams. This involves a detailed look at the format and use of the Real-time Transport Protocol (RTP) and its sister protocol, Real-time Transport Control Protocol (RTCP). Also in this chapter you will see how other functionality introduced in the previous chapter applies to real call flows. This functionality includes dual-tone multiple frequency (DTMF) relay and the transmission of Voice Band Data such as facsimile and modem transmissions. Finally, the chapter covers pertinent configuration in Cisco PacketCable products, including the BTS 10200 and Cisco media gateways. The chapter also revisits some MTA configuration file parameters associated with these topics.

RTP Protocol Stack

The Real-time Transport Protocol (RTP) is used for the transmission of the bearer packets in PacketCable. The PacketCable 1.x specification for CODECs identifies RFC 1889 as the compliant specification for RTP (and RTCP).

NOTE RFC 3550 actually obsoletes RFC 1889; the differences between these two documents are minimal, and in fact, these differences are small enough that they are beyond the level of detail covered in this book. This chapter addresses the key points you need to be familiar with from these two specifications.

In PacketCable, RTP is encapsulated inside the UDP transport layer protocol. Recall that the UDP protocol includes a checksum that provides error detection on packets, but it does not provide any sort of retransmission scheme. So if a voice packet is lost, it is truly lost. In the previous chapter you learned about Packet Loss Concealment (PLC) techniques that mitigate the effects of lost packets. These techniques are implemented in the CODECs and occur after the payload is extracted from the RTP packets.

RTP and RTCP port numbers are from the random range of UDP ports (1,024 through 65,535). Typically, RTP uses even port numbers, and its corresponding RTCP stream uses the next odd port number.

NOTE The RFC 3550 specification removes this restriction if the RTP/RTCP port pairs are explicitly specified.

RTP Message Format

Like any other protocol RTP packets comprise a payload portion and a packet header. In PacketCable voice the payload consists of either digital samples of the voice stream or coded parameters characterizing the voice stream.

Figure 11-1 depicts the format of the RTP header.

Figure 11-1 *RTP Header Format*

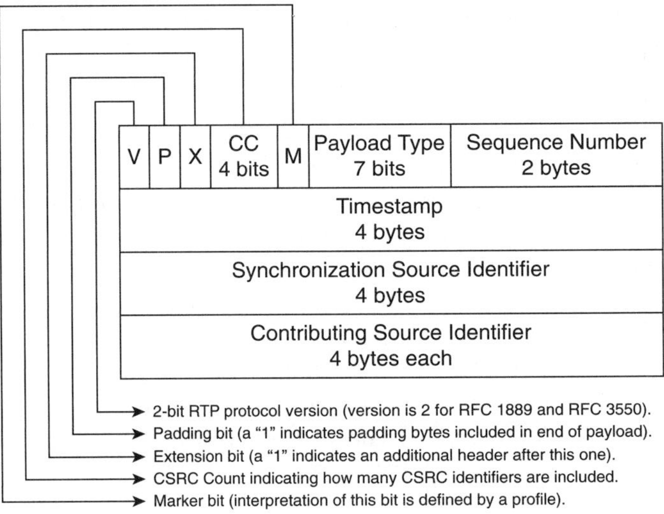

Some fields to note include the Payload Type, Sequence Number, Timestamp, and Synchronization Source Identifier. The sections that follow cover these fields in more detail.

NOTE The Marker bit is typically set only in the first packet after a period of silence suppression.

Payloads and Payload Types

The format of RTP payloads is designed to be flexible; therefore, it is not defined in the RTP specification. Instead, a separate profile document defines this for a given application. RFC 1890 defines such a profile for audio and video communications.

NOTE RFC 3551 obsoletes RFC 1890, but as was the case with RFC 1889 and RFC 3550, the differences between RFC 1890 and RFC 3551 are minimal.

Logically, the Payload Type field in the RTP header identifies what the payload is. You have two categories of payload types:

- **Static**—Static payload types are predefined in the aforementioned RFCs. The static range of payload types is defined as 0–95.
- **Dynamic**—Dynamic payload types are defined during the call setup. The dynamic range of payload types is defined as 96–127.

Table 11-1 shows some of the static payload type mappings defined in RFC 1890. Most of these are not relevant in PacketCable, so they are grayed out.

Table 11-1 *Static Payload Type Mappings*

Payload Type	Encoding Name	Audio(A)/Video(V)
0	PCMU (G.711 μ-law)	A
3	GSM	A
4	G.723	A
5, 6	DVI4	A
7	LPC	A
8	PCMA (G.711 a-law)	A
9	G.722	A
10,11	L16	A
12	QCELP	A
13	Comfort Noise (CN)	A
14	MPA	A
15	G.728	A
16,17	DVI4	A
18	G.729	A
25	CelB	V

continues

Table 11-1 *Static Payload Type Mappings (Continued)*

Payload Type	Encoding Name	Audio(A)/Video(V)
26	JPEG	V
28	nv	V
31	H261	V
32	MPV	V
33	MP2T	AV
34	H263	V

So you can see the G.711 variants are defined by payload types 0 and 8, and G.728 is defined by payload type 15. G.729E, BV16, and iLBC are not statically defined, so they must be defined using the dynamic method.

The definition of a dynamic payload type is done using another protocol. As you might have guessed, this is a protocol you have already learned about, the Session Description Protocol (SDP). Basically, SDP parameters map the dynamic payload type to a CODEC. Remember the Session Attribute SDP parameter (a), which is used to extend SDP to an application. Well, dynamic payload type definition is one of those applications. Specifically the attribute rtpmap is used. The general form of this parameter, as defined in RFC 2327, is as follows:

a=rtpmap:*payload type encoding name*/*clock rate*[/*encoding parameters*]

The *payload type* is the dynamic payload type number identified in the media descriptor. The *encoding name* is the registered name of the encoding format. The *clock rate* is the RTP timestamp clock rate and typically equates to the number of samples taken in 1 second (that is, 8000 for narrowband voice). The optional parameter, *encoding parameters*, defines the number of channels for audio streams and is meaningless for video streams. If not included, the default of one channel is assumed.

When the call agent sends a CRCX or MDCX with a Local Connection Option (LCO) containing a CODEC without a defined static payload type, it uses this parameter to assign a dynamic value and returns this to the call agent in the acknowledgement message.

For example, suppose a CRCX is sent with the following LCO:

```
L: mp:20; 20; 20, a:G729E;PCMU;PCMA
```

The 200 - OK response would contain a mapping for the G.729E CODEC as follows:

```
m=audio 60294 RTP/AVP 96 0 8
     a=mptime:20 20 20
a=rtpmap:96 G729E/8000/1
```

The order of CODECs listed in the SDP media announcement parameter (m) is the same as the order specified in the previous LCO. Thus, payload type 96 corresponds to G729E, 0 to

PCMU, and 8 to PCMA. The 0 and 8 values are well-defined static values, whereas the 96 value is dynamic, so it is further specified with the rtpmap parameter.

Because the iLBC CODEC is defined dynamically, the rtpmap attribute is also needed, as demonstrated here:

```
a=rtpmap:97 iLBC/8000
```

If the 20 ms block/frame length is used with iLBC, another attribute is needed as well. This is the format-specific parameter fmtp:

```
a=fmtp:97 mode=20
```

The parameter mode identifies the CODEC as using the 20 ms block length; if not specified, the default block length of 30 ms is assumed.

Similarly, if the BV16 CODEC is used, the payload type is defined dynamically, and the rtpmap attribute is required. A sample definition is as follows:

```
a=rtpmap:98 BV16/8000
```

Please refer to Chapter 6, "Signaling Interfaces and MGCP Overview," for more information about SDP.

The RTP payload format for G.711 packets is among those defined in RFC 3551. The format is fairly straightforward and is simply the 8-bit encoding of each sample with the sign bit as the most significant bit. If you are using a 20 ms packetization period, you then have 160 samples according to the Nyquist criteria (8,000 samples/sec × 20 ms). A G.711 sample is 8 bits, so you have 160 8-bit samples corresponding to 160 bytes of RTP payload.

The RTP payload format for G.728 packets is also defined in RFC 3551. Remember from Chapter 10, "Audio CODECs," that the G.728 frame size is 5 bytes comprised of four 10-bit codepoint vectors. So if RTP packets are sent out every 10 ms, 4 G.728 frames are in each RTP packet.

The RTP payload format for G.729E packets is also defined in RFC 3551. In Chapter 10 you learned that G.729E works by translating samples from a 10 ms period into algebraic codewords and then transmits these parameters as well as other variables such as gain parameters. All these parameters result in a frame size of 118 bits, which is rounded up to 120 bits (15 bytes) to make a frame an even number of bytes. Each RTP packet could consist of multiple G.729E frames. Note: Each RTP packet could also optionally include a G.729B frame used for silence suppression.

The RTP payload format for iLBC consists of a single iLBC frame of 304 bits (38 bytes) if the 20 ms block length is used and a single iLBC frame of 400 bits (50 bytes) if the 30 ms block length is used.

The RTP payload for BV16 consists of some number of BV16 frames, with each frame being 80 bits (10 bytes) long.

Another RTP payload format worth mentioning is the one used for comfort noise packets when silence suppression is implemented. These packets are sent when you have no actual voice packets to transmit to reassure the other devices that the endpoint is still functional. The static payload type 13 is used for these packets. You can find more information on silence suppression and RTP later in this chapter.

Sequence Number

The sequence number parameter increments by one for each RTP packet sent. So logically, the receiver can examine this field to detect missing and out-of-order packets. For security reasons, the initial sequence number should be randomly generated.

Timestamp

The timestamp field refers to the instant the first byte in the payload is sampled. Just as with the sequence number field, the initial value of this field should be randomly generated. For audio streams with fixed sampling rates this parameter increments by one for each sampling interval. For example, the sampling interval for narrowband voice is 125 microseconds (1/8,000). Thus, if a 20-ms packetization period is used, this parameter increments by 160 for each RTP packet, and if a 10-ms packetization period is used, this parameter increments by 80 for each RTP packet.

Synchronization Source (SSRC)

The Synchronization Source (SSRC) parameter is a randomly generated 32-bit number that identifies the source of the RTP packets (that is, the gateway to which the speaking user is connected). In most cases the gateway sourcing the RTP stream generates this parameter. The timing and sequence number information is relative for all RTP packets with the same SSRC identifier. Thus, the gateway to which the listening user is connected can correlate RTP packets based on the SSRC field.

Another field in the RTP header is the Contributing Source (CSRC) parameter, also a 32-bit number. With RTP a separate device known as an RTP mixer can be used to combine multiple RTP streams, such as in a conference call scenario. This field identifies those contributing RTP sources by listing their SSRC identifiers. The RTP mixer creates a new SSRC parameter identifying itself, and the SSRC parameters of the contributing RTP sources comprise its CSRC parameter.

Because the CSRC field can contain any number of contributing sources (from 0 to 15), it is of variable length. Recall that the number of contributing sources is identified in the CC field depicted in Figure 11-1. Also as these contributing sources enter and leave the conference, this field changes to reflect that. This in turn causes the length of the RTP header to vary. Because of this variability, this field can cause a problem in PacketCable.

As you will see in Part V of this book, having fixed-size packets is desirable for admission control purposes. Consequently, the transmission of this parameter by PacketCable endpoints is currently optional and usually avoided. However, the ability to receive this parameter is mandatory.

RTP Packet Trace Analysis

Now it's time to put what you've learned about CODECs and RTP to use and examine some packet traces. Figure 11-2 is an Ethereal capture of the RTP stream of one direction of a call using the G.711 μ–law CODEC.

Figure 11-2 *G.711 Packet Trace*

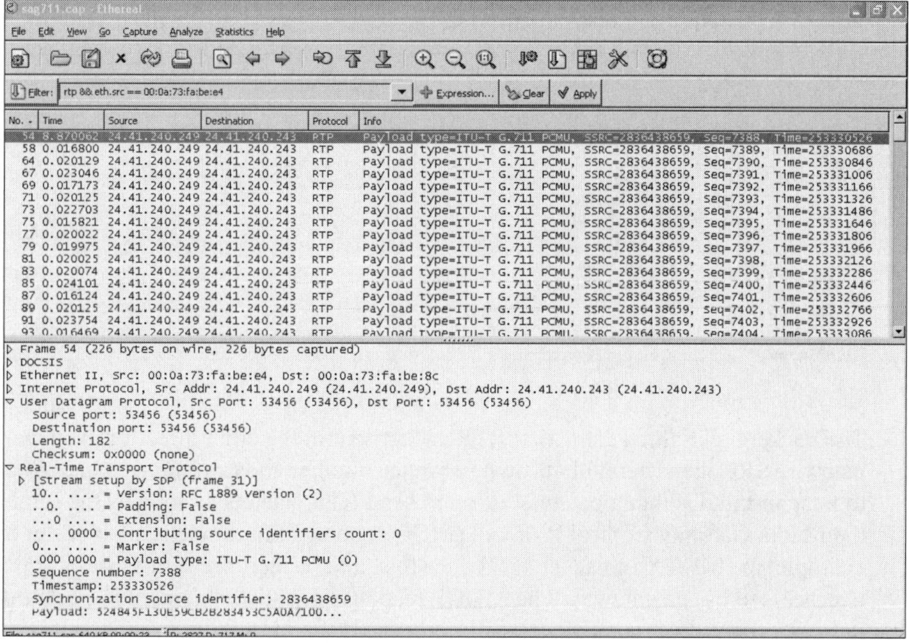

The first transmitted RTP packet in this call is the one decoded at the bottom of the figure. Notice that both the sequence number and the timestamp values are randomly generated. The Synchronization Source identifier is also random and is used by the other end of the call to correlate packets. As you can see from the view of all the packets, the sequence numbers are incrementing by 1, and the timestamp values are incrementing by 160. Recall that this is the expected value for a call using a packetization period of 20 ms. The time display format was set to "time since previous packet," so you can see this is also around 20 ms as expected. Finally, notice the RTP payload type is 0, which is the statically defined value for the G.711 μ–law CODEC.

Figure 11-3 shows the point in the call where the other end begins transmitting RTP packets.

Figure 11-3 *Bidirectional G.711 Packet Trace*

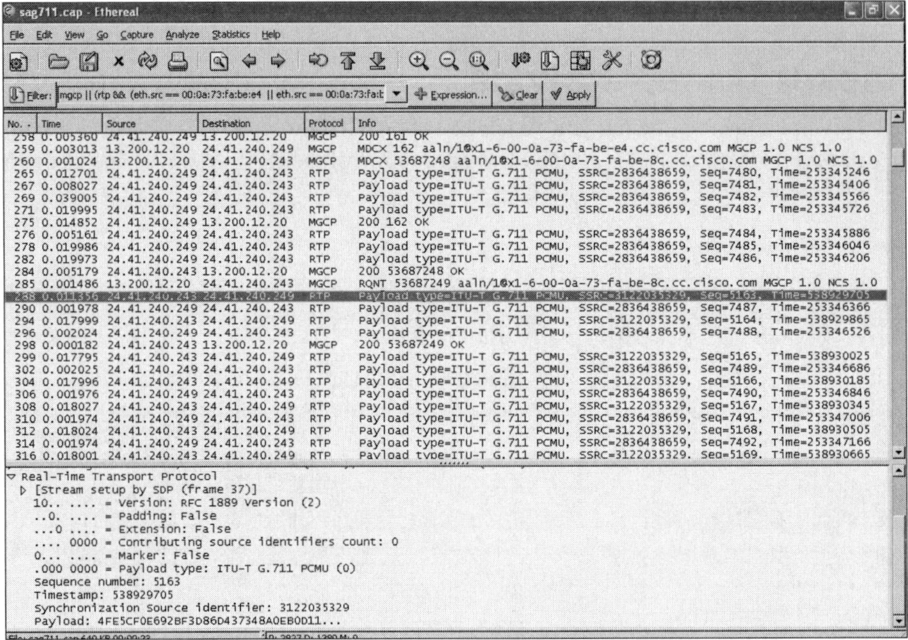

The packet highlighted is the first RTP packet sent in the other direction. You can see it has its own SSRC identifier with its own sequence number and timestamp space. Another thing to keep in mind is that a gateway starts to send RTP packets as soon as the send option in the Media Gateway Control Protocol (MGCP) connection mode is enabled. In this example, the MDCX in packet #260 does this, and packets are sent as soon as this MDCX is processed by the gateway. The MGCP response acknowledgement in packet #284 indicates the gateway has successfully processed the MDCX.

Figure 11-4 shows a packet trace of a call using the G.728 CODEC. Notice that the payload type is now the static value of 15 reserved for the G.728 CODEC. Another thing to note is the packet length. The size of the RTP payload is 20 bytes. You can figure this out if you take the UDP length and then subtract the UDP header (8 bytes) and RTP header (12 bytes). Remember that a G.728 frame is 5 bytes (which comprises four codewords), so an RTP payload of 20 bytes contains 4 G.728 frames.

Figure 11-4 *G.728 Packet Trace*

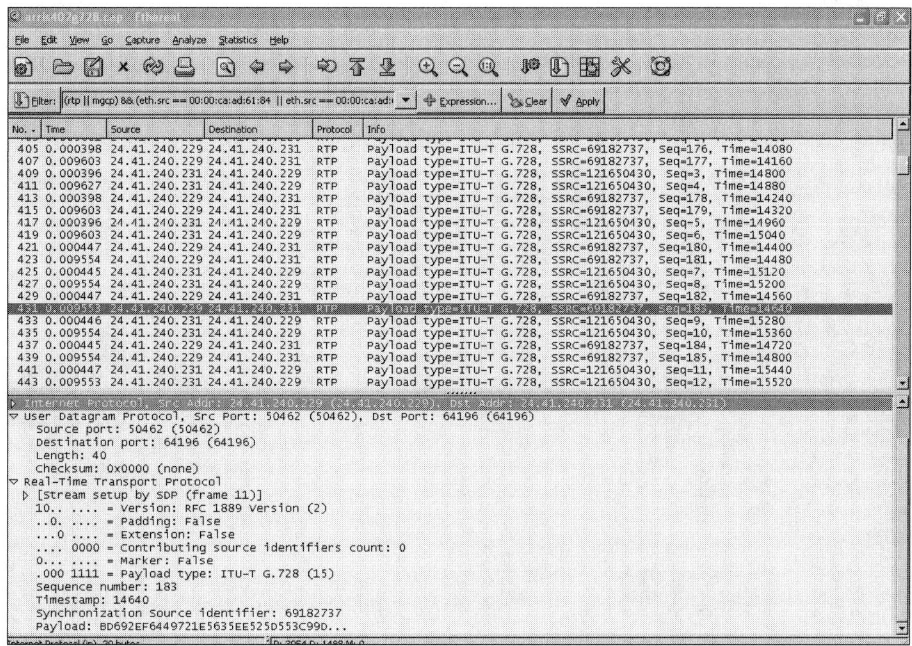

Similarly, Figure 11-5 shows a packet trace of a call using the G.729E CODEC. Notice that in this case the payload type is the dynamic value 96, which was previously defined as G.729E. Because a dynamic payload type is used in this case, Ethereal is unable to recognize the payload as G.729E and fully decode it.

The UDP packet length is 35, meaning the RTP payload size is 15 bytes. This turns out to be the size needed to hold one G.729E frame. One G.729E frame is 118 bits, which is 15 when rounded up to the nearest whole number of bytes.

Figure 11-5 *G.729 Packet Trace*

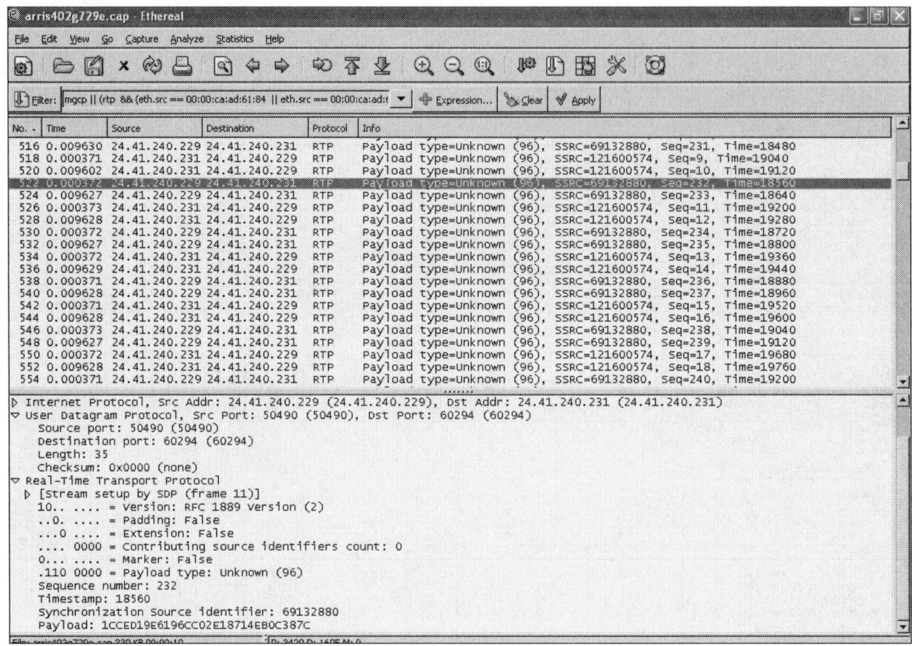

Packet Size Comparison

Now that you know how all these CODECs work, it is useful to examine the effective bandwidth requirements of each.

NOTE At the time of the writing of this book no MTAs were available that supported either Global IP Sound's iLBC or Broadcom's BV16 royalty free CODEC; therefore, no packet traces can be examined. Also the 4.5 version of the BTS 10200 does not support these CODECs.

To summarize what you saw in the example packet traces:

The G.711 CODEC causes an RTP payload of 160 bytes when used with a 20-ms packetization rate (it would be 80 bytes with a 10-ms packetization rate).

The G.728 CODEC causes an RTP payload of 20 bytes when used with a 10-ms packetization rate.

The G.729E CODEC causes an RTP payload of 15 bytes when used with a 10-ms packetization rate.

You also know that each packet has a 12-byte RTP header, an 8-byte UDP header, and a 20-byte IP header. Layer 2 overhead (such as DOCSIS) is also associated with each packet, which you will learn more about later.

Table 11-2 describes the resulting bandwidth constraints of these various CODECs:

Table 11-2 *CODEC Bandwidth Constraints*

CODEC	CODEC Bit Rate	Pack. Period	RTP Payload	RTP/UDP/IP Header	Total Packet Size	Packets per Second	Resulting Bit Rate of One Call
G.711	64 Kbps	10 ms	80 bytes	40 bytes	120 bytes	100	96 Kbps
G.711	64 Kbps	20 ms	160 bytes	40 bytes	200 bytes	50	80 Kbps
G.711	64 Kbps	30 ms	240 bytes	40 bytes	280 bytes	33 1/3	~74.67 Kbps
G.728	16 Kbps	10 ms	20 bytes	40 bytes	60 bytes	100	48 Kbps
G.728	16 Kbps	20 ms	40 bytes	40 bytes	80 bytes	50	32 Kbps
G.728	16 Kbps	30 ms	60 bytes	40 bytes	100 bytes	33 1/3	~26.67 Kbps
G.729E	11.8 Kbps	10 ms	15 bytes	40 bytes	55 bytes	100	44 Kbps
G.729E	11.8 Kbps	20 ms	30 bytes	40 bytes	70 bytes	50	28 Kbps
G.729E	11.8 Kbps	30 ms	45 bytes	40 bytes	85 bytes	33 1/3	~22.67 Kbps
iLBC	15.2 Kbps	20 ms	38 bytes	40 bytes	78 bytes	50	31.2 Kbps
iLBC	13.33 Kbps	30 ms	50 bytes	40 bytes	90 bytes	33 1/3	24 Kbps
BV16	16 Kbps	10 ms	20 bytes	40 bytes	60 bytes	100	48 Kbps
BV16	16 Kbps	20 ms	40 bytes	40 bytes	80 bytes	50	32 Kbps
BV16	16 Kbps	30 ms	60 bytes	40 bytes	100 bytes	33 1/3	~26.67 Kbps

You can use these numbers to traffic engineer your PacketCable network, which is one of the topics of Chapter 16, "PacketCable Network Design Considerations." Remember from Chapter 10 that bandwidth constraints should just be one of the considerations when choosing a CODEC; you must also take other metrics, such as delay, voice quality, and robustness to packet loss, into consideration. Also other factors can affect bandwidth requirements not addressed in this table. This includes silence suppression, payload header suppression, and security overhead. Payload header suppression is covered in Part V of the book, and security is covered in Chapter 16. You will learn more about silence suppression next.

Silence Suppression (also Known as VAD)

In a typical phone conversation, only one party is talking at a time; therefore, bandwidth is wasted by continuously sending packets in both directions. If no voice packets are sent when a subscriber is silent, then this bandwidth can be used for other calls or other applications.

As it turns out, instead of not transmitting any RTP packets at all, it is beneficial to send another type of RTP packet known as a *comfort noise packet*. A couple of reasons exist as to why this is a better approach. One, if a person doesn't hear anything at all in a telephone call, he or she might think that person has been disconnected or he or she has a problem with the phone or the phone network. People are used to always hearing something, even if it is just noise as heard in the traditional PSTN network. This noise is known as comfort noise; thus, the packets providing this are known as comfort noise packets. Now you might ask yourself, "If you are still transmitting packets, how is bandwidth saved?" The answer is that these packets are much smaller than normal voice packets and are sent much less frequently; consequently, the bandwidth constraints are much less. The other reason sending comfort noise is a good idea is that the receipt of these packets assures intermediate devices that the RTP stream is still active and does not inadvertently tear down these calls. Devices such as the CMTS can monitor the activity of calls, and if no packets are received, in a given amount of time the call is considered dead and torn down.

RFC 3389 defines the comfort noise payload format and designates the static payload type 13 for its RTP packets. The first byte in the payload defines the comfort noise level, and the remaining bytes are optional spectral information. This spectral information is of various lengths, but a typical RTP payload size for comfort noise packets is three bytes. The packet rate of comfort noise packets is implementation-specific.

For information on how silence suppression relates to DOCSIS and DQoS, see Chapter 13, "Analyzing, Implementing, and Troubleshooting DQoS."

DTMF Relay

As covered in Chapter 10, DTMF relay is another application that uses dynamic RTP payload types. RFC 2833 defines a payload type called a Named Telephony Event (NTE) to identify DTMF digits.

An important thing to keep in mind is that even though named telephony events have a different payload type than regular voice packets, they are all part of the same audio stream. That is, the RTP SSRC identifier is the same for both packet types. Consequently, the RTP sequence numbers and timestamp parameters are of the same number space. This is needed so that the receiving gateway can correlate the DTMF digit tones with the regular audio stream.

Figure 11-6 illustrates the NTE payload format.

Figure 11-6 *RFC 2833 NTE Payload Format*

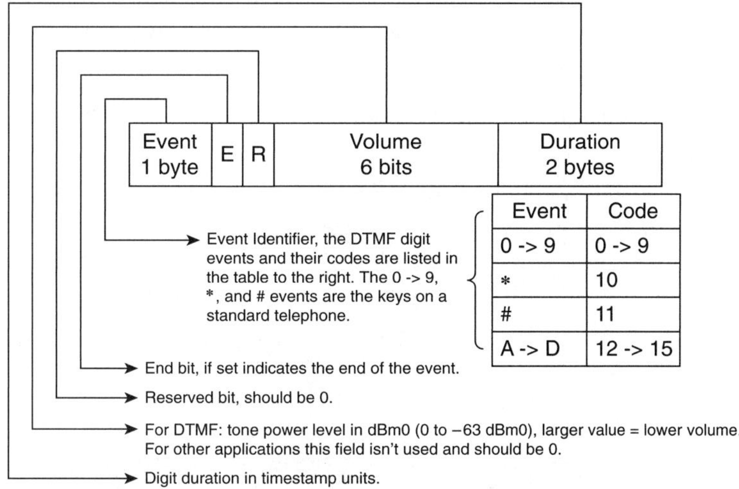

Figure 11-7 depicts the SDP information used by the MTA to identify the dynamic payload type it uses for DTMF relay.

Figure 11-7 *Dynamic Payload Type Definition*

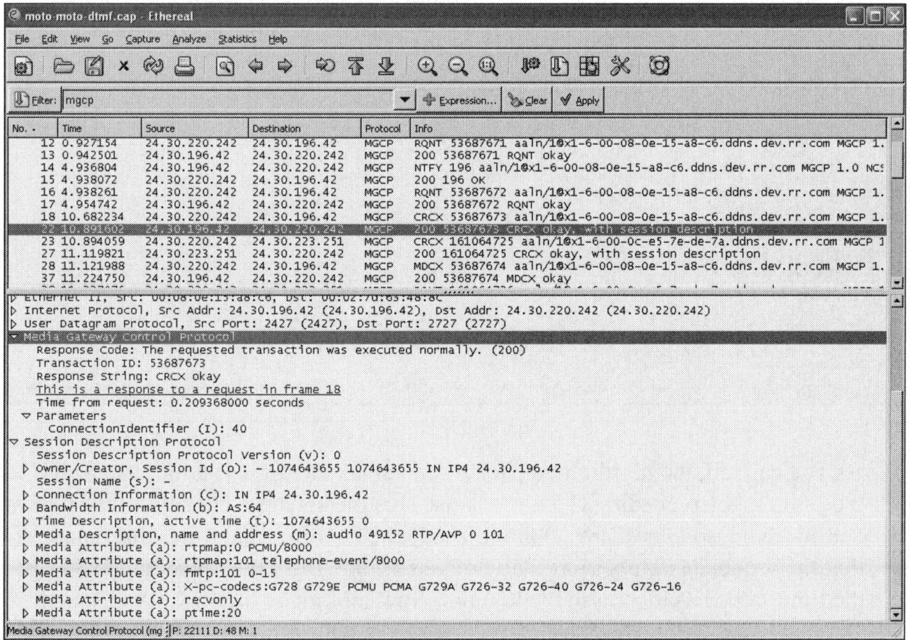

From the trace you can see that payload type 101 is designated for use for RFC 2833 NTE packets. The SDP media attribute line a=rtpmap:101 telephone-event/8000 signifies this. Also notice the next SDP media attribute line, a=fmtp:101 0-15. This is the format-specific attribute and has the following format:

a=fmtp:*format format specific parameters*

In this case the format is 101, the dynamic payload type used for NTE packets. The format-specific parameters identify the events that can be carried in these packets. These numbers refer to the same event codes used in the packets; refer to Figure 11-6 for this information. Note: All gateways implementing RFC 2833 are required to support event codes 0–15, so the inclusion of them in SDP is optional.

Next, Figure 11-8 shows a trace of what happens after the originating party presses a DTMF digit on their telephone.

Figure 11-8 *Start of DTMF Relay Trace*

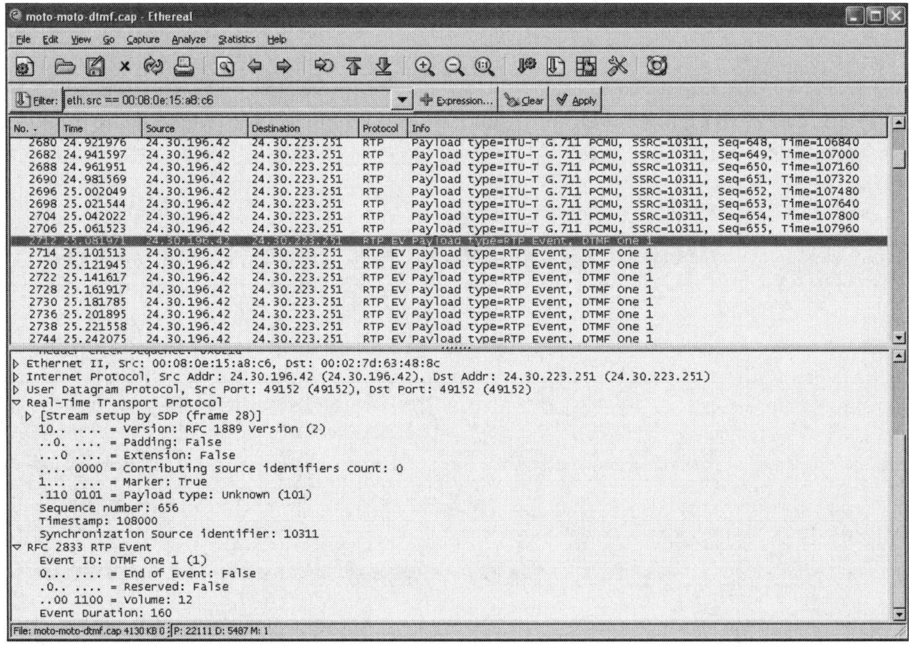

As you can tell, the highlighted packet represents the digit 1; notice the SSRC identifier is the same as the preceding G.711 packets. This means these packets are all part of the same audio stream, so they share a common sequence number and timestamp space. You can see that the sequence number has incremented by one from the preceding G.711 packet, and the timestamp has increased by 40 units. Now looking at the RFC 2833 specific fields, you can tell that this event is the digit 1 with a volume level of –12 dBm0. Note that the event duration is 160 timestamp units (that is, 20 milliseconds).

If you were to look at the next packet, you would see its duration is 320, the one after that is 480, the one after that is 640, and so on. So what does this mean? Well, if you look at the timestamps of each packet, you see that the interpacket time is approximately 20 ms, the same as the G.711 packets. This is because NTE packets are sent at the same rate as the regular audio stream. Remember that for a 20-ms packetization rate the RTP timestamp field in regular audio packets increases by 160 (20 ms × 8000 samples/sec). This is also true for the event duration field in NTE packets.

An interesting aspect of this call is that the RTP timestamp is constant for every packet of this event; this is because the caller is holding down the digit for a lot longer than 20 ms. If you were to look at the event duration for the last packet of the event, you could figure out the approximate total time the digit was pressed. Just divide this number by 8,000 to get the time. Figure 11-9 depicts the last event packet in this example.

Figure 11-9 *End of DTMF Relay Trace*

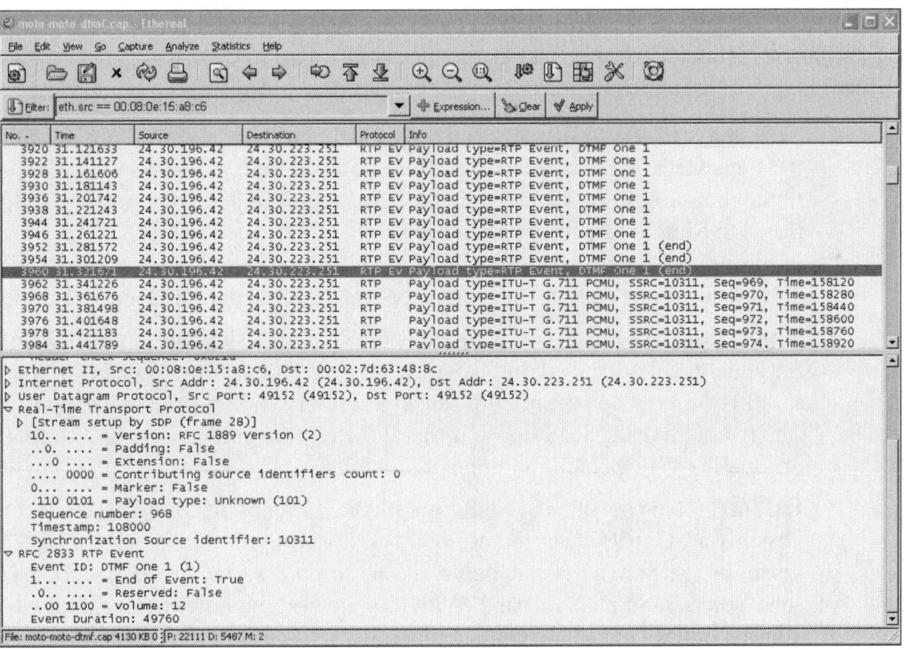

So if you divide the event duration of 49,760 by 8,000, you get 6.22 seconds for the digit duration. You can see another interesting observation in this trace. Notice that three packets indicate the end of the event. This is done for reliability. If one of these packets is lost, the terminating gateway can still detect the end of the event.

RFC 2833 specifies that the gateway can play out digit tones either based on the event duration (1) or continuously until stopped by certain criteria being met (2). This criterion consists of the receipt of an "end-of-event," the receipt of a mutually exclusive event, the receipt of an audio packet, or timeout expiry because of no packets being received. PacketCable specifies that the second method must be used. In the preceding example, tone playout stopped because of the receipt of an "end-of-event."

Fax Relay (T.38)

In the previous chapter you learned that the T.38 standard for fax relay is the preferred method of facsimile transmission in a PacketCable network. To support T.38 the MGCP Fax package (package name "FXR") is used by the call agent (CMS/MGC) and gateways (MTA/MG). This package is defined in both the PacketCable 1.5 NCS specification and the PacketCable 1.5 TGCP specification, and its support is a requirement for PacketCable 1.5. To illustrate how this package can be used, this section examines messaging from an actual T.38 call.

The call agent can send a prioritized list of fax handling procedures to the gateway in the fax LCO option (L: fxr/fx). This parameter is included in the CRCX and/or MDCX message. The possible values of this parameter are as follows:

- **T.38 Strict (t38)**—Call agent controlled fax using T.38 with either UDPTL or TCP as the transport layer protocol. T.38 is used only if both gateways indicate they support it; the remote endpoint must indicate T.38 support in its SDP parameters. Specifically, the SDP Simple Capability Declaration feature defined in RFC 3407 is used for this purpose; you'll see an example in a little bit. Consequently, you face one risk in using T.38 Strict. If the remote gateway supports T.38 but not RFC 3407, T.38 fails.
- **T.38 Loose (t38-loose)**—Call agent controlled fax using the same procedures as T.38 Strict are invoked; except indication of support from remote side is not required. This should be used when the remote gateway supports T.38 but doesn't support RFC 3407. The risk with using T.38 Loose is that if the remote gateway doesn't support T.38, the originating gateway still attempts it and fails.
- **Off (off)**—No special procedures are invoked except disabling echo cancellation and possibly a CODEC up-speed. Remember, if a compression-based CODEC is used for voice, the gateways must switch to a non-compression CODEC (such as G.711) to send Voice Band Data in-band. With this method the gateways generate a nopfax(start) event to the call agent when fax tone is detected.
- **Gateway (gw)**—The gateway is permitted to control fax handling without involving the call agent. Gateway methods are not defined in PacketCable and are left to the discretion of the gateway vendors. If the gateway does do something special, it still notifies the call agent with a gwfax(start) event so the call agent knows not to do anything that might jeopardize the gateway's capability to handle the fax

transmission. Gateway mode is also the default mode if nothing is specified in the CRCX message. When the gateway detects the fax is over, it generates a gwfax(stop) event to the call agent, and if it detects a failure, it generates a gwfax(failure) event to the call agent.

For example, a CRCX indicating the T.38-Loose method is to be used for fax handling is as follows:

```
CRCX 161155926 aaln/2@x1-6-00-00-ca-dd-a7-ec.cc.cisco.com MGCP 1.0 NCS 1.0
C: 3bb
L: mp:20;20;20,  a:G729E;PCMU;PCMA,  fxr/fx:t38-loose,  e:off,  s:off,  t:a0
M: inactive
K: 161155923
```

Notice that this message also indicates either the G.729E, G.711μ, or G.711a CODECs are to be used for voice with the G.729E CODEC preferred. The acknowledgement to this message is as follows:

```
200 161155926 OK
I: 7

v=0
o=- 722410 722410 IN IP4 24.34.240.246
s=-
c=IN IP4 24.34.240.246
t=0 0
m=audio 61730 RTP/AVP 96 0 8
b=AS:81
a=mptime:20 20 20
a=rtpmap:96 G729E/8000/1
a=sqn: 0
a=cdsc: 1 audio RTP/AVP 0 8 96 100
a=cpar: a=rtpmap:100 telephone-event/8000/1
a=cpar: a=fmtp:100 0-15,144,149,159
a=cdsc: 5 image udptl t38
a=cpar: a=T38FaxVersion:0
a=cpar: a=T38FaxRateManagement:transferredTCF
a=cpar: a=T38FaxMaxDatagram:160
a=cpar: a=T38FaxUdpEC:t38UDPRedundancy
```

Notice the SDP attribute parameters beginning with the a=sqn: 0 line. This the RFC 3407 capability information used to define a complete capability set of media attributes supported by the endpoint; among these attributes are the gateway's fax capabilities. This information also lets the call agent know that, although the gateway endpoint supports the use of T.38, it cannot use it at the same time it is also using one of the voice CODECs.

The capability set begins with a sequence number, a=sqn: 0, and is followed by one or more capability descriptors. The first capability set always uses sequence number 0, and if there are other capability sets, this number is incremented. A capability descriptor is denoted by the cdsc attribute name. In this case, two of these attributes indicate the media formats the endpoint supports at a given time.

The first one, a=cdsc: 1 audio RTP/AVP 0 8 96 100, identifies the RTP voice payload formats supported by the endpoint. This attribute begins with the capability number (in this casc 1), followed by the media type (in this case audio), the transport used (in this case

RTP), and a format list. In this case the format list identifies the supported RTP payload types. The first three are G.711µ (0), G.711a (8), and the dynamically defined G.729E (96). The last one is also something you have seen before; it is the payload type used for RFC 2833 DTMF relay. Notice the rtpmap and fmtp attributes defined next as capability parameters cpar that further define DTMF relay.

The next capability descriptor, a=cdsc: 5 image udptl t38, indicates support of T.38 fax relay using the UDPTL protocol for transport. The capability number is 5 because four media formats were denoted in the previous capability descriptor. The capability parameters after this capability descriptor further describe the endpoints use of T.38. The T38FaxVersion attribute identifies the T.38 version number and is equal to 0. The T38FaxRateManagement attribute is equal to transferredTCF, which is the expected value when UDPTL is used for transport. The T38FaxMaxDatagram identifies the maximum UDP datagram size, and in this case it is set to the minimum value of 160 bytes. Finally, the T38FaxUdpEC attribute identifies the error correction method, which in this case is t38UDPRedundancy.

The fax LCO option and the SDP information from the originating endpoint (including the fax parameters) are then sent in a CRCX message to the terminating gateway endpoint. Similarly, in the response message the endpoint includes its fax capabilities in the SDP parameters.

As you've seen before, an MDCX is then sent to the originating endpoint to reserve dynamic quality of service (DQoS) resources and to pass along this SDP data. These steps are no different than the regular call examples you've seen before. DQoS resources are reserved at both legs of the call, ring tone and ring back tone are provided, at some point the call is answered, and DQoS resources are committed.

At this point RTP voice packets are sent and received by both gateways. This is because the media announcement parameter (m) in the CRCX and MDCX messages indicates a media type of audio. You'll see later the call agent changes this to image for the sending of T.38 packets. One aspect interesting to this call flow is the CODEC used by both gateways. Recall that the compression CODEC G.729E was identified as preferred for both gateways. The terminating gateway is sending G.729E packets, but the originating gateway is sending G.711µ packets instead. This is because this gateway knows it is sending facsimile data instead of voice by the detection of the fax tone. So it automatically up-speeds to the G.711µ CODEC to avoid corrupting the fax tone signal caused by using a compression CODEC. This allows the terminating gateway to properly identify this tone as fax tone.

Remember in the basic call flows the call agent requested notification of the on-hook and hook flash events as soon as the call was active. In this case, the call agent also requests notification of T.38 events in the RQNT message:

```
RQNT 161155935 aaln/2@x1-6-00-00-ca-dd-a7-e4.cc.cisco.com MGCP 1.0 NCS 1.0
X: 1FA00000016
R: L/hu(N),L/hf(N),FXR/t38(N)
Q: process, loop
K: 161155933
```

The originating gateway then notifies the call agent that it is ready to start using T.38 for the fax transmission. First, it mutes the audio stream, and then it sends the T.38 start event to the call agent as soon as it detects an event such as the T.30 CNG (calling fax) tone or the V.21 fax preamble:

```
NTFY 30 aaln/2@x1-6-00-00-CA-DD-A7-EC.cc.cisco.com MGCP 1.0 NCS 1.0
X: 1FC00000023
O: fxr/t38(start)
```

After receiving this message, the call agent instructs the originating gateway to start transmitting T.38 fax packets instead of audio packets. This is done with an MDCX as follows:

```
MDCX 161155939 aaln/2@x1-6-00-00-ca-dd-a7-ec.cc.cisco.com MGCP 1.0 NCS 1.0
I: 7
C: 3bb
L: mp:20,  a:image/t38
K: 161155937
```

The LCO attribute a:image/t38 indicates image or fax packets should now be sent using the T.38 protocol. The response to this message is as follows:

```
100 161155939 MDCX Response Pending

v=0
o=- 722410 722412 IN IP4 24.34.240.246
s=-
c=IN IP4 24.34.240.246
t=0 0
m=image 61730 udptl t38
b=AS:47
a=mptime:20
a=T38FaxVersion:0
a=T38FaxRateManagement:transferredTCF
a=T38FaxMaxDatagram:160
a=T38FaxUdpEC:t38UDPRedundancy
a=sqn: 0
a=cdsc: 1 audio RTP/AVP 0 8 96 100
a=cpar: a=rtpmap:96 G729E/8000/1
a=cpar: a=rtpmap:100 telephone-event/8000/1
a=cpar: a=fmtp:100 0-15,144,149,159
a=cdsc: 5 image udptl t38
a=cpar: a=T38FaxVersion:0
a=cpar: a=T38FaxRateManagement:transferredTCF
a=cpar: a=T38FaxMaxDatagram:160
a=cpar: a=T38FaxUdpEC:t38UDPRedundancy
```

The capability sets are the same that you saw in the initial CRCX response; however, notice the media announcement parameter, m=image 61730 udptl t38, has changed to indicate T.38 image packets are now being used utilizing the UDPTL protocol and port number 61730. This is the same port number that was used for RTP packets; this is done so that classifiers used to identify packets receiving DQoS do not have to be modified when T.38 is invoked. Also notice the bandwidth SDP attribute, b=AS:47, has decreased from the value in the CRCX response, b=AS:81. The amount of bandwidth needed by the MTA to support voice for the worst case is 81 kbps (that is, most bandwidth-intensive CODEC) from the possible CODEC list. In this case, the G.711 CODEC requires the most bandwidth. T.38 requires less bandwidth resources than G.711; thus, in this step the MTA also contacts the CMTS to decrease the amount of committed DQoS to the amount needed to handle T.38.

An MDCX is then sent to the terminating endpoint to tell it to start sending T.38 image packets, and the updated SDP information from the originating endpoint is also sent. The response to this message contains updated SDP information (such as T.38 transport protocol and port number) for the terminating endpoint. Likewise, bandwidth parameters are adjusted, and the terminating MTA contacts the CMTS to modify DQoS resources to accommodate T.38.

This updated information for the terminating endpoint is then sent to the originating endpoint in another MDCX message. The call agent also requests notification from the gateways for further T.38 events.

At this point the facsimile pages transmission occurs using the T.38 protocol. Figure 11-10 shows some of these messages. For the details on the use and format of these messages please refer to the T.38 specification.

Figure 11-10 *T.38 Packet Trace*

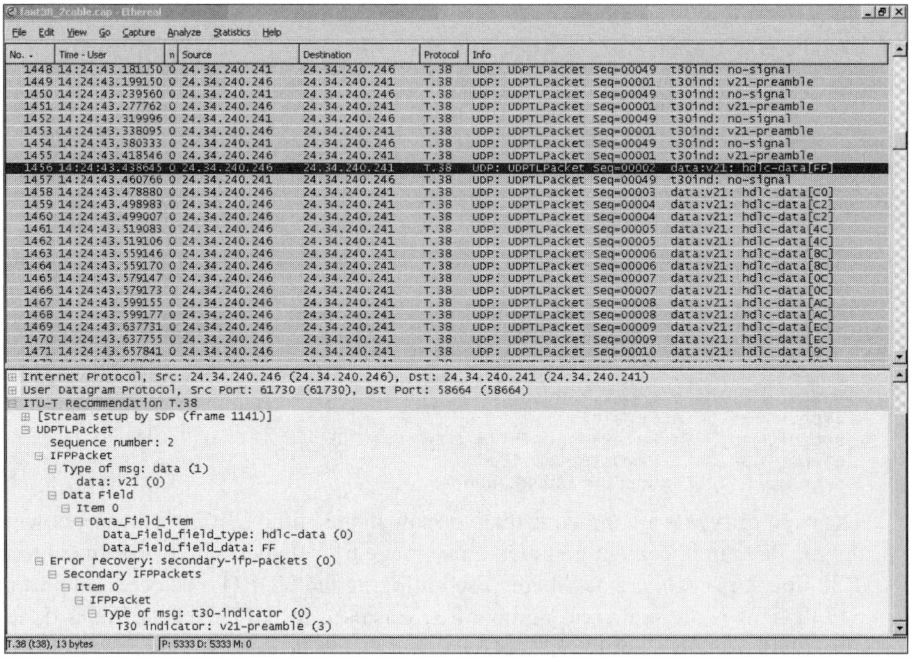

```
faxt38_2cable.cap - Ethereal
File Edit View Go Capture Analyze Statistics Help
No. Time - User n Source Destination Protocol Info
1448 14:24:43.181150 0 24.34.240.241 24.34.240.246 T.38 UDP: UDPTLPacket Seq=00049 t30ind: no-signal
1449 14:24:43.199150 0 24.34.240.246 24.34.240.241 T.38 UDP: UDPTLPacket Seq=00001 t30ind: v21-preamble
1450 14:24:43.239560 0 24.34.240.241 24.34.240.246 T.38 UDP: UDPTLPacket Seq=00049 t30ind: no-signal
1451 14:24:43.277762 0 24.34.240.246 24.34.240.241 T.38 UDP: UDPTLPacket Seq=00001 t30ind: v21-preamble
1452 14:24:43.319996 0 24.34.240.241 24.34.240.246 T.38 UDP: UDPTLPacket Seq=00049 t30ind: no-signal
1453 14:24:43.338095 0 24.34.240.246 24.34.240.241 T.38 UDP: UDPTLPacket Seq=00001 t30ind: v21-preamble
1454 14:24:43.380333 0 24.34.240.241 24.34.240.246 T.38 UDP: UDPTLPacket Seq=00049 t30ind: no-signal
1455 14:24:43.418546 0 24.34.240.246 24.34.240.241 T.38 UDP: UDPTLPacket Seq=00001 t30ind: v21-preamble
1456 14:24:43.438645 0 24.34.240.246 24.34.240.241 T.38 UDP: UDPTLPacket Seq=00002 data:v21: hdlc-data[FF]
1457 14:24:43.460766 0 24.34.240.241 24.34.240.246 T.38 UDP: UDPTLPacket Seq=00049 t30ind: no-signal
1458 14:24:43.478880 0 24.34.240.246 24.34.240.241 T.38 UDP: UDPTLPacket Seq=00003 data:v21: hdlc-data[C0]
1459 14:24:43.498983 0 24.34.240.246 24.34.240.241 T.38 UDP: UDPTLPacket Seq=00004 data:v21: hdlc-data[C2]
1460 14:24:43.499007 0 24.34.240.246 24.34.240.241 T.38 UDP: UDPTLPacket Seq=00004 data:v21: hdlc-data[C2]
1461 14:24:43.519083 0 24.34.240.246 24.34.240.241 T.38 UDP: UDPTLPacket Seq=00005 data:v21: hdlc-data[4C]
1462 14:24:43.519106 0 24.34.240.246 24.34.240.241 T.38 UDP: UDPTLPacket Seq=00005 data:v21: hdlc-data[4C]
1463 14:24:43.559146 0 24.34.240.246 24.34.240.241 T.38 UDP: UDPTLPacket Seq=00006 data:v21: hdlc-data[8C]
1464 14:24:43.559170 0 24.34.240.246 24.34.240.241 T.38 UDP: UDPTLPacket Seq=00006 data:v21: hdlc-data[8C]
1465 14:24:43.579147 0 24.34.240.246 24.34.240.241 T.38 UDP: UDPTLPacket Seq=00007 data:v21: hdlc-data[0C]
1466 14:24:43.579173 0 24.34.240.246 24.34.240.241 T.38 UDP: UDPTLPacket Seq=00007 data:v21: hdlc-data[0C]
1467 14:24:43.599155 0 24.34.240.246 24.34.240.241 T.38 UDP: UDPTLPacket Seq=00008 data:v21: hdlc-data[AC]
1468 14:24:43.599177 0 24.34.240.246 24.34.240.241 T.38 UDP: UDPTLPacket Seq=00008 data:v21: hdlc-data[AC]
1469 14:24:43.637731 0 24.34.240.246 24.34.240.241 T.38 UDP: UDPTLPacket Seq=00009 data:v21: hdlc-data[EC]
1470 14:24:43.637755 0 24.34.240.246 24.34.240.241 T.38 UDP: UDPTLPacket Seq=00009 data:v21: hdlc-data[EC]
1471 14:24:43.657841 0 24.34.240.246 24.34.240.241 T.38 UDP: UDPTLPacket Seq=00010 data:v21: hdlc-data[9C]

Internet Protocol, Src: 24.34.240.246 (24.34.240.246), Dst: 24.34.240.241 (24.34.240.241)
User Datagram Protocol, Src Port: 61730 (61730), Dst Port: 58664 (58664)
ITU-T Recommendation T.38
  [Stream setup by SDP (frame 1141)]
  UDPTLPacket
    Sequence number: 2
    IFPPacket
      Type of msg: data (1)
        data: v21 (0)
      Data Field
        Item 0
          Data_Field_item
            Data_Field_field_type: hdlc-data (0)
            Data_Field_field_data: FF
    Error recovery: secondary-ifp-packets (0)
      Secondary IFPPackets
        Item 0
          IFPPacket
            Type of msg: t30-indicator (0)
              T30 indicator: v21-preamble (3)
T.38 (t38), 13 bytes    P: 5333 D: 5333 M: 0
```

When the facsimile transmission is finished, the terminating gateway notifies the call agent by sending the T.38 stop event as shown:

```
NTFY 15 aaln/2@x1-6-00-00-CA-DD-A7-E4.cc.cisco.com MGCP 1.0 NCS 1.0
 X: 1FA00000018
 O: fxr/t38(stop)
```

This triggers the call agent to instruct the endpoints to stop sending T.38 image packets and go back to sending audio RTP packets. An MDCX is used for this purpose as shown:

```
MDCX 161155945 aaln/2@x1-6-00-00-ca-dd-a7-e4.cc.cisco.com MGCP 1.0 NCS 1.0
 I: 7
 C: 3bb
 L: mp:20,  a:G729E
 K: 161155943
```

Similar to before, this causes the endpoints to send updated SDP information, which is passed along from one gateway to the other by the call agent.

Finally, the call is disconnected, triggering on-hook notification to the call agent, which in turn triggers DLCX messages to tear down the connections. The 250 - Connection Deleted messages returned in this case are a little different than what you've seen for normal voice calls:

```
250 161155951 Connection Deleted
P: FXR/PGS=0, FXR/PGR=1, PS=1309, OS=32439, PR=2033, OR=94772, PL=0, JI=0, LA=7, PC/
  RPS=2007, PC/ROS=93940, PC/RPL=0, PC/RJI=0
```

Notice the inclusion of the FXR package parameters, FXR/PGS and FXR/PGR. These indicate the number of fax pages sent (in this case 0) and received (in this case 1). Logically, this is the response message sent by the terminating gateway.

RTCP Overview

Real-time Transport Control Protocol (RTCP) works with RTP and provides four basic functions:

- Provides feedback of quality of associated RTP sessions.
- Transports the canonical name (CNAME) parameter used to associate multiple RTP streams. It is with this parameter that separate audio and video RTP streams can be correlated.
- Negotiates the rate at which RTCP packets are sent.
- Conveys a minimal set of session control information.

There are five defined types of RTCP packets:

- Sender reports (SR) for transmission and reception statistics from active senders (that is, endpoints actively transmitting RTP packets)
- Receiver reports (RR) for reception statistics from nonactive senders (that is, endpoints receiving RTP packets but not transmitting any)
- Source description (SDES) items that include parameters such as CNAME
- The end of participation indicator (BYE) sent to indicate the end of an RTP stream
- Application-specific functions (APP)

Multiple RTCP packets are concatenated to form what is referred to as a *compound* RTCP packet. The first packet in a compound RTCP packet is either an SR or an RR. Also every compound RTCP packet must contain an SDES packet containing the CNAME parameter. Thus, in PacketCable, you typically see a compound RTCP packet consisting of an SR followed by an SDES with the CNAME parameter. The last RTCP packet sent also ends with the BYE packet type.

PacketCable endpoints can start transmitting RTCP packets as soon as they have received the SDP information from the remote endpoint. Recall from the NCS portion of the book that this information is referred to as a Remote Connection Descriptor. Also when a PacketCable endpoint starts receiving RTCP packets, it must send them as well.

It is suggested that the amount of bandwidth consumed by RTCP be limited to 5 percent of the associated RTP bandwidth. Another thing to note is that the RTCP packets do not traverse the DOCSIS network on the Unsolicited Grant Service (UGS) service flow set up by DQoS for the RTP packets. By default, they are sent on the default Best Effort service flow across the DOCSIS network; consequently, no guaranteed bandwidth exists for these packets. You can, however, create a separate service flow for RTCP packets. This is discussed in more detail in Chapter 16, along with the optional setting up of DQoS for NCS signaling packets.

RTCP Message Format

The RTCP packet header is similar to the RTP packet header in that it begins with a 2-bit version field followed by a padding bit. The remaining fields depend on the type of RTCP packet.

For the SR and RR packet types the next five bits identify the number of reports contained in the packet (this is typically equal to 1 in PacketCable). For the SDES and BYE packet types this 5-bit field identifies the number of chunks of data in the packet. One chunk is available for each SSRC or CSRC. This value is also typically equal to 1 for PacketCable endpoints. The APP packet type is not used in PacketCable, so it is not discussed in any more detail.

The next field is the 1-byte RTCP packet type identifier. Table 11-3 shows the defined packet type values.

Table 11-3 *Defined Packet Types*

Abbreviation	Packet Type	Value
SR	Sender Report	200
RR	Receiver Report	201
SDES	Source Description	202
BYE	Goodbye	203
APP	Application Defined	204

The remainder of the fields differs depending on the packet type. The next two figures show traces of the three RTCP packet types most commonly seen in PacketCable, starting with the sender report (SR) packet depicted in Figure 11-11.

Figure 11-11 *RTCP SR Packet Trace*

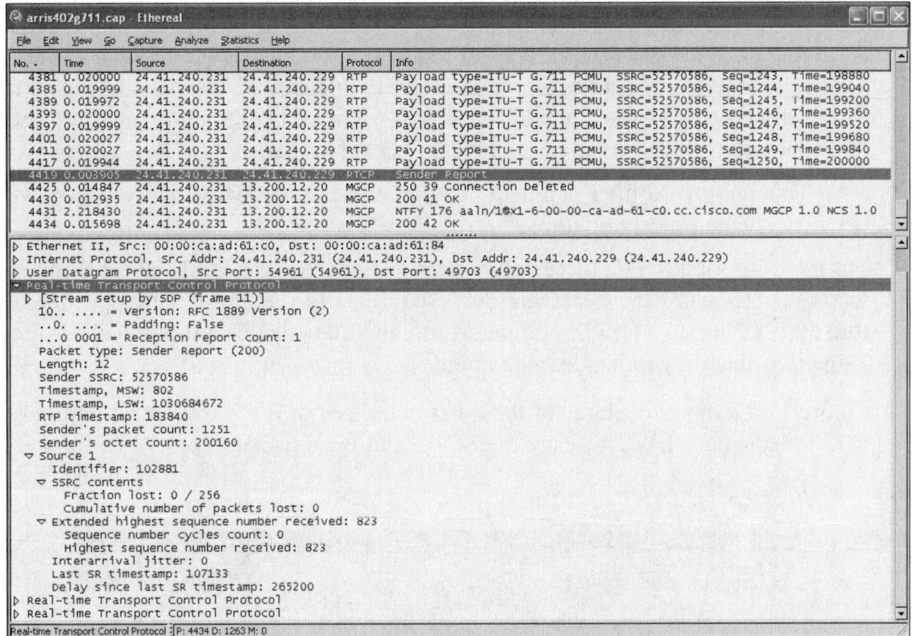

As you can tell, this is a compound RTCP packet consisting of three RTCP packet types beginning with a sender report. After the sender report packet type field is a length field. This length field is a little confusing; it is the size of the RTCP packet including the header in 32-bit (4-byte) words minus 1. For example, in this case the length is 12, meaning the RTCP packet size is 52 bytes, $(12 + 1) \times 4$. You see the next field is the SSRC of the sender, followed by NTP and RTP timestamps referring to the time the report packet was sent. The next two fields are the sender's packet count and octet count. These values represent the total number of packets and payload bytes sent by the sender since transmission began.

The rest of this packet is a receiver report block containing statistics about received RTP packets:

- It begins with the SSRC of the sender, which would be the SSRC of the other end of the call.
- Next is the fraction lost parameter, defined as the number of packets lost divided by the number of packets expected since the last receiver report.

- The cumulative number of packets lost is the number of packets expected minus the number of packets received since the start of the conversation. You might be wondering how the number of expected packets is calculated. This is done by subtracting the initial sequence number from the sequence number of the last packet received.
- Next are some fields that identify the highest sequence number in a received RTP packet. Note all of these receiver report block statistics are relative to the time referenced by the timestamp parameter in this packet. It takes some time for the gateway to generate the receiver report, so in all likelihood many more packets have been received by the time this RTCP packet is sent.
- The interarrival jitter parameter estimates the statistical variance of received RTP packets in timestamp units.

For more information on these parameters please refer to RFC 1889. The important thing to know here is how these statistics are used. Recall from the signaling portion of the book that the response to a DLCX command contains statistical information about the call. This is the information you just learned about.

Figure 11-12 shows captures of the other two types of RTCP packets commonly seen in a PacketCable network. These are the SDES and BYE packet types.

Figure 11-12 *RTCP SDES and BYE Packet Traces*

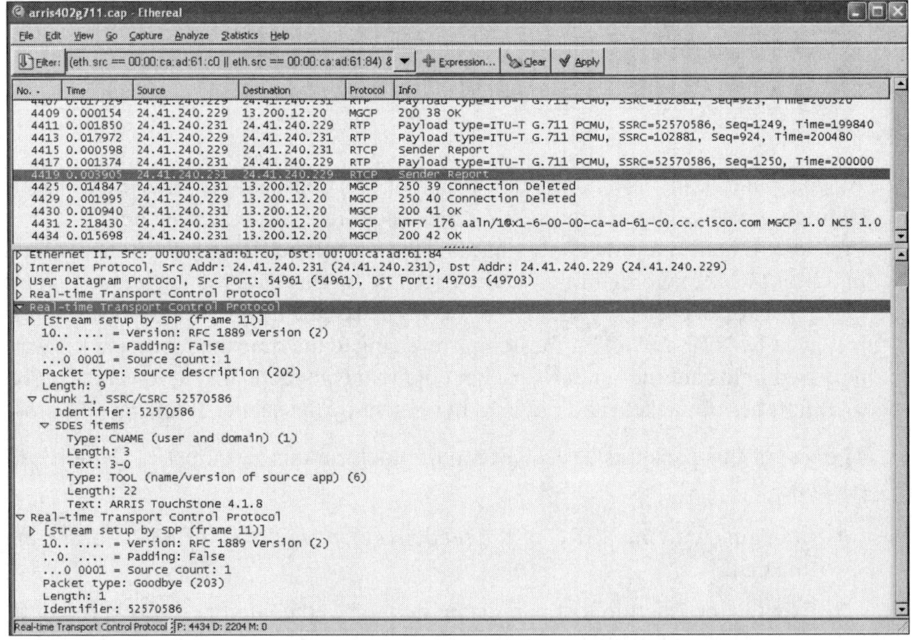

The SDES packet type consists of constructs known as *chunks*. Each chunk refers to a single SSRC or CSRC; thus, in PacketCable you usually have only one chunk in an SDES packet. A chunk begins with the SSRC or CSRC identifier, followed by sub-parameters expressed in Tag/Length/Value (TLV) format with a 1-byte type field and a 1-byte length field. The CNAME parameter (type 1) is the only mandatory parameter. In PacketCable the CNAME field is cryptographic and a random value so that it does not convey the identity of the endpoint.

The BYE packet type signifies the end of a conversation, and the only data it contains is the SSRC or CSRC identifier of the device leaving the conversation.

The call metrics covered so far can provide valuable information about call quality, but additional information such as Mean Opinion Score (MOS) calculations, delay measurements, and signal level measurements would be extremely helpful as well. This information can be obtained using the MGCP voice over IP (VoIP) metrics package (package name "XRM") on the call agent (CMS/MGC) and gateways (MTA/MG). This package is defined in both the PacketCable 1.5 NCS specification and the PacketCable 1.5 TGCP specification, and its support is a requirement for PacketCable 1.5.

This data is based on the VoIP metric block defined in IETF RFC 3611 "RTP Control Protocol Extended Reports (RTCP-XR)." RTCP-XR data is concatenated with RTCP SR and RR payloads. The use of the VoIP metrics block has three facets:

- The *collection* function of accumulating locally calculated data as well as data received from remote endpoints.
- The *responding* function of sending the accumulated local data to remote endpoints. The request for this information is sent between endpoints in the SDP parameters.
- The *reporting* function of sending the accumulated local and remote data to the call agent.

The call agent can optionally enable or disable the *reporting* function on a given MGCP endpoint for a particular connection. If this is enabled, the *collection* and *responding* functions are also enabled. When enabled, an endpoint sends in its SDP information a request for the remote endpoint to transmit XR blocks. By default, an endpoint responds to a remote endpoint if signaled to do so. If the call agent disables *reporting* on an endpoint, it also disables *responding* to remote endpoints.

Table 11-4 describes the individual metrics that are "collected."

Table 11-4 *VoIP Extended Metrics*

Metric	Description	Value Range
Network Packet Loss Rate (NLR)	Proportion of packets lost since the start of transmission	0 to 0.996
Jitter Buffer Discard Rate (JDR)	Proportion of packets discarded because of late arrival since the start of transmission	0 to 0.996
Burst Loss Density (BLD)	Average proportion of packets lost and discarded during burst periods	0 to 0.996
Gap Loss Density (GLD)	Average proportion of packets lost and discarded during gap periods	0 to 0.996
Burst Duration (BD)	Average length of burst periods (in milliseconds)	0 to 65,535
Gap Duration (GD)	Average length of gap periods (in milliseconds)	0 to 65,535
Minimum Gap Threshold (GMN)	Gap/burst transition threshold	16 for PacketCable systems
Round Trip Delay (RTD)	Delay between RTP endpoints (in milliseconds)	0 to 65,535
End System Delay (ESD)	Delay including voice encoding/decoding, packetization, and jitter buffer (in milliseconds)	0 to 65,535
Signal Level (SL)	Ratio of average signal level to a 0 dBm0 reference during active speech periods (in dB)	-30 to +3 (typical values from -15 to -20)
Noise Level (NL)	Ratio of average background noise level to a 0 dBm0 reference during silent periods (in dB)	-40 to -70
Residual Echo Return Loss (ERL)	Estimate of line echo after echo cancellation (in dB)	0 to 80
R-factor (NSR)	Conversational transmission quality rating	0 to 100
External R Factor (XSR)	R-factor for an attached external network	0 to 100
MOS-LQ (MLQ)	Estimated MOS Listening Quality (LQ) (times 10)	10 – 50
MOS-CQ (MCQ)	Estimated MOS Conversational Quality (CQ) (times 10)	10 – 50
Packet Loss Concealment Type (PLC)	Type of Packet Loss Concealment algorithm	0 – Unspecified 1 – Disabled 2 – Enhanced 3 - Standard

Table 11-4 *VoIP Extended Metrics (Continued)*

Metric	Description	Value Range
Jitter Buffer Adaptive (JBA)	Adaptability of the jitter buffer	0 - Unknown 1 - Reserved 2 - Non-adaptive 3 - Adaptive
Jitter Buffer Rate (JBR)	Rate of adjustment of an adaptive jitter buffer	0 to 15
Jitter Buffer - Nominal Delay (JBN)	Nominal delay applied to received packets arriving exactly on time (in milliseconds)	0 to 65,535
Jitter Buffer - Maximum Delay (JBM)	Maximum delay applied to received packets, corresponds to the delay of the earliest received packet that would not be discarded (in milliseconds)	0 to 65,535
Jitter Buffer - Absolute Max Delay (JBS)	Maximum delay size that an adaptive jitter buffer can reach in worst case conditions (in milliseconds)	0 to 65,535
Inter-Arrival Jitter (IAJ)	Mean deviation off the difference in packet spacing at the receiver compared to the sender (in milliseconds)	Same as the parameter in RTCP SR

The RTCP-XR separates a call into two different periods:

- **Bursts**—A *burst* is a period when packet loss and discard rate is high enough to cause a noticeable decrease in voice quality.
- **Gaps**—A *gap* is a period when packet loss and discards are infrequent and have a negligible effect on voice quality.

Notice the Minimum Gap Threshold (GMN) parameter in Table 11-4 is set to a value of 16 for PacketCable systems. This defines what gap and burst periods are.

A burst is defined as the longest packet sequence that starts with a lost or discarded packet, does not contain any sequence of GMN (in this case 16) or more packets, and ends with a lost or discarded packet.

A gap is defined as one of the following: (1) the period from the start of the RTP session to the first burst, (2) the period from the last burst to the time of the report or end of the RTP session, or (3) the period between bursts.

A call agent enables extended metric collection through the use of the xrm/mcr attribute in the LCO parameter list. Typically, this would be included in the CRCX message, but it could also be included in the MDCX message. The possible values of this attribute are off (disable collecting, responding, and reporting), on (enable collecting, responding, and reporting), and negotiate (allow collecting and responding if requested, but disable reporting). For a CRCX if this parameter is not included, a default of negotiate is assumed; if this parameter is not included in an MDCX, the previous value is assumed. An example of what this message might look like is as follows:

```
CRCX 161096008 aaln/1@x1-6-00-13-10-50-38-02.cc.cisco.com MGCP 1.0 NCS 1.0
C: 284
L: mp:20;20;20,  a:PCMU;PCMA;G728,  e:off,  s:off,  t:a0, xrm/mcr:on
M: inactive
K: 161095965
```

The gateway then includes the SDP attribute, a=rtcp-xr:voip-metrics, in its response message. This indicates it has enabled VoIP metric collecting, responding, and reporting:

```
200 161096008 OK
I: 575

v=0
o=- 161096008 1397 IN IP4 24.34.240.245
s=-
c=IN IP4 24.34.240.245
b=AS:82
t=0 0
m=audio 53456 RTP/AVP 0 8 15
a=mptime:20 20 20
a=rtcp-xr:voip-metrics
```

This SDP information is then passed along to the terminating gateway endpoint in a CRCX message. If the LCO option, xrm/mcr:on, is also included, collecting, responding, and reporting is enabled for that endpoint as well. If the LCO option is set to negotiate or if it is not included at all, only collecting and responding is enabled on the remote endpoint.

When the call is deleted, the DLCX response message reports the extended metrics. A sample of what this exchange might look like is as follows:

```
DLCX 161096021 aaln/1@x1-6-00-13-10-50-38-02.cc.cisco.com MGCP 1.0 NCS 1.0
I: 575
C: 284
K: 161096017
250 161096021 Connection was deleted
P: PS=579, OS=93798, PR=706, OR=112960, PL=1, JI=197, LA=49, PC/RPS=609,
  PC/ROS=98658, PC/RPL=2, PC/RJI=0
XRM/LVM: NLR=128, JDR=40, BLD=33, GLD=10, BD=55, GD=45, RTD=1000, ESD=1000, SL=-10,
  NL=2, RERL=23, GMN=80, RF=19, XRF=19, MLQ=19, MCQ=22, PLC=1, JBA=2, JBR=8, JPN=40,
  JBM=80, JBS=120, IAJ=10
XRM/RVM: NLR=128, JDR=40, BLD=33, GLD=10, BD=55, GD=45, RTD=1000, ESD=1000, SL=-10,
  NL=2, RERL=23, GMN=80, RF=19, XRF=19, MLQ=19, MCQ=22, PLC=1, JBA=0, JBR=8, JPN=30,
  JBM=60, JBS=100, IAJ=15
```

The XRM/LVM parameter details the local "collected" metrics and the XRM/RVM parameter details the remote "responded" metrics. Caution—this message is not from an actual call, so the parameter values shown might not be typical.

Component Configuration and Troubleshooting

The last part of this chapter covers configurating and troubleshooting information relevant to the topics of this chapter for various PacketCable components. These components are the CMS/MGC (specifically Cisco's BTS 10200), the MG (specifically Cisco's VXSM and VISM cards), and the MTA.

The Cisco CMS: BTS 10200

In the "MTA Configuration Tasks" section of Chapter 7 you learned how to configure an MTA on the BTS 10200 for PacketCable operation. That section focused on key parameters relating to the NCS protocol. In the "Trunking Media Gateway Configuration Tasks" section of Chapter 8 you learned how to configure a media gateway on the BTS 10200 for PacketCable operation. In that section, parameters pertinent to the TGCP protocol were the focus. Two of the tables covered in both of these sections also contain parameters related to the topics covered in these past two chapters. These are the media gateway profile (MGW-PROFILE) table and the quality of service (QOS) table.

The relevant media gateway profile parameters are as follows:

- **CODEC-NEG-SUPP**—Designates whether the endpoint supports CODEC negotiation. This field works with the CODEC identified in the QOS table as well as with what the endpoint returns in the AUEP response. Refer to Chapter 7 for more detail.
- **DTMF_OOB_SUPP**—Whether or not the gateway supports mid-call notification of DTMF digits by sending a NTFY command.
- **EC_SUPP**—Specifies whether the BTS should signal the gateway to turn on or off its echo cancellers.
- **FAX_INBAND_METHOD**—If set to GW_SPECIFIED, the BTS is not involved in how the gateways handle in-band fax transmissions; if set to T38_FXR_OFF, the BTS explicitly instructs the gateways not to do T.38 fax relay, regardless of whether the gateway supports the gateway controlled method of T.38 fax relay.
- **T38_FXR_LOOSE_SUPP**—Identifies whether the gateway supports call agent controlled T.38 Loose mode fax.

The relevant quality of service parameters are as follows:

- **CODEC_TYPE**—If CODEC negotiation is supported, this field identifies the preferred CODEC. If CODEC negotiation is not supported this field identifies the CODEC used. Values used in PacketCable are PCMU (G.711 μ-law), PCMA (G.711 a-law), G.728, and G.729E. BTS version 4.5 does not support iLBC and BV16.
- **DTMF_PREF_MODE**—Identifies the preferred method of DTMF relay. Values used in PacketCable are DTMF-INBAND and DTMF-GW-NTE (RFC 2833). Remember, DTMF-INBAND should be used only if G.711 is the CODEC.

- **FAX_T38_ENABLED**—Whether or not T.38 is used for facsimile transmission.
- **HPTIME**—Indicates the higher supported packetization period in a range for the gateway (units are ms). Valid values in PacketCable are 10, 20, and 30.
- **LPTIME**—Indicates the lower supported packetization period in a range for the gateway (units are ms). Valid values in PacketCable are 10, 20, and 30.
- **SILENT_SUPPRESSION**—Indicates whether or not silence suppression is enabled on the media gateway.

The Cisco MG: VXSM

In Chapter 8 you also learned the basic steps of how to configure Cisco's media gateways, such as the VXSM and VISM cards. These cards have several configurable parameters relating to the topics of these past two chapters that are worth a closer look. The sections that follow cover the commands to view and adjust these parameters.

NOTE Some of these you have already seen, and some will be new to you. Remember that you can use the **?** command if you don't recall a command's exact syntax.

Configuring CODEC-Related Options

You can view the current CODEC configurations using the **dspcodecparam(s)** commands as demonstrated here:

```
MGX8800.5.VXSM.a > dspcodecparams
==========================================================
       Display All Codecs Entries (dspcodecparams)
==========================================================
Type       Codec     Pref V PP Jtr Mode  Max   Nom   Min   DTMF   Payload
==== ============ ==== ==== ======== ===== ===== ===== ======= =======
aal5        G.729a  2    20  adapt     100   30    2       n/a     18
aal5      G.726-32  3    20  adapt     100   30    2       n/a      2
aal5        G.711u  7    20  fixed      40    2    0       n/a      0
aal5        G.711a  8    20  adapt     100   30    2       n/a      8
aal5       G.729ab  1    20  adapt     100   30    2       n/a     99
aal5 clearChannel  9    20  fixed     135   70    0       n/a    125
aal2        G.729a  0     *  adapt      60   20    2    enable      0
aal2      G.726-32  0     *  adapt      60   20    2   disable      0
aal2        G.711u  0     *  adapt      60   20    2   disable      0
aal2        G.711a  0     *  adapt      60   20    2   disable      0
aal2       G.729ab  0     *  adapt      60   20    2    enable      0
aal2 clearChannel  0     *  fixed     135   70    0   disable      0
```

The only CODECs shown in the preceding example that are relevant to a PacketCable deployment are the G.711 variants with the AAL5 ATM adaptation type. The preference parameter prioritizes the use of one CODEC over another, where the lower number has the higher preference. The value 0 indicates this CODEC is not used; notice all the AAL2 entries have this value. Notice the DTMF option here is not applicable for the AAL5 ATM adaptation type.

Jitter parameters are also shown with this command. For example, you can see that the G.711 μ-law variant is set to use a fixed jitter buffer, and the G.711 a-law variant is set to use an adaptive jitter buffer. Also notice that you can see the RTP payload types with this command. These parameters can be modified using the **cnfcodecparam** command.

You can view how CODEC negotiation is handled using the **dspcodecnego** command:

```
MGX8800.5.VXSM.a > dspcodecnego
======================================================
         Codec Negotiation Option (dspcodecnego)
======================================================
Adaptation Type                  :        aal5
Codec Negotiation Option         :        lcoRcdLcl
```

This setting determines how the VXSM card determines the CODEC to use for a call. In this case, the value sent from in the call agent in the Local Connection Options (LCO) is used first. If nothing is specified, the SDP information from the Remote Connection Descriptor (RCD) of the remote endpoint is used. If that is also not available, the Local CODEC List (LCL) is used. This order can be adjusted with the **cnfcodecnego** command.

Configuring DTMF Relay

DTMF relay is enabled/disabled with the **cnfxgcpdtmf** command. You can view the current status of DTMF relay using the **dspxgcpdtmf** command, as demonstrated here:

```
MGX8800.5.VXSM.a > dspxgcpdtmf
======================================================
              Advertising xGCP DTMF Relay
======================================================
Advertising xGCP DTMF Relay (Originating GW):   enable
```

You can adjust the manner in which DTMF relay packets are transferred with the **cnfdsp -dtmf** command and verify the operation with the **dspdsp -dtmf** command, as demonstrated here:

```
MGX8800.5.VXSM.a > dspdsp -dtmf 1.2.1.1 1 -enc 1
==============================================================================
                  Display       DTMF       Parameters
==============================================================================
DS0       PacketCount    RepeatInterval(mSec)    KeepAlive
==============================================================================
2017      3              5                       50
```

The parameters of the command include the DS1 line number and the packet encapsulation type (1 means RTP). The configuration in this case indicates that when a DTMF digit is detected, three DTMF relay packets are initially sent 5 milliseconds apart. If the tone is still active, additional packets are sent in 50 millisecond intervals.

You can also use commands that enable you to see DTMF relay packets sent and received. These commands are **cnfdtmfrxsnoop** for received packets and **cnfdtmftxsnoop** for transmitted packets. You can use the **dspdtmfsnoop** command to see the current status of these commands.

Adjusting the Echo Parameters

You can modify echo cancellation attributes for voice interfaces with the **cnfvifec** command and display current settings using the **dspvifec(s)** commands, as demonstrated here:

```
MGX8800.5.VXSM.a > dspvifecs
=======================================================
       Echo Canceller in all Voice Interfaces
=======================================================
  DS1/E1    DS0 Group    ECAN      ECAN Coverage
========== =========  =======  ===============
 1.3.2.1.1       0      enable          128
```

In this case you can see that echo cancellation has been enabled and the coverage is the default of 128 milliseconds. You can also adjust echo cancellation attributes used by the VXSM DSP resources with the **cnfdsp -ec** command and display the status using the **dspdsp -ec** command.

Adjusting Silence Suppression Parameters

You can configure silence suppression or voice activity detection (VAD) parameters for a particular voice interface with the **cnfvifvad** command and display the status with the **dspvifvad(s)** commands, as demonstrated here:

```
MGX8800.5.VXSM.a > dspvifvads
=======================================================
        VAD in all Voice Interfaces
=======================================================
  DS1/E1     DS0 Group     VAD    VAD Timer
==========  =========  ======= =========
 1.3.2.1.1       0       enable       250
```

In this case, VAD is enabled, and the VAD timer is set to 250 milliseconds; both of these are default values. The VAD timer indicates the length of time after voice inactivity is detected before silence suppression is activated. You can also use the **cnfvifparam** command to specify whether comfort noise is sent when silence suppression is activated. You can also adjust VAD attributes used by the VXSM DSP resources with the **cnfdsp -vad** command and display the status using the **dspdsp -vad** command.

Configuring Tone Parameters

If not specified in the call signaling, tone timeout values default to values defined on the VXSM card. You can customize these timeout values using the **cnfxgcpprofttone** command and display the values using the **dspxgcpprofttone** command, as demonstrated here:

```
MGX8800.5.VXSM.a > dspxgcpprofttone 0
======================================================
    Tone Timeout in Profile (dspxgcpprofttone)
======================================================
Index                         :      0
Mwi Timeout (ms)              :      16
Rt  Timeout (s)               :      180
Rbk Timeout (s)               :      180
Cg  Timeout (s)               :      180
Bz  Timeout (s)               :      30
Dl  Timeout (s)               :      16
Sl  Timeout (s)               :      16
Rg  Timeout (s)               :      180
Ro  Timeout (s)               :      30
```

The individual tones are message waiting indication (mwi), ring back toward the PSTN (rt), ring back toward the VoIP network (rbk), congestion (cg), busy (bz), dial tone (dl), slutter dial tone (sl), ring tone (rg), and reorder tone (ro).

The VXSM card has the capability to help you debug connectivity issues by letting you manually generate a tone. The command to do this is **cnftonegen**. The syntax for this is as follows:

```
MGX8800.5.VXSM.a > cnftonegen
Err: Not enough arguments.
Syntax: cnftonegen <LineNum> <ds0>
    -gen <generate> -tn <tone> -amp <amplitude> -dir <direction>
   LineNum -- (bay.line.path.vtg.vt or bay.line.path.ds1)
                  bay {1 - upper}
                  line (range=1..4)
                  path (range=1..3)
                  vtg (range=1..7)
                  vt (range=1..4)(ds1) (range=1..3)(e1)
                  ds1 (range=1..28)
   ds0 -- ds0 index
                  T1: (range=1..24)
                  E1: (range=1..31)
   gen -- generate: {1 ¦ 2}
                  1 - enable
                  2 - disable
   tn -- tone: (range=404..2804Hz)
   amp -- amplitude: (range=-800..30 (0.1db)) (increments of 5)
   dir -- direction: {1 ¦ 2}
                  1 - towards TDM
                  2 - towards network
```

Notice that with this command you can generate tones out the TDM connection to the PSTN or out the network connection to the VoIP network.

Setting RTCP Parameters

You can use the **cnfdspparam** command to set RTCP-related parameters and use the corresponding **dspdspparam** command to display current settings, as demonstrated here:

```
MGX8800.5.VXSM.a > dspdspparam
=========================================================
              List DSP Parameters
=========================================================
SID Payload Type                :           decimal
RTCP Control                    :           true
RTCP Interval(milliseconds)     :           5000
RTCP Interval Multiplier        :           5
VAD Adaptive                    :           true
DTMF Power Level (0.1 dBm)      :           -120
DTMF Power Twist (0.1 dB)       :           0
RTCP Timer Control              :           startRtcpPktRcvd
```

The transmission and reception of RTCP packets can be enabled or disabled here (in this case it's enabled). The RTCP interval (for sending RTCP report packets) is equal to 5000 ms, and the RTCP interval multiplier equal to 5 (indicating the number of times RTCP reports can fail before exception processing is performed). The RTCP timer control parameter indicates when the RTCP receive timer is started (in this case it starts as soon as the first RTCP packet is received). In later versions of VXSM code, this command can also set parameters related to RTCP Extended Reports (RTCP-XR) and the VoIP metric block.

Configuring Voice Band Data (VBD) Options

On the VXSM the voice band event mapping feature determines how VBD events are handled. You can use the **cnfeventmapping** command to change the current settings and use the **dspeventmappings** command to display the current settings, as demonstrated here:

```
MGX8800.5.VXSM.a > dspeventmappings
=========================================================
    Voice Band Data Event Mapping Configurations
=========================================================
Evt. Map. Index  Evt. Index  Evt. Handle Type  Evt. Prof. Index  Evt. Handle Mode
===============  ==========  ================  ================  ================
1                ced         vbd               1                 nse
1                v21Tone     vbd               1                 nse
1                v21Modem    vbd               1                 nse
1                bellModem   vbd               1                 nse
1                v8bis       vbd               1                 nse
1                v23Modem    vbd               1                 nse
1                v18aTone    tty               1                 none
```

The event mapping index is always 1 for VoIP switching. The event index identifies the type of mapping: ced (fax/modem pass-through), v21Tone (T.38 fax relay), and v18aTone (TTY pass-through), and the rest are for low-speed modem tones. The event handle type is either none, vbd, fax, or tty. The value none means handling of this event has been disabled. The event mapping profile index refers to handling attributes specific to the event type. If the handle type is vbd, a VBD profile is referenced; if the handle type is fax, a FAX profile is referenced; and if the handle type is tty, a TTY profile is referenced. How these profiles are defined is shown next. The event handling mode nse means the VXSM detects tones from both the TDM side and IP side and sends an NSE event to the peer upon detection from the TDM side. The mode none means the VXSM detects tones from both the TDM side and IP side but does not send an NSE event to the peer upon detection from the TDM side.

NOTE The NSE protocol is Cisco proprietary and might not be understood by third-party devices.

VBD profiles determine the handling of fax/modem pass-through when it is detected. Here, information such as the up-speed CODEC and jitter buffer parameters are defined. You can add these profiles with the **addvbdprof** command, configure them with the **cnfvbdprof** command, and display the current settings with the **dspvbdprof(s)** commands, as demonstrated here:

```
MGX8800.5.VXSM.a > dspvbdprof 1
=================================================================
              Voice Band Data Configuration Profile
=================================================================
VBD Profile Index               :      1
VBD Upspeed CODEC Type          :      G.711u
VBD Jitter Buffer Delay Mode    :      fixed
VBD Jitter Buffer Maximum Delay :      135
VBD Jitter Buffer Nominal Delay :      70
VBD Jitter Buffer Minimum Delay :      0
VBD Inactivity Timeout          :      4
VBD Locally Disable VAD         :      enable
VBD Packet Period Control       :      enable
VBD Packet Period               :      10
```

In this example the up-speed CODEC is set to G.711 μ-law, and a fixed jitter buffer (with a nominal size of 70 ms and a maximum size of 135 ms) is used. The inactivity timeout means that calls are torn down after 4 seconds of inactivity. The packet period parameters indicate that the VXSM switches the packetization period to 10 ms during up-speeding.

Similarly, TTY profiles determine the handling of TTY pass-through when it is detected. The parameters are very similar to the VBD profile parameters. You can add these profiles with the **addttyprof** command, configure them with the **cnfttyprof** command, and display the current settings with the **dspttyprof(s)** commands as demonstrated here:

```
MGX8800.5.VXSM.a > dspttyprof 1
=================================================================
              TTY Configuration Profile
=================================================================
TTY Profile Index               :      1
TTY CODEC Type                  :      G.711u
TTY Jitter Buffer Delay Mode    :      fixed
TTY Jitter Buffer Maximum Delay :      135
TTY Jitter Buffer Nominal Delay :      70
TTY Jitter Buffer Minimum Delay :      0
TTY Locally Disable VAD         :      enable
TTY Packet Period Control       :      enable
TTY Packet Period               :      10
```

T.38 fax relay profiles determine the handling of T.38 fax relay when a V21 tone is detected. You can add these profiles with the **addfaxprof** command, configure them with the **cnffaxprof** command, and display current settings with the **dspfaxprof(s)** commands, as demonstrated here:

```
MGX8800.5.VXSM.a > dspfaxprof 1
======================================================
             Fax Relay Configuration Profile
======================================================
Fax Profile Index                :      1
Control Mode                     :      t38FallbackPt
Varient of T.38 Standered        :      0
Transport Protocol               :      udp
Method Used for TCF              :      network
Maximum Transmission Rate        :      14400
High-Speed Data Packet Rate      :      30
Low-Speed Data Redundancy        :      5
High-Speed Data Redundancy       :      1
Named-Signal-Event Timeout       :      1000
Inactivity Timeout               :      10
Nominal Delay                    :      200
Error-Correcting-Mode Status     :      enable
NSF Code Override Status         :      enable
NSF Manufacturing Country Code   :      173
NSF Vendor Code                  :      0051
```

The control mode t38FallbackPt means that when the V.21 preamble is detected, T.38 is attempted; if this fails, pass-through is attempted. Alternatively, this can be set to t38Only, in which case pass-through is not attempted if the T.38 attempt fails. You can see UDP is used for the T.38 transport protocol, and the maximum fax transmission rate is set to 14.4 Kbps. The inactivity timeout means that if no fax activity is detected for 10 seconds, the call switches back to voice mode.

Remember from Chapter 8 the voice interfaces (VIFs) used to apply a group of configurable parameters to a group of DS0 trunk endpoints. Event mappings are an example of these configurable parameters. You can use the **dspvifeventmapping(s)** commands to display current mappings and use the **cnfvifeventmapping** command to create a mapping. For example, to associate the event mapping defined by index 1 with the VIF used, you would use the following command:

```
MGX8800.5.VXSM.a > cnfvifeventmapping 1.3.2.1.1 0 1
```

BERT Testing

You can verify DS1 circuit quality by performing bit error rate testing (BERT). You can perform this testing directly on a VXSM card using the following commands:

```
MGX8800.5.VXSM.a > ? bert
    Available commands
    ------------------
    addbert                     Add DS1 Bert
    cnfbert                     Config DS1 Bert
    delbert                     Delete DS1 Bert
    dspbert                     List a DS1 Bert
```

```
dspberts                      List all DS1 BERTs
startbert                     Start DS1 Bert
stopbert                      Stop DS1 Bert
```

Bearer Tracing

Another powerful troubleshooting feature available on the VXSM cards is the bearer tracing tool. With this tool, valuable information about the content of bearer streams can be transmitted real time to an external file server for analysis. Probes are available at various points in the bearer path—such as the inputs and outputs from both the TDM network and the VoIP network, the output from the echo canceller, and the output from the jitter buffer—to collect this information. The external file server can be connected either via the VXSM control PVC or through the Ethernet port on the VXSM back card.

A summary of the configuration steps is as follows:

Step 1 Specify the DS0 endpoint that needs to be traced using the **addbearertraceendpt** command. With this command you can specify what trace probes are activated.

Step 2 Configure the external file server using the **addbearertracesrvprof** command. Either the TFTP or FTP protocol can be used to communicate with the file server.

Step 3 Associate the traced endpoint with the external file server using the **addbearertrace** command.

Step 4 If TFTP is used, make sure the proper write permissions exist on the file server.

Step 5 Start tracing on the endpoint using the **bearertracestart** command.

Step 6 Disconnect the call to stop tracing on the endpoint or use **bearertracestop** to stop tracing.

Trace configurations can be deleted and displayed by replacing **add** in the previous commands with **del** or **dsp**.

The Cisco MG: VISM

This section examines VISM configuration commands and parameters. Recall that the operating mode, as well as the current configuration of several parameters, can be verified with the **dspvismparam** command, as shown in the following example:

```
MGX8800.1.3.VISM8.a > dspvismparam
  VISM mode:                    voipSwitching/voipTrunking
  VISM features Bit Map:        0x5bc
  FunctionModuleType:           VISM-PR-8T1
  CAC flag:                     enable
```

```
DS0s available:              192
Template number:             2
Percent of functional DSPs:  100
IP address:                  13.13.4.132
Subnet mask:                 255.255.255.255
Bearer IP address:           13.13.4.152
Bearer Subnet mask:          255.255.255.255
Bearer IP ping:              disable
RTCP report interval:        5000 msec
RTCP receive multiplier:     3
RTP receive timer:           disable
ControlPrecedence/Tos:       0x60
BearerPrecedence/Tos:        0xa0
Aal2 muxing status:          disable
Tftp Server Dn               TFTPDOMAIN
Aggregate Clipping           enable
Aggregate Svc Bandwidth      0
Codec negotiation option     1
Profile negotiation option   1
VAD Duty Cycle               61
VAD Tolerance                100
VISM Initiated NW COT        Off
VISM CO4 Timer               1000 msec
CALEA flag                   enable
Oam Loopback CLP Bit         1
Continuity Check Cell        1
SupportdModuleType:          VISM-PR-8T1.
VismNSAP:                    47009181000000000059a3fa172000001071bff00
```

You already know that the VISM mode of voipSwitching/voipTrunking is what is required for a PacketCable VoIP network. Several of these other parameters are worth a discussion, so keep this command output in mind because it will be referred to often.

Configuring CODEC-Related Options

From the results of the **dspvismparam** command in the preceding section, you can see that this card is using template #2, which supports only the uncompressed CODECs (clear channel and G.711 variants). At the moment these are the only PacketCable identified CODECs supported by the VISM card. However, if you needed to change to this template from another one or to another one, you would use the **cnfcodectmpl** command. To see a list of the CODECs supported by each template, use the **dspcodectmpls** command.

NOTE Keep in mind that changing the CODEC template requires a card reset.

If you do plan on supporting multiple CODECs on your VISM card, you might need to alter how CODEC negotiation is handled. Notice from the **dspvismparam** command output that the CODEC negotiation option is set to 1. This means that first the VISM card picks the CODEC based on the Local Connection Option parameter it receives from the MGC. You can alter this setting to instead make the VISM pick the CODEC based on the CODEC list from the other endpoint (remember this is part of the MGCP Remote Connection Descriptor parameter). You can also make the VISM pick CODECs based on a predefined list. This list (referred to as the Local MIB) can be viewed with the **dspcodecparams** command and can be modified with the **cnfcodecparams** command. The preference scale is from 1 to 13, with 1 being the most preferred (a value of 0 removes that CODEC from consideration). To change the way the CODEC is selected, use the command **cnfcodecneg**. In addition to using the **dspvismparam** command, you can obtain the current CODEC selection with the command **dspcodecneg**. The following is sample output from these commands.

```
MGX8800.1.3.VISM8.a > dspcodecparams
                                Packet                          IANA
Codec            Codec String   Period (ms)   Preference   Type
---------        ------------   -----------   ----------   ----
 1 - G.711u            PCMU       20             1          0
 2 - G.711a            PCMA       10             8          8
 3 - G.726-32K       G726-32       10            5          2
 4 - G.729a           G729a       10             2         96
 5 - G.729ab          G729ab      10             1         96
 6 - clr chan          CCD        10             9         96
 7 - G.726-16K       G726-16       10            3         96
 8 - G.726-24K       G726-24       10            4         96
 9 - G.726-40K       G726-40       10            6         96
11 - G.723.1-H         G723H      30            10         96
12 - G.723.1a-H       G723AH      30            11         96
13 - G.723.1-L         G723L      30            12         96
14 - G.723.1a-L       G723AL      30            13         96
15 - Lossless        LOSSLESS        10            14          96
MGX8800.1.3.VISM8.a > cnfcodecneg
Syntax : cnfcodecneg "codecNegotiationOption"
          codecNegOpt -- Acceptable values are:
             1 = LCO, RCD, Local Mib,
             2 = LCO, Local Mib, RCD,
             3 = RCD, LCO, Local Mib,
             4 = RCD, Local Mib, LCO,
             5 = Local Mib, LCO, RCD,
             6 = Local Mib, RCD, LCO
MGX8800.1.3.VISM8.a > dspcodecneg
Codec Negotiation Option = 1
```

Adjusting the Jitter Buffer Parameters

For CODECs such as G.711, where the latency jitter is relatively constant, a fixed-size jitter buffer is recommended, and for other CODECs an adaptive jitter buffer is recommended. You can change the jitter mode, fixed or adaptive, with the command **cnfjtrmode**. You can modify the size of the jitter buffer or the initial size in the adaptive case on a per-CODEC basis with the command **cnfcodecjtrdelay**; for facsimile transmission the command **cnffaxjtrdelay** is used instead, which allows you to define a separate treatment for fax calls. You can display current settings with the **dspcodecjtrdelays** command.

NOTE The **cnfjtrmode** command changes the settings for all CODECs supported by the current CODEC template. If you just want to change the values for a specific CODEC, use the **cnfcodecjtrdelay** command.

Negotiating Dynamic RTP Payload Types

Whether RTP payload types are negotiated or not depends on the setting of the dynamic payload, which you can display using the **dspdynamicpayload** command. You can change this setting using the **cnfdynamicpayload** command. When enabled, RTP payload types are negotiated using the dynamic range 96 to 127. If disabled, the statically defined payload type values are used. For CODECs these are the values displayed in the output of the **dspcodecparams** command.

Configuring DTMF Relay

To enable DTMF relay on the VISM card, use the **cnfvoiptransparams** command. You can verify these values with the **dspvoipparams** command and adjust the on/off times for DTMF digit out-pulsing using the **cnfendptcadence** command.

Adjusting the Echo Parameters

Each DS0 on a VISM card is assigned a DSP for echo cancellation. You can enable or disable echo cancellation on a per-line basis using the **cnfecanenable** command. Also you can adjust the maximum tail length using the **cnfecantail** command. The default value is 32 ms. You can use the **cnferl** command to adjust the echo return loss applied by the echo canceller DSPs (the default is –6 dB). Additionally, you can configure residual echo control on a per-line basis using the **cnfecanrec** command. With this command, each line can be instructed to cancel echo only, to suppress residual echo, or to generate comfort noise. You can display the current echo setting using the **dsplndsp** command. For an active call, you can obtain echo information using the **dspmonecanendpt** command.

Configuring Tone Plans

Each country has its own definition of what frequencies constitute a tone, such as ring back tone or busy tone. This is known as a tone plan and must be configured on each T1 or E1 on the VISM card using the command **cnflntoneplan**. You can display the current settings with the **dsplntoneplan** command and display a list of all supported tone plans with the **dsptoneplans** command. If the tone plan you need does not exist, you can upload one to the VISM card using the **addtoneplan** command. Also you can configure the amount of time each type of tone is played out with the **cnflntonetimeout** command and display the settings with the **dsplntonetimeout** command.

Setting RTCP Parameters

The **cnfrtcpprepint** command configures the interval for sending RTCP packets. If you refer to the **dspvismparam** command output at the beginning of the "The Cisco MG: VISM" section, you can see a field called RTCP report interval equal to 5000 msec and a field called RTCP receive multiplier equal to 3. These are the values configured with this command. Multiply these two values together to get the interval between RTCP packets (in this case 15 seconds).

You can use the **cnfrtprxtimer** command to set up the VISM card to tear down connections when no RTP packets have been received for 5 seconds. The 5-second time limit is not configurable. Be cautious when using this option because call waiting scenarios and calls utilizing silence suppression could inadvertently get torn down. You can verify the current setting with the **dspvismparam** command.

Configuring T.38 Facsimile Parameters

The **cnft38fxlco** command configures on a per-line basis how the VISM card handles a fax call. Each line can be configured to try to pass the fax via in-band first or via T.38. The **cnft38nsettimeout** command enables you to configure the amount of time to wait for an acknowledgement to the T.38 offer NSE on each line. The configuration of several other T.38 parameters is done with the **cnft38params** command. You can verify the current settings with the commands **dspt38fxlco(s)**, **dspt38nsettimeout(s)**, and **dspt38params**.

Detecting Voice Band Data

The **cnfvbdpol** command defines what the VISM card should do in the event the admission control check for a fax or modem up-speed fails or if carrier is lost. The **cnfvbdcodec** command defines the up-speed CODEC in the event Voice Band Data is detected. This CODEC should be G.711. You can verify current settings with the **dspconvbdpol** and **dspvbdcodec** commands.

Adjusting Silence Suppression Parameters

The **cnfvoiptimerparam** command configures the time before VAD or silence suppression goes into effect; its default value is 250 ms. You can verify the current value with the **dspvoipparams** command. You can adjust the parameters the VISM uses when making admission control decisions on calls utilizing VAD with the **cnfcacparams** command.

Monitoring Active Calls

You can use the **dsprtpcnt** command to give active RTP statistics, such as packets and bytes sent and received on a particular endpoint.

BERT Testing

For a VISM card, resources on the SRM cards do BERT for verifying DS1 circuit quality. The commands are executed from the active PXM card; the following is a list of these commands:

```
MGX8800.7.PXM.a > ? bert
    Available commands
    ------------------
    cnfbert
    delbert
    dspbert
    dspbertcap
```

MTA Configuration File

In Part II of this book, you learned that the MTA gets its configuration during the provisioning process. Specifically, an MTA configuration file containing several SNMP objects is transferred and loaded by the MTA. In this MTA configuration file is information such as the Kerberos realm name and the DNS name of the CMS. This configuration file could also contain some MIBs that are relevant to the topics of these past two chapters. Examples of these MIBs are

- **CODEC MIBs: "pktcSigDevCodecIndex", "pktcSigDevCodecType", and "pktcSigDevCodecMax"**—These MIB objects contain the supported CODEC types for the MTA. The "pktcSigDevCodecType" object has an integer value, and among the possible values are G.729E (5), PCMU (6), G.728 (8), PCMA (9), ILBC (13), and BV16 (14). The "pktcSigDevCodecMax" object indicates the maximum number of simultaneous calls for that CODEC that the MTA can support at a time.
- **DTMF MIB: "pktcEnNcsMinimumDtmf"**—This object defines the minimum play-out time for DTMF digits at the egress gateway when RFC 2833 DTMF relay is used.

- **Echo MIB: "pktcSigDevEchoCancellation"**—This object specifies whether the MTA supports echo cancellation.
- **VAD MIB: "pktcSigDevSilenceSuppression"**—This object specifies whether the MTA supports silence suppression (VAD).
- **Voice Band Data MIBs: "pktcSigDevConnectionMode" and "pktcEnNcsEndPntFaxDetection"**—The object "pktcSigDevConnectionMode" is a bit map that identifies the types of bearer traffic it supports (voice – bit 0, fax – bit 1, modem – bit 2). The object "pktcEnNcsEndPntFaxDetection" enables/disables the detection of distinctive fax calling tone (CNG); the default value is false (disabled) .

Chapter Summary

PacketCable bearer packets are transported using the Real-time Transport Protocol (RTP). Working with RTP is RTCP (Real-time Transport Control Protocol) providing diagnostic information of the associated RTP session as well as other functionality. These two protocols are defined in RFC 1889.

RTP packets are encapsulated inside UDP packets. The format of the payload portion of a RTP packet varies depending on the payload type. These payload types can be statically defined in a separate document or defined dynamically via another protocol. RFC 1890 defines some static payload types for CODECs such as G.711 and G.728. Dynamic payload types are in the range 96 to 127 and are defined in PacketCable using the Session Description Protocol (SDP). For audio CODECs the RTP payload is either sample-based or frame-based, depending on the nature of the CODEC. RTP comfort noise packets are utilized when silence suppression is implemented to save bandwidth and reassure devices that the network is operating properly. Other fields of the RTP header, such as sequence number, timestamp, and synchronization source, should be randomly generated for improved security. These fields are used by the remote gateway to correlate packets.

Assuming no Payload Header Suppression is used, each RTP packet has 40 bytes of RTP/UDP/IP header overhead. Using this information along with the CODEC and packet rate you can figure out the bandwidth requirements of a single call. These numbers are then used when traffic engineering your PacketCable network. However, other factors, such as voice quality and toleration to impairments such as packet loss, should also be taken into consideration when choosing a CODEC.

DTMF relay as defined under RFC 2833 can be used as an alternative to passing these tones in-band. Passing tones in-band can be a problem when using a CODEC that employs a compression algorithm. Also in-band transmission provides no protection against dropped packets. DTMF relay uses a separate payload type known as a Named Telephony Event to transmit these digits.

The fax package defined in the PacketCable 1.5 NCS and TGCP specifications is used for T.38 fax relay in PacketCable. The recommended method of fax relay in PacketCable is call agent controlled. Either T.38 Strict mode or T.38 Loose mode can be used in this case; T.38 Loose mode should be used when one of the gateway endpoints does not support RFC 3407 for indicating T.38 fax capabilities.

RTCP provides feedback to gateways on the quality of associated RTP sessions. This feedback is sent in sender and receiver report blocks. The type of this information in these report blocks includes data on octets and packets sent/received as well as estimates of packet loss and jitter. The VoIP extended metrics package defined in the PacketCable 1.5 NCS and TGCP specifications can be used to provide much more information on VoIP call quality.

The PacketCable components that typically have configurable parameters relating to the bearer path are the CMS, MTA, and MG. If you are using Cisco's CMS, the BTS 10200, the media gateway profile (MGW-PROFILE) table and the quality of service (QoS) table contain these types of parameters. Cisco's VXSM and VISM media gateways have several configuration and show commands relating to these topics. These topics include CODEC negotiation, echo cancellation, Voice Band Data transmission, DTMF relay, and silence suppression. The VXSM card also has troubleshooting features such as manual tone generation, BERT testing, and bearer tracing.

The MTA configuration file can also contain SNMP MIB objects that define bearer path behavior.

Chapter Review

1 What are the standards that define the RTP and RTCP protocols as used in PacketCable?

Answer: RTP and RTCP are defined under RFC 1889.

2 What transport protocol and port numbers are typically used for RTP and RTCP packets?

Answer: Typically, random UDP ports are used with the even port numbers for RTP and the next odd port number for RTCP.

3 What type of RTP payload do "waveform encoder" CODECs generate? Give an example.

Answer: They generate sample-based payloads. G.711 is an example where 8 bits represent each sample.

4 What type of RTP payload do "vocoder" CODECs generate? Give an example.

Answer: They generate frame-based payloads. G.728 is an example where 10-bit codepoint vectors are used. G.729E, iLBC, and BV16 are also examples.

5 What is the CSRC field in a RTP packet and why is it not typically used in PacketCable?

Answer: It is the Contributing Source Identifier used to reference sources when RTP stream mixing is used. This can result in variable-length RTP packets, which is bad for DQoS admission control.

6 What SDP attributes define a dynamic payload type?

Answer: The rtpmap attribute, which defines the payload and clock rate, and the fmtp attribute, which defines parameters of the dynamic payload

7 If the G.729E CODEC operates on 10-ms samples, how can it have an output packet rate of 20 ms or 30 ms?

Answer: Frame-based CODECs typically concatenate multiple frames together in a single packet to achieve a different packet rate. This improves bandwidth efficiency by eliminating the number of packet headers.

8 Why is the transmission of comfort noise packets beneficial when implementing VAD?

Answer: It assures the listener that the call is still up and keeps intermediate devices from thinking the subscriber is down

9 If a 30-ms packet rate is used, by how much would you expect the timestamp field to increment for each packet?

Answer: 240 (30 ms × 8000 samples/sec)

10 True or False: The SSRC field is the same for comfort noise packets and regular audio packets from the same source?

Answer: True. This is needed so that the receiver can correlate periods of silence with periods of activity.

11 What are the three types of RTCP packet commonly seen in PacketCable?

Answer: SR (Sender Report), SDES (Source Description), and BYE

12 What kind of information is provided in an SR packet?

Answer: Bytes and octets sent, bytes and packets received, packets lost, and jitter

13 How would you configure the BTS to instruct the gateway to use a packet rate of either 20 ms or 30 ms?

Answer: In the QOS table, set LPTIME to 20 and HPTIME to 30.

14 What VISM command would you use to give the CODEC order in the Remote Connection Descriptor preference over the Local Connection Options?

Answer: cnfcodecneg

15 How would you configure the VISM card to generate RTCP packets every 20 seconds?

Answer: Using the **cnfrtcprepint** command, set the RTCP interval to 10,000 and the multiplier to 2.

16 What VISM command would you use to look at the current echo configuration on a T1 or E1?

Answer: dsplndsp

17 How does a gateway indicate it can support audio CODECs and T.38, but not at the same time?

Answer: Using RFC 3407 to define a capability set with separate capability descriptors for audio and fax

18 What are the three facets to the use of the VoIP metrics block?

Answer: Collecting of local data, responding to remote endpoints, and reporting to a call agent

19 What VXSM command verifies whether DTMF relay is enabled?

Answer: dspxgcpdtmf

20 What VXSM command specifies the external file server for collecting bearer trace data?

Answer: addbearertracesrvprof

PART V

Quality of Service

CHAPTER 12

DQoS Architecture and Framework

A cable operator's ability to offer quality of service (QoS) treatment for multimedia applications gives that cable operator a significant edge over other service providers that have no control over the Multiple Systems Operator (MSO)'s transport network. This ability allows the cable MSO to be more than just a raw bit provider. Not only can an MSO offer a service like telephony that makes use of their DQoS resources, but they can also provide dynamic quality of service (DQoS) resources for services operated by third-party providers. Whether operated in-house or externally, these QoS-enabled services can open up new sources of revenue. The concepts presented in this chapter, as well as in Chapter 13, "Analyzing, Implementing, and Troubleshooting DQoS," and Chapter 14, "Multimedia Applications," detail how DQoS is provided and how these resources are accounted for.

Throughout this book, you have seen numerous discussions on the importance of a QoS mechanism to manage Data-Over-Cable Service Interface Specification (DOCSIS) resources. In Chapter 1, "PacketCable Overview," you learned that this QoS requirement ensures that voice calls are of sufficient quality and provides a mechanism to authorize and control DOCSIS resources—an essential part of PacketCable. Chapter 2, "PacketCable Functional Components," covered the multimedia terminal adapter (MTA), cable modem (CM), cable modem termination system (CMTS), call management server (CMS), and Record Keeping Server (RKS) PacketCable components, all of which contribute to providing this QoS. Chapter 7, "NCS," covered some parameters within the Network-based Call Signaling (NCS) protocol that communicate DQoS information between the CMS and MTA. Finally, in Chapter 10, "Audio CODECs," and Chapter 11, "RTP and RTCP," you learned how voice is transmitted in a PacketCable VoIP network.

This chapter covers DQoS in greater detail, including how this functionality is provided and managed and what the protocols are that make it work. The DOCSIS 1.1 protocol, the NCS protocol, and the Common Open Policy Service (COPS) protocol all contribute to PacketCable DQoS functionality. Chapter 13 builds on the technical information covered here by examining the implementation of these technologies in providing quality of service for PacketCable telephony. Chapter 14 goes into detail on the theory and technical aspects behind PacketCable Multimedia (PCMM). The PCMM specifications detail a generalized framework for providing dynamic QoS for any type of multimedia application. The topics covered in Chapter 12 are the foundation for many of the topics covered in Chapter 14.

DQoS Overview

When it comes to QoS for a PacketCable call, the call is separated into three pieces:

- The access network between the originating subscriber and the backbone
- The access network between the backbone and the terminating subscriber
- The backbone itself

QoS across the backbone can be provided using Differentiated Services Code Point (DSCP) marking or Multiprotocol Label Switching (MPLS). How backbone QoS is provided is not explicitly defined by PacketCable; however, Chapter 16, "PacketCable Network Design Considerations," introduces some QoS mechanisms that can be used.

If the originating or terminating subscriber is a PacketCable customer, the access network contains the DOCSIS network between the CMTS and the MTA. QoS across the DOCSIS access network is provided via PacketCable DQoS, as described in this chapter.

NOTE QoS discussed in this chapter is only for bearer packets; QoS for signaling packets (NCS) is also an important consideration and is covered in Chapter 16.

It is no secret that DOCSIS resources are a valuable commodity; typically, hundreds of users are competing for this bandwidth across a DOCSIS MAC domain. Remember, a DOCSIS MAC domain is the collection of users connected to a common set of downstream and upstream DOCSIS channels. Normally, high-speed data users and PacketCable customers utilize this same bandwidth pipe. PacketCable DQoS ensures that voice customers are given preferential treatment in receiving the amount of bandwidth and at the intervals needed for quality voice communication.

Figure 12-1 illustrates the PacketCable components and protocols involved in providing DQoS functionality.

Figure 12-1 *DQoS Architecture*

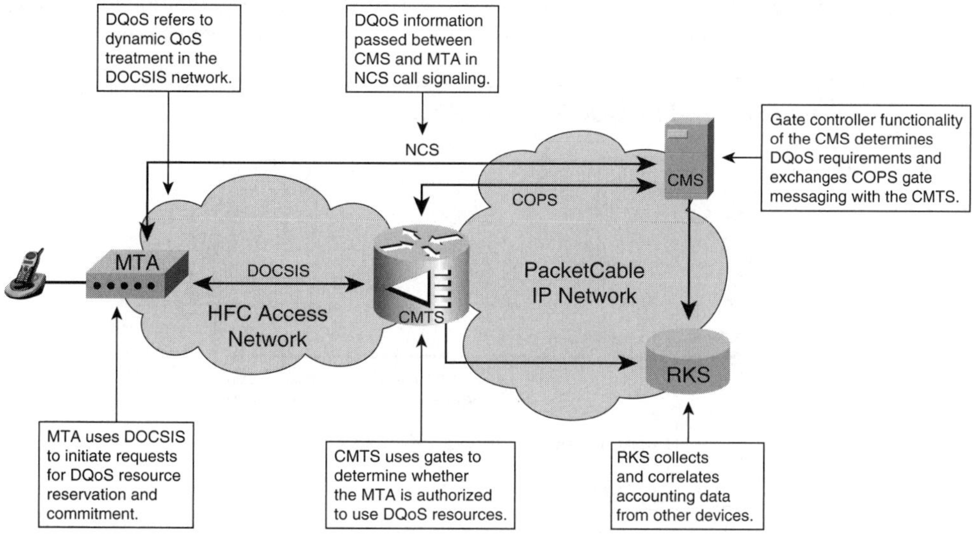

As you can see, four PacketCable components are involved in providing DQoS functionality:

- **Embedded MTA (eMTA)**—DQoS information is exchanged between an eMTA and the CMS using the NCS protocol. An embedded MTA internally signals its cable modem component to request QoS using the DOCSIS protocol. This is done through the use of DOCSIS 1.1 dynamic service flow messages exchanged between the CMTS and eMTA.
- **CMTS**—DOCSIS bandwidth is authorized, reserved, and committed by the CMTS. Access to this premium bandwidth is identified at the CMTS by a DQoS gate. What a gate is and how it works is covered later—for now, think of a gate as the key to premium DOCSIS resources. Because the CMTS is where DOCSIS resources are controlled, it is also referred to as the Policy Enforcement Point (PEP) in the DQoS architecture.
- **CMS**—Gates are created on the CMTS under the direction of the CMS. Thus, the CMS also functions as a gate controller. The CMS directs the CMTS to set up DQoS gates and then relays this information to the MTA using the NCS protocol. Because the CMS is controlling the manipulation of DQoS gates, it is referred to as the Policy Decision Point (PDP) in the DQoS architecture.
- **RKS**—The value of premium DQoS resources in a DOCSIS network has already been discussed. Because these resources are so valuable, they must be accurately accounted for. The CMTS uses event messaging to record DQoS usage per subscriber

and passes this information to an RKS for accounting purposes. The identity of the RKS(s) is relayed to the CMTS by the CMS. Keep in mind that the RKS has nothing do with providing DQoS functionality; it simply records DQoS usage.

Figure 12-2 shows a high-level overview of the messaging between the MTA, CMS, and CMTS used in providing DQoS functionality.

Figure 12-2 *DQoS Messaging*

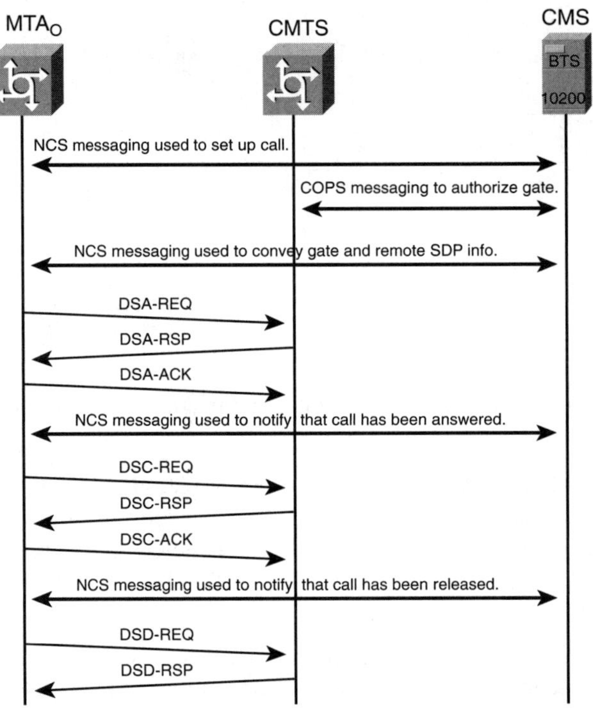

The details of the COPS messaging between the CMS and CMTS components and the DOCSIS messaging between the MTA and CMTS components are the focus for much of this chapter.

Gates

DQoS gates control access to premium DOCSIS resources between an MTA and a CMTS. A 4-byte gate identifier assigned by the CMTS references these gates. For PacketCable 1.x the gate ID typically refers to two gates—one for the upstream direction and one for the downstream direction. Thus, a single gate ID references the DQoS resources for a single leg in an NCS call.

Although the gate ID can reference a bidirectional flow, the gates themselves are unidirectional. In fact, in Chapter 14 you'll learn that PCMM upstream and downstream gates are generally created separately and have separate gate IDs.

Gates are allocated and scheduled on the CMTS under the direction of the CMS. A gate consists of the following:

- Classification components used to identify packets permitted to use the premium resources. The classifiers consist of a direction (upstream or downstream), the IP protocol, the source and destination IP addresses, and port numbers.
- Policing components ensuring only resources allocated for the gate are used. These components describe the traffic needing DQoS resources.

Other information is often associated with a gate. An example is accounting components ensuring resources are accurately billed. This information includes a billing identifier and the location of the RKS(s).

Gate States

A PacketCable 1.x gate transitions through four states or phases:

- Allocated
- Authorized
- Reserved
- Committed

Before an MTA can request DQoS resources, the allocation and authorization phases need to complete. In the allocation phase the CMS tells the CMTS (using the COPS protocol) to allocate a pair of gates for a particular MTA. The messaging here includes the maximum number of gates the MTA is allowed to have at once; the CMTS checks to see whether adding two additional gates violates this maximum. Assuming the MTA is permitted to receive two additional gates, the CMTS assigns a 4-byte gate identifier for the newly allocated gate pair and returns this value to the CMS.

After the CMS learns what type of resources the MTA requires and the classification parameters needed to identify packets permitted to use those resources, it sends this information to the CMTS in a COPS "Gate Set" message. The CMTS then determines whether it is capable of providing these resources for the MTA. The CMTS responds to the CMS with an indication of whether the authorization was successful. At this point the gate is in the *authorized state*. Remember, no resources have actually been reserved at this point.

The allocation and authorization phases can occur in separate steps, or they can occur simultaneously; this is up to the CMS vendor implementation. If separate steps are used, a *Gate Alloc* message is first sent from the CMS to the CMTS. If the CMTS was able to allocate the gates, a *Gate Alloc Ack* message is sent to the CMS containing the gate ID. If the CMTS was unable to allocate the gates, a *Gate Alloc Err* message, indicating the reason for the error, is sent to the CMS instead.

In the authorization phase the CMS sends a *Gate Set* message to the CMTS containing the DQoS resource and classification parameters. If the CMTS is able to authorize the request a *Gate Set Ack* message is returned; if the CMTS was unable to authorize the request a *Gate Set Error* message is returned, indicating the reason for the error. In a single-step implementation only the *Gate Set* message exchange occurs; when the CMTS receives the *Gate Set*, it does the allocation check and gate ID assignment in addition to the authorization check. The 4.x and later versions of the BTS 10200 use a single-step implementation.

After resources are authorized, the MTA requests the CMTS (using DOCSIS 1.1) for all or some of these resources to be reserved for use. The MTA is directed to reserve resources in NCS messaging from the CMS. This usually occurs right before the CMS requests ring tone and ring back tone be provided to the call parties. The reserved resources envelope is always less than or equal to the authorized resources envelope.

In the reservation phase, resources are held for an MTA, ensuring they are available when needed. Until they are needed, these resources can be used for ordinary best effort data service. Remember, reserved resources cannot be used by other DQoS customers because the same resources cannot be reserved multiple times; they can be used only by Best Effort services. This protects against the possibility of a MTA attempting to commit DQoS resources that aren't available because they were assigned to other users.

In the commit phase, resources are made available for use by the MTA, and no one else can use them. This usually happens at the point the phone call is answered. As with the reservation phase, NCS messaging from the CMS tells the MTA to commit resources, and DOCSIS 1.1 messaging signals the CMTS to commit resources.

Usually, DQoS resources are provided to an MTA in two separate phases, the reservation phase and the commitment phase. This dual-phase model ensures that resources are available when needed and does not tie up resources that are not needed. For example, you don't need to commit resources for a phone call if the terminating party isn't there to answer the phone. On the other hand, if quality resources cannot be provided, there is no point in allowing the call to go through. Another benefit of the dual-phase approach is that the billing of resources doesn't begin until the commit phase, where the voice call is actually active.

NOTE The CMS might decide to reserve and commit resources at the same time, but this is generally discouraged. One situation in which this might occur is for OFFNET calls where ring-back tone is supplied from the PSTN. In this case, DQoS resources in the direction from the PSTN to the MTA might be committed immediately to let the ring back signal use DQoS resources.

Keep in mind the amount of resources committed can be different from the amount reserved, and the amount of resources reserved can be different from the amount of resources authorized, although in most PacketCable 1.x deployments they are typically the same. The committed resources envelope is always less than or equal to the reserved resources envelope, which is always less than or equal to the authorized resources envelope. Figure 12-3 illustrates the relationship between these resource categories.

Figure 12-3 *DQoS Resources*

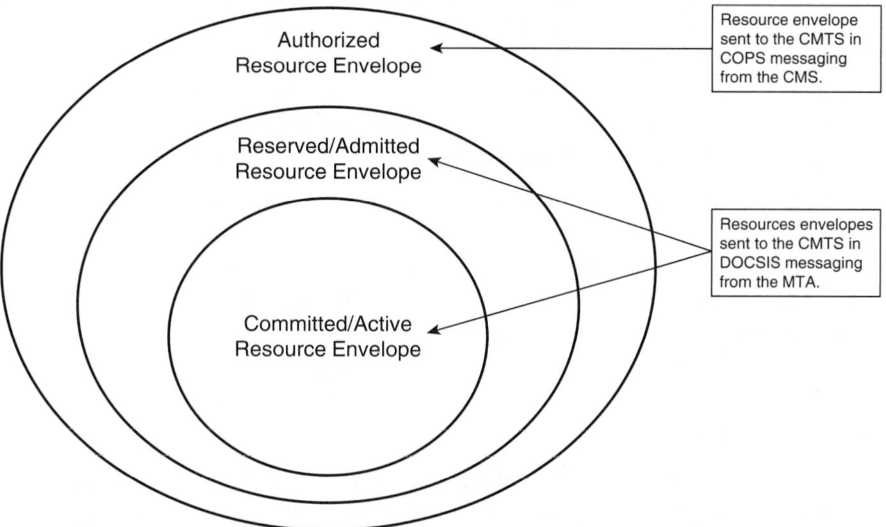

As you can tell in Figure 12-3, reserved resources are also referred to as *admitted resources*, and committed resources are also referred to as *active resources*. The *admitted* and *active* terms are used in the DOCSIS specification to refer to the status of DOCSIS service flows. A DOCSIS service flow that is statically defined in the cable modem configuration file is classified as *provisioned*. A DOCSIS service flow (either static or dynamic) that the CMTS accepts and is able to service is classified as *admitted*. And a DOCSIS service flow (either static or dynamic) that the CMTS allows traffic to be sent on is classified as *active*. As you can see, the DOCSIS concepts of *admitted* and *active* service flows map directly to the PacketCable concepts of *reserved* and *committed* DQoS resources.

Earlier it was mentioned that in PacketCable 1.x the *authorized*, *reserved*, and *committed* resource envelopes are typically equal. That's because a majority of today's deployments use only the G.711 CODEC; consequently, the CMS knows from the beginning the CODEC the call is going to use so it can authorize the exact amount of DQoS resources necessary to support it. Even if Voice Band Data such as fax is detected and the MTA doesn't support a relay mechanism, no CODEC change is required because G.711 supports Voice Band Data in-band. Conversely, if CODEC negotiation results in multiple possible

CODECs, the CMS does not know what CODEC is going to be used for the call. So the CMS generates an authorization envelope big enough to handle any of the possible CODECs; this is also known as the least upper bound (LUB) constraint. For example, if the CODEC list contains G.711 and G.729E, the CMS authorizes enough resources to support G.711 even though the actual call might end up using G.729E, which has a much smaller bandwidth requirement.

Now, you might be wondering in what situations the reserved and committed resource envelopes would be different. One example is safeguarding for CODEC changes because of Voice Band Data detection. For example, if voice calls are set up to use G.729E, but the MTA wants to be sure a CODEC up-speed to G.711 to support fax transmission gets the necessary resources, it can reserve the G.711 envelope but only commit the G.729E envelope. Then if fax tone is detected requiring a CODEC change, the MTA is assured the process of committing additional resources will be successful.

Another point to remember about reserved and committed resources is that they are dynamic and can be changed during the life of a call. For example, if a CODEC change is necessary to support fax transmission (usually triggered by NCS signaling from the CMS), the MTA sends DOCSIS messaging to the CMTS to change the committed resource envelope (and possibly the reserved as well).

COPS Protocol

DQoS gate information is communicated between the CMS and the CMTS using the Common Open Policy Service (COPS) protocol. COPS is a server/client protocol that runs over TCP, the COPS server is also referred to as the Policy Decision Point (PDP), and the COPS client is also referred to as the Policy Enforcement Point (PEP). For PacketCable 1.x, the CMS/GC is the COPS server/PDP, and the CMTS is the COPS client/PEP. Each COPS message consists of a common header followed by a variable number of objects. Figure 12-4 details the format of this header.

Figure 12-4 *COPS Common Header Format*

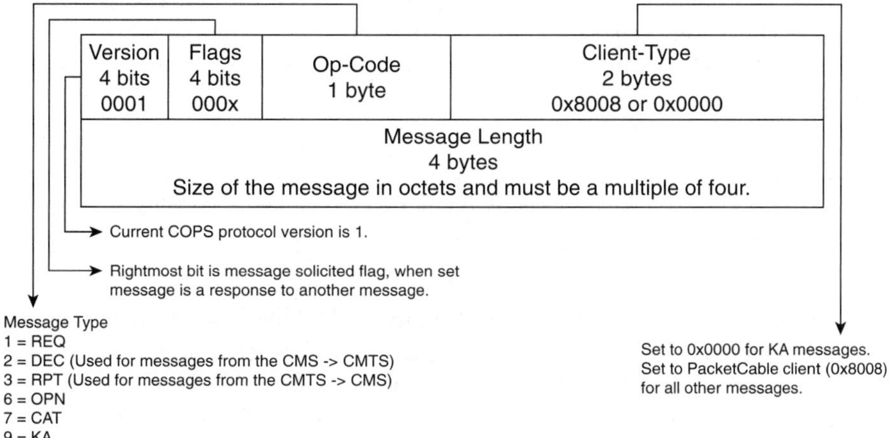

Figure 12-5 illustrates the format for the variable objects.

Figure 12-5 *COPS Object Format*

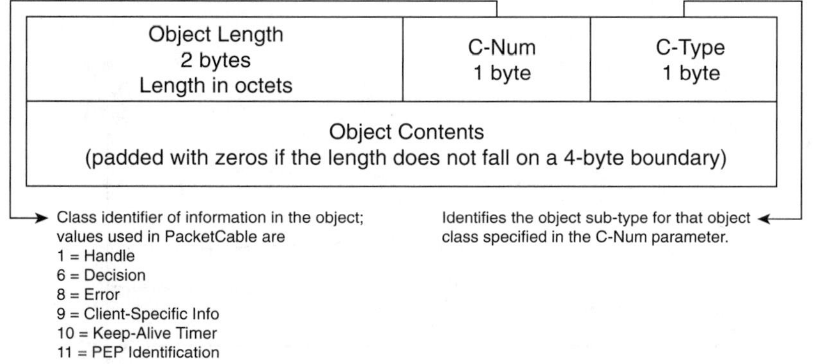

For messages going from the CMS to the CMTS in a Decision Message, the C-Num is 6 (Decision) and the C-Type is 4 (Client-Specific Decision Data) for PacketCable objects.

For messages going from the CMTS to the CMS in a Report-State Message, the C-Num is 9 (Client-Specific Info) and the C-Type is 1 (Signaled Client SI) for PacketCable objects.

COPS Initialization

After the CMTS is configured for PacketCable operation and the CMTS is configured and activated on the CMS, the COPS initialization sequence begins. First, the CMS opens the TCP connection used for COPS on TCP port 2126. Notice that for this connection to be established, the CMS must be configured with the location of the CMTS; however, the CMTS does not need to be configured with the location of the CMS. This is assuming that IP Security (IPsec) is not being used for the COPS connection. If IPsec is being used, both the CMS and CMTS must be explicitly configured to have a security association between them.

NOTE Refer to Chapter 7 for how to configure a CMTS on the Cisco BTS 10200.

After this TCP connection is established, the CMTS identifies itself by sending a COPS Client-Open (message type 6 - OPN) to the CMS. This identification consists of a PEP-ID (Policy Enforcement Point ID), typically the configured FQDN of the CMTS. The CMS responds by sending a Client-Accept (message type 7 - CAT) to the CMTS. In this message is the value of the keepalive timer to be used by the CMTS. The CMTS completes the initialization process by sending a Request (message type 1 - REQ) to the CMTS. This message defines the Handle object used in future messages between the CMS and CMTS. Figure 12-6 illustrates this initialization sequence as well.

Figure 12-6 *COPS Initialization Sequence*

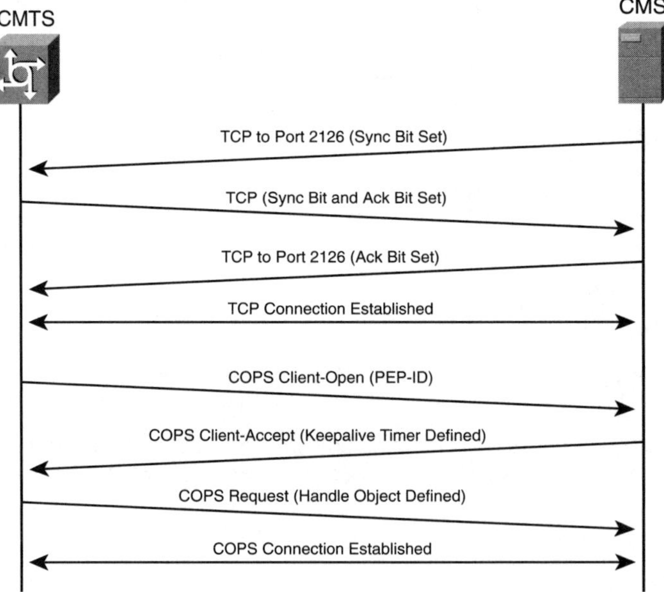

Figure 12-7 shows an example of this initialization sequence on a Cisco uBR7246VXR. IOS debugs were enabled to show the messages. Please be very cautious when enabling debugs such as the **debug cops detail** command because it causes quite a bit of output.

Figure 12-7 *COPS Initialization on a Cisco CMTS*

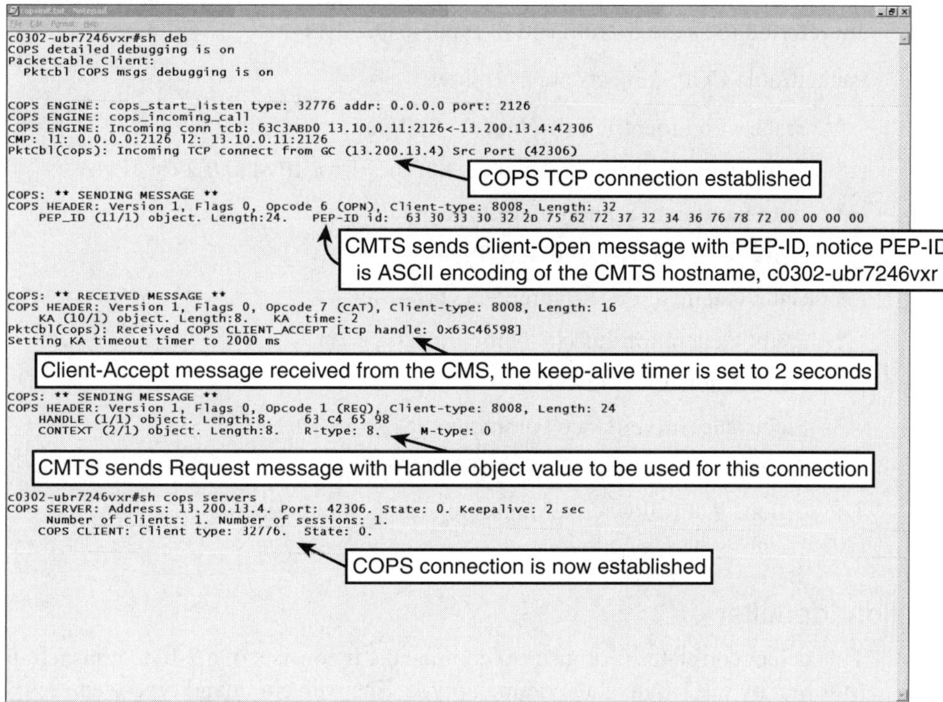

CMTS to CMS Heartbeats

Keepalive messages (message type 9 - KA) are sent from the CMTS to the CMS periodically. The CMS responds to these keepalive messages by sending a keepalive message back to the CMTS. If the CMS fails to receive a keepalive message from the CMTS by the keepalive timer value, it assumes the connection has been lost. The CMS then attempts to reestablish the TCP connection that if successful causes the CMTS to reinitialize the COPS connection. Similarly, if the CMTS fails to receive the keepalive echo from the CMS, it assumes the connection has been lost and listens for a new TCP connection.

PacketCable Objects in COPS

As shown in Figure 12-5, PacketCable objects are included in either Client Specific Decision Data objects (C-Num 6, C-Type 4) or Signaled Client SI objects (C-Num 9, C-Type 1), depending on the direction of the message. These objects begin with a 2-byte length field followed by another 1-byte object number and 1-byte subtype definition. These are referred to as the S-Num and S-Type, respectively.

PacketCable COPS objects are as follows:

- Transaction Identifier (S-Num=1, S-Type=1)
- Subscriber Identifier (S-Num=2, S-Type=1 for IPv4 and 2 for IPv6)
- Gate Identifier (S-Num=3, S-Type=1)
- Activity Count (S-Num=4, S-Type=1)
- Gate Specification (S-Num=5, S-Type=1)
- Event Generation Info (S-Num=7, S-Type=1)
- PacketCable Error (S-Num=9, S-Type=1)
- Electronic Surveillance Parameters (S-Num=10, S-Type=1)
- PacketCable Reason (S-Num=13, S-Type=1)

The sections that follow cover these objects in greater detail.

Transaction Identifier

This object correlates responses to commands. It consists of a 2-byte transaction identifier followed by the 2-byte gate command type. The gate command type identifies the DQoS message type. Table 12-1 lists the possible values.

Table 12-1 *DQoS Message Types*

Gate Command Type	Gate Command	Message Direction
1	GATE-ALLOC	CMS → CMTS
2	GATE-ALLOC-ACK	CMTS → CMS
3	GATE-ALLOC-ERR	CMTS → CMS
4	GATE-SET	CMS → CMTS
5	GATE-SET-ACK	CMTS → CMS
6	GATE-SET-ERR	CMTS →CMS
7	GATE-INFO	CMS → CMTS
8	GATE-INFO-ACK	CMTS → CMS
9	GATE-INFO-ERR	CMTS → CMS

Table 12-1 *DQoS Message Types (Continued)*

Gate Command Type	Gate Command	Message Direction
10	GATE-DELETE	CMS → CMTS
11	GATE-DELETE-ACK	CMTS → CMS
12	GATE-DELETE-ERR	CMTS → CMS
13	GATE-OPEN	CMTS → CMS
14	GATE-CLOSE	CMTS → CMS

You've already learned how the GATE-ALLOC and GATE-SET commands are used in the gate allocation and authorization phases. The GATE-INFO commands are used by the CMS to query information about a gate. The GATE-OPEN command is used by the CMTS to inform the CMS that DQoS resources have been committed. The GATE-CLOSE command is used by the CMTS to inform the CMS that a gate has been deleted because of MTA interaction or inactivity. The GATE-DELETE commands are typically seen only in error scenarios where the CMS needs to tear down a gate.

Subscriber Identifier

The Subscriber ID object is the IP address of the subscriber, used to associate QoS information with a particular subscriber to help prevent denial of service attacks.

Gate Identifier

The Gate ID is the 4-byte gate identifier assigned by the CMTS.

Activity Count

When sent in a GATE-ALLOC or GATE-SET message from the CMS, this field indicates the maximum amount of simultaneous sessions permitted for the subscriber. When sent in a GATE-ALLOC-ACK or GATE-SET-ACK message from the CMTS, this field indicates the number of gates currently assigned to a subscriber. This field is also used to help prevent denial of service attacks.

Gate Specification

This field describes the gate parameters authorized for a particular subscriber. Figure 12-8 illustrates the format for this field.

Figure 12-8 *Gate Specification Format*

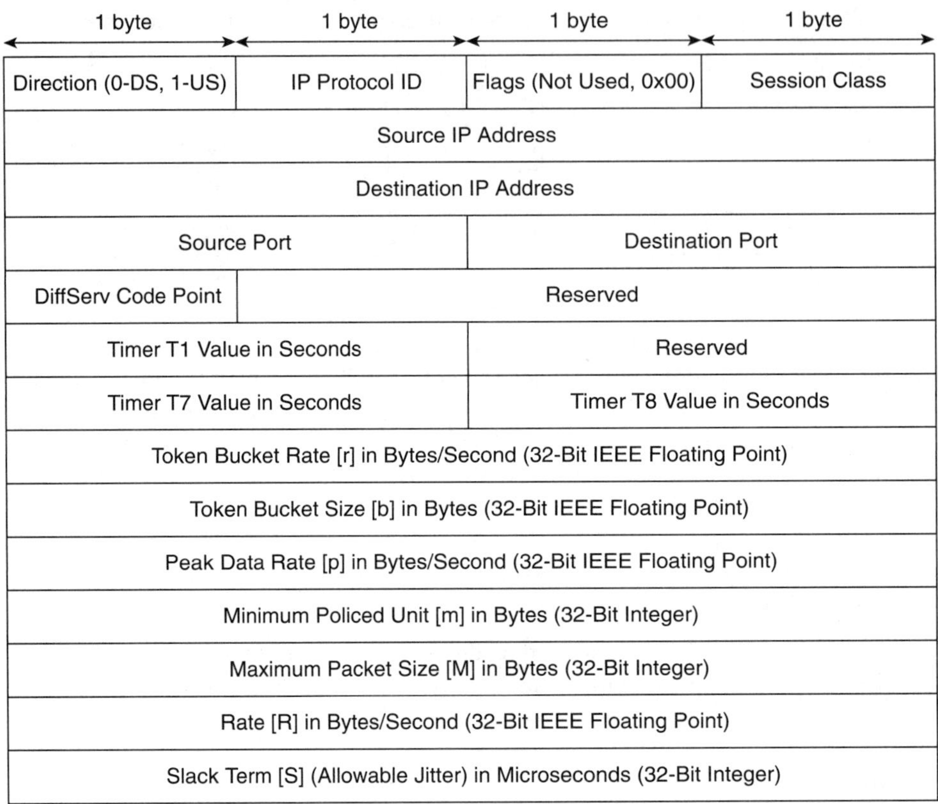

The Session Class field identifies the gate as belonging to a normal VoIP session (0x01), a high-priority VoIP session (0x02), or an unspecified session (0x00). The high-priority session class could be used for emergency 911 calls so that they receive preferential treatment over ordinary phone calls. For example, if an attempt to reserve resources for a 911 call is made and no bandwidth is available, ordinary calls could be dropped to create the necessary resources for this high-priority session. Another possible implementation is to pre-designate a certain amount of bandwidth explicitly for high-priority calls.

Event Generation Information

This object passes along RKS information to the CMTS. This includes primary and secondary RKS server IP addresses and port numbers, a billing correlation identifier, and the mode in which the CMTS is to send event messages. The CMTS can either send event messages in real time or accumulate them and send them at periodic time intervals. Please refer to Chapter 15, "Event Messaging and Lawful Intercept," for more information on how event messaging is used in PacketCable.

PacketCable Error

This object consists of a 2-byte error code and a 2-byte error sub-code. Table 12-2 lists the currently defined error codes.

Table 12-2 *Error Codes*

Error Code	Definition
1	No gates currently available
2	Unknown Gate ID
3	Illegal session class value
4	Subscriber exceeded gate limit
5	Gate already set
6	Missing required object
7	Invalid object
8	Illegal DSCP field value
127	Other, unspecified error

The error sub-code conveys further information about the error. For example, for error codes 6 and 7 this field contains the S-Num and S-Type values of the object missing or invalid.

Electronic Surveillance Parameters

This object contains all the information needed to support electronic surveillance or lawful intercept. This includes the lawful intercept Delivery Function (DF) IP addresses and port numbers used for both "call detail" and "call content" taps, the call content correlation identifier, the billing correlation identifier, and the flags indicating whether the CMTS should duplicate "call detail" and/or "call content" information. Please refer to Chapter 15 for more information on how electronic surveillance is used in PacketCable.

PacketCable Reason

This object communicates why a gate is being deleted or closed. It consists of a 2-byte reason code and a 2-byte reason sub-code. The reason code identifies the message as a

GATE-DELETE operation (0) or a GATE-CLOSE operation (1). Gate Delete messages are sent from the CMS to the CMTS, whereas Gate Close messages are sent from the CMTS to the CMS. Table 12-3 defines the values of the reason sub-code for a Gate Delete operation.

Table 12-3 *Gate Delete Reason Sub-Codes*

Reason Sub-Code	Definition
0	Normal operation
1	Local gate-coordination not completed
2	Remote gate-coordination not completed
3	Authorization revoked
4	Unexpected GATE-OPEN
5	Local GATE-CLOSE failure
127	Other, unspecified error

Table 12-4 defines the values of the reason sub-code for a Gate Close operation.

Table 12-4 *Gate Close Reason Sub-Codes*

Reason Sub-Code	Definition
0	Client initiated release (normal operation)
1	Reservation reassignment (e.g., for priority session)
2	Lack of reservation maintenance
3	Lack of DOCSIS MAC-layer responses
4	Timer T0 expiration (no GATE-SET from CMS)
5	Timer T1 expiration (no COMMIT from MTA)
6	Timer T7 expiration (Service Flow reservation timeout)
7	Timer T8 expiration (Upstream service flow activity timeout)
127	Other, unspecified error

DQoS Timers

Timers used in PacketCable 1.x DQoS include

- **T0**—The amount of time a gate can be in the allocated state. Timer starts when a gate is allocated and stops when the gate is authorized. This timer is defined on the CMTS. The recommended value of this timer is 30.
- **T1**—The amount of time a gate can be in the authorized state. Timer starts when a gate is authorized and stops when the gate is committed. Note: The timer does not stop when the gate is put into the reserved state. This timer is typically set in the GATE-SET message from the CMS, but if it is absent, the provisioned value on the CMTS is used. The recommended value of this timer is between 200 and 300 seconds.
- **T5**—The amount of allowable time between when a CMS signals an MTA to delete a connection and the time the resulting GATE-CLOSE command is received from the CMTS. If the GATE-CLOSE command is not received within this time period, the CMS issues a GATE-DELETE to the CMTS to delete the gate. The recommended value of this timer is 5 seconds.
- **T7**—The amount of time reserved but not committed resources are to be held. This timer is specified by the CMS and is sent to the CMTS in a GATE-SET message. The CMTS then maps this parameter to the DOCSIS Timeout for Admitted QoS parameter, which is sent to the MTA in a DSA-RSP or DSC-RSP message. The recommended value of this parameter is 200 seconds.
- **T8**—The amount of time committed resources can remain unused. This is also referred to as the inactivity timer. Like T7, this timer is specified by the CMS and is sent to the CMTS in a GATE-SET message. The CMTS maps this parameter to the DOCSIS Timeout for Active QoS parameter, which is sent to the MTA in a DSA-RSP or DSC-RSP message. The default value of this timer is 0, meaning the CMTS should not poll the service flow for activity. It is recommended that a non-zero value be used for this parameter to ensure that "hung" resources can be reclaimed.

Authorizing DQoS Resources

The CMS determines what resources the MTA is authorized to use and signals this information to the CMTS. In the gate specification parameter in the Gate Set message, DQoS resource envelopes are defined using Resource Reservation Protocol (RSVP) flow specifications. The values of the RSVP FlowSpec are determined by information the CMS knows about the MTA DQoS requirements for a call.

You might be asking yourself, how does the CMS know what amount of DQoS resources a MTA requires? The answer is from information obtained from the NCS call signaling. As you might recall, in NCS the Local Connection Descriptor and Remote Connection Descriptor parameters describe the endpoint's bearer information. This information is encoded using the Session Description Protocol (SDP). SDP was introduced in Chapter 6, "Signaling Interfaces and MGCP Overview," and its use in the NCS protocol was covered in Chapter 7. Please refer to these chapters for more information on how the NCS and SDP protocols function. The specific SDP fields used are the media announcement parameter (m), the bandwidth parameter (b), and the "packetization period" attributes (either mptime or ptime) as demonstrated in the following example:

```
200 161096008 OK
I: 575

v=0
o=- 161096008 1397 IN IP4 24.34.240.245
s=-
c=IN IP4 24.34.240.245
b=AS:82      <- Bandwidth requirement in kbps of the media sessions
t=0 0
m=audio 53456 RTP/AVP 0 8 15    <- Possible CODECs used by the MTA endpoint
a=X-pc-secret:base64:<hex digit string omitted>
a=X-pc-csuites-rtp:62/51 64/51 60/51 60/50
a=X-pc-csuites-rtcp:81/71 81/70 80/70
a=X-pc-nrekey:0
a=mptime:20 20 20     <- Corresponding packetization periods used with CODECs
```

NOTE In Chapter 10 you learned that this CODEC list is negotiated between the gateway endpoints. In a majority of today's deployments only the G.711 CODEC is used; therefore, the CODEC list consists of this single CODEC.

Summary of RSVP Flow Specifications

An RSVP flow specification consists of two parts: the traffic description (TSpec) and the resource description (RSpec). Within RSVP you have two types of services: controlled load and guaranteed. The guaranteed type of service is used for applications where delay is an issue, such as voice.

A TSpec of a guaranteed type of service consists of the following parameters:

- **(b)**—Token bucket depth in bytes, which represents a reservoir of credit correlating to the maximum amount of data permitted to be sent in a given time interval
- **(r)**—Token bucket rate in bytes/second, which specifies how fast credits are added to this reservoir
- **(p)**—Peak rate in bytes/second that the source sends
- **(m)**—Minimum policed unit in bytes, which is typically the smallest packet size the source sends
- **(M)**—Maximum packet size in bytes that the source sends

An RSpec of a guaranteed type of service consists of the following parameters:

- **(R)**—Reserved rate in bytes/second, which is the amount of bandwidth allocated to the flow
- **(S)**—Slack term in microseconds (PacketCable uses this term to define maximum allowable packet jitter)

In PacketCable, voice packets in a call are sent one at a time and are of a constant size. Thus, the (b), (m), and (M) parameters must be equal to each other. These values equate to the size of a single digitized voice packet, including the IP and higher layer headers. The MTA transmits these voice packets at a constant rate; consequently, the (r), (p), and (R) parameters must also be equal to each other. The (S) parameter is defined on the CMS; by default, it has a value of 800 microseconds for the upstream and 0 for the downstream.

Mapping SDP Information into RSVP Flow Specifications

Now that you have some understanding of how these protocols are formatted it's time to examine how the CMS converts SDP information into the RSVP flow specifications used in the COPS PacketCable Gate Specification object.

The CMS uses the CODEC and the packetization period obtained from SDP to create the RSVP flow specifications. The easiest way to illustrate how this works is with an example.

Assume the MTA has a SDP descriptor with the following parameters:

```
m=audio 49182 RTP/AVP 0
a=ptime:20
```

RTP payload type 0 is the G.711 μ-law CODEC; the packetization period is 20 ms. In Chapter 11, you saw the resulting IP packet size in this case is 200 bytes. This is the value of the (b), (m), and (M) parameters in the TSpec portion of the RSVP flow specification. The (r), (p), and (R) parameters in the RSVP FlowSpec correspond to the rate at which the MTA sends data. In this example, a 200-byte packet is sent every 20 ms, equating to a rate of 10,000 bytes/sec (200 byte/20 ms). The only other remaining RSVP FlowSpec parameter is the slack term that is either defined on the CMS or the default value is used.

Figure 12-9 illustrates this process; also for more detail refer to Part IV of this book.

Figure 12-9 *Mapping SDP Parameters into RSVP FlowSpecs*

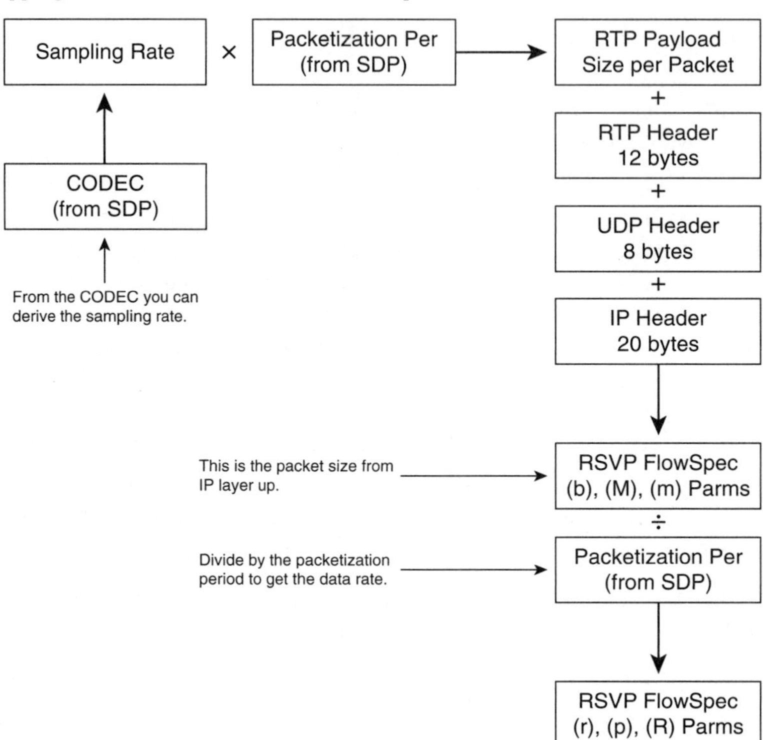

Table 11-2 in Chapter 11 shows the bandwidth constraints of all the possible CODEC and packetization period combinations for PacketCable. Table 12-5 shows what the corresponding Flow Specification parameters would be for each of these CODECs. Neither this table nor the formula in Figure 12-9 shows the impact of bearer stream security or payload header suppression on these parameters. How payload header suppression works and affects bearer traffic is covered later in this chapter. Bearer stream security works by encapsulating the RTP packets inside Encapsulation Security Payload (ESP) frames used for IP security; the net effect of this is packet sizes are increased by four bytes. More on security is covered in Chapter 16.

Table 12-5 *CODEC Flow Specification Parameters*

CODEC	Pack. Period	Total Packet Size	FlowSpec (b),(M),(m) Parameters	Resulting Bit Rate of One Call	FlowSpec (r),(p),(R) Parameters
G.711	10 ms	120 bytes	120 bytes	96 Kbps	12,000 bytes/sec
G.711	20 ms	200 bytes	200 bytes	80 Kbps	10,000 bytes/sec
G.711	30 ms	280 bytes	280 bytes	~74.67 Kbps	9,334 bytes/sec

Table 12-5 *CODEC Flow Specification Parameters (Continued)*

CODEC	Pack. Period	Total Packet Size	FlowSpec (b),(M),(m) Parameters	Resulting Bit Rate of One Call	FlowSpec (r),(p),(R) Parameters
G.728	10 ms	60 bytes	60 bytes	48 Kbps	6,000 bytes/sec
G.728	20 ms	80 bytes	80 bytes	32 Kbps	4,000 bytes/sec
G.728	30 ms	100 bytes	100 bytes	~26.67 Kbps	3,334 bytes/sec
G.729E	10 ms	55 bytes	55 bytes	44 Kbps	5,500 bytes/sec
G.729E	20 ms	70 bytes	70 bytes	28 Kbps	3,500 bytes/sec
G.729E	30 ms	85 bytes	85 bytes	~22.67 Kbps	2,834 bytes/sec
iLBC	20 ms	78 bytes	78 bytes	31.2 Kbps	3,900 bytes/sec
iLBC	30 ms	90 bytes	90 bytes	24 Kbps	3,000 bytes/sec
BV16	10 ms	60 bytes	60 bytes	48 Kbps	6,000 bytes/sec
BV16	20 ms	80 bytes	80 bytes	32 Kbps	4,000 bytes/sec
BV16	30 ms	100 bytes	100 bytes	~26.67 Kbps	3,334 bytes/sec

Reserving and Committing DQoS Resources

So far you have seen how an authorized resource envelope is created. Next, you will learn how reserved and committed resource envelopes are created. Remember, NCS call signaling parameters trigger an embedded MTA to initiate requests for DQoS resource reservation and commitment. DOCSIS mechanisms are used by the MTA; thus, a review of DOCSIS aspects pertinent to QoS is warranted.

Reviewing DOCSIS 1.1 QoS

In Chapter 1 you learned that the DOCSIS 1.1 RFI specification added support for quality of service mechanisms in DOCSIS. PacketCable makes use of these mechanisms to provide DQoS premium services. The key aspects of DOCSIS used by PacketCable DQoS are summarized here. For more detail please refer to the DOCSIS 1.1 RFI specification, which can be obtained at http://www.cablemodem.com.

DOCSIS Headers

Some parameters in DOCSIS headers convey information on how voice packets use DQoS resources. Consequently, a review of how DOCSIS headers are used is necessary. Also being familiar with how DOCSIS messages are formatted can help you in the capacity planning discussion in Chapter 16. Figure 12-10 illustrates the format of the DOCSIS header and extended headers typically used for PacketCable voice packets.

Figure 12-10 *DOCSIS Headers and Extended Headers*

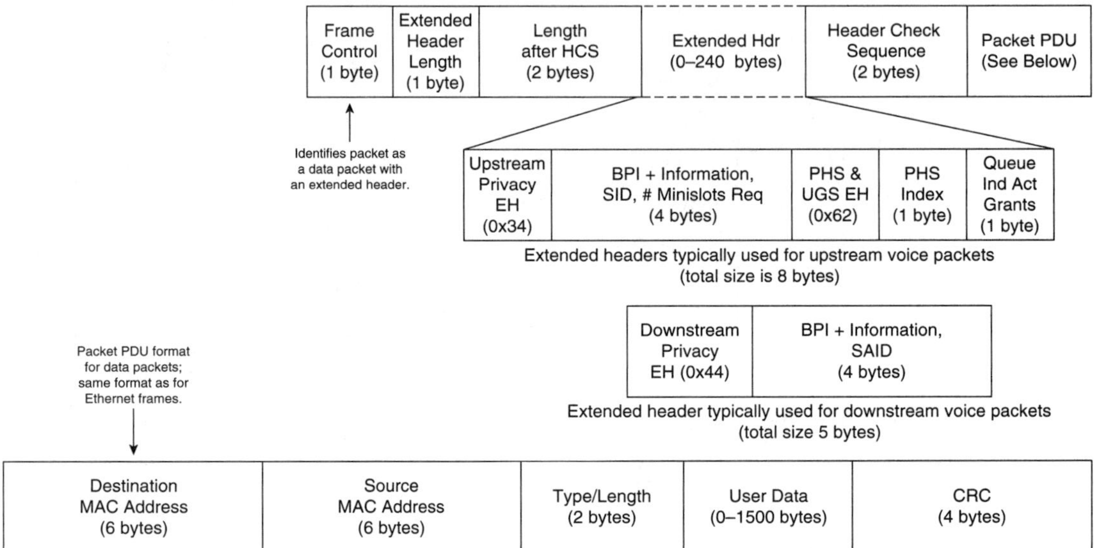

All DOCSIS messages have at least a 6-byte header. This header defines whether the packet is a DOCSIS management message or an actual data packet. This header also contains the length of the packet and a header checksum. Additionally, in the header is a flag that indicates the presence of a DOCSIS extended header. If this flag is set, the length of the extended header is also included in the DOCSIS header. If DOCSIS extended headers are present, they are included between the two bytes in the header used for length and the header checksum. DOCSIS extended headers begin with a 4-bit type field and a 4-bit length field followed by the variable length data.

Some uses of the DOCSIS extended header include piggybacking data grant requests and providing packet fragmentation functionality. In PacketCable, extended headers convey information about Baseline Privacy, Payload Header Suppression, and Unsolicited Grant Service (UGS) grant usage.

The top of Figure 12-10 shows the general format of the DOCSIS header used for IP data packets. Below this are the formats of the extended headers typically used in PacketCable. First the extended header used for upstream (MTA → CMTS) packets is shown, and then the extended header used for downstream (CMTS → MTA) packets. In both cases, a 5-byte Baseline Privacy extended header identifies whether BPI is enabled, the SID or SAID the packet is associated with, and piggyback minislot requests.

NOTE For PacketCable, voice piggybacks are disallowed, so this value is zero. Also SIDs are valid only for upstream packets; downstream packets use a Security Association Identifier (SAID) that is typically the same value as the upstream SID.

Another extended header (EH) used in upstream packets is the 3-byte UGS EH (type 6, length 2). It contains a queue indicator (QI) that lets the CMTS know whether the modem has packets waiting in the queue to be sent and the number of active grants used by the service flow. For PacketCable 1.0 the number of active grants is always one unless silence suppression is used; in this case the number of active grants is equal to zero during periods of silence. In PacketCable 1.5, this parameter can be set to values greater than 1. This can be used to allow multiple calls to share the same set of DQoS resources; the value of this parameter then indicates the number of active calls or sub-flows. If the QI bit is set, this typically indicates the MTA is missing UGS opportunities. This can be because of noise in the DOCSIS plant and/or a malfunctioning CMTS or MTA. If packets are stuck in the queue for too long, jitter problems can result, causing poor voice quality. If the party a PacketCable MTA is talking to complains about hearing garbled voice, look for set QI bits. This extended header also contains the Payload Header Suppression Index (PHSI) if one exists. Payload Header Suppression (PHS) can also be active on a downstream service flow; in this case an additional 2-byte extended header is included (type 5, length 1). PHS is covered later in the chapter.

Dynamic Service Flows

In DOCSIS 1.1 all packets traversing the Hybrid Fiber Coaxial (HFC) network do so via unidirectional service flows; associated with a service flow are parameters describing the traffic. For a typical high-speed data user two Best Effort service flows are defined in the modem's configuration file (one in the upstream direction and one in the downstream direction). All traffic to and from the modem and CPE devices behind the modem use these primary service flows.

For a PacketCable user, voice traffic is separated from the rest of the traffic by using dedicated service flows better suited for voice traffic. There is one upstream service flow and one downstream service flow for voice. These service flows are created when needed by DOCSIS management messages introduced in DOCSIS 1.1. These messages allow for the dynamic creation, modification, and deletion of service flows.

Dynamic service flow creation can be initiated by either the cable modem or the CMTS. For PacketCable DQoS, as described in this chapter, dynamic service flow creation is initiated from the cable modem. The NCS messaging sent from the CMS to the embedded MTA actually triggers the service flow creation by the cable modem, as demonstrated here:

```
MDCX 161096010 aaln/1@x1-6-00-13-10-50-38-02.cc.cisco.com MGCP 1.0 NCS 1.0
I: 575
C: 284
L: dq-gi:1463,  dq-rr:snrcresv          ← DQoS parameters
M: recvonly
K: 161096008

v=0
o=- 161096009 29086 IN IP4 24.34.240.247
s=-
c=IN IP4 24.34.240.247
b=AS:82
t=0 0
m=audio 53456 RTP/AVP 0 8 15
a=X-pc-secret:base64:<hex digit string omitted>
a=X-pc-csuites-rtp:62/51 64/51 60/51 60/50
a=X-pc-csuites-rtcp:81/71 81/70 80/70
a=X-pc-nrekey:0
a=mptime:20 20 20
a=ptime:20
```

This message is Step #7 in the basic call example in Chapter 7. Notice the Local Connection Option parameters dg-gi (DQoS gate identifier) and dq-rr (DQoS resource reservation) in this message. The gate identifier tells the MTA how to correlate the DQoS resources it is trying to reserve to the DQoS resource previously authorized on the CMTS. The resource reservation parameter tells the MTA that it needs to initiate the request for DQoS resource reservation for both the upstream (send) and downstream (receive) directions. Also notice the remote SDP information in this message. The MTA uses this information to identify the amount of resources required and to build the classifiers used to associate IP packets to these resources.

Assuming the CMTS can successfully reserve resources, it assigns a resource identifier and returns this value to the MTA. The MTA sends this value to the CMS in the response to the previous DQoS request. The response for the request in the example is as follows:

```
200 161096010 OK
K:
DQ-RI: 2BF
```

For more information on NCS, please refer to Chapter 7.

The DOCSIS message used to create a dynamic service flow is a Dynamic Service Addition Request (DSA-REQ) message. The CMTS then either accepts or permits this service flow creation and sends the result to the cable modem using the Dynamic Service Addition Response (DSA-RSP) message. The CM acknowledges the receipt of this message with a Dynamic Service Addition Acknowledgement (DSA-ACK) message.

These service flows can also be modified using DOCSIS messaging. Just as with dynamic service flow creation, a 3-way handshake is used to modify a service flow. These messages are the DSC-REQ, DSC-RSP, and DSC-ACK referring to a Dynamic Service Change Request, Response, and Acknowledgement message.

The dynamic service change message exchange is typically used to commit DQoS resources. Again, parameters in the NCS messaging sent from the CMS to the MTA tell it when to initiate the request for resource commitment:

```
MDCX 161096016 aaln/1@x1-6-00-13-10-50-38-02.cc.cisco.com MGCP 1.0 NCS 1.0
I: 575
C: 284
L: dq-gi:1463,  dq-rr:snrccomt,  dq-ri:2BF
M: sendrecv
K: 161096013
```

This is Step #14 from the basic call flow example in Chapter 7. As you can see, the DQoS resource reservation parameter is now set to commit for both the send and receive directions. The DQoS resource identifier (dq-ri) parameter, which identifies the resources previously reserved on the CMTS, is also included.

Logically, these service flows are also deleted using DOCSIS messages. The messages used for this are the DSD-REQ and DSD-RSP referring to a Dynamic Service Deletion Request and Response.

Authorization Block Usage

According to the DOCSIS 1.1 RFI, the Authorization Block parameter contains an authorization "hint" to be passed to the authorization function of the CMTS. The parameter can be in a CM initiated DSA-REQ or DSC-REQ and a CMTS initiated DSA-RSP or DSC-RSP.

PacketCable uses this parameter to transfer DQoS gate and resource information between the CM and CMTS. In a CM-initiated DSA-REQ or DSC-REQ the Authorization Block parameter is used to pass the gate identifier the MTA/CM wants to use to the CMTS. The CMTS then verifies whether the embedded MTA is permitted to use the resources authorized by that gate. In a CMTS-initiated DSA-RSP or DSC-RSP the authorization block parameter is used to provide the embedded MTA with the DQoS resource identifier that references the premium DOCSIS resources to be used by the MTA. This information can then be passed along to the CMS, where decisions can be made on whether these resources can be shared. For example, in a call-waiting scenario resources can be shared among multiple gates, but these resources can be committed for only one of the gates at a time. In this type of call, the resource identifier parameter would also be included in the CM initiated DSA-REQ or DSC-REQ to notify the CMTS it wants to reuse existing resources.

The Authorization Block parameter is encoded in a Type/Length/Value (TLV) format. Type 1 indicates a PacketCable authorization block; its value is further broken down into TLV pairs. Subtype 1 indicates a DQoS gate ID that is 4 bytes long. Subtype 2 indicates a DQoS resource ID that is also 4 bytes long.

In PacketCable 1.5, subtype 3 is added to the specification. This parameter is the DQoS "sub-flow" status and is valid only when the maximum number of UGS grants per interval parameter is greater than one. It is used when multiple gates share the same DOCSIS service flows. It has four possible values: 0 (Admitted), 1 (Active), 2 (Deleted), and 3 (Move).

Figure 12-11 illustrates an example decode of the Authorization Block TLV.

Figure 12-11 *Authorization Block Encoding*

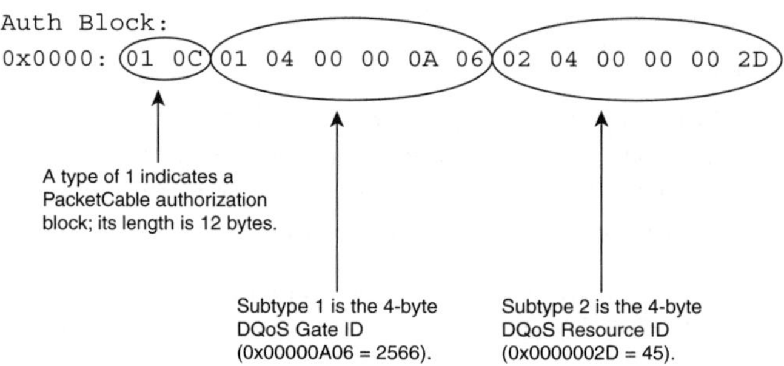

Generic Service Flow Parameters

Some of the service flow parameters are applicable only to an upstream service flow, while some of them are applicable only for a downstream service flow. A few apply to both directions.

NOTE Several of the parameters described in the downstream section can also be applied to an upstream Best Effort service flow. However, only parameters relevant to the service flows used for PacketCable 1.x are covered here.

These service flow parameters are

- **Service Flow Reference**—Used to associate a classifier with a new service flow.
- **Service Flow Identifier**—4-byte identifier used by the CMTS that uniquely identifies a service flow.
- **QoS Parameter Set Type**—This 3-bit field defines whether the specified QoS parameters apply to a provisioned service flow (LSB - bit 0), an admitted service flow (bit 1), or an active service flow (MSB - bit 2). These bits can be set individually or in conjunction. In PacketCable this field is one of three values. If this field is set to 010 (2), the QoS parameters apply to a service flow in the admitted state. This setting corresponds to DQoS resources in the reserved state in the two-phase activation model. If the field is set to 100 (4), the QoS parameters apply to a service flow in the

active state equating to the committed state in the two-phase activation model. If the field were set to 110 (6), the QoS parameters would be both admitted and activated (that is, reserved and committed), indicating a single-phase activation.

- **Timeout for Admitted QoS Parameters**—This field indicates the maximum length of time (in seconds) admitted QoS resources in excess of the active QoS resources can be held. If this timer expires, these unused excess resources are released. In PacketCable, this field cannot be populated by the MTA but is typically set by the CMTS in the DSA-RSP or DSC-RSP. The CMTS sets this timer to the DQoS timer T7 received in the Gate Set message from the CMS. If unspecified, a default value of 0 indicating an infinite timeout is used. Also note that in most situations the admitted (reserved) and active (committed) resource envelopes are the same.
- **Timeout for Active QoS Parameters**—This field indicates the maximum length of time (in seconds) resources on an active service flow can remain unused. If this timer expires, the active and admitted QoS resources are released. In PacketCable, this field cannot be populated by the MTA; it is set by the CMTS in the DSA-RSP or DSC-RSP message. The CMTS sets this timer to the DQoS timer T8 received in the Gate Set message from the CMS. If unspecified, a default value of 0 indicating an infinite timeout is used. Note: This field should be used by the CMTS as it ensures that "hung" resources are eventually recovered.

Upstream Service Flow Parameters

In Chapter 1 you learned that DOCSIS 1.1 provides for service scheduling types better suited for real-time applications that are more sensitive to delay and packet drops than the traditional Best Effort service type. Two of these service types, the Unsolicited Grant Service (UGS) and the Unsolicited Grant Service with Activity Detection (UGS-AD) are especially useful for providing voice services.

A UGS service provides fixed size periodic intervals for a modem to send traffic in the upstream direction. Unlike Best Effort intervals, these intervals are dedicated for a particular user. Also no delay is involved in requesting this bandwidth as the grants are provided without an explicit request from the modem—hence, the grants are unsolicited. By now you are aware that a gateway digitizes voice using a CODEC and a packetization period. This results in the periodic generation of fixed size packets, ideal for the UGS service.

The UGS-AD service is very similar to the UGS service except it is designed for calls using silence suppression or VAD to conserve bandwidth. When the voice communication is active, it is identical to the UGS service; however, when there are pauses or silence in the conversation, you need not waste grants so the service switches to a polling service. This is the DOCSIS Real-Time Polling Service (RTPS), where the modem is given periodic unicast opportunities to request bandwidth. When the CMTS detects UGS grants are not being used, it switches to the RTPS service. When conversation starts again, the modem uses one of these bandwidth requests (by sending a DOCSIS request packet), which signals the CMTS to start providing UGS grants again. Keep in mind that the CMTS is determining when to switch between UGS and RTPS; no DOCSIS DSX messaging is exchanged with the MTA in this case.

The service flow parameters used to define a UGS or UGS-AD service are as follows:

- **Service Identifier**—The 14-bit service identifier is assigned by the CMTS to reference the upstream service flow. It is used in the bandwidth allocation messages to identify the user of a data or request opportunity. It is also used in the DOCSIS extended header to reference the device sending the packet.
- **Service Flow Scheduling Type**—This field identifies the scheduling service for upstream data and request transmissions. This is where the service flow is identified as a UGS (6) or UGS-AD (5) service. Other possible values include: Best Effort (2), Non Real-Time Polling (3), and Real-Time Polling (4).
- **Request/Transmission Policy**—This field is a bit map that controls when and how the service flow can transmit data and data requests. A bit value of "1" indicates true, and a bit value of "0" indicates false. For PacketCable UGS and UGS-AD, service flows bits 0 through 6, and 8 are always set; hence, the value is either 0x000001ff or 0x0000017f depending upon whether or not PHS is allowed. Table 12-6 defines these lower nine bit positions.

Table 12-6 *Request/Transmission Policy Bits*

Bit Position	Value	Value for UGS/UGS-AD
0	Use of broadcast request opportunities disallowed	1 (service flow has dedicated data grants, so no need to make bandwidth requests)
1	Use of priority multicast request opportunities disallowed	1 (service flow has dedicated data grants, so no need to make bandwidth requests)
2	Use of Request/Data opportunities for requests disallowed	1 (service flow has dedicated data grants, so no need to make bandwidth requests)
3	Use of Request/Data opportunities for data disallowed	1 (service flow has dedicated data grants, so no need to use these contention data slots)
4	Piggybacking of requests with data disallowed	1 (service flow has periodic dedicated data grants, so no need to piggyback requests)
5	Concatenation disallowed	1 (concatenation of voice packets not allowed)
6	Fragmentation disallowed	1 (fragmentation of voice packets not allowed)
7	Payload Header Suppression disallowed	If set to 0, payload header suppression can be used to increase data efficiency
8	Packets that do not fit the UGS size are dropped	1 (packets larger than the UGS size are dropped)

- **Nominal Polling Interval**—Interval in microseconds for dedicated request opportunities. Used in the UGS-AD service during silent periods to give the MTA uncontested opportunities to signal the CMTS when voice starts again. For PacketCable this parameter must be an integer multiple of the Nominal Grant Interval.
- **Tolerated Poll Jitter**—Amount of time in microseconds the polling interval is permitted to vary from the Nominal Polling Interval. This parameter is used for the UGS-AD service type.
- **Unsolicited Grant Size**—The size of each digitized voice packet in bytes. This size is from the DOCSIS Header Frame Control byte to the CRC at the end of the packet. In other words, the "Ethernet" portion of the DOCSIS overhead is included as well as the general 6-byte DOCSIS header and any DOCSIS extended headers.
- **Nominal Grant Interval**—Interval in microseconds for dedicated data opportunities. Used by the UGS and UGS-AD services to give the MTA uncontested opportunities to transmit voice packets. This field directly corresponds to the MTA packetization period and can only have a value of 10 ms, 20 ms, or 30 ms.
- **Tolerated Grant Jitter**—Amount of time in microseconds the grant interval is permitted to vary from the Nominal Grant Interval. This parameter defines the permissible jitter for the UGS and UGS-AD service types. In PacketCable this parameter is specified by the CMS; if it is left unspecified, a default value of 800 μs is used.
- **Grants per Interval**—Number of dedicated data grants given to the MTA during each grant interval. For PacketCable 1.0, this value is always equal to "1." In PacketCable 1.5 more flexibility is permitted with this parameter. It can be set to larger values to allow multiple calls to share the same DQoS resources. For example with this set to "2", each call gets 1 grant per interval, creating sub-flows within the main DQoS flow. Obviously, for this to be permitted the characteristics of these sub-flows needs to be identical (i.e., same CODEC, packetization, use of silence suppression, etc.).

Downstream Service Flow Parameters

In addition to the service flow for the premium treatment of voice packets in the upstream, a service flow for voice packets is also created for the downstream. Remember, all downstream traffic is Best Effort. However, you have ways to prioritize voice traffic over ordinary data traffic and to reserve bandwidth for this traffic, as you will see. The key parameters used to define the downstream service flow are as follows:

- **Traffic Priority**—Defines the service flow priority. If multiple service flows exist that are identical in all QoS parameters besides priority, the higher priority service flow is given preference. If not included, the default priority of "0" is used. For PacketCable 1.x downstream flows, this is set to the value "5."

- **Maximum Sustained Traffic Rate**—The token bucket (R) parameter for rate limiting packets expressed in bits per second. The packet size is from after the DOCSIS Header Check Sequence to the CRC at the end of the packet. In other words, the "Ethernet" portion of the DOCSIS overhead is included, but the general 6-byte DOCSIS header and any DOCSIS extended headers are not.
- **Maximum Traffic Burst**—The token bucket size (B) parameter for rate limiting packets expressed in bytes. Again, this parameter accounts for the total packet size in the same way as the Maximum Sustained Traffic Rate parameter. The minimum value is 1,522 bytes, and if unspecified, a default value of 3,044 bytes is implied. This parameter is meaningful only if the Maximum Sustained Traffic Rate parameter is non-zero.
- **Minimum Reserved Traffic Rate**—The minimum rate in bits per second reserved for the service flow. Packet size is calculated the same as in the previous parameters.
- **Minimum Reserved Rate Packet Size**—The assumed minimum packet size in bytes for which the Minimum Reserved Traffic Rate is applied. Packet size is calculated the same as in the previous parameters.

Classifying Packets

Packets are assigned to service flows by matching packet classifiers. Some of the things a packet classifier can match packets upon include source and/or destination IP addresses, source and/or destination UDP/TCP port numbers, IP protocol type, and IP TOS bits. You have upstream packet classifiers used to assign packets to upstream service flows and downstream packet classifiers used to assign packets to downstream service flows. A cable modem can be associated with multiple upstream and downstream classifiers; in this case a classifier priority is used to determine the order in which classifiers are applied.

NOTE If a packet fails to match any classifier, it is sent out the primary service flow. Remember, the primary service flows are the first upstream and downstream service flows defined in the cable modem configuration file. Also keep in mind these classifiers can change during the life of a call, for example, with a call transfer scenario.

Some fields associated with DOCSIS classifiers used in PacketCable are

- **Classifier Reference/Classifier Identifier**—If this is the first time the classifier is introduced in the call, the Classifier Reference field is used and populated with a unique value by the MTA. If this classifier already exists, the Classifier Identifier field is used, populated with the value assigned by the CMTS.

- **Service Flow Reference/Service Flow Identifier**—One of these fields is used to correlate the Classifier with its corresponding service flow. If the service flow already exists, the Service Flow Identifier is used; otherwise, the Service Flow Reference is used.
- **Rule Priority**—This field is used to identify the order classifiers are applied. Higher priority classifiers are applied first. For PacketCable voice service flows, you have only one classifier for each service flow, so this field really doesn't matter. In this case, the priority is always the default of 128.
- **Classifier Activation State**—This field determines whether or not this classifier should be used for selecting packets. A value of inactive (0) is typically used for admitted service flows. Whether or not a classifier is used also depends on the state of the service flow; an active classifier still is not used if its associated service flow is not active.
- **Dynamic Service Change Action**—Action to be taken with this classifier. Values are 0 - Add Classifier, 1 - Replace Classifier, and 2 - Delete Classifier. This parameter is used in PacketCable 1.5 for the addition and deletion of sub-flows when multiple calls share the same DQoS resources.

Payload Header Suppression

In Chapter 11, you learned that voice packets are encapsulated using the RTP protocol. These RTP packets are encapsulated in UDP packets, which are encapsulated in IP packets, which are encapsulated in DOCSIS frames. Figure 12-12 shows the packet format.

Figure 12-12 *Voice Packet Header Format*

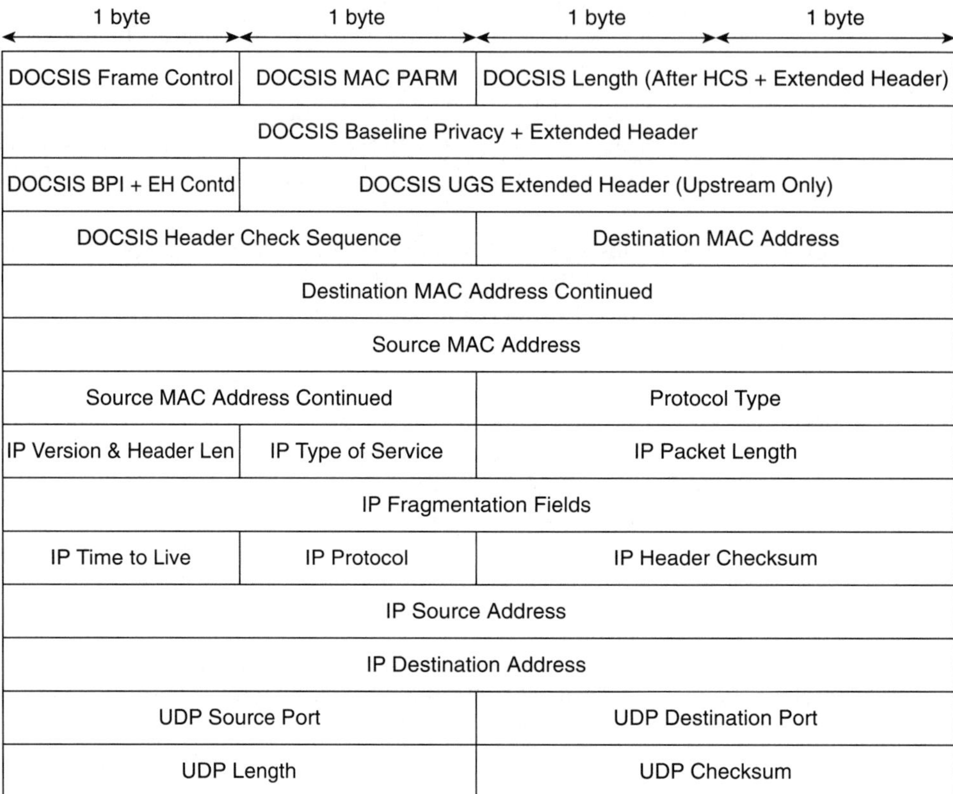

This is 56 bytes of overhead before getting to the RTP portion of the packet! Actually, the RTP protocol adds another 12 bytes of overhead before getting to the payload. Are all of these fields really needed in every packet? A closer look at these fields reveals that several are not used and several have the same value throughout the life of the call.

Payload Header Suppression (PHS) functionality eliminates the need to send these repetitive fields. Instead, a rule is created where the rule index represents these repetitive fields; thus, only the rule needs to be transmitted in these packets.

A PHS Rule accomplishes this using a number of parameters including

- **PHS Field (PHSF)**—String of bytes representing the header portion of a packet in which one or more bytes are suppressed. This is the uncompressed header that is kept in memory by the receiving end and then reinserted back into the packet before forwarding. For upstream packets the PHSF begins with the Destination MAC address. For downstream packets the MAC addresses cannot be suppressed because downstream packets are sent to all devices, and the MAC addresses are needed to determine the packet destination. Therefore, the PHSF begins with the protocol type field.

- **PHS Index (PHSI)**—A byte value used in the DOCSIS extended header to reference a PHS Rule.
- **PHS Mask (PHSM)**—A bit map that specifies which bytes in the PHSF are to be suppressed. Note: One bit refers to one byte in the PHSF. A bit value of "1" means to suppress the corresponding byte, whereas a bit value of "0" means the corresponding byte is not suppressed.
- **PHS Size (PHSS)**—Length of the PHSF in bytes and also the number of bits in the PHSM.
- **PHS Verify (PHSV)**—A flag that determines whether the sending entity should verify that the uncompressed header bytes actually match the PHSF bytes before suppressing them.

As with most things, the easiest way to understand how PHS works is through an example. Figure 12-13 shows an Ethereal capture of a DSC-REQ that adds PHS rules to both the upstream and downstream voice service flows.

Figure 12-13 *Payload Header Suppression Example*

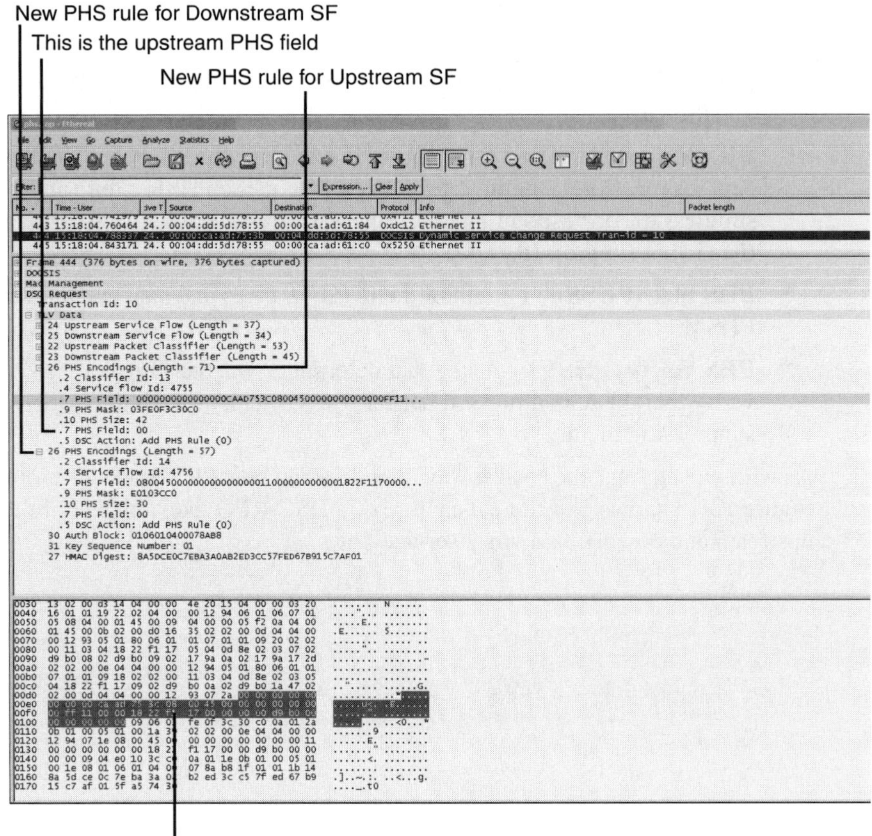

The first set of PHS encodings is for voice upstream service flow 4755 using classifier 13. Because this is an upstream service flow the PHSF is 42-bytes long, note this matches the PHSS. Remember the PHSF for an upstream service flow begins with the Destination MAC address and continues to the end of the UDP header. The PHS Mask field, 0x03FE0F3C30C0, indicates which of these bytes are to be suppressed. Notice, the PHSM is 6 bytes or 48 bits long. Each bit corresponds to 1 byte in the PHSF, which is 42 bytes long, so the last 6 bits are just zero padding. The second PHSF field shown in the Figure 12-13 output is a bug; this field is actually the PHSV field. A value of "0" means that the sending entity must verify that the values of the fields to be suppressed in the actual packets match the corresponding values in the PHSF field. If this verification fails, then no suppression occurs.

The impact of PHS on an upstream packet is illustrated in Figure 12-14. The fields that are shaded are the ones that are going to be suppressed. To give you a little help, look at the first 12 bits of the PHSM (0x03F = 0000 0011 1111). The first 6 bits of 0 indicate the first 6 bytes of the PHSF (Destination MAC Address) are not to be suppressed. The second 6 bits of 1 indicate the next 6 bytes of the PHSF (Source MAC Address) are to be suppressed.

Figure 12-14 *Resulting Packet Header after PHS*

<table>
<tr><td>1 byte</td><td>1 byte</td><td>1 byte</td><td>1 byte</td></tr>
<tr><td>DOCSIS Frame Control</td><td>DOCSIS MAC PARM</td><td colspan="2">DOCSIS Length (After HCS + Extended Header)</td></tr>
<tr><td colspan="4">DOCSIS Baseline Privacy + Extended Header</td></tr>
<tr><td>DOCSIS BPI+ EH Contd</td><td colspan="3">DOCSIS UGS Extended Header</td></tr>
<tr><td colspan="2">DOCSIS Header Check Sequence</td><td colspan="2">CMTS MAC Address (0x00 00)</td></tr>
<tr><td colspan="4">CMTS MAC Address Continued (0x00 00 00 00)</td></tr>
<tr><td colspan="4">MTA MAC Address (0x00 00 ca ad)</td></tr>
<tr><td colspan="2">MTA MAC Address Continued (0x75 3c)</td><td colspan="2">Protocol Type (0x0800 for IP)</td></tr>
<tr><td>0x45</td><td>IP ToS (0x00)</td><td colspan="2">IP Packet Length (0x00 00)</td></tr>
<tr><td colspan="2">IP Identification (0x00 00)</td><td colspan="2">IP Fragmentation Flags & Offset (0x00 00)</td></tr>
<tr><td>IP Time to Live (0xff)</td><td>IP Protocol (0x11 for UDP)</td><td colspan="2">IP Header Checksum (0x00 00)</td></tr>
<tr><td colspan="4">MTA IP Address (0x18 22 f1 17)</td></tr>
<tr><td colspan="4">Remote Gateway IP Address (0x00 00 00 00)</td></tr>
<tr><td colspan="2">UDP/RTP Source Port (0xd9 b0)</td><td colspan="2">UDP/RTP Destination Port (0x00 00)</td></tr>
<tr><td colspan="2">UDP Length (0x00 00)</td><td colspan="2">UDP Checksum (0x00 00)</td></tr>
</table>

Mapping DOCSIS Service Flows into RSVP Flow Specifications

When the CMTS receives a DSA-REQ or DSC-REQ from an embedded MTA with service flow parameters for an admitted QoS parameter set, it compares the received DOCSIS parameters against the RSVP flow specifications that defined the authorized resource envelope. Remember, the reserved resource envelope must be less than or equal to the authorized resource envelope, or else the request is denied. The gate ID in the authorization block parameter in the DSX-REQ enables the CMTS to correlate DOCSIS requests to gates. To make the comparison the CMTS needs to convert the Layer 2 DOCSIS parameters into the equivalent Layer 3 RSVP flow specification parameters.

Figure 12-15 illustrates the mapping logic for an upstream service flow.

Figure 12-15 *DOCSIS Upstream Service Flows to RSVP Mapping*

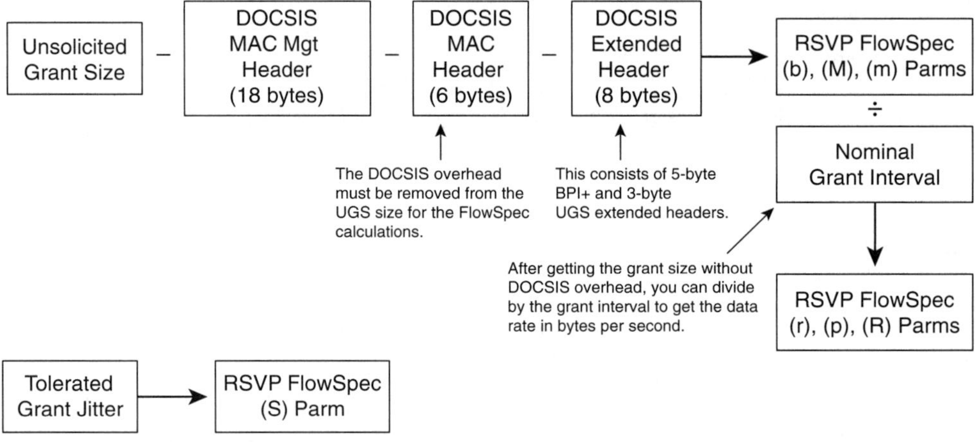

The following example further illustrates this conversion. On the Cisco CMTS a couple of debugs were enabled that show the DOCSIS messages and the resulting RSVP flow specifications.

The output from these debugs has been modified just to illustrate the pertinent parts and is shown in Figure 12-16. Debugging DQoS on the Cisco CMTS is covered in detail in Chapter 13.

Figure 12-16 *Example Upstream Mapping*

```
Found Upstream Service Flow TLV
        Service Flow Reference : 1
        QoS Parameter Set Type : 0x2 <---- Service Flow Is in the Admitted Phase
        Scheduling Type : 6 <---- Unsolicited Grant Service
        Request/Transmission Policy : 0x17F
        Unsolicited Grant Size : 232 <---- Subtracting the 32 bytes of
        Nominal Grant Interval : 20000     DOCSIS header yields
        Tolerated Grant Jitter : 800
        Grants Per Interval : 1

PktCbl(r2d): REQ - r/b/p/m/M 1176256512 1128792064 1176256512 200 200 (R/S)
1176256512 (800) <---- Grant Jitter
```

10 KBps in IEEE Floating-Point Format

This upstream service flow is from a DSA-REQ message. The QoS Parameter Set is 0x2, which means the QoS resources are to be placed in the admitted state or reserved state. Scheduling Type 6 is the UGS service. The Request Transmission Policy sets all the bits except for bit 7. This bit allows for Payload Header Suppression; the other bits do things such as disallow the use of contention request and data opportunities, disallow concatenation, and disallow fragmentation for this service flow. The UGS size is 232 bytes, the expected value for the G.711 CODEC with a 20 ms packetization.

So, if you subtract the 32 bytes of DOCSIS overhead from the UGS size, you are left with 200 bytes. You can see from the RSVP flow specification, the (m) and (M) parameters have values of 200. The (b) parameter as well as the (r), (p), and (R) parameters are expressed in single precision IEEE floating-point format, which is a 32-bit field.

NOTE The format of IEEE floating-point numbers is defined in the IEEE-754 specification. In summary, the 32-bit field is broken up into three sub-fields: a Sign field, an Exponent field, and a Significand field. The Sign field is the most significant bit (Bit 31) and is equal to zero (positive) for all the flow specification parameters. The next 8 bits comprise the Exponent field. If you convert this 8-bit field to decimal and subtract by 127, you get the value of the exponent. The remaining bits comprise the Significand. These bits represent fractions of base 2. The most significant bit (MSB) represents the value 1/2, the next MSB represents the value 1/4, the next the value 1/8, the next 1/16, etc. To compute the Significand, add up the values where the bits are set (equal to 1). To get the decimal value of the floating-point number multiply the Significand value plus 1 by the number 2 raised to the exponent value. Figure 12-17 shows some examples. If you do a search on the Internet, you can find several utilities to do this conversion.

Figure 12-17 *Sample Floating-Point-to-Decimal Conversions*

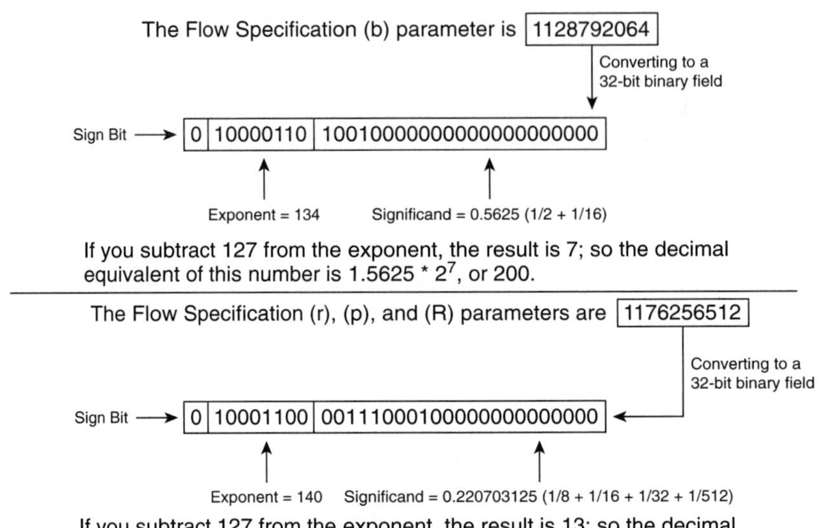

If you now return to the example, you would see that if you were to do the conversion, the (b) value is also 200 bytes. Taking this value and dividing by the grant interval, which is 20,000 microseconds or 20 milliseconds, yields the data rate. This rate computes to be 10,000 bits per second. Converting this to IEEE floating-point notation yields the value 1,176,256,512 shown in the debug. The (S) parameter is taken from the value of the Tolerated Grant Jitter and is 800 microseconds.

Figure 12-18 illustrates the logic for mapping the downstream service flow parameters.

In Figure 12-19, an example is used to help you understand the conversion. This example is from the same call as the previous example, and information from the same debugs is collected.

Figure 12-18 *DOCSIS Downstream Service Flows-to-RSVP Mapping*

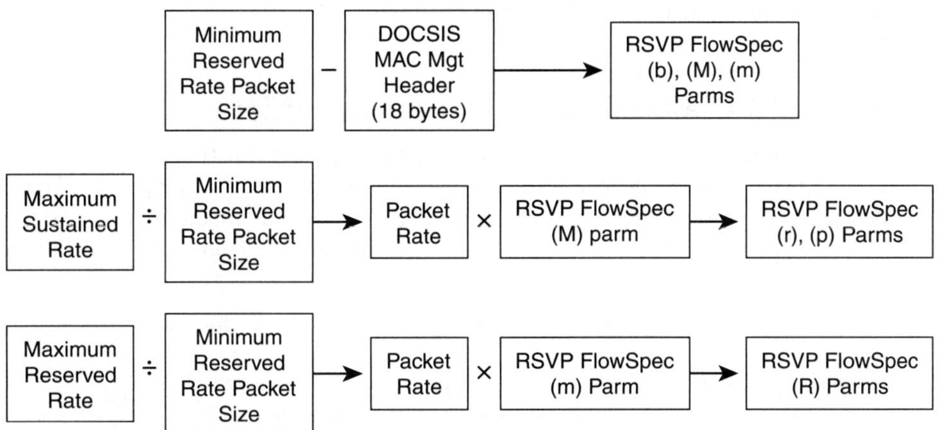

Figure 12-19 *Sample Downstream Mapping*

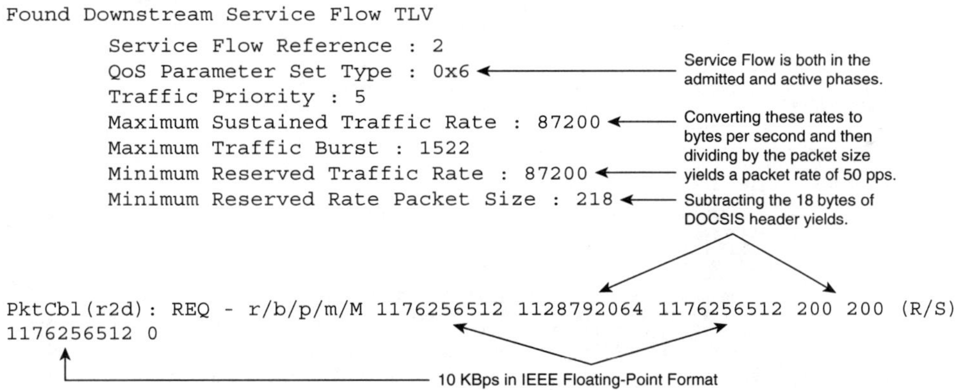

This time, the QoS parameter set is 0x6, meaning QoS resources are to be admitted and activated. The traffic priority is "5," making it a higher priority than any service flows left at the default of "0". The Minimum Reserved Rate Packet Size is 218 bytes. Subtracting the 18 bytes of DOCSIS MAC Management header yields 200 bytes, which is the value of the (b), (m), and (M) RSVP flow specification parameters. Remember, the (b) parameter is expressed in 32-bit IEEE floating-point format. The Maximum Sustained Traffic Rate is 87,200 bits per second, dividing that by 8 yields 10,900 bytes per second. If you then divide that by the Minimum Reserved Rate Packet Size, you get a packet rate of 50 packets per second. Taking this value and then multiplying by the previously calculated packet size

(200 bytes) yields a rate of 10,000 bytes per second; putting this value in 32-bit IEEE floating-point format results in the values shown in the debug.

Mapping DOCSIS Classifiers into Gate Classifiers

The mapping of DOCSIS classifiers into gate classifiers is fairly straightforward. The IP protocol type and source and destination IP address are all direct mappings. The DOCSIS destination port start and end should be equal, and that is mapped to the gate destination port. The gate source port is often "wildcarded" or removed from the classification process by setting it to a value of zero.

Figure 12-20 shows the classifier mappings from the same example using the same debugs.

Figure 12-20 *Mapping DOCSIS Classifiers to Gate Classifiers*

```
Found Upstream Packet Classifier TLV
        Classifier Reference : 1
        Service-Flow Reference : 1
        Rule Priority : 128
        Activation State : 0
        Found IP Packet Classifier Sub-TLV
                Protocol : 17
                Source Address : 24.37.241.2
                Destination Address : 13.3.22.92
                Source Port Start : 1086
                Source Port End : 1086
                Destination Port Start : 49294
                Destination Port End : 49294

PktCbl(r2d): cfr: proto=17 src=24.37.241.2 dst=13.3.22.92 sport=1086 dport=49294
PktCbl(r2d): flowspec cfr: proto=17 src=24.37.241.2 dst=13.3.22.92 sport=0
dport=49294


Found Downstream Packet Classifier TLV
        Classifier Reference : 2
        Service-Flow Reference : 2
        Rule Priority : 128
        Activation State : 1
        Found IP Packet Classifier Sub-TLV
                Protocol : 17
                Source Address : 13.3.22.92
                Destination Address : 24.37.241.2
                Destination Port Start : 1086
                Destination Port End : 1086

PktCbl(r2d): cfr: proto=17 src=13.3.22.92 dst=24.37.241.2 sport=0 dport=1086
PktCbl(r2d): flowspec cfr: proto=17 src=13.3.22.92 dst=24.37.241.2 sport=0
dport=1086
```

As you can see, the IP packet classifier mappings are straightforward. Notice the use of the "activation state" parameter; it has a value of 0 (inactive) for the upstream classifier because this classifier is not yet used because its associated service flow is in the admitted state. The service flow associated with the downstream is admitted and activated at once so this classifier is put into the active state (value is 1) .

Revisiting CMTS Authorization

So after a reserved QoS parameter envelope (indicated by an admitted service flow in a DSA-REQ) or a committed QoS parameter envelope (indicated by an active service flow in a DSC-REQ or DSA-REQ) are converted into RSVP flow specifications, the authorization process can occur.

If the reserved envelope is less than or equal to the authorized envelope and those DOCSIS resources are currently available, the CMTS accepts the DSA-REQ by indicating a positive confirmation code in the DSA-RSP sent back to the MTA. Also included in the DSA-RSP are Service Flow Identifiers assigned to these new service flows. A Service Identifier is also assigned to the upstream service flow. Values for the Timeout for Active parameters and Timeout for Admitted parameter timers can also be included. In this case, only the Timeout for Admitted parameter makes sense at this point and indicates how long the CMTS waits for the CM to activate the resources that have been reserved. The CMTS also assigns Classifier Identifiers to both classifiers. Finally, in the authorization block parameter the DQoS resource identifier, which can be used by the CMS to manage HFC resources and possibly reuse resources, is returned.

Likewise, if the committed envelope is less than or equal to the reserved envelope, the CMTS activates those DOCSIS resources. Notice, the CMTS does not need to check to see if these resources are available as this is assured because of the previous reservation process. The Timeout for Active parameter is included in the response message in this case indicating how long the CMTS allows the service flows to remain idle.

Chapter Summary

PacketCable DQoS enables high-voice quality across the DOCSIS network between a MTA and a CMTS. The embedded MTA, CMTS, CMS, and RKS components all contribute to providing dynamic quality of service. The NCS protocol is used to communicate QoS information between the eMTA and CMS, the COPS protocol is used to communicate QoS information between the CMS and CMTS, and the DOCSIS MAC management protocol is used to communicate QoS information between the eMTA and CMTS.

The fundamental component of PacketCable DQoS is a gate, which is the key to premium DOCSIS resources. A gate contains a policing component for managing resources and a classification component for identifying packets. Other information such as accounting components for tracking and billing resources can also be associated with a gate. PacketCable gates transition through four states: allocation, authorization, reservation, and commitment. The allocation and authorization states occur via COPS-based messaging between the CMS and CMTS. The reservation and commitment states occur via DOCSIS messaging between the embedded MTA and CMTS. These transitions are triggered by NCS call signaling parameters sent from the CMS. Having separate reservation and commit stages provides two-phase activation. This ensures resources are available when needed and that resources are not wasted when not needed.

The CMS communicates DQoS information with the CMTS using the COPS protocol. Additional COPS objects are defined for PacketCable usage. Among these objects are parameters that define gate messages, define gate specifications, identify subscribers, provide billing information, and provide wiretapping information. The COPS protocol uses TCP as its transport layer protocol. DQoS timers minimize the chance resources are wasted or possibly stuck.

DQoS resources are characterized using RSVP flow specifications. The authorized resource envelope is determined by the CMS by interpreting a call's SDP information. Reserved and committed resource envelopes are determined by mapping DOCSIS service flow parameters into RSVP flow specs. The committed resources are always less than or equal to the reserved resources, which in turn are always less than or equal to the authorized resources.

DOCSIS 1.1 provides QoS mechanisms used by PacketCable. Among these mechanisms are dynamic service flows, premium categories of service flows, classifiers, and Payload Header Suppression. The Unsolicited Grant Service (UGS) is the premium type of service flow used by PacketCable. Payload Header Suppression increases DOCSIS bandwidth efficiency by eliminating the repetitive transmission of useless and redundant fields in packet headers.

Chapter Review

1 List the PacketCable components that contribute to providing PacketCable DQoS:

Answer: MTA, CM, CMTS, CMS, RKS

2 What are the four states a gate can pass through? Identify the components involved in putting the gate into each state.

Answer: Allocated (CMS and CMTS), Authorized (CMS and CMTS), Reserved (CMS, eMTA and CMTS), Committed (CMS, eMTA and CMTS)

3 Why is a two-phase activation model important?

Answer: It ensures that resources aren't committed until needed and ensures that resources are available when needed.

4 How is an authorized DQoS resource envelope determined?

Answer: By the SDP information included in the NCS call setup signaling, which is translated into RSVP FlowSpec parameters sent in the COPS signaling to the CMTS

5 How is a reserved DQoS resource envelope determined?

Answer: By DOCSIS service flow parameters

6 How is a committed DQoS resource envelope determined?

Answer: By DOCSIS service flow parameters

7 What DOCSIS extended headers are used in PacketCable?

Answer: BPI+, UGS information, Payload Header Suppression

8 How are the DOCSIS Admitted QoS timeout and Active QoS timeout parameters determined?

Answer: They are set to the DQoS timers T7 and T8, which are set by the CMS in a GATE-SET message.

9 What are the two conditions that must be true for a DOCSIS classifier to be used?

Answer: Classifier state must be active, and associated service flow must be active.

10 What is the expected UGS size if the G.711 codec and a 10-ms packetization period are being used?

Answer: 152 bytes (80 bytes RTP payload, 40 bytes RTP/UDP/IP header, 32 bytes DOCSIS header)

11 What DOCSIS parameter is used to pass DQoS gate information between the eMTA and CMTS?

Answer: Authorization Block

12 What field of a PacketCable COPS gate specification can be used to identify an emergency 911 call?

Answer: Session class

13 How does the CMTS figure out the timer used to send the COPS keepalive message?

Answer: The CMS sends this as a parameter in the COPS Client-Accept (CAT) message during initialization.

14 What is the QI bit in the UGS extended header used for?

Answer: It is used by the MTA to let the CMTS know it has a backlog of voice packets in its transmit queue and needs extra grants.

15 Match the DQoS timers (T0, T1, T5, T7, T8) to their definitions:

A: Time a gate can remain in the authorized state before moving to the reserved state

B: Time a gate can remain in the allocated state before moving to the authorized state

C: Time a gate can remain in the authorized state before moving to the committed state

D: Time between a NCS DLCX is sent to a MTA and the resulting GATE-CLOSE message is received from the CMTS

E: Time committed resources can remain unused before they are released

F: Time reserved resources can remain so before they are released

Answer: T0 - B, T1 - C, T5 - D, T7- F, T8 - E

16 True or False: QoS for signaling packets as well as bearer packets is covered by PacketCable DQoS.

Answer: False. PacketCable DQoS covers only the media stream.

17 True or False: Although a gate ID often refers to two service flows, a DQoS gate is unidirectional.

Answer: True

18 True or False: DOCSIS Packet concatenation and fragmentation are permitted options for voice packets.

Answer: False

19 True or False: The DOCSIS admitted and active states correspond to the DQoS reserved and committed states.

Answer: True

20 True or False: The PHSF field for an upstream packet and a downstream packet are composed of the same header fields.

Answer: False. Upstream PHSF field begins with the Destination MAC address, and downstream PHSF field begins with the Ethernet Type field.

21 True or False: The CMTS can also be referred to as a COPS Policy Decision Point (PDP).

Answer: False. The CMTS is the COPS Policy Enforcement Point (PEP); the CMS is the PDP.

22 True or False: GATE-OPEN and GATE-CLOSE commands are sent from the CMTS to the CMS to ensure proper gate coordination.

Answer: True

23 True or False: PacketCable COPS connections are initiated by the CMTS.

Answer: True

CHAPTER 13

Analyzing, Implementing, and Troubleshooting DQoS

The previous chapter introduced you to the concepts and technologies used in guaranteeing premium quality of service (QoS) for PacketCable subscribers. You learned what a DQoS gate is and its possible states. You learned how the Common Open Policy Service (COPS) protocol is used with respect to distributed quality of service (DQoS). You also learned about the Data-Over-Cable Service Interface Specification (DOCSIS) QoS mechanisms used in PacketCable. This chapter builds upon this knowledge by illustrating how these concepts are applied in the context of actual calls. Chapter 7, "NCS (Network-based Call Signaling)," examines the manner in which the Network-based Call Signaling (NCS) protocol provides call feature functionality. This chapter provides a closer look at how the DQoS protocols interact with NCS. The chapter also examines DQoS configuration on the Cisco cable modem termination system (CMTS) and call management server (CMS). Finally, the chapter provides guidance on troubleshooting both DQoS and the Cisco components.

Protocol Message Flows and Operation

Chapter 7, "NCS (Network-based Call Signaling)," examined call flows with a specific focus on NCS messages and parameters. The call flows in this chapter build on those by illustrating how PacketCable DQoS functionality is incorporated in NCS. Because NCS was covered previously, it's not examined in as much detail in this chapter; instead, DQoS is the focus. Because the DQoS covered thus far addresses only the DOCSIS access portion of the PacketCable network, only ONNET-to-ONNET calls between MTAs are examined.

If you are interested in calls where either the originating or terminating party resides in the public switched telephone network (PSTN), refer to Chapter 8, "TGCP and the PSTN Interconnect," for the messaging for that side of the call. The messaging for the ONNET portion of the call is, for the most part, similar to what is described in this chapter. One possible exception occurs when a PacketCable subscriber calls a user in the PSTN. In Chapter 8 you learned that ring back tone is often supplied in the voice path from the PSTN in this case. Consequently, DQoS resources for the direction from the PSTN interconnect to the multimedia terminal adapter (MTA) are typically committed at the ringing phase instead of by waiting for the call to be answered. This is so the ring back tone can use DQoS resources. For ONNET-to-ONNET calls, ring back tone is typically supplied locally by the MTA, so you don't need to commit resources during the ringing phase.

Some points to keep in mind about these example call flows are as follows:

- These traces are for calls between MTAs controlled by the same call agent or CMS. Calls between endpoints controlled by different CMSs require the use of the CMS signaling protocol (Call Management Server Signaling [CMSS]) covered in Chapter 9, "Call Management Server Signaling Protocol (CMSS)." As far as DQoS is concerned, this really doesn't matter because it applies only to the Hybrid Fiber Coaxial (HFC) access network.
- All the MTAs involved are connected to the same CMTS; as far as DQoS is concerned, it doesn't matter whether the MTAs are on the same CMTS or on different ones.
- The PacketCable specifications do not dictate how messages are used to formulate calls; this is up to the vendors of the various components. As long as the specifications are not violated, you might have many ways to accomplish the same task. All the examples in this text utilize the Cisco BTS 10200 (running version 4.5) as the softswitch or CMS and either the Cisco uBR7246vxr or uBR10k (running Cisco IOS Software Release 12.3[17a]BC2) as the CMTS.

Basic ONNET Call

The first message flow covered should be somewhat familiar to you; this is the same call between Oscar (the originating party) and Tina (the terminating party) first described in Chapter 7. They are both PacketCable customers, and their MTAs are referred to as MTA_o and MTA_t. Just as in Chapter 7, the examination of their phone call is broken up into three stages:

- Stage 1 details the messaging used from when Oscar picks up his phone until the point where Tina's phone is ringing.
- Stage 2 shows the messaging from there up to the point where Oscar and Tina have bidirectional voice communication.
- Stage 3 shows the messaging resulting from Oscar and Tina hanging up their phones.

This example uses Ethereal traces to illustrate the messages and their parameters. These traces were obtained from two places:

- Using an RF sniffer on the HFC network to capture the DOCSIS management messages, NCS messages, and the relationship between them. Remember, DOCSIS messages can also be captured on a Cisco CMTS using the cable monitor utility described in Appendix B, "Cable Monitor and Ethereal."
- Off of the wide area network (WAN) between the CMTS and the CMS. This is needed so that the COPS messages can be captured relative to the NCS messages.

Basic ONNET Call: Stage 1

Figure 13-1 illustrates the DQoS and NCS signaling used for Stage 1.

Figure 13-1 *Basic ONNET Call Flow (Stage 1)*

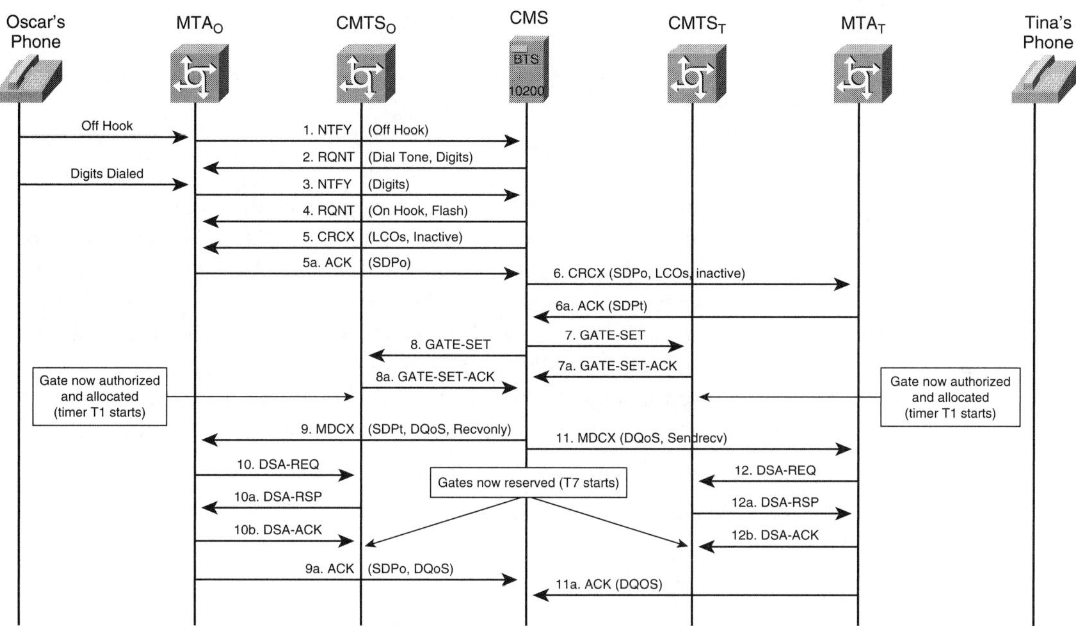

Notice a number identifies each command and its associated responses. Messages between an MTA and the CMS are NCS, messages between a CMTS and the CMS are COPS, and messages between an MTA and a CMTS are DOCSIS. Some of the NCS responses are left out because they were covered in detail in Chapter 7. The following list examines this call flow one message at a time:

1 Oscar picks up his phone, triggering an NTFY message to be sent from his MTA to the CMS. The Observed Event of hook detection (O: hd) is included in this message.

2 The CMS requests the MTA to provide Oscar with dial tone and to be notified as soon as digits are dialed or if Oscar hangs up his phone. Refer to Chapter 7 for more detail on the format of this or any other NCS message.

3 Oscar dials Tina's number, triggering another NTFY message to be sent to the CMS.

4 The CMS issues a new list of requested events consisting of only the on hook and flash hook events.

5 Now that the CMS has all the digits it needs to route the call, a connection is initiated on Oscar's MTA between his phone and the VoIP PacketCable network. This CRCX is as follows:

```
CRCX 107374314 aaln/1@x1-6-00-13-10-50-38-02.cc.cisco.com MGCP 1.0
 NCS 1.0
C: 21
L: mp:20;20;20, a:PCMU;PCMA;G728, e:off, s:off, t:a0, sc-rtp:60/50,
 sc-rtcp:80/70
M: inactive
K: 107374313
```

The connection is put in the inactive state, meaning the media stream is not passed between the analog endpoint and the IP network. The CMS also includes the Local Connection Parameter option to influence how Oscar's MTA handles the connection. Among these parameters are the a: and mp: parameters, identifying which CODECs and packetization periods can be used for the call. The security parameters are included to explicitly tell the MTA to use only the NULL encryption and authentication algorithms for RTP and RTCP security. Without these parameters, the MTA can request more bandwidth to account for bearer security overhead.

The resulting response from Oscar's MTA is as follows:

```
200 107374314 OK
I: 40EA

v=0
o=- 107374314 16618 IN IP4 24.34.240.235
s=-
c=IN IP4 24.34.240.235
b=AS:80  <- Bandwidth required for multimedia session in kbps
m=audio 53456 RTP/AVP 0 8 15  <- Prioritized CODEC list
a=X-pc-csuites-rtp:60/50
a=X-pc-csuites-rtcp:80/70
a=X-pc-nrekey:0
a=mptime:20 20 20 <- Packetization periods for use with each CODEC
```

In the response to this message is Session Description Protocol (SDP) information, detailing the IP address and Real-time Transport Protocol (RTP) port number used to reach this endpoint as well as its supported CODECs and packetization periods. Also notice the inclusion of the SDP bandwidth parameter (b). In Chapter 12, "DQoS Architecture and Framework," you learned that a G.711 call using a 20-ms packetization period requires 80 kbps of bandwidth (from a Layer 3 perspective).

6 The CMS then requests the creation of a connection on Tina's MTA and provides information on how to reach Oscar's MTA. Parameters in this CRCX are very similar to the parameters in Oscar's CRCX. Oscar's SDP information is also included so that Tina's MTA can know how and where to forward voice packets. The response to this CRCX is as follows:

```
200 107374315 OK
I: 4FF2

v=0
o=- 107374315 20466 IN IP4 24.34.240.247
s=-
c=IN IP4 24.34.240.247
b=AS:80  <- Bandwidth required for multimedia session in kbps
t=0 0
m=audio 53456 RTP/AVP 0 8 15  <- Prioritized CODEC list
a=X-pc-csuites-rtp:60/50
a=X-pc-csuites-rtcp:80/70
a=X-pc-nrekey:0
a=mptime:20 20 20  <- Packetization periods for use with each CODEC
```

In the response to this CRCX are the IP address and RTP port number used to reach Tina's MTA endpoint. Also in the response are the supported CODECs and packetization periods that could be used for this call. Tina's MTA uses the intersection of the values designated in the Local Connection Options (LCOs) with the values Oscar's MTA supports (identified in his SDP information).

7 Up to now, these steps have been purely NCS and identical to those in Chapter 7. However, it is at this point that DQoS functionality begins. Because the CMS now knows the possible CODECs and packetization periods that can be used for the call, it knows how much DOCSIS bandwidth is required. In the SDP response in the previous step, notice the three possible CODEC/packetization period parameter pairs for the connection. Each combination has different bandwidth requirements, and any one of these could conceivably be used for the connection. So how does the CMS know how much DOCSIS bandwidth is required? The answer is, it doesn't. But if resources for the most bandwidth-intensive option were authorized, then any of the less stringent options would work as well. Thus, the CMS uses the Least Upper Bound (LUB) resource requirement when setting up DQoS resources on the CMTS.

   Remember that in a majority of today's deployments, only the noncompression CODEC G.711 is used; therefore, only a single CODEC/packetization pair is set in the LCO, eliminating the need of the CMS to compute LUB resource requirements. The resource requirements in this case are just the requirements of that single CODEC/packetization pair.

Because the G.711 CODEC does not use compression, it usually requires the most bandwidth and, hence, comprises the LUB anyway. Thus, whether only this single CODEC were specified or a CODEC list including this CODEC, the resource requirements of the G.711/20 ms pair are used by the CMS to create gates on the CMTS in either case.

The first gates are created for Tina's MTA. Ethereal traces of this Gate-Set message are shown in Figures 13-2 through 13-4. Another benefit of using Ethereal is that it fully decodes the PacketCable COPS objects. Figure 13-2 depicts much of the Gate-Set message, while Figures 13-3 and 13-4 show the decoded upstream and downstream gate specification objects within this Gate-Set message.

Figure 13-2 *Ethereal Trace of Gate-Set Message*

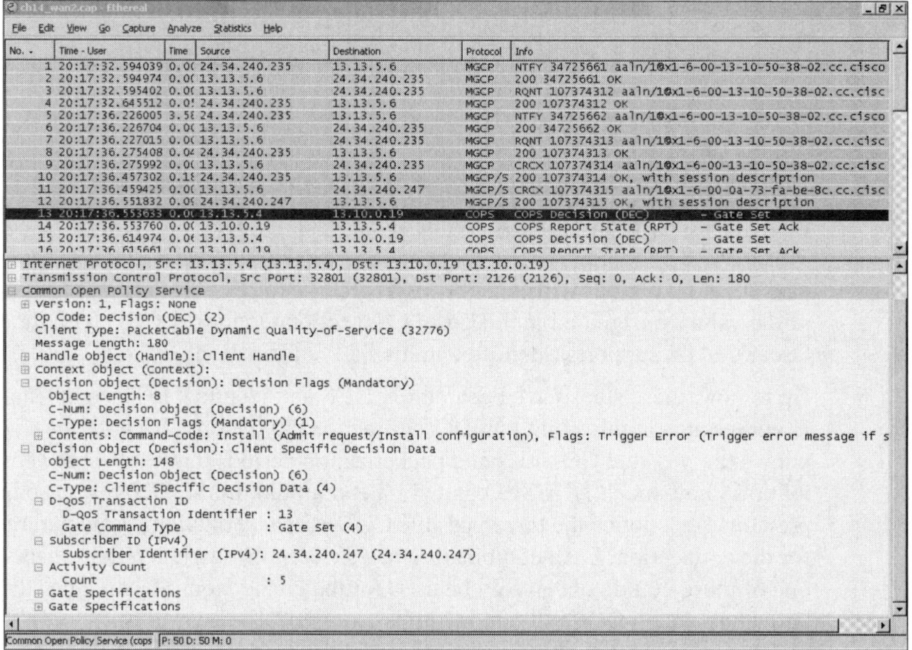

In Figure 13-2, you can see that a Gate-Set message is sent to the CMTS to authorize and allocate upstream and downstream gates for Tina's MTA. Because a Gate-Set message is sent from the CMS to the CMTS, a COPS Decision message is used, and the PacketCable objects are specified in Client Specific Decision Data objects. The CMS is pushing this policy to the CMTS to tell it how to handle the QoS for Tina's end of the phone call.

The Subscriber ID is the IP address of Tina's MTA. The Activity Count parameter triggers the allocation check. In this instance, it is set to 5, meaning that Tina's MTA is not permitted to have more than five active gates at one time. As you can see, the BTS combines the allocation and authorization phases into a single step with this Gate-Set message; hence, no Gate-Allocation message exchange is required.

The gate specifications contain the parameters used to define the authorization envelope for the DQoS resources. Figure 13-3 shows the upstream gate specification in greater detail.

Figure 13-3 *Ethereal Trace of Upstream Gate Specification*

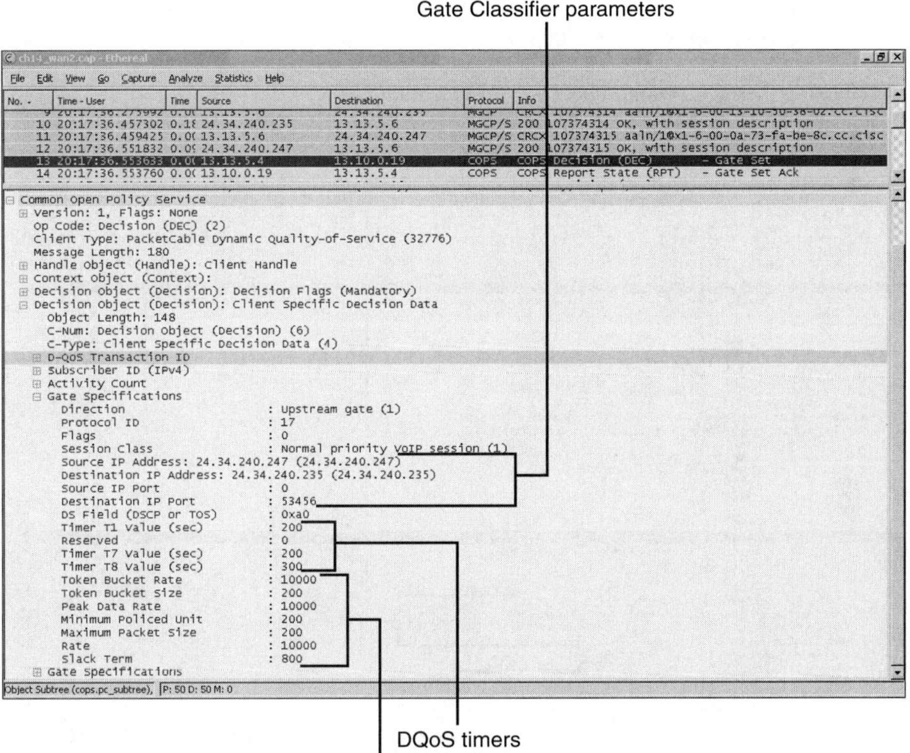

From the Direction field you can tell this gate specification is for an upstream gate. The Session Class parameter indicates this is an ordinary call; you will see use of this parameter later in this chapter. The next four fields comprise the gate classifier and are the source and destination IP addresses and RTP port numbers used for the voice flow. Notice that the source IP port number is 0, meaning all source ports match this gate classifier. The next field is used for overwriting DSCP bits or IP precedence bits and is also covered in greater detail later in this chapter. Next are the timers; recall from

Chapter 12, "DQoS Architecture and Framework," that timer T1 starts as soon as a gate is authorized and stops when gate resources are committed. If resources are not committed within this interval, the gate is closed. Timers T7 and T8 control the amount of time unused resources are permitted to be in the admitted and active states.

The remaining parameters comprise the gate flow specification and are determined from the CODEC and packetization period. In Table 12-5 of Chapter 12 you saw that for G.711 with a 20 ms results in the (b), (M), and (m) Flow Specification parameters equal to 200 bytes and the (r), (p), and (R) FlowSpec parameters equal to 10,000 bytes/sec. The Slack Term values are statically configured on the BTS, which you'll learn about in the section, "BTS Configuration." For more details on how these FlowSpec parameters are determined, see Chapter 12. Figure 13-4 shows the downstream gate specification.

Figure 13-4 *Ethereal Trace of Downstream Gate Specification*

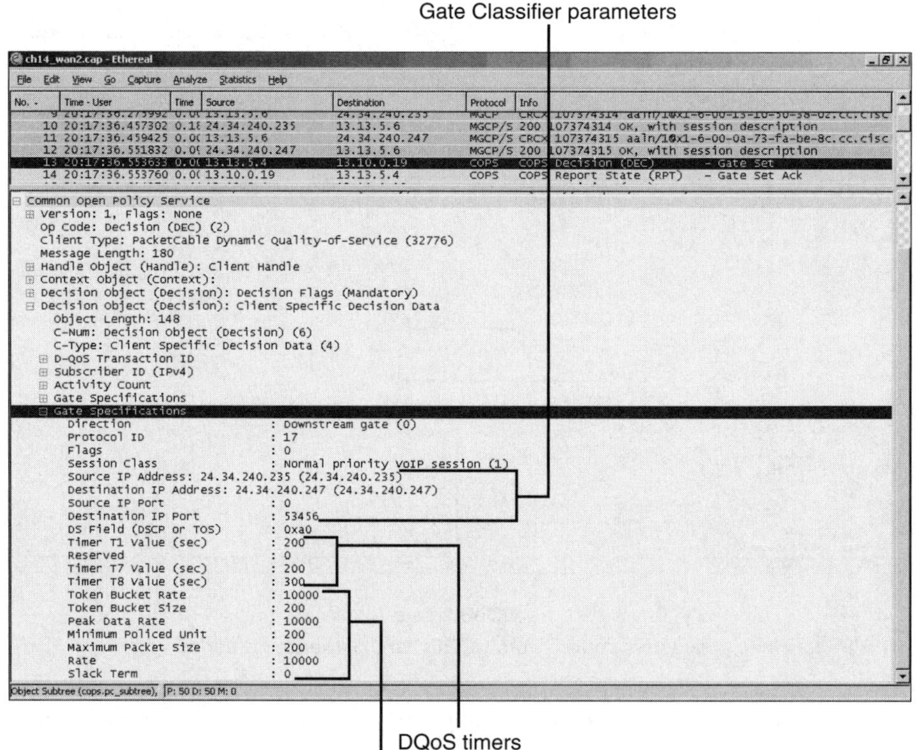

The parameters here are very similar to those in the upstream gate with a few exceptions. Notice that the source and destination IP addresses in the Gate Classifier are reversed, and the destination RTP port number refers to the port on Tina's MTA.

At this point the CMTS performs the allocation check and authorizes the creation of the gates. A Gate-Set-Ack message, illustrated in Figure 13-5, is sent to the CMS in response.

Figure 13-5 *Ethereal Trace of Gate-Set-Ack*

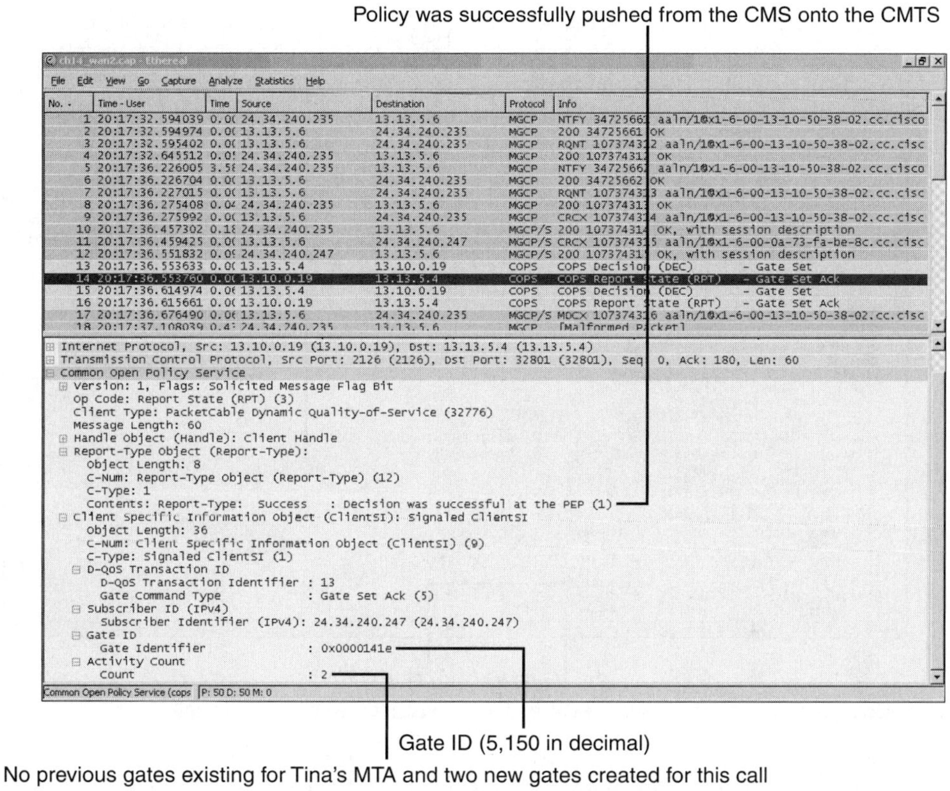

Policy was successfully pushed from the CMS onto the CMTS

Gate ID (5,150 in decimal)

No previous gates existing for Tina's MTA and two new gates created for this call

Two things to note about the Gate-Set-Ack message are the Gate ID assigned by the CMTS and the activity count. The activity count of 2 indicates that gates have been allocated for this new call for both the upstream and downstream directions and no previous gates existed. Also note timer T1 is started at this time. In this case, T1 is set to 200 seconds. So if these resources are not committed in this time, the gate is closed by the CMTS.

The Gate ID is used by the MTA to reference this gate, as you will see in a bit.

8. Similarly, the CMS signals the CMTS to create gates for Oscar's MTA. The manner in which this is done is the same as in the previous step. Gate specifications are defined for both an upstream and a downstream gate. The CMTS performs the allocation check for Oscar's MTA and creates the gates and assigns a gate identifier.

Figure 13-6 illustrates the resulting Gate-Set-Ack message.

Figure 13-6 *Ethereal Trace of Gate-Set-Ack for Oscar's MTA*

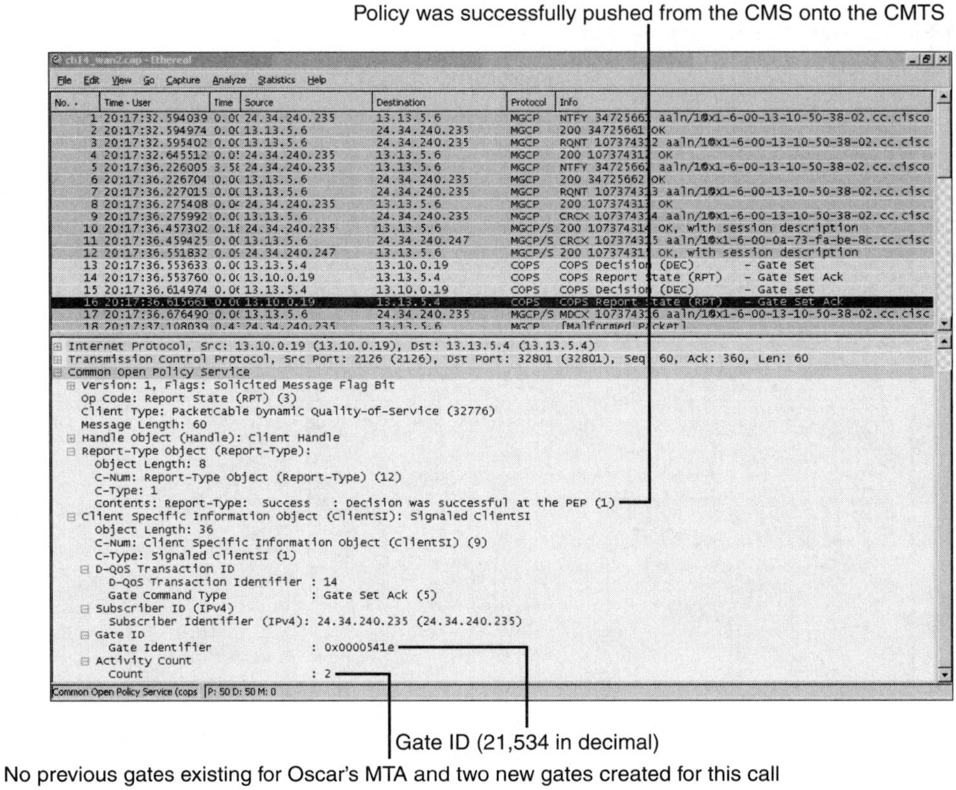

9 At this point, DQoS gates for both Oscar's MTA and Tina's MTA are in the authorized state. Both of these MTAs are informed of the existence of these gates by the CMS. The CMS includes this gate information in an NCS MDCX command, as follows:

```
MDCX 107374316 aaln/1@x1-6-00-13-10-50-38-02.cc.cisco.com MGCP 1.0
 NCS 1.0
I: 40EA
C: 21
L: dq-gi:541e,  dq-rr:snrcresv  <- DQoS parameters
M: recvonly  <- Endpoint connection mode set to receive only
K: 107374314
```

```
v=0
o=- 107374315 20466 IN IP4 24.34.240.247
c=IN IP4 24.34.240.247
b=AS:80
t=0 0
m=audio 53456 RTP/AVP 0 8 15
a=X-pc-csuites-rtp:60/50
a=X-pc-csuites-rtcp:80/70
a=X-pc-nrekey:0
a=mptime:20 20 20
a=ptime:20
```

This MDCX is sent to Oscar's MTA. In the MDCX you see that Tina's SDP information is included, so Oscar can know how and where to forward voice packets. Also the mode of the connection has been changed to recvonly, allowing packets from the VoIP network to be heard by Oscar. This is done in case ring back tone is provided from the far end.

This information is also covered in Chapter 7. Looking at the DQoS parameters you see the gate ID parameter, dq-gi:541e, which matches the gate authorized in the previous step. The resource reservation parameter, dq-rr:snrcresv, indicates resources should be reserved only by the MTA at this point. If Tina is not able to answer the call, tying up valuable resources that could be used by other customers is pointless.

10 After Oscar's MTA receives the MDCX with DQoS gate information described in Step 9, it signals the CMTS to reserve resources. Recall from Chapter 12 that an embedded MTA does this using DOCSIS Dynamic Service Flow messages. In this case, a Dynamic Service Addition Request (DSA-REQ) is sent to the CMTS. In this message are classifiers and service flow descriptors for an upstream and a downstream service flow for voice traffic. Figure 13-7 illustrates the trace of the upstream information.

Figure 13-7 *Upstream Information in DSA-REQ*

```
ch14_rf2.cap - Ethereal
File Edit View Go Capture Analyze Statistics Help
No. . Time - User       Time Source             Destination        Protocol Info
   14 16:36:12.631119 0.0( 13.13.5.6           24.34.240.235      MGCP     200 34725662 OK
   15 16:36:12.631152 0.0( 13.13.5.6           24.34.240.235      MGCP     RQNT 107374313 aaln/1@x1-6-00-13-10-50-38-02.cc.cisc
   16 16:36:12.679261 0.04 24.34.240.235       13.13.5.6          MGCP     200 107374313 OK
   17 16:36:12.680405 0.0( 13.13.5.6           24.34.240.235      MGCP     CRCX 107374314 aaln/1@x1-6-00-13-10-50-38-02.cc.cisc
   18 16:36:12.860539 0.1E 24.34.240.235       13.13.5.6          MGCP/S   200 107374314 OK, with session description
   19 16:36:12.863863 0.0( 13.13.5.6           24.34.240.247      MGCP/S   CRCX 107374315 aaln/1@x1-6-00-0a-73-fa-be-8c.cc.cisc
   22 16:36:12.955050 0.0( 24.34.240.247       13.13.5.6          MGCP/S   200 107374315 OK, with session description
   23 16:36:13.080299 0.1: 13.13.5.6           24.34.240.235      MGCP/S   MDCX 107374316 aaln/1@x1-6-00-13-10-50-38-02.cc.cisc
   24 16:36:13.152058 0.0; 00:13:10:50:38:01   00:0e:39:d2:d4:8c  DOCSIS   Dynamic Service Addition Request Tran-id = 18
   25 16:36:13.152713 0.0( 00:0e:39:d2:d4:8c   00:13:10:50:38:01  DOCSIS   Dynamic Service Add Response ID = 18 (okay/success)
   26 16:36:13.250363 0.0( 00:13:10:50:38:01   00:0e:39:d2:d4:8c  DOCSIS   Dynamic Service Add Ack ID = 18 (okay/success)

DOCSIS
Mac Management
DSA Request
  Transaction Id: 18
  TLV Data
    24 Upstream Service Flow (Length = 35)
      .1 Service Flow Ref: 1
      .6 QOS Parameter Set: Perform admission control add apply to addmitted set (0x02)
      .15 Scheduling Type: Unsolicited Grant Service (0x00000006)
      .16 Request/Transmission Policy: 0x0000017f
      .19 Unsolicited Grant Size (bytes): 232
      .20 Nominal Grant Interval (usec): 20000
      .21 Tolerated Grant Jitter (usec): 800
      .22 Grants Per Interval: 1
    22 Upstream Packet Classifier (Length = 47)
      .1 Classifier Ref: 1
      .3 Service Flow Ref: 1
      .5 Rule Priority: 128
      .6 Activation State: Inactive
      9 IP Classifier (Length = 32)
        ..2 IP Protocol: 17
        ..3 Source Address: 24.34.240.235 (24.34.240.235)
        ..4 Destination Address: 24.34.240.247 (24.34.240.247)
        ..7 Source Port Start: 53456
        ..8 Source Port End: 53456
        ..9 Dest Port Start: 53456
        ..10 Dest Port End: 53456
    25 Downstream Service Flow (Length = 32)
    23 Downstream Packet Classifier (Length = 39)
    30 Auth Block: 010601040000541E
    31 Key Sequence Number: 02
    27 HMAC Digest: F21FB9BA85983D2B14C819696854B4ABCCFF611E
DOCSIS Mac Management (docsis   P: 1323 D: 62 M: 0
```

Gate ID 0x0000541E (21,534 decimal)

From the upstream service flow encodings you can see that an Unsolicited Grant Service (UGS) flow is requested. Note that the QoS parameter set indicates that this service flow is only to be admitted and not yet activated. This equates to the reserved state for DQoS resources. The grant interval is equal to 20,000 μs, or 20 ms, which is the packetization interval. The UGS size is equal to 232 bytes, which is what you get after you add the DOCSIS MAC-layer overhead to the 200-byte RTP packets. Recall from the previous chapter that the DOCSIS MAC Management Header is 18 bytes, the regular DOCSIS Header is 6 bytes, and you have 8 bytes of DOCSIS Extended Headers (thus, you have 32 bytes of DOCSIS MAC-layer overhead). Consequently, in this case the reserved resource envelope is equal to the authorized resource envelope.

Also note the classifier for this service flow. It is set to match on source and destination IP address and User Datagram Protocol (UDP)/RTP port number.

The classifier is also in the "inactive" state, meaning it is not yet used because the service flow is not ready for use. Figure 13-8 depicts the downstream service flow and classifier encodings.

Figure 13-8 *Downstream Information in DSA-REQ*

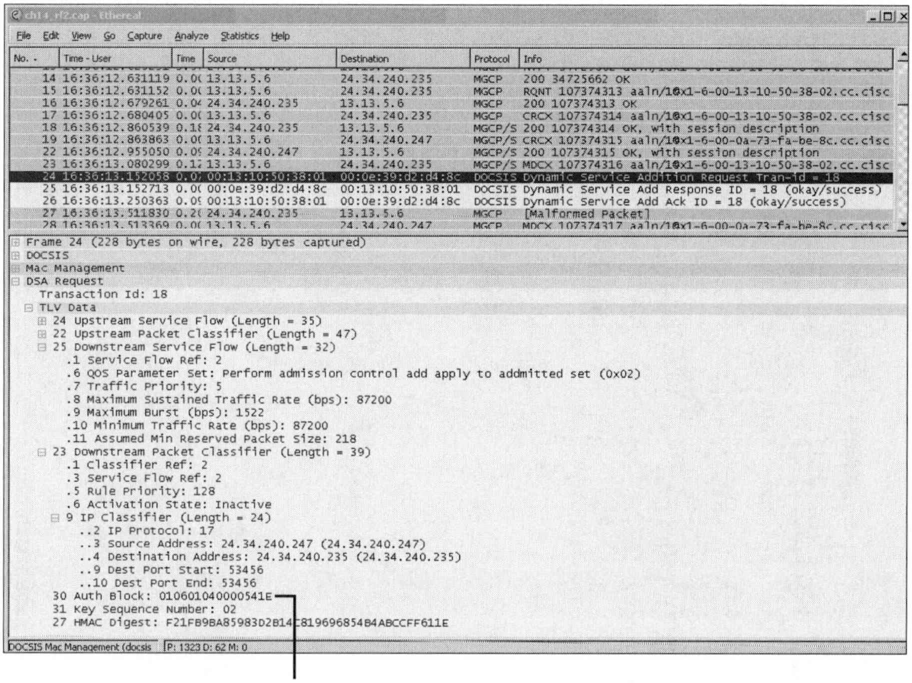

Notice the last parameter in the DSA-REQ, the authorization block. Within this parameter is the DQoS Gate ID, which identifies the gate associated with these DOCSIS service flows. Chapter 12 explains the format of this parameter in greater detail. The CMTS searches for an authorized gate matching this gate identifier. Assuming one exists, the CMTS then converts the DOCSIS classifiers and service flow parameters into gate classifiers and gate flow specifications. The resulting gate flow specifications comprise the reserved resource envelope. For more details on how these conversions are done, refer to Chapter 12. If the authorized and reserved gate classifiers match up and the reserved resource envelope is less than the authorized resource envelope, the CMTS then attempts to create the dynamic service flows.

Because reserved resources are guaranteed available when needed, the CMTS checks to see whether bandwidth exists to support these new service flows. If resources are available, the service flows are accepted, and service flow identifiers are assigned as shown in Figure 13-9. If resources are not currently available, the service flows are rejected. Also remember that these reserved resources can be used by Best Effort service flows until they are committed. The CMTS then issues a DSA-RSP to the MTA, shown in Figure 13-9.

Figure 13-9 *Ethereal Trace of DSA-RSP*

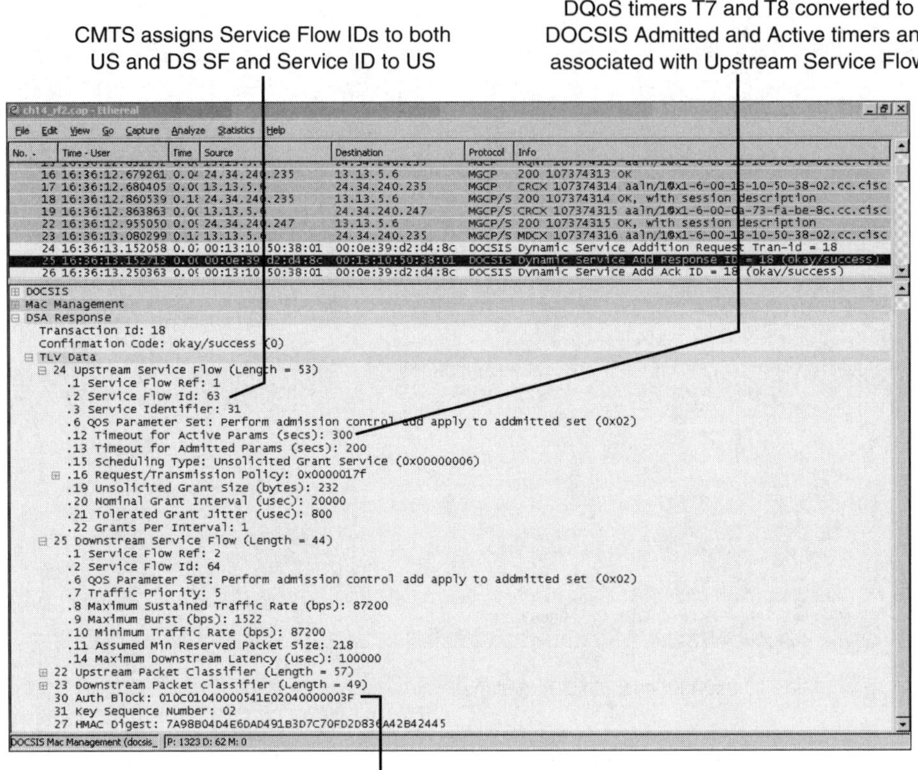

So in looking at the DSA-RSP you can see that the DSA-REQ is accepted, noted by the okay/success confirmation code. You can also see that a service flow identifier (63) is assigned to the upstream service flow as well as a DOCSIS SID (31).

A service flow identifier was also assigned to the downstream service flow (64), and classifier identifiers were assigned to both classifiers.

Looking at the authorization block, you might notice two parameters are now included. The first one is the gate identifier you've seen before; the second one is the DQoS resource identifier. This field uniquely identifies the DQoS resources used by these service flows. This information then can be relayed to the CMS where decisions can be made about sharing these DQoS resources among multiple service flows and gates. For example, in a call waiting scenario only one of the calls is active at a time, so the DQoS resources could be shared between both calls. Oscar's MTA acknowledges the receipt of the DSA-RSP by sending a DSA-ACK back to the CMTS. The CMTS also starts timer T7 as soon as the resources are reserved.

Timer T7 indicates the length of time these resources can remain reserved without being committed. If this timer expires, the CMTS deletes this service flow with a DSD-REQ and closes the DQoS gate with a Gate-Close message to the CMS. As you can see in Figure 13-9, the DQoS timers T7 and T8 were mapped to the DOCSIS admitted and active timers and included with the upstream service flow. Also notice the CMTS has assigned a maximum downstream latency parameter to the downstream service flow. This parameter specifies the maximum amount of delay that occurs on the CMTS between receiving a voice packet on its WAN interface and forwarding it out the cable interface.

Finally, the DQoS resource identifier parameter is relayed to the CMS in the acknowledgement to the MDCX from Step 9 as follows:

```
200 107374316 OK
DQ-RI: 3F
```

11 An MDCX command is also sent to Tina's MTA:

```
MDCX 107374317 aaln/1@x1-6-00-0a-73-fa-be-8c.cc.cisco.com MGCP 1.0
 NCS 1.0
I: 4FF2
C: 21
L: dq-gi:141e,  dq-rr:snrcresv
M: sendrecv  <- Endpoint connection mode is set to send/receive
K: 107374315
```

In this command is the gate identifier referencing the authorized DQoS resources to be used by Tina's MTA. Just as it did with Oscar's MTA, the CMS requests Tina's MTA to reserve these resources using the resource reservation parameter, dq-rr:snrcresv. How Tina's MTA reserves these resources is covered in Step 12. The MDCX also changes the NCS connection mode to sendrecv. This enables the exchange of bearer data between the PacketCable VoIP network and Tina's phone. Tina's MTA now generates RTP packets to Oscar's MTA, although at this point they would be sent Best Effort because DQoS resources have not yet been committed.

12 After the receipt of the MDCX, Tina's MTA reserves her DQoS resources. This process is similar to Step 10. Tina's MTA sends a DSA-REQ for the creation of service flows using her authorized gate. The CMTS checks the validity of the request and, assuming resources are available, transitions her gate into the reserved state. Service flow identifiers, classifier identifiers, an upstream service identifier, and a DQoS resource identifier are all returned in the DSA-RSP. Timer T7 is started for Tina's reserved resources, and in the acknowledgement to the MDCX the DQoS resource identifier assigned by the CMTS is returned.

This concludes the first stage of this basic call.

Basic ONNET Call: Stage 2

In Stage 2, the messaging used up to the point where the call is active is examined. Figure 13-10 shows this part of the call.

Figure 13-10 *Basic ONNET Call Flow (Stage 2)*

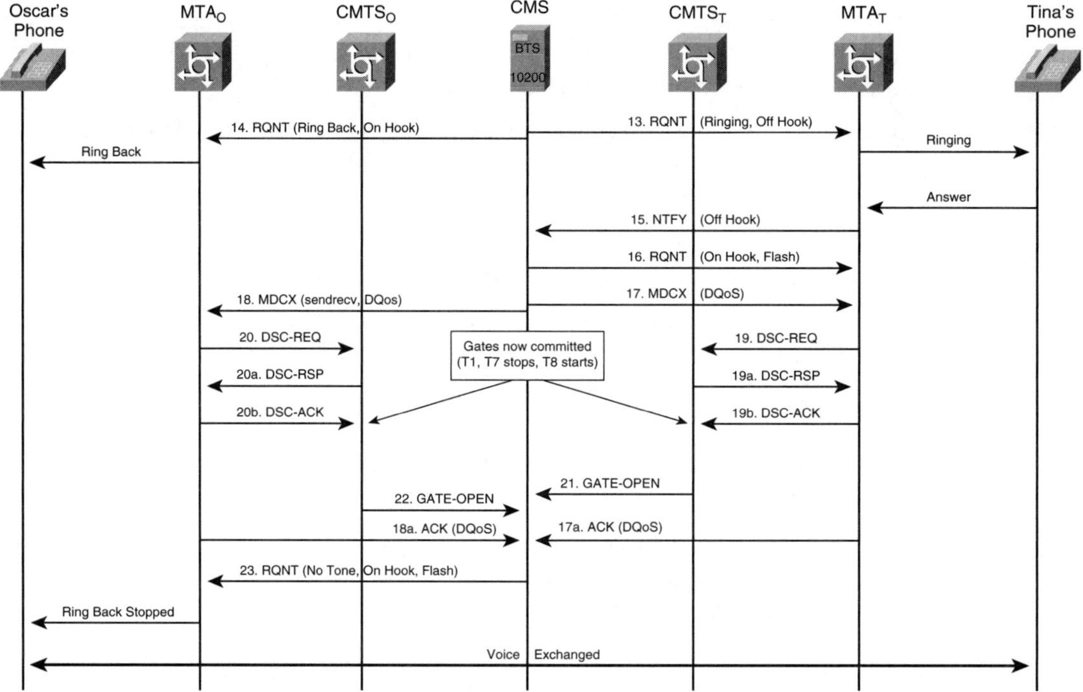

13 At this point the CMS knows DQoS resources are available and have been reserved for both ends of the call. So it sends an NCS RQNT to Tina's MTA telling it to ring her phone and to be notified when Tina answers.

14 The CMS also alerts Oscar's MTA that Tina's phone is ringing by instructing Oscar's MTA to provide ring back tone using an RQNT command.

15 Eventually, Tina answers her phone, triggering an NTFY to be sent from her MTA to the CMS.

16 The CMS then requests to be notified if Tina hangs up her phone or performs a hook flash.

17 Now that Tina has answered her phone, the quality of service resources on the DOCSIS network can be committed for use. An MDCX is sent to Tina's MTA that contains DQoS parameters that instruct her MTA to commit her DQoS resources. This MDCX is as follows:

```
MDCX 107374321 aaln/1@x1-6-00-0a-73-fa-be-8c.cc.cisco.com MGCP 1.0
 NCS 1.0
I: 4FF2
C: 21
L: dq-gi:141e,  dq-rr:snrccomt,  dq-ri:41  <- DQoS parameters
K: 107374318
```

As you can see, three DQoS parameters are included in the NCS LCO parameter. The first one (dq-gi) is the gate identifier referring to the same gate used by Tina throughout this call. The second parameter (dq-rr) is the DQoS resource reservation parameter. It is set to snrccomt, which means the MTA is to commit resources for both the send and receive directions. The particular resources to be committed are noted by the next parameter (dq-ri), which is the DQoS resource identifier. This refers to the same resources reserved by the CMTS in Step 12. How Tina's MTA commits DQoS resources is covered in Step 19.

18 In much the same fashion, the CMS instructs Oscar's MTA to commit his DQoS resources. This MDCX is as follows:

```
MDCX 107374322 aaln/1@x1-6-00-13-10-50-38-02.cc.cisco.com MGCP 1.0
 NCS 1.0
I: 40EA
C: 21
L: dq-gi:541e,  dq-rr:snrccomt,  dq-ri:3F  <- DQoS parameters
M: sendrecv  <- Endpoint connection mode changed to send/receive
K: 107374319
```

In the MDCX used in this step, Oscar's MTA is also instructed to change its connection state to send/receive. This permits the exchange of packets between the telephone port on Oscar's MTA and the PacketCable VoIP network.

19 Now that Tina's MTA has received the MDCX with DQoS gate information (Step 17), it signals the CMTS to commit the resources. A Dynamic Service Change Request (DSC-REQ) is sent to the CMTS to activate the service flows previously reserved. In this message are the classifiers and service flow descriptors for an upstream and a downstream service flow used for voice traffic. Figure 13-11 illustrates the trace of the upstream information.

Figure 13-11 *Activation of the Upstream Service Flow*

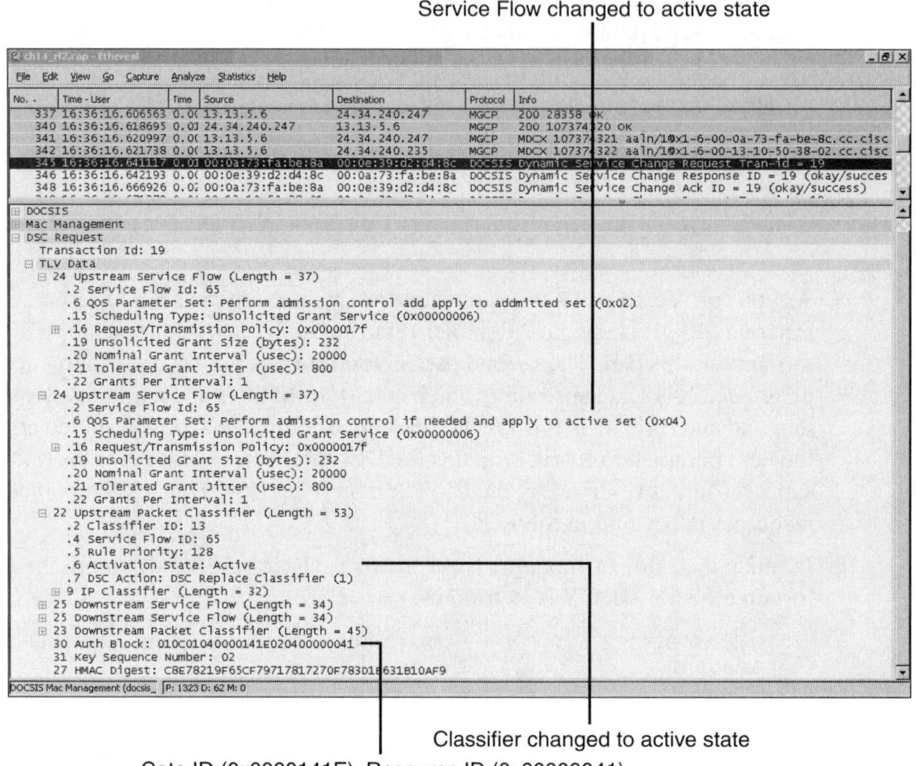

Notice the two sets of upstream service flow encodings that have nearly identical parameters. The only difference is in the QoS Parameter Set where one is equal to "admitted" and the other is equal to "active." The "active" SF encoding indicates what DQoS resources are to be committed, which in this case is all the reserved resources. Conceivably, only a subset of the reserved resources could have been activated, in which case the "admitted" and "active" QoS parameters would have been different.

The classifier's state is now "active"; that enables the classifier for use in classifying packets. The authorization block parameter identifying the DQoS gate and resources associated with these service flows is also included.

Figure 13-12 illustrates information relating to the downstream service flow.

Figure 13-12 *Activation of the Downstream Service Flow*

Service Flow changed to active state

Classifier changed to active state

Gate ID (0x0000141E), Resource ID (0x00000041)

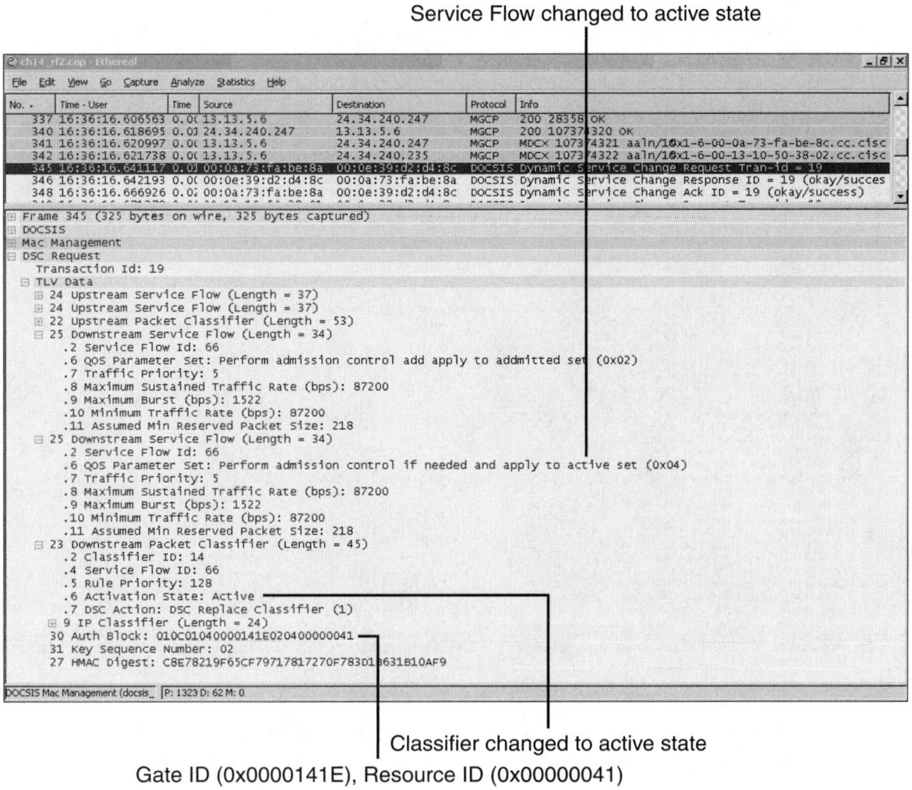

As you can see, the downstream service flow and classifiers are also activated.

At this point, the CMTS converts the DOCSIS parameters into gate specifications to create the committed resource envelope. If this envelope is less than or equal to the reserve resource envelope, the DSC-REQ is successful. Remember, you do not need to do another bandwidth check because this was already performed when the resources were reserved.

Figure 13-13 shows a trace of the resulting DSC-RSP.

Figure 13-13 *Ethereal Trace of the DSC-RSP*

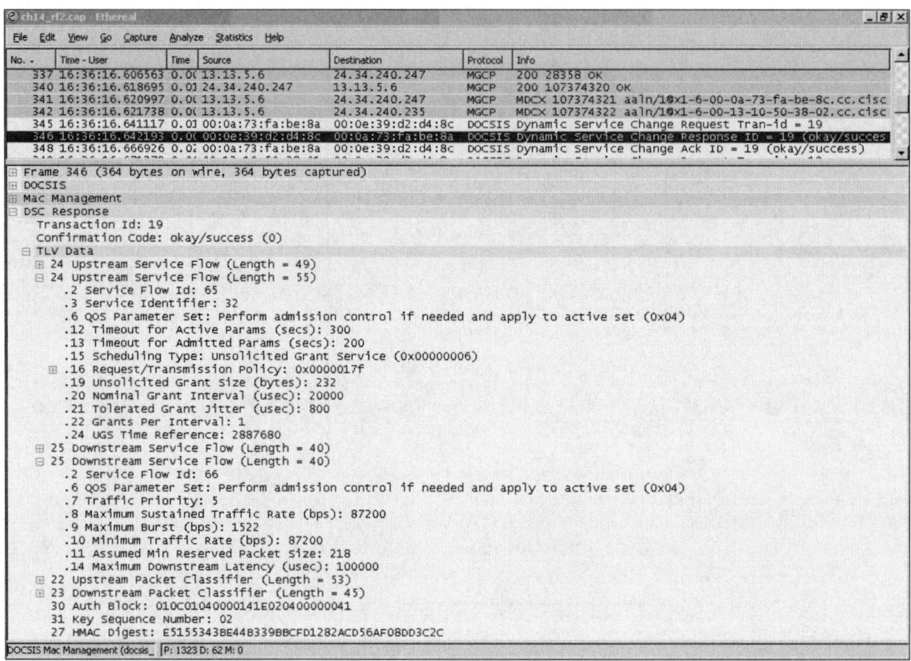

The Confirmation Code parameter indicates that the DSC-REQ was successful. Note the inclusion of the active and admitted timers in the service flow encodings. The values of these timers are taken from the DQoS timers T7 and T8, respectively. In this case, the admitted timer is meaningless because all reserved resources have been committed. However, if not all the reserved resources were committed, this timer would apply to the uncommitted resources. The active timer is set 300 seconds, meaning if no packets are received on this service flow after this length of time the CMTS tears down both the upstream and the downstream service flows and closes the gate.

NOTE The default value of timer T8 is 0 seconds, which means even if the DQoS resources were never used, they would remain available. This can cause "hung" service flows on your CMTS, which takes away bandwidth from other users. This timer is set on the CMS and should always be set to a nonzero value.

One final observation about the upstream service flow encoding is the inclusion of the UGS Time Reference parameter. This parameter contains a timestamp alerting the MTA of when its UGS grants occur. Finally, Tina's MTA acknowledges the receipt of the DSC-RSP by sending a DSC-ACK to the CMTS and alerts the CMS of the successful activation of resources in the MDCX acknowledgement.

20 This same process occurs on Oscar's MTA resulting from the MDCX in Step 18. Oscar's service flows are activated using a three-message DSC exchange very similar to the process described in the previous step.

21 The CMTS also notifies the CMS that these gates are now active. This is part of DQoS gate coordination that ensures the CMTS and CMS are in synchronization on the status of gates. First, the CMTS sends a Gate-Open message to the CMS, alerting it that Tina's gate is now active. This message is fairly straightforward and is shown in Figure 13-14.

Figure 13-14 *Ethereal Trace of Gate-Open Message*

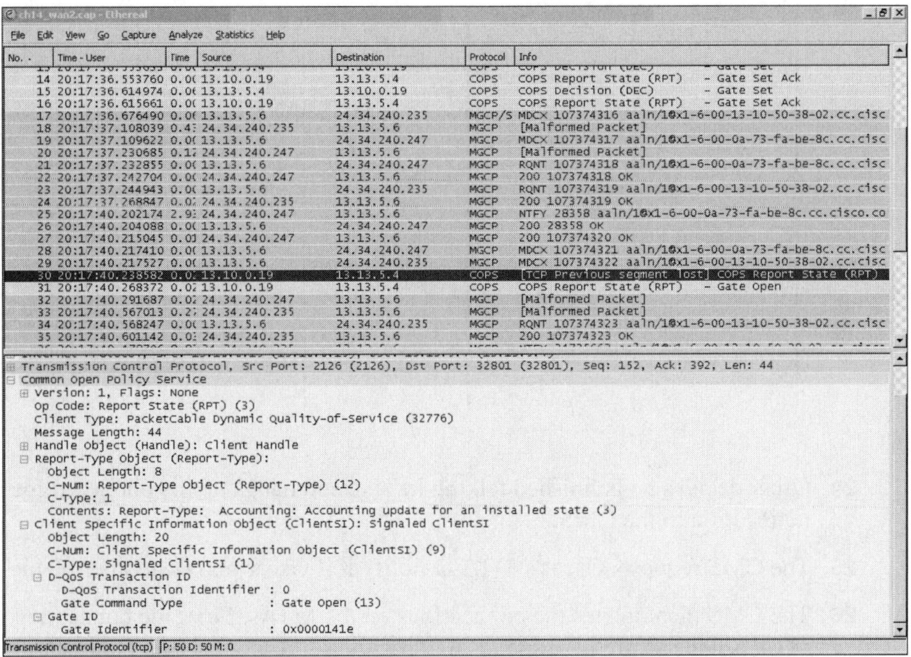

22 The CMTS sends a similar Gate-Open to the CMS about the activation of Oscar's gate.

23 Because Tina has answered the call, the CMS requests Oscar's MTA to stop providing ring back tone. Also notification of an on hook or flash hook event is requested by the CMS.

At this point, Tina and Oscar are engaged in an active voice call with guaranteed quality of service across the DOCSIS network.

Basic ONNET Call: Stage 3

Stage 3 of the call examines the NCS and DQoS signaling used to disconnect the call and is illustrated in Figure 13-15.

Figure 13-15 *Basic ONNET Call Flow (Stage 3)*

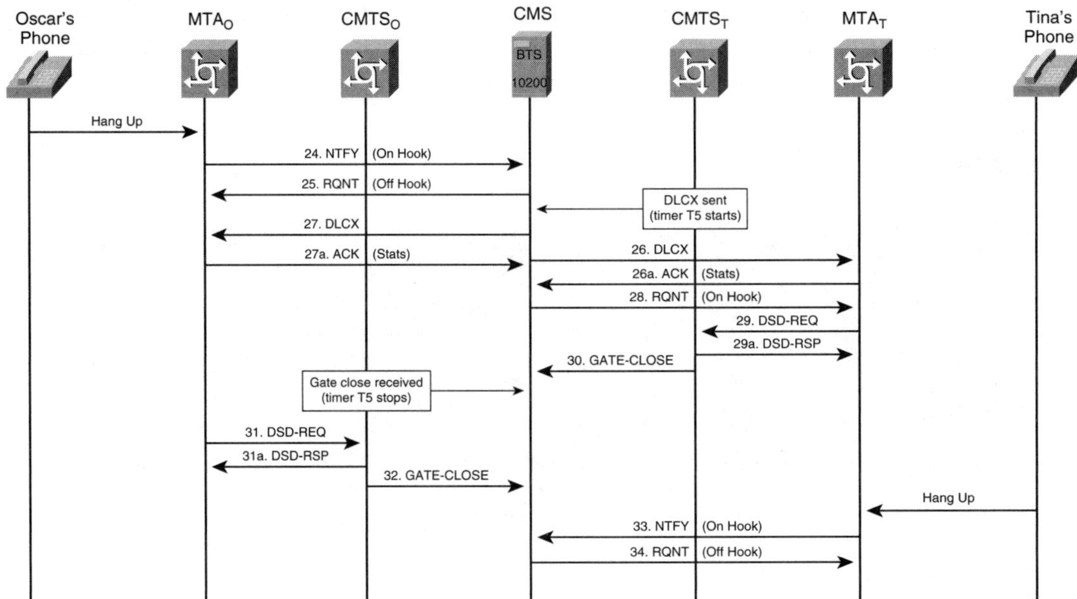

24 Oscar decides he is finished talking to Tina and hangs up his phone, triggering a notification to the CMS.

25 The CMS requests Oscar's MTA to notify it if Oscar makes another phone call.

26 The CMS then deletes the connections on the MTAs. First, the connection on Tina's MTA is deleted. In the response to the DLCX, the Connection Parameter (P:) is included, providing the CMS with statistics about the call. This includes packets sent/received, jitter, and latency information. Remember, the information received from Oscar's MTA in RTCP reports is included in the response as well. If these MTAs supported the VoIP metrics package as described in the PacketCable 1.5 CODEC specification, these additional quality metrics would be included here as well.

27 The connection on Oscar's MTA is also deleted. Call statistic information as seen from Oscar's MTA is reported to the CMS as well as information received from RTCP reports originated from Tina's MTA.

28 The CMS then requests to be notified after Tina hangs up her phone.

29 As soon as Tina's MTA deletes the NCS connection as requested by the CMS in Step 26, it then signals the CMTS for deletion of the DOCSIS service flows. This is done by sending a Dynamic Service Deletion Request (DSD-REQ) to the CMTS, as depicted in Figure 13-16.

Figure 13-16 *Ethereal Trace of DSD-REQ*

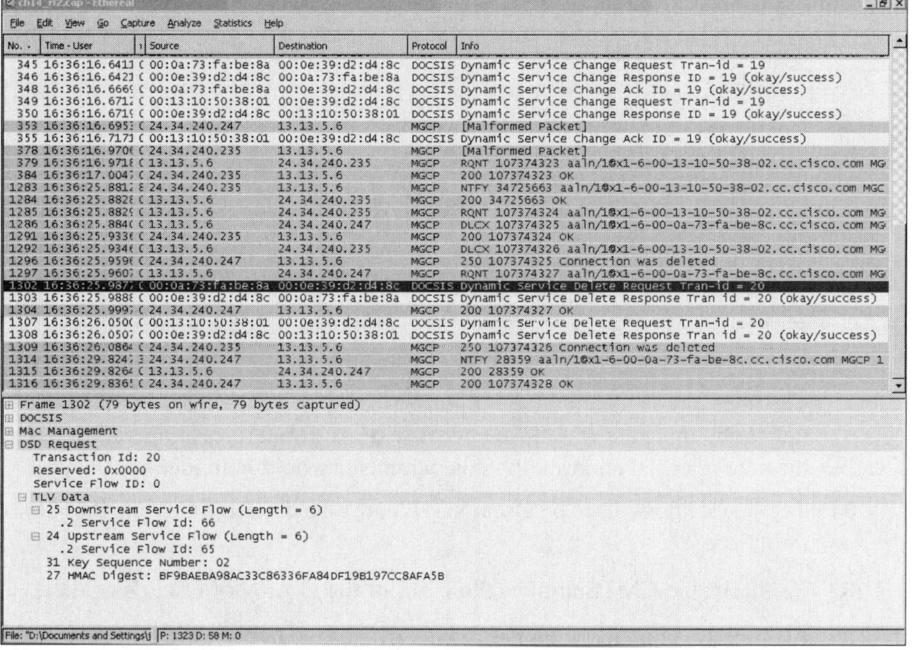

The service flow identifiers of the upstream and downstream service flows to be deleted are included in the TLV portion of the message. Note that if service flow IDs are included in the TLV part of the message, the service flow parameter in the beginning of the message is set to zero. The CMTS acknowledges the successful deletion of these service flows by sending a DSD-RSP back to Tina's MTA with a positive confirmation code.

30 To ensure the CMTS and the CMS are synchronized in their knowledge of active gates, the CMTS notifies the CMS about the deletion of these gates. First, a Gate-Close message is sent referring to Tina's gate. Figure 13-17 depicts this message.

Figure 13-17 *Ethereal Trace of Gate-Close*

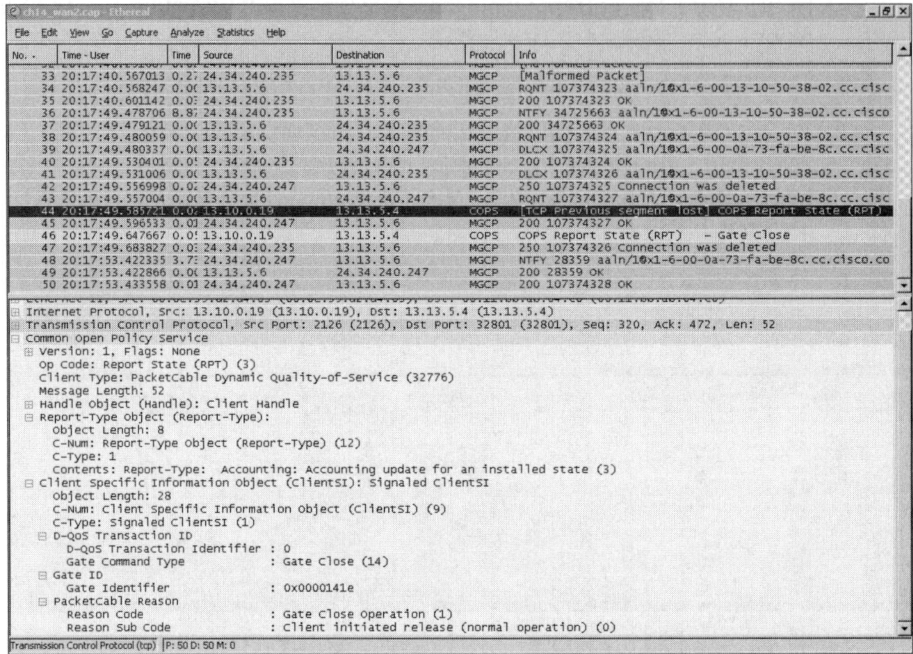

This message is also pretty straightforward. Notice the Reason Sub Code parameter indicating this gate was closed because of a normal release operation by the MTA. If the gate released abnormally, this parameter would help identify the probable cause.

31 The service flows used by Oscar's MTA are also deleted following the same procedure as in Step 29.

32 Similarly, the CMTS notifies the CMS of the closing of Oscar's gate.

33 Eventually, Tina hangs up her phone, triggering an on-hook notification to the CMS.

34 Finally, the CMS requests to be notified when she takes her phone off hook again.

VAD and UGS-AD

Now that you know how a basic call works, this section covers some variations that you might encounter. As you know, the UGS scheduling service works by providing dedicated fixed size opportunities to transmit data at periodic intervals. Because these opportunities are dedicated to specific devices, no other device can use this bandwidth. So if the device associated with the UGS flow does not use these intervals, these intervals are wasted. In a typical phone call, only one person is talking at a time. This means that approximately 50 percent of the time, these intervals are in fact wasted. The MTA still sends packets when its user is not speaking; however, these packets just contain silence with the exception of some ambient noise detected by the phone.

Silence suppression or voice activity detection (VAD) provides a means for taking advantage of this fact. This is done by simply not sending RTP packets during periods of silence. During these silence periods normal RTP packets are not sent, but comfort noise packets are sent instead to keep the other end of the call from thinking their call has been disconnected.

In PacketCable, VAD can be used to improve bandwidth efficiency in a DOCSIS upstream. The UGS-AD scheduling service introduced in Chapter 12 allows DOCSIS to realize the bandwidth savings VAD provides. As you recall, the UGS-AD service works like a normal UGS flow while the MTA user is talking; however, it switches to a Real-Time Polling Service (RTPS) when its user is silent. The best way to see how this works is with an example. So now you can revisit the call between Oscar and Tina, only this time both of their MTAs have silence suppression enabled.

The only difference in NCS signaling relative to the previous example is a single parameter in the create connections:

```
CRCX 161122188 aaln/1@x1-6-00-0c-e5-11-17-d0.company-y.cisco.com MGCP 1.0 NCS 1.0
C: 1ab
L: mp:20; 10; 10; 10,  a:PCMU;PCMA;G729E;G729,  e:off,  s:on  <- VAD enabled
M: inactive
K: 161122187
```

Looking at the Local Connection Option parameters, notice the parameter s: is now set to on. This parameter identifies whether or not silence suppression is enabled. This alerts the MTA that the UGS-AD DOCSIS scheduling service should be used for the call, as illustrated in Figure 13-18.

Figure 13-18 *Ethereal Trace of DSA-REQ Using UGS-AD*

```
ugsadtest - Ethereal
File  Edit  View  Go  Capture  Analyze  Statistics  Help
Filter: docsis.mgmt.type >= 15 || mgcp     Expression...  Clear  Apply

No.   Time      Source             Destination        Protocol Info
16670 8.918443  24.34.241.104      13.200.12.20       MGCP     200 161122187 OK
16671 8.920673  13.200.12.20       24.34.241.104      MGCP     CRCX 161122188 aaln/1@x1-6-00-0c-e5-11-17-d0.company-y
17227 9.214443  24.34.241.104      13.200.12.20       MGCP     200 161122188 OK, with session description
17236 9.219400  13.200.12.20       24.34.240.252      MGCP     CRCX 107441380 aaln/1@x1-6-00-0c-e5-11-18-18.company-y
17922 9.588967  24.34.240.252      13.200.12.20       MGCP     200 107441380 OK, with session description
18031 9.646664  13.200.12.20       24.34.241.104      MGCP     MDCX 161122189 aaln/1@x1-6-00-0c-e5-11-17-d0.company-y
18389 9.836442  00:0c:e5:11:17:ce  00:04:dd:5d:78:55  DOCSIS   Dynamic Service Addition Request Tran-id = 8
18390 9.837488  00:04:dd:5d:78:55  00:0c:e5:11:17:ce  DOCSIS   Dynamic Service Add Response ID = 8 (okay/success)
18488 9.889018  00:0c:e5:11:17:ce  00:04:dd:5d:78:55  DOCSIS   Dynamic Service Add Ack ID = 8 (okay/success)
18634 9.964443  24.34.241.104      13.200.12.20       MGCP     200 161122189 OK, with session description
18643 9.968533  13.200.12.20       24.34.240.252      MGCP     MDCX 107441381 aaln/1@x1-6-00-0c-e5-11-18-18.company-y

    Header check sequence: 0xf731
  ▷ Mac Management
  ▽ DSA Request
      Transaction Id: 8
    ▽ TLV Data
      ▽ 24 Upstream Service Flow (Length = 47)
          .1 Service Flow Ref: 1
          .6 QOS Parameter Set: Perform admission control add apply to addmitted set (0x02)
          .15 Scheduling Type: Unsolicited Grant Service w/Activity Detection (0x00000005)
        ▷ .16 Request/Transmission Policy: 0x0000017f
          .17 Nominal Polling Interval(usec): 40000
          .18 Tolerated Poll Jitter (usec): 20000
          .19 Unsolicited Grant Size (bytes): 231
          .20 Nominal Grant Interval (usec): 10000
          .21 Tolerated Grant Jitter (usec): 800
          .22 Grants Per Interval: 1
      ▷ 22 Upstream Packet Classifier (Length = 47)
      ▷ 25 Downstream Service Flow (Length = 32)
      ▷ 23 Downstream Packet Classifier (Length = 39)
        30 Auth Block: 0106010400084AE9
Scheduling Type (docsis.tlv.sflow.sd  P: 61330 D: 46 M: 0
```

Notice the DOCSIS Scheduling Type is now the UGS-AD service. Two additional parameters are needed for this service type than are needed with the regular UGS service. They are the Nominal Polling Interval, which defines how often to provide dedicated request opportunities to the MTA when it is operating in the Real-Time Polling Service (RTPS) mode, and the Tolerated Poll Jitter, which defines the acceptable jitter time for these request opportunities.

Figure 13-19 illustrates how the transition between talking and silence or UGS and RTPS works with a UGS-AD service flow.

Figure 13-19 *Ethereal Trace of How UGS-AD Works*

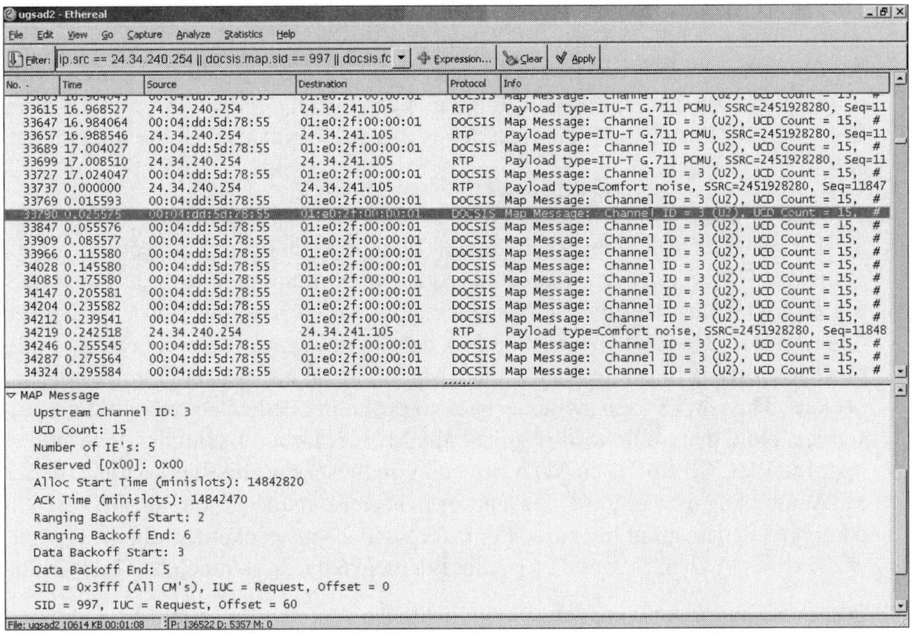

```
ugsad2 - Ethereal
File  Edit  View  Go  Capture  Analyze  Statistics  Help
Filter: ip.src == 24.34.240.254 || docsis.map.sid == 997 || docsis.fc   Expression...  Clear  Apply
No.   Time       Source             Destination        Protocol Info
33615 16.968527  24.34.240.254      24.34.241.105      RTP    Payload type=ITU-T G.711 PCMU, SSRC=2451928280, Seq=11
33647 16.984064  00:04:dd:5d:78:55  01:e0:2f:00:00:01  DOCSIS Map Message:  Channel ID = 3 (U2), UCD Count = 15,  #
33657 16.988546  24.34.240.254      24.34.241.105      RTP    Payload type=ITU-T G.711 PCMU, SSRC=2451928280, Seq=11
33689 17.004027  00:04:dd:5d:78:55  01:e0:2f:00:00:01  DOCSIS Map Message:  Channel ID = 3 (U2), UCD Count = 15,  #
33699 17.008510  24.34.240.254      24.34.241.105      RTP    Payload type=ITU-T G.711 PCMU, SSRC=2451928280, Seq=11
33727 17.024047  00:04:dd:5d:78:55  01:e0:2f:00:00:01  DOCSIS Map Message:  Channel ID = 3 (U2), UCD Count = 15,  #
33737 0.000000   24.34.240.254      24.34.241.105      RTP    Payload type=Comfort noise, SSRC=2451928280, Seq=11847
33769 0.015593   00:04:dd:5d:78:55  01:e0:2f:00:00:01  DOCSIS Map Message:  Channel ID = 3 (U2), UCD Count = 15,  #
33790 0.025575   00:04:dd:5d:78:55  01:e0:2f:00:00:01  DOCSIS Map Message:  Channel ID = 3 (U2), UCD Count = 15,  #
33847 0.055576   00:04:dd:5d:78:55  01:e0:2f:00:00:01  DOCSIS Map Message:  Channel ID = 3 (U2), UCD Count = 15,  #
33909 0.085577   00:04:dd:5d:78:55  01:e0:2f:00:00:01  DOCSIS Map Message:  Channel ID = 3 (U2), UCD Count = 15,  #
33966 0.115580   00:04:dd:5d:78:55  01:e0:2f:00:00:01  DOCSIS Map Message:  Channel ID = 3 (U2), UCD Count = 15,  #
34028 0.145580   00:04:dd:5d:78:55  01:e0:2f:00:00:01  DOCSIS Map Message:  Channel ID = 3 (U2), UCD Count = 15,  #
34085 0.175580   00:04:dd:5d:78:55  01:e0:2f:00:00:01  DOCSIS Map Message:  Channel ID = 3 (U2), UCD Count = 15,  #
34147 0.205581   00:04:dd:5d:78:55  01:e0:2f:00:00:01  DOCSIS Map Message:  Channel ID = 3 (U2), UCD Count = 15,  #
34204 0.235582   00:04:dd:5d:78:55  01:e0:2f:00:00:01  DOCSIS Map Message:  Channel ID = 3 (U2), UCD Count = 15,  #
34212 0.239541   00:04:dd:5d:78:55  01:e0:2f:00:00:01  DOCSIS Map Message:  Channel ID = 3 (U2), UCD Count = 15,  #
34219 0.242518   24.34.240.254      24.34.241.105      RTP    Payload type=Comfort noise, SSRC=2451928280, Seq=11848
34246 0.255545   00:04:dd:5d:78:55  01:e0:2f:00:00:01  DOCSIS Map Message:  Channel ID = 3 (U2), UCD Count = 15,  #
34287 0.275564   00:04:dd:5d:78:55  01:e0:2f:00:00:01  DOCSIS Map Message:  Channel ID = 3 (U2), UCD Count = 15,  #
34324 0.295584   00:04:dd:5d:78:55  01:e0:2f:00:00:01  DOCSIS Map Message:  Channel ID = 3 (U2), UCD Count = 15,  #

MAP Message
  Upstream Channel ID: 3
  UCD Count: 15
  Number of IE's: 5
  Reserved [0x00]: 0x00
  Alloc Start Time (minislots): 14842820
  ACK Time (minislots): 14842470
  Ranging Backoff Start: 2
  Ranging Backoff End: 6
  Data Backoff Start: 3
  Data Backoff End: 5
  SID = 0x3fff (All CM's), IUC = Request, Offset = 0
  SID = 997, IUC = Request, Offset = 60
File: ugsad2 10614 KB 00:01:08    P: 136522 D: 5357 M: 0
```

In this capture of DOCSIS packets, a filter was applied to display only the data packets and the DOCSIS MAP packets for an active call utilizing silence suppression. Remember, a DOCSIS MAP message is sent to all the cable modems on a particular downstream and identifies the DOCSIS timeslots each device is permitted to use and for what purpose. These are sent for each upstream channel associated with the downstream channel. These purposes include transmission of data packets, transmission of data request packets, and transmission of DOCSIS ranging messages. Each of these intervals denoted in the MAP messages also identify a particular SID authorized to use these intervals. The SID used by the MTA in this example is 997. Refer to the DOCSIS RF specification for more information on DOCSIS bandwidth allocation and the MAP message.

If you look at the beginning of the capture, you see a sequence of MAP packets each followed by an RTP packet. These MAP packets contain the dedicated UGS data grants the MTA uses for sending RTP packets. Thus, at this point in the call the MTA user is actively speaking. Notice that the RTP payload type for these packets is G.711 PCMU, which is the CODEC used for transmitting the voice stream. The user then stops speaking, causing VAD to kick in at the MTA. The MTA then stops transmitting G.711 RTP packets and begins to transmit comfort noise RTP packets. An RTP payload type indicating comfort noise identifies these packets, as seen in Figure 13-19. So if the MTA is still generating RTP packets, what is thc bandwidth saving? The answer is that these packets are sent much less frequently than the normal voice packets. Looking at the first comfort noise packet in the trace you can see that the time stamp is set to zero; this was done so you can see the delta time between these packets. Looking at the next comfort noise packet, you see that it comes approximately 242 ms afterward. Recall that the G.711 packets occur every 20 ms, so the comfort noise packets are sent much less regularly. Note another advantage of the MTA sending comfort noise packets is that they keep the DQoS activity timer T8 from expiring. Logically, for this to work T8 must be set to a value greater than the interval between comfort noise packets.

So as soon as the CMTS detects that these data grants provided every 20 ms are not being used, it switches over to the Real-Time Polling Service. Instead of providing dedicated data grants the CMTS provides dedicated request opportunities. The highlighted message in Figure 13-19 illustrates one of these messages. Notice, the Interval Usage Code or IUC for this interval is Request. When the MTA detects that voice is again present it uses one of these request opportunities to tell the CMTS it again needs data grants to transmit voice packets. The CMTS then switches back to providing dedicated data grants as in the UGS service. Note that until voice resumes, the MTA is also not using its dedicated request opportunities. So why is an MTA not using request opportunities so much better than an MTA not using data grants? The answer is because request intervals are very small compared to data grant intervals. For G.711 and 20-ms packetization, the typical grant size is 232 bytes. A DOCSIS request packet is a mere 6 bytes, so much less bandwidth is wasted.

One more important consideration with silence suppression should be addressed before moving on to the next topic. Are the reserved and/or committed resource envelopes different for a UGS service and a UGS-AD service? DQoS reserved resources for a UGS-AD service are going to be the same as a UGS service because enough resources have to be available for when the user is speaking. Therefore, implementing VAD is not going to allow you to have more calls on a DOCSIS upstream because the same amount of resources are reserved. Then the question is, "Will implementing VAD allow more Best Effort traffic to be transmitted?" This relates to whether the committed resource envelopes are different for a UGS and UGS-AD service. Remember, Best Effort traffic can utilize unused reserved resources, but it cannot utilize unused committed resources. So if the CMTS is not adjusting the committed resource envelope during periods of silence, Best Effort utilization is not affected either. If this is the case, then no benefit exists from a DOCSIS perspective of implementing VAD.

As an example, look at the UGS-AD service flow throughput during periods of talking:

```
C0103-7246VXR#show cable modem 24.34.240.247 qos
Sfid  Dir Curr  Sid   Sched  Prio MaxSusRate  MaxBrst    MinRsvRate  Throughput
          State       Type
31    US  act   15    BE     0    384000      3044       0           305
73    US  act   36    UGS_A  0    0           0          0           87199
32    DS  act   N/A   BE     0    3000000     3044       0           202
74    DS  act   N/A   BE     5    87200       1522       87200       86996
```

The results of the CMTS **show** command indicate that the throughput on the UGS-AD flow is 87,199 bps, which is approximately the expected value. The DOCSIS header and extended headers are not accounted for in this throughput number; therefore, the packet size is 218 bytes (200 byte IP packet + 18 byte DOCSIS MAC management header). So...

218 bytes/packet * 8 bits/byte * 50 packets/sec = 87,200 bits/sec.

As soon as silence is detected, you can tell the UGS-AD service flow throughput drops dramatically. The CMTS **show** command that follows indicates that the throughput has dropped to 2,536 bps:

```
C0103-7246VXR#show cable modem 24.34.240.247 qos
Sfid  Dir Curr  Sid   Sched  Prio MaxSusRate  MaxBrst   MinRsvRate  Throughput
          State       Type
31    US  act   15    BE     0    384000      3044      0           304
73    US  act   36    UGS_A  0    0           0         0           2536
32    DS  act   N/A   BE     0    3000000     3044      0           175
74    DS  act   N/A   BE     5    87200       1522      87200       85693
```

Admission Control of Service Flows

Because upstream DOCSIS bandwidth is typically the bottleneck in a PacketCable network, it is desirable to control how much upstream bandwidth is used for voice so that bandwidth is available for other applications, namely high-speed data. In conjunction with controlling voice bandwidth, it is also desirable to control this bandwidth according to the type of voice application. For example, you probably want to make sure some bandwidth is available for emergency calls separate from the bandwidth reserved for everyday calls.

The way to do this is by implementing an admission control policy on the CMTS upstream channels. Based on how your network is traffic engineered you can limit the percentage of bandwidth available for both voice and data traffic. You can also specify some small amount of bandwidth explicitly for the handling of emergency calls.

On a Cisco CMTS (with Cisco IOS Software Release 12.3[13]BC and later), admission control policies can be defined restricting the amount of bandwidth allocated to different DOCSIS service types (UGS, RTPS, BE, and so on) or to different defined service class names (for example, a service class can be defined for G.711 voice using a 20-ms packetization period). Not only can admission control policies be defined based on upstream bandwidth utilization, but they can also be defined based on downstream bandwidth utilization, CPU utilization, and memory resources. Admission control checks happen when cable modems register and when dynamic service creation is attempted (for example, when PacketCable voice flows are admitted) .

For each admission control policy enabled, three thresholds are defined:

- Minor
- Major
- Critical

For upstream and downstream bandwidth utilization policies, these thresholds represent the percent utilization at which these thresholds are met. When a minor or major threshold is reached, service flows are still admitted; however, an alarm is generated (via SNMP) to a preconfigured collection device. This alerts a system operator that bandwidth resources are becoming scarce. When a critical threshold is reached, service requests are gracefully declined. Also the critical thresholds for bandwidth resources can be defined based on exclusive resource amounts and non-exclusive resource amounts. Exclusive resources are reserved 100 percent for that particular scheduling type or service class, whereas nonexclusive resources can be shared among multiple scheduling types or service classes.

The configuration details of admission control on a Cisco CMTS are covered in Chapter 16, "PacketCable Network Design Considerations."

Messaging in an Emergency 911 Call

Because PacketCable voice services can be used to provide primary line voice service, the support for all the call feature functionality with a traditional telephone line must be available. One of these features is the ability to get in touch with the police, fire department, or medical personnel in the event of an emergency. As you know, in the United States dialing 911 accomplishes this.

As the PacketCable network is a distributed architecture where the call control element (that is, CMS) is separate from the bearer network (that is, CMTS), a means of distinguishing 911 calls throughout the network is required. Why is this needed? You find a couple of answers for this. One is for admission control; the other reason is that in the event bandwidth is fully utilized, you might want a means to give preference to emergency calls over ordinary phone calls. This second reason is known as *911 preemption* and can take place regardless of the existence of any admission control policies. How an emergency call works in terms of DQoS is illustrated next.

In terms of DQoS and NCS messaging, the difference is in the Gate Specifications parameter in the Gate-Set message. A trace of this message is shown in Figure 13-20.

The only difference in this Gate-Set message as opposed to the ones in the basic call example is the Session Class parameter. Notice that in this case it is set to High priority VoIP session instead of Normal priority VoIP session. This is how the CMS communicates to the CMTS to treat this call separately from an ordinary call.

Figure 13-20 *Ethereal Trace of Gate-Set for a 911 Call*

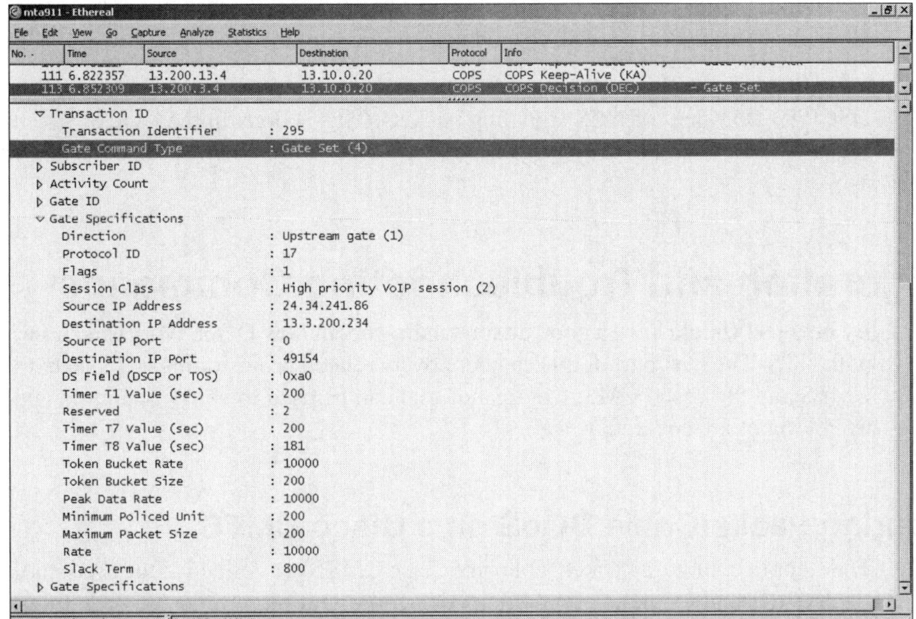

Tying DOCSIS QoS to Backbone QoS

Even though the quality of service mechanisms you learned about so far apply only to the DOCSIS network, these mechanisms can be manipulated to influence quality of service for the backbone network as well. This is also done through a parameter in the Gate-Set message. If you look at the Gate-Set message in Figure 13-20 used for the previous example, you can see a Gate Specification parameter called DS Field. The DS stands for Differentiated Services, and the field's value is the Differentiated Services Code Point (DSCP) or IP Type of Service, value to be applied to voice packets after they leave the CMTS. Usually, this field is relevant only for the upstream gate as these packets traverse the backbone IP network after they leave the CMTS. Packets passing through the downstream gate are going directly to the MTA from the CMTS, so there is no IP backbone requiring QoS.

As you might guess, this parameter is configured on the CMS so that it can be sent to the CMTS in the Gate-Set message. Later on you will see how to set this parameter on the Cisco BTS. When this parameter is set, the CMTS overwrites the DSCP or IP TOS byte with the value in the DS Field before sending it out the WAN interface. In this example, the DS Field is set to 0xa0 or in binary 1010 0000. This example also happens to be for an IP TOS interpretation of this field that starts with a 3-bit IP precedence value followed by five

IP type-of-service bits. So the IP precedence value is 101, which equates to a precedence of critical. This is commonly used for voice packets. All the remaining bits are set to zero. How this information is used by the backbone to provide QoS is addressed in Chapter 16.

Keep in mind that this technique addresses only bearer packets. You can also overwrite the DSCP/IP TOS field of NCS signaling packets. This is done in the DOCSIS configuration file in the definitions of the signaling service flows. More on this is discussed in Chapter 16 as well.

Configuration and Troubleshooting Commands

By now you should have a good understanding about how DQoS works and how it interacts with NCS. The next part of this chapter covers related provisioning steps on both the Cisco CMTS and the Cisco CMS. Commands that can be used to verify operation and troubleshoot are covered as well.

Configuring PacketCable DQoS on a Cisco CMTS

The steps to configure PacketCable on a Cisco CMTS are defined in the sections that follow. Each of these tasks includes recommended configuration settings; these settings should be fine for the general case but might require some adjusting depending on the environment.

Enabling PacketCable Operation on the CMTS

The most basic step, and the only essential one, is to enable PacketCable operation on the CMTS:

```
Router(config)#packetcable
```

Without this command the CMTS cannot respond or generate DQoS COPS commands, and PacketCable operation cannot exist. After this command is entered, the CMTS attempts to establish a COPS connection on the TCP connection created by the CMS.

Configure the Event Messaging Element ID

To configure the Event Messaging Element ID (optional), enter the following command:

```
Router(config)#packetcable element-id n
```

This is used to identify the CMTS in event messaging records; if this is not specified, the CMTS defaults to a random value between 0 and 99,999.

Specifying the Maximum Number of CMTS Allocated Gates

To specify the maximum number of gates allocated on the CMTS (Optional), enter the following command:

```
Router(config)#packetcable gate maxcount n
```

The range of values for *n* is 1 to 2,097,152, with 2,097,152 being the default value. This command refers to the total number of PacketCable 1.x gates plus the total number of PacketCable Multimedia gates. Keep that in mind if you have both services running on the CMTS.

NOTE Each PacketCable 1.x gate ID typically refers to an upstream and downstream gate, so divide this number by two to get the maximum number of PacketCable 1.x gates possible on the CMTS. By setting this parameter you are limiting the total number of allocated gates on the CMTS as a whole, not on a per-upstream basis. To limit gate creation on a per-upstream basis follow the admission control techniques covered earlier.

Specifying the PacketCable DQoS Timer Values

To specify PacketCable DQoS timer values (optional), enter the following commands:

```
Router(config)#packetcable timer T0 timer-value
Router(config)#packetcable timer T1 timer-value
```

Table 13-1 defines these timers.

Table 13-1 *PacketCable DQoS Timers*

Timer	Range (in milliseconds)	Default Value (in milliseconds)	Description
T0	1–1,000,000,000	30,000	Maximum amount of time between when a gate is allocated and authorized
T1	1–1,000,000,000	300,000	Maximum amount of time between when a gate is authorized and committed

Remember, timer T1 is typically set by the CMS in the Gate-Set message. The value specified here is used only if this parameter is not specified by the CMS. Also note the BTS in releases 4.x and later does the allocation and authorization steps together, so timer T0 would not come into play.

Configuring UGS-AD Flows to Switch to RTPS Scheduling

If you are using Unsolicited Grant Service with Activity Detection (UGS-AD) service flows in conjunction with voice activity detection, you can define the length of time inactivity must occur before the CMTS switches the service flow to RTPS. The command to do this is as follows and, if unspecified, the default value of 10 seconds is used:

```
Router(config)#cable service flow inactivity-threshold time
```

If you want to prevent UGS-AD flows from ever switching to RTPS scheduling, you can use the command **no cable service flow inactivity-threshold**. However, by doing this you are basically making UGS-AD equivalent to UGS, so no bandwidth savings occur during periods of silence.

Timing Out a UGS Flow

If you happen to be providing voice service using dynamically created UGS flows without PacketCable DQoS, you can get "hung" UGS service flows. This is because without DQoS you have no way to set timer T8, which maps to the DOCSIS activity timer. If the cable modem does not explicitly set this timer, the default value of 0 is assumed, which means "never time out." So a UGS flow without traffic continues to stay up indefinitely, consuming valuable DOCSIS resources that could be used by other customers. To protect against this, use the following configuration command to override the default value of 0:

```
Router(config)#cable service flow activity-timeout time
```

Verifying and Troubleshooting DQoS on a Cisco CMTS

You can verify DQoS COPS sessions are established on the CMTS with the following command:

```
Router#show cops servers
```

A sample output of this command is as follows:

```
C0103-7246VXR#show cops servers
COPS SERVER: Address: 13.13.6.196. Port: 57531. State: 0. Keepalive: 2 sec
     Number of clients: 1. Number of sessions: 1.
    COPS CLIENT: Client type: 32776.  State: 0.
COPS SERVER: Address: 13.13.5.4. Port: 32801. State: 0. Keepalive: 2 sec
     Number of clients: 1. Number of sessions: 1.
    COPS CLIENT: Client type: 32776.  State: 0.
COPS SERVER: Address: 172.18.98.62. Port: 36196. State: 0. Keepalive: 60 sec
     Number of clients: 1. Number of sessions: 1.
    COPS CLIENT: Client type: 32778.  State: 0.
```

In this example you have two CMSs with active COPS sessions to this CMTS. These are the first two COPS servers listed. You can tell they are PacketCable 1.x CMSs, based on the client type of 32776. In hexadecimal, this is 0x8008, which is the value used for COPS in PacketCable 1.x. The third COPS server listed is a PCMM policy server. Client type 32778

(0x800A) is used for COPS in PCMM. Also notice the CMS/Policy Server IP address and TCP port number are shown as well as the value of the COPS keepalive timer.

Recall that if COPS security is not enabled, the CMTS automatically initiates the COPS connection after the CMS (or PCMM policy server) initiates the TCP connection. So if the COPS connection is not showing up on the CMTS, you most likely have a misconfiguration on the CMS or a routing issue between these devices. To see the TCP connections on the CMTS (including those used for COPS) use the **show tcp brief** command as demonstrated here:

```
C0103-7246VXR#show tcp brief
TCB       Local Address          Foreign Address        (state)
649984B0  13.13.0.34.23          172.18.98.229.2402     ESTAB
63DE8DEC  13.10.0.19.2126        13.14.6.12.64503       ESTAB
6436017C  13.10.0.19.2126        13.13.6.196.57531      ESTAB
649BFE88  13.10.0.19.3918        172.18.98.62.36196     ESTAB
63F7CADC  13.10.0.19.2126        13.13.83.116.49879     ESTAB
6494FCB0  13.10.0.19.2126        13.13.5.4.32801        ESTAB
```

Remember, PacketCable 1.x DQoS sessions use port 2126 on the CMTS (PCMM sessions use port 3918).

Displaying Call History

The CMTS maintains a history of active and recent phone calls on a per cable interface basis. The CMTS also distinguishes between high priority (emergency 911) calls and ordinary calls. The command **show cable calls** displays this information as demonstrated here:

```
C0103-7246VXR#show cable calls
Interface   ActiveHiPriCalls  ActiveAllCalls  PostHiPriCallCMs  RecentHiPriCMs
Cable3/0    0                 0               0                 0
Cable4/0    0                 0               0                 0
Cable5/0    0                 2               0                 0
Cable6/0    0                 0               0                 0
Total       0                 2               0                 0
```

The category PostHiPriCallCMs details the number of regular active calls made after a recent emergency 911 call. The category RecentHiPriCMs details the number of recent emergency 911 calls made. In this example, you can see two regular active calls on Cable interface 5/0.

The length of time during which the CMTS keeps records of emergency 911 calls can be configured with the following global configuration command:

```
Router(config)#cable high-priority-call-window minutes
```

The *minutes* value can be from 0–10,080 minutes (10,080 minutes = 1 week), and 0 is the default.

A similar command to the **show cable calls** command is the **show cable modem calls** command, which lists all the cable modems with active calls or recent high priority calls, as demonstrated in the following output:

```
C0103-7246VXR#show cable modem calls
Cable Modem Call Status Flags:
H: Active high priority calls
R: Recent high priority calls
V: Active voice calls (including high priority)
MAC Address     IP Address      I/F       Prim  CMCallStatus  LatestHiPriCall
                                          Sid                    (min:sec)
0013.1050.3801 13.34.14.122    C5/0/U0    2     V                     -
000a.73fa.be8a 13.34.14.78     C5/0/U0    15    V                     -
```

Verifying Current PacketCable Settings

Current PacketCable settings can be verified using the following command:

```
Router#show packetcable global
```

The following example demonstrates sample output from this command:

```
C0103-7246VXR#show packetcable global
Packet Cable Global configuration:
Packetcable DQOS Enabled   : Yes
Packetcable Multimedia Enabled   : Yes
Element ID: 78819
Max Gates : 2097152
Not Allow non-PacketCable UGS
Intercept Hold Content: Enabled
Default Multimedia Timer value -
  T1              : 200000 msec
  Persistent gate : 0 hour
  Volume Limit    : STOPPED
Default DQoS Timer value -
  T0      : 30000 msec
  T1      : 300000 msec
```

In this example, you can see that PacketCable operation has been enabled (as well as PacketCable Multimedia operation). PCMM configuration is covered in Chapter 14, "Multimedia Applications." You also see the default values of timer T0 and T1, as well as the maximum number of allocated gates permitted. The element ID is used for Event Messaging records; more on this parameter is covered in Chapter 15, "Lawful Intercept." The Not Allow non-PacketCable UGS option is default and means that UGS service flows that do not have a correlated authorization block parameter or DQoS gate ID are not permitted. If you want to allow UGS service flows to exist for non-PacketCable applications, this option can be changed with the following command:

```
Router(config)#packetcable authorize vanilla-docsis-mta
```

Displaying Information About Active DQoS Gates

You can display information about active DQoS gates by using the following command:

```
C0103-7246VXR#show packetcable gate ?
  <0-4294967295>  gate id
  counter         gate counter
  downstream      downstream gates
  dqos            Packetcable dqos gate
  multimedia      Packetcable multimedia gate
  summary         gate summary
  upstream        upstream gates
```

Several options are associated with this command. The following paragraphs look at these options one at a time.

The **counter** option enables you to see the number of gates committed since the CMTS was last reset or since the counter was cleared, as demonstrated in the following output:

```
C0103-7246VXR#show packetcable gate counter commit
Total gates committed(since bootup or clear counter) = 20
```

This command shows all gates committed, including PCMM gates. If you want to see just PacketCable 1.x DQoS gates or PCMM gates, you can use the keyword **dqos** or **multimedia** at the end of the command.

You can reset this counter using the following command:

```
Router#clear packetcable gate counter commit
```

The **summary** option displays all the currently active gates and their associated states. These states include ALLOC (allocated), AUTH (authorized), RSVD (reserved), and COMMIT (committed). Two other possible states, INVLD (invalid) and UNKWN (unknown), indicate error situations. A sample output of this command is as follows:

```
C0103-7246VXR#show packetcable gate summary
GateID    i/f      SubscriberID    GC-Addr        State     Type  SFID(us) SFID(ds)
5166      Ca5/0    24.34.240.247   13.13.5.4      COMMIT    DQoS  81       82
21550     Ca5/0    24.34.240.235   13.13.5.4      COMMIT    DQoS  79       80
Total number of  gates = 2
Total Gates committed(since bootup or clear counter) = 20
```

In this example, two active gates are in the committed state. Their gate identifiers, service flow identifiers, MTA IP addresses, CMS IP addresses, and the cable linecard to which the MTA is connected are all provided.

The remaining options enable you to view more detailed information about a DQoS gate. By specifying the gate ID you can see all the information that was included within the Gate-Set message for that gate, as demonstrated here:

```
C0103-7246VXR#show packetcable gate 5166
GateID                  : 5166
    Subscriber ID       : 24.34.240.247
    COPS connection     :
        server handle   : 0x63F35F90
        server address  : 13.13.5.4
        server port     : 32801
        client address  : 13.10.0.19
```

```
State              : COMMIT
CALEA Version      :  -
Gate specs [UPSTREAM]
  Gate classifier  :  protocol 17,
                      src addr/port 24.34.240.247/0,
                     dest addr/port 24.34.240.235/53456
  diffserv dscp    : 0xA0
  timer t1(s)     : 200
  timer t7(s)     : 200
  timer t8(s)     : 300
  commit flags     : 0x0
  session class    : 0x1
  flowspec # 1     : [r/b/p/m/M 10000/200/10000/200/200] [R/S: 10000/800]
Gate specs [DOWNSTREAM]
  Gate classifier  :  protocol 17,
                      src addr/port 24.34.240.235/0,
                     dest addr/port 24.34.240.247/53456
  diffserv dscp    : 0xA0
  timer t1(s)     : 200
  timer t7(s)     : 200
  timer t8(s)     : 300
  commit flags     : 0x0
  session class    : 0x1
  flowspec # 1     : [r/b/p/m/M 10000/200/10000/200/200] [R/S: 10000/0]
```

In this example, you see all the DQoS information associated with both the upstream and downstream gates using this gate identifier. To see only the upstream or only the downstream gate information, you can use the keywords **upstream** or **downstream** before the gate identifier.

Displaying Service Flow Information

In the **show packetcable gate summary** output in the previous section, you might notice that the cable interface and the service flow identifiers associated with the gate are listed. You can obtain information pertaining to specific service flows with the **show interface cable** *x/y* **service-flow** *z* command as demonstrated in the following examples:

```
C0103-7246VXR#show interface cable 5/0 service-flow 81
Sfid  Sid   Mac Address     QoS Param Index    Type     Dir     Curr   Active
                            Prov  Adm  Act                      State  Time
81    40    000a.73fa.be8a   0      20   20    dyn(S)   US      act    06:56
C0103-7246VXR#show interface cable 5/0 service-flow 82
Sfid  Sid   Mac Address     QoS Param Index    Type     Dir     Curr   Active
                            Prov  Adm  Act                      State  Time
82    N/A   000a.73fa.be8a   0      23   23    dyn(S)   DS      act    50:32
```

First, the upstream service flow for the MTA is listed, and then the downstream service flow is listed. Notice they are both currently in the act (Active) state. The length of time they have been active is also indicated. The reason the upstream gate indicates such a smaller active time is that the MTA is sending DSC refresh messages for the upstream service flow, which cause the active timer to get reset.

Also notice the QoS parameter set indices for the provisioned, admitted, and active resources. These indices define these resource envelopes. Because these service flows are dynamic, the provisioned index is always 0, and most of the time the admitted and active indices are identical. An exception to this is if the MTA commits DQoS resources using a compression CODEC for voice but reserves DQoS resources for G.711 in case voice band data is detected. You can display the DOCSIS parameters defining these resources with the commands demonstrated in the following examples:

```
C0103-7246VXR#show interfaces cable 5/0 qos paramset 20 verbose
Index:                                    20
Name:
Direction:                                Upstream
Minimum Packet Size                       0 bytes
Admitted QoS Timeout                      200 seconds
Active QoS Timeout                        300 seconds
Scheduling Type:                          Unsolicited Grant Service
Request/Transmission Policy:              0x17F
Unsolicited Grant Size:                   232 bytes
Nominal Grant Interval:                   20000 usecs
Tolerated Grant Jitter:                   800 usecs
Grants per Interval:                      1
IP ToS Overwrite [AND-mask,OR-mask]:      0xFF,0x0
Parameter Presence Bitfield:              {0xC00, 0x3CC000}
C0103-7246VXR#show interfaces cable 5/0 qos paramset 23 verbose
Index:                                    23
Name:
Direction:                                Downstream
Traffic Priority:                         5
Maximum Sustained Rate:                   87200 bits/sec
Max Burst:                                1522 bytes
Minimum Reserved Rate:                    87200 bits/sec
Minimum Packet Size                       218 bytes
Admitted QoS Timeout                      0 seconds
Active QoS Timeout                        0 seconds
Scheduling Type:                          Reserved
Max Latency:                              100000 usecs
Parameter Presence Bitfield:              {0x13E0, 0x0}
```

You can append the keywords **classifiers verbose** to the **show interface cable** *x/y* **service-flow** *z* command to see all the classifiers and their parameters associated with that service flow, as demonstrated in the following example:

```
C0103-7246VXR#show interface cable 5/0 service-flow 81 classifiers verbose
Sfid                                : 81
Classifiers:
Classifier Id                       : 21
Service Flow Id                     : 81
CM Mac Address                      : 000a.73fa.be8a
Direction                           : upstream
Activation State                    : active
Classifier Matching Priority        : 128
PHSI                                : 0
Number of matches                   : -
IP Classification Parameters:
        IP Source Address           : 24.34.240.247
        Source IP Address Mask      : 255.255.255.255
        Destination IP Address      : 24.34.240.235
        Destination IP Address Mask : 255.255.255.255
        IP Protocol Type            : 17
        Source Port Low             : 53456
        Source Port High            : 53456
        Destination Port Low        : 53456
        Destination Port High       : 53456
```

Similarly, you can see the Payload Header Suppression parameters associated with a service flow by appending the keywords **phs verbose** to the previous command instead of the keywords **classifiers verbose**.

By appending the keywords **counters verbose** to the **show interface cable** *x/y* **service-flow** *z* command, you can see the current throughput rates of the service flow in terms of bits per second (bps) and packets per second. You can also see the total number of packets and bytes that have traversed the service flow and if you have had any rate limit delayed or dropped packets, as demonstrated in the following example:

```
C0103-7246VXR#show interface cable 5/0 service-flow 81 counters verbose
Sfid                    : 81
Packets                 : 238110
Octets                  : 51907980
RateLimit Delayed Pkts  : 0
RateLimit Dropped Pkts  : 0
Bits/sec                : 87200
Packets/Sec             : 49
```

Remember that a 20-ms packetization rate implies 50 packets are sent per second, and for the G.711 CODEC this results in a bit rate of 87,200 bps (218 bytes/packet * 8 bits/byte * 50 packets/sec = 87,200 bps). For UGS service flows you should never have any rate limited packets because these flows have dedicated transmit opportunities. For downstream service flows you can have rate limited packets if the downstream is very heavily congested. However, these downstream flows are also prioritized, so this shouldn't occur either. If you notice these counters have nonzero values for voice service flows, this most likely indicates a problem with the CMTS. These types of problems should immediately be investigated because rate limited packets can definitely result in voice quality issues.

By simply using the **verbose** option on the **show interface cable** *x/y* **service-flow** *z* command, you can see all the information previously discussed (that is, classifiers, phs, and counters information).

Examining DOCSIS SID Information in Upstream Service Flows

You can also obtain information for upstream service flows by looking at DOCSIS SID (Service ID) information, as demonstrated in the following output:

```
C0103-7246VXR#show interface cable 5/0 sid 40
Sid  Prim  MAC Address     IP Address      Type Age     Admin   Sched  Sfid
                                                        State   Type
40   15    000a.73fa.be8a 13.34.14.78      dyn  1h30m   enable  UGS    81
```

Notice the command output indicates this is a dynamically created UGS service, and the length of time it has been active is displayed. Unlike with the service flow, output this time is not re-sent when DSC refresh messages are received.

You can see the DOCSIS parameters as well as the current throughput by appending the keywords **qos verbose** to the **show interface cable** *x/y* **sid** *z* command, as demonstrated here:

```
C0103-7246VXR#show interface cable 5/0 sid 40 qos verbose
Sid                                      : 40
Minimum Packet Size                      : 0 bytes
Admitted QoS Timeout                     : 200 seconds
Active QoS Timeout                       : 300 seconds
Scheduling Type                          : Unsolicited Grant Service
Unsolicited Grant Size                   : 232 bytes
Nominal Grant Interval                   : 20000 usecs
Grants per interval                      : 1
Tolerated Grant Jitter                   : 800 usecs
Request/Transmission policy              : 0x17F
IP ToS Overwrite[AND-mask, OR-mask]      : 0xFF, 0x0
Current Throughput                       : 87200 bits/sec, 49 packets/sec
```

You can also obtain counter information with this command by appending the keywords **counters verbose** as demonstrated here:

```
C0103-7246VXR#show interface cable 5/0 sid 40 counters verbose
Sid                             : 40
Request polls issued            : 0
BWReqs {Cont,Pigg,RPoll,Other}  : 0, 0, 0, 0
No grant buf BW request drops   : 0
Rate exceeded BW request drops  : 0
Grants issued                   : 285712
Packets received                : 285509
Bytes received                  : 62240962
Fragment reassembly completed   : 0
Fragment reassembly incomplete  : 0
Concatenated packets received   : 0
Queue-indicator bit statistics  : 21601 set, 180 granted
Good Codewords rx               : 0
Corrected Codewords rx          : 0
Uncorrectable Codewords rx      : 0
Concatenated headers received   : 0
Fragmentation headers received  : 0
Fragmentation headers discarded: 0
```

A great deal of valuable troubleshooting information is available with the output of this command. The number of contention and piggyback bandwidth requests made by this SID is shown; recall these types of requests are not allowable by PacketCable MTAs because they have dedicated opportunities to send data. Consequently, these counters should always be equal to 0. The number of bandwidth request drops because of no upstream bandwidth being available or because the modem exceeds its allotted rate are also shown. These should also always be 0 for PacketCable MTAs because their resources have been guaranteed. The Queue-indicator bit statistics is a field that is very useful for voice troubleshooting.

Recall that in the UGS extended header is a QI bit that a modem can set when it detects a backlog of UGS packets waiting to be sent. A modem continues to set this bit until the backlog is cleared. Per DOCSIS, a CMTS should grant up to 1 percent of additional bandwidth to compensate for clock rate mismatches. Thus, as you can tell in this example, a little less than 1 percent of these requests have been granted. Another good place for diagnosing problems with UGS flows comes from looking at the Grants issued and Packets received counters. These should be approximately equal as the CMTS is providing exactly the number of data grants to handle the RTP packets sent by the MTA. In the preceding example, notice these parameters differ by approximately the number of QI requests granted. Thus, for some reason, the MTA is missing grants and trying to catch up using the QI bit indicator. This most likely indicates a problem with the MTA, possibly related to synchronizing its CODEC clock with the DOCSIS clock.

Much of the information obtained from service flow and SID show commands can also be seen with **show cable modem** <*Modem IP* or *MAC*> command output. By appending the keyword calls you can see whether the modem has any active or recent emergency call activity. You can use the keyword classifiers to see classifier parameters and the current number of matches. Use the cpe keyword to see the IP address and MAC address of the embedded MTA or other CPE devices behind the cable modem. Finally, you can use the qos keyword to display all service flow parameters associated with that modem and their current throughputs.

Displaying Upstream Statistics

You can view upstream statistics, such as the total number of received frames with correctable and uncorrectable Forward Error Correction (FEC) errors on, with the **show interface cable** *x/y* **upstream** *z* command as demonstrated here:

```
C0103-7246VXR#show interface cable 5/0 upstream 0
Cable5/0: Upstream 0 is up
     Received 22338 broadcasts, 39894 multicasts, 3287890 unicasts
     0 discards, 2625 errors, 0 unknown protocol
     3350122 packets input
     Codewords: 3238801 good 29 corrected 21 uncorrectable
     76 noise, 0 microreflections
     Total Modems On This Upstream Channel : 12 (12 active)
```

By adding the keywords **ugs statistics** to the end of the **show interface cable** *x/y* **upstream** *z* command, you can see the number of active UGS flows on the upstream as well as the maximum, minimum, and average number of UGS flows on the upstream in the last hour and in the last 5 minutes, as demonstrated in the following output:

```
C0103-7246VXR#show interface cable 5/0 upstream 0 ugs statistics
UGS Statistics for Upstream 0
# of Active UGS on the Upstream : 4
                      UGS Allocation Statistics
                     max          min          avg
Last 1 Hour          4            0            2
Last 5 Min           4            3            4
```

Displaying Upstream Capacity Issues

A useful command for diagnosing upstream capacity issues is **show interface cable** *x/y* **mac-scheduler** *z*, where *z* is the Upstream Channel ID, as demonstrated in the following output:

```
C0103-7246VXR#show interface cable 5/0 mac-scheduler 0
     DOCSIS 1.1 MAC scheduler for Cable5/0/U0
     Queue[Rng Polls] 0/128, 0 drops, max 1
     Queue[CIR Grants] 0/64, 0 drops, max 1
     Queue[BE(7) Grants] 0/64, 0 drops, max 1
     Queue[BE(6) Grants] 0/64, 0 drops, max 0
     Queue[BE(5) Grants] 0/64, 0 drops, max 0
     Queue[BE(4) Grants] 0/64, 0 drops, max 2
     Queue[BE(3) Grants] 0/64, 0 drops, max 2
     Queue[BE(2) Grants] 0/64, 0 drops, max 0
     Queue[BE(1) Grants] 0/64, 0 drops, max 2
     Queue[BE(0) Grants] 0/64, 0 drops, max 2
     Req Slots 1146301987, Req/Data Slots 0
     Init Mtn Slots 15354859, Stn Mtn Slots 1105561
     Short Grant Slots 36410, Long Grant Slots 1671233
     ATDMA Short Grant Slots 0, ATDMA Long Grant Slots 0
     ATDMA UGS Grant Slots 0
     Awacs Slots 0
     Fragmentation count 295
     Fragmentation test disabled
     Avg upstream channel utilization : 3%
     Avg percent contention slots : 94%
     Avg percent initial ranging slots : 3%
     Avg percent minislots lost on late MAPs : 0%
     Sched Table Rsv-state: Grants 0, Reqpolls 0
     Sched Table Adm-State: Grants 4, Reqpolls 0, Util 5%
     UGS    : 4 SIDs, Reservation-level in bps 370400  <- 4 active UGS flows
     UGS-AD : 0 SIDs, Reservation-level in bps 0
     RTPS   : 0 SIDs, Reservation-level in bps 0
     NRTPS  : 0 SIDs, Reservation-level in bps 0
     BE     : 18 SIDs, Reservation-level in bps 0
```

As you can see, in this case you have four active UGS flows on this DOCSIS upstream, and 370,400 bps of bandwidth have been reserved to accommodate these flows. With this command you can get all sorts of information concerning QoS on this upstream. You can also see that the four UGS grants consume approximately 5 percent of the upstream bandwidth.

Displaying Downstream Capacity Issues

You can obtain information for diagnosing downstream capacity issues using the **show interface cable** *x/y* **downstream** command, as demonstrated in the following output:

```
C0103-7246VXR#show interface cable 5/0 downstream
Cable5/0: Downstream is up
     661071205 packets output, 3161281948 bytes, 0 discarded
     0 output errors
     21 total active devices, 12 active modems
        Total downstream bandwidth: 42884 Kbps
        Total downstream reserved/reservable bandwidth: 352/38595 Kbps
        Estimated usable bandwidth: 37500 Kbps
        Worst case latency for low latency queue: 1280 usecs
```

```
      Current Upper limit for worst case latency: 100000 usecs
      Ranging lists counts
   Maintainence list 12, max 0
   Continous list 0, max 0
   Pending list 0, max 0
```

As you can see, 352 kbps of downstream bandwidth has been reserved. In this case, the only reserved bandwidth is for the four active calls (notice the amount of reserved downstream bandwidth is roughly the same as the amount of reserved upstream bandwidth).

Displaying Statistical Information on DOCSIS DSX Messages Sent on the Downstream Channel

To obtain statistical information on DOCSIS DSX messages sent on the downstream channel, use the **show controller cable** *x/y* **downstream** command, as demonstrated in the following output:

```
C0103-7246VXR#show controller cable 5/0 downstream
 Cable5/0 Downstream is up
  Frequency 741.0000 MHz, Channel Width 6 MHz, 256-QAM, Symbol Rate 5.360537 Msps
  FEC ITU-T J.83 Annex B, R/S Interleave I=32, J=4
  Downstream channel ID: 4
  Dynamic Services Stats:
  DSA: 0  REQs  84 RSPs  0 ACKs
  0 Successful DSAs  0 DSA Failures
  DSC: 0  REQs  274 RSPs  0 ACKs
  0 Successful DSCs  0 DSC Failures
  DSD:  7 REQs  72 RSPs
  6 Successful DSDs  1 DSD Failures
  DCC: 0  REQs  0 RSPs  0 ACKs
  0 Successful DCCs  0 DCC Failures
  DCC end of transaction counts:
  DCC unknown cause(0) offline(0) if down(0) no cm(0)
  DCC no resource(0) no retries(0) reject(0) unknown state (0)
  DCC rebuild err (0) T15 timeout(0) reinit MAC (0) dcc succeeds(0)
```

In a PacketCable 1.x environment you see Response (RSP) hits only for the DSA and DSC dynamic service messages. This is because in PacketCable 1.x DSA and DSC transactions are always initiated from the MTA; therefore, the only downstream messages in the three-way exchange are the responses. DSD message exchanges are also usually initiated by the MTA; however, in error situations and cases such as activity timer expiry these exchanges are initiated from the CMTS. As you can see in this example, 72 DSD-RSPs have been sent on the downstream while only 7 DSD-REQs have been sent (of these 7 CMTS-initiated DSD exchanges, 6 were successful) .

Displaying General Information for the Cable Line Card

To obtain overall information on the cable linecard, use the **show interface cable** *x/y* command, as demonstrated in the following output:

```
C0103-7246VXR#show interface cable 5/0
Cable5/0 is up, line protocol is up
```

```
Hardware is BCM3210 ASIC, address is 000e.39d2.d48c (bia 000e.39d2.d48c)
MTU 1500 bytes, BW 37500 Kbit, DLY 1000 usec,
   reliability 255/255, txload 1/255, rxload 1/255
Encapsulation MCNS, loopback not set
Keepalive not set
ARP type: ARPA, ARP Timeout 08:00:00
Last input 00:00:00, output 00:00:00, output hang never
Last clearing of "show interface" counters never
Input queue: 0/1024/0/0 (size/max/drops/flushes); Total output drops: 0
Queueing strategy: weighted fair
Output queue: 0/1024/64/0 (size/max total/threshold/drops)
   Conversations  0/1/256 (active/max active/max total)
   Reserved Conversations 9/9 (allocated/max allocated)
   Available Bandwidth 28125 kilobits/sec
30 second input rate 247000 bits/sec, 200 packets/sec
30 second output rate 247000 bits/sec, 200 packets/sec
   2142149 packets input, 394418703 bytes, 1 no buffer
   Received 62269 broadcasts, 0 runts, 0 giants, 0 throttles
   1 input errors, 1 CRC, 0 frame, 0 overrun, 0 ignored, 0 abort
   2157552 packets output, 385305856 bytes, 0 underruns
   0 output errors, 0 collisions, 4 interface resets
   0 output buffer failures, 0 output buffers swapped out
```

As you know, a call using a 20-ms packetization period sends 50 packets per second. As you can see, the highlighted output indicates that currently 200 packets/sec are being sent and received on this cable interface. This is because the only active traffic is four phone calls using a 20-ms packet rate.

Debugging PacketCable DQoS

This section covers some **debug** commands that you can use to see DQoS-related messages. As a rule of thumb, try to troubleshoot problems first by using the **show** commands or by a protocol sniffer because **debug** commands can output quite a bit of data, so you should exercise caution when using them in a live production environment.

Debugging DOCSIS Dynamic Service Flow Messages During a Call

The first step in debugging DOCSIS dynamic service flow messages during a call is to enable a debug on the particular cable modems involved in the call. The syntax of this command is as follows:

```
Router#debug cable mac-address modem-mac-address [verbose]
```

The *modem-mac-address* is the cable modem MAC address of the MTA used in the example. Remember this is the cable modem MAC address, not the MTA MAC address. The **verbose** option gives more detail in the output. Without the **verbose** option the details of each message and parameter are not provided.

Identifying the modems to debug is the first step; next, turn on the protocol debugs of interest. To see the DOCSIS dynamic service flow messages, enable the following debug:

```
Router#debug cable dynsrv
```

Recall that in DOCSIS messages, most of the parameters are encoded in Tag/Length/Value, or TLV, format. Enable the following **debug** as well to see these parameters decoded. Without this **debug** enabled, the dynamic service messages are shown only in hex format.

```
Router#debug cable tlvs
```

The following output shows key parts of the resulting output from turning on the previous three debugs during a call. Note, only the originating device was traced, yet the amount of output is still quite large (the highlighted comments point out key parts of the **debug** output).

```
C0103-7246VXR#show debug
CMTS:
  CMTS dynsrv debugging is on
  CMTS TLV encodings debugging is on
CMTS specific:
  Debugging is on for Address 0013.1050.3801, Mask ffff.ffff.ffff (verbose)
Aug  9 19:28:49.792: DSA-REQ-RECD: OrgMac->0013.1050.3801 OrgId->89
Aug  9 19:28:49.792: DSA-STATE-CREATED: OrgMac->0013.1050.3801 OrgId->89
Aug  9 19:28:49.796: Found Upstream Service Flow TLV
Aug  9 19:28:49.796:     Service Flow Reference : 1
Aug  9 19:28:49.796:     QoS Parameter Set Type : 0x2      <- Admit service flow only
Aug  9 19:28:49.796:     Scheduling Type : 6             <- UGS upstream scheduling type
Aug  9 19:28:49.796:     Request/Transmission Policy : 0x17F
Aug  9 19:28:49.796:     Unsolicited Grant Size : 232  <- Standard UGS size for G.711/
  20ms
Aug  9 19:28:49.796:     Nominal Grant Interval : 20000     <- 20 ms grant interval
Aug  9 19:28:49.796:     Tolerated Grant Jitter : 800
Aug  9 19:28:49.796:     Grants Per Interval : 1
Aug  9 19:28:49.796: Found Upstream Packet Classifier TLV
Aug  9 19:28:49.796:     Classifier Reference : 1
Aug  9 19:28:49.796:     Service-Flow Reference : 1
Aug  9 19:28:49.796:     Rule Priority : 128
Aug  9 19:28:49.796:     Activation State : 0              <- Classifier not active yet
Aug  9 19:28:49.796:     Found IP Packet Classifier Sub-TLV
Aug  9 19:28:49.796:             Protocol : 17
Aug  9 19:28:49.796:             Source Address : 24.34.240.235
Aug  9 19:28:49.796:             Destination Address : 24.34.240.247
Aug  9 19:28:49.796:             Source Port Start : 53456         <- RTP port numbers
Aug  9 19:28:49.796:             Source Port End : 53456
Aug  9 19:28:49.796:             Destination Port Start : 53456
Aug  9 19:28:49.796:             Destination Port End : 53456
Aug  9 19:28:49.796: Found Downstream Service Flow TLV
Aug  9 19:28:49.796:     Service Flow Reference : 2
Aug  9 19:28:49.796:     QoS Parameter Set Type : 0x2      <- Admit service flow only
Aug  9 19:28:49.796:     Traffic Priority : 5   <- DS voice given relatively high
  priority
Aug  9 19:28:49.796:     Maximum Sustained Traffic Rate : 87200
Aug  9 19:28:49.796:     Maximum Traffic Burst : 1522
Aug  9 19:28:49.796:     Minimum Reserved Traffic Rate : 87200
Aug  9 19:28:49.796:     Minimum Reserved Rate Packet Size : 218
Aug  9 19:28:49.796: Found Downstream Packet Classifier TLV
Aug  9 19:28:49.796:     Classifier Reference : 2
Aug  9 19:28:49.796:     Service-Flow Reference : 2
Aug  9 19:28:49.796:     Rule Priority : 128
Aug  9 19:28:49.796:     Activation State : 0
Aug  9 19:28:49.796:     Found IP Packet Classifier Sub-TLV
Aug  9 19:28:49.796:             Protocol : 17
Aug  9 19:28:49.796:             Source Address : 24.34.240.247
Aug  9 19:28:49.796:             Destination Address : 24.34.240.235
Aug  9 19:28:49.796:             Destination Port Start : 53456
```

```
Aug  9 19:28:49.796:           Destination Port End : 53456
Aug  9 19:28:49.796: Auth Block:
Aug  9 19:28:49.796: 0x0000: 01 06 01 04 00 00 14 3E    <- DQoS Gate ID contained here
Aug  9 19:28:49.796: Sfref = 1, SFID = 103        <- Service Flow IDs assigned by CMTS
Aug  9 19:28:49.796: Sfref = 2, SFID = 104
Aug  9 19:28:49.796: Cfr-ref = 1, CFID = 33, SF-ref 1, SFID 103<- Classifier IDs
 assigned
Aug  9 19:28:49.796: Cfr-ref = 2, CFID = 34, SF-ref 2, SFID 104
Aug  9 19:28:49.796: Added Auth Block(633A2828) len=14    <- DQoS parms added
Aug  9 19:28:49.796: DSA-RSP-SENT: CM->0013.1050.3801 TranscId->89 ConfCode->0
Aug  9 19:28:49.896: DSA-ACK-RECD: OrgMac->0013.1050.3801 OrgId->89 ConfCode->0
Aug  9 19:28:50.196: DSA-REQ End. Transaction over-T8 timer expired. OrgMac-
 >0013.1050.3801 OrgId->89
Aug  9 19:28:50.196: DYN-SRV-STATE-DESTROYED : OrgMac->0013.1050.3801 OrgId->89
Aug  9 19:28:54.996: DSC-REQ-RECD: OrgMac->0013.1050.3801 OrgId->90
Aug  9 19:28:54.996: DSC-STATE-CREATED: OrgMac->0013.1050.3801 OrgId->90
Aug  9 19:28:54.996: Found Upstream Service Flow TLV
Aug  9 19:28:54.996:    Service Flow Id : 103
Aug  9 19:28:54.996:    QoS Parameter Set Type : 0x2
Aug  9 19:28:54.996:    Scheduling Type : 6
Aug  9 19:28:54.996:    Request/Transmission Policy : 0x17F
Aug  9 19:28:54.996:    Unsolicited Grant Size : 232
Aug  9 19:28:54.996:    Nominal Grant Interval : 20000
Aug  9 19:28:54.996:    Tolerated Grant Jitter : 800
Aug  9 19:28:54.996:    Grants Per Interval : 1
Aug  9 19:28:54.996: Found Upstream Service Flow TLV
Aug  9 19:28:54.996:    Service Flow Id : 103
Aug  9 19:28:54.996:    QoS Parameter Set Type : 0x4     <- Activate these resources
Aug  9 19:28:54.996:    Scheduling Type : 6
Aug  9 19:28:54.996:    Request/Transmission Policy : 0x17F
Aug  9 19:28:54.996:    Unsolicited Grant Size : 232
Aug  9 19:28:54.996:    Nominal Grant Interval : 20000
Aug  9 19:28:54.996:    Tolerated Grant Jitter : 800
Aug  9 19:28:54.996:    Grants Per Interval : 1
Aug  9 19:28:54.996: Found Upstream Packet Classifier TLV
Aug  9 19:28:54.996:    Classifier Id : 33
Aug  9 19:28:54.996:    Service Flow Id : 103
Aug  9 19:28:54.996:    Rule Priority : 128
Aug  9 19:28:54.996:    Activation State : 1              <- Activate classifier
Aug  9 19:28:54.996:    DSC Action : 1
Aug  9 19:28:54.996:    Found IP Packet Classifier Sub-TLV
Aug  9 19:28:54.996:           Protocol : 17
Aug  9 19:28:54.996:           Source Address : 24.34.240.235
Aug  9 19:28:54.996:           Destination Address : 24.34.240.247
Aug  9 19:28:54.996:           Source Port Start : 53456
Aug  9 19:28:54.996:           Source Port End : 53456
Aug  9 19:28:54.996:           Destination Port Start : 53456
Aug  9 19:28:54.996:           Destination Port End : 53456
Aug  9 19:28:54.996: Found Downstream Service Flow TLV
Aug  9 19:28:54.996:    Service Flow Id : 104
Aug  9 19:28:54.996:    QoS Parameter Set Type : 0x2
Aug  9 19:28:54.996:    Traffic Priority : 5
Aug  9 19:28:54.996:    Maximum Sustained Traffic Rate : 87200
Aug  9 19:28:54.996:    Maximum Traffic Burst : 1522
Aug  9 19:28:54.996:    Minimum Reserved Traffic Rate : 87200
Aug  9 19:28:54.996:    Minimum Reserved Rate Packet Size : 218
Aug  9 19:28:54.996: Found Downstream Service Flow TLV
Aug  9 19:28:54.996:    Service Flow Id : 104
Aug  9 19:28:54.996:    QoS Parameter Set Type : 0x4
Aug  9 19:28:54.996:    Traffic Priority : 5
Aug  9 19:28:54.996:    Maximum Sustained Traffic Rate : 87200
Aug  9 19:28:54.996:    Maximum Traffic Burst : 1522
Aug  9 19:28:54.996:    Minimum Reserved Traffic Rate : 87200
Aug  9 19:28:54.996:    Minimum Reserved Rate Packet Size : 218
```

```
Aug  9 19:28:54.996: Found Downstream Packet Classifier TLV
Aug  9 19:28:54.996:    Classifier Id : 34
Aug  9 19:28:54.996:    Service Flow Id : 104
Aug  9 19:28:54.996:    Rule Priority : 128
Aug  9 19:28:54.996:    Activation State : 1
Aug  9 19:28:54.996:    DSC Action : 1
Aug  9 19:28:54.996:    Found IP Packet Classifier Sub-TLV
Aug  9 19:28:54.996:            Protocol : 17
Aug  9 19:28:54.996:            Source Address : 24.34.240.247
Aug  9 19:28:54.996:            Destination Address : 24.34.240.235
Aug  9 19:28:54.996:            Destination Port Start : 53456
Aug  9 19:28:54.996:            Destination Port End : 53456
Aug  9 19:28:54.996: Auth Block:
Aug  9 19:28:54.996: 0x0000: 01 0C 01 04 00 00 14 3E 02 04 00 00 00 67   <- DQoS parms
Aug  9 19:28:55.000: DSC-RSP-SENT: CM->0013.1050.3801 TranscId->90 ConfCode->0
Aug  9 19:28:55.048: DSC-ACK-RECD: OrgMac->0013.1050.3801 OrgId->90 ConfCode->0
Aug  9 19:28:55.348: DSC-REQ End. Transaction over-T8 timer expired. OrgMac-
  >0013.1050.3801 OrgId->90
Aug  9 19:28:55.348: DYN-SRV-STATE-DESTROYED : OrgMac->0013.1050.3801 OrgId->90
Aug  9 19:29:08.812: DSD-REQ-RECD: I/F Cable5/0/U0: OrgMac->0013.1050.3801 SFID = 0
Aug  9 19:29:08.812: Found Downstream Service Flow TLV
Aug  9 19:29:08.812:    Service Flow Id : 104
Aug  9 19:29:08.812: Found Upstream Service Flow TLV
Aug  9 19:29:08.812:    Service Flow Id : 103
Aug  9 19:29:08.812: DSD-RSP-SENT: To CM->0013.1050.3801 TranscId->91
Aug  9 19:29:11.813: DSD-REQ End. T10 timer expired. OrgMac->0013.1050.3801 OrgId-
  >91
Aug  9 19:29:11.813: DYN-SRV-STATE-DESTROYED : OrgMac->0013.1050.3801 OrgId->91
```

Debugging to Find Admission Control Information

Another debug that can help provide some information relating to admission control is the following:

```
Router#debug cable mac-scheduler [admission-control]
```

Caution: The use of this debug can also generate quite a lot of output. These debugs should be used only in a controlled environment with a relatively small number of cable modems.

Debugging COPS Messaging with DQoS

So far the debugs shown can be used to provide information on the DOCSIS messaging involved with DQoS. The next set of debug commands can be used to provide information on the COPS messaging involved with DQoS. These debugs are as follows:

```
C0103-7246VXR#debug packetcable ?
  cops        Debug pktcbl COPS messages
  gate        Debug gate messages
  ipc         Debug pktcbl IPC messages
  subscriber  Debug on a particular subscriber
```

Just as the first step to debugging DOCSIS is to enable debugs on the cable modems of interest, the first step to debugging COPS is to enable debugs on the MTAs of interest. The following command is used:

```
Router#debug packetcable subscriber MTA-ip-address [verbose]
```

Remember, the IP address here is of the MTA, not the cable modem. Again, the **verbose** option provides more detail in the output.

The **cops** option can be used to debug problems with the COPS protocol itself, such as invalid messages or parameters. However, it is not very useful in looking at call flows. Another useful debug for diagnosing COPS issues is **debug cops** [**detail**].

The **ipc** option is useful for debugging interprocess communication on a CMTS where more than one processor is involved in the handling of the call. For example, on the uBR10012 CMTS messaging between the PRE module and the cable linecards would be shown. This command can also be used on the uBR7246VXR CMTS to show messaging when more advanced linecards, such as the MC28u, are utilized.

The **gate** option is the best one to use for looking at the COPS messages themselves. As you can see, several options are contained within this option:

```
C0103-7246VXR#debug packetcable gate ?
  commit           Debug commit messages
  control          Debug gate control messages
  database         Debug changes to gate database
  docsis-mapping   Debug docsis mapping messages
  events           Debug event messages
```

Logically, the **commit** suboption shows information pertaining to when a gate is committed and alone is not that interesting.

Probably the most useful of these options is the **control** suboption that shows the DQoS gate commands. The output of this command when used in conjunction with the **debug packetcable subscriber** command for the life of a call is as follows:

```
C0103-7246VXR#show debug
PacketCable Client:
  Pktcbl gate control msgs debugging is on
PacketCable specific:
  Debugging is on for Subscriber 24.34.240.235, Mask 255.255.255.255 (verbose)
Aug  9 20:01:16.546: Pktcbl(gcp): Received GATE SET message, tid=0x37
Aug  9 20:01:16.546: --- Pktcbl(gcp): Received GCP message ------
Aug  9 20:01:16.546: TRANSACTION ID        : Object.[snum/stype/len 1/1/8]
Aug  9 20:01:16.546:     transaction id    : 0x37
Aug  9 20:01:16.546:     gcp cmd           : 4 (GATE SET)
Aug  9 20:01:16.546: SUBSCRIBER ID (IPV4)  : Object.[snum/stype/len 2/1/8]
Aug  9 20:01:16.546:     Addr              : 24.34.240.235
Aug  9 20:01:16.546: ACTIVITY COUNT        : Object.[snum/stype/len 4/1/8]
Aug  9 20:01:16.546:     Count             : 5
Aug  9 20:01:16.546: GATE SPEC             : Object.[snum/stype/len 5/1/60]
Aug  9 20:01:16.546:     direction         : 1 (UPSTREAM)
Aug  9 20:01:16.546:     protocol id       : 17
Aug  9 20:01:16.546:     commit flag       : 0x0
Aug  9 20:01:16.546:     session class     : 1
Aug  9 20:01:16.546:     source            : 24.34.240.235
Aug  9 20:01:16.546:     dest              : 24.34.240.247
```

```
Aug  9 20:01:16.546:     src port          : 0 (0x0)
Aug  9 20:01:16.546:     dest port         : 53458 (0xD0D2)
Aug  9 20:01:16.546:     dscp              : 0xA0
Aug  9 20:01:16.546:     timer t1          : 200
Aug  9 20:01:16.546:     timer t7          : 200
Aug  9 20:01:16.546:     timer t8          : 300
Aug  9 20:01:16.546:     flowspec # 1      : [r/b/p/m/M 10000/200/10000/200/200] [R/
  S: 10000/800]
Aug  9 20:01:16.546: GATE SPEC             : Object.[snum/stype/len 5/1/60]
Aug  9 20:01:16.546:     direction         : 0 (DOWNSTREAM)
Aug  9 20:01:16.546:     protocol id       : 17
Aug  9 20:01:16.546:     commit flag       : 0x0
Aug  9 20:01:16.546:     session class     : 1
Aug  9 20:01:16.546:     source            : 24.34.240.247
Aug  9 20:01:16.546:     dest              : 24.34.240.235
Aug  9 20:01:16.546:     src port          : 0 (0x0)
Aug  9 20:01:16.546:     dest port         : 53456 (0xD0D0)
Aug  9 20:01:16.546:     dscp              : 0xA0
Aug  9 20:01:16.546:     timer t1          : 200
Aug  9 20:01:16.546:     timer t7          : 200
Aug  9 20:01:16.546:     timer t8          : 300
Aug  9 20:01:16.546:     flowspec # 1      : [r/b/p/m/M 10000/200/10000/200/200] [R/
  S: 10000/0]
Aug  9 20:01:16.546: ------------------------------------
Aug  9 20:01:16.546: Pktcbl(gcp): Building GCP message, added obj TRANSACTION ID
  ; len:8 padding:0
Aug  9 20:01:16.546: Pktcbl(gcp): Building GCP message, added obj SUBSCRIBER ID
  (IPV4); len:8 padding:0
Aug  9 20:01:16.546: Pktcbl(gcp): Building GCP message, added obj GATE ID          ;
  len:8 padding:0
Aug  9 20:01:16.546: Pktcbl(gcp): Building GCP message, added obj ACTIVITY COUNT
  ; len:8 padding:0
Aug  9 20:01:16.546: Pktcbl(gcp): Built GCP message, GATE SET ACK        , length:
  32, copsLen 60
Aug  9 20:01:16.546:  --- Pktcbl: Sending GCP message ------
Aug  9 20:01:16.546: TRANSACTION ID        : Object.[snum/stype/len 1/1/8]
Aug  9 20:01:16.546:     transaction id    : 0x37
Aug  9 20:01:16.546:     gcp cmd           : 5 (GATE SET ACK)
Aug  9 20:01:16.546: SUBSCRIBER ID (IPV4)  : Object.[snum/stype/len 2/1/8]
Aug  9 20:01:16.546:     Addr              : 24.34.240.235
Aug  9 20:01:16.546: GATE ID               : Object.[snum/stype/len 3/1/8]
Aug  9 20:01:16.546:     GateID            : 21606 (0x5466)
Aug  9 20:01:16.546: ACTIVITY COUNT        : Object.[snum/stype/len 4/1/8]
Aug  9 20:01:16.546:     Count             : 2
Aug  9 20:01:16.546:  ------------------------------------
Aug  9 20:01:21.014: Pktcbl(gcp): Building GCP message, added obj TRANSACTION ID
  ; len:8 padding:0
Aug  9 20:01:21.014: Pktcbl(gcp): Building GCP message, added obj GATE ID          ;
  len:8 padding:0
Aug  9 20:01:21.014: Pktcbl(gcp): Built GCP message, GATE OPEN          , length:
  16, copsLen 44
Aug  9 20:01:21.014:  --- Pktcbl: Sending GCP message ------
Aug  9 20:01:21.014: TRANSACTION ID        : Object.[snum/stype/len 1/1/8]
Aug  9 20:01:21.014:     transaction id    : 0x0
Aug  9 20:01:21.014:     gcp cmd           : 13 (GATE OPEN)
Aug  9 20:01:21.014: GATE ID               : Object.[snum/stype/len 3/1/8]
Aug  9 20:01:21.014:     GateID            : 21606 (0x5466)
Aug  9 20:01:21.014:  ------------------------------------
Aug  9 20:01:31.126: Pktcbl(gcp): Building GCP message, added obj TRANSACTION ID
  ; len:8 padding:0
Aug  9 20:01:31.126: Pktcbl(gcp): Building GCP message, added obj GATE ID          ;
  len:8 padding:0
Aug  9 20:01:31.126: Pktcbl(gcp): Building GCP message, added obj PKTCBL REASON
  ; len:8 padding:0
```

```
Aug  9 20:01:31.126: Pktcbl(gcp): Built GCP message, GATE CLOSE        , length:
  24, copsLen 52
Aug  9 20:01:31.126:  --- Pktcbl: Sending GCP message ------
Aug  9 20:01:31.126: TRANSACTION ID       : Object.[snum/stype/len 1/1/8]
Aug  9 20:01:31.126:     transaction id   : 0x0
Aug  9 20:01:31.126:     gcp cmd          : 14 (GATE CLOSE)
Aug  9 20:01:31.126: GATE ID              : Object.[snum/stype/len 3/1/8]
Aug  9 20:01:31.126:     GateID           : 21606 (0x5466)
Aug  9 20:01:31.126: PKTCBL REASON        : Object.[snum/stype/len 13/1/8]
Aug  9 20:01:31.126:     reason           : [code/subcode: 0x1/0x0]
Aug  9 20:01:31.126:  -------------------------------------
```

In this debug you can see all the DQoS gate messages and their associated parameters.

The **database** suboption also provides useful information concerning gate creation, deletion, and state transitions, as illustrated in the following output:

```
Aug  9 20:06:12.488: Pktcbl(gdb): Allocating gate entry 0/0. Instance cnt: 26,
  gate_count is now 2
Aug  9 20:06:12.488: Pktcbl(gdb): Created new Subscriber IE for 24.34.240.235
Aug  9 20:06:12.488: Pktcbl(gdb): Updated subscriber IE [subs addr: 24.34.240.235,
  gate: 5226]
Aug  9 20:06:12.488: Pktcbl(gdb): Created gate IE on Cable5/0, gateid = 5226
Aug  9 20:06:12.488: Pktcbl(gdb): Started gate [id 5226] timer t1 [200000 msec]
Aug  9 20:06:12.608: Pktcbl(gdb): handle=5226 dir=0 type=3 (get gate flowspec)
Aug  9 20:06:12.608: r/b/p/m/M 1176256512 1128792064 1176256512 200 200 (R/S)
  1176256512 0
Aug  9 20:06:12.608: Pktcbl(gdb): handle=5226 dir=1 type=2 (get gate flowspec)
Aug  9 20:06:12.608: r/b/p/m/M 1176256512 1128792064 1176256512 200 200 (R/S)
  1176256512 800
Aug  9 20:06:12.608: Pktcbl(gdb): handle=5226 dir=0 type=2 (get gate flowspec)
Aug  9 20:06:12.608: r/b/p/m/M 1176256512 1128792064 1176256512 200 200 (R/S)
  1176256512 0
Aug  9 20:06:12.608: Pktcbl(gdb): handle=5226 dir=1 type=3 spec=6344AF58 (set gate
  flowspec)
Aug  9 20:06:12.608: Pktcbl(gdb): handle=5226 dir=0 type=3 spec=6344AF74 (set gate
  flowspec)
Aug  9 20:06:12.608: Pktcbl: params_enc : auth str=0x633A2828 len=8
Aug  9 20:06:12.708: Pktcbl(gdb): Get gate spec info, handle=5226 dir=1 spec=6344AFC0
Aug  9 20:06:12.708: Pktcbl(gdb): Get gate spec info, handle=5226 dir=0 spec=6344AFE8
Aug  9 20:06:20.072: Pktcbl(gdb): TOS Overwrite gate spec info,gate_id=5226 dir=1
  gie=6489C71C
Aug  9 20:06:20.072: Pktcbl(gdb): TOS Overwrite Gate=5226 DSCP=0xA0 mask=0x0
Aug  9 20:06:20.072: Pktcbl(gdb): Get gate spec info, handle=5226 dir=0 spec=6344AFD0
Aug  9 20:06:20.072: Pktcbl(gdb): Get gate spec info, handle=5226 dir=1 spec=6344AFD0
Aug  9 20:06:20.072: Pktcbl(gdb): handle=5226 dir=0 type=4 (get gate flowspec)
Aug  9 20:06:20.072: r/b/p/m/M 1176256512 1128792064 1176256512 200 200 (R/S)
  1176256512 0
Aug  9 20:06:20.072: Pktcbl(gdb): handle=5226 dir=1 type=2 (get gate flowspec)
Aug  9 20:06:20.072: r/b/p/m/M 1176256512 1128792064 1176256512 200 200 (R/S)
  1176256512 800
Aug  9 20:06:20.072: Pktcbl(gdb): handle=5226 dir=0 type=2 (get gate flowspec)
Aug  9 20:06:20.072: r/b/p/m/M 1176256512 1128792064 1176256512 200 200 (R/S)
  1176256512 0
Aug  9 20:06:20.072: Pktcbl(gdb): handle=5226 dir=1 type=4 spec=6344AF40 (set gate
  flowspec)
Aug  9 20:06:20.072: Pktcbl(gdb): handle=5226 dir=0 type=4 spec=6344AF5C (set gate
  flowspec)
Aug  9 20:06:20.072: Pktcbl(gdb): Cancelled gate [id 5226] timer t1
Aug  9 20:06:20.072: Pktcbl: params_enc : auth str=0x633A2828 len=14
```

The **docsis-mapping** suboption decodes the DQoS parameters from the DOCSIS authorization block parameter. It also shows debug messaging pertaining to the RSVP Flow

Specification-to-DOCSIS parameter mapping used to determine whether a service flow's QoS parameters fall with the authorized resource envelope, as demonstrated in the following output:

```
Aug  9 20:09:40.958: PktCbl(d2r): extract id: gate=21614, resource=0
Aug  9 20:09:40.958: PktCbl(d2r): extract id: gate=21614, resource=0
Aug  9 20:09:40.958: PktCbl(d2r): DSA-REQ received, gateid: 21614
Aug  9 20:09:40.958: PktCbl(r2d): cfr: proto=17 src=24.34.240.235 dst=24.34.240.247
  sport=53456 dport=53458
Aug  9 20:09:40.958: PktCbl(r2d): flowspec cfr: proto=17 src=24.34.240.235
  dst=24.34.240.247 sport=0 dport=53458
Aug  9 20:09:40.958: PktCbl(r2d): REQ - r/b/p/m/M 1176256512 1128792064 1176256512
  200 200 (R/S) 1176256512 800
Aug  9 20:09:40.958: PktCbl(r2d): cfr: proto=17 src=24.34.240.247 dst=24.34.240.235
  sport=0 dport=53456
Aug  9 20:09:40.958: PktCbl(r2d): flowspec cfr: proto=17 src=24.34.240.247
  dst=24.34.240.235 sport=0 dport=53456
Aug  9 20:09:40.962: PktCbl(r2d): REQ - r/b/p/m/M 1176256512 1128792064 1176256512
  200 200 (R/S) 1176256512 0
Aug  9 20:09:41.058: PktCbl(d2r): DSA-ACK received, gateid: 21614
Aug  9 20:09:41.350: PktCbl(d2r): DSA-ACK received, gateid: 5230
Aug  9 20:09:47.874: PktCbl(d2r): extract id: gate=21614, resource=151
Aug  9 20:09:47.874: PktCbl(d2r): extract id: gate=21614, resource=151
Aug  9 20:09:47.874: PktCbl(d2r): extract id: gate=21614, resource=151
Aug  9 20:09:47.874: PktCbl(d2r): DSC-REQ received, gateid(21614) flow handle(21614)
Aug  9 20:09:47.874: PktCbl(d2r): max_fs_type us=2, ds=2 qos_set_type us=6 ds=6
Aug  9 20:09:47.874: PktCbl(r2d): cfr: proto=17 src=24.34.240.235 dst=24.34.240.247
  sport=53456 dport=53458
Aug  9 20:09:47.874: PktCbl(r2d): flowspec cfr: proto=17 src=24.34.240.235
  dst=24.34.240.247 sport=0 dport=53458
Aug  9 20:09:47.874: PktCbl(r2d): REQ - r/b/p/m/M 1176256512 1128792064 1176256512
  200 200 (R/S) 1176256512 800
Aug  9 20:09:47.874: PktCbl(r2d): cfr: proto=17 src=24.34.240.247 dst=24.34.240.235
  sport=0 dport=53456
Aug  9 20:09:47.874: PktCbl(r2d): flowspec cfr: proto=17 src=24.34.240.247
  dst=24.34.240.235 sport=0 dport=53456
Aug  9 20:09:47.874: PktCbl(r2d): REQ - r/b/p/m/M 1176256512 1128792064 1176256512
  200 200 (R/S) 1176256512 0
Aug  9 20:09:57.774: PktCbl(d2r): DSD-REQ received, gateid: 5230
Aug  9 20:09:57.846: PktCbl(d2r): DSD-REQ received, gateid: 21614
```

Finally, the **event** suboption decodes PacketCable Event Messages that are exchanged with the Record Keeping Server (RKS).

Testing PacketCable DQoS Functionality

The Cisco CMTS also has several test commands that can be used to test PacketCable DQoS functionality. These commands are executed from EXEC configuration mode. The following commands simulate COPS messaging sent to the CMTS. The CMTS responds to these messages just as if they were received from the actual gate controller (such as the PacketCable CMS). For these commands to work, a PacketCable COPS connection to the gate controller specified in the command must be operational. The list that follows provides a summary of these commands:

- **test packetcable gc client-accept** *GC IP Address*—Executing this command simulates a COPS Client-Accept (CAT) message sent to the CMTS from the gate controller IP address specified. Recall that this message is used when a COPS connection is first established. The CMTS responds to this message by sending a COPS Request (REQ) message to the GC.
- **test packetcable gc gate-alloc** *GC IP Sub IP*—Executing this command simulates a PacketCable Gate-Alloc message sent to the CMTS from the gate controller IP address specified for the subscriber specified with an Activity Count of "1." The CMTS responds to this message with a Gate-Alloc-Ack message if it can honor the allocation and a Gate-Alloc-Err message if it cannot. Remember, in this message is an error code indicating the reason for the error.
- **test packetcable gc gate-delete** *GC IP Sub IP Gate ID Reason*—Executing this command simulates a PacketCable Gate-Delete message sent to the CMTS from the gate controller IP address specified for the subscriber specified for the Gate ID specified with the Reason subcode specified. The CMTS responds to this message with a Gate-Delete-Ack message if it can honor the deletion and with a Gate-Delete-Err message if it cannot. Again, in this message is an error code indicating the reason for the error.
- **test packetcable gc gate-info** *GC IP Sub IP Gate ID*—Executing this command simulates a PacketCable Gate-Info message sent to the CMTS from the gate controller IP address specified for the subscriber specified for the Gate ID specified. The CMTS responds to this message with a Gate-Info-Ack message if it can honor the deletion and a Gate-Info-Err message if it cannot. Again, in this message is an error code indicating the reason for the error.
- **test packetcable gc gate-set** *GC IP* **[dqos | multimedia]** *Sub IP Gate ID*—Executing this command simulates a PacketCable Gate-Set message sent to the CMTS from the gate controller IP address specified for the subscriber specified for the Gate ID specified. For this message to succeed the Gate ID specified must already be allocated. The CMTS responds to this message with a Gate-Set-Ack message if it can honor the request and with a Gate-Set-Err message if it cannot. Again, in this message is an error code indicating the reason for the error. The Gate-Set parameters resulting from this Gate-Set are as follows:

```
C0103-7246VXR#show packetcable gate 37902
GateID                  : 37902
    Subscriber ID       : 24.34.240.235
    COPS connection     :
        server handle   : 0x63F35F90
        server address  : 13.13.5.4
        server port     : 32801
        client address  : 13.10.0.19
    State               : AUTH
    CALEA Version       : I03
```

```
    Gate specs [UPSTREAM]
      Gate classifier  :  protocol 17,
                          src addr/port 5.5.1.4/0,
                         dest addr/port 5.5.1.5/0
      diffserv dscp     : 0x60
      timer t1(s)      : 37856
      timer t7(s)      : 120
      timer t8(s)      : 120
      commit flags      : 0x1
      session class     : 0x2
      flowspec # 1      : [r/b/p/m/M 9750/195/9750/195/195] [R/S: 9750/0]
    Gate specs [DOWNSTREAM]
      Gate classifier  :  protocol 17,
                          src addr/port 5.5.1.5/0,
                         dest addr/port 5.5.1.4/0
      diffserv dscp     : 0x60
      timer t1(s)      : 37856
      timer t7(s)      : 120
      timer t8(s)      : 120
      commit flags      : 0x1
      session class     : 0x2
      flowspec # 1      : [r/b/p/m/M 9750/195/9750/195/195] [R/S: 9750/0]
    Billing Info
      primary RKS       : [addr/port 1.10.90.1/43690]
      secondary RKS     : [addr/port 1.10.90.2/56797]
      flags             : 0
      billing corr ID   : [31 32 33 34 31 32 33 34 31 32 33 34 31 32 33 34 00 00
00 00 00 00 00 00 ]
```

- **test packetcable gc initiate** [**dqos** | **multimedia**] *GC IP*—Executing this command simulates the opening of a PacketCable COPS connection to the gate controller IP address specified. The CMTS sends a COPS Client-Open (OPN) message to the gate controller from port 2126 if the **dqos** option is used and from port 3918 if the **multimedia** option is used. The gate controller should respond with a COPS Client-Accept (CAT) message, which the CMTS responds to with a COPS Request (REQ) message.
- **test packetcable gate create** *Sub IP* [**dqos** | **multimedia**] *# of gates*—Executing this command simulates the allocation of the number of gates specified for the subscriber specified. The gate controller associated with these gates has an IP address of 0.0.0.0. No COPS messages are sent from the CMTS as a result of this command. The gates created with this command can be used with other test commands to verify functionality.

BTS Configuration

Chapter 7 covers much of the BTS 10200 configuration needed for PacketCable operation. The steps relevant to DQoS operation are revisited here.

For the BTS to function as a PacketCable CMS the call-agent-profile table must be defined indicating it supports DQoS. This is done by setting the DQOS SUPP field to Y as demonstrated here:

```
CLI>show call_agent_profile
ID=CA104
CMS_ID=104
MGC_ID=104
FEID=CA104
DQOS_SUPP=Y
CDB_BILLING_SUPP=Y
EM_BILLING_SUPP=N
GTD_SUPP=N
PCMM_SUPP=N
AMID=99999
PCMM_VERSION_MAJOR=1
PCMM_VERSION_MINOR=0
Reply : Success: Entry 1 of 1 returned.
```

DQoS timers and variables are among the parameters configured in the CA-CONFIG table. The configuration used in the call flow examples is as follows:

```
CLI>show ca-config type=DQOS%
TYPE=DQOS-T1-TIMER
DATATYPE=INTEGER
VALUE=200
TYPE=DQOS-T8-TIMER
DATATYPE=INTEGER
VALUE=300
Reply : Success:  Entries 1-2 of 2 returned.
```

You have more configurable DQoS variables than what is shown here; if values are not specified in the CA-CONFIG table, the values in the CA-CONFIG-BASE table are used. The DQoS parameters in this table are as follows:

```
CLI>show ca-config-base type=DQOS%; display=type,default_value
TYPE=DQOS-DS-SLACK-TERM
DEFAULT_VALUE=0
TYPE=DQOS-GATE-TIMER
DEFAULT_VALUE=2
TYPE=DQOS-T1-TIMER
DEFAULT_VALUE=200
TYPE=DQOS-T5-TIMER
DEFAULT_VALUE=5
TYPE=DQOS-T7-TIMER
DEFAULT_VALUE=200
TYPE=DQOS-T8-TIMER
DEFAULT_VALUE=0
TYPE=DQOS-US-SLACK-TERM
DEFAULT_VALUE=800
Reply : Success:  Entries 1-7 of 7 returned.
```

Timers T1, T7, and T8 and the slack term parameters are sent to the CMTS in Gate-Set messages. The GATE-TIMER parameter is the maximum length of time for a response from the CMTS to be received after a message is sent to the CMTS; for example, the tolerable amount of time between a Gate-Set and Gate-Set-ACK.

Table 13-2 defines these timers and parameters.

Table 13-2 *CA-CONFIG Table DQoS Variables*

Variable Name	Data Type and Value Range	Default/ Recommended Value and Comments	Description
DQOS-DS-SLACK-TERM-TIMER	Integer 0–60000 microseconds	0 microseconds	Tolerated latency for downstream service flows (0 indicates no restriction)
DQOS-GATE-TIMER	Integer 0–10 seconds	2 seconds	CMTS gate operation response timer
DQOS-T1-TIMER	Integer 0–500 seconds	200 seconds	Amount of time a gate can remain in the authorized state
DQOS-T5-TIMER	Integer 1–60 seconds	5 seconds	Amount of time the CMTS must hold resources in the Admitted QoS set in excess of the Active QoS set
DQOS-T7-TIMER	Integer 0–300 seconds	200 seconds	Amount of time the CMTS must hold resources in the Admitted/Reserved QoS set in excess of the Active/ Committed QoS set
DQOS-T8-TIMER	Integer 0–300 seconds	0 seconds Keeping this parameter at the default of 0 seconds provides no protection from "hung" dynamic service flows that can tie up bandwidth. It is recommended that you set this to a nonzero value. 200–300 seconds should be adequate.	Amount of time Committed/Active CMTS resources can remain unused (0 indicates forever)
DQOS-US-SLACK-TERM-TIMER	Integer 0–60000 microseconds	800 microseconds	Tolerated grant jitter for upstream service flows

Recall from Chapter 7 that CMTS network elements are configured in the AGGR table. The field DQOS SUPP must be set to Y for PacketCable DQoS to work. Also notice the COPS keepalive timer is configured in this table. The configuration used in the call flow examples is as follows:

```
CLI>show aggr
ID=PC-CMTS
TSAP_ADDR=PC-CMTS.cc.cisco.com
ES_SUPP=Y
DQOS_SUPP=Y
KA_TIMER=2
ACK_TIMEOUT=1000
ES_EVENT_SUPP=Y
IPSEC_SA_ESP_CS=3DES-MD5,3DES-SHA1,NULL-MD5,NULL-SHA1
IPSEC_SA_LIFETIME=86400
IPSEC_SA_GRACE_PERIOD=21600
IPSEC_ULP_NAME=IP
IKE_GROUP=2
IKE_SA_LIFETIME=86400
IKE_CS=3DES-MD5,3DES-SHA1
TYPE=CMTS
Reply : Success: Entry 1 of 1 returned.
```

Also in Chapter 7, you learned about the QoS table where many parameters that dictate how QoS for a subscriber is handled. Table 13-3 includes the key fields and fields relating to DQoS.

Table 13-3 *QoS Table Fields Pertinent to DQoS*

Field	Default Value	Recommended Value	Description
CLIENT_TYPE	NONE	DQOS For a PacketCable Multimedia subscriber this should be set to a value of MM_COPS	Identifies type of client requiring QoS.
CODEC-TYPE	N/A	PCMU Optionally you might want to create an entry for a compression CODEC such as G.728 or G.729E	Preferred CODEC.
DQOS_CMTS_DSCP_TOS	160 (10100000 in binary)	160 (10100000 in binary)	DSCP/IP ToS byte value sent to the CMTS in COPS Gate-Set messages; the CMTS overwrites RTP packets with this value before sending them out the WAN interface.

continues

Table 13-3 *QoS Table Fields Pertinent to DQoS (Continued)*

Field	Default Value	Recommended Value	Description
MAX_DQOS_AUTH_BANDWIDTH	HIGH	HIGH	Maximum authorized amount of DQoS bandwidth (High = G.711, Low = G.72x).
MAX_DQOS_SESSIONS	5	5	Maximum number of DQoS session permitted for a subscriber.
SILENT_SUPPRESSION	N	N If you want to utilize the UGS-AD service to free up DOCSIS bandwidth, set this parameter to Y	Enable silence suppression (a.k.a. VAD). Determines the value of MGCP Local Connection Options parameter s:.

The QOS table configuration used for all the examples in this chapter is as follows:

```
CLI>show qos id=G711u_20ms
ID=G711u_20ms
CODEC_TYPE=PCMU
RSVP_REQUIRED=N
RESOURCE_RESERVATION=BE
IPTOS_RTP_LOWDELAY=N
IPTOS_RTP_THROUGHPUT=N
IPTOS_RTP_RELIABILITY=N
IPTOS_RTP_PRECEDENCE=CRITICAL
MAX_DQOS_SESSIONS=5
MAX_DQOS_AUTH_BANDWIDTH=HIGH
DTMF_PREF_MODE=DTMF_INBAND
DQOS_CMTS_DSCP_TOS=160
LBW=0
HBW=0
AUTO_GAIN_CONTROL=N
GAIN_VALUE=0
SILENT_SUPPRESSION=N
LPTIME=20
HPTIME=20
VIDEO_CODEC_TYPE=NONE
CLIENT_TYPE=DQOS
DQOS_DSCP_TOS_BITMASK=224
DOCSIS_DSCP_TOS=160
DOCSIS_DSCP_TOS_BITMASK=224
FAX_T38_ENABLED=Y
Reply : Success: Entry 1 of 1 returned.
```

Verifying and Troubleshooting DQoS on the BTS

Recall from Chapter 7 that the status of a CMTS or aggregation device can be verified by issuing a **status** command as demonstrated here:

```
CLI>status aggr id=PC-CMTS
AGGR ID -> PC-CMTS
OPER STATE -> AGGR IN Service
RESULT -> ADM configure result in success
REASON -> ADM executed successfully
Reply : Success:
```

If your CMTS does not indicate it is in service, you are going to have a problem establishing the COPS session. Some things to check for include proper provisioning on the CMTS and IP connectivity between the CMS and CMTS. Also verify that no firewalls or other security mechanisms are in place blocking the TCP ports and IP addresses used for the COPS session. If secure COPS is not being used, the BTS attempts to open the TCP connection to the CMTS as soon as the aggregation entry is configured, and the CMTS should automatically initiate the opening of the COPS connection.

Measurements on the number of attempted and successful Gate-Set messages sent from the BTS to each CMTS can be obtained as well. Also information on the number of gates successfully committed is recorded. Tracking the Gate-Open messages received from the CMTS when a gate is committed does this. A sample report on these measurements is as follows:

```
CLI>report measurement_dqos_summary aggr_id=PC-CMTS;start_time=2006-08-09 16:00:00
TIMESTAMP                                   2006-08-09 16:00:00
CALL_AGENT_ID                               CA104
CONDITION                                   Normal
AGGR_ID                                     PC-CMTS
DQOS_GATESET_ATTMP                          18
DQOS_GATESET_SUCC                           18
DQOS_GATE_COMMIT                            16
TIMESTAMP                                   2006-08-09 16:15:00
CALL_AGENT_ID                               CA104
CONDITION                                   Normal
AGGR_ID                                     PC-CMTS
DQOS_GATESET_ATTMP                          8
DQOS_GATESET_SUCC                           8
DQOS_GATE_COMMIT                            8
TIMESTAMP                                   2006-08-09 16:30:00
CALL_AGENT_ID                               CA104
CONDITION                                   Normal
AGGR_ID                                     PC-CMTS
DQOS_GATESET_ATTMP                          2
DQOS_GATESET_SUCC                           2
DQOS_GATE_COMMIT                            2
TIMESTAMP                                   2006-08-09 16:45:00
CALL_AGENT_ID                               CA104
CONDITION                                   Normal
AGGR_ID                                     PC-CMTS
DQOS_GATESET_ATTMP                          0
DQOS_GATESET_SUCC                           0
DQOS_GATE_COMMIT                            0
Reply : Success:  Entries 1-4 of 4 returned.
```

In this example you can see that from 16:00:00 to 16:15:00, 18 DQoS calls were attempted and completed on CMTS PC-CMTS, and 16 of the calls were answered. Eight more calls were made in the next 15-minute interval, 2 in the interval after that, and none were made in the next 15-minute intervals. Remember, if the DQOS_GATE_COMMIT and DQOS_GATESET_ATTMP counters are equal, this indicates all calls that were attempted were answered.

Troubleshooting Tips

The last part of this chapter contains some troubleshooting tips that might help you diagnose and fix problems relating to DQoS.

COPS Protocol Issues

A problem in the COPS protocol results in error messages being sent from the CMTS to the CMS. These are the Gate-Set-ERR, GATE-ALLOC-ERR, GATE-INFO-ERR, and Gate-Delete-ERR message types. A PacketCable Error parameter is included in this message that identifies the reason for the error, such as an invalid or missing parameter. In these cases the offending parameter is also identified. Other causes of these error messages include no gates being available and the subscriber exceeding his or her gate limit. If you receive the no gates available indication, check the CMTS configuration for the maximum number of gates setting. If it is set correctly and you still have a problem, you have probably exceeded the capacity of the CMTS. If you are receiving the subscriber exceeding his or her limit problem, verify the setting of the MAX-DQOS-SESSION parameter in the QoS table on the BTS. If this is correct, you might have a misbehaving MTA attempting to sabotage your network by tying up DQoS resources.

Unexpected Gate Deletion from a Gate-Close Message

In a normal call the PacketCable gate is deleted by a Gate-Close message being sent from the CMTS to the CMS. In these Gate-Close messages are the PacketCable reason code and reason subcode parameters that indicate why the gate is to be deleted. A reason code of 1 is used for Gate-Close messages; 0 is used for Gate-Delete messages, which you'll learn more about in a little bit. The reason subcode identifies the cause of the gate deletion and normally has a value of 0, indicating a normal client release. You might recall that this was the value seen in the examples.

If the gate were abnormally deleted by the CMTS, this reason subcode tells why. The case where a normal call was preempted to make room for an emergency 911 call would be identified here. Also if the cable modem lost RF connectivity to the CMTS, an unexpected Gate-Close would result. Remember that the CMTS maintains a heartbeat with each cable modem via the station maintenance mechanism in DOCSIS. If enough of these heartbeats are missed, the CMTS knows the cable modem is no longer functional, so any DQoS gates associated with it are removed. A CMTS debug of a Gate-Close generated from this scenario is as follows:

```
Aug  9 21:43:06.828:  --- Pktcbl: Sending GCP message ------
Aug  9 21:43:06.828: TRANSACTION ID        : Object.[snum/stype/len 1/1/8]
Aug  9 21:43:06.828:     transaction id    : 0x0
Aug  9 21:43:06.828:     gcp cmd           : 14 (GATE CLOSE)
Aug  9 21:43:06.828: GATE ID               : Object.[snum/stype/len 3/1/8]
Aug  9 21:43:06.828:     GateID            : 21626 (0x547A)
Aug  9 21:43:06.828: PKTCBL REASON         : Object.[snum/stype/len 13/1/8]
Aug  9 21:43:06.828:     reason            : [code/subcode: 0x1/0x3]
```

If you refer to Chapter 12, you can see that subcode 3 equates to a lack of DOCSIS MAC-layer responses.

Other reasons for abnormal Gate-Close messages are timer expirations. The expiration of timer T0 means that a gate was allocated, but no Gate-Set message authorizing the gate was received in this time. If this timer is expiring, it is either set too short on the CMTS, or a problem exists with the CMS. Recall that the recommended value is 30 seconds.

The expiration of timer T8 means that committed resources are not being used. This is the activity timer, and its expiration is typically indicative of a malfunctioning MTA. In this case you want these committed resources to be released so that they can be reused. This is why setting this timer to 0 can be a bad idea. So how about the case where voice activity detection (VAD) is deployed? When the subscriber is not talking, resources are not being used. Couldn't this cause the activity timer to expire and delete a functional gate? The answer is yes if it is set too low. But remember, while the MTA is detecting silence, it periodically transmits comfort noise RTP packets. If timer T8 is set greater than this interpacket time, it should not expire. Also, note it is a somewhat common misconception that the regular transmission of RTCP packets keeps this timer from expiring. This is not the case; RTCP port numbers are one off from the RTP port numbers, so they do not match the classifiers for the voice service flow. RTCP packets are sent across the Best Effort service flow.

So what about timers T1 and T7? Recall that timer T1 is the time between when a gate is authorized and committed. Timer T7 is the amount of time resources are allowed to be reserved without being committed. Notice both timers stop when a gate is committed. So under what scenarios are resources authorized or reserved but not committed? The answer is when a call is not answered. So an unexpected Gate-Close indicating the expiry of one of these timers is probably because of a call not being answered. Whichever timer expires first would be the cause of the gate closure. A sample debug of a Gate-Close message because of a call not answering is as follows:

```
Aug  9 21:56:29.364:  --- Pktcbl: Sending GCP message ------
Aug  9 21:56:29.364: TRANSACTION ID        : Object.[snum/stype/len 1/1/8]
Aug  9 21:56:29.364:     transaction id    : 0x0
Aug  9 21:56:29.364:     gcp cmd           : 14 (GATE CLOSE)
Aug  9 21:56:29.364: GATE ID               : Object.[snum/stype/len 3/1/8]
Aug  9 21:56:29.364:     GateID            : 5126 (0x1406)
Aug  9 21:56:29.364: PKTCBL REASON         : Object.[snum/stype/len 13/1/8]
```

```
Aug  9 21:56:29.364:     reason          : [code/subcode: 0x1/0x5]
Aug  9 21:56:29.364:   -------------------------------------
```

In this case the Gate-Close results from timer T1 expiring. Remember that on the BTS both timers can be defined. In this case they were both set to 200 seconds, so because timer T1 starts first, it expires first. Also note you also have call-signaling no-answer timers. If this timer expires first, the CMS initiates the call deletion.

Unexpected Gate Deletion from a Gate-Delete Message

Gates can also be deleted by the CMS. However, this is not normal and typically indicates a problem. Usually this problem is related to gate coordination. Remember that timer T5 indicates the maximum amount of time allowed between when the CMS signals the MTA to delete a connection and the time the CMTS sends out the corresponding Gate-Close. If this Gate-Close is not received, this indicates that the MTA did not issue a DSD-REQ to delete the service flows after receiving the DLCX command. This might be caused by a malfunctioning MTA or a misbehaving MTA attempting to steal service. The CMS ensures this does not occur by sending a Gate-Delete to the CMTS when timer T5 expires.

DOCSIS Protocol Issues

Problems can also occur in the DOCSIS protocol with the dynamic service flow messages sent between the embedded MTA and CMTS. Recall from the examples in this book that in each dynamic service flow response message the confirmation code parameter is included. In those cases this parameter had a value of zero, indicating a normal or successful operation. This parameter also identifies problems in the DOCSIS message exchange when they occur. Appendix C of the DOCSIS RF specification defines the values of this parameter. These errors can be problems with the service flow, classifier, and/or payload header parameters. This type of problem is indicative of a malfunctioning MTA.

More common problems that you might encounter here are authorization failures and resource problems. If you are getting authorization failures, you might have a configuration problem. For example, if your MTA is not configured for DQoS operation on the CMS, no authorization block TLV is sent, causing the CMTS to reject the dynamic service flow request with this reason. If you are getting codes indicating a temporary resource problem, this typically indicates that upstream resources are not currently available for this service flow. In this case, the DSA-REQ gets rejected, indicating a confirmation code of Reject temporary, and the MTA responds to the MDCX (or CRCX) causing the DQoS reservation with an error code of 526 - No bandwidth. Recheck your network traffic engineering in this case. If you are getting a permanent resource problem, this might be because of your admission control policy, for example, if on the CMTS no bandwidth is made available for voice calls.

Chapter Summary

Quality of service for a PacketCable call is dynamically created as calls are placed and received. This functionality, known as PacketCable DQoS, is provided using a collection of protocols. NCS, DOCSIS, and COPS all contribute to meeting this requirement. Chapters 6, 7, and 12 detail these protocol specifics, while this chapter delves into the mechanics of their interaction in providing guaranteed QoS.

The easiest way to understand this protocol interaction is through the examination of real call flows. Messages can be obtained via external protocol traces or if a Cisco CMTS is used by the cable monitor utility and/or Cisco IOS debug commands. As seen in these traces, DQoS-specific parameters are passed between protocols to communicate QoS information between the embedded MTA, CMTS, and CMS PacketCable components. The creation, deletion, authorization, and management of DQoS resources are all topics of this chapter.

Some other functions of DQoS include the use of timers to ensure proper coordination of resources and to ensure resources are not wasted. The ability to influence quality of service across the IP backbone of the PacketCable network is covered, as well the ability to reserve bandwidth for various classes of call types. This latter function can be used to prioritize emergency calls when bandwidth is limited.

Voice activity detection or silence suppression can be used to increase bandwidth efficiency by greatly reducing the number of packets sent when a PacketCable subscriber is not speaking. The DOCSIS UGS-AD service flow type is used for this purpose.

If a Cisco CMTS, such as the uBR7246VXR or the uBR10012, are used in your PacketCable network, you have a few required configuration steps and some optional ones that can be manipulated to govern how DQoS is handled. Also you have a number of **show** and **debug** commands in Cisco IOS that can be used to view the current configuration and performance and assist you in troubleshooting problems.

Similarly, if the Cisco BTS 10200 is your CMS, you have some configuration steps that are required for PacketCable DQoS operation and some parameters that can be adjusted. Statistics on DQoS usage can also be obtained on this product.

Return codes in the COPS and DOCSIS messaging, as well as the NCS messaging during a call, can help you identify problems. Unexpected Gate-Close messages could indicate timer expiries or loss of DOCSIS station maintenance. A Gate-Delete message could indicate a problem with gate coordination. Failures in the DOCSIS messaging typically indicate a problem with reserving DOCSIS resources.

Chapter Review

1 If both the originating and terminating MTAs in a basic call support multiple CODECs, each with different bandwidth requirements, how is the DQoS authorized resource envelope determined?

Answer: The CMS uses the Least Upper Bound requirement when creating the DQoS gate flow specifications.

2 Can voice RTP packets be sent between MTAs if their associated gate resources have not been committed?

Answer: Yes. If the NCS connection mode permits the capability to send packets, the MTA sends packets; however, they are sent over the Best Effort service flow.

3 What does the CMTS do with the information in the Activity Count parameter sent by the CMS in a GATE-ALLOC or Gate-Set message?

Answer: It performs an allocation check to see whether by creating new gates for a new call the number of active gates exceeds the number specified in this parameter.

4 What does a port number of zero indicate in a gate classifier?

Answer: No restriction, all port numbers match the classifier.

5 Can PacketCable DQoS be used to influence backbone quality of service for NCS signaling packets?

Answer: No. However, it can be used to influence backbone QoS for RTP bearer packets.

6 In a Cisco PacketCable network, is the DQoS timer T1 configured on the BTS or on the CMTS?

Answer: Both. However, the value sent by the BTS has precedence over the value defined on the CMTS.

7 How does the MTA learn about the DQoS gate identifier created on the CMTS under the direction of the CMS?

Answer: The CMS relays this information to the MTA in the NCS LCO parameter dq-gi.

8 Why is setting the DQoS timer T8 to the value 0 a bad idea?

Answer: It doesn’t ensure that hung resources are eventually reusable.

9 Will the implementation of VAD alone allow you to fit more calls on a DOCSIS upstream channel?

Answer: No, but it allows more Best Effort traffic to pass.

10 How can the CMTS differentiate an emergency 911 call from an ordinary phone call?

Answer: Using the session class value in the gate specification parameter

11 What is the only required step for enabling PacketCable functionality on a Cisco CMTS?

Answer: The configuration command **packetcable**

12 What IOS **show** command can be used to look at the number of active committed gates?

Answer: show packetcable gate summary

13 How do you define the tolerable grant jitter values for a DQoS gate on the BTS 10200?

Answer: By defining the DQOS-US-SLACK-TERM and DQOS-DS-SLACK-TERM parameters in the CA-CONFIG table

14 What command on the BTS 10200 can be used to see the number of DQoS gates committed on a CMTS in the past hour?

Answer: report measurement-dqos-summary

15 If you are receiving Gate-Set-ERR messages when trying to authorize gates on a CMTS, what are some likely causes?

Answer: No available gates and subscriber exceeding gate limit

16 What events indicate the closure of a gate because of no answer?

Answer: Expiring of timers T1 and T7

17 What **show** command on a Cisco router can be used to see whether the QI bit has been set for an active voice call?

Answer: show interface cable *x/y* **sid** *z* **counters verbose** (for the UGS SID)

CHAPTER 14

Multimedia Applications

Now that you know how distributed quality of service (DQoS) provides premium service for telephone calls, the focus shifts to other applications for DQoS and PacketCable. Back in the very first chapter you learned that cable providers are expanding their service offerings to include applications such as videoconferencing, streaming media, interactive gaming, and telephony. Telephony is just the beginning; PacketCable can be used to provide a guaranteed and reliable network for any of these service offerings.

The architecture and protocols used to provide these services are defined under the PacketCable Multimedia (PCMM) specifications. PCMM provides the ability to provide quality of service (QoS) over a Data-Over-Cable Service Interface Specification (DOCSIS) network for different types of multimedia applications. PCMM is unaware of the application specifics requiring QoS. Thus, it does not specify the details of other aspects of the multimedia application, such as signaling and provisioning. This is different from PacketCable 1.x, where these application details are defined. PacketCable 2.0 addresses these additional architecture mechanisms for offering multimedia applications. Technologies such as Session Initiation Protocol (SIP) and IP Multimedia Subsystem (IMS), as well as PCMM for providing QoS over the DOCSIS network, are all used in PacketCable 2.0.

This chapter begins with an overview of how PacketCable multimedia works. This includes an in-depth look at how PacketCable Multimedia varies from PacketCable telephony. Differences in the Common Open Policy Service (COPS) and DOCSIS protocols and their messaging are compared to what you learned previously. Also call flows are examined to help illustrate PCMM functionality. The use of security and event messaging in PCMM is also discussed. Finally, the chapter addresses relevant configuration and troubleshooting of PCMM on a Cisco cable modem termination system (CMTS).

PCMM Overview

PacketCable Multimedia defines a framework for providing QoS, security, and resource accounting for any type of service in a DOCSIS network requiring this functionality. The methods of providing these mechanisms in PCMM are based on mechanisms defined in PacketCable 1.x. As you will see, PacketCable 1.x components, such as the multimedia terminal adapter (MTA) and call management server (CMS), are similar to a PCMM client and application server.

The main goal of PCMM is to provide a framework where reliable DOCSIS resources can be reserved for a variety of applications, which include videoconferencing, interactive gaming, streaming media, and so on. Voice and video applications tend to be very sensitive to network delay and interpacket delay (jitter); therefore, they can greatly benefit from the reliable resources provided by PCMM. Applications that are not real-time in nature can benefit from PCMM as well. For example, high-speed data users can be given boosts to their Best Effort service to allow for faster file transfers. PCMM can be used to give subscribers temporary speed increases in their high-speed data (HSD) service to entice them to move to one of these higher-speed service offerings. Another anticipated use of PCMM is with Deep Packet Inspection (DPI) devices. Either these devices can sit inline in the PacketCable network so they can inspect all network traffic, or traffic can be replicated to these devices. A DPI device can then recognize certain type of traffic and initiate a request for QoS to the PCMM devices in the network. For example, they could identify Real-time Transport Protocol (RTP) voice streams, and then initiate the request for DOCSIS Unsolicited Grant Service (UGS) service flows. Also a DPI device can be used to identify misbehaving subscribers or viruses and initiate requests to limit or cut off the QoS assigned to these subscribers.

The PCMM framework allows for client devices to use PCMM resources without having to be tightly integrated in the resource reservation and commitment processes. This is different from PacketCable 1.x, where embedded MTAs are required to fully understand and implement DQoS resource manipulation. PCMM also separates resource policy decisions from application-specific decisions, which increases flexibility in providing services. Finally, the PCMM architecture is such that a third-party service provider can manipulate QoS on the Multiple Systems Operator (MSO) network.

PCMM Components

The first step in understanding how PCMM works is to look at what the various contributing components are. Figure 14-1 is a high-level illustration of the devices involved in providing PacketCable Multimedia services.

Figure 14-1 *PacketCable Multimedia Components*

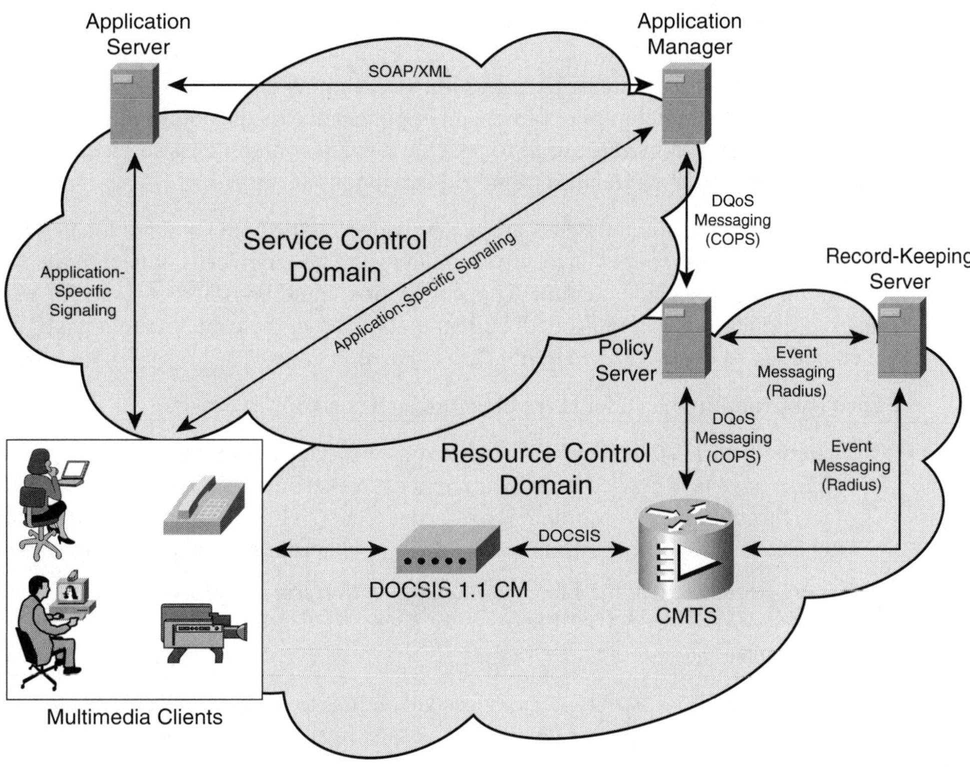

As you can see from the diagram, seven components are used to provide PCMM services:

- Multimedia clients
- Application Server
- Application Manager
- Policy Server
- Cable modem termination system (CMTS)
- Cable modem (CM)
- Record Keeping Server (RKS)

The following sections describe these components in further detail.

PacketCable Multimedia Clients

Logically, the multimedia client is the device using a multimedia service. One example would be a software application on a personal computer providing streaming audio or video, such as Internet radio or perhaps a movie. Another example is a gaming console connected to the Internet to play against an opponent across the country. Another example is a video camera and monitor used to provide a video conferencing service. And yes, another example is an MTA used to provide telephony service.

Usually, multimedia client devices are not provided by the MSO and are located at the customer's premises, so they are untrusted devices. These devices connect to the MSO network through the cable modem. This connection could be Ethernet, USB, or wireless. The exception to this is a multimedia client embedded in the cable modem itself, such as the embedded MTA you are no doubt familiar with by now.

PacketCable multimedia clients are categorized as one of three types:

- **Client Type 1**—Client type 1 is what most of today's multimedia clients are. These clients have no knowledge or awareness of PacketCable, DOCSIS, or even QoS. In fact, this client type is commonly referred to as *QoS-unaware*. These clients are generic IP networking applications such as PC applications and gaming consoles.

  These devices are at the mercy of the application manager to set up quality of service resources. You'll learn more about application managers later in the chapter.

- **Client Type 2**—Client type 2 devices are similar to the embedded MTA devices used in PacketCable telephony. These devices are aware of PacketCable, DOCSIS, and QoS but they also require an application manager to do the QoS resource allocation and part of the authorization. However, this type of client still does much of the QoS resource manipulation. This type of client is commonly referred to as *Push model*.
- **Client Type 3**—The third type of multimedia client is also PacketCable QoS–aware. The difference between it and a type 2 device is that these devices do not need an application manager to set up QoS resources. These devices use the Resource Reservation Protocol (RSVP) to communicate directly with the CMTS to set up QoS. This type of client is commonly referred to as *Pull model*.

Currently, only client type 1 devices have been fully defined in the PCMM specifications. Consequently, they are the focus of this chapter.

Application Server (AS)

In many cases, it is beneficial to have another device handle the multimedia session requests on behalf of the client. This device is known as the Application Server (AS). You just learned that a majority of today's PCMM clients are of type 1, which know nothing about QoS or PacketCable. So for these clients, an AS is typically used to initiate multimedia sessions. The AS might or might not be managed by the MSO. The means in which session requests are communicated from the client to the AS is unspecified by PCMM.

Application Manager (AM)

The device that handles PCMM requests from clients (either directly or via an AS) is the Application Manager (AM). In addition to coordinating client application requests, an AM maintains application-level state and is the device responsible for applying service policies. Because clients and AS devices are typically untrusted, the AM also authenticates incoming requests.

Just as the communication protocol between the client and AS is unspecified in PCMM, the communication protocol between the client and AM is also unspecified. The protocol used between the AS and AM is specified and is the PacketCable Multimedia Web Service Interface Specification (PKT-SP-MM-WS). This specification defines the use of a SOAP/ extensible markup language (SOAP/XML)–based web services interface for AS to AM communication. Please refer to this specification for the details on how this communication is done.

One example of an AM is a call agent. A call agent communicates with clients (in this case, gateways) using a signaling protocol such as Media Gateway Control Protocol (MGCP) or SIP. Remember, in PCMM the protocol used between the AM and clients is undefined. When a gateway wants to make a phone call, it contacts the call agent using the signaling protocol. The call agent then determines whether the gateway is authorized to make the call and figures out how to route the call. The call agent can also determine the QoS resources required for the call and signal this information to another device. Recall this is done in PacketCable 1.x by converting SDP information into RSVP flow specifications sent in COPS messages. In PCMM this other device is the policy server, and the protocol used to send this policy information is COPS.

Another sample AM is a simple web server. Suppose this server receives HTML requests from a client for an application such as streaming media. After the server determines whether the client is authorized to receive the media, it figures out the QoS resources required and sends this information to the policy server.

Yet another example might be a gaming console that signals a server to initiate a game against an opponent in another geographic area. In this case, a proprietary protocol is usually used between the gaming console and the server. The gaming server authorizes the gaming console, determines the required QoS resources, and sends this information to the policy server.

In all these cases, if everything is successful, a gate identifier referring to the QoS resources is eventually returned to the AM. It's important to remember that the gate identifier is not sent to PCMM client type 1 devices from the AM. These clients are "QoS-unaware," so they wouldn't know what to do with this information. Instead, the AM reserves and commits resources on behalf of the client. The signaling protocol between the client and the application manager can be used to determine when the AM should reserve or commit resources, however.

The AM might not know the entire details on how to signal the policy server for QoS for an application. In this case, it might simply use a Service Name in the request to the Policy Server (PS) and let the PS handle converting the Service Name into the various QoS parameters. Obviously, for this to work common service names need to be configured on the AM and PS components.

Policy Server (PS)

The Policy Server is the device responsible for taking the service requests from the AM and converting them into resource policy requests used to set up quality of service resources on a CMTS.

The PS also applies policy rules to requests received from the AM. Sample policies include limits on the number of gates allocated to a subscriber, limits on the types of QoS available to a subscriber, limits on which applications a policy server accepts, and limits on the impact of service on a particular CMTS. Assuming all checks pass, the PS figures out where that subscriber is and "pushes" a policy to the CMTS where the subscriber's cable modem is connected. So as you can see, the AM doesn't need to know what CMTS the subscriber's modem is attached to; the PS figures this out instead.

The PS also functions as a proxy between the AM and CMTS as it sees all the QoS messages to and from these devices. Because of this, an MSO might elect to deploy PS devices in a hierarchical fashion to satisfy scalability and redundancy concerns. In this case, some policy servers can be used to make per-subscriber policy decisions, others to make per-CMTS policy decisions, others to make per-region policy decisions, and so on.

Policy servers can also be classified as "stateful" or "stateless," similar to the SIP proxy servers discussed in Chapter 9, "Call Management Server Signaling Protocol (CMSS)." A stateful PS keeps track of gates and resources and can be used to notify you when resources become scarce. Stateless policy servers do not have the ability to track resources, but as a tradeoff they are simpler to implement.

So what component is like the Policy Server in PacketCable 1.x? The answer is the Gateway Controller (GC) component of the CMS. This is the COPS server that controls gate creation on a CMTS. The call agent component of the CMS that controls telephony call states is similar to the AM.

Cable Modem Termination System (CMTS)

The CMTS is the COPS Policy Enforcement Point (PEP) for premium DOCSIS resources in PacketCable Multimedia, just as it is in PacketCable telephony. It communicates policy information with one or more Policy Servers for the authorization of resources. This information can either be "pushed" to the CMTS by the PS or "pulled" to the PS from the CMTS, depending on the client type. Currently, only client type 1 devices are supported, so policy information is always "pushed" to the CMTS in PCMM. Also with client type 1 devices, DOCSIS DSX messages will always initiate from the CMTS as these messages are triggered by the policy information "pushed" by the PS. This is a key difference from PacketCable 1.x, where the DSX messaging initiates from the MTA. The CMTS can be instructed to track DOCSIS resource use based on time requirements and traffic volume requirements.

Cable Modem (CM)

The requirement of a cable modem in PacketCable Multimedia is the same as it is in PacketCable telephony; it must operate in at least DOCSIS 1.1 mode. The clients are generally assumed to be non-embedded (as opposed to the embedded MTA defined in PacketCable 1.x that interfaces directly with the DOCSIS layer). Thus, all client communication is done through the CM using protocols over ordinary IP.

RKS

The RKS component also serves a similar function as it did in PacketCable telephony. It uses event messaging to keep track of QoS resources. The difference here is that messages come from policy servers as well as CMTSs. Also event messaging in PCMM is generalized because several of the messages and parameters previously defined are meaningful only for the telephony application.

Service and Resource Control Domains

Many of the concepts in this chapter are related to the concepts introduced in Chapter 12, "DQoS Architecture and Framework." One of these concepts was the COPS terminology of Policy Decision Points (PDPs) and Policy Enforcement Points (PEPs).

A PDP is the COPS server requesting or denying a policy creation, and the PEP is the COPS client where the policy is applied. Remember that in PacketCable 1.x the CMTS is the PEP, and the gate controller function of the CMS is the PDP. In PCMM, COPS is used in two places; hence, you have two sets of PDP/PEP relationships. The CMTS is a PEP in its communication with the PS, and the PS in the PDP in its communication with the CMTS. But the PS is a PEP in its communication with the AM, and the AM is the PDP.

Because policy servers perform both COPS functions, the multimedia specification separates the policy domains into a resource control domain (RCD) and a service control domain (SCD). Figure 14-1 illustrates how these different domains apply to PCMM.

The RCD is the logical grouping of elements that provide connectivity (that is, CMTS, CM, Hybrid Fiber Coaxial [HFC] network) and network resource–level policy management (that is, Policy Server) along the packet forwarding paths to and from a client. In the RCD, the CMTS is the PEP and the PS is the PDP.

The SCD is the logical grouping of elements that offer applications and content (that is, Application Servers and Application Managers) to service subscribers (clients). The SCD is associated more with the higher-level concept of services than with physical DOCSIS resources. In the SCD, the PS is the PEP and the AM is the PDP. Multiple SCDs can operate in a single RCD, and a single SCD can operate in multiple RCDs. For example, in a particular cable division, both a gaming application and a videoconferencing application might be available. Likewise, the gaming application might be available in multiple cable divisions.

For example, assume you have a type 1 multimedia client wanting to use a particular application. It signals this to the AM (either directly or via the AS). The AM then makes decisions based on SCD policies on whether or not the client is authorized to use the application. If it is, the AM figures out what the required QoS resources are for the application and issues a request to a PS. The PS then makes a decision based on RCD policies on whether or not the client is authorized to use the proposed QoS resources. If it is, the request is sent to the CMTS where the client is connected. Then, if the CMTS has the physical resources to grant the request, it does so, and it sends the gate identifier to the PS. The PS then in turn forwards this information back to the AM. Thus, the PS proxies QoS requests between the AM and CMTS. In fact, the PS translates SCD parameters into RCD parameters, and vice versa.

PCMM DOCSIS Usage Considerations

The use of the DOCSIS 1.1 protocol in PCMM is virtually the same as it is in PacketCable telephony. You have authorized, reserved, and committed resource envelopes, and gates are used to correlate DOCSIS service flow requests to these resource envelopes.

The main difference is that upstream scheduling types in addition to the Unsolicited Grant Service (UGS) and Unsolicited Grant Service with Activity Detection (UGS-AD) types discussed previously can also be used. Specifically, the Real-Time Polling Service (RTPS) and Non Real-Time Polling Service (nRTPS) are used. Also note that the Best Effort (BE) upstream scheduling type can be used to provide premium QoS services through the use of DOCSIS parameters such as traffic priority and minimum reserved rate. In other words, you can guarantee, restrict, or prioritize bandwidth on a BE flow using QoS.

Remember that all downstream service flows are Best Effort, but these same DOCSIS parameters can be used to provide QoS. So, refer to Chapter 12 or the DOCSIS specifications for the details on those parameters.

The RTPS service is designed for real-time applications that generate variable-sized data at periodic intervals. It works by providing periodic dedicated request opportunities. The modem then uses these request opportunities to specify the size of the data grants it requires. This service type is well-suited for applications such as MPEG video. The nRTPS service is designed for non real-time applications requiring variable-sized data grants on a regular basis. These applications are ensured request opportunities even during congestion periods. One difference between nRTPS and RTPS is that the jitter between successive requests does not matter for nRTPS because its applications are not sensitive to delay variations. Also the polling interval used in nRTPS flows is typically much greater (that is, less frequent) than it is with RTPS.

Service flow parameters used to define RTPS, nRTPS, and BE services include

- **Service Identifier**—The 14-bit service identifier is assigned by the CMTS to reference the upstream service flow. It is used in the bandwidth allocation messages to identify the user of a data or request opportunity. It is also used in the DOCSIS extended header to reference the device sending the packet.
- **Service Flow Scheduling Type**—This field identifies the scheduling service for upstream data and request transmissions. This is where the service flow is identified as a Best Effort (2), Non Real-Time Polling (3), Real-Time Polling (4), UGS-AD (5), or UGS (6) service.
- **Request/Transmission Policy**—This field is a bitmap that controls when and how the service flow can transmit data and data requests. A bit value of 1 indicates true, and a bit value of 0 indicates false. The lower nine bit positions are defined as shown in Table 14-1.

Table 14-1 *Request/Transmission Policy Bits*

Bit Position	Meaning	Description and Typical Value
0	Use of broadcast request opportunities disallowed.	1 for RTPS because service flow has dedicated unsolicited request opportunities, 0 for BE, and usually 0 for nRTPS because they have no or very little unsolicited request opportunities, respectively.
1	Use of priority multicast request opportunities disallowed.	1 for RTPS because service flow has dedicated unsolicited request opportunities, 0 for BE, and usually 0 for nRTPS because they have no or very little unsolicited request opportunities, respectively.
2	Use of Request/Data opportunities for requests disallowed.	1 for RTPS because service flow has dedicated unsolicited request opportunities, 0 for BE, and usually 0 for nRTPS because they have no or very little unsolicited request opportunities, respectively.

continues

Table 14-1 *Request/Transmission Policy Bits (Continued)*

Bit Position	Meaning	Description and Typical Value
3	Use of Request/Data opportunities for data disallowed.	1 for RTPS because service flow has dedicated request opportunities; for BE and nRTPS these intervals can be used for data transmissions.
4	Piggybacking of requests with data disallowed.	1 for RTPS because service flow has periodic request opportunities, so no need to piggyback requests; usually 0 for BE and nRTPS service flows because piggybacking requests increase efficiency.
5	Concatenation disallowed.	RTPS, nRTPS, and BE service flow types can be configured to use concatenation if desired.
6	Fragmentation disallowed.	RTPS, nRTPS, and BE service flow types can be configured to use fragmentation if desired.
7	Payload Header Suppression disallowed.	RTPS, nRTPS, and BE service flow types can be configured to use PHS if desired.
8	Packets that do not fit the UGS size are dropped.	0; this field not meaningful for RTPS, nRTPS, and BE service flows.

- **Nominal Polling Interval**—Interval in microseconds for dedicated request opportunities; used in the RTPS and nRTPS services to give the CM uncontested opportunities to signal the CMTS.
- **Tolerated Poll Jitter**—Amount of time in microseconds the polling interval is permitted to vary from the Nominal Polling Interval. This parameter is used for the RTPS service type.
- **Minimum Reserved Traffic Rate**—Can be used for BE, RTPS, and nRTPS service flows; the minimum rate in bits per second reserved for the service flow. The packet size is from after the DOCSIS Header Check Sequence to the cyclic redundancy check (CRC) at the end of the packet. In other words, the "Ethernet" portion of the DOCSIS overhead is included, but the general 6-byte DOCSIS header and any DOCSIS extended headers are not.
- **Maximum Sustained Traffic Rate**—Can be used for BE, RTPS, and nRTPS service flows. The token bucket (R) parameter for rate-limiting packets is expressed in bits per second. Packet size is calculated the same as in the previous parameter.
- **Traffic Priority**—Used for nRTPS and BE service flows; defines the service flow priority. If multiple service flows exist that are identical in all QoS parameters besides priority, the higher priority service flow is given preference. If not included, the default priority of 0 is used. For example, if you have 10 residential users configured

with a Best Effort service of priority 0 and 10 commercial services users configured with a Best Effort service of priority 3, the CMTS services data grants from the commercial services users before the residential users.

In PCMM, the use of service class names to define dynamic flows is also permitted. You'll see more about how this is done later in the chapter. The concept and use of service class names is defined in the DOCSIS RFI specification. A service class name can be defined in the DOCSIS modem configuration file instead of all the class of service or service flow QoS parameters. The modem then sends this name in the DOCSIS registration request to the CMTS. On the CMTS, this service class name is also defined along with its corresponding QoS parameters. After the CMTS receives the registration request, it looks up the service class name and returns its parameters to the modem in the registration response message. Thus, a cable operator can simplify its modem provisioning by just using service class names in configuration files corresponding to the customer's level of service; for example, you might have a "Bronze," "Silver," and "Gold" service. The details of these service levels are defined on each CMTS and could be different on different CMTSs if desired.

Not only can static flows make the use of service class names, but dynamic flows can as well. In this case, the service class name is included in the DOCSIS dynamic service flow messages.

NOTE When a flow is defined with a service class name, additional QoS parameters can be defined as well. If the same parameters are defined on the CMTS and with the service class name, the parameters defined with the name take precedence. For example, if a Max Sustained Rate of 512 kbps is defined in a modem configuration file along with a service class name of "Bronze," and in the CMTS definition of service class name "Bronze" the Max Sustained Rate is defined as 256 kbps, the modem's Max Sustained Rate is 512 kbps.

PCMM COPS Usage Considerations

The COPS protocol is used in two places in PCMM:

- Between the application manager and policy server
- Between the policy server and CMTS

The use of the COPS protocol in PCMM is a bit different than the use of it in PacketCable telephony. First of all, remember that a PacketCable 1.x gate referred to two gates—one upstream and one downstream—sharing the same gate identifier. In PCMM, gates are truly unidirectional, and a gate ID refers to either an upstream gate or a downstream gate. Also previously a gate was used to define the authorized envelope of resources for an embedded MTA; in PCMM, gates can be used to define the reserved and committed resource envelopes as well as the authorized envelope. Remember, the committed or active envelope is always less than or equal to the reserved or admitted envelope, which is in turn always less than or equal to the authorized envelope.

The common header for each COPS packet is similar to the format used in PacketCable telephony; the only difference is that the client type is set to a different value, as shown in Figure 14-2.

Figure 14-2 *PCMM COPS Header Format*

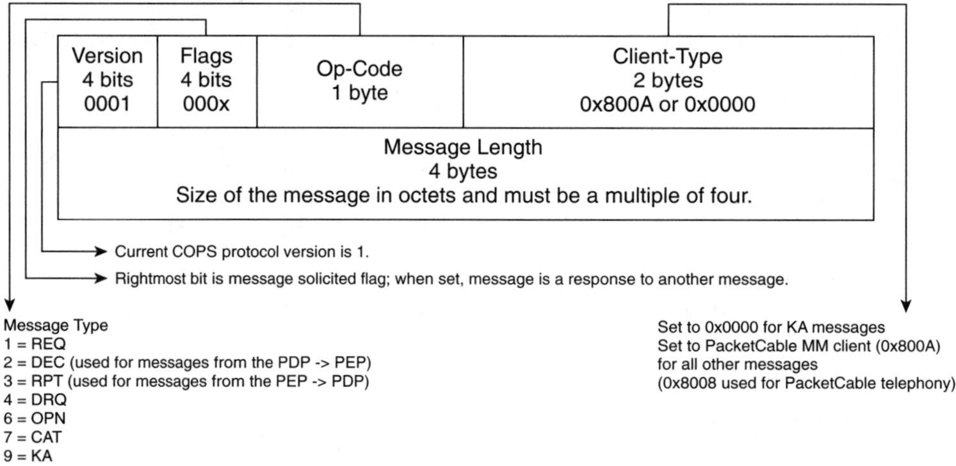

After this header are a number of objects in the form shown in Figure 14-3.

Figure 14-3 *PCMM COPS Object Format*

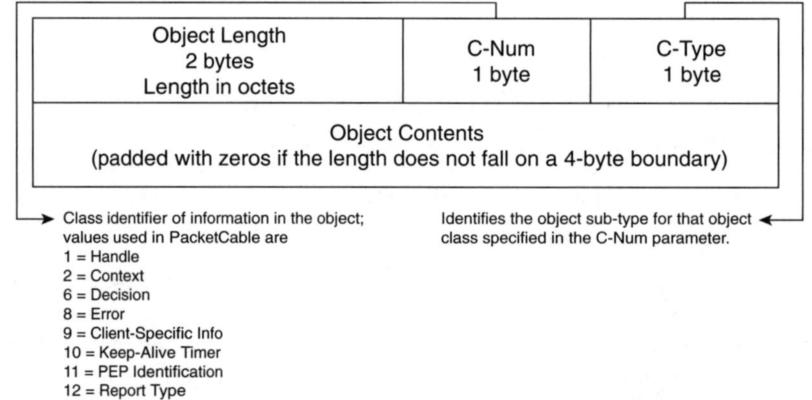

For messages going from the PDP to the PEP in a Decision Message, the C-Num is 6 (Decision) and the C-Type is 4 (Client-Specific Decision Data) for PacketCable objects.

For messages going from the PEP to the PDP in a Report-State Message, the C-Num is 9 (Client-Specific Info) and the C-Type is 1 (Signaled Client SI) for PacketCable objects.

COPS Initialization

The initialization process in PCMM is almost the same as in PacketCable telephony. One difference is that the PEP (CMTS or PS) listens on TCP port 3918 as opposed to port 2126 in telephony.

NOTE This allows a CMTS to support PacketCable telephony and PCMM simultaneously without interfering with each other.

After the TCP session is established, the PEP sends a COPS Client-Open (OPN) message to the PDP; in this message, the PEP identifies itself and includes the Version Info object. The PDP then responds with a Client-Accept (CAT) message containing the value of the Keep-Alive timer. The PEP then responds to this message with a Request (REQ) message to the PDP.

As is the case with PacketCable telephony, the PDP must be configured with the location of the PEP; however, the PEP does not need to be configured with the location of the PDP for the COPS connections to be established. This is assuming that IP Security (IPsec) is not being used for the COPS connection. If IPsec is being used, both the PDP and PEP need to be explicitly configured to have a security association between them.

PEP to PDP Heartbeats

As with PacketCable 1.x, keepalive messages (message type 9—KA) are sent from the PEP to the PDP periodically. The PDP responds to these keepalive messages by sending a keepalive message back to the PEP. If the PDP fails to receive a keepalive message from the PEP by the keepalive timer value, it assumes the connection has been lost. It then attempts to reestablish the TCP connection, which if successful causes the PEP to reinitialize the COPS connection. Similarly, if the PEP fails to receive the keepalive echo from the PDP, it assumes the connection has been lost and listens for a new TCP connection.

PacketCable Multimedia Objects in COPS

Remember that in PacketCable 1.x PacketCable gate messages sent from the CMS to the CMTS are sent in COPS Decision (DEC) messages, and PacketCable gate messages sent from the CMTS to the CMS are sent in COPS Report State (RPT) messages. In this case, the CMS is the PDP and the CMTS is the PEP. The same rules apply in PCMM—that is, PCMM gate messages sent from the PDP to the PEP are sent in COPS DEC messages, and PCMM gate messages sent from the PEP to the PDP are sent in COPS RPT messages.

As shown in Figure 14-3, PacketCable objects sent in DEC messages are included in Client Specific Decision Data objects (C-Num 6, C-Type 4) and PacketCable objects sent in RPT messages are included in Signaled Client SI objects (C-Num 9, C-Type 1). In either case, these objects begin with a 2-byte length field followed by another 1-byte object number and 1-byte subtype definition. These are referred to as the S-Num and S-Type, respectively.

PacketCable Multimedia COPS objects are as follows:

- Transaction Identifier (S-Num = 1, S-Type = 1)
- Application Manager Identifier (AMID) (S-Num = 2, S-Type = 1)
- Subscriber Identifier (S-Num = 3, S-Type = 1)
- Gate Identifier (S-Num = 4, S-Type = 1)
- Gate Specification (S-Num = 5, S-Type = 1)
- Classifier/Extended Classifier (S-Num = 6, S-Type = 1, 2)
- Traffic Profile (S-Num = 7, S-Type = 1, 2, 3, 4, 5, 6, 7, 8)
- Event-Generation-Info (S-Num = 8, S-Type = 1)
- Volume-Based Usage Limit (S-Num = 9, S-Type = 1)
- Time-Based Usage Limit (S-Num = 10, S-Type = 1)
- Opaque Data (S-Num = 11, S-Type = 1)
- Gate Time Info (S-Num = 12, S-Type = 1)
- Gate Usage Info (S-Num = 13, S-Type = 1)
- PacketCable Error (S-Num = 14, S-Type = 1)
- Gate State (S-Num = 15, S-Type = 1)
- Version Info (S-Num = 16, S-Type = 1)
- Policy Server Identifier (PSID) (S-Num = 17, S-Type = 1)
- Synch Options (S-Num = 18, S-Type = 1)
- Msg Receipt Key (S-Num = 19, S-Type = 1)

The following sections describe these PacketCable Multimedia COPS objects in greater detail.

Transaction ID

This object correlates responses to commands. It consists of a 2-byte transaction identifier followed by the 2-byte gate command type. The gate command type identifies the DQoS message type. Table 14-2 shows the possible values.

Table 14-2 *PCMM DQoS Message Types*

Gate Command Type	Gate Command	Message Direction
4	GATE-SET	PDP -> PEP
5	GATE-SET-ACK	PEP -> PDP
6	GATE-SET-ERR	PEP -> PDP
7	GATE-INFO	PDP -> PEP
8	GATE-INFO-ACK	PEP -> PDP
9	GATE-INFO-ERR	PEP -> PDP
10	GATE-DELETE	PDP -> PEP
11	GATE-DELETE-ACK	PEP -> PDP
12	GATE-DELETE-ERR	PEP -> PDP
15	GATE-REPORT-STATE	PEP -> PDP
16	GATE-CMD-ERR	PEP -> PDP
17	PDP-CONFIG	PDP -> PEP
18	PDP-CONFIG-ACK	PEP -> PDP
19	PDP-CONFIG-ERR	PEP -> PDP
20	SYNCH-REQUEST	PDP -> PEP
21	SYNCH-REPORT	PEP -> PDP
22	SYNCH-COMPLETE	PEP -> PDP
23	MSG-RECEIPT	PDP -> PEP

As you can see, several of the commands are the same as in PacketCable 1.x. The Gate Set message is used to initialize and modify gates as well as to control and update the state of a gate on the CMTS. The Gate Info message is used to query gate status, and the Gate Delete message is used to delete specific gates and their associated DOCSIS service flows. In PCMM, DOCSIS service flows and their associated gates on the CMTS are deleted via the receipt of a Gate Delete message from the PS. This is different from PC 1.x, where service flows and gates are usually deleted based on DOCSIS signaling from the MTA, and the Gate Delete message is used only under error conditions.

Notice that the Gate Allocation messages have been removed, as well as the Gate Open and Gate Close messages. Remember, the Gate Open and Gate Close messages were used by the CMTS to notify the CMS when a gate was committed or deleted. In PCMM, this functionality is rolled into the Gate Report State message, which notifies the PDP of gate changes that are not the result of a Gate Set message. One example: If a type 2 or 3 multimedia client commits its resources, the CMTS sends out a Gate Report State message to the policy server. Another example when this message would be sent is the expiration of a timer. Another use of the Gate Report State message is for notification of a gate reaching its traffic volume limit.

The PDP-Config message can be used when the COPS connection is first established. If the PDP is an Application Manager, this message contains the AMID(s) that are used by the PDP, and if the PDP is a Policy Server, this message contains the PSID(s) that are used by the PDP. This information allows a PEP to associate gate states for a PDP when the previous COPS connection is lost.

A PDP can request synchronization between its gate database and a PEP gate database by sending a Synch-Request message. Parameters such as the AMID, PSID, and Subscriber Identifier are specified in the Synch-Request as filtering criteria indicating which gate's information should be returned. A Synch-Report message is returned to the PDP for each gate matching the criteria. As you might imagine, the response to a synchronization request can consist of many responses; the PEP indicates it has finished sending responses by sending a Sync-Complete message. You find two types of synchronization requests: full and incremental. If it receives a full synchronization request, the PEP returns information for all active gates matching the filtering criteria. An incremental request returns only the information for gates whose state has recently changed. This is used in situations where the COPS connection was recently lost. When it receives an incremental synchronization request, the PEP returns information for all gates with an outstanding unconfirmed Gate-Report-State message.

The Msg-Receipt message is used by the PDP to tell the PEP that it has successfully received a message that the PEP indicated required acknowledgement. The PEP indicates it requires an acknowledgement by including the Message Receipt Key object in the message it sends to the PDP.

Application Manager Identifier (AMID)

This 4-byte identifier is defined on the Application Manager and is unique within a service provider. Because a single AM can be used in multiple service provider networks, the AM would have an AMID for each service provider. The AMID consists of a 2-byte Application Manager Tag followed by a 2-byte Application Type. Logically, the tag identifies the AM, and the type identifies the type of application for which the gate is used. The Application Type values used are up to the service provider; the only exception is the value zero, which is reserved to indicate no defined application association. The policy server uses the AMID to associate Gate messages received from the CMTS with a particular AM and application type.

Subscriber Identifier

The IP address representing the subscriber could be the IP address of the client CPE device or of the cable modem. If private addresses are used on the client CPE device and NAT is performed, the CM address is used instead of the CPE address. This parameter associates QoS information with a particular subscriber to help prevent denial-of-service attacks.

Gate Identifier

The 4-byte gate identifier is assigned by the CMTS; remember, in PCMM this refers a single unidirectional gate.

Gate Specification

The PCMM gate specification is quite different than the one used in telephony. The parameters related to packet classification and traffic characteristics have been removed and put into other objects, namely the classifier and traffic profile objects. The gate specification object format is shown in Figure 14-4.

Figure 14-4 *PCMM Gate Specification Format*

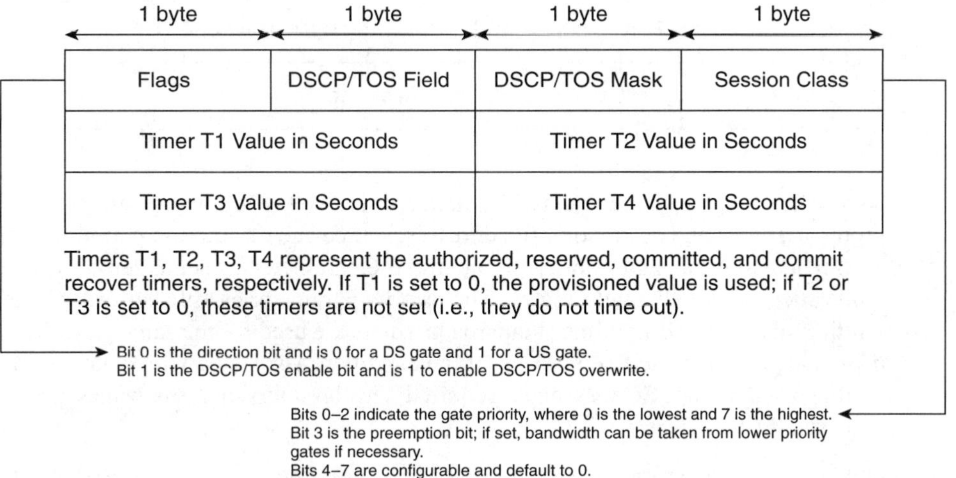

The Session Class field identifies the relative priority of the gate. Higher priority gates could preempt lesser priority gates if resources were congested, much like how emergency

911 calls can receive preferential treatment over ordinary phone calls in PacketCable telephony. The authorized, reserved, and committed timers work the same way as the T1, T7, and T8 timers in PacketCable telephony. The committed recovery timer is the amount of time a gate can remain in the inactive state after the AM or PS is notified.

Classifier/Extended Classifier

One or more classifiers identify the packets that utilize the gate. Two types of classifiers are currently defined in the PCMM specification, the standard classifier and the extended classifier. Figure 14-5 depicts the format of the standard classifier object.

Figure 14-5 *PCMM Standard Classifier Format*

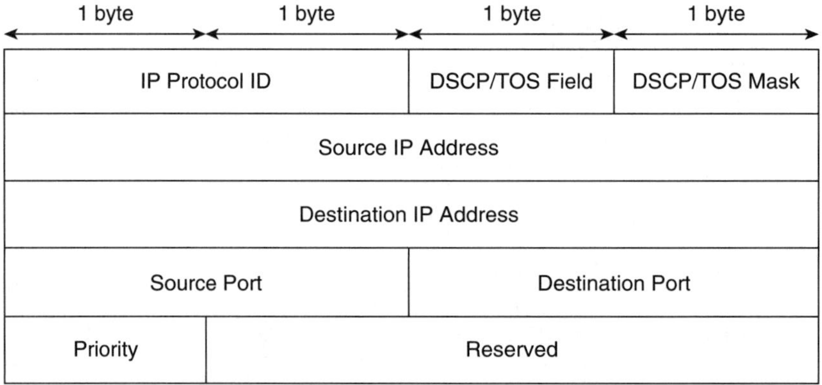

As you can see, the standard classifier field consists of the following parameters: IP protocol, source and destination IP addresses, source and destination port numbers, DSCP/TOS field and mask, and priority. If one of these parameters is set to zero, it is considered wildcarded, where all values match. The priority parameter establishes order in applying multiple classifiers. If multiple standard classifiers are used for the same gate, they must all be included in the Gate Set message. If the classifiers associated with a gate need to be modified, a new Gate Set message is required; the classifiers in this new message overwrite any existing classifiers.

Unlike standard classifiers, extended classifiers have operation actions associated with them so individual classifiers can be added, deleted, activated, and inactivated. Extended classifiers also allow for ranges of IP addresses and port numbers to be specified. The use of extended classifiers over standard classifiers is preferred. Figure 14-6 illustrates the format of the Extended Classifier object.

Figure 14-6 *PCMM Extended Classifier Format*

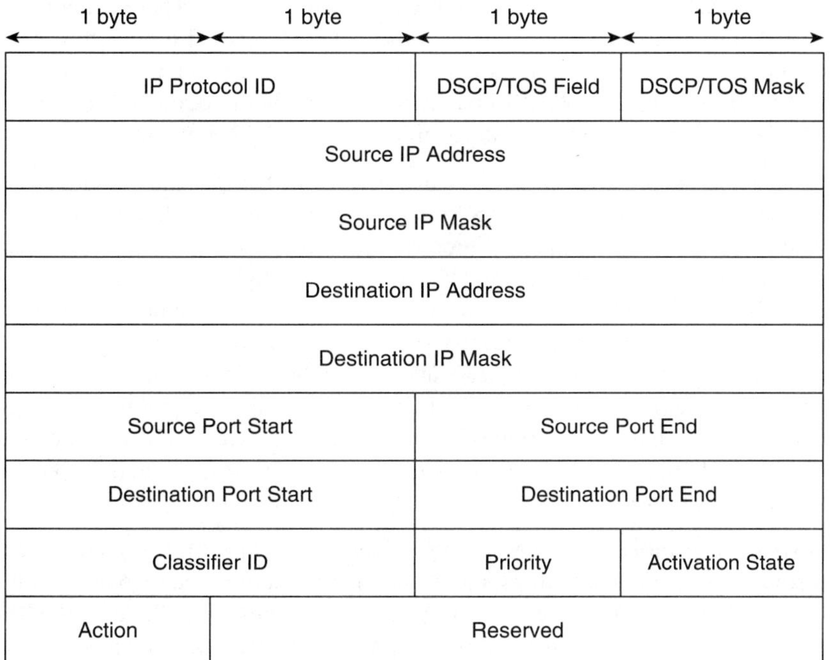

If the Activation State field is equal to 0 (Inactive), the classifier must not be used to map traffic, and if it is equal to 1 (Active), it must be used. The Action field has the following possible values: 0 (Add), 1 (Replace), 2 (Delete), and 3 (No change).

Traffic Profile

The traffic characteristics of a gate can be defined in one of three ways in PCMM: by an RSVP flow specification (S-Type = 1), by a DOCSIS service class name (S-Type = 2), or by DOCSIS specific parameterization (S-Type = 3).

The RSVP flow specification method is similar to the method used in PacketCable telephony. The second method is simply a named DOCSIS service class. The traffic profile parameters are defined on the CMTS under the same service class name. One key difference exists here compared to how service class names are used in DOCSIS. In DOCSIS, additional parameters can be defined with the service class name overriding the parameters defined on the CMTS. This is not permitted in PCMM. The third method is where the traffic profile is defined directly by DOCSIS Tag/Length/Value (TLV) parameters.

In most situations, any one of these methods can be used. An exception is if you want to use the UGS-AD or nRTPS DOCSIS service scheduling types. These cannot be created using RSVP flow specs; therefore, you must use either the named service flow or DOCSIS TLV method. As you can see, using the RSVP flow spec method has some inherent limitations in its ability to map to DOCSIS parameters. The other two methods do not have these limitations because they are also specified using DOCSIS parameters.

A benefit of using the service class name method is that QoS information is logged in to the ServiceFlowLogTable in the QoS Management Information Base (MIB). This occurs only when the named service class method is used. The service class name method is also useful in situations where a particular service might vary slightly from location to location. For example, if a "boost" service in one market uses a maximum downstream rate of 5 Mbps and a "boost" service in another market uses a maximum downstream rate of 4 Mbps, a named service class called "boost" can exist on the CMTS devices in each market with the appropriate definition. This way, the PCMM devices do not need to be configured for these market differences. Also service class names can be used for the communication between the AM and PS, so the AM doesn't need to know the specific details on how to create QoS resource envelopes.

Up to three resource envelopes can be defined in the traffic profile object: one for authorized resources, one for reserved resources, and one for committed resources. A bit field, called the Envelope field, indicates which resource envelopes are included—the LSB is for authorized resources, the next bit is for reserved resources, and the next bit is for committed resources. Of course, if resources are to be activated immediately, then all three of these resource envelopes would be included and the same.

The presence of a reserved resource envelope causes the CMTS to initiate DOCSIS DSX messaging to place these resources in the admitted state. The presence of a committed resource envelope causes the CMTS to initiate DOCSIS DSX messaging to place these resources in the active state.

A limitation with the service class name method is that the same name must be used for all three resource envelopes. Therefore, this method can be used only when all three envelopes share a common set of parameters.

Remember from Chapter 12 that two types of services are defined by RSVP flows specs: controlled load and guaranteed. The guaranteed service is for applications that are delay sensitive and is defined by a traffic spec (TSpec) and resource spec (RSpec), and the controlled load service is for applications that require some amount of bandwidth but are not delay sensitive and are defined by only a TSpec. Chapter 12 defined the TSpec and RSpec parameters. PacketCable telephony is defined using the guaranteed service. Streaming video and interactive gaming are also sensitive to delay and would also be defined as a guaranteed service. An example of an application using a controlled load service is a web-based bandwidth on demand service.

If your AM defines a guaranteed service using RSVP flow specs, this gets converted to either a UGS or RTPS DOCSIS service flow. A controlled load service gets converted to a BE DOCSIS service flow.

NOTE Whether or not a UGS or RTPS service flow results depends on whether the Reserved Rate (R) and Bucket Rate (r) parameters are equal. If they are, that means the bandwidth rate is constant; hence, a UGS flow results. Otherwise, an RTPS flow results. For more details on how RSVP flow specifications are mapped into DOCSIS parameters, please refer to section 9 of the PacketCable Multimedia specification.

Event-Generation-Info

This object works the same as in telephony (see Chapter 12) and passes along RKS information to the CMTS or policy server. This includes primary and secondary RKS server IP addresses and port number, as well as a billing correlation identifier.

Volume-Based Usage Limit

This object specifies the maximum amount of data in kilobytes that can traverse a gate. Bytes are counted from after the DOCSIS MAC header HCS to the end of the CRC—that is, the "Ethernet" portion of the DOCSIS header counts, but the DOCSIS header and extended headers do not.

NOTE The CMTS is not responsible for enforcing this limit, but it does send a message to the PS/AM when this limit is reached.

Time-Based Usage Limit

This object specifies the maximum amount of time gate resources can remain committed. It is in units of seconds.

NOTE The CMTS is not responsible for enforcing this limit, however.

Opaque Data

This object contains arbitrary data the PS or AM wants to associate to a gate. The CMTS stores this information but does nothing else with it.

Gate Time Info

This object contains the total amount of time (in seconds) the gate was committed. This information can be queried by the PS and/or AM.

Gate Usage Info

This object contains an octet counter representing the number of kilobytes that have traversed the gate. Again, bytes are counted from the DOCSIS header HCS to the end of the CRC. This information can be queried by the PS and/or AM.

PacketCable Error

This object consists of a 2-byte error code and a 2-byte error subcode. Currently defined error codes are shown in Table 14-3.

Table 14-3 *Error Codes*

Error Code	Definition
1	Insufficient resources
2	Unknown Gate ID
6	Missing required object
7	Invalid object
8	Volume-based usage limit exceeded
9	Time-based usage limit exceeded
10	Session Class limit exceeded
11	Undefined Service Class name
12	Incompatible envelope
13	Invalid subscriber identifier
14	Unauthorized AMID
15	Number of classifiers not supported
16	Policy exception

Table 14-3 *Error Codes (Continued)*

Error Code	Definition
17	Invalid field value in object
18	Transport error
19	Unknown gate command
20	DOCSIS 1.0 CM
21	Number of SIDs exceeded in CM
22	Number of SIDs exceeded in CMTS
23	Unauthorized PSID
24	No state for PDP
25	Unsupported Synch Type
26	State data incomplete
127	Other, unspecified error

The error subcode conveys further information about the error. For example, for error codes 6, 7, and 17 this field contains the S-Num and S-Type values of the object missing or invalid.

Gate State

This object conveys the current state of the gate and consists of a 2-byte state field and a 2-byte reason field. This parameter is sent in a Gate Report State message. The values of the state field are 1 (Idle/Closed), 2 (Authorized), 3 (Reserved), 4 (Committed), and 5 (Committed Recovery). Logically, the reason field contains the reason the gate is in this state and has the values shown in Table 14-4.

Table 14-4 *Reason Codes*

Reason Value	Description
1	Close initiated by CMTS because of reservation reassignment
2	Close initiated by CMTS because of lack of DOCSIS responses
3	Close initiated by CMTS because of timer T1 expiry
4	Close initiated by CMTS because of timer T2 expiry
5	Inactivity timer (T3) expired
6	Close initiated by CMTS because of a lack of reservation maintenance
7	Gate state unchanged, but volume limit reached

continues

Table 14-4 *Reason Codes (Continued)*

Reason Value	Description
8	Close initiated by CMTS because of timer T4 expiry
9	Gate state unchanged, but timer T2 expiry caused reservation reduction
10	Gate state unchanged, but time limit reached
11	Close initiated by PS or CMTS; volume limit reached
12	Close initiated by PS or CMTS; time limit reached
13	Close initiated by CMTS, other
65,535	Other

Version Info

This object contains the major and minor version numbers of PacketCable multimedia the application is using. Both fields are 2-byte integers. The current specification has a major version number of 2 and a minor version number of 0.

PSID

This object is a 4-byte integer that uniquely identifies the Policy Server.

Synch Options

A PDP can ensure its database is synchronized with a PEP by issuing a synchronization request to it. This object consists of a 1-byte Report Type followed by a 1-byte Synch Type. The Report Type can be either 0, telling the PEP to return standard report data, or 1, telling the PEP to return complete gate data. The Synch Type can be 0, indicating full synchronization, or 1, indicating incremental synchronization.

Msg Receipt Key

This object is a 32-bit integer assigned by the PEP. When it is included in a message, it tells the PDP to confirm receipt of the message.

PCMM Message Flow Template

Okay, now that you know the basics of PCMM, it's time to look at an example of what the messaging might look like in a PCMM session. For the purposes of this discussion a

PCMM session starts when a subscriber initiates an application requiring QoS and finishes when the subscriber's application is finished. For example, in PacketCable telephony a multimedia session equates to a phone call.

For this example it is assumed that the multimedia application is not PacketCable QoS–aware, so you know this means the multimedia client is a legacy type 1 device. Again, examples of these client types include applications running on a personal computer such as streaming audio and video, gaming consoles, and IP phones. This client device would be connected to a cable modem via the modem's LAN interface (Ethernet, USB, wireless, and so on). Figure 14-7 depicts what this messaging might look like, and the list that follows details the nine steps depicted in this message flow.

Figure 14-7 *PCMM Session Message Flow Example*

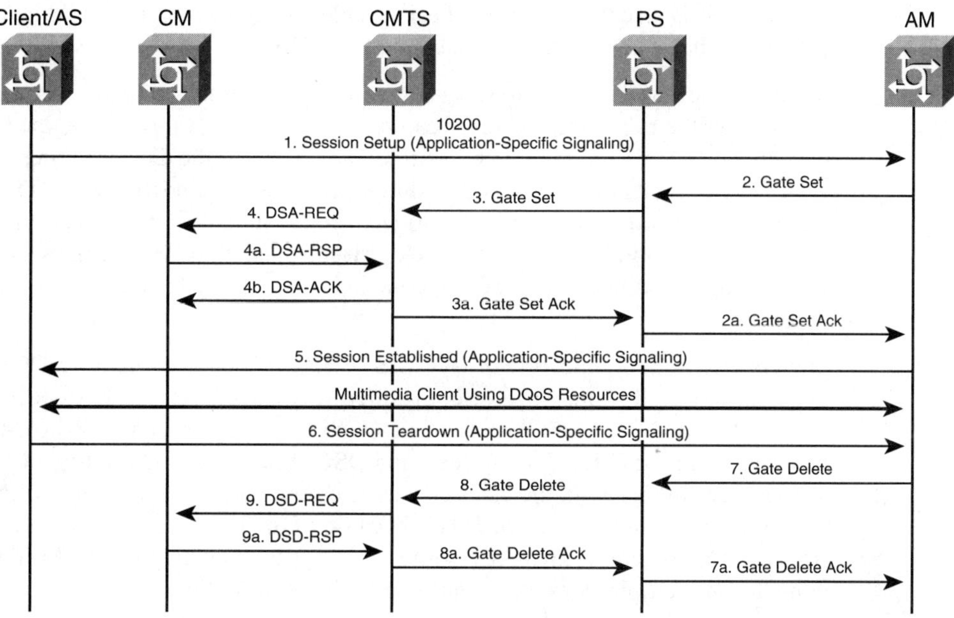

1 In Step 1 the client (either directly or via an application server) and the AM exchange messaging, which indicates the request to create a new session. Remember, this signaling is out of the scope of PCMM, but some possible protocols include MGCP and SIP. These protocols can be used to create audio and/or video multimedia connections on the multimedia client. Another possible application is HTML, where the client simply opens a web browser to the server, requesting some kind of QoS on-demand service.

2 At some point the AM figures out what the required QoS parameters are and signals a policy server for their creation. This is done by sending a Gate Set message, which identifies the subscriber and contains the classifier, gate specification, and traffic profile parameters needed to achieve the desired QoS.

3 When the PS receives the Gate Set, it checks to see whether that AM is authorized to make such a request. It does this by applying predefined policies. Sample policies include limits on the number of gates allocated to a subscriber, limits on the types of QoS available to a subscriber, limits on which applications a policy server accepts, and limits on the impact of service on a particular CMTS. Assuming all checks pass, the PS figures out where that subscriber is and initiates a Gate Set to the CMTS where the subscriber's cable modem is connected. So as you can see, the AM doesn't need to know which CMTS the subscriber's modem is attached to; the PS figures this out instead. This is an enhancement to PacketCable telephony where the CMTS must be defined on the CMS (aggregation table on the BTS).

4 Similarly, the CMTS checks to see whether that PS is authorized and, if so, checks to see whether the requested resources can be granted. If all of this is true, the CMTS then creates a gate, assigns a gate identifier, and converts the COPS parameters to DOCSIS parameters. If a reserved resource envelope is included, the CMTS signals the service flow creation to the cable modem concerning the admission of these resources. If a committed resource envelope is included, the CMTS signals the service flow creation/modification to the cable modem concerning the activation of these resources.

   The creation and modification of DOCSIS resources is done using the DOCSIS dynamic service flow messages. So, the three-way handshake of a DSA-REQ, DSA-RSP, and DSA-ACK is used to create a DOCSIS service flow, and the three-way handshake of a DSC-REQ, DSC-RSP, and DSC-ACK is used to modify a DOCSIS service flow. In this case, the gates and service flows are created, reserved, and committed all in the same step. This is because of the initial Gate Set message sent from the AM containing all three resource envelopes; alternatively, the AM could have done this in multiple steps using multiple Gate Set messages.

   If the service flow and gate creation process are successful, the CMTS notifies the PS of this in the Gate Set Acknowledgement message. If this process is unsuccessful, a Gate Set Error message indicating the reason for failure is returned instead. The PS then relays this message to the AM.

5 At this point, the multimedia session's QoS resources are active, so the AM relays this information back to the client device. Again, the signaling used to accomplish this is out of the scope of PCMM.

Note that it is entirely possible no coordination exists between the application signaling and the QoS setup. Therefore, the client could have started sending data traffic before QoS was available; in this case, the traffic would have traversed the default BE service flows before the QoS was committed.

6. When the client is finished with the multimedia session, the client device signals this information to the AM.
7. The AM then requests the deletion of QoS resources by sending a Gate Delete message to the PS.
8. The PS in turn forwards this information to the appropriate CMTS in a Gate Delete message.
9. The CMTS receives this message and immediately requests deletion of the DOCSIS service flows by sending a DSD-REQ to the CM. The CM acknowledges the deletion of the service flow(s) by sending a DSD-RSP back to the CMTS.

NOTE If the CMTS receives a DSD-REQ from the CM before this point, the service flows are deleted, but the gate still remains active until the PS/AM signals for gate deletion.

Now that the CMTS knows the service flows are successfully deleted, it deletes the gate and sends a Gate Delete Acknowledgement to the PS confirming this. If there was a problem in the gate deletion, a Gate Delete Error message is sent instead.

Finally, the PS forwards this message back to the AM.

PCMM Message Flow Example

The previous call flow template is a fairly good indication of what a PCMM session might look like. This section shows a real-world example of the life of a PCMM session. In this case, the application is a bandwidth-on-demand or Speed Preview service where a subscriber is temporarily given an increase in QoS for their high-speed data (HSD) service. The subscriber activates the Speed Preview service via a web page managed by the service provider. Traces of messages are examined to help you gain a greater understanding of how PCMM works. These message traces can also help you see how PCMM differs from PacketCable 1.x. Figure 14-8 illustrates the messaging used for the creation of the PCMM session, and the list that follows describes each step of the message flow.

Figure 14-8 *PCMM Session Message Flow Example Part I*

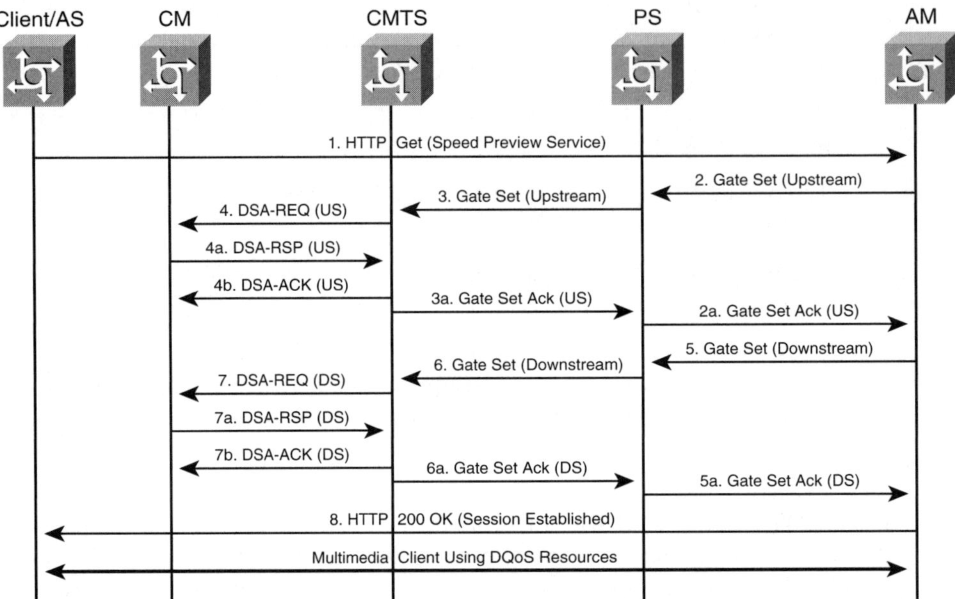

1 The subscriber initiates a request for the Speed Preview service by selecting an option on a service provider web page. This web server is the application manager for the Speed Preview service. In addition to initiating the PCMM session, the web server can also collect billing information so that the subscriber is properly charged for the session. The message used to initiate the request from the subscriber's web browser to the service provider's web server is an HTTP Get request. Figure 14-9 shows a trace of this message.

Figure 14-9 *HTTP Get Request*

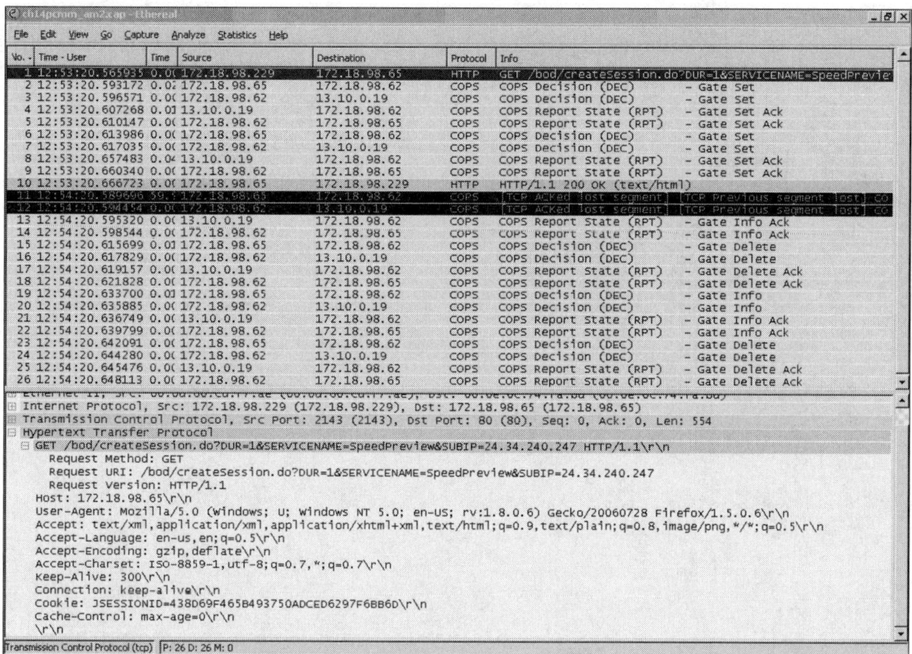

If you look at the Request URI, you can see some parameters that identify details of the application request. First of all, you can see the service name of Speed Preview is specified, as well as the IP address of the subscriber. In this example, the multimedia session is set up to be active for 1 minute (DUR=1). This message might be sent directly from the subscriber, or it might be sent on behalf of the subscriber from another device. The latter is the case in this example. Notice that the source IP address of the HTTP packet is different from the subscriber IP address. The device sending the HTTP packet is the Application Server. Recall that the Application Server is an intermediary device between PCMM clients and the AM.

2 As soon as the AM receives the HTTP request, it verifies that this particular subscriber is authorized for the service and determines what the required QoS parameters are for the Speed Preview service. In this case, the service corresponds to the creation of an upstream gate as well as a downstream gate. First, the AM signals the policy server for the creation of the upstream gate using a Gate Set message. Figure 14-10 shows this message.

Figure 14-10 *Gate Set (Upstream)*

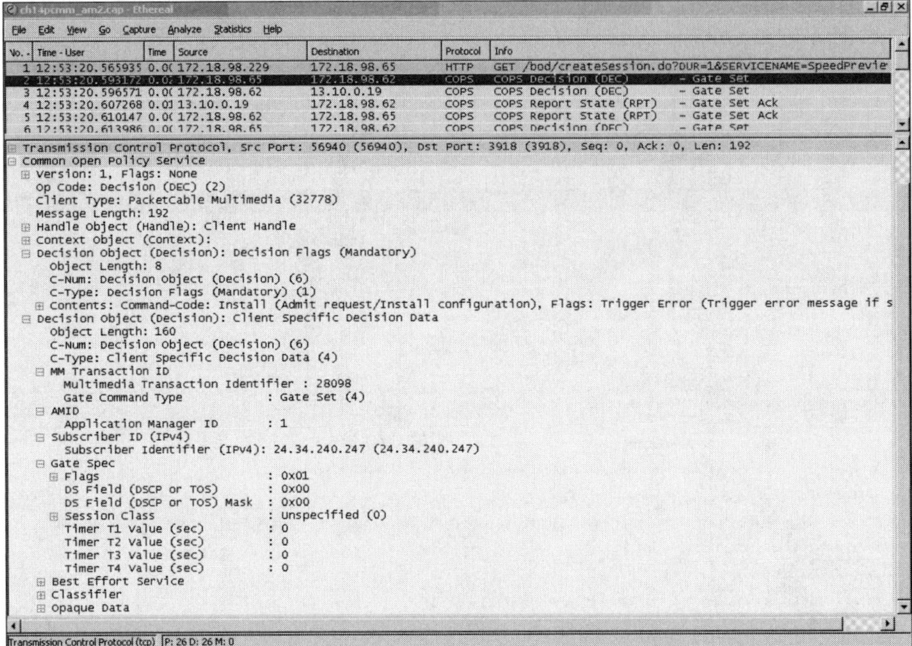

Notice that this message is sent to port 3918 on the policy server, and the COPS client type indicates PacketCable Multimedia. The PacketCable Multimedia parameters are contained with COPS Client Specific Decision Data objects. The first parameter is the Transaction ID, which identifies this message as a Gate Set. This parameter is followed by the Application Manager ID referencing the AM and application type and the subscriber ID referencing the user of the PCMM session. The next parameter is the Gate Specification. The Flags subparameter identifies this gate as for the upstream direction (remember, if the least significant bit in this field is set, this indicates an upstream gate). The session class parameter indicates the gate is of the lowest priority. All the timer values are set to 0, meaning these timers never expire. As you'll soon see, all resources for this gate are immediately activated; therefore, timers T1 and T2 do not apply. Recall from the previous step that one of the parameters used when creating the gate was duration. This multimedia session is designed to be set up for this time period, regardless of whether the subscriber actually uses it; therefore, timer T3 is also set to 0. Because timer T3 isn't used, timer T4 isn't needed either, so it is also set to 0. The traffic profile type used in this case is the DOCSIS parameters for an upstream Best Effort Service. Figure 14-11 shows the details of this parameter.

Figure 14-11 *Traffic Profile Gate Set (Upstream)*

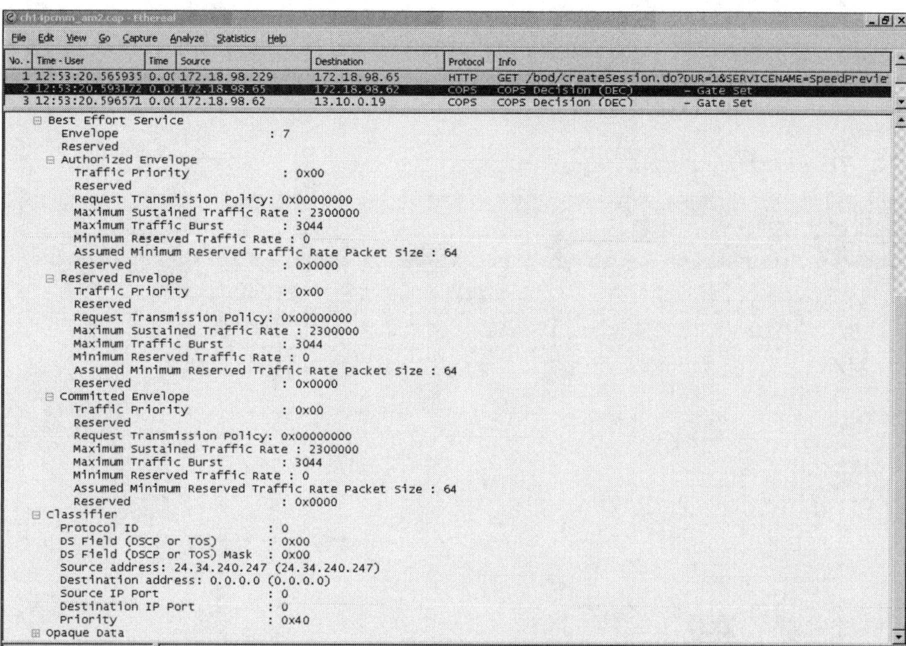

The envelope parameter is equal to 7, meaning authorized, reserved, and committed resource envelopes are all included. As you can tell, the parameters of all three of these resource envelopes are identical. This tells the PS and, ultimately, the CMTS that DOCSIS resources are to be immediately admitted and activated. After the traffic profile parameter is the classifier parameter. In this case, a standard classifier is used that simply states to match all traffic with a source IP address matching the subscriber (that is, CPE) IP address. The last parameter is the opaque data parameter, which specifies arbitrary data the AM wants to be associated with the gate.

3 When the PS receives the Gate Set, it checks to see whether that AM is authorized to make such a request. It then applies any predefined policies, such as making sure the creation of this gate does not violate any limits identified for this subscriber. The PS also does a lookup for this subscriber IP to determine what CMTS the subscriber is connected to. After it has this information, the PS sends a Gate Set message to the CMTS. In this case, this message is nearly identical to the Gate Set message sent from the AM. Remember, though, that this is not necessarily always the case.

4 Upon receiving the Gate Set message, the CMTS checks to see whether the requested resources can be granted. In this case this check is successful. Because the requested resources are to be immediately reserved and activated, the CMTS initiates the DSA message exchange to the cable modem where the subscriber is located. Figure 14-12 shows the DSA-REQ message.

Figure 14-12 *DSA-REQ (Upstream)*

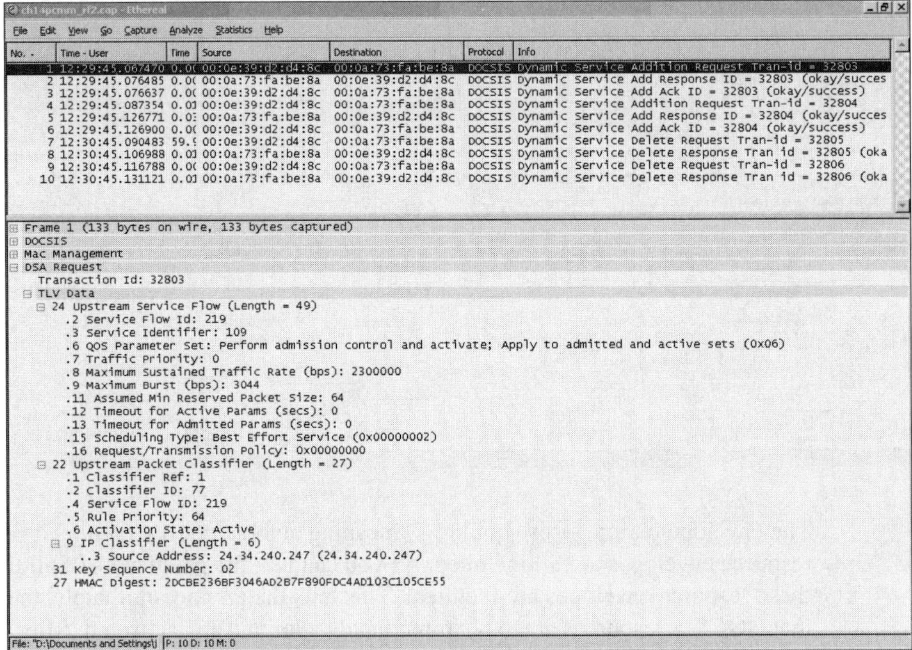

Because the DSA-REQ initiates from the CMTS, you already have a Service Flow ID, Service ID, and Classification ID for the upstream resources. As you can see, the QoS parameter set indicates these DOCSIS resources are admitted and activated and the classifier state is active. Finally, do you notice something else different compared to a DSA-REQ in PacketCable 1.x? Because this message originates from the CMTS and the client is QoS-unaware, sending an authorization block containing the gate ID to the cable modem is not necessary. The cable modem responds to the DSA-REQ with a DSA-RSP message indicating the request was successful. The CMTS then completes the three-way handshake by sending a DSA-ACK message.

The CMTS then lets the PS know the service flow and gate creation was successful by sending a Gate Set Acknowledgement message, as shown in Figure 14-13.

Figure 14-13 *Gate-Set Ack (Upstream)*

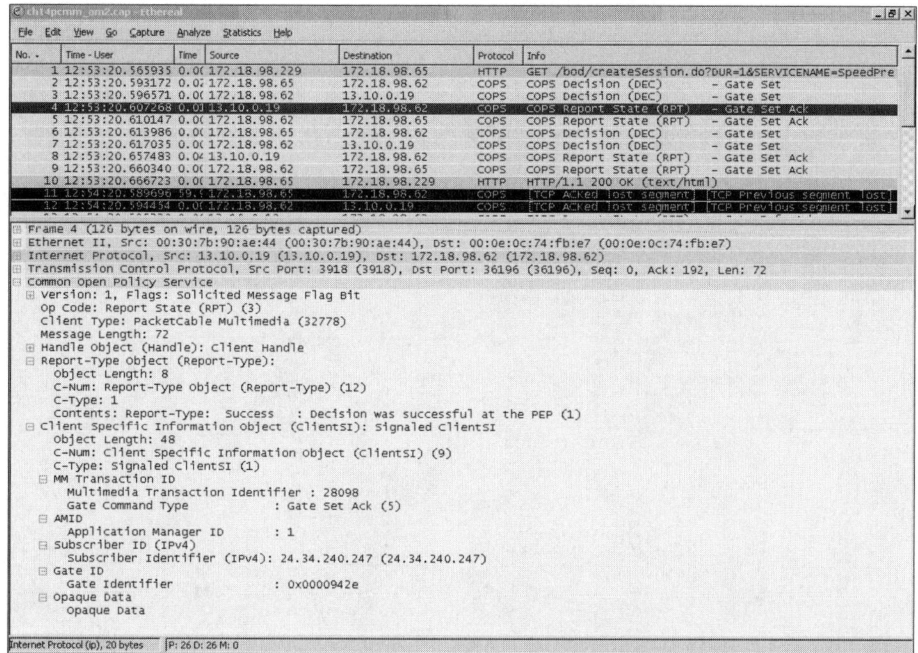

As you can see, this message indicates the policy was successfully installed on the CMTS, and the gate identifier for the upstream QoS is returned in the Gate Set Ack message. This is so the AM and PS can refer to this gate in future messages. Upon receiving this message, the PS sends a Gate Set Ack of its own to the AM.

5 The AM then starts a similar process for the creation of downstream QoS resources for the client. Figure 14-14 shows the resulting Gate Set message.

Figure 14-14 *Gate Set (Downstream)*

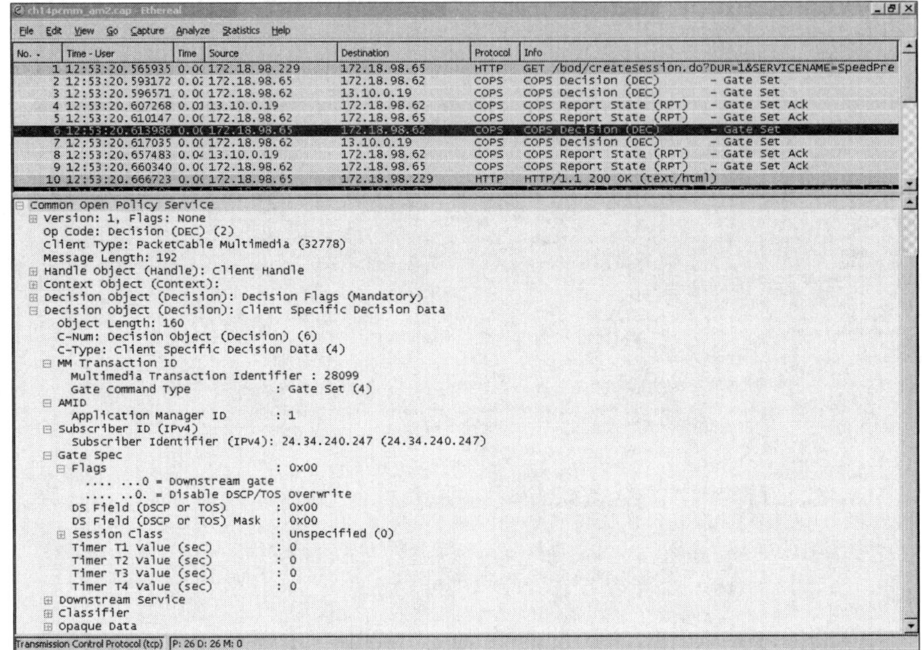

As you can see, in this case the Flags field indicates this is a Downstream Gate. The traffic profile in this case is a DOCSIS-based definition for a Downstream Service Flow. Figure 14-15 shows the details of this parameter.

Figure 14-15 *Traffic Profile, Gate Set (Downstream)*

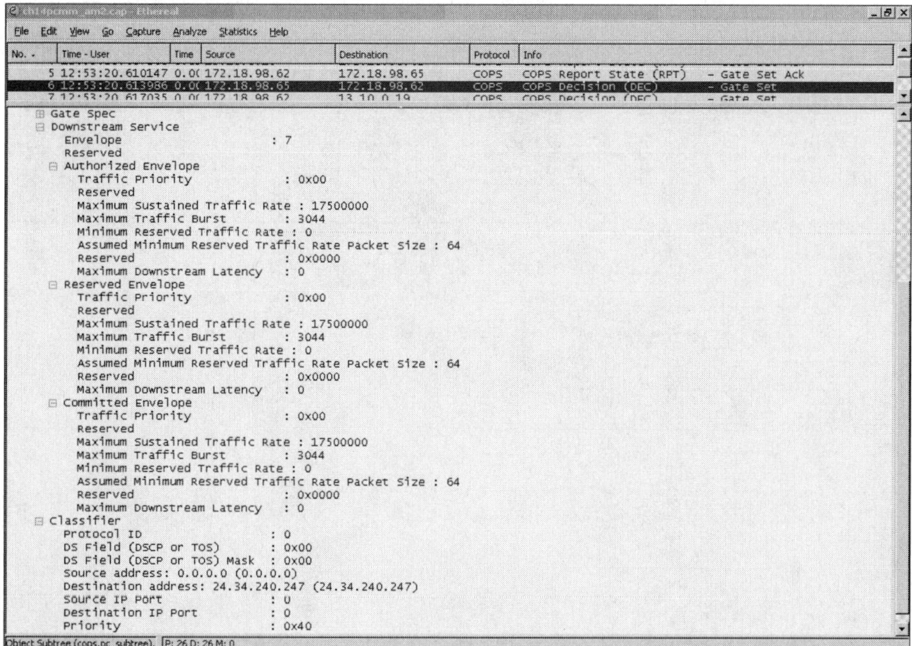

As was the case with the upstream traffic profile definition, identical authorized, reserved, and committed resource envelopes are included. A standard classifier is set up to match all packets destined for the subscriber's IP address.

6 Upon receiving this message, the PS performs any predefined policy checks and, if successful, generates a Gate Set message to the CMTS for the creation of the downstream gate. In this case, this message is nearly identical to the Gate Set message in the previous step.

7 As it did with the upstream request, the CMTS authorizes the required downstream resources, and then admits and activates the corresponding DOCSIS resources. Figure 14-16 shows the resulting DSA-REQ sent to the cable modem.

Figure 14-16 *DSA-REQ (Downstream)*

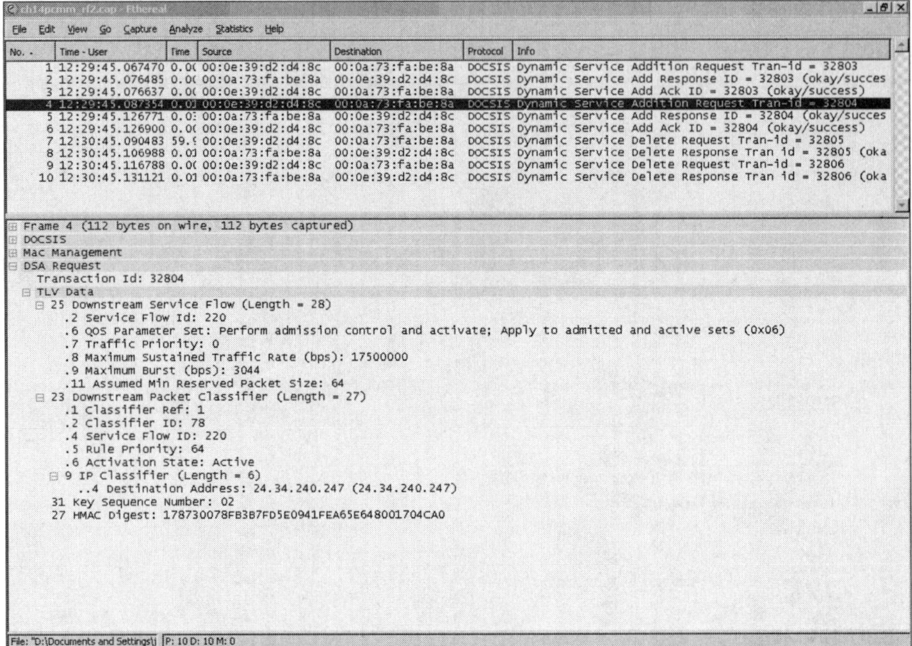

As expected, the QoS parameter set specifies that the resources are admitted and activated and the classifier state is activated. Also a service flow ID and classifier ID are assigned to this QoS. The modem then sends a DSA-RSP message indicating this request is successful, and the CMTS responds to this message with a DSA-ACK. Knowing that the Gate Set was successful, the CMTS sends a Gate Set Acknowledge message to the PS containing the gate identifier. The PS in turn forwards this information to the AM.

8 At this point, the Speed Preview QoS resources are active in both directions, so the AM relays this information back to the client device. An HTTP 200 - OK response to the original HTTP Get Request accomplishes this.

The multimedia session is now active, and the subscriber is enjoying the use of the enhanced QoS resources. However, eventually the duration time allocated to the subscriber expires, and the multimedia session must be deleted. Figure 14-17 illustrates the messaging used to accomplish this.

Figure 14-17 *PCMM Session Message Flow Example, Part II*

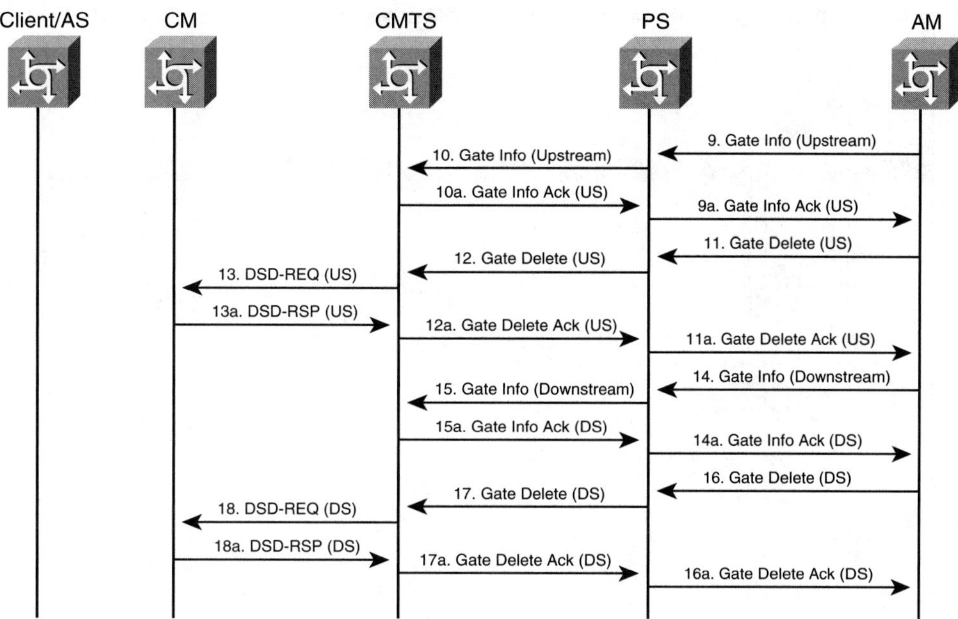

9 The first thing the AM does is request the PS for the current status of the upstream gate by sending a Gate Info message as shown in Figure 14-18.

Figure 14-18 *Gate Info (Upstream)*

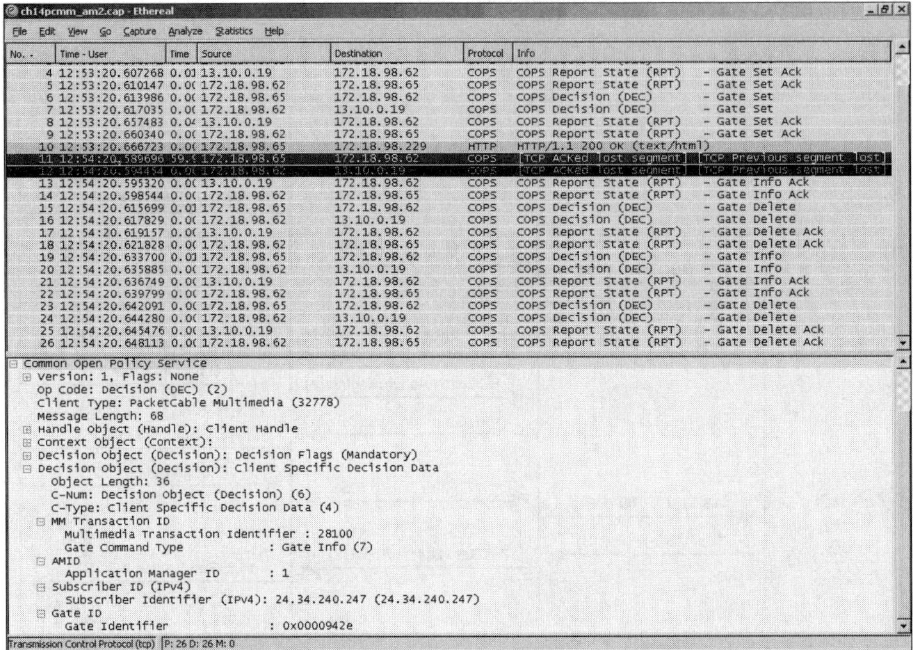

Notice the Gate ID is the value returned by the CMTS in the Gate Set Acknowledge message at the end of Step 3.

10 To retrieve all the gate information for the AM, the PS needs to issue a Gate Info message of its own to the CMTS. This parameter looks nearly identical to the previous Gate Info message.

The CMTS then collects all the required information and returns it to the PS in a Gate Info Ack message, as shown in Figure 14-19.

Figure 14-19 *Gate Info Ack (Upstream)*

```
ch14pcmm_am2.cap - Ethereal
File Edit View Go Capture Analyze Statistics Help
No. - Time - User       Time  Source          Destination      Protocol  Info
   9 12:53:20.660340 0.0( 172.18.98.62    172.18.98.65     COPS      COPS Report State (RPT)   - Gate Set Ack
  10 12:53:20.666723 0.0( 172.18.98.65    172.18.98.229    HTTP      HTTP/1.1 200 OK (text/html)
  11 12:54:20.589696 59.( 172.18.98.65    172.18.98.62     COPS      [TCP ACKed lost segment] [TCP Previous segment lost]
  12 12:54:20.594454 0.0( 172.18.98.62    13.10.0.19       COPS      [TCP ACKed lost segment] [TCP Previous segment lost]
  13 12:54:20.595320 0.0( 13.10.0.19      172.18.98.62     COPS      COPS Report State (RPT)   - Gate Info Ack
  14 12:54:20.598544 0.0( 172.18.98.62    172.18.98.65     COPS      COPS Report State (RPT)   - Gate Info Ack
  15 12:54:20.615699 0.01 172.18.98.65    172.18.98.62     COPS      COPS Decision (DEC)       - Gate Delete

Common Open Policy Service
  Version: 1, Flags: Solicited Message Flag Bit
  Op Code: Report State (RPT) (3)
  Client Type: PacketCable Multimedia (32778)
  Message Length: 220
  Handle Object (Handle): Client Handle
  Report-Type Object (Report-Type):
  Client Specific Information Object (ClientSI): Signaled ClientSI
    Object Length: 196
    C-Num: Client Specific Information Object (ClientSI) (9)
    C-Type: Signaled ClientSI (1)
    MM Transaction ID
      Multimedia Transaction Identifier : 28100
      Gate Command Type                 : Gate Info Ack (8)
    AMID
      Application Manager ID            : 1
    Subscriber ID (IPv4)
      Subscriber Identifier (IPv4): 24.34.240.247 (24.34.240.247)
    Gate ID
      Gate Identifier                   : 0x0000942e
    Gate Spec
    Classifier
    Best Effort Service
      Envelope                          : 7
      Reserved
      Authorized Envelope
      Reserved Envelope
      Committed Envelope
    Gate State
      State                             : Committed (4)
      Reason                            : Other (65535)
    Gate Time Info
      Time Committed                    : 59
    Gate Usage Info
      Octet Count                       : 0
    Opaque Data
Common Open Policy Service (cops  P: 26 D: 26 M: 0
```

As you can see, all the parameters from the Gate Set used for the creation of this gate, as well as some additional parameters, are included in this message. The Gate State parameter indicates the gate is currently committed. The Gate Time Info message indicates the gate has been committed for 59 seconds (remember, the application was initially set up for a duration of 1 minute). The Gate Usage Info message indicates no traffic has used this gate. Upon receiving this information, the PS creates a similar message and sends it to the AM.

11 At this point, the AM knows the time for the upstream gate has expired, so the AM requests the deletion of the upstream QoS resources by sending a Gate Delete message to the PS. Figure 14-20 shows this message.

Figure 14-20 *Gate Delete (Upstream)*

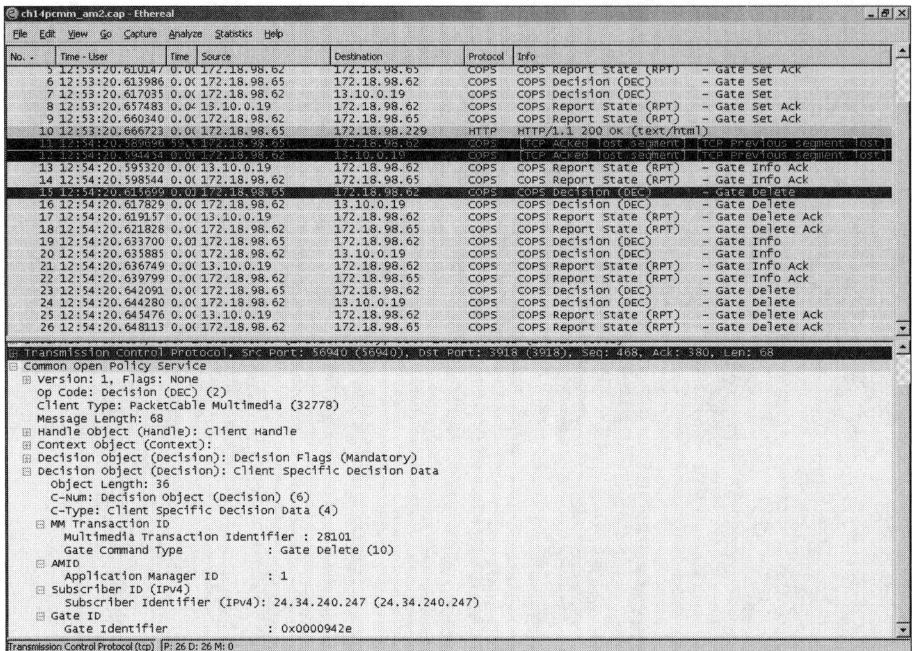

12 The PS then in turn passes this information to the CMTS in a Gate Delete message.

13 The CMTS receives this message and immediately requests deletion of the upstream DOCSIS service flow by sending a DSD-REQ to the CM. Figure 14-21 shows this message.

Because you have only a single service flow to delete in this request, the Service Flow ID is put in that field instead of in the TLV data. Recall that with PacketCable 1.x, both the upstream and the downstream service flows are deleted with the same message; therefore, the service flow IDs are put in the TLV data. The CM acknowledges the deletion of the service flow by sending a DSD-RSP back to the CMTS.

Now that the CMTS knows the service flow is successfully deleted, it deletes the gate and sends a Gate Delete Acknowledgement to the PS confirming this. The PS then forwards this message back to the AM.

Figure 14-21 *DSD-REQ (Upstream)*

```
ch14pcmm_rf2.cap - Ethereal
File Edit View Go Capture Analyze Statistics Help
No. - Time - User        Time Source            Destination       Protocol Info
   1 12:29:45.067470 0.0( 00:0e:39:d2:d4:8c 00:0a:73:fa:be:8a DOCSIS Dynamic Service Addition Request Tran-id = 32803
   2 12:29:45.076485 0.0( 00:0a:73:fa:be:8a 00:0e:39:d2:d4:8c DOCSIS Dynamic Service Add Response ID = 32803 (okay/succes
   3 12:29:45.076637 0.0( 00:0e:39:d2:d4:8c 00:0a:73:fa:be:8a DOCSIS Dynamic Service Add Ack ID = 32803 (okay/success)
   4 12:29:45.087354 0.01 00:0e:39:d2:d4:8c 00:0a:73:fa:be:8a DOCSIS Dynamic Service Addition Request Tran-id = 32804
   5 12:29:45.126771 0.0: 00:0a:73:fa:be:8a 00:0e:39:d2:d4:8c DOCSIS Dynamic Service Add Response ID = 32804 (okay/succes
   6 12:29:45.126900 0.0( 00:0e:39:d2:d4:8c 00:0a:73:fa:be:8a DOCSIS Dynamic Service Add Ack ID = 32804 (okay/success)
   7 12:30:45.090483 59.( 00:0e:39:d2:d4:8c 00:0a:73:fa:be:8a DOCSIS Dynamic Service Delete Request Tran-id = 32805
   8 12:30:45.106988 0.01 00:0a:73:fa:be:8a 00:0e:39:d2:d4:8c DOCSIS Dynamic Service Delete Response Tran id = 32805 (oka
   9 12:30:45.116788 0.0( 00:0e:39:d2:d4:8c 00:0a:73:fa:be:8a DOCSIS Dynamic Service Delete Request Tran-id = 32806
  10 12:30:45.131121 0.01 00:0a:73:fa:be:8a 00:0e:39:d2:d4:8c DOCSIS Dynamic Service Delete Response Tran id = 32806 (oka

+ Frame 7 (60 bytes on wire, 60 bytes captured)
+ DOCSIS
+ Mac Management
- DSD Request
    Transaction Id: 32805
    Reserved: 0x0000
    Service Flow ID: 219
  - TLV Data
      31 Key Sequence Number: 02
      27 HMAC Digest: 3C0B701393026D23B5AF2AF96E6AB51DDDC5CD2A

DOCSIS Dynamic Service Delete Re | P: 10 D: 10 M: 0
```

14 A similar process then occurs for the downstream multimedia gate. First, the AM queries the PS for gate information by sending a Gate Info message.

15 To get this information the PS initiates a Gate Info message of its own to the CMTS. The CMTS honors this request by putting all the downstream gate information in a Gate Info Ack message and sends this to the PS. Among the information in this message are the Gate Time Info and Gate Usage Info fields, indicating how long the gate has been committed and the amount of traffic that has traversed it. After receiving this information, the PS sends the data to the AM in a Gate Info Ack message.

16 Knowing that the time setup for the gate has expired, the AM sends a Gate Delete message to the PS requesting the removal of this gate.

17 The PS then services this request by sending a Gate Delete message to the CMTS.

18 This triggers the CMTS to send a DSD-REQ message to the cable modem concerning the deletion of the downstream service flow. The cable modem responds to this message with a DSD-RSP message, indicating the service flow has been deleted. This triggers a Gate Delete Ack message to be sent to the PS, which in turn triggers a Gate Delete Ack message to be sent to the AM.

Event Messaging

As you know, tracking QoS resources is a vital element of PacketCable and just like with PacketCable, telephony event messaging to a Record Keeping Server (RKS) is used with PCMM. PacketCable event messaging uses the Accounting-Request and Accounting-Response messages defined in the IETF RADIUS protocol. In PCMM, both the CMTS and the Policy Servers send event messages to the RKS. The main differences between event messaging in PCMM and event messaging in PacketCable telephony are that the messages have been generalized and specifics to telephony have been classified as optional.

As you'll learn in Chapter 15, "Lawful Intercept," PacketCable telephony uses 14 event messages. Most of these are optional in PCMM, with only QoS Reserve, QoS Commit, QoS Release, and Time Change being mandatory. The first three messages are generated by the CMTS, and the Time Change message can be sent by either a CMTS or PS. Also three new messages have been added for use by policy servers: Policy Request, Policy Delete, and Policy Update.

Logically, the PS sends a Policy Request message whenever a new policy is created. This is whenever it receives the CMTS response to a Gate Set message. The PS sends a Policy Delete when it learns of a policy deletion; that is, a Gate Delete from the AM, a Gate Delete Acknowledge from the CMTS, or a Gate Report State from the CMTS indicating the release of QoS resources. The Policy Update message is sent when the AM modifies gate policy parameters such as traffic profile or classifiers.

Security

Just like with PacketCable telephony, PacketCable Multimedia also defines security between components. In Chapter 16, "PacketCable Network Design Considerations," you'll learn all about the security mechanisms used in PacketCable telephony, and many of these same protocols and concepts apply to PCMM. Therefore, only a very high-level description of these mechanisms is mentioned here.

The following interfaces should all be protected using the IPsec Encapsulation Security Payload (ESP) protocol:

- PS-CMTS
- AM-PS
- CMTS-RKS
- PS-RKS

All of these interfaces must support the Internet Key Exchange (IKE) key management protocol utilizing preshared keys. Optionally, IKE with certificates and Kerberized IPsec can be implemented.

PCMM Configuration on a Cisco CMTS

By now you should have a good understanding about how PCMM works. The next part of this chapter covers related provisioning steps on a Cisco CMTS. Chapter 13, "Analyzing, Implementing, and Troubleshooting DQoS," covered the process of configuring PacketCable telephony on Cisco CMTSs in detail. You'll see that many of the commands used in PCMM are similar.

The most basic step, and the only essential one, is to enable PacketCable Multimedia operation on the CMTS:

```
Router(config)#packetcable multimedia
```

Without this command, the CMTS does not respond to or generate PCMM COPS commands, and PacketCable multimedia operation does not exist. As soon as this command is entered, the CMTS attempts to establish a COPS connection on the TCP connection created by the PCMM policy server.

To specify PCMM timer values (optional), use the following command:

```
Router(config)#packetcable timer multimedia T1 timer-value
```

Timer T1 is defined in milliseconds, and you might recall it is the amount of time resources can remain in the authorized state. This timer is typically set by the PS in the GATE-SET message. The value specified here is used only if this parameter is not specified (that is, set to 0) by the PS. If unspecified on the CMTS, the default value of 200 seconds is assumed.

Recall that one of the ways to define traffic profiles is by using DOCSIS service class names. When a device (such as a PS or CMTS) receives a request containing a service class name, it looks up the name in its configuration database to resolve it into the specific DOCSIS service flow parameters. Consequently, this name and its mapping must exist on the device receiving the request (PS or CMTS). On a Cisco CMTS, service class names are configured in global configuration mode using the **cable service class** command. Configuration for the Speed Preview service flows used in the call flow example is as follows:

```
cable service class 100 name SpeedPreview_US
cable service class 100 upstream
cable service class 100 max-rate 2300000
cable service class 100 min-packet-size 64
cable service class 100 admission-timeout 0
cable service class 100 activity-timeout 0
cable service class 101 name SpeedPreview_DS
cable service class 101 downstream
cable service class 101 max-rate 17500000
cable service class 101 min-packet-size 64
```

You can determine the values of these service classes with the **show cable service-class** *ID* **verbose** command as demonstrated here:

```
C0103-7246VXR#show cable service-class 100 verbose
Index:                              100
Name:                               SpeedPreview_US
Direction:                          Upstream
```

```
Traffic Priority:                          0
Maximum Sustained Rate:                    2300000 bits/sec
Max Burst:                                 3044 bytes
Minimum Reserved Rate:                     0 bits/sec
Minimum Packet Size                        64 bytes
Admitted QoS Timeout                       0 seconds
Active QoS Timeout                         0 seconds
Maximum Concatenated Burst:                1522 bytes
Scheduling Type:                           Best Effort
Request/Transmission Policy:               0x0
IP ToS Overwrite [AND-mask,OR-mask]:       0xFF,0x0
Parameter Presence Bitfield:               {0xE48, 0x0}
C0103-7246VXR#sh cable service-class 101 v
Index:                                     101
Name:                                      SpeedPreview_DS
Direction:                                 Downstream
Traffic Priority:                          0
Maximum Sustained Rate:                    17500000 bits/sec
Max Burst:                                 3044 bytes
Minimum Reserved Rate:                     0 bits/sec
Minimum Packet Size                        64 bytes
Admitted QoS Timeout                       200 seconds
Active QoS Timeout                         0 seconds
Scheduling Type:                           Undefined
Max Latency:                               0 usecs
Parameter Presence Bitfield:               {0x248, 0x0}
```

Verifying and Troubleshooting PCMM on a Cisco CMTS

The commands used to verify and troubleshoot PCMM on a Cisco CMTS for the most part are the exact same commands used to verify and troubleshoot PacketCable telephony. These commands were covered in detail in Chapter 13, so please refer to that chapter for more information.

- Use the **show cops servers** command to see that the COPS connection to the PS is established (the client type should be 32778).
- If the COPS connection is not there, you can check to see whether the PS has opened the TCP connection (to port 3918 on the CMTS) using the **show tcp brief** command.
- The **show packetcable global** command tells you whether PCMM is enabled and the value of timer T1.
- The **show packetcable gate** command shows you information on PCMM gates, just as it did with PC 1.x gates. The **multimedia** keyword can be used to just show PCMM gates if you have both types on the CMTS. The following is some sample output of PCMM gates:

```
C0103-7246VXR#show packetcable gate multimedia summary
GateID    i/f     SubscriberID    GC-Addr       State     Type  SFID(us) SFID(ds)
37938     Ca5/0   24.34.240.247  172.18.98.62   COMMIT    MM             222
54322     Ca5/0   24.34.240.247  172.18.98.62   COMMIT    MM     221
Total number of Multimedia gates = 2
Total Gates committed(since bootup or clear counter) = 92
```

```
C0103-7246VXR#show packetcable gate 54322   <- Detailed information for Upstream
 PCMM gate
GateID                    : 54322
    Subscriber ID         : 24.34.240.247
    COPS server handle    : 0x644754F8
    State                 : COMMIT
      Upstream
      classifier          :
        protocol          : 0
        priority          : 64
        src addr/port     : 24.34.240.247/0
        dest addr/port    : 0.0.0.0/0
        dscp-tos/mask     : 0x0 0x0
      gate spec           : flags 0x1
      dscp-tos/mask       : 0x0 0x0
      timer t1(s)         : 200
      timer t2(s)         : 0
      timer t3(s)         : 0
      timer t4(s)         : 0
      session class       : 0x0
      Traffic Profile     : Best Effort.
      Authorized : <- Below is the Authorized Resource Envelope
Traffic Priority 0x0
Req Transmission Policy 0x0
      Max Sustain Rate 2300000
          Max Burst        3044
      Min Rate         0
      Min pkt size     64
      Reserved : <- Below is the Reserved Resource Envelope
Traffic Priority 0x0
Req Transmission Policy 0x0
      Max Sustain Rate 2300000
          Max Burst        3044
      Min Rate         0
      Min pkt size     64
      Committed : <- Below is the Committed Resource Envelope
Traffic Priority 0x0
Req Transmission Policy 0x0
      Max Sustain Rate 2300000
          Max Burst        3044
      Min Rate         0
      Min pkt size     64
C0103-7246VXR#show packetcable gate 37938 <- Detailed information for Downstream
 PCMM gate
GateID                    : 37938
    Subscriber ID         : 24.34.240.247
    COPS server handle    : 0x644754F8
```

```
State               : COMMIT
  Downstream
  classifier        :
    protocol        : 0
    priority        : 64
    src addr/port   : 0.0.0.0/0
    dest addr/port  : 24.34.240.247/0
    dscp-tos/mask   : 0x0 0x0
  gate spec         : flags 0x0
  dscp-tos/mask     : 0x0 0x0
  timer t1(s)       : 200
  timer t2(s)       : 0
  timer t3(s)       : 0
  timer t4(s)       : 0
  session class     : 0x0
  Traffic Profile   : Downstream.
  Authorized : <- Below is the Authorized Resource Envelope
  Traffic priority 0
  Max Sustain Rate 17500000
  Max Burst        3044
  Min Rate         0
  Min pkt size     64
  DS Latency       0
  Reserved : <- Below is the Reserved Resource Envelope
  Traffic priority 0
  Max Sustain Rate 17500000
  Max Burst        3044
  Min Rate         0
  Min pkt size     64
  DS Latency       0
  Committed : <- Below is the Committed Resource Envelope
  Traffic priority 0
  Max Sustain Rate 17500000
  Max Burst        3044
  Min Rate         0
  Min pkt size     64
  DS Latency       0
```

- The **show packetcable gate counter multimedia** command tells you the number of PCMM gates committed since the CMTS was last reset or the counter was cleared (this can be done with the command **clear packetcable gate counter commit multimedia**).
- DOCSIS Service Flows, Classifiers, and SIDs created as a result of PCMM can be seen using the same commands described in the last chapter.

- Use the **show controller cable** *x/y* **downstream** command to view statistics on DSX messages. For PCMM, you see hits on the Request and Acknowledgement counters for DSA and DSC messages because these always originate from the CMTS for PCMM client type 1 devices.
- The debugging of DOCSIS service flow creation as a result of PCMM can be seen with the same commands that are used for PC 1.x, as described in Chapter 13.
- Likewise, the debugging of COPS gate messaging for PCMM can be seen with the same commands that are used for PC 1.x DQoS. The following is an example of the **debug packetcable gate control** command output used with the **debug packetcable subscriber** *sub IP* **verbose** command:

```
Aug 10 20:22:06.984: Pktcbl(mm): Received GATE SET message, tid=0x6DCE
Aug 10 20:22:06.984: --- Pktcbl(mm): Received GCP message ------
Aug 10 20:22:06.984: TRANSACTION ID       : Object.[snum/stype/len 1/1/8]
Aug 10 20:22:06.984:     transaction id   : 0x6DCE
Aug 10 20:22:06.984:     gcp cmd          : 4 (GATE SET)
Aug 10 20:22:06.984: AM ID                : Object.[snum/stype/len 2/1/8]
Aug 10 20:22:06.984:     AM ID            : 1
Aug 10 20:22:06.984: SUBSCRIBER ID (IPV4) : Object.[snum/stype/len 3/1/8]
Aug 10 20:22:06.984:     Addr             : 24.34.240.247
Aug 10 20:22:06.984: GATE SPEC            : Object.[snum/stype/len 5/1/16]
Aug 10 20:22:06.984:     flag             : 0x1
Aug 10 20:22:06.984:     dscp             : 0x0
Aug 10 20:22:06.984:     dscp tos mask    : 0x0
Aug 10 20:22:06.984:     Timers  t1       : 0,  t2 : 0
Aug 10 20:22:06.984:             t3       : 0,  t4 : 0
Aug 10 20:22:06.984:     session class    : 0x0
Aug 10 20:22:06.984: TRAFFIC PROFI
C0103-7246VXR#LE      : Object.[snum/stype/len 7/3/80]
Aug 10 20:22:06.984:     envelope         : 0x7
Aug 10 20:22:06.984:     service number   : 0x0
Aug 10 20:22:06.984:     Authorized :
Aug 10 20:22:06.984: Traffic Priority   : 0
Aug 10 20:22:06.984: Request Xmit Policy: 0x0
Aug 10 20:22:06.984: Max Sustained Rate : 2300000
Aug 10 20:22:06.984: Max Burst          : 3044
Aug 10 20:22:06.984: Min Reserved Rate  : 0
Aug 10 20:22:06.984: Min Pkt Size       : 64
Aug 10 20:22:06.984:     Reserved   :
Aug 10 20:22:06.984: Traffic Priority   : 0
Aug 10 20:22:06.984: Request Xmit Policy: 0x0
Aug 10 20:22:06.984: Max Sustained Rate : 2300000
Aug 10 20:22:06.984: Max Burst          : 3044
Aug 10 20:22:06.984: Min Reserved Rate  : 0
Aug 10 20:22:06.984: Min Pkt Size       : 64
Aug 10 20:22:06.984:     Committed  :
```

```
Aug 10 20:22:06.984: Traffic Priority   : 0
Aug 10 20:22:06.984: Request Xmit Policy: 0x0
Aug 10 20:22:06.984: Max Sustained Rate : 2300000
Aug 10 20:22:06.984: Max Burst          : 3044
Aug 10 20:22:06.984: Min Reserved Rate  : 0
Aug 10 20:22:06.984: Min Pkt Size       : 64
Aug 10 20:22:06.984: CLASSIFIER           : Object.[snum/stype/len 6/1/24]
Aug 10 20:22:06.984:     protocol         : 0x0
Aug 10 20:22:06.984:     dscp             : 0x0
Aug 10 20:22:06.984:     dscp tos mask    : 0x0
Aug 10 20:22:06.984:     src/port         : 24.34.240.247 0
Aug 10 20:22:06.984:     dest/port        : 0.0.0.0 0
Aug 10 20:22:06.984:     priority         : 64
Aug 10 20:22:06.984: OPAQUE               : Object.[snum/stype/len 11/1/12]
Aug 10 20:22:06.984:     data             : [33 36 34 31 39 31 38 00 ]
Aug 10 20:22:06.984: --------------------------------------
Aug 10 20:22:06.984: Pktcbl(mm): Change profile 0 qos 0
Aug 10 20:22:06.996: Pktcbl(mm): Building GCP message, added obj TRANSACTION ID
 ; len:8 padding:0
Aug 10 20:22:06.996: Pktcbl(mm): Building GCP message, added obj AM ID
 ; len:8 padding:0
Aug 10 20:22:06.996: Pktcbl(mm): Building GCP message, added obj SUBSCRIBER ID
 (IPV4); len:8 padding:0
Aug 10 20:22:06.996: Pktcbl(mm): Building GCP message, added obj GATE ID
 ; len:8 padding:0
Aug 10 20:22:06.996: Pktcbl(mm): Building GCP message, added obj OPAQUE
 ; len:12 padding:0
Aug 10 20:22:06.996: Pktcbl(mm): Built GCP message, GATE SET ACK       , length:
 44, copsLen 72
Aug 10 20:22:06.996:  --- Pktcbl: Sending GCP message ------
Aug 10 20:22:06.996: TRANSACTION ID       : Object.[snum/stype/len 1/1/8]
Aug 10 20:22:06.996:     transaction id   : 0x6DCE
Aug 10 20:22:06.996:     gcp cmd          : 5 (GATE SET ACK)
Aug 10 20:22:06.996: AM ID                : Object.[snum/stype/len 2/1/8]
Aug 10 20:22:06.996:     AM ID            : 1
Aug 10 20:22:06.996: SUBSCRIBER ID (IPV4) : Object.[snum/stype/len 3/1/8]
Aug 10 20:22:06.996:     Addr             : 24.34.240.247
Aug 10 20:22:06.996: GATE ID              : Object.[snum/stype/len 4/1/8]
Aug 10 20:22:06.996:     GateID           : 37942 (0x9436)
Aug 10 20:22:06.996: OPAQUE               : Object.[snum/stype/len 11/1/12]
Aug 10 20:22:06.996:     data             : [33 36 34 31 39 31 38 00 ]
Aug 10 20:22:06.996:  --------------------------------------
Aug 10 20:22:07.016: Pktcbl(mm): Received GATE SET message, tid=0x6DCF
Aug 10 20:22:07.016: --- Pktcbl(mm): Received GCP message ------
Aug 10 20:22:07.016: TRANSACTION ID       : Object.[snum/stype/len 1/1/8]
Aug 10 20:22:07.016:     transaction id   : 0x6DCF
Aug 10 20:22:07.016:     gcp cmd          : 4 (GATE SET)
Aug 10 20:22:07.016: AM ID                : Object.[snum/stype/len 2/1/8]
```

```
Aug 10 20:22:07.016:     AM ID            : 1
Aug 10 20:22:07.016: SUBSCRIBER ID (IPV4) : Object.[snum/stype/len 3/1/8]
Aug 10 20:22:07.016:     Addr             : 24.34.240.247
Aug 10 20:22:07.016: GATE SPEC            : Object.[snum/stype/len 5/1/16]
Aug 10 20:22:07.016:     flag             : 0x0
Aug 10 20:22:07.016:     dscp             : 0x0
Aug 10 20:22:07.016:     dscp tos mask    : 0x0
Aug 10 20:22:07.016:     Timers  t1       : 0,  t2 : 0
Aug 10 20:22:07.016:             t3       : 0,  t4 : 0
Aug 10 20:22:07.016:     session class    : 0x0
Aug 10 20:22:07.016: TRAFFIC PROFILE      : Object.[snum/stype/len 7/8/80]
Aug 10 20:22:07.016:     envelope         : 0x7
Aug 10 20:22:07.016:     service number   : 0x0
Aug 10 20:22:07.016:     Authorized :
Aug 10 20:22:07.016: Traffic Priority   : 0
Aug 10 20:22:07.016: Max Sustained Rate : 17500000
Aug 10 20:22:07.016: Max Burst          : 3044
Aug 10 20:22:07.016: Min Reserved Rate  : 0
Aug 10 20:22:07.016: Min Pkt Size       : 64
Aug 10 20:22:07.016: Max Latency        : 0
Aug 10 20:22:07.016:     Reserved   :
Aug 10 20:22:07.016: Traffic Priority   : 0
Aug 10 20:22:07.016: Max Sustained Rate : 17500000
Aug 10 20:22:07.016: Max Burst          : 3044
Aug 10 20:22:07.016: Min Reserved Rate  : 0
Aug 10 20:22:07.016: Min Pkt Size       : 64
Aug 10 20:22:07.016: Max Latency        : 0
Aug 10 20:22:07.016:     Committed   :
Aug 10 20:22:07.016: Traffic Priority   : 0
Aug 10 20:22:07.016: Max Sustained Rate : 17500000
Aug 10 20:22:07.016: Max Burst          : 3044
Aug 10 20:22:07.016: Min Reserved Rate  : 0
Aug 10 20:22:07.016: Min Pkt Size       : 64
Aug 10 20:22:07.016: Max Latency        : 0
Aug 10 20:22:07.016: CLASSIFIER           : Object.[snum/stype/len 6/1/24]
Aug 10 20:22:07.016:     protocol         : 0x0
Aug 10 20:22:07.016:     dscp             : 0x0
Aug 10 20:22:07.016:     dscp tos mask    : 0x0
Aug 10 20:22:07.016:     src/port         : 0.0.0.0 0
Aug 10 20:22:07.016:     dest/port        : 24.34.240.247 0
Aug 10 20:22:07.016:     priority         : 64
Aug 10 20:22:07.016: OPAQUE               : Object.[snum/stype/len 11/1/12]
Aug 10 20:22:07.016:     data             : [33 36 34 31 39 31 38 00 ]
Aug 10 20:22:07.016: --------------------------------------
Aug 10 20:22:07.016: Pktcbl(mm): Change profile 0 qos 0
Aug 10 20:22:07.040: Pktcbl(mm): Building GCP message, added obj TRANSACTION ID
 ; len:8 padding:0
```

```
Aug 10 20:22:07.040: Pktcbl(mm): Building GCP message, added obj AM ID
 ; len:8 padding:0
Aug 10 20:22:07.040: Pktcbl(mm): Building GCP message, added obj SUBSCRIBER ID
 (IPV4); len:8 padding:0
Aug 10 20:22:07.040: Pktcbl(mm): Building GCP message, added obj GATE ID
 ; len:8 padding:0
Aug 10 20:22:07.040: Pktcbl(mm): Building GCP message, added obj OPAQUE
 ; len:12 padding:0
Aug 10 20:22:07.040: Pktcbl(mm): Built GCP message, GATE SET ACK      , length:
 44, copsLen 72
Aug 10 20:22:07.040:  --- Pktcbl: Sending GCP message ------
Aug 10 20:22:07.040: TRANSACTION ID       : Object.[snum/stype/len 1/1/8]
Aug 10 20:22:07.044:     transaction id   : 0x6DCF
Aug 10 20:22:07.044:     gcp cmd          : 5 (GATE SET ACK)
Aug 10 20:22:07.044: AM ID                : Object.[snum/stype/len 2/1/8]
Aug 10 20:22:07.044:     AM ID            : 1
Aug 10 20:22:07.044: SUBSCRIBER ID (IPV4) : Object.[snum/stype/len 3/1/8]
Aug 10 20:22:07.044:     Addr             : 24.34.240.247
Aug 10 20:22:07.044: GATE ID              : Object.[snum/stype/len 4/1/8]
Aug 10 20:22:07.044:     GateID           : 54326 (0xD436)
Aug 10 20:22:07.044: OPAQUE               : Object.[snum/stype/len 11/1/12]
Aug 10 20:22:07.044:     data             : [33 36 34 31 39 31 38 00 ]
Aug 10 20:22:07.044:  -------------------------------------
```

- Finally, many of the test commands covered in Chapter 13 can also be used to troubleshoot the COPS connectivity between the CMTS and PS.

Chapter Summary

The PacketCable Multimedia (PCMM) specifications provide a generalized framework for enabling DOCSIS QoS services for a wide range of multimedia applications. PCMM is unaware of the application specifics; hence, it does not specify application details such as signaling and provisioning. The possible range of PCMM applications includes real-time applications such as videoconferencing, interactive gaming, streaming media, and telephony, as well as non real-time applications such as web-based bandwidth on demand and Speed Preview.

There are seven components in the PCMM architecture: clients, Application Server (AS), Application Manager (AM), Policy Server (PS), CMTS, CM, and RKS. Three types of multimedia clients exist:

- Type 1 is referred to as "QoS-unaware" because it knows nothing about the PacketCable network it is using.
- Type 2 is referred to as "Push model" because it is QoS-aware but relies on other components such as the AM and PS to push QoS policies to the CMTS.
- Type 3 is referred to as "Pull model" because it is QoS-aware and able to trigger the CMTS to pull QoS policies from the PS.

Currently, only type 1 clients are fully defined in PCMM. The AS is an intermediary device that initiates multimedia sessions on behalf of clients. The Application Manager figures out the resources needed by a client and conveys this information to a Policy Server. The Policy Server applies rules to requests from the AM, such as checking CMTS usage and subscriber QoS restrictions. The CMTS, cable modem, and RKS event messaging server perform similar functions in PCMM that they do in PacketCable 1.x. However, significant differences in how these functions are performed do exist. For example, in PCMM dynamic DOCSIS service flow creation is initiated from the CMTS, unlike in PacketCable 1.x, where they are initiated from the embedded MTA.

COPS is used in two places in PCMM—between the AM and PS and between the PS and CMTS. Consequently, a PCMM network is split into service control domains (SCDs) and resource control domains (RCDs). An SCD is a logical grouping of elements that provide an application, and an RCD is the logical grouping of elements providing connecting to QoS resources.

Like PacketCable telephony, the DOCSIS network must be at least DOCSIS 1.1 capable in PCMM. However, the use of DOCSIS in PCMM varies from its previous use. In addition to the unsolicited grant services (UGS and UGS-AD), the polling services are also permitted (RTPS and nRTPS). The Real-Time Polling Service is used for jitter-sensitive applications that transmit variable-sized packets. The Non Real-Time Polling Service is used for applications that are sensitive to packet loss and require regular transmission opportunities but can tolerate packet jitter. Also through the manipulation of certain DOCSIS service flow parameters, the Best Effort scheduling type can also be used in PCMM to provide QoS.

The COPS messaging and parameters used in PCMM are significantly different than those in PacketCable telephony. The COPS protocol is still used for resource allocation and authorization, but in PCMM it is also used for resource reservation and commitment. Among the objects defined in COPS for PCMM are gate specifications, classifiers, and traffic profiles. PCMM gate specifications contain timer values as well as parameters such as gate priority and DSCP overwrite settings, and classifiers determine the type of packets permitted to use a PCMM gate. A traffic profile defines the QoS resources required for the application and can be defined by RSVP flow specifications, DOCSIS TLV parameters, or DOCSIS service class names. Traffic profiles can contain authorized resource envelopes, reserved resource envelopes, and committed resource envelopes. The presence of reserved and committed resource envelopes triggers DOCSIS messages to the cable modem connected to the PCMM subscriber. Also objects exist that can be used to track gate bandwidth use and the time in which PCMM gates are active. The use of event messaging for real-time billing and packet encryption for added security also exist in PCMM, just as they did in telephony.

PCMM configuration and troubleshooting commands on a Cisco CMTS are similar to the corresponding commands used for PacketCable telephony. Show commands exist, which output the current status and details of PCMM gates and their DOCSIS service flows. Debug commands exist, which output the COPS and DOCSIS messaging used to create QoS for PCMM sessions.

Chapter Review

1 True or False: The Application Manager doesn't need to know the CMTS to which its clients are physically connected.

Answer: True. It relies on the Policy Server to figure this out.

2 True or False: PCMM defines the protocol used between the AM and PS devices.

Answer: True. The COPS protocol is used.

3 True or False: PCMM defines the protocol used between the AM or AS and the client devices.

Answer: False. This is out of the scope of PCMM.

4 True or False: If a PCMM gate is set up using a DOCSIS service class name defined on the CMTS, certain QoS parameters can also be overwritten using DOCSIS TLV parameters.

Answer: False

5 True or False: The way in which PCMM is configured, verified, and debugged on a Cisco CMTS is nearly identical to the way these things are done for PacketCable telephony.

Answer: True

6 True or False: A PCMM gate identifier refers to the combination of an upstream gate and a downstream gate.

Answer: False. In PCMM, gates are always unidirectional.

7 True or False: The types of event messages used in PCMM are different than those used in PC telephony.

Answer: True. Messages specific to the telephony application have been removed and messages sent from the PS have been added.

8 True or False: The CMTS initiates the deletion of a gate that exceeds its configured volume usage limit or exceeds its time usage limit.

Answer: False. The CMTS is not responsible for enforcing these limits.

9 True or False: If the CMTS receives a PCMM gate specification with a timer T1 value of 0, this means the resources remain authorized until torn down by the PS.

Answer: False. This means the CMTS provisioned value is used.

10 What type of multimedia client uses RSVP to set up its QoS?

Answer: Type 3

11 For each of the following possible PCMM applications, identify the DOCSIS scheduling type (UGS, UGS-AD, RTPS, nRTPS, or BE) that makes the most sense:

a Video telephony

b On-demand minimal bandwidth guaranteed service

c Standalone telephony MTA implementing VAD

d Interactive gaming

Answer: a) RTPS, b) BE, c) UGS-AD, d) RTPS

12 When is a Policy Server a COPS PDP and when is it a COPS PEP?

Answer: The PS is a Policy Decision Point for the resource control domain in its communication with the CMTS and a Policy Enforcement Point for the service control domain in its communication with the AM.

13 What is the purpose of the Gate Report State message?

Answer: It is sent by the CMTS to alert the PS of a change in the resource status of a gate because of events like client messaging and timer expiry.

14 What is the TCP port used for PCMM, and why is it different than the one used for PC telephony?

Answer: 3918. So both applications can simultaneously exist without interfering with each other.

15 What could a Deep Packet Inspection (DPI) device be used for in a PCMM network?

Answer: To initiate PCMM session requests based on inspected packets (such as voice/video RTP streams)

16 Under what situation would an incremental synchronization request be sent by a PDP?

Answer: To recover gate information after the COPS connection is lost

17 What is a reason for using extended classifiers over standard classifiers?

Answer: So individual classifiers can be added, activated, or deleted without having to resend the entire list of classifiers, or they could be used to specify a range of IP addresses and port numbers.

18 Why can't traffic profiles using RSVP flow specs be used to trigger the creation of UGS-AD and nRTPS service flows?

Answer: Because of limitations in the ability to map flow spec parameters into DOCSIS parameters

19 What Cisco IOS command can be used to see the active list of PCMM gates?

Answer: show packetcable gate multimedia summary

Part VI

Network Considerations

CHAPTER 15

Event Messaging and Lawful Intercept

This chapter covers both the use of Event Messaging in a PacketCable network and the lawful intercept requirement that is mandatory for service providers.

In a PacketCable 1.x network, Event Messaging can be used to collect accounting information on phone calls. In a PacketCable Multimedia (PCMM) network, Event Messaging can be used to track distributed quality of service (DQoS) resource use for any type of multimedia session. In both cases, PacketCable Event Messaging enables the service provider to accurately bill customers for services.

This chapter discusses the reasons why PacketCable Event Messaging is needed. The chapter begins with an overview of the transport layer protocol used for Event Messaging—the Remote Authentication Dial-In User Service (RADIUS). Then the chapter examines the different types of event messages, followed by coverage of call flows and message traces, to gain a better understanding of how these types of event messages are used. Next, you learn about the relevant configuration and troubleshooting tips on the Cisco BTS 10200 and Cisco CMTS platforms.

After this point, the chapter's focus turns to lawful intercept. Lawful intercept is the ability for a law enforcement agency, such as the police department or the FBI, to conduct a wiretap on a subscriber's phone calls. This wiretap could be collecting information about who the tapped subscriber called or was called by, or it could be collecting the actual voice conversations.

For a voice over IP (VoIP) service provider, such as a PacketCable provider, special equipment and techniques are required to perform wiretapping services. This chapter provides an overview of this equipment and associated techniques. The chapter also covers how this equipment interfaces with various PacketCable components, such as the call management server (CMS), cable modem termination system (CMTS), and Media Gateway (MG). Like other chapters, this chapter provides analysis of real call flows with a focus on the configuration and troubleshooting relevant to lawful intercept.

Keep in mind that the bulk of this chapter relates to lawful intercept as it applies in the United States. Different countries might have different requirements than those described here.

Event Messaging Motivation and Background

As you know, in a VoIP network, such as in PacketCable, many components are involved in the handling of a phone call or PCMM multimedia session. Each one of these components is aware of accounting information. For example, the CMS and Media Gateway Controller (MGC) components are aware of information such as calling and called party identities, call feature use, and call state; this is because these components are responsible for call signaling. The CMTS component, on the other hand, is in the bearer path, so it can provide information such as when DQoS resources are reserved and committed. The CMS, MGC, and CMTS are all trusted components and are the only components that can generate event messages in PacketCable 1.x. In PCMM, the Policy Server and the CMTS can generate event messages. The fact that there are typically multiple components with specific accounting information is the reason event messages are used in PacketCable. Each one of these devices can independently send information to the Record Keeping Server (RKS). The RKS can then correlate this data into call detail record (CDR) records.

The PacketCable Event Messaging specification (PKT-SP-EM) details a transport layer–independent attribute format for event message Tag/Length/Value (TLV) pairs. Currently, only the RADIUS transport layer is required for the transport of these TLVs. The RADIUS protocol is used in an IP network to provide authentication, authorization, and accounting (AAA) functions. A description of these functions is as follows:

- **Authentication**—Method of identifying users including the login and password dialog.
- **Authorization**—Method for remote access control to a particular service or device. It works by assembling a set of attributes that describe what the user is authorized to perform. These attributes are compared to the information contained in a database for a given user, and the result is returned to AAA to determine the user's actual capabilities and restrictions.
- **Accounting**—Method for collecting and sending security server information used for billing, auditing, and reporting, such as user identities, start and stop times, executed commands, number of packets, and number of bytes. Enables you to track the services users are accessing, as well as the amount of network resources they are consuming. When activated, the network access server (NAS) reports user activity to the security server in the form of accounting records. Each accounting record is composed of accounting attribute value (AV) pairs and is stored on the access control server.

PacketCable Event Messaging uses the RADIUS accounting function. RADIUS is defined in IETF RFC 2865, and its use for accounting is specified in RFC 2866. The RADIUS protocol is a client/server protocol where requests are sent from the client to the server and responses are returned from the server to the client. The accounting aspect of the RADIUS protocol consists of only two messages:

- **Accounting-Request**—Contains accounting Tag/Length/Value (TLV) data to be stored on the server
- **Accounting-Response**—Used to tell the client the TLV data has successfully been stored

The RADIUS protocol uses User Datagram Protocol (UDP) (specifically, port 1813) for transport, which does not guarantee delivery. This is why Accounting-Response messages are required to acknowledge the successful receipt of Accounting-Request messages. If the RADIUS client (that is, CMS/MGC/CMTS) does not receive an Accounting-Response within a specified time interval, it resends the Accounting-Request until a configured maximum amount of retries is reached.

The collection of TLVs associated with a single multimedia session is used by the RKS to generate a complete billing record known as a call detail record (CDR). These CDRs are then typically transferred to a back-office application such as a billing system. The interface between the RKS and back-office billing system is not specified by PacketCable.

Event Messaging has the advantage that records can be generated in real time. Consequently, the MSO knows which users are using what resources at any given moment. This enables services such as prepaid calling cards. Also the Multiple Systems Operator (MSO) can more quickly detect whether a subscriber is fraudulently using unauthorized resources and react accordingly. Event messages can also be stored on PacketCable components for a period of time and then "batched" together before being sent to the RKS. "Batch" processing provides a more efficient use of network resources.

In PacketCable 1.x, event messages are separated into messages for the originating leg of a call and messages for the terminating leg of a call. A Billing Correlation Identifier (BCID) is generated by the originating call agent (CMS or MGC) to identify event messages for the originating call leg, and another BCID is generated by the terminating call agent (CMS or MGC) to identify event messages for the terminating call leg. A CMS generates a BCID after receiving an NCS off-hook notification from a multimedia terminal adapter (MTA). An MGC generates a BCID on receipt of a Signaling System 7 (SS7) IAM message or on the receipt of a Trunk Gateway Control Protocol (TGCP) NTFY with digits in the case of an incoming operator services call on a channel-associated signaling (CAS) trunk. The CMS sends the BCID for a call, as well as the location of the RKS servers, to the CMTS in the COPS Gate Set message. Each CMS and MGC is also assigned a Financial Entity Identifier (FEID).

In Chapter 9, "Call Management Server Signaling Protocol (CMSS)," you learned that a PacketCable zone is defined as the set of MTAs in one or more Data-Over-Cable Service Interface Specification (DOCSIS) networks under the control of a single CMS system. The collection of PacketCable zones under the administrative control of the same service provider is referred to as a PacketCable domain. Thus, calls between PacketCable zones in the same domain are referred to as *intra-domain* calls, and calls between PacketCable zones in different domains are referred to as *inter-domain* calls. When calls are made between PacketCable zones, the originating and terminating call agent exchange BCIDs and FEIDs.

Event Message Types

Logically, event messages (EM) are triggered by an event. Sample events include the start and stop of call signaling and the activation and deactivation of DQoS resources. The following is a list of the event messages currently defined. The Event Message Identifier (EMID) used to identify the message in RADIUS packets is also listed in parentheses.

- **Call Answer (EMID 15)**—Sent by CMS/MGC when the media stream is started. This is when the connection is answered (that is, NTFY off-hook received, CMSS 200 OK response to INVITE received, Integrated Services User Part [ISUP] ANM sent/received).
- **Call Disconnect (EMID 16)**—Sent by CMS/MGC when the media stream is stopped. This is when the connection is being disconnected (that is, DLCX sent, ISUP REL or RLC received, CMSS 200 OK response to BYE sent).
- **Conference Party Change (EMID 21**)—Sent by CMS when a party is added, placed on hold, or retrieved from hold in a conference call. This EM is used only for lawful intercept.
- **Database Query (EMID 3)**—Sent by CMS when an SS7 TCAP query for a service such as local number portability (LNP) is made.
- **Interconnect Start (EMID 13)**—Sent by MGC when messages for public switched telephone network (PSTN) interconnect call setup are sent or received (that is, address complete message [ACM] sent/received).
- **Interconnect Stop (EMID 14)**—Sent by MGC when messages for PSTN interconnect call teardown are sent or received (that is, RLC sent/received).
- **Media Alive (EMID 20)**—Sent by CMS/MGC/CMTS to indicate the call is still active. The PacketCable components are preconfigured with a media alive generation time (the default value is midnight). At this time the components check for any active calls equal to or older than 1440 minutes (24 hours). If one of these calls exists, this event message is generated.
- **Media Report (EMID 11)**—Sent by CMS/MGC when a change in media stream information occurs. This EM is used only for lawful intercept.
- **Media Statistics (EMID 22)**—Sent by CMS/MGC when media stream statistics defined by the VoIP metrics NCS/TGCP package are reported by the gateway. The receipt of standard Real-time Transport Control Protocol (RTCP) statistics in a DLCX 250 response does not trigger this EM.

- **QoS_Commit (EMID 19)**—Sent by CMTS when quality of service (QoS) is committed. This is when the DSA-ACK or DSC-ACK confirming the successful activation of QoS resources is sent (as in PCMM) or received (as in PC 1.x).
- **QoS Release (EMID 8)**—Sent by CMTS when QoS is released; for example, when a DSD-RSP confirming the successful deletion of QoS resources is sent.
- **QoS_Reserve (EMID 7)**—Sent by CMTS when QoS is reserved. This is when the DSA-ACK or DSC-ACK confirming the successful admission of QoS resources is sent (as in PCMM) or received (as in PC 1.x).
- **Redirection (EMID 24)**—Sent by CMS/MGC to indicate that a call involving a lawful intercept tapped subscriber has been redirected. Logically, this EM is used only for lawful intercept.
- **Service Activation (9)**—Sent by CMS when a call feature (such as Automatic Callback) is activated. It is sent at the time the vertical service code (VSC) for feature activation is dialed (the VSC is typically of the *XX format).
- **Service Deactivation (10)**—Sent by CMS when a call feature (such as Automatic Callback) is deactivated. It is sent at the time the vertical service code (VSC) for feature deactivation is dialed (the VSC is typically of the *XX format).
- **Service Instance (EMID 6)**—Sent by CMS when a call feature (such as Call Waiting, Call Forwarding, etc.) is invoked. It is sent at the time a call is forwarded, put on hold, transferred, and so on.
- **Signal Instance (EMID 12)**—Sent by CMS when an endpoint is instructed to play a signal (such as dial tone, ring tone, ring back tone, and so on). This EM is used only for lawful intercept.
- **Signaling Start (EMID 1)**—Sent by CMS/MGC when signaling is started to support a call. This occurs after routable digits are received from the originating party (that is, NTFY with called #, SS7 IAM, TGCP NTFY with digits, CMSS INVITE).
- **Signaling Stop (EMID 2)**—Sent by CMS/MGC when signaling is stopped, usually when a DLCX is sent to an endpoint. It could also be triggered by the transmission of a CMSS BYE or CANCEL message or the transmission/receipt of an SS7 RLC message.
- **Surveillance Stop (EMID 23)**—Sent by CMS/MGC to indicate the end of a call content and/or call data lawful intercept wiretap. Generally sent as a result of a call for a tapped subscriber ending but is also used when call content and/or call data can no longer be intercepts (for example, because of call forwarding). Logically, this EM is used only for lawful intercept.
- **Time Change (EMID 17)**—Sent by CMS/MGC/CMTS when a time change on the component by more than 200 milliseconds occurs (for example, when a transition into or out of Daylight Savings Time occurs).

Chapter 14, "Multimedia Applications," discussed the topic of Event Messaging for use in PacketCable Multimedia. In PCMM, support for most of the previously mentioned event messages is optional because most of them are very specific to the application of telephony and not useful for PCMM. Only the messages that can be sourced from the CMTS are mandatory: QoS Reserve, QoS Commit, QoS Release, and Time Change. In PCMM, event messages can also be sent from the Policy Server. The Time Change message can be sent from the PS as well as the following three new messages, which have been added specifically for use by Policy Servers:

- **Policy Request (EMID 31)**—Sent whenever a new policy is created. This is whenever it receives the CMTS response to a Gate Set message.
- **Policy Delete (EMID 32)**—Sent by PS when it learns of a policy deletion—that is, a Gate Delete from the Application Manager (AM), a Gate Delete Acknowledge from the CMTS, or a Gate Report State from the CMTS indicating the release of QoS resources.
- **Policy Update (EMID 33)**—Sent when the AM modifies gate policy parameters such as traffic profile or classifiers.

The PCMM specific attributes used for the event messages are defined in section 7 of the PacketCable Multimedia specification.

Event Messaging Call Flow

To help illustrate how Event Messaging is used in PacketCable telephony, this section examines messages from a real call. This particular example is an ONNET-to-OFFNET call flow. The Network-based Call Signaling (NCS), Trunk Gateway Control Protocol (TGCP), and COPS messages, as well as the RADIUS messages from the call, were captured using Ethereal. Because call signaling and DQoS were already covered in detail, they are not covered in detail here. The CMS and MGC components in this example are integrated into a single device, a BTS 10200 system. The call flow is broken into three parts or stages, with the first part illustrated in Figure 15-1.

The first event messaging packets occur after the CMS has received the digits dialed by Oscar (the originating party). This triggers the CMS to generate a Signaling Start Event Message to the RKS. Figure 15-2 illustrates this message.

Figure 15-1 *Call Flow with Event Messaging: Part I*

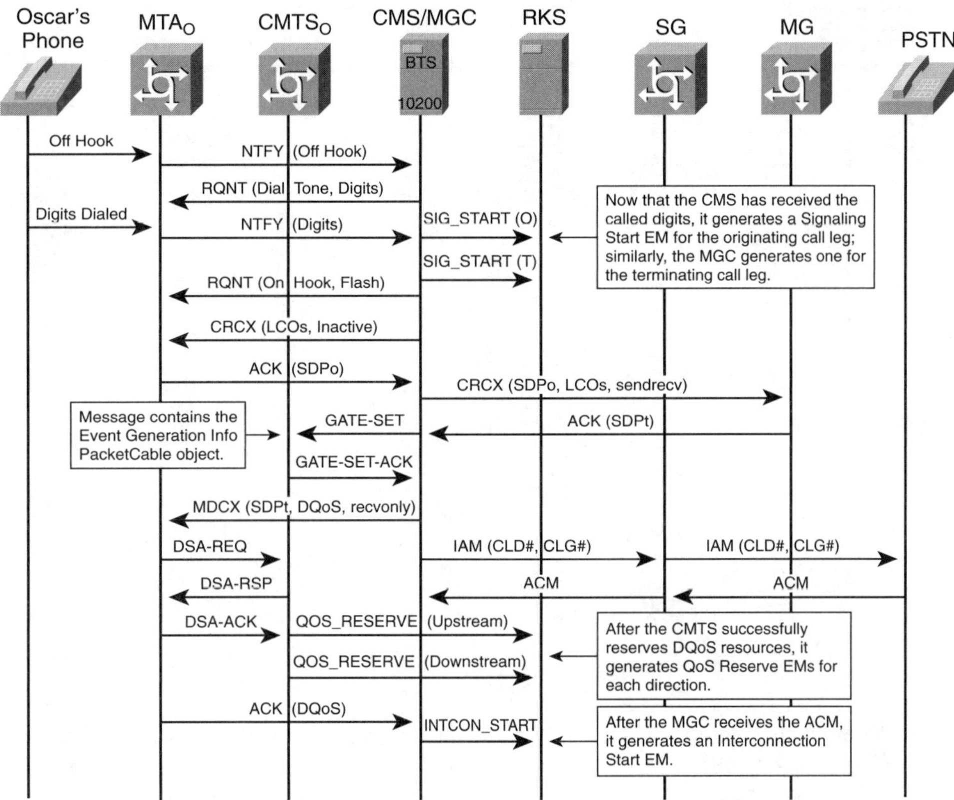

Notice this message is a RADIUS Accounting-Request (identified by the Code parameter equal to 4). The Identifier parameter associates Accounting-Request and Accounting-Response messages. The length of this message is 274 bytes. The next field is the 16-byte Authenticator, which authenticates the reply from the RADIUS server.

RADIUS attributes then follow, expressed as attribute value pairs (AVP). The first two attributes are always the NAS-IP-Address and Acct-Status-Type mandatory standard attributes defined by RADIUS. The NAS-IP-Address attribute contains the IP address of the PacketCable component originating the message (in this case, the CMS). The Acct-Status-Type attribute always contains the value Interim-Update in PacketCable Event Messaging.

Figure 15-2 *Signaling Start EM Example*

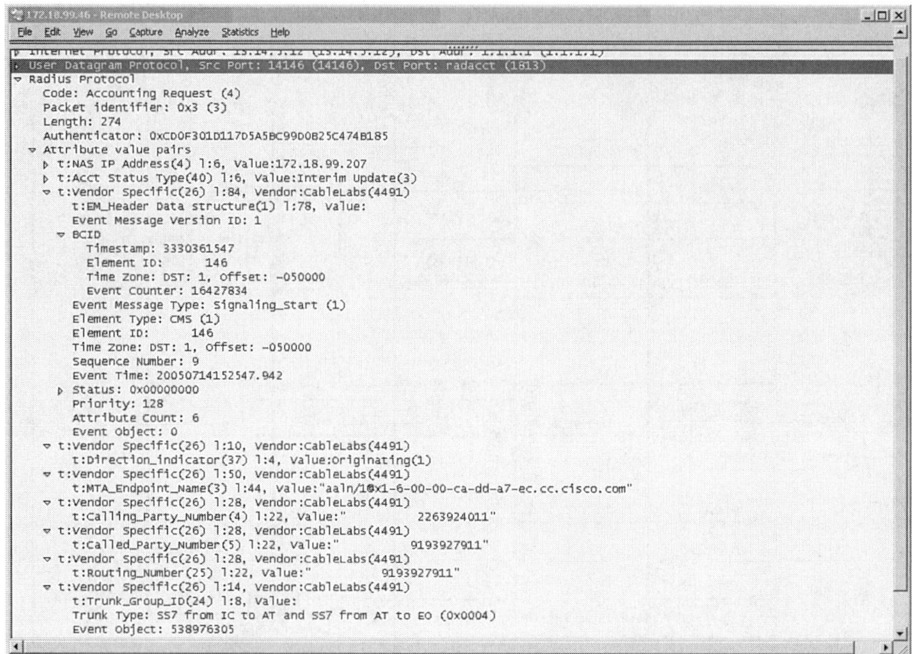

After this is where PacketCable specific attributes are defined. These always begin with the type 26 (indicating vendor specific), a 1-byte length field, and the vendor ID of 4491 (indicating CableLabs). The next byte identifies the first PacketCable EM attribute. PacketCable EM attributes are defined in the PacketCable Event Messaging specification. PacketCable attribute types are 1 byte in length and are followed by a 1-byte length field, which is the length of the PacketCable attribute value plus 2.

In Figure 15-2 you can see the first PacketCable attribute is type 1, which is the Event Messaging (EM) header attribute. The length field is equal to 78, which means the attribute value is 76 bytes in length. The EM header attribute is the first PacketCable attribute in every PacketCable Event Message. This attribute is comprised of a number of parameters, including

- Version ID (in this case, PacketCable 1.0)
- BCID (which itself is comprised of a number of parameters)
- Event Message Type (in this case, a Signaling Start EM)
- Sending Element Type (in this case, CMS)
- Element ID (this is statically configured)
- Time Zone (Daylight Savings Time or not and offset from UTC)
- Sequence Number
- Event Time

- Status
- Priority (importance relative to other EMs)
- Attribute Count (# of following attributes, in this case 6)
- Event Object (either accounting or PCES [lawful intercept])

The remaining PacketCable attributes in this message include Direction Indicator, MTA Endpoint Name, Calling Party Number, Called Party Number, Routing Number, and Trunk Group ID. The Direction Indicator attribute indicates this message is for an originating call leg. The MTA Endpoint Name is logically the NCS endpoint name. The Calling and Called Party Numbers are self-explanatory. The Routing Number specifies the number of the terminating party. The Routing Number can be different from the Called Number if the number was translated on the CMS as a result of a LNP query, for example. The Trunk Group ID identifies the trunk group type (that is, SS7, CAS) and contains the ASCII identifier for the trunk group configured on the MGC. For more details on the format or meaning of any of these PacketCable attributes refer to the PacketCable Event Messaging specification.

Because this is an integrated CMS/MGC, the MGC component of the BTS generates a Signaling Start EM for the terminating leg of the call. After this, an NCS connection is created on the MTA by the CMS, and a TGCP connection is created on the MG by the MGC containing the MTA endpoint's Session Description Protocol (SDP) information. The CRCX response from the MG contains its endpoint's SDP information. At this point the CMS knows how many DQoS resources need to be authorized, so it sends a COPS Gate Set message to the CMTS for this purpose (see Figure 15-3).

Figure 15-3 *COPS Gate Set Message*

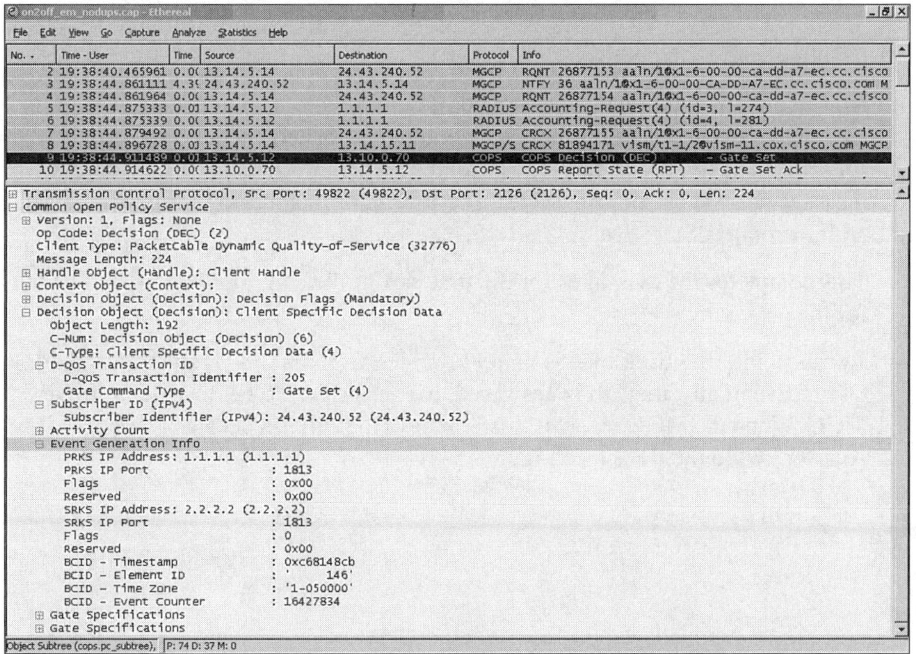

Most of these parameters should be familiar to you, with the possible exception of the Event Generation Info. This parameter contains the IP address and port number used to reach the Primary RKS (in this case, 1.1.1.1 and UDP port 1813). In this case, an IP address and port number for a Secondary RKS is also included. Event messages are sent only to the Primary RKS unless the sending entity detects it is unavailable. The sending entity detects this by a failure to receive Accounting Response messages. The least significant bit of the Flags parameter in the Event Generation Info parameter tells the CMTS whether to send EMs real time or to accumulate records into batch files that are sent at periodic intervals. In this case, this bit is set to "0," which means to send EMs real time. The other bits in the Flags parameter are currently undefined. The final parameter in the Event Generation Info object is the BCID. This parameter contains a Timestamp, the Element Identifier of the CMS, the Time Zone, and an Event Counter. The Timestamp comprises the high order bits of the Network Time Protocol (NTP) reference. The Time Zone specifies whether Daylight Savings Time (DST) is active (in this case it is) and the time offset from Universal Time (formally Greenwich Mean Time). In this case the offset is –5 hours. The Event Counter is an integer that increases for each transaction.

The CMTS then assigns a gate identifier for the authorized DQoS resources and returns this to the CMS in the Gate Set Acknowledgement message. The CMS then relays the DQoS information to the MTA as well as the SDP information of the MG. The CMS also changes the connection mode on the MTA so it can receive ring back tone from the PSTN. This MDCX also triggers the MTA to reserve DQoS resources. As you know, this causes a DSA-REQ/DSA-RSP/DSA-ACK exchange to occur between the MTA and CMTS, the net effect of which is the reservation of DQoS resources. The completion of this causes the CMTS to generate QoS Reserve EMs to the RKS specified in the previous Gate Set message. Separate QoS Reserve messages are sent for each direction (upstream/downstream) of DQoS resources.

During this time the MGC component of the BTS has sent an ISUP IAM to the PSTN via the signaling gateway. This message contains the called and calling party phone numbers. As soon as the PSTN realizes it can route this call, it sends an ISUP address complete message (ACM) back to the MGC. The receipt of this message triggers the MGC to send an Interconnect Start EM to the RKS.

This completes the messages for the first part of the call. Figure 15-4 shows the messages for the next part.

At this point, ring back tone is being provided from the PSTN to the originating party, Oscar. Eventually, the call is answered, triggering the PSTN to send an ISUP ANM to the MGC. When the MGC receives this, it knows the call has been answered, so it sends a Call Answer EM to the RKS (see Figure 15-5).

Figure 15-4 *Call Flow with Event Messaging: Part II*

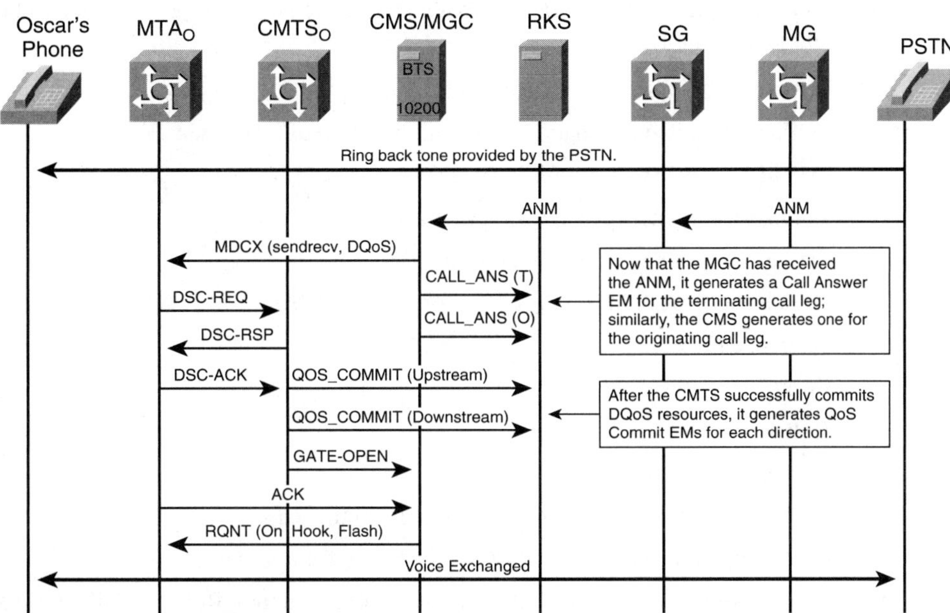

Figure 15-5 *Call Answer EM Example*

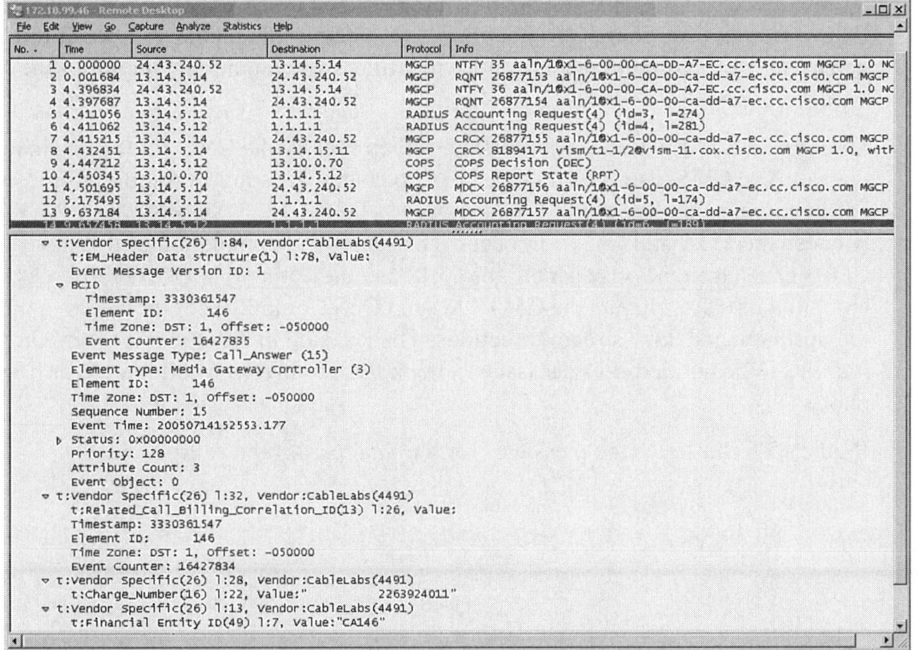

As with all PacketCable event messages, the first PacketCable attribute is type 1, which is the Event Messaging (EM) header attribute. This attribute is comprised of a number of parameters, including

- Version ID (in this case, PacketCable 1.0)
- BCID (which itself is made up of a number of parameters)
- Event Message Type (in this case, a Call Answer EM)
- Sending Element Type (in this case, MGC)
- Element ID (this is statically configured)
- Time Zone (Daylight Savings Time or not and offset from UTC)
- Sequence Number
- Event Time
- Status
- Priority (importance relative to other EMs)
- Attribute Count (# of following attributes, in this case 3)
- Event Object (either accounting or PCES [lawful intercept])

The remaining PacketCable attributes in this message include Related Call BCID, Charge Number, and FEID. The Related Call BCID attribute contains the BCID used by the CMS for the originating leg of the call—remember, this message is sent from the MGC for the terminating leg of the call. The format of the Related Call BCID attribute is the same as the BCD attribute. The Charge Number attribute specifies the phone number of the billable party. The FEID is the Financial Entity Identifier, which uniquely identifies the MSO in billing records.

Similarly, a Call Answer event message is then sent by the CMS for the originating leg of the call. The CMS also instructs the MTA to change its connection to send/receive mode and to commit DQoS resources. This triggers a DSC-REQ/DSC-RSP/DSC-ACK exchange between the MTA and CMTS to occur. The successful completion of this also triggers a COPS Gate Open message sent to the CMS and indicating that DQoS resources have been committed. It also triggers the CMTS to send QoS Commit event messages to the RKS for the upstream and downstream directions. The last step in this stage is where the CMS requests to be notified if Oscar issues a hook flash to invoke a call feature or hangs up his phone.

Figure 15-6 illustrates the messages for the final part of the call.

Figure 15-6 *Call Flow with Event Messaging: Part III*

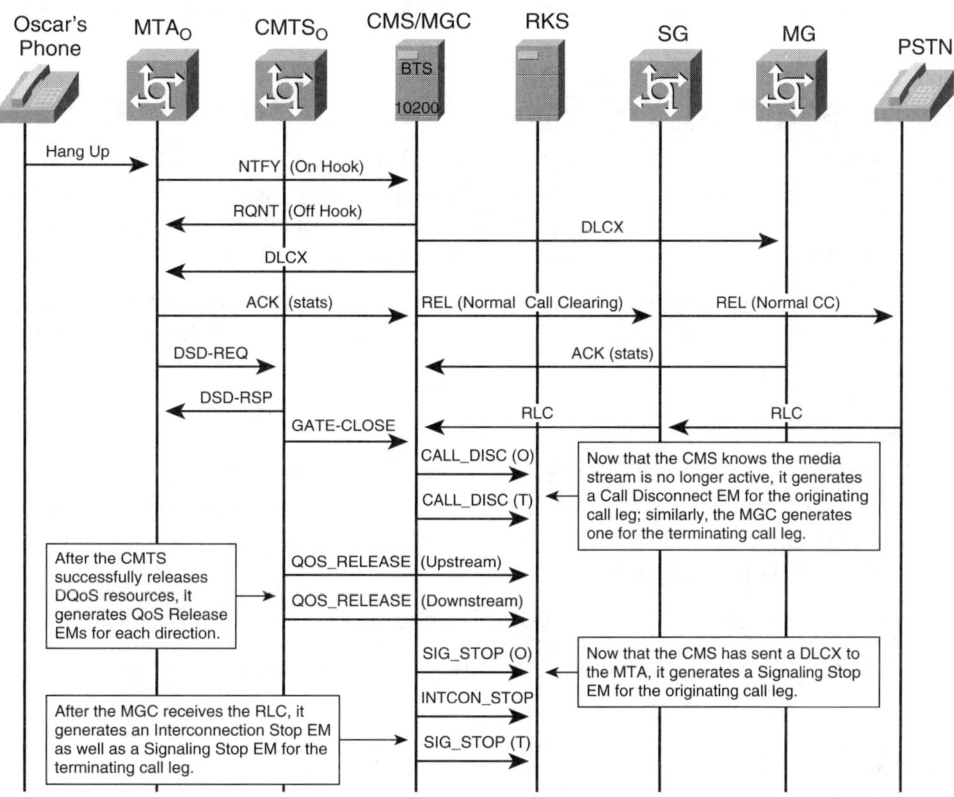

This stage begins with Oscar hanging up his phone, triggering an NTFY sent to the CMS and an RQNT returned asking to be notified when the phone is taken off hook for the next call. The MGC then deletes the connection on the MG, and similarly the CMS deletes the connection on the MTA. Sending DLCX messages to tear down connections triggers the CMS to send a Call Disconnect EM to the RKS for the originating leg and the MGC to send a Call Disconnect EM to the RKS for the terminating leg. The sending of the DLCX by the CMS also triggers the CMS to send a Signaling Stop EM to the RKS for the originating leg.

The DLCX to the MTA also triggers the MTA to initiate the DSD-REQ/DSD-RSP exchange with the CMTS to release DQoS resources. This in turn causes the CMS to send a COPS Gate Close message to the CMS, as well as QoS Release EMs to the RKS for the upstream and downstream DQoS resources.

Along with sending the DLCX to the MG to tear down the call, the MGC also sends an ISUP release (REL) message to the PSTN. The PSTN acknowledges the successful call teardown by sending an ISUP RLC to the MGC. This triggers the MGC to send both an Interconnect Stop EM and a Signaling Stop EM for the terminating leg to the RKS.

Event Messaging Configuration and Troubleshooting

Now that you know how PacketCable Event Messaging works, it's time to learn how to configure and troubleshoot it on Cisco equipment. Much of this configuration is on the CMS/MGC component (that is, BTS 10200). Also a small amount of configuration is required on the Cisco CMTS for Event Messaging generation.

Problems related to Event Messaging are most likely either configuration issues or network issues (such as IP routing between the BTS, CMTS, and RKS). A few other troubleshooting techniques are mentioned in this section.

Configuring Event Messaging on a BTS 10200

Chapter 7, "NCS (Network-based Call Signaling)," covered much of the BTS 10200 configuration needed for PacketCable operation. The specific steps relevant to Event Messaging operation are as follows.

Step 1 Define the RKS(s) (RADIUS Profile table).

Step 2 (Optional) Modify the Event Messaging variables (Call Agent Configuration table).

Step 3 Enable the sending of EMs (Call Agent profile table).

Step 4 (Optional) Configure a billing type for subscribers (Subscriber table).

Step 5 (Optional) Configure Media Alive EM generation (Activity table).

The following sections cover each of these steps in greater detail.

Step 1: Define the RKS

An RKS is defined in the RADIUS-PROFILE table. The syntax to add an entry to this table as follows:

```
CLI>add radius-profile id=RKS id; tsap-addr=RKS IP address; acc-req-retransmit=# of
  retransmission; acc-rsp-timer=time to wait for response;
```

The configuration of this table used for the call flow examples is as follows:

```
CLI>show radius_profile
ID=prirks
TSAP_ADDR=1.1.1.1
ENCRYPTION_KEY=0000000000000000
ACC_REQ_RETRANSMIT=3
ACC_RSP_TIMER=2
```

```
IPSEC_SA_ESP_CS=3DES-MD5,3DES-SHA1,NULL-MD5,NULL-SHA1
IPSEC_SA_LIFETIME=86400
IPSEC_SA_GRACE_PERIOD=21600
IPSEC_ULP_NAME=IP
IKE_GROUP=2
IKE_SA_LIFETIME=86400
IKE_CS=3DES-MD5,3DES-SHA1
SERVER_TYPE=OTHER
ID=secrks
TSAP_ADDR=2.2.2.2
ENCRYPTION_KEY=0000000000000000
ACC_REQ_RETRANSMIT=3
ACC_RSP_TIMER=2
IPSEC_SA_ESP_CS=3DES-MD5,3DES-SHA1,NULL-MD5,NULL-SHA1
IPSEC_SA_LIFETIME=86400
IPSEC_SA_GRACE_PERIOD=21600
IPSEC_ULP_NAME=IP
IKE_GROUP=2
IKE_SA_LIFETIME=86400
IKE_CS=3DES-MD5,3DES-SHA1
SERVER_TYPE=OTHER
Reply : Success:  Entries 1-2 of 2 returned.
```

The field ACC-REQ-RETRANSMIT indicates how many retransmission attempts of Accounting-Request messages are made to the RKS if no Accounting-Response message is received; in the sample configuration, three retransmission attempts are made. The field ACC-RSP-TIMER defines how long the BTS waits for an Accounting-Response message after an Accounting-Request message is sent; in the example the BTS waits two seconds. The remaining fields in this table relate to configuring IP security for the BTS-to-RKS communication.

Step 2: (Optional) Modify Event Messaging Variables

Recall that the CALL-AGENT-CONFIG table defines a variety of variables that affect how the BTS processes calls. Some of these variables relate to Event Messaging.

A sample configuration of these variables is as follows:

```
CLI>show ca-config
TYPE=RETRY-PRI-RKS-TIMER
DATATYPE=INTEGER
VALUE=10
TYPE=EM-FILE-OPEN-TIME
DATATYPE=INTEGER
VALUE=600
TYPE=EM-FILE-SIZE
DATATYPE=INTEGER
VALUE=20
TYPE=BATCH-LATENCY
DATATYPE=INTEGER
VALUE=60
TYPE=BATCH-MODE-SUPP
DATATYPE=BOOLEAN
VALUE=N
TYPE=RADIUS-DSCP-TOS
DATATYPE=INTEGER
VALUE=96
```

The first three variables specify how EMs are stored in files on the call agent when loss of communication with the RKSs prevent EMs from being transferred. An open EM file does not close automatically when RKS communication is restored; the file closes automatically when the open time reaches the EM-FILE-OPEN-TIME value or the file size reaches the EM-FILE-SIZE value, whichever occurs first. The BATCH-MODE-SUPP variable specifies whether EMs are sent real time or sent in batches, and the BATCH-LATENCY variable specifies the time interval between batches if the BATCH-MODE-SUPP variable is set. The RADIUS-DSCP-TOS variable defines the setting of the IP type of service (ToS) byte in Accounting-Request messages sent by the BTS.

Remember that if variables do not show up in the CA-CONFIG table, the default values in the CA-CONFIG-BASE table are used. You can always add these variables to the CA-CONFIG table and modify their values from the defaults if you want.

Step 3: Enable the Sending of EMs

The BTS 10200 has the capability to generate either call detail blocks (CDB), which is the traditional post-call billing data, or PacketCable event messages. In a PacketCable network, you should have the BTS configured to generate event messages. As you have seen, the BTS 10200 is both a CMS and an MGC, and this is also true with respect to Event Messaging.

For the BTS to send PacketCable event messages, the CALL-AGENT-PROFILE table must be provisioned similar to the following (the highlighted fields are related to Event Messaging) :

```
CLI>show call_agent_profile
ID=CA104
CMS_ID=104
MGC_ID=104
FEID=CA104
DQOS_SUPP=Y
CDB_BILLING_SUPP=N
EM_BILLING_SUPP=Y
PRI_RKS_PROFILE_ID=prirks
SEC_RKS_PROFILE_ID=secrks
GTD_SUPP=N
PCMM_SUPP=N
AMID=99999
PCMM_VERSION_MAJOR=1
PCMM_VERSION_MINOR=0
Reply : Success: Entry 1 of 1 returned.
```

The EM-BILLING-SUPP field must be set to Y to enable the BTS generation of event messages; consequently, the CDB-BILLING-SUPP field should be set to N to disable the generation of CDBs. You can force both parameters set to Y; however, this is strongly discouraged because it can impact BTS performance. The PRI-RKS-PROFILE-ID and SEC-RKS-PROFILE-ID fields identify the RKS components defined in the first step. The CMS-ID and MGC-ID fields specify the values used for the Element ID attribute in messages originated by the CMS and MGC components of the BTS, respectively.

Logically, the FEID field specifies the value used for the Financial Entity Identification attribute.

Step 4: (Optional) Configure a Billing Type for Subscribers

For each PacketCable MTA subscriber, you can specify a billing type (flat rate or measured rate) and an account ID. This is done by modifying the corresponding entry in the SUBSCRIBER table as demonstrated here:

```
CLI>change subscriber id=sub1@jriddel_lsys1;billing_type=MR1;account_id=123456789;
```

Step 5: (Optional) Configure Media Alive EM Generation

The Media Alive EM verifies that calls that have been active for at least 24 hours are still active. On the BTS this is enabled by adding an entry to the ACTIVITY table as demonstrated here:

```
CLI>show activity
ID=MEDIA-ALIVE-EM
FREQ=DAILY
DAY_OF_MONTH=NA
DAY_OF_WEEK=NA
START_TIME=0000
FIXED_TIME_INTERVAL=N
ENABLED=Y
SO_ENABLED=N
RESTART_ENABLED=N
LAST_CHANGED=2006-08-13 22:04:53
```

In this example Media Alive EMs are generated every day at midnight (00:00).

Recovering Billing Files

If communication is lost between the BTS and RKS devices, alarms are generated by the BTS, and billing data files are written locally on the BTS call agent machine. Specifically, they are put in the directory /opt/BTSem, and they begin with the name PKT_EM followed by a timestamp. These files can then be manually transferred to the RKS using FTP.

Retrieving Event Messaging Measurements

You can obtain measurements on the number of event messages logged, acknowledged, and retransmitted using the **report measurement_em_summary** command as demonstrated here:

```
CLI>report measurement_em_summary
TIMESTAMP                                2006-08-13 22:15:00
CALL_AGENT_ID                            CA104
CONDITION                                Normal
```

```
BILLING_EM_ACKED                    0
BILLING_EM_LOGGED                   0
BILLING_EM_RETRANS                  0
```

Configuring and Troubleshooting Event Messaging on a Cisco CMTS

The steps to configure PacketCable Event Messaging on a Cisco CMTS are as follows (only the EM-related steps are covered; for general PacketCable configuration steps refer to Chapter 13, "Analyzing, Implementing, and Troubleshooting DQoS"):

Step 1 (Optional) Configure the Event Messaging Element ID:

```
Router(config)#packetcable element-id n
```

This defines the Element ID attribute for the CMTS in outgoing event messages; if this is not specified, the CMTS defaults to a random value between 0 and 99,999.

Step 2 Enable the AAA access control model:

```
Router(config)#aaa new-model
```

This enables AAA services for authorization, authentication, and accounting. Remember, PacketCable Event Messaging uses RADIUS accounting for transport. Consequently, this command must be configured for Event Messaging records to be generated.

Because this command enables all AAA services, you want to make sure the authorization and authentication services are properly defined. For example, you should have commands defined that enable these AAA services to interface with the appropriate RADIUS and/or Terminal Access Controller Access Control System Plus (TACACS+) servers.

Step 3 Configure the CMTS to recognize and use vendor-specific attributes:

```
Router(config)#radius-server vsa send accounting
```

PacketCable Event Messaging uses vendor-specific attributes specified by CableLabs; therefore, this command must be entered to enable the generation of these attributes.

Step 4 Debug Event Messaging on the CMTS.

Remember in Chapter 13 that you were introduced to the **debug packetcable gate** series of commands. One of the options for this command is **event**, which when used with the **debug packetcable subscriber** command decodes the PacketCable Event Messaging sent to the RKS.

Also you can verify that the Event Generation Info object is set properly by looking at gates or debugging the gate control messages as described in Chapter 13.

Step 5 Debug RADIUS accounting.

You can display RADIUS statistical information with the **show radius statistics** command. The command output tells you the number of RADIUS Accounting-Request packets sent and the number that did and did not receive a corresponding Accounting-Response packet. It also gives metrics on the average and maximum delay between sending a request and receiving a response.

PacketCable Lawful Intercept Motivation and Background

In the United States telecommunication carriers are required to adhere to the Communications Assistance for Law Enforcement Act, more commonly referred to as CALEA. Under this act, telecom carriers must provide certain capabilities to law enforcement agencies provided they have the proper court authorization. MSOs offering telephony service fall under the classification of telecommunication carriers; therefore, they must also adhere to CALEA. That is, if an agency such as the FBI has a valid warrant to place one of an MSO's PacketCable customers under surveillance, the MSO must be able to provide them with the information identified under CALEA. An ANSI standard already exists describing the CALEA requirements for traditional service providers; however, because the PacketCable environment introduces new challenges, a PacketCable specification also exists for lawful intercept.

These "new challenges" are primarily because of the distributed nature of a PacketCable network. For traditional circuit-switched telephony service providers, lawful intercept processing is done on the telephony switch, so it is relatively easy to control. In a PacketCable network, the lawful intercept data comes from distributed components and needs to be formatted into the standard interfaces used by law enforcement agencies. Consequently, lower-level protocols are specified by PacketCable that allow components to be interoperable and for enough data to be made available to conform to the CALEA standard at the higher level.

This specification is under the PacketCable 1.5 series of specifications and is often referred to as PacketCable Electronic Surveillance (PCES).

Historically, only voice communication set up by the CMS is a requirement for lawful intercept. Other types of communication offered by the MSO, such as e-mail and Internet access, are categorized as information services and did not fall under the domain of CALEA. However, recent court rulings have changed this, and information services are now covered and must be implemented by May 2007. Service providers must also respond to national security wiretaps and supply the same information as for CALEA.

Surveillance Types

Three types of surveillance are performed by law enforcement agencies (LEAs):

- Pen register
- Trap and trace
- Interception

The first two are both for the recording of call identifying information similar to what is captured in a call billing record. The difference is the type of calls they apply to; *pen register* is for calls originated by the subject under surveillance, and *trap and trace* is for calls received by the subject under surveillance. Today, these two types of surveillance are always requested together, which is known as *pen and trace*. The *interception* type includes capturing the call content information (that is, phone conversations) in addition to capturing the call identifying data. Interception traces are authorized only when other evidence, like that obtained from pen and trace, can establish probable cause to actually listen to the content of protected conversations.

Lawful Intercept Terminology

Before getting into too much detail about the interworkings of lawful intercept in a PacketCable network, you need to be familiar with some more terminology. The PacketCable components involved in surveillance or wiretapping are known as Intercept Access Points (IAPs). PacketCable CMSs and MGCs are the sources (IAPs) for call identifying information. In most cases, MGC information is redundant and can be discarded. PacketCable CMTSs and MGs are the sources (IAPs) for call content information. Because the call agent components are not in the path of the voice stream, they cannot be call content IAPs. You might be wondering why the MTA itself is not an IAP for lawful intercept. The reason is because of security; the MTA is an untrusted device because it resides on the customer premises. It is a CALEA requirement that the end user must not be able to tell that they are under surveillance. So no signaling or content differences between a regular call and one under surveillance can exist from the MTA standpoint.

Later you'll see under what situations each of these PacketCable devices are used as an IAP.

The process of electronic surveillance is also generally separated into four distinct functions:

- Access
- Delivery
- Collection
- Administration

IAPs perform the Access Function (AF) of lawful intercept because they have access to the subscriber's call information; thus, IAPs are also known as AF devices. Both call identifying information (also known as *call detail*) and call content information are provided to LEAs via a device that performs the Delivery Function (DF). The DF device is a mediation server provided by the PacketCable MSO that processes the data from the AFs and passes this data along to collection devices under the control of the LEA. The information collected by the MSO DF is sent to an LEA device, which performs the Collection Function (CF). The administrative function refers to the user interface and control of these other three functions. The administrative function of the AF and DF devices is the responsibility of the service provider, whereas the administrative function of the CF is the responsibility of the LEA. As you can see, between the DF and CF devices you have a boundary or demarcation point between the service provider and the LEA responsibilities. Figure 15-7 illustrates these functional components.

Figure 15-7 *Lawful Intercept Functional Components*

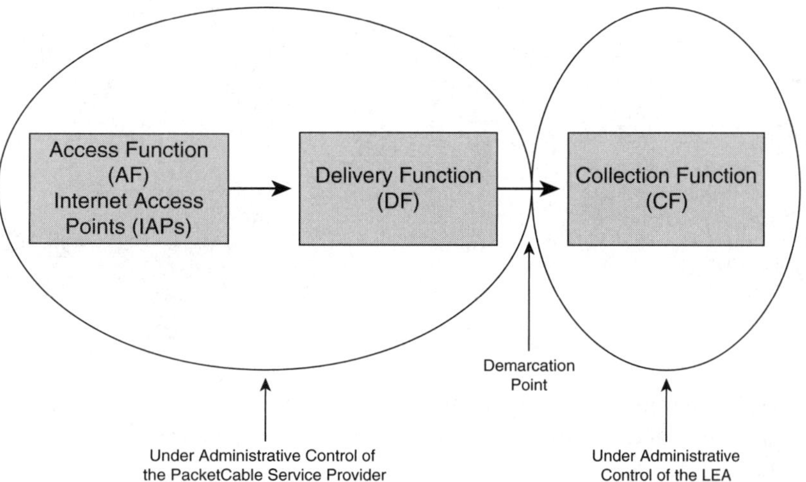

The communication channels between the DF and CF in which call detail and call content information are provided to the LEA are known as the call data connection (CDC) and call content connection (CCC), respectively. It is important to note that these communication channels are one way; the CF cannot request anything from the DF. You'll learn much more about how these connections function later.

In many cases, the CMS and CMTS components are the only IAPs required for intercepting both call data and call content information. An exception to this is if a subscriber under surveillance has forwarded his or her number to one serviced by the legacy PSTN. In this case, a call coming from the PSTN to the suspect is forwarded back to the PSTN; thus, no CMTS is in the call path and, hence, able to collect call content information. Consequently, the MG component becomes the call content IAP in this case. If you are using the BTS as the CMS/MGC, then the BTS tries to set up the call content intercept at the entry point in the network. So if the suspect originates a call, the entry point is the CMTS, and content is intercepted there. If a call to the suspect is coming from the PSTN, the entry point is the MG, and the BTS tries to set up the content intercept there. If the MG doesn't support interception, the BTS sets up the intercept at the exit point (that is, the CMTS). If neither the entry point nor the exit point supports interception, the BTS informs the DF of this and the DF can try another method of interception.

Figure 15-8 is an illustration of the lawful intercept architecture in a PacketCable network.

Figure 15-8 *PacketCable Lawful Intercept Architecture*

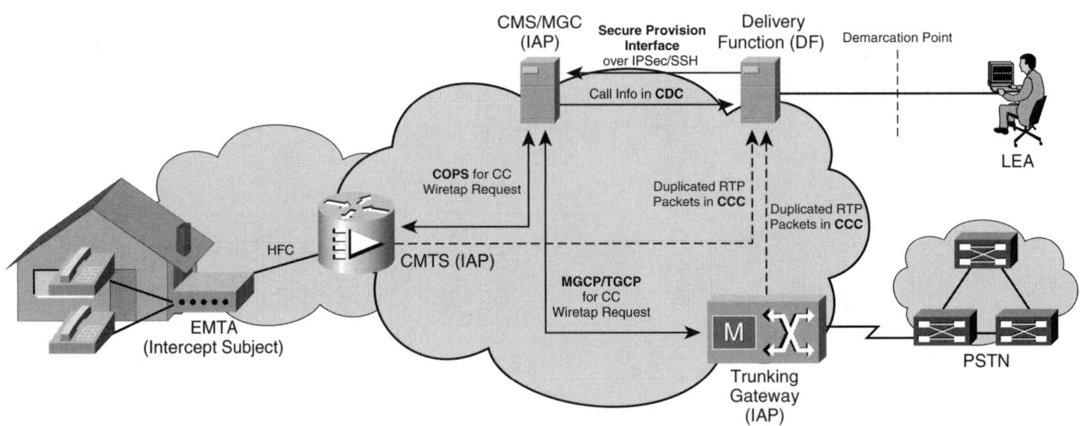

Note that multiple LEAs (up to a maximum of five) can monitor the same subject at the same time, requesting different information or the same information, and must not be able to detect each other.

AF-DF Communication

The first type of AF-to-DF communication covered is for the administration function of initiating, modifying, or removing a surveillance instance or wiretap. The PacketCable lawful intercept specification does not dictate how this communication is done. However, a typical implementation is for the PacketCable administrator to manually configure surveillance information on the mediation server or DF device. Defined on the DF device are the identities of the provisioning interfaces to the AF devices including IP addresses and login and password information. For example, with the Cisco BTS10200 CMS, the login and password for the CALEA user would be provisioned, which would allow the DF to access the CLI of the BTS EMS. Also defined on the DF are the identities and connectivity information about the LEA CF devices.

When the administrator needs to initiate a new intercept, it configures this on the DF. The identity of the tapped subscriber is defined, along with information such as the type of intercept and the warrant data authorizing the intercept. Also specified is the CMS IAP device the subscriber is associated with. This information is used by the DF to initiate the communication channel to the CMS. At this point the DF can perform the necessary provisioning on the CMS. For security reasons, this connection is established using the secure shell TCP/IP protocol. Again, remember the administration interface is not specified by PacketCable; the previous steps describe how BTS provisioning on the DF is typically handled.

The next type of AF-to-DF communication covered is for the access function of sending call detail and call data from the PacketCable components to the DF. After the surveillance is set up on the CMS, the IAPs need to provide the DF with the suspect's call information when a call is made or received. Earlier, you learned that call detail information is similar to the type of information you see in a call billing record. So what protocol is logical for the application of sending call detail data? The answer is the same protocol you learned about at the beginning of this chapter, the RADIUS Accounting protocol with PacketCable Event Messaging attributes.

The CMS sends Event Messaging packets to the DF device in real time. When the DF device receives one of these messages, it acknowledges this by sending an Accounting Response message. Note: For lawful intercept, an IAP always sends Event Messaging packets in real time. The format of these packets is nearly identical to those used in basic Event Messaging, except for the addition of a few parameters relevant to lawful intercept. Figure 15-9 shows an Ethereal trace of one of these packets.

Figure 15-9 *Lawful Intercept Event Messaging Trace*

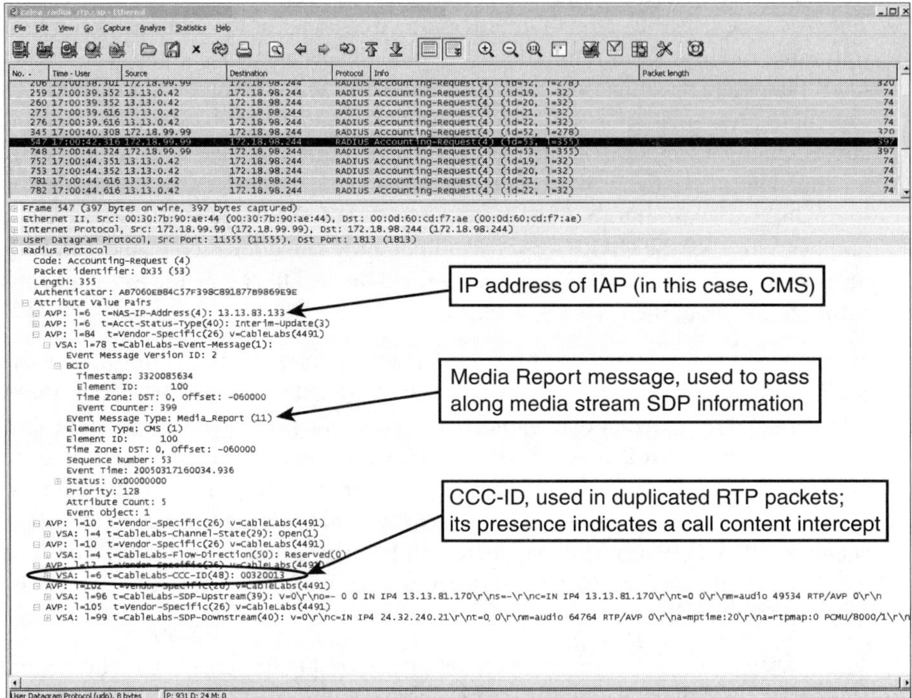

The following event message types are used for lawful intercept:

- Service Instance
- Signaling Start
- Signaling Stop
- Call Answer
- Call Disconnect
- QoS Reserve
- QoS Release
- QoS Commit
- Media Report
- Signal Instance
- Conference Party Change
- Surveillance Stop
- Redirection

The last five EM types listed are used only for lawful intercept. For more details on how these event messages are used for lawful intercept and the attributes contained within them, please refer to Appendix A of the PacketCable Event Messaging specification, which details the use of Event Messaging in lawful intercept.

Remember that the CMTS can also send lawful intercept Event Messaging packets to the DF device (these are the QoS_Reserve, QoS_Commit, and QoS_Release EM types). However, the CMTS is usually configured not to send them because the information in these EMs is no longer needed with the CMS sending the Media_Report EMs. Also remember that the CMTS and MG components become IAPs when a call content intercept is required. So if other PacketCable devices need to send call identifying and call content data to the DF, how are these devices informed of where and how they need to forward intercepted information? Does the DF need to communicate with these components as well? The answer is no because in a call content intercept information from the CMS is still required, the CMS can inform the other components about the DF device. Keep in mind that the CMS triggers call-content intercepts on the CMTS and MG components on a per-call basis. Also triggering intercepts in the middle of call is supported.

Communicating Lawful Intercept Information to the CMTS

So how does the CMS communicate lawful intercept data to the CMTS? The answer was briefly covered in Chapter 13, which dealt with PacketCable DQoS. Electronic surveillance parameters are sent in a PacketCable COPS object in a Gate Set message. The following is a sample trace of one of these messages on a Cisco CMTS obtained via IOS **debug** commands:

```
Aug 14 20:18:42.853: Pktcbl(gcp): Received GATE SET message, tid=0x20
Aug 14 20:18:42.853: --- Pktcbl(gcp): Received GCP message ------
Aug 14 20:18:42.853: TRANSACTION ID        : Object.[snum/stype/len 1/1/8]
Aug 14 20:18:42.853:     transaction id    : 0x20
Aug 14 20:18:42.853:     gcp cmd           : 4 (GATE SET)
Aug 14 20:18:42.853: SUBSCRIBER ID (IPV4)  : Object.[snum/stype/len 2/1/8]
Aug 14 20:18:42.853:     Addr              : 24.34.240.246
Aug 14 20:18:42.853: ACTIVITY COUNT        : Object.[snum/stype/len 4/1/8]
Aug 14 20:18:42.853:     Count             : 5
Aug 14 20:18:42.853: LAWFUL INTERCEPT      : Object.[snum/stype/len 10/1/48]
Aug 14 20:18:42.853:     df_cdc_addr       : 0.0.0.0
Aug 14 20:18:42.853:     df_cdc_port       : 0
Aug 14 20:18:42.853:     flags             : 0x2
Aug 14 20:18:42.853:     df_ccc_addr       : 172.18.99.11
Aug 14 20:18:42.853:     df_ccc_port       : 65002
Aug 14 20:18:42.853:     ccc_id            : 3407914
Aug 14 20:18:42.853:     billing corr ID  : [C8 8B 59 22 20 20 20 20 20 31 30 34 31
  2D 30 34 30 30 30 30 00 00 00 C4 ]
Aug 14 20:18:42.853: GATE SPEC             : Object.[snum/stype/len 5/1/60]
Aug 14 20:18:42.853:     direction         : 1 (UPSTREAM)
Aug 14 20:18:42.853:     protocol id       : 17
Aug 14 20:18:42.853:     commit flag       : 0x0
Aug 14 20:18:42.853:     session class     : 1
Aug 14 20:18:42.853:     source            : 24.34.240.246
Aug 14 20:18:42.853:     dest              : 24.34.240.235
Aug 14 20:18:42.853:     src port          : 0 (0x0)
```

```
Aug 14 20:18:42.853:      dest port         : 53456 (0xD0D0)
Aug 14 20:18:42.853:      dscp              : 0xA0
Aug 14 20:18:42.853:      timer t1          : 200
Aug 14 20:18:42.853:      timer t7          : 200
Aug 14 20:18:42.853:      timer t8          : 300
Aug 14 20:18:42.853:      flowspec # 1      : [r/b/p/m/M 10000/200/10000/200/200] [R/
  S: 10000/800]
Aug 14 20:18:42.853: GATE SPEC              : Object.[snum/stype/len 5/1/60]
Aug 14 20:18:42.853:      direction         : 0 (DOWNSTREAM)
Aug 14 20:18:42.853:      protocol id       : 17
Aug 14 20:18:42.853:      commit flag       : 0x0
Aug 14 20:18:42.853:      session class     : 1
Aug 14 20:18:42.853:      source            : 24.34.240.235
Aug 14 20:18:42.853:      dest              : 24.34.240.246
Aug 14 20:18:42.853:      src port          : 0 (0x0)
Aug 14 20:18:42.853:      dest port         : 61674 (0xF0EA)
Aug 14 20:18:42.853:      dscp              : 0xA0
Aug 14 20:18:42.853:      timer t1          : 200
Aug 14 20:18:42.853:      timer t7          : 200
Aug 14 20:18:42.853:      timer t8          : 300
Aug 14 20:18:42.853:      flowspec # 1      : [r/b/p/m/M 10000/200/10000/200/200] [R/
  S: 10000/0]
Aug 14 20:18:42.853: -------------------------------------
```

Included in the Gate Set message are the DF IP address and port number to which the CMTS needs to send call content (CCC) information. The flags field consists of two flags; the least significant bit is the DUP-EVENT flag, and the next bit is the DUP-CONTENT flag. So in the example this value is 0x2, meaning only the DUP-CONTENT flag is set. A set DUP-EVENT flag tells the CMTS to send duplicated Event Messaging packets to the DF identified for CDC, and the DUP-CONTENT flag tells the CMTS to send duplicated Real-time Transport Protocol (RTP) packets to the DF identified for CCC. The DUP-EVENT flag is usually not sent as the QoS_xxx event messages are not used anymore, for the reasons discussed earlier. The CCC-ID is the identifier used in the header of these duplicated RTP packets. Finally, so information from other IAPs (namely the CMS) can be correlated, the Billing-Correlation-ID field is also included.

So what does a duplicated RTP packet look like? Figure 15-10 is an Ethereal packet trace of a call content surveillance.

As you can see, the duplicated RTP packets are simply encapsulated in UDP and prepended with the CCC-ID.

Figure 15-10 *Call Content Messaging Trace*

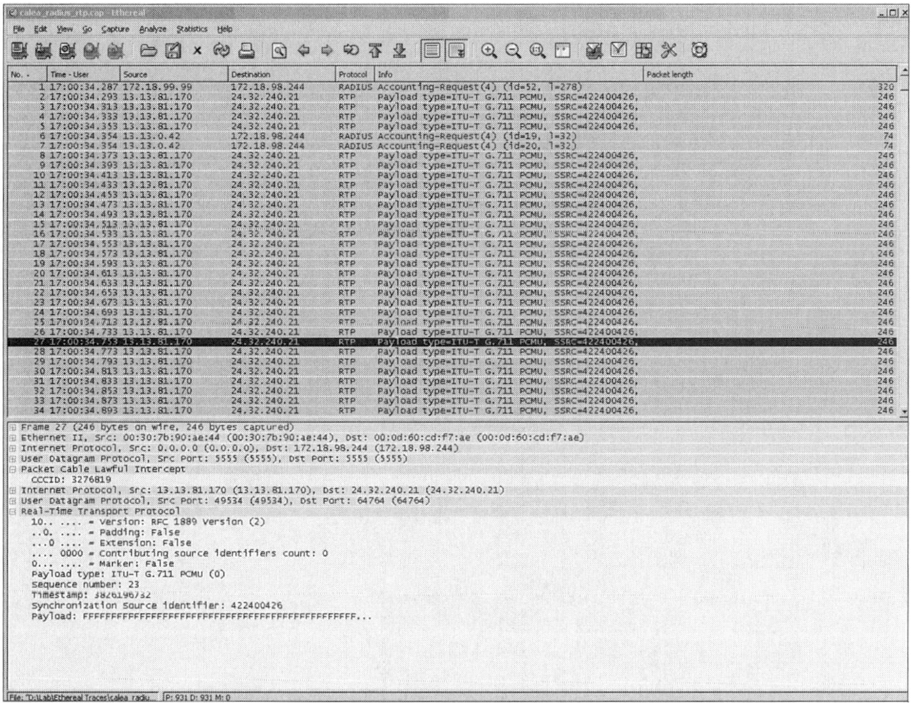

Communicating Lawful Intercept Information to the MG

So what about the case where the trunking MG needs to perform the packet replication? Remember, this typically happens in scenarios where the CMTS is not in the voice path of a tapped subscriber call, such as in a call forwarding scenario. If you are using the BTS 10200, this also happens when calls to tapped subscribers originate from the PSTN.

In this case, lawful intercept parameters are inserted into the TGCP messaging sent from the MGC to the MG. You were introduced to these parameters in Chapter 8. These parameters are as follows:

- **Call Content Connection Identifier (es-cci)**—Because multiple IAPs can contribute to lawful intercept, in this case the CMS/MGC and the MG, the Call Content Connection Identifier is a 32-bit value used to correlate all this information. As you'll see shortly, this field is added to the header of all intercepted voice packets before they are sent to the electronic surveillance delivery function.
- **Call Content Destination (es-ccd)**—This field identifies the IP address and UDP port number of the delivery function that is collecting the intercepted voice packets.

Figure 15-11 depicts a sample trace of TGCP messaging used for a call content trace.

Figure 15-11 *Lawful Intercept TGCP Messaging Trace*

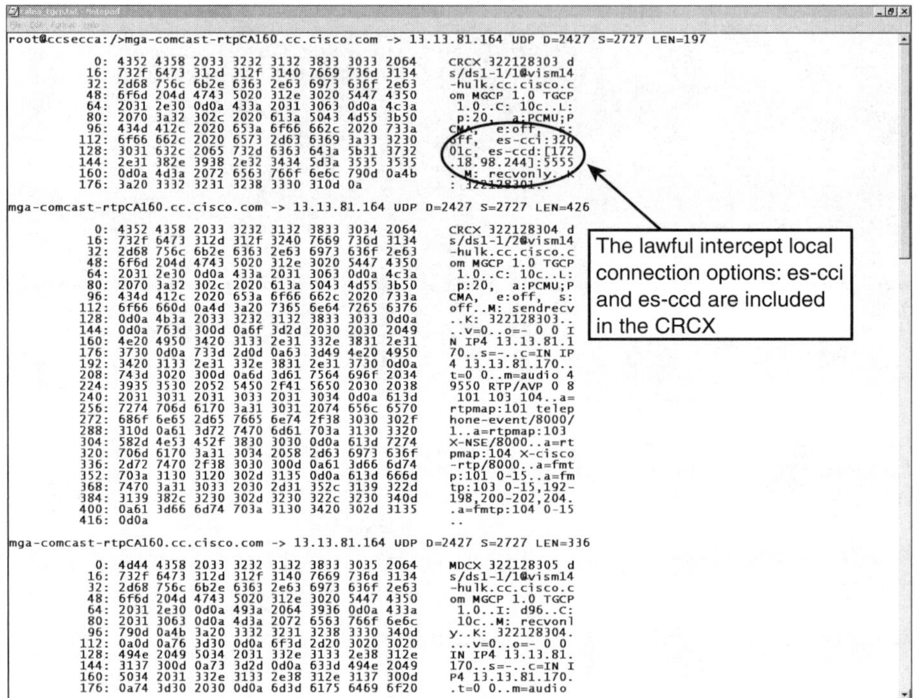

Communicating Lawful Intercept Information Across CMSS

The previous discussion assumed the CMS and MGC components are bundled into the same device. If this is not the case and CMSS is used as the signaling protocol between the CMS and MGC components, the lawful intercept information is carried in the CMSS signaling.

In Chapter 9, you learned about CMSS extensions to SIP used to pass additional information. Two of these extensions are used for lawful intercept:

- The P-DCS-LAES extension contains information needed to support lawful intercept for CMS–to-CMS call types. Basically, it is used when the responsibility for lawful intercept moves from one CMS to another. This is required in situations where a subscriber under surveillance has forwarded his or her phone.
- The P-DCS-REDIRECT extension contains information specific to redirected or forwarded calls. For lawful intercept, the REFER method transports information.

Support of these headers is required for CMSS compliance.

DF-CF Communication

As you know, the DF is under the domain of the PacketCable service provider, and the CF is under the domain of the LEA; the interface between these two is known as the *demarcation point*. The demarcation point is physically connected and colocated to the DF. You also learned that CDC and CCC channels transfer intercept information to the CF, but how?

Capturing Call Data and PCESP

A specialized protocol known as the PacketCable Electronic Surveillance Protocol (PCESP) delivers call detail information in CDC channels from the DF to the demarcation point. PCESP uses TCP as its transport layer protocol and the ASN.1 language to define its messages.

Abstract Syntax Notation One (ASN.1) is a stream-lined encoding language that can apply to a wide range of applications. The ASN.1 language is an international standard, and its basic notation is defined under the ITU-T recommendation X.680. This standard is also adopted by the International Standards Organization (ISO) and is defined in the document ISO/IEC 8824-1.

In Chapter 5, "The MTA Configuration File," you learned about one well-known application that uses the ASN.1 language syntax; the TCP/IP network management protocol Simple Network Management Protocol (SNMP). ASN.1 utilizes a Tag/Length/Value (TLV) format that allows complex data types to be defined out of a small number of predefined simple data types, other complex data types, or both. Examples of these simple data types are integers, strings, and Boolean or true/false variables. Complex data types can be created by defining lists, structures, choices, etc. made up of these simple types or other complex types. After you see how ASN.1 is used in the PCESP protocol, this becomes clearer.

So what does one of these messages look like? Figure 15-12 is an Ethereal trace of a PCESP message.

Figure 15-12 *PCESP ASN.1 Trace #1*

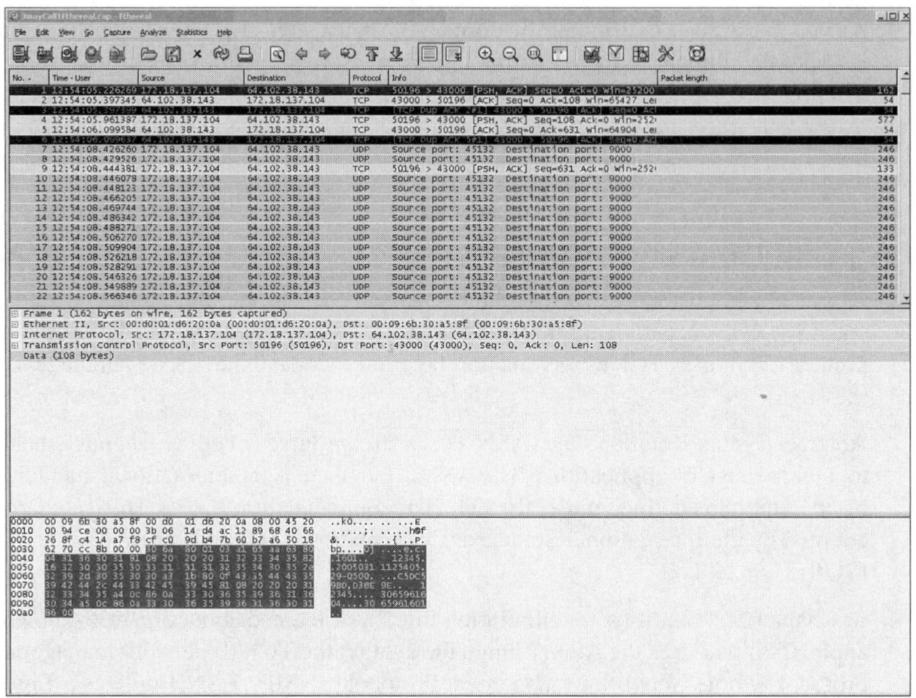

The PCESP portion of the packet is highlighted in the raw hex portion of the Ethereal capture. Before that, you can see that the packet is using TCP and is sent from the DF device to the CF device. After the TCP header is the ASN.1 encoded TLV data. The tag is always a single byte in length and is composed of three fields: the two most significant bits identify the tag class; the next bit identifies the tag format; and the final five bits identify the tag number.

So looking at the first tag in the example, you see it has a value of 0x30 (0011 0000 in binary), which is a tag class of 00, a tag format of 1, and a tag number of 10000. Tag class 00 is categorized as a universal class and means this is a well-defined tag in the ASN.1 standard. Tag format 1 means this tag is constructed out of multiple other tag types. The tag number field is meaningful with respect to the tag class; for the universal class, 10000 identifies the value is a sequence. So now you know the value is a sequence of other tag types. The next byte is the start of the length parameter. The most significant bit (MSB) of this byte is an indicator if the length is larger than 127 bytes. If it is not set, the remaining 7 bits in this byte represent the length of the parameter. If it is set, the remaining 7 bits identify the number of length bytes following this byte. In this case this byte is set to 0x6a. So, the indicator isn't set, meaning the length of this sequence tag is 106 bytes, which is the length of the rest of the packet. Hopefully, now you're getting the hang of decoding ASN.1.

The next tag byte is equal to 0x80. This is a tag class of 10, which is the context-specific class. The tag format of 0 means this tag is a primitive or single value. So now you know that this tag is a single value specific to the PCESP application. In the PCESP protocol, tag number 0 identifies the protocol version, and tag number 1 identifies the message type. So the TLV 0x800103 equates to PCESP protocol version 3. The next tag, 0xa1, identifies the next value as a PCESP message and identifies the value as being constructed of multiple tags. Within the PCESP protocol you have several types of messages, each used to convey different information that occurs during a tapped subscriber's call. Table 15-1 lists the messages you are most likely to see, along with a brief description and its tag number.

Table 15-1 *PCESP Messages*

PCESP Message	Tag Number	Description
Answer	1	Sent after call answered; indicates bidirectional voice communication now established
CCChange	12	Change in call content delivery; indicates a change in the call SDP parameters (used only in call content intercepts)
CCClose	2	End of call content delivery (used only in call content intercepts)
CCOpen	3	Start of call content delivery (used only in call content intercepts)
Origination	5	Suspect has attempted to place a call
Redirection	7	Call has been redirected, such as with a call transfer or call forwarding scenario
Release	8	Suspect's call is being torn down
Termination Attempt	10	Someone is calling the suspect
Media Report	18	SDP information has been exchanged (used only in call data intercepts)
Network Signal	16	Signal, such as a tone, is applied toward the suspect
Service Instance	19	Event such as a hook flash occurred, such as with a call waiting scenario
Subject Signal	17	Signaling information such as dialed digits

The next tag, 0xaa, is the message type, which is also a constructed data type (tag class 10 = context specific, format 1 = constructed data, tag number 01010 = 10 [message type]). From the previous table you can see the tag number 10 indicates a termination attempt message. The remaining tags are all parameters of this message, they are all context-specific, and their tag numbers identify the specific parameter for a termination attempt message.

You can see from Figure 15-12 that among these parameters are the lawful intercept case/warrant ID; the accessing element ID, which identifies the IAP; the timestamp of the event; the calling and called party phone numbers; and the call identifier. Note: The timestamp parameter identifies the time the IAP detected the event triggering the message, in this case when the CMS detected a call coming in destined for the tapped subscriber. For more details on these parameters, please refer to the PacketCable specification.

Figure 15-13 illustrates another example for your reference.

Figure 15-13 *PCESP ASN.1 Trace #2*

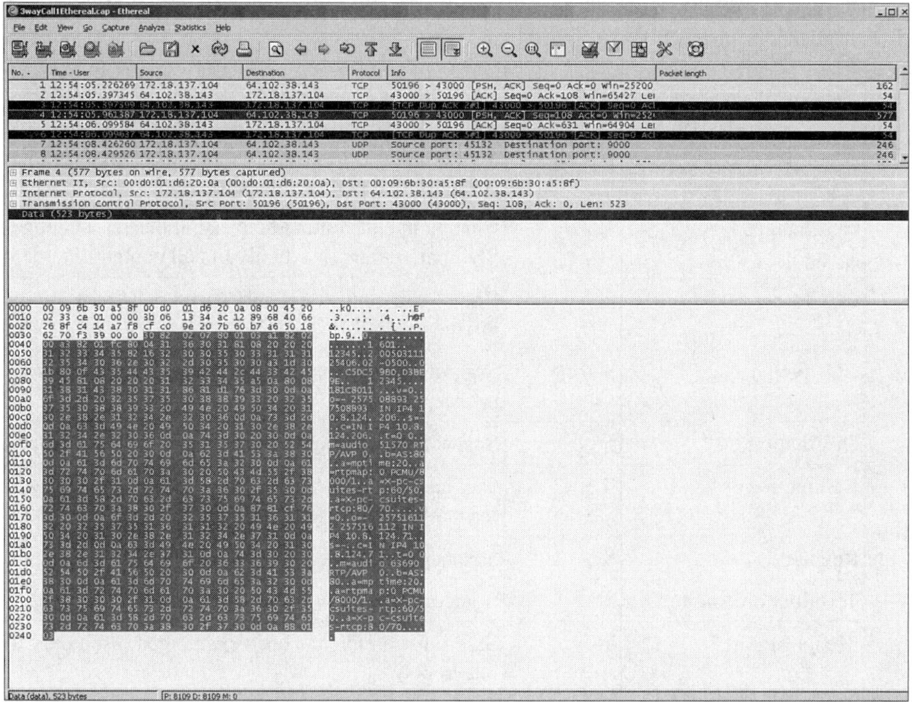

As you can see, this PCESP message is much bigger. If you look at the first TLV, you'll see it begins with 0x30820207. You know the 0x30 identifies the tag as a sequence of other tags, but pay attention to the next byte, 0x82. Here the length indicator bit mentioned earlier is set. This means that the following two bytes denote the length of this tag, in this case 0x0207, which in decimal equates to 519 bytes. The next tag, 0xa3, is the message type, which is also a constructed data type (tag class 10 = context specific, format 1 = constructed data, tag number 00011 = 3 [message type]). Message tag number 3 identifies a CCOpen message. Like the previous example the Case ID, Accessing Element ID, and Event Time parameters are included with this message. Also included is the CCC ID, which is the call content identifier used in the CCC channel packets you'll learn about in the next section. Finally, the SDP information from both gateways is included, as well as the direction of the traffic flow. The value 3 means traffic is flowing in both directions.

Lawful Intercept Call Flows

The call flows in this section illustrate how lawful intercept messaging is used between the IAPs and the DF, as well as between the DF and CF when a PacketCable subscriber is under surveillance. To keep the call flows manageable, some of the messaging not directly related to lawful intercept functionality (such as DOCSIS) is not shown. Also in these examples, the target under surveillance and the parties with which it is communicating are under control of the same CMS; consequently, no CMSS signaling is required.

Also only basic calls are covered here; more complicated call scenarios, such as call waiting, 3-way calling, call forwarding, etc., require additional messaging and are outside the scope of this book. Please consult the PacketCable specification for this information.

Suspect with Call Data Tap Originates Call

The next two figures show the lawful intercept messaging used when a subscriber under a call identifying intercept originates a call. This would be for the pen register and/or trap and trace surveillance types. Messaging from the point the lawful intercept target takes his or her phone off hook to the point the terminating party answers is shown in Figure 15-14.

Figure 15-14 *Call Identifying Intercept for Originating Party: Part I*

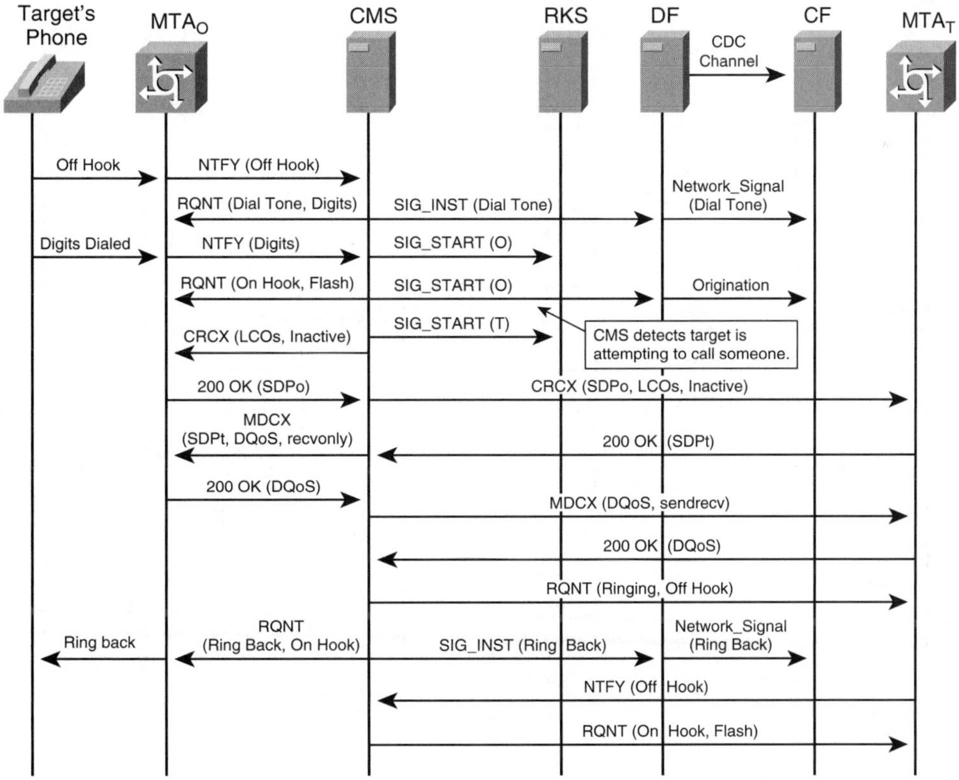

Target's Phone
MTA_O
CMS
RKS
DF
CDC Channel
CF
MTA_T
Off Hook
NTFY (Off Hook)
RQNT (Dial Tone, Digits)
SIG_INST (Dial Tone)
Network_Signal (Dial Tone)
Digits Dialed
NTFY (Digits)
SIG_START (O)
RQNT (On Hook, Flash)
SIG_START (O)
Origination
SIG_START (T)
CRCX (LCOs, Inactive)
CMS detects target is attempting to call someone.
200 OK (SDPo)
CRCX (SDPo, LCOs, Inactive)
MDCX (SDPt, DQoS, recvonly)
200 OK (SDPt)
200 OK (DQoS)
MDCX (DQoS, sendrecv)
200 OK (DQoS)
RQNT (Ringing, Off Hook)
Ring back
RQNT (Ring Back, On Hook)
SIG_INST (Ring Back)
Network_Signal (Ring Back)
NTFY (Off Hook)
RQNT (On Hook, Flash)

When the CMS requests the target's MTA to provide the dial tone signal, it also generates a Signaling Instance EM to the DF telling it that dial tone is being provided to the target. The DF then relays this information to the CF in a Network Signal PCESP message. After receiving a dial tone, the target dials the called party's phone number. This triggers an NCS NTFY message containing the called number digits to the CMS, which in turn triggers a Signaling Start EM for the originating call leg to be sent to the RKS. Inside this message are attributes such as the calling and called party phone numbers. This event message is also sent to the DF, which causes an Origination PCESP message to be sent to the CF. The call then proceeds as you've seen before: Connections are created on the gateways, and SDP information is exchanged; DQoS gates are authorized on the CMTSs; and the gateways are instructed to reserve these DQoS resources. After this, the CMS requests ring tone be provided to the called party and ring back tone be provided to the target. This triggers a Signaling Instance EM to the DF telling it that ring back tone is being provided to the target. In the last steps in this figure, the call is answered, triggering a NTFY off hook message sent to the CMS.

Figure 15-15 illustrates the next part of the call flow.

Figure 15-15 *Call Identifying Intercept for Originating Party: Part II*

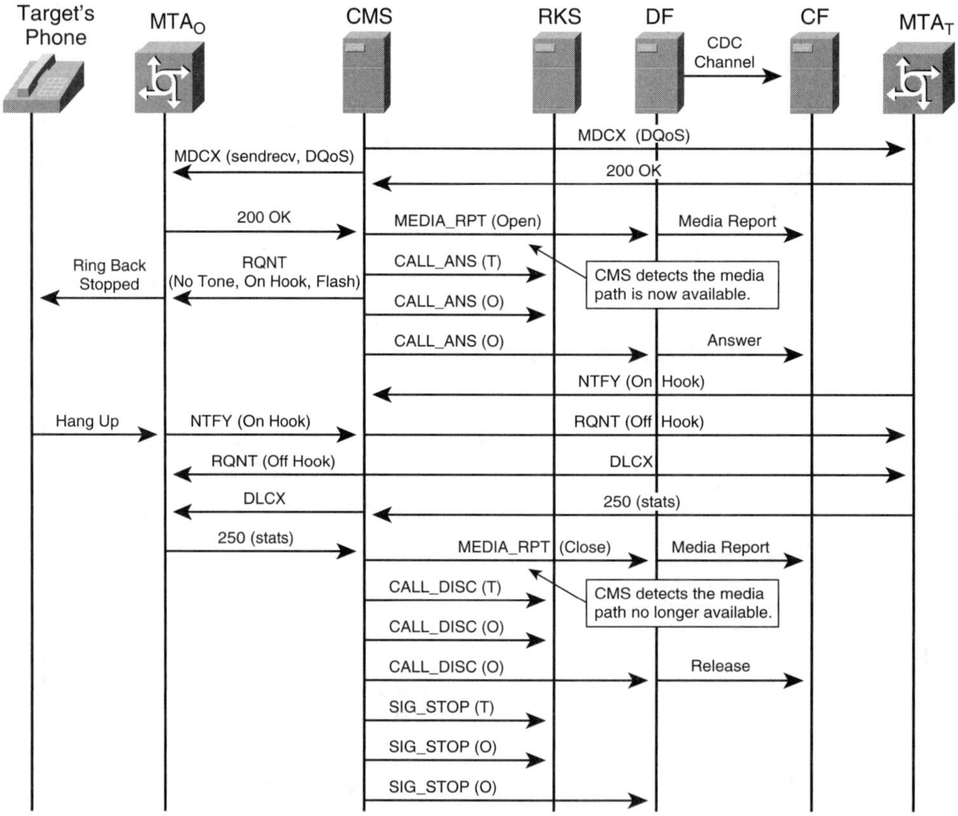

Because the call has been answered, the CMS instructs both gateways to commit their DQoS resources. When the CMS learns the target's resources have been successfully committed (via the receipt of the MDCX ACK), it sends a Media Report EM to the DF indicating the target's multimedia session is now open. Inside this message is the SDP information used by the gateways to exchange RTP traffic. The receipt of this message causes the DF to send a Media Report PCESP message to the CF. A Call Answer EM for the target's leg of the call is also forwarded to the DF, resulting in an Answer PCESP message to the CF. Eventually, both parties finish talking and hang up their phones. This triggers DLCX messages to the MTAs to delete the connections, which in turn triggers DOCSIS messaging to release DQoS resources. As soon as the CMS receives the DLCX ACK response packet from the target's MTA it sends another Media Report EM to the DF. This message indicates that the target's multimedia session has been closed. The DF then sends a Media Report PCESP indicating this to the CF. The call teardown process also triggers a Call Disconnect EM and a Signaling Stop EM sent to the DF. These messages indicate the reason for call teardown. The DF then sends this information to the CF in a Release PCESP message.

One final comment about this call flow is that the Media Report messages in this example indicate the target's multimedia session is opened and closed for both the upstream and downstream directions. Alternatively, separate Media Reports could have been used, each indicating the status of media for only the upstream or only the downstream direction.

Suspect with Call Data Tap Receives Call

Next, take a closer look at the case where this suspect under call identity surveillance receives a call. Again, two figures illustrate this messaging, with Figure 15-16 containing the first part.

Figure 15-16 *Call Identifying Intercept for Terminating Party: Part I*

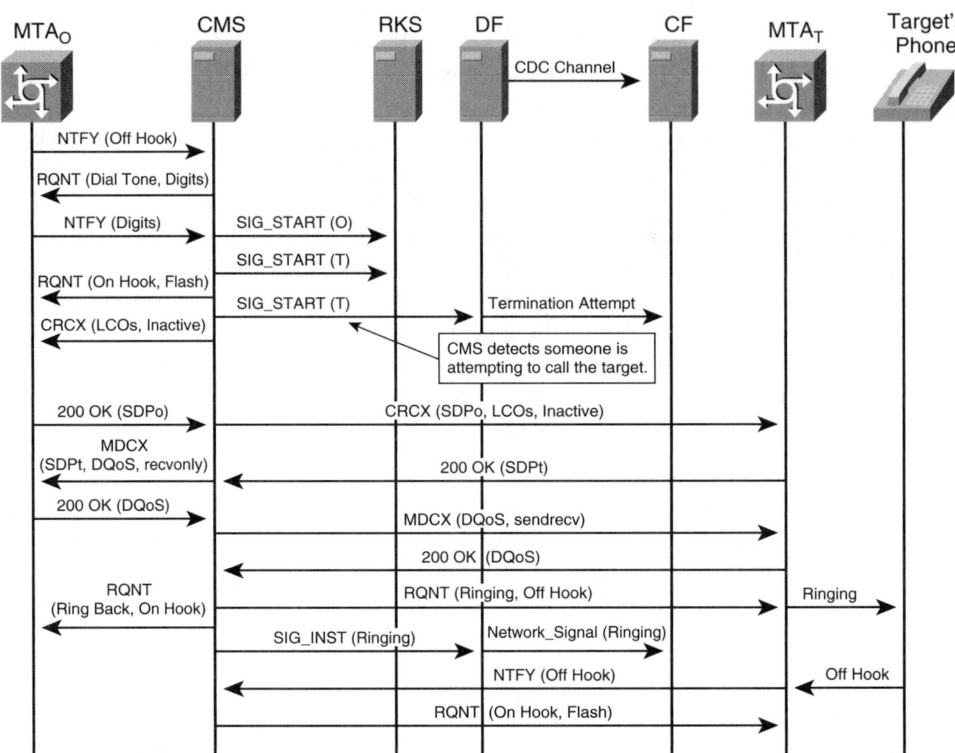

The main difference here is in the first PCESP message. In this case the Termination Attempt message is sent instead of the Origination message. This is because the Signaling Start EM received by the DF indicates the lawful intercept target is the terminating party. Also a Signaling Instance EM and Network Signal PCESP message are generated for ring tone being provided to the target.

Figure 15-17 shows the messaging for the remainder of the call.

As you can see, the messaging used here is nearly identical to the messaging shown in the second half of the previous call flow.

Also the same uses of the Media Report message(s) apply as in the previous example.

Figure 15-17 *Call Identifying Intercept for Terminating Party: Part II*

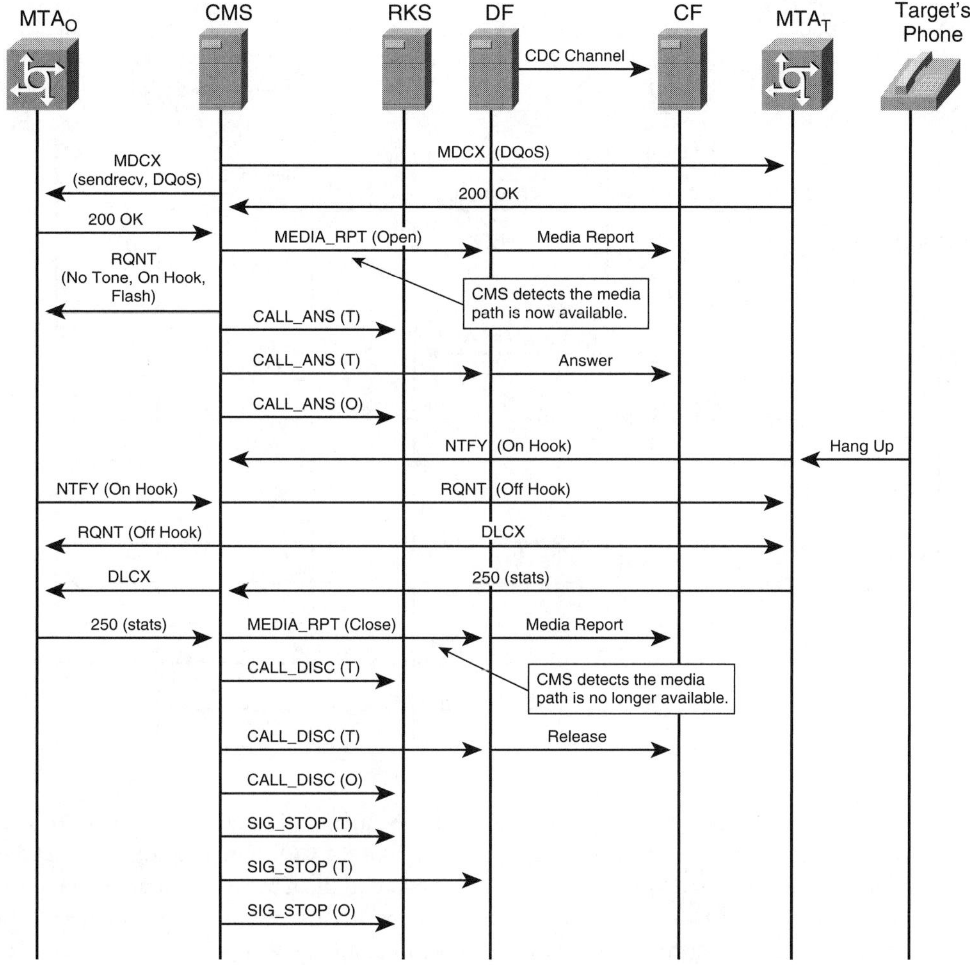

Suspect with Call Content Tap Originates Call

Now take a look at what happens if the subscriber is under a call content or "interception" type of surveillance. The lawful intercept messages seen in this case are split into three figures, with Figure 15-18 containing the first part.

Figure 15-18 *Call Content Intercept for Originating Party: Part I*

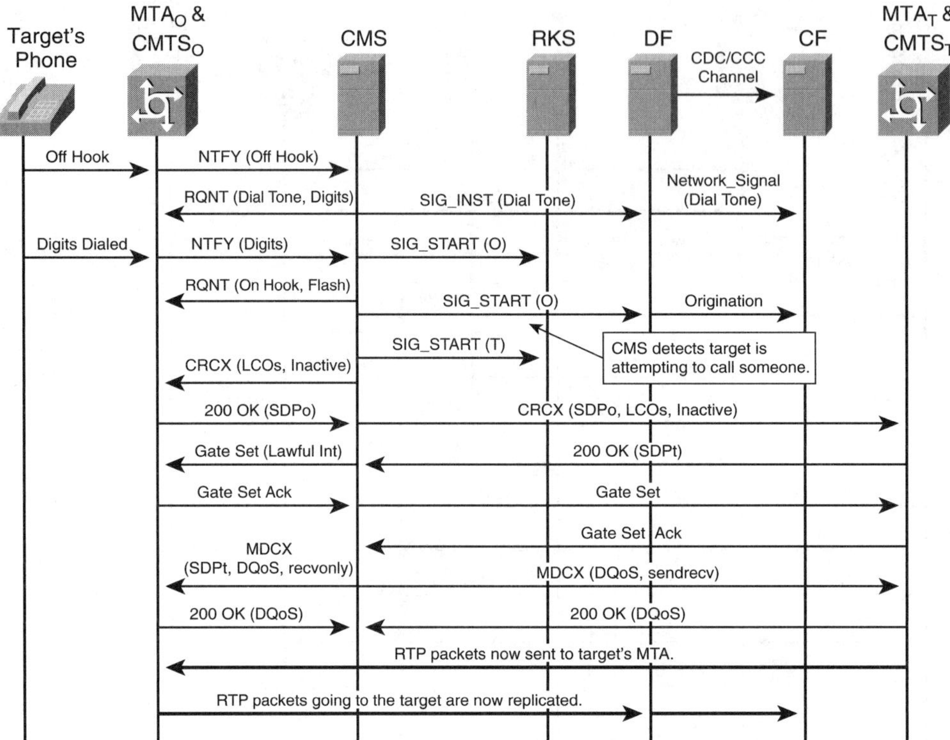

A majority of this messaging is the same as the call detail intercept example for the originating party. One exception is the inclusion of the COPS messaging between the CMS and CMTS devices. Notice the Gate Set message to the CMTS to which the target's MTA is connected includes the lawful intercept object. In this case, this object contains the IP address and UDP port number used to send call content packets, as well as the call content connection identifier (CCC-ID). The request to the CMTS to send call content is indicated by the DUP_CONTENT flag's being set. So at this point the CMTS has all the information it needs to send call content data to the DF. Recall that the call content data between an AF and a DF is simply replicated RTP packets prepended with the CCC-ID. Toward the end of Figure 15-18 you can see that the connection on the called party's MTA is set to send/receive. This enables packets sent by the called party's MTA to be received by the target's MTA. Because a call content intercept is active on the CMTS, the CMTS replicates these packets to the DF. The DF then prepends a timestamp to these packets, indicating the time received, and sends them to the CF.

Figure 15-19 illustrates the messaging for the next phase.

Figure 15-19 *Call Content Intercept for Originating Party: Part II*

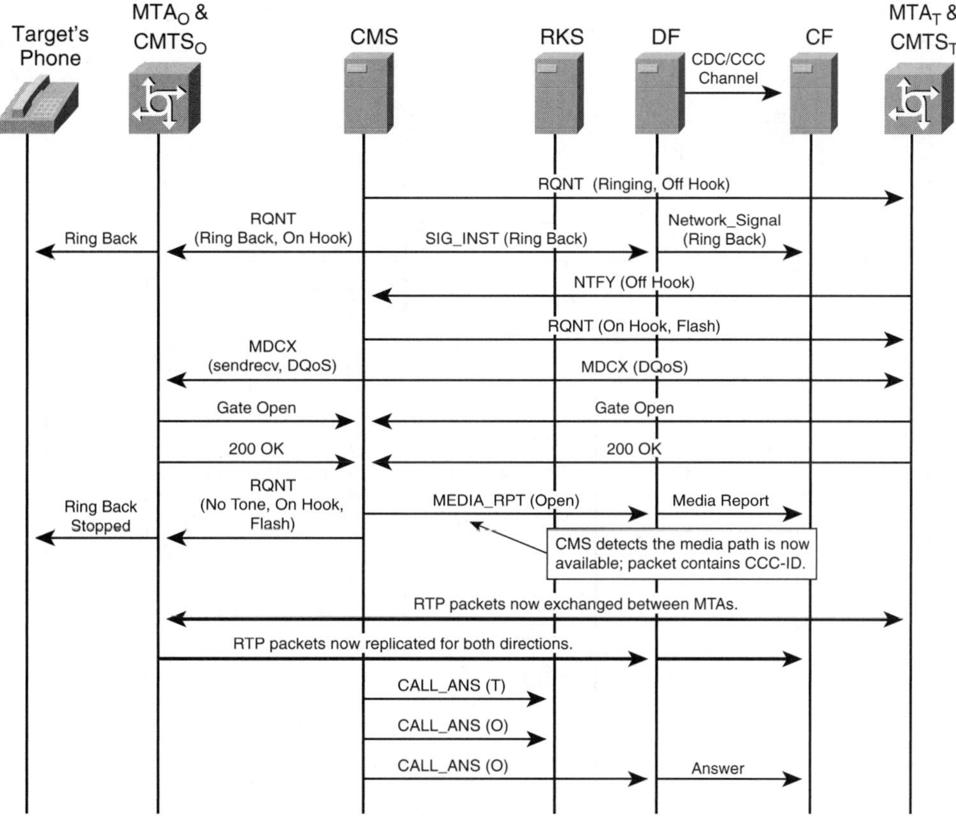

Again, you have seen much of this messaging before. When the media stream resources for the target have been committed, indicated by the acknowledgement of the MDCX packet to commit DQoS resources and change the connection mode to send/receive, a Media Report (Open) EM is sent from the CMS to the DF. You can see one attribute in this message that wasn't in this message in the call detail examples. This is the CCC-ID, which allows the DF (and ultimately the CF) to be able to correlate call detail data with call content data. The DF forwards this identifier to the CF in the Media Report PCESP message it sends. As soon as the media stream is activated in both directions, RTP packets are exchanged between the target's MTA and the called party's MTA. The CMTS replicates both directions of this RTP traffic to the DF, which sends this information to the CF in the CCC channel.

Figure 15-20 illustrates the final part of this messaging.

Figure 15-20 *Call Content Intercept for Originating Party: Part III*

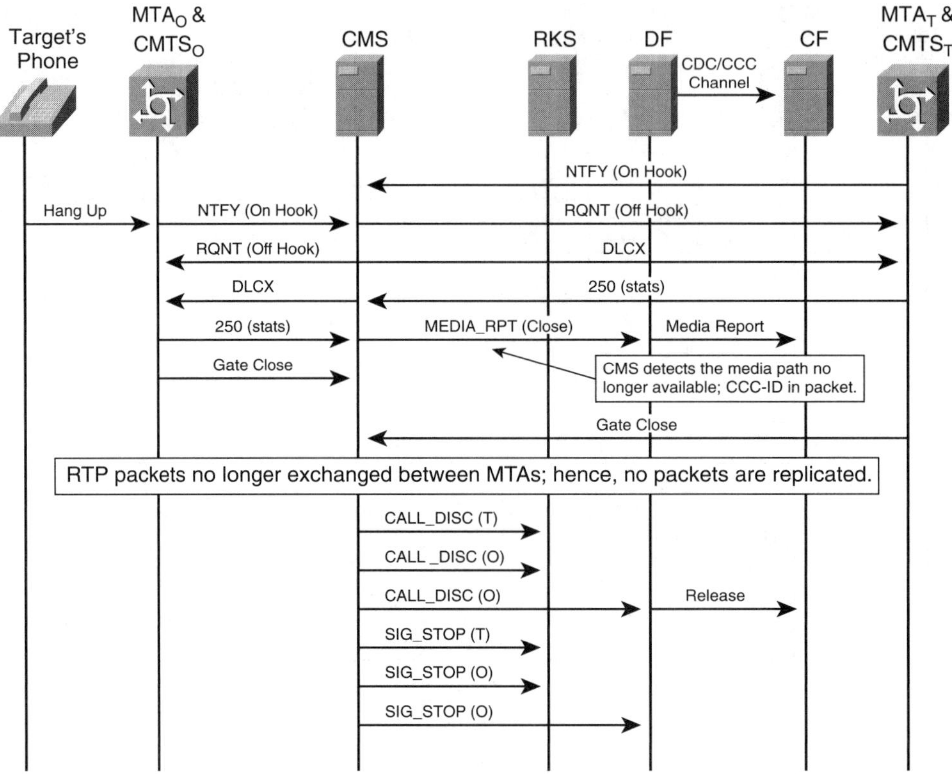

The messaging in this figure should be self-explanatory.

Suspect with Call Content Tap Receives Call

The final CDC call flow example is when the same suspect as in the previous example receives a call. Figures 15-21, 15-22, and 15-23 depict this call flow.

Figure 15-21 *Call Content Intercept for Terminating Party: Part I*

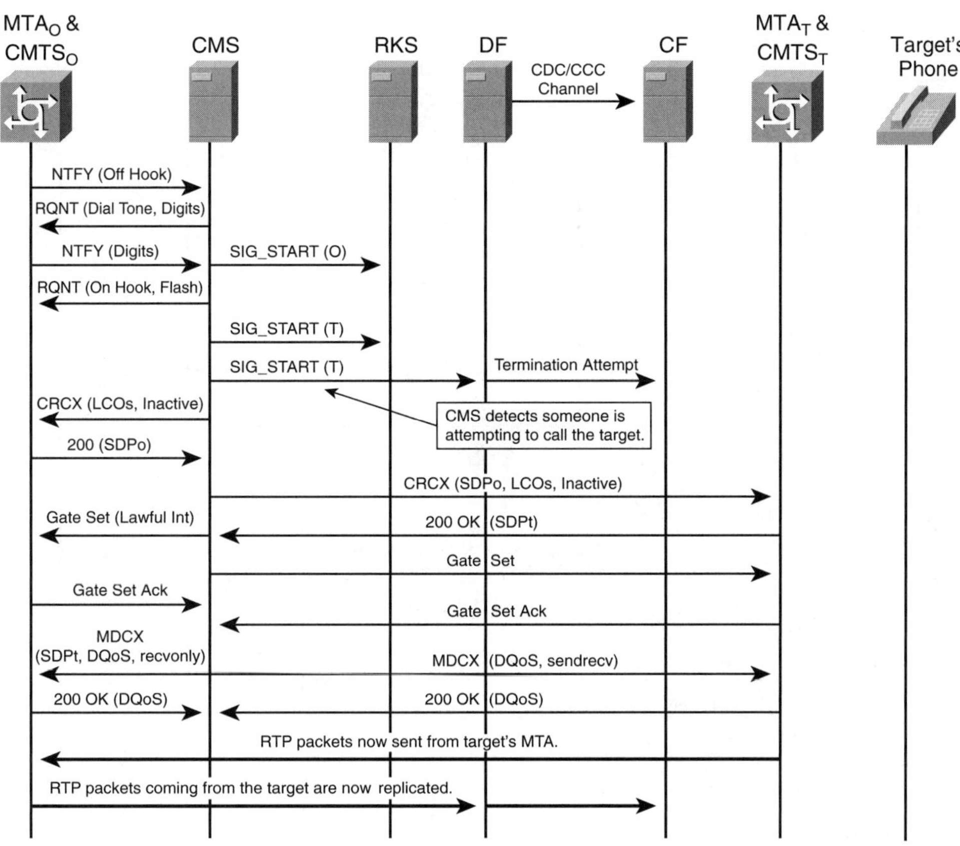

Figure 15-22 *Call Content Intercept for Terminating Party: Part II*

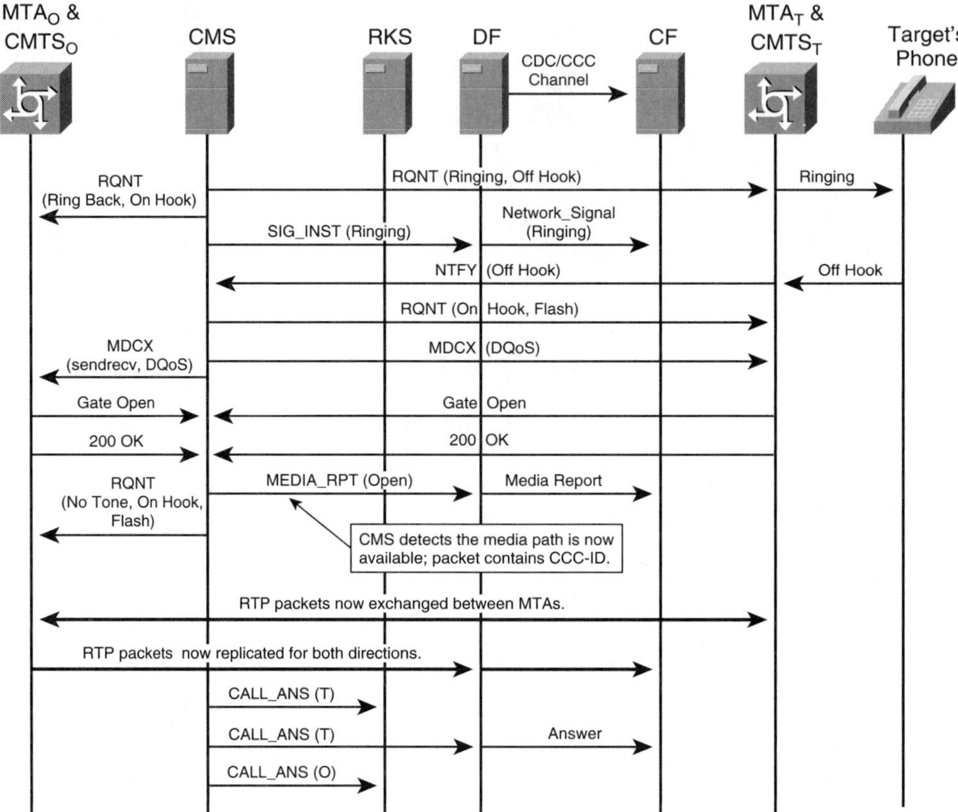

By now the figures should be self-explanatory. One thing to note is where the call content intercept is made. Notice it is made at the CMTS to which the originating party is connected, not the CMTS to which the target is connected. In other words, the Gate Set message to authorize resources for the originating party has the lawful intercept object, not the Gate Set message to authorize resources for the target. This is because the BTS sets up the call content intercept at the entry point in the network, which in this case is the CMTS to which the originating party's MTA is connected.

Figure 15-23 *Call Content Intercept for Terminating Party: Part III*

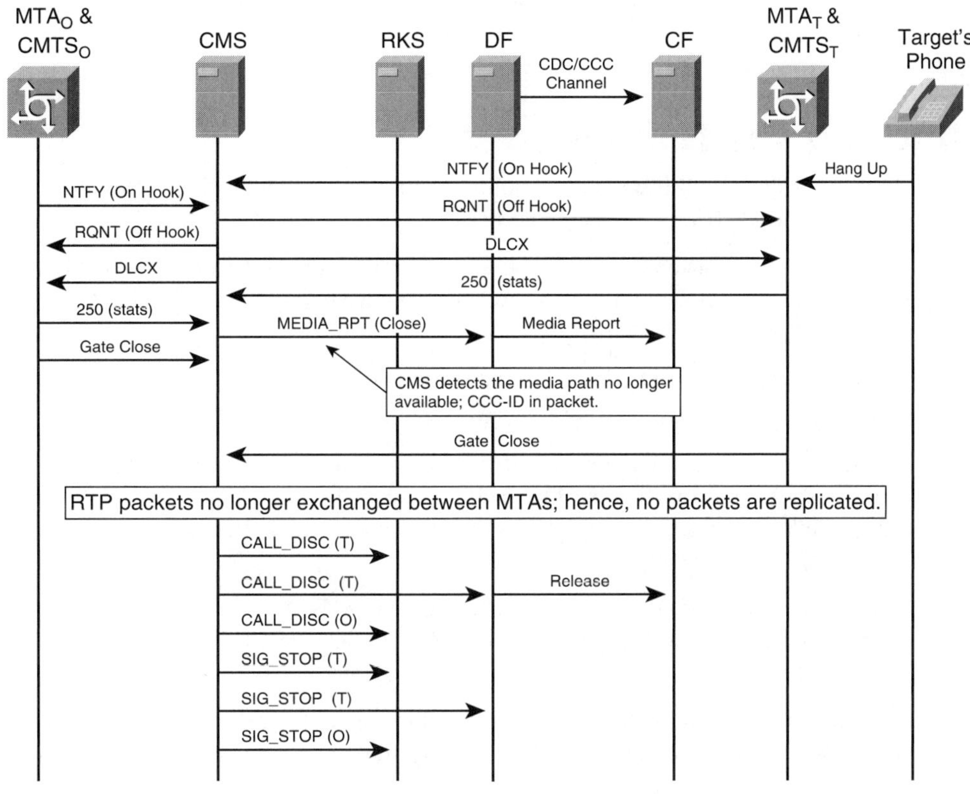

Capturing Call Content

The format of the packets sent from the DF device to the CF device in the CCC channel is rather simple. The RTP packets from the tapped subscriber's conversation are prepended with an identifier and a timestamp and then sent in UDP/IP packets.

NOTE For the CCC channel, UDP is the transport layer protocol, as opposed to the CDC channel, which uses TCP.

The identifier is the 4-byte CCC identifier mentioned earlier. Remember, this identifier is also included in the Media Report PCESP packets, as shown in the previous examples. The CCC-ID is generated by the DF device, and it is important to remember that the CCC identifier is the same for all packets associated with a conversation, regardless of whether they are sent from the target or to the target. The timestamp field is an 8-byte NTP timestamp that reflects the time at which the corresponding information was detected by the DF.

Configuring Lawful Intercept

Hopefully, at this point you have a solid understanding about how electronic surveillance works in a PacketCable network.

The next part of this chapter covers the configuration on the Cisco BTS10200 related to electronic surveillance or lawful intercept. You have two types of BTS lawful intercept configuration:

- Configuration of DF devices done manually
- Dynamic configuration of subscriber taps

For the latter type, the DF device initiates the provisioning interface to the CMS provisioning platform (that is, BTS EMS) securely. For the Cisco BTS10200 this is done using the Secure Shell protocol. After the DF device establishes this session, it enters the necessary BTS provisioning.

The configuration of DF devices is performed on the BTS under a different user than normal provisioning; by default, the CALEA user is the only user with access to the lawful intercept tables.

The Electronic Surveillance Server (ESS) table contains the definition of the DF devices. Records must be added to this table for the destination ports of call detail and call content information as follows:

```
CLI>add ess cdc_df_address=IP address of DF device; cdc_df_port=UDP port of DF
  device; use-packet-cable-iap=y;
```

Table 15-2 lists the key fields.

Table 15-2 *ESS Fields*

Field	Default Value	Recommended Value	Description
CDC-DF-ADDRESS	N/A	N/A	FQDN or IP address of the DF device where call detail data is sent
ACC-REQ-RETRANSMIT	3	3	Number of retransmissions of unacknowledged RADIUS Accounting Request packets
ACC-RSP-TIMEOUT	2 secs	2 secs	Length of time the BTS waits for an Accounting Response packet before it considers the Accounting Request unacknowledged and retransmits

Table 15-2 *ESS Fields (Continued)*

Field	Default Value	Recommended Value	Description
CDC-DF-PORT	1813	N/A	UDP port on the DF device where call detail data is sent
USE-PACKETCABLE-IAP	N	Y	Whether the not the BTS figures out and signals a PacketCable IAP when call content needs to be duplicated

Some other aspects about BTS configuration deserve some discussion. On the definition of the CMTS (the AGGR-ID table), make sure the ES-SUPP and ES-EVENT-SUPP fields are set properly. The ES-SUPP field indicates whether the CMTS supports electronic surveillance, and the ES-EVENT-SUPP field indicates whether the CMTS supports PacketCable Event Messaging for electronic surveillance (this is usually disabled). Another area to watch out for is TGCP/MGCP hair pinning on the MG component. Hair pinning is when a call originates and terminates on the same gateway; in this case, the call agent can signal the gateway to internally connect the two endpoints, eliminating the need for converting the media streams to and from RTP packets. As you might guess, in this case the gateway cannot duplicate RTP packets because none exist to duplicate. Set the MGW-PROFILE table field MGCP_HAIRPIN_SUPP to N to protect against this possibility.

As for the configuration of other Cisco components, you won't find much to it because they are given lawful intercept information via call signaling packets. For the Cisco VXSM and VISM cards, the software version itself tells you whether CALEA functionality is supported. For the VXSM card you can verify whether CALEA is enabled using the **dspxgcpcalea** command, as shown (if this command doesn't work, this indicates you are running a software version that does not support CALEA):

```
MGX8800.6.VXSM.a > dspxgcpcalea
======================================================
                  CALEA Status
======================================================
CALEA                  :  enable
```

You can enable CALEA functionality using the **cnfxgcpcalea** command, and you can display information about any active CALEA calls using the **dspxgcpcaleacalls** command.

This can be verified using the **version** command on the VISM card; CALEA functionality must also be enabled using the **cnfcalea** command. The following is an example of VISM lawful intercept verification:

```
MGX8800.1.3.VISM8.a > version
***** Cisco Systems. MGX VISM Card *****
    Firmware Version    = 003.053.025.201 (WITH CALEA)
    Backup Boot version = VI8_BT_3.2.00
    Xilinx Firmware version = 10/ 2/1998
    DSPCOM FPGA version     =  4/24/2005
<Remainder of output omitted>
MGX8800.1.3.VISM8.a > dspcalea
CALEA : enable
```

Troubleshooting Tips

The last part of this chapter contains some troubleshooting tips that might help you in diagnosing and fixing problems relating to lawful intercept.

As you've seen throughout the book, one of the best ways to troubleshoot is to simply look at the messaging. Of course, if these messages are encrypted, you have to find some other means to troubleshoot, such as examining call agent log files.

If you aren't seeing any messages between an IAP and the DF, you most likely have a problem with IP routing or DNS resolution. Check routing tables, DNS configuration, and firewalls. If you are using the BTS, make sure the CDC_DF_ADDRESS configured in the ESS table is resolvable and reachable. Also make sure this address and the UDP port specified with the CDC_DF_PORT entry are not blocked. If you see RADIUS accounting request messages sent out but no accounting response messages coming back, more than likely a problem exists on the DF device, or you have a network connectivity issue. Check with your documentation for that device or the manufacturer for assistance there.

If you are having a problem with call content delivery from an IAP, debugging the COPS or TGCP messaging can help here. If you are using a Cisco CMTS, the **debug packetcable gate control** command shows you the COPS Gate Set message. You should see the lawful intercept object in this message. Refer to Chapter 13 for more detail on these **debug** commands. If you don't see this object and the Cisco BTS is your CMS, check that you have configured for lawful intercept support on the AGGR-ID and that the USE_PACKETCABLE_IAP option is set to Y on your ESS table entry. If you aren't seeing the lawful intercept parameters in the TGCP or MGCP messaging and you are using the Cisco VXSM or VISM card, check the commands mentioned in the configuration section. The BTS determines whether an MG supports lawful intercept by sending an AUEP to the gateway for capabilities. Consequently, if the gateway is not configured to support lawful intercept (or if the BTS has not performed the AUEP exchange with it), the BTS will not attempt to capture call content on the gateway.

Problems in the DF-to-CF communication are a little trickier to troubleshoot. Make sure everything is configured properly on the DF. If this looks correct and you still have issues, you probably need to work with the LEA responsible for the CF device.

Finally, if you are having issues dynamically creating taps on your CMS, check the CMS provisioning configuration on the DF device. For example, with the BTS CMS make sure secure shell is configured with the right IP address and login information.

Chapter Summary

The distributed nature of the PacketCable network makes the accounting of resources difficult because no one device used to establish multimedia sessions has both the call signaling information and the bearer path data. This is one of the motivating factors behind the development of PacketCable Event Messaging. With Event Messaging, different devices independently send accounting information to a Record Keeping Server (RKS). The RKS correlates this information and generates call detail records (CDR), which can then be transferred to a billing mediation device.

PacketCable Event Messaging uses the RADIUS Accounting protocol for the transport of event messages (EM). PacketCable-specific information is encoded as vendor-specific attributes in RADIUS Accounting-Request messages. Event messages are created independently for the originating and terminating legs of a call. An attribute known as the Billing Correlation Identifier (BCID) correlates event messages for a call. Logically, event messages are triggered by events such as the start and stop of call signaling and the activation and deactivation of a media stream.

The identity of the RKS is configured on the CMS and MGC components, such as the Cisco BTS 10200, and sent to the CMTS component during the DQoS authorization process. Consequently, a majority of the configuration tasks related to Event Messaging occur on the CMS/MGC components. Troubleshooting event messages should begin with verifying component configuration/operation and network connectivity.

In most countries, telecommunications service providers are required to provide capabilities allowing authorized law enforcement agencies (LEA) to initiate surveillances of their subscribers. This requirement also exists in PacketCable. A PacketCable deployment must be able to support both call identity and call content surveillance types. Call identity surveillance types capture the details about a suspect's call, whereas the call content surveillance types involve capturing the voice media streams and call detail data.

The PacketCable components involved in electronic surveillance are known as Intercept Access Points (IAP). The CMS and MGC are IAPs for call identifying information, whereas the CMTS and MG are IAPs for call content surveillances. It is common practice to classify the functions of lawful intercept into four categories, and the IAPs perform the access function (AF). The delivery function (DF) is performed by a separate device that collects all the surveillance data generated by the IAPs. This device is controlled by the PacketCable MSO and is a mediation server between the MSO and the LEAs requesting surveillances. The LEAs use a device that collects data from the MSO's DF devices; this device is known as the Collection Function (CF) device. The fourth function of lawful intercept is the administrative function and involves the provisioning and control of the other devices.

When a new intercept is required, it is provisioned on the DF device. The DF then initiates a provisioning connection to the PacketCable CMS provisioning interface and sets up the intercept. If the CMS needs to involve either the CMTS or the MG component in the intercept, it does so in the COPS or TGCP messaging. Call identifying information is sent from the IAPs to the DF device using PacketCable Event Messaging. Call content information is sent from the IAPs to the DF by encapsulating duplicated RTP packets inside UDP packets appended with a call content connection identifier (CCC-ID).

As the DF device is collecting information, it is also forwarding this information on to the appropriate CF device. Call identifying or call detail information is sent in a CDC channel to the CF device. The CDC channel is a TCP connection in which call detail information is sent using the PacketCable Electronic Surveillance Protocol (PCESP). These packets consist of Tag/Length/Value parameters encoded using the ASN.1 syntax. Call content information is sent in a CCC channel to the CF device. The CCC channel is a UDP connection, and its packets are of the same form as the call content packets coming from the IAPs, except for the addition of a timestamp field.

Lawful intercept configuration on the Cisco BTS 10200 CMS is relatively simple. Much of the work in lawful intercept configuration is on the DF device. Cisco does not manufacture the DF device but instead works with third-party companies. Troubleshooting tasks are similar to Event Messaging troubleshooting; verifying connectivity and analyzing packet traces are good places to start.

Chapter Review

1 True or False: The Direction Indicator attribute is the first CableLabs attribute in every Event Messaging packet.

Answer: False. The EM Header attribute is the first CableLabs attribute in EM packets.

2 True or False: The Service Instance EM attribute is used when a *XX VSC code is dialed to activate a call feature.

Answer: False. The Service Activation EM is used for this.

3 True or False: It is strongly discouraged not to configure the BTS for the simultaneous generation of EMs and call detail blocks (CDBs).

Answer: True. Doing so can impact performance.

4 True or False: The PacketCable MSO is not responsible for providing or managing the CF device.

Answer: True. This is the responsibility of the LEA.

5 True or False: The Event Messaging sent between an IAP and a DF can be sent real time or in bulk fashion, as with Event Messaging to an RKS server.

Answer: False. Call identifying event messaging must be sent real time.

6 True or False: The ASN.1 language is a standardized flexible protocol for defining complex data types.

Answer: True

7 True or False: In a call content surveillance the PCESP message Media Report is never used in the DF-CF communication channel.

Answer: False. Nowadays this message is typically used instead of the CCOpen, CCChange, and CCClose.

8 True or False: On the Cisco BTS 10200 CMS any user can provision the lawful intercept tables.

Answer: False. Only specially privileged users have access to these tables.

9 What are the three functions provided by the RADIUS protocol?

Answer: Authorization, authentication, accounting

10 Identify an application where the use of real-time Event Messaging is especially useful.

Answer: Fraud detection, debit-based calling card

11 List three event message types that are used only for lawful intercept.

Answer: Conference Party Change, Media Report, Redirection, Signal Instance, Surveillance Stop

12 What is the terminology for the three types of surveillance? Describe each.

Answer: Pen register for call identifying data on calls originated by suspect; "Trap and trace" for call identifying data on calls received by suspect; and "Interception" for call content on all suspect calls

13 Why is the PacketCable MTA not a valid device for capturing call content?

Answer: The MTA is an untrusted device.

14 Describe a scenario where call content is not captured by a CMTS.

Answer: The case where a suspect has a phone forwarded to an OFFNET number and a call destined for the suspect comes in from the PSTN—in this case, no CMTS is in the voice path, and the MG component becomes an IAP—or if using the BTS and a call originates from the PSTN.

15 How is an electronic surveillance configured on a PacketCable CMS?

Answer: Dynamically by the DF device

16 What is the function of the two flags in the lawful intercept object of a Gate Set message?

Answer: One indicates whether event messages are duplicated to a DF, and the other indicates whether RTP packets are duplicated to a DF.

17 Suppose you have a PCESP object that is 500 bytes in length and consists of a sequence of other objects. Excluding the data portion, what would this TLV look like in hexadecimal format?

Answer: 0x308201F4 (500 in hex is 0x1f4)

18 If the MGC is not sending lawful intercept options to a VISM or VXSM card when it needs to, what is the probable cause?

Answer: The software version does not support CALEA, or CALEA functionality is not enabled.

CHAPTER 16

PacketCable Network Design Considerations

At this point you have been introduced to all the major components, protocols, and technologies associated with a PacketCable deployment. However, you should be familiar with some information not yet addressed. This chapter serves two primary purposes:

- To serve as a catchall for these concepts
- To provide you with some important considerations when you are designing your PacketCable network

Examples of these design considerations include security, traffic engineering, quality of service (QoS), and high availability requirements. The security portion of this chapter covers the mechanisms used by PacketCable to provide data confidentiality and data integrity for the protocols used to communicate between components. The traffic engineering portion covers how to determine voice call capacity throughout a PacketCable network. Much of this focus is on the typical bottleneck in the network, the Data-Over-Cable Service Interface Specification (DOCSIS) upstream channel. The QoS considerations address the treatment of all the different types of packets present in a PacketCable network. How these packets are treated can greatly affect the quality of service perceived by your customers. Similarly, ensuring your network is constantly available and operational is also vital in keeping your customers happy. Redundancy techniques used to ensure high availability throughout the network address these concerns. Finally, at the end of this chapter you find a brief discussion about the future direction of PacketCable. This includes developments in DOCSIS as well as some of the key features that will be available with PacketCable Release 2.

Security Considerations

The PacketCable security specification defines security requirements for every signaling and media communication channel in the PacketCable network. Figure 16-1 is a diagram illustrating all the signaling and media channels discussed in this book.

Figure 16-1 *PacketCable Signaling and Media Channels*

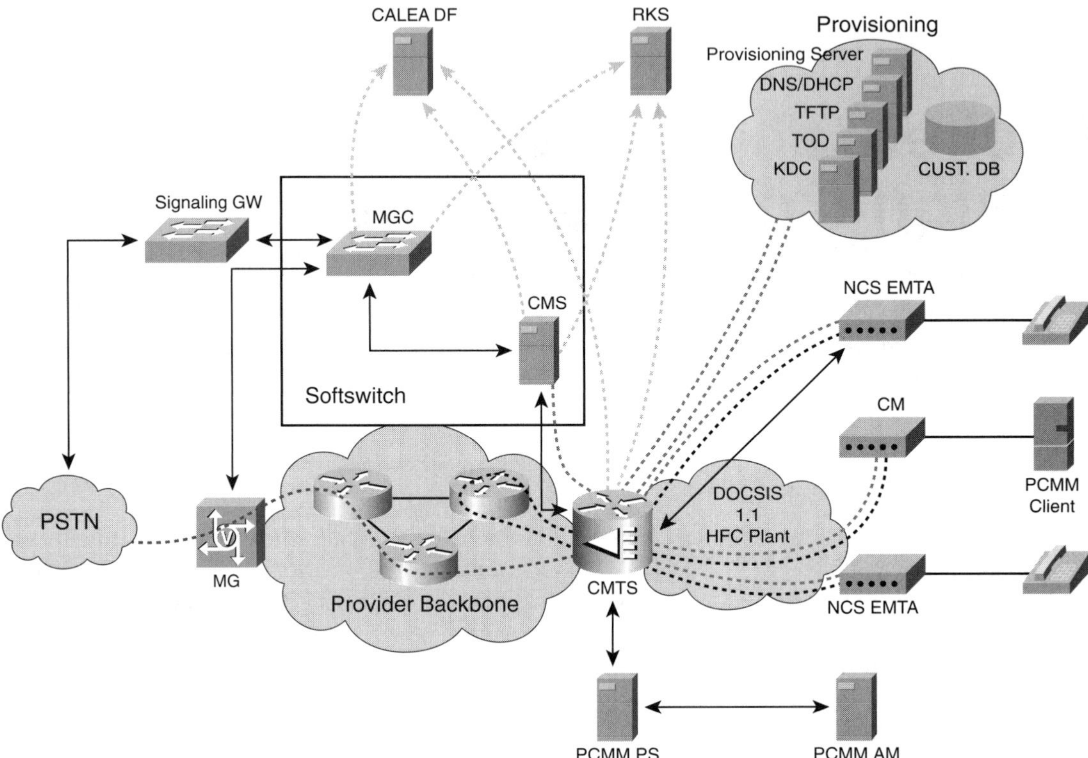

The PacketCable security requirements include

- Baseline Privacy Initialization Plus (BPI+) encryption for Data-Over-Cable Service Interface Specification (DOCSIS) packets between the multimedia terminal adapter (MTA) and cable modem termination system (CMTS)
- Kerberos-based Simple Network Management Protocol version 3 (SNMPv3) for MTA provisioning packets
- Kerberos-based IPsec for MTA-to-call management server (CMS) call signaling packets (Network-based Call Signaling [NCS])
- Internet Key Exchange (IKE) or Kerberos-based IPsec for Media Gateway (MG) to Media Gateway Controller (MGC) call signaling packets (Trunk Gateway Control Protocol [TGCP])
- Session Initiation Protocol (SIP) Transport Layer Security (TLS) for CMS-to-CMS call signaling packets (Call Management Server Signaling [CMSS])

- IPsec for Real-time Transport Protocol (RTP) and Real-time Transport Control Protocol (RTCP) media packets sent and received by the MTA and MG components (security keys in this case are communicated by the CMS/MGC in call signaling messages)
- IKE or Kerberos-based IPsec for Common Open Policy Service (COPS) packets between the CMTS and CMS, CMTS and PacketCable Multimedia (PCMM) policy server, and PCMM policy server and application manager
- IKE or Kerberos-based IPsec for Event Messaging packets to and from the Record Keeping Server (RKS)
- IKE or Kerberos-based IPsec for Event Messaging packets to and from the lawful intercept Delivery Function (DF)

NOTE In reality, a majority of today's deployments implement very few of these requirements. Typically, only BPI+ and provisioning security are implemented. This is primarily because of the Multiple Systems Operator (MSO) choosing to live with the security risks in exchange for the reduced network complexity.

Types of Security Requirements

PacketCable security requirements protect against a number of potential threats, such as

- Theft of network services
- Theft of bearer channel information
- Theft of signaling channel information
- Service disruption threats
- Repudiation

Theft of Network Services

In Chapter 3, "Provisioning Overview," and Chapter 4, "Provisioning Flows," you learned how the secure method of PacketCable provisioning protects against the possibility of MTA clones stealing service by masquerading as legitimate MTAs. The use of digital certificates and the authentication capabilities provided with the use of Kerberos and SNMPv3 greatly minimize this risk. Conceivably, other devices such as the Key Distribution Center (KDC), SNMP provisioning entity, and CMS could also be cloned in an attempt to steal service.

As with the MTA cloning case, using the secure method of PacketCable provisioning helps immensely in minimizing the chances of the provisioning elements (KDC, SNMP provisioning entity, and so on) from being cloned.

A PacketCable MTA responds only to NCS messages from call agents configured in its configuration file. Thus, an MTA does not respond to messages from unknown CMS devices. However, an MTA could potentially respond to a CMS masquerading as a legitimate (for example, configured) CMS. Mechanisms such as IPsec-based secure NCS protect against this possibility.

Recall that the cable modem and MTA provisioning processes include configuration file hash checks used to verify that the contents of these files have not been tampered with. The CMTS also does a hash check when a modem registers to protect against rogue cable modems trying to manipulate their configuration. The use of PacketCable distributed quality of service (DQoS) ensures that only DOCSIS service flow requests that have associated gates are successful. These gates also make sure that these requests cannot be for more QoS than what is authorized by the CMS or PCMM policy server. All of these checks ensure that users are only getting the services authorized.

By simply not paying his or her bill, a customer can in a sense steal network resources. Consequently, PacketCable devices (such as the CMS and provisioning system) are typically designed so that it is very simple to disable services for nonpaying customers. Also you can implement "pay as you go" services to make sure services are properly paid for.

Theft of Bearer Channel Information

MTA clones can potentially be used to listen in on other user's conversations by diverting traffic intended for the actual MTA. Again, the use of the secure method of provisioning greatly reduces the risk of this happening.

Recall that the DOCSIS network is a shared network, so potentially one cable modem could be manipulated to see the traffic of another. The use of BPI+ encryption across the DOCSIS network makes user traffic meaningless to a device that does not have the decryption mechanisms; therefore, you are encouraged to use BPI+ for all MTAs. BPI+ encrypts the packets only across the DOCSIS network; after they leave the CMTS, this encryption is gone. Consequently, other mechanisms such as IP Security encryption of RTP packets must be used to secure bearer information end to end.

Theft of Signaling Channel Information

A certain amount of information can be obtained from call signaling packets (that is, NCS, TGCP, CMSS). Information such as called and calling party phone numbers and calling party name can potentially be seen by unwanted parties (such as telemarketers). Encrypting these packets greatly minimizes this possibility.

Service Disruption Threats

These types of threats include denial-of-service (DoS) attacks. For example, a malicious user could attempt to flood a DOCSIS upstream with traffic, making it difficult for legitimate traffic to pass. Mechanisms such as PacketCable DQoS help ensure that only authorized users can use DOCSIS resources.

A user with access and enough knowledge might also issue call signaling messages to prematurely end active sessions. Again, encrypting call signaling packets and DQoS packets protects against this possibility.

Repudiation

Repudiation is the refusal of one party to fulfill a contractual obligation to another party because they feel their service contract has been falsified. For example, if MTA cloning were common in a PacketCable network, customers might refuse to pay, claiming their service has been compromised. The provisioning, security, and call signaling mechanisms in PacketCable make it very difficult for someone to compromise MTA identity, thus greatly reducing the risk and the viability of repudiation claims.

Usage of the IP Security Architecture (IPsec)

As you can see, a majority of the PacketCable security mechanisms utilize the IP Security architecture. IPsec provides security for the protocols that are within the IP layer payload (that is, TCP and User Datagram Protocol [UDP] layers and above).

The security services provided by IPsec include

- **Data Confidentiality**—The IPsec sender can encrypt packets before transmitting them across a network.
- **Data Integrity**—The IPsec receiver can authenticate packets sent by the IPsec sender to ensure that the data has not been altered during transmission.
- **Data Origin Authentication**—The IPsec receiver can authenticate the source of the IPsec packets sent. This service is dependent upon the data integrity service.
- **Anti-Reply**—The IPsec receiver can detect and reject replayed packets. It is a security service where the receiver can reject old or duplicate packets to protect itself against reply attacks. IPsec provides this optional service by use of a sequence number combined with the use of data authentication.

IPsec uses two protocols to provide security services:

- **Authentication Header (AH)**—Provides an authentication service to the entire IP packet, including the IP header.

- **Encapsulation Security Payload (ESP)**—Provides a data encryption service to IP payload, which includes the upper-layer protocols (that is, TCP, UDP). ESP packets are encapsulated in IP packets, and new IP headers are created (using IP protocol 50). ESP also provides an authentication service to the ESP header and encrypted packet, but not for the IP header.

Only the ESP protocol is used with IP Security in PacketCable because use of the AH protocol isn't deemed to provide much additional security for a PacketCable network.

IPsec operates in two modes:

- **Transport**—Transport mode is used when the end hosts are encrypting/decrypting the packets. In this case, only the IP payload is encrypted, and the original header is left intact. Only transport mode is used with IP Security PacketCable.
- **Tunnel**—Tunnel mode is used when intermediary devices are encrypting/decrypting the packets. The entire original packets are encrypted and new IP headers are added.

IPsec enables the establishment of Security Associations (SA) between entities (that is, PacketCable devices). An SA is unidirectional and contains the set of information enabling secure communication between entities; it includes a transform identifier and the shared secret encryption and authentication keys used for protecting the traffic. A transform identifier specifies the operations done on a dataflow to provide data authentication, data confidentiality (encryption), and optionally data compression. Several transform identifiers are identified in the PacketCable security specification, but only Triple DES and Null encryption with the ESP protocol must be supported.

PacketCable also mandates two authentication algorithms: Hashed Message Authentication Code (HMAC) Message Digest 5 (HMAC-MD5-96) and Hashed Message Authentication Code (HMAC) Secure Hash Algorithm 1 (HMAC-SHA-1-96). These restrictions apply to the signaling communication channels in a PacketCable network. For RTP security the Advanced Encryption Standard (AES) algorithm must be supported as well as the Multilinear Modular Hash (MMH) authentication algorithms MMH2 and MMH4. It is important to note that the encryption algorithms used for RTP to do not impose any additional overhead on packet size (the payload is simply encrypted). The authentication algorithms add a small amount of overhead because of the addition of a Message Authentication Code (MAC). This is 2 bytes if MMH2 is used and 4 bytes if MMH4 is used. For RTCP security support of the AES Cipher Block Chaining (AES-CBC) encryption algorithm and the HMAC-SHA1-96 authentication algorithm is mandated.

In PacketCable, SAs are established dynamically via a key exchange protocol (such as IKE or Kerberos) or in call signaling messages (this is used for RTP/RTCP security). As you learned in the provisioning chapters, the Kerberos protocol uses a third-party device known as the Key Distribution Center (KDC) to distribute security keys to other devices. The Internet Key Exchange (IKE) protocol was also briefly described in Chapter 3. In IKE, key management is handled directly between the parties involved. This is possible because the parties know the identity of each other well in advance. Also typically a small number of security associations are required on each party, permitting the computationally intense security algorithms to reside directly in the devices involved. IKE has two phases of key negotiation. Phase 1 negotiates a shared secret between two IKE peers. The shared secret negotiated in phase 1 enables IKE peers to communicate securely in phase 2. In phase 2 negotiation, IKE establishes keys (SAs) for other applications such as IPsec. The IKE protocol uses UDP port 500 to negotiate keys.

In the call signaling chapters for NCS (Chapter 7, "NCS [Network-based Call Signaling]") and TGCP (Chapter 8, "TGCP and the PSTN Interconnect"), you learned about local connection options used for the establishment of bearer channel security; specifically, the RTP ciphersuite (sc-rtp) and RTCP ciphersuite (sc-rtcp) parameters. These parameters are specified by the call agent (CMS or MGC) and define the set of security encryption and authentication algorithms to be used on the RTP and RTCP packets sent by the gateway. These chapters also identified additional SDP attributes used in PacketCable. Four of these are used in providing RTP/RTCP security:

- **X-pc-secret**—Contains the end-to-end encryption key information used to establish RTP and RTCP security
- **X-pc-nrekey**—Contains a counter for the number of rekey events, used with RTP security
- **X-pc-csuites-rtp**—Contains the list of supported ciphersuites for RTP security
- **X-pc-csuites-rtcp**—Contains the list of supported ciphersuites for RTCP security

When a PacketCable gateway is instructed to secure RTP and/or RTCP via the security Local Connection Options, it returns these parameters. As you can see, one of these parameters, X-pc-secret, is the encryption key used by the IP Security process on the gateways to encrypt and authenticate RTP and/or RTCP packets.

NOTE The PacketCable security specification references RFC 2401 (which defines the IP Security architecture) and RFC 2406 (which defines the ESP protocol). These RFCs have been declared obsolete by the IETF. RFC 4301 obsoletes RFC 2401 and RFC 4301 obsoletes RFC 2406.

Configuring and Troubleshooting PacketCable Security

Because PacketCable security applies to all PacketCable components, security configuration applies to all of these components as well.

Broadband Access Center (BAC) Configuration

In the chapters in Part II, "MTA Provisioning," you learned how provisioning security and NCS signaling security is provided and configured for the MTA component. Provisioning security was covered at length in those chapters, so refer to them for the details.

In Chapter 5, "The MTA Configuration File," you learned NCS signaling security is enabled for an MTA in its configuration file by setting the SNMP MIB pktcMtaDevCmsIpsecCtrl to a value of true or 1. Recall that the index for this MIB is the ASCII representation of the CMS Fully Qualified Domain Name (FQDN) used for NCS. In the examples shown in this section, the CMS FQDN is mgcp-104ca104.roseville.cc.cisco.com.

Enabling this MIB triggers the MTA to initiate the secure MTA to CMS initialization steps (SEC-1 through SEC-10) covered in Chapter 4. Steps SEC-5 and SEC-6 are the AS-REQ/AS-RSP exchange with the telephony KDC for the service key used to establish IPsec SAs for NCS communication with the CMS. Thus, this key must be configured on the KDC for these steps to succeed. In BAC, this key is configured using the **keygen** utility in the <BAC_HOME>/kdc directory, as demonstrated here:

```
root@ss-nms37# ./keygen -c mgcp-104ca104.roseville.cc.cisco.com COMPANY-Y.CISCO.COM
  changeme
The following CMS service key have been written to the keys directory [/opt/CSCObpr/
  kdc/solaris/keys]
cms,mgcp-104ca104.roseville.cc.cisco.com@COMPANY-Y.CISCO.COM
```

NOTE After generating a new service key, it is necessary to restart the KDC; otherwise, the KDC responds with an error indicating an unknown application service request (AS-REQ).

After the MTA receives the service key in the AS-RSP, it initiates an AP-REQ message to the CMS for the application security keys used to encrypt NCS messages with the IPsec ESP protocol. The AP-REQ contains an encrypted form of the service key that the CMS uses to determine whether this particular MTA is authorized. Consequently, the CMS also needs to know this service key. How this information is communicated from the KDC to the CMS is not explicitly defined by PacketCable.

If you are using the BTS 10200 as the CMS, you must manually configure the service key. The BTS configuration is covered in the next section. Logically, you need to know the service key to manually enter it. In the output of the **keygen** example you can see that it specifies the location of the service key generated by the KDC. This key is in raw hexadecimal format. The UNIX Octal Dump utility (**od**) can be used to view the key by using the **-x** (interpret as hex) option as shown in the following example:

```
root@ss-nms37# od -x cms,mgcp-104ca104.roseville.cc.cisco.com@COMPANY-Y.CISCO.COM
0000000 0000 1c97 5785 a0c9 8d9a 02f4 4bb0 5cce
0000020 e75f 3712 c238 6051 737d
0000032
```

As you can see, this file is 13 words (equal to 26 bytes) in length. The first word (0x0000) is the service key version and always has this value. The remaining 12 words (24 bytes/48 hexadecimal characters) are the service key. These are the values that must be configured on the CMS as shown in the next section.

Troubleshooting the AS-REQ/AS-RSP exchange was covered in detail in Chapter 4, so refer to that chapter for assistance.

BTS 10200 Configuration and Troubleshooting

As you can tell from Figure 16-1, several of the signaling communication channels in PacketCable involve the CMS and MGC components. In fact, these components also control the establishment of bearer communication channels on the MTA and MG components. Therefore, a majority of security configuration is performed on the CMS and MGC components as well.

Version 4.5 of the Cisco BTS 10200 supports security configuration for the following communication channels:

- IPsec using Kerberos for NCS to/from the MTA
- IPsec using IKE for TGCP to/from the MG
- IPsec using IKE for COPS to/from the CMTS
- IPsec using IKE for Remote Authentication Dial-In User Service (RADIUS)–based Event Messaging to/from the RKS
- IPsec using IKE for RADIUS-based Event Messaging to/from the Communications Assistance for Law Enforcement Act (CALEA) Delivery Function (DF)
- IPsec for RTP/RTCP between MTA and MG gateways

Configuring NCS Signaling Security on the BTS 10200

In the BAC configuration section, you learned about some of the steps required to configure NCS signaling security. Logically, configuration is also needed on the CMS.

The steps for configuring NCS signaling security on the BTS 10200 are as follows:

Step 1 Configure Kerberos parameters in the IPSEC-KERBEROS table. In this table the telephony Kerberos Realm name is specified, along with the CMS FQDN used by the KDC. Also in this table is the Kerberos service key used in the AP-REQ/AP-RSP exchange with the MTA to establish IPsec keys. After this key is entered, the BTS automatically encrypts it. This table also contains other various Kerberos parameters; for more details on the parameters in this table or any other table, please refer to the BTS Command Line Interface Reference Guide. An example of this table is as follows:

```
CLI>show ipsec-kerberos
KRB_FQDN=mgcp-104ca104.roseville.cc.cisco.com
KRB_REALM=COMPANY-Y.CISCO.COM
KRB_SRV_KEY=bb946e46c00582c432350b79e153afa06c68f806db9e7587d30c7aedf3
  a99305f314032564e2cb4cfc507a629697fe58
SRV_KEY_VERSION=1
KRB_SRV_KEY_COMP_FLAG=N
KRB_MAX_OLD_SRV_KEYS=32
KRB_ACK_FLAG=Y
KRB_TIMEOUT=3
KRB_MAX_RETRY=10
KRB_EXP_RETRY_TIME=2
KRB_MAX_RETRY_TIME=20
KRB_REEST_SA_ACK_FLAG=Y
```

As you can see, the CMS FQDN (KRB_FQDN) and Kerberos Realm name (KRB_REALM) are the same values used to generate the service key in the BAC configuration example. The service key (KRB_SRV_KEY) is an encrypted form of the service key generated in that example.

NOTE Old Kerberos service keys are stored in the IPSEC-KERBEROS-KEYS table. A service key cannot be added to the IPSEC-KERBEROS table that exists in this table.

Step 2 The next step is defining policies that define the traffic to which IP security is applied. This is done by defining entries in the IPSEC-POLICY table that associates source and destination IP address ranges and UDP/TCP port numbers with a particular "action." The action field can be one of four values: Permit, Apply, IPsec, or Bypass. If the Permit action is specified, matching inbound traffic is accepted only if it is secured using IPsec. If the Apply action is specified, matching outbound traffic is secured using IPsec. The IPsec option is the combination of the Permit and Apply options. The Bypass option specifies traffic that is allowed to be sent and received without IP security. For example, the following configuration can be used to require an MTA to communicate NCS securely with the BTS:

```
CLI>show ipsec-policy
ID=MTA_out
```

```
SRC_IPADDR=13.13.5.38
DEST_IPADDR=24.34.240.241
ACTION=IPSEC
ID=MTA_in
SRC_IPADDR=13.13.5.6
DEST_IPADDR=24.34.240.241
ACTION=IPSEC
ID=passAP
SRC_PORT=1293
DEST_PORT=1293
ACTION=BYPASS
```

The BTS uses two interfaces with different IP addresses for call signaling traffic, necessitating the need for two entries with the IPsec action. In these entries, the CMS IP is the source, and the MTA IP is the destination. These entries alone require all traffic to and from the MTA to use IP Security, even the AP-REQ/AP-RSP exchange used to establish security. Consequently, the last entry must enable AP-REQ/AP-RSP messages to pass without IPsec. Recall that UDP port 1293 is used for Kerberos AP-REQ/AP-RSP messages.

On the active BTS call agent machine, you can use the UNIX ipsecconf utility to view these configured policies, as demonstrated here:

```
bts-prica# ipsecconf
#INDEX 22                   <- Entry to bypass security on incoming AS-REQs
{  sport 1293 dport 1293 } bypass { dir in }
#INDEX 24                   <- Entry to bypass security on outgoing AS-RSPs
{  sport 1293 dport 1293 } bypass { dir out }
#INDEX 29        <- Entry to require security on incoming messages from MTA
{ saddr 24.34.240.241 daddr 13.13.5.38 } permit {  encr_auth_algs any
  encr_algs any }
#INDEX 30           <- Entry to apply security on outgoing messages to MTA
{ saddr 13.13.5.38 daddr 24.34.240.241 } apply {  encr_auth_algs any
  encr_algs any sa shared }
#INDEX 31        <- Entry to require security on incoming messages from MTA
{ saddr 24.34.240.241 daddr 13.13.5.6 } permit {  encr_auth_algs any
  encr_algs any }
#INDEX 32         <- Entry to apply security on outgoing messages to MTA
{ saddr 13.13.5.6 daddr 24.34.240.241 } apply {  encr_auth_algs any
   encr_algs any sa shared }
```

Step 3 The next step is to configure the security parameters in the MGW-PROFILE table entry used by the MTA. Recall that this table defines a template of common parameters for a group of gateways. The use of this table was detailed in the BTS configuration section of Chapter 7. Some of these parameters apply to secure NCS call signaling. Table 16-1 shows these parameters (as you can see, the default values tend to be the recommended values).

Table 16-1 *MGW-PROFILE Table Fields Related to NCS Security*

Field	Default Value	Recommended Value	Description
KRB_RESET_FLAG	Y	Y	If enabled, Kerberos SA is automatically reestablished upon expiration
IPSEC_SA_ESP_CS	3DES-MD5, 3DES-SHA1, NULL-MD5, NULL-SHA1	3DES-MD5, 3DES-SHA1, NULL-MD5, NULL-SHA1	List of supported IPsec ESP ciphersuites for authentication and encryption in priority order
IPSEC_SA_GRACE_PERIOD	21,600 seconds	21,600 seconds	IPsec key expiration grace time; must be less than or equal to 25 percent of the IPSEC_SA_LIFETIME value
IPSEC_SA_LIFETIME	86,400 seconds	86,400 seconds	Lifetime of IPsec security association
IPSEC_ULP_NAME	IP	IP	IPsec SA Upper Layer Protocol name; indicates what type of traffic the SA is created for

On the active BTS call agent machine, you can use the UNIX **ipseckey dump** utility to view the established security associations, as demonstrated here:

```
bts-prica#ipseckey dump
Base message (version 2) type DUMP, SA type ESP. <- SA for traffic from
  CMS to MTA
Message length 232 bytes, seq=1, pid=25381.
SA: SADB_ASSOC spi=0x402, replay=64, state=MATURE
SA: Authentication algorithm = hmac-md5
SA: Encryption algorithm = 3des-cbc
SA: flags=0x80000000 < X_USED >
SRC: Source address (proto=0/<unspecified>)
SRC: AF_INET: port 0, 13.13.5.6 (mgcp-104CA104.roseville.cc.cisco.com).
DST: Destination address (proto=0/<unspecified>)
DST: AF_INET: port 0, 24.34.240.241 (x1-6-00-00-CA-DD-A7-
  E4.cc.cisco.com).
AKY: Authentication key.
AKY: 2fc9ea06f903930362a133a9d8f7620f/128
EKY: Encryption key.
```

```
EKY: 46f862cbba6d7f29373d5d38a840fe7cc43be6042a196179/192
 LT: Lifetime information
CLT: 33144 bytes protected, 0 allocations used.
CLT: SA added at time Wed Aug 16 13:07:56 2006
CLT: SA first used at time Wed Aug 16 13:07:56 2006
CLT: Time now is Wed Aug 16 15:17:59 2006
SLT: Soft lifetime information:  9223372036854775807 bytes of lifetime,
  2147483647 allocations.
SLT: 64800 seconds of post-add lifetime.
SLT: 64800 seconds of post-use lifetime.
SLT: 9223372036854742663 more bytes can be protected.
SLT: Soft expiration occurs in 56997 seconds, at Thu Aug 17 07:07:56
  2006.
HLT: Hard lifetime information:  9223372036854775807 bytes of lifetime,
  2147483647 allocations.
HLT: 86400 seconds of post-add lifetime.
HLT: 86400 seconds of post-use lifetime.
HLT: 9223372036854742663 more bytes can be protected.
HLT: Hard expiration occurs in 78597 seconds, at Thu Aug 17 13:07:56
  2006.
Base message (version 2) type DUMP, SA type ESP.     <- SA for traffic
  from MTA to CMS
Message length 232 bytes, seq=1, pid=25381.
SA: SADB_ASSOC spi=0xfabf9e59, replay=64, state=MATURE
SA: Authentication algorithm = hmac-md5
SA: Encryption algorithm = 3des-cbc
SA: flags=0x80000000 < X_USED >
SRC: Source address (proto=0/<unspecified>)
SRC: AF_INET: port 0, 24.34.240.241 (x1-6-00-00-CA-DD-A7-
  E4.cc.cisco.com).
DST: Destination address (proto=0/<unspecified>)
DST: AF_INET: port 0, 13.13.5.6 (mgcp-104CA104.roseville.cc.cisco.com).
AKY: Authentication key.
AKY: 5ce39409caec75d897ace12fc7208720/128
EKY: Encryption key.
EKY: 318a70ab2970e33438ade98668a18c628a86c2f2e5f22057/192
 LT: Lifetime information
CLT: 18560 bytes protected, 0 allocations used.
CLT: SA added at time Wed Aug 16 13:07:56 2006
CLT: SA first used at time Wed Aug 16 13:07:56 2006
CLT: Time now is Wed Aug 16 15:17:59 2006
SLT: Soft lifetime information:  9223372036854775807 bytes of lifetime,
  2147483647 allocations.
SLT: 86400 seconds of post-add lifetime.
SLT: 86400 seconds of post-use lifetime.
SLT: 9223372036854757247 more bytes can be protected.
```

```
SLT: Soft expiration occurs in 78597 seconds, at Thu Aug 17 13:07:56
 2006.
HLT: Hard lifetime information:  9223372036854775807 bytes of lifetime,
 2147483647 allocations.
HLT: 108000 seconds of post-add lifetime.
HLT: 108000 seconds of post-use lifetime.
HLT: 9223372036854757247 more bytes can be protected.
HLT: Hard expiration occurs in 100197 seconds, at Thu Aug 17 19:07:56
 2006.
Dump succeeded for SA type 0.
```

Configuring TGCP, COPS, and RADIUS Signaling Security on the BTS 10200

Security for TGCP call signaling to a trunking MG, COPS (DQoS) signaling to a CMTS, RADIUS (Event Messaging) signaling to an RKS, and RADIUS (Event Messaging) signaling to a CALEA DF all use the IKE protocol for the security key exchange needed to establish IPsec SAs. Consequently, it stands to reason that the configurations for these various interfaces are similar. The only difference is in the BTS tables that are configured. The steps for configuring security on the BTS 10200 for these interfaces are as follows:

Step 1 As in the secure NCS case, policies that define the traffic to which IP Security is applied must be configured in the IPSEC-POLICY table. For example, the following configuration can be used to require a particular MG to communicate TGCP securely with the BTS:

```
CLI>show ipsec-policy
ID=MG_in
SRC-IPADDR=13.13.81.164
SRC-PORT=2427
DEST-FQDN=MGA-RTPCA160.CC.CISCO.COM
ULP-NAME=UDP
ACTION=PERMIT
ID=MG_out
SRC-FQDN=MGA-RTPCA160.CC.CISCO.COM
DEST-IPADDR=13.13.81.164
DEST-PORT=2427
ULP-NAME=UDP
ACTION=APPLY
```

Notice in this example that only TGCP traffic sent and received from the MG must be secured; other traffic, such as ICMP pings, can pass nonsecurely. Because the BTS uses multiple interfaces with different IP address for call signaling traffic, the FQDN that resolves to these multiple IP addresses is specified. A sample configuration requiring a CMTS to communicate COPS securely with the BTS is as follows:

```
CLI>show ipsec-policy
ID=CMTS
SRC-FQDN=SGA-RTPCA160.CC.CISCO.COM
DEST-FQDN=PC-CMTS.CC.CISCO.COM
ACTION=IPSEC
ID=passIKE
SRC_PORT=500
DEST_PORT=500
ACTION=BYPASS
```

The first entry here causes the BTS to secure all packets sent to the CMTS, and it requires all packets received from the CMTS to be secure. This causes the IKE packets used to set up security to be dropped because they are not secured. Consequently, the second entry is needed, which permits IKE packets (IKE uses UDP port 500) to be sent and received nonsecurely. Likewise, you must perform configurations requiring an RKS and a CALEA DF to communicate RADIUS securely with the BTS, as demonstrated here (for these configurations to work, the pass IKE entry is needed as well):

```
CLI>show ipsec-policy
ID=RKS
SRC-FQDN=SGA-RTPCA160.CC.CISCO.COM
DEST-FQDN=PC-RKS.CC.CISCO.COM
ACTION=IPSEC
ID=CALEA_DF
SRC-FQDN=SGA-RTPCA160.CC.CISCO.COM
DEST-FQDN=PC-DF.CC.CISCO.COM
ACTION=IPSEC
```

Step 2 The next step is to configure the security parameters for the particular device (that is, MG, CMTS, RKS, or CALEA DF). Different BTS tables are used for each device type. For a trunking MG, this is the MGW-PROFILE table entry used by the MG. Remember this same table was used to configure security parameters for the MTA/NCS case. The difference here is that this time the IKE fields are relevant, and the Kerberos fields are not. For a CMTS, the AGGR table is used; recall that this is the table where you define a CMTS on the BTS. The use of this table was also detailed in the "Configuring the CMTS" section in Chapter 7. The RADIUS-PROFILE table defines an RKS on the BTS, and this is also where you define its security parameters. The use of this table was detailed in the "Defining the RKS" section of Chapter 15, "Event Messaging and Lawful Intercept." Also in Chapter 15 you learned that a CALEA DF is defined in the ESS table by logging in to the BTS EMS platform using the special CALEA user. This table is also where security parameters for the BTS-to-DF communication are defined.

The security parameters are defined in the same fields for all of these tables. These fields also have the same value ranges and default values in all the aforementioned tables. Table 16-2 shows these fields (as you can see, the default values tend to be the recommended values).

Table 16-2 *Table Fields Related to TGCP, COPS, and RADIUS Security*

Field	Default Value	Recommended Value	Description
IKE_CS	3DES-MD5, 3DES-SHA1	3DES-MD5, 3DES-SHA1	List of supported ciphersuites for authentication and encryption in priority order
IKE_GROUP	2	2	Diffie-Helman exchange group (1 or 2)
IKE_KEY	N/A	Same value as remote device	Preshared key for security association between BTS and MG, CMTS, RKS, or CALEA DF
IKE_SA_LIFETIME	86,400 seconds	86,400 seconds	Lifetime of IKE security association
IPSEC_SA_ESP_CS	3DES-MD5, 3DES-SHA1, NULL-MD5, NULL-SHA1	3DES-MD5, 3DES-SHA1, NULL-MD5, NULL-SHA1	List of supported IPsec ESP ciphersuites for authentication and encryption in priority order
IPSEC_SA_GRACE_PERIOD	21,600 seconds	21,600 seconds	IPsec key expiration grace time; must be less than or equal to 25 percent of the IPSEC_SA_LIFETIME value
IPSEC_SA_LIFETIME	86,400 seconds	86,400 seconds	Lifetime of IPsec security association
IPSEC_ULP_NAME	IP	IP	IPsec SA Upper Layer Protocol name; indicates what type of traffic the SA is created for

NOTE The IKE-KEY field value used to establish security is encrypted automatically by the BTS after it is entered. You have no way of displaying the original value entered.

Configuring RTP/RTCP on the BTS 10200

The steps for configuring secure media (RTP/RTCP) on the BTS 10200 are as follows:

Step 1 Configure the SECURED-MEDIA-ONLY variable in the CA-CONFIG table appropriately. If this is set to Y, calls fail if they cannot establish media security as configured; if set to N, the BTS uses media security only if it determines that both gateways support the configured media security parameters. The default value of this parameter is N.

Step 2 Earlier you learned that the call agent uses the **sc-rtp** and **sc-rtcp** Local Connection Option parameters to tell a gateway how RTP and RTCP security need to be handled. These parameters are sent in an NCS (for MTAs) or a TGCP (for MGs) CRCX or MDCX command. These options contain prioritized lists of ciphersuites (recall that a ciphersuite is the combination of a security authentication and encryption algorithm). On the BTS, you must create an entry in the CIPHERSUITE-PROFILE table that refers to an RTP/RTCP ciphersuite pair. The only fields for this table are an ID and optional description, as demonstrated here:

```
CLI>show ciphersuite_profile
ID=noBearSec
DESCRIPTION=No RTP or RTCP security
```

Step 3 Next, define prioritized ciphersuite lists for RTP and RTCP using this CIPHERSUITE-PROFILE ID in the CIPHERSUITE table, as demonstrated here:

```
CLI>show ciphersuite
ID=noBearSec
PROTO_TYPE=RTCP
AUTH_ALGO=RTCP_NULL
ENCRYPT_ALGO=RTCP_NULL
PRIORITY=1
ID=noBearSec
PROTO_TYPE=RTP
AUTH_ALGO=RTP_NULL
ENCRYPT_ALGO=RTP_NULL
PRIORITY=1
```

In this case, only a single entry is defined for RTP, and only a single entry is defined for RTCP. The PROTO_TYPE field indicates whether the entry is for RTP or RTCP. The PRIORITY field specifies the relative priority if there is more than one entry in the table for a protocol. The value of this field can be from 1 to 32, where the lower number indicates the higher priority. In this example, authentication and encryption have been disabled (set to NULL) for both RTP and RTCP. Table 16-3 shows the relationship between the authentication/encryption algorithms and the ciphersuite ID for the algorithms currently supported on the BTS.

Table 16-3 *BTS Supported RTP and RTCP Ciphersuites*

RTP/RTCP	AUTH/ENCR	Algorithm	ID
RTP	ENCR	RTP_NULL	50
RTP	ENCR	RTP_AES	51
RTP	ENCR	RTP_XDESX_CBC	53
RTP	ENCR	RTP_DES_CBC_PAD	54
RTP	ENCR	RTP_3DES_CBC	56
RTP	ENCR	RTP_RC4	57
RTP	AUTH	RTP_NULL	60
RTP	AUTH	RTP_MMH_2	62
RTP	AUTH	RTP_MMH_4	64
RTCP	ENCR	RTCP_NULL	70
RTCP	ENCR	RTP_AES_CBC	71
RTCP	ENCR	RTCP_XDESX_CBC	72
RTCP	ENCR	RTP_DES_CBC_PAD	73
RTCP	ENCR	RTP_3DES_CBC	74
RTCP	AUTH	RTCP_NULL	80
RTCP	AUTH	RTCP_HMAC_SHA1_96	81
RTCP	AUTH	RTCP_HMAC_MD5_96	82

Step 4 The final step is to associate subscribers and/or trunk groups with these ciphersuite lists. This is done using the QOS table. This table was covered in detail in Chapters 7, 8, and 11. It is used to influence how connections are set up. An entry in this table defines parameters such as preferred CODEC, DSCP/IP type of service (ToS) settings, use of silence suppression, and use of T.38 for fax. RTP and RTCP ciphersuite lists can also be specified by adding the CIPHERSUITE-PROFILE-ID field, which indexes the previously created ciphersuite lists. MTA subscribers can be associated with a QoS table entry either directly in the SUBSCRIBER table or indirectly via the SUBSCRIBER-PROFILE table. PSTN trunk groups can be associated with a QoS table entry directly in the TRUNK-GRP table.

VXSM Configuration

Much of the VXSM configuration required for PacketCable operation is covered in Chapter 8; only the configuration related to security is covered here. Cisco VXSM media gateways support both TGCP call signaling security and RTP/RTCP bearer security.

The security keys for TGCP call signaling are exchanged with the call agent (MGC) using the IKE protocol. You already saw how this is configured on the Cisco MGC (BTS 10200). The steps to configure this on the VXSM card are as follows:

Step 1 Just as with the BTS 10200, the first step is to create policies that define the traffic to which IP security is applied. The **addipsecsp** command is used for this purpose, and its syntax is as follows:

```
MGX8800.5.VXSM8.a > addipsecsp SP Index Src Addr Prefix Len Src Addr Dst
  Addr Prefix Len Dst Addr SP Protocol SP Src Port SP Dst Port SP
  Directionality SP Mirroring SP Action
```

The *SP Protocol* can be 0 (any protocol), 1 (TCP), 2 (UDP), or 3 (Raw). The *SP Directionality* is either 1 for inbound traffic or 2 for outbound traffic. The *SP Mirroring* is either 1 (mirrored) or 2 (nonmirrored). If it is set to 1, this means the policy applies to traffic in both directions (that is, source and destination IP addresses and port numbers are swapped and the *SP Directionality* value is meaningless). The *SP Action* is either 1 (secure—apply IP Security), 2 (bypass—let traffic pass nonsecurely), or 3 (discard—drop matching traffic).

Step 2 Next, the transform set used by the IPsec ESP protocol is defined with the **addipsecxform** command:

```
MGX8800.5.VXSM8.a > addipsecxform IPsec Transform Index IPsec ESP
  Encryption IPsec ESP Authentication IPsec Encapsulation Mode
  IPsec Transform Index -- IPsec Transform Index: 1..20
  IPsec ESP Encryption -- Encapsulation Payload Encryption Algorithm
    (3 - 3DES, 8 - NULL)
  IPsec ESP Authentication -- Encapsulation Payload Authentication
    Algorithm (2 - SHA1-96, 3 - MD5-96)
  IPsec Encapsulation Mode -- Encapsulation Mode (1 - transport)
```

Step 3 IP Security parameters, like Diffie-Hellman group, are defined by creating an IPsec proposal using the **addipsecprop** command:

```
MGX8800.5.VXSM.a > addipsecprop Proposal Index End Point IP Address [-
  antiReplay Anti Replay][-pfsEnable Pfs Enable][-dhgrp Diffie Hellman
  Group][-unitoftime Unit of Time][-hardlifetime Hard Lifetime][-select
  Use Select]
```

Notice the *End Point IP Address* parameter, which is the MGC IP address used for TGCP. Thus, if the MGC uses multiple IP addresses, this command must be added multiple times.

Step 4 Next, the IPsec transform set defined in Step 2 and the proposal defined in Step 3 are associated together using the **addipsecprxfassoc** command:

```
MGX8800.5.VXSM.a > addipsecprxfassoc Proposal Index IPsec Transform
   Index
```

Step 5 You then need to create an entry in the Policy Database Key Manager Association table, which contains mappings between the security policy table and the key management table. Next, the IPsec transform set defined in Step 2 and the proposal defined in Step 3 are associated together using the **addipsecprxfassoc** command:

```
MGX8800.5.VXSM.a > addipsecspkm SP Index SP Key Mgr Assoc Key Type SP
   Key Mgr Assoc Key Index
```

The key type is always 1, indicating IKE.

Step 6 Transform sets for the IKE key management protocol need to be created using the **addipsecikexform** command:

```
MGX8800.5.VXSM.a > addipsecikexform IKE Transform Id IKE Encr Algo IKE
  Auth Algo
   IKE Transform Id -- IKE Transform Index: 1..10
   IKE Enc Algo -- IKE Transform Encryption Algorithm (1 - DES, 2 - 3DES)
   IKE Auth Algo -- IKE Transform Authentication Algorithm (1 - MD5,
   2 - SHA)
```

Step 7 You must define the Diffie-Hellman group number and key lifetime in an IKE proposal using the **addipsecikeprop** command:

```
MGX8800.5.VXSM.a > addipsecikeprop Index [-dhgrp Diffie Hellman
  Group][-lifetimeunit Lifetime Unit][-lifetime Lifetime]
   Index -- IKE  Proposal Index: 1..10
   dhgrp -- Diffie Hellman Group (1 - dhGroup1(default), 2 - dhGroup2)
   lifetimeunit -- IKE Proposal Lifetime Unit (1 - seconds(default), 2 -
     minutes, 3 - hours)
   lifetime -- IKE Proposal Lifetime (60..86400 (seconds), 1..1440
    (minutes), 1..24 (hours), default:28800 (seconds))
```

Step 8 You then need to associate the IKE transform set and proposal defined in the previous two steps using the **addipsecikeprxfassoc** command:

```
MGX8800.5.VXSM.a > addipsecikeprxfassoc Index IKE Transform Idx
```

Step 9 Next, define the remote IPsec peer (in this case, the MGC) with the **addipsecikepeer** command:

```
MGX8800.5.VXSM.a > addipsecikepeer IP Config Index Remote IP Index
   Remote IP Addr Type Remote IP Addr Auth Pre Shared Key Proposal Index
   Auth Method
```

The authentication method is a value of 1, indicating preshared keys. Logically, this command must be repeated for each MGC IP address.

Step 10 The final step is to configure the use of IPsec on a network interface using the **addipsecnwif** command:

```
MGX8800.5.VXSM.a > addipsecnwif IP Config Index [-dfbit DF Bit][-pmtuage
  PMTU age]
   IP Config Index -- IP index (range=1..16)
   dfbit -- Network Interface DF Bit (1 - clear(default), 2 - set,
     3 - copy)
   pmtuage -- Network IF PMTU Age: 1..65535 default: 10
```

As you know, the security keys and parameters used for bearer security are sent between gateways via the call signaling messages; hence, no key exchange protocol configuration is needed on the VXSM. You can configure what RTP and RTCP ciphersuites are enabled on the VXSM card, though—you can even define their relative priority. However, keep in mind that the priority specified in the Local Connection Options sent by the MGC overrides this priority. But if the MGC does not specify these options, the priority list defined on the VXSM is used. The authentication and encryption algorithms for RTP and RTCP currently supported on the VXSM card are listed in Table 16-4.

Table 16-4 *VXSM Supported RTP and RTCP Ciphersuites*

RTP/RTCP	AUTH/ENCR	Algorithm	ID
RTP	ENCR	RTP_NULL	50
RTP	ENCR	RTP_AES	51
RTP	AUTH	RTP_NULL	60
RTP	AUTH	RTP_MMH_2	62
RTP	AUTH	RTP_MMH_4	64
RTCP	ENCR	RTCP_NULL	70
RTCP	ENCR	RTP_AES_CBC	71
RTCP	AUTH	RTCP_NULL	80
RTCP	AUTH	RTCP_HMAC_SHA1_96	81

The **cnfciphersuite** command can be used to specify which of these algorithms can be used and their relative preference, and the **cnfbearersec** command is used to enable/disable bearer security.

Cisco CMTS Configuration

Configuring IP Security for COPS traffic on a Cisco CMTS is just like configuring IP Security in general on any IOS-based router. The only difference here is that only the security options specified by PacketCable for secure COPS are used. Also Cisco IOS does not currently support Kerberos for security key establishment; therefore, the IKE method must be used on the CMTS.

NOTE For more details of security configuration on Cisco IOS platforms, consult the following URL:

http://www.cisco.com/en/US/partner/products/sw/iosswrel/ps1831/products_configuration_guide_chapter09186a0080347983.html

A sample configuration follows, with comments that should help you understand the steps required:

Step 1 Configure security policies for IKE.

The first step is defining security parameters used by the key management protocol, IKE. This includes defining the encryption algorithm (encr), authentication algorithm (hash), Diffie-Hellman group number (group), and authentication method (authentication). These parameters need to match the IKE parameters configured on the CMS (on the BTS, these are the IKE_ parameters in the AGGR table). A sample configuration is as follows:

```
crypto isakmp policy 1
 encr 3des
 hash md5
 authentication pre-share
 group 2
```

As you can see, in this case the Triple DES encryption algorithm and MD5 authentication algorithm are specified, as are Diffie-Hellman group 2. The authentication method is preshared keys. This means a common key is defined on the CMTS and the CMS (on the BTS, this is the IKE_KEY field in the AGGR table). If these keys do not match, IKE fails. Use the following command on the CMTS to view the IKE policy:

```
C0103-7246VXR#sh crypto isakmp policy
Global IKE policy
Protection suite of priority 1
        encryption algorithm:   Three key triple DES
        hash algorithm:         Message Digest 5
        authentication method:  Pre-Shared Key
        Diffie-Hellman group:   #2 (1024 bit)
        lifetime:               86400 seconds, no volume limit
Default protection suite
```

```
encryption algorithm:   DES - Data Encryption Standard (56 bit
 keys).
hash algorithm:          Secure Hash Standard
authentication method:   Rivest-Shamir-Adleman Signature
Diffie-Hellman group:    #1 (768 bit)
lifetime:                86400 seconds, no volume limit
```

Step 2 Configure the preshared secret key for IKE.

This step is where you define the preshared key on the CMTS. In this command, the IP address on the CMS used for COPS signaling is specified. Consequently, if you have multiple interface addresses on multiple CMS machines, you need to enter this command multiple times. The example that follows shows where PACKETCABLE is used as the preshared key:

```
crypto isakmp key PACKETCABLE address 13.13.5.36
crypto isakmp key PACKETCABLE address 13.13.5.37
crypto isakmp key PACKETCABLE address 13.13.5.4
crypto isakmp key PACKETCABLE address 13.13.5.5
```

To view the current settings, use the following command:

```
C0103-7246VXR#sh crypto isakmp key
Keyring             Hostname/Address                   Preshared Key
default             13.13.5.36                         PACKETCABLE
                    13.13.5.37                         PACKETCABLE
                    13.13.5.4                          PACKETCABLE
                    13.13.5.5                          PACKETCABLE
```

Step 3 Configure IPsec ESP parameters.

Next, you need to define the security parameters used by the IPsec ESP protocol. This includes specifying the security association lifetime as well as the transform set. Recall that in PacketCable the ESP protocol is always used, and the mode of the transform set is always transport. The following sample configuration is for the Triple DES encryption algorithm and both the MD5-HMAC and SHA-HMAC authentication algorithms:

```
crypto ipsec security-association lifetime seconds 86400

crypto ipsec transform-set 3DES_MD5 esp-3des esp-md5-hmac
 mode transport
crypto ipsec transform-set 3DES_SHA1 esp-3des esp-sha-hmac
 mode transport
```

You can view the current configuration by entering the following:

```
C0103-7246VXR#sh crypto ipsec transform-set
```

```
Transform set 3DES_MD5: { esp-3des esp-md5-hmac  }
   will negotiate = { Transport,  },

Transform set 3DES-SHA1: { esp-3des esp-sha-hmac  }
   will negotiate = { Transport,  },
```

Step 4 Define traffic requiring IP security.

The outgoing traffic that is protected and the incoming traffic that must be protected are identified on the CMTS using access lists. You need to define separate access lists for each CMTS/CMS IP address combination. You'll see the reason for this in the next step. The following example defines four extended access lists, each mapping all traffic to and from a particular CMTS/CMS IP address pair:

```
ip access-list extended COPS1
 deny udp any eq isakmp any eq isakmp
 permit ip host 13.13.0.34 host 13.13.5.36
 permit ip host 13.13.5.36 host 13.13.0.34
ip access-list extended COPS2
 deny udp any eq isakmp any eq isakmp
 permit ip host 13.13.0.34 host 13.13.5.37
 permit ip host 13.13.5.37 host 13.13.0.34
ip access-list extended COPS3
 deny udp any eq isakmp any eq isakmp
 permit ip host 13.13.0.34 host 13.13.5.4
 permit ip host 13.13.5.4 host 13.13.0.34
ip access-list extended COPS4
 deny udp any eq isakmp any eq isakmp
 permit ip host 13.13.0.34 host 13.13.5.5
 permit ip host 13.13.5.5 host 13.13.0.34
```

Notice the first line in each access list is required to allow the IKE packets needed to establish SAs to pass unsecured.

Step 5 Configure IPsec ESP policies.

This step is where the information specified in the previous steps is tied together. An IPsec policy is defined that specifies the key exchange protocol used (in this case IKE **ipsec-isakmp** as defined in Steps 1 and 2), the remote IPsec peer, the transform sets listed in order of priority (defined in Step 3), and the access list identifying IPsec traffic (defined in Step 4). The remote peer identifies where secured packets are sent, in this case the CMS. Because in this example the CMS can use multiple IP addresses, there are multiple entries in the policy (one for each CMS IP address). A sample configuration is as follows:

```
crypto map COPS 1 ipsec-isakmp
 set peer 13.13.5.36
 set transform-set 3DES_MD5 3DES_SHA1
match address COPS1
crypto map COPS 2 ipsec-isakmp
 set peer 13.13.5.37
 set transform-set 3DES_MD5 3DES_SHA1
match address COPS2
crypto map COPS 3 ipsec-isakmp
 set peer 13.13.5.4
 set transform-set 3DES_MD5 3DES_SHA1
match address COPS3
crypto map COPS 4 ipsec-isakmp
 set peer 13.13.5.5
 set transform-set 3DES_MD5 3DES_SHA1
match address COPS4
```

Step 6 Apply the IP security policy to the interface.

The final step is to apply the IPsec policy to the appropriate CMTS interface(s). Again, because transport mode is used, this is on the CMTS interface(s) with the IP address(es) used for COPS:

```
interface GigabitEthernet 0/3
  ip address 13.13.0.34 255.255.255.252
  crypto map COPS
```

At this point, the IP security policy is applied to outbound traffic. If the traffic matches a Permit action in one of the specified access lists, the CMTS looks to see if security associations have been established. If they already have, the packet is secured with the ESP protocol and sent. If the SAs have yet to be established, the CMTS initiates the IKE messaging necessary to set them up.

Keep in mind that traffic not matching access list permits is sent normally (that is, unprotected). This means that COPS connections for other devices fail to match the crypt map and are created nonsecurely. To protect against this and any other unwanted traffic interface, you should create access lists. Also you can use the global configuration command **cops listener access-list** *ACL number or name* to specify an access list indicating the permitted COPS listeners (that is, CMS devices and/or PCMM policy servers).

You can see if the IKE associations have been established using the **show crypto isakmp sa** [**detail**] command as demonstrated here:

```
C0103-7246VXR#sh crypto isakmp sa det
Codes: C - IKE configuration mode, D - Dead Peer Detection
       K - Keepalives, N - NAT-traversal
       X - IKE Extended Authentication
       psk - Preshared key, rsig - RSA signature
       renc - RSA encryption
C-id  Local           Remote          I-VRF    Encr Hash Auth DH Lifetime Cap.
2     13.13.0.34      13.13.5.5                3des md5  psk  2  04:24:19
1     13.13.0.34      13.13.5.5                3des md5  psk  2  03:44:20
```

You can view statistics on the number of packets encrypted and decrypted by using the following command:

```
C0103-7246VXR#sh crypto engine connections active
  ID Interface              IP-Address      State  Algorithm             Encrypt  Decrypt
2004 GigabitEthernet0/3     13.13.0.34       set    HMAC_MD5+3DES_56_C          0   140297
2005 GigabitEthernet0/3     13.13.0.34       set    HMAC_MD5+3DES_56_C     140296        0
```

Use the **show crypto ipsec sa** command to display detailed information about IPsec security associations, as demonstrated here:

```
C0103-7246VXR#sh crypto ipsec sa
interface: GigabitEthernet0/3
    Crypto map tag: COPS, local addr. 13.13.0.34
   protected vrf:
   local  ident (addr/mask/prot/port): (13.13.0.34/255.255.255.255/0/0)
   remote ident (addr/mask/prot/port): (13.13.5.5/255.255.255.255/0/0)
   current_peer: 13.13.5.5:500
     PERMIT, flags={origin_is_acl,}
    #pkts encaps: 145532, #pkts encrypt: 145532, #pkts digest 145532
    #pkts decaps: 145530, #pkts decrypt: 145530, #pkts verify 145530
    #pkts compressed: 0, #pkts decompressed: 0
    #pkts not compressed: 0, #pkts compr. failed: 0
    #pkts not decompressed: 0, #pkts decompress failed: 0
    #send errors 41, #recv errors 0
     local crypto endpt.: 13.13.0.34, remote crypto endpt.: 13.13.5.5
     path mtu 1500, ip mtu 1500, ip mtu idb GigabitEthernet0/3
     current outbound spi: A9720B9
     inbound esp sas:
      spi: 0x1D5C4919(492587289)
        transform: esp-3des esp-md5-hmac ,
        in use settings ={Transport, }
        slot: 0, conn id: 2004, flow_id: 5, crypto map: COPS
        sa timing: remaining key lifetime (k/sec): (4493418/15217)
        IV size: 8 bytes
        replay detection support: Y
     inbound ah sas:
     inbound pcp sas:
     outbound esp sas:
      spi: 0xA9720B9(177676473)
        transform: esp-3des esp-md5-hmac ,
        in use settings ={Transport, }
        slot: 0, conn id: 2005, flow_id: 6, crypto map: COPS
        sa timing: remaining key lifetime (k/sec): (4493418/15217)
        IV size: 8 bytes
        replay detection support: Y
     outbound ah sas:
     outbound pcp sas:
```

Certain configuration changes take effect only during the SA negotiation process; if you want to make changes take effect immediately, you can manually clear SAs using the **clear crypto sa** command. Entering this command as shown clears all SAs; you can identify specific SAs by appending the **map**, **peer**, or **spi** optional keywords.

As a last resort, you can use **debug** commands to help troubleshoot. The output from **debug crypto engine**, **debug crypto isakmp**, and **debug crypto ipsec** shows valuable information related to the security establishment process.

This section is a very high-level review of the use of security in PacketCable. Please consult the PacketCable security specification (PKT-SP-SEC) for more detailed information.

QoS Considerations

To ensure your PacketCable network is operationally sound and your customers are achieving a high-quality level of service, you can use some key metrics. Throughout this book, many of these metrics have already been discussed. The following is a recap of some of them:

- **Maximizing voice quality**—A mean opinion score (MOS) of at least 3.5 should be present for all calls, or customer satisfaction undoubtedly suffers. In fact, an MOS of at least 4.0, which is considered "toll quality," should be your goal.
- **Minimizing delay**—The ITU recommendation (G.114) states that the acceptable one-way delay for telephony is 150 ms; if this delay is greater than around 200 ms, voice quality suffers as callers are unsure their connections are operational and begin to talk over each other. The PacketCable 1.x architecture framework technical reports specify that the one-way delay for local IP access and IP egress (meaning the end-to-end delay excluding the IP backbone) is less than 45 ms.
- **Minimizing packet loss**—The occasional dropped packet is undetectable for a voice conversation, but a grouping of dropped packets definitely impacts voice quality. Also dropped packets are much more damaging to voice-band modem and facsimile calls. A 3 percent packet loss is regarded as the maximum tolerable amount for quality voice communications. Often, a stricter limit of 1 percent is used as the tolerable threshold.
- **Minimizing jitter variations**—If the receipt of voice packets is not periodic and the inter-packet delivery times vary greatly, large jitter buffers are needed to compensate for this disparity. This increases delay and could potentially adversely affect call quality. If jitter buffers are set too small, they can overflow, and packets can be lost. Adaptive jitter buffers are generally optimal.
- **Ensuring high availability**—The standard for the availability of telephony services is 99.94 percent. If the components in your network are unreliable, this standard is not met, and your customers will be dissatisfied with the quality of their telephony service.

To meet these metrics you can implement several design considerations in a PacketCable network. One is QoS treatment for the various types of IP packets, ensuring they reliably reach their destination in a timely manner. Another consideration is proper traffic engineering, assuring your network is designed to handle the volume of traffic it receives. The final major consideration area is high availability. If your network is stable and reliable, you should have minimal downtime impacting your customer's ability to make and receive phone calls. The remainder of this chapter addresses these topics in detail.

As you know, the QoS considerations in the majority of this book have dealt with the voice packets only. However, ensuring the reliable delivery of these RTP packets should not be your only QoS concern. Other types of PacketCable messaging, such as NCS call signaling packets and COPS gate control packets, can also benefit from QoS treatment.

Characterizing Call Signaling Traffic

So what are the traffic characteristics of call signaling? To help figure this out, the following four tables show the typical call signaling packets sent and received by an MTA during the life of a basic call. In these tables the size of each packet, as well as the time sent relative to the first packet being sent, is shown. Remember, in DOCSIS, packet size is from the byte after the standard DOCSIS header and extended headers to the end of the CRC. For example, a 200-byte IP packet would have a packet size of 218 bytes. Also in this discussion call signaling traffic for the upstream is defined as the NCS messaging and the DOCSIS dynamic service flow (that is, DSX) messaging, whereas call signaling traffic for the downstream is just the NCS messaging. You'll learn more about this in the section, "Figuring out the Classifiers."

Table 16-5 is for packets sent by an MTA when it is the originator of a call.

Table 16-5 *Packets Sent by Originating MTA*

Packet Description	Packet Size in Bytes	Time Transmitted in Seconds Relative to First Packet
NTFY off hook	190	0.000
RQNT for dial tone	63	0.052
NTFY digits	207	3.632
RQNT for on hook/flash acknowledgement	63	3.682
CRCX acknowledgement with SDP	283	3.863
DSA-REQ to reserve DQoS	222	4.155
DSA-ACK	52	4.253
MDCX to reserve DQoS acknowledgement	73	4.514
RQNT for ring back tone acknowledgement	63	4.675
DSC-REQ to commit DQoS	319	7.674
DSC-ACK	52	7.720
MDX to commit DQoS acknowledgement	73	7.973
RQNT to disable ring back acknowledgement	63	8.007
NTFY on hook (hang up)	190	16.884
RQNT for future off hook acknowledgment	63	16.936
DSD-REQ to release DQoS	73	17.053
DLCX acknowledgement with call statistics	191	17.089

As you can see in this call, it took the subscriber ~3.5 seconds to dial the called party number, ~3 seconds for the terminating party to answer, and a talk time of ~9 seconds. If you were to add up all the packet sizes and divide by the total call time, you would get a result of ~1,049 bps. However, it is more interesting to look at the highest traffic rate during any 1-second interval. It is common practice in traffic characterization not to compute rates on intervals less than 1 second to account for the spike caused by one or two messages. As you can see, the largest amount of traffic sent in any 1-second time period is between the NTFY with dialed digits and the MDCX to reserve DQoS. In this case, because the RQNT for ring back tone comes almost immediately after the MDCX, that is included in the analysis as well. So during this time there are 963 bytes sent in 1.043 seconds, resulting in a rate of 7,386 bps.

Table 16-6 is for packets sent by an MTA when it is the recipient of a call.

Table 16-6 *Packets Sent by Terminating MTA*

Packet Description	Packet Size in Bytes	Time Transmitted in Seconds Relative to First Packet
CRCX acknowledgement with SDP	283	0.000
DSA-REQ to reserve DQoS	222	0.577
DSA-ACK	52	0.613
MDCX to reserve DQoS acknowledgement	73	0.679
RQNT for ring tone acknowledgement	63	0.691
NTFY off hook (answer)	187	3.650
RQNT for on-hook/flash acknowledgement	63	3.664
DSC-REQ to commit DQoS	319	3.686
DSC-ACK	52	3.712
MDX to commit DQoS acknowledgement	73	3.740
DLCX acknowledgement with call statistics	186	13.004
DSD-REQ to release DQoS	73	13.032
RQNT for on-hook acknowledgement	63	13.045
NTFY on hook (hang up)	187	16.870
RQNT for future off-hook acknowledgment	63	16.881

In this case, the busiest period is from the answer (NTFY off hook) to the activation of voice path (MDCX to commit DQoS). The rate in this period comes out to 5,552 bps. Thus, in this example the largest rate of upstream call signaling traffic is 7,386 bps, which occurs in the dialing stage for a call origination. Most likely the biggest contributor here is the SDP information returned in the CRCX ACK. If the MTA supports several CODECs as well as DTMF relay, T.38, RTP/RTCP encryption, etc., the SDP can cause this number to be a bit larger. However, capping upstream call signaling traffic at 15–20 kbps is generally adequate.

Now the downstream call signaling traffic is examined. Table 16-7 is for packets received by an MTA when it is the originator.

Table 16-7 *Packets Received by Originating MTA*

Packet Description	Packet Size in Bytes	Time Transmitted in Seconds Relative to First Packet
NTFY off hook acknowledgement	64	0.000
RQNT for dial tone	353	0.000
NTFY digits acknowledgement	64	3.632
RQNT for on hook or flash	172	3.632
CRCX	245	3.681
MDCX to reserve DQoS; also has SDP	417	4.081
RQNT for ring back tone	189	4.649
MDCX to commit DQoS	208	7.622
RQNT to disable ring back	177	7.972
NTFY on hook (hang up) acknowledgment	64	16.883
RQNT for future off hook	164	16.883
DLCX	152	16.935

In this case, the busiest 1-second interval is between receiving digits and sending ring back tone (a rate of 8,696 bps).

The final table, Table 16-8, is for packets received by an MTA when it is the recipient of a call.

Table 16-8 *Packets Received by Terminating MTA*

Packet Description	Packet Size in Bytes	Time Transmitted in Seconds Relative to First Packet
CRCX with SDP	465	0.000
MDCX to reserve DQoS	197	0.650
RQNT for ring tone	179	0.773
NTFY off hook (answer) acknowledgement and RQNT for on hook/flash	175	3.743
MDCX to commit DQoS	195	3.757
DLCX	152	13.020
RQNT for on hook	169	13.097
NTFY on hook (hang up) acknowledgement and RQNT for future off hook	187	16.963

In this case, the busiest period is from connection creation to the sending of the RQNT for ringing. It computes to a rate of 6,728 bps. Thus, the worst case overall for downstream signaling is 8,696 bps.

QoS for NCS Packets on the DOCSIS Network

When a PacketCable MTA needs to send an NCS call signaling message to the CMS, such as an RSIP or NTFY, it uses whatever service flow has been set up to handle this traffic as defined in the cable modem configuration file. In contrast to the voice RTP packets, no QoS-enabled path is dynamically set up when these packets need to be sent. Similarly, downstream NCS messages sent from the CMS to the MTA utilize whatever DOCSIS service flow has been configured to handle these packets.

As with any other type of traffic sent from a DOCSIS 1.1 cable modem, you can define dedicated service flows to provide a certain level of service for NCS signaling packets. You can define these service flows in the modem's configuration file, along with packet classifiers used to identify the traffic authorized to utilize these service flows.

Using the Primary Service Flows

At a minimum, every DOCSIS 1.1 configured cable modem needs two service flows defined: one for the upstream direction and one for the downstream direction. These service flows are also known as the primary service flows. For ordinary high-speed data users, these are typically the only service flows defined and contain parameters such as maximum transmission rates and maximum burst sizes. No packet classifiers are needed to identify packets for these flows because they are the default service flows. That is, when a packet fails to match any defined classifier, it is sent on these default primary service flows. If multiple upstream and downstream service flows are statically defined in a modem's configuration file, the first upstream flow defined and the first downstream flow defined become the primary service flows.

If a PacketCable MTA has only the required primary service flows defined in its modem configuration file, then these flows are also used for call signaling traffic. This configuration probably works fine if the network load is relatively light. But if the MTA signaling traffic has to compete for DOCSIS bandwidth with high-speed data users, then this configuration has the potential for problems.

Because the primary upstream service flow utilizes a Best Effort DOCSIS scheduling service, each packet the MTA needs to send must have a dedicated data grant. The cable modem portion of the embedded MTA (eMTA) obtains these grants by first sending bandwidth requests in contention intervals. If the DOCSIS upstream is heavily utilized, it is entirely possible other modems are also transmitting during these intervals. If multiple devices are transmitting, a collision occurs, and these requests are not detected by the CMTS. At this point, the modems wait some random period before retransmitting the request. If another collision occurs, the modem waits a longer period and then retransmits. This process continues until the request transmission is successful or the maximum number of retries is reached. A modem knows its request packet was received by the CMTS when it sees a unicast data grant or grant pending information element with its Service Identifier (SID) in the subsequent DOCSIS bandwidth allocation map messages.

Just because the CMTS receives the bandwidth request is no guarantee that the MTA or modem receives a data grant. If the upstream channel is heavily utilized, bandwidth might not be available; it is in this situation that a grant pending information element is sent instead of the data grant. This lets the modem or MTA know that the bandwidth request was received, but it cannot be honored at this time. The modem does not retransmit the request. As soon as bandwidth is available, the CMTS provides the data grant to the MTA, but at this point it might be too late.

Using a Premium Best Effort Service Flow

One way to alleviate these potential problems is by defining a separate Best Effort service flow for upstream signaling information. Service flow parameters such as Minimum Reserved Rate and Traffic Priority can be used to provide a measure of QoS for Best Effort flows.

By defining a Minimum Reserved Rate for the service flow, you have an assurance that some bandwidth is available when needed; this is also commonly referred to as a committed information rate (CIR) service. The CMTS reserves these resources and does not give them to another QoS flow even if all other DOCSIS resources are totally utilized. Of course, this can also be a drawback because when the MTA is idle, these resources cannot be used by any other QoS-enabled service. Additionally, if the aggregated Minimum Reserved Rates of services flows on a US or DS channel reaches the channel capacity, the admission of additional modems with QoS guarantees is rejected because the CMTS cannot reserve more total bandwidth than it has on a given channel. As you learned in the DQoS portion of the book, unused reserved resources can be used for ordinary Best Effort data. This is because this utilization is only temporary and ceases as soon as the reserved resources are needed by an authorized device.

The Traffic Priority parameter can also be used to differentiate signaling service flows. The CMTS uses this parameter to determine which service flows should be serviced first, assuming other QoS parameters are similar, such as Scheduling Type and Minimum Reserved Rate. If these other parameters are similar, the CMTS uses a priority queuing scheme when deciding which requests to honor. For example, if the signaling service flows are defined with a priority of 2, and all the primary Best Effort flows are defined with a priority of 1, then all requests from signaling flows are honored before those of the primary flows. Of course, this is assuming that no other QoS restraints, such as Maximum Sustained Rate or Maximum Burst Size, have been violated.

A sample configuration is as follows:

```
24 (Upstream Service Flow Block)
    S01 (Flow Reference) = 2
    S06 (QoS Parameter Set Type) = 7
    S07 (Traffic Priority) = 3                    <- Higher priority than HSD traffic
    S08 (Max Sustained Traffic Rate) = 20000     <- Rate capped at 20 kbps
    S10 (Min Reserved Traffic Rate) = 0           <- No bandwidth reserved
    S11 (Assumed Min Reserved Rate Packet Size) = 64
    S15 (Service Flow Scheduling Type) = 2        <- "2" means Best Effort
    S16 (Request/Transmission Policy) = 0
```

In this example, the traffic priority is set to 3 to ensure call signaling bandwidth requests are serviced before high-speed data users (which are assumed to have the default priority of 0). A minimum reserved rate of 0 is used because the increased priority is sufficient to differentiate call signaling traffic from high-speed data (HSD) traffic. Because you could potentially have several MTAs on an upstream channel, you do not want to tie up a lot of bandwidth for MTAs that are going to be idle most of the time. The maximum sustained rate is set to a value that should be large enough to handle the highest anticipated volume of call signaling traffic but low enough so that other applications cannot steal this preferential bandwidth.

Using an nRTPS Service Flow

Defining the Minimum Reserved Rate and/or Traffic Priority parameters only helps if request packets are received by the CMTS. Remember that these requests are still sent in contention intervals, and you have no guarantee that they will be successful.

Also in periods where utilization is high and when many QoS enabled flows, such as voice Unsolicited Grant Service (UGS) flows, are present, the number of contention intervals offered by the CMTS decreases dramatically. Thus, if several devices are competing for this small number of requests, the odds of a collision occurring is pretty good. The way around this problem is by offering the MTAs dedicated request opportunities to send signaling packets.

This is done by defining the signaling flow using the Non Real-Time Polling Service (nRTPS) instead of the Best Effort service. The nRTPS service works by issuing dedicated request opportunities on a regular basis. These opportunities are typically not periodic as compared to the Real-Time Polling Service (RTPS) and much less frequent than for an RTPS flow. Note: These types of QoS flows are also discussed in the PCMM section of the book.

The following is a sample configuration:

```
24 (Upstream Service Flow Block)
    S01 (Flow Reference) = 2
    S06 (QoS Parameter Set Type) = 7
    S07 (Traffic Priority) = 3                    <- Higher priority than HSD traffic
    S08 (Max Sustained Traffic Rate) = 20000     <- Rate capped at 20 kbps
    S10 (Min Reserved Traffic Rate) = 0           <- No bandwidth reserved
    S11 (Assumed Min Reserved Rate Packet Size) = 64
```

```
S15 (Service Flow Scheduling Type) = 3          <- "3" means nRTPS
S16 (Request/Transmission Policy) = 0
S17 (Nominal Polling Interval) = 100000         <- Polls sent every 100 ms
S23 (IP ToS Overwrite) = 0.104
```

In this case, the Nominal Polling Interval is set to 100 ms; this is often enough such that each MTA is given a sufficient amount of dedicated request opportunities but not so often that it affects other upstream traffic. The flow is still given traffic priority over other types of traffic. The Request/Transmission policy is configured so that this flow is permitted to use contention intervals and piggyback requests if it needs additional bandwidth. Also note the inclusion of the IP ToS Overwrite parameter; this assigns designated DSCP or IP Precedence values to call signaling packets so that they are given proper QoS treatment throughout the network. Backbone QoS is covered in greater detail later in this chapter.

Just as with most things, there is a potential downside to using nRTPS service flows for call signaling. Depending on the implementation of the CMTS upstream scheduler, the scheduling of these dedicated request intervals might influence the number of other QoS service flows an upstream can support. For example, the maximum number of UGS flows available for voice traffic might be impacted.

TIP As of Cisco IOS Software Release 12.3(13a)BC, the Cisco CMTS supports the Low Latency Queuing (LLQ) method of upstream scheduling. The LLQ method can be enabled on a per-upstream basis for UGS, RTPS, and/or nRTPS traffic. By enabling this mode, you change the behavior of the upstream scheduler for that particular service scheduling type. Instead of dedicating periodic data or request intervals at the time the service flow is created, you create an LLQ for the flow. When the normal data/request interval time for the flow approaches, the CMTS "pops" a data/request from the LLQ by scheduling bandwidth for the flow. The net effect of this is that more QoS flows can coexist on an upstream channel. The drawback to enabling LLQ is it can introduce additional jitter. Consequently, it is recommended you enable LLQ only for applications that are not real-time—such as nRTPS for call signaling. It is NOT recommended to enable LLQ for PacketCable UGS flows. For more information on upstream scheduling on the Cisco CMTS refer to http://www.cisco.com/en/US/partner/tech/tk86/tk804/technologies_white_paper09186a0080672b3a.shtml.

Designing the Downstream Service Flow

Downstream call signaling packets can also encounter problems if no QoS is enabled for them. If the primary DS service flow is used for call signaling, then the CMTS has no way of differentiating them from other packets. So when the DS flow is exceeding its maximum transmission rate, call signaling packets can be delayed or dropped just like any other packet. Also if the overall downstream channel is congested, these packets can be delayed or dropped if they aren't designated as preferential over other traffic. The following is a sample configuration for a downstream call signaling service flow:

```
25 (Downstream Service Flow Block)
    S01 (Flow Reference) = 4
    S06 (QoS Parameter Set Type) = 7              <- Higher priority than HSD traffic
    S07 (Traffic Priority) = 7
    S08 (Max Sustained Traffic Rate) = 20000  <- Rate capped at 20 kbps
    S10 (Min Reserved Traffic Rate) = 0          <- No bandwidth reserved
    S11 (Assumed Min Reserved Rate Packet Size) = 64
```

This flow has the maximum traffic priority value. The way in which the traffic priority parameter works is different for downstream flows than it is for upstream flows. With a Cisco CMTS this parameter is used as a weighting factor when deciding how much bandwidth is made available to multiple flows with different priorities. For example, suppose two flows each attempt to send 2 Mbps worth of traffic that does not violate any other QoS parameter: One flow has a priority of 1, and the other a priority of 7. If you have only 2 Mbps of downstream bandwidth, the CMTS allocates 1/(1+7) of the bandwidth for the priority one flow and 7/(1+7) of the bandwidth for the priority seven flow. Consequently, this is why the maximum priority value is used in the example instead of just a value greater than HSD traffic. This flow is also defined with a maximum sustained rate of 20 kbps. As with the earlier examples, this is designed to handle the largest volume of anticipated call signaling traffic but is low enough that other applications cannot steal this service and benefit from the preferential QoS treatment.

Figuring Out the Classifiers

Now that the service flows have been created, classifiers must also be created to make sure the signaling traffic is routed to these flows. This is a bit more complicated than it first appears.

Using source and/or destination IP addresses is one option. However, there are some limitations with this approach. Remember, the MTA IP address is determined via DHCP during the provisioning process, and even if it is known ahead of time, it could change. Also using this parameter would require each MTA in the network to use a different classifier and, hence, a different modem configuration file. The CMS IP address(s) could be used because these are relatively static. Of course, markets with different CMS devices require different classifiers and configuration files.

Another option is using the embedded MTA MAC address. Unlike an IP address, this is static. The need to use different classifiers and modem configuration files still exists, but with a provisioning system like Cisco's BAC that can generate dynamic configuration files, this can be achieved without too much difficulty. In Chapter 5 you learned how variables can be used with BAC templates to dynamically generate configuration files. This is a perfect situation to use these capabilities. Just define a variable for the MTA MAC address and define its value with each MTA device. This can be done by the front office provisioning system when MTAs are added. The modem template file then references this variable in the classifier description. On a Cisco CMTS this works only for upstream classifiers because downstream MAC classifiers are not currently supported. Also note that this implementation maps other types of traffic, such as DHCP, SNMP, ARP, DNS, and RTCP to the upstream service flow; consequently, the maximum sustained rate in the service flow examples discussed earlier might need to be adjusted.

Another option is matching packets based on the IP precedence/DSCP codepoint values. These values can be set up properly using methods talked about later in this chapter, but the problem is controlling other traffic from using these same values. For example, an application behind the LAN on the MTA can potentially utilize the upstream signaling flow by simply using the same IP Precedence value. Remember, setting maximum sustained rates in the service flows provides some protection against this theft of service.

A better option is to use UDP port numbers, possibly in conjunction with IP Precedence/DSCP and CMS IP address(es). Remember that, by default, the NCS signaling protocol uses port 2427 on the MTA and port 2727 on the CMS. However, these values are just defaults, and other values can be used. For example, the MTA and CMS ports can be changed by setting certain SNMP MIBs in the MTA configuration file. Please refer to the provisioning portion of the book for additional information. Also keep in mind the CMS port is just the initial port where the MTA sends the RSIP message. If you recall from Chapter 7, the CMS supplies the NCS Notified Entity parameter that can instruct the MTA to change what port it is sending messages. Also many CMS vendors use different source ports to send different messages. An example of this is the auditing procedure with the Cisco BTS 10200. In this instance a separate port is used to send AUEP messages. It is because of these variables that it is often wise to use just the MTA port number as the classifier; this corresponds to a source port classifier for the upstream and a destination port classifier for the downstream, as shown here:

```
22 (Upstream Packet Classification Block)
    S01 (Classifier Reference) = 1
    S03 (Flow Reference) = 2
    S05 (Rule Priority) = 1
    S06 (Activation State) = 1
    S09 (IP Packet Classification)
        T02 (IP Protocol) = 17                <- IP protocol 17 is UDP
        T07 (Source Port Start) = 2427
        T08 (Source Port End) = 2427
22 (Upstream Packet Classification Block)
    S01 (Classifier Reference) = 2
    S03 (Flow Reference) = 2
    S05 (Rule Priority) = 64
    S06 (Activation State) = 1
    S10 (Ethernet LLC Packet Classification)
        T03 (Enet/DSAP/MAC Type) = 3.15.22  <- Maps DOCSIS DSX messages
```

Notice the two upstream classifiers in the example. The first one is for mapping NCS call signaling packets as discussed previously. The second one maps the DOCSIS dynamic service flow messages used to create, modify, and delete the voice service flows. To ensure these messages get through to the CMTS, it is a good idea to map them to the call signaling flow as well. For downstream classification only, mapping the NCS packets is sufficient because the DSA and DSC (and usually DSD) transactions are all initiated from the MTA. On a Cisco CMTS the responses are handled directly by the cable linecard and not subject to traffic shaping. The following is a sample downstream classifier:

```
23 (Downstream Packet Classification Block)
    S01 (Classifier Reference) = 3
    S03 (Flow Reference) = 4
    S05 (Rule Priority) = 1
    S06 (Activation State) = 1
    S09 (IP Packet Classification)
        T02 (IP Protocol) = 17
        T09 (Destination Port Start) = 2427
        T10 (Destination Port End) = 2427
```

Backbone QoS Overview

Now that you know how to ensure QoS for packets on the DOCSIS portion ofthe network, you can turn your focus to QoS treatment for packets across your backbone network. For NCS and COPS packets this is between the wide-area network (WAN) interface(s) of the CMTS and the CMS; for RTP packets this is between the WAN interface(s) of the CMTS and the VoIP interface of the remote gateway; for TGCP packets this is between the MG and the MGC; and for SIP/CMSS packets this between the two CMS devices.

Backbone QoS is typically provided by marking packets with different priorities so that each intermediary device can differentiate between them and process them accordingly using a defined policy. This packet marking can be done using Differentiated Services Code Point (DSCP) codepoints or IP Precedence bits in the ToS byte of the IP header.

The six most significant bits in the IP header ToS byte comprise the DSCP codepoint value. The two least significant bits in the IP header ToS byte are currently unused and should be set to zero. The main idea behind DSCP is that boundary devices (such as PacketCable components) evaluate packets and assign DSCP codepoint values. Inside the network these values determine how packets are treated. This decision-making process is performed on each device in the core network that the packet traverses; hence, this QoS treatment is also referred to as per hop behavior, or PHB.

Because 6 bits are used for DSCP codepoints, there are 64 possible PHB values. The three most significant bits define the DSCP level; thus, there are eight DSCP levels. The PHB commonly used for RTP traffic is 46, also known as the Expedited Forwarding (EF) PHB. The PHB commonly used for voice signaling traffic is 26, which is also known as the Assured Forwarding 31 (AF31) PHB.

Table 16-9 lists the DSCP codepoints. With each table entry the corresponding value of the IP ToS header byte is shown in decimal, hexadecimal, and binary format. Also the legacy interpretation of this field is shown, where the IP ToS header byte is broken down into a 3-bit IP Precedence value followed by bit flags for minimizing delay (IP D), maximizing throughput (IP T), and maximizing reliability (IP R). The final 2 bits are unused by DSCP and, hence, always equal to 00.

Table 16-9 *DSCP to IPv4 TOS Conversions*

DSCP Value	DSCP Level	Dec Value	Hex Value	Binary Value	IP Prec	IP D	IP T	IP R	Unused Bits
0	0	0	00	00000000	000	0	0	0	00
1	0	4	04	00000100	000	0	0	1	00
2	0	8	08	00001000	000	0	1	0	00
3	0	12	0C	00001100	000	0	1	1	00
4	0	16	10	00010000	000	1	0	0	00
5	0	20	14	00010100	000	1	0	1	00
6	0	24	18	00011000	000	1	1	0	00
7	0	28	1C	00011100	000	1	1	1	00
8	1(cs1)	32	20	00100000	001	0	0	0	00
9	1	36	24	00100100	001	0	0	1	00
10	1(af11)	40	28	00101000	001	0	1	0	00
11	1	44	2C	00101100	001	0	1	1	00
12	1(af12)	48	30	00110000	001	1	0	0	00
13	1	52	34	00110100	001	1	0	1	00
14	1(af13)	56	38	00111000	001	1	1	0	00
15	1	60	3C	00111100	001	1	1	1	00
16	2(cs2)	64	40	01000000	010	0	0	0	00
17	2	68	44	01000100	010	0	0	1	00
18	2(af21)	72	48	01001000	010	0	1	0	00
19	2	76	4C	01001100	010	0	1	1	00
20	2(af22)	80	50	01010000	010	1	0	0	00
21	2	84	54	01010100	010	1	0	1	00
22	2(af23)	88	58	01011000	010	1	1	0	00
23	2	92	5C	01011100	010	1	1	1	00
24	3(cs3)	96	60	01100000	011	0	0	0	00

Table 16-9 *DSCP to IPv4 TOS Conversions (Continued)*

DSCP Value	DSCP Level	Dec Value	Hex Value	Binary Value	IP Prec	IP D	IP T	IP R	Unused Bits
25	3	100	64	01100100	011	0	0	1	00
26	3(af31)	104	68	01101000	011	0	1	0	00
27	3	108	6C	01101100	011	0	1	1	00
28	3(af32)	112	70	01110000	011	1	0	0	00
29	3	116	74	01110100	011	1	0	1	00
30	3(af33)	120	78	01111000	011	1	1	0	00
31	3	124	7C	01111100	011	1	1	1	00
32	4(cs4)	128	80	10000000	100	0	0	0	00
33	4	132	84	10000100	100	0	0	1	00
34	4	136	88	10001000	100	0	1	0	00
35	4	140	8C	10001100	100	0	1	1	00
36	4	144	90	10010000	100	1	0	0	00
37	4	148	94	10010100	100	1	0	1	00
38	4	152	98	10011000	100	1	1	0	00
39	4	156	9C	10011100	100	1	1	1	00
40	5(cs5)	160	A0	10100000	101	0	0	0	00
41	5	164	A4	10100100	101	0	0	1	00
42	5	168	A8	10101000	101	0	1	0	00
43	5	172	AC	10101100	101	0	1	1	00
44	5	176	B0	10110000	101	1	0	0	00
45	5	180	B4	10110100	101	1	0	1	00
46	5(ef)	184	B8	10111000	101	1	1	0	00
47	5	188	BC	10111100	101	1	1	1	00
48	6(cs6)	192	C0	11000000	110	0	0	0	00

continues

Table 16-9 *DSCP to IPv4 TOS Conversions (Continued)*

DSCP Value	DSCP Level	Dec Value	Hex Value	Binary Value	IP Prec	IP D	IP T	IP R	Unused Bits
49	6	196	C4	11000100	110	0	0	1	00
50	6	200	C8	11001000	110	0	1	0	00
51	6	204	CC	11001100	110	0	1	1	00
52	6	208	D0	11010000	110	1	0	0	00
53	6	212	D4	11010100	110	1	0	1	00
54	6	216	D8	11011000	110	1	1	0	00
55	6	220	DC	11011100	110	1	1	1	00
56	7(cs7)	224	E0	11100000	111	0	0	0	00
57	7	228	E4	11100100	111	0	0	1	00
58	7	232	E8	11101000	111	0	1	0	00
59	7	236	EC	11101100	111	0	1	1	00
60	7	240	F0	11110000	111	1	0	0	00
61	7	244	F4	11110100	111	1	0	1	00
62	7	248	F8	11111000	111	1	1	0	00
63	7	252	FC	11111100	111	1	1	1	00

In a PacketCable network, DSCP codepoints are usually set at the point where packets originate. One exception to this is with packets originated by the MTA.

Remember that the MTA is an untrusted network entity, so DSCP codepoints for these packets are typically overwritten by the CMTS instead of trusting the MTA to properly set them. Similarly, PCMM clients are also typically untrusted, so DSCP codepoints are typically overwritten at the CMTS.

After your network is set up to properly mark packets, core devices must be configured so that each class of packets receives the proper QoS. A recommended way to do this with Cisco routers is by utilizing class-based weighted fair queuing (CBWFQ) along with Low Latency Queuing (LLQ) for the prioritized treatment of RTP packets. With CBWFQ you can define multiple traffic classes based on various packet matching criteria such as IP address, TCP/UDP port numbers, and DSCP values. This is similar to how DOCSIS classifiers are defined. After these classes are defined, different levels of QoS can be assigned to each class, such as reserved bandwidth and prioritization. Again, this is similar to how DOCSIS service flows are defined. The LLQ feature provides a strict priority queue within CBWFQ. When LLQ is configured for voice over IP (VoIP) delay, sensitive RTP packets are mapped to the strict priority LLQ queue and are sent before packets in other queues. LLQ is appropriate only for delay-sensitive RTP packets. For VoIP signaling packets it is recommended you use CBWFQ with bandwidth reservation. CBWFQ derives the weight for the packets belonging to the corresponding class from the bandwidth specified. Then this weight is used to ensure that the queue for the corresponding class is serviced fairly.

Setting DSCP Codepoints

This section details how DSCP codepoints can be set on the various devices in a PacketCable network.

Packets Originating from the MTA

If you are confident that your MTAs can be trusted to properly set DSCP values, you can instruct the MTA to set these values by defining the following SNMP MIBs in the MTA configuration file:

- **pktcSigTosFormatSelector**—Determines whether the IPv4 TOS field is defined as a DSCP codepoint. Recommended value is 2, indicating this field is interpreted as a DSCP codepoint.
- **pktcSigDefCallSigTos**—Specifies the DSCP setting to be used for NCS call signaling packets. Recommended value is PHB 26 (0x68).
- **pktcSigDefMediaStreamTos**—Specifies the DSCP setting to be used for RTP packets. Recommended value is PHB 46 (0xB8).

Packets Originating from the CMTS

If you are not confident in trusting the MTA to properly set DSCP values, you can set up your network to overwrite these values at the CMTS. Back in Chapter 13, "Analyzing, Implementing, and Troubleshooting DQoS," it was discussed how the CMTS, in conjunction with the CMS, can overwrite DSCP values for RTP bearer packets. In summary, in the GATE-SET message sent from the CMS to the CMTS, the COPS PacketCable object DS Field is included, which is the value with which the IP ToS byte on RTP voice packets leaving the CMTS are overwritten. The decision on whether or not this object is included and what the value of this parameter will be is configurable on the CMS.

If you are using the Cisco BTS 10200 as your CMS, the value of the DS Field GATE-SET object is defined in the DQOS-CMTS-DSCP field in the QOS table entry used by the MTA subscriber. The format of this field is the DSCP value in decimal (so refer to Table 16-9 for the necessary conversions).

Additionally, you can configure the CMS to instruct the MTA on what IP ToS byte value to use for outgoing RTP packets. The CMS does this by including the type of service LCO, t, in a CRCX or MDCX message, as demonstrated here:

```
CRCX 161096008 aaln/1@x1-6-00-13-10-50-38-02.cc.cisco.com MGCP 1.0 NCS 1.0
C: 284
L: mp:20;20;20,  a:PCMU;PCMA;G728,  e:off,  s:off,  t:a0
M: inactive
K: 161095965
```

To make the BTS send the t parameter you need to set the IPTOS-RTP-SUPP field in the MGW-PROFILE table entry used by the MTA to Y. The value of the t parameter is specified by fields in the IPTOS-RTP-xxx fields in the QOS table. Specifically, the following fields must be set:

- **IPTOS-RTP-PRECEDENCE**—This field has one of the following values: NETCONTROL (IP Precedence 7–111), INTERNETCONTROL (IP precedence 6–110), CRITICAL (IP Precedence 5–101), FLASHOVERRIDE (IP Precedence 4–100), FLASH (IP Precedence 3–011), IMMEDIATE (IP precedence 2–010), PRIORITY (IP Precedence 1–001), or ROUTINE (IP Precedence 0–000). The typical value used is CRITICAL.
- **IPTOS-RTP-LOWDELAY**—T field indicates whether the Delay bit is set in the IP ToS byte.
- **IPTOS-RTP-THROUGHPUT**—T field indicates whether the Throughput bit is set in the IP ToS byte.
- **IPTOS-RTP-RELIABILITY**—T field indicates whether the Reliability bit is set in the IP ToS byte.

To achieve a certain DSCP value, just convert it to the equivalent IP ToS settings. For example to set the t parameter to DSCP 46 (0xB8) you need to set IPTOS-RTP-PRECEDENCE to CRITICAL, IPTOS-RTP-LOWDELAY to Y, IPTOS-RTP-THROUGPUT to Y, and IPTOS-RTP-RELIABILITY to N.

To overwrite the DSCP values of NCS call signaling packets, include the DOCSIS service flow parameter, IP ToS Overwrite, in the signaling service flow definition in the cable modem configuration file. This parameter is defined as a 2-byte field where the first byte is the TOS-AND-MASK and the second byte is the TOS-OR-MASK. How this parameter works is a little confusing, so an example is included that should help explain it. Suppose this parameter is equal to 0.104, as it is in the nRTPS upstream service flow example used earlier. If this parameter is converted to hexadecimal, a TOS-AND-MASK of 0x00 and a TOS-OR-MASK of 0x68 result. First, the IP ToS byte of all upstream packets are binary AND'ed with the TOS-AND-MASK; because anything AND'ed with 0 is 0, this nulls out the current setting. Next, this result, 0x00, is binary OR'ed with the TOS-OR-MASK; because anything OR'ed with 0 is unchanged, this results in setting the IP ToS byte to 0x68.

Recall that 0x68 is DSCP codepoint 26, more commonly known as AF31, which is generally used for call signaling.

Another type of PacketCable packets that can benefit from QoS treatment that actually originate from the CMTS are COPS DQoS packets. For a Cisco CMTS this value can be set using the following command-line interface (CLI) command:

```
CMTS(config)#cops ip dscp value
```

The DSCP value can be entered as an integer from 0–63 or by using the DSCP name, such as af31. The recommended value is to use the same DSCP for COPS that you are using for the MGCP protocols, which is typically af31. The default value is 0.

NOTE You might also want to consider setting DSCP values for event messaging and lawful intercept packets coming from the CMTS.

Packets Originating from the CMS

Several types of PacketCable packets originate from the CMS that can benefit from QoS treatment across the backbone. Among these are NCS, TGCP, COPS, SIP/CMSS, EM, and lawful intercept packets.

If you are using the Cisco BTS 10200 as your CMS, the way to define these DSCP codepoint values is described next.

To define the IP TOS bytes used for MGCP packets originating from the BTS (this includes the NCS and TGCP PacketCable variants), use the following parameters defined in the call agent configuration table (CA-CONFIG):

- **MGCP-SIG-TOS-PRECEDENCE**—This field has a value from 0–7 and maps to the IP Precedence bits in the IP ToS byte. The typical value used is 3.
- **MGCP-SIG-TOS-LOWDELAY**—This field indicates whether the Delay bit is set in the IP ToS byte.
- **MGCP-SIG-TOS-THROUGHPUT**—This field indicates whether the Throughput bit is set in the IP ToS byte.
- **MGCP-SIG-TOS-RELIABILITY**—This field indicates whether the Reliability bit is set in the IP ToS byte.

Again, to achieve a certain DSCP codepoint value, just enter the IP ToS byte parameter interpretation equivalents. For example, to set NCS/TGCP packets to AF31 (0x68), set the MGCP-SIG-TOS-PRECEDENCE to 3, MGCP-SIG-TOS-LOWDELAY to N, MGCP-SIG-TOS-THROUGPUT to Y, and MGCP-SIG-TOS-RELIABILITY to N.

Similarly, to define the IP TOS bytes used for SIP/CMSS packets originating from the BTS, use the following parameters defined in the softswitch trunk group profile table (SOFTSW-TG-PROFILE):

- **SIP-SIG-TOS-PRECEDENCE**—This field has one of the following values: NETCONTROL (IP Precedence 7–111), INTERNETCONTROL (IP Precedence 6–110), CRITICAL (IP Precedence 5–101), FLASHOVERRIDE (IP Precedence 4–100), FLASH (IP Precedence 3–011), IMMEDIATE (IP Precedence 2–010), PRIORITY (IP Precedence 1–001), or ROUTINE (IP Precedence 0–000). The typical value used is CRITICAL.
- **SIP-SIG-TOS-LOWDELAY**—This field indicates whether the Delay bit is set in the IP ToS byte.
- **SIP-SIG-TOS-THROUGHPUT**—This field indicates whether the Throughput bit is set in the IP ToS byte.
- **SIP-SIG-TOS-RELIABILITY**—This field indicates whether the Reliability bit is set in the IP ToS byte.

To define the IP TOS bytes used for outgoing COPS packets, use the COPS-DSCP-TOS parameter. To define the IP TOS byte used for outgoing RADIUS packets, use the RADIUS-DSCP-TOS parameter; both are defined in the call agent configuration table (CA-CONFIG). The value of these fields is the DSCP codepoint value in decimal; for example, if you wanted to use AF31 (PHB 26), you would set this parameter to 104.

NOTE If the values of parameters defined in the CA-CONFIG table are changed, they won't take effect until a system switchover occurs. You can use the **control call_agent** command from an EMS CLI session to force a system switchover, as demonstrated in the following example:

```
CLI>status call-agent id=CA104
APPLICATION INSTANCE -> Call Agent [CA104]
PRIMARY STATUS -> STANDBY
SECONDARY STATUS -> ACTIVE
Reply : Success:
CLI>control call_agent id=CA104;target_state=active-standby;
APPLICATION INSTANCE -> Call Agent [CA104]
REASON -> Application instance is reconfigured successfully
Reply : Success:
CLI>status call-agent
CALL AGENT STATUS IS... ->
APPLICATION INSTANCE -> Call Agent [CA104]
PRIMARY STATUS -> ACTIVE
SECONDARY STATUS -> STANDBY
Reply : Success:
```

In this example, the initial state was with the secondary call agent machine active and the primary call agent machine standby. This is referred to as the standby-active state. So a target state of active-standby is needed to force a failover. To switch back, a target state of standby-active is used with the **control** command. A platform stop/start from the UNIX prompt also causes the CA-CONFIG table changes to take effect. However, it is recommended to do the system switchover method instead because it is less service-impacting.

Packets Originating from the MG

Similar to the MTA gateway case, call signaling and bearer packets coming from your trunking media gateways can be given QoS treatment to improve service. DSCP codepoints or IP Precedence bits for RTP voice packets can be specified via MGCP/TGCP using the type of service Local Connection Option. If you are using the Cisco BTS 10200 as your CMS, refer to the section, "Packets Originating from the CMTS," for the details on how to configure this functionality.

If you are using the Cisco VXSM card within an MGX8800 multiservice switch as your trunking gateway, you can define IP Precedence values for bearer and signaling packets using the following command:

```
MGX8800.5.VXSM.a > cnfiptos ControlTOS BearerTOS
```

The *ControlTOS* field is the DSCP codepoint value (specified in decimal) for signaling traffic, and the *BearerTOS* field is the DSCP codepoint value (specified in decimal) for bearer traffic.

You can view current settings with the **dspiptos** command:

```
MGX8800.5.VXSM.a > dspiptos
=========================================================
                VoIP Service Types
=========================================================
Control Service Type      :              104
Bearer Service Type       :              184
```

In this case, signaling is set to the default of 104 (AF31), and bearer is set to the default of 184 (EF).

If you are using the Cisco VISM card within an MGX8800 multiservice switch as your trunking gateway, IP Precedence values for RTP and signaling packets can be defined using the following command:

```
MGX.1.11.VISM8.a > cnfdeftos ctrprec ctrtos [berprec bertos]
```

The parameters *ctrprec* and *berprec* should be configured with the desired IP precedence values (0–7) for control/signaling packets and bearer/RTP packets, respectively. The parameters *ctrtos* and *bertos* are for the settings of the various IP types of service bits in the IP ToS byte for control/signaling packets and bearer/RTP packets, respectively. The possible values for these parameters are 0, 1, 2, 4, and 8, where 0 indicates no ToS bits are set, 1 indicates the cost bit is set, 2 indicates the reliability bit is set, 4 indicates the throughput bit is set, and 8 indicates the delay bit is set. Note: The bearer parameters are optional and are used only if the call agent does not dictate values in the MGCP messaging.

Another related QoS consideration for both VXSM and VISM cards is the use of separate IP addresses and separate ATM connections for bearer traffic and control traffic. Keeping these separate does provide some measure of security, but the real benefit is it doesn't allow one type of traffic to influence the other. If you have a lot of call traffic going through a card, call signaling traffic can get significantly delayed if both types of traffic share the same ATM virtual circuit. It is recommended to keep these connections separate. Configuring the connections is covered in Chapter 8.

Configuring Backbone QoS Policies

Modular QOS CLI (MQC) is used for configuring backbone QoS on Cisco IOS–based equipment. MQC utilizes a three-step QoS configuration mechanism, where the steps are as follows:

Step 1 **Define traffic classes**—This defines what type of traffic constitutes a given class (RTP, NCS, and so on). For example:

```
Router(config)#class-map VoIP_Bearer
Router(config-cmap)#match ip dscp ef
Router(config-cmap)#exit
Router(config)#class-map VoIP_Signaling
Router(config-cmap)#match ip dscp af31
```

In this example, classes are defined only for PacketCable bearer and call signaling traffic; other traffic classes can easily be defined as well.

Step 2 **Associate policies with each traffic class**—These policies specify how traffic flows are treated for each class using QoS parameters. Examples of these QoS parameters include minimum reserved bandwidth, queuing strategy, and prioritization. For example:

```
Router(config)#policy-map PacketCable
    Router(config-pmap)#class VoIP_Bearer
    Router(config-pmap-c)#priority 91000
    Router(config-pmap-c)#exit
    Router(config-pmap)#class VoIP_Signaling
    Router(config-pmap-c)#bandwidth 5000
    Router(config-pmap)#class class-default
    Router(config-pmap-c)#fair-queue
```

The priority statement assigns the bearer traffic class to an LLQ, which is a strict priority queue, meaning 91 kbps are reserved for these packets and the packets are serviced before any other type of traffic. The bandwidth statement reserves 5 kbps for call signaling traffic. This means that at least 5 kbps of bandwidth are guaranteed to be available for call signaling traffic at all times. The default class is needed to handle traffic not matching any previously defined traffic class.

NOTE This configuration is just an example. In a deployment MQC configuration, all the various types of traffic should be taken into consideration for the network.

Step 3 **Attach the policies to the backbone interfaces**—This step activates the QoS policy on the backbone interfaces. For example:

```
Router(config)#interface GigabitEthernet 1/0/0
Router(config-if)#service-policy output PacketCable
```

Use of MPLS in Backbone

Multiprotocol Label Switching (MPLS) is a way of speeding up backbone packet traversal by utilizing labels for switching packets instead of performing more time- and processor-intensive routing table lookups. These labels are prepended to packets on the ingress edge routers in the network and stripped off on the egress edge routers. Routers in the core of the network then switch packets based on these labels; hence, they are known as label switch routers (LSR). The label information is populated on LSRs using a label distribution protocol (LDP). It is beyond the scope of this book to go into the specific details of how MPLS works. You can find numerous books on this subject, and if you are interested in learning more, you should consult these resources.

For IPv4 networks the MPLS label field is 4 bytes long and inserted between the MAC layer header and the IP header. Figure 16-2 shows the format of this label.

Figure 16-2 *MPLS Label Format*

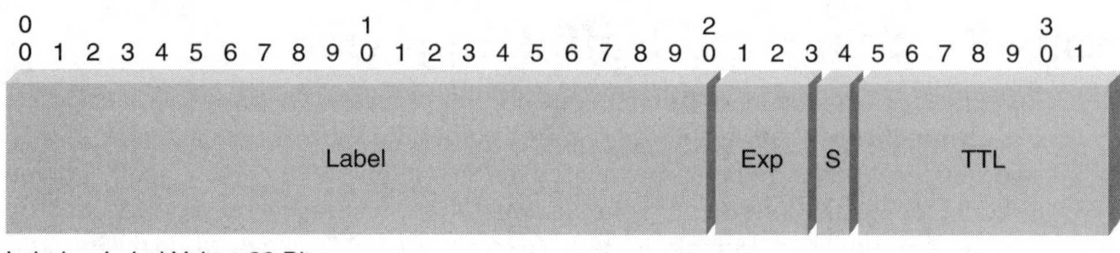

Label Label Value, 20 Bits
Exp Experimental Use, 3 Bits
S Bottom of Stack, 1 Bit
TTL Time to Live, 8 Bits

The actual tag field used to switch packets is the 20-bit label field. The 3-bit experimental field is used to convey IP Precedence/DSCP values. This is required so that MPLS label switch routers in the core can do packet classification by only examining the MPLS label. By default, the three IP Precedence bits get copied directly into the EXP field at the edge router. If DSCP codepoints are used in the network and because the EXP field contains just three bits, the entire IP DSCP value of six bits cannot be directly mapped into the EXP field. As a result, entire DSCP classes are mapped to EXP values based on their DSCP level; for example, the EF PHB maps to an EXP value of 5, and the AF31 PHB maps to an EXP value of 3.

An advantage of using MPLS in a service provider network is that it can provide security through the use of virtual private networks (VPN). The service provider edge routers, also known as PE routers, receive and classify routing information from directly connected customer edge (CE) routers. These CE routers are then logically tied in to a specific VPN. Information is exchanged between CE routers only if they are members of the same VPN. This means that the routing information within a VPN is isolated from other VPNs; thus, routes are secure, and IP addresses can even be reused. The LSRs in the core are referred to as provider (P) routers and don't have any VPN awareness; they simply switch packets based on their MPLS labels. If MPLS VPNs are used, a second MPLS label is prepended to each packet. One label is used for switching within the VPN, and the other is used for switching in the core network.

For a cable network, the CMTS can act as an edge LSR. VPNs can then be defined, and cable customers can be assigned to a specific VPN. A common application for this is where an MSO provides its customers with the option of choosing a specific Internet service provider (ISP) for its HSD services. Different ISPs are assigned to different VPNs, and their routing information is completely isolated from each other.

For a PacketCable service provider it might also be desirable to have its telephony network mapped to a specific MPLS VPN. This way, the PacketCable network is secured from other services (such as HSD) that the MSO might be providing.

Capacity Planning and Traffic Engineering

Proper capacity planning or traffic engineering is crucial in ensuring your customers' calls can obtain the necessary resources required across the PacketCable network. If this is not done properly, calls can be of low quality, or they might not complete at all. Obviously, if either of these situations exists, your customer's satisfaction decreases dramatically. Thus, you must design your network properly from end to end. Some of these design considerations are covered in this section.

Determining the Call Capacity of a DOCSIS Upstream

In all likelihood the bottleneck of your network is going to be the HFC leg; specifically, the DOCSIS upstream. It is important to realize that calls attempted that are unable to receive DOCSIS bandwidth do not even receive an announcement indicating that the call cannot currently be completed because of a lack of network resources; connecting to an announcement server takes bandwidth on the CMTS just as for an ordinary call. Typically, a tone such as fast busy is applied locally by the MTA to indicate failure to the user. This scenario is obviously extremely undesirable because it results in unhappy customers. Upstream capacity is partially dependent upon the options chosen by the MSO for voice deployment (such as CODEC, packetization, PHS, and security) and partially dependent upon the DOCSIS channel characteristics. The impact on the choices for all these variables is now covered in detail.

Initially, in this discussion it is assumed that your customers are connected in sparse mode; in other words, the MTAs on each fiber node or group of fiber nodes are physically connected to a single DOCSIS upstream only. Logically, one way you can provide more upstream bandwidth to your customers is to provide them physical access to multiple DOCSIS upstream channels. This connectivity method is known as dense mode. With dense mode combining, more bandwidth is available, and techniques such as dynamic load balancing can be used to optimize bandwidth usage. You learn more about dynamic load balancing later in this chapter. For now, the simpler case of sparse mode combining using a single upstream is the logical place to start.

Before you can figure out how many simultaneous calls can exist on a single upstream, you must first know the amount of bandwidth consumed by a single call. Chapter 11, "RTP and RTCP," covers how to determine the IP packet size based on CODEC and packetization period. In Chapter 12, "DQoS Architecture and Framework," you learned each upstream voice packet has an 18-byte DOCSIS MAC management header, a 6-byte DOCSIS header, and typically 8 bytes of DOCSIS extended headers (assuming BPI+ is enabled). Table 16-10 shows the net result of the 32 bytes of DOCSIS header overhead.

Table 16-10 *Single Call—Upstream DOCSIS Bandwidth Constraints*

CODEC	Pack. Period	Packets per Second	RTP Payload	IP Packet Size	DOCSIS Frame Size (UGS Size)
G.711	10 ms	100	80 bytes	120 bytes	152 bytes
G.711	20 ms	50	160 bytes	200 bytes	232 bytes
G.711	30 ms	33 1/3	240 bytes	280 bytes	312 bytes
G.728	10 ms	100	20 bytes	60 bytes	92 bytes
G.728	20 ms	50	40 bytes	80 bytes	112 bytes
G.728	30 ms	33 1/3	60 bytes	100 bytes	132 bytes
G.729E	10 ms	100	15 bytes	55 bytes	87 bytes

continues

Table 16-10 *Single Call—Upstream DOCSIS Bandwidth Constraints (Continued)*

CODEC	Pack. Period	Packets per Second	RTP Payload	IP Packet Size	DOCSIS Frame Size (UGS Size)
G.729E	20 ms	50	30 bytes	70 bytes	102 bytes
G.729E	30 ms	33 1/3	45 bytes	85 bytes	117 bytes
iLBC	20 ms	50	38 bytes	78 bytes	110 bytes
iLBC	30 ms	33 1/3	50 bytes	90 bytes	122 bytes
BV16	10 ms	100	20 bytes	60 bytes	92 bytes
BV16	20 ms	50	40 bytes	80 bytes	112 bytes
BV16	30 ms	33 1/3	60 bytes	100 bytes	132 bytes

Table 16-10 shows the resulting IP packet size, DOCSIS frame size, and packet rate for every possible CODEC/packetization period combination in PacketCable. This table does not take into account the effects of Payload Header Suppression (PHS) or secure RTP, which you'll learn more about later.

So now you probably think you know exactly how much upstream bandwidth each voice packet consumes. Well, not yet—you also need to take into consideration the lower-layer DOCSIS overhead. First, an error correction mechanism exists, adding overhead. This mechanism is Forward Error Correction (FEC), and it works by dividing each packet into constructs known as codewords and then appending parity bytes to each codeword. FEC parity bytes provide some measure of protection against bit errors caused by RF impairments. The more parity bytes used per codeword, the better protection against errors you have, but the more overhead you have.

NOTE In DOCSIS, the symbol k represents the codeword size, and the symbol T represents the number of parity bytes, where the number of parity bytes per codeword is 2*T. A general guideline for how much FEC coverage should be used is 7 percent, where FEC coverage = $2T/(2T + k)$.

The other types of lower-layer DOCSIS overhead are preamble and guard time. Remember that the DOCSIS upstream is an asynchronous transmission media, meaning additional signaling is required so that the other end (that is, the CMTS) knows when it is receiving a DOCSIS packet. This is done by prepending the upstream transmission with a well-known bit pattern, the preamble. Also to ensure the roll-off from successive upstream transmissions does not interfere with the previous transmission, you have a guard time period included between upstream transmissions. This value is defined to be at least as long as five symbols plus the maximum timing error between a CM and the CMTS. A typical guard time value is 8 symbols.

At this point, all this information might be a bit confusing, so some examples can hopefully clear things up for you. Assume the G.711 CODEC is used with a 20-ms packetization period. This equates to 160 bytes of RTP payload. Adding the 40 bytes of IP/RTP/UDP header and 32 bytes of DOCSIS MAC layer overhead yields a total packet size of 232 bytes. It is from this 232-byte packet that the codewords are made. So where are these codeword and other DOCSIS layer parameters defined? Well, they are defined on the CMTS and then sent to the cable modems in the Upstream Channel Descriptor (UCD) DOCSIS messages.

For a Cisco CMTS these parameters are defined under each cable interface in configuration mode. Figure 16-3 depicts this configuration on an MC5x20S cable linecard in a Cisco uBR10K CMTS.

Figure 16-3 *Sample MC5x20 Configuration*

Here, the DOCSIS parameters for the first upstream channel are defined. Parameters pertinent to capacity planning are channel-width, minislot-size, and modulation-profile.

```
C0701-10012-PRE-A#show running-config interface cable 7/0/0
Load for five secs: 2%/0%; one minute: 1%; five minutes: 1%
Time source is NTP, .20:39:47.154 EST Tue Nov 14 2006

Building configuration...

Current configuration : 1510 bytes
!
interface Cable7/0/0
 no ip address
 load-interval 30
 cable tftp-enforce mark-only
 no cable packet-cache
 cable bundle 1
 cable downstream annex B
 cable downstream modulation 256qam
 cable downstream interleave-depth 32
 cable downstream frequency 609000000
 cable downstream channel-id 0
 no cable downstream rf-shutdown
 cable downstream rf-power 60
 cable upstream 0 connector 0
 cable upstream 0 frequency 22000000
 cable upstream 0 channel-width 3200000 1600000
 cable upstream 0 minislot-size 2
 cable upstream 0 power-level 0
 cable upstream 0 range-backoff 3 6
 cable upstream 0 modulation-profile 29
 no cable upstream 0 shutdown
 cable upstream 1 connector 1
 cable upstream 1 frequency 22000000
 cable upstream 1 channel-width 3200000 1600000
 cable upstream 1 minislot-size 2
 cable upstream 1 power-level 0
 cable upstream 1 range-backoff 3 6
 cable upstream 1 modulation-profile 29
 no cable upstream 1 shutdown
 cable upstream 2 connector 2
 cable upstream 2 frequency 22000000
 cable upstream 2 channel-width 3200000 1600000
 cable upstream 2 minislot-size 2
 cable upstream 2 power-level 0
 cable upstream 2 range-backoff 3 6
 cable upstream 2 modulation-profile 29
 no cable upstream 2 shutdown
 cable upstream 3 connector 3
 cable upstream 3 frequency 22000000
 cable upstream 3 channel-width 3200000 1600000
 cable upstream 3 minislot-size 2
 cable upstream 3 power-level 0
 cable upstream 3 range-backoff 3 6
 cable upstream 3 modulation-profile 29
 no cable upstream 3 shutdown
```

Three upstream parameters defined here are relevant to capacity planning. One is the channel width parameter, which defines the size of the upstream RF pipe; in this case, it is set to the maximum possible value in DOCSIS 1.1, which is 3.2 MHz. Logically, the larger the upstream channel, the greater its capacity for handling user traffic. The second pertinent parameter is minislot size; in DOCSIS, bandwidth is defined in minislots, not bytes. You'll see how to convert from bytes to minislots in a bit. The third capacity-related parameter is the cable modulation profile. The modulation profile defines all the other upstream parameters. In this case, modulation profile #29 is used, which is defined as the default mix type of profile for the Cisco MC5x20 cable linecard (the global configuration command **cable modulation-profile 29 mix** was used to create it).

You might be wondering, what is a mix type of modulation profile? Well, as you know, you have several different types of DOCSIS messages, such as data packets (used to carry RTP, MGCP, and other TCP/IP protocols) and management messages (such as the ranging messages a cable modem sends and receives to stay online with the CMTS). These messages can all use the same upstream modulation scheme (for example, QPSK, 16-QAM), or they can use different ones. Remember, QPSK modulation uses 2 bits to represent each symbol, whereas 16-QAM modulation uses 4 bits to represent each symbol. Thus, with 16-QAM, twice as much data is sent in the same time interval than is sent with QPSK, but as a tradeoff, 16-QAM is more sensitive to RF impairments or noise. In a mix type of modulation scheme, data packets use 16-QAM modulation (so more user data can be transmitted) and ranging messages use QPSK modulation (to lessen the possibility of RF impairments causing a modem to drop offline). Figure 16-4 shows the details of modulation profile 29.

Figure 16-4 *Default Mix Modulation Profile*

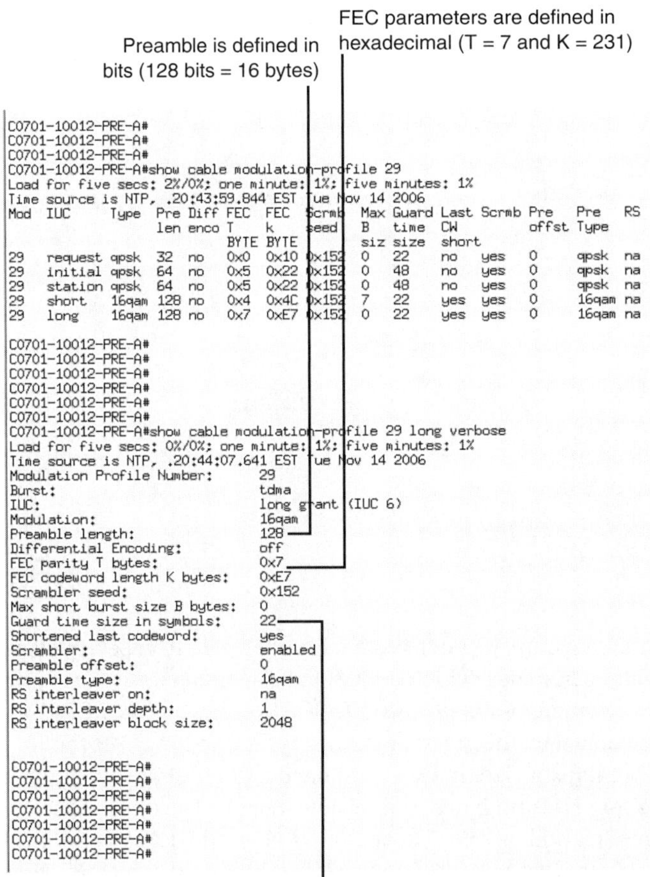

Notice in Figure 16-4 two types of data grants are defined: short and long. The distinction as to whether short or long data grants are used is based on the configured maximum burst size for short data grants. If the size of the transmission the modem wants to make is less than or equal to this value, it uses short data grants; otherwise, it must use long data grants. You can see from the figure that the maximum burst size for short data grants is equal to 7 minislots, where one minislot equals 6.25 μs. But what is this size in terms of bytes? Because of DOCSIS requirements and scheduler efficiency it is desirable to have one minislot equal to 16 bytes. To verify that one minislot equals 16 bytes use the following formula:

$$\frac{\#\,bytes}{\min islot} = \frac{\#\,timeticks}{\min islot} * \frac{6.25\ u\,\text{sec}}{timetick} * \frac{\#\,symbols}{\text{sec}} * \frac{\#\,bits}{symbol} * \frac{1\ byte}{8\ bits}$$

For 16-QAM modulation, 4 bits are used to represent a symbol. The channel symbol rate is deduced from the channel width: for a 3.2 MHz channel this is 2,560 kilosymbols per second, and for a 1.6 MHz channel this is 1,280 kilosymbols per second. So in this example the formula simplifies to 16 bytes/minislot as shown here:

$$\frac{16\ bytes}{\min islot} = \frac{2\ timeticks}{\min islot} * \frac{6.25\ u\,\text{sec}}{timetick} * \frac{2{,}560\ ksymbols}{\text{sec}} * \frac{4\ bits}{symbol} * \frac{1\ byte}{8\ bits}$$

Thus, the maximum burst size for a short data grant is 112 bytes (7 minislots * 16 bytes/minislot). Because voice packets are larger than 112 bytes, large data grants are used.

NOTE The value of 112 bytes corresponds to the size required for the transmission of a 64-byte packet. TCP clients such as web browsers regularly send 64-byte acknowledgement packets upstream, so short data grants are optimized for these packets.

Getting back to the example, now that you know long data grants will be used, you can figure out the total DOCSIS voice packet size. As you can see from Figure 16-4, the codeword parity parameter T is equal to 0x7, and the codeword length parameter k is equal to 0xE7 (231 in decimal). Thus, two codewords are needed: one of size 231 and one of size 1. Because shortened last keyword is enabled, you have no need to pad this last codeword to 231 bytes; however, it must be padded to the minimum size of 16 bytes.

NOTE Shortened last codeword permits the last codeword to be less than the codeword size; otherwise, zero fill bytes are required to make the last codeword k bytes long. In most cases, shortened last codeword should be enabled unless you have older cable modems on your network that do not support this functionality.

For each of these codewords, 14 parity bytes are inserted. So in total, a 231-byte codeword is transmitted followed by 14 parity bytes, and then a 16-byte codeword is transmitted followed by 14 parity bytes, for a total of 275 bytes.

Finally, the preamble and guard time overhead needs to be accounted for. From the figure, you can see the preamble length for the long data grant is 128 bits or 16 bytes. The guard time is given in symbols and is equal to 22. With 16-QAM modulation there are 4 bits per symbol, so the guard time equates to 88 bits or 11 bytes.

So the total amount of DOCSIS upstream capacity required to transmit a single voice packet is 302 bytes (275 + 16 + 11). But DOCSIS allocates upstream data in terms of minislots not bytes, and only integer numbers of minislots can be granted. So any fractional amount must be rounded up to the next integer. This is why it's important to keep the minislot size to the minimum value of 16 bytes, which minimizes the number of wasted bytes in a minislot. In this case, 19 minislots (304 bytes) are required per voice packet. As you can see, this is a lot of overhead considering you have only 160 bytes of voice payload.

The modulation profile can be altered from the default, which is much more efficient. Figure 16-5 illustrates an optimized modulation profile.

Figure 16-5 *Optimized Mix Modulation Profile*

FEC parameters have been modified (T = 9 and K = 232)

```
C0701-10012-PRE-A#
C0701-10012-PRE-A#
C0701-10012-PRE-A#
C0701-10012-PRE-A#
C0701-10012-PRE-A#
C0701-10012-PRE-A#sh cable modulation-profile 30
Load for five secs: 0%/0%; one minute: 1%; five minutes: 1%
Time source is NTP, .16:32:07.178 EST Wed Nov 15 2006
Mod  IUC     Type  Pre Diff FEC  FEC  Scrmb Max Guard Last  Scrmb Pre   Pre   RS
                   len enco T    k    seed  B   time  CW          offst Type
                            BYTE BYTE       siz size  short
30   request qpsk  32  no   0x0  0x10 0x152 0   22    no    yes   0     qpsk  na
30   initial qpsk  64  no   0x5  0x22 0x152 0   48    no    yes   0     qpsk  na
30   station qpsk  64  no   0x5  0x22 0x152 0   48    no    yes   0     qpsk  na
30   short   16qam 128 no   0x4  0x4C 0x152 7   22    yes   yes   0     16qam na
30   long    16qam 128 no   0x9  0xE8 0x152 0   22    yes   yes   0     16qam na

C0701-10012-PRE-A#
C0701-10012-PRE-A#
C0701-10012-PRE-A#
C0701-10012-PRE-A#sh cable modulation-profile 30 long verbose
Load for five secs: 2%/0%; one minute: 1%; five minutes: 1%
Time source is NTP, .16:32:12.930 EST Wed Nov 15 2006
Modulation Profile Number:     30
Burst:                         tdma
IUC:                           long grant (IUC 6)
Modulation:                    16qam
Preamble length:               128
Differential Encoding:         off
FEC parity T bytes:            0x9
FEC codeword length K bytes:   0xE8
Scrambler seed:                0x152
Max short burst size B bytes:  0
Guard time size in symbols:    22
Shortened last codeword:       yes
Scrambler:                     enabled
Preamble offset:               0
Preamble type:                 16qam
RS interleaver on:             na
RS interleaver depth:          0
RS interleaver block size:     0

C0701-10012-PRE-A#
C0701-10012-PRE-A#
C0701-10012-PRE-A#
C0701-10012-PRE-A#
C0701-10012-PRE-A#
C0701-10012-PRE-A#
C0701-10012-PRE-A#
C0701-10012-PRE-A#
C0701-10012-PRE-A#
```

With this modulation profile long data grants are also used for voice packets because the maximum burst size for the short data grant interval is still 7 minislots, or 112 bytes. The FEC codeword size has been increased to 0xE8, or 232 bytes. This number should look familiar because it is the size of one voice packet. As a result, only a single FEC codeword is required and consequently one set of FEC parity bytes. The number of FEC parity bytes per codeword has also been increased to two times 0x9 bytes, or 18 bytes. This permits greater protection against errors caused by RF noise. Before, the FEC coverage was 14 bytes for every 231-byte codeword for a coverage of 14/(231 + 14) or ~5.7 percent; with this optimized modulation profile the FEC coverage is now 18 bytes for every 232-byte codeword, for a coverage of 18/(232 + 18), or ~7.2 percent. Coverage of around 7 percent is generally regarded as optimum. With this optimized profile, each voice packet consumes 277 bytes (232-byte codeword + 18 bytes FEC parity + 16 bytes preamble + 11 bytes guard time) of DOCSIS upstream channel capacity. Again, because DOCSIS schedules bandwidth in minislots, not in bytes, this is rounded up to 18 minislots (288 bytes) of upstream DOCSIS capacity used per voice packet.

This optimized modulation profile does not exist by default on the CMTS. It can be created by entering the commands shown in Figure 16-6 from global configuration mode.

Figure 16-6 *Modulation Profile Creation*

```
C0701-10012-PRE-A#
C0701-10012-PRE-A#conf t
Enter configuration commands, one per line.  End with CNTL/Z.
C0701-10012-PRE-A(config)#
C0701-10012-PRE-A(config)#cable modulation-profile ?
  <1-10>      DOCSIS 1.X Modulation Profile Group
  <121-130>   DOCSIS 1.X/2.0 Mixed Modulation Profile Group for MC520 Line Card
  <21-30>     DOCSIS 1.X Modulation Profile Group for MC520 Line Card
  <221-230>   DOCSIS 2.0 Only ATDMA Modulation Profile Group for MC520 Line Card

C0701-10012-PRE-A(config)#cable modulation-profile 30 ?
  initial     Initial Ranging Burst
  long        Long Grant Burst
  mix         Create default QPSK/QAM-16 mix modulation profile
  qam-16      Create default QAM-16 modulation profile
  qpsk        Create default QPSK modulation profile
  reqdata     Request/data Burst
  request     Request Burst
  robust-mix  Create robust QPSK/QAM-16 mix modulation profile
  short       Short Grant Burst
  station     Station Ranging Burst

C0701-10012-PRE-A(config)#cable modulation-profile 30 mix
C0701-10012-PRE-A(config)#cable modulation-profile 30 long 9 232 0 22 16qam scrambler 152 no-diff 128 shortened
C0701-10012-PRE-A(config)#end
C0701-10012-PRE-A#
C0701-10012-PRE-A#
C0701-10012-PRE-A#
001990: .Nov 15 16:30:30.020 EST: %SYS-5-CONFIG_I: Configured from console by cisco on consolesh cab
C0701-10012-PRE-A#sh cable modu
C0701-10012-PRE-A#sh cable modulation-profile 30
Load for five secs: 3%/0%; one minute: 1%; five minutes: 1%
Time source is NTP, .16:30:34.993 EST Wed Nov 15 2006
Mod IUC     Type  Pre Diff FEC  FEC  Scrmb Max Guard Last Scrmb Pre   Pre   RS
                  len enco T    k    seed  B   time  CW         offst Type
                           BYTE BYTE       siz size  short
30  request qpsk  32  no   0x0  0x10 0x152 0   22    no   yes   0     qpsk  na
30  initial qpsk  64  no   0x5  0x22 0x152 0   48    no   yes   0     qpsk  na
30  station qpsk  64  no   0x5  0x22 0x152 0   48    no   yes   0     qpsk  na
30  short   16qam 128 no   0x4  0x4C 0x152 7   22    yes  yes   0     16qam na
30  long    16qam 128 no   0x9  0xE8 0x152 0   22    yes  yes   0     16qam na

C0701-10012-PRE-A#
C0701-10012-PRE-A#
C0701-10012-PRE-A#
C0701-10012-PRE-A#
C0701-10012-PRE-A#
C0701-10012-PRE-A#
```

So with the optimized modulation profile, each call consumes 18 minislots for every packet. With the G.711 CODEC and a 20-ms packetization interval, 50 packets are sent every second. Thus, the upstream DOCSIS capacity constraint of a single call is 115.2 kbps (18 minislots/packet * 16 bytes/minislot * 8 bits/byte * 50 packets/second). Because you now know how much DOCSIS capacity one call takes up, you can figure out approximately how many calls can simultaneously exist on the entire upstream channel.

The theoretical throughput capacity of a 16-QAM, 3.2 MHz channel is 10.24 Mbps (2,560 kilosymbols/sec * 4 bits/sec). However, not all this capacity is available for voice traffic. For example, DOCSIS ranging messages must be sent to keep modems online as well as permit new modems to come online. Another example of other traffic existing on the DOCSIS upstream is signaling traffic required to set up each voice call. A good rule of thumb is that 80 percent of the upstream capacity can be used for voice traffic (in this case, 8.192 Mbps). Of course, using these design criteria means a minimal amount of bandwidth is available for other types of traffic such as high-speed data users. If you intend to support other services on the upstream channels, you must use the percentage of bandwidth allocated for voice traffic in this calculation. For this discussion, it's assumed that 80 percent of the upstream capacity is available for voice traffic. Thus, in this case a single DOCSIS upstream can support a maximum of 71 simultaneous calls (8.192 Mbps/115.2 kbps). For information on how to control how much bandwidth is used for voice, please see the admission control discussion near the end of this chapter.

Earlier it was mentioned that several variables affect voice packet size and DOCSIS upstream capacity. This next section looks at those variables one at a time and their impact on what you've already learned.

The Effect of Changing Packetization Interval

The packetization intervals supported in PacketCable are 10 ms, 20 ms, and 30 ms. In the previous example a 20-ms packetization interval was used along with the G.711 CODEC.

Now you can examine what happens when the packetization interval is changed. First, take a look at the 10 ms case. As you might guess, the only thing that is going to change in the voice packets is the size of the RTP payload. Instead of 160 bytes of payload in a single packet, each packet now contains only 80 bytes of payload. As a tradeoff, twice as many packets must be sent; you'll see the benefits and drawbacks of this in a bit. The decrease of 80 bytes in the RTP payload results in a decrease of 80 bytes in the required UGS size; in other words, the UGS size of upstream voice packets will now be 152 bytes instead of 232 bytes.

If the optimized modulation profile from the previous example is used, a single shortened codeword is required. The total physical DOCSIS layer overhead is the same as before: 18 FEC parity bytes, 16 preamble bytes, and 11 guard time bytes. Thus, the total packet size is 197 bytes, which would require 13 minislots per packet. As you might guess, because this modulation profile was optimized for a 20-ms packetization interval, it is not optimal for a 10-ms packetization interval. To optimize the profile, change the FEC codeword size parameter k to the UGS size, and then modify the FEC parity byte T until a ~7 percent coverage results. In this case, change k to 152 (0x98) and T to 6. With this new profile only

12 FEC parity bytes are needed; thus, the total packet size is 191 bytes, meaning 12 minislots are required per packet.

So what effect did changing the packetization interval have on the number of calls possible on a single upstream channel? Each call now sends 100 packets per second instead of 50, and each one of these packets is 12 minislots instead of 18. Thus, the upstream capacity constraint of a single call is 153.6 kbps (12 minislots/packet * 16 bytes/minislot * 8 bits/byte * 100 packets/second) compared to 115.2 kbps before. Using the 80 percent of upstream bandwidth available for voice guideline means a single DOCSIS upstream can support a maximum of 53 simultaneous calls (8.192 Mbps / 153.6 kbps).

As you can see, by sending packets twice as often, the negative effects of all the packet overhead have increased. Only 53 calls per upstream are now possible instead of 71. So why decrease the packetization interval? By sending the packets more often, the amount of delay in sending packets end-to-end has decreased. If delay is an issue in your network, decreasing the packetization interval is one way to improve call quality. Also because a fewer number of voice samples reside in each packet, the loss of a single packet is not as damaging. It is for these reasons that voice-band modem and facsimile calls, which are much more sensitive to delay and packet loss, benefit from a smaller packetization interval.

Logically, if decreasing the packetization interval reduces call capacity but improves end-to-end delay, then increasing the packetization interval has the reverse effect. Suppose an interval of 30 ms is used. Now the RTP payload size per packet would be 240 bytes, and the UGS size would be 312 bytes. There would now be 33 1/3 packets sent per second.

Again, the modulation profile can be modified so that the FEC codeword size is 312 bytes, and the T parameter is set to the maximum of 10. Note: In this case, the FEC coverage is ~6 percent, which is the maximum possible for this size codeword. Each RTP packet now requires 23 minislots (368 bytes) for a per call bandwidth of ~98.13 kbps. By using the same method as before, 83 calls are possible on a single upstream. One way to still achieve the 7 percent FEC coverage guideline is to make each voice packet two codewords. In other words, set the codeword size to half the UGS size (156 bytes) and the T parameter to 6. If this is done, each RTP packet still requires 23 minislots, so the per-call bandwidth is unchanged, yet the FEC coverage has increased. Because of the delay consequences, it is generally not recommended to use a 30-ms packetization interval.

The Effect of Changing CODEC

Thus far, all the examples have assumed that the noncompression G.711 CODEC is used. However, one of the major benefits of digitally encoding voice is that compression techniques can be implemented, allowing more calls to utilize the same bandwidth capacity. As you learned in Chapter 11, four compression-based CODECs are supported in PacketCable. Two of these are the ITU G.728 and G.729 Annex E CODECs, and the other two are the royalty-free iLBC and BV16 CODECs, which must be supported in

PacketCable 1.5. The net effect of these CODECs is smaller packet sizes, as shown in Table 16-9.

To see how the use of one of these CODECs changes upstream call capacity, take a look at an example of using G.729E with a 20-ms packet rate. This results in a UGS size of 102 bytes. So an optimized modulation profile has a FEC codeword size of 0x66 (102) bytes and an FEC parity parameter of 4 (resulting in an FEC coverage of ~7.3 percent). The amount of DOCSIS upstream capacity required to send one packet would be 137 bytes; rounding to the nearest minislot boundary results in nine minislots (144 bytes) required. So the amount of upstream capacity consumed by one call is 57.6 kbps. This results in an upstream capacity of 142 calls, quite a difference from the G.711 case. Similar results would be seen with the other compression CODECs as well.

As with most things, you face a tradeoff with these bandwidth savings. Voice quality is slightly less using a compression CODEC; however, in most cases the difference is negligible. The exception is with fax and voice-band modem calls; in these cases the CODEC is typically changed to the G.711 CODEC after these tones are detected. Also dropped packets are usually more damaging when a compression CODEC is utilized. You must take extra care to mitigate the possibility of dropped packets when using compression CODECs.

The Effect of Implementing PHS

Another way to increase the amount of simultaneous calls possible on a DOCSIS upstream is by implementing Payload Header Suppression (PHS). PHS was introduced and covered in detail back in Chapter 12, so refer to that chapter for more detail. In summary, PHS works by substituting an index for the redundant parts of the packet header overhead. For voice communication a majority of the fields contained in the UDP header, IP header, and DOCSIS MAC management header are identical for each packet, so transmitting this redundant information is pointless. Instead, when the DOCSIS dynamic service flows used for voice are created, a PHS index is also set up representing this redundant information. In theory, for upstream packets the entire UDP header (8 bytes), IP header (20 bytes), and DOCSIS MAC management header (14 bytes) can be replaced by a PHS index. Because the PHS index is sent within DOCSIS headers, which are already being sent anyway, a savings of 42 bytes per packet can result. For downstream packets the source and destination MAC addresses cannot be suppressed because they are needed by the MTAs to identify packets, so at most the PHS savings is 28 bytes per packet (30 bytes of data can be suppressed, but a 2-byte DOCSIS extended header must be added that contains the PHS index). The number of bytes suppressed is entirely dependent on the MTA implementation of PHS.

By using the initial G.711/20 ms example and assuming PHS is implemented optimally, you can readily see the impact of PHS.

Recall that without PHS, the UGS size in this case is 232 bytes. If PHS is implemented so that 42 bytes are suppressed in the upstream, the resulting UGS size drops to 190 bytes. An optimized modulation profile would have a codeword size of 0xBE (190) bytes, a codeword parity byte parameter of 7 (resulting in a coverage of 6.9 percent). Thus, the total amount of upstream capacity required to transmit one voice packet is 231 bytes, which when rounded up to the nearest minislot boundary results in 15 minislots required (240 bytes). The upstream capacity consumed by one call would then be 96 kbps, resulting in a total upstream capacity of 85 calls.

So the upstream call capacity has increased from 71 to 85 by implementing PHS. Also keep in mind the savings in the downstream direction as well. Obviously, because PHS is a DOCSIS feature, it affects capacity only through the DOCSIS portion of your network. So what is the downside of implementing PHS? In theory it should be very minimal, but really it depends on the MTA and CMTS equipment vendors. The process of replacing packet headers with a PHS index and vice versa should be seamless because it is done in hardware.

Also note that PHS can be implemented in conjunction with using a compression CODEC to drastically increase upstream call capacity. For example, if the G.728 CODEC is also used, the resulting UGS size is now 70 bytes. This is small enough so that short data grants are now used, and only six minislots (= 96 bytes) are required per packet! As a result, 213 calls are now theoretically possible in this same amount of upstream bandwidth.

The Effect of Implementing VAD

The topic of implementing silence suppression or voice activity detection (VAD) has also been addressed throughout this book. VAD works by not sending packets or by sending smaller and less frequent comfort noise packets during periods where the subscriber is silent. However, even when the subscriber is silent and DOCSIS bandwidth is not active, it is still admitted and, hence, unavailable for other QoS services. Consequently, VAD has no effect on upstream call capacity.

The Effect of Implementing Security

So far, all the variables discussed except for VAD have resulted in increasing upstream call capacity. One important consideration can decrease call capacity if implemented. This is PacketCable security for voice packets, which was discussed earlier in this chapter. However, the impact is not as great as you might first think. Remember that RTP packet encryption results in the payload being encrypted, but the header is unaffected. Thus, encryption alone does not change packet size. If authentication is also enabled, a small trailer known as the Message Authentication Check (MAC) is appended to the packet. For RTP, the MMH_2 and MMH_4 authentication algorithms are supported, which impose a 2-byte and a 4-byte MAC, respectively.

So for the G.711/20 ms case the UGS size increases to 236 bytes if the MMH_4 security authentication algorithm is used. In this case, 18 minislots are still required to transmit a single voice packet (just as in the case without security); so capacity is the same!

Differences Among Cable Linecards

You might also encounter capacity differences relating to the cable hardware in your CMTS. For example, if you are using a Cisco CMTS in your network, differences in the architecture of the individual cable linecards can affect call capacity. Namely, advanced cards such as the MC28U and MC5x20 series have functionality that can detect and remove certain types of RF impairments. As a tradeoff for this advanced functionality, larger preamble patterns and guard times are often needed compared to legacy cable linecards such as the MC16C. You might have noticed when creating a cable modulation profile in Cisco IOS specific number ranges are reserved depending on the type of cable linecard to account for these differences.

Line Usage and Traffic Pattern Considerations

Now that you know how to calculate upstream call capacity, you need to also figure out how many MTAs should be allowed on a per-upstream basis. This decision is based on the combined probability each MTA phone line is active in a given time interval.

A common metric used in computations relating to traffic engineering is the erlang, named after the Dutch mathematician Agner Krarup Erlang. The erlang represents the utilization of a telephony line or a group of lines. For a single line, it has a value between 0 and 1 that correlates to the percentage of time that a line is active or "off hook." So if one line has an erlang value of 0.5, it is active half of the time and idle the other half. If a group of six lines is each active 50 percent of the time, then the erlang value for the group is 3.0 (6 times 0.5).

On an end-office switch in the PSTN, calls either come into a switch on lines and leave the switch on trunks or come into the switch on trunks and leave the switch on lines. Remember, trunks are DS0 time-division multiplexed circuits that exist on a T1 or an E1 carrier connected to another switch. If the number of lines on the switch equates to the number of trunks on the switch, then it is certain that you have a trunk available for each active line, even if all the lines on the switch are active at the same time. Having this much capacity is not practical because the odds of all lines being active at the same time is extremely rare. If instead the number of trunks is equal to the erlang value of all the lines, then these trunks would be sufficient and constantly used if the line usage for the group was uniform in time. In other words, using the example of six lines, each with an erlang value of 0.5, if at all points in time only half of the lines were in use, then three trunks would suffice. However, the odds of such uniformity in time are also extremely rare.

As you might guess, a value in between these two extremes is optimal. Of course, with the number of trunks less than the number of lines, you always face the possibility that calls will not go through. This variable is referred to in the PSTN as the blocking probability, or grade of service (GOS). The grade of service represents the tolerable amount of failed calls; a common value used for this parameter is 1 percent. This means that out of every 100 calls attempted, one fails or gets blocked because of a lack of trunk resources. Note: A GOS of 1 percent is typical for an ONNET-to-ONNET call; a value of 0.5 percent is typical for a call going through the PSTN. The relationship between the grade of service, combined line erlang value, and number of required trunks is defined in the erlang B formula. Figure 16-7 shows the erlang B formula.

Figure 16-7 *Erlang B Formula*

$$P_r = \frac{\frac{A^N}{N!}}{\sum_{J=0}^{N} \frac{A^j}{j!}}$$

- A = Offered Traffic Load in Erlangs
- N = Number of Trunks
- P_r = Probability of Blockage (GoS)

This formula is well known, and tables and spreadsheets exist that figure out one variable given the other two variables. By simply doing an Internet search, you can find a web site that can figure out one of these variables given the other two. For example, suppose you can accept a grade of service of 1 percent and have a total traffic load of 50 erlangs—this formula then returns the value 64 for the number of trunks needed.

When the total number of erlangs exceeds a certain threshold, it has been observed that a much simpler relationship exists between erlangs and trunks; this is the 90 percent rule, where the number of trunks required is simply the erlang traffic load divided by 90 percent or 0.9. The erlang traffic load upon which one should switch to this simpler formula is dependent upon the grade of service. For a grade of service of 1 percent, this demarcation point is 181 erlangs. So if the traffic load is greater than 181 erlangs and the grade of service is 1 percent, the 90 percent formula should be used instead of the erlang B formula.

Some other points about the erlang-B formula are worth mentioning. One is that the erlang B formula does not take into account retries. In other words, it assumes that when a customer attempts a call and it fails, that customer does not hang up and try again. Another formula exists that takes retries into account. This is the extended erlang B formula, and as you might guess, it is an enhancement to the erlang B formula. In addition to the parameters previously mentioned, it also uses the percentage of failures that retry. Programs and web sites that compute using this formula can also be found on the Internet. If you don't have an accurate number for the percentage of call retries, a value of 50 percent is a reasonable guess. Another aspect of the erlang B formula worth mentioning is that it assumes a rather large number of traffic sources. For situations where the number of lines to the number of trunks ratio is less than 8 to 1, another formula called the Engset formula has been proven to give more accurate results.

The erlang value for a single line is always between 0 and 1. Erlang values are not tied to a specific time interval; it is a common misconception that erlang usage defines line usage over a one-hour period. A value of one erlang for a single line means this line is continuously utilized for any time period. If the time period is 1 hour, then the usage could be for a single 60-minute call or ten 6-minute calls or any other combination that adds up to 1 hour. Because individual line utilization is always between 0 and 1, a different metric is often used for per-line utilization, which has a much larger scale. The Centrum Call Second (CCS) metric accomplishes this by measuring line utilization in terms of number of intervals active in an hour period. This interval is 100 seconds, and one CCS means a line is active for one Centrum Second per hour. Because one erlang indicates the line is active for the entire hour (3600 seconds), you have 36 CCS in one erlang. A typical value used to measure the utilization of a line in the PSTN is 3–6 CCS. Table 16-11 shows the CCS to erlang conversion for these values.

Table 16-11 *CCS-Erlang Conversion*

3 CCS	0.083 erlang
4 CCS	0.111 erlang
5 CCS	0.139 erlang
6 CCS	0.167 erlang

Another term traditionally used for traffic engineering *is Busy Hour utilization*, which equates to the utilization for the busiest hour of the day. Voice networks are often designed to handle the load during the busiest hour on the busiest day of the year; this term is referred to as the High Day Busy Hour (HDBH). If the network can handle the traffic volume during this time, it can handle the load at all other times during the year. Other factors involved in traffic engineering include average call hold time and average call completion rate. Common values for these parameters include 8 minutes for residential calls, 2.5 minutes for business calls, and a 75 percent call completion rate.

In a PacketCable network the traffic engineering techniques discussed so far can be used to decide how much OFFNET trunking gateway capacity is needed to handle the call loads generated by the line side PacketCable MTAs. This is covered in more detail later. Additionally, these techniques can also be used to figure out how many MTAs should be allowed to exist on a single DOCSIS upstream. In this case, the line side part of the equation is still the phone lines on the MTAs, but the trunk side is the number of simultaneous calls supported by the DOCSIS upstream. In other words, the trunk side is the number of available UGS service flows on the DOCSIS upstream. This parameter was calculated in the previous section.

Another important assumption before moving on is the number of voice lines supported on a single MTA. PacketCable MTAs typically support two telephony lines; however, it is entirely up to the MSO to decide how many lines to enable per MTA. Obviously, if it is known that only one line per MTA is utilized instead of two, more MTAs can coexist on a DOCSIS upstream channel.

So to figure out the number of MTAs that should be allowed on a single DOCSIS upstream, take the number of trunk resources (also known as upstream UGS service flows) and the acceptable grade of service or blocking probability to figure out the traffic load in erlangs. Using a grade of service of 1 percent as an example and the maximum number of calls supported in the first example, 71, and plugging these into the erlang B formula yields an erlang value of 57. Recall that this was using the G.711 CODEC, a 20 ms packetization interval, no PHS, the MC5x20 cable linecard, no security, and 80 percent of the upstream bandwidth. As you have seen, a change in any one of these variables changes the supported erlang load on a single upstream. For example, if PHS is optimally implemented, 85 calls can exist, and the erlang value becomes 70. This means the busiest load that one upstream can handle for an hour is a total MTA line utilization of 57 erlangs in the nonPHS case and 70 erlangs in the PHS case.

So, now that the total erlang capacity is known, the per-MTA erlang usage can be used to figure out the total number of MTAs. For this example, assume both MTA lines are enabled for all customers and that the utilization for line one is 5 CCS (0.139 erlang) and the utilization value of line two is 3 CCS (0.083 erlang). Note: A per-line utilization of 5 CCS is a typical value seen in the PSTN. In this example it assumed that all line ones share the same utilization pattern as do all line twos. If a more accurate separation of utilization is known, this can be used in the calculations as well. For example, if you have a mixture of residential and business subscribers on a single upstream, each with different utilization values, you can use the percentages of each in the sample calculations that follow. Similarly, if both MTA lines are not enabled for all customers this variable can be incorporated into the following formula as well:

MTAs * (line1 erlang + line2 erlang) = Total erlang

Therefore, the # MTAs is equal to the total erlang value divided by the erlang value of a single MTA. So, if the maximum total traffic use is 57 erlangs and each MTA contributes 0.222 erlangs of traffic, then at most you could have 256 MTAs (57/0.222) on a single upstream. If PHS were implemented, then at most you could have 315 MTAs (70/0.222).

Another number worth looking at as a sanity check is the total number of lines to the total number of trunks ratios. For the non-PHS case this would be 7.21 (512/71), and for the PHS case this would be 7.41 (630/85).

Traffic engineering using the erlang B formula is an excellent way of providing initial estimates of required network resources. However, as actual call patterns are unpredictable, this method is not foolproof. For example, this formula assumes a random (Poisson) distribution of call arrivals and does not take into account events that trigger a sudden burst of call activity, such as a certain television advertisement or a natural disaster. For example, in the United States the reality television program *American Idol* causes huge bursts of traffic previously not seen in the PSTN. It is for this reason that monitoring trunk or, in this case, DOCSIS upstream use is critical. If upstream use is consistently high because of a high volume of voice calls, the traffic engineering numbers must be adjusted and network engineering must be modified accordingly.

CMTS Considerations Beyond the Upstream Limitations

A critical topic to keep in mind is that all of this discussion so far has focused on the limitations on an individual upstream channel. Other CMTS limitations might exist, impacting your upstream capacity and traffic engineering numbers. These limitations can be on a per-DOCSIS MAC domain, cable linecard, or on the chassis level.

For example, a single downstream channel typically supports four upstream channels on a Cisco CMTS. The combination of downstream and upstream channels available to a group of users is known as a DOCSIS MAC domain. If the downstream channel uses 256-QAM modulation, the resulting downstream capacity is around 38 Mbps. If each upstream channel uses a 3.2 MHz channel and 16-QAM modulation, then it has a capacity of around 10 Mbps. So the total upstream bandwidth is roughly equal to the downstream bandwidth in the MAC domain. But if the downstream channel uses 64-QAM modulation, then its capacity is around 27 Mbps and cannot support as much bandwidth as its four upstream channels; consequently, in this scenario the downstream becomes the bottleneck.

If you have DOCSIS 2.0 upstream channels that support much more capacity, reaching downstream channel capacity is a very real possibility. On a Cisco CMTS the virtual interface bundling feature can be used to map upstream channels to downstream channels to achieve an optimal association. For example, with an MC28U card on a Cisco uBR7246VXR, three 3.2 MHz 16-QAM upstream channels could be associated with a 64-QAM downstream channel to provide a voice service, and the remaining five upstream channels could be associated to the other downstream channel, providing a high-speed data service.

Another consideration for a MAC domain is the DOCSIS limitation of 8,191 unicast SIDs. Remember, a DOCSIS SID is required for each upstream service flow. If these SIDs are spread out evenly across the four upstream channels, then 2,047 SIDs are available per upstream. Each MTA typically has at least 2 SIDs: one for primary Best Effort and one for call signaling traffic. Each active call also utilizes an SID for its UGS traffic. In the earlier example at most 71 simultaneous calls can be active, and the traffic engineering calculation identifies 256 MTAs as the maximum on an upstream. Thus, 583 SIDs are consumed for these MTAs. So in this case, the SID limitation is not an issue. However, if techniques are employed where more calls are possible on an upstream and if the utilization is low enough so that more MTAs can coexist, this might become an issue—not to mention that newer services imply more sophisticated CM configurations (that is, more service flows and thus more SIDs). Also, PCMM can trigger the creation of upstream service flows that require SIDs as well. Finally, if DOCSIS set-top boxes are used on the CMTS, they can eat into the pool of available SIDs rapidly.

A logical way to figure out the maximum number of active voice calls possible on a MAC domain is to take the maximum number of calls per upstream and multiply it by the total number of upstream channels. So for the basic G.711/20-ms example, this is 71 times 4 for a total of 284 calls. Similarly, to get the maximum number per linecard multiply this by the number of MAC domains on a linecard. If the CMTS is a Cisco uBR10k and MC5x20 cable linecards are used, this would be 1,420 (284 * 5) because you have five MAC domains on the MC5x20S card. Finally, to get the maximum number per chassis simply add up the totals for each cable linecard. For a Cisco uBR10k this could be 8 MC5x20 cards, although 7 is more likely as one card is typically used for redundancy. Note: Linecard redundancy is covered later in this chapter. So using the same example, this would be 9,940 calls per chassis!

However, this type of analysis might not be accurate because it doesn't take into account several other factors. For example, with this many calls active it is important that you have enough WAN-side capacity to handle the traffic. For example, if all these calls are with subscribers not on the same CMTS chassis, then all the RTP traffic would need to leave the box. So for G.711/20-ms calls, each packet leaving the CMTS would be 218 bytes long (assuming an Ethernet WAN interface), and each call would generate 50 packets per second. So for 9,940 calls the WAN interface traffic would be close to 900 Mbps. Given that Ethernet is not 100 percent efficient, this RTP traffic alone would in all likelihood exhaust a Gigabit uplink. Other important aspects to keep in mind include CPU utilization, memory resources, and packet switching limitations. Note: With a Cisco CMTS it is recommended you keep CPU utilization below 80 percent. Thus, the traffic engineering numbers discussed earlier might be fine for a single upstream but would not scale to covering an entire chassis. Also keep in mind that not all CMTS processors (that is, the NPEs for the uBR7246VXRs and the PREs for the uBR10012s) have the same performance, which might not scale to the chassis level.

Capacity Planning on the CMS

Capacity issues relating to the maximum number of simultaneous calls are typically not an issue when it comes to the CMS. After a call is active, you don't have much involvement with the CMS unless some kind of mid-call feature is invoked, such as call waiting. However, capacity issues in terms of call ramp-up rate can be an issue. The number of calls possible in a single second interval is a common metric used to evaluate CMS performance. This metric is usually directly related to both the number of CPUs used on the CMS as well as the performance or speed of these CPUs. Also keep in mind the type of call can make a big difference in the impact on the CMS. For example, a simple ONNET-to-ONNET call is typically far less processor-intensive than an ONNET-to-OFFNET call performing a 1-800 TCAP query or a call using Class V features such as three-way calling and call forwarding.

Also you often face limitations because of the memory allocated to the database used to store configuration data. For example, the number of total provisioned subscribers is typically limited by the amount of memory available on the CMS.

The Cisco BTS 10200 CMS runs on four SUN servers with two of the servers used as a redundant pair of call agents and the other two servers used as a redundant pair of element management systems used for provisioning and billing. Currently, three SUN server platforms are supported for use with the Cisco BTS 10200 software: the Netra 240, Netra 440, and Netra 1280 server platforms. Which hardware configuration you use depends on the size capacity of the network. For small deployments the Netra 240 servers can be used with 2 processors and 8 GB of memory on each. For larger deployments either the Netra 440 servers or Netra 1280 servers can be used. Use Netra 440 servers with 4 processors and 8 GB of memory. The Netra 1280 server platform can run in three different hardware configurations depending on the network size: 4 CPUs/8 GB of memory, 8 CPUs/16 GB of memory, or 12 CPUs/24 GB of memory.

As for traffic engineering on the CMS, you can use the erlang techniques discussed earlier. Based on the total line utilization of your subscribers, you can estimate the OFFNET trunk capacity required. Obviously, one additional piece of information you need is the percentage of calls that are off network. As an example, suppose your network has a total line utilization of 1,000 erlangs and that 60 percent of the calls are either ONNET-to-OFFNET or OFFNET-to-ONNET. Thus, you have 600 erlangs of PSTN traffic in your network. This amount of traffic is far above the threshold on which the simple 90 percent rule mentioned earlier can be used instead of the more complex erlang B formula. Using this formula, the result in 667 trunks required. Remembering that one trunk equates to one DS0 and that you have 24 DS0s in a T1, you can conclude that 28 T1s are needed. This is equal to one T3. So roughly for every 600 erlangs of OFFNET traffic in your network you need a T3 connection to the PSTN.

Capacity Planning on the Trunking Gateway

Capacity planning on trunking gateways is a bit simpler than it is in other places in your network. Basically, you just have to make sure that the gateway's IP connectivity to the network core can handle the maximum amount of traffic generated by all the gateway's endpoints. A majority of the IP traffic entering and exiting the trunking gateway is RTP traffic, but you also have some signaling traffic and some RTCP traffic as well.

If you are using Cisco VXSM or VISM cards in an MGX 8800 chassis as your trunking gateway, you need to be concerned about capacity in two places. One is on the ATM virtual circuits between the card and either the PXM or RPM cards. Here you need to define your circuits to handle the total capacity of a single card. Please refer to Chapter 8 for more detail relating to ATM capacity in the Cisco MGX.

The other place where capacity is a concern is in considering the traffic to and from the MGX as a whole. Here you must be sure you have enough capacity to the PacketCable backbone to handle all the traffic generated by and sent to all the VXSM and VISM cards in the MGX. Typically, you want to have at least two RPM-XF modules in the MGX as well as two Gigabit Ethernet interfaces to the PacketCable backbone. Of course, if you have several VXSM cards in the chassis (especially the OC-3 variety), you might need additional RPM-XF modules to handle the capacity.

Redundancy, Redundancy

A major challenge in implementing a successful voice over IP network is to simulate a level of service availability as close to the legacy PSTN networks as possible.

Manufacturers of telephony equipment deployed in circuit-switched networks have had decades of experience in providing an infrastructure with minimal downtime. The Telcordia standards organization mandates an availability of 99.94 percent for an end-to-end PSTN network and an availability of 99.99 percent for an individual network element. Therefore, to replace a legacy PSTN network a PacketCable network must also have an end-to-end availability of 99.94 percent.

You might be asking yourself how this availability number is calculated. You have two standard methods for calculating end-to-end high availability—the multiplication method and the addition method. The multiplication method is recommended for most cases, whereas the addition method is effective only for cases where availability is extremely high.

When you use the multiplication method, end-to-end availability (E2EA-M) is the product of the availabilities of the individual network components (AN1, AN2, ... ANx), as shown in the following formula:

E2EA-M = AN1 * AN2 * ...ANx

When you use the addition method end-to-end, availability (E2EA-A) is the sum of the unavailabilities of the individual network components (UN1, UN2, ... UNx) subtracted from 1, as shown in the following formula:

E2EA-A = 1 - (UN1 + UN2 + ...UNx)

As an example, suppose the network is made up of two components, each with an availability of four 9s (99.99 percent). Then, using the multiplication method, the end-to-end availability is 0.9999 * 0.9999 = 0.99980001. If the addition method is used, end-to-end availability is 1 - (0.0001 + 0.0001) = 0.9998. As you might imagine, if an individual component has a poor availability, the availability of the entire network suffers; hence, an end-to-end system is only as available as its weakest link.

Figure 16-8 depicts an example for a PacketCable environment. In this case, the availability is for a single zone PacketCable network to provide service for an ONNET-to-ONNET call between two subscribers connected to the same CMTS. Availability is calculated using the addition method, and the individual unavailabilities for each component are listed as well.

Figure 16-8 *High Availability Example*

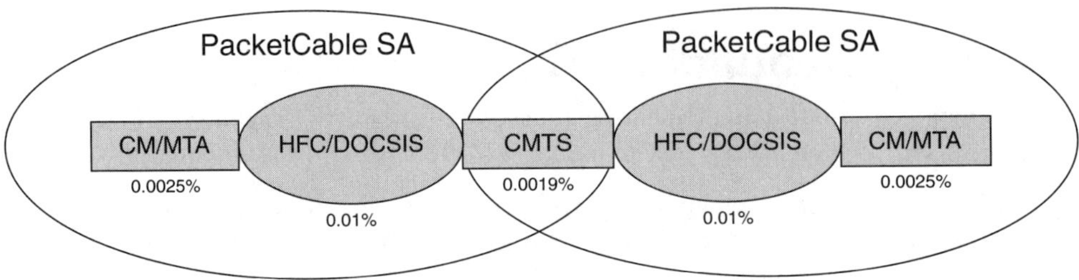

This example is a bit oversimplified because it doesn't account for components like the CMS responsible for call signaling and the provisioning components.

To achieve a high end-to-end availability for a PacketCable network, you must have high availability exist for all the network components. This includes the CMS, CMTS, MTA, MG, and SG components, as well as the OSS servers and backbone devices. The next part of this chapter covers high availability for several of these network components.

CMS Availability

Because the CMS component is the "brains" in a PacketCable network, ensuring it's highly available is a primary concern. Typically, the CMS is deployed on redundant sets of hardware, where one unit is operating in active mode, and one or more other units are operating in standby mode. These machines then use some sort of heartbeat mechanism to ensure the availability of the other. If a standby machine detects the loss of the active machine, it takes over the active role, minimizing the impact to the network. Also if the active machine detects an event harmful enough, it can initiate a failover to the standby machine; for example, if the active CMS detects the loss of Signaling System 7 (SS7) connectivity, it can decide to initiate a switchover. During periods where the network is stable, the multiple machines synchronize configuration changes and the status of its gateways.

Basic redundancy considerations should be implemented for the call agent devices similar to what you will soon learn about concerning the CMTS devices. For example, multiple power sources and cooling fans should be available so a single component failure does not bring down the system.

The Cisco BTS 10200 CMS uses the type of redundant architecture mentioned thus far. The individual components are deployed in pairs on separate SUN hardware platforms. Specifically, SUN Netra 240, 440, or 1280 servers are used. These SUN servers support a pair of either AC or DC power supplies. It is recommended that these power sources be uninterruptible. Note: Using these platforms the BTS 10200 CMS is Network Equipment Building System (NEBS) compliant for environmental capability, it conforms to the FCC standard for its Electromagnetic Interface (EMI), and it conforms to the Underwriters Laboratories (UL) standard for information technology equipment.

Recall that the Cisco BTS 10200 call agent runs three different main components. One is the call agent process responsible for basic call processing, and the other two are feature server processes responsible for the handling of a variety of call features such as local number portability (LNP) queries and call forwarding. Typically, the two feature server processes and the call agent process are installed on the same machines, but separate hardware platforms can also be used. Also a pair of servers called Element Management Systems (EMS) is included in the BTS architecture. The EMS consists of two different components: one responsible for system configuration and the other for call billing. Both the call agent and EMS function of the BTS operate in a dual active/standby pair for redundancy using multiple SUN servers. It's important to note that these devices do not operate in a load-sharing fashion where both are simultaneously actively processing calls. One device always assumes the active role, and one is idle in the standby role.

Periodically during the life of a call, call state information is synchronized between the active and standby machines. The two machines are also continually exchanging keepalive messages. If the standby machine detects the active machine is no longer operational because of missing keepalive messages, it immediately assumes the active role and can continue processing active calls. Thus, active calls are not lost during the switchover period; however, the ability to process new calls might not be possible during the switchover period. When the failed call agent returns to service, it assumes the standby role.

The call agent and EMS machines communicate with each other and with external devices via multiple network interfaces. On each CA machine you have two Ethernet interfaces used for EMS communication, redundancy, and management and two Ethernet interfaces used for call signaling. On each EMS machine, two Ethernet interfaces are used for call agent communication, billing collection, redundancy, and management. This BTS configuration is logically referred to as a 4/2 configuration.

All the Ethernet cables from these machines are wired into a pair of Cisco series switches dedicated for the BTS platform. Each of these switches has a management VLAN and a call signaling VLAN. Theses switches are also connected to a pair of upstream routers used for call signaling purposes and one or two upstream routers used for management purposes. Thus, as you can see there is redundancy everywhere in this configuration. There is redundant call signaling interfaces on each call agent machine, and there is redundant management interfaces on both the call agent and EMS machines to protect against a cut cable or bad network interface card (NIC). Also multiple series switches and multiple upstream routers protect against a switch failure. Figure 16-9 is an illustration of this configuration.

Figure 16-9 *BTS High Availability Architecture*

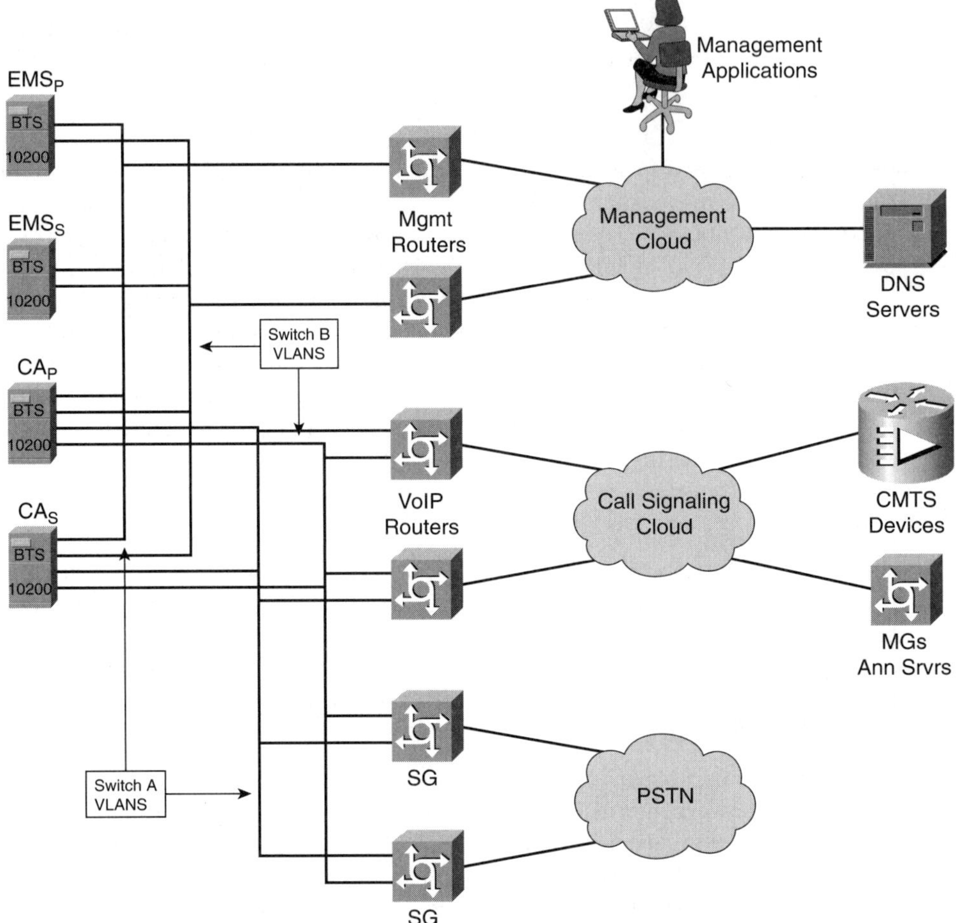

IP routing within the management cloud and within the call signaling cloud should have multiple paths through different routers to reach the edge devices shown in Figure 16-9. Also the IP feature, IRDP, should be enabled on the four interfaces on the VoIP routers connected to the switches and on the two interfaces on the management routers connected to the switches, with the VoIP router interfaces having priority. IRDP enables the BTS devices to dynamically discover their default gateway routes.

The use of DNS is also very important to the health of the BTS; therefore, redundant DNS servers should be in place. These servers should be placed behind a device configured for server load balancing. This way, a single IP address can be configured on the BTS, and the SLB device can handle spreading the DNS load between multiple servers.

High availability for call signaling is also achieved through the use of logical IP addresses for the call signaling interfaces of the BTS. Each call agent device has two physical interfaces dedicated to call signaling, for a total of four interfaces and four IP addresses. But because only one call agent device is active at any given time, only two of these interfaces are functional. To prevent devices from contacting interfaces that are not functional, virtual IP addresses are used that always represent the active call agent. A process known as IP Manager handles the migration of logical IP addresses to the standby device during a failover.

Another suggestion to improve availability and overall network health is related to the way OFFNET routing is configured. To increase the availability of OFFNET PSTN routing, you can use multiple trunking media gateways to connect to the PSTN. Also if possible, implement connectivity to multiple places within the PSTN. Assuming these routes are equal in terms of cost, you can then set up call routing to load share across these gateways to prevent a single gateway from being overloaded.

CMTS Availability

To ensure the CMTS devices in your network have a minimal amount of downtime, consider a couple of areas:

- System redundancy
- Processor redundancy
- HFC redundancy
- WAN redundancy

The following sections cover each of these areas in greater detail.

System Redundancy

On any network, device cooling systems are needed to dissipate the heat generated by the circuits in the various components. They control temperature and prevent the device from overheating and potentially failing. Having backup cooling systems can prevent this problem if one fails. If you are using a Cisco CMTS, the cooling system consists of a fan assembly module with four fans that work at two different speeds: low and high. The operating speed for the fans is determined by the temperature of the fan module at the module air outlet. The fan module provides built-in redundancy against a single fan failure. If a single fan failure occurs, the fan assembly can still cool the chassis using the three functioning fans. On the chassis, the SINGLE FAN FAILURE LED turns yellow, indicating a failure of one of the fans. It is recommended you replace the fan module as soon as possible, even though the three fans are sufficient to cool the system. If you have multiple fan failures or the chassis temperature gets too high, the system automatically shuts down.

Redundant power supplies are also recommended on the CMTS. The Cisco uBR10k and uBR7246VXR both support both DC power entry modules and AC power entry modules and can have either two internal DC power entry modules or two internal AC power entry modules, but not a combination. A single power entry module can provide sufficient power for a fully loaded chassis if one power entry module fails. This happens automatically without impacting system operation.

If you are using the Cisco uBR10k, you also have at least one Timing, Communication and Control Plus (TCC+) card that performs the following functions:

- Generates and distributes a 10.24 MHz clock references to each cable interface linecard
- Generates and distributes 32-bit time stamp references to each cable interface linecard
- Allows software to independently power off any or all cable interface linecards
- Monitors the supply power usage of the chassis
- Provides two redundant RJ-45 ports for external timing clock reference inputs such as a Global Positioning System (GPS) or BITS clock

It is recommended to install two TCC+ cards where one is active and the other is standby. If the TCC+ card in the first slot is working at system power-up, it automatically becomes the active card, and the TCC+ card in the second slot becomes the standby card. The TCC+ cards monitor each other so that if the active card fails, the standby card becomes active without loss of data. The two RJ-45 connecters used for external timing are labeled primary and secondary and can be used to get a Stratum 3 external clock reference source that is traceable to a Stratum 1 clock source. This permits the Cisco uBR10012 reference clock to be synchronized to a Stratum 1 clock source, providing a free-running DOCSIS-quality clock reference and time stamp to the cable interface linecards. If present, the primary external clock reference on the active TCC+ card is used. If it is lost, the secondary clock reference on the active TCC+ card is used. If the active TCC+ card stops functioning, control is transferred to the standby TCC+ card, which uses its primary and secondary clock reference sources. If neither card has a valid clock reference, the active TCC+ card uses its own internal clock to provide the DOCSIS-quality clock reference and time stamp. Recall an external clock source is recommended to increase the quality of facsimile and voice-band data calls.

Note: Similarly, the Cisco uBR7246VXR can be equipped with a cable clock card that can provide an external T1 clock. Two DB-15 interfaces are available on the card to connect to a primary and secondary external clock. However, no card redundancy exists because the chassis supports only a single clock card.

Processor Redundancy

If the processor on your CMTS stops functioning, obviously it can no longer service your customers. Having redundant processors can greatly lessen the chances of this occurring.

On the Cisco uBR10k CMTS the main processor module is called the Performance Routing Engine (PRE). Currently, two different Performance Routing Engines are available, PRE1 and PRE2. The PRE performs all Layer 2 and Layer 3 packet routing and forwarding by using an advanced parallel processing architecture called Parallel eXpress Forwarding (PXF). This process separates the control plane from the data plane. The PXF architecture supports high-performance throughput with IP services enabled on every port. A general-purpose RISC processor supports high-performance Layer 3 forwarding of traffic.

Two PRE modules can be installed to provide processor-level redundancy. If the primary PRE fails, the secondary PRE automatically takes over operation of the chassis. Because all Cisco uBR10012 linecards are physically connected to both the primary and the secondary PRE modules, a switchover of PRE modules does not require human intervention to reset the interface cards because they automatically fail over to the redundant PRE. Also note that the PRE module is hot-swappable if you have a redundant PRE module in the chassis.

Two different configuration options are available with router processor redundancy:

- RPR
- RPR+

With the RPR option and two PRE modules under normal operating conditions, one PRE module is active and the other one is standby. When the active PRE module fails, the system automatically performs a switchover, and the standby PRE module takes control. Under normal operating conditions, the standby PRE module is only partially initialized. During a switchover, the standby PRE module completes its initialization procedure, including loading the Cisco IOS software, verifying the physical components of the chassis, and parsing the startup configuration file. Switchover involves resetting all installed and enabled linecards and respective interfaces. When the cable linecards are reset, all the previously registered modems go offline and are forced to reregister.

With the RPR+ option the standby PRE is fully initialized and configured. This significantly reduces the switchover time compared to RPR. The running configuration is continuously synchronized between the two PREs, and as a result, linecards don't get reset during the switchover. As a result, modems stay online and retain their Service Identifiers (SID) as well during the switchover.

RPR+ works by having the configuration information synchronized from the active PRE to the standby PRE at startup and whenever changes to the active PRE configuration occur. This synchronization occurs in two separate phases. First, when a standby PRE first comes online, configuration information is synchronized in bulk from the active RP to the standby RP. When a system with RPR+ is initialized, the active PRE performs a chassis discovery (discovery of the number and type of linecards and fabric cards in the system) and parses the startup configuration file. The active PRE then synchronizes this data to the standby PRE and instructs the standby PRE to complete its initialization. This method ensures that both PREs contain the same configuration information. Even though the standby PRE is fully initialized, it interacts only with the active PRE to receive incremental changes to the configuration files as they occur. CLI commands on the standby PRE are not supported. Incremental synchronizations contain either the modifications to the shelf configuration or the trigger that caused the modification. These changes include synchronization of startup configuration every time the running configuration is changed to startup configuration, synchronization of running configuration every time a CLI command is entered, and synchronization of any changes to the chassis state because of the linecard insertion/ linecard removal, etc.

When RPR+ is configured and two PREs are installed, two types of route processor switchover can occur:

- **Automatic switchover**—Switchover initiated by the standby PRE when it determines that the active PRE has failed.
- **Manual switchover**—Switchover initiated by CLI command used by an operator to force a failover from the active PRE to the standby PRE.

The default redundancy mode on the Cisco uBR10012 router is standard RPR redundancy. The following configuration is required to enable RPR+ on the CMTS:

```
router(config)#redundancy
router(config-red)#main-cpu
router(config-r-mc)#auto-sync standard
```

NOTE The Cisco uBR7246VXR CMTS uses a single processor card known as the Network Processing Engine (NPE).

HFC Redundancy

You usually provide redundancy in the HFC network by designating a single cable linecard or even an entire CMTS chassis as backup if an operational card fails.

The Cisco uBR7246VXR and uBR10K CMTS platforms use a proprietary protocol known as Hot Standby Connection to Connection Protocol (HCCP) to provide linecard and chassis redundancy. With the uBR7246vxr, HCCP provides 1+1 chassis redundancy for its cable linecards between two uBR7246 CMTS systems. This has the drawback that an entire CMTS chassis sits idle as a standby or "protect" CMTS. Consequently, this type of redundancy is often deemed cost prohibitive by MSOs. The uBR10k uses HCCP to provide individual linecard redundancy. Both types of redundancy require the use of a separate device known as an RF switch. By using a Cisco RF switch and the uBR10k you can have N+1 redundancy for your cable linecards where one backup or protect card provides redundancy for N active or working cards. There are two types of N+1 redundancy configuration:

- **8+1 (7+1)**—An eight-card redundancy scheme in which seven working cable linecards are protected by a single card. This is the default N+1 configuration for the Cisco uBR10k, which as you might recall supports eight cable linecards. This redundancy scheme is also referred to as 7+1 redundancy, which is the more physically accurate term.
- **4+1**—A four-card redundancy scheme in which four working cable linecards are protected by a single card.

The Cisco uBR3x10 RF switch is a rack-mount unit comprised of RF combiners and splitters. It has an Ethernet interface used to provide SNMP control for automatic redundancy channel selection. Each slot is comprised of ten upstream switch modules that provide 75-ohm termination and automatic upstream redundancy switching from 5 to 70 MHz and three downstream switch modules that provide 75-ohm termination and automatic downstream redundancy switching from 54 to 860 MHz. So, to connect to an MC5x20 card with 5 downstream channels and 20 upstream channels, two RF switch slots are needed. Eight slots (W1–W8) are designated for working or active cable linecards, two slots (P1–P2) are designated for protect cable linecards, and eight slots are designated for cable plant connectivity. The CMTS instructs the RF switch to switch the relays between active and protect slot modules using the SNMP protocol.

The 7+1 configuration is designed for a fully populated uBR10k chassis with seven active linecards and one protect linecard. This configuration can also be used for a chassis with five or six active linecards. If the number of active linecards is four or less, use the 4+1 configuration.

If you are trying to deploy 7+1 cable linecard redundancy for an uBR10k with MC5x20 cable linecards, you require two Cisco RF switches. These RF switches should be rack mounted directly above the uBR10k. Also specialized coaxial cables with MCX connectors are needed for interconnecting an MC5x20 to an RF switch. Figure 16-10 shows a picture of one of these cables.

Figure 16-10 *RF Switch Cables*

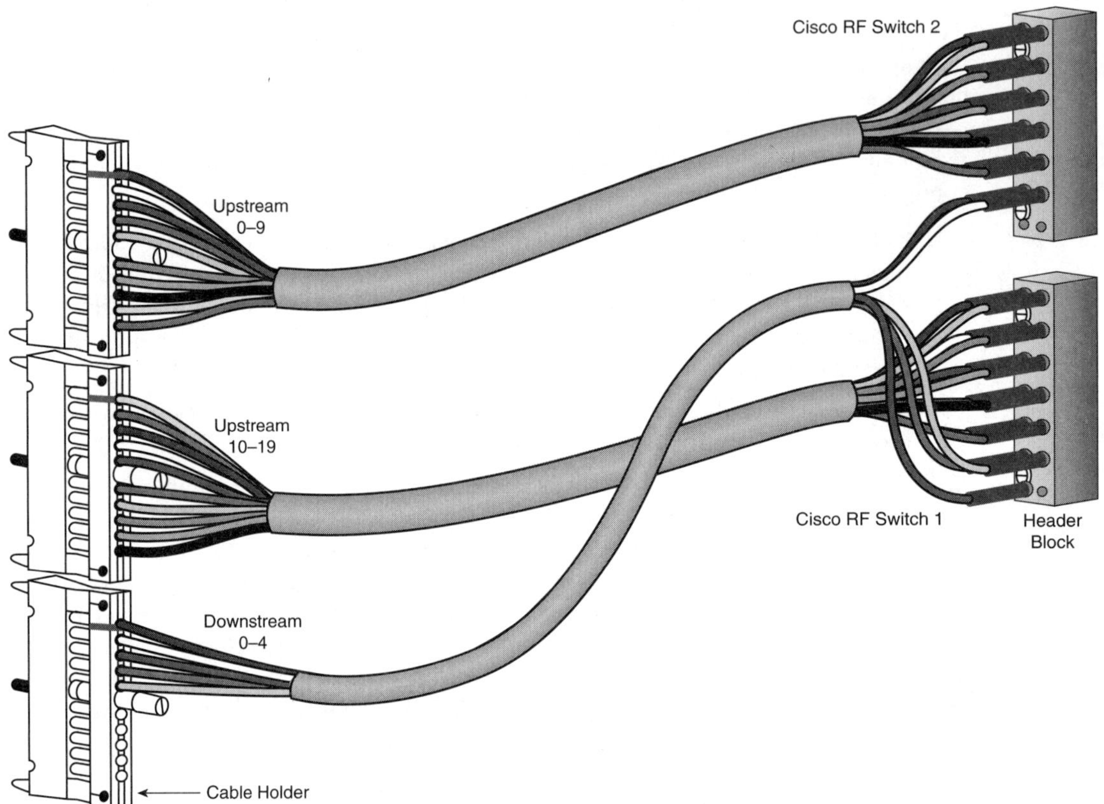

Looking at this figure, you can see that the physical connection to the RF switch is made with a piece of equipment known as a header block. The ports on the header block are A through N, and the RF cables are color-coded. Figure 16-11 illustrates the recommended mapping of cables to header block ports.

Figure 16-11 *Header Block Layout*

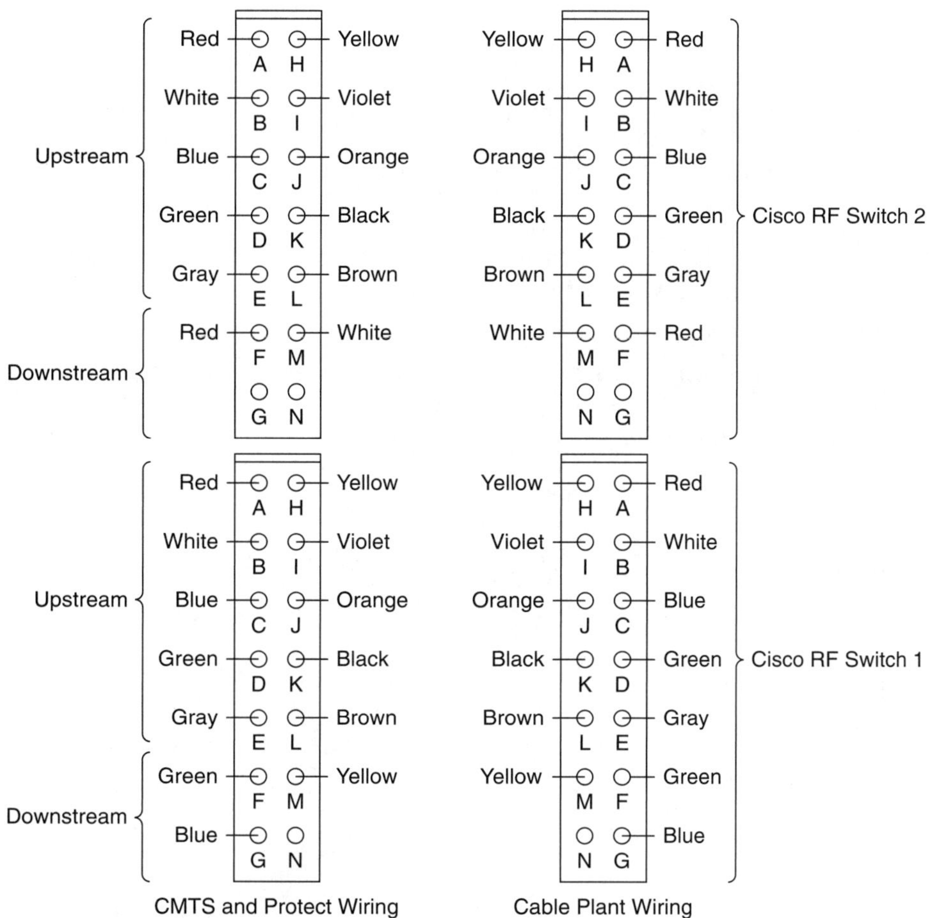

In this figure, the two connectors at the bottom half are for connectivity to RF switch #1, and the two at the top half are for connectivity to RF switch #2. The left half of this figure illustrates how the cabling to an MCx20 cable linecard should be done. The ports A–E and H–L on these two header blocks are for the 20 upstream channels on an MC5x20 linecard, and the ports F and M on the top header block and G, F, and M on the bottom header block are for the five downstream channels in numerical order.

Table 16-12 shows the recommended mapping of slots between the uBR10k and RF switches. Notice that slot 5/1 is designated as the protect linecard. Also note that the RF SW slots refer to the slots on both RF switches.

Table 16-12 *8+1 Redundancy Mapping Recommendation*

uBR10k MC5x20 Slot	8/0	8/1	7/0	7/1	6/0	6/1	5/0	5/1
RF SW Slot	W1	W2	W3	W4	W5	W6	W7	P1

For example, the connectivity matrix for the working MC5x20 card in slot 8/0 and the protect MC5x20 card in slot 5/1 are shown in Table 16-13.

Table 16-13 *Sample MC5x20-to-RF Switch Connectivity Mapping*

Working Linecard 8/0 Ports	**RF Switch—Port**	**Protect Linecard 5/1 Ports**	**RF Switch — Port**	**Cable Color**
US0	SW2-1A	US0	SW2-P1A	Red
US1	SW2-1B	US1	SW2-P1B	White
US2	SW2-1C	US2	SW2-P1C	Blue
US3	SW2-1D	US3	SW2-P1D	Green
US4	SW2-1H	US4	SW2-P1H	Yellow
US5	SW2-1I	US5	SW2-P1I	Violet
US6	SW2-1J	US6	SW2-P1J	Orange
US7	SW2-1K	US7	SW2-P1K	Black
US8	SW2-1E	US8	SW2-P1E	Gray
US9	SW2-1L	US9	SW2-P1L	Brown
US10	SW1-1E	US0	SW1-P1E	Gray
US11	SW1-1L	US1	SW1-P1L	Brown
US12	SW1-1A	US2	SW1-P1A	Red
US13	SW1-1B	US3	SW1-P1B	White
US14	SW1-1C	US4	SW1-P1C	Blue
US15	SW1-1D	US5	SW1-P1D	Green
US16	SW1-1H	US6	SW1-P1H	Yellow
US17	SW1-1I	US7	SW1-P1I	Violet
US18	SW1-1J	US8	SW1-P1J	Orange
US19	SW1-1K	US9	SW1-P1K	Black

Table 16-13 *Sample MC5x20-to-RF Switch Connectivity Mapping (Continued)*

Working Linecard 8/0 Ports	RF Switch— Port	Protect Linecard 5/1 Ports	RF Switch — Port	Cable Color
DS0	SW2-1F	DS0	SW2-P1F	Red
DS1	SW2-1M	DS1	SW2-P1M	White
DS2	SW1-1G	DS2	SW1-P1G	Blue
DS3	SW1-1F	DS3	SW1-P1F	Green
DS4	SW1-1M	DS4	SW1-P1M	Yellow

If your network design calls for 4+1 redundancy, only one RF switch is required. Table 16-14 shows the recommended mapping of slots between the uBR10k and the RF switch. As it was for the 8+8 case, slot 5/1 is designated as the protect linecard. Also it is assumed that the working linecards are in slots 8/0, 8/1, 7/0, and 7/1.

Table 16-14 *4+1 Redundancy Mapping Recommendation*

uBR10k MC5x20 Slot	8/0	8/1	7/0	7/1	5/1
RF SW Slot	W1, W5	W2, W6	W3, W7	W4, W8	P1, P2

When a protect linecard takes over for a failed working card, all the configuration is automatically copied between cards except for three configuration options: the downstream modulation, downstream interleave depth, and the DOCSIS annex mode. Manual configuration of these options is required on the protect linecard.

As of Cisco IOS Software Release 12.3(13a)BC, simplified HCCP configuration is supported for the 8+1 redundancy scheme on the uBR10k. As of Cisco IOS Software Release 12.3(17a)BC, simplified HCCP configuration is also supported for the 4+1 redundancy scheme on the uBR10k. These enhancements make HCCP configuration much easier on the cable operator. In the simplified configuration the assumption is that if one upstream or downstream channel fails, causing a single MAC domain to failover, the entire linecard should failover. This assumption is realistic because if the failover is because of a malfunction on the MC5x20 card, the card must be replaced; therefore, you do not want any customer traffic on this card. In this configuration, the IP address(es) of the RF switches must be manually specified and by default are assigned to the host name rfsw-1 and rfsw-2. The redundancy configuration consists of defining a linecard group in redundancy configuration mode (the same place where PRE RPR+ redundancy was defined) and then identifying the chassis slots used for working and protect cable linecards. The following is a sample configuration for a 4+1 redundancy scheme where the linecard is slot 8/0 is protected by the linecard in slot 5/1:

```
router(config)#ip host rfsw-1 192.168.101.1
router(config)#redundancy
router(config-red)#linecard-group 1 cable
```

```
router(config-red-lc)#rf-switch protection-mode 4+1
router(config-red-lc)#member subslot 5/1 protect
router(config-red-lc)#member subslot 8/0 working rfsw-slot 2 6
```

By default, the configuration assumes working linecards are connected to the RF switch(es) according to Table 16-12 for 8+1 and Table 16-14 for 4+1. If they are not, you can specify this by adding the **rfsw-slot** keyword; its options are the RF switch slot# connected to the top half of the MC5x20 card (DS 1-2, US 0-9) and the RF switch slot# connected to the bottom half of the MC5x20 card (DS 3-5, US 10-19). Notice in the sample configuration that slot 8/0 is using RF switch slots 2 and 6 instead of the standard slots 1 and 5.

You can display the configuration by using the following command:

```
route#sh redundancy linecard all
 Redundancy Mode: 4 + 1
 Interface Redundancy Detail
 ---------------------------
Interface Config   Grp Mbr RfSw-Name        RfSw-IP-Addr    RfSw-Slot Bitmap
Ca5/1/0   Protect  1   80  rfsw-1           192.168.101.1    2         0xFFFFFFFF
Ca5/1/1   Protect  2   80  rfsw-1           192.168.101.1    2         0xFFFFFFFF
Ca5/1/2   Protect  3   80  rfsw-1           192.168.101.1    2         0xFFFFFFFF
Ca5/1/3   Protect  4   80  rfsw-1           192.168.101.1    2         0xFFFFFFFF
Ca5/1/4   Protect  5   80  rfsw-1           192.168.101.1    2         0xFFFFFFFF
Ca8/0/0   Working  1   80  rfsw-1           192.168.101.1    2         0xFFFFFFFF
Ca8/0/1   Working  2   80  rfsw-1           192.168.101.1    2         0xFFFFFFFF
Ca8/0/2   Working  3   80  rfsw-1           192.168.101.1    2         0xFFFFFFFF
Ca8/0/3   Working  4   80  rfsw-1           192.168.101.1    2         0xFFFFFFFF
Ca8/0/4   Working  5   80  rfsw-1           192.168.101.1    2         0xFFFFFFFF
NOTE:
   In 4+1 Global Config Mode, the 8 port RF Switch is logically
   partitioned into 2 switches, each logical switch consists of 4 ports.
   Hence, 'RfSw-Slot' column (above) shows a value between 1 & 4.
 Subslot Summary
--------------------
                                RF Switch Slot*
   Subslot         Status      Current          Default
 -----------------------------------------------------------
    5/0            Unknown        -, -             -, -
    5/1            Protect       P, P             P, P
    6/0            Unknown        -, -             -, -
    6/1            Unknown        -, -             -, -
    7/0            Unknown        -, -             7, 3
    7/1            Unknown        -, -             8, 4
    8/0            Working       2, 6             5, 1
    8/1            Unknown        -, -             6, 2
 ----------------------------------------------------------------
 NOTE:
 Legend: - = Unused; P= Protect
 *    Indicates the slot numbers as seen on the front panel
      of RF Switch and maybe different from'Rfsw-Slot' column.
```

The configuration, by default, assumes the RF switch host names are rfsw-1 and rfsw-2, and the SNMP community name used to talk to the RF switches is private. These can be changed using the linecard redundancy configuration commands **rf-switch name** and **rf-switch snmp-community**. Also you can use some global commands to manually manipulate linecard redundancy, as demonstrated here:

```
router#redundancy linecard-group ?
  lockout     Lockout protection channel
  revertback  Specify LC you are reverting back to
```

```
switchover  Request an LC switch
unlockout   Unlockout protection channel
```

The **switchover** option enables you to initiate a failover on a specified working linecard, and the **revertback** option enables you to switch from the protect card back to the previous working card. The **lockout** option can be used to keep a working linecard from switching over when doing routing maintenance that would cause a temporary signal loss.

If you are running older code on the uBR10k or do not want a single MAC domain failover to cause an entire linecard failover, you can use the traditional configuration methods. In this type of HCCP configuration bit maps are used to convey to the CMTS how the RF switch cabling is set up. These bit maps are defined for each DOCSIS MAC domain. The bit map field is a 32-bit field where each bit corresponds to a port on the header block. A value of 1 for a given port means that port is in use for that MAC domain, and a value of 0 means that port is not in use. Figure 16-12 illustrates how these bit maps are determined, assuming the default combining of upstream channels to downstream channels.

Figure 16-12 *Bit Maps for 8+1 Redundancy*

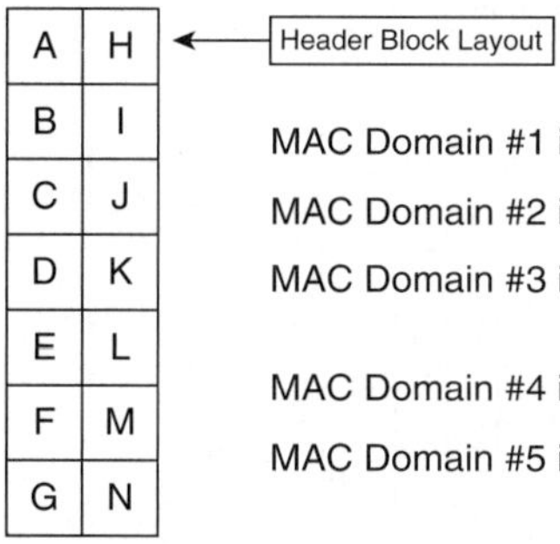

Header Block Layout

A	H
B	I
C	J
D	K
E	L
F	M
G	N

MAC Domain #1 is DS#0 (F) and US#0–3 (A, B, C, D) for SW#2

MAC Domain #2 is DS#1 (M) and US#4–7 (H, I, J, K) for SW#2

MAC Domain #3 is US#8–9 (E, L) for SW#2 and DS#2 (G) and US#10–11 (E, L) for SW#1

MAC Domain #4 is DS#3 (F) and US#12–15 (A, B, C, D) for SW#1

MAC Domain #5 is DS#4 (M) and US#16–19 (H, I, J, K) for SW#1

BIT MAPS

	32	31	30	29	28	27	26	25	24	23	22	21	20	19	18	17	16	15	14	13	12	11	10	9	8	7	6	5	4	3	2	1
	A	H	B	I	C	J	D	K	E	L	F	M	G	N																		
1	1	0	1	0	1	0	1	0	0	0	1	0	0	0	0	0	0	0	0	0	0	0	0	0	0	0	0	0	0	0	0	0
2	0	1	0	1	0	1	0	1	0	0	0	1	0	0	0	0	0	0	0	0	0	0	0	0	0	0	0	0	0	0	0	0
3	0	0	0	0	0	0	0	0	1	1	0	0	0	0	0	0	0	0	0	0	0	0	0	0	0	0	0	0	0	0	0	0
3	0	0	0	0	0	0	0	0	1	1	0	0	1	0	0	0	0	0	0	0	0	0	0	0	0	0	0	0	0	0	0	0
4	1	0	1	0	1	0	1	0	0	0	1	0	0	0	0	0	0	0	0	0	0	0	0	0	0	0	0	0	0	0	0	0
5	0	1	0	1	0	1	0	1	0	0	0	1	0	0	0	0	0	0	0	0	0	0	0	0	0	0	0	0	0	0	0	0

Bit Map for MAC Domains #1 and #4 is 0xAA200000
Bit Map for MAC Domains #2 and #5 is 0x55100000
Bit Map for MAC Domain #3 is 0x00C0 for SW#2 and 0x00C8 for SW#1

From this figure you can see how these bit maps are determined. Notice that the bit maps for MAC domains #1 and #4 are identical, as are the maps for domains #2 and #5; the only difference is that different RF switches are used. MAC domain #3 is the special case; it has two upstream channels connected to each RF switch; therefore, two bit maps are needed for this interface. A bit map of 0xFFFFFFFF indicates all MAC domains; therefore, a failover on a cable interface configured with this bit map causes the RF switch to failover the relays for all upstream and downstream connections on the linecard when one of them fails. This is the bit map assumed when using the simple configuration option.

Figure 16-13 illustrates how the bit maps are determined in the 4+1 redundancy case, assuming the default combining of upstream channels to downstream channels.

Figure 16-13 *Bit Maps for 4+1 Redundancy*

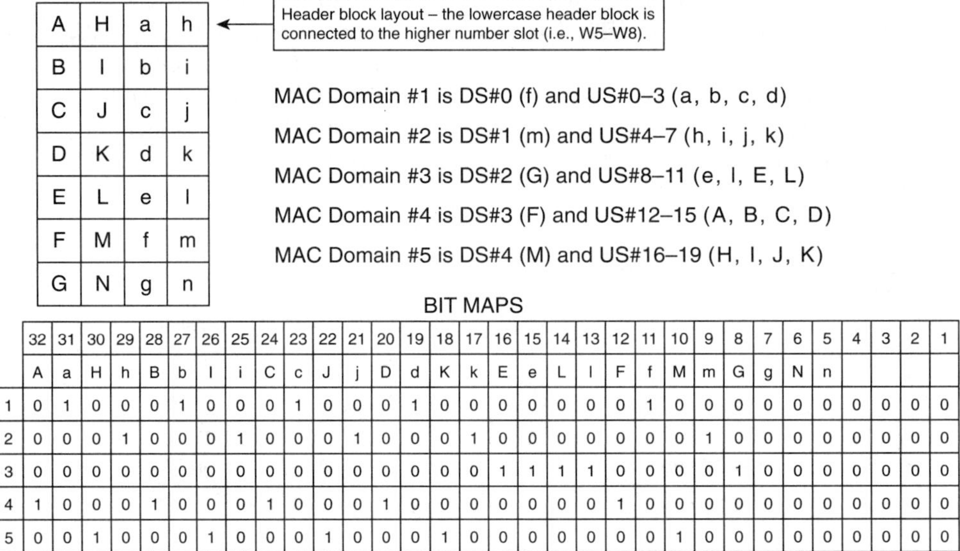

	32	31	30	29	28	27	26	25	24	23	22	21	20	19	18	17	16	15	14	13	12	11	10	9	8	7	6	5	4	3	2	1
	A	a	H	h	B	b	I	i	C	c	J	j	D	d	K	k	E	e	L	l	F	f	M	m	G	g	N	n				
1	0	1	0	0	0	1	0	0	0	1	0	0	0	1	0	0	0	0	0	0	0	1	0	0	0	0	0	0	0	0	0	0
2	0	0	0	1	0	0	0	1	0	0	0	1	0	0	0	1	0	0	0	0	0	0	0	1	0	0	0	0	0	0	0	0
3	0	0	0	0	0	0	0	0	0	0	0	0	0	0	0	0	1	1	1	1	0	0	0	0	1	0	0	0	0	0	0	0
4	1	0	0	0	1	0	0	0	1	0	0	0	1	0	0	0	0	0	0	0	1	0	0	0	0	0	0	0	0	0	0	0
5	0	0	1	0	0	0	1	0	0	0	1	0	0	0	1	0	0	0	0	0	0	0	1	0	0	0	0	0	0	0	0	0

The specific Cisco IOS commands used to configure HCCP are as follows (all commands are issued under the cable interface configuration):

- **hccp** *group* **working** *member-id*—Identifies the interface as a working interface for the specified group.
- **hccp** *group* **protect** *working-member-id ipaddr*—Identifies the interface as a protect interface for the specified group. The IP address specified is where redundancy status messages are transmitted and received.
- **hccp** *group* **channel-switch** *member-id switch-name* **rfswitch-group** *rfswitch-ip-address module-bitmap position*—Configures the RF switch as a hot standby HCCP member in a particular HCCP group.

It is recommended you set the *keepalive* value as low as possible on working cable interfaces so that switchover occurs as fast as possible. If no messages are received on any of the interface's upstream channels within three *keepalive* periods, the interface is declared down, and switchover occurs. The default value is 10 seconds, but it can be set to the minimum value of 1 second using the following interface command: **keepalive 1**.

It might also be a good idea to disable keepalive on protect interfaces to prevent them from switching back; this can be achieved using the **no keepalive** command on the protect cable interface. If you don't want a failover to occur because of a failure to receive upstream DOCSIS messages, disable keepalives on all working interfaces.

Another HCCP option that you might want to alter from the default is revert time. This is the time that a previously working interface automatically switches back to operational mode so that the protect interface can service other working interfaces. The default value is 30 minutes. You can disable this functionality by using the **no hccp** *group#* **revertime** command on the protect interface. Revert time is usually disabled to give the cable operator enough time to troubleshoot the reason for failover. As soon as operation is restored, you can also manually switch back by issuing the global command, **hccp** *group#* **switch** *member#*.

The **show hccp** command can be used to obtain information relating to HCCP.

Some basic configuration is also required on the RF switch; these configuration commands are as follows:

- **set ip address** *ip-address netmask*—Defines the IP address and netmask for the Ethernet interface
- **set snmp community read-write** *string*—Defines the SNMP read/write community string
- **set protection {4|8}**—Defines the protection scheme, 4+1 or 8+1
- **set switchover-group** *group-name module-bitmap*—Adds switchover groups with corresponding bitmaps

The following is a sample configuration:

```
rfswitch>sh config
IP addr: 192.168.101.1
Subnet mask: 255.255.255.0
MAC addr: 00-03-8F-01-04-87
Gateway IP: 192.168.101.3
TFTP host IP: 192.168.101.3
ARP timeout: 14400 secs
DHCP lease time: infinite
TELNET inactivity timeout: 600 secs
Password: *****
SNMP Community: private
SNMP Traps: Enabled
SNMP Trap Interval: 300 sec(s)
SNMP Trap Hosts: none
Card Protect Mode: 4+1
Protect Mode Reset: Disabled
```

```
Slot Config: 0x03ff 0x1c00 (13 cards)
Watchdog Timeout: 20 sec(s)
Group definitions: 1
  ALL         0xffffffff
```

As you know, the CMTS and RF switch communicate via SNMP. To prevent outside SNMP attacks, you should define an access list restricting access to the CMTS interface used to communicate with the RF switch. Ideally, a dedicated subnet is used for the RF switch communication, which is isolated from the rest of the network. For example, the Fast Ethernet interface is often used for this purpose. The following example shows this configuration, where the CMTS interface to the RF switch has an IP address of 10.10.10.1:

```
CMTS(config)#access-list 99 permit 10.10.10.1 0.0.0.0
CMTS(config)#snmp-server community private rw 99
```

NOTE Please be aware that some measure of HFC availability can also be achieved by providing your customers with physical connectivity to multiple upstream and downstream channels. As mentioned in the capacity portion of this chapter, wiring fiber nodes in dense mode provides this access. Recall that dense mode is when more than one upstream channel controlled by a downstream channel is accessible by the end devices. When one upstream channel is no longer operational, the devices automatically try to use another upstream channel associated with their downstream channel. Multiple downstream channels can also be made available to your end devices. These downstream channels can be on the same CMTS device or entirely different devices. If different CMTS devices are used, then even an entire CMTS chassis failure would only temporarily disrupt service. When a cable modem detects downstream failure, it continues to scan the DOCSIS spectrum in search of another downstream channel. As you might guess, moving upstream and downstream channels is still disruptive to active calls. Also note this does not provide any protection from physical problems in the distribution portion of the HFC network such as a fiber cut.

WAN Redundancy

Depending on the volume and types of traffic running on your CMTS devices, a single WAN uplink might be sufficient in terms of bandwidth. However, from a high-availability standpoint this is very bad because it is a single point of failure. Having multiple WAN uplinks is a much better design. Also, if possible, have these links connect to different upstream routers so if one of these routers fails, your network is still operational.

If your CMTS is the Cisco uBR10k, you can have up to four full height Gigabit Ethernet linecards in a single chassis to achieve this WAN side redundancy. Figure 16-14 depicts this setup, where the gray cloud represents the rest of your VoIP network.

Figure 16-14 *CMTS WAN Side Availability*

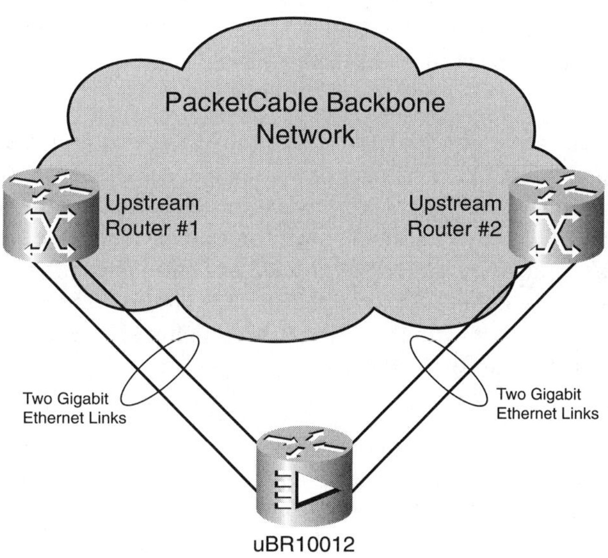

NOTE Because there is more than one path the CMS can use to reach the CMTS, define a loopback interface on the CMTS and define that interface for the COPS DQoS connection.

Another advantage of having multiple WAN uplinks is that they can be used in a load sharing fashion to prevent one part of your network from getting too heavily utilized. IP routing protocols such as Open Shortest Path First (OSPF), Intermediate System-to-Intermediate System (IS-IS), or Enhanced Interior Gateway Routing Protocol (EIGRP) can be used to automatically load balance traffic across these multiple links.

If your CMTS is the uBR7246VXR using an NPE-G1 processor, there are three Gigabit Ethernet links available for redundancy WAN connectivity. If you have one of the older processor cards, multiple WAN linecards can be inserted into the chassis to achieve this redundancy.

NOTE Both cable linecards and WAN linecards are hot-swappable on both Cisco CMTS platforms, and on the uBR10k, processor cards are as well, provided you have a redundant card in the chassis.

MTA Availability

A PacketCable service provider cannot do a great deal to ensure the high availability of the MTA component because it is physically located at the customer's premises and not easily accessible.

One thing that can be done, however, is to use MTAs that have a backup power supply in the event of a failure in the primary power source. Another procedure the service provider can utilize to improve MTA availability is to remotely monitor the status/health of the MTA power supply. This way you should have some advanced notice when a backup power supply is in use or when it becomes drained or faulty. This information can be transmitted to OSS back office applications via SNMP traps and/or SYSLOG messaging. The PacketCable 1.5 specification CL-SP-MIB-BB defines these battery monitoring MIBs, so refer to this document for the details.

MG Availability

Ensuring the high availability of your trunking media gateways assures that your subscribers can communicate with subscribers residing on the PSTN. Examples of ways to achieve high availability with a media gateway are similar to those discussed for the CMTS case. Redundant power sources and cooling fans, redundant WAN connections, and redundancy with respect to hardware components are all applicable for the media gateway case as well.

If you are using Cisco VISM cards within an MGX 8800 ATM series switch, the following sections cover what you can do to achieve high media gateway availability.

PXM Redundancy

The MGX 8800 switch should be equipped with two PXM processor cards to achieve 1+1 redundancy. One of these cards operates in active mode, and the other operates in standby mode. The standby card regularly synchronizes its database with the primary PXM through an internal mechanism. If the standby PXM detects the active PXM is no longer operational, it takes over the active role. If you are also using the PXM back cards to connect to the core of your network, you can utilize Y-cables to connect to both PXM cards. These cables take as input the copper or fiber connection to the rest of your network and then split it into both PXM back cards.

SRM Redundancy

A redundant set of SRM cards should also be used to achieve 1+1 redundancy. These cards are used for bulk distribution from the SRM T3 back card connections to various T1 trunks configured on VISM cards and also provide N+1 redundancy for VISM cards. One pair of

SRM cards should be used in the upper shelf slots 15 and 16, and one pair of SRM cards should be used in the lower shelf slots 31 and 32. Remember, upper shelf SRM cards work with VISM cards in the upper shelf (slots 1–6, 9–14), and lower shelf SRM cards work with VISM cards in the lower shelf (slots 17–22, 25–30).

If you are using the SRM back cards for DS3 connectivity to the PSTN, Y-cables can be used here as well so that this PSTN connection is connected to a pair of SRM cards for protection against a single card failure.

VXSM and VISM Redundancy

For VXSM cards, a variety of redundancy schemes are possible at both the card level and the line level. The card redundancy technique is 1:1, where an adjacent card operates in standby mode for an active card. For VXSM back cards operating in 1:1 redundancy, the PSTN time-division multiplexing (TDM) lines are often split using Y-cables with an output going to a back card interface for each card. This type of redundancy protects against a back card failure as well as a front card failure. Optionally, you might choose to provide only front card redundancy; in this case, a special type of back card is used for the redundancy slot, which has no physical connections. If the active front card fails, the standby front card becomes active, and the lines on the back card are automatically connected through the redundancy back card.

If you are using the OC-3 variety of VXSM cards, you can also have Synchronous Optical Network (SONET) Automatic Protection Switching (APS) 1+1 and 1:1 line redundancy. Recall that an OC-3 VXSM card supports four OC-3s on a single height back card. APS line redundancy is provided by designating a line on one back card as working and a line on another back card as protect. If these two back cards are for a pair of redundant VXSM cards, you have 1:1 APS line redundancy. Because you can have two single height back cards associated with a VXSM front card, you can also have intracard linecard redundancy by using OC-3 back cards in both slots; this is known as 1+1 APS line redundancy.

N+1 VISM card redundancy is supported via the SRM cards. In each shelf on the MGX chassis, a VISM card can be designated as a protect card that can take over if any of the other VISM cards in that shelf fail. Please note that active calls are lost during the switchover process.

By executing the CLI command **dspcds** on the active PXM card, you can tell if N+1 redundancy is in place and active. The card operating in protect mode has a card state of standby, whereas the working cards show a status of active. Also if N+1 redundancy is currently active, where x represents the slot number of the active card that failed and y represents the slot number of the protect card now active, you can tell this in the redundancy column output of this command. You see "Covering slot x" for the now active protect card and "Covered by slot y" for the previously working card.

To configure redundancy use the following command from the PXM CLI:

```
mgx.1.8.PXM.a > addred redPrimarySlotNum redSecondarySlotNum redType
```

For a VXSM or VISM card, the slot numbers would be from 1–6, 9–14, 17–22, and 25–30. The redundancy type would be 2 for 1:N.

APS redundancy on an OC-3 VXSM card is configured using the **addapsln** command on the VXSM CLI.

ATM VC Redundancy

Redundant ATM virtual circuits (VC) establish multiple paths VXSM and VISM traffic can use to reach the rest of your network. These VCs should be connected to the core of your network through different physical devices. For example, if you are using the RPM-XF cards in the MGX chassis, you should use two cards with one VC to each card. If you are using the PXM back cards to reach your network core, you should use different ports with upstream connectivity to different devices. The QoS portion of this chapter recommended using different VCs for bearer and control traffic. It is also recommended that you use a pair of VCs for each of these types of traffic. Thus, for each card you should have four VCs built, two bearer VCs and two control VCs; this configuration is referred to as 2B+2C. In Chapter 8 you learned how to create VCs. These VCs can be configured to protect each other using the **cnfconprotect** command. The syntax for a VXSM card is as follows:

```
MGX888000.5.VXSM.a > cnfconprotect ifNum vpi vci protect fallbackPort fallbackVpi
fallbackVci
```

The *ifNum* and *fallbackPort* are always 1, the *vpi* and *vci* identify the active PVC, and the *fallbackVpi* and *fallbackVci* identify the standby PVC. The *protect* field should be set to 1 meaning protected; 2 means unprotected.

Note that a VXSM or VISM card only sends traffic through its active VCs, but it can receive traffic from both its active VC and its standby VC.

As mentioned in the traffic engineering section, if you are using the PXM back cards to connect to the network core, you need two ATM OC-3 connections to support a fully populated shelf. To achieve full redundancy you should use four ATM OC-3 connections, connected to multiple upstream routers. Note: The OC-3 PXM back card supports four connections. It also spreads out the VISM active and standby PVCs on the chassis evenly across these four ATM OC-3 circuits. For example, assuming you have the maximum of 22 VISM cards in the chassis, configure five active VCs through two of these OC-3 pipes and six active VCs on the other two OC-3 pipes. Similarly, configure the standby VCs in the reverse manner so that there are 11 VCs on each OC-3. Note: If you are utilizing a 2B+2C configuration as previously mentioned, you would have 22 VCs on each OC-3.

If you are using the Gigabit Ethernet interfaces on the RPM-XF back cards for your core network connectivity, use all four interfaces for redundant connectivity to at least a pair of upstream routers. As in the earlier case, split up the active and standby VXSM and VISM

VCs so that half of them connect to one RPM-XF card and half of them connect to the other. Also, build a VC between the two RPM-XF cards so that they can route between each other to reach the individual cards. This is especially needed if you are summarizing routes on the RPM cards as external routers do not know which RPM card has operational VCs to which cards.

On the router configuration for the VC, it is also recommended that you use the option **oam-pvc manage** so that the router marks the VC as down if OAM heartbeat cells are no longer received. Otherwise, the router might think a VC is operational when it actually is not.

Other High-Availability Topics

The call agent FQDN used for MGCP communication can be defined to be resolved locally or externally via a DNS server. As it is not anticipated that the FQDN IP address resolution changes very often, it is recommended you define this resolution locally. Another option is to use local resolution first and then, if this fails, to use external DNS resolution second. This configuration for Cisco VXSM and VISM cards is covered in detail in Chapter 8.

To improve the clock source for PSTN circuits, it is recommended you configure an external BITS clock source. This can improve voice quality, especially for voice-band modem and facsimile call types. Also a secondary clock source should be available and defined if the primary clock source fails. On the Cisco MGX this is configured on the PXM module using the **cnfclksrc** command.

SG Availability

Chapter 8 covered the use of SS7 Signaling Gateways in detail. This network component provides call signaling connectivity to the PSTN via the SS7 protocol. These devices are critical to the ability of PacketCable subscribers to be able to communicate with PSTN subscribers; therefore, this component should also have a high degree of availability.

Again, the physical redundancy considerations of multiple power supplies and cooling fans apply to these gateways as well. Typically, this component is deployed in mated pairs. These pairs can share a point code with the CMS and provide SS7 A link connectivity to the PSTN STPs, or they can have unique point codes and provide SS7 D link connectivity to the PSTN STPs. A fairly common way of deploying a pair of SGs connected to a mated pair of STPs is by having an SS7 signaling link from each SG to each STP for a total of four links. Also a C link is typically configured between the SG pair. This provides redundancy for the TDM side of the SG devices. The SIGTRAN protocol used for connectivity to the CMS/MGC on the IP side of the SG devices also has built-in redundancy considerations in the protocol. If you are using Cisco routers as your SGs and the Cisco BTS 10200 as your CMS/MGC, the common way of providing IP redundancy is by connecting both SGs to both call signaling VLANs. This type of configuration was depicted in Figure 16-9.

OSS Server Availability

Redundancy ensuring high availability for critical back office OSS/NMS devices can also prove useful. Applications such as the DHCP and TFTP servers used for modem and MTA registration, the Kerberos server used for MTA registration, and the DNS servers used by the CMS are all critical applications that need to be highly available. If you are using Cisco BACC for your PacketCable network, there is a great deal of redundancy built in to the product. For example, multiple DHCP and DNS servers are available as well as redundant Device Provisioning Engine (DPE) devices. Please refer to the provisioning portion of the book for more information on BACC.

Also as mentioned in the CMS availability section, DNS availability can be achieved through the use of server load balancing (SLB). The recommended configuration is to have SLB running on a pair of devices (such as Cisco 7600 switches) one hop away from the DNS servers. The SLB devices should also be assigned a single virtual IP address for which devices such as the CMS are configured. The Hot Standby Routing Protocol (HSRP) can be used to run between the two Cisco switches configured for SLB so either one of them can assume this virtual IP address if one switch goes down.

Note: Other devices, such as Communications Assistance for Law Enforcement Act (CALEA), Delivery Function (DF) servers, Kerberos servers, and announcement servers, can also benefit from DNS high availability with SLB.

It is also a good idea to permit MTAs to reuse IP addresses when possible. This limits the amount of dynamic DNS updates that need to be made by the DHCP servers and also reduces the chances the secondary DNS server's database is inaccurate.

Other CMTS Design Considerations

If you are using the Cisco CMTS in your network, some other design considerations that have not previously been discussed can affect PacketCable performance.

Load Balancing

One of these considerations is the use of load balancing, which allows traffic to be optimally split across upstream and downstream channels on a cable linecard. As mentioned earlier, load balancing applies only when cable modems have physical access to these multiple upstream and downstream channels. By default, the CMTS attempts to equally split modems across upstream channels when they register; however, given the types of traffic generated by different modems, this might not be optimal. One of the options that can be configured is dynamic load balancing.

Dynamic load balancing works by moving online modems to different upstream channels based on defined criteria, such as when the load difference between channels exceeds a predefined threshold. Modems are instructed to change upstream channels via DOCSIS

Upstream Channel Change (UCC) messages. If you are providing both PacketCable services and high-speed data services on your network, you might want to be able to move modems around to provide them with access to more DOCSIS bandwidth. The steps and commands needed to configure dynamic load balancing are as follows:

Step 1 Define a load balance group:

```
CMTS(config)#cable load-balance group n method {modems ¦ service-flows
   ¦ utilization}
```

The *n* parameter is the group number from 1–80. If the **modems** method is used, load balancing decisions are based on the number of modems; if the **service-flows** method is used, decisions are based on the total number of service flows; and if the **utilization** method is used, decisions are based on channel utilization. The utilization method is probably the most useful in a PacketCable environment. Note: The utilization method won't start until utilization reaches 25 percent of total bandwidth.

Step 2 Define how often the CMTS checks load conditions (Optional).

```
CMTS(config)#cable load-balance group n interval seconds
```

The *n* parameter is the group number from 1–80. The **interval** value is the minimum amount of time between which a single modem can be moved. The range is from 1 to 1000 seconds, and the default value is 10 seconds.

Step 3 Define load balancing thresholds (Optional).

```
CMTS(config)#cable load-balance group n threshold load {value1 [enforce
   value2] ¦ minimum number} ¦ pcmm value3 ¦ stability value4 ¦ ugs
   value35
```

The **load** value is the maximum tolerable load difference that can exist between interfaces before load balancing during modem registration begins (default value is 10 percent). The **enforce** value is the maximum tolerable load difference that can exist between interfaces before dynamic load balancing of online modems begins (if unspecified, this does not occur). The **minimum** number is used with the modems and service flows methods and is the amount after which load balancing begins. The **stability** value specifies the minimal acceptable percentage of good periodic ranging requests. If this percentage falls below this number, the CMTS begins moving modems (default value is 50 percent). The **ugs** option indicates that the CMTS should move modems with active UGS flows after the interface UGS usage reaches the defined percentage (default value is 70 percent). This option is valid only when the following command is also configured.

Step 4 Permit the CMTS to move modems with active UGS flows (Optional).

```
CMTS(config)#cable load-balance group n policy ugs
```

In most cases you would not want MTAs to move that have active voice calls because this causes a temporary disruption in service. A possible exception is if the difference in channel utilization is so great that moving active MTAs is the only way to alleviate this difference.

Step 5 Assign the load balancing group to interfaces:

```
CMTS(config-if)#cable load-balance group n
```

or

```
CMTS(config-if)#cable upstream us# load-balance group n
```

The first command is for assigning downstream channels to a load balance group, and the second is for assigning upstream channels.

Step 6 The current load balancing configuration can be viewed with the following command:

```
CMTS#show cable load-balance [group n] [all | load | pending |
  statistics | target]
```

The **group** option specifies information only for a certain group. The **load** option shows interface load and group assignments. The **pending** option shows all modems in the process of moving. The **statistics** option shows cumulative load balancing statistics. The **target** option shows the current target interfaces being used for load balancing.

Admission Control

Admission control works by monitoring upstream and downstream bandwidth utilization, CPU utilization, and memory resources. Admission control checks happen when cable modems register and when dynamic service creation is attempted.

When admission control based on monitoring CPU utilization is configured and this utilization gets close to 100 percent, the CMTS drops Best Effort traffic.

For a PacketCable 1.x deployment, admission control can be used to limit the amount of upstream and/or downstream capacity consumed by PacketCable voice traffic. In a PacketCable Multimedia deployment, admission control can be used to limit the amount of upstream and/or downstream capacity consumed by the PCMM applications running on a CMTS.

The steps to configure admission control are as follows:

Step 1 Define the events that trigger admission control:

```
CMTS(config)#cable admission-control event {cm-registration | dynamic-
  service}
```

If the **cm-registration** option is enabled, admission checks are made when modems register; if the **dynamic-service** option is enabled, the attempt is made to dynamically create a voice call.

Step 2 To configure admission control for CPU utilization:

```
CMTS(config)#cable admission-control {cpu-5sec | cpu-avg} minor num1
   major num2 critical num3
```

If the **cpu-5sec** option is enabled, CPU checks are for 5-second periods, and if the **cpu-avg** option is enabled, CPU checks are for 1-minute periods. *num1*, *num2*, and *num3* are the threshold values for the minor, major, and critical alarms, respectively.

Step 3 To configure admission control for memory resources:

```
CMTS(config)#cable admission-control {io-mem | proc-mem | total-mem}
   minor num1 major num2 critical num3
```

If the **io-mem** option is enabled, memory checks are for router Input/Output memory; if the **proc-mem** option is enabled, memory checks are for router process memory; and if the **total-mem** option is enabled, memory checks are for combined router I/O and process memory. *num1*, *num2*, and *num3* are the threshold values for the minor, major, and critical alarms, respectively.

Step 4 To configure admission control for downstream bandwidth:

```
CMTS(config)#cable admission-control ds-bandwidth {data | voice} minor
   num1 major num2 exclusive num3 [non-exclusive num4]
```

or

```
CMTS(config-if)#cable admission-control ds-bandwidth {data | voice}
   minor num1 major num2 exclusive num3 [non-exclusive num4]
```

If this command is configured globally, it applies to all cable interfaces on the router; if it is configured on a specific interface, it applies only to that interface. If the **data** option is enabled, the thresholds apply to downstream data traffic; if the **voice** option is enabled, the thresholds apply to downstream voice traffic. *num1* and *num2* are the threshold values for the minor and major alarms, respectively. *num3* represents the percentage of throughput reserved for this class (voice or data). *num4* represents the percentage of throughput above the exclusive amount that can be used for this class (voice or data).

Step 5 To configure admission control for upstream bandwidth:

```
CMTS(config)#cable admission-control us-bandwidth {scheduling-type
   sched-type | service-class classname} minor num1 major num2 exclusive
   num3 [non-exclusive num4]
```

or

```
CMTS(config-if)# cable admission-control us-bandwidth {scheduling-type
   sched-type ¦ service-class classname} minor num1 major num2 exclusive
   num3 [non-exclusive num4]
```

or

```
CMTS(config-if)# cable upstream us# admission-control us-bandwidth
   {scheduling-type sched-type ¦ service-class classname} minor num1
   major num2 exclusive num3 [non-exclusive num4]
```

If this command is configured globally, it applies to all upstream cable interfaces on the router; if it is configured on a specific interface, it applies only to upstream channels on that interface, and if it applies to a specific upstream, it applies only to that upstream channel. The *sched-type* defines the type of traffic the thresholds apply to; valid values are BE, NRTPS, RTPS, AD-UGS, and UGS. The *classname* defines the service class that the thresholds apply to, this class name must be defined on the CMTS. *num1* and *num2* are the threshold values for the minor and major alarms, respectively. *num3* represents the percentage of throughput reserved for this class. *num4* represents the percentage of throughput above the exclusive amount that can be used for this class.

Step 6 To troubleshoot and check the current configuration of admission control use the following commands:

```
CMTS#show cable admission-control {global ¦ interface {Cable intf# ¦
  all}}
CMTS#debug cable admission-control {cpu ¦ ds-bandwidth ¦ events ¦ memory
   ¦ us-bandwidth}
```

As an example of how to use the admission control feature, look at the following:

```
router#sh cable admission-control global
CM registration event enabled
Dynamic service event enabled
Resource  Minor #times Major #times Critical #times Current
 Resource - Upstream Bandwidth
 -----------------------------
Sched       Class       Minor          Major         Exclusive          Non-Excls
Type        Name        Level          Level         Level              Level
   UGS        -          30             40             50                 50
 Resource - Downstream Bandwidth
 -------------------------------
Flow                    Minor          Major         Exclusive          Non-Excls
Type                    Level          Level         Level              Level
voice                    30             40             50                 50
```

In this example, both upstream and downstream admission control is defined for voice traffic. These commands were entered globally, so they apply to all cable interfaces on the chassis. When the voice utilization on one of these interfaces hits 30 percent, a minor alarm is generated, and an SNMP trap is sent to configured collection devices. When voice utilization reaches 40 percent, a major alarm is generated, and an SNMP trap is sent to configured collection devices. The exclusive level of 50 means that 50 percent of upstream and downstream capacity has been reserved for voice, and no other application can reserve QoS from these resources. The sum of exclusive levels for all types of traffic cannot exceed 100 percent. The nonexclusive level of 50 means the other 50 percent of upstream and downstream capacity can be used for voice if it is available. However, other applications can reserve QoS from these resources. The addition of the exclusive and nonexclusive levels represents the total amount of bandwidth capacity that can be used by an application. So, in this case, voice is permitted to potentially use 100 percent of upstream and downstream resources. The admission control software uses the 80 percent of raw channel capacity rule to determine total upstream capacity. This is done to account for upstream physical layer overhead such as preambles and guard time intervals needed to transmit an upstream packet.

For example, recall for a 3.2 MHz channel using 16-QAM modulation the raw channel capacity is 10.240 Mbps, and 80 percent of this is 8.192 Mbps. So, if 50 percent of this capacity is reserved for UGS, this is 4.096 Mbps. If basic (no PHS or security) G.711/20 ms voice is used, the UGS grant size is 232 bytes, and 50 grants are issued each second. So one call consumes 92,800 bps of upstream bandwidth as far as admission control is concerned. This means it is guaranteed that 44 (4,096 kbps/92.8 kbps) calls are always supported on an upstream.

Spectrum Management

The spectrum management feature available with some cable linecards allows the CMTS to make intelligent decisions in choosing upstream frequencies. If an upstream channel is defined with bands of frequencies from which to choose an upstream frequency instead of a single frequency, the CMTS can avoid noisy channels, improving modem performance. When noise detected in a certain frequency range is bad enough, the upstream channel automatically moves to a cleaner frequency. In a PacketCable environment this can help move MTAs from noisy channels, but care should be taken because during the frequency transition a few voice packets will be lost.

This frequency hopping can happen when a configurable threshold amount of station maintenance ranging request messages is missed. Missed ranging request messages indicate RF noise and/or impairments in the upstream channel. Frequency hopping can also be configured to occur at specified times.

When defining spectrum groups, keep the following guidelines in mind:

- Avoid frequencies with known ingress noise problems, such as those used for amateur radio.
- Avoid frequencies below 20 MHz.

The steps to configure spectrum groups are as follows:

Step 1 Define fixed frequencies and/or frequency bands to be used for the upstream channel in a spectrum group. Each of these configuration commands can be repeated multiple times.

```
CMTS(config)#cable spectrum-group group# [time day hh:mm:ss] frequency
   up-freq [power-level]
```

or

```
CMTS(config)#cable spectrum-group group# [time day hh:mm:ss] band up-
   freq up-freq2 [power-level]
```

The *group#* parameter is a value from 1–32. The **time** options can be used to specify a frequency move at the specified time. For the frequency option, the *up-freq* parameter specifies a specific center upstream frequency (in Hz), and the optional *power-level* parameter denotes what the received input power level should be for this frequency (default is 0 dBmV). For the **band** option the two frequency parameters specify a range where the CMTS can search for an appropriate upstream center frequency.

Step 2 Specify the minimum time between successive frequency hops; valid range is 1 to 300 seconds:

```
CMTS(config)#cable spectrum-group group# hop period seconds
```

Step 3 Specify the frequency hop threshold (percentage of missing station maintenance opportunities needed to trigger a frequency hop). Default is 20 percent:

```
CMTS(config)#cable spectrum-group group# hop threshold [percent]
```

Step 4 Identify whether or not upstream channels in a spectrum group share the same frequency (Optional):

```
CMTS(config)#cable spectrum-group group# shared
```

Step 5 Assign the spectrum group to one or more upstream channels:

```
CMTS(config-if)#cable spectrum-group group#
```

or

```
CMTS(config-if)#cable upstream us# spectrum-group group#
```

The first option assigns the spectrum group to all upstream channels on the interface, whereas the second assigns the spectrum group only to that specific upstream channel.

Step 6 To verify spectrum group configuration and operation:

```
CMTS#show cable spectrum-group [group#] [detail]
```

Interface Queues

By default, the WAN and cable interfaces have an input hold-queue depth of 75. When a packet is received on an interface, they are temporarily stored in this queue. Holding packets in a queue prevents a single interface from flooding the network with too many packets. Packets sent in excess of the queue capacity are dropped. To better accommodate packets that are delay sensitive, such as voice packets, it is recommended that you increase the queue size to 1024. This can be done using the following command under the interfaces:

```
CMTS(config-if)#hold-queue in 1024
```

Similarly, the output hold queue depths should also be modified on the cable interfaces from the default of 40. These queues contain packets waiting to be sent out an interface. If more packets are waiting to be sent than the interface can handle, these packets are dropped. To lessen the chance voice packets are dropped, increase this queue size to 1024 as well.

```
CMTS(config-if)#hold-queue out 1024
```

Limiting Burst Sizes

If you have a mixed environment where there are DOCSIS 1.0 devices on your network, you can have an issue with regard to maximum burst size. DOCSIS 1.0 modems do not have the ability to fragment large bursts, and if they are provisioned with the ability to transmit in large bursts, this can cause voice packets to be significantly delayed behind these transmissions. It is recommended to limit the maximum burst of DOCSIS 1.0 modems to 1600 bytes, which is large enough to handle the maximum sized Ethernet packet. As a safeguard to protect against modems defined with unlimited burst sizes, you can define a default maximum burst size on the cable interface, where the burst size can be from 0–4096 bytes.

```
CMTS(config-if)#cable default-phy-burst maxburst
```

PacketCable Future

The cable field is a fast-developing field, with advancements in technologies such as DOCSIS enabling support for a wide range of services. The bulk of this book has assumed the HFC portion of the network is operating in DOCSIS 1.1 mode. However, this is only the minimum, and more advanced versions of DOCSIS can increase the performance characteristics of a PacketCable network.

For example, the DOCSIS 2.0 series of specifications include several enhancements to the physical layer of DOCSIS. For TDMA channels modulation schemes are supported in the DOCSIS upstream greater than 16-QAM, specifically 32-QAM and 64-QAM. Synchronous Code Division Multiple Access (S-CDMA) upstream channels are also supported with DOCSIS 2.0. With S-CDMA different devices can transmit on the same frequency at the same time by using different code values that do not interfere with each other.

Also with DOCSIS 2.0, upstream interleaving is now supported, which helps protect against problems caused by impulse noise, and the range of the FEC T parameter has been increased from 10 to 16.

DOCSIS 3.0 is currently in the development stages, but its main feature is the capability to combine upstream and downstream channels to achieve far greater bandwidth. DOCSIS 3.0 decouples the physical layer of DOCSIS from the MAC layer. This allows for channel bonding, where multiple physical upstream and downstream channels can be combined to form a larger logical channel. This technology, also known as Modular CMTS (M-CMTS), enables the dynamic allocation of edge (RF) resources so you can respond more flexibly to changing demands from applications (including VoIP).

Another key feature of DOCSIS 3.0 is the support of IPv6 and the migration from IPv4. Also DOCSIS 3.0 is backward compatible with DOCSIS 2.0, 1.1, and 1.0 networks, meaning devices running different versions of DOCSIS can coexist on the same MAC domain.

This book has dealt with the features, protocols, and devices associated with the PacketCable 1.0 and 1.5 series of specifications. Recall that PacketCable 1.5 is an enhancement to PacketCable 1.0; everything required for a PacketCable 1.0 network is also required for a PacketCable 1.5 network. New functionality with PacketCable 1.5 includes the royalty-free CODECs, T.38 fax relay, inter-domain call capability using CMSS, and lawful intercept.

The PacketCable 2.0 specifications are geared for more advanced applications and rely on the PacketCable Multimedia framework for access network QoS. The PacketCable 2.0 series of specifications are also in the development stages but include the following topics:

- Support for SIP-based endpoints that might or might not be on the cable network. Examples of these endpoints include SIP clients embedded in phones, set-top boxes, and PC software. Methods of addressing QoS for these endpoints as well as issues relating to authorization, registration, and NAT/firewall traversal are addressed as well.
- Support and migration strategies for the implementation of IPv6 in a PacketCable network.
- Wireless integration so that customers can use the relatively low-cost PacketCable network when they are in range but can seamlessly switch to a cellular network when they are not.
- Cross-platform integration; for example, a set-top box can be used in conjunction with an MTA to provide a service like caller ID on the television.
- Commercial services extensions, where an application like PBX-based VoIP services can be provided across a PacketCable network.

- SIP applications like "Presence," where the location of a customer can be determined dynamically and the appropriate phone notified. This is also referred to as a "find me, follow me feature."
- Another anticipated application is "Click to dial," where web pages can be integrated with the telephony network to provide various services.

Chapter Summary

Security, capacity planning, ensuring high-availability, and implementing a networkwide QoS policy are all very important design considerations for any type of network but especially for a PacketCable network. To provide reliable and high-quality telephony services to PacketCable customers similar or better than possible in the traditional PSTN, these design considerations are crucial. Minimizing packet loss and jitter, maximizing availability, and maximizing call quality are all dependent upon these considerations.

PacketCable security covers requirements for all the signaling and media communication channels in a PacketCable network. These requirements provide protection against several types of threats, including theft of network services, theft of bearer channel information, theft of signaling channel information, service disruption, and repudiation. A majority of the communication channels are protected using the IP Security (IPsec) architecture. Specifically, the Encapsulation Security Payload (ESP) protocol is used in transport mode. Security keys are established via the IKE protocol, Kerberos, or in call signaling messages. The latter method is how keys are exchanged for bearer channel security. The CMS dictates what security ciphersuites are to be used in the NCS/TGCP local connection option parameters and the gateways exchange security information in their SDP parameters. Security-related configuration tasks consist of defining security parameters such as transform sets and keys and identifying the traffic requiring protection.

A large percentage of the QoS discussion in the PacketCable specifications is concerned with voice RTP packets. Other types of packets, such as call signaling packets, can also benefit from QoS treatment. For NCS signaling packets, QoS can be achieved via the cable modem configuration file. For upstream packets, a premium Best Effort flow can be set up, or a Non Real-Time Polling Service can be set up, depending on the degree of anticipated congestion in the upstream channel. For downstream packets, a premium Best Effort flow can be set up, giving these signaling packets a preferential treatment over other types of traffic. The QoS parameters minimum reserved rate and traffic priority can be used to ensure some measure of QoS treatment for these flows. Also required in the cable modem configuration are classifiers that direct call signaling traffic to these premium service flows.

Differentiated Service Code Point marking is the recommended way to implement QoS in the backbone of the network. DSCP codepoints are set on the edge devices in the network, and then intermediary devices treat traffic based on these codepoint markings. Each intermediary device treats these packets independently; hence, DSCP codepoints are also referred to as Per Hop Behaviors (PHBs). It is recommended that you treat voice RTP packets a strict priority LLQ throughout the network to give these packets preference and minimize end-to-end delay. Call signaling packets can be given a waited queuing preference and/or guaranteed some minimum amount of backbone bandwidth. DSCP markings should be made on the MTA, CMTS, CMS, and MG PacketCable components to achieve QoS for voice bearer and signaling packets. Another option for the backbone network is the use of Multiprotocol Label Switching (MPLS). MPLS can be used to speed up backbone traversal by switching packets based on labels in the packet header instead of performing routing lookups. MPLS can also be used to provide security in the network through the use of virtual private networks (VPNs).

Capacity planning in a PacketCable network should begin at the DOCSIS upstream channel because this is typically the most congested point in the network. CMTS upstream capacity is partially dependent upon the options chosen for voice deployment (CODEC, packetization period, PHS, security, etc.) and partially dependent upon the physical attributes of the upstream channel (modulation profile, channel width, FEC coverage, etc.). The size of the payload portion of RTP packets is dependent upon the CODEC used and on the packetization period used. Payload Header Suppression (PHS) can be used to mitigate the negative effects of the DOCSIS, IP, and UDP protocol headers. DOCSIS header overhead for upstream packets include the Ethernet-like MAC management portion, the standard 6-byte header, extended headers, FEC parity bytes, preamble overhead, and guard time overhead. Depending on the variable chosen for deployment, the DOCSIS parameters might need to be modified to optimize upstream performance. With a Cisco CMTS this involves changes to the upstream modulation profile. A common misconception is that implementing silence suppression increases upstream call capacity; this is not true because admission control decisions are based on reserved bandwidth, not committed bandwidth.

The traffic engineering metric of erlangs is commonly used to measure line usage and to calculate required trunk capacity. The total traffic load in erlangs, the required trunk capacity, and the blocking probability are related using the erlang B formula. Blocking probability or grade of service refers to the percentage of calls that fail because of a lack of trunk resources. This formula can be used on a PacketCable CMS/MGC to relate MTA lines to PSTN trunks. It can also be used on a PacketCable CMTS to relate MTA lines on an upstream to the call capacity of the upstream channel. When the total erlang traffic value exceeds a certain number, a simpler relationship exists between lines and trunks. This is the 90 percent rule, where the number of trunks required equals the number of lines divided by 0.9. Individual line utilization is also commonly expressed in Centrum Call Seconds (CCS), where the typical utilization of a phone line is between 3 and 6 CCS. Traffic engineering using models such as the erlang B formula is best used as an initial estimate. Because of limitations in this formula and the complexity of real-world traffic patterns, monitoring actual traffic characteristics and adapting traffic engineering design accordingly is recommended.

On a CMTS, limitations beyond the DOCSIS upstream can affect capacity analysis as well. This includes downstream bandwidth limitations as well as CPU and memory restrictions.

Proper testing is recommended in any network design to determine the true capacity and the interaction of all variables.

The capacity of a PacketCable CMS is typically more dependent on metrics such as call setup rate, call complexity, and the number of provisioned subscribers than on the number of simultaneous calls. The number and performance of CPUs on the CMS and the amount of database memory available determine these capacity limitations.

Capacity planning on a PacketCable MG involves having enough backbone IP bandwidth to handle the maximum trunk load to the PSTN. If you are using Cisco VXSM and/or VISM cards in a Cisco MGX 8800 series switch, you also need to calculate bandwidth requirements in the terms of ATM cells. ATM connections then need to be built from each VXSM/VISM card to the network backbone through either the PXM back card or a RPM card.

The availability requirement for a PacketCable network is 99.94 percent. This end-to-end availability can be calculated using either the multiplication method or the addition method. The multiplication method uses the product of individual component availabilities and the addition method uses the sum of individual component unavailabilities (subtracted from one) to compute end-to-end availability. Basic considerations, such as power supply redundancy, cooling system redundancy, and network connectivity redundancy, should be applied to all PacketCable network components if possible.

CMS availability is achieved by using pairs of devices with redundant network interfaces. Some sort of heartbeat mechanism is used between these devices to ensure that the other is operational. If the standby machine detects the active machine is no longer working, it can assume the active role and immediately take over with minimal impact on the network. During normal operation, call state information is typically transferred from the active machine to the standby machine. In addition to the use of multiple network interfaces, multiple upstream devices with multiple IP paths to network components should be implemented to maximize availability. Also multiple off network routing paths should be configured to maximize the ability to reach PSTN subscribers.

One of the ways high availability can be achieved with the Cisco uBR10k CMTS is by using redundant processors. The standby processor can detect a failure in the active processor and immediately take over with an minimal amount of disruption in the network. If the RPR+ feature is enabled, cable modems won't even have to reinitialize on a processor switchover. Thus, previously active calls won't be lost.

HFC high availability is achieved on the uBR10k CMTS by the use of a separate Cisco device, the uBR3x10 RF switch. With the use of an RF switch N+1 redundancy can be configured for your cable linecards. That means one cable linecard can act as a protect card for the remaining working cards. Two types of N+1 configuration are possible. One is 8+1 redundancy where a single cable linecard is protecting all of the remaining cable linecards in an uBR10K chassis. The maximum total number of cards in a chassis is eight, so the largest number of working cards is seven. This configuration requires a pair of RF switches.

If you have four or less working cards, 4+1 redundancy can be configured; this requires only a single RF switch. The Hot Standby Connection to Connection Protocol (HCCP) is used to provide this redundancy.

Some measure of MTA availability can be provided by using MTAs with backup power sources in the event you have a house power failure. Also the MSO should be able to remotely monitor the health of this power supply to get some advanced notification when a replacement unit might be required.

If you are using Cisco VXSM and/or VISM cards in a Cisco MGC 8800 series switch as your trunking gateway, you have several ways in which to configure high availability. Both the PXM and SRM modules can be installed in redundant pairs where one unit acts in the standby role, can detect the failure of the active unit, and can take over this role. 1:1 redundancy is available for VXSM cards as well as TDM lines. N+1 redundancy is available for VISM cards. Redundant virtual circuits should also be built on each VXSM and VISM card so that you have two paths to the network IP core.

Signaling gateway high availability is built into the Sigtran protocol used for IP communication to the MGC. It is also common for SG devices to be installed in pairs with multiple connections to pairs of PSTN STPs.

High availability for OSS servers should also be configured in your network. If you are using Cisco BACC, this is generally part of the architecture as multiple DPEs are usually installed, as well as multiple DHCP and DNS servers.

If you are using a Cisco CMTS and are providing multiple upstream and/or downstream channels to your customers, dynamic load balancing can be configured to improve DOCSIS utilization. Also with a Cisco CMTS admission control can be configured to base the amount of devices permitted on a particular DOCSIS channel contingent upon CPU utilization thresholds, memory availability thresholds, or available DOCSIS bandwidth. Upstream bandwidth restrictions can be applied based upon the type of upstream traffic. In other words, bandwidth can be allocated based upon the upstream scheduling type (UGS, RTPS, etc.). Upstream channels can also be defined using spectrum groups so that the CMTS can dynamically find a good upstream frequency. Other Cisco CMTS design considerations include changing interface queue depths and limiting the burst size of DOCSIS 1.0 modems.

Implementing DOCSIS 2.0 in your network can provide much greater upstream bandwidth amounts in the same channel bands used in a DOCSIS 1.x upstream. DOCSIS 3.0 further increases DOCSIS bandwidth by allowing multiple upstream and downstream channels to be combined together. DOCSIS 3.0 also addresses the implementation of IPv6 in the DOCSIS network. The PacketCable 2.0 specifications address advanced applications of PacketCable using a PacketCable Multimedia infrastructure for achieving DOCSIS QoS. Among these applications are the integration of wireless networks with the PacketCable network, SIP-based clients and applications, and cross-platform integration of DOCSIS devices.

Chapter Review

1 Why is using the secure method of PacketCable provisioning important?

Answer: It protects against possible theft of network services via the cloning of PacketCable components. It also protects against thefts of bearer and signaling channel information.

2 Why is using PacketCable DQoS important?

Answer: It ensures only authorized devices are using DOCSIS QoS services and prevents the misuse and possible service disruption of DOCSIS resources.

3 Briefly describe the Anti-Reply service of IP Security.

Answer: It enables devices to reject old or duplicated traffic that could be an indication of a denial-of-service attack.

4 Describe the amount of overhead imposed by implementing RTP security.

Answer: The only overhead is because of the appending of a small (2- or 4-byte) Message Authentication Code.

5 How is the NCS call signaling service key communicated between the BAC KDC and BTS?

Answer: It must be manually entered in the IPSEC-KERBEROS table.

6 What BTS table associates RTP and RTCP ciphersuites to MTA subscribers?

Answer: QOS table

7 True or false: Packets not matching a crypto map access list on a Cisco CMTS interface are dropped.

Answer: False. They are permitted without security.

8 What problems can occur if jitter buffers in a VoIP network are configured too large or too small?

Answer: If they are too large, unnecessary delay is inserted into the network, potentially affecting voice quality, and if they are too small, they can overflow, and packets can be lost.

9 How can problems occur if no DOCSIS QoS is set up for call signaling packets?

Answer: They can be delayed or even lost if they have to compete with ordinary Best Effort data traffic for bandwidth.

10 How does the CMTS let a cable modem know it has received its bandwidth request but cannot currently accommodate the request?

Answer: It issues a grant pending information element for that DOCSIS SID in the bandwidth allocation messages.

11 In addition to directing NCS packets to premium call signaling service flows, what other types of packets, if any, should you also direct to use these flows?

Answer: The DOCSIS dynamic service flow messages should also utilize the flows for the upstream direction.

12 Where in an IP packet is the DSCP codepoint value defined, and how many possible values can it have?

Answer: In the most significant 6 bits of the IP ToS header byte; thus, it can have 64 values.

13 What PHBs are typically used for voice bearer and signaling traffic?

Answer: For bearer traffic the Expedited Forwarding (EF) PHB is typically used, and for signaling traffic the Assured Forwarding 31 (AF31) PHB is typically used.

14 On Cisco routers how can policies, based on PHB markings, be implemented?

Answer: Using the Modular QoS CLI (MQC) feature

15 What are the ways in which DSCP codepoints are set on outgoing packets from a PacketCable MTA?

Answer: They can be set via SNMP MIBs in the MTA configuration file. Bearer packets can also be overwritten at either the MTA or CMTS under direction of the CMS. Signaling packets can also be overwritten on the CMTS because of configuration parameters in the cable modem configuration file.

16 When you are using the Cisco BTS 10200, some types of packets can be configured using only the IP Precedence interpretation of the IP ToS header byte. So is the configuration of DSCP codepoints for these packets possible? If so, how?

Answer: Yes. By doing the conversion of the desired DSCP codepoint into the IP Precedence interpretation values, DSCP codepoints can be configured.

17 If you are using the Cisco VXSM or VISM media gateway on a Cisco MGC 8800 ATM series switch, why is it a good idea to use separate ATM connections for call signaling and bearer/RTP traffic?

Answer: So these traffic types can be treated independently without influencing each other

18 For IPv4 packets, where is the MPLS label located?

Answer: In between the IP header and the MAC (layer 2) header

19 If the CMTS cannot allocate bandwidth for a subscriber's call, will the subscriber hear a network announcement indicating this?

Answer: No. A fast busy tone is typically used because DOCSIS bandwidth is also required to connect to an announcement server.

20 What are the two extended headers used in DOCSIS upstream packets?

Answer: The 5-byte BPI+ and 3-byte UGS extended headers

21 If PHS is available so that the maximum number of bytes are suppressed in both the upstream and downstream directions, what is the bandwidth savings in terms of bytes per packet?

Answer: 42 bytes for upstream packets and 28 for downstream packets

22 How is the determination made as to whether a short data grant or long data grant is used by a cable modem to send an IP packet?

Answer: By the configured maximum burst size for the short data grant. If the upstream transmission size is less than or equal to this value, short data grants are used; otherwise, long data grants are needed.

23 What are the benefits and drawbacks from increasing or decreasing the packetization period?

Answer: A smaller packetization period decreases capacity because of the amount of header overhead required, but it decreases end-to-end delay. A larger packetization period has the reverse effect.

24 True or false: The erlang metric represents utilization over a 1-hour period.

Answer: False. The erlang metric represents a utilization percentage and is time independent.

25 True or false: Call retries are accounted for in the erlang B formula.

Answer: False. The extended erlang B formula should be used to account for retries.

26 What are the three types of SUN servers currently supported for a Cisco BTS 10200 softswitch?

Answer: Netra 240, Netra 440, and Netra 1280 SUN servers

27 As a general rule how many erlangs of OFFNET traffic are supported on a T3 PSTN connection?

Answer: ~600 erlangs

28 True or false: Even if both BTS 10200 call agent devices are operational, only one process calls.

Answer: True. No load sharing occurs. One unit is active, and the other is idle in standby mode.

29 What is the 4/2 configuration with respect to the Cisco BTS 10200 call agent?

Answer: Configuration where four network interfaces are used on the two call agent machines, and two network interfaces are used on the two Element Management System machines to achieve high availability

30 What is the recommended way to configure high availability for DNS resolution?

Answer: Redundant DNS machines should be available behind a device or pair of devices configured with server load balancing (SLB).

31 Why are logical IP addresses used for the call signaling interfaces on the BTS 10200?

Answer: So gateways don't waste time contacting the standby call agent, which does not process calls and so does not respond to call signaling packets

32 How can an external clock source be provided to a Cisco CMTS?

Answer: On a uBR7246VXR via the cable clock card and on a uBR10k via the TCC+ cards

33 Other than through the use of a RF switch, how else can HFC high availability be achieved?

Answer: By providing cable modems with physical access to multiple upstream and downstream channels. This way if a single channel is unavailable, the modem can use an alternate channel.

34 How does an individual cable interface know what physical cables should be considered when determining whether or not a failover is needed?

Answer: Bit maps are defined on the interface indicating what upstream and downstream channels are used on that interface.

35 What type of hardware is required to use the SRM cards for redundant DS3 PSTN connectivity?

Answer: Y-cables are needed on the SRM back cards.

36 What does the 2B+2C VXSM or VISM configuration refer to?

Answer: Using redundant and separate virtual circuits for both the voice bearer traffic and call signaling traffic generated from a card

37 What three criteria can be used as the basis of dynamic load balancing decisions?

Answer: Number of modems, number of service flows, and RF channel utilization

38 What would be the command syntax be to reserve 50 percent of upstream bandwidth for voice UGS traffic and to let an additional 20 percent be used if available? Also a minor notification should be sent when 30 percent of bandwidth is used, and a major notification should be sent when 40 percent of bandwidth is used for voice UGS flows.

Answer: cable upstream *n* admission-control us-bandwidth scheduling-type UGS minor 30 major 40 exclusive 50 non-exclusive 20

39 In addition to traditional TDMA upstream channels, what other type of upstream channel is now available with DOCSIS 2.0?

Answer: S-CDMA

40 List two applications anticipated relating to PacketCable 2.0.

Answer: "Presence" and "Click to dial"

Part VII

Appendixes

APPENDIX **A**

Standards and Specifications

This appendix lists the standards and specifications referred to in this book. These standards and specifications can be obtained from the following sources:

- The CableLabs (that is, DOCSIS and PacketCable) specifications can be obtained at http://www.cablelabs.com.
- You can find all IETF RFCs online at http://www.ietf.org/rfc/rfc*xxxx*.txt, where *xxxx* is the number of the RFC. If you do not know the number of the RFC, you can try searching by topic at http://www.rfc-editor.org/rfcsearch.html.
- SCTE standards can be obtained at http://www.scte.org/standards/index.cfm?pID=59.
- ITU recommendations can be obtained at http://www.itu.int/rec/T-REC/en.
- ETSI standards can be obtained at http://www.etsi.org/services_products/freestandard/home.htm.

NOTE For the PacketCable specifications the corresponding SCTE, ITU, and ETSI specifications are listed when applicable.

Table A-1 *IETF RFCs*

RFC ID	Description
RFC 0768	User Datagram Protocol
RFC 0791	Internet Protocol
RFC 0792	Internet Control Message Protocol
RFC 0868	Time Protocol
RFC 0951	Bootstrap Protocol
RFC 1034	Domain Names—Concepts and Facilities
RFC 1035	Domain names—implementation and specification

continues

Table A-1 *IETF RFCs (Continued)*

RFC ID	Description
RFC 1157	Simple Network Management Protocol (SNMP)
RFC 1305	Network Time Protocol (Version 3) Specification, Implementation, and Analysis
RFC 1321	MD5 Message-Digest Algorithm
RFC 1350	The TFTP Protocol (Revision 2)
RFC 1825	Security Architecture for the Internet Protocol
RFC 1826	IP Authentication Header
RFC 1827	IP Encapsulating Security Payload (ESP)
RFC 1889	RTP: A Transport Protocol for Real-Time Applications
RFC 1890	RTP Profile for Audio and Video Conferences with Minimal Control
RFC 1945	Hypertext Transfer Protocol—HTTP/1.0
RFC 2011	SNMPv2 MIB for the Internet Protocol using SMIv2
RFC 2045	Multipurpose Internet Mail Extensions (MIME) Part One
RFC 2068	Hypertext Transfer Protocol—HTTP/1.1
RFC 2104	HMAC: Keyed-Hashing for Message Authentication
RFC 2113	IP Router Alert Option
RFC 2131	Dynamic Host Configuration Protocol (DHCP)
RFC 2132	DHCP Options and BOOTP Vendor Extensions
RFC 2136	Dynamic Updates in the Domain Name System (DNS UPDATE)
RFC 2198	RTP Payload for Redundant Audio Data
RFC 2205	Resource ReSerVation Protocol (RSVP)—Version 1 Functional Specification
RFC 2210	The use of RSVP with IETF Integrated Services
RFC 2211	Specification of the Controlled-Load Network Element Service
RFC 2212	Specification of the Guaranteed Quality of Service
RFC 2234	Augmented BNF for Syntax Specifications: ABNF
RFC 2246	The TLS Protocol Version 1.0
RFC 2326	Real Time Streaming Protocol (RTSP)
RFC 2327	SDP: Session Description Protocol
RFC 2367	PF_KEY Key Management API, Version 2
RFC 2396	Uniform Resource Identifiers (URI): Generic Syntax

Table A-1 *IETF RFCs (Continued)*

RFC ID	Description
RFC 2397	The "data" URL scheme
RFC 2401	Security Architecture for the Internet Protocol
RFC 2403	The Use of HMAC-MD5-96 within ESP and AH
RFC 2404	The Use of HMAC-SHA-1-96 within ESP and AH
RFC 2406	IP Encapsulating Security Payload (ESP)
RFC 2407	The Internet IP Security Domain of Interpretation for ISAKMP
RFC 2408	Internet Security Association and Key Management Protocol (ISAKMP)
RFC 2409	The Internet Key Exchange (IKE)
RFC 2437	PKCS #1: RSA Cryptography Specifications Version 2.0
RFC 2451	The ESP CBC-Mode Cipher Algorithms
RFC 2459	Internet X.509 Public Key Infrastructure—Certificate and CRL Profile
RFC 2474	Definition of the Differentiated Serviced Field (DS Field) in the IPv4 and IPv6 Headers
RFC 2475	An Architecture for Differentiated Services
RFC 2571	An Architecture for Describing SNMP Management Frameworks
RFC 2572	Message Processing and Dispatching for SNMP
RFC 2573	SNMP Applications
RFC 2574	User-based Security Model (USM) for SNMPv3
RFC 2575	View-based Access Control Model (VACM) for SNMP
RFC 2597	Assured Forwarding PHB Group
RFC 2598	An Expedited Forwarding PHB
RFC 2616	Hypertext Transfer Protocol -- HTTP/1.1
RFC 2630	Cryptographic Message Syntax
RFC 2638	A Two-bit Differentiated Services Architecture for the Internet
RFC 2669	DOCSIS Cable Device MIB for DOCSIS compliant CMs and CMTSs
RFC 2670	Radio Frequency (RF) Interface MIB for MCNS/DOCSIS compliant RF interfaces
RFC 2702	Requirements for Traffic Engineering Over MPLS
RFC 2705	Media Gateway Control Protocol (MGCP) Version 1.0
RFC 2747	RSVP Cryptographic Authentication
RFC 2748	The COPS (Common Open Policy Service) Protocol

continues

Table A-1 *IETF RFCs (Continued)*

RFC ID	Description
RFC 2749	COPS usage for RSVP
RFC 2750	RSVP Extensions for Policy Control
RFC 2751	Signaled Preemption Priority Policy Element
RFC 2753	A Framework for Policy-based Admission Control
RFC 2793	RTP Payload for Text Conversation
RFC 2833	RTP Payload for DTMF Digits, Telephony Tones and Telephony Signals
RFC 2863	The Interfaces Group MIB
RFC 2865	Remote Authentication Dial In User Service (RADIUS)
RFC 2866	RADIUS Accounting
RFC 2960	Stream Control Transmission Protocol
RFC 2961	RSVP Refresh Overhead Reduction Extensions
RFC 2996	Format of the RSVP DCLASS Object
RFC 3006	Integrated Services in the Presence of Compressible Flows
RFC 3046	DHCP Relay Agent Information Option
RFC 3084	COPS Usage for Policy Provisioning (COPS-PR)
RFC 3164	The BSD Syslog Protocol
RFC 3168	The Addition of Explicit Congestion Notification (ECN) to IP
RFC 3261	SIP: Session Initiation Protocol
RFC 3262	Reliability of Provisional Responses in the Session Initiation Protocol (SIP)
RFC 3263	Session Initiation Protocol (SIP): Locating SIP Servers
RFC 3265	Session Initiation Protocol (SIP)–Specific Event Notification
RFC 3286	An Introduction to the Stream Control Transmission Protocol (SCTP)
RFC 3311	The Session Initiation Protocol (SIP) UPDATE Method
RFC 3312	Integration of Resource Management and Session Initiation Protocol (SIP)
RFC 3323	A Privacy Mechanism for the Session Initiation Protocol (SIP)
RFC 3325	Privacy Extensions to the Session Initiation Protocol (SIP) for Asserted Identity within Trusted Networks
RFC 3332	Signaling System 7 (SS7) Message Transfer Part 3 (MTP3)—User Adaptation Layer (M3UA)
RFC 3372	Session Initiation Protocol for Telephones (SIP-T): Context and Architectures
RFC 3389	Comfort noise RTP payload type

Table A-1 *IETF RFCs (Continued)*

RFC ID	Description
RFC 3398	Integrated Services Digital Network (ISDN) User Part (ISUP) to Session Initiation Protocol (SIP) Mapping
RFC 3407	Session Description Protocol (SDP) Simple Capability Declaration
RFC 3420	Internet Media Type message/sipfrag
RFC 3435	Media Gateway Control Protocol (MGCP) Version 1.0
RFC 3436	Transport Layer Security over Stream Control Transmission Protocol
RFC 3495	Dynamic Host Configuration Protocol (DHCP) Option for CableLabs Client Configuration
RFC 3515	The Session Initiation Protocol (SIP) Refer Method
RFC 3550	RTP: A Transport Protocol for Real-Time Applications
RFC 3551	RTP Profile for Audio and Video Conferences with Minimal Control
RFC 3594	PacketCable Security Ticket Control Sub-Option for the DHCP CableLabs Client Configuration (CCC) Option
RFC 3603	Private Session Initiation Protocol (SIP) Proxy-to-Proxy Extensions for Supporting the PacketCable Distributed Call Signaling Architecture
RFC 3611	RTP Control Protocol Extended Reports (RTCP XR)
RFC 3661	MGCP Return Code Usage
RFC 3842	A Message Summary and Message Waiting Indication Event Package for the Session Initiation Protocol (SIP)
RFC 3868	Signalling Connection Control Part User Adaptation Layer (SUA)
RFC 3966	The tel URI for Telephone Numbers
RFC 4103	RTP Payload for Text Conversation
RFC 4120	The Kerberos Network Authentication Service (V5)
RFC 4301	Security Architecture for the Internet Protocol
RFC 4306	Internet Key Exchange (IKE) protocol
RFC 4346	The Transport Layer Security (TLS) Protocol Version 1.1
RFC 4366	Transport Layer Security (TLS) Extensions

Table A-2 *ANSI, Telcordia, ITU, and Other Specifications*

ID	Description
IANA RTP Parameter Assignments	Listing of RTP parameter values (such as payload type). Refer to the following URL: http://www.iana.org/assignments/rtp-parameters
ANSI J-STD-025-A	Lawfully Authorized Electronic Surveillance
ANSI T1.111	Signaling System No. 7 (SS7)—Message Transfer Part (MTP)
ANSI T1.521-1999	A Packet Loss Concealment Technique for use with ITU-T Recommendation G.711
FIPS-180-1	Secure Hash Algorithm, Department of Commerce, NIST
FIPS-186-2	Digital Signature Standard, Department of Commerce, NIST
FIPS-197	Advanced Encryption Standard (AES), Department of Commerce, NIST
FIPS-81	Federal Information Processing Standard: DES Modes of Operation
IMTC VoIP Forum	Service Operability Implementation Agreement 1.0
ISO 639-2	Codes for the Representation of Names of Languages
ISO/IEC 8802-3	Telecommunications and information exchange between systems
ITU-T, E.123	Notation for national and international telephone numbers, e-mail addresses, and Web addresses
ITU-T, E.164	The international public telecommunications numbering plan
ITU-T, G.107	The E-model, a computational model for use in transmission planning
ITU-T, G.114	One-way transmission time
ITU-T, G.165	Echo cancellers
ITU-T, G.168	Digital network echo cancellers
ITU-T, G.711	Pulse Code Modulation (PCM) of voice frequencies
ITU-T, G.726	Adaptive differential pulse code modulation (ADPCM)
ITU-T, G.728	Coding of speech at 16 kbit/s using low-delay code-excited linear prediction
ITU-T, G.729-Annex E	Coding of speech at 8 kbit/s using CS-ACELP
ITU-T, H.225	Call signaling protocols and media stream packetization for packet-based multimedia communications systems
ITU-T, H.245	Control protocol for multimedia communication
ITU-T, H.323	Packet-based multimedia communications systems
ITU-T, J.83	Digital multi-program systems for television, sound and data services for cable distribution
ITU-T, M.3010	Principles for a telecommunications management network

Table A-2 *ANSI, Telcordia, ITU, and Other Specifications (Continued)*

ID	Description
ITU-T, M.3100	Generic network information model
ITU-T, P.861	Objective quality measurement of telephone-band (300–3400 Hz) speech codecs
ITU-T, P.862	Perceptual evaluation of speech quality (PESQ): An objective method for end-to-end speech quality assessment of narrow-band telephone networks and speech codecs
ITU-T, Q.706	Specifications of SS7—Signaling performance
ITU-T, Q.709	Specifications of SS7—Hypothetical signaling reference connection
ITU-T, Q.714	Specifications of SS7—Signaling connection control part
ITU-T, Q.761	Functional description of the ISDN User Part of SS7
ITU-T, Q.762	General function of messages and signals of the ISDN User Part of SS7
ITU-T, Q.763	Signaling System No. 7—ISDN User Part formats and codes
ITU-T, Q.764	Signaling System No. 7—ISDN User Part signaling procedures
ITU-T, Q.767	Application of the ISDN User Part of CCITT signaling system No. 7 for international ISDN interconnections
ITU-T, Q.921	ISDN user-network interface—Data link layer specification
ITU-T, Q.931	ISDN user-network interface layer 3 specification for basic call control
ITU-T, T.4	Standardization of Group 3 facsimile terminals for document transmission
ITU-T, T.30	Procedures for document facsimile transmission in the general switched telephone network
ITU-T, T.37	Procedures for the transfer of facsimile data via store-and-forward on the Internet
ITU-T, T.38	Procedures for real-time Group 3 facsimile communication over IP networks
ITU-T, V.18	Operational and interworking requirements for DCEs operating in the text telephone mode
ITU-T, V.151	Procedures for end-to-end connection of analog PSTN text telephones over an IP network utilizing text relay
ITU-T, V.152	Procedures for supporting voice-band data over IP networks
ITU-T, X.509	OSI—The Directory: Authentication framework
ITU-T, X.680	Information technology—Abstract Syntax Notation One (ASN.1): Specification of basic notation

continues

Table A-2 *ANSI, Telcordia, ITU, and Other Specifications (Continued)*

ID	Description
ITU-T, X.690	ASN.1 encoding rules: Specification of Basic Encoding Rules (BER), Canonical Encoding Rules (CER), and Distinguished Encoding Rules (DER)
ITU-T, X.733	OSI—System Management: Alarm reporting function
Proceeding of the 4th Workshop on Fast Software Encryption	MMH: Software Message Authentication in Gbit/sec rates
RSA Labs, PKCS #7	Cryptographic Message Syntax Standard
Telcordia	Bellcore, Local Exchange Routing Guide (LERG)
Telcordia BR-751-100-112	Bellcore, Common Language General Codes
Telcordia GR-1089-CORE	Bellcore, Generic Requirements for Electronic Equipment Cabinets
Telcordia GR-1176-CORE	Bellcore, OSSGR: Custom Call Handling Features (FSD 80 Series)
Telcordia GR-11-CORE	Bellcore Automatic Message Accounting Format (BAF) Requirements and Definitions
Telcordia GR-1364-CORE	Bellcore Generic Requirements
Telcordia GR-246	Bell Communications Research Specification of Signaling System Number 7
Telcordia GR-317-CORE	Bellcore, LSSGR: Switching System Generic Requirements for Call Control using ISDN User Part
Telcordia GR-474	Bellcore, Network Maintenance: Alarm and Control for Network Elements
Telcordia GR-499-CORE	Bellcore, Transport Systems Generic Requirements
Telcordia GR-506-CORE	Bellcore, LSSGR: Signaling for Analog Interfaces
Telcordia GR-517-CORE	Bellcore, LEC Traffic Environment Characteristics
Telcordia LSSGR	Switch Processing Time Generic Requirements
Telcordia SR-2275	Bellcore Notes on the Networks

Table A-2 *ANSI, Telcordia, ITU, and Other Specifications (Continued)*

ID	Description
Telcordia TA-NWT-000909	Bellcore, Generic Requirements and Objectives for Fiber in the Loop (FITL) Systems
Telcordia TR-NPL-000258	Bellcore, Compatibility Information for Feature Group D Switched Access Service
Telcordia TR-NWT-000303	Bellcore, Integrated DLC System Generic Requirements
Telcordia TR-TSY-000531	Bellcore, Interoffice LSSGR: Verification Connections
TIA and ANSI J-STD-025	Lawfully Authorized Electronic Surveillance

Table A-3 *PacketCable 1.0 Specifications*

PacketCable Specification/Technical Report	SCTE ID	ITU ID	ETSI ID
PacketCable Audio/Video Codecs Specification PKT-SP-CODEC-I06-050812	ANSI/SCTE 24-2 2006	J.161	TS 101 909-03
PacketCable Dynamic Quality-of-Service Specification PKT-SP-DQOS-I12-050812	ANSI/SCTE 24-4 2006	J.163	TS 101 909-05
PacketCable Network-Based Call Signaling Protocol Specification PKT-SP-EC-MGCP-I11-050812	ANSI/SCTE 24-3 2006	J.162	TS 101 909-04
PacketCable Event Message Specification PKT-SP-EM-I12-050812	ANSI/SCTE 24-9 2006	J.164	TS 101 909-10
PacketCable Internet Signaling Transport Protocol (ISTP) Specification PKT-SP-ISTP-I02-011221	ANSI/SCTE 24-11 2006	J.165	TS 101 909-12
PacketCable MIBs Framework Specification PKT-SP-MIBS-I10-050812	ANSI/SCTE 24-6 2006	J.166	TS 101 909-07
PacketCable MTA MIB Specification PKT-SP-MIB-MTA-I10-0508012	ANSI/SCTE 24-7 2006	J.168	TS 101 909-08
PacketCable Signaling MIB Specification PKT-SP-MIB-SIG-I09-050812	ANSI/SCTE 24-8 2006	J.169	TS 101 909-09
PacketCable MTA Device Provisioning Specification PKT-SP-PROV-I11-050812	ANSI/SCTE 24-5 2006	J.167	TS 101 909-06

continues

Table A-3 *PacketCable 1.0 Specifications (Continued)*

PacketCable Specification/Technical Report	SCTE ID	ITU ID	ETSI ID
PacketCable Security Specification PKT-SP-SEC-I12-050812	ANSI/SCTE 24-10 2006	J.170	TS 101 909-11
PacketCable PSTN Gateway Call Signaling Protocol Specification PKT-SP-TGCP-I10-050812	ANSI/SCTE 24-12 2006	J.171	TS 101 909-13
CableLabs→ Definition MIB Specification CL-SP-MIB-CLABDEF-I05-050408	N/A	N/A	N/A
PacketCable NCS Basic Packages Technical Report PKT-TR-MGCP-PKG-V01-020315	N/A	N/A	N/A
PacketCable Architecture Call Flows Technical Report—On-Net MTA to On-Net MTA PKT-TR-CF-ON-ON-V02-030815	N/A	N/A	N/A
PacketCable Architecture Call Flows Technical Report—On-Net MTA to PSTN Telephone PKT-TR-CF-ON-PSTN-V02-030815	N/A	N/A	N/A
PacketCable Architecture Call Flows Technical Report—PSTN Telephone to On-Net MTA PKT-TR-CF-PSTN-ON-V02-030815	N/A	N/A	N/A
PacketCable 1.0 Architecture Framework Technical Report PKT-TR-ARCH-V01-991201	ANSI/SCTE 24-1 2006	J.160	TS 101 909-2
PacketCable OSS Overview Technical Report PKT-TR-OSS-V02-991201	N/A	N/A	N/A

Table A-4 *PacketCable 1.5 Specifications*

PacketCable Specification/Technical Report	SCTE ID	ITU ID	ETSI ID
PacketCable Analog Trunking for PBX Specification PKT-SP-ATPBX1.5-I01-060419	N/A	N/A	N/A
PacketCable Embedded MTA Analog Interface and Powering Specification PKT-SP-AIP1.5-I01-050128	ANSI/SCTE 24-14 2002	J.173	TS 101 909-18
PacketCable Audio Server Protocol Specification PKT-SP-ASP1.5-I01-050128	ANSI/SCTE 24-17 2002	J.175	TS 101 909-19

Table A-4 *PacketCable 1.5 Specifications (Continued)*

PacketCable Specification/Technical Report	SCTE ID	ITU ID	ETSI ID
PacketCable CMS Subscriber Provisioning Specification PKT-SP-CMSPROV1.5-I01-050128	ANSI/SCTE 24-19 2004	J.177	N/A
PacketCable Call Management Server Signaling Specification PKT-SP-CMSS1.5-I02-050812	ANSI/SCTE 24-18 2004	J.178	N/A
PacketCable Audio/Video Codecs Specification PKT-SP-CODEC1.5-I01-050128	N/A	N/A	N/A
BroadVoice→16 Speech Codec Specification PKT-SP-BV16-CODEC1.5-I01-031030	ANSI/SCTE 24-21 2006	N/A	N/A
PacketCable Dynamic Quality-of-Service Specification PKT-SP-DQOS1.5-I02-050812	N/A	N/A	N/A
PacketCable Event Message Specification PKT-SP-EM1.5-I02-050812	N/A	N/A	N/A
PacketCable Electronic Surveillance Specification PKT-SP-ESP1.5-I01-050128	ANSI/SCTE 24-13 2006	N/A	TS 101 909-20-01
PacketCable Management Event MIB Specification PKT-SP-EVEMIB1.5-I02-050812	N/A	J.176	TS 101 909-10
PacketCable Management Event Mechanism PKT-SP-MEM1.5-I02-050812	ANSI/SCTE 24-16 2002	J.172	TS 101 909-16
PacketCable MTA Extension MIB Specification PKT-SP-MIB-EXMTA1.5-I01-050128	N/A	N/A	N/A
PacketCable Signaling Extension MIB Specification PKT-SP-MIB-EXSIG1.5-I02-050812	N/A	N/A	N/A
PacketCable MTA MIB Specification PKT-SP-MIB-MTA1.5-I01-050128	N/A	N/A	N/A
PacketCable MIBs Framework Specification PKT-SP-MIBS1.5-I01-050128	N/A	N/A	N/A
PacketCable Signaling MIB Specification PKT-SP-MIB-SIG1.5-I01-050128	N/A	N/A	N/A
PacketCable Network-Based Call Signaling Protocol Specification PKT-SP-NCS1.5-I02-050812	N/A	N/A	N/A

continues

Table A-4 *PacketCable 1.5 Specifications (Continued)*

PacketCable Specification/Technical Report	SCTE ID	ITU ID	ETSI ID
PacketCable MTA Device Provisioning Specification PKT-SP-PROV1.5-I02-050812	N/A	N/A	N/A
PacketCable Security Specification PKT-SP-SEC1.5-I01-050128	N/A	N/A	N/A
PacketCable PSTN Gateway Call Signaling Protocol Specification PKT-SP-TGCP1.5-I02-050812	N/A	N/A	N/A
CableLabs→ Battery Backup MIB Specification CL-SP-MIB-BB-I01-050128	N/A	N/A	N/A
CableLabs→ Definition MIB Specification CL-SP-MIB-CLABDEF-I05-050408	N/A	N/A	N/A
PacketCable 1.5 Architecture Framework Technical Report PKT-TR-ARCH1.5-V01-050128	N/A	N/A	TS 101 909-23

A majority of the PacketCable 1.0 specifications exist with enhancements for PacketCable 1.5, as shown in this table.

Table A-5 *PacketCable Multimedia Specifications*

PacketCable Specification/Technical Report	SCTE ID	ITU ID	ETSI ID
PacketCable Multimedia Specification PKT-SP-MM-I03-051221	N/A	J.179	N/A
PacketCable Multimedia Web Service Interface Specification PKT-SP-MM-WS-I01-051221	N/A	N/A	N/A
PacketCable Multimedia Architecture Framework Technical Report PKT-TR-MM-ARCH-V02-051221	N/A	N/A	N/A

APPENDIX **B**

Cable Monitor and Ethereal

Throughout the book, you often saw packet traces used to illustrate how the various PacketCable protocols work and interrelate. Obtaining these traces can also provide great insight and assistance in troubleshooting issues you might encounter in a PacketCable deployment. The Ethereal packet sniffer utility is excellent at providing this functionality; embedded in this utility are several enhancements for decoding the protocols related to PacketCable. One of these is the capability to decode Data-Over-Cable Service Interface Specification (DOCSIS) frames and to decipher DOCSIS messages and parameters. These DOCSIS packet traces can be obtained through the use of an external radio frequency (RF) packet sniffer or via features on the cable modem termination system (CMTS). With a Cisco CMTS, the cable monitor feature provides this functionality, eliminating the need to purchase an external RF analyzer.

Ethereal Overview

The Ethereal packet sniffer utility is opensource software. This means that anyone can use it without having to buy any kind of license or pay any kind of fee. You can obtain the Ethereal software from their website at http://www.ethereal.com. Also not only is the application freely available, but the source code is openly available as well. This allows anyone to add functionality to the software, such as for new and evolving protocols. As a result of this, Ethereal is often ahead of other protocol analyzers in terms of emerging technologies. Ethereal can be installed on Windows-based machines as well as UNIX- and Linux-based machines.

NOTE Before installing the Ethereal software you need to download and install the packet capture software, libpcap, which is also available on the Ethereal website.

Using Cable Monitor to Decode DOCSIS Packets

With a Cisco CMTS, two features are used for the replication of DOCSIS packets: cable intercept and cable monitor. Neither of these features has any impact on the treatment of the original packets that are replicated.

The cable intercept feature is intended for use in providing lawful intercept functionality. This feature is not intended for use as a troubleshooting tool and shouldn't be used that way. It replicates packets from a specified MAC address and forwards that data to a specific IP address and UDP port number residing on a lawful intercept delivery function device. In summary, a suspect's packets are encapsulated into UDP packets by the CMTS and then sent to a Delivery Function (DF) mediation device. This tap can either be directly configured on the CMTS or initiated remotely. For more information about lawful intercept refer to Chapter 15, "Lawful Intercept."

The cable monitor feature is intended for use in analyzing DOCSIS packets for troubleshooting purposes. It works by replicating DOCSIS packets out an Ethernet interface (can be 10 Mbps, 100 Mbps, or 1 Gbps). An important distinction of cable monitor compared to cable intercept is that these replicated packets are not encapsulated in IP frames; they are replicated and sent out the Ethernet interface as they are sent and received on the cable interface. Thus, the Ethernet interface must be connected either directly or through a switch to the machine running the Ethereal sniffer utility. Because cable monitor packets are typically not valid Ethernet frames, they cannot be routed over an IP network; this is why the protocol analyzer machine must be connected on the same subnet as the CMTS. It's important to remember that the CMTS Ethernet interface should be dedicated for the cable monitor application; no other type of traffic should utilize this interface. Figure B-1 shows a high-level illustration of how the cable monitor feature is used.

Figure B-1 *Cable Monitor Diagram*

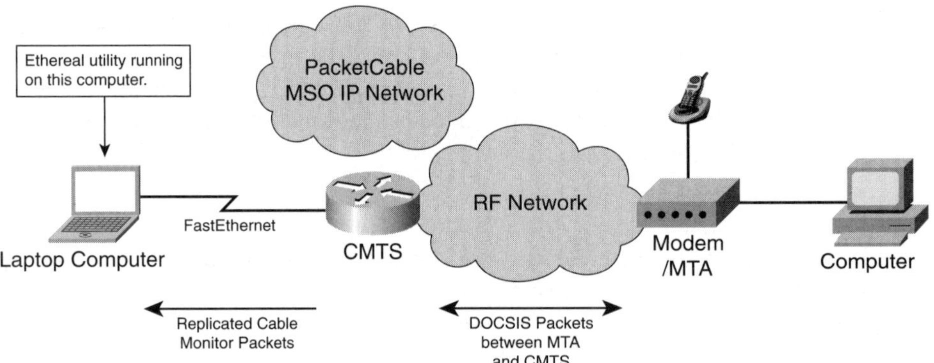

The following parameters can be used to define what packets are captured using cable monitor:

- Packet direction (incoming/upstream or outbound/downstream or both)
- Specific MAC address(es)
- Specific DOCSIS service identifier(s) (SIDs)
- Packets matching a defined access list

Cable monitor is configured on a cable interface using the following command:

```
CMTS(config-if)#cable monitor {incoming ¦ outbound} {timestamp} interface
  WAN_interface [access-list ACL ¦ mac-address MAC ¦ sid SID] {packet-type [data
  docsis ¦ data ethernet ¦ mac {type [dsa ¦ dsc ¦ dsd ¦ map-grant ¦ map-req]}]}
```

- If the optional timestamp parameter is included, a 4-byte timestamp (in hundredths of a second) is appended to packets.
- The access-list option can be defined either as a numbered access list or as a named access list. Packets that are permitted in the access list are monitored, and packets that are denied are not.
- The mac-address option identifies a particular device you want to monitor. This can be a cable modem or a host device such as a MTA or PC. This can also be the broadcast address or a multicast address.
- The sid option specifies a specific upstream service flow to monitor.
- The packet-type option can be used to capture only a specific kind of DOCSIS packet. If you want to capture data packets only from a specific device, you would use the data option here with either an Ethernet or a DOCSIS encapsulation. With DOCSIS encapsulation the standard 6-byte DOCSIS header and any extended headers are captured, as well as the payload, whereas with Ethernet encapsulation, the header information is not captured (thus, the packets look like normal Ethernet packets).

  An important aspect to keep in mind when using DOCSIS encapsulation is that you might be unable to capture large packets. Remember, the maximum size of an Ethernet packet is 1518 bytes. So if you are doing DOCSIS encapsulation, the DOCSIS headers can make a frame bigger than this maximum size; consequently, these packets are lost.

- The mac option can be used to specify only DOCSIS MAC management messages to be captured. If the sid option is used with the mac option, you can denote specific types of MAC messages to capture. These options include the dynamic service flow messages for addition (dsa), modification (dsc), and deletion (dsd), as well as two options for capturing bandwidth allocation (MAP) messages. If the map-grant option is used, any MAP message with a grant interval for that SID is captured, and if the map-request option is used, any MAP message with a request interval for that SID is captured.

NOTE If no packet-type option is specified, both data packets and DOCSIS MAC messages are captured, and the data packets are captured using the DOCSIS encapsulation format.

Multiple cable monitor instances can be simultaneously defined to capture multiple types of traffic. Individual cable monitor instances can be removed by prefixing the keyword no to the corresponding configuration command; also all cable monitor instances can be removed from an interface using the **no cable monitor all** interface command.

Because the CMTS Ethernet interface should be dedicated for cable monitor, disabling other types of packets from being sent over this interface is recommended. Configuration commands you can use to get rid of non-cable monitor traffic are as follows:

- To prevent the Ethernet heartbeat keep-alive messages from being sent out this interface, enter the following command:.

  ```
  CMTS(config-if)#no keepalive
  ```

- To prevent Cisco Discovery Protocol packets from being sent out this interface, enter the following command.

  ```
  CMTS(config-if)#no cdp enable
  ```

You can verify your cable monitor configuration using the following command:

```
CMTS#show interface cable x/y monitor
```

Figure B-2 shows an example of the steps required to configure and verify the cable monitor feature. In this case, all types of traffic from both the cable modem component and MTA component of a PacketCable MTA are captured to a PC running Ethereal connected to interface GigabitEthernet 0/1.

Using an External RF Analyzer

If you are using another vendor's CMTS in your network or simply do not want to use the cable monitor feature (perhaps because of the Ethernet message size limitation), RF traces can also be obtained with an external RF protocol analyzer. This device typically has both a downstream and an upstream RF interface and is located near the CMTS. The downstream feed from the CMTS is split with one connection going to the analyzer and the other connection(s) going to the various cable plants. The downstream analyzer connection typically also requires ~50 dBmV of attenuation so that the input level is around 0 dBmV. Similarly, the upstream feed(s) from a cable plant is split with one connection going to the analyzer and the other connection going to the CMTS. This way, the upstream interface can capture the upstream traffic from all modems in that cable plant.

Figure B-2 *Cable Monitor Configuration Example*

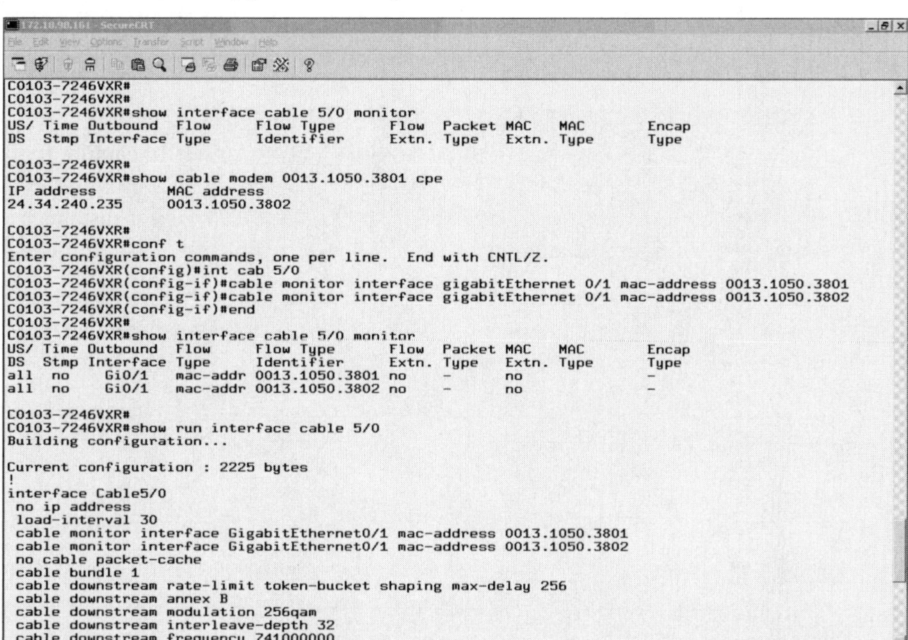

The analyzer locks on to the downstream and upstream channels and captures DOCSIS packets. Usually, specific filters can be set up to capture only particular types of messages or only messages from certain MAC addresses. These captured packets are then imported into a sniffer utility such as Ethereal, where they can be analyzed.

You find some key differences in the data obtained from capturing RF packets via an external analyzer as compared to via the cable monitor utility. For example, if Baseline Privacy Initialization Plus (BPI+) is enabled on your network (which it should be for a PacketCable MTA), you cannot decipher the DOCSIS payload if the capture was obtained via an external analyzer. Using the cable monitor feature, you can decipher the DOCSIS payload because it displays packets after the BPI+ decryption process has occurred for upstream packets and before the BPI+ encryption process has occurred for downstream packets. Another difference is that with an external RF analyzer you can collect DOCSIS Synchronization messages and DOCSIS header CRC bytes. This information is not available using the cable monitor feature because these functions are handled in hardware before the cable monitor code is invoked.

Configuring Ethereal to Decode DOCSIS Packets

By default, Ethereal interprets frames as ordinary Ethernet frames and thus does not understand DOCSIS headers. With the Preferences option under the Edit menu you can change this behavior. In the Preferences menu, under Protocols, select the Frame option and enable the Treat All Frames As DOCSIS Frames checkbox, as shown in Figure B-3.

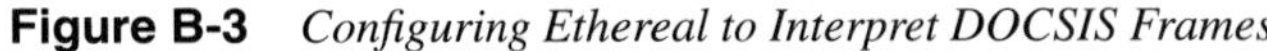

Figure B-3 *Configuring Ethereal to Interpret DOCSIS Frames*

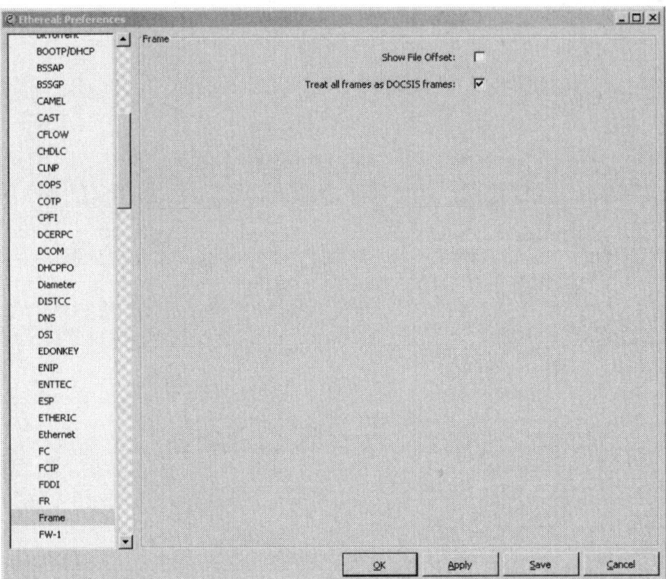

Now Ethereal expects each frame to contain a DOCSIS header.

Using Ethereal to Examine MTA Provisioning

Following are some examples that illustrate how cable monitor and Ethereal can be used in a PacketCable network. First, you will see how these tools can be used to collect a trace of the provisioning process of a PacketCable MTA. The cable monitor configuration is similar to the one in Figure B-1, except that the broadcast MAC address FFFF.FFFF.FFFF is also configured. This is so broadcast DHCP packets are captured; without this option you cannot see the DHCP Offer and Acknowledgement packets. Figure B-4 shows part of the modem provisioning process with a Ranging Response message highlighted. As you can see, all the DOCSIS parameters in this message are decoded by Ethereal. In this case you can see the MTA is being instructed to change its DOCSIS timing offset.

Figure B-4 *Trace of Cable Modem Provisioning*

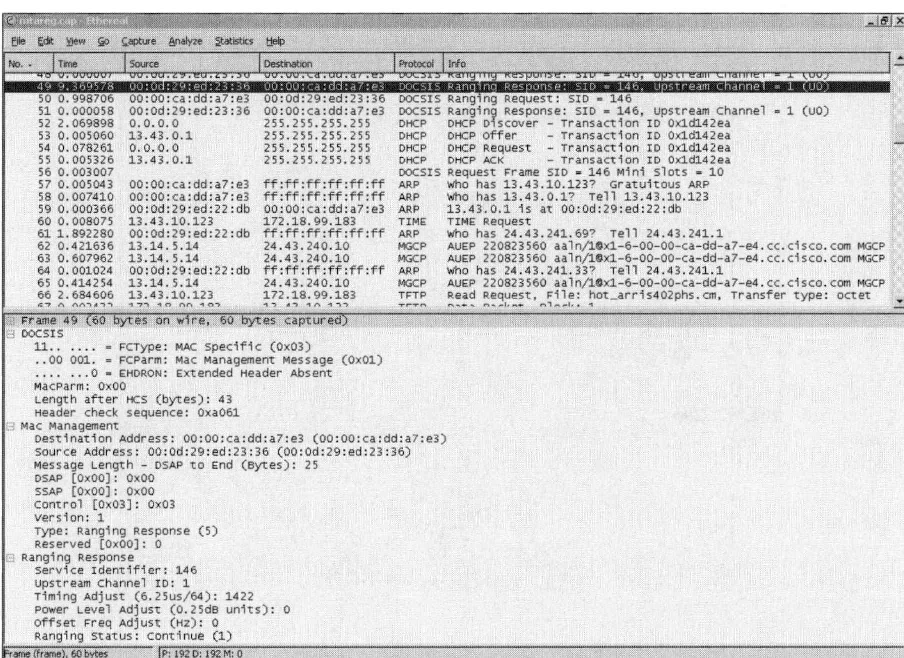

Filters can be set up on DOCSIS parameters just like with any other protocol decoded by Ethereal. For example, if you only wanted to see DOCSIS MAC Management messages, you can apply the filter docsis_mgmt. Filters could also be set up to look at only certain DOCSIS messages or messages for a specific SID. In the packet details plane, if you highlight the parameter you want to filter on and click the right mouse button you can use the Apply As Filter or Prepare As Filter options to have Ethereal automatically create the right packet filter.

NOTE Filtering is especially useful if you are capturing DOCSIS bandwidth allocation (MAP) messages because of the sheer number of packets without filtering.

Another Ethereal feature related to the PacketCable provisioning process is that Ethereal fully decodes option #122, the PacketCable DHCP option. If you look at the BOOTP/DHCP option in the Protocols menu, as shown in Figure B-5, you can see that option #122 has been defined as this option (this is the default). If you have a MTA that uses the older option #177, you can set this value here as well and select the standard used to define the format of the PacketCable DHCP option.

Figure B-5 *Defining the PacketCable DHCP Option*

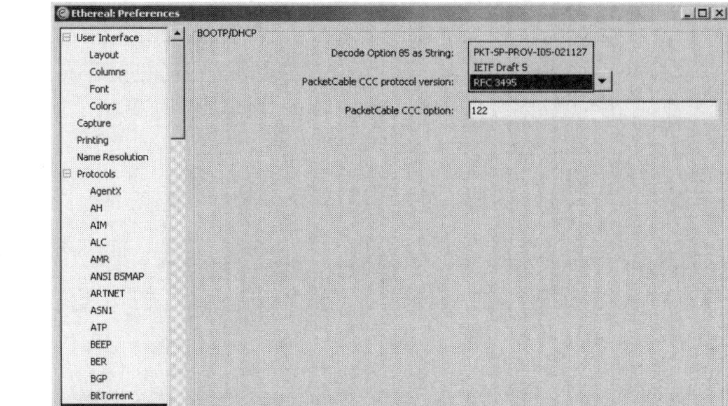

Figure B-6 shows a portion of the MTA registration process.

This trace illustrates the steps from the MTA sending out a DHCP Discover message up to the point where it is downloading its configuration file. Pay attention to the highlighted packet; it shows the Kerberos AP Request message sent from the MTA to the provisioning server to get security keying information about the CMS. Notice Ethereal automatically identifies this message as a PacketCable message. For more information on PacketCable provisioning refer to those sections of the book. But as you can see, if you are having failures in the MTA provisioning process, having a packet trace of this process can help immensely.

Figure B-6 *Trace of MTA Provisioning*

```
No.  Time     Source             Destination        Protocol Info
 93 1.515975  0.0.0.0            255.255.255.255    DHCP   DHCP Discover  - Transaction ID 0x1a12f58
 94 0.004209  13.43.0.1          24.43.240.10       DHCP   DHCP Offer     - Transaction ID 0x1a12f58
 95 0.007839  0.0.0.0            255.255.255.255    DHCP   DHCP Request   - Transaction ID 0x1a12f58
 96 0.003932  13.43.0.1          24.43.240.10       DHCP   DHCP ACK       - Transaction ID 0x1a12f58
 97 0.008301  00:00:ca:dd:a7:e4  ff:ff:ff:ff:ff:ff  ARP    Who has 24.43.240.10?  Gratuitous ARP
 98 0.008418  00:00:ca:dd:a7:e4  ff:ff:ff:ff:ff:ff  ARP    Who has 24.43.240.1?  Tell 24.43.240.10
 99 0.000484  00:0d:29:ed:22:db  00:00:ca:dd:a7:e4  ARP    24.43.240.1 is at 00:0d:29:ed:22:db
100 0.003093  24.43.240.10       172.18.99.184      DNS    Standard query A ss-nms32.cisco.com
101 0.001415  172.18.99.184      24.43.240.10       DNS    Standard query response A 172.18.99.182
102 0.007662                                        DOCSIS Request Frame SID = 146 Mini Slots = 91
103 0.004828  24.43.240.10       172.18.99.182      PKTC   AP Request (SNMPv3)
104 0.015460  172.18.99.182      24.43.240.10       PKTC   AP Reply (SNMPv3)
105 0.014230  24.43.240.10       172.18.99.182      SNMP   INFORM iso.3.6.1.2.1.1.3.0 iso.3.6.1.6.3.1.1.4.1.0 iso.3.6.1.2
106 0.002084  172.18.99.182      24.43.240.10       SNMP   RESPONSE iso.3.6.1.2.1.1.3.0 iso.3.6.1.6.3.1.1.4.1.0 iso.3.6.1
107 0.004019  172.18.99.182      24.43.240.10       SNMP   SET iso.3.6.1.4.1.4491.2.2.1.1.2.5.0 iso.3.6.1.4.1.4491.2.2.1.
108 0.008684  24.43.240.10       172.18.99.182      SNMP   RESPONSE iso.3.6.1.4.1.4491.2.2.1.1.2.5.0 iso.3.6.1.4.1.4491.2
109 0.003709  24.43.240.10       172.18.99.182      TFTP   Read Request, File: bpr01060000cadda7e40001, Transfer type: oc
110 0.002116  172.18.99.182      24.43.240.10       TFTP   Data Packet, Block: 1
111 0.005794  24.43.240.10       172.18.99.182      TFTP   Acknowledgement, Block: 1
112 0.000993  172.18.99.182      24.43.240.10       TFTP   Data Packet, Block: 2
113 0.005022  24.43.240.10       172.18.99.182      TFTP   Acknowledgement, Block: 2
114 0.000996  172.18.99.182      24.43.240.10       TFTP   Data Packet, Block: 3

Frame 103 (662 bytes on wire, 662 bytes captured)
DOCSIS
Ethernet II, Src: 00:00:ca:dd:a7:e4, Dst: 00:0d:29:ed:22:db
Internet Protocol, Src Addr: 24.43.240.10 (24.43.240.10), Dst Addr: 172.18.99.182 (172.18.99.182)
User Datagram Protocol, Src Port: 1293 (1293), Dst Port: 1293 (1293)
PacketCable
  Key Management Message ID: AP Request (0x02)
  Domain of Interpretation: SNMPv3 (2)
  Version: 1.0
  Kerberos AP-REQ
    Pvno: 5
    MSG Type: AP-REQ (14)
    Padding: 0
    APOptions: 20000000 (Mutual required)
    Ticket
    Authenticator des3-cbc-sha
  Server Nonce: 0x00000000
  Application Specific Data
  List of Ciphersuites (2): HMAC-MD5/DES HMAC-MD5/NULL
  Re-establish Flag: False
  SHA-1 HMAC: 45F1A7B2EDBD4269E05F6311AAF01228BE3E0F5F
```

Using Ethereal to Examine a Phone Call

The example in this section examines a cable monitor capture of the messages that occur during the life of a basic call. Again, in the cable monitor configuration both the MAC address of the cable modem and of the MTA need to be included to get the full message trace. This is because the Network-based Call Signaling (NCS) and Real-time Transport Protocol (RTP) packets are sent to and from the MTA address, but the DOCSIS packets used to reserve and activate quality of service (QoS) resources are sent to and from the cable modem MAC address. Because this example is an ONNET-to-ONNET call between two MTAs in the same DOCSIS MAC domain, the MAC addresses of both MTA are monitored. Figure B-7 shows the beginning of the call flow, from where the originating MTA issues a bandwidth request packet to send the off-hook notification until the point where the originating MTA issues the DSA-REQ message to reserve bandwidth for the call.

Figure B-7 *Trace of a Basic Call: Part I*

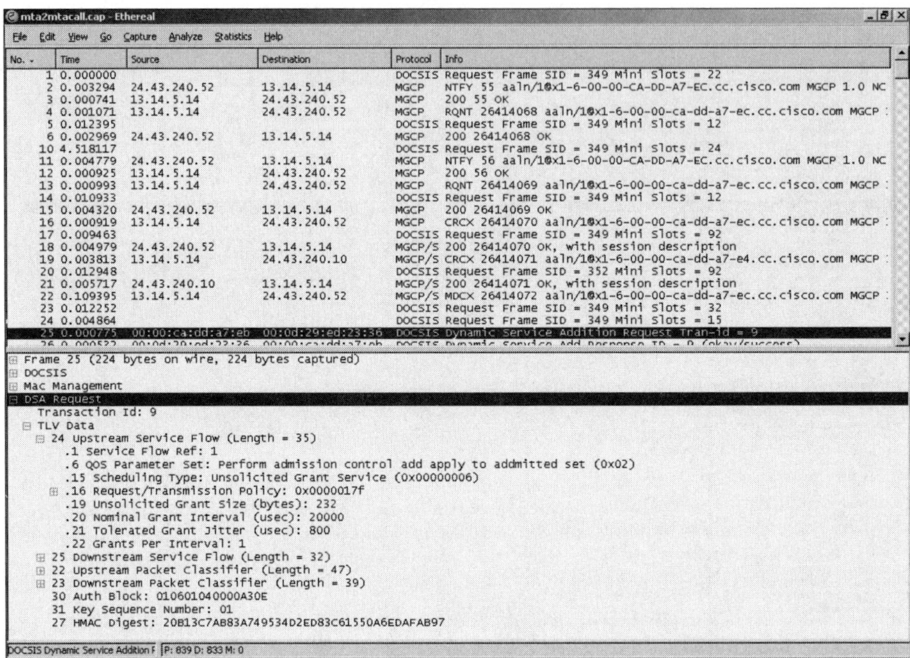

The highlighted packet is the DOCSIS DSA-REQ from the originating MTA; the details of the UGS upstream service flow are shown in this figure. Figure B-8 is a packet trace of the call at the point where the call is answered. In this capture, you can see the NCS and DOCSIS messages used to answer the call and activate the service flows. At the end of this figure is the flow of voice RTP packets between the two MTAs.

Using Ethereal to Generate Call Flow Schematics

Ethereal also has the ability to identify VoIP calls in captured traces and analyze them. Under the Statistics option select VoIP Calls..., and Ethereal detects any MGCP and/or SIP calls that are present in the captured packets. For example, Figure B-9 is the output of this process for the basic ONNET-to-ONNET call captured in the previous section.

Figure B-8 *Trace of a Basic Call: Part II*

File Edit View Go Capture Analyze Statistics Help

No.	Time	Source	Destination	Protocol	Info
288	0.009921			DOCSIS	Request Frame SID = 352 Mini Slots = 22
289	0.004182	24.43.240.10	13.14.5.14	MGCP	NTFY 23 aaln/1@x1-6-00-00-CA-DD-A7-E4.cc.cisco.com MGCP 1.0 NC
290	0.000564	13.14.5.14	24.43.240.10	MGCP	200 23 OK
291	0.000886	13.14.5.14	24.43.240.10	MGCP	RQNT 26414076 aaln/1@x1-6-00-00-ca-dd-a7-e4.cc.cisco.com MGCP
292	0.004139	24.43.240.10	24.43.240.52	RTP	Payload type=ITU-T G.711 PCMU, SSRC=28695344, Seq=113, Time=18
293	0.000871	24.43.240.10	24.43.240.52	RTP	Payload type=ITU-T G.711 PCMU, SSRC=28695344, Seq=113, Time=18
294	0.005880			DOCSIS	Request Frame SID = 352 Mini Slots = 12
295	0.005991	24.43.240.10	13.14.5.14	MGCP	200 26414076 OK
296	0.002223	13.14.5.14	24.43.240.10	MGCP	MDCX 26414077 aaln/1@x1-6-00-00-ca-dd-a7-e4.cc.cisco.com MGCP
297	0.001109	13.14.5.14	24.43.240.52	MGCP	MDCX 26414078 aaln/1@x1-6-00-00-ca-dd-a7-ec.cc.cisco.com MGCP
298	0.003898	24.43.240.10	24.43.240.52	RTP	Payload type=ITU-T G.711 PCMU, SSRC=28695344, Seq=114, Time=18
299	0.000869	24.43.240.10	24.43.240.52	RTP	Payload type=ITU-T G.711 PCMU, SSRC=28695344, Seq=114, Time=18
300	0.002386			DOCSIS	Request Frame SID = 352 Mini Slots = 37
301	0.000661			DOCSIS	Request Frame SID = 349 Mini Slots = 37
302	0.004128	00:00:ca:dd:a7:e3	00:0d:29:ed:23:36	DOCSIS	Dynamic Service Change Request Tran-id = 10
303	0.000497	00:00:ca:dd:a7:eb	00:0d:29:ed:23:36	DOCSIS	Dynamic Service Change Request Tran-id = 10
304	0.000002			DOCSIS	Request Frame SID = 352 Mini Slots = 15
305	0.000004			DOCSIS	Request Frame SID = 349 Mini Slots = 15
306	0.001053	00:0d:29:ed:23:36	00:00:ca:dd:a7:e3	DOCSIS	Dynamic Service Change Response ID = 10 (okay/success)
307	0.000002	00:0d:29:ed:23:36	00:00:ca:dd:a7:e3	DOCSIS	Dynamic Service Change Response ID = 10 (okay/success)
308	0.001520	00:0d:29:ed:23:36	00:00:ca:dd:a7:eb	DOCSIS	Dynamic Service Change Response ID = 10 (okay/success)
309	0.000002	00:0d:29:ed:23:36	00:00:ca:dd:a7:eb	DOCSIS	Dynamic Service Change Response ID = 10 (okay/success)
310	0.000611	24.43.240.10	13.14.5.14	MGCP	100 26414077 MDCX Response Pending
311	0.000128	24.43.240.52	13.14.5.14	MGCP	100 26414078 MDCX Response Pending
312	0.003680			DOCSIS	Request Frame SID = 352 Mini Slots = 10
313	0.000423			DOCSIS	Request Frame SID = 349 Mini Slots = 10
314	0.003695	24.43.240.10	24.43.240.52	RTP	Payload type=ITU-T G.711 PCMU, SSRC=28695344, Seq=115, Time=18
315	0.000002	00:00:ca:dd:a7:e3	00:0d:29:ed:23:36	DOCSIS	Dynamic Service Change Ack ID = 10 (okay/success)
316	0.000005	00:00:ca:dd:a7:eb	00:0d:29:ed:23:36	DOCSIS	Dynamic Service Change Ack ID = 10 (okay/success)
317	0.005759			DOCSIS	Request Frame SID = 349 Mini Slots = 15
318	0.000127			DOCSIS	Request Frame SID = 352 Mini Slots = 15
319	0.004097	24.43.240.52	13.14.5.14	MGCP	[Malformed Packet]
320	0.000250	24.43.240.10	13.14.5.14	MGCP	[Malformed Packet]
321	0.000482	13.14.5.14	24.43.240.52	MGCP	000 26414078
322	0.000149	13.14.5.14	24.43.240.52	MGCP	RQNT 26414079 aaln/1@x1-6-00-00-ca-dd-a7-ec.cc.cisco.com MGCP
323	0.000230	13.14.5.14	24.43.240.10	MGCP	000 26414077
324	0.009244	24.43.240.10	24.43.240.52	RTP	Payload type=ITU-T G.711 PCMU, SSRC=28695344, Seq=116, Time=18
325	0.002444			DOCSIS	Request Frame SID = 349 Mini Slots = 12
326	0.003209	24.43.240.52	13.14.5.14	MGCP	200 26414079 OK
327	0.011767	24.43.240.52	24.43.240.10	RTP	Payload type=ITU-T G.711 PCMU, SSRC=28694509, Seq=0, Time=1920
328	0.002708	24.43.240.10	24.43.240.52	RTP	Payload type=ITU-T G.711 PCMU, SSRC=28695344, Seq=117, Time=18
329	0.017267	24.43.240.52	24.43.240.10	RTP	Payload type=ITU-T G.711 PCMU, SSRC=28694509, Seq=1, Time=1936
330	0.002705	24.43.240.10	24.43.240.52	RTP	Payload type=ITU-T G.711 PCMU, SSRC=28695344, Seq=118, Time=18
331	0.017395	24.43.240.52	24.43.240.10	RTP	Payload type=ITU-T G.711 PCMU, SSRC=28694509, Seq=2, Time=1952
332	0.002704	24.43.240.10	24.43.240.52	RTP	Payload type=ITU-T G.711 PCMU, SSRC=28695344, Seq=119, Time=19
333	0.017158	24.43.240.52	24.43.240.10	RTP	Payload type=ITU-T G.711 PCMU, SSRC=28694509, Seq=3, Time=1968

File: mta2mtacall.cap 183 KB 00:0(P: 839 D: 833 M: 0

Figure B-9 *Using Ethereal to Detect VoIP Calls*

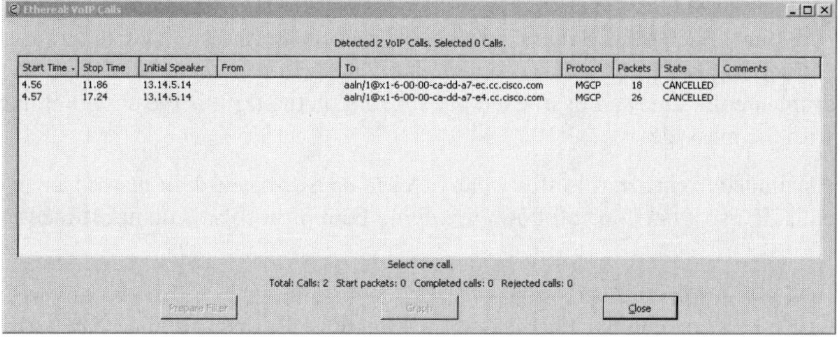

In this case, Ethereal detected two VoIP call legs; one to each MTA endpoint. The Initial Speaker in this case is an IP address of the CMS, and the To field shows the MGCP endpoint names of the MTAs involved in this call. Ethereal can also generate diagrams illustrating these detected VoIP calls. To do this, select the VoIP call legs you want graphed and click the Graph option. Figure B-10 shows the result of this operation.

Figure B-10 *Using Ethereal to Generate Call Flows*

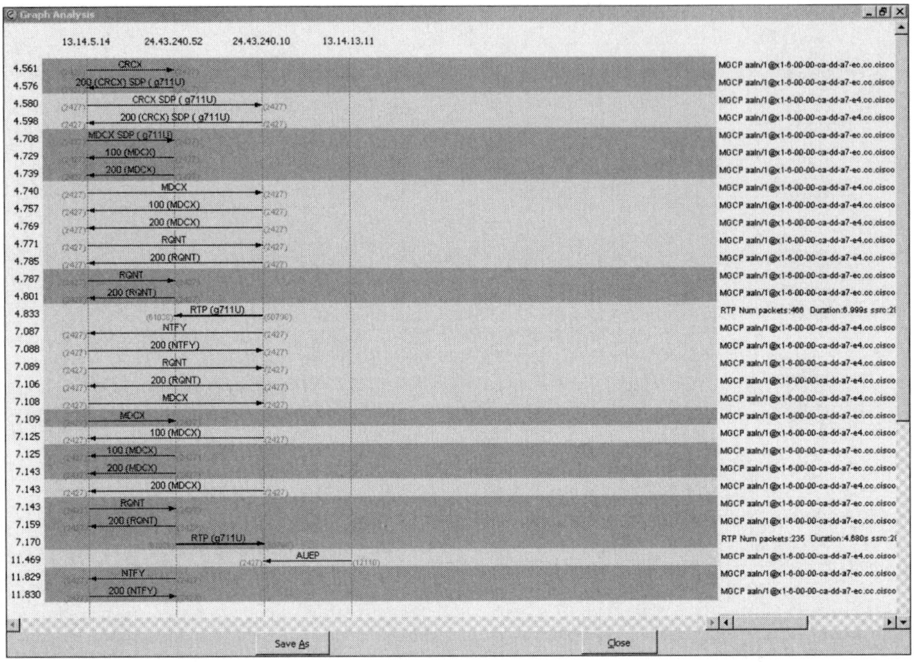

In this case, the vertical line on the left represents the CMS, and the next two vertical lines are the two MTAs; the next vertical line is another CMS IP address used to audit the MTA endpoints. Each MGCP message and its direction are shown in the diagram, and each VoIP call leg is given its own color to help distinguish it from the others. Also on the left of the graph are the timestamps of each packet, and on the right is the endpoint name associated with the message.

As alluded to earlier, this utility can be used on SIP-based calls as well as on MGCP-based calls. It also works on call flows involving both protocols as demonstrated in Figure B-11.

NOTE One limitation of this feature is that Ethereal does not detect that a MGCP call leg is present until it sees the CRCX message; thus, the dialing stage of the call is not detected.

Figure B-11 *Call Flow for an MGCP and SIP Call*

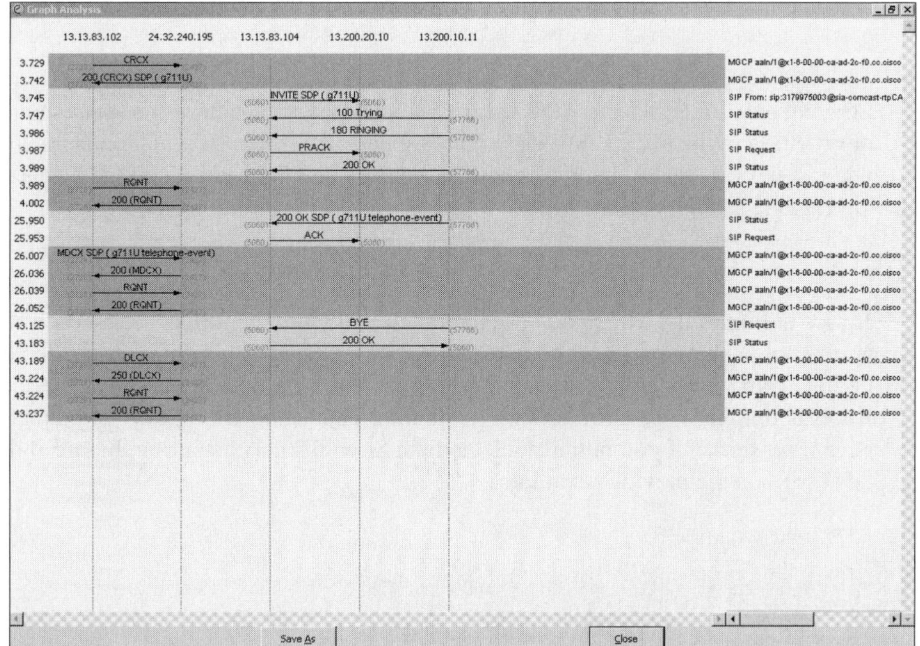

Using Ethereal to Examine RTP Streams

Ethereal also has the capability to gather information about the RTP media streams it captures. This utility is also accessed under the Statistics menu. If you select the RTP option, you see two sub-options: Show All Streams and Stream Analysis. The Show All Streams option triggers Ethereal to detect all the RTP media streams present in the capture file. Figure B-12 illustrates the result of this operation on the previous basic ONNET-to-ONNET call example.

Figure B-12 *RTP Stream Detection*

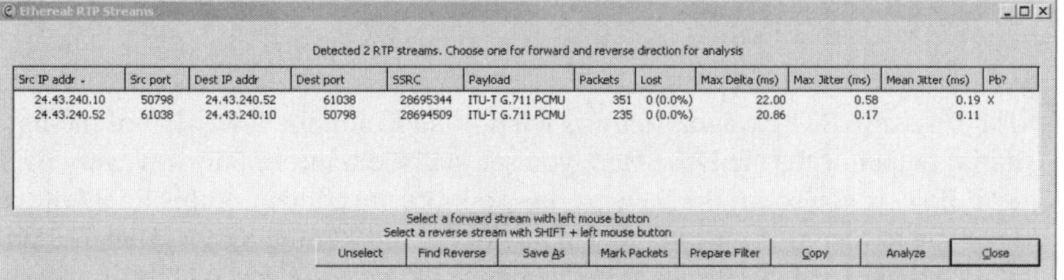

Detected 2 RTP streams. Choose one for forward and reverse direction for analysis

Src IP addr	Src port	Dest IP addr	Dest port	SSRC	Payload	Packets	Lost	Max Delta (ms)	Max Jitter (ms)	Mean Jitter (ms)	Pb?
24.43.240.10	50798	24.43.240.52	61038	28695344	ITU-T G.711 PCMU	351	0 (0.0%)	22.00	0.58	0.19	X
24.43.240.52	61038	24.43.240.10	50798	28694509	ITU-T G.711 PCMU	235	0 (0.0%)	20.86	0.17	0.11	

Select a forward stream with left mouse button
Select a reverse stream with SHIFT + left mouse button

Unselect | Find Reverse | Save As | Mark Packets | Prepare Filter | Copy | Analyze | Close

As you can see, in this case two RTP media streams were detected—one from the originating MTA to the terminating MTA and one in the reverse direction.

TIP Ethereal only identifies the UDP streams as RTP if it also captures the corresponding SDP information in the MGCP/SIP call signaling packets. If this information is not available, you can tell Ethereal to interpret these streams as RTP by right-clicking one and selecting the Decode as RTP option.

Notice in Figure B-12 that you have data such as the number of packets in each stream, the number of packets missing, and packet jitter statistics. This information can be very useful in diagnosing voice quality issues. If you select both of these streams and then select the Analyze option, more information is available. This is the same information that would have been shown if you initially selected the Stream Analysis option. Figure B-13 shows this result on the previous example.

Figure B-13 *RTP Stream Analysis*

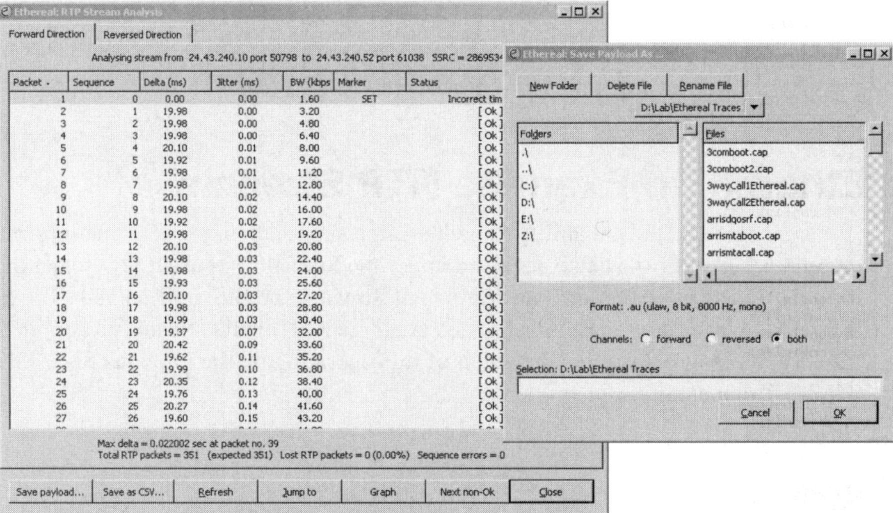

The Ethereal: RTP Stream Analysis window shown in the lower left of the figure is the initial output of the analysis. Here you see statistical information on every packet in this RTP flow. If you select the Save payload option, the content of this RTP flow can be saved in an audio file, as shown in the window in the right of the figure. This file can then be played back using any standard media player to listen to the actual voice stream. Thus, you can then use your own ear to assess the call voice quality. Of course, in a production environment this should be done only to troubleshoot network quality. Note: If security is enabled in your network so that the RTP packets are encrypted, this type of analysis is not possible.

Examining PacketCable COPS DQoS Packets

So far the use of Ethereal has focused on the communication between the MTA and other network components such as the CMTS and CMS. Packet sniffer utilities can also provide value in other places in a PacketCable network. For example, one use is in troubleshooting the Common Open Policy Service (COPS) packets sent between the CMS and CMTS used to provide DQoS.

If a Cisco switch is in the path between the CMS and CMTS components, such as a 7600, 4500, or 2900 series device, the port monitor feature can be used to replicate packets to a machine running Ethereal.

Figure B-14 shows the steps required to configure port monitoring and the show commands used to verify its operation. In this example the CMTS is connected to interface GigabitEthernet 0/1, and the Ethereal machine is connected to interface GigabitEthernet 0/3.

Figure B-14 *Port Monitor Configuration*

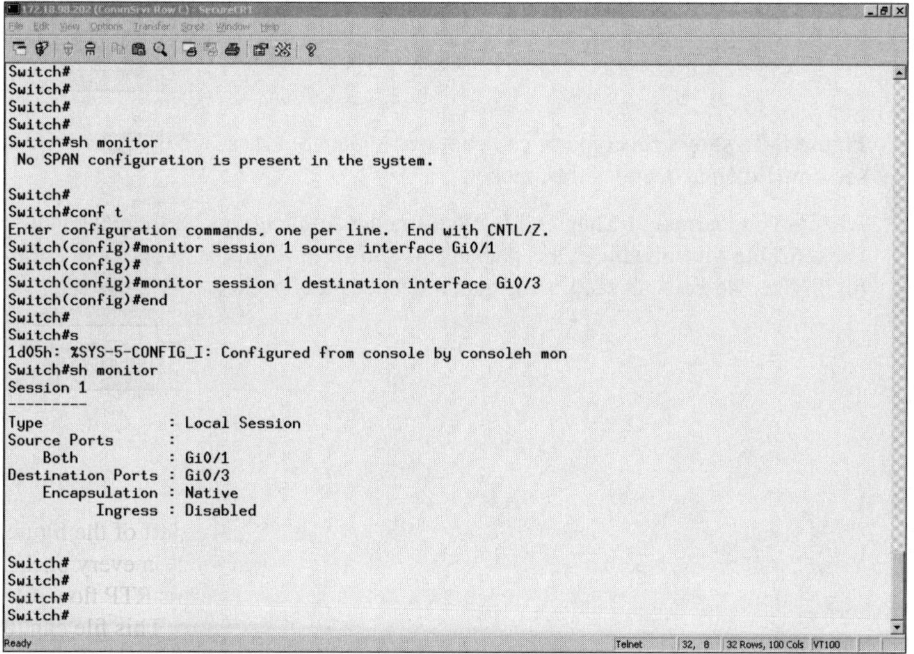

Ethereal also needs to be told what TCP port is used for PacketCable COPS packets. For PacketCable telephony this should be port 2126. To check this configuration, look at the COPS option in the Edit > Preferences > Protocols menu, as depicted in Figure B-15.

Figure B-15 *PacketCable COPS Option*

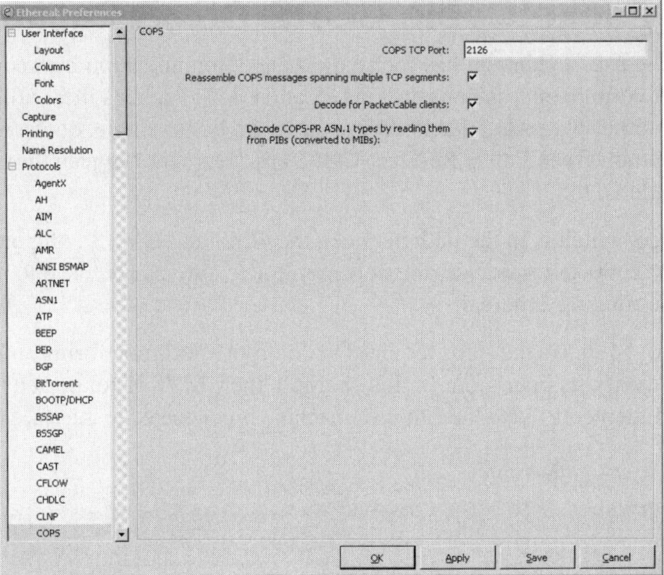

Figure B-16 shows this type of call capture. A Gate Set message is highlighted so you can see how Ethereal decodes this packet.

Also as you learned in Chapter 14, "Multimedia Applications," Ethereal can also decode PacketCable Multimedia COPS packets, as shown in Figure B-17. A Gate Set message is highlighted so you can see how Ethereal decodes this packet.

Figure B-16 *PacketCable COPS Example*

```
mta2mta41.cap - Ethereal
File Edit View Go Capture Analyze Statistics Help
No. -  Time      Source          Destination     Protocol Info
 70 0.000416  13.200.12.20    24.34.241.84    MGCP     RQNT 53757727 aaln/1@x1-6-00-00-ca-ad-2c-f0.company-y.cisco.co
 71 0.013836  24.34.241.84    13.200.12.20    MGCP     200 53757727 OK
 72 0.015844  13.200.12.20    24.34.241.84    MGCP     CRCX 53757728 aaln/1@x1-6-00-00-ca-ad-2c-f0.company-y.cisco.co
 73 0.012398  172.18.98.148   13.10.0.20      COPS     COPS Keep-Alive (KA)
 74 0.002409  24.34.241.84    13.200.12.20    MGCP/S   200 53757728 OK, with session description
 75 0.003650  13.200.12.20    24.34.240.86    MGCP/S   CRCX 36106 aaln/1@x1-6-00-00-ca-ad-27-38.company-y.cisco.com M
 76 0.018481  24.34.240.86    13.200.12.20    MGCP/S   200 36106 OK, with session description
 77 0.003067  13.200.13.4     13.10.0.20      COPS     COPS Decision (DEC)
 78 0.001510  13.10.0.20      13.200.13.4     COPS     COPS Report State (RPT)
 79 0.051291  13.200.13.4     13.10.0.20      COPS     COPS Decision (DEC)
 80 0.001457  13.10.0.20      13.200.13.4     COPS     COPS Report State (RPT)
 81 0.051508  13.200.12.20    24.34.241.84    MGCP/S   MDCX 53757729 aaln/1@x1-6-00-00-ca-ad-2c-f0.company-y.cisco.co
 82 0.028442  24.34.241.84    13.200.12.20    MGCP     200 53757729 OK
 83 0.001519  13.200.12.20    24.34.240.86    MGCP     MDCX 36107 aaln/1@x1-6-00-00-ca-ad-27-38.company-y.cisco.com M
 84 0.028462  24.34.240.86    13.200.12.20    MGCP     200 36107 OK
 85 0.003199  13.200.12.20    24.34.240.86    MGCP     RQNT 36108 aaln/1@x1-6-00-00-ca-ad-27-38.company-y.cisco.com M
Frame 77 (278 bytes on wire, 278 bytes captured)
Ethernet II, Src: 00:07:84:a4:ee:0a, Dst: 00:04:dd:5d:78:03
Internet Protocol, Src Addr: 13.200.13.4 (13.200.13.4), Dst Addr: 13.10.0.20 (13.10.0.20)
Transmission Control Protocol, Src Port: 50984 (50984), Dst Port: 2126 (2126), Seq: 80, Ack: 80, Len: 224
Common Open Policy Service
  Version: 1, Flags: None
  Op Code: Decision (DEC) (2)
  Client Type: PacketCable Dynamic Quality-of-Service (32776)
  Message Length: 224
  Handle Object (Handle): Client Handle
  Context Object (Context):
  Decision Object (Decision): Decision Flags (Mandatory)
  Decision Object (Decision): Client Specific Decision Data
    Object Length: 192
    C-Num: Decision Object (Decision) (6)
    C-Type: Client Specific Decision Data (4)
    D-QoS Transaction ID
      D-QoS Transaction Identifier : 3
      Gate Command Type            : Gate Set (4)
    Subscriber ID (IPv4)
      Subscriber Identifier (IPv4): 24.34.240.86 (24.34.240.86)
    Activity Count
      Count                        : 5
    Event Generation Info
    Gate Specifications
    Gate Specifications
Transmission Control Protocol (tcp)  P: 172 D: 172 M: 0
```

Figure B-17 *PacketCable Multimedia COPS Example*

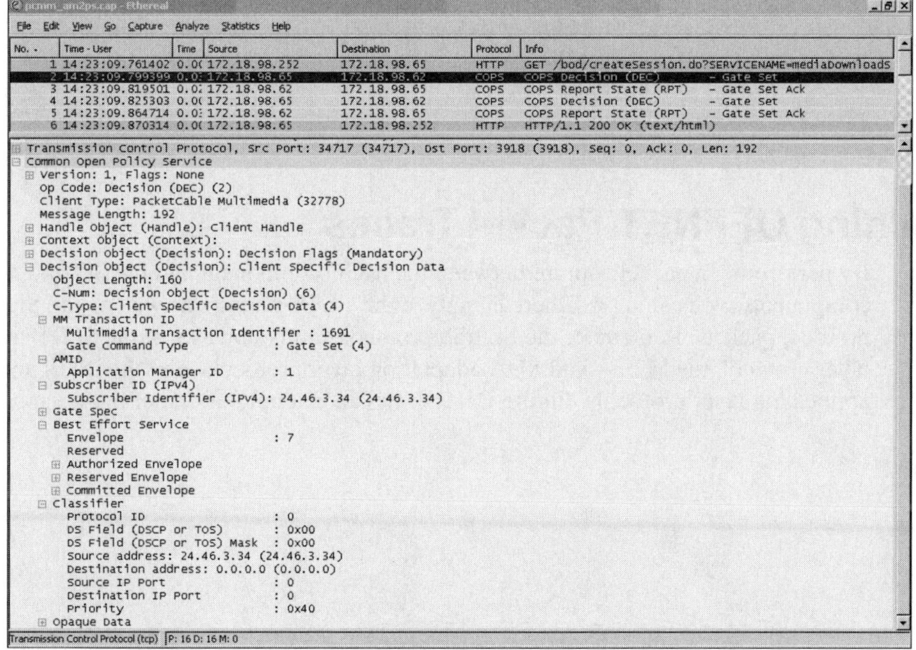

Examining PacketCable Event Messaging Packets

Ethereal also automatically decodes the PacketCable Tag/Length/Value (TLV) attributes used for Event Messaging. No special configuration is needed to enable this functionality. Figure B-18 shows an example of one of these packet captures. In this case a Signaling Start message is depicted.

Figure B-18 *PacketCable Event Messaging Option*

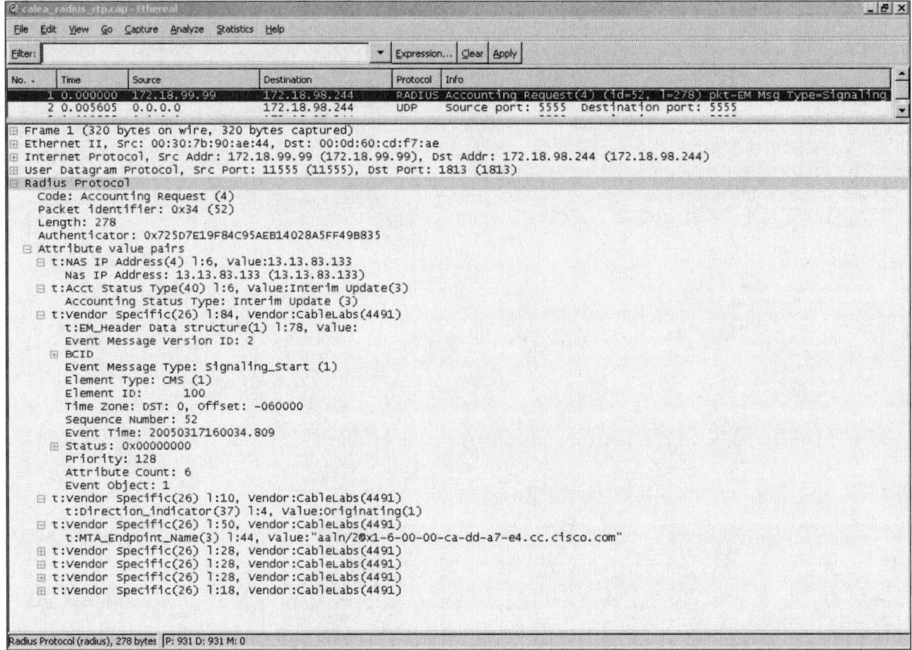

Examining OFFNET Packet Traces

By performing a packet capture between the MGC component and the SG and MG components, you can have Ethereal analyze the TGCP packets as well as the Sigtran protocol packets. Remember the Sigtran protocol family consists of the SCTP transport layer protocol, the M3UA and SUA adaptation layer protocols, and the ISUP and TCAP application layer protocols. Figure B-19 shows an example of one of these captures.

Figure B-19 *OFFNET Call Trace*

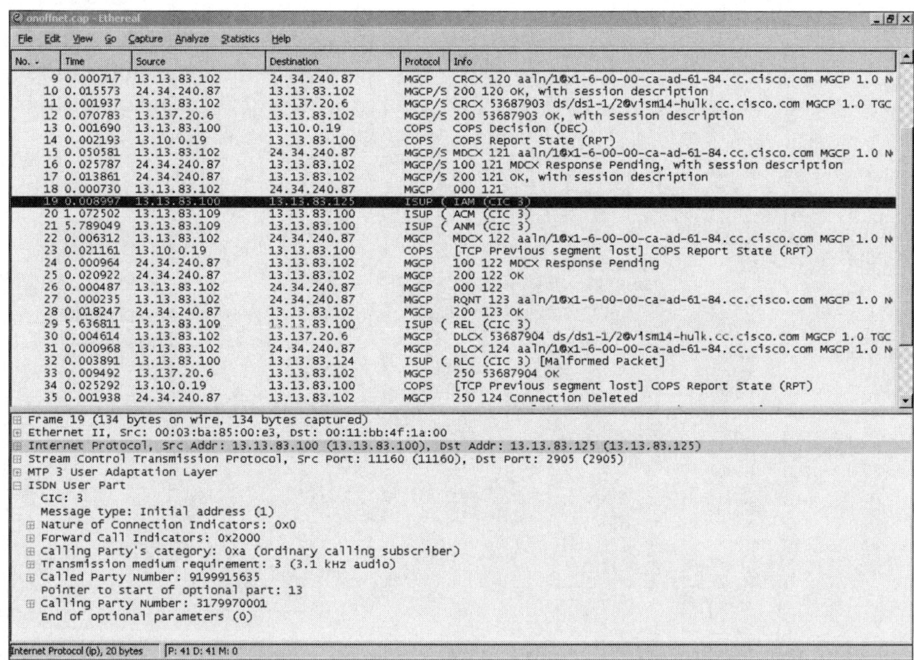

This trace is of an ONNET-to-OFFNET call; thus, as you can see, ISUP messages are present and fully decoded by Ethereal.

Examining PacketCable Lawful Intercept Packets

Finally, Ethereal also has the capability to decode the call content lawful intercept packets sent to a CALEA DF mediation device. To do this you need to configure the DF UDP port so that Ethereal knows how to identify these packets. This is done with the PacketCable Lawful Intercept (PCLI) option from the Edit > Preferences > Protocols menu, as shown in Figure B-20.

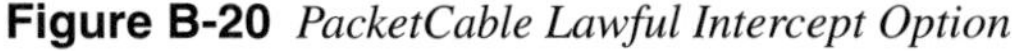

Figure B-20 *PacketCable Lawful Intercept Option*

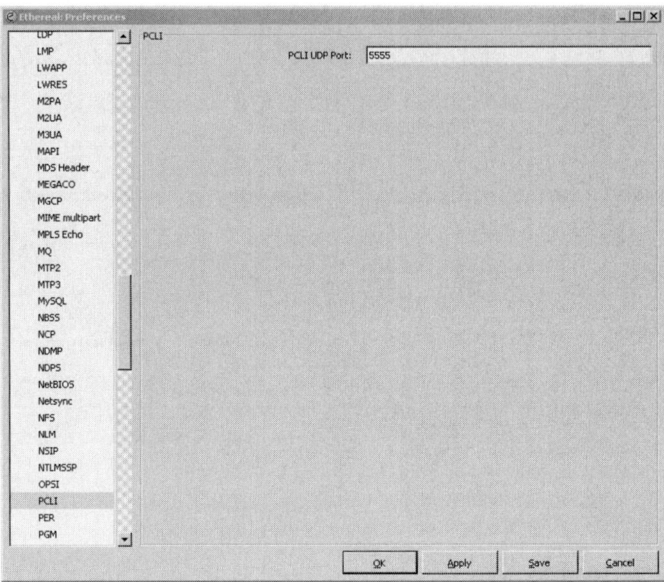

As you can see, in this case UDP port 5555 is being used for call content delivery. From the packet details pane you can see Ethereal has identified a packet as a PacketCable Lawful Intercept packet and has decoded the call content identifier parameter (CCC-ID).

Appendix C

Complete Call Flows

Throughout this text you have seen call flows that focused on certain aspects of a call. For example, in Chapter 7, "NCS (Network-based Call Signaling)," the focus of the call flows was on the Network-based Call Signaling (NCS) call signaling packets between the MTA and the CMS, whereas in Chapter 8, "TGCP and the PSTN Interconnect," the focus was on the Trunk Gateway Control Protocol (TGCP), Sigtran, and Integrated Services User Part (ISUP) protocols, and in Chapter 14, "Multimedia Applications," the focus was on the Common Open Policy Service (COPS) and Data-Over-Cable Service Interface Specification (DOCSIS) protocols. This appendix illustrates the whole picture of the call types most common in a PacketCable deployment.

As stated throughout this book, the PacketCable specifications do not define explicit call flows used to provide telephony services. As long as the rules governed by the protocols involved in the call are not violated, you might have many ways to reach the same goal. Of course, some ways are better than others. The examples shown in this appendix are taken with the Cisco BTS as the call management server/Media Gateway Controller (CMS/MGC), the Cisco uBR10k as the cable modem termination system (CMTS), the Cisco MGX8880/VISM as the Media Gateway (MG), and the Cisco 7206 ITP as the Signaling Gateway (SG).

Also note that in the call flows shown, packet acknowledgements are generally not depicted unless they contain key information. This is done in an attempt to keep the call flows manageable and easier to comprehend. For example, MGCP message acknowledgements with Session Description Protocol (SDP) and/or gate information are shown but a basic "200 - Okay" is typically not shown. Also the event messaging packets encapsulated in RADIUS Accounting Request packets are shown, but the corresponding RADIUS Accounting Response packets are not shown.

Basic ONNET-to-ONNET Call

This section begins with the most basic of call types, a simple ONNET-to-ONNET call. This is a call between two PacketCable subscribers controlled by the same CMS. The subscribers might or might not be physically connected to the same CMTS. The call flows here show two separate CMTSs; however, if only one CMTS is involved, all the same messages are sent to that one CMTS. The number of messages and their sequencing remains the same. The only PacketCable components involved in this case are the CMS, CMTS, Record Keeping Server (RKS), and the two multimedia terminal adapters (MTAs). In this call, Oscar, the originating party, places a call to Tina, the terminating party.

Figures C-1 through C-3 illustrate an implementation of this type of call.

Figure C-1 *Basic ONNET-to-ONNET Call: Part I*

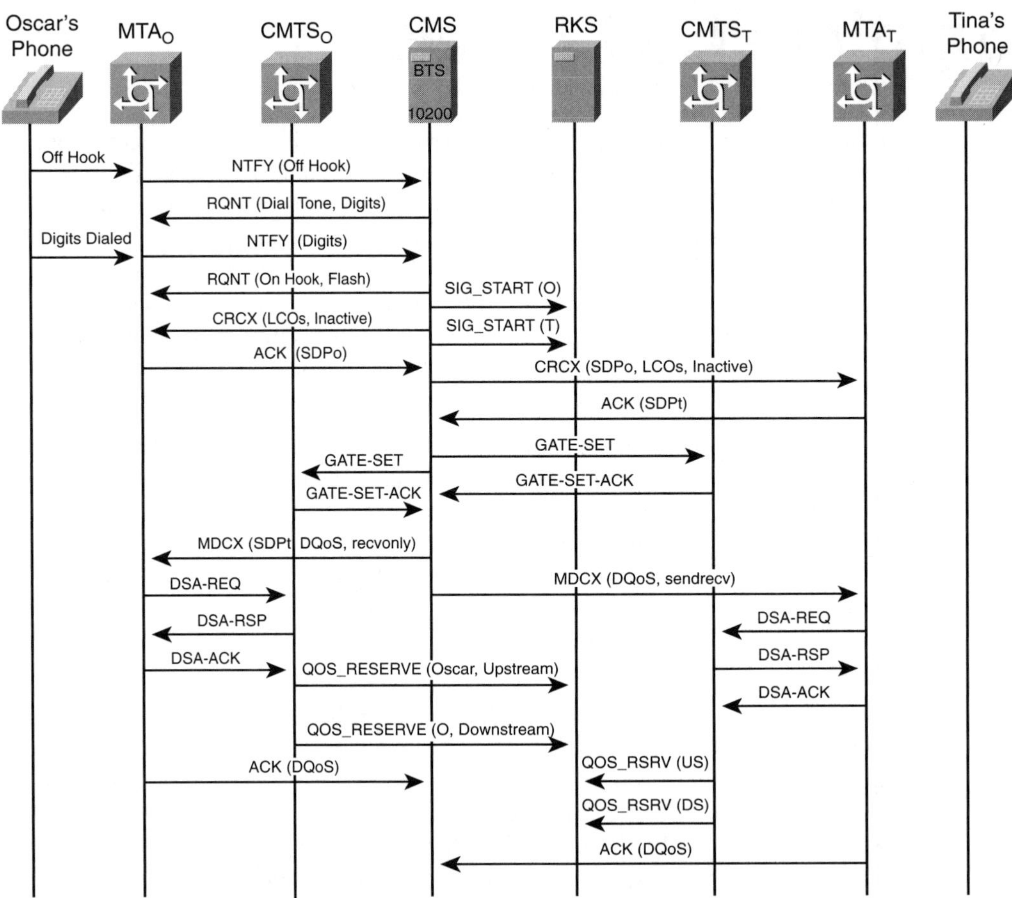

Figure C-2 *Basic ONNET-to-ONNET Call: Part II*

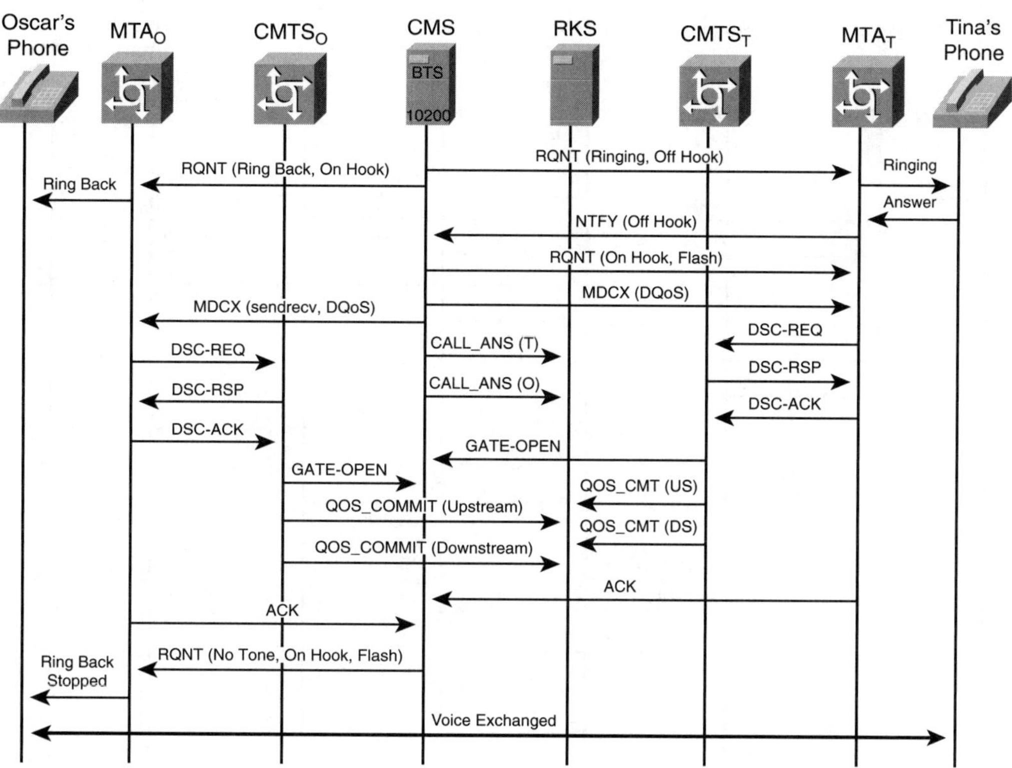

Figure C-3 *Basic ONNET-to-ONNET Call: Part III*

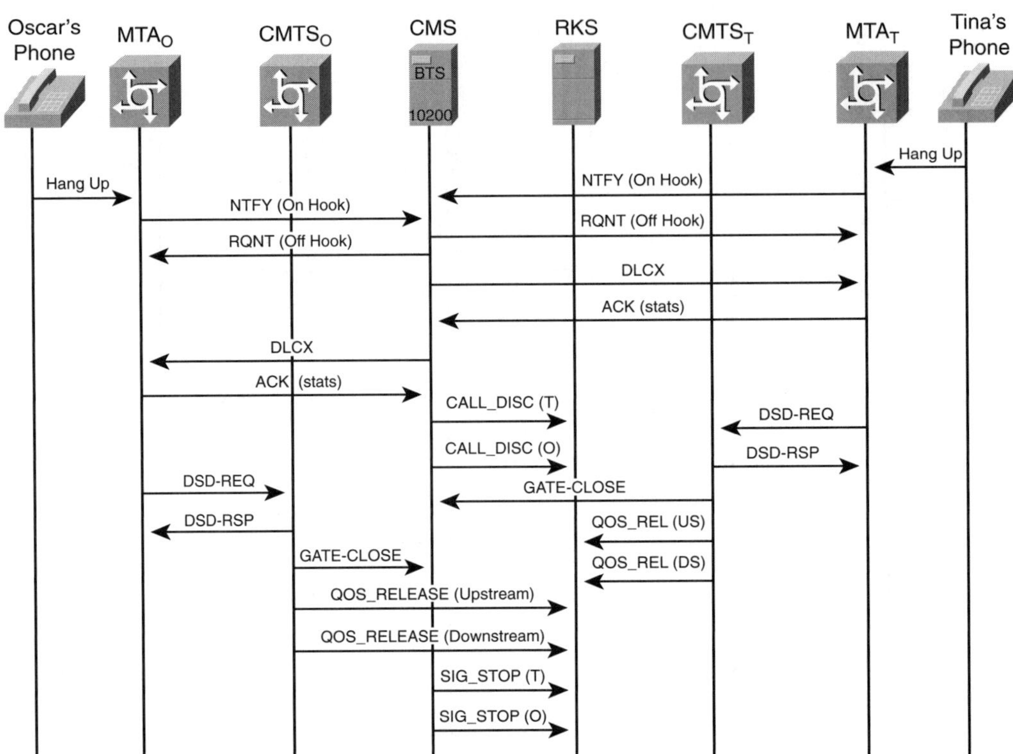

Basic ONNET-to-OFFNET Call

The next example is a basic ONNET-to-OFFNET call. This is when a PacketCable subscriber calls someone whose telephony service resides in the domain of the legacy PSTN. In this case, a trunking MG, SG, and an MGC are also required to interface with the PSTN. Because the BTS is both a PacketCable CMS and a PacketCable MGC, it is shown only once in the call flow traces. The protocol used for communication between the MGC and the SG is the Sigtran family of protocols. Sigtran allows Signaling System 7 (SS7) messages to be sent over an IP-based network. Also in these traces, only the upper application layer protocol, Integrated Services User Part (ISUP), is shown in the traces. Note that the ISUP messages shown between the MGC and SG utilize the Sigtran protocols as the lower layers, whereas the ISUP messages shown between the SG and the PSTN utilize the traditional SS7 protocol stack.

Figures C-4 through C-6 illustrate an implementation of this type of call.

Figure C-4 *Basic ONNET-to-OFFNET Call: Part I*

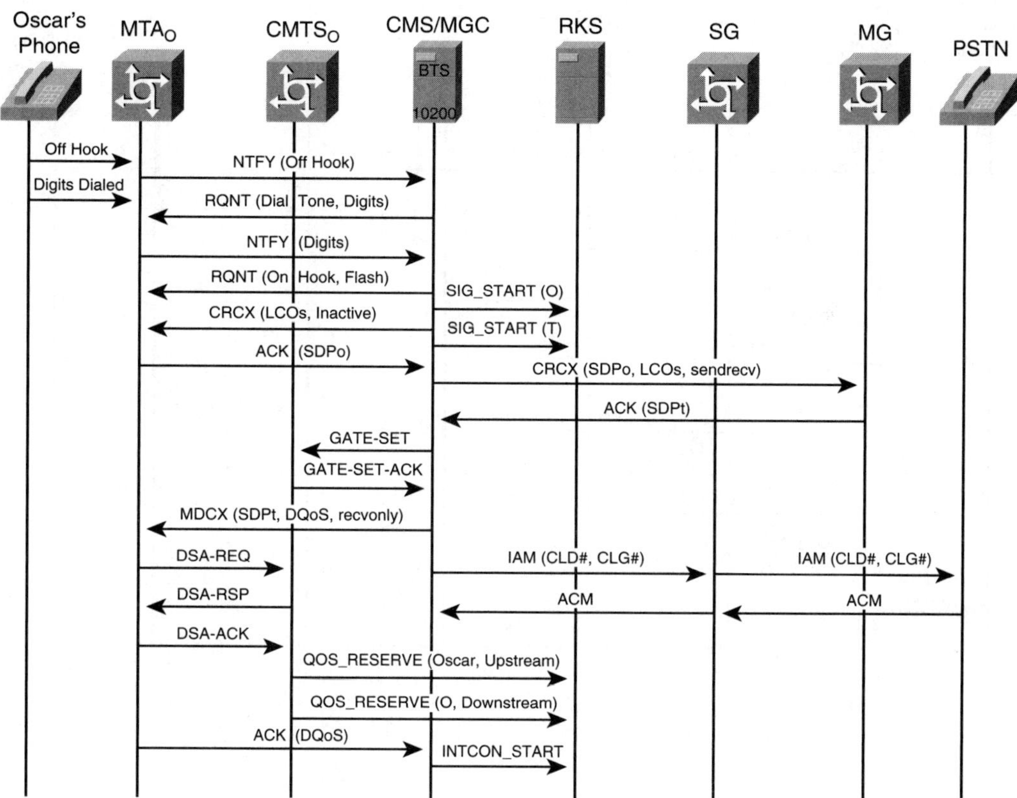

Figure C-5 *Basic ONNET-to-OFFNET Call: Part II*

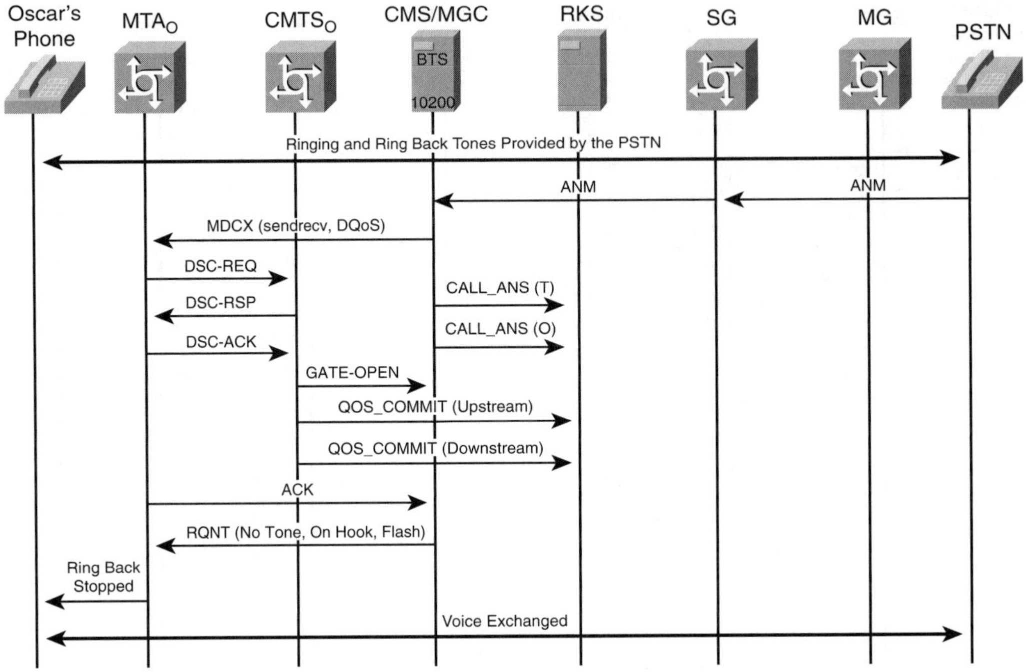

Oscar's Phone
MTA_O
$CMTS_O$
CMS/MGC
BTS 10200
RKS
SG
MG
PSTN

Ringing and Ring Back Tones Provided by the PSTN
ANM
ANM
MDCX (sendrecv, DQoS)
DSC-REQ
CALL_ANS (T)
DSC-RSP
CALL_ANS (O)
DSC-ACK
GATE-OPEN
QOS_COMMIT (Upstream)
QOS_COMMIT (Downstream)
ACK
RQNT (No Tone, On Hook, Flash)
Ring Back Stopped
Voice Exchanged

Figure C-6 *Basic ONNET-to-OFFNET Call: Part III*

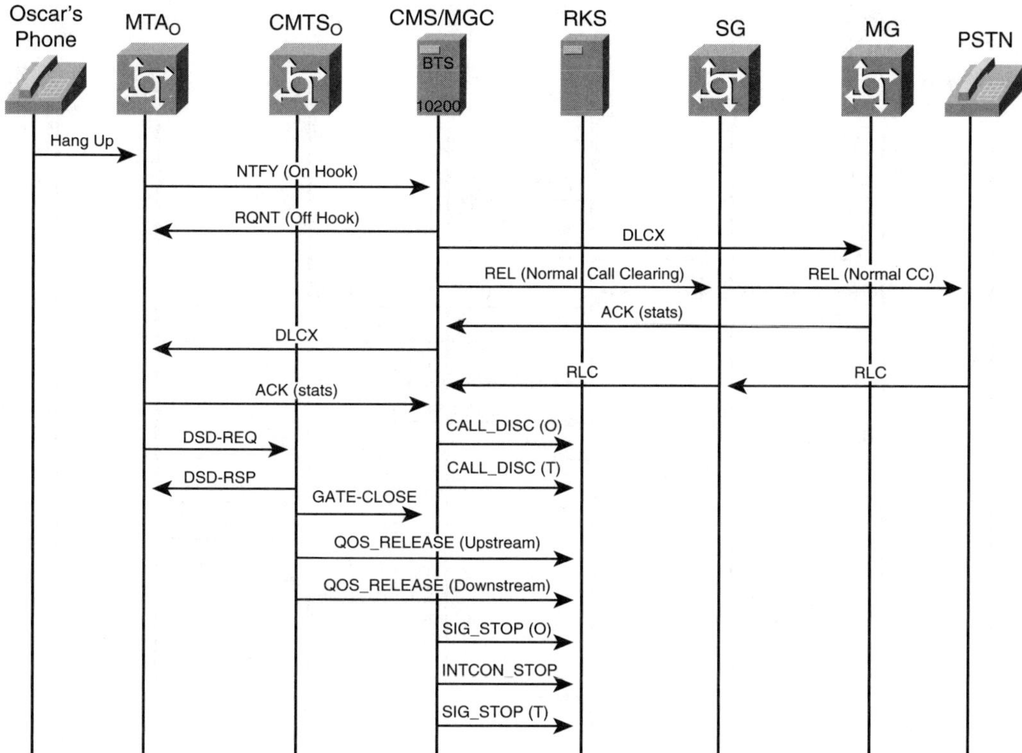

Basic OFFNET-to-ONNET Call

This is the reverse case of the previous example. This time the PSTN subscriber is calling the PacketCable subscriber. The same components are involved in the call.

Figures C-7 through C-9 illustrate an implementation of this type of call.

Figure C-7 *Basic OFFNET-to-ONNET Call: Part I*

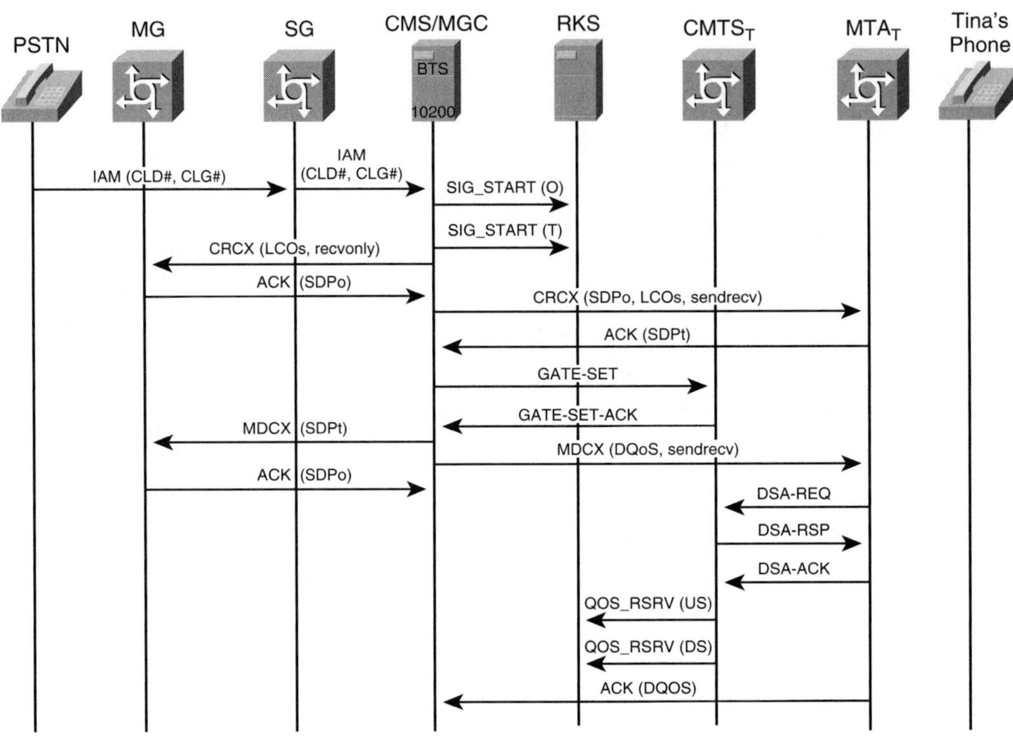

Figure C-8 *Basic OFFNET-to-ONNET Call: Part II*

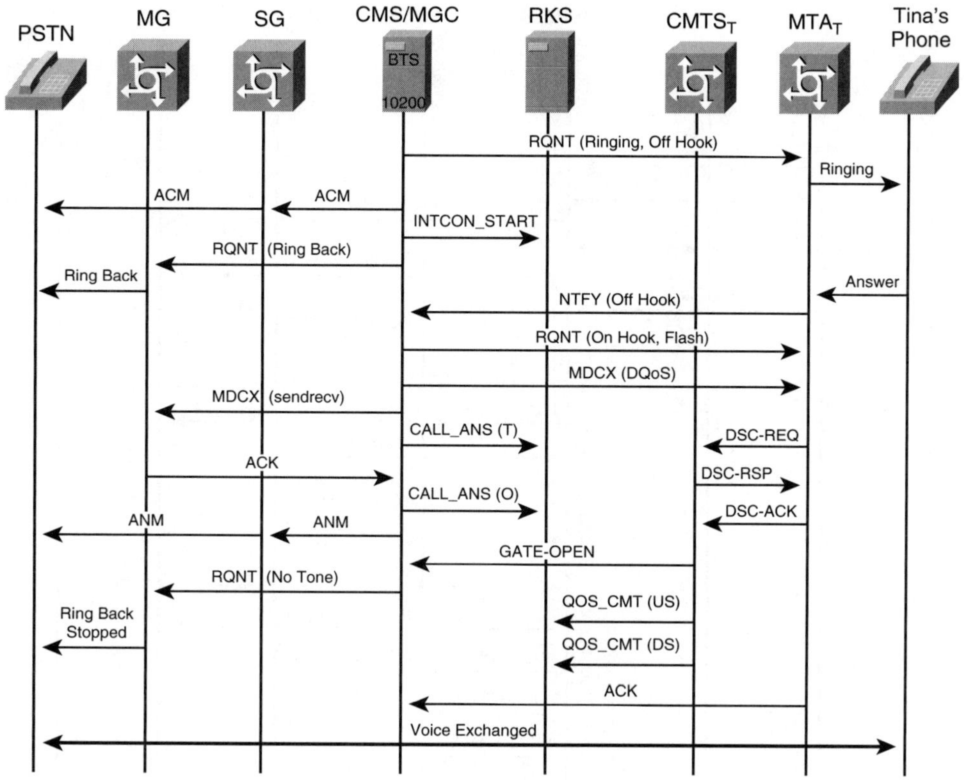

PSTN
MG
SG
CMS/MGC
BTS 10200
RKS
$CMTS_T$
MTA_T
Tina's Phone
RQNT (Ringing, Off Hook)
Ringing
ACM
ACM
INTCON_START
RQNT (Ring Back)
Ring Back
Answer
NTFY (Off Hook)
RQNT (On Hook, Flash)
MDCX (DQoS)
MDCX (sendrecv)
CALL_ANS (T)
DSC-REQ
ACK
DSC-RSP
CALL_ANS (O)
DSC-ACK
ANM
ANM
GATE-OPEN
RQNT (No Tone)
QOS_CMT (US)
Ring Back Stopped
QOS_CMT (DS)
ACK
Voice Exchanged

Figure C-9 *Basic OFFNET-to-ONNET Call: Part III*

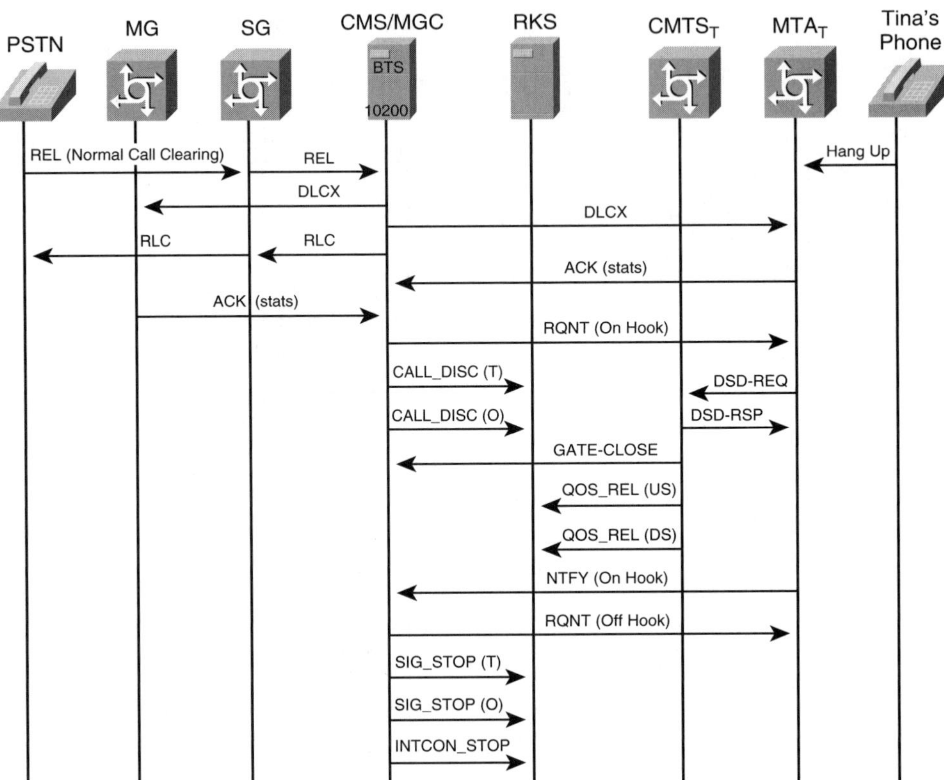

Basic ONNET-to-SIP/CMSS Call

One way in which a PacketCable provider can reduce costs is to route calls between softswitches over a QoS-enabled IP network instead of the PSTN. In this case, no trunking MG or SG is required. The next call flow example shows the messaging when a PacketCable subscriber calls another subscriber connected to a remote softswitch. If this remote switch is a PacketCable CMS, the Call Management Server Signaling (CMSS) protocol is used between switches; otherwise, the standard Session Initiation Protocol (SIP) protocol can be used. Because CMSS is simply a variant of SIP, the messaging is very similar. Key differences you might encounter between the two include the provisional acknowledgement packets (PRACK) sent to acknowledge the receipt of 1xx response packets and UPDATE packets sent to alert the remote CMS of dynamic quality of service (DQoS) resource reservation.

Figures C-10 through C-12 illustrate an implementation of this type of call where SIP is used between soft switches.

Figure C-10 *Basic ONNET-to-SIP/CMSS Call: Part I*

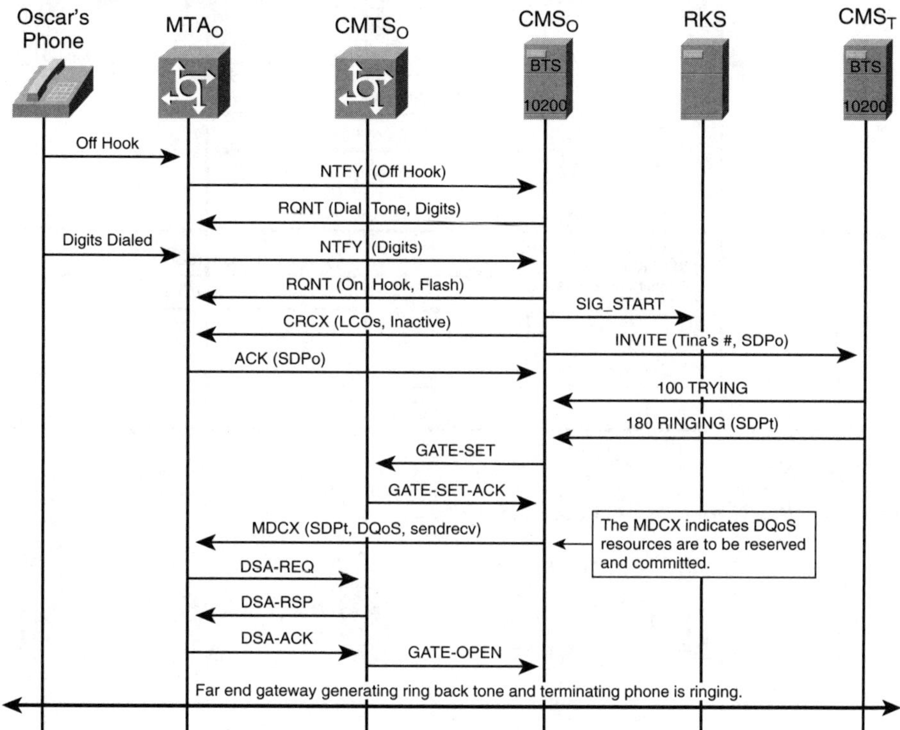

Figure C-11 *Basic ONNET-to-SIP/CMSS Call: Part II*

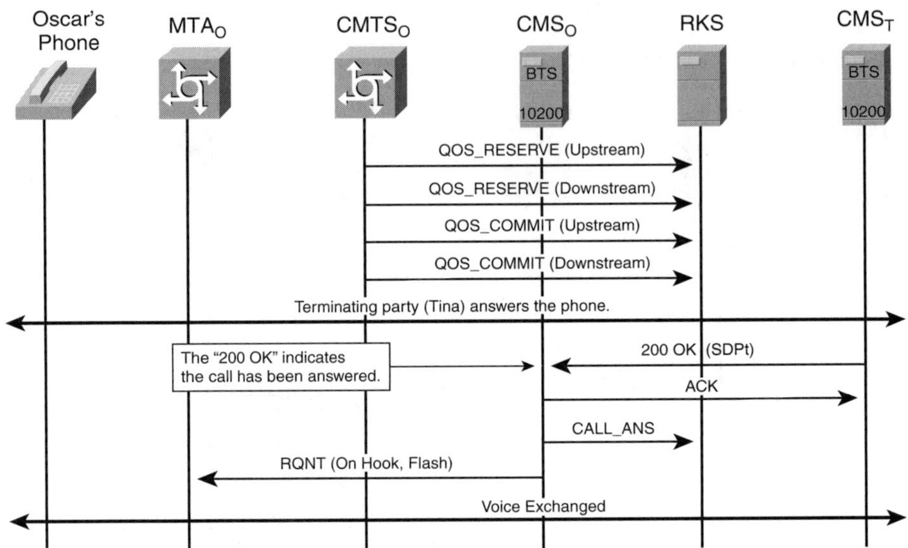

Figure C-12 *Basic ONNET-to-SIP/CMSS Call: Part III*

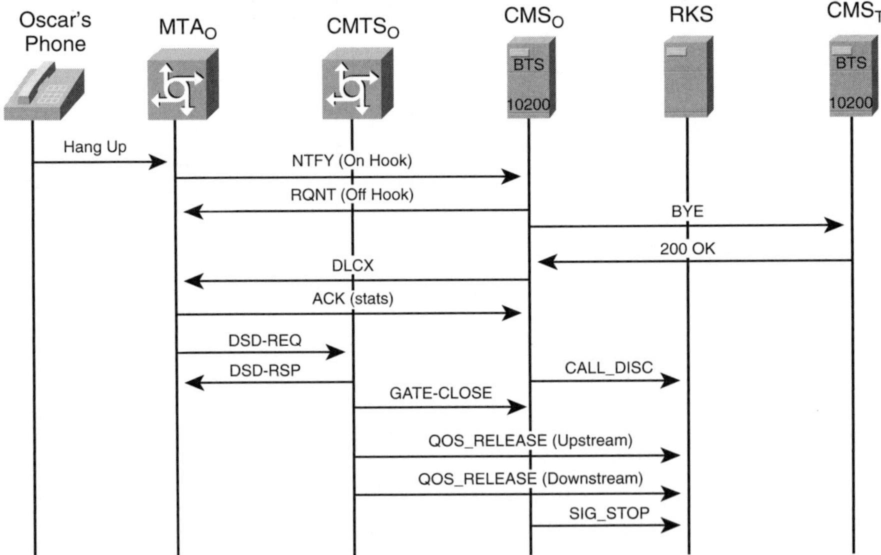

Basic SIP/CMSS-to-ONNET Call

This example is the reverse of the previous one. In this case, a subscriber controlled by a remote softswitch calls a local PacketCable subscriber.

Figures C-13 through C-15 illustrate an implementation of this type of call, where SIP is used between softswitches.

Figure C-13 *Basic SIP/CMSS-to-ONNET Call: Part I*

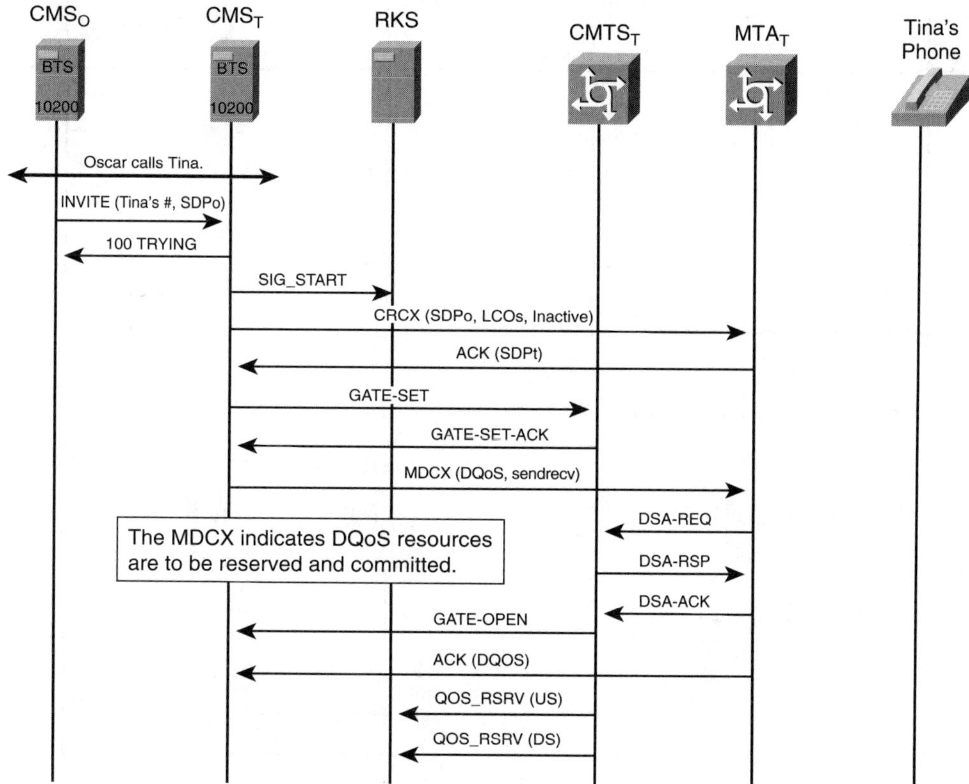

Figure C-14 *Basic SIP/CMSS-to-ONNET Call: Part II*

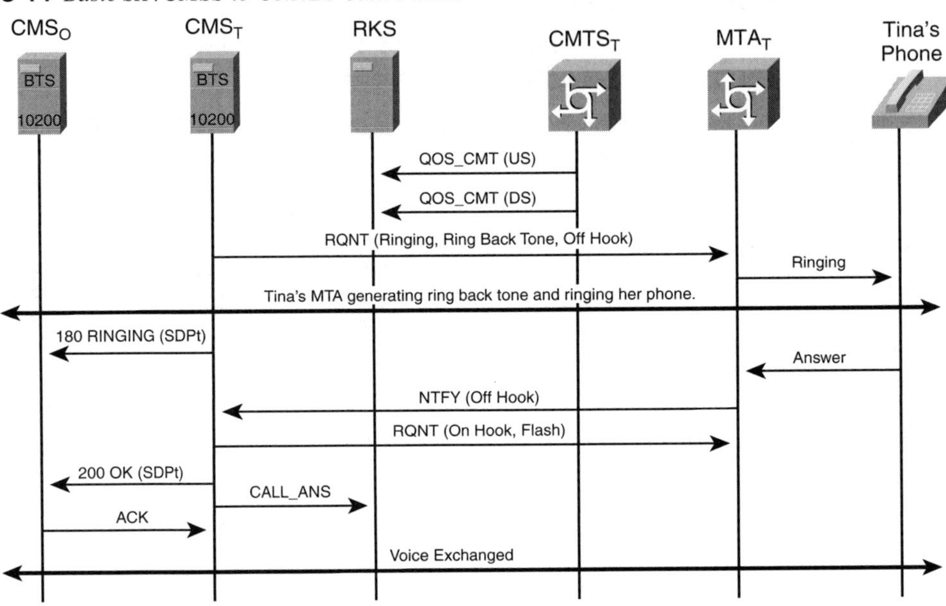

Figure C-15 *Basic SIP/CMSS-to-ONNET Call: Part III*

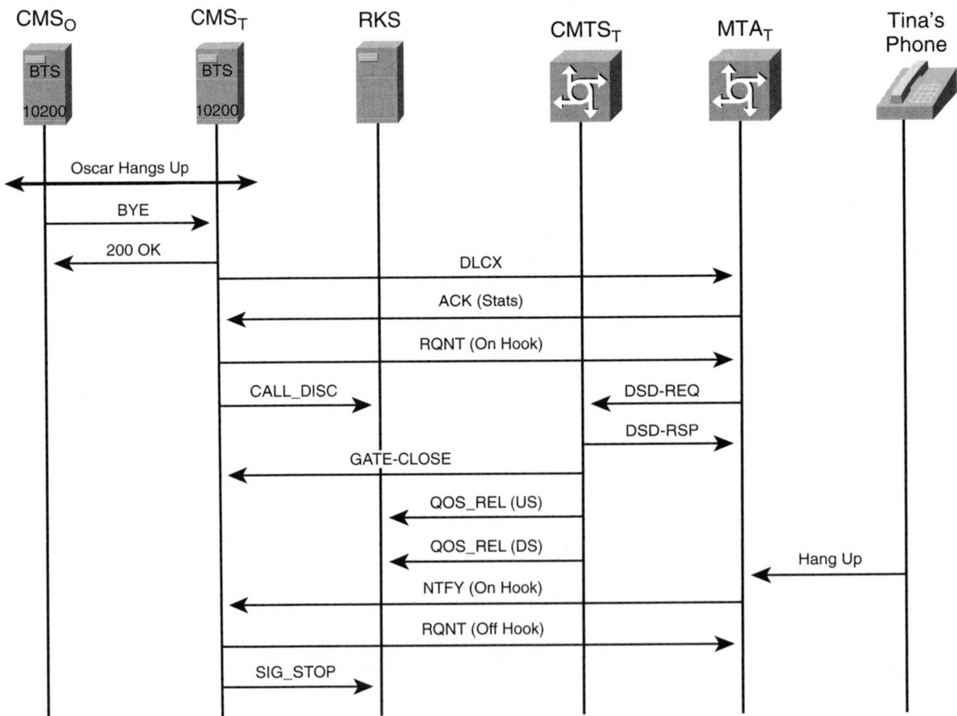

Call Waiting

These next three examples show the messaging involved when certain Custom Local Area Signaling Service (CLASS) call features are invoked. Specifically, the call waiting, call forwarding, and three-way calling features are examined. These CLASS features are included because they are the most interesting when it comes to messaging. Note: In an attempt to keep the call flows manageable all MTAs are assumed to be under the control of the same CMS and physically connected to the same CMTS. As you will soon see, these call flows are rather lengthy and complex.

First, a call waiting messaging flow is shown. In this case two PacketCable subscribers first establish a basic ONNET-to-ONNET call; then an OFFNET subscriber calls one of these PacketCable subscribers. Because three parties are involved in this call, they are referred to as Alan (A), Bob (B), and Cathy . Alan and Bob are the two PacketCable subscribers, and Cathy is reachable via the PSTN.

Figures C-16 through C-26 illustrate the messaging used in this call flow.

Figure C-16 *Call Waiting Feature: Part I*

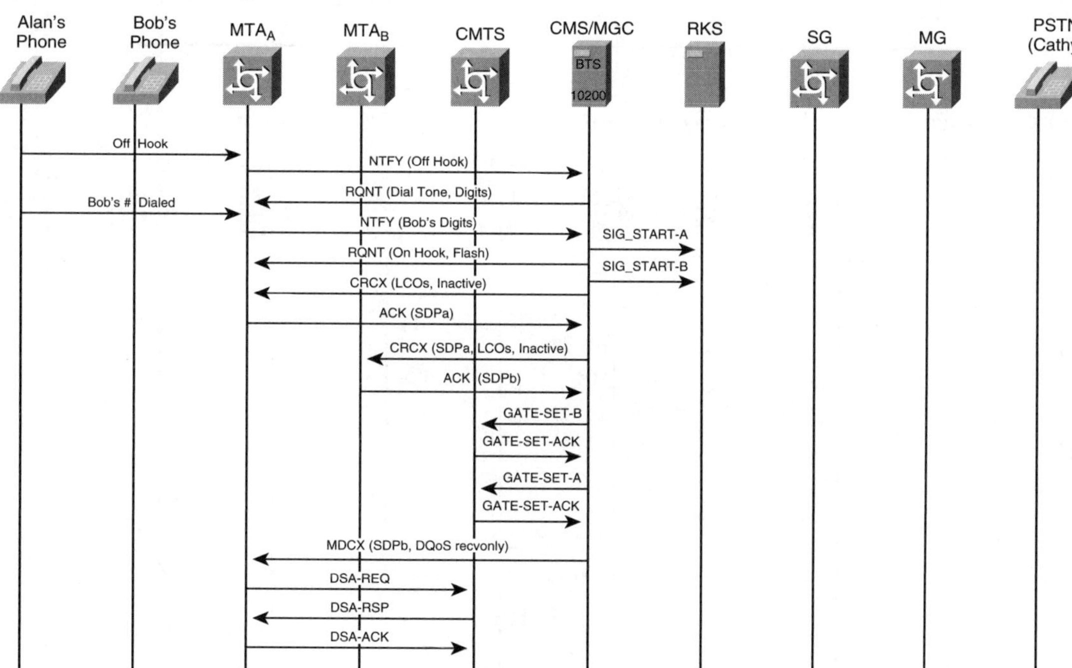

Figure C-17 *Figure C-17Call Waiting Feature: Part II*

Alan's Phone
Bob's Phone
MTA_A
MTA_B
CMTS
CMS/MGC
BTS
10200
RKS
SG
MG
PSTN (Cathy)
QOS_RESERVE (A, Upstream)
QOS_RESERVE (A, DS)
ACK (DQoS)
MDCX (DQoS, sendrecv)
DSA-REQ
DSA-RSP
DSA-ACK
QOS_RESERVE (B, Upstream)
QOS_RESERVE (B, DS)
ACK (DQoS)
RQNT (Ringing)
Ringing
RQNT (Ring Back)
Bob's phone is ringing and Alan is hearing ring back tone.
Ring Back

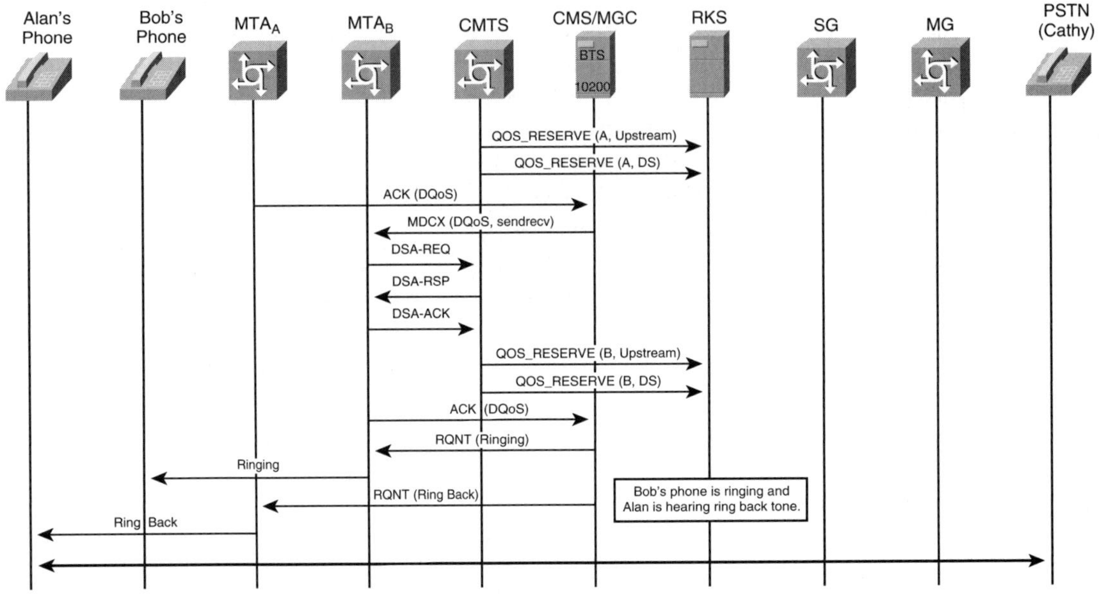

Figure C-18 *Figure C-18Call Waiting Feature: Part III*

Alan's Phone
Bob's Phone
MTA_A
MTA_B
CMTS
CMS/MGC
BTS
10200
RKS
SG
MG
PSTN (Cathy)
Answer
NTFY (Off Hook)
CALL_ANS-A
RQNT (On Hook, Flash)
CALL_ANS-B
MDCX (DQoS)
MDCX (DQoS, sendrecv)
DSC-REQ
DSC-RSP
DSC-ACK
GATE-OPEN-B
QOS_COMMIT (B, Upstream)
QOS_COMMIT (B, DS)
ACK (DQoS)
DSC-REQ
DSC-RSP
DSC-ACK
GATE-OPEN-A

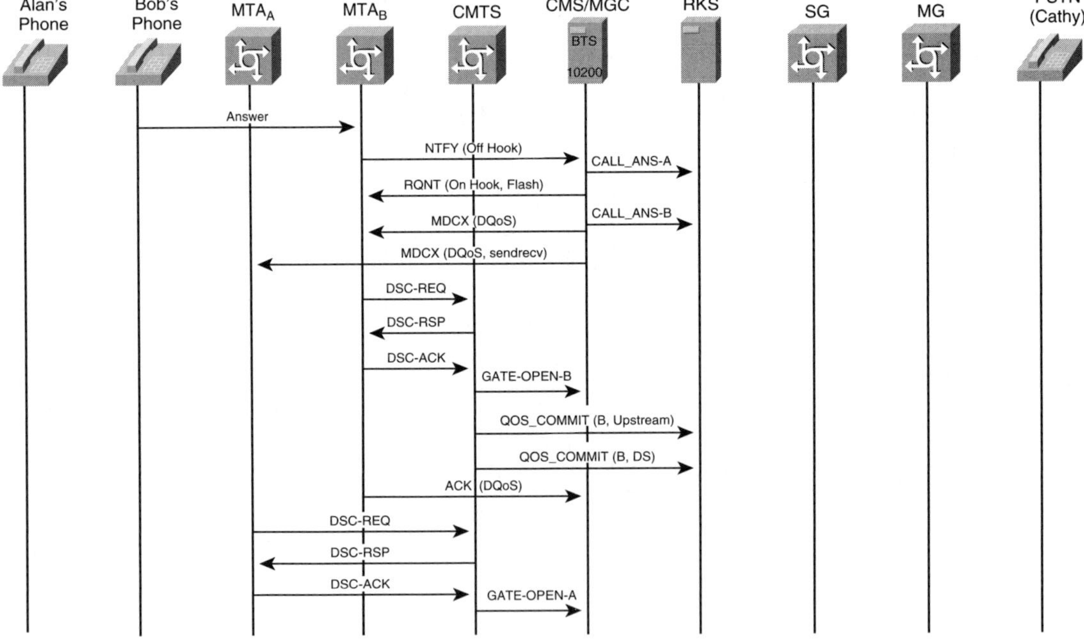

Figure C-19 *Call Waiting Feature: Part IV*

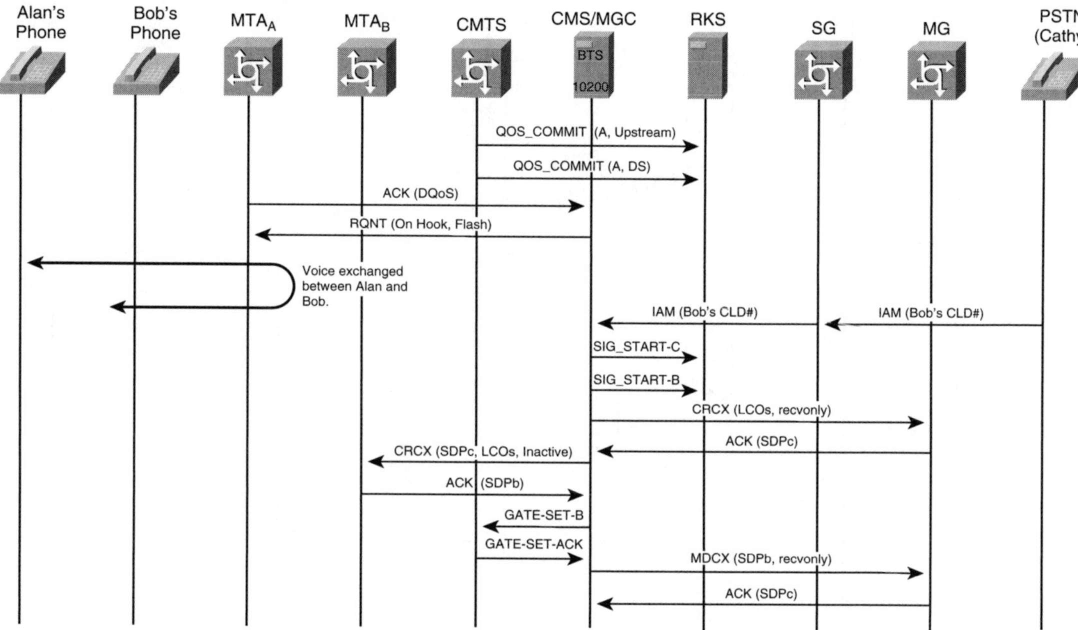

Figure C-20 *Call Waiting Feature: Part V*

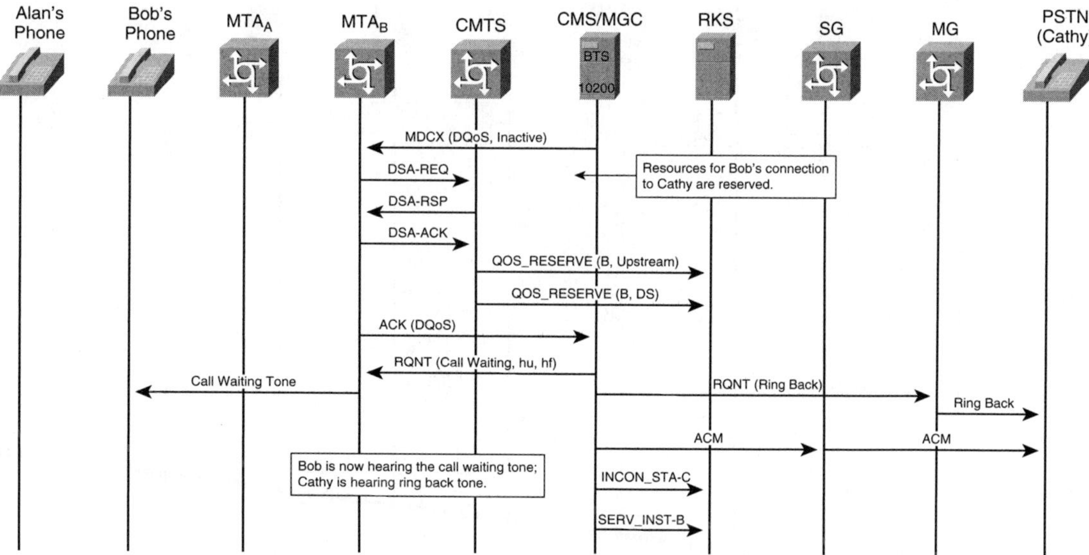

Figure C-21 *Call Waiting Feature: Part VI*

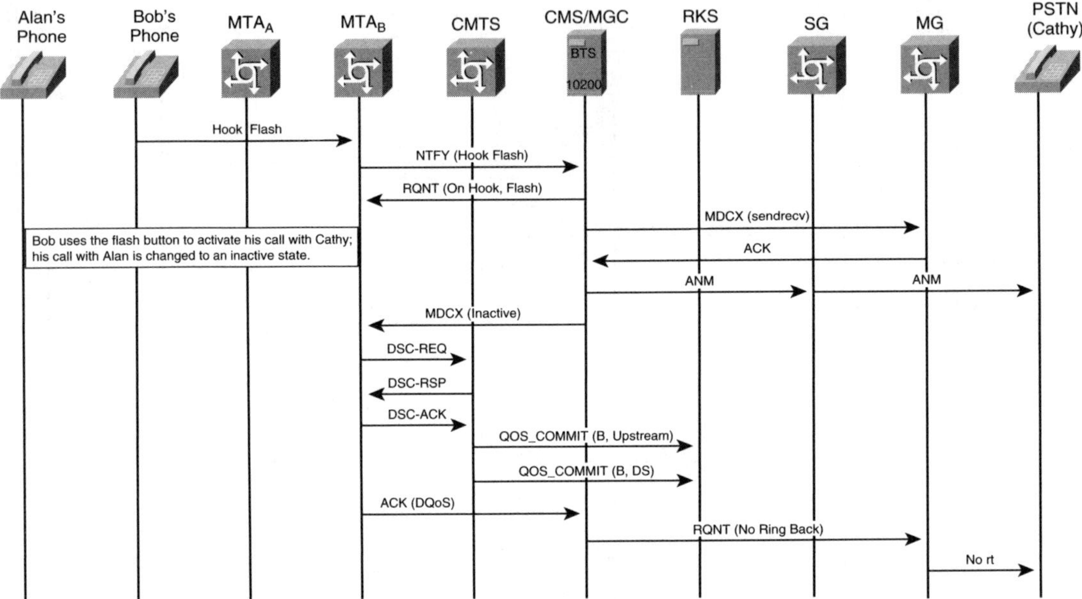

Figure C-22 *Call Waiting Feature: Part VII*

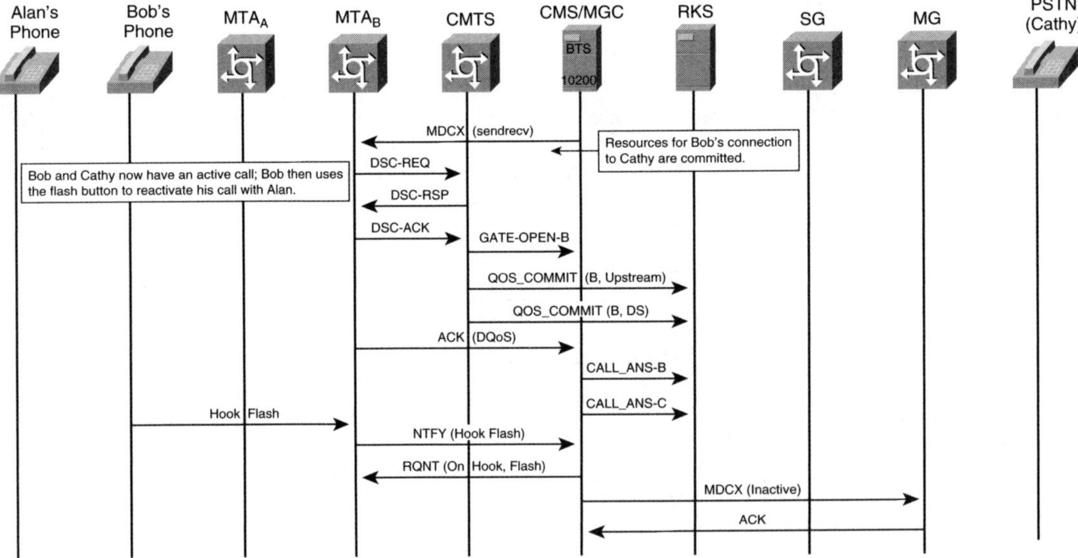

Figure C-23 *Call Waiting Feature: Part VIII*

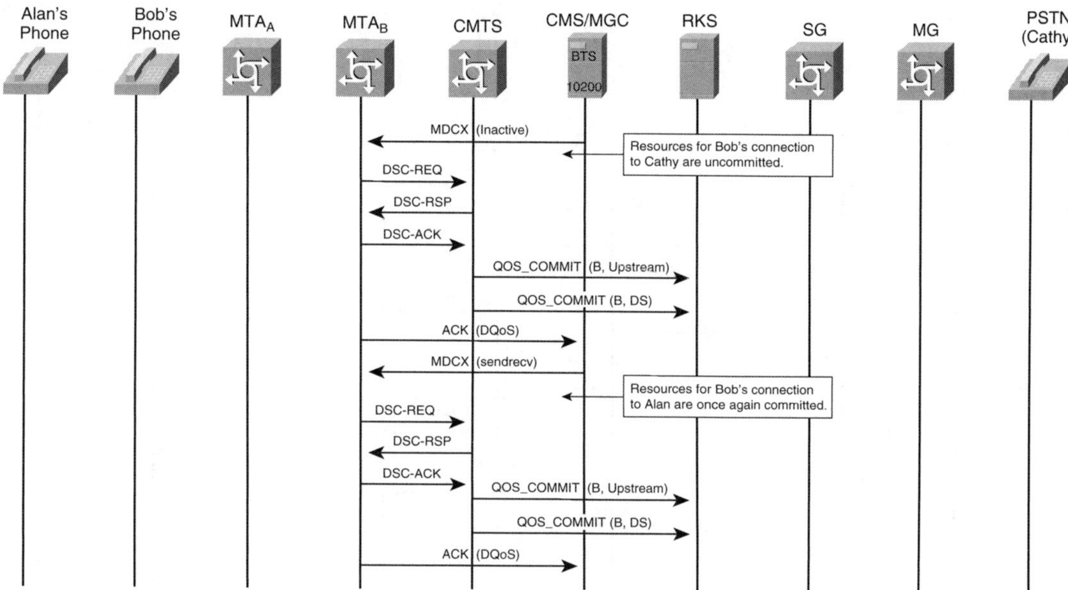

Figure C-24 *Call Waiting Feature: Part IX*

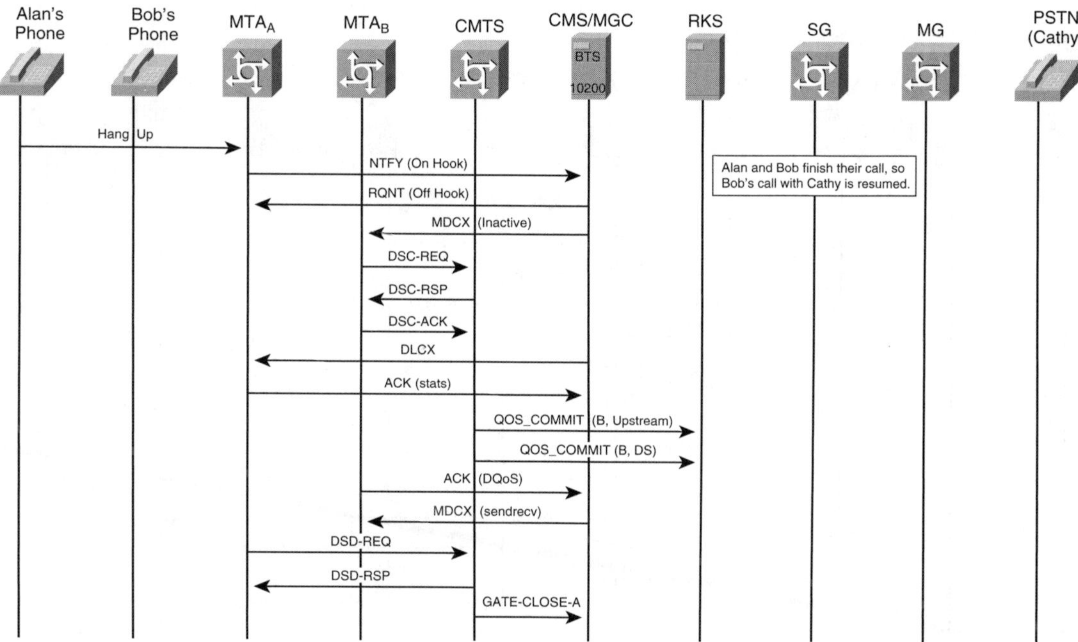

Figure C-25 *Call Waiting Feature: Part X*

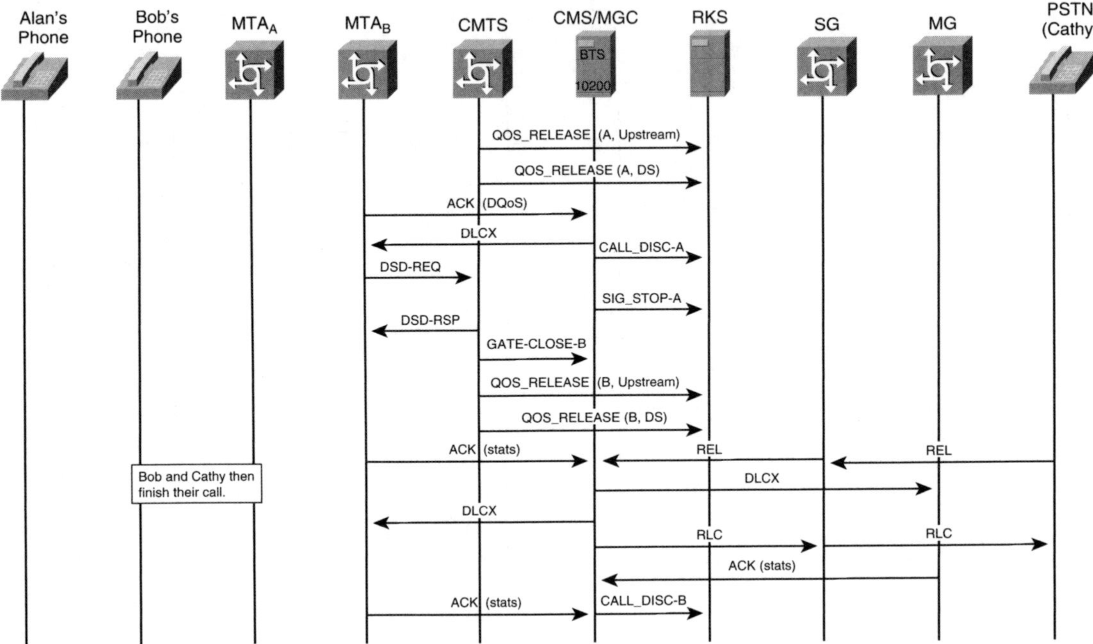

Figure C-26 *Call Waiting Feature: Part XI*

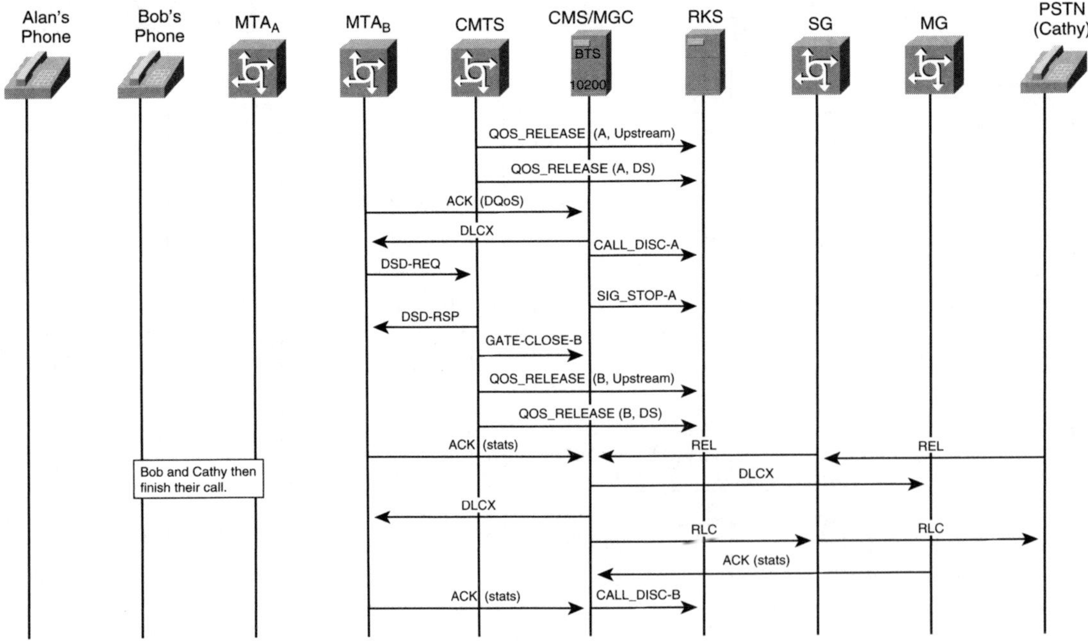

Call Forwarding

The next example involves the call forwarding feature. In this case a PacketCable subscriber forwards his or her phone to an OFFNET number such as his or her cell phone.

The call forwarding feature can work in multiple ways. For example, a subscriber can have his or her phone forwarded only when it is busy or only after a number of rings have first been attempted. In this case, unconditional call forwarding has occurred; that is, incoming calls are immediately forwarded regardless of the current status of the line. In this case, Bob uses the digit string *72 to activate unconditional call forwarding to his cell phone.

Figures C-27 through C-32 illustrate the messaging used in this call flow.

Figure C-27 *Call Forwarding Feature: Part I*

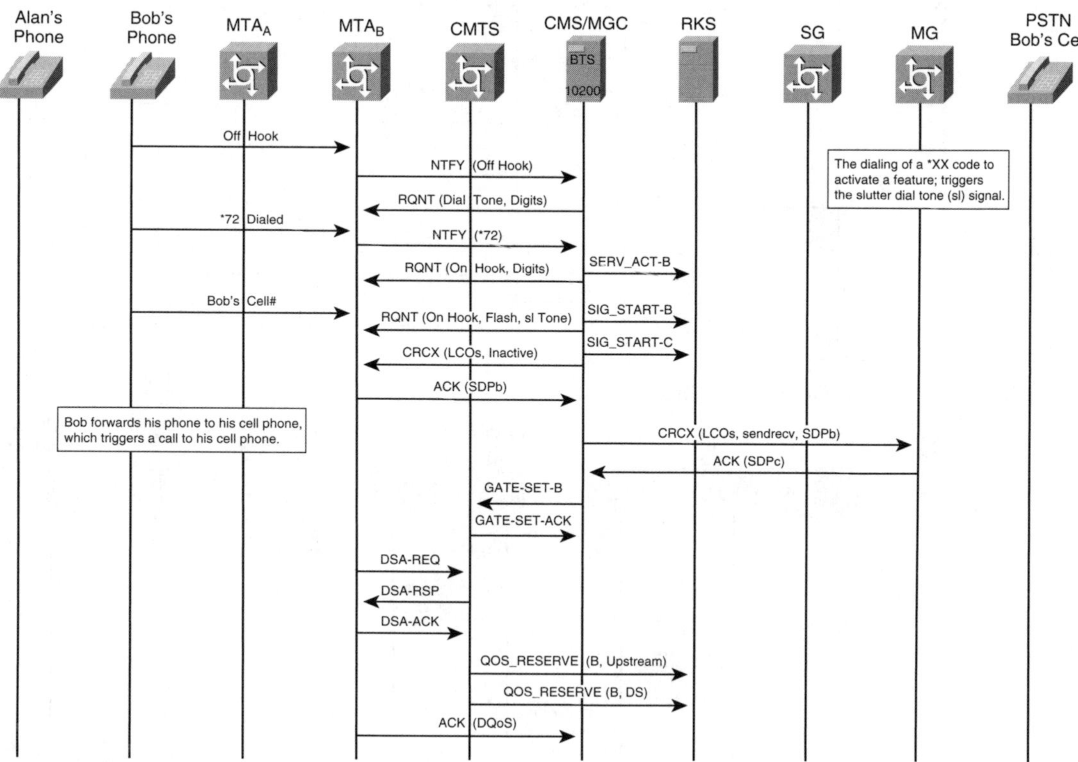

Figure C-28 *Call Forwarding Feature: Part II*

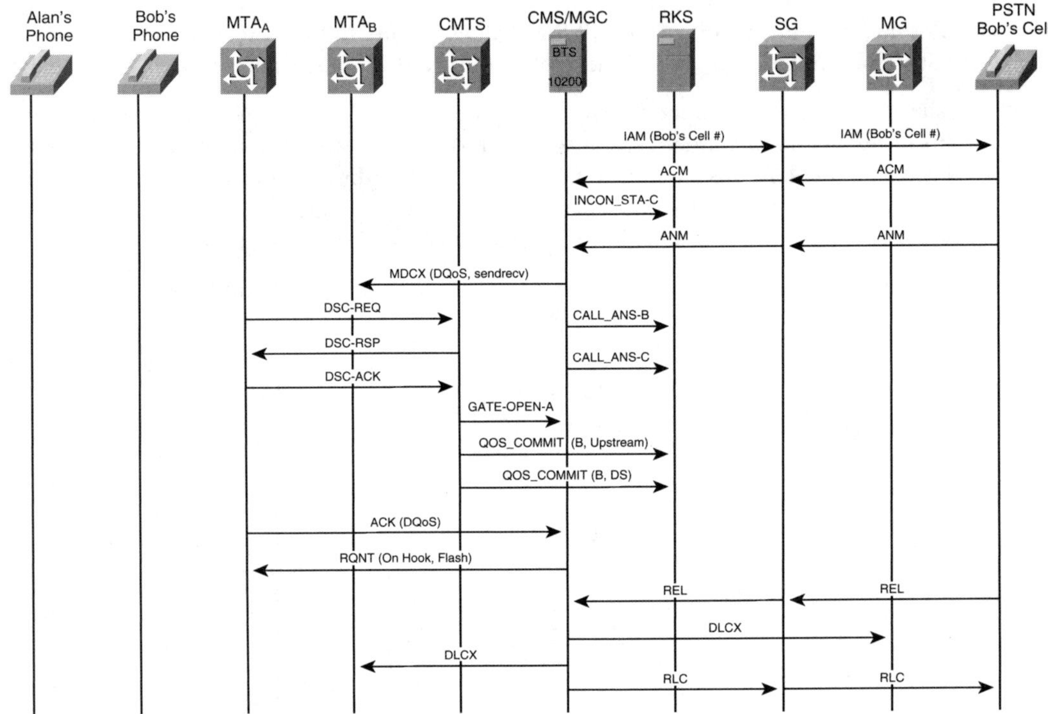

Figure C-29 *Call Forwarding Feature: Part III*

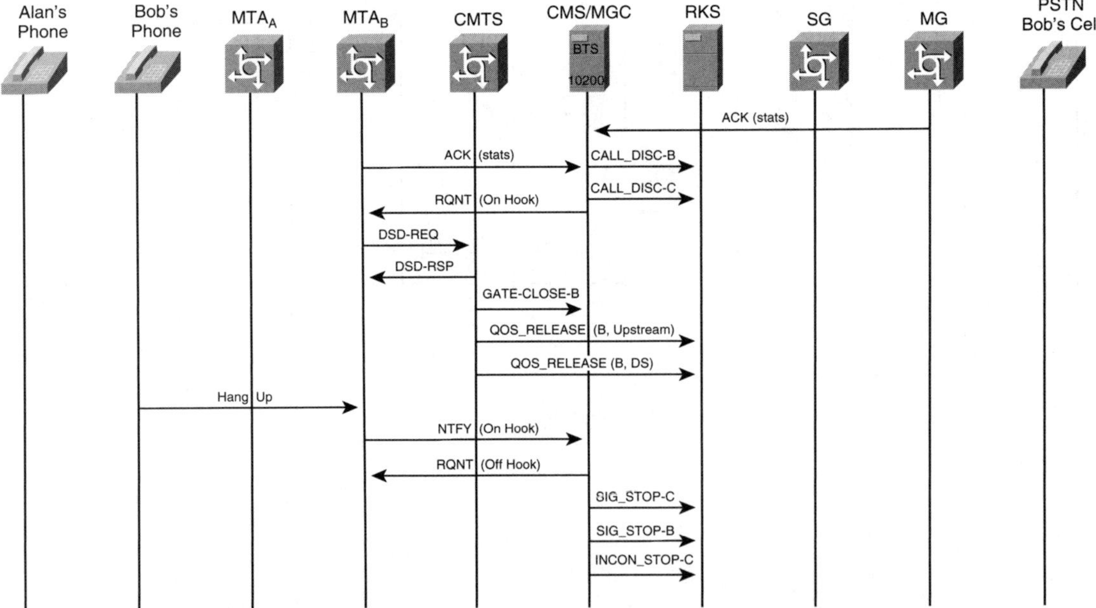

Figure C-30 *Call Forwarding Feature: Part IV*

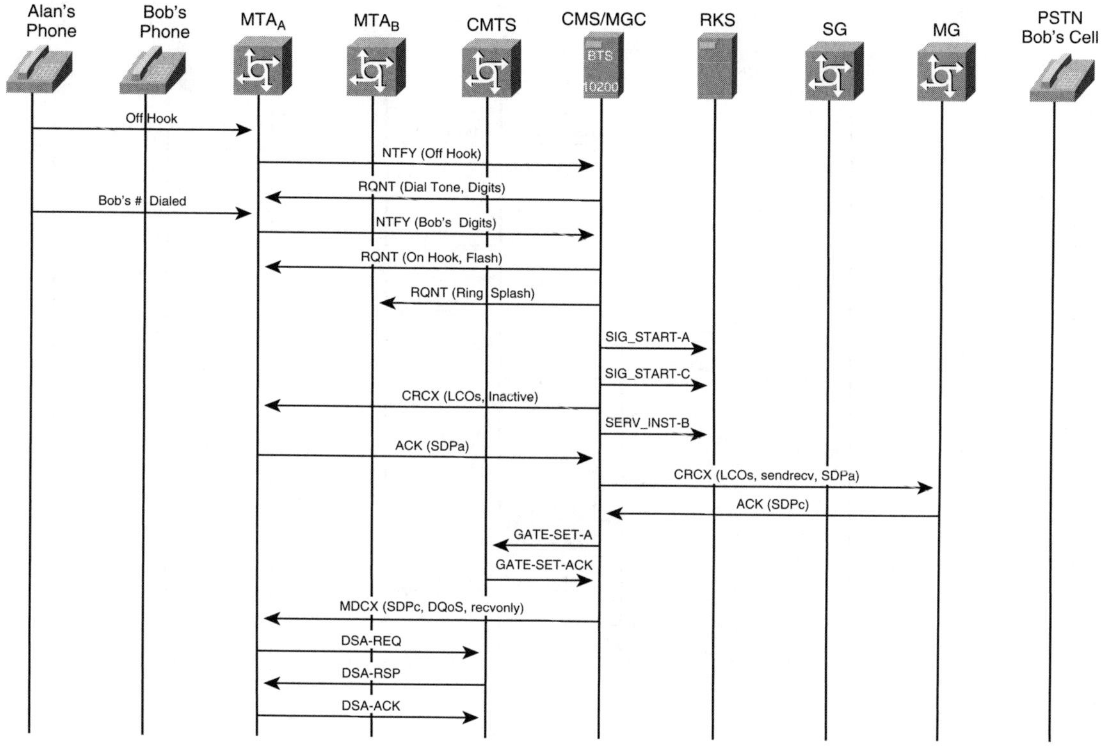

Alan's Phone | Bob's Phone | MTA_A | MTA_B | CMTS | CMS/MGC | RKS | SG | MG | PSTN Bob's Cell

BTS 10200

Off Hook
NTFY (Off Hook)
RQNT (Dial Tone, Digits)
Bob's # Dialed
NTFY (Bob's Digits)
RQNT (On Hook, Flash)
RQNT (Ring Splash)
SIG_START-A
SIG_START-C
CRCX (LCOs, Inactive)
SERV_INST-B
ACK (SDPa)
CRCX (LCOs, sendrecv, SDPa)
ACK (SDPc)
GATE-SET-A
GATE-SET-ACK
MDCX (SDPc, DQoS, recvonly)
DSA-REQ
DSA-RSP
DSA-ACK

Figure C-31 *Call Forwarding Feature: Part V*

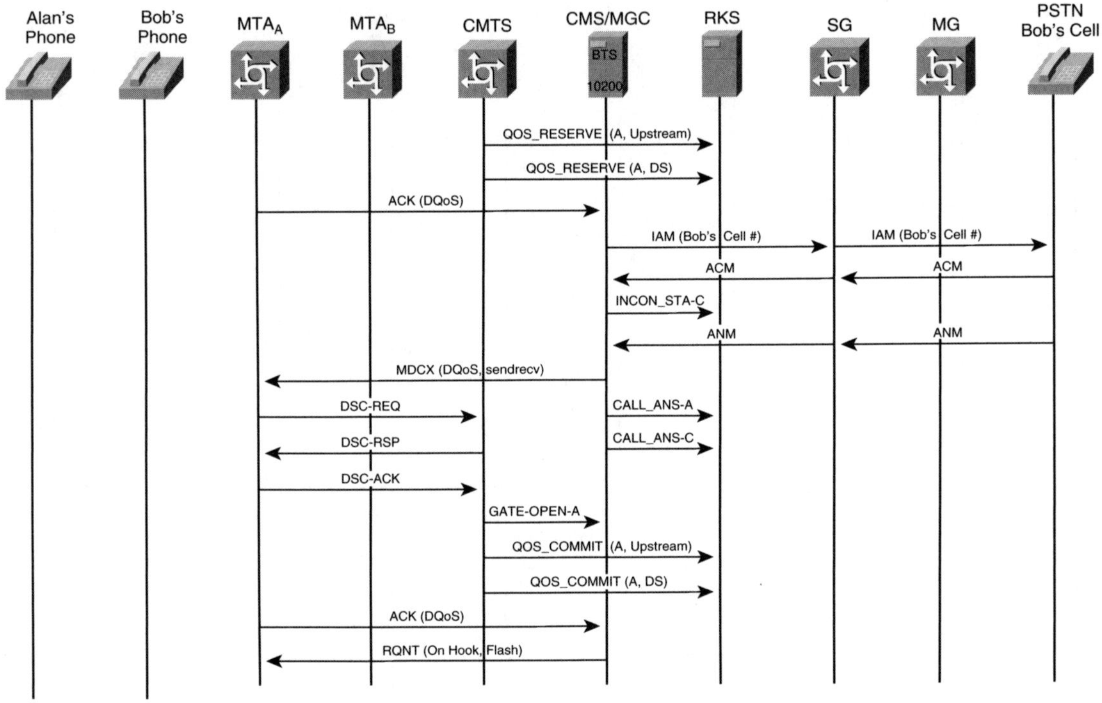

Alan's Phone | Bob's Phone | MTA_A | MTA_B | CMTS | CMS/MGC (BTS 10200) | RKS | SG | MG | PSTN Bob's Cell

QOS_RESERVE (A, Upstream)
QOS_RESERVE (A, DS)
ACK (DQoS)
IAM (Bob's Cell #)
IAM (Bob's Cell #)
ACM
ACM
INCON_STA-C
ANM
ANM
MDCX (DQoS, sendrecv)
DSC-REQ
CALL_ANS-A
DSC-RSP
CALL_ANS-C
DSC-ACK
GATE-OPEN-A
QOS_COMMIT (A, Upstream)
QOS_COMMIT (A, DS)
ACK (DQoS)
RQNT (On Hook, Flash)

Figure C-32 *Call Forwarding Feature: Part VI*

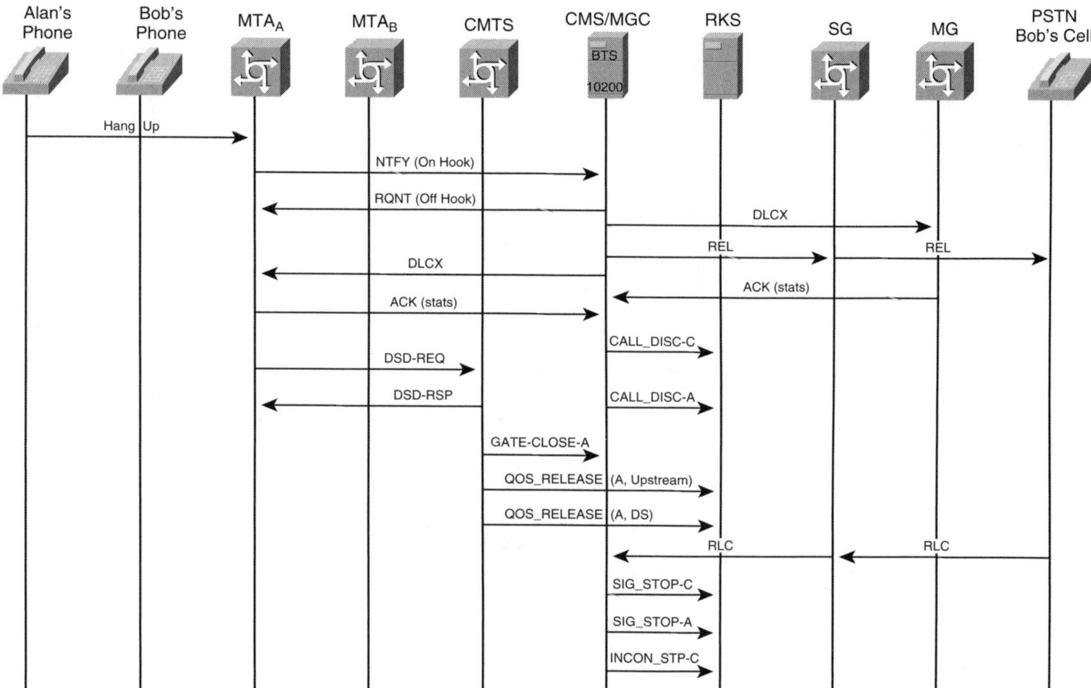

Three-Way Calling

The final call feature examined is three-way calling. In this example, a basic ONNET-to-ONNET call first occurs between a pair of PacketCable subscribers. This is followed by one of these subscribers conferencing an OFFNET subscriber into the conversation.

Figures C-33 through C-41 illustrate the messaging used in this call flow.

Figure C-33 *Three-way Calling Feature: Part I*

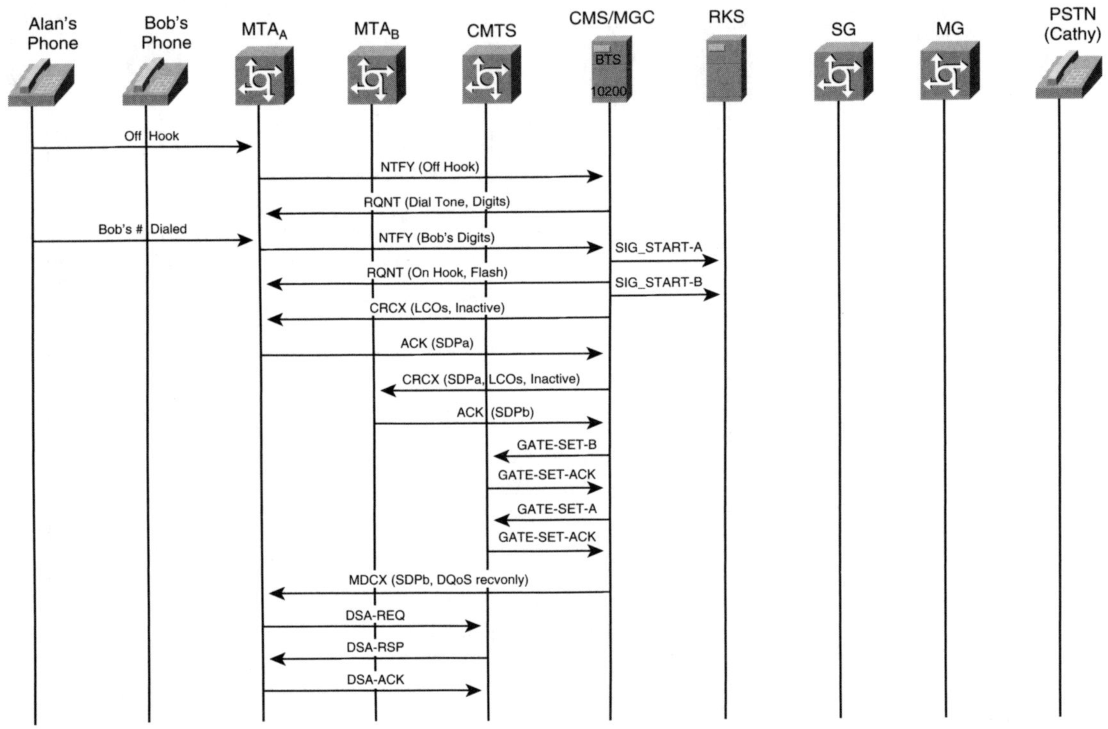

Alan's Phone
Bob's Phone
MTA_A
MTA_B
CMTS
CMS/MGC
BTS 10200
RKS
SG
MG
PSTN (Cathy)

Off Hook
NTFY (Off Hook)
RQNT (Dial Tone, Digits)
Bob's # Dialed
NTFY (Bob's Digits)
SIG_START-A
RQNT (On Hook, Flash)
SIG_START-B
CRCX (LCOs, Inactive)
ACK (SDPa)
CRCX (SDPa, LCOs, Inactive)
ACK (SDPb)
GATE-SET-B
GATE-SET-ACK
GATE-SET-A
GATE-SET-ACK
MDCX (SDPb, DQoS recvonly)
DSA-REQ
DSA-RSP
DSA-ACK

Figure C-34 *Three-way Calling Feature: Part II*

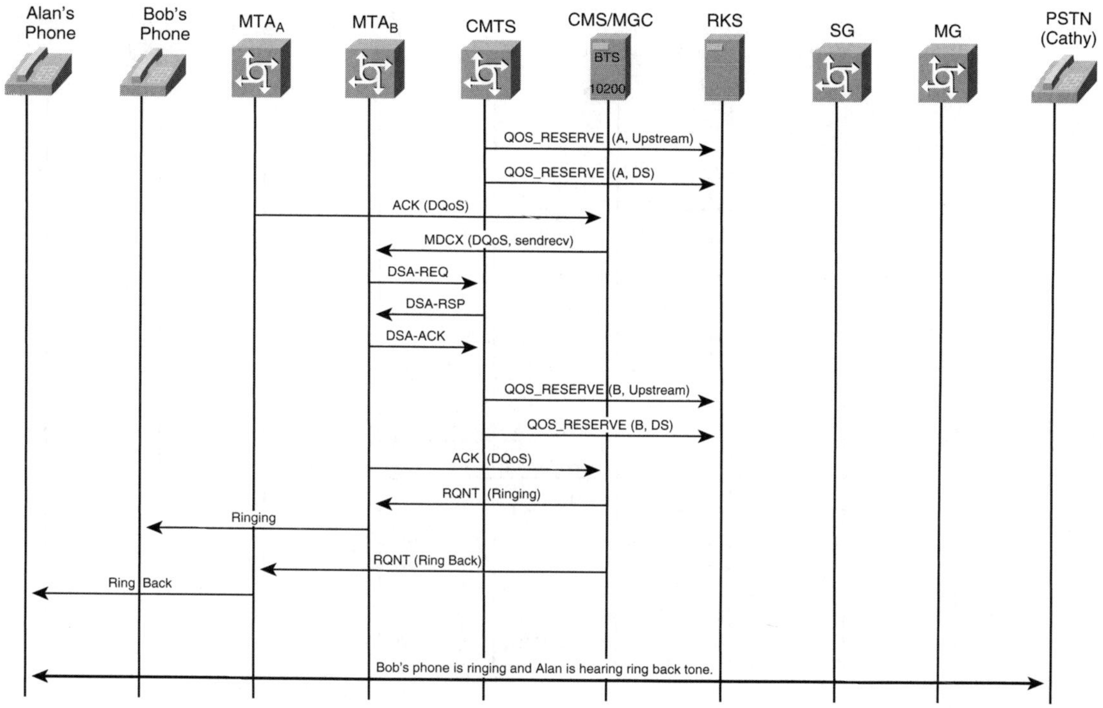

Alan's Phone
Bob's Phone
MTA_A
MTA_B
CMTS
CMS/MGC
BTS
10200
RKS
SG
MG
PSTN (Cathy)
QOS_RESERVE (A, Upstream)
QOS_RESERVE (A, DS)
ACK (DQoS)
MDCX (DQoS, sendrecv)
DSA-REQ
DSA-RSP
DSA-ACK
QOS_RESERVE (B, Upstream)
QOS_RESERVE (B, DS)
ACK (DQoS)
RQNT (Ringing)
Ringing
RQNT (Ring Back)
Ring Back
Bob's phone is ringing and Alan is hearing ring back tone.

Figure C-35 *Three-way Calling Feature: Part III*

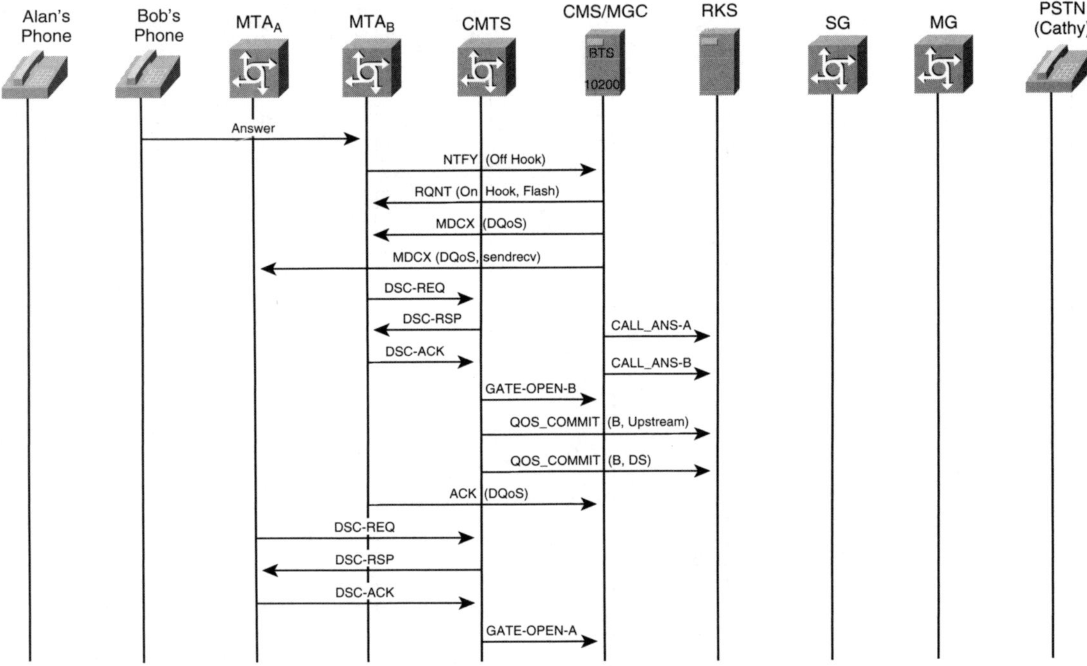

Figure C-36 *Three-way Calling Feature: Part IV*

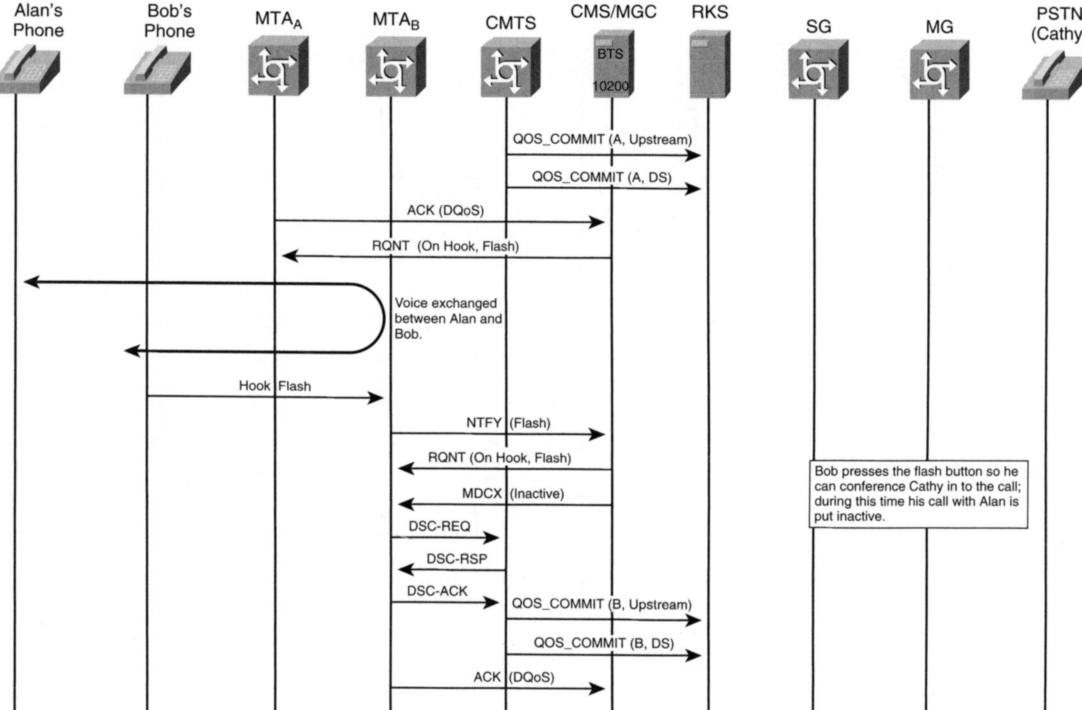

Figure C-37 *Three-way Calling Feature: Part V*

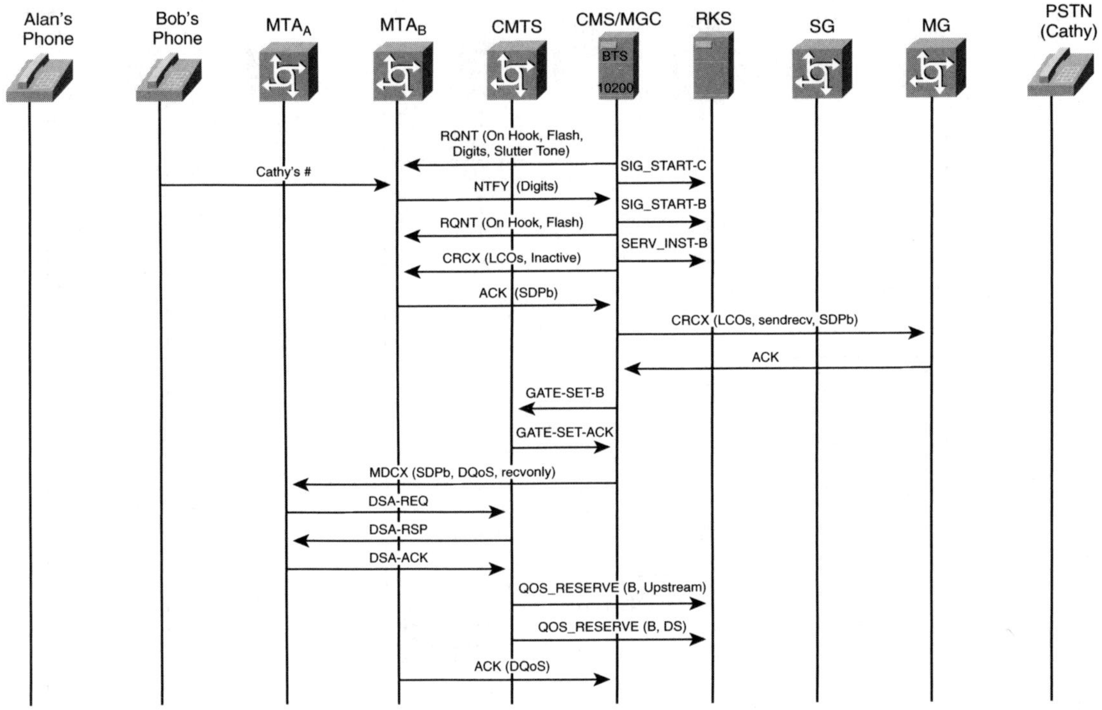

Figure C-38 *Three-way Calling Feature: Part VI*

Alan's Phone
Bob's Phone
MTAA
MTAB
CMTS
CMS/MGC
BTS
10200
RKS
SG
MG
PSTN (Cathy)
IAM (Cathy's CLD#)
IAM (Cathy's #)
ACM
ACM
INCON_STA-C
ANM
ANM
MDCX (sendrecv)
DSC-REQ
DSC-RSP
DSC-ACK
GATE-OPEN-B
QOS_COMMIT (B, Upstream)
QOS_COMMIT (B, DS)
ACK (DQoS)
CALL_ANS-B
CALL_ANS-C

Figure C-39 *Three-way Calling Feature: Part VII*

Alan's Phone
Bob's Phone
MTAA
MTAB
CMTS
CMS/MGC
BTS
10200
RKS
SG
MG
PSTN (Cathy)
Hook Flash
NTFY (Flash)
RQNT (On Hook, Flash)
MDCX (Conference)
Now that Bob and Cathy are in an active call, Bob presses the flash button again to bring Alan into the conversation.
DSC-REQ
DSC-RSP
DSC-ACK
QOS_COMMIT (B, Upstream)
QOS_COMMIT (B, DS)
ACK (DQoS)
MDCX (Conference)
ACK
Hang Up
NTFY (On Hook)
RQNT (Off Hook)

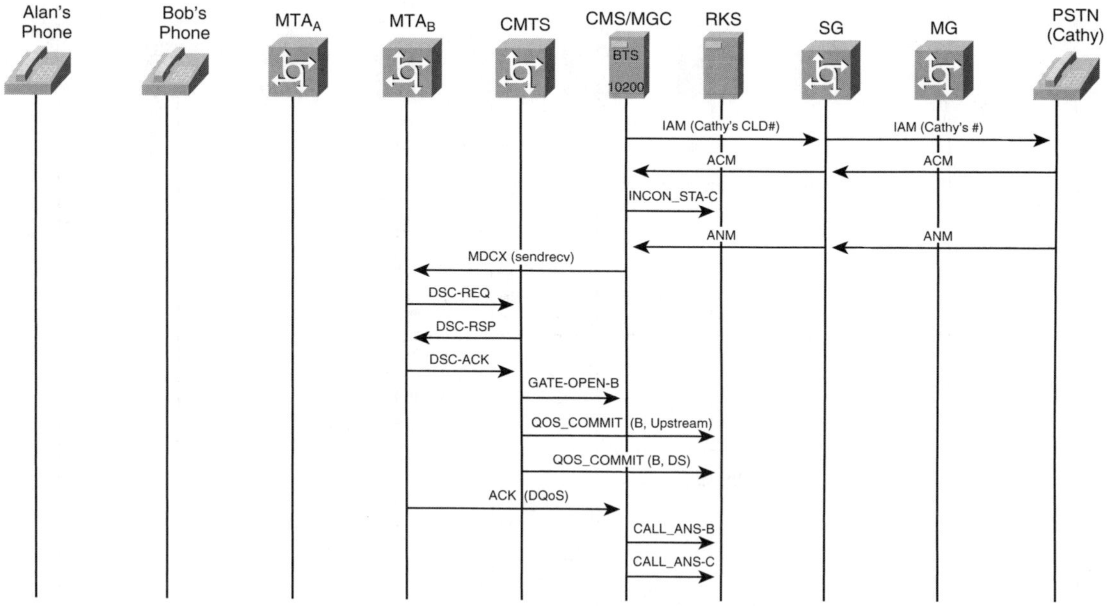

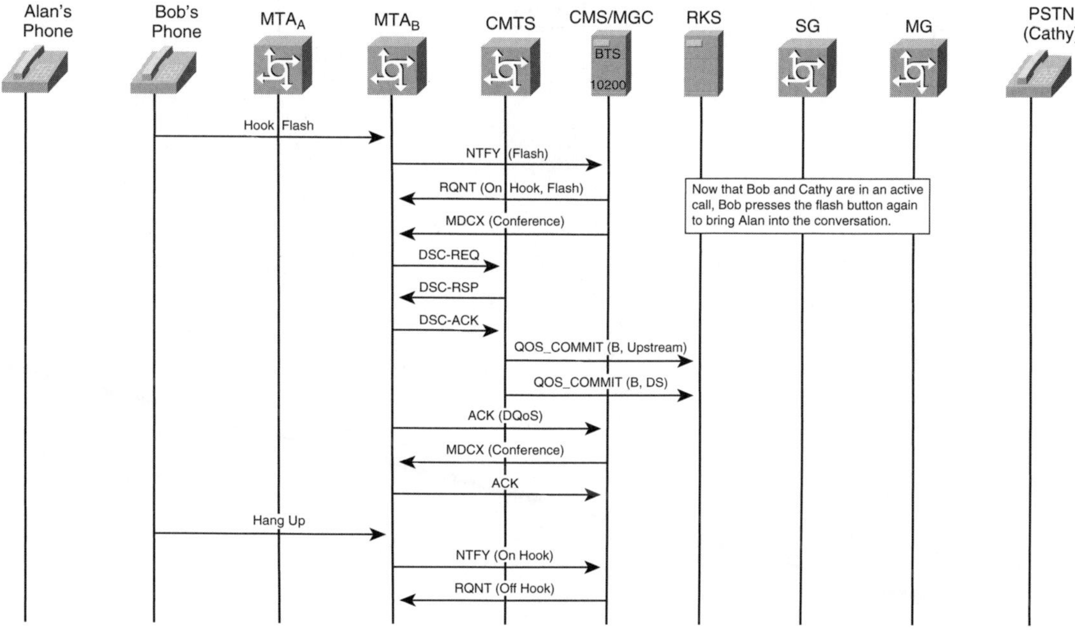

Figure C-40 *Three-way Calling Feature: Part VIII*

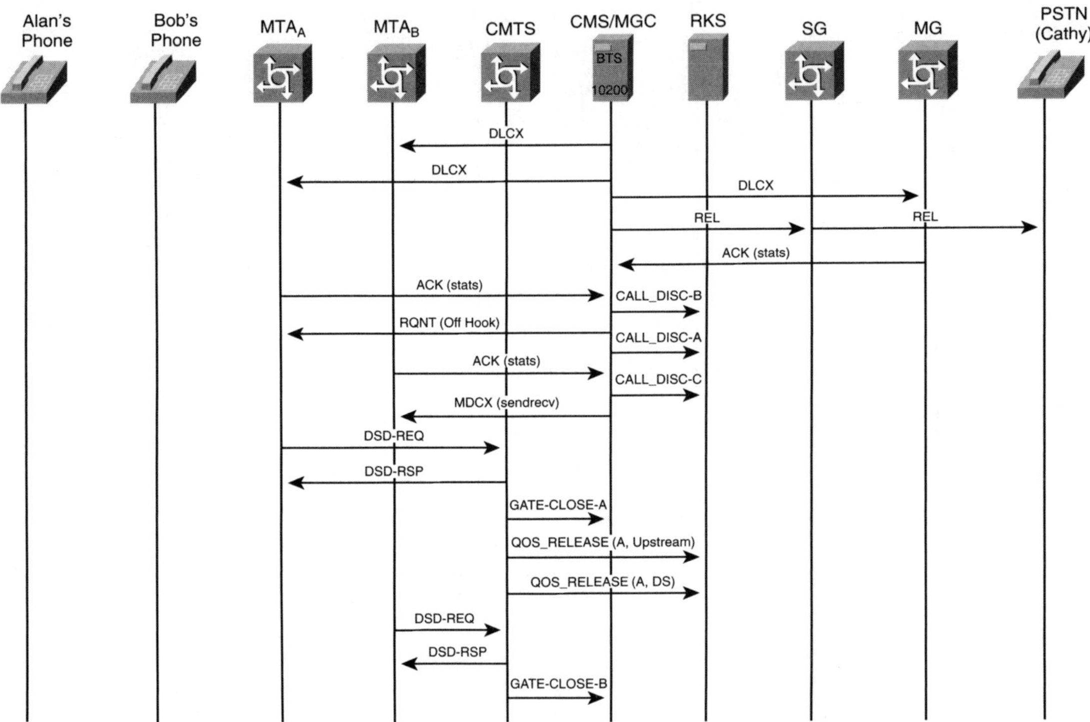

Figure C-41 *Three-way Calling Feature: Part IX*

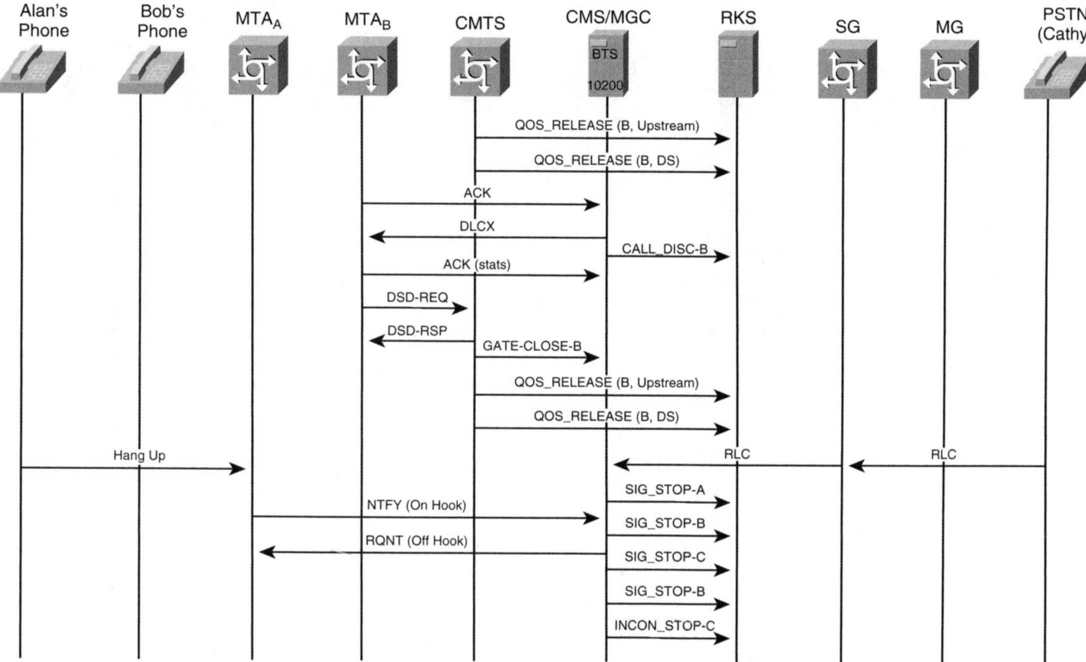

No Bandwidth Available

The final call flow shown is an example of the messaging that occurs when DQoS bandwidth reservation fails. Remember in this scenario no announcement can be provided to your customers as playing an announcement also requires DQoS bandwidth.

Figures C-42 and C-43 illustrate the messaging used in this call flow.

Figure C-42 *No Bandwidth Call: Part I*

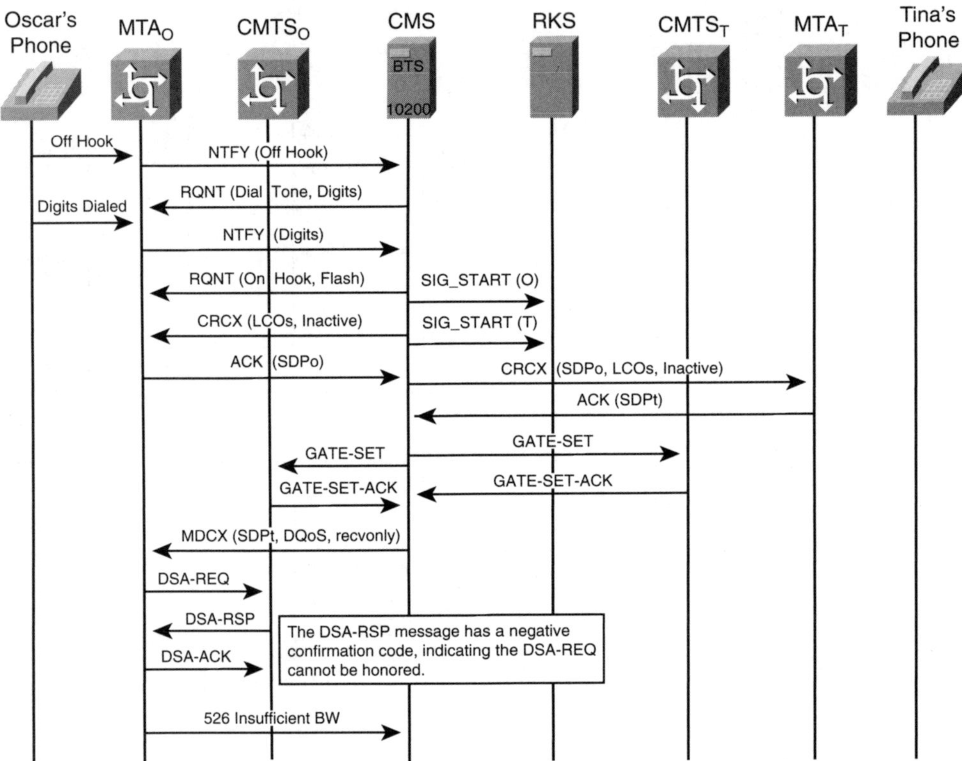

Figure C-43 *No Bandwidth Call: Part II*

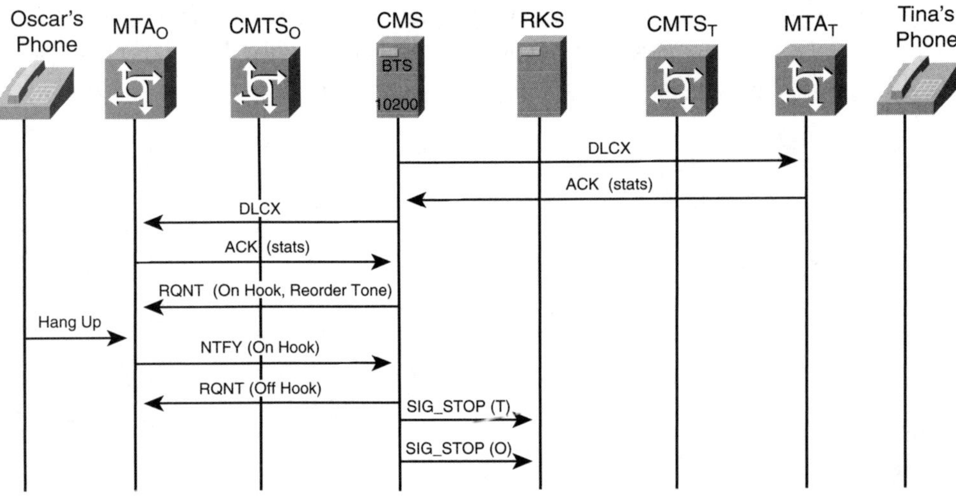

Numbers

A

B

C

D

E

F

G

H

I

J-K

L

M

O

P

Q

R

S

T

U

V

W

X-Y-Z

Cisco Press

FUNDAMENTALS SERIES

ESSENTIAL EXPLANATIONS AND SOLUTIONS

1-57870-168-6

When you need an authoritative introduction to a key networking topic, **reach for a Cisco Press Fundamentals book**. Learn about network topologies, deployment concepts, protocols, and management techniques and **master essential networking concepts and solutions.**

Look for Fundamentals titles at your favorite bookseller

802.11 Wireless LAN Fundamentals
ISBN: 1-58705-077-3

Cisco CallManager Fundamentals: A Cisco AVVID Solution
ISBN: 1-58705-008-0

Cisco LAN Switching Fundamentals
ISBN: 1-58705-089-7

Cisco Unity Fundamentals
ISBN: 1-58705-098-6

Data Center Fundamentals
ISBN: 1-58705-023-4

IP Addressing Fundamentals
ISBN: 1-58705-067-6

IP Routing Fundamentals
ISBN: 1-57870-071-X

Network Security Fundamentals
ISBN: 1-58705-167-2

Storage Networking Fundamentals
ISBN: 1-58705-162-1

Voice over IP Fundamentals
ISBN: 1-57870-168-6

Coming in Fall 2005

Cisco CallManager Fundamentals: A Cisco AVVID Solution, Second Edition
ISBN: 1-58705-192-3

Visit **www.ciscopress.com/series** for details about the Fundamentals series and a complete list of titles.

Safari
BOOKS ONLINE
ENABLED